Organic selenium compounds: their chemistry and biology

This is a volume in the series

THE CHEMISTRY OF ORGANOMETALLIC COMPOUNDS

THE CHEMISTRY OF ORGANOMETALLIC COMPOUNDS
A Series of Monographs

Dietmar Seyferth, *editor*

Department of Chemistry,
Massachusetts Institute of Technology
Cambridge, Massachusetts

PUBLISHED:

Chemistry of the iron group metallocenes

Part One; ferrocene, ruthenocene, osmocene
By Myron Rosenblum

The organic compounds of lead

By Hymin Shapiro and F. W. Frey

The organic chemistry of tin

By Wilhelm P. Neumann

The organic compounds of germanium

By M. Lesbre, P. Mazerolles and J. Satge

Organometallic compounds of arsenic, antimony, and bismuth

By G. O. Doak and Leon D. Freedman

Organic selenium compounds: their chemistry and biology

Edited by Daniel L. Klayman and Wolfgang H. H. Günther

Organic selenium compounds: their chemistry and biology

EDITED BY

DANIEL L. KLAYMAN

Walter Reed Army Institute of Research
Division of Medicinal Chemistry
Washington, D.C.

WOLFGANG H. H. GÜNTHER

Xerox Corporation
Rochester Corporate Research Center
Chemistry Research Laboratory
Webster, New York

Wiley-Interscience
a Division of
JOHN WILEY & SONS, INC.

NEW YORK · LONDON · SYDNEY · TORONTO

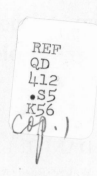

Library of Congress Cataloging in Publication Data:

Klayman, Daniel L.
  Organic selenium compounds.

  (The Chemistry of organometallic compounds)
  Bibliography: p.
  1. Organoselenium compounds. I. Günther, Wolfgang
H. H., joint author. II. Title. [DNLM: 1. Selenium.
QD 412.S5 068 1972]

QD412.S5K56   546′.724′2   72-5448
ISBN 0-471-49032-6

Contributors

Lars-Börge Agenäs
Joseph F. Alicino
Leslie G. S. Brooker
Ehrenfried Bulka
Richard G. Crystal
John A. Ford, Jr.
Arne Fredga
Wolfgang H. H. Günther
L. Henriksen
K. A. Jensen
C. K. Jørgensen
Márton Kajtár
Isabella L. Karle
Jerome Karle
Daniel L. Klayman
John A. Kowald
James E. Kuder
Marcel A. Lardon
Natalia N. Magdesieva
John L. Martin

Anna Markowska
Henry G. Mautner
Jan Michalski
Luigi Mortillaro
P. H. Nielsen
Roland Paetzold
Manfred Reichenbächer
J. Roy
Mario Russo
Milton L. Scott
Jay R. Shapiro
Henry J. Shine
Robert J. Shine
Alex Shrift
Richard B. Silverman
Günther Snatzke
Ulla Svanholm
Earl J. Van Lare
Roderich Walter
Göran Zdansky
Nicolai S. Zefirov

Foreword

A century and a half has passed since Jöns Jacob Berzelius discovered the element selenium in lead chamber deposits at Gripsholm in Sweden. He had invested money in a sulfuric acid plant, and the discovery of selenium appears to have been the only benefit he ever derived from his investment. The industry was not profitable, but at length his money was returned when the insured plant was destroyed by fire. Berzelius realized at once the strong similarity between sulfur and selenium; the new element was just like "a new kind of sulfur." He also pointed out that it had some metallic properties; that is, it was something intermediate between the metals and the metalloids.

For about a century, organoselenium chemistry was a neglected field. In the 1920's the entire literature consisted of about 200 papers. Now and then a sulfur chemist made a casual digression into the selenium domain, prepared a few compounds, and found them similar to the sulfur analogs, though less stable. The sudden appearance of red smears instead of the expected product was (and still is) a common and disconcerting experience. The highly unpleasant odor of many selenium compounds can also be discouraging. Organosulfur compounds have a bad reputation in this respect, but the selenium analogs are worse. Nevertheless, in the last three or four decades interest in organoselenium chemistry has increased appreciably, and results of some practical importance have emerged.

There have been analytical problems. For a long time the determination of carbon and hydrogen in the presence of selenium presented difficulties, but they have been overcome. In the determination of selenium itself, the destruction of the organic substance without loss of selenium has proved to be a problem. Studies of selenium in biological materials called for routine methods, applicable even to very small amounts of the element. Here, neutron activation analysis utilizing the radioactive nuclide ^{75}Se has rendered very good service. This isotope is also used in metabolic studies.

Selenium compounds generally react rapidly and quantitatively, especially in oxidation-reduction processes. In organosulfur chemistry the corresponding reactions are more sluggish, allowing for the more frequent isolation of intermediates. The Se–Se bond is distinctly less stable than the S–S bond and is easily cleaved in various reversible reactions. Selenium compounds can often be used as model substances for studies in organosulfur chemistry and *vice versa*. This aspect has been found fruitful in polarographic research, for example. Comparison of u.v. or i.r. spectra of analogous sulfur and selenium compounds has also yielded valuable information.

The C–Se bond is less stable than the C–S bond. Often selenium compounds are colored red by the free element when they are exposed to light. In certain cases they are also decomposed by X-rays, which may cause trouble in crystal structure work or high-resolution electron spectroscopy (ESCA).

Natural selenium has six stable isotopes with the mass numbers 74, 76, 77, 78, 80, and 82. All except ^{74}Se are present in quantities sufficient to give distinct lines in ordinary mass spectra of selenium compounds. This makes the spectra very rich in lines, a feature which might be a disadvantage. On the other hand, the five lines form a characteristic group, making it easy to identify selenium-containing fragments. The isotope ^{77}Se has a nonzero magnetic moment, and its coupling with protons has been found useful in elucidating the constitution and conformation of selenium compounds by n.m.r.

For stereochemists it was frustrating that no optically active selenoxides could be obtained. Of the different explanations that have been discussed, two may be mentioned: reversible addition of water, giving the nonchiral $>$Se(OH)$_2$ group, and rapidly established redox equilibria. In both cases chirality in hydroxyl-free media should be possible. Recently some promising results have been reported.

It is interesting to compare the element pairs, sulfur-selenium and phosphorus-arsenic. Compounds of sulfur and selenium are often isomorphic, as are compounds of phosphorus and arsenic. Sulfur and phosphorus are typical metalloids, while selenium and arsenic have some metallic properties. Sulfur and phosphorus may expand their electron shells, but do so rather reluctantly; selenium and arsenic do so readily. This is a very important difference. Monoesters of sulfuric and phosphoric acid are rather stable to hydrolysis in neutral aqueous solutions but lend themselves readily to all kinds of enzymatic reactions—a necessary condition for their fundamental roles in physiological processes. The esters of selenic and arsenic acid (and other oxygen acids of these elements) hydrolyze spontaneously and rapidly. Sulfur and phosphorus are normal constituents of the living organism and are present in considerable quantities. Selenium and arsenic are highly toxic but may play a beneficial role in very low concentrations.

The peculiar irritating effect of hydrogen selenide on the mucous membranes of the nose was already observed by Berzelius, who gave a vivid description of the unpleasant experience. A chemist who has received a dose of this gas into his nostrils will certainly never forget it. The effect may be due in part to the strong acidity of hydrogen selenide. Its pK_a is 3.73; this means that it is comparable to formic acid and about 2000 times stronger than hydrogen sulfide. In addition, hydrogen selenide is a very powerful reducing agent whose effect on the biological redox systems may be quite destructive.

For many years selenium was regarded as rather uninteresting from the biological point of view. Around 1930, however, attention was drawn to the toxic properties of this element, which is very widespread, although its overall concentration in the earth's crust is only $10^{-5}\%$. It was discovered that selenium, if present in the soil even in minute amounts, is taken up by the vegetation and may cause damage to animals grazing on the toxic plants. Alkali disease and blind staggers, diseases of livestock known for a long time, could now be attributed to chronic selenium poisoning. Certain plant species were found to accumulate selenium in spectacular quantities. Some of them are dependent on selenium for their normal growth and never occur on selenium-free soils. Although such "indicator plants" are highly toxic to most animals, they are capable of supporting parasites, insect larvae with very high selenium tolerance. The relationship between selenium and vegetation presents many interesting problems.

In the 1930's it was also shown that the feeding of selenium-containing cereals to hens may cause malformations, monstrosities, or death in the chickens. This was one of the first cases in which a chemical agent was found to affect the embryo in a detrimental way without apparent injury to the mother.

In the 1950's it was discovered that some selenium compounds, when present in very small quantities, can have positive effects in preventing lethal liver necrosis in laboratory animals kept on an artificial diet ("factor 3" effect). Chronic diseases and cases of ill-thrift in domestic animals in various parts of the world were found to be selenium-responsive. There is strong indication, therefore, that selenium is really an essential element. Moreover, in certain cases, selenium and vitamin E can replace each other.

To what extent can selenium compounds replace their sulfur analogs in biological systems? No general answer can be given. The esters of selenic acids have already been touched upon. Selenocystine and its derivatives are rather toxic to animals, but they play an important role in selenium-accumulating plants. In the case of thioctic acid, the selenium analog is said to be a powerful antagonist. On the other hand, selenocoenzyme A has a certain catalytic activity and competes with the sulfur compound in enzymatic reactions.

Like many other elements, selenium has been suspected of being a carcinogen but without rigorous proof. Conversely, a number of its compounds have been tested as antitumor agents; however, to my knowledge, no results of practical importance have been obtained.

Organoselenium chemistry continues to be of interest to scientists in various other fields—biochemistry, physiology, toxicology, and botany. Perhaps it will become important in medicine one day. In recent years the biological interest has been a powerful incentive to the study of organoselenium compounds, but as yet we know rather little about their biological function and there is no lack of interesting problems. Just now organosulfur chemistry is said to be in a period of explosive development, and it is probable that research in organoselenium chemistry will be further stimulated by the many new ideas and findings forthcoming in this closely related field.

ARNE FREDGA

University of Uppsala
Institute of Chemistry
Uppsala, Sweden
September 1971

Preface

The organic chemistry of selenium has undergone major expansion during the past two decades. Nevertheless, the topic is not covered in general organic textbooks and is mentioned only occasionally in works devoted to organo-sulfur chemistry. The parallel chemical and biological behavior of sulfur and selenium compounds has been frequently stressed; however, scrutiny of the literature reveals a fascinating individuality in the two series of chalcogens. There is a definite need, therefore, to discuss not only their similarities but also their differences.

The last significant review of the field, that made by Heinrich Rheinboldt (in *Houben-Weyl Methoden der Organischen Chemie*, Vol. IX, Stuttgart, 1955), gave excellent coverage to many aspects of synthetic selenium chemistry. It failed, however, to include such important classes as organic selenium-oxygen (R-O-Se type) and selenium-phosphorus compounds, as well as selenium-containing heterocycles, dyes, and amino acids. The physical chemistry of organoselenium compounds was also generally omitted.

We were of the opinion that, in view of the recent progress in the chemistry as well as in the biology of organically bound selenium, there was a need for a new and truly comprehensive monograph. When we initially contemplated undertaking such a project it soon became apparent that we would not do justice to topics in which we had limited personal research experience. Eventually, a large number of researchers and specialists in organoselenium chemistry and biology were enlisted to bring this book to fruition. As editors we are grateful to the contributors for their willingness to undertake the task of preparing critical surveys of the pertinent literature and for their under-standing when limitations of space resulted in the exercise of editorial prerogative.

The goal of this collaboration has been to provide a complete review of the chemistry and biology of organic selenium compounds. It is addressed to researchers who are interested in organochalcogen chemistry, to biochemists investigating the intriguing and complex metabolic fate of selenium in plants and animals, and to future investigators whose enthusiasm for a new and fertile field of research is yet to be awakened.

About one half of the volume is devoted to the organic chemistry of the various classes of compounds bearing a selenium functionality. The discussions of the place of selenium in a biological and biochemical context are divided according to the systems with which they interact, whereas the physicochemical and analytical aspects of selenium are categorized according to the appropriate methodology. Numerous areas of organoselenium chemistry and biology have been reviewed for the first time in this book. A definite effort has been made to cite and discuss all factual data pertinent to an understanding of the field and to compare, where possible, the properties of organic compounds of selenium with those of sulfur and, occasionally, of tellurium. Thus, emphasis has been placed on imparting a comprehension of the various aspects of the field. Tabulations of the vast number of organoselenium compounds could not be accommodated in the book without adding substantially to its cost.

We are grateful to many of our contributors who served in a dual capacity as authors and critics. In addition, we were fortunate to receive the generous cooperation of the following individuals who undertook the time-consuming task of reviewing the manuscripts which were in the area of their personal expertise: Monte Blau, N. P. Buu-Hoi, Robert B. Fox, Douglas V. Frost, Howard E. Ganther, James Z. Ginos, T. Scott Griffin, Ronald D. Grigsby, Erhard Gross, Kurt J. Irgolic, Erwin Klingsberg, Orville A. Levander, Mario Mammi, W. McFarlane, George W. A. Milne, J. Morel, S. N. Nigam, C. Paulmier, Peter J. Peterson, Marcel Renson, Frank Saeva, Klaus Schwarz, John S. Thayer, Joseph E. Tomaszewski, Philip W. West, Gayl H. Wiegand, Raymond F. X. Williams, and Kenneth J. Wynne.

The production of the book was aided by a generous financial grant from the Selenium-Tellurium Development Association. The support of this organization is gratefully acknowledged.

Special thanks go also to the Walter Reed Army Institute of Research and the Xerox Corporation for making their facilities available to us. In addition, appreciation is extended to Fatima N. Johnson, Ann A. Norton, Veronica A. Nelson, Lois Eason, William L. Callahan, and Robert Presswood for editorial assistance, and to Theodore P. Hoffman and other staff members of John Wiley and Sons for their efforts in bringing out this volume.

Finally, we express our gratitude to our wives, Dora and Helge, for their patience and encouragement during their "book widowhood."

DANIEL L. KLAYMAN
WOLFGANG H. H. GÜNTHER

Washington, D.C.
Webster, New York
March 1972

Contents

Abbreviations

The following abbreviations have been used:

c.d.	circular dichroism
DMF	N,N-dimethylformamide
DMSO	dimethyl sulfoxide
e.s.r.	electron spin resonance
g.l.c.	gas-liquid chromatography
i.p.	intraperitoneal
i.r.	infrared
n.m.r.	nuclear magnetic resonance
o.r.d.	optical rotatory dispersion
ppm	parts per million
THF	tetrahydrofuran
t.l.c.	thin-layer chromatography
u.v.	ultraviolet

All temperatures are given in degrees Celsius unless specified otherwise.

Organic selenium compounds: their chemistry and biology

I Nomenclature of organic selenium compounds

WOLFGANG H. H. GÜNTHER

Xerox Corporation
Chemistry Research Laboratory
Webster, New York

A. INTRODUCTION

A thorough knowledge of chemical nomenclature is essential to the organic chemist both to recall information from the literature and to communicate effectively the results of his own research. This chapter presents and discusses only the rules of nomenclature that pertain specifically to organic selenium moieties. A general knowledge with regard to the naming of organic compounds is assumed.

In addition to citing the practices followed by *Chemical Abstracts* (*C.A.*) and the definitive nomenclature rules of the International Union of Pure and Applied Chemistry (IUPAC), an attempt has been made to examine special problem areas, including the use of established and new trivial names.

B. THE NAMING AND INDEXING OF SELENIUM COMPOUNDS BY *CHEMICAL ABSTRACTS*

In the interest of reliable information retrieval it appears important to discuss the naming and indexing system of *Chemical Abstracts* for the selenium field. A summary of the principles being followed by *C.A.* is periodically provided at the beginning of *C.A.* Index volumes.[1]

Historically, *C.A.* has not imposed any nomenclature rules, but has followed existing usage wherever this could be reconciled with sound indexing principles. With the development of definitive rules these were incorporated, and excellent use has been made of cross references to assure that a particular topic or compound can be located with a minimum amount of work and little chance for error.

In agreement with IUPAC rules (see below), *C.A.* has named selenium compounds in analogy to and by extrapolation from sulfur compounds. The specific cases cited in the following are illustrative of current *C.A.* usage. No attempt has been made to extend the listing in anticipation of future practices.

Selenium Analogs of Alcohols and Phenols.   Species RSeH are named by adding the suffix -selenol to the name of the parent compound. Example:

CH_3CH_2SeH	Ethaneselenol
C_6H_5SeH	Benzeneselenol

Selenocyanates.   Compounds RSeCN are indexed in the same manner as are thiocyanates under "selenocyanic acid" and as selenocyanato derivatives:

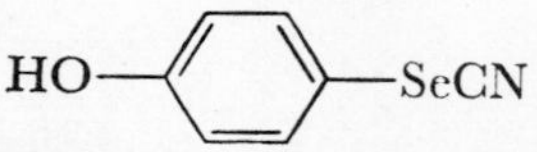

HO—⟨benzene ring⟩—SeCN   Selenocyanic acid, *p*-hydroxyphenyl ester
                                                           Phenol, *p*-selenocyanato-

Isoselenocyanates, R—NCSe, are treated analogously to selenocyanates.

Selenium-Oxygen Acids.   There are three classes of selenium-oxygen acids: RSeOH, selenenic acid; $RSeO_2H$, seleninic acid; and $RSeO_3H$, selenonic acid. They and their simple derivatives are named according to the organic group attached to the selenium, using modifying terms as in derivatives of carboxylic acids, for example:

$C_6H_5SeO_3H$	Benzeneselenonic acid
$2\text{-}C_{10}H_7SeO_2CH_3$	2-Naphthaleneseleninic acid, methyl ester
C_6H_5SeBr	Benzeneselenenyl bromide
$C_6H_5SeONH_2$	Benzeneseleninamide

Selenides and Diselenides.   Simple, unsubstituted species of the general formula R_2Se are indexed under names such as:

$(CH_3)_2Se$	Methyl selenide
$(C_6H_5Se—)_2$	Phenyl diselenide

Unsymmetrical or substituted compounds either are entered under "selenide" or "diselenide," as in:

$C_2H_5SeCH_3$ Selenide, ethyl methyl
CF_3SeCF_3 Selenide, bis(trifluoromethyl)

or an alkylselenyl (arylselenyl) prefix is applied to indicate a derivative of a selenol:

$C_6H_5SeCH_2COOH$ Acetic acid, (phenylselenyl)-

Tri-, tetra-, and polyselenides are treated analogously.

Selenium Oxides. Compounds R_2SeO are named selenoxides. When symmetrical and unsubstituted, they are entered under their own headings, as in

$(CH_3)_2SeO$ Methyl selenoxide

Unsymmetrical or substituted species are given an inverted entry, as

$\left(O_2N-\!\!\left\langle\underset{}{\bigcirc}\right\rangle-\!\!\right)_2 SeO$ Selenoxide, bis(*p*-nitrophenyl)

$CH_3-\!\!\underset{\underset{O}{\|}}{Se}-\!\!\left\langle\underset{}{\bigcirc}\right\rangle$ Selenoxide, methyl phenyl

Correspondingly, compounds R_2SeO_2 are named selenones and have entries similar to those of the selenoxides.

Selenonium Compounds. Compounds of the general structure $R_3Se^+X^-$ are indexed with uninverted names as selenonium compounds, such as

$(C_6H_5)_3Se^+Cl^-$ triphenylselenonium chloride.

Compounds in which the significant element selenium is a ring member are given appropriate cyclic names.

Selenium Halides and Related Species. Compounds of the formulas. $RSeX_3$ and R_2SeX_2 were formerly indexed at "selenium compounds." They are now indexed as inverted entries under "selenium."

EXAMPLES

$(C_2H_5)_2SeBr_2$ Selenium, dibromodiethyl-
CH_3SeBr_3 Selenium, tribromomethyl-
$(CH_3)_2Se(OH)_2$ Selenium, dihydroxydimethyl-

$CH_3CONH-\!\!\left\langle\underset{}{\bigcirc}\right\rangle-\!\!SeBr_3$ Selenium, (*p*-acetamidophenyl) tribromo-

However, dihalides of unsymmetrical selenyl derivatives, RSeR′, are named in *C.A.* by adding "*Se,Se*-dihalide" to the modification of the index entry, as in:

$C_6H_5SeBr_2CH_2COOH$ Acetic acid, (phenylselenyl)-, *Se, Se*-dibromide

Note further the alkyl(aryl)selenenyl halide designation for species RSeX, which are now considered to be derivatives of alkyl(aryl)selenenic acids.

Selenium Analogs of Carbonyl Compounds. Species R–C(=Se)R′ are indexed in *C.A.* as ketones, with the selenium indicated in the inverted part of the name by "seleno-." General studies are indexed under "ketones" with modifications beginning "seleno-."

EXAMPLES

$CH_3C(=Se)CH_3$ 2-Propanone, seleno-
$C_6H_5C(=Se)CH_3$ Acetophenone, seleno-
$C_6H_{11}C(=Se)CH_3$ Ketone, cyclohexyl methyl seleno-

Selenium analogs of aldehydes do not receive a special listing in the *C.A.* nomenclature section.[1] Entries analogous to those of thioaldehydes using the ending -selenal present an alternative to the ketone rules above.

Selenium Analogs of Carboxylic Acids. Seleno derivatives of carboxylic acids have been named by means of the prefix seleno-(Se replacing O). No characteristic prefix names have been used to distinguish between oxo and hydroxy substitution. Thus, the name acetic acid, seleno- will represent both $CH_3C(=Se)OH$ and $CH_3C(=O)SeH$. Derivatives containing one of the tautomers in stabilized form are indexed under a more specific name, such as

$CH_3C(=O)SeCH_3$ Acetic acid, seleno, *Se*-methyl ester

Substitution of both oxygen atoms by selenium leads to

$C_6H_5C(=Se)SeH$ Benzoic acid, diseleno-

The above is used in cases in which the oxygen acid has a trivial name. Otherwise, the ending -selenoic acid applies, as in

$CH_3(CH_2)_6C(=Se)SeCH_3$ Octanediselenoic acid, *Se*-methyl ester.

Amides and Ureas. Replacement of a key oxygen atom by selenium is generally indicated through use of the prefix seleno-. Thus, carboxamides and ureas are entered under the name of the oxygen derivative, as in:

$C_6H_5—C(=Se)NH_2$ Benzamide, seleno-
$CH_3NH—C(=Se)NH_2$ Urea, 1-methyl-2-seleno-

C. IUPAC RULES OF NOMENCLATURE

In recognition of the similarities between the elements and of the fact that many selenium derivatives were prepared specifically as analogs of known sulfur compounds, definitive rules for characteristic groups containing selenium have been formulated by the Commission on Nomenclature of the International Union of Pure and Applied Chemistry (IUPAC)[2] as follows:*

RULE C-701-1: Organic compounds of selenium are named as far as possible analogously to the corresponding sulfur compounds. The prefixes and suffixes used, with examples for their application, are shown in Table I-1. Structures not listed in the table are named by placing the syllables seleno before the name of the corresponding oxygen compound.

EXAMPLES

1	Selenourea	$Se=C(NH_2)_2$
2	Ethaneselenol	CH_3-CH_2-SeH
3	3-(Hydroseleno)propionic acid	$HSe-CH_2-CH_2-COOH$
4	Diethyl selenide	$CH_3-CH_2-Se-CH_2-CH_3$
5	(Ethylseleno)acetic acid	$CH_3-CH_2-Se-CH_2-COOH$

6 Selenacyclopentane⎫
7 Tetrahydroselenole⎭

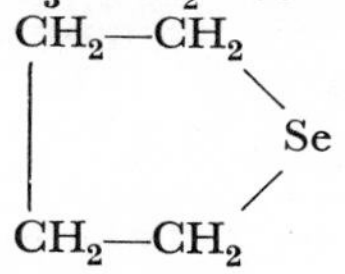

8 1,4-Episeleno-1,4-dihydronaphthalene

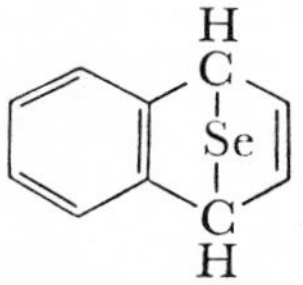

9 Hexaneselenal⎫
10 1-Pentanecarboselenaldehyde⎭

$$CH_3-[CH_2]_4-CH=Se$$

11 *p*-Selenoformylbenzoic acid

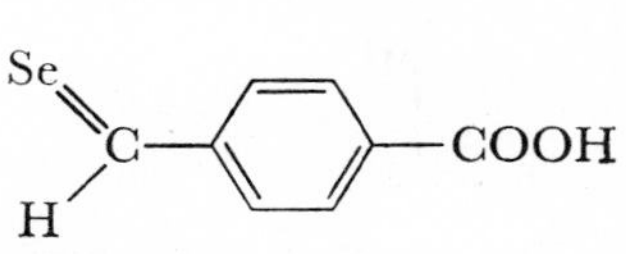

12 3-Selenoxohexane⎫
13 Hexane-3-selone⎭

$$CH_3-CH_2-CH_2-C-CH_2-CH_3$$
$$\|$$
$$Se$$

14 Selenobenzoic acid

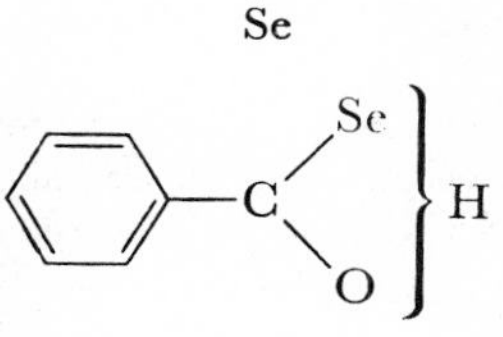

* Excerpts from IUPAC nomenclature rules are reproduced with permission.

Table I-1　Prefixes and suffixes for use with compounds containing selenium

GROUP	PREFIX	SUFFIX OR FUNCTIONAL-CLASS NAME	SEE EXAMPLES ABOVE
—SeH	hydroseleno-	-selenol	2, 3
—Se—	seleno-	selenide	4, 5
—Se—	selena-		6
—Se—⎤ bridge ⎦	episeleno-		8
—Se—Se—⎤ bridge　⎦	epidiseleno-		
—C⟨Se|H (selenal structure)		-selenal	9
—C(=Se)⟨H	selenoformyl-	-carboselenaldehyde	10, 11
＼C=Se ／	selenoxo-	-selone	12, 13
—C⟨Se／⟩H⟨O／		-carboselenoic acid	14
—C⟨Se／⟩H⟨S／		-carboselenothioic acid	15
＼SeO ／	seleninyl-	selenoxide	16, 17
＼SeO_2 ／	selenonyl-	selenone	
—SeO_3H	selenono-	-selenonic acid	
—SeO_2H	selenino-	-seleninic acid	
—SeOH	seleneno-	-selenenic acid	

15   Cyclohexanecarboselenothioic acid

16   Ethyl phenyl selenoxide
17   (Ethylseleninyl)benzene

$$C_6H_5\text{---}Se\text{---}CH_2\text{---}CH_3$$
$$\overset{\|}{O}$$

Heterocyclic compounds containing selenium are named by following the definitive IUPAC nomenclature guidelines, Sections A and B.[3,4]

RULE B-1-1: Monocyclic compounds containing one or more hetero atoms in a three- to ten-membered ring are named by combining the appropriate prefix or prefixes— (here: selena-)—with a stem from Table I-2, eliding "a" where necessary. The state of hydrogenation is indicated either in the stem, as shown in Table I-2, or by the prefixes dihydro-, tetrahydro-, *etc.*

EXAMPLES

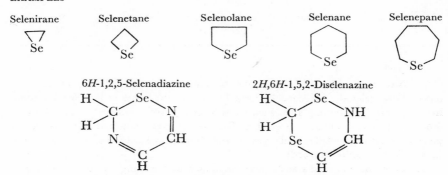

Selenirane   Selenetane   Selenolane   Selenane   Selenepane

6H-1,2,5-Selenadiazine   2H,6H-1,5,2-Diselenazine

According to Section B-2 of the IUPAC rules, it is permissible to form trivial and semitrivial names of selenium heterocycles in analogy to the trivial names of the corresponding sulfur compounds, wherever the latter would be applicable, including their use in fusion names.

EXAMPLES

Selenophene   Selenanthrene   Phenoselenazine

Where a heteroatom replaces a carbon atom, the "a" nomenclature is applied and the following provisions pertain.

Table I-2   Stems indicating ring size

NO. OF MEMBERS IN RING	RINGS CONTAINING NITROGEN		RINGS CONTAINING NO NITROGEN	
	Unsaturation	Saturation	Unsaturation	Saturation
3	-irine	-iridine	-irene	-irane
4	-ete	-etidine	-ete	-etane
5	-ole	-olidine	-ole	-olane
6	-ine	Use prefix perhydro-	-ine	-ane
7	-epine	Use prefix perhydro-	-epine	-epane
8	-ocine	Use prefix perhydro-	-ocin	-ocane
9	-onine	Use prefix perhydro-	-onin	-onane
10	-ecine	Use prefix perhydro-	-ecin	-ecane

RULE B-4

4.1. Names of monocyclic hetero compounds may be formed by prefixing "a" terms (here: selena-), preceded by their locants, to the name of the corresponding hydrocarbon. Numbering is assigned so as to give lowest numbers in the following order: first to hetero atoms (as given in Rule B-1.1), next to hetero atoms as a complete set, next to multiple bonds, next to substituents as a complete set, and then to substituents in alphabetic order.

4.2. Fused heterocyclic systems may be named by prefixing "a" terms, preceded by their locants, to the name of the corresponding hydrocarbon. The numbering of the corresponding hydrocarbon is retained, irrespective of the position of the hetero atoms; where there is a choice, low numbers are assigned in the following order: first to hetero atoms as a complete set, next to hetero atoms in order of Rule A-11.3, and then to multiple bonds in the heterocyclic compound according to the principles of Rule A-11.3. These principles are applied in one of two ways, as follows:

(*a*) When the corresponding hydrocarbon does not contain the maximum number of noncumulative double bonds and can be named without the use of hydro-prefixes, as for cyclohexane and indan, then the hydrocarbon is named in that state of hydrogenation.

(*b*) When the two conditions of paragraph (*a*) are not fulfilled, positions in the skeleton of the corresponding hydrocarbon that are occupied by hetero atoms are denoted by "a" prefixes, and the parent heterocyclic compound is considered to be that which contains the maximum number of conjugated or isolated double bonds, but the corresponding hydrocarbon is named in the form in which it contains the maximum number of noncumulative double bonds. Hydrogen additional to that present in the parent heterocyclic compound is named by hydro- prefixes and/or as *H* in front of the "a" terms.

EXAMPLES

Selenacyclopentane   8-Selenapurine

1,4-Diselenanaphthalene   2,3-Diselena-1,5-diazaindane

RULE B-6: According to the "a" nomenclature, heterocyclic compounds containing cationic hetero atoms are named in conformity with the preceding rules by replacing "oxa-," "thia-," "aza-," etc., by "oxonia-," "thionia-," "azonia-," etc., the anion being designated in the usual way.

EXAMPLES AS APPLIED TO SELENIUM CATIONS

1-Selenoniaanthracene chloride

1-Selenoniabicyclo[2.2.1]heptane chloride

$$\begin{array}{c} CH_2 - Se^+ - CH_2 \\ |6 \quad 1| \quad 2| \\ {}^7CH_2 \quad \\ |5 \quad 4| \quad 3| \\ CH_2 - CH - CH_2 \end{array} \quad Cl^-$$

D. TRIVIAL NAMES OF SELENIUM COMPOUNDS AND SPECIAL PROBLEM AREAS

Historically, trivial names have been applied to all chemical entities for which exact structural information was not available, and this process continues today mainly in the areas of biological and macromolecular chemistry. In addition, authors are frequently faced with the dilemma that adherence to the precise chemical nomenclature may result in names that are extremely complex, unwieldy, and difficult to remember. In the interest of effective communication, the retention of some trivial names and the coining of new shorthand terms cannot be avoided. It is estimated that *C.A.* continues to use more than 12,000 trivial names. In this context the IUPAC rules (Section B-2) permit the formation of new terms for selenium heterocycles on the basis of existing trivial names for sulfur heterocycles.

The latter style of emphasizing the relationship between sulfur and selenium is certainly desirable, and the prefix seleno- has been used widely to indicate selenium analogs of sulfur compounds. Such names as selenocysteamine, selenocystine, selenotaurine, selenoglutathione, selenocoenzyme A, selenooxytocin, selenoxanthogen, and selenolipoic acid are common in the literature.

A distinct conflict arises, however, when Section C of the IUPAC rules is considered. There the prefix seleno- is uniquely defined as denoting the replacement of oxygen by selenium. Since any definitive system of nomenclature would break down if more than one interpretation per term were possible, all "seleno-" names generated from common names of organic sulfur compounds are trivial names and should be so designated upon first usage. An author using the seleno- prefix to indicate the replacement of sulfur may also find it difficult to discover his compound in *C.A.* indices since systematically correct terms are now generally entered without cross-indexing to new trivial names.

For use within a systematic frame of nomenclature Sections A and B of the IUPAC Rules appear to apply well to selenium and tellurium heterocycles, and few problems are anticipated in their application. By contrast, Section C is given in less detail and its interpretation may, at times, be difficult.

In attempting to name selenium compounds analogously to the corresponding sulfur compounds, the single root selen- had long to suffice as a potential replacement for thi-, mercapt-, sulf-, and sult-. A *thione* denotes $C=S$, a *sulfone* denotes the $-SO_2-$ group, and a *sultone* is $-SO_2-O-$ as part of a ring. Attached to the root selen-, the ending -one is exhausted in the single $-SeO_2-$.

Thus, the $C=Se$ moiety had to be named by the somewhat cumbersome prefix selenoxo- or selenocarbonyl-. The recent[2] addition of the -selone ending to denote $C=Se$ and the adoption of the root sel- by the IUPAC Inorganic Nomenclature Commission to distinguish, for example, diselane, H_2Se_2, from diselenane, the analog of dioxane, have brought some welcome relief. It may be expected that numerous extensions can be built around the root sel-, such as application to analogs of sultones, sultams, and others.

A multiplicity of choices exists in the naming of organic selenium-halogen compounds. By analogy to IUPAC Rule C-6.2 dealing with sulfur compounds, the names phenylselenium chloride (C_6H_5SeCl), phenylselenium trichloride ($C_6H_5SeCl_3$), and diphenylselenium dichloride [$(C_6H_5)_2SeCl_2$] should apply. However, there is no ambiguity or overlap if the relationship to other organoselenium compounds or selenium oxidation states is indicated by such names as selenenyl halide, RSeX, or selenide dihalide, R_2SeX_2. Furthermore, a distinct acid halide like $RSe(O)Cl$ should be called alkyl-(aryl)seleninyl chloride since the selenin- portion already indicates the oxidation state.

The matter becomes more confusing if rules for organometallic chemistry (IUPAC Section D; publication expected late in 1972) are applied. The foregoing examples then might alternatively be named as coordination compounds, that is, chloro(phenyl)selenium, trichloro(phenyl)selenium, and dichloro(diphenyl)selenium, giving ligands in alphabetical order and introducing parentheses to avoid ambiguity. However, this writer feels that the latter style places undue emphasis on a "central" selenium atom, which is clearly contraindicated in numerous complex organic species where selenium does not dominate the structure.

Compound classes based on selenium-phosphorus moieties present another problem area. A discussion of some of the complexities in this field and some relevant references can be found in Chapter X.

An area for which no general rules appear to have been developed concerns species containing intercalated linear sequences of sulfur and selenium atoms. A number of compounds in this class are now known (see Chapter IV) with these general topological characteristics: (1) the chalcogen sequence contains carbon substituents at the terminal atoms; (2) each chain link may occur in a divalent or multivalent form, the latter usually represented by oxygen substitution; (3) to date branched chains have been postulated but have not been proved to exist. A variety of names exists in the literature, and some of these are cited in Chapter IV.

To this author, it appears that a general class designation derived from a -thiaselane root may be the least ambiguous alternative. Thus, the species $C_6H_5SSeCH_3$ may be named 1-phenyl-2-methylthiaselane with sulfur having precedence over selenium according to IUPAC Rule B-1. The related compound $CH_3SSeC_6H_5$ would then be 1-methyl-2-phenylthiaselane. Structure **90** (Chapter IV) would be 1,5-diphenyl-1,5-dithia-2,3,4-triselane-1,1,5,5-tetroxide:

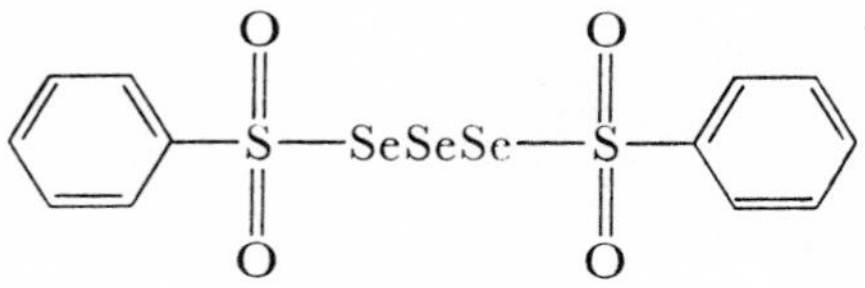

90

Secondary substituents on the organic moieties may be designated by primed numbers and placed in parentheses.

The foregoing proposal appears to meet this author's criteria of clarity, ease of use, and lack of ambiguity. However, while these suggestions appear practical for the sulfur-selenium case, the scheme cannot readily be applied to chains of random heteroatoms, singly and in combination. A new term would have to be coined to provide a total heteroatom count with "a" terms then giving the number and location of each element. Since future discoveries and unexpected structural features cannot readily be anticipated, the fashioning of a self-consistent chemical nomenclature remains an evolutionary process, with all workers in this field contributing to the framework.

REFERENCES

1. *Chem. Abstr.*, **56,** Subject Index A–H, p. in ff. January–June 1962; reprints available.
2. "IUPAC—Nomenclature of Organic Chemistry, Section C (Characteristic Groups Containing Carbon, Hydrogen, Oxygen, Nitrogen, Halogen, Sulfur, Selenium and/or Tellurium)," Butterworths, London, 1965; 3rd ed., 1971.
3. "IUPAC—Nomenclature of Organic Chemistry, Section A (Hydrocarbons) and Section B (Fundamental Heterocyclic Systems)," Butterworths, London, 1958; 3rd ed., 1971.
4. Extracts of the IUPAC Rules (Sections A and B) are reprinted in "Handbook of Chemistry and Physics," Section C, 52nd ed., The Chemical Rubber Co., Cleveland, Ohio, 1971.

II Elemental selenium: structure and properties

RICHARD G. CRYSTAL

Xerox Corporation
Chemistry Research Laboratory
Webster, New York

A. ATOMIC STRUCTURE

1. Historical Notes

The formal discovery of selenium dates back to the work of two Swedish chemists, J. J. Berzelius and J. G. Gahn,[1] who noticed a curious residual slime during the oxidation of sulfur dioxide from copper pyrites. As tellurium

Table II-1   Portion of the periodic table showing the location of selenium

PERIOD	ELECTRON SHELLS	GROUP			
		IVB	VB	VIB	VIIB
		C	N	O	F
		6	7	8	9
2	K	2	2	2	2
	L	4	5	6	7
		Si	P	S	Cl
		14	15	16	17
3	K	2	2	2	2
	L	8	8	8	8
	M	4	5	6	7
		Ge	As	Se	Br
		32	33	34	35
4	K	2	2	2	2
	L	8	8	8	8
	M	18	18	18	18
	N	4	5	6	7
		Sn	Sb	Te	I
		50	51	52	53
	K	2	2	2	2
	L	8	8	8	8
5	M	18	18	18	18
	N	18	18	18	18
	O	4	5	6	7
		Pb	Bi	Po	At
		82	83	84	85
	K	2	2	2	2
	L	8	8	8	8
6	M	18	18	18	18
	N	32	32	32	32
	O	18	18	18	18
	P	4	5	6	7

had been discovered some 35 years before, the investigators thought that they had found a new route to tellurium refinement. Further work indicated, however, that this was indeed a new element that was similar to tellurium. The actual discovery may, however, date to the fourteenth century and the work of Arnold of Villanova, who recorded in his "Rosarius Philoso-phorum" that a red sulfur deposit (*sulfur rubeum*) had formed on oven chamber walls after vapor from crude sulfur had been condensed.

2. Classification in the Periodic Table and Atomic Structure

Selenium is located between sulfur and tellurium in Group VIb in the series oxygen, sulfur, selenium, tellurium, and polonium. Its atomic number is 34, which ranks it in period 4 of the periodic table, between arsenic and bromine (*cf.* Table II-1). Selenium has been classified both as a metal and as a nonmetal. It is interesting that it lies between the nonmetals oxygen and sulfur and the metals tellurium and polonium by group and between arsenic, a metal, and bromine, a nonmetal, by period. Selenium has an atomic weight of 78.96. Six stable isotopes have been isolated[2] and are summarized in Table II-2 along with their relative abundance in nature. In addition,

Table II-2 Natural occurrence of stable isotopes

Mass:	74	76	77	78	80	82
Ratio of natural occurrence:	1	10	9	27	57	11

several unstable isotopes, including 70, 72, 73, 75, 79, 81, 83, 85, 86, and 87, have been identified. The outer electron shell of selenium is $4s^24p^4$, and its inner shells (2, 8, 18) are completely filled. It is markedly similar to sulfur in its chemistry, its primary oxidation states being -2, 0, $+2$, $+4$, and $+6$. Reported atomic radius values vary in the literature from 1.17 Å[3] to 1.40 Å[4].

3. Occurrence

In order of abundance, selenium ranks seventieth among the elements and comprises approximately $10^{-5}\%$ of the earth's crust.[5] The element is widely distributed in igneous-type rocks and even sea water; however, the principal sources for commercial applications are copper-bearing ores and sulfur deposits. The production of selenium, unlike that of most common elements, is always a by-product of the mining of other elements, primarily copper, zinc, nickel, and silver. It is also recovered from sulfur deposits and uranium purification. Selenium plays a major role in some plant-life cycles and can be found in concentrations as high as 1.5%.[6] Although recovery from seleniferous plants has been contemplated, no commercial process has been developed. Selenium is known to form natural compounds with at least 16 other elements and is a major component in at least 39 principal mineral species.[7]

High purity is essential for utility in the electrical industry. As impurities play a crucial role in electrical performance, manufacturers add controlled amounts of desired impurities and thus demand initial selenium levels exceeding 99.99%. On an industrial scale, purification is achieved by treating the crude selenium with boiling sodium sulfite solution, during which time the selenium is dissolved by the following reversible reaction:

$$\mathrm{Na_2SO_3 + Se \rightleftharpoons Na_2SeSO_3}$$

After approximately 2 hours, the hot solution is filtered to remove insoluble impurities and then cooled to deposit the now purified selenium, which is separated, allowing reuse of the sodium sulfite.[8] A final distillation of the element then completes the purification.

B. ALLOTROPIC FORMS

1. Ring-Chain Equilibrium

Like sulfur, selenium exhibits several allotropic forms, all of which are stable at room temperature. On a molecular level, selenium can form Se, Se_2, Se_4, Se_6, Se_8, and Se_n aggregates,[2] but only Se_8 rings and Se_n polymeric chains are known to exist at room temperature.[9] Eisenberg and Tobolsky[9] have studied the Se_8 ring-Se_n chain equilibrium in liquid selenium and have developed a theoretical model for equilibrium polymerization of the eight-membered rings to chains. The theoretical floor temperature (*i.e.*, the lowest temperature at which Se chains exist in the melt) was calculated to be 83°. At higher temperatures, the equilibrium was calculated to shift strongly toward chains with an approximate average chain length of 10,000 Se atoms at the melting point. Briegleb,[10] in an attempt to determine the relative ring/chain ratios, extracted selenium glasses quenched from various temperatures with CS_2, thus separating the insoluble Se_n chains from the soluble Se_8 rings. His data, summarized in Table II-3, may be a good approximation of relative ring concentrations at the various selenium liquid temperatures indicated.

Because the polymeric form is insoluble in any solvent system, thus making solution characterization impossible, little is known about the true average molecular weights or chain lengths of polymeric selenium. Unlike organic polymers, selenium can readily change molecular weight by simple

Table II-3 CS$_2$ *extraction data*[a]

TEMPERATURE (°K)	% CS$_2$ SOLUBLE SELENIUM (Se$_8$ RINGS)
393	55
493	41
573	30
673	23
773	18
873	15
923	14

[a] Data from Briegleb.[10]

chain scission, radical formation, and radical recombination. The constant presence of radicals at either chain end hinders the stabilization of a particular chain length. Shirai[11] *et al.* monitored the viscosity of selenium melts to which small amounts of chlorine had been added. As the chlorine concentration was increased, more of the free selenium chain ends were capped with halogen until a concentration was reached at which any additional chlorine resulted in chain scission. As shorter polymer chain lengths result in lower melt viscosities, these workers were able to determine a point where plots of viscosity *vs.* chlorine content took an abrupt change and thus where the chain scission process had begun. Through stoichiometry, they were able to deduce that the average Se$_n$ chain at 230° contained 7.2 × 10^3 Se atoms, a number in good agreement with the Eisenberg and Tobolsky[9] calculation previously mentioned.

2. Amorphous Selenium (α-Selenium)

In the amorphous state, selenium behavior is very similar to that of organic glasses. Above approximately 230°, selenium is a free-flowing liquid. As the liquid is cooled, however, its viscosity rapidly increases to about 80°, where it then falls—a phenomenon very similar to that observed for amorphous sulfur. This can be explained partially by the fact that the ring-chain equilibrium favors rings at the lower temperatures, the ring form exhibiting lower viscosity. At the higher end of this temperature range, where higher Se$_n$ concentration is favored, selenium viscosity behavior follows typical polymeric behavior, where viscosity increases logarithmically with decreasing temperature. Below 31°,[12] however, amorphous selenium undergoes a transition to a hard, brittle glass. Thus selenium is glassy below 31°, vitreous between 31° and 230°, and liquid above 230°.

Colloidal selenium, a particulate form of amorphous selenium, can be prepared by chemically reducing cold aqueous solutions of selenious acid with hydrazine, sulfur dioxide, hydrogen selenide, and so on. The red particles give the suspension a reddish yellow tinge at low concentrations and a red color at higher concentrations. When heated above 60°, colloidal amorphous selenium crystallizes into colloidal hexagonal selenium.[13]

3. Crystalline Selenium

Selenium is known to occur in three crystal forms—α-monoclinic, β-monoclinic, and hexagonal (trigonal) crystals. The monoclinic forms, composed wholly of puckered Se_8 rings, are sometimes referred to as red (α-monoclinic) and dark red (β-monoclinic) selenium. The hexagonal form, comprised of helical Se_n chains, is usually referred to as β-selenium or gray selenium. Table II-4 summarizes the principal lattice parameters obtained from the various crystal forms.

Table II-4   Lattice parameters for crystalline selenium[a]

| | UNIT CELL PARAMETER | | | | | | | |
CRYSTALLINE MODIFICATION	a_0 (Å)	b_0 (Å)	c_0 (Å)	Angle	Volume (Å^3)	Z	Density (g/cm^3)	SPACE GROUP
α-Monoclinic	9.05	9.05	11.61	90°46′	982.85	32	4.46	$C_{2h}^5(P2_1/n)$
β-Monoclinic	12.85	8.07	9.31	93° 8′	963.99	32	4.50	$C_{2h}^5(P2_1/a)$
Hexagonal	4.35		4.95		93.86	3	4.82	$D_3^4(C3_121)$

[a] Data from Abdullayev et al.[14]

α-Monoclinic selenium can be prepared by evaporating saturated CS_2 solutions to give, usually, mixtures of both α- and β-forms. The α-crystals are red and take on flat hexagonal and sometimes polygonal shapes. The β-form crystals are needle-like or prismatic and deep red. Separation of the two forms is usually achieved manually.[15]

Hexagonal selenium has been referred to as β-selenium, gray selenium, black selenium, metallic selenium, and trigonal selenium. Depending on the method of preparation, it can take on any one of several appearances. The crystal form is comprised entirely of spiral Se_n chains oriented along the c crystallographic axis. The chains are arranged in a 3/1 helix such that each Se atom is positioned directly above and below its third nearest neighbors along the chain. The hexagonal form is the most stable; when heated above 110°, both monoclinic forms transform to hexagonal selenium. The hexagonal form is also prepared by crystallization of amorphous selenium in the temperature range 70–210°.[16,17] Polycrystalline aggregates are usually formed by this method. Recently, large single hexagonal crystals have been grown under high-pressure crystallization[18] and by Czochralski crystal-pulling techniques.[19]

The phase diagram for selenium shown in Fig. II-1 summarizes the forms of the element listed above.

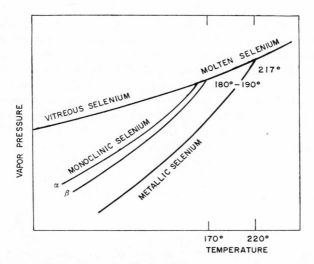

Fig. II-1   Pressure-temperature diagram indicating phases associated with selenium.

4. Crystallization of Amorphous Selenium

Selenium readily crystallizes from the molten amorphous state into the most stable polymeric hexagonal form. If it is assumed that the unordered state is composed of a mixture of Se_8 rings and Se_n chains, then any crystallization act resulting in the most stable polymeric crystalline form must be associated with opening of Se_8 rings, followed by polymerization of Se_n chains. Krebs and Morsch[20] theorized that during the crystallization act Se_8 rings open, forming Se_8 chains with free-radical ends, which then polymerize during crystallization. Selenium crystallization rates are markedly accelerated by the addition of impurities such as alkali metals, thallium, and halogens.[21,22] Such phenomena can easily be rationalized by assuming that these elements react with Se_8 to form both metallic ions and selenium ions capable of inducing further Se_8 ring openings. This mechanism should lower energy barriers to crystallization.

Because of its polymeric nature, trigonal selenium crystallization is markedly similar to more classical organic polymer crystallization. Unlike organic polymers, however, selenium polymer chains can readily change molecular weight by chain scission, resulting in radical formation and radical recombination. Selenium readily crystallizes into morphologies similar to those of organic polymers, which are characterized by long, thin lamellae or fibrous entities growing outward in a radial direction from a central nucleating source into a characteristic spherulitic structure.

Amorphous selenium readily crystallizes in the 70–200° temperature range. Kinetic measurements[17,23] indicate that maximum crystallization rates are achieved in the vicinity of 140°. Above that temperature, crystallization is governed primarily by thermodynamics; that is, greater supercooling below the 219.2° equilibrium melting point[16] leads to a faster crystallization rate. Below 140°, crystallization is apparently diffusion controlled so that increases in melt viscosity with lower temperatures inhibit motion of chains to the crystal growth face, resulting in decreases in crystallization rates with decreasing temperature. Kinetic data suggest that below 100° higher concentrations of Se_8 rings in amorphous selenium function as impurities and contribute to lower crystallization rates.[17]

C. PHYSICAL PROPERTIES

1. Vapor Properties

Selenium vapors are yellow in color and are paramagnetic.[24] Vapor in equilibrium with the liquid is composed of primarily paramagnetic associations of Se_2 and Se_6 having partial pressures of 310 and 450 mm Hg, respectively.[8] Small amounts of Se_4 and Se_8 moieties have also been detected in the vapor. Monatomic selenium exists at higher temperatures. The latent heat of vaporization of liquid selenium at 684° is 211.5 cal/g. The atomic spacing between atoms in Se_2 has been reported to be 2.15–2.19 Å, and the dissociation energy is 62.5 kcal/mole.[25] The equilibrium between Se_2 and Se_6 and the liquid Se_8-Se_n equilibrium are temperature dependent.[26] The concentration of Se_2 is believed to increase with increasing temperature and has been estimated to be 12 % at 185° and 25 % at 684°. The temperature variation in vapor molecular weight has been approximated.[14] The vapor pressures over solid selenium[28,29] and over liquid selenium[27,30,31] have been reported.

2. Thermal Properties

As would be expected, values of selenium heat capacities, densities, and coefficients of expansion depend on both allotropic form and temperature

Table II-5   Average values for heat capacity (C_p), *density* (ρ), *and thermal expansion coefficients* (α) *of selenium allotropic forms*

References are in parentheses.

ALLOTROPIC FORM	C_p (cal/g °C)	ρ (g/cm³)	$\alpha \times 10^5$
Amorphous (below T_g)	0.095 (32,35)	4.28 (13)	3.24 (13)
Amorphous (above T_g)	0.114 (32)	4.06 (36)	3.79 (13)
Trigonal	0.084 (32,34)	4.82 (14)	$\begin{cases} 7.4\text{-} \perp c\text{-axis (38)} \\ 1.7\text{-} \parallel c\text{-axis (38)} \end{cases}$
Monoclinic	0.082 (33)	4.50 (14)	
Liquid	0.118 (33)	3.95 (36,37)	

of measurement. Table II-5 summarizes average values from several sources. From a statistical-mechanics point of view, trigonal selenium heat capacity follows the Debye theory for monatomic crystals at low temperature, the calculated Debye temperature being 52°K. Thus, low-temperature heat capacity varies as the cube of the temperature and is dependent only on a single vibrational mode.

Crystal[16] has determined that the equilibrium melting point of trigonal selenium is 219.2°. The α-monoclinic form melts at 170°, while the β-monoclinic melts over a broad range, 144–180°.[8]

3.   Self-Diffusion and Diffusion of Impurities

The radioactive ^{75}Se isotope has been used in the measurement of selenium diffusion coefficients. Abdullayev and Bashaliev[39] have reported self-diffusion coefficients for amorphous and crystalline (trigonal) selenium and their temperature dependences. Additions of even trace amounts of impurities result in sharp changes. The diffusion of various impurities has received considerable attention, undoubtedly because of interest in electrical properties.[40–42] It is interesting to note that sulfur, which undergoes similar allotropic transformations, renders the highest coefficient, whereas tellurium renders coefficients similar to those of other elements. It is believed that S atoms can diffuse through and replace selenium in chains and rings, whereas the metallic elements can only move through interstitial points within the crystal lattice.[8,40,42]

4.   Mechanical Properties

The mechanical properties of selenium are very similar to those of visco-elastic polymeric materials. Because of its dynamic molecular weight,

Table II-6 Properties of selenium (References are in parentheses)

(Atomic number: 34; atomic weight: 78.96; electron configuration: $1s^2\ 2s^2\ 2p^6\ 3s^2\ 3p^6\ 3d^{10}\ 4s^2\ 4p^4$)

PROPERTY	LIQUID	AMORPHOUS	TRIGONAL	α-MONOCLINIC
Density (g/cm³)	3.99 at 220°C	4.28 (13)	4.80 (14)	4.50 (14)
Melting point (°C)		$T_g = 31$ (43)	219.2 (16)	170 (8)
Boiling point (°C)	684			
Thermal conductivity (cal/cm² °K)	1.72×10^{-3} (24,25)	1.72×10^{-3}	$\sim 2 \times 10^{-2}$	
Heat capacity (cal/g °C)	0.114 (32)	0.095 (32)	0.084 (32)	
Thermal expansion coefficient	3.79×10^{-5} (13)	3.24×10^{-5} (13)	7.5×10^{-5} (see text)	
Self-diffusion coefficient at 35°K (cm²/sec)		7.7×10^{-12}	3.8×10^{-12} (39)	
Self-diffusion coefficient at 50°K (cm²/sec)		$\sim 900 \times 10^{-12}$	6.3×10^{-12} (39)	

Property				
Absorption edge (Å) (at $\alpha = 10^5$ cm^{-1})	~6200 at 220°C	5400	~6200	~5000
Optical band gap (eV)	~2.0 at 220°C	2.3	~2.0	~2.5
Photoresponse edge (Å)		~4600	~6200	
Dielectric constant	6.1	6.3		
Reflectivity (short λ)		22%		
Resistivity at 20°C (Ω cm)	10^{15}–10^{16} extrap. to 20°C	10^{13}–10^{16}	10^4–10^6–10^{12}	
Thermal band gap (eV)	2.2–2.3		1.2–1.6	
Majority carrier	holes		holes	
Carrier drift mobility (cm²/V sec)				
Electrons		5×10^{-3} $\epsilon = 0.25$–0.285 eV		2 $\epsilon = 0.25$ eV 1.3 at 105°C
Holes		1.4×10^{-1} $\epsilon = 0.14$ eV		$\epsilon = 0.3$ eV
Carrier range (cm²/V)				
Electrons		2×10^{-7}		
Holes		4×10^{-8}		

however, bond interchange, that is, cleavage of bonds followed by interchain healing, is feasible. Eisenberg and Teter[43] indicate that such bond interchange is not significant at or near room temperature. Several researchers[11,43,44] have reported data on amorphous and liquid selenium viscosity as a function of both temperature and halogen content. Shirai et al.[11] have derived an empirical expression which relates the viscosity of liquid selenium to the average number of Se atoms in a polymer chain—the chain lengths being controlled by the addition of monovalent chlorine, which apparently end-caps Se_n chains at lower compositions and subsequently cleaves these chains at higher concentrations where end-capping is stoichiometrically completed. Small amounts of chain end-capping by halogens results in more classical polymeric mechanical behavior in selenium, perhaps because end-capping and termination of chain-end radicals lead to more stable molecular weights and less bond interchange during measurements.

Eisenberg and Teter[43] indicated that the glass transition and also stress relaxation properties are strongly dependent on iodine content. The decrease in the glass transition temperature (T_g) is similar to the decreases in T_g of polymers observed upon lowering molecular weights.

5. Optical Properties

The optical properties of selenium have received considerable attention[46-49] because of optical filter and photoelectric applications. Both crystalline and amorphous selenium have sharply defined bands in transmissivity and reflection spectra in the u.v. to far i.r. wavelength region. As would be expected with anisotropic crystalline materials, the bands are strongly dependent on light polarization.

6. Electrical Properties

Because of direct applications of the electrical properties of selenium to many commercial devices, the literature in this area is voluminous and a detailed review would exceed the scope of this book. Key references will be cited, however, to direct the interested reader to detailed descriptions.

The thermoelectric properties of selenium have been determined by several investigators.[50,52,53] Typical values are 0.805 mV/°C against lead, 1.0 mV/°C against copper, and 1.1134 mV/°C against platinum. The resistivity of selenium has been measured as varying between 10^2 to 10^{12} ohm-cm, depending on crystallinity, impurities, temperature and thermal history, field intensity, frequency, and illumination. As a general rule, metallic impurities decrease while nonmetallics increase selenium conductivity. The former effect is probably caused by the low solubility of metallics in selenium. In a given subgroup in the periodic chart, larger-atom metallics have a greater effect; for example, silver decreases conductivity more than copper, mercury more than cadmium, and thallium more than indium. Larger quantities of impurities, however, may reverse the effect and increase conductivity.[54,55] Henkels and Maczuk[56] give an encompassing account of the effects of many impurities on crystalline selenium conductivity. For pure amorphous and liquid selenium, the temperature dependence of specific conductivity is described by Eckart.[57]

Although selenium can be categorized as a semiconductor with low carrier mobility, excitation with electromagnetic radiation wavelengths below the absorption edge causes marked increases in conductivity, that is, photoconductivity. Such properties have led to the use of selenium and its compounds in photocell devices as well as in xerography. The photoelectric properties of selenium and their relation to optical properties have been summarized by several authors.[59-64]

7. Summary of Key Properties

Table II-6 summarizes the key physical properties of the various forms of selenium.

REFERENCES

1. J. J. Berzelius, *Acad. Handl. Stockholm*, **39,** 13 (1818).
2. R. C. Brasted, "Comprehensive Inorganic Chemistry," Vol. VIII, Van Nostrand, Princeton, N.J., 1961, p. 2.
3. W. C. Bragg, *Phys. Mag. (Phil. Mag.)*, **6,** 1960 (1920).
4. E. N. Gapon, *Z. Phys.*, **44,** 535 (1927).
5. D. M. Chizhikov and V. P. Shchastlivyi, "Selenium and the Selenides," Collets Ltd., London, 1968.
6. I. Rosenfeld and O. A. Beath, "Selenium," Academic Press, New York, N.Y., 1964, p. 279.
7. B. Mason, "Principles of Geochemistry," 3rd ed., John Wiley, New York, N.Y., 1966.
8. D. M. Chizhikov and V. P. Shchastlivyi, "Selenium and the Selenides," Collets Ltd., London, 1968, p. 107.
9. A. Eisenberg and A. V. Tobolsky, *J. Polymer Sci.*, **46,** 19 (1960).
10. G. Briegleb, *Z. Phys. Chem.*, **144A,** 321 (1929).
11. T. Shirai, S. Hamada, and K. Kobayashi, *J. Chem. Soc. Japan, Pure Chem. Sect.*, **84,** 968 (1963).
12. A. Eisenberg, *J. Polymer Sci. Lett.*, **1,** 177 (1963).

13. J. W. Mellor, "Inorganic and Theoretical Chemistry," Vol. X, Longmans, London, 1930, p. 693.

14. G. B. Abdullayev, Y. G. Asadov, and K. P. Mamedov, "The Physics of Selenium and Tellurium," W. C. Cooper, Ed., Pergamon Press, London, 1969, p. 179.

15. S. Iizima, J. Tayani, and M. Nicolet, "The Physics of Selenium and Tellurium," W. C. Cooper, Ed., Pergamon Press, London, 1969, p. 199.

16. R. G. Crystal, *J. Polymer Sci., Part A-2*, **8,** 1755 (1970).

17. R. G. Crystal, *J. Polymer Sci., Part A-2*, **8,** 2153 (1970).

18. J. D. Harrison, *J. Appl. Phys.*, **39,** 3672 (1968).

19. R. C. Keezer and M. W. Bailey, *Mater. Res. Bull.*, **2,** 185 (1967).

20. H. Krebs and W. Morsch, *Z. Anorg. Allg. Chem.*, **263,** 305 (1960).

21. F. Eckart, *Monatsber. Deut. Akad. Rad. Wiss. Berlin*, **3,** 183 (1961).

22. V. G. Sidyakin, *Izv. Vysskikh Uchebn. Zavedenii, Fiz.*, **2,** 25 (1962).

23. S. V. Dzhalilov, *J. Phys. Chem. (USSR)*, **39,** 1376 (1965).

24. Yu. A. Tresnikov, *Izv. Vysskikh Uchebn. Zavedenii, Khim. Tekhnol.*, **3,** 440 (1960).

25. J. Thouvenin and A. Rauch, *C. R. Acad. Sci., Paris*, **255,** 866 (1962).

26. A. P. Saunders, *J. Phys. Chem.*, **4,** 423 (1900).

27. K. Niwa, *J. Fac. Sci. Hokkaido Univ.*, **3,** 75 (1940).

28. D. J. Fourie and C. M. Wolf, *Z. Phys.*, **169,** 326 (1962).

29. T. Sato and H. Kenaco, *Technol. Rept. Tohoku Univ.*, **16,** 18 (1958).

30. L. E. Dodd, *J. Amer. Chem. Soc.*, **42,** 1579 (1920).

31. L. E. Hedman, *Ark. Fysik*, **21,** 161 (1962).

32. A. Bettendorf and A. Wullner, *Pogg. Ann.*, **133,** 203 (1868).

33. M. P. Moudein-Monval, *C. R. Acad. Sci., Paris*, **182,** 1465 (1926).

34. J. Dewar, *Proc. Roy. Soc., Ser. A*, **76,** 525 (1906).

35. P. Chaudari, P. Beardmore, and M. B. Bever, *Phys. Chem. Glasses*, **7,** 157 (1966).

36. P. W. Bridgman, *Proc. Amer. Acad. Arts Sci.*, **74,** 21 (1940).

37. B. I. Boltaksand and B. T. Plachenov, *Zh. Tekhn. Fiz.*, **27,** 2228 (1957).

38. M. Stravmanis, *Z. Krist.*, **102,** 432 (1940).

39. G. B. Abdullayev and A. A. Bashaliev, *Zh. Tekhn. Fiz.*, **27,** 1971 (1957).

40. Z. L. Amarova and G. B. Abdullayev, *Tr. Tashkentsk. Konf. po Mirnomu Ispotliz. At. Energii. Arak Nauk Uz., SSR*, **1,** 256 (1959).

41. A. A. Bashaliev, *Fiz. Tverd. Tela*, **1,** 348 (1959).

42. N. E. Aliev and N. I. Ibragemov, *Tr. Inst. Fiz. Akad. Nauk Azerb. SSR*, **10,** 30 (1960).

43. A. Eisenberg and L. Teter, *J. Amer. Chem. Soc.*, **87,** 2108 (1965).

44. S. Hamada, T. Sato, and T. Shirai, *Bull. Chem. Soc. Japan*, **41,** 135 (1968).

45. M. L. Williams, R. F. Landel, and J. D. Ferry, *J. Amer. Chem. Soc.*, **77,** 3701 (1955).

46. R. Frerichs, *J. Opt. Soc. Amer.*, **43,** 1153 (1953).

47. H. Gelbie and J. G. Cannon, *J. Opt. Soc. Amer.*, **45,** 489 (1955).

48. G. Lucovsky, "The Physics of Selenium and Tellurium," W. C. Cooper, Ed., Pergamon Press, London, 1969, p. 255.

49. R. S. Caldwell and H. Y. Fan, *Phys. Rev.*, **114,** 664 (1959).

50. I. P. Shapiro, *Zh. Tekh. Fiz.*, **21,** 717 (1951).

51. H. W. Henkels, *Phys. Rev.*, **76,** 1737 (1949).

52. R. M. Holmes, *Phys. Rev.*, **25,** 826 (1925).

53. W. Progser, *Czech. J. Phys.*, **B10,** 306 (1960).

54. B. D. Cullity, *Metals Technol.*, **15,** 1 (1948).

55. F. Neiland, *Philips Res. Rep.*, **9,** 259 (1954).

56. H. W. Henkels and J. Maczuk, *J. Appl. Phys.*, **25,** 11 (1954).

57. F. Eckart, *Ann. Physik*, **14,** 233 (1954).

58. H. P. Lanyen, *Phys. Rev.*, **130,** 134 (1963).

59. E. M. Pell, "Xerography and Related Processes," J. H. Dessauer and H. E. Clark, Eds., Focal Press, New York, N.Y., 1965, p. 65.

60. T. S. Moss, "Optical Properties of Semiconductors," Butterworths, London, 1959.

61. H. W. Henkels and J. Maczuk, *J. Appl. Phys.*, **24,** 1056 (1953).

62. W. E. Spear, *J. Phys. Chem. Solids*, **21,** 110 (1961).

63. J. L. Hartke, *Phys. Rev.*, **125,** 1177 (1962).

64. W. E. Spear, *Proc. Phys. Soc.*, *B*, **76,** 826 (1960).

III Reagents and methods for the introduction of selenium into organic molecules

WOLFGANG H. H. GÜNTHER

Xerox Corporation
Chemistry Research Laboratory
Webster, New York

This chapter summarizes the reagents commonly used for the introduction of selenium into organic compounds. Reactions directly from the element or those occurring with simple inorganic or organic selenium derivatives are discussed in terms of their mechanisms and their general utility. In addition, some individual conversions of special interest are cited, even though their scope may yet remain to be demonstrated. While older literature sources are quoted chiefly for historical reasons, an attempt has been made to present references to well-documented recent work. As selenium and its simple derivatives are now sold by a variety of chemical supply houses, individual commercial sources are not given.

A. ELEMENTAL SELENIUM

The element is available commercially as a gray to black powder (97–99.99999 % Se) which serves as an excellent starting material for most synthetic purposes. The low solubility of selenium contraindicates the use of coarser lumps or fused rods, which require extended reaction times. The main advantage claimed for commercial red selenium is its very finely divided state, but the gray material is quite satisfactory for preparative work; in addition, most samples of the red allotropic forms have a comparatively short shelf-life and quickly convert to a darker powder, especially in the presence of light, moisture, strong alkalies, and acids. Since most organic derivatives of selenium are purified more readily than the element, the 97 % Se powder is entirely satisfactory for preparative purposes. Care must be taken, however, to avoid contamination with sulfur or tellurium because of the similarity of organic compounds of these elements to those of selenium. Boiling points and solubilities are frequently quite close, and crystal packing may be identical between sulfur and selenium compounds, allowing formation of mixed crystals.

Although, in general, a close analogy exists between reactions of sulfur and selenium isologs, this similarity does not always extend to the elements. Frequently, reactions with elemental selenium return unchanged starting materials, unlike numerous well-documented conversions with elemental sulfur.[1] In spite of this apparent lack of reactivity of the element, a number of valuable synthetic methods lead from selenium to well-defined organic derivatives in a single step.

1. Thermal Reactions

Aromatic monoselenides are formed by heating symmetrical diaryl sulfones with elemental selenium[2-7] to temperatures in the 200–300° range. This procedure (equation 1), while it represents the initial discovery of the

$$(1) \qquad Ar\text{—}SO_2\text{—}Ar + Se \xrightarrow[300°]{\Delta} Ar\text{—}Se\text{—}Ar + SO_2$$

aromatic selenide class of compounds, has found only limited application.

Monoselenides incorporate an additional atom of selenium to yield diselenides, which are, however, contaminated with varying quantities of polyselenides (equation 2).[3] The latter split off red selenium under a variety

$$(2) \qquad C_6H_5\text{—}Se\text{—}C_6H_5 + Se_x \xrightarrow{\Delta} (C_6H_5Se\text{—})_2 + (C_6H_5Se\text{—})_2Se_n$$

of conditions, such as treatment with solvents, mild heating, or contact with oxidizing or reducing agents, and this has led to the proposal of a branched diselenide structure for a β-naphthyl derivative thus prepared.[8] A later investigation[9] established that this assignment had been erroneous and that incorporation of selenium into monoselenides does indeed lead to the normal, unbranched diselenide chain.

When a similar incorporation was carried out[10] on the unsymmetrical 2-benzylselenoethylamine hydrochloride,[11] the reaction appeared to follow scheme 3. The two main reaction products, after equilibration of the initially

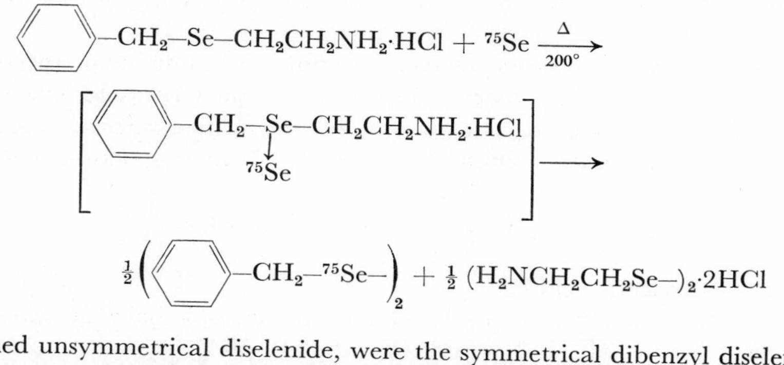

formed unsymmetrical diselenide, were the symmetrical dibenzyl diselenide and bis(2-aminoethyl) diselenide dihydrochloride. In addition, a small amount of dibenzyl monoselenide was formed. When radioactive ^{75}Se was used in this experiment, radioactivity was found associated exclusively with the dibenzyl diselenide and monoselenide fractions. The incorporation is, thus, not random but follows a well-defined course suggesting an initial attachment of the tracer to the monoselenide to form an unstable branched intermediate, which then rearranges by migration of the benzyl group.

The rearrangement of a branched into a linear structure is known in other systems, such as the Meisenheimer rearrangement,[12,13] which involves an N → O shift of substituents to give a hydroxylamine derivative from a tertiary amine-oxide. A reverse-reaction type, namely, the conversion of a sulfenate into a sulfoxide, has recently been described.[14]

It has been postulated that the Meisenheimer, as well as the Mislow,[14] rearrangement involves the formation of free radicals which recombine to form the thermodynamically most favored product, which may be either the linear or the branched form. In the selenium case highest stability may be expected for the linear diselenide. The correct choice of substituents should, then, make the incorporation of elemental selenium into monoselenides a versatile tool, especially for the synthesis of radioactively labeled selenium compounds.

In a similar fashion elemental selenium is incorporated into the thia-cyclobutane ring to form a 1-thia-2-selenacyclopentane system in 75% yield[15] (equation 4). The utility of a trace of potassium cyanide in this

(4)

reaction indicates that one of the attacking species is the highly nucleophilic selenocyanate ion. It is noteworthy that an attempt to incorporate an additional S atom by the same procedure resulted only in polymer formation.[15]

The following represents further examples of individual reactions. Hexafluorodimethyl diselenide is formed in satisfactory yield when the mercury salt of trifluoroacetic acid is heated with elemental selenium (equation 5).[16] A similar conversion occurs with the corresponding mercury

$$(5) \qquad (CF_3COO)_2Hg + Se_x \rightarrow CF_3SeSeCF_3$$

salt of difluoroacetic acid.[17]

Elemental selenium also reacts with trifluoroiodomethane at 260–285° to yield $(CF_3Se-)_2$ (10–15%) and $(CF_3-)_2Se$ (45–50%).[18]

Heating 2,3-diiodohexafluoro-2-butene with elemental selenium at 180°

under autogenous pressure gives an 8% yield of tetra(trifluoromethyl)-p-diselenin (equation 6).[19] Under these conditions the corresponding sulfur

$$(6) \qquad CF_3-\overset{I}{C}=\overset{I}{C}-CF_3 + Se \xrightarrow{180°}$$

analog loses one S atom and contracts to form tetra(trifluoromethyl)-thiophene, indicating that the diselenin ring system possesses the greater thermal stability.

Reaction of selenium with 2,2'-biphenylylene mercury at 330° gives a 91 % yield of 2,2'-biphenylylene selenide.[20]

Similar reactions have been applied for the synthesis of diphenyl selenide,[3] di-o-tolyl selenide,[21] di-p-tolyl selenide,[21] and 1,1-dinaphthyl selenide.[22]

The excellent yields in these reactions indicate a procedure of general utility, depending only on the availability of the mercury compounds, but requiring a high stability of the selenium-organic product. Thermal displacement reactions on elements other than carbon allow the mutual transformations in equation 7.[23]

$$(7) \qquad (Et_3M)_2Te \underset{+Se}{\overset{-Te}{\longrightarrow}} (Et_3M)_2Se \underset{+S}{\overset{-Se}{\longrightarrow}} (Et_3M)_2S$$

$$\underset{+S}{\overset{-Te}{\longleftarrow}}$$

$$(M = Si, Ge, or Sn)$$

Vigorous reaction conditions are required with temperatures in the 200–300° range. Reaction times of 7–90 hours gave moderate to good yields of products.[23]

2. Addition Reactions under Mild Conditions

Solutions of isonitriles in solvents boiling generally below 100° (petroleum ether, ether, benzene, chloroform, *etc.*) react readily with powdered elemental selenium to give the corresponding isoselenocyanates (equation 8)

$$(8) \qquad R-NC + Se \rightarrow R-N=C=Se$$

in good to excellent yields.[24–29] This general reaction has found wide application since it provides a convenient route to substituted selenoureas and selenosemicarbazides and, furthermore, to a variety of selenium-substituted heterocycles.

Derivatives of trivalent phosphorus react smoothly with selenium to form substituted phosphine selenides (equation 9), selenophosphites, and

$$(9) \qquad R_3P + Se \rightarrow R_3P=Se$$

selenophosphates in excellent yields. These reactions on phosphorus are general and represent one of the prime sources for such materials (*cf.* Chapter X on selenium-phosphorus compounds).

3. Reactions with Metal-Organic Reagents

Grignard reagents, organolithium, organosodium, and organopotassium derivatives containing a potential carbanion in relatively polar solvents (diethyl ether, tetrahydrofuran, liquid ammonia) react vigorously with dry selenium at room temperature to form metal salts of the corresponding selenols.

Thus, benzeneselenol is formed from phenylmagnesium bromide (equation 10),[30-32] and 2-naphthylselenol is derived from 2-naphthylmagnesium

$$(10) \qquad C_6H_5MgBr + Se \rightarrow C_6H_5SeMgBr \xrightarrow{H_2O} C_6H_5SeH$$

iodide.[33] Numerous other examples have been described.[34,35] Since aliphatic selenium derivatives are usually accessible from halides by direct nucleophilic substitution, the formation of an intermediate Grignard reagent appears unnecessary in many cases, and application of the above type of reaction has been restricted largely to aromatic molecules. One well-documented route to benzylselenol,[36] however, employs the Grignard intermediate for comparatively large-scale syntheses of this useful reagent. Lithium-organic intermediates, conveniently accessible *via* exchange reactions with commercial *n*-butyllithium, react in ether, but not in hydrocarbon solvents, to give lithium salts of the selenols. In the example given here[37] (scheme 11), the free selenol is unstable and alkylation or acylation reactions

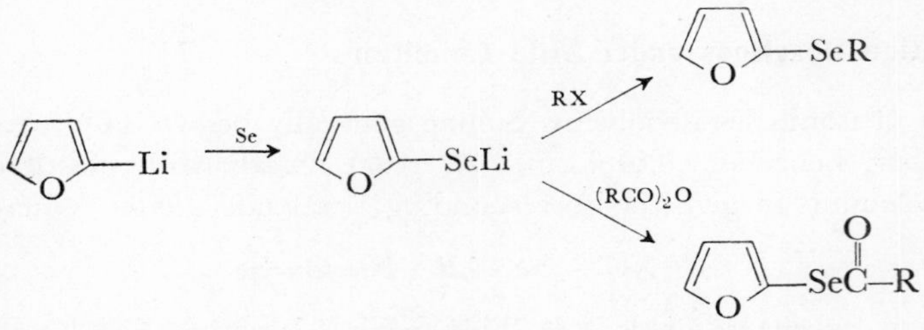

were carried out without isolation on the intermediate selenol anion. Other examples of this type of reaction have been described,[38,39] and the procedure appears to offer a convenient general route to selenols and selenol derivatives.

Alkali-metal acetylides in liquid ammonia reacted with elemental selenium to form alkali alkynylselenides[40-43] (scheme 12), which were immediately

$$(12) \qquad R—C{\equiv}C—Na \xrightarrow{Se} R—C{\equiv}C—SeNa \xrightarrow{R'X} R—C{\equiv}C—SeR'$$

alkylated to unsymmetrical acetylenic selenides, since the acetylenic selenols appear to be highly unstable. This type of reaction may also be carried out at room temperature in ether or acetonitrile,[44] followed by trapping of the intermediate with carbon disulfide (equation 13). The structure of the product

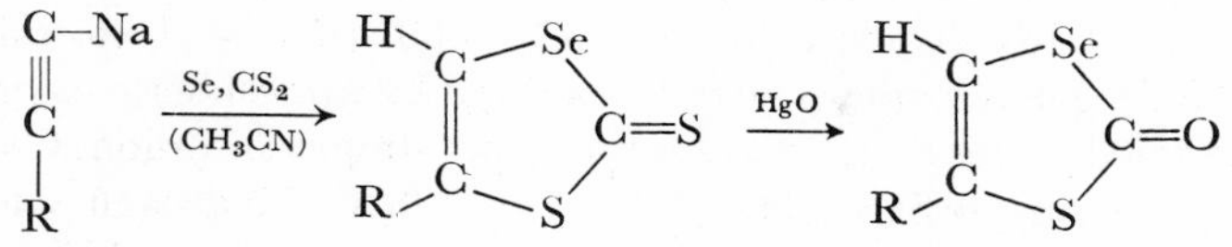

and proof for the direction of addition are clearly shown by the further conversion, indicated in scheme 13, to give a unique heterocyclic compound containing oxygen, sulfur, and selenium ligands at a single carbon atom.

4. Reactions of Selenium Atoms

Flash photolysis of carbon diselenide and of carbon oxyselenide yields triplet- and singlet-state Se atoms, respectively, and these have been shown to undergo addition and insertion reactions with olefins and aliphatic hydrocarbons.[45–48]

Reaction kinetics and mass spectrometric determinations indicated a marked similarity between excited S and Se atoms. Although relatively stable episulfides may be generated by this process, the corresponding selenium products are much less stable and the transient formation of episelenides has been postulated in reaction of olefins with atomic selenium. The half-lives of the species thus formed appear to be very short (30 msec to a few seconds), and products of the thermal sequence reaction were not identified. Singlet selenium is reported to insert into C-H bonds, giving such novel structures as *tert*-butylselenol, cyclopropylselenol, cyclobutylselenol, and methylsilylselenol.[48] Although the flash photolysis systems did not allow application to preparative-scale synthetic procedures, the principles demonstrated here may perhaps be usefully employed once a convenient source of atomic selenium becomes available. In this connection it might be interesting to re-examine the numerous thermal reactions of selenium with organic compounds which either have given unidentified selenium-organic materials or have led to "catalyzed" rearrangement reactions, presumably *via* unstable selenium derivatives.

B. HYDROGEN SELENIDE AND ALKALI SELENIDES

Hydrogen selenide, H_2Se, mol. wt. 80.98, is a colorless gas at room temperature; b.p. $-41.4°$, d. (liquid at $-42°$) 2.12; m.p. $-65.7°$, d.

(solid at $-170°$) 2.45; critical temperature $+141°$, critical pressure 91 atm. It liquefies at $0°$ under 6.6 atm, at $18°$ under 8.6 atm, at $52°$ under 21.5 atm, at $100°$ under 47.1 atm. Gas weight per liter at $0°/760$ mm Hg is 3.6643 g; it is readily soluble in water (377 ml in 100 ml H_2O at $4°$, 270 ml in 100 ml H_2O at $22.5°$) and in most organic solvents. In aqueous solution the compound is a fairly strong acid with pK_1 3.88, $pK_2 \sim 11$ (formic acid, pK 3.75). Hydrogen selenide is readily oxidized by atmospheric oxygen and must be handled under an inert atmosphere of nitrogen or argon. Oxidation potential [$H_2Se(aq) \rightarrow Se + 2H^+ + 2e^-$] = 0.36 V. Hydrogen selenide is extremely toxic (6 ppm in air is lethal to guinea pigs) and should be treated with great caution. The penetrating and disagreeable odor, frequently described as reminiscent of rotting radishes, generally provides sufficient warning to the experimenter upon first exposure. The acute effect of a single whiff on mucous membranes may take the form of the symptoms of a heavy head cold, lasting as long as a week. This observation by Berzelius has been substantiated repeatedly in the author's experience; however, no lasting ill effects have resulted. For details on toxicity see Chapter XIIIC.

Hydrogen selenide is now commercially available as a high-purity, compressed gas in 1-lb lecture bottles and larger cylinders, and this source is recommended for routine applications. The compound may be prepared through direct combination of the elements[49,50] by passing the vapors over pumice fragments heated to 350–400°, trapping impurities successively at $-20°$ and $-40°$, and condensing the product in liquid nitrogen.

For occasional use, the production of gaseous H_2Se by hydrolysis of aluminum selenide in water or dilute acid under an atmosphere of nitrogen may be recommended.[49,50] Aluminum selenide is readily prepared by igniting a stoichiometric mixture of aluminum powder and elemental selenium by means of a burning magnesium wire. A carbon crucible with a loosely fitting asbestos lid is suitable for this vigorous reaction.[50] The material is also available commercially. The generally low purity of Al_2Se_3 presents some problems when accurate quantitative control of hydrogen selenide concentrations is desired. In these cases the gas should be dried over P_2O_5, trapped at low temperature, and redistilled in a slow stream of N_2.

The principal modes of reaction with organic molecules for hydrogen selenide and the alkali selenides are addition to unsaturated moieties and nucleophilic displacement reactions at suitably activated centers.

1. Addition to Multiple Bonds

a. Isolated and Activated Carbon-Carbon Double Bonds. To the author's knowledge no interactions of hydrogen selenide with isolated C-C double bonds in aliphatic systems have been reported. However, suitably activated

C-C double bonds react readily with hydrogen selenide. Thus 2 molecules of benzalacetophenone condense with 1 molecule of H_2Se to form a symmetrical monoselenide (**1**),[51] and dibenzal acetone yields 1-selena-2,6-diphenyl-cyclohexane-4-one (**2**).[51]

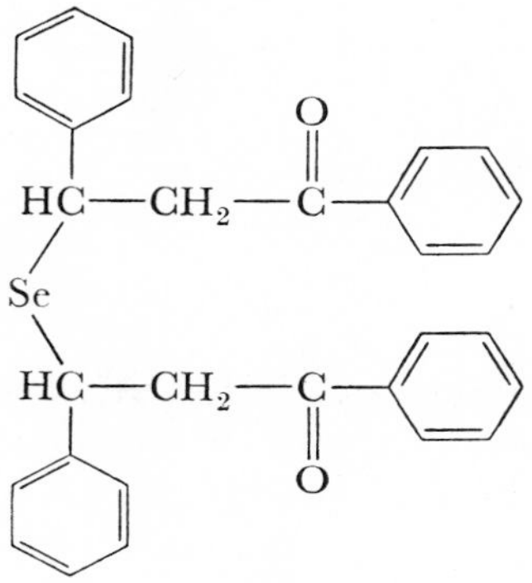

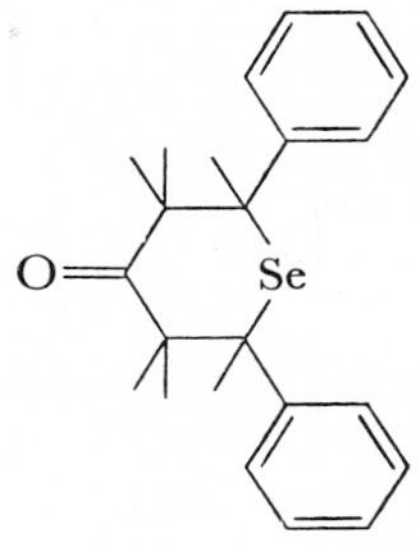

In a similar manner 1-selena-2,6-dimethylcyclohexan-4-one-3,5-di(ethyl-carboxylate) forms, in a multiple-step reaction from acetaldehyde, acetone di(ethylcarboxylate) and hydrogen selenide at $-10°$ in ethanol in the presence of 1,4-diazabicyclo[2.2.2]octane. The simple reaction conditions[52] compensate for the relatively poor yield of 10%.

For further examples of hydrogen selenide addition to activated C-C double bonds, see Ashby et al.[53]

b.   Addition and Displacement Reactions on Imino Functions.   As in the corresponding sulfur cases this procedure represents a very general method for the preparation of selenocarbonyl groups under mild conditions. Thus, 2-selenophthalide and a variety of analogs were prepared via the iminoester hydrochlorides (scheme 14).[54]

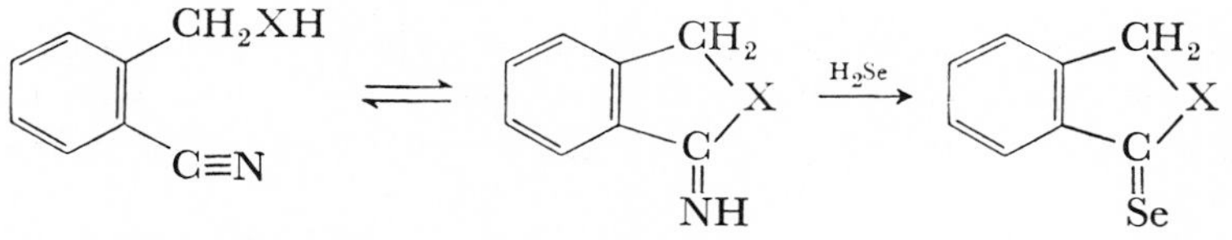

O-Methyl selenobenzoate[55] was derived from the corresponding imino-benzoate hydrochloride with H_2Se in 36% yield in a procedure analogous to that used for the thiocarbonyl analog.[56]

The selenium analog of dithizone was prepared[57] from 3-nitro-1,5-diphenylformazan and H_2Se in ethanolic ammonia. The reaction presumably involves the reduction of the nitro function by H_2Se and addition of H_2Se to the unsaturated system, followed by displacement of ammonia and controlled oxidation to avoid formation of the diselenide (scheme 15).

(15)

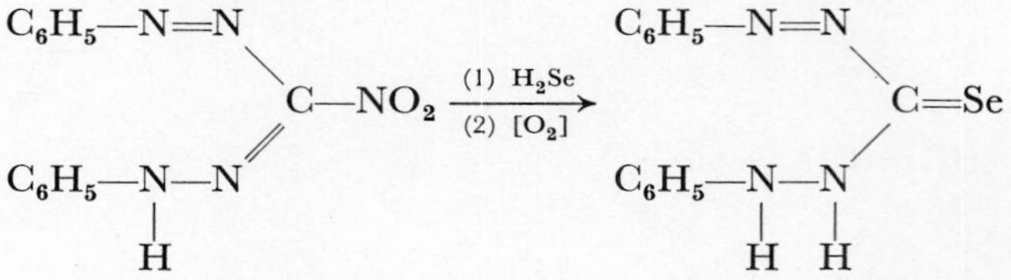

$$
\begin{array}{c}
C_6H_5-N=N \\
\diagdown \\
\qquad\qquad C-NO_2 \\
\diagup \\
C_6H_5-N-N \\
| \\
H
\end{array}
\xrightarrow[\text{(2) } [O_2]]{\text{(1) } H_2Se}
\begin{array}{c}
C_6H_5-N=N \\
\diagdown \\
\qquad\qquad C=Se \\
\diagup \\
C_6H_5-N-N \\
| \quad | \\
H \quad H
\end{array}
$$

In a related type of reaction the addition of hydrogen selenide to an imino function resulted in displacement of an associated methylmercaptide rather than the nitrogen moiety (scheme 16).[58,59]

(16)

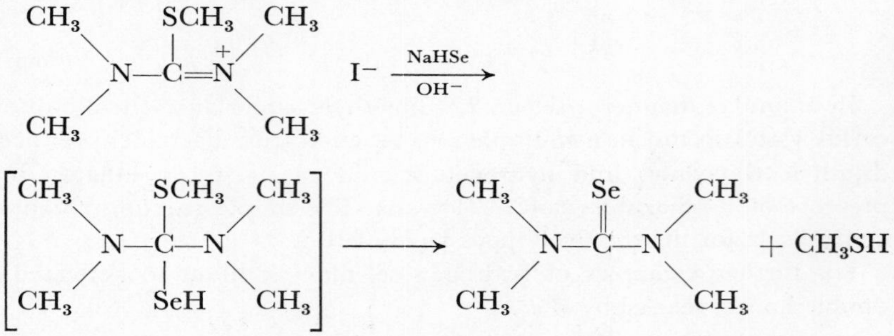

$$
\begin{array}{c}
CH_3 \quad\quad SCH_3 \quad CH_3 \\
\diagdown \qquad | \quad \overset{+}{} \diagup \\
\qquad N-C=N \\
\diagup \qquad\qquad\qquad \diagdown \\
CH_3 \qquad\qquad\qquad CH_3
\end{array}
\quad I^- \xrightarrow[\text{OH}^-]{\text{NaHSe}}
$$

$$
\left[
\begin{array}{c}
CH_3 \quad\quad SCH_3 \quad CH_3 \\
\diagdown \qquad | \qquad \diagup \\
\qquad N-C-N \\
\diagup \qquad | \qquad \diagdown \\
CH_3 \qquad SeH \quad CH_3
\end{array}
\right]
\longrightarrow
\begin{array}{c}
CH_3 \quad\quad Se \quad\quad CH_3 \\
\diagdown \qquad || \qquad \diagup \\
\qquad N-C-N \\
\diagup \qquad\qquad \diagdown \\
CH_3 \qquad\qquad CH_3
\end{array}
+ CH_3SH
$$

Hydrogen selenide also adds smoothly to aliphatic or aromatic carbodiimides in ether or benzene solution to yield symmetrically disubstituted selenoureas according to equation 17.[60]

(17)
$$
R-N=C=N-R + H_2Se \longrightarrow \underset{\underset{Se}{||}}{RHN-C-NHR}
$$

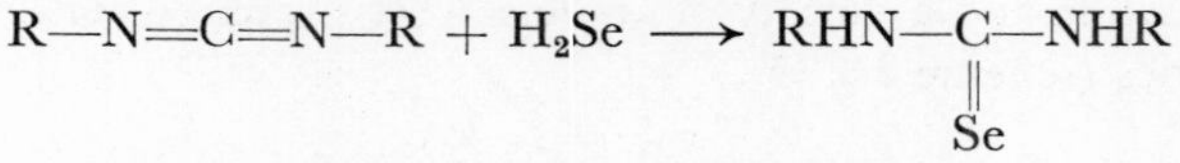

Without isolation of the intermediate carbodiimide, it is possible to start from the corresponding thiourea, desulfurize it by reaction with mercuric oxide or lead oxide, and treat the crude product with hydrogen selenide.[61]

c.   Addition to Carbon-Nitrogen Triple Bonds.

Addition to cyano functions readily occurs to yield terminal selenoamide groups. Reaction 18 illustrates

(18)
$$
H_2N-C\equiv N + H_2Se \xrightarrow{\text{catalyst}} \underset{\underset{Se}{||}}{H_2N-C-NH_2}
$$

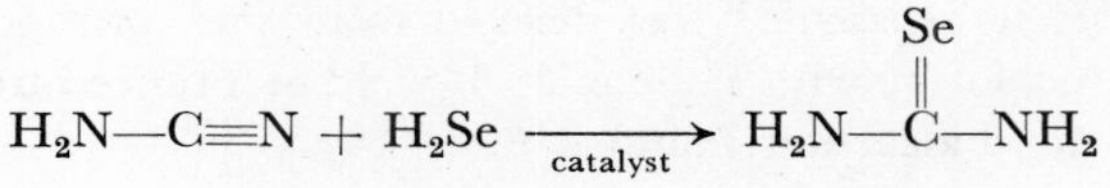

the preparation of selenourea from cyanamide and hydrogen selenide in methanol containing sodium methoxide[62] or in ether in the presence of either ammonia[63] or a catalytic amount of aqueous HCl.[64,65]

Correspondingly, alkyl cyanates yield O-alkyl selenocarbamates[66] (scheme 19), and the general process appears useful for any moiety containing CN groups.

$$(19) \qquad RO\text{—}CN + H_2Se \xrightarrow{\text{dry ether}} RO\text{—}\overset{\displaystyle Se}{\overset{\|}{C}}\text{—}NH_2$$

Further pertinent examples are syntheses of selenoamides from nitriles, usually in ethanol in the presence of alcoholic ammonia[67-70] or sodium ethoxide.[71,72]

d. Addition to Conjugated Diacetylenes. Suitably substituted diacetylenes, readily accessible by oxidation of terminal acetylenes, react with hydrogen selenide in ethanol containing sodium ethoxide and catalytic amounts of Ag^+ or Cu^+ ions to yield substituted selenophenes (equation 20).[73] Product yields appear to be good to excellent, and the procedure should provide a general route to selenophenes. The corresponding thiophenes may be prepared by an analogous procedure with hydrogen sulfide.[74]

$$(20) \qquad \begin{array}{c} R\text{—}C\equiv C \\ | \\ R'\text{—}C\equiv C \end{array} \xrightarrow[\text{C}_2\text{H}_5\text{ONa, Ag}^+]{\text{H}_2\text{Se}} \quad \underset{R \quad \ \ Se \quad \ \ R'}{\boxed{}}$$

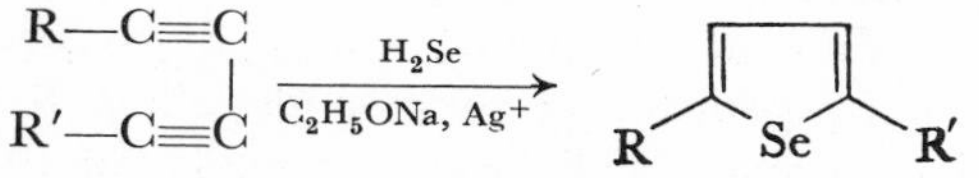

e. Addition and Displacement on Carbonyl Groups. Replacement of a carbonyl oxygen atom by selenium has been observed under a variety of reaction conditions. Thus, cyclic oligomers and polymers of selenoformaldehyde are obtained by treating formaldehyde with H_2Se in the presence of aqueous hydrochloric acid.[75]

The corresponding reaction with various ketones is accompanied by a reduction yielding diselenides,[76] not dimeric selenoketones as originally assumed.[77]

The interaction of ketones, their cyanohydrins, ammonia, and hydrogen selenide yields 2,2,5,5-tetraalkyl-4-selenoxoimidazolidines (equation 21).[78]

$$(21) \qquad \begin{array}{c} R \qquad OH \\ \diagdown \ / \\ C \\ / \ \diagdown \\ R' \qquad C\equiv N \end{array} + NH_4SeH + \begin{array}{c} R \\ \diagdown \\ C\!=\!O \\ / \\ R' \end{array} \longrightarrow$$

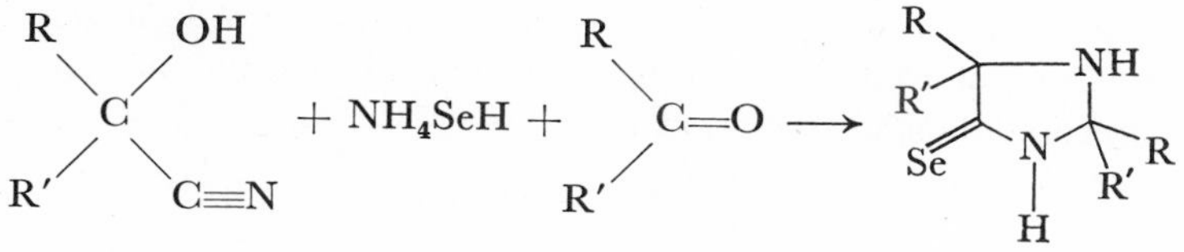

Identical products are obtained when these reactions are carried out with the α-aminonitriles corresponding to the cyanohydrins.[79]

The direct replacement of carbonyl oxygen by selenium has also been reported for one example of a 1,3-dicarbonyl compound. Pentane-2,4-dione in methanolic hydrogen chloride in the presence of nickel chloride, when treated with hydrogen selenide, deposited the monomeric complex bis(4-selenopent-3-ene-2-selenoxo)nickel(II) in 36% yield (equation 22).[80]

(22)

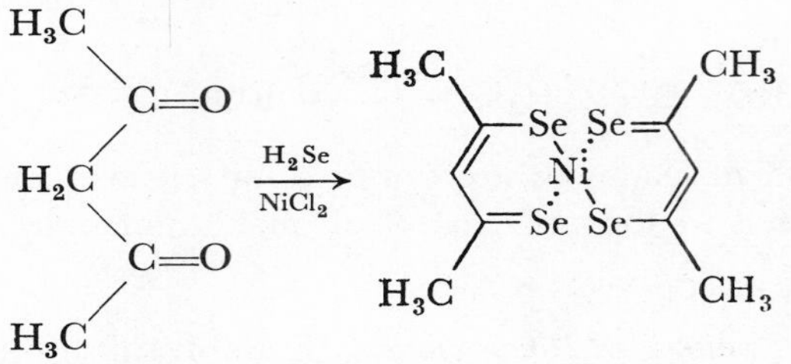

It must be assumed that the stability of the metal complex prevents further reactions of the monomeric selenocarbonyl ligand to dimeric and polymeric species.

2. Displacement Reactions

a. Ring-Opening Reactions. A few specific cases are known in which hydrogen selenide or an alkali selenide was used to effect a heteroatom displacement reaction in small ring compounds. Thus, the hydride reacts with ethylene oxide to give 2-hydroxyethylselenol and the corresponding diselenide, the latter possibly by air oxidation of the intermediate selenol (equation 23).[81]

$$(23) \quad \overset{\triangle}{O} + H_2Se \longrightarrow HOCH_2CH_2SeH \xrightarrow{[O]} \tfrac{1}{2}(HOCH_2CH_2Se-)_2$$

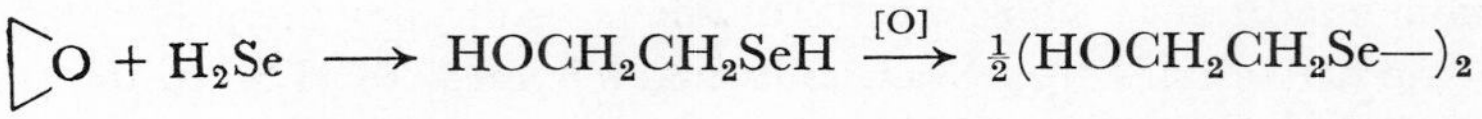

By contrast, reaction of ethylenimine with hydrogen selenide in ethanol yields only the 2:1 reaction product bis(2-aminoethyl) selenide (scheme 24), isolated as the picrate.[82]

$$(24) \quad 2\,\overset{\triangle}{NH} + H_2Se \longrightarrow (H_2NCH_2CH_2-)_2Se$$

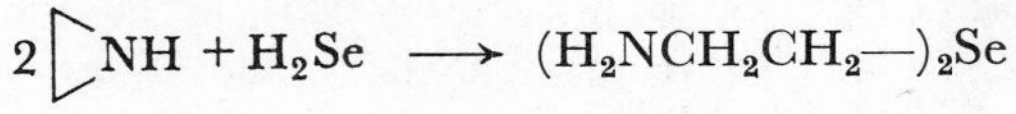

Dropwise addition of β-propiolactone to a solution of sodium selenide in water at $0°$, followed by acidification and air oxidation yields β,β'-diselenodipropionic acid (equation 25).[83]

(25)

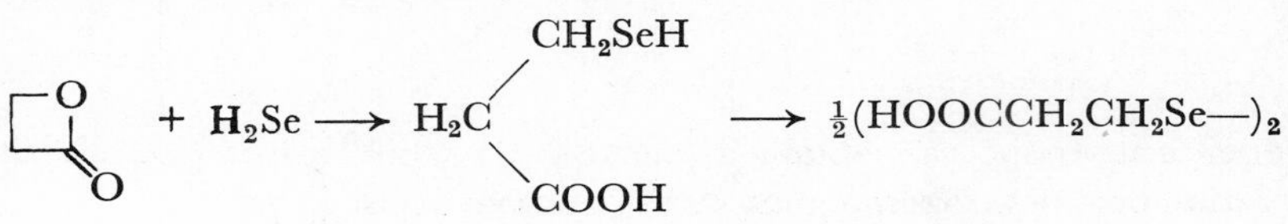

b. *Halide Displacement Reactions.* The highly nucleophilic hydroselenide anion readily displaces aliphatic halides to give selenols, the derived diselenides, and monoselenides.

Although many selenium reactions require separate generation of hydrogen selenide, two excellent procedures[84,84a] avoid the handling of this gas completely. In the first method, elemental selenium is reacted with sodium in liquid ammonia to form disodium monoselenide or disodium diselenide, depending on the Se/Na stoichiometry employed.[85,86] Direct alkylation in liquid ammonia can be carried out to give relatively high (80%) yields of aliphatic monoselenides and symmetrical diselenides, respectively. The sequence of diselenide anion, dialkyl diselenide, further sodium reduction to alkylselenol, and repeat of the alkylation step then gives ready access to unsymmetrically substituted monoselenides without isolation of intermediates (scheme 26). The general process is applicable also to corresponding sulfur and tellurium derivatives.[87]

$$(26) \quad 2Se + 2Na \xrightarrow{\ NH_3\ } Na_2Se_2 \xrightarrow{\ 2RX\ }$$

$$RSeSeR \xrightarrow{\ 2Na/NH_3\ } 2RSeNa \xrightarrow{\ 2R'X\ } 2RSeR'$$

Klayman and Griffin[84a] have recently developed a very convenient method for the production of ethanolic hydroselenide ion from selenium in which the element is reduced by 1 mole of sodium borohydride under nitrogen. The exothermic reaction, accompanied by vigorous hydrogen evolution, is complete in a few minutes to give a colorless solution of sodium hydrogen selenide and triethyl borate according to equation 27a. The solution thus

$$(27a) \quad Se + NaBH_4 + 3C_2H_5OH \rightarrow NaHSe + B(OC_2H_5)_3 + 3H_2$$

formed has been used directly for the synthesis of selenides, selenoureas, and selenols in good yield.

Aqueous sodium hydrogen selenide can be formed similarly by the reduction of selenium with 2 moles of sodium borohydride. Hydrogen and sodium tetraborate are the by-products of the reaction, as shown in equation 27b.

$$(27b) \quad 2Se + 4NaBH_4 + 7H_2O \rightarrow 2NaHSe + Na_2B_4O_7 + 14H_2$$

Sodium hydrogen selenide, so generated, was converted, for example, into dibenzyl diselenide in 86% yield. Neither of the boron-containing by-products indicated in equations 27a and 27b interferes with the subsequent reactions of the hydroselenide ion and work-up procedures.[84a]

C. POTASSIUM SELENOCYANATE AND RELATED COMPOUNDS

The potassium salt of selenocyanic acid is one of the best-established reagents for the introduction of selenium into organic molecules. The compound is readily prepared (equation 28) by fusion of an intimate mixture of selenium

$$(28) \qquad\qquad KCN + Se \rightarrow KSeCN$$

powder with potassium cyanide in a porcelain vessel.[88,208] In an alternative preparation[89] the powdered ingredients are simple refluxed in ethanol until sufficient selenium has dissolved. Excess KCN is converted to $KHCO_3$ with gaseous CO_2, and pure KSeCN (mol. wt. 144.08; d. 2.347) is then obtained as a white, granular, crystal mass from ethanol and acetone. The compound is very hygroscopic and decomposes slowly on contact with acidic gases and moisture, that is, the common laboratory atmosphere.* The salt appears to be stable for many years in well-sealed containers. Potassium selenocyanate is soluble in water, the lower aliphatic alcohols, acetone, methyl ethyl ketone, DMF, acetonitrile, and other polar organic solvents. This high solubility in organic solvent systems earns KSeCN a unique position among the readily accessible inorganic selenium derivatives.

1. Nucleophilic Displacement Reactions

The major mode of application for KSeCN takes advantage of the high nucleophilicity of the selenocyanate anion towards aliphatic halides and aromatic diazonium salts, as illustrated in equations 29 and 30.

$$(29) \qquad\qquad RX + KSeCN \rightarrow R\!-\!SeCN + KX$$

$$(30) \qquad Ar\!-\!N_2{}^+X^- + KSeCN \rightarrow Ar\!-\!SeCN + KX + N_2$$

In the first case the components are simply heated under reflux in a suitable organic solvent (usually alcohol or acetone) until the calculated amount of potassium halide has been formed. The diazonium replacement

---

* A point of caution: selenocyanic acid is very unstable and decomposes readily to the poisonous hydrogen cyanide. In contact with skin or clothing KSeCN generates red stains of selenium which are difficult to remove by ordinary washing procedures. Chemical conversion of the element to colorless, soluble species is required and may be accomplished by the application of aqueous solutions of potassium sulfite, sodium formaldehyde sulfoxylate, or sodium dithionite. Ammonium or sodium sulfide has been suggested, but these compounds have additional staining possibilities with traces of heavy metal ions and are likely to result in a lingering smell. If organic solvents are preferred or if the garment is to be dry cleaned, a solution of triphenylphosphine in trichloroethylene appears safe and effective. In all cases the bleach resistance of the fabric dye should be tested by application of the reagent to an inconspicuous spot.

reaction is carried out without catalysts in aqueous solutions buffered to pH 5.5 or above by adding an aqueous solution of KSeCN slowly, with stirring, to the diazonium salt. Evolution of nitrogen is a convenient indicator of the progress of this reaction, which may have to be completed by heating the mixture on a steam bath.

Reaction products are, in nearly every case, "normal" selenocyanates resulting from attack of a selenium nucleophile, and numerous products are cited in Chapters IV and VIII. However, chlorotriphenylmethane reacts to form the isoselenocyanate,[90] indicating that selenocyanate ion may also exist as a nitrogen nucleophile, unless a radical process via triphenylmethyl is involved in this single reported case.

Reactions of vicinal aliphatic dibromides with potassium selenocyanate have been observed to lead to olefin formation, rather than substitution.[209,210]

2. Electrophilic Substitution via Selenocyanogen and Related Species

Starting from potassium selenocyanate or the corresponding lead or silver salt, the selenocyanate ion may be converted into selenium dicyanide $Se(CN)_2$, dicyanodiselenide (selenocyanogen) $(SeCN)_2$, and dicyanotriselenide $Se(SeCN)_2$.

Of these, selenium dicyanide is known to undergo a single conversion with an organic species (scheme 31)[91] and does not appear to be a very practical reagent for the introduction of selenium into organic molecules.

(31) $(C_6H_5)_3Bi + Se(CN)_2 \rightarrow (C_6H_5)_2BiCN + C_6H_5SeCN$

Dicyanotriselenide,[91,93] pale yellow leaflets from $CHCl_3$ or orange needles from benzene (both forms melt at 133–134°), is prepared from aqueous KSeCN by oxidation with chlorine[92] or with NO_2 and nitric acid.[94] The compound has been well characterized by X-ray methods,[95] and a few reactions with organic molecules are known. Thus, aniline and substituted anilines react according to equation 32[91,96] by shaking the components in

(32)

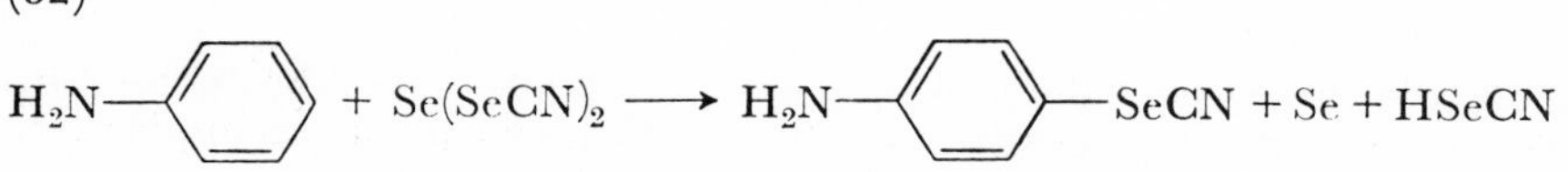

ether at room temperature for 18–20 hours. However, the overall process from KSeCN to the selenium-substituted aromatic moiety is very inefficient and does not constitute a practical synthetic method.

Selenocyanogen,[93] a very labile compound prepared earlier from KSeCN and iodine[97] or lead tetraacetate[98] in, presumably, impure form, has recently[99] been generated by reaction of KSeCN with a stoichiometric quantity of bromine at $-60°$ in methanol. Its electrophilic substitution reactions with suitable organic species have been investigated, and numerous products were prepared[99-103] without intermediate isolation of the pure reagent.

It was shown that the stoichiometry of scheme 33 pertains and that these reactions occur in a fairly general fashion with aromatic amines, indoles, indolines, and pyrroles. Yields are generally good, and further extensions of this procedure appear likely.

$$(33) \qquad 2\,KSeCN + Br_2 \xrightarrow[\text{MeOH}]{-60°} (SeCN)_2 + 2\,KBr$$

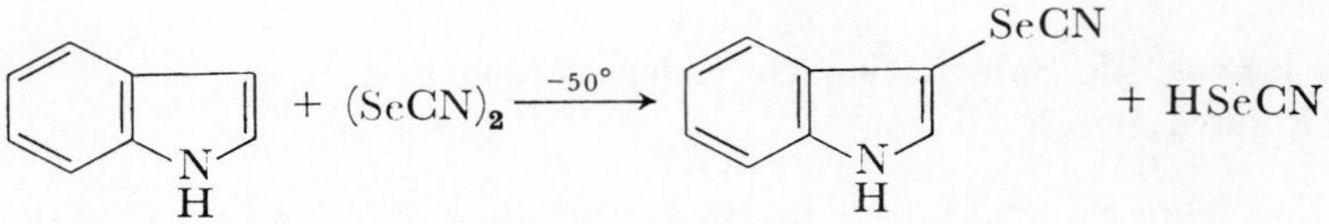

$$HSeCN \longrightarrow Se + HCN$$

Syntheses and properties of selenocyanates and their further conversion reactions are discussed in Chapter IV.

D. PHOSPHORUS PENTASELENIDE

The possible existence of phosphorus pentasulfide was suggested by Berzelius[104] as early as 1834, and claims to its preparation[105] date back to 1862. It appears much less reactive towards organic species than the analogous phosphorus pentasulfide, which is widely used for the preparation of organosulfur compounds and sulfur-containing organophosphorus materials. Since little is known concerning the detailed mechanism by which these species react, rational extensions of the known synthetic applications are extremely difficult.

Phosphorus pentaselenide was prepared[106] from red phosphorus and gray selenium, which were heated to 450° for 12 hours in an atmosphere of nitrogen in a sealed, heavy-walled ampoule. The insoluble, black-purple solid, which showed no evidence of crystallinity, was then powdered in a ball mill under an inert atmosphere and stored over P_2O_5. Older procedures[107] were reported[106] to give a much less active product. Since there are no specific ways of characterizing P_2Se_5, variations in reagent quality may have contributed much to the confusion that exists in regard to its use.

1. Reactions with Alcohols

Kudchadker *et al.*[106] report a number of reactions of phosphorus pentaselenide with primary alcohols; these yielded *O,O'*-dialkyl diselenophosphates (as the potassium salts), as well as bis(*O,O'*-dialkyl diselenophosphate) and an analog containing a sequence of three Se atoms linking phosphorus, namely, bis(*O,O'*-diethyl diselenophosphate) selenium(II). For details of preparation and properties of selenium-phosphorus compounds, see Chapter X.

2. Reactions with Carbonyl Compounds

The direct conversion of amide and heterocyclic carbonyls to the corresponding thiocarbonyl compounds constitutes one of the major applications for phosphorus pentasulfide. Numerous attempts have been made to generate a simple and reliable general method by which selenium can be introduced in place of double-bonded oxygen; however, to the author's knowledge, no such method exists at this time, and the examples cited below represent specific cases only.

Tetramethylselenourea was prepared[108] in 5.6% yield from tetramethylurea and P_2Se_5 by heating the components in benzene for 72 hours. An analogous experiment with tetraethylurea failed to give a pure product, although the infrared spectrum indicated that the crude oil contained tetraethylselenourea.[108]

In order to circumvent the preparation of phosphorus pentaselenide, the preparation of selenolactams from lactams[109] was undertaken by refluxing the starting materials with red phosphorus and gray selenium in xylene for 24 hours (equation 34). Although most of the products were obtained in

(34)

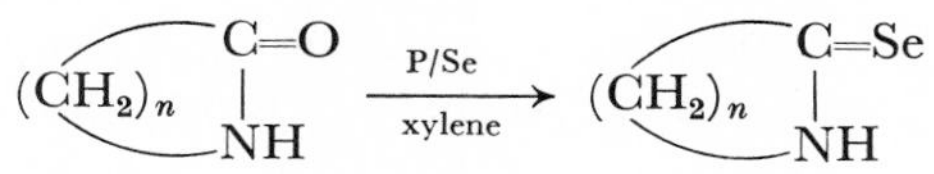

yields in the range of 1%, the compound with $n = 11$ gave a 40% yield of pure selenolactam.

A quantitative conversion of an epidithiopentadienal with phosphorus selenide to the resonance-stabilized 2,4-diphenyl-(1,2)-thiaselenolo[5,1-b]-(1,2)dithiole was reported[110,111] to occur on boiling in chlorobenzene for 3.5 hours (equation 35).

(35)

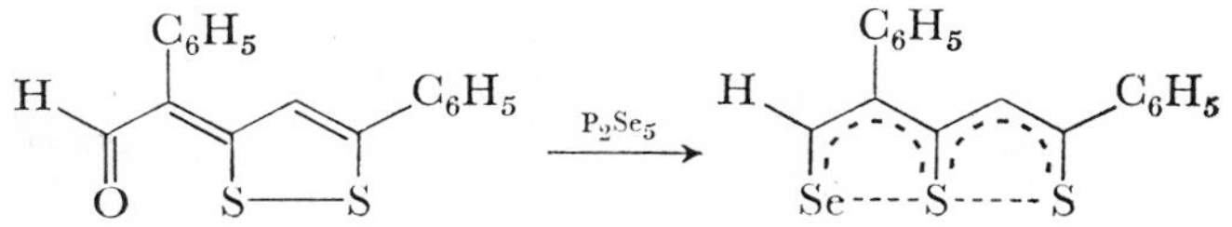

Although these examples indicate the possibility of a general method for carbonyl oxygen *vs.* selenium exchange, a systematic study of reaction conditions, solvents, temperatures, and potential catalysts is lacking. It may be significant that the highest yields were generated at rather high temperatures and that the product stability under these forced conditions may be rather low. Slow reaction rates with the insoluble phosphorus pentaselenide contribute further to difficulties in attempts to optimize reaction conditions.

E. SELENIUM DIOXIDE, SELENIOUS ACID, AND SELENIUM TRIOXIDE

Numerous oxidations of organic compounds by selenium dioxide and its hydrate, selenious acid, have been reported.[112] Since such work centered on organic derivatives that did not contain selenium, relatively little is known about the selenium-containing intermediates and by-products that are formed in many cases. Thus, selenium dioxide is not generally considered a useful reagent for the formation of new C-Se bonds. However, as a stimulus towards further work, several cases will be cited in which well-defined organoselenium compounds have been generated by this route.

1. Substitution at Carbon

Indole reacts with SeO_2 in benzene to give 3,3'-diindolyl selenide in 23 % yield.[113] The product structure was further confirmed by deselenation and isolation of 3,3'-diindolyl.[114] Various methyl-substituted indoles, when subjected to the same treatment, rendered similarly low yields of indolyl di- and triselenides,[113] with traces of the tetraselenide observed by mass spectrometry. In all cases the selenium was introduced into the 3-position of indole. The main reaction products had resulted from coupling of oxidized indole species and did not contain selenium.[113] Because of the lack of consistency in product structure, even with these closely related starting materials, and in the absence of a mechanistic interpretation, the reaction offers little possibility of general application at this point.

The reaction of selenium dioxide with olefins offers an attractive route to the introduction of allylic substituents, although an allylic rearrangement of the double bond appears to occur fairly regularly[115] and must be taken into consideration. A mechanistic investigation[115] of the reaction in 99 % acetic acid indicated that at least two pathways account for the observed product formation, the minor one of which involves intermediates containing C-Se bonds. The proposed pathway (scheme 36) was supported by isolation of a colorless, crystalline material, the composition and properties of which were consistent with the selenoxide depicted in scheme 36.

(36) $Ph—CH_2—CH=CH—Ph + HOSe=O^+ \longrightarrow$

$$O=Se—OH$$
$$|$$
$$Ph—CH_2—\overset{}{C}H—CH—Ph \xrightarrow[-H^+]{HOAc}$$
$$+$$

$$O=Se—OH$$
$$|$$
$$Ph—CH_2—\overset{}{C}H—CH—Ph$$
$$|$$
$$OAc$$

$Ph—CH_2—CH=CH—Ph$

$$OAc$$
$$|$$
$$Ph—CH_2—CH—\overset{}{C}H—Ph$$
$$|$$
$$Se=O$$
$$|$$
$$Ph—CH_2—\overset{}{C}H—CH—Ph$$
$$|$$
$$OAc$$

$\xrightarrow[\text{low yield}]{300°}$ $Ph—CH=CH—CH—Ph$
$$|$$
$$OAc$$

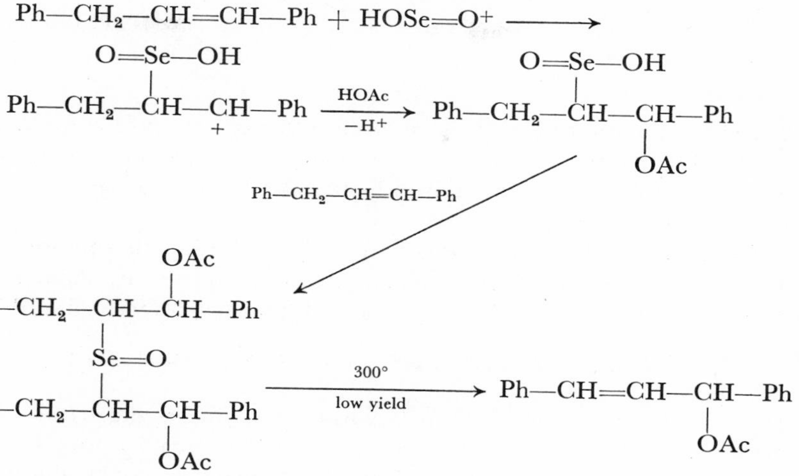

In a similar fashion the reactions of selenium dioxide in acetic acid with ethylene[116] and with cyclohexene[117] were found to generate selenium-containing products to which similar structures were assigned. These cases appear related to a number of olefin addition reactions reported for arylselenenyl acetates in the presence of pyridine.[118] Reaction scheme 37[118]

(37) $RSe—OCOCH_3 + H_2C=CHR' \rightarrow RSe—CH_2—CHR'—OCOCH_3$

appears rather similar to the initial steps in scheme 36 and suggests that a combination of selenious acid, acetic anhydride, and pyridine may, *via* an intermediate mixed anhydride of acetic and selenious acid, constitute a new reagent for the specific introduction of selenium into aliphatic unsaturated molecules. Product stability should be enhanced by starting with a terminal olefin and by reducing the selenoxide to a selenide before isolation.

These speculations aside, it appears feasible, then, to generate some well-defined organoselenium compounds, starting with selenium dioxide, and further work along these lines is indicated.

2. Reactions at Heteroatoms in Organic Molecules

The following examples of organic selenium compounds formed from selenious acid involve bonds other than C–Se. Aromatic orthodiamines undergo a general reaction with Se(IV) in aqueous and alcoholic solutions to give derivatives of condensed 2,1,3-selenadiazoles (equation 38). Reactions are

(38)

$$\text{(o-diaminobenzene)} + SeO_2 \longrightarrow \text{(benzoselenadiazole)} Se + 2H_2O$$

generally quantitative and have been utilized extensively for accurate spectrophotometric and fluorometric analysis of selenium. Details of the analytical uses are discussed in Chapter XVI, and the general class is treated in Chapter XIC.

The reaction of selenious acid with sulfhydryl reagents conforms to the general stoichiometry given in reaction 39, as proposed by Painter[119] and recently reconfirmed by Ganther.[120] The dithioselenides formed in this very

(39) $4RSH + H_2SeO_3 \rightarrow RS-Se-SR + RSSR + 3H_2O$

general reaction are difficult to separate from the symmetrical disulfide by-products. Thus, Stekol[121] assigned a tetrathioselenide structure to the selenium derivatives of cysteine and of glutathione obtained by this method, although these products were probably mixtures, according to reaction 39. The rapid interaction of selenite with sulfhydryl groups under physiological conditions (pH near neutrality to acidic) must be taken into account whenever the metabolic fate of administered selenium in intact animals, tissue preparations or biological extracts is investigated.

The conversion of Se(IV) to selenide, as given above, is, however, only one of the potential reactions of this system. It has recently[121] been found by the author that 2-(dimethylamino)ethanethiol hydrochloride in methanol, when added slowly and with rapid stirring to an excess of selenious acid in chilled methanol, yields a crystalline precipitate with a stoichiometry of 2 moles of thiol to 1 mole of selenium. The product is stable under acidic conditions in aqueous and alcoholic solvents and can be recrystallized from boiling ethanol or methanol. The empirical formula of $C_8H_{22}Cl_2N_2OS_2Se$ suggests an S,S-dialkyl dithioselenite structure (3) for this material. However, the

$$\begin{array}{c} CH_3 \\ \diagdown \\ N-CH_2CH_2S-Se-SCH_2CH_2-N \\ \diagup \quad\quad\quad \| \quad\quad\quad \diagdown \\ CH_3 \quad\quad\quad O \quad\quad\quad CH_3 \end{array} \cdot 2HCl$$

CH_3 (upper left), CH_3 (upper right), CH_3 (lower left), CH_3 (lower right)

3

reaction could not readily be extended to form analogous derivatives of other thiols.

3. Electrophilic Substitution with Selenium Trioxide

Reactions of aromatic ring systems with sulfuric acid, sulfur trioxide, and chlorosulfonic acid are well-known procedures for the introduction of sulfur into organic compounds. By contrast, the corresponding reactions in the selenium series have not been well investigated. It has been observed, however, that selenium trioxide in liquid sulfur dioxide reacts smoothly with benzene and with chloro- and bromobenzene according to equation 40.[122]

$$(40) \qquad C_6H_6 + SeO_3 \xrightarrow{\text{liq. } SO_2} \underset{\substack{85\% \\ \text{m.p. } 64°}}{C_6H_5SeO_3H} + \underset{3-5\%}{C_6H_5SeO_2C_6H_5}$$

Lower yields of selenonic acids and increased production of the selenones were observed when the selenation was carried out with 20% SeO_3 in H_2SeO_4. The reaction of SeO_3 with benzene in sulfur dioxide may also result in a product (m.p. 142°) that was characterized[124] as the acidium salt between phenylseleninic and phenylselenonic acid (**4**) after an erroneous assignment[123] as the selenonic acid had been made.

$$\left[C_6H_5Se \diagup[\text{OH}]{}^{\diagdown}\text{OH} \right]^+ \quad [C_6H_5SeO_3]^-$$

4

F. GENERAL NUCLEOPHILIC SELENIUM REAGENTS

This section describes reagents containing highly nucleophilic selenium anions, all capable of nucleophilic attack on carbon, displacing aliphatic halide and sulfonic ester substituents, as well as ring opening of epoxides and small-ring lactones. All are characterized by reagent preparation from elemental selenium and by use without isolation of the pure inorganic selenium compound. Aqueous solvents are common to this group, and while the primary products may differ in constitution, they are all readily convertible into aliphatic diselenides. Relatively mild reaction conditions and good to excellent yields prevail throughout. This class of reagents appears most useful for the formation of new C-Se bonds.

1. Potassium Selenosulfate

Elemental selenium dissolves readily in aqueous solutions of potassium sulfite at 80–100° to yield potassium selenosulfate (equation 41).

$$(41) \qquad K_2SO_3 + Se \xrightarrow{\text{aq.}} KSeSO_3K$$

As the rate of selenium dissolution depends on the sulfite concentration, as well as on pH and temperature, the much less soluble sodium sulfite is not recommended. The solubility of selenium in aqueous solutions of Na_2SO_3 and K_2SO_3 has been determined[205] from 0 to 100° and in a sealed tube from 100 to 152°. It is possible to isolate pure potassium selenosulfate,[125] for which a shelf-life of at least 6 years has been observed.[206] Recent methods[126,127] describe the preparation of the reagent and its use *in situ* as an excellent selenium nucleophile for displacement reactions on aliphatic halides (equation 42) or ring opening of aziridines (equation 43) to give well-characterized alkyl selenosulfates (seleno-Bunte salts). The isolation of selenosulfates is not required for the preparation of diselenides.

(42) $BrCH_2CH_2NH_3^+Br^- + KSeSO_3K$

$$H_3\overset{+}{N}CH_2CH_2SeSO_3^-$$

(43)

$$\begin{array}{c} H_2C\text{——}CH_2 \\ \diagdown \quad N \quad \diagup \\ \overset{+}{} \\ H \qquad H \end{array} + SeSO_3^{2-} + H^+$$

Both acidic and alkaline hydrolysis conditions yield the symmetrical diselenides. Since selenosulfuric acid is unstable[207] and decomposes into elemental selenium, water, and sulfur dioxide, the alkaline hydrolysis of crude alkyl selenosulfates contaminated by K_2SeSO_3 generally yields cleaner diselenides. If desired, diselenides can be prepared directly from alkyl halides or sulfonates by reaction with potassium selenosulfate at pH 12–13, obviating a separate hydrolysis step. A further variant, which has given excellent results in the hands of the author, consists in forming the alkyl selenosulfate at pH 9–10, freeing the solution from unreacted organic starting material and traces of selenium, and then raising the pH until precipitation of the diselenide appears complete.

2.   Selenium Solutions in Rongalite

Elemental selenium dissolves in aqueous solutions of sodium formaldehyde sulfoxylate (Rongalite, $NaSO_2CH_2OH$) in the presence of sodium hydroxide[128] to yield a reagent containing nucleophilic selenium anions. When a stoichiometry of 1 Se:5.5 NaOH:2.4 Rongalite is used, the reaction with alkyl halides yields mainly symmetrical selenides.[129]

When 1 g-atom of selenium reacts with 1 mole equivalent of Rongalite,[129,130] the brown solution contains chiefly the diselenide dianion.

Reactions with alkyl halides, alkyl sulfonates, and other species subject to nucleophilic displacement reactions proceed at room temperature or on heating to yield symmetrical diselenides in one step. The process is extremely convenient and has been widely applied. However, all reagents containing the diselenide dianion as the dominant species are likely to be contaminated by tri- and oligoselenide dianions, and the resulting alkyl-substituted products reflect this distribution. Alkyl polyselenides are unstable to light and storage in the dry state, and diselenides prepared by this route will generally turn red from deposited elemental selenium. Attempts at purification by crystallizing the aged products meet with little success since the finely divided element will partly redissolve in diselenide solutions.

In the author's experience the best purification is based on a partial reduction of the contaminated diselenide with, for instance, hypophosphorous acid[131] which converts the central Se atoms into hydrogen selenide. Crystallization in the absence of oxygen will then yield pure diselenides. An excess of reducing agent may, however, lower the recovery yield by also generating the corresponding selenols, which may remain in solution. Volatile diselenides may be freed from all polyselenide impurities by distillation.

It should be pointed out further that many older literature references claim instability of diselenides. Frequently, the appearance of elemental selenium in these cases was simply a consequence of the polyselenide contamination. A detailed discussion of diselenides is given in Chapter IV.

3.   Alkali Polyselenides

Aqueous sodium or potassium hydroxides in concentrations above 15 % react with powdered selenium to yield deep brown solutions that contain highly reactive selenide anions. The nature of this disproportionation reaction has been investigated on a number of occasions,[132–136] but no definite stoichiometry can yet be assigned to the products. Although the presence of Se_n^{2-} is postulated by all investigators, the oxidized species are variously quoted as SeO_3^{2-} and $Se_2O_3^{2-}$. Since the methodology, in every case, involved analysis by chemical reactions likely to disturb the redox equilibrium, final clarification will have to come through spectroscopic techniques applicable to the original solutions. For reactions of such polyselenide solutions with alkyl halides[137,138] similar arguments apply regarding the distribution of species, as given in Section III.F.2, and only very impure diselenides are isolated from the initial mixture. Because the reaction is carried out in a highly alkaline medium, further complications arise from competing hydrolyses. However, the simplicity of the procedure is such that further variants incorporating chemical or electrochemical reduction steps to yield a "clean" diselenide reagent might profitably be investigated.

4. Bis(methoxymagnesium) Diselenide

It has recently been observed[139] that elemental selenium may enter the methoxide-forming reaction between Grignard magnesium and absolute methanol to give a deep brown solution containing diselenide anions (scheme 44). The reagent may be obtained in crystalline form. Its com-

$$(44) \qquad 6CH_3OH + 2Mg + 2Se \rightarrow (CH_3OMgSe)_2 \cdot 4CH_3OH$$

position and reactions have led to a structure assignment as bis(methoxymagnesium) diselenide, $(CH_3OMgSe)_2$. The reagent in methanol solution contains selenium anions in a highly reactive form, allowing nucleophilic displacement, not only on alkyl halides but even in ring-opening reactions, by attack on the γ-carbon of lactones. Benzaldehyde was cleanly converted to dibenzyl diselenide by the reagent only in the presence of morpholine.

The advantages of this new reagent are the use of a nonaqueous solvent for a single-step diselenide formation and the derived absence of hydrolytic side reactions. The disadvantages involve possible tri- and oligoselenide formation, as discussed in Section III.F.2.

G. SELENIUM HALIDES AND OXYHALIDES

The use of sulfur halides and oxyhalides as reagents for the synthesis of organic sulfur and/or halogen compounds is well established. By comparison, the chemistry of the corresponding selenium compounds has been exploited to a much lesser degree. Table III-1 lists a number of selenium halogen derivatives that have found some utility in organic syntheses. Of these, the oxyhalides are derived from selenium dioxide by treatment with hydrogen halide and concentrated sulfuric acid. The simple halides are prepared by combining the calculated quantities of the elements. However, maintenance of the desired stoichiometry during preparation and storage and in the course of chemical reactions presents some problems since exchange reactions take place readily. Thus, selenium monochloride and selenium tetrachloride equilibrate according to reactions 45 and 46.[140,141] Similar observations were made with respect to selenium bromides.[142,144]

$$(45) \qquad\qquad Se_2Cl_2 \rightleftharpoons SeCl_2 + Se$$

$$(46) \qquad\qquad SeCl_4 + Se \rightleftharpoons 2SeCl_2$$

The halides are further compatible with elemental selenium, forming numerous nonstoichiometric "alloys." As a consequence, some uncertainty

Table III-1 Selenium halides and oxyhalides[a]

NAME	FORMULA	APPEARANCE	MELTING POINT (BOILING POINT)	DENSITY
Selenium monochloride	Se_2Cl_2	Red-brown liquid	$-85°$ (130° dec.)	d_4^{24} 2.77
Selenium monobromide[b]	Se_2Br_2	Dark red liquid	(225–230° dec.)	d_4^{15} 3.604
Selenium tetrachloride	$SeCl_4$	Colorless to yellow cubic crystals	305° sealed tube, 196° subl., 288° dec.	d 3.80
Selenium tetrabromide[b]	$SeBr_4$	Yellow-orange crystals	75° dec.	
Selenium oxychloride	$SeOCl_2$	Pale yellow liquid	8.5° (176.4°)	d^{22} 2.42
Selenium oxybromide	$SeOBr_2$	Red-yellow crystals	41.6° (217°/740 mm dec.)	d^{50} 3.38

[a] Data from Feher,[143] unless otherwise indicated.
[b] Ref. 144.

exists with regard to reaction conditions and yields when following reports from the older literature.

1. Addition Reactions to Multiple Bonds

Selenium monochloride reacts with isolated C-C double bonds according to scheme 47.[145–150] Depending on the reactivity of the olefin, the reaction

(47)

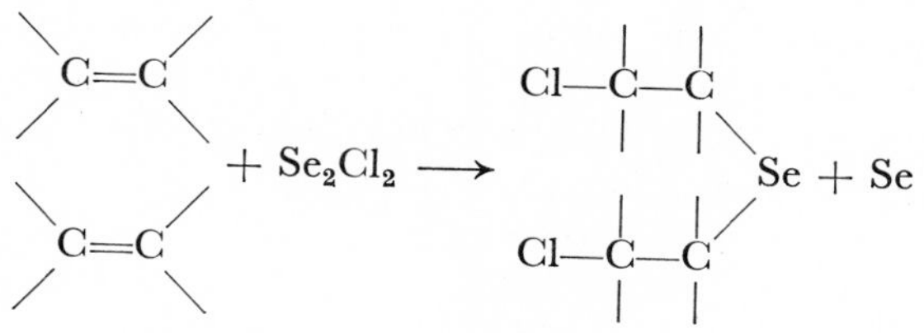

temperatures may range from $-40°$ to the reflux temperature in halogenated hydrocarbons or benzene. It appears important to avoid an excess of Se_2Cl_2 in these reactions since chlorination of the product may otherwise occur to give a bis(2-chloroalkyl) selenide dichloride. Olefins containing two double bonds in suitable positions are converted to cyclic selenides containing chlorine substituents in both β-positions.[145]

Selenium monobromide appears to react in a different fashion, leading to diselenides according to equation 48.[151] However, this represents a single example, and further studies seem indicated.

(48) $Se_2Br_2 + F_2C{=}CF_2 \rightarrow (BrF_2C{-}CF_2Se{-})_2$

Selenium tetrachloride adds smoothly to double bonds to yield bis(2-chloroalkyl) selenide dichlorides according to equation 49.[152,153] It is

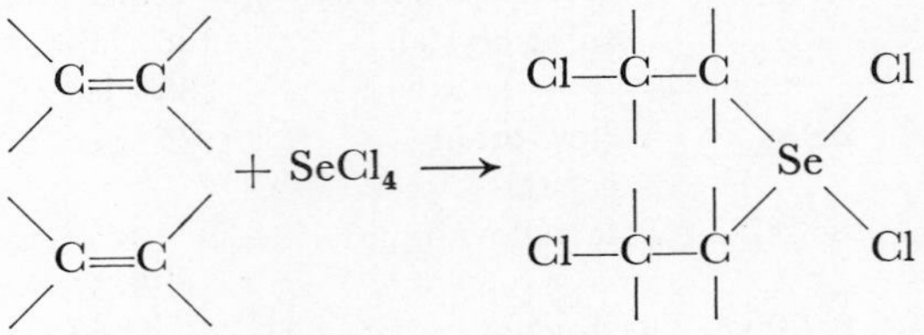

interesting that reaction type 49 can also be carried out by combining selenium dioxide, saturated aqueous HCl (or HBr), and the desired alkenes.[154] By this method it was also possible to prepare bis(2,2'-dibromodivinyl)-selenium dibromide from acetylene.[154]

Further examples and specific products are discussed in Chapter V.

2. Electrophilic Substitutions

A variety of reactions are known in which selenium is introduced in place of hydrogen by reaction of suitably substituted aliphatic and aromatic species with either selenium tetrachloride or selenium oxychloride. These are summarized in this section, although mechanistic details are, generally, lacking.

Thus, the condensation of acetone[155] and of methyl aryl ketones[156] with $SeCl_4$ occurs readily in ether, according to scheme 50. The reaction with selenium oxychloride[157] occurs under similar mild conditions, again yielding selenide dichlorides (scheme 50).

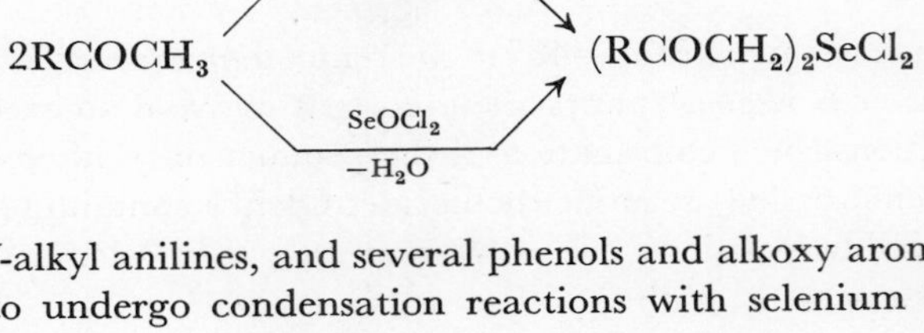

Aniline, *N*-alkyl anilines, and several phenols and alkoxy aromatic species are known to undergo condensation reactions with selenium halides and oxyhalides. Thus 4,4′-bis(dimethylamino)phenyl selenide is formed according to equation 51 by combining the reactants in petroleum ether.

(51) 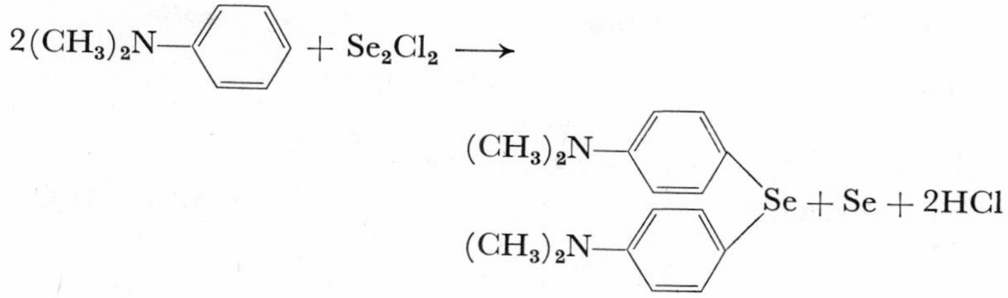 $2(CH_3)_2N$—⟨ ⟩ + Se_2Cl_2 ⟶

$(CH_3)_2N$—⟨ ⟩

$(CH_3)_2N$—⟨ ⟩ Se + Se + 2HCl

The reaction of dimethylaniline and $SeOCl_2$ in ether gives the same product as reaction 51 in 65 % yield.[159] However, the mechanism of oxygen or halide elimination is not known.

In a similar fashion an ingenious, recent synthesis of polycondensed selenophenes[160–162] involves an eventual loss of a substituent on selenium by an unknown process (equation 52). The proposed reaction mechanism[162]

(52) 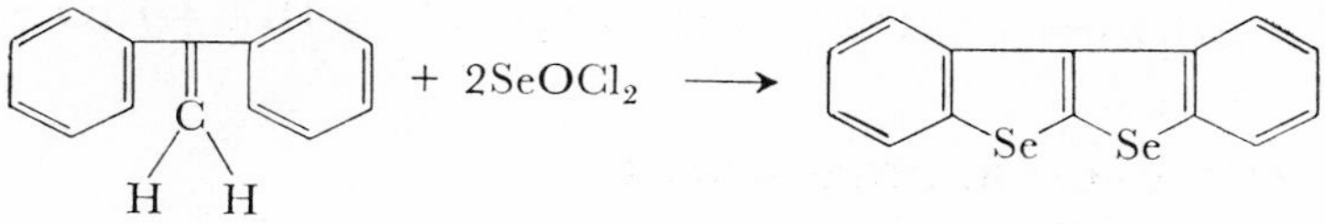 ⟨diarylethylene⟩ $\overset{H}{\underset{H}{C}}$ + $2SeOCl_2$ ⟶ ⟨polycondensed structure⟩ Se Se

involves attack on the terminal double bond of the diarylethylene by $SeOCl_2$, possibly by an addition-elimination sequence, until both aliphatic hydrogens are replaced by an —SeOCl group. Ring closure then takes place at the sterically favored *ortho* position of the aromatic moiety with elimination of hydrogen chloride, presumably through electrophilic attack by the —SeO+ group. The product should now be the diselenoxide, but elimination of oxygen takes place somehow and the only type of product isolated is the polycondensed aromatic species.

In a similar manner electrophilic cyclization reactions of aryl biphenyl selenide dichlorides occur readily, as shown in equation 53.[163]

(53) 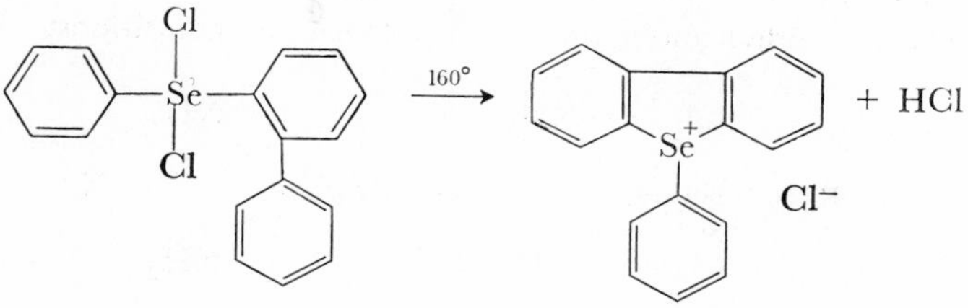 ⟨structure with Se, Cl, Cl⟩ $\xrightarrow{160°}$ ⟨cyclized structure⟩ $\overset{+}{Se}$ + HCl

Cl⁻

Naphthols and naphthol ethers, as well as alkoxybenzenes (and presumably a variety of related aromatics), react according to equation 54.[155,164]

(54)

However, the formation of selenide dichlorides has been reported,[165,166] under essentially similar conditions, for phenyl ethers, and the case appears to require some careful reinvestigation. The reaction with phenol itself, as well as with resorcinol and *o*- and *p*-cresol, is reported to yield triaryl selenonium halides according to equation 55.[155]. References to numerous other attempted conversions are given by Rheinboldt.[159]

$$(55) \qquad 3HOArH + SeOCl_2 \rightarrow (HOAr)_3Se^+Cl^- + HCl + H_2O$$

H. SIMPLE ORGANIC INTERMEDIATES

This section is intended mainly to draw attention to a few organic derivatives of selenium that have general applicability for the formation of new C-Se bonds. In these cases the organic moiety acts as a protecting group for the selenium in the initial reaction and in any further synthetic sequence, but it is then removed by a suitable procedure.

1. Selenourea

While all selenium nucleophilic reagents cited in the preceding sections contain a selenium anion and are generally used in alkaline media, the neutral selenourea is readily alkylated to give isoselenouronium salts (selenopseudoureas) (reaction 56) in organic solvents. Hydrolytic side reactions are avoided under the very mild reaction conditions.

Reactions are carried out in refluxing ethanol, 2-propanol, or acetone, and applications of the technique range from reactions with simple halides[167] and more complex halogeno sugars[168] to steroid *p*-toluenesulfonates,[169,170] to cite only a few examples.

On occasion, the isoselenouronium salt cannot be isolated and the reaction products represent direct replacement of halogen by selenium. Instances are, especially, the preparation of seleno-substituted nitrogen heterocycles, such as 6-selenopurines.[171,172] In most cases, the isoselenouronium salts are readily hydrolyzed in base to the selenol. If desired, additional alkylation or acylation reactions on the selenol may be performed *in situ*.

Additional synthetic applications of selenourea involving reactions on the N atoms are described in Chapters VIIIB, XIC, and XID.

2. Benzylselenol

In the synthesis of complex aliphatic selenols it is frequently necessary to introduce the selenium function in such a fashion that sequence reactions may be carried out on the product without adversely affecting the sensitive selenium substituent. Protection by formation of easily reducible diselenides (see Chapter IV) introduces some problems, since further reactions must then be applied to a symmetrical, difunctional compound. A better solution for transient blocking of selenols is the formation of selenol esters[173–178] (RSeCOR′), from which the acyl groups are readily removed by aminolysis under very mild conditions. However, this sensitivity towards hydrolytic and aminolytic scission does not commend acyl protection for more than one or two sequential steps. Similar arguments apply to most other substituent groups that may be introduced *via* reactions with monofunctional selenium reagents. An exception is the benzylselenium moiety. Benzylselenoalkanes are stable to a wide variety of reagents and reaction conditions, and the protecting benzyl group may be removed very cleanly by reduction with sodium in liquid ammonia.[179]

Benzylselenol can be prepared by reaction of benzylmagnesium bromide with elemental selenium[179] or, better, by reduction of dibenzyl diselenide with lithium aluminum hydride.[181] The diselenide, in turn, can be generated from benzyl chloride by any of the diselenide reagents cited in Section III.F, and constitutes a convenient storage form.

The reactions of benzylselenol and its anion are typically those encountered with other selenium nucleophiles, except that the reactions appear to proceed even more smoothly and that the resulting benzyl alkyl mono-selenides are rather stable and readily isolable.

Addition to double bonds[182–184] can be illustrated by a portion of a selenocystine synthesis (scheme 57).[181,182]

(57) 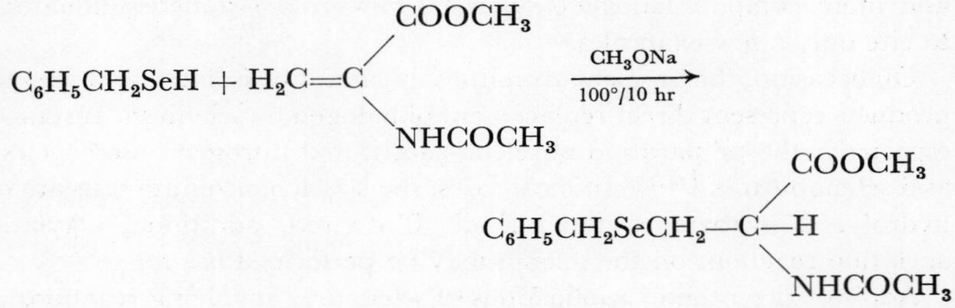 $C_6H_5CH_2SeH + H_2C=C\begin{subarray}{l} \\ \text{COOCH}_3 \\ \\ \text{NHCOCH}_3 \end{subarray}$ $\xrightarrow[\text{100°/10 hr}]{\text{CH}_3\text{ONa}}$

$$C_6H_5CH_2SeCH_2-\underset{\underset{\text{NHCOCH}_3}{|}}{\overset{\overset{\text{COOCH}_3}{|}}{C}}-H$$

Nucleophilic ring-opening reactions occur readily with propiolactone,[185] γ-butyrolactone,[185] (equation 58), and substituted butyrolactones, including phthalide,[187] by selenide attack on the ω-carbon as previously observed[188] for sodium methyl selenide in its reaction with 2-amino-γ-butyrolactone.

(58) 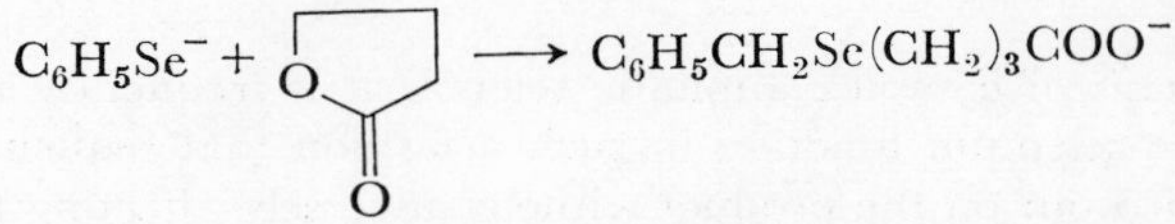 $C_6H_5Se^- + \underset{\underset{O}{\parallel}}{O}\!\!\diagdown \!\!\!\square \longrightarrow C_6H_5CH_2Se(CH_2)_3COO^-$

The high nucleophilicity of the species is further illustrated by its reaction (equation 59) with gramine on heating without additional reagents.[189]

(59)

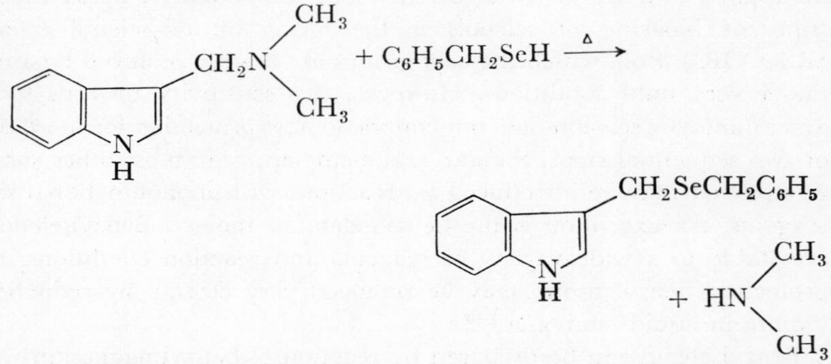

Displacement by the benzyl selenide anion of p-toluenesulfonate ion from serine and homoserine in peptides[190–192] represents another recent use for benzylselenol. This reaction, which constitutes a well-controlled three-step replacement of oxygen by selenium (reaction 60), should be extremely valuable and applicable to many other systems:

(60) $R-OH \xrightarrow{\text{CH}_3\text{C}_6\text{H}_4\text{SO}_2\text{Cl}} R-O-SO_2C_6H_4CH_3 \xrightarrow{\text{C}_6\text{H}_5\text{CH}_2\text{Se}^-}$

$$R-SeCH_2C_6H_5 \xrightarrow{\text{Na/NH}_3} RSeH$$

3. Carbon Diselenide

Reactions with CSe_2 are discussed here although many of the organic reaction products of this compound are derived without formation of new C-Se bonds.

Carbon diselenide has been prepared by reaction of carbon tetrachloride with phosphorus selenide,[193,194] cadmium selenide,[195] or hydrogen selenide,[193,196] The most satisfactory method involves passing methylene chloride vapor in dry nitrogen over molten selenium kept at 600°[197,198] and trapping the effluent gases at low temperature. Yields of product (b.p. 45.5–46°/ 50 mm Hg) are relatively low (*i.e.*, in the 30 % range).

Carbon diselenide and undefined by-products convey one of the worst odors this author has ever experienced. Regular updraft hoods are rather ineffective in moving the heavy vapors beyond the laboratory building, and complaints from the surrounding populace add to the discomfort, even if the working area stays fairly clear. Hence, it is advisable to work in a closed system with efficient traps. Charcoal, in combination with ethanolic potassium hydroxide or primary aliphatic amines, such as ethanolamine, may be used for effective removal of the carbon diselenide. Alkaline hydrogen peroxide or permanganate solutions have also been found useful to "deactivate" flasks and to rinse off accidental spills.

The chemical reactions of carbon diselenide are somewhat analogous to those of carbon disulfide, although significant differences have been noted by various authors. Thus, CSe_2 reacts with secondary amines to form *N*,*N*-dialkyldiselenocarbamic acids.[198]

The reaction with primary amines[192,200] proceeds with formation of symmetrical selenoureas. A proposed mechanism[199] for this condensation involves an isoselenocyanate intermediate. The formation of analogous cyclic compounds[199] (reaction 61) supports this mechanism.

(61)

(A = NH, O, or S)

For the synthetic process it is important to avoid an excess of CSe_2 during the reaction,[199] since polymerization of the selenide may be initiated by amines[198] as well as by the analogous phosphines.[201] Structure **5** has been proposed for the polymer,[201] but no detailed studies have been undertaken to determine polymer properties.

5

Interaction of carbon diselenide with active methylene groups occurs readily[202] by the general process in scheme 62, which also illustrates a trapping method for the 1,1-diselenolate anions:

(62)

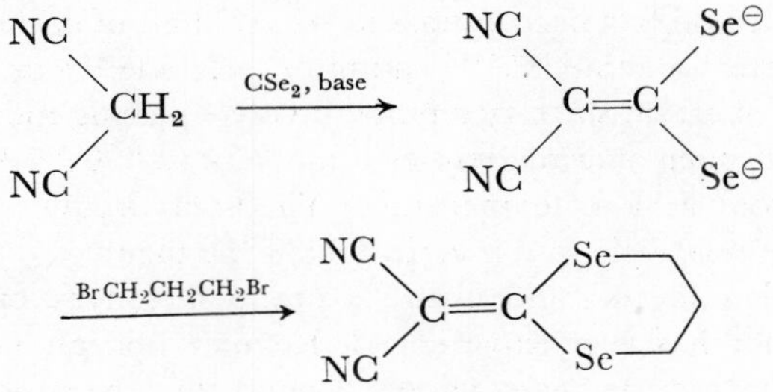

A number of analogous reactions have been summarized.[202] The preparation of some diselenoxanthogenates and their sequence reactions with phosphorus trichloride have been described.[203,204]

REFERENCES

1. W. A. Pryor, "Mechanisms of Sulfur Reactions," McGraw-Hill, New York. N.Y., 1962, pp. 162 ff.

2. F. Krafft and W. Vorster, *Ber.*, **26**, 2817 (1893).

3. F. Krafft and R. E. Lyons, *Ber.*, **27**, 1761 (1894).

4. J. Loevenich, H. Fremdling, and M. Föhr, *Ber.*, **62**, 2864 (1929).

5. F. Krafft and A. Kaschau, *Ber.*, **29**, 443 (1896).

6. N. M. Cullinane, A. G. Rees, and C. A. J. Plummer, *J. Chem. Soc.*, 153 (1939).

7. N. M. Cullinane, N. M. E. Morgan, and C. A. J. Plummer, *Rec. Trav. Chim. Pays-Bas*, **56**, 629 (1937).

8. J. Loevenich, H. Fremdling, and M. Föhr, *Ber.*, **62**, 2856 (1929).

9. G. Bergson, *Ark. Kemi*, **10**, 127 (1956).

10. W. H. H. Günther, unpublished observation.

11. W. H. H. Günther and H. G. Mautner, *J. Amer. Chem. Soc.*, **87**, 2708 (1965).

12. J. Meisenheimer, *Ber.*, **52**, 1667 (1919).

13. J. Meisenheimer, H. Greeske, and A. Willmersdorf, *Ber.*, **55**, 513 (1922).

14. E. G. Miller, D. R. Rayner, and K. Mislow, *J. Amer. Chem. Soc.*, **88**, 3139 (1966).

15. A. Biezais and G. Bergson, *Acta Chem. Scand.*, **18**, 815 (1964).

16. N. N. Yarovenko, V. N. Shemanina, and G. B. Gazieva, *Zh. Obshch. Khim.*, **29**, 942 (1959).

17. R. N. Haszeldine, *J. Chem. Soc.*, 4259 (1952).

18. J. W. Dale, H. J. Emeléus, and R. N. Haszeldine, *J. Chem. Soc.*, 2939 (1958).

19. C. G. Krespan, *J. Amer. Chem. Soc.*, **83**, 3434 (1961).

20. D. Hellwinkel and G. Fahrbach, *Justus Liebigs Ann. Chem.*, **715**, 68 (1968).

21. F. Zeiser, *Ber.*, **28**, 1671 (1895).

22. R. E. Lyons and G. C. Bush, *J. Amer. Chem. Soc.*, **30,** 831 (1908).

23. N. S. Vyazankin, M. N. Bochkarev, and L. P. Sanina, *Zh. Obshch. Khim.*, **38,** 414 (1968).

24. K. A. Jensen and E. Frederiksen, *Z. Anorg. Allg. Chem.*, **230,** 285 (1936).

25. R. Huls and M. Renson, *Bull. Soc. Chim. Belges*, **65,** 684 (1956).

26. C. Collard-Charon, R. Huls, and M. Renson, *Bull. Soc. Chim. Belges*, **71,** 531, 541 (1962).

27. C. T. Pedersen, *Acta Chem. Scand.*, **17,** 1459 (1963).

28. E. Bulka and K.-D. Ahlers, *Z. Chem.*, **3,** 348 (1963).

29. W. J. Franklin and R. L. Werner, *Tetrahedron Lett.*, 3003 (1965).

30. F. Taboury, *Bull. Soc. Chim. Fr.*, **29,** 761 (1903); **35,** 668 (1906).

31. G. T. Morgan and W. H. Porritt, *J. Chem. Soc.*, 1755 (1925).

32. L. D. Pettit, A. Royston, C. Sherrington, and R. J. Whewell, *J. Chem. Soc., B*, 558 (1968).

33. B. Sjöberg and S. Herdevall, *Acta Chem. Scand.*, **12,** 1347 (1958).

34. D. G. Foster, *J. Amer. Chem. Soc.*, **61,** 2973 (1939).

35. H. J. Backer and J. B. G. Hurenkamp, *Rec. Trav. Chim. Pays-Bas*, **61,** 802 (1942).

36. E. P. Painter, *J. Amer. Chem. Soc.*, **69,** 232 (1947).

37. E. Niwa, H. Aoki, H. Tanaka, K. Munakata, and M. Namiki, *Chem. Ber.*, **99,** 3215 (1966).

38. Ya. L. Goldfarb, V. P. Litvinov, and S. A. Ozolin', *Akad. Nauk SSSR, Izv. Ser. Khim.*, Nr. 6, 1347 (1968).

39. E. S. Gould and J. D. McCullough, *J. Amer. Chem. Soc.*, **73,** 1109 (1951).

40. L. Brandsma, H. E. Wijers, and J. F. Arens, *Rec. Trav. Chim. Pays-Bas*, **81,** 583 (1962).

41. L. Brandsma, H. E. Wijers, and C. Jonker, *Rec. Trav. Chim. Pays-Bas*, **83,** 208 (1963).

42. A. A. Petrov, S. I. Radschenko, K. S. Mingaleva, I. G. Savich, and V. B. Lebedev, *Zh. Obshch. Khim.*, **34,** 1899 (1964); *Chem. Abstr.*, **61,** 8179g (1964).

43. Yu. A. Boiko, B. S. Kupin, and A. A. Petrov, *Zh. Org. Khim.*, **4,** 1307 (1968).

44. R. Mayer and K. Müller, *Z. Chem.*, **4,** 384 (1964).

45. A. B. Callear and W. J. R. Tyerman, *Proc. Chem. Soc.*, 296 (1964).

46. A. B. Callear and W. J. R. Tyerman, *Trans. Faraday Soc.*, **61,** 2395 (1965); **62,** 1, 371 (1966).

47. W. J. R. Tyerman, W. B. O'Callaghan, P. Kebarle, O. P. Strausz, and H. E. Gunning, *J. Amer. Chem. Soc.*, **88,** 4277 (1966).

48. O. P. Strausz and H. E. Gunning, "The Chemistry of Sulfides." A. V. Tobolsky, Ed., Interscience Publishers, New York, N.Y., 1968, p. 23.

49. F. Feher, "Handbook of Preparative Inorganic Chemistry," Vol. 1, G. Brauer, Ed., Academic Press, New York, N.Y., 1963, p. 418.

50. H. Rheinboldt, "Houben-Weyl Methoden der Organischen Chemie," Vol. IX, E. Müller, Ed., Georg Thieme Verlag, Stuttgart, 1955, p. 953.

51. J. Gosselck and E. Wolters, *Z. Naturforsch.*, **17,** 131 (1962).

52. W. Hansel and R. Haller, *Naturwissenschaften*, **55,** 83 (1968).

53. J. Ashby, L. A. Cort, J. A. Elvidge, and U. Eisner, *J. Chem. Soc., C*, 2311 (1968).

54. M. Renson and R. Collienne, *Bull. Soc. Chim. Belges*, **73,** 491 (1964).

55. R. Mayer, S. Scheithauer, and D. Kunz, *Chem. Ber.*, **99,** 1393 (1966).

56. U. Schmidt, E. Heymann, and K. Kabitzke, *Chem. Ber.*, **96,** 1478 (1963).

57. R. S. Ramakrishna and H. M. N. H. Irving, *Chem. Ind.* (London), 325 (1969).

58. D. L. Klayman and R. J. Shine, *Chem. Commun.*, 372 (1968).

59. D. L. Klayman and R. J. Shine, *J. Org. Chem.*, **34**, 3549 (1969).

60. F. Zetzschke and H. Pinske, *Ber.*, **74**, 1022 (1941).

61. R. A. Zingaro, F. C. Bennett, Jr., and G. W. Hammar, *J. Org. Chem.*, **18**, 294 (1953).

62. H. J. Backer and H. Bos, *Rec. Trav. Chim. Pays-Bas*, **62**, 580 (1943).

63. A. Verneuil, *Bull. Soc. Chim. Fr.*, **41**, 599 (1884).

64. L. C. King and R. J. Hlavacek, *J. Amer. Chem. Soc.*, **73**, 1864 (1951).

65. R. E. Dunbar and E. P. Painter, *J. Amer. Chem. Soc.*, **69**, 1833 (1947).

66. K. A. Jensen, M. Due, A. Holm, and C. Wentrup, *Acta Chem. Scand.*, **20**, 2091 (1966).

67. A. Baroni, *Atti Accad. Italia, Rend. Classe Sci., Fis., Mat., Nat* [7], **1**, 543 (1940); *Chem. Abstr.*, **37**, 869 (1943).

68. F. von Dechend, *Ber.*, **7**, 1273 (1874).

69. G. Hofmann, *Justus Liebigs Ann. Chem.*, **250**, 314 (1889).

70. W. Becker, *Ber.*, **37**, 2551 (1904).

71. K. Kindler, *Justus Liebigs Ann. Chem.*, **431**, 189, 206, 207 (1923).

72. L. G. S. Brooker, G. H. Keyes, and F. L. White, *J. Amer. Chem. Soc.*, **57**, 2494 (1935).

73. R. F. Curtis, S. N. Hasnain, and J. A. Taylor, *Chem. Commun.*, 365 (1968).

74. K. E. Schulte, J. Reisch, W. Herrman, and G. Bohn, *Arch. Pharm.*, **296**, 456 (1963).

75. H. J. Bridger and R. W. Pittman, *J. Chem. Soc.*, 1371 (1950).

76. D. S. Margolis and R. W. Pittman, *J. Chem. Soc.*, 799 (1957).

77. R. E. Lyons and W. E. Brandt, *Ber.*, **60**, 824 (1927).

78. F. Asinger, H. Berding, and H. Offermanns, *Monatsh. Chem.*, **99**, 2072 (1968).

79. C. G. Barraclough, R. L. Martin, and I. M. Stewart, *Australian J. Chem.*, **22**, 891 (1969).

80. G. A. Heath, I. M. Stewart, and R. L. Martin, *Nucl. Inorg. Lett.* **5**, 169 (1969).

81. N. N. Yarovenko and M. A. Raksha, *Zh. Obshch. Khim.*, **30**, 6064 (1960).

82. H. G. Mautner and M. Hecht, personal communication.

83. D. L. MacPeek and W. H. Rauscher, U.S. Pat. 2,729,676 (1956).

84. L. Brandsma and H. E. Wijers, *Rec. Trav. Chim. Pays-Bas*, **82**, 68 (1963).

84a. D. L. Klayman and T. S. Griffin, *J. Amer. Chem. Soc.*, in press (1973).

85. F. Feher, "Handbook of Preparative Inorganic Chemistry," Vol. 1, G. Brauer, Ed., Academic Press, New York, N.Y., 1963, p. 421.

86. W. Klemm, H. Sodomann, and P. Langmesser, *Z. Anorg. Allg. Chem.*, **241**, 281 (1939).

87. Yu. N. Shlyk, G. M. Bogolyubov, and A. A. Petrov, *Zh. Obshch. Khim.*, **38**, 1199 (1968).

88. G. R. Waitkins and R. Shutt, *Inorg. Syn.*, **2**, 186 (1946).

89. W. Muthmann and E. Schroeder, *Ber.*, **33**, 1766 (1900).

90. T. Tarantelli and C. Pecile, *Ann. Chim.* (Rome), **52**, 75 (1962).

91. F. Challenger, A. T. Peters, and J. Halévy, *J. Chem. Soc.*, 1648 (1926).

92. A. Verneuil, *Bull. Soc. Chim. Fr.*, [2], **46**, 193 (1886).

93. A. Verneuil, *Ann. Chim.* (Paris), [6], **9**, 328 (1886).

94. W. Muthmann and E. Schroeder, *Ber.*, **33**, 1766 (1900).

95. O. Aksnes and O. Foss, *Acta Chem. Scand.*, **8**, 702, 1787 (1954).

96. T. W. Campbell and M. T. Rogers, *J. Amer. Chem. Soc.*, **70,** 1029 (1948).

97. L. Birkenbach and K. Kellermann, *Ber.*, **58,** 790 (1925).

98. H. P. Kaufmann and F. Kogler, *Ber.*, **59,** 178, 185 (1926).

99. L.-B. Agenäs, *Acta Chem. Scand.*, **17,** 268 (1963).

100. L.-B. Agenäs, *Ark. Kemi*, **23,** 145, 155 (1965).

101. L.-B. Agenäs, *Ark. Kemi*, **28,** 145 (1968).

102. L.-B. Agenäs, *Ark. Kemi*, **30,** 417 (1969).

103. L.-B. Agenäs, Inaugural Dissertation, University of Uppsala, Sweden, 1969.

104. See Gmelin's "Handbuch der Anorganischen Chemie," Sys. Nr. 9–12, Verlag Chemie, Weinheim/Bergstr., 1942, p. 244.

105. L. Carius and W. Bogen, *Justus Liebigs Ann. Chem.*, **124,** 57 (1862).

106. M. V. Kudchadker, R. A. Zingaro, and K. J. Irgolic, *Can. J. Chem.*, **46,** 1415 (1968).

107. W. Muthmann and A. Clever, *Z. Anorg. Chem.*, **13,** 191 (1897).

108. K. A. Jensen, G. Felbert, and B. Kägi, *Acta Chem. Scand.*, **20,** 281 (1965).

109. H. E. Hallam and C. M. Jones, *J. Chem. Soc., B*, 1033 (1969).

110. J. H. van den Hende and E. Klingsberg, *J. Amer. Chem. Soc.*, **88,** 5045 (1966).

111. E. Klingsberg, U.S. Pat. 3,407,213 (1968).

112. For relevant reviews see: N. Rabjohn, *Org. Reactions*, **5,** 331 (1949); E. N. Trachtenberg, "Oxidation," R. L. Augustine, Ed., Marcel Dekker, New York, N.Y., 1969.

113. J. F. K. Wilshire, *Australian J. Chem.*, **20,** 359 (1967).

114. J. Bergman, *Acta Chem. Scand.*, **22,** 1883 (1968).

115. J. P. Schaefer, B. Horvath, and H. P. Klein, *J. Org. Chem.*, **33,** 2647 (1968).

116. D. H. Olson, *Tetrahedron Lett.*, 3053 (1966).

117. K. Wiberg and S. D. Nielsen, *J. Org. Chem.*, **29,** 3353 (1964).

118. G. Hölzle and W. Jenny, *Helv. Chim. Acta*, **41,** 331, 593 (1958).

119. E. P. Painter, *Chem. Rev.*, **28,** 179 (1941).

120. H. E. Ganther, *Biochemistry*, **7,** 2898 (1968).

121. W. H. H. Günther, unpublished data.

122. K. Dostál, P. Mošna, and Z. Žák, *Z. Chem.*, **6,** 153 (1966).

123. M. Schmidt and I. Wilhelm, *Chem. Ber.*, **97,** 872 (1964).

124. R. Paetzold and D. Lienig, *Z. Chem.*, **4,** 186 (1964).

125. F. Foerster, F. Lange, O. Drossbach, and W. Seidel, *Z. Anorg. Allg. Chem.*, **128,** 312 (1928).

126. W. H. H. Günther and H. G. Mautner, *J. Med. Chem.*, **7,** 229 (1964).

127. D. L. Klayman, *J. Org. Chem.*, **30,** 2454 (1965).

128. L. Tschugaeff and W. Chlopin, *Ber.*, **47,** 1269 (1914); *Zh. Russ. Fiz. Khim. Obshch.*, **47,** 364 (1915); *Chem. Zentr.*, **1,** 825 (1916).

129. M. L. Bird and F. Challenger, *J. Chem. Soc.*, 570 (1942.)

130. A. Fredga, *Acta Chem. Scand.*, **17,** S51 (1963).

131. W. H. H. Günther, *J. Org. Chem.*, **31,** 1202 (1966).

132. E. Schulek and E. Körös, *J. Inorg. Nucl. Chem.*, **13,** 58 (1960).

133. P. Adami, *Atti Mem. R, Acad. Sci. Lett. Arti Padova, Mem., Classe Sci. Fis-Mat.*, **51,** 13 (1934/35).

134. G. Calcagni, *Gazz. Chim. Ital.*, **53**, 114 (1923).

135. R. L. Espil, *Bull. Soc. Chim. Fr.*, **7**, 155 (1910).

136. J. J. Berzelius, "Lehrbuch der Chemie," Vol. 3, 3rd ed., Dresden-Leipzig, 1834, p. 26.

137. T. W. Campbell, H. G. Walker, and G. M. Coppinger, *Chem. Rev.*, **50**, 279 (1950).

138. H. Rheinboldt, "Houben-Weyl Methoden der Organischen Chemie," Vol. IX, E. Müller, Ed., Georg Thieme Verlag, Stuttgart, 1955, p. 1088.

139. W. H. H. Günther, *J. Org. Chem.*, **32**, 3929 (1967).

140. D. M. Yost and C. E. Kircher, *J. Amer. Chem. Soc.*, **52**, 4680 (1930).

141. H. Brintzinger, K. Pfannstiel, and H. Vogel, *Z. Anorg. Allg. Chem.*, **256**, 75 (1948).

142. N. W. Tideswell and J. D. McCullough, *J. Amer. Chem. Soc.*, **78**, 3026 (1956).

143. For preparative details see F. Feher, "Handbook of Preparative Inorganic Chemistry," Vol. 1, G. Brauer, Ed., Academic Press, New York, N.Y., 1963, p. 341 ff.

144. For a recent study see: N. Katsaros and J. W. George, *Chem. Commun.*, 662 (1968).

145. F. Lautenschlaeger, *J. Org. Chem.*, **34**, 4002 (1969).

146. C. E. Boord and C. F. Cope, *J. Amer. Chem. Soc.*, **44**, 395 (1922).

147. E. S. Gould and J. D. McCullough, *J. Amer. Chem. Soc.*, **73**, 1105 (1951).

148. F. H. Heath and W. L. Semon, *J. Ind. Eng. Chem.*, **12**, 1101 (1920).

149. H. W. Bausor, C. S. Gibson, and W. J. Pope, *J. Chem. Soc.*, **117**, 1453 (1920).

150. E. H. Harvey and H. A. Schuette, *J. Amer. Chem. Soc.*, **50**, 2837 (1928).

151. N. N. Yarovenko, M. A. Raksha, and V. N. Shemanina, *Zh. Obshch. Khim.*, **30**, 4069 (1960).

152. R. F. Riley, J. Flato, and D. Bengels, *J. Org. Chem.*, **27**, 2651 (1962).

153. T. Smedslund, Dissertation, Helsingfors Universitet, Helsinki, Finland, 1932; *Finska Kemistsamfundets Medd.*, **41**, 13 (1932); *Chem. Abstr.*, **26**, 5905 (1932).

154. H. Funk and W. Papenroth, *J. Prakt. Chem.*, **8**, 256 (1959).

155. A. Michaelis and F. Kunckell, *Ber.*, **30**, 2826 (1897).

156. F. Kunckell and R. Zimmermann, *Justus Liebigs Ann. Chem.*, **314**, 281 (1901).

157. R. E. Nelson and R. N. Jones, *J. Amer. Chem. Soc.*, **52**, 1588 (1930).

158. O. Behaghel and K. Hofmann, *Ber.*, **72**, 591 (1939).

159. H. Rheinboldt, "Houben-Weyl Methoden der Organischen Chemie," Vol. IX, E. Müller, Ed., Georg Thieme Verlag, Stuttgart, 1955, p. 991, ref. 2.

160. S. Patai, M. Sokolovsky, and A. Friedlander, *Proc. Chem. Soc.*, 181 (1960).

161. S. Patai and A. K. Muszkat, *Bull. Res. Council Israel*, **10A**, 73 (1961).

162. S. Patai, K. A. Muszkat, and M. Sokolovsky, *J. Chem. Soc.*, 734 (1962).

163. B. Greenberg, E. S. Gould, and W. Burlant, *J. Amer. Chem. Soc.*, **78**, 4028 (1956).

164. G. T. Morgan and F. H. Burstall, *J. Chem. Soc.*, 3260 (1928).

165. G. V. Boyd, M. Doughty, and J. Kenyon, *J. Chem. Soc.*, 2196 (1949).

166. F. Kunckell, *Ber.*, **28**, 609 (1895).

167. S.-H. Chu and H. G. Mautner, *J. Org. Chem.*, **27**, 2899 (1962).

168. G. Wagner and P. Nuhn, *Arch. Pharm.*, **297**, 461 (1964).

169. A. Segaloff and R. B. Gabbard, U.S. Pat. 3,372,173 (1968).

170. A. Segaloff and R. B. Gabbard, *Steroids*, **5**, 219 (1965).

171. H. G. Mautner, *J. Amer. Chem. Soc.*, **78**, 5292 (1956).

172. E. Dyer and C. E. Minnier, *J. Med. Chem.*, **11,** 1232 (1968).

173. H. G. Mautner and W. H. H. Günther, *J. Amer. Chem. Soc.*, **83,** 3342 (1961).

174. H. G. Mautner, S.-H. Chu, and W. H. H. Günther, *J. Amer. Chem. Soc.*, **85,** 3458 (1963).

175. J. Gosselck, H. Barth, and L. Beress, *Justus Liebigs Ann. Chem.*, **671,** 1 (1964).

176. H.-D. Jakubke, *Chem. Ber.*, **97,** 2816 (1964).

177. W. H. H. Günther and H. G. Mautner, *J. Med. Chem.*, **7,** 229 (1964).

178. W. H. H. Günther and H. G. Mautner, *J. Amer. Chem. Soc.*, **87,** 2708 (1965).

179. E. P. Painter, *J. Amer. Chem. Soc.*, **69,** 232 (1947).

180. H. J. Klosterman and E. P. Painter, *J. Amer. Chem. Soc.*, **69,** 2009 (1947).

181. S.-H. Chu, W. H. H. Günther, and H. G. Mautner, *Biochem. Prep.*, **10,** 154 (1963).

182. G. Zdansky, *Ark. Kemi*, **17,** 273 (1960).

183. L.-B. Agenäs, *Ark. Kemi*, **23,** 463 (1965).

184. G. Zdansky, *Ark. Kemi*, **19,** 559 (1962).

185. L.-B. Agenäs and B. Persson, *Acta Chem. Scand.*, **21,** 837 (1967).

186. L.-B. Agenäs, *Ark. Kemi*, **24,** 415 (1965).

187. L.-B. Agenäs and B. Persson, *Acta Chem. Scand.*, **21,** 835 (1967).

188. H. Plieninger, *Chem. Ber.*, **83,** 265 (1950).

189. L.-B. Agenäs, *Ark. Kemi*, **31,** 31 (1969).

190. D. Theodoropoulos, I. L. Schwartz, and R. Walter, *Tetrahedron Lett.*, 241 (1967).

191. D. Theodoropoulos, I. L. Schwartz, and R. Walter, *Biochemistry*, **6,** 3927 (1967).

192. C. S. Pande, J. Rudick, and R. Walter, *J. Org. Chem.*, **35,** 1440 (1970).

193. B. Rathke, *Justus Liebigs Ann. Chem.*, **152,** 181, 199 (1869).

194. B. Rathke, *Ber.*, **36,** 600 (1903).

195. E. Bartal, *Chem. Ztg.*, **30,** 1044 (1906).

196. H. G. Grimm and H. Metzger, *Ber.*, **69B,** 1356 (1936).

197. D. J. G. Ives, R. W. Pittman, and W. Wardlaw, *J. Chem. Soc.*, 1080 (1947).

198. D. Barnard and D. T. Woodbridge, *J. Chem. Soc.*, 2922 (1961).

199. J. S. Warner, *J. Org. Chem.*, **28,** 1642 (1963).

200. J. S. Warner and T. F. Page, Jr., *J. Org. Chem.*, **31,** 606 (1966).

201. K. A. Jensen and P. H. Nielsen, *Acta Chem. Scand.*, **17,** 549 (1963).

202. K. A. Jensen, *Quart. Rept. Sulfur Chem.*, **5,** 45 (1970).

203. A. Rosenbaum, H. Kirchberg, and E. Leibnitz, *J. Prakt. Chem.*, **19,** 1 (1963).

204. A. Rosenbaum, *J. Prakt. Chem.*, **37,** 200 (1968).

205. G. S. Klebanov and N. A. Ostapkevich, *Zh. Priklad. Khim.*, **33,** 1957 (1960); *Chem. Abstr.*, **55,** 3171g (1961).

206. V. Zelionkaite, J. Janickis, and R. Liksiene, *Lietuvos TSR Aukstuja Mokykln Mokslo Darbai, Chem. ir Chem. Tech.*, **1,** 98 (1961); *Chem. Abstr.*, **58,** 6444c (1963).

207. F. Foerster, F. Lange, O. Drossbach, and W. Seidel, *Z. Anorg. Allg. Chem.*, **128,** 245 (1923).

208. M. Gina and R. Bianco, *Gazz. Chim. Ital.*, **89,** 693 (1959).

209. T. van Es, *J. S. African Chem. Inst.*, **21,** 82 (1968).

210. T. van Es, *Carbohydrate Res.*, **5,** 282 (1967).

IV Selenols and their derivatives (excluding selenides)

DANIEL L. KLAYMAN

Walter Reed Army Institute of Research
Division of Medicinal Chemistry
Washington, D.C.

A. SELENOLS

1. Aliphatic Selenols

Selenols, which are also known by the cumbersome name of selenomercaptans, are selenium analogs of alcohols and thiols. They have a notorious reputation for possessing odors described variously as "vile," "highly disgusting," "quite intolerable," and "nauseating beyond description." Simple aliphatic selenols are water-insoluble liquids whose boiling points tend to be somewhat closer to the boiling points of their analogous thiols than to those of tellurols (cf. Table IV-1). Like thiols, selenols are soluble in aqueous alkali

Table IV-1   Comparison of the boiling points of thiols, selenols, and tellurols

R	RSH, B.P. (°C)	REF.	RSeH, B.P. (°C)	REF.	RTeH, B.P. (°C)	REF.
CH_3	5.8–6.2	1	25.5	5	57.0	7
C_2H_5	36.2	2	53.5	6	90.0	7
$n\text{-}C_3H_7$	67.5	3	83	6	121.0	7
$n\text{-}C_4H_9$	97–98	4	113.5	6	151.0	7

because of the high ionizability of the selenium proton. Selenols are more acidic than thiols, just as hydrogen selenide ($pK_a = 3.73$) has a greater acid strength than hydrogen sulfide ($pK_a = 6.94$). 2-Aminoethaneselenol has been found to be completely in zwitterionic form at pH 7–10, in contrast to 2-aminoethanethiol, which is only 60% zwitterionic at pH 10.[9]

Selenols are readily oxidized and must be maintained in an inert atmosphere to prevent their conversion to diselenides by atmospheric oxygen. They form yellow salts with heavy metal ions. As active hydrogen donors, selenols have been shown to have the ability to act as catalysts in the reduction of disulfides, a sulfoxide, and an aromatic azo compound.[10]

PREPARATION

a. From Alkyl Sulfates.   The first synthesis of a selenol was reported in 1847 by Siemens,[11] who distilled a mixture of potassium selenide and potassium ethyl sulfate to give ethaneselenol. A few years later Joy[12] used diethyl sulfate in a similar procedure, while Wöhler and Dean[13] applied the sulfate displacement reaction (equation 1) to the preparation of methaneselenol.

(1) $(CH_3)_2OSO_3 + 2KHSe \rightarrow 2CH_3SeH + K_2SO_4$

Siemens' method was refined by Shaw and Reid,[14] who devised an elaborate apparatus for the generation of ethaneselenol.

b. From Alkyl Halides. Methaneselenol,[15] as well as the ethane,[16] n-propane,[16] n-butane,[16] n-decane,[17] cyclopentylmethane,[18] 2-ethylhexane,[18] and phenylethane[19] congeners, has been made by displacing bromide or iodide from the appropriate alkyl halide with an alkali hydrogen selenide (equation 2). A series of selenoglycerols (**1, 2, 3**) was obtained from the

(2) $$RX + HSe^- \rightarrow RSeH + X^-$$

requisite bromopropanes by the action of sodium hydrogen selenide.[20] Com-

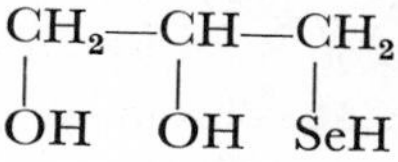

$$\begin{array}{ccc}
\text{CH}_2\text{—CH—CH}_2 & \text{CH}_2\text{—CH—CH}_2 & \text{CH}_2\text{—CH—CH}_2 \\
| \quad\; | \quad\; | & | \quad\; | \quad\; | & | \quad\; | \quad\; | \\
\text{OH} \;\; \text{OH} \;\; \text{SeH} & \text{SeH} \;\; \text{OH} \;\; \text{SeH} & \text{SeH} \;\; \text{SeH} \;\; \text{SeH} \\
\mathbf{1} & \mathbf{2} & \mathbf{3}
\end{array}$$

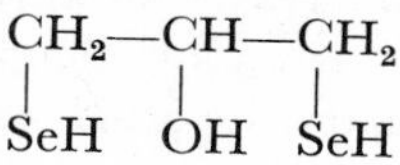

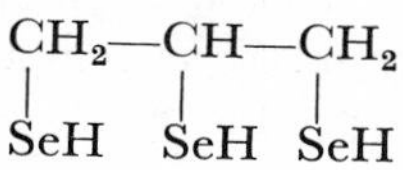

pounds **1** and **2** are water soluble, whereas **3** is not. The selenol, **4**, was derived

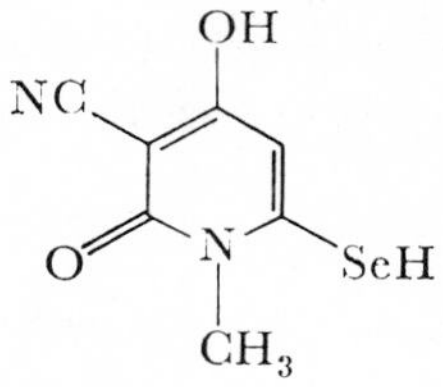

4

from chlororicinic acid by hydrogen selenide displacement.[21]

c. From Selenocyanates. o-Cyanobenzyl selenocyanate, on hydrolysis with concentrated sulfuric acid, was reported[22] to give o-cyanobenzylselenol (equation 3). Selenols have also been formed when aliphatic selenocyanates

(3)

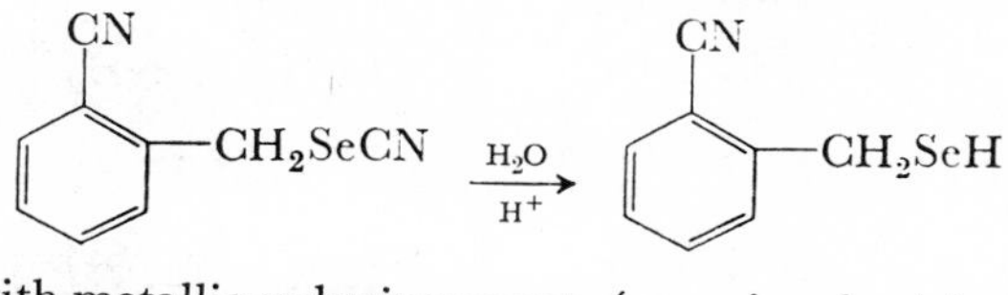

were treated with metallic reducing agents (*e.g.*, zinc dust) in acid media.[23,24]

d. From a Grignard Reagent. Elemental selenium, like elemental sulfur, reacts with Grignard reagents;[25] however, the production of aromatic selenols by this method is more satisfactory than that of aliphatic selenols. Cyclohexylmagnesium chloride has been reported to react with selenium in anhydrous ether to yield a selenium-containing intermediate which on hydrolysis gave cyclohexaneselenol (equation 4).[26] Benzylselenol[27] and

(4)

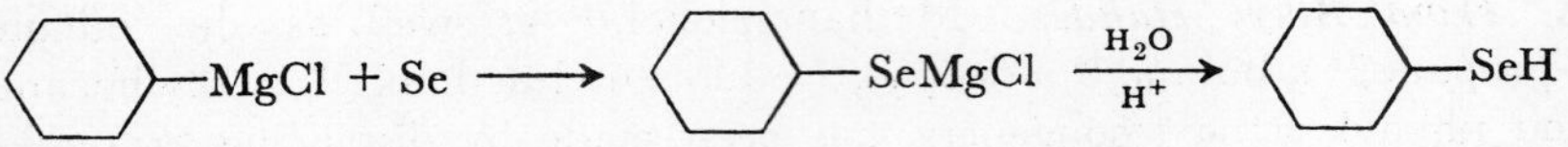

n-butylselenol[27a] have also been made *via* a Grignard reagent. *tert*-Butyl-mercaptan[28] has been successfully synthesized by this method; however, *tert*-butylselenol* and tritylselenol[29] have not.

e. From a Selenopseudourea.†

Mercaptans can be readily prepared by the alkali treatment of thiopseudoureas;[30] however, the first reported attempt to make methaneselenol from *Se*-methylselenopseudourea failed to give the desired product.[31] It was later reported that α-acetobromoglucose, on reaction with selenourea, gave the selenopseudourea, which, on hydrolysis with potassium hydroxide solution, yielded α-acetoselenoglucose.[32] Tosylate groups in two steroids have been displaced by selenourea to ultimately give 3β-selenol derivatives of pregnenolone and dehydroepiandrosterone.[33] *Se*-Alkylselenopseudoureas (free bases) were heated in air to yield diselenides, which were then reduced to selenols (equation 5).[18] Aminoalkylated

$$(5) \qquad 2RSe\overset{\overset{\displaystyle NH}{\|}}{C}-NH_2 \xrightarrow{\Delta} RSeSeR \xrightarrow{LiAlH_4} 2RSeH$$

selenoureas undergo rearrangement (transguanylation) in aqueous solution to give guanidinoselenols *via* an intermediate *gem*-2,2-diaminoselenazolidine (equation 6).[34,35] It has been noted that 2-amino-2-selenazoline, like

(6)

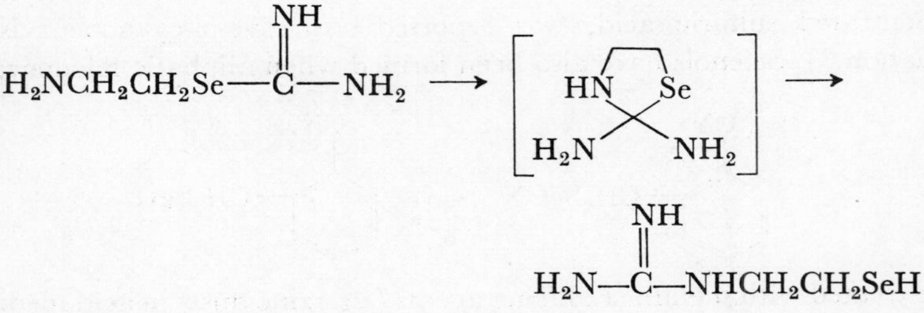

2-amino-2-thiazoline, is ring-opened when treated with alkali (equation 7)[35]

---

* The first synthesis of *tert*-butylselenol was achieved by the reaction of isobutane and selenium under photolytic conditions (O. P. Strausz and H. E. Gunning, "The Chemistry of Sulfides," A. V. Tobolsky, Ed., Interscience Publishers, New York, N.Y., 1966, p. 42).

† Selenopseudoureas are discussed further in Chapter VIIIB.

or hydrogen sulfide (equation 8).[36] The products, 2-ureido- and 2-thioureido-

(7)

$$H_2N\!-\!\underset{\substack{\|\\O}}{C}\!-\!NHCH_2CH_2SeH$$

via reaction of the cyclic $N{=}\!Se$ / NH_2 compound with H_2O / HO^-

(8)

$$H_2N\!-\!\underset{\substack{\|\\S}}{C}\!-\!NHCH_2CH_2SeH$$

via reaction of the cyclic $N{=}\!Se$ / NH_2 compound with H_2O / HS^-

ethaneselenol, respectively, were not isolated but were characterized as their diselenides.

f. By Reduction of Diselenides. Dialkyl diselenides have been cleaved by sodium in alcohol to give selenols;[37] however, the latter are produced in a superior manner by using sodium in liquid ammonia (equation 9).[5,38,39]

(9) $\qquad (RSe\!-\!)_2 \xrightarrow[\text{liq. NH}_3]{2Na} 2RSeNa \xrightarrow{2H^+} 2RSeH$

Sodium borohydride was applied to the reduction of dibenzyl diselenide to benzylselenol by Zdansky[39a] and to the preparation of 2-aminoethylselenols as intermediates by Günther and Mautner.[40] 2-Aminoethylselenol (selenocysteamine) itself was later isolated from the borohydride reduction of bis(2-aminoethyl) diselenide (equation 10) by Klayman[41] and others.[9]

(10) $\qquad (H_2NCH_2CH_2Se\!-\!)_2 + \tfrac{1}{2}NaBH_4 \rightarrow 2H_2NCH_2CH_2SeH$

Sodium borohydride has also been used to generate sodium selenoglycolate from the corresponding diselenide.[42] The first example of a *gem*-diselenol was prepared by reducing with sodium borohydride the polydiselenide resulting from the reaction of methylene bromide with bis(methoxymagnesium) diselenide (equation 11).[43] The *gem*-diselenol, while not

(11) $\quad (CH_3OMgSe\!-\!)_2 + CH_2Br_2 \longrightarrow$

$$(CH_{2.4}Se_{3.6})_n \xrightarrow[\text{NaOH}]{NaBH_4} [HSeCH_2SeH]$$

isolated, was reasonably stable in alkaline solution and was characterized as its dibenzoyl derivative. Lithium aluminum hydride has been used to reduce dibenzyl diselenide in ether solution to benzylselenol.[44]

Hypophosphorous acid, which reduces many selenium-containing functional groups, has been used successfully to cleave aliphatic diselenides to selenols.[10,45,46] It is of interest that disulfides and sulfoxides were found to be unaffected by hypophosphorous acid unless catalytic quantities of diselenide were added.[10]

Many aliphatic diselenides have been found to be readily reduced to selenols with dithiothreitol (Cleland's reagent) in aqueous solution (equation 12).[47]

$$(12) \quad (RSe—)_2 + \underset{\substack{| \\ SH}}{CH_2}—\underset{\substack{| \\ OH}}{CH}—\underset{\substack{| \\ OH}}{CH}—\underset{\substack{| \\ SH}}{CH_2} \longrightarrow 2RSeH + \overset{OH\,OH}{\underset{S—S}{\diamondsuit}}$$

g. Ring Opening by Hydrogen Selenide. Ethylenimine[9] and N-substituted aziridines[48] have been opened by hydrogen selenide (equation 13) to form

$$(13) \quad \underset{H}{\overset{\triangledown}{N}} + H_2Se \longrightarrow H_2NCH_2CH_2SeH \overset{\overset{\triangledown}{\underset{H}{N}}}{\longrightarrow} (H_2NCH_2CH_2—)_2Se$$

2-aminoethylselenols, which, in the case of the N-substituted aziridines, were immediately converted to selenazolidines by condensing them with aldehydes or ketones. If the quantity of hydrogen selenide present is limited, the aminoselenol will attack another molecule of the aziridine to form the bis(2-aminoethyl) selenide.

Ethylene oxide reacted with hydrogen selenide for 100 hours under pressure to give a low yield of 2-hydroxyethylselenol (equation 14), which

$$(14) \quad \overset{\triangledown}{O} + H_2Se \longrightarrow HOCH_2CH_2SeH$$

was contaminated with the corresponding diselenide.[49]

h. Miscellaneous Methods. Aluminum selenide has been reported to react with an alcohol at 250–350° in a sealed tube to give the selenol mixed with diselenide, as illustrated by equation 15.[50]

$$(15) \quad Al_2Se_3 + i\text{-}C_3H_7OH \rightarrow \underset{60\%}{i\text{-}C_3H_7SeH} + \underset{20\%}{(i\text{-}C_3H_7Se—)_2}$$

Triethylaluminum reacts with selenium at 80° to give ethylselenol and ethane (equation 16).[51]

$$(16) \quad Et_3Al + Se \overset{80°}{\longrightarrow} Et_2AlSeEt \overset{H_2O}{\longrightarrow} EtSeH + 2C_2H_6$$

Hydrogen selenide adds to ethylene at elevated temperatures and pressures in the presence of MoS_2 to give ethaneselenol and diethyl selenide (equation 17).[52] The presence of water decreases the yield of both products.

$$(17) \quad CH_2{=}CH_2 + H_2Se \rightarrow CH_3CH_2SeH + (CH_3CH_2)_2Se$$
$$\qquad\qquad\qquad\qquad\qquad\quad 14\% \qquad\qquad 66\%$$

Cyclohexanone reacts with the hydrogen selenide in DMF in the presence of pyridine or piperidine to give an unstable product having some of the characteristics of cyclohexaneselenol.[53] This material gives the same derivative with oxalyl chloride as authentic cyclohexaneselenol.

The compound bis(trifluoromethylseleno)mercury, when heated in a sealed tube with hydrochloric acid at 50–100°, gives trifluoromethaneselenol in low yield (equation 18).[54] This selenol, on treatment with aqueous

$$(18) \quad (F_3CSe)_2Hg + HCl \xrightarrow[\text{tube}]{\text{sealed}} F_3CSeH + (F_3CSe{-})_2 + F_3CSeHgCl$$

base, decomposes to give fluoride ion quantitatively.

Aminolysis of selenol esters is a good general method by which selenols may be formed. It should be pointed out that the aminolysis of N,Se-dibenzoylselenocysteamine (**5**) with n-butylamine occurs over 100 times more rapidly than that of its sulfur analog to give N-benzoylselenocysteamine (**6**) (equation 19).[55] γ-Selenolbutyrolactone was opened with n-butylamine

$$(19) \quad C_6H_5CONHCH_2CH_2SeCOC_6H_5 + n\text{-}C_4H_9NH_2 \rightarrow$$
$$\textbf{5}$$
$$C_6H_5CONHCH_2CH_2SeH + C_6H_5CONHC_4H_9$$
$$\textbf{6}$$

to obtain an amidobutaneselenol, which was then oxidized to the diselenide.[10]

The introduction of selenium into an organic molecule by benzylselenol is frequently utilized in selenoamino acid syntheses (cf. Chapter XIIA). Debenzylation by sodium in liquid ammonia leaves a residual selenol moiety.

REACTIONS

a. Oxidation. Selenols must be handled in an air-free atmosphere because of the great ease with which they can be oxidized to diselenides. Rapid conversion to diselenides, if desired, can be effected by a variety of oxidizing agents (cf. Section IV.B). Ethaneselenol was oxidized with nitric acid beyond the diselenide stage to ethylseleninic acid hydronitrate salt.[14] Ordinarily, however, selenols do not serve as immediate precursors of seleninic acids.

b. With Alkyl Halides. Selenols, as their alkali-metal salts, react with alkyl halides to give selenides, the synthesis of which is discussed in detail in Chapter V.

c. Acylation.   A selenol, in the presence of base, can be acylated with an acyl halide (equation 20)[33,40,47,53,56] or an acid anhydride (equation 21)[47]

$$(20) \qquad RSeH + R'-\overset{\overset{O}{\|}}{C}-X \xrightarrow{OH^-} RSe-\overset{\overset{O}{\|}}{C}-R'$$

$$(21) \quad RSeH + R'-\overset{\overset{O}{\|}}{C}-O-\overset{\overset{O}{\|}}{C}-R' \xrightarrow{OH^-} RSe-\overset{\overset{O}{\|}}{C}-R' + R'COOH$$

to give a selenol ester. Thionobenzoyl chloride, on reaction with N,N-dimethylaminoethylselenol, yielded the thionoselenobenzoate **7** (equation 22).[57] The first known selenolactone, γ-selenolbutyrolactone (**8**), was

$$(22) \qquad \overset{CH_3}{\underset{CH_3}{\diagdown}}N-CH_2CH_2SeH + C_6H_5-\overset{\overset{S}{\|}}{C}-Cl \xrightarrow{OH^-}$$

$$\overset{CH_3}{\underset{CH_3}{\diagup}}N-CH_2CH_2-Se-\overset{\overset{S}{\|}}{C}-C_6H_5$$
$$\mathbf{7}$$

prepared by distilling 4-selenolbutyric acid at 160° under reduced pressure (equation 23).[10]

$$(23) \qquad HOOCCH_2CH_2CH_2SeH \xrightarrow{160°} \underset{\mathbf{8}}{O=\!\!\!\overset{}{\underset{Se}{\bigcirc}}} + H_2O$$

d. Ring opening.   Ethylene oxide was opened by methaneselenol to give 2-methylselenoethanol (equation 24).[10] The reactions of other epoxides with

$$(24) \qquad CH_3SeH + \triangle\!\!\!O \longrightarrow CH_3SeCH_2CH_2OH$$

selenols in the presence of a basic catalyst or under u.v. irradiation have been investigated.[58,59] Isobutylene oxide gave the normal addition product **9a**, but use of catalytic amounts of boron trifluoride-etherate at 20° gave 1:2.5 normal/abnormal (**9b**) addition products (equation 25).[59]

$$(25) \quad C_3H_7SeH + H_3C-\overset{CH_3}{\underset{O}{\diagup\!\!\!\!\triangle}} \xrightarrow[Et_2O]{BF_3}$$

$$H_3C-\overset{\overset{CH_3}{|}}{\underset{\underset{OH}{|}}{C}}-CH_2SeC_3H_7 + H_3C-\overset{\overset{CH_3}{|}}{\underset{\underset{SeC_3H_7}{|}}{C}}-CH_2OH$$
$$\qquad\qquad \mathbf{9a} \qquad\qquad\qquad\qquad \mathbf{9b}$$

1-Propaneselenol has been added to α-epoxyalkenes, styrene oxide, butadiene monoepoxide, and isoproprene monoepoxide.[58]

Benzylselenol reacted with ethylenimine to give 2-benzylselenoethylamine (equation 26).[60]

(26) $\qquad$ $C_6H_5CH_2SeH$ +
$\longrightarrow$ $C_6H_5CH_2SeCH_2CH_2NH_2$

The lactone, α-amino-γ-butyrolactone, has been opened with sodium methyl selenolate to give the selenium analog of methionine (equation 27).[61]

(27)
+ CH_3SeNa $\longrightarrow$
$\longrightarrow$

$$CH_3SeCH_2CH_2\underset{\underset{NH_2}{|}}{CH}\text{---}COONa$$

β-Propiolactone, certain γ- and δ-lactones,[63,64] and phthalide[65] combined with sodium benzyl selenolate in DMF to give benzylseleno derivatives of carboxylic acids; however, steric hindrance was held responsible for the failure of other lactones to open.

The opening of sultones by methaneselenol has been reported to give methylselenoalkyl sulfonates (equation 27a).[66]

(27a) CH_3Se^- +
$\longrightarrow$ $CH_3Se\text{---}\overset{\overset{R}{|}}{CH}(CH_2)_nSO_3^-$

e.   Addition to Double or Triple Bonds.

Benzylselenol has been added to double bonds activated by aldehyde,[39a] keto (equation 28),[67] or carboxyl[68,69]

(28) $C_6H_5CH_2SeH$ + $CH_2{=}CH\text{---}\overset{\overset{O}{\|}}{C}\text{---}CH_3$ $\longrightarrow$

$$C_6H_5CH_2SeCH_2CH_2\text{---}\overset{\overset{O}{\|}}{C}\text{---}CH_3$$

groups. Debenzylation of the products was effected by sodium in liquid ammonia to give selenols. Aliphatic and aromatic selenols add to alkynes

(29) $C_6H_5CH_2SeH$ + $HC{\equiv}C\text{---}\overset{\overset{O}{\|}}{C}C_6H_5$ $\longrightarrow$

$$C_6H_5CH_2Se\text{---}CH{=}CH\text{---}\overset{\overset{O}{\|}}{C}\text{---}C_6H_5$$

(equation 29) in the presence of catalytic amounts of piperidine or, in some instances, by simply combining the reactants at room temperature.[70]

f. With Aldehydes and Ketones. 2-Substituted selenazolidines[71] and 2,2-disubstituted selenazolidines[48] have been reported to be formed by the reaction of 2-aminoethylselenols with aldehydes and ketones, respectively. Ethaneselenol was condensed with several aliphatic ketones to give seleno-ketals.[14]

By saturating a solution of paraformaldehyde in benzylselenol with dry hydrogen chloride, it was possible to prepare benzyl chloromethyl selenide (equation 30),[72,73] a valuable intermediate for the synthesis of selenoamino acids.

$$(30) \quad C_6H_5CH_2SeH + (HCHO)_x \xrightarrow{HCl} C_6H_5CH_2SeCH_2Cl + H_2O$$

g. Heavy Metal Salts. Methaneselenol reacts with mercuric cyanide (equation 31), lead acetate, silver nitrate, cupric acetate, or bismuth tri-

$$(31) \quad CH_3SeH + Hg(CN)_2 \rightarrow (CH_3Se)_2Hg$$

chloride to give selenides as yellow crystalline powders.[15] Recently, the cationoid complexes of 2-aminoethylselenol and related compounds with nickel, platinum, copper, cobalt, zinc, palladium, and cadmium salts have been studied.[74,74b] The stability constants of many of these complexes have been calculated.[74a]

h. With Grignard Reagents and Other Organometallics. Tschugaeff[16] found that selenols react with methylmagnesium iodide to give methane (equation 32), which could be measured quantitatively (Zerevitinov method).

$$(32) \quad RSeH + CH_3MgI \rightarrow RSeMgI + CH_4$$

In another investigation, ethylmagnesium bromide, freed of solvent and cooled with liquid nitrogen, was treated with butaneselenol to give ethane and butylselenomagnesium bromide, $C_4H_9SeMgBr$.[75] The magnesium bromide moiety of the latter compound could be displaced by cadmium chloride in THF to yield the cadmium complex, $C_4H_9SeCdCl \cdot CdCl_2$. Similar displacements could be performed by triethylgermanium bromide and mercuric chloride.[75]

Nelson[76] reported that selenols and thiols react somewhat selectively with diethylmercury, tetramethyllead and triethylbismuth. Trimethylgallium combines with methaneselenol with elimination of methane to give a dimeric product, tetramethyl-μ,μ'-bismethylselenodigallium (**10**).[77] The reactions of selenols with di-π-cyclopentadienyltitanium(IV) dichloride,[78] bis(triethyl-germanyl)mercury, and analogous organometallics of cadmium and zinc[79] have also been described.

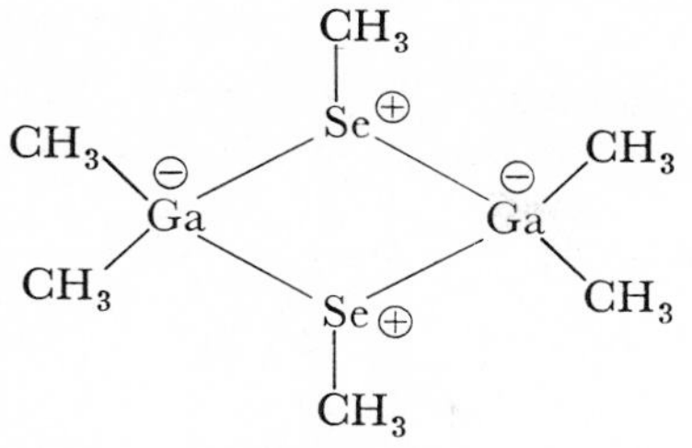

$$\begin{array}{c}
CH_3 \\
| \\
Se^{\oplus} \\
CH_3\diagdown \quad \diagup \quad \diagdown \quad \diagup CH_3 \\
Ga^{\ominus} \qquad\qquad Ga^{\ominus} \\
CH_3\diagup \quad \diagdown \quad \diagup \quad \diagdown CH_3 \\
Se^{\oplus} \\
| \\
CH_3
\end{array}$$

10

Ethaneselenol reacted with boron tribromide in a 1:1 ratio to give bromoethylselenoborane ($Br_2BSeC_2H_5$) and in a 2:1 ratio to yield the unstable $BrB(SeC_2H_5)_2$, which disproportionated at room temperature to bromoethylselenoborane and tris(ethylseleno)borane (**11**).[80] The latter compound could also be prepared by combining the above reactants in a 3:1 ratio (equation 33).[81]

(33) $\qquad\qquad 3C_2H_5SeH + BBr_3 \rightarrow B(SeC_2H_5)_3 + 3HBr$

11

MISCELLANEOUS*

The observed boiling points of fourteen selenols and those calculated by the Kinney equation have been found to be in close agreement.[82] The microscopic dissociation constant for the SeH group in selenoglycolic acids has been determined.[42]

Sodium, calcium, barium, and zinc salts of selenols are alleged to impart antioxidant, detergent, and anticorrosive properties to lubricating oils.[83] Aralkyl selenols were reported to be superior to mercaptans for the modification of the high-temperature polymerization of butadiene-vinyl compounds.[84]

2. Aromatic Selenols

Selenophenol (also known as benzeneselenol and phenylselenol) was first prepared in 1894.[85] It is an extremely foul-smelling, light-sensitive oil, a drop of which on the skin has been reported to form a blister that itches intensely.[86] Its boiling point is 182°,[87] as compared to 169.5° for thiophenol.[88] (N.B. Tellurophenol is unknown.)

The selenophenoxide anion is more nucleophilic than thiophenoxide towards carbon in 2-bromo-5-nitrothiophene ($k_{C_6H_5Se^-}/k_{C_6H_5S^-} = 4.6$)[89] and towards sulfur and selenium in a number of compounds.[90]

* A discussion of the spectral properties of aliphatic selenols can be found in Chapter XV.

PREPARATION

a. From Grignard Reagents. The reaction of elemental selenium with aromatic Grignard reagents (equation 34) was discovered independently

$$(34) \qquad C_6H_5MgX + Se \longrightarrow C_6H_5SeMgX \xrightarrow{H^+} C_6H_5SeH$$

in 1903 by Taboury[91] and by Wuyts and Cosyns.[92] The former investigator was also interested in making thiols in a similar manner.[93,94] Numerous aromatic selenols[25,87,95,96] were subsequently prepared, and it was noted that in each case the selenophenol obtained was accompanied by the corresponding diaryl selenide and diselenide.[25,97] Improved procedures for the preparation of selenophenols by this method were later developed,[98,99] including a modification in which the reaction was run in a stream of dry hydrogen.[100,101] Selenophenols bearing methyl,[56,100,102,103] phenyl,[104] methoxy,[56,105] bromo,[102] chloro,[106] and fluoro[29,107] groups on the ring have been made *via* the Grignard route. 2-Naphthylselenol was prepared similarly but not isolated.[108]

b. From an Aryllithium Compound. A method which has been reported to work when the Grignard method (discussed above) fails is the reaction of an aryllithium with selenium, followed by acid hydrolysis of the intermediate lithium selenide (equation 35).[109] Yields are low, and the method has been

$$(35) \qquad ArLi + Se \xrightarrow{Et_2O} ArSeLi \xrightarrow{H_2O} ArSeH + LiOH$$

found to have only limited application.[110,111] A modification of this procedure has led, however, to the synthesis of 3,4-methylenedioxyselenophenol.[112]

c. By Reduction of Diselenides. The earliest preparation of selenophenol was accomplished by Krafft and Lyons,[85] who reduced diphenyl diselenide with sodium in alcohol in a carbon dioxide atmosphere (equation 36).

$$(36) \qquad (C_6H_5Se{-})_2 \xrightarrow[\substack{(1)\ Na \\ C_2H_5OH \\ (2)\ H^+}]{} 2C_6H_5SeH$$

Bis(2-aminophenyl) diselenide has also been reduced in this manner.[113]

Raney nickel has been used to simultaneously cleave a nitro-substituted aromatic diselenide to a selenol and reduce the nitro group to an amine function.[114] Bis(2,2'-dinaphthyl) diselenide was reduced in boiling benzene-ethanol solution by sodium borohydride.[108]

Other methods of diselenide reduction involve the use of zinc dust in either acetic acid[115-118] or alkali solution.[119,120] The selenophenols, reduced in the acid media, were obtained in the form of their zinc salts, which can be used directly for numerous transformations (*vide infra*).

d. From Selenocyanates. Aromatic selenocyanates have been converted to selenols by treatment with zinc dust in acid solution (equation 37),[121,122,]

$$(37) \qquad\qquad ArSeCN \xrightarrow[H^+]{Zn} ArSeH$$

[124–126] with aqueous or alcoholic alkali,[108,123,127–131] with alkaline dextrose solution (equation 38),[132,133] and with alkaline sodium hydrosulfite solution

(38)

Selenohydroquinone

(equation 39).[113,134,135] 8-Selenocyanatoquinoline (**12**) was reduced to

(39)

selenoxine (**13**) with hypophosphorous acid (equation 40).[136]

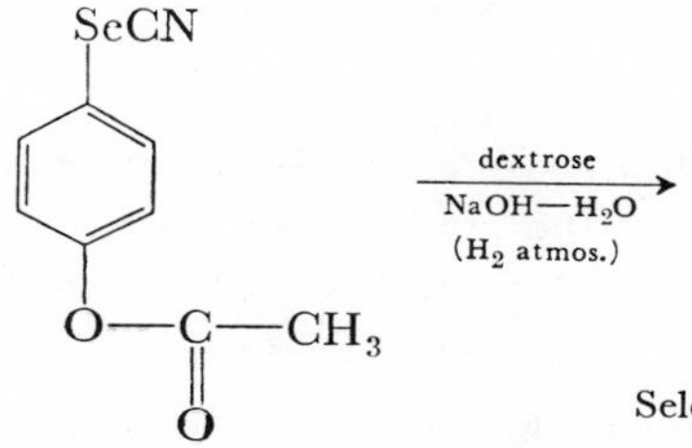

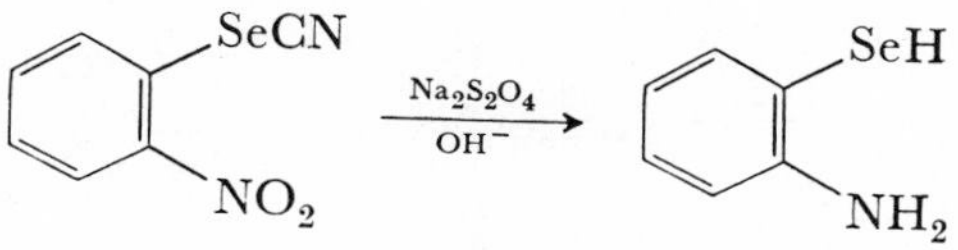

(40)

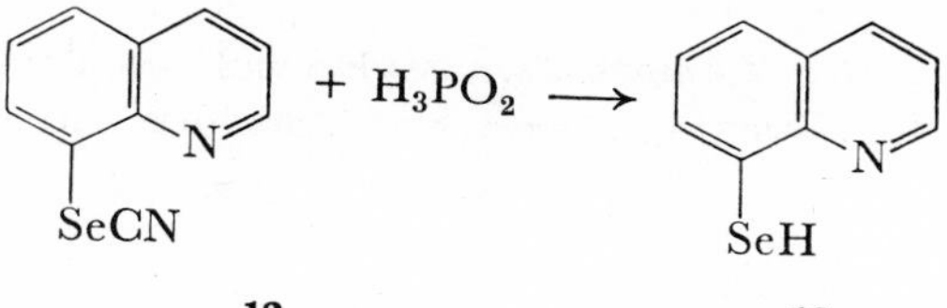

12   **13**

e. Aromatic Nucleophilic Substitution. *o*-Chloronitrobenzene,[127] 1-chloro-2-aminoanthraquinone,[137] 2-bromopyridine,[138] 2-bromopyridine-*N*-oxide,[138] and 4-chloropyridine[139] were converted into selenols on treatment with selenide ion. An attempt to make 2-selenopyridine *via* the selenopseudourea failed.[138]

f. Miscellaneous Methods. Benzeneselenonic acid was reduced to selenophenol by agents such as hydrogen sulfide, sulfur dioxide, and zinc dust in hydrochloric acid.[86] Sodium borohydride has been used to convert phenylene-1,4-diseleninic acid to phenyl-1,4-diselenol.[43]

In the late nineteenth century, Chabrié claimed that the reaction of selenium tetrachloride with benzene in the presence of anhydrous aluminum chloride gave, among other products, selenophenol.[140,141] His reported melting point of 60° for the latter compound, however, aroused the suspicion of Krafft and Kaschau,[142] who reinvestigated the reaction and found that Chabrié's "selenophenol" was, in actuality, diphenyl diselenide. Bradt and Green[143] verified Krafft and Kaschau's findings and identified triphenyl-selenonium chloride as an additional product.

REACTIONS

Certain products derived from aromatic selenols are discussed in detail in other parts of this book, and the reader is referred to them: selenoacetals and selenoketals (Section IV.A.3), diselenides (Section IV.B), selenides (Chapter V), and selenol esters (Chapter VIIIA).

a. With Diazonium Compounds. Keimatsu and his co-workers[105,129,133,144,145] studied the reaction of benzeneselenols, used generally as their sodium salts, with diazonium compounds (equation 41). The products are diphenyl

(41)

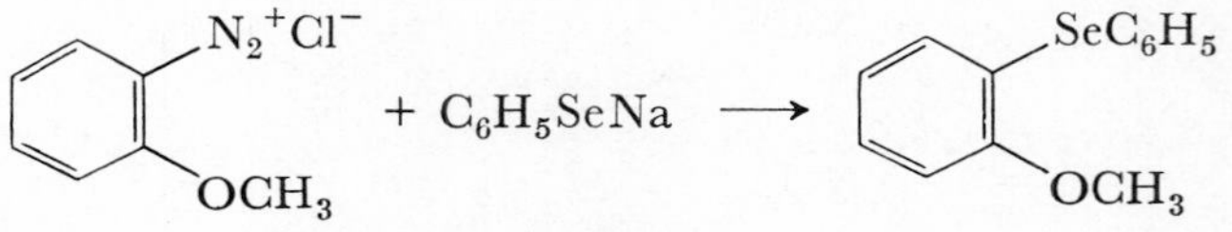

selenides, some of which can be further transformed into selenanthrenes.[129,145] Phenyl *o*-nitrophenyl selenide has been prepared by the Keimatsu method.[146]

b. With Unsaturated Compounds. Arylselenols condense readily with chalcones by 1,4-addition to give γ-keto selenides (equation 42).[147] No

(42)

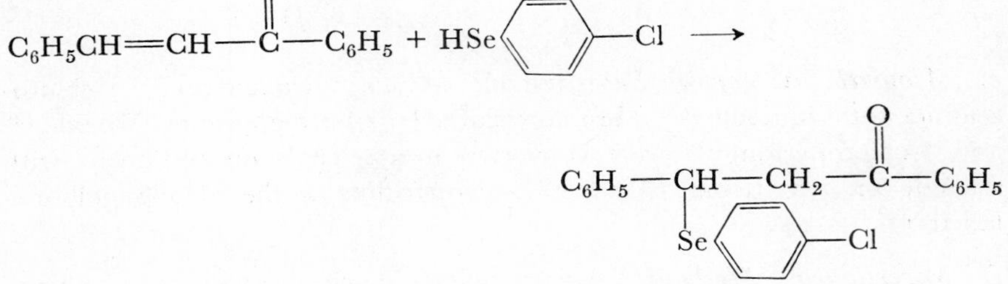

addition takes place if the chalcone is *ortho*-substituted. Similarly, seleno-phenol has been shown to condense with α,β-unsaturated acids and esters (equation 43).[148,149]

$$
\text{(43)} \quad C_6H_5SeH + \overset{\overset{\displaystyle CH_3}{|}}{CH}=CH-COOC_2H_5 \longrightarrow
$$

$$
C_6H_5Se-\overset{\overset{\displaystyle CH_3}{|}}{CH}-CH_2COOC_2H_5
$$

Selenophenol adds mainly in a 1,4-manner to 1,3-dienes (equation 44).[150]

$$
\text{(44)} \quad C_6H_5SeH + CH_2=CH-CH=CH_2 \rightarrow C_6H_5SeCH_2CH=CH-CH_3
$$

The n.m.r. and i.r. spectra indicated that a mixture of *cis* and *trans* isomers had been obtained, as well as some 1,2-addition products. Other olefins which have been found to react with selenophenol include ethylene, propylene, isobutylene, and cyclohexene.[151]

Arylselenols also add to alkynes, generally in the presence of a base catalyst. The addition of selenophenol to various acetylenes[70] and other alkynes[152] to give α,β-unsaturated selenides (equation 45) has been investigated.

$$
\text{(45)} \quad C_6H_5SeH + RC{\equiv}CR' \longrightarrow C_6H_5Se-\overset{\overset{\displaystyle R}{|}}{C}=CH-R'
$$

Propiolic acid also combines with arylselenols (equation 46) to give mixtures

$$
\text{(46)} \quad ArSeH + HC{\equiv}C-COOH \rightarrow ArSeCH=CHCOOH
$$

of *cis* and *trans* β-arylselenoacrylic acids.[153-155] Recently, Azerbaev *et al.*[156,157] reported that the nucleophilic addition of selenophenol to acetylenic alcohols in the presence of a base catalyst gave the corresponding phenylselenovinyl carbinols.

c. Ring-Opening Reactions. Aromatic selenols behave very much like aliphatic selenols in ring-opening reactions (*vide supra*). Thus, the addition of alkyl- and arylselenols to epoxides has been found to yield hydroxy selenides under free-radical or ionic conditions (equation 47).[58,59,158]

$$
\text{(47)} \quad C_6H_5SeH + \underset{O}{\triangle}-CH=CH_2 \longrightarrow C_6H_5SeCH_2-\underset{\underset{\displaystyle OH}{|}}{CH}-CH=CH_2
$$

Selenophenol opens diketene in the presence of *p*-toluenesulfonic acid catalyst to give the acetyl acetate, **14**, which rearranges further to a selenochromone (equation 48).[159]

(48)

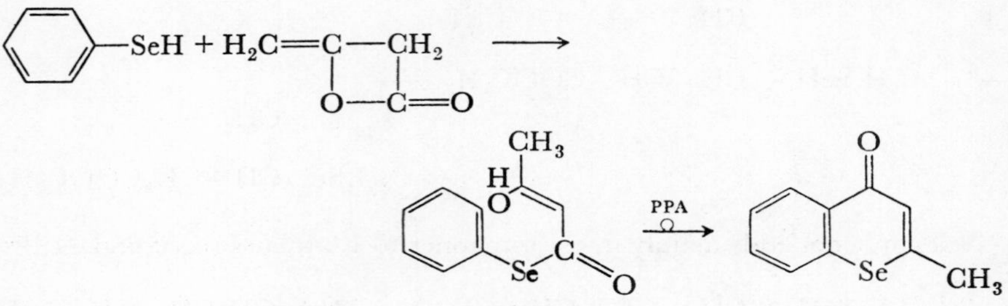

14

Phthalide has been opened by selenophenol in the presence of sodium ethoxide to give the selenide, **15** (equation 49).[160] The latter can be cyclized

(49)

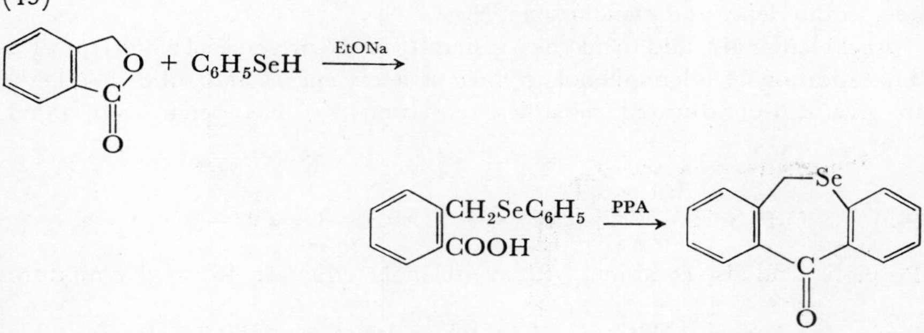

15   **16**

with polyphosphoric acid (PPA) to a dibenzoselenepin, **16**.

d. Cyclization. 1-Selenochromones have been synthesized by the condensation of selenophenols with β-keto esters in the presence of polyphosphoric acid (equation 50)[161,162] by the Simonis chromone cyclization.

(50)

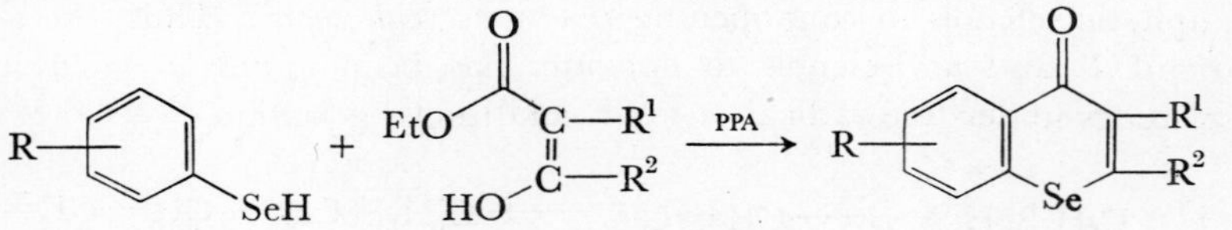

Intramolecular condensations of appropriately *ortho*-substituted selenophenols can lead to selenanaphthene (**17**) formation. For example, *o*-selenocinnamic acid was oxidized *via* the selenenic acid to **17**,[130] while *o*-ethylbenzeneselenol was dehydrocyclized to also give **17** in low yield (equation 51).[163]

(51)

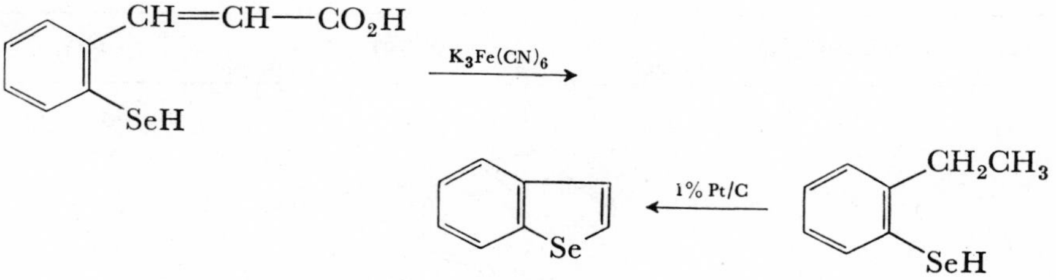

17

o-Aminoselenophenol as its zinc salt is a particularly versatile reagent for the construction of numerous heterocyclic systems. A summary of its utility is given in Table IV-2. Further information pertaining to selenium heterocycles will be found in Chapter XI.

Table IV-2 Condensations of the zinc salt of 2-aminoselenophenol with various unsaturated and carbonyl-containing substrates leading to heterocyclic compounds

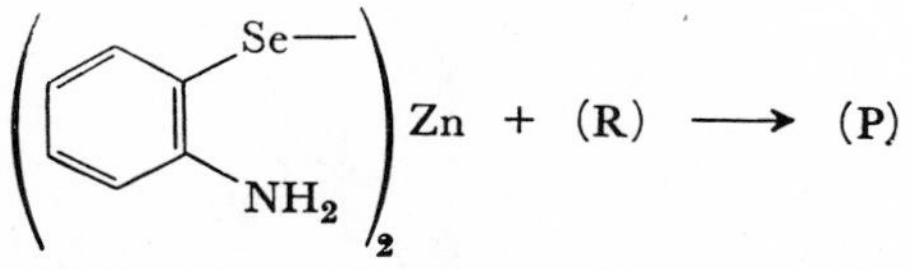

TYPE OF REACTANT (R)	PRODUCT (P)	REF.
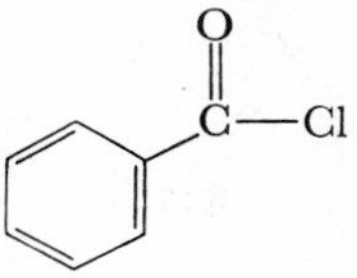	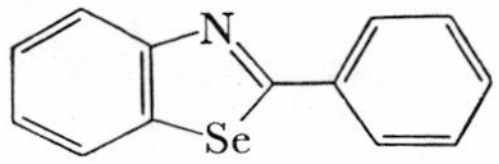	115, 116, 118, 127, 164–166
HCOOH	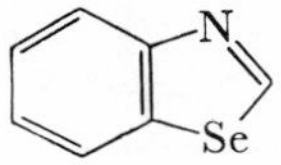	164
CH—CH_3 ‖ CH | COOH	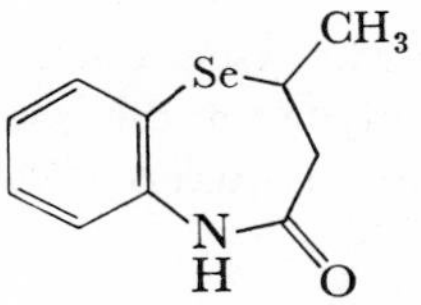	167
C—COOH ≡ C—COOH	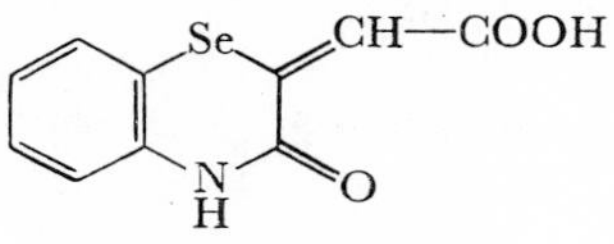	167

Table IV-2   (*Continued*)

TYPE OF REACTANT (R)	PRODUCT (P)	REF.
$\begin{array}{c} CH_3 \\ | \\ C-C_2H_5 \\ || \\ CH \\ | \\ COC_2H_5 \end{array}$		168
		168
$\begin{array}{c} CH_3 \quad O \\ | \quad\quad || \\ CH_3-C=CH-C-CH_3 \end{array}$		169, 170
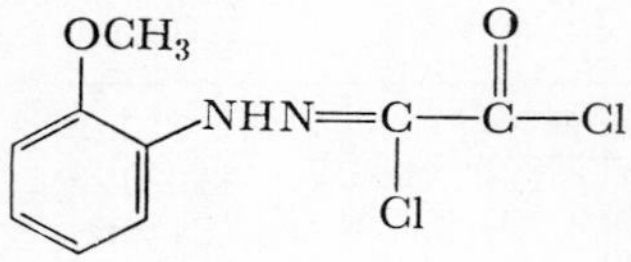	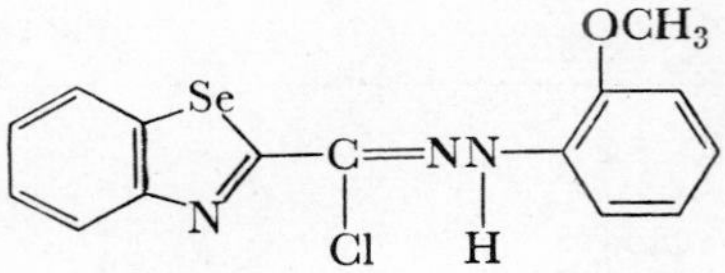	171
$\begin{array}{c} NH \\ || \\ H_2N-C-NH-C\equiv N \end{array}$		117

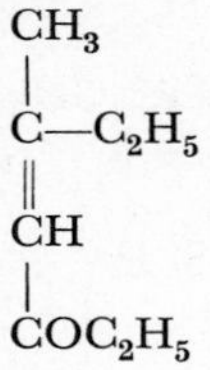

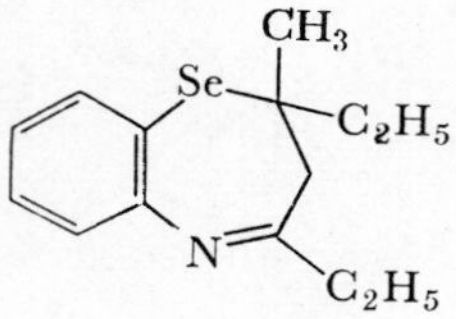

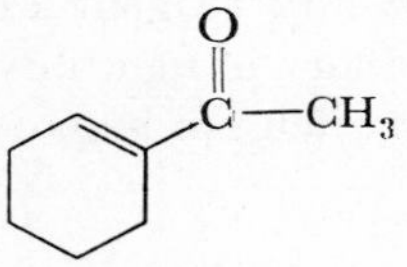

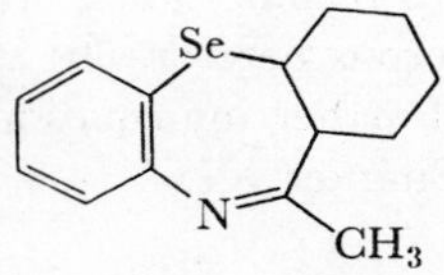

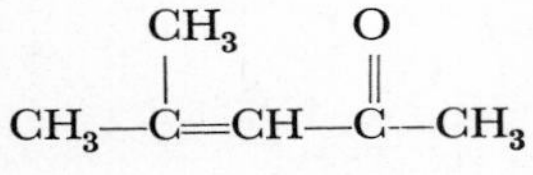

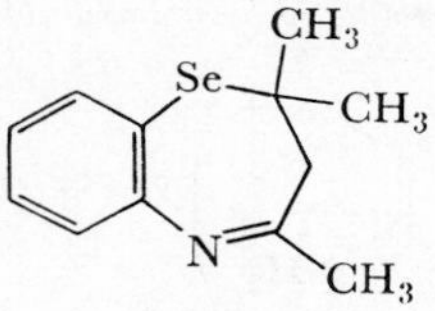

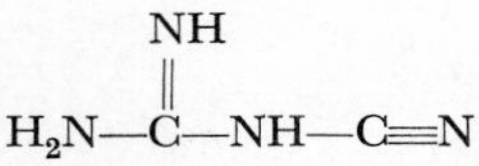

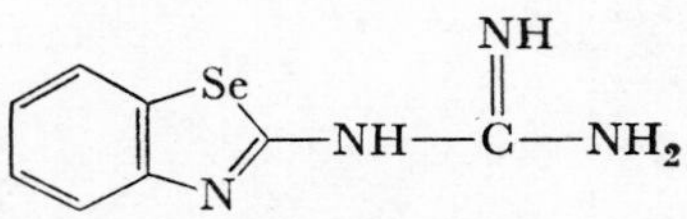

e. Mannich Reaction. The aminomethylation of selenophenol by condensing it with formaldehyde in the presence of primary or secondary aliphatic or aromatic amines (equations 52, 53) has been studied by Pollack

(52)

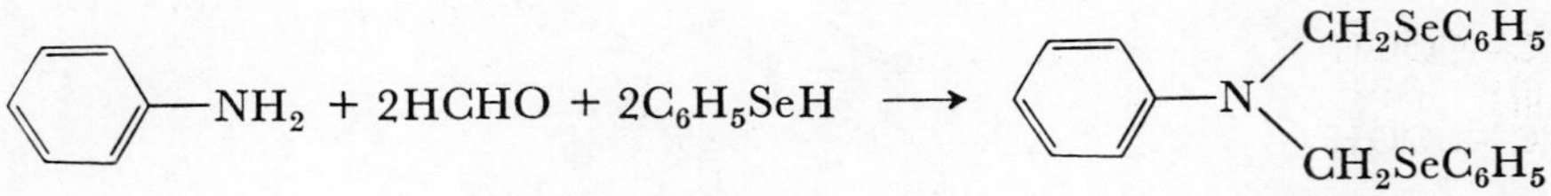

(53)

$$\text{C}_6\text{H}_5\!-\!\text{NHR} + \text{HCHO} + \text{C}_6\text{H}_5\text{SeH} \longrightarrow \underset{\overset{|}{\text{R}'}}{\text{C}_6\text{H}_5\!-\!\text{N}}\!-\!\text{CH}_2\text{SeC}_6\text{H}_5$$

and Grillot.[172–174] It is interesting that in acid solution aromatic substitution (equation 54), rather than aminomethylation, occurs.[172]

(54)

$$\text{H}_2\text{N}\!-\!\text{C}_6\text{H}_5 + \text{HCHO} + \text{C}_6\text{H}_5\text{SeH} \xrightarrow{\text{H}^+}$$

$$\text{H}_2\text{N}\!-\!\text{C}_6\text{H}_4\!-\!\text{CH}_2\text{SeC}_6\text{H}_5$$

f. **Synthesis of Selenoaryl Esters of Amino Acids.** Certain esters of selenophenol[175,176] and α-selenonaphthol,[177] prepared by the method of Wieland,[178] are used as active *N*-acylating reagents for the synthesis of linear and cyclic peptides. To the *N*-protected amino acid in THF is added ethyl chloroformate at −10° and then the aromatic selenol (equation 55).

(55)

$$\text{C}_6\text{H}_5\!-\!\underset{\overset{\|}{\text{O}}}{\text{OC}}\text{NHCH}_2\text{COOH} + \text{C}_6\text{H}_5\text{SeH} \xrightarrow[\text{(C}_2\text{H}_5)_3\text{N}]{\overset{\overset{\text{O}}{\|}}{\text{Cl}-\text{C}-\text{OC}_2\text{H}_5}}$$

$$\text{C}_6\text{H}_5\!-\!\underset{\overset{\|}{\text{O}}}{\text{OC}}\text{NHCH}_2\underset{\overset{\|}{\text{O}}}{\text{C}}\text{SeC}_6\text{H}_5$$

g. **With Metal Compounds.** In addition to forming salts of alkali metals and zinc, aromatic selenols react with ethylmercuric chloride (equation 56),[121] diarylmercury (equation 57),[179] silicon chloride, germanium chloride,

(56) $\underset{\text{COOH}}{\overset{\text{SeNa}}{\text{C}_6\text{H}_4}} + \text{C}_2\text{H}_5\text{HgCl} \longrightarrow \underset{\text{COOH}}{\overset{\text{SeHgC}_2\text{H}_5}{\text{C}_6\text{H}_4}}$

(57) $\text{C}_6\text{H}_5\text{SeH} + (\text{C}_6\text{H}_5)_2\text{Hg} \rightarrow (\text{C}_6\text{H}_5\text{Se}\!-\!)_2\text{Hg} + \text{C}_6\text{H}_5\text{HgC}_6\text{H}_5$

and stannic chloride (equations 58, 59, 60),[180] and di-π-cyclopentadienyl-

(58) $4\text{C}_6\text{H}_5\text{SeH} + \text{SiCl}_4 \rightarrow \text{Si}(\text{SeC}_6\text{H}_5)_4$

(59) $4\text{C}_6\text{H}_5\text{SeH} + \text{GeCl}_4 \rightarrow \text{Ge}(\text{SeC}_6\text{H}_5)_4$

(60) $4\text{C}_6\text{H}_5\text{SeH} + \text{SnCl}_4 \rightarrow \text{Sn}(\text{SeC}_6\text{H}_5)_4$

titanium(IV) dichloride (equation 61).[78]

$$(61)\quad C_6H_5SeH + (C_5H_5)_2TiCl_2 \xrightarrow{2N(C_2H_5)_3}$$
$$(C_5H_5)_2Ti(SeC_6H_5)_2 + 2(C_2H_5)_3N\cdot HCl$$

h. Demethylation of Quaternary Ammonium Salts of Alkaloids. On the basis of the earlier work of Shamma *et al.*[181] on the sodium thiophenolate demethylation of certain quaternary alkaloids, Šimánek and Klasek[182] found that demethylation with selenophenolate occurs at a greater rate. The latter workers noted that *N*-demethylation proceeds very readily in alkaloids whose N atom is part of one ring only, whereas alkaloids whose N atom is shared by two rings of a bicyclic system are demethylated at a slower rate or not at all.

MISCELLANEOUS*

A use suggested for aryl selenols is in the prevention of darkening of aralkyl-amines during storage.[183]

3. Selenoacetals and Selenoketals

Selenoacetals and selenoketals are generally derived from the condensation of an aldehyde or ketone with a selenol. They are not obtained, however, from a selenoaldehyde or selenoketone reaction with an alcohol, for in this case the final product would, in principle, be a simple acetal or ketal.

PREPARATION

a. From Aldehydes or Ketones. Aliphatic aldehydes react with 2 equiv-alents of a selenol in the presence of hydrogen chloride to give selenoacetals (equation 62).[132] Selenoketals can be made from ketones in a similar manner

$$(62)\quad 2RSeH + R'CHO \xrightarrow{HCl} R'\!-\!\overset{\displaystyle H}{\underset{\displaystyle SeR}{\overset{|}{\underset{\diagdown}{C}}}}\overset{\diagup SeR}{} + H_2O$$

(equation 63).[14,132]

$$(63)\quad 2RSeH + \underset{R''}{\overset{R'}{\diagdown}}C=O \longrightarrow \underset{R''}{\overset{R'}{\diagdown}}C\underset{\diagdown SeR}{\overset{\diagup SeR}{}} + H_2O$$

* A discussion of the spectral properties of aromatic selenols will be found in Chapter XV.

b. From α-Chloroethers and Zinc. The reaction of α-chlorinated ethers with zinc generates an aldehyde that condenses with alkyl- or arylselenols *in situ*.[184–186] The mechanism for the formation of the resultant selenoacetals is shown in scheme 64. When an α-dichloro ether is the precursor, a triseleno-

(64)
$$CH_3OCH_2Cl + Zn \rightarrow CH_3OCH_2\cdot + ZnCl\cdot$$
$$CH_3OCH_2\cdot \rightarrow HCHO + CH_3\cdot$$
$$2CH_3\cdot \rightarrow C_2H_6$$
$$ZnCl\cdot + CH_3OCH_2Cl \rightarrow CH_3OCH_2\cdot + ZnCl_2$$
$$2RSeH + HCHO \rightarrow RSeCH_2SeR$$

orthoformate is obtained, as indicated in scheme 65.[187]

(65)
$$2CH_3OCHCl_2 + Zn \rightarrow 2ClCHO + C_2H_6 + ZnCl_2$$
$$ClCHO + 3RSeH \rightarrow HC(SeR)_3 + H_2O + HCl$$

c. From Diazomethane and Diselenides. Petragnani and Schill[188] recently found that diazomethane cleaves diphenyl or dibenzyl diselenides by a photochemical reaction to give selenoacetals in quantitative yield (equation 66).

(66)
$$(C_6H_5Se\text{—})_2 + CH_2N_2 \rightarrow C_6H_5SeCH_2SeC_6H_5 + N_2$$

REACTION

Shaw and Reid[14] oxidized selenoketals derived from ethaneselenol with dilute (1:4) nitric acid in the hope of obtaining the selenium analog of sulfonal [2,2-bis(ethylsulfonyl)propane]. Instead, ethylseleninic acid was formed as its nitric acid salt (equation 67). It is interesting that these workers

(67)

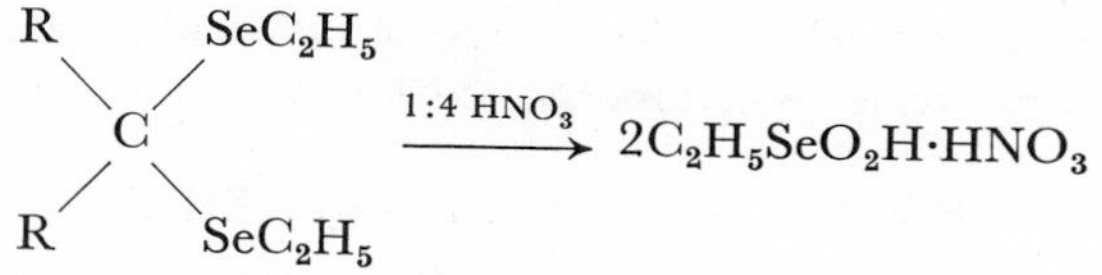

$$
\begin{array}{c}
R \qquad SeC_2H_5 \\
\diagdown \qquad \diagup \\
C \xrightarrow{\;1:4\ HNO_3\;} 2C_2H_5SeO_2H\cdot HNO_3 \\
\diagup \qquad \diagdown \\
R \qquad SeC_2H_5
\end{array}
$$

observed ethaneselenol to be less susceptible than the selenoketals to conversion to the seleninic acid.

B. DI- AND TRISELENIDES

1. Diselenides

Simple dialkyl diselenides are generally yellow to orange oils, whereas diaryl diselenides are yellow solids. The disappearance of the color in the course of a reduction or an oxidation is a convenient indicator of the progress of the reaction. The coloration of diselenides has been observed to diminish as the temperature is lowered and to intensify at higher temperatures.[189-193] It was suggested, but not proved, by Campbell[192] that this thermochromic effect was due to the dissociation of diselenides into free radicals. Kuder and Lardon[193] have shown, however, that a free-radical process is not responsible for this phenomenon. Rather, it was found that the thermochromic behavior is due to a single molecular species whose chromophore absorbs at the edge of the visible region and the population of whose vibrational energy levels is very temperature dependent.

The odors of diselenides are somewhat less offensive than those of the comparable, more volatile selenols. Many of the former compounds have been reported to be skin irritants.

Dimethyl diselenide has been isolated from a natural source, namely, the plant *Astragalus racemosus*, whose unpleasant odor is attributed to the release of this and other volatile selenium compounds.[194] On the basis of i.r. studies of dimethyl diselenide, it has been suggested that diselenides have dihedral angles approximating 90°, an indication of the importance of the interaction between unshared pairs of $4p\pi$ electrons of adjacent Se atoms.[195,196]

Selenenyl selenocyanates may be considered to be unsymmetrical diselenides. This group of compounds is discussed in Section IV.F.4.b, following the subsection (4a) devoted to the closely related selenenyl thiocyanates.

PREPARATION

a. From Selenols.   Diselenides are formed with such ease by air oxidation of selenols that only by the meticulous exclusion of atmospheric oxygen can the transformation be prevented. For practical purposes, the intentional conversion of selenols to diselenides may be accelerated by bubbling air through a solution of a selenol[63] or by the use of oxidizing agents such as dilute hydrogen peroxide (equation 68).[19,131,136] Bergson and Biezais[68] used

$$(68) \qquad 2RSeH + H_2O_2 \rightarrow (RSe-)_2 + 2H_2O$$

ferric chloride as both an indicator and a catalyst in the simultaneous air oxidation and cyclization of a compound bearing thiol and selenol groups. The use of chloramine and dimethylchloramine as coupling agents for the formation of symmetrical and unsymmetrical diselenides has been described by Sisler and Kotia.[27a]

b. From Alkyl Halides.   Displacement of halogen from an alkyl halide by an alkali diselenide constitutes a direct method for the synthesis of diselenides (equation 69). Sodium and potassium diselenide are prepared by dissolving

$$(69) \qquad 2RX + Na_2Se_2 \rightarrow (RSe-)_2 + 2NaX$$

elemental selenium in an aqueous solution of the alkali-metal selenide.[116,197,198] A solution of selenium in sodium *sulfide* has also been used[198,200] but is less satisfactory, as it frequently leads to mixtures of diselenides and disulfides.

The reagents more commonly used today are derived by adding selenium to either liquid ammonia in which sodium[38,201,202] or lithium[201] has been dissolved, or to a solution of sodium formaldehyde sulfoxylate (Rongalite C) and sodium or ammonium hydroxides.[203–207] Diselenides made by the latter method may be contaminated by traces of mono- and polyselenides.

An interesting reagent, prepared by the reaction of magnesium turnings with selenium in methanol and tentatively described as bis(methoxymagnesium) diselenide, has been developed by Günther.[208] Nucleophilic attack of this compound on alkyl halides (as well as lactones) gives diselenides in good yield.

Recently, it was found by Klayman and Griffin[208a] that selenium is acted upon very rapidly by aqueous sodium borohydride (1 molar equivalent) to generate sodium diselenide in a highly convenient manner (equation 69a). The sodium hydrogen selenide solution, which formed initially at room

$$(69a) \qquad 2Se + 2NaBH_4 + 6H_2O \rightarrow Na_2Se_2 + 3H_3BO_3 + 7H_2$$

temperature, was heated on a steam bath for 1 hour to dissolve the elemental selenium completely.

Ethanolic sodium selenide could be made by combining 3 moles of selenium with 2 moles of sodium borohydride in ethanol. The resultant mixture, after the evolution of hydrogen ceased, was heated to complete the dissolution of the selenium and expel the mole of hydrogen selenide formed according to equation 69b.

(69b)
$$3Se + 2NaBH_4 + 6C_2H_5OH \rightarrow Na_2Se_2 + H_2Se + 2B(OC_2H_5)_3 + 6H_2$$

The halide displacement method has been used to prepare simple dialkyl diselenides,[38,196,201,202,207] bis(hydroxylalkyl) diselenides,[203] dibenzyl diselenides,[197,200,208–210] diselenide dicarboxylic acids,[211–213] *tert*-aminoalkyl diselenides,[213a] cyclic diselenides,[204,214] bis[β-(3-indolylethyl)] diselenide,[206] and diselenide sugars.[215] The action of potassium diselenide on diethyl sulfate has served to form diethyl diselenide.[216]

Bogert and his co-workers[115,116] reported that the aromatic nucleophilic displacement reaction of sodium diselenide on *o*-chloronitrobenzene gave *o,o'*-dinitrodiphenyl diselenide. When this technique was applied by Baker and Moffitt[217] to *p*-chloronitrobenzene, however, the monoselenide rather than the diselenide was obtained.

c. From Selenocyanates.   One of the most important means by which diselenides can be obtained is by alkaline hydrolysis of selenocyanates. Although no clear-cut mechanism for this reaction has been established, one
has been suggested by Challenger *et al.*[218] which implicates selenols as the
initial products of the hydrolysis (scheme 70). In another proposed mech-

$$(70) \qquad 2ArSeCN + H_2O \longrightarrow 2ArSeH + 2HCNO$$

$$2ArSeH \xrightarrow{[O]} (ArSe—)_2 + H_2O$$

anism, the reaction is viewed as proceeding *via* a selenenic acid intermediate
(scheme 71).[219]

$$(71) \qquad RSeCN + H_2O \rightarrow RSeOH + HCN$$

$$RSeOH + RSeCN \rightarrow (RSe—)_2 + HCNO$$

When care has been taken to minimize exposure of the hydrolysis products
to atmospheric oxygen, it has been possible to isolate aryl selenols.[131,132,135]
The presence of alkylating agents during the course of the hydrolysis of
selenocyanates has led to the direct formation of selenides (equation
72),[132,217,220–223] further suggesting that selenols are the initial products of

$$(72) \qquad ArSeCN + H_2O \xrightarrow[OH^-]{CH_3I} ArSeCH_3$$

the reaction, and that diselenides result from their air oxidation. Recently,
Agenäs,[224] in a review of this subject, reported finding cyanide but no
cyanate ion in the alkaline hydrolysis of aryl selenocyanates.

A mechanism that accounts for most of the observations cited above is
submitted in scheme 73. Product formation may be a kinetically controlled

$$(73) \qquad OH^- + RSeCN \xrightarrow{k_1} RSe^- + HOCN$$

$$RSe^- + RSeCN \xrightarrow{k_2} (RSe—)_2 + CN^-$$

process that is dependent on two second-order rate constants.

Base hydrolysis of the appropriate aliphatic selenocyanate has led to the
formation of simple alkyl,[207,225,226] phthalimidoethyl,[227,228] benzamido-
ethyl,[229] cyclic,[230,234] and β-keto[235] diselenides. It is of interest that hydrolysis
of trifluoromethyl selenocyanate with 20% aqueous alkali caused fluoride
ion to be formed quantitatively.[54]

The alkaline hydrolysis of aromatic selenocyanates, first described by Bauer,[127] has been applied to the formation of various
phenyl,[118,129,134,135,236–238] biphenyl,[239,241] and indolyl[242–244] diselenides.

Recently, Biezais-Zirnis and Fredga[244a] synthesized the unusual bis-
diselenide **17a** containing a 14-membered ring by treating 1,8-bis(selenocyanatomethyl)naphthalene with methanolic potassium hydroxide (equation
73a).

(73a)

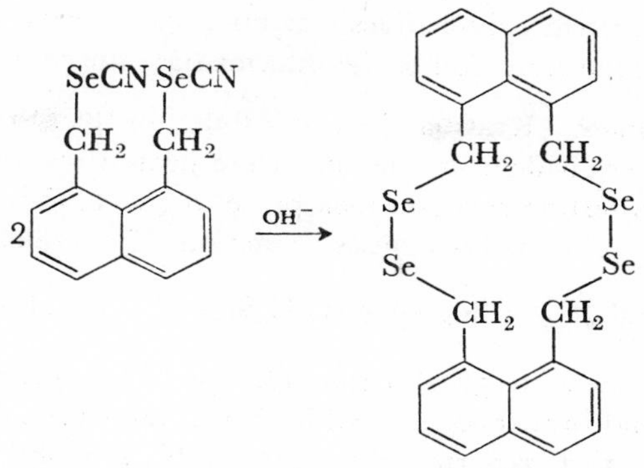

17a

Like the hydrolysis of selenocyanates under alkaline conditions, hydrolysis by mineral acids appears to proceed *via* selenol intermediates, which in a few instances have actually been isolated.[22,136,245] In most cases, however, air oxidation is permitted to take the products to the diselenide stage. Several selenocyanatocarboxylic acids have been converted to the corresponding diselenides by acid treatment (equation 74).[221,246,247] Pyrolysis of 1,1-

(74) $$NCSeCH_2COOH + H_2O \xrightarrow[\text{[O]}]{H^+} (-SeCH_2COOH)_2$$

bis(selenocyanatomethyl)cyclohexane at 175–185° resulted in the formation of 2,3-diselenaspiro[4.5]decane (equation 75).[248]

(75)

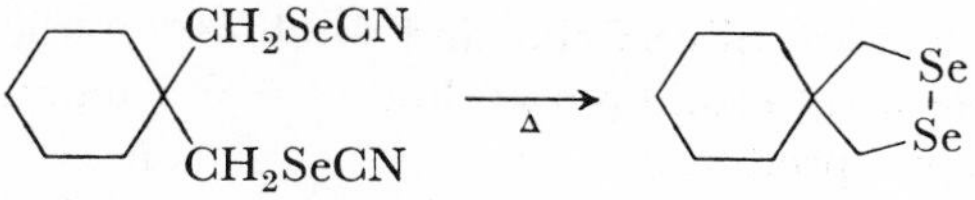

Sodium *o*-bromobenzoate, when heated with an aqueous solution of potassium selenocyanate and copper bronze in a sealed tube at 150°, gave bis(*o*-carboxyphenyl) diselenide.[249]

Phenylselenenyl selenocyanate, while insensitive to simple hydrolysis, is converted to diphenyl diselenide on treatment with an ethanolic or benzene solution of ammonia or alcoholic bromine.[250] Tin and hydrochloric acid transformed *m*-nitrophenylselenocyanate into bis(*m*-aminophenyl) diselenide.[275]

In a reaction analogous to the preparation of selenenyl sulfides from aryl selenocyanates and aryl thiols,[251] Nakasaki[252] treated aryl selenocyanates with arylselenols in an attempt to make unsymmetrical diselenides (equation 76). Although hydrogen cyanide was evolved, as anticipated, only a mixture

(76) $RSeCN + R'SeH \rightarrow HCN + [RSeSeR'] \rightarrow$

$$\tfrac{1}{2}(RSe-)_2 + \tfrac{1}{2}(R'Se-)_2$$

of the two symmetrical diselenides was obtained in each case, presumably because of disproportionation of the intermediate unsymmetrical diselenide.

d. From Selenides.   Krafft and Lyons[85] found in 1894 that, when diphenyl selenide and 1 equivalent of selenium were heated together at about 320° for 2 days, an insertion reaction took place to give a 35% yield of diphenyl diselenide and some polyselenides (reaction 77), which were noted to

$$(77) \qquad C_6H_5SeC_6H_5 + Se_x \xrightarrow{\Delta} (C_6H_5Se-)_2 + (C_6H_5Se-)_2Se_x$$

extrude selenium with great ease. The possible existence of branched diselenides[253] and branched triselenides[254] was later proposed.

Loevenich et al.[132] reported that heating bis(2-naphthyl) selenide with selenium gave two products, bis(2-naphthyl) diselenide (**18**), m.p. 126–127°, and the so-called branched diselenide **19**, m.p. 112–114°. Unlike **18**, **19** was

18   **19**

oxidizable to the selenoxide and selenium; boiling ethanol also caused selenium extrusion. Bergson[255] re-examined this reaction and concluded that the low-melting "branched" diselenide **19** was, in actuality, a mixture of the mono- and diselenides. Silverwood and Orchin[256] studied the reaction of diphenyl selenide with selenium and showed that a mixture of diphenyl di- and triselenides was produced.

Distillation of diphenyl diselenide with copper powder has been found to result in the abstraction of a Se atom to give diphenyl selenide.[85]

e. From Diazonium Salts.   Lesser and Weiss[119] found that the addition of sodium hydrogen selenide to diazotized anthranilic acid gave a 50% yield of diselenosalicylic acid (equation 78) and a small quantity of the monoselenide.

Schoeller[257] tried to obviate the use of hydrogen selenide required for the preparation of sodium hydrogen selenide. Instead, he heated a mixture of elemental selenium with potassium hydroxide to 140° and added it as an ice-cooled aqueous polyselenide solution to diazotized anthranilic acid.

An 85% yield of diselenosalicylic acid was reported, but application of this reaction to aniline gave an 80% yield of diphenyl selenide and only a small quantity of the diselenide. The Lesser and Weiss method has been used to make bis(2-naphthyl) diselenide[132] but has been reported to fail with diazotized *m*-aminobenzoic acid.[258]

*f. **From Grignard Reagents.*** The addition of selenium to a Grignard reagent is a method for the preparation of aromatic selenols (*cf.* Section IV.A.2). There are reports[104,258] of diselenide formation by this route, probably resulting from the air oxidation of the intermediate selenol.

Strecker and Willing[259] noted that the addition of selenium monochloride to phenylmagnesium bromide gave a mixture of diphenyl selenide and diselenide. With benzylmagnesium bromide and selenium monochloride, a mixture of dibenzyl diselenide and dibenzyl selenide dichloride was obtained (equation 79). Selenium monobromide is reported to give diphenyl

(79) $C_6H_5CH_2MgBr + Se_2Cl_2 \rightarrow (C_6H_5CH_2Se-)_2 + (C_6H_5CH_2)_2Se \cdot Cl_2$

diselenide with phenylmagnesium bromide.[260]

Gould and McCullough[109] found that, when a Grignard reagent could not be formed, diaryl diselenides could be obtained (albeit in low yields) by the reaction of an aryllithium compound with elemental selenium in ether, followed by hydrolysis and air oxidation of the product (scheme 80).

(80) $\qquad$ ArLi $+$ Se $\longrightarrow$ ArSeLi

$\qquad$ ArSeLi $+$ H$_2$O $\longrightarrow$ ArSeH $+$ LiOH

$\qquad$ 2ArSeH $\xrightarrow{\text{[O]}}$ (ArSe$-$)$_2$ $+$ H$_2$O

*g. **From Selenium Monohalides.*** In the paragraph above the preceding, mention is made of selenium monohalides as precursors of diselenides. In addition to reacting with Grignard reagents, selenium monobromide reacts with many aromatic hydrocarbons under Friedel-Crafts conditions to give mixtures of di- and monoselenides (equation 81).[261]

(81) $C_6H_6 + Se_2Br_2 + AlBr_3 \xrightarrow[\text{(then H}^+\text{)}]{\text{CS}_2}$

$\qquad\qquad\qquad\qquad\qquad$ (C$_6$H$_5$Se$-$)$_2$ $+$ (C$_6$H$_5$)$_2$Se $+$ C$_6$H$_5$Br

Pentafluorophenyllithium reacts with selenium monochloride to give bis(pentafluorophenyl) diselenide (equation 82), which can also be made by

(82)

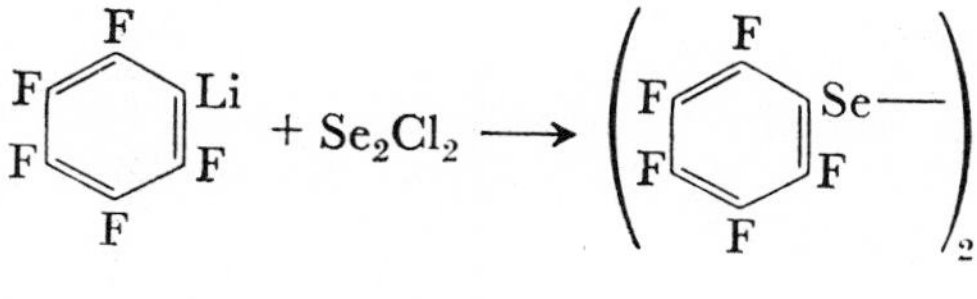

the reaction of selenium with either pentafluorophenylmercuric chloride or iodopentafluorobenzene.[262]

Yarovenko *et al.*[49] caused selenium monobromide to add to the double bond of perfluoroethylene to give a low yield of the halogenated diselenide (equation 83).

$$(83) \qquad CF_2{=}CF_2 + Se_2Br_2 \xrightarrow[110^\circ]{\text{sealed tube}} \left(F_2{-}\overset{\overset{\displaystyle Br}{|}}{C}{-}\overset{\overset{\displaystyle F_2}{|}}{C}{-}Se{-} \right)_2$$

h. *From Selenosulfates.* Diselenide formation from organic selenosulfates can be effected with ease by acid[10,40,74a] or alkaline[41,43,74a] hydrolysis. In some instances, conversion to the diselenide takes place by simply heating an aqueous solution of the selenosulfate until sulfur dioxide is no longer evolved. 2-Aminoethaneselenosulfuric acid, for example, is unstable above 60° in aqueous solution, giving bis(2-aminoethyl) diselenide.[41] The mechanism proposed for this hydrolysis is shown in scheme 84. 2-Aminoethaneseleno-

$$(84) \qquad \overset{\displaystyle H}{\underset{\displaystyle \cdot\cdot}{HO}}{:} + \overset{\overset{\displaystyle O^-}{|}}{SO_2}SeCH_2CH_2\overset{+}{N}H_3 \longrightarrow H_3\overset{+}{N}CH_2CH_2Se^- + H_2SO_4$$

$$H_3\overset{+}{N}CH_2CH_2Se^- + \overset{\overset{\displaystyle SO_3^-}{|}}{Se}CH_2CH_2\overset{+}{N}H_3 \longrightarrow (H_3\overset{+}{N}CH_2CH_2Se{-})_2 + SO_3{}^{2-}$$

$$H_2SO_4 + SO_3{}^{2-} \longrightarrow SO_2 + SO_4{}^{2-} + H_2O$$

sulfuric acid, when pyrolyzed at 145–150°, gives bis(2-aminoethyl) diselenide quantitatively as the sulfuric acid salt.[41] Another method recently found to rapidly convert 2-aminoethaneselenosulfuric acid to the diselenide consists of treating it with aqueous hydrogen sulfide. The diselenide is then obtained as the thiosulfuric acid salt (equation 85).[263]

$$(85) \quad 2H_2NCH_2CH_2SeSO_3H + 3H_2S \rightarrow$$
$$(H_2NCH_2CH_2Se{-})_2{\cdot}H_2S_2O_3 + 3S + 3H_2O$$

Selenoctic acid was made *via* the corresponding diselenosulfate, which was hydrolyzed *in situ* to give the diselenide directly.[264]

m- and *p*-Nitrobenzyl selenosulfates were converted electrolytically to the corresponding diselenides, but the related nitrobenzyl disulfides could not be made in this manner.[265] Failure was similarly encountered in the attempt to convert *o*-nitrobenzyl selenosulfate to the diselenide.

Oxidation of selenosulfates with hydrogen peroxide [266,267] or iodine[190,198,265] also gives diselenides.

i. *From Selenium Dioxide.* The preparation of diselenides by the use of selenium dioxide is not a reliable procedure, but instances in which this has been achieved will be cited.

The reaction of benzene with selenium dioxide under Friedel-Crafts conditions gives a 3:1 mixture of diphenyl selenide and diphenyl diselenide.[268]

Florey et al.[269,270] treated several steroids with selenium dioxide, incorporating a Se atom into them. Later, Baran[271] oxidized testosterone in *tert*-butyl alcohol with selenious acid and obtained seleno-1-dehydrotestosterone, which had been described earlier but not identified.[269] He proposed that the Se atom was part of the $\Delta^{1,4}$-3-keto moiety, as shown in structure **20**.

20

Selenium dioxide treatment of 6,7-dihydro-5*H*-benzocycloheptene[272] also led to diselenide formation, while mono-, di- and triselenides were observed as products in the reaction of indoles with SeO_2.[273] The reaction of selenium dioxide with trifluoroacetic anhydride gave some bis(trifluoromethyl) diselenide.[274]

j. Reduction of Seleninic and Selenenic Acids.   The numerous agents which have been used to reduce aromatic seleninic acids ($ArSeO_2H$) to diselenides include aqueous sodium bisulfite solution,[275] hydrazine,[276] hypophosphorous and phosphorous acids,[277] mercaptans,[278] and alkaline glucose.[279,280] An aromatic seleninic acid anhydride, similarly, gave a diselenide on treatment with sodium metabisulfite.[281] 2-Aminoethaneselenenic acid was reduced to bis(2-aminoethyl) diselenide by means of hydrazine.[282]

k. From Selenenyl Halides and Selenium Trihalides.   Phenylselenenyl bromides have been reduced with zinc dust to diselenides,[283,284] whereas tin has been used to reduce trifluoromethyl-[49] and trichloromethylselenenyl chlorides.[285] Phenylselenium tribromide has also been reduced by zinc dust to the diselenide.[283]

l. Miscellaneous Reactions.   A series of unsymmetrical diselenides was prepared by the reaction of an aromatic selenenyl bromide with an aromatic selenol in chloroform (reaction 86).[191]

Heating silver trifluoroacetate with selenium in a sealed tube resulted in a mixture of bis(trifluoromethyl) mono- and diselenides with loss of carbon dioxide.[274] Silver heptafluorobutyrate, similarly, gave bis(heptafluoropropyl) mono- and diselenides.[286] The mercury salts of trifluoroacetic[287] and difluoroacetic[288] acids, when treated with selenium, behaved in a comparable manner, giving hexafluorodimethyl and tetrafluorodimethyl diselenides, respectively.

Recently, Bacon and Pande[289] reported that 2-bromonitrobenzene, when heated with selenium in pyridine for 24 hours in the presence of cuprous oxide, gave bis(2-nitrophenyl) diselenide contaminated with a small quantity of the corresponding selenide.

Virtually all attempts to make selenoaldehydes or selenoketones have led to diselenide or polyselenide formation (*cf.* Chapter VII). For example, treatment of benzophenone or acetone in ethanolic hydrogen chloride solution with hydrogen selenide gave dibenzhydryl diselenide (equation 87)

$$(87) \qquad C_6H_5\underset{\underset{O}{\|}}{C}C_6H_5 + H_2Se \longrightarrow \tfrac{1}{2}\left(\underset{C_6H_5}{\overset{C_6H_5}{\diagup}}CH\!-\!Se\!-\!\right)_2$$

or diisopropyl diselenide, respectively.[19] The reaction of benzaldehyde with bis(methoxymagnesium) diselenide in the presence of morpholine yielded dibenzyl diselenide.[208]

Organomercury selenides such as **21** generate the corresponding diselenide on heating in a sealed tube with iodine (equation 88)[286,290] or dry hydrogen chloride.[290]

$$(88) \qquad (C_2F_5Se\!-\!)_2Hg + I_2 \xrightarrow[60°, 120\ hr]{sealed\ tube} (C_2F_5Se\!-\!)_2 + HgI_2$$
21

The unsymmetrical methyl phenyl diselenide was prepared by mixing equimolar quantities of selenophenol and dimethyl diselenide at room temperature (equation 89).[201] The volatile methaneselenol was removed

$$(89) \qquad C_6H_5SeH + (CH_3Se\!-\!)_2 \rightleftharpoons C_6H_5SeSeCH_3 + CH_3SeH$$

under reduced pressure, but attempts to distill the unsymmetrical diselenide resulted in its disproportionation.

REACTIONS

a. **Reduction.** Diselenides have been reduced to selenols in acidic, neutral, or basic media by a variety of reagents which include sodium in alcohol[37,85,292] or liquid ammonia,[38] zinc in sodium hydroxide solution[284] or glacial acetic acid,[115,117,118] sodium borohydride,[40,41,108] lithium aluminum hydride,[18,44] and hypophosphorous acid[10,45,46,136] (*cf.* Section IV.A).

The thiol, reduced glutathione, has been found to be only slightly effective in the reduction of diselenides.[47] Also, 2-mercaptoethanol, while capable of reducing the disulfide linkage of coenzyme A and isocoenzyme A, was reported by Günther and Mautner[60] to be unable to cleave diselenides. Later studies by Walter *et al.*,[293] however, indicated that Se-Se bond cleavage in selenocystine occurs with 2-mercaptoethanol over a wide pH range. The dithiol, dithiothreitol, readily reduces most diselenides to selenols.[47]

Dickson and Tappel[294] found that selenocystine is completely reduced to selenocysteine by cysteine or glutathione if either of the latter is present in concentrations about a thousand-fold that of the diselenide.

b. Oxidation. Diselenides are oxidized most readily to seleninic acids (*cf.* Section IV.D); however, by the correct choice of reaction conditions and oxidizing agent it is also possible to obtain selenenic acids (*cf.* Section IV.C), seleninic anhydrides (*cf.* Section IV.D), or selenonic acids (*cf.* Section IV.E).

Woodbridge[295] found that at room temperature dialkyl diselenides react with a stoichiometric quantity of *tert*-butyl peroxide in benzene to give the corresponding seleninic anhydride.[296] At 75°, however, 1 mole of diselenide catalytically decomposes at least 50 moles of hydroperoxide (initial concentration 0.5 *M*). By decomposing the hydroperoxide with a seleninic acid or anhydride, the induction period due to the conversion of the diselenide to the anhydride can be obviated. Caldwell and Tappel[297] observed that diselenodicarboxylic acids are over 24 times more effective than dithiodipropionic acid in the reduction of hydrogen peroxide. Selenocystine was found by these investigators[298] to be superior to cystine in reducing hydroperoxides.

The initial oxidation products of disulfides are thiolsulfinates (RSOSR) and thiolsulfonates (RSSO$_2$R).[296]. Although the analogous diselenide oxidation products are unknown, they have been postulated to be intermediates in the formation of seleninic anhydrides[296] and seleninic acids.[298] The appearance of an intermediate product in the oxidation of 1,2-diselenolane-4-carboxylic acid having a u.v. maximum at 440 nm was attributed by Bergson[204,299] to a transient selenolseleninate **22**.

22

c. Halogenolysis. Treatment of diselenides by chlorine, thionyl, or sulfuryl chlorides in inert solvents gives selenenyl chlorides (RSeCl).

Brominolysis, similarly, results in selenenyl bromide formation (*cf.* Section IV.C). Excess reagent causes further halogenation to take place, giving organic selenium trihalides ($RSeX_3$) (*cf.* Section IV.D).

d. Other Cleavage Reactions.   The effect of strongly alkaline solutions on diselenides has been studied by numerous workers who have found that, generally, more than one product is obtained. In 1908 Frerichs and Wildt[300] reported that diselenodiglycolic anilide, on heating a short time with 25% sodium hydroxide, gave the corresponding selenide, selenoxanilide, sodium selenoglycolate, and sodium selenide. Fredga[301] and Behaghel and Seibert[283] proposed that diselenides are hydrolyzed to the corresponding selenol and seleninic acid according to equation 90. Painter *et al.*[302] also studied the

$$(90) \qquad 2RSeSeR \underset{}{\overset{2H_2O}{\rightleftharpoons}} 3RSeH + RSeO_2H$$

decomposition of diselenides in boiling 5% aqueous sodium hydroxide and noted that some form of inorganic selenium was generated in most cases.

Rheinboldt and Giesbrecht[191] investigated the alkaline hydrolysis of unsymmetrical aromatic diselenides and obtained selenols, symmetrical diselenides, and seleninic acids according to scheme 91.

$$(91) \qquad 3ArSeSeAr' + 3H_2O \rightarrow 3ArSeOH + 3Ar'SeH$$
$$2ArSeOH \qquad\qquad \rightarrow ArSeO_2H + ArSeH$$
$$ArSeOH + ArSeH \quad \rightarrow ArSeSeAr + H_2O$$

$$3ArSeSeAr' + 2H_2O \rightarrow 3Ar'SeH + ArSeSeAr + ArSeO_2H$$

Bird and Challenger[37] cleaved dimethyl and diethyl diselenides with mercuric cyanide, giving methyl and ethyl selenocyanates, respectively. Potassium cyanide treatment of bis(chloromethyl) diselenide gave dissimilar results in that bis(chloromethyl) selenide and potassium selenocyanate were reported to be formed.[303]

Phenyllithium on reaction with diphenyl diselenide gave diphenyl selenide and, presumably, selenophenol as the lithium salt.[304]

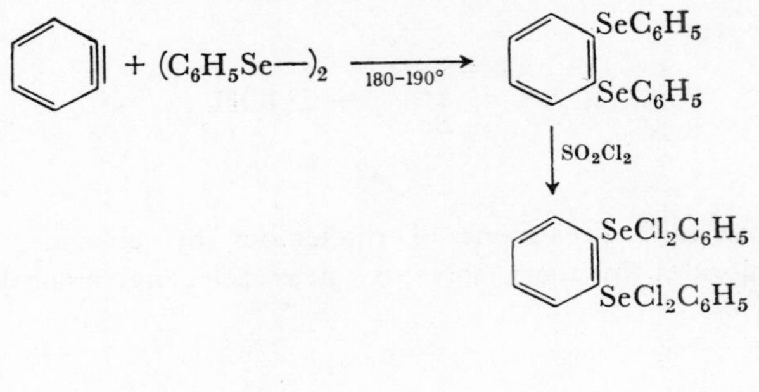

Diphenyl diselenide is cleaved in 57% yield by benzyne at 180–190° in *o*-dichlorobenzene (equation 92).[305] Under the identical conditions diphenyl ditelluride was cleaved to the extent of 97%, whereas diphenyl disulfide was unaffected.

Diazomethane reacted with diphenyl and dibenzyl diselenides to give selenoacetals quantitatively (equation 93).[188] The reaction was found

$$(93) \qquad C_6H_5SeSeC_6H_5 + CH_2N_2 \rightarrow C_6H_5SeCH_2SeC_6H_5 + N_2$$

applicable to ditellurides as well as to disulfides and was suggested to proceed *via* one of several possible photochemical pathways.

Unsymmetrical selenides can be made by the reaction of diselenides with Grignard reagents (equation 94).[306]

$$(94) \qquad (RSe—)_2 + R'MgBr \rightarrow RSeR' + RSeMgBr$$

The rate of cleavage of various diphenyl diselenides with sodium bis-(phenylsulfonyl)methane (equation 95) was found to be most rapid when

the phenyl ring of the diselenide was substituted with a nitro group.[307]

e. **Thermal Decomposition.** Morgan and Burstall[231-233] noted that, when cyclic diselenides were heated strongly, elemental selenium was extruded with concomitant ring contraction (equation 96). Bis(chloromethyl)

diselenide, similarly, lost one Se atom when it was heated, giving bis-(chloromethyl) selenide.[303] The pyrolysis of dibenzhydryl diselenide at 200° under aspiration gave a selenium-free product, 1,1,2,2-tetraphenylethane (equation 97),[19] while diisopropyl diselenide, under these conditions,

yielded an equimolar mixture of propane and propene.

f. Exchange Reactions. Dimethyl diselenide, on treatment with seleno-phenol, gave the unsymmetrical methyl phenyl diselenide with elimination of methaneselenol.[291] Bis(pentafluorophenyl) disulfide and diselenide reacted with one another to give unsymmetrical bis(pentafluorophenyl)-selenenyl sulfide, which was separated from the starting materials by t.l.c. on kieselguhr.[308]

g. Transhydrogenation. The catalytic action of diselenides in transhydro-genation has been studied by Morimoto,[226,309] who found that cinnamic acid was reduced to 2-phenylpropionic acid when heated with diphenyl diselenide in tetralin. Morimoto suggested that in the rate-determining step tetralin is first dehydrogenated to naphthalene with the formation of 4 equivalents of selenol (equation 98). The selenol then provides the H atoms

$$(98) \quad \text{[structure]} + 2(C_6H_5Se{-})_2 \longrightarrow \text{[structure]} + 4C_6H_5SeH$$

for the reduction of the cinnamic acid and reverts to the diselenide (equation 99).

$$(99) \quad C_6H_5CH{=}CH{-}COOH + 2C_6H_5SeH \rightarrow$$
$$C_6H_5CH_2CH_2COOH + (C_6H_5Se{-})_2$$

It was noted by Silverwood and Orchin[256] that aromatic diselenides can serve as dehydrogenation agents. Inasmuch as the results obtained at 290° resemble those obtained with elemental selenium, it was suggested that the diaryl diselenides probably decompose to elemental selenium and the monoselenide. Pyrolysis of diphenyl diselenide, in actuality, gave diphenyl selenide, a second material tentatively identified as diphenyl triselenide, and selenium.

h. Action of Raney Nickel. Diphenyl diselenide, on treatment with Raney nickel which had been degassed at 200°, gave biphenyl or diphenyl selenide, or both, depending on the reaction temperatures.[310,311] Deselenization of bis(o-biphenylyl) diselenide to biphenyl was achieved with Raney nickel, but deselenization did not occur with diphenyl diselenide.[312]

i. With Mercury and Its Compounds. Fredga[313] reported that, when an aqueous solution of the disodium salt of diselenodiglycolic acid was shaken at room temperature with mercury, the diselenide was cleaved to give a mercury derivative (equation 100). The same reaction was reported by Margolis and Pittman[19] to occur with other diselenides. The diselenodiglycolic acids could be restored on treatment of the mercury derivatives with I_2/KI solution.[313]

$$(100) \qquad (\text{—SeCH}_2\text{COONa})_2 + \text{Hg} \xrightarrow[\text{H}^+]{\text{then}} \text{Hg} \underset{\text{SeCH}_2\text{COOH}}{\overset{\text{SeCH}_2\text{COOH}}{<}}$$

The reaction of diphenyl diselenide with mercuric chloride was reported by Bradt and Green[314] to give a complex, $(\text{C}_6\text{H}_5\text{Se}-)_2\cdot2\text{HgCl}_2$. Gould and Amendola[315] could obtain this material only in extremely poor yield and suggested that selenophenol, which was probably present in the original diphenyl diselenide, reacted to form chloromercuriselenophenolate (equation 101). Diselenido dicarboxylic acids were reported by Fredga[356] to combine

$$(101) \qquad \text{C}_6\text{H}_5\text{SeH} + \text{HgCl}_2 \rightarrow \text{C}_6\text{H}_5\text{SeHgCl} + \text{HCl}$$

with mercuric chloride according to equation 101a. Both the mercury salt

$$(101a) \qquad 2\text{RSeSeR} + 3\text{HgCl}_2 + 2\text{H}_2\text{O} \rightleftharpoons 3\text{RSeHgCl} + \text{RSeO}_2\text{H} + 3\text{HCl}$$

and the seleninic acid could be restored to the diselenide state by treatment with hydriodic acid. Okamoto and Yano[315a] have recently found that diphenyl diselenide reacts with dibutyl mercury at room temperature in dioxane to give butyl phenyl selenide and $(\text{C}_6\text{H}_5\text{Se})_2\text{Hg}$. At reflux temperature there was obtained an 80% yield of the unsymmetrical selenide and elemental mercury (equation 101b).

$$(101b) \qquad (\text{C}_6\text{H}_5\text{Se}\text{—})_2 + (\text{C}_4\text{H}_9)_2\text{Hg} \rightarrow 2\text{C}_6\text{H}_5\text{SeC}_4\text{H}_9 + \text{Hg}$$

j. With Metal Carbonyls. Welcman and Rot[316] found that dimethyl, diethyl, and dipropyl diselenides and their perfluoro derivatives react with manganese pentacarbonyl hydride to give dimeric products (equation 102).

$$(102) \qquad \text{Mn(CO)}_5\text{H} + (\text{F}_3\text{CSe}\text{—})_2 \rightarrow [\text{Mn(CO)}_4\text{SeCF}_3]_2$$

Cyclopentadienyliron carbonyls were observed by Welcman and Rozenbuch[317] to react with diselenides to give mononuclear complexes of the type $\text{RSeFe(CO)}_2\text{C}_5\text{H}_5$, where R = CF_3, C_2H_5, or C_3H_7. Cyclopentadienylmolybdenum tricarbonyl gives several complex products with diphenyl diselenide.[318]

MISCELLANEOUS*

The Kinney equation has been applied to the prediction of the boiling points of three diselenides.[82] It has been observed that mixture melting points of related diselenides, disulfides, and selenenyl sulfides are frequently undepressed.[90,251]

Aliphatic diselenides[319] and diphenyl diselenide[320] are claimed to be effective oxidation inhibitors for lubricating fluids. Other diaryl diselenides

---

* A discussion of the spectral properties of diselenides will be found in Chapter XV.

have been found to be useful as antifog and antibronzing agents for photographic emulsions.[321]

2. Triselenides

Only a limited number of triselenides are known. In general, they are difficult to obtain in good yield and in high purity. Aromatic triselenides are moderately stable yellow or orange-yellow materials that melt lower than the corresponding diselenides. At elevated temperatures triselenides readily extrude selenium, leaving the diselenide. The i.r. spectra of related di- and triselenides do not vary appreciably.

PREPARATION

The first triselenides were reported by Levi and Baroni,[254] who applied a method[322] developed earlier for the synthesis of trisulfides. These workers combined carbon disulfide solutions of ethaneselenol and selenium oxychloride, resulting in an exothermic reaction and the formation of a mixture of diethyl di- and triselenides (equation 103).

(103) $4C_2H_5SeH + SeOCl_2 \rightarrow$

$$C_2H_5SeSeSeC_2H_5 + C_2H_5SeSeC_2H_5 + 2HCl + H_2O$$

Rheinboldt and Giesbrecht[323] found that aromatic selenenyl selenocyanates in benzene react quantitatively with thiols, resulting in the evolution of hydrogen cyanide and the formation of products which include the diaryl triselenide (equation 104). These investigators speculated that the

(104) $2ArSeSeCN + 2RSH \rightarrow ArSeSeSeAr + (RSe-)_2 + Se + 2HCN$

tetraselenide was formed initially, followed by loss of selenium and formation of the triselenide.

Silverwood and Orchin,[256,324] on pyrolyzing diphenyl diselenide, obtained selenium, diphenyl *mono*selenide, and a yellow crystalline material that analyzed for a mixture of the di- and triselenide. On dissolving in any solvent, the latter material readily decomposed to selenium and the diselenide. On standing for 7 days, the compound was transformed into a red solid which on exposure to u.v. light reverted to the yellow form. When diphenyl diselenide was heated with selenium *in vacuo* at 290° for 24 hours and the products were chromatographed in the dark, one of the fractions appeared to be identical to the yellow material mentioned above (scheme105).

(105) $2ArSeSeAr \rightleftharpoons ArSeSeSeAr + ArSeAr$

$$\Updownarrow$$

$$ArSeSeAr + Se$$

Wilshire[273] found that heating 2-methylindole with selenium dioxide in benzene gave under apparently identical conditions either 3,3'-diindolyl di- or triselenide. The latter compound decomposed rapidly in methanol, whereas the diselenide was stable. The di- and triselenides exhibited virtually identical i.r. spectra.

REACTIONS

The ease with which many triselenides lose selenium on heating and in solution has already been mentioned. Diaryl triselenides have been found by Rheinboldt and Giesbrecht[323] to decompose in potassium hydroxide solution to give the corresponding diaryl diselenide and aryl seleninate (equation 106). These workers also noted that, although diaryl triselenides

$$(106) \quad 3ArSeSeSeAr + 8KOH \rightarrow$$

$$3K_2Se + 2ArSeSeAr + 2ArSeO_2K + 4H_2O$$

are unaffected by hydrochloric acid, they are cleaved by hydrobromic[323] (equation 107) and hydriodic acids[323,325] (equation 108) to give different

$$(107) \qquad ArSeSeSeAr + 2HBr \longrightarrow 2ArSeBr + H_2Se$$

$$(108) \qquad ArSeSeSeAr \xrightarrow{HI} (ArSe{-})_2 + Se$$

types of products. Halogen treatment of triselenides affords the aryl selenenyl halide and the selenium monohalide (equation 109).[323]

$$(109) \qquad 2ArSeSeSeAr + 3X_2 \rightarrow 4ArSeX + Se_2X_2$$

C. SELENENIC ACIDS AND DERIVATIVES

1. Selenenic Acids

Selenenic acids, RSeOH, contain selenium in the bivalent state and are amphoteric in nature. Aromatic selenenic acids, which constitute virtually all known examples of the class, range in color from yellow to purple, intensification of the color occurring in alkaline solution.

PREPARATION

a. From Selenenyl Halides. Arylselenenic acids were first implicated by Behaghel and Siebert[283] as intermediates in the hydrolysis of selenenyl halides to seleninic acids and diselenides (scheme 110). Later, when Behaghel

$$(110) \qquad C_6H_5SeBr + H_2O \rightarrow C_6H_5SeOH + HBr$$

$$2C_6H_5SeOH \rightarrow C_6H_5SeO_2H + C_6H_5SeH$$

$$C_6H_5SeH + C_6H_5SeOH \rightarrow (C_6H_5Se{-})_2 + H_2O$$

and Müller[326] heated *o*-nitrophenylselenenyl bromide in water for 1 hour, they obtained the corresponding selenenic acid (**23**) (equation 111), as

(111)

23

orange-red needles. The *m*-nitro isomer gave only the corresponding diselenide when subjected to the same treatment. Jenny[327] found that anthraquinone-1-selenenyl bromide, on refluxing with silver oxide in dioxane, yielded anthraquinone-1-selenenic acid.

b. From Selenocystamine. The only known aliphatic selenenic acid is apparently stabilized by having a cationic group present in the molecule. 2-Aminoethaneselenenic acid hydrochloride (**24**) is prepared quantitatively by the oxidation of selenocystamine dihydrochloride with *m*-chloroperbenzoic acid or hydrogen peroxide (equation 112).[282] Similarly, oxidation

$$(112) \quad (HCl \cdot H_2NCH_2CH_2Se-)_2 \xrightarrow{[O]} 2HCl \cdot H_2NCH_2CH_2SeOH$$
$$\mathbf{24}$$

of selenocystamine free base with peroxide gives 2-aminoethaneselenenic acid, which is less stable than its hydrochloride salt, **24**. The analogous sulfur compound, 2-aminoethanesulfenic acid, appears to be unknown.

c. From Seleninic Acids. Seleninic acids can be reduced with hypophosphorous to the selenenic acid stage (equation 113), and if sufficient hypo-

$$(113) \quad ArSeO_2H + H_3PO_2 \rightarrow ArSeOH + H_3PO_3$$

phosphorous acid is present the reduction proceeds to give diselenide (equation 114).[277] Similarly, phosphorous acid, depending on the quantity

$$(114) \quad 2ArSeO_2H + 3H_3PO_2 \rightarrow (ArSe-)_2 + 3H_3PO_3 + 2H_2O$$

used, can also give either the selenenic acid or diselenide.[277] Other agents that have been employed in the reduction of seleninic acids to selenenic acids are ethyl and *tert*-butyl mercaptans, *p*-thiocresol, thiophenol (equation 115), selenophenol,[278] hydrazine and its salts, phenylhydrazine, semi-

$$(115) \quad ArSeO_2H + 2C_6H_5SH \rightarrow ArSeOH + (C_6H_5S-)_2 + H_2O$$

carbazide hydrochloride, and hydroxylamine hydrochloride.[276] The last three compounds gave impure products, while hydrazine, like the other reducing agents mentioned above, led, in some instances, to diselenide formation. Hydrazine gave a quantitative yield of 2-benzoylbenzeneselenenic acid from the corresponding seleninic acid.[328]

REACTIONS

2-Aminoethaneselenenic acid can be oxidized with hydrogen peroxide to 2-aminoethaneseleninic acid.[282] Conversely, the aminoselenenic acid hydrochloride can be reduced to selenocystamine dihydrochloride by means of hydrazine (equation 116), whereas sodium borohydride takes the compound to the selenol stage.[282]

$$(116) \quad 2HCl \cdot H_2NCH_2CH_2SeOH + H_2NNH_2 \rightarrow$$
$$(HCl \cdot H_2NCH_2CH_2Se—)_2 + N_2 + 2H_2O$$

2-Benzoylbenzeneselenenic acid (**25**) was oxidized with hydrogen per-

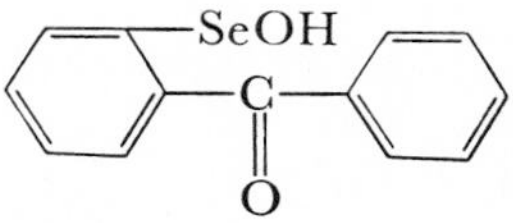

25

oxide in acetone to the corresponding seleninic acid.[328] Hydrobromic acid converted **25** to 2-benzoylbenzeneselenenyl bromide, while pyrolysis at *ca.* 140° led to the corresponding diselenide. The latter could also be extracted from a 2-month-old sample of **25**, indicating that disproportionation had slowly taken place.

2. Selenenic Esters (RSeOR′)

The selenium counterparts of sulfenic esters are selenenic esters, for which an alternative name, selenenyl alkoxides,[329] has been suggested.

PREPARATION

Selenenic esters have not been prepared by the direct esterification of selenenic acids.

Behaghel and Müller[326] heated *o*-nitrobenzeneselenenyl bromide with methanolic silver acetate and mistakenly reported that the product was the selenenyl acetate **26**. Cook and Donia,[330] working with 2,4-dinitrobenzene-

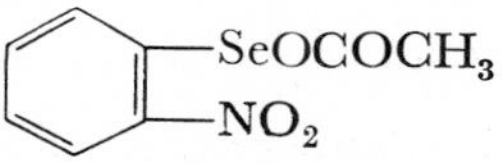

26

selenenyl bromide, and Hölzle and Jenny[331] demonstrated that the product obtained in this type of reaction is, in actuality, the selenenic ester **27**. The

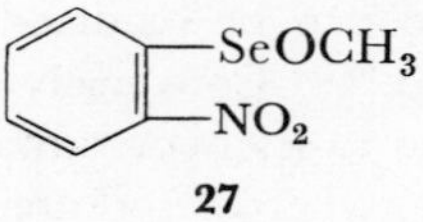

$$\text{27}$$

latter authors suggested, however, that the acetate is an intermediate in the reaction; indeed, Jenny[238,332,333] was able, in some instances, to isolate acetates before their ultimate conversion to selenenic esters.

The direct conversion of anthraquinone-1-selenenyl bromide to the methyl ester of anthraquinone-1-selenenic acid was achieved by heating the former in methanol in the presence of moist silver oxide (equation 117).[327]

(117)

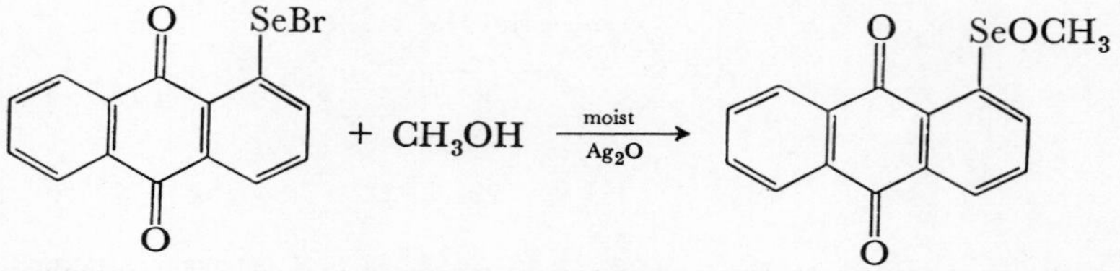

Certain selenocyanates, such as *o*-nitrobenzeneselenocyanate, can be used as precursors for the preparation of selenenic esters. Displacement on selenium occurs in the presence of either silver acetate, silver carbonate, silver hydroxide, or cupric acetate and catalytic amounts of pyridine (equation 118).[331]

(118)

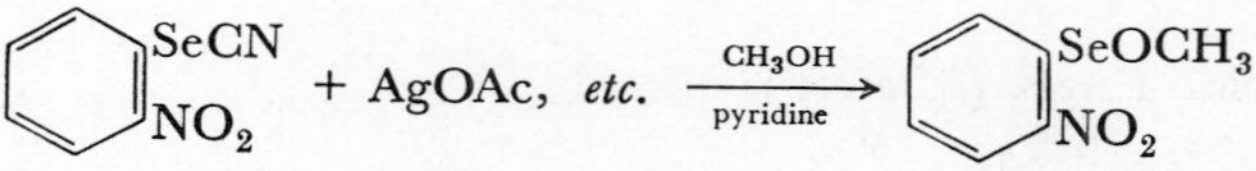

This method fails, however, when there is no electron-withdrawing group on the ring or when it is applied to aromatic thiocyanates.

REACTIONS

a. **With Acetic Acid.** Selenenyl esters have been converted to acetates by treatment with acetic acid (equation 119).[238,334] The original ester can be

(119) $\qquad \text{RSeOR}' + \text{HOAc} \rightleftharpoons \text{RSeO—C—CH}_3 + \text{R}'\text{OH}$
$$\overset{\|}{O}$$

regenerated by heating the acetate with the appropriate alcohol.

b. **With Thiosulfate and Other Nucleophiles.** Foss and Svendsen[329] found that the reaction of selenenyl esters with thiosulfate ion to give selenenyl thiosulfates (equation 120) could be utilized to quantitate the esters. The

(120) $\qquad \text{ArSeOR} + \text{S}_2\text{O}_3{}^{2-} + \text{H}^+ \rightarrow \text{ArSeSSO}_3{}^- + \text{ROH}$

excess thiosulfate ion could be back-titrated inasmuch as the selenenyl thiosulfates are unreactive towards iodine. These workers also found that the alkoxide group can be displaced, as indicated by the equilibria in equations 121 and 122.

(121) $ArSeOC_2H_5 + CH_3OH \rightleftharpoons ArSeOCH_3 + C_2H_5OH$

(122) $ArSeOR + NH_3 \rightleftharpoons ArSeNH_2 + ROH$

c.   *Oxidation.*   Ethyl 2,4-dinitrobenzeneselenenate (**28**) was oxidized by

(123)

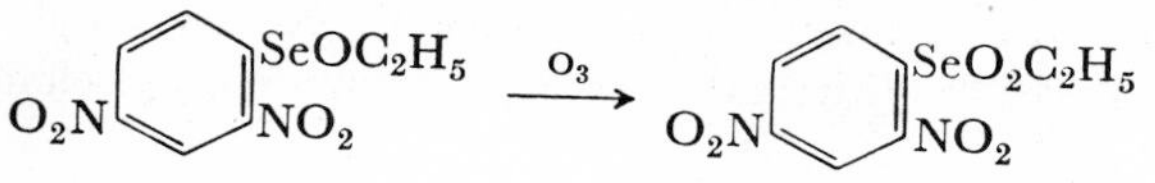

28

ozone to the seleninic ester (equation 123),[296] which was readily hydrolyzable to the seleninic acid.

REACTIONS OF SELENENYL ACETATES

Selenenyl acetates are hydrolyzed by water to selenenic acids (equation 124)[238,333] and are capable of adding to various olefinic systems such as

(124) $RSe{-}O{-}\overset{\displaystyle O}{\underset{\| }{C}}{-}CH_3 + H_2O \longrightarrow RSeOH + CH_3COOH$

cyclohexene (equation 125) and styrene.[334] In the latter reaction, alter-

(125)

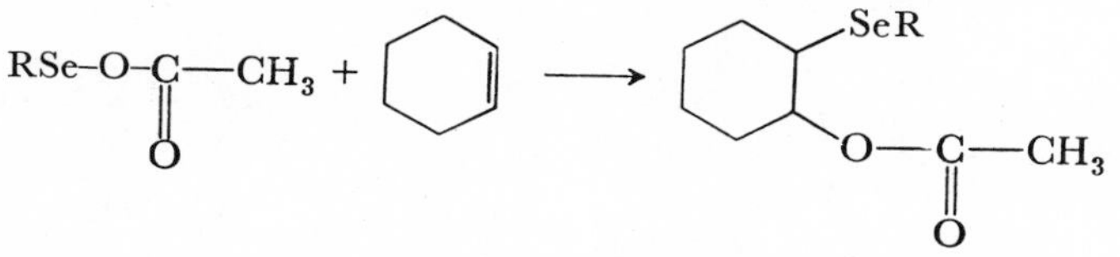

native starting materials may be either the corresponding selenenic ester or the selenenyl bromide, which are heated with the olefin in the presence of acetic acid. The acetoxy groups of the products can be hydrolyzed with aqueous-alcoholic potassium hydroxide to give hydroxy selenides.

If the olefin in the above reaction is polarized due to the presence of an electron donating group, as in 2,3-dihydropyran, substitution rather than addition takes place (equation 125a).[334a] Substitution reactions of this type

(125a)

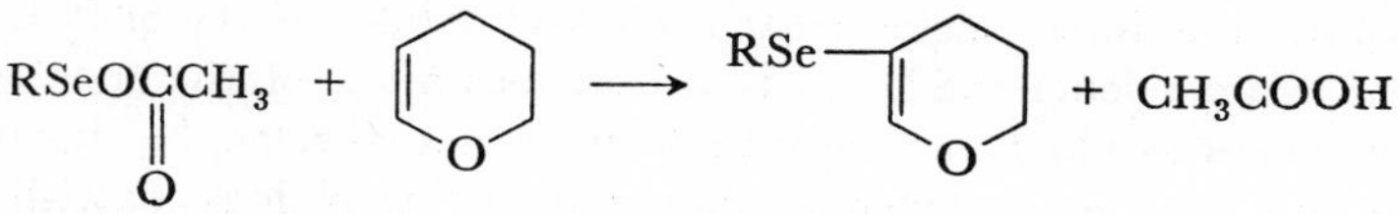

have not been observed to occur with sulfenic esters.

3. Selenenamides

Only a very limited number of examples of this class are known. The Se-N bond in selenenamides is hydrolyzed slowly at room temperature and rapidly by acids or at elevated temperatures. Selenenamides can be assayed with thiosulfate ion by the method of Foss and Svendsen.[329]

PREPARATIONS

a. From Selenenyl Halides. A number of perfluoroalkylselenenyl halides, on treatment with amines, have afforded perfluoroalkylselenenamides (equation 126a).[286,290] With methylamine, pentafluoroethylselenenyl chloride

$$(126a) \quad CF_3CF_2CF_2SeX + NH(CH_3)_2 \rightarrow CF_3CF_2CF_2SeN(CH_3)_2$$

gave N-methylpentafluoroethylselenenamide ($F_5C_2SeNHCH_3$) or N-methyl-bispentafluoroethylselenenimide [$F_5C_2SeN(CH_3)SeC_2F_5$], depending on the ratio of the reactants.[290]

2,4-Dinitrobenzeneselenenyl bromide reacts with substituted anilines to give N-arylselenenamides (equation 126b)[330] With 1- and 2-naphthyl-

$$(126b) \quad \underset{O_2N}{\underset{}{\bigcirc}}\overset{SeBr}{\underset{NO_2}{}} + 2ArNH_2 \longrightarrow \underset{O_2N}{\underset{}{\bigcirc}}\overset{SeNHAr}{\underset{NO_2}{}} + Ar\overset{+}{N}H_3Br^-$$

amines, nuclear substitution took place to give **29** and **30**, respectively. In

29   30

contrast, Pitombo[335] found that 2-naphthylselenenyl chloride reacts with 2-naphthylamine to give the expected selenenamide.

Aromatic substitution, when it occurs, takes place *para* to the amine function of an aniline if that position is free.[335-338] Yagupol'skii and Voloshchuk,[339,340] in studying the reaction of trifluoromethylselenenyl chloride with various anilines, suggested that the low stability of the Se-N bond in the intermediate selenenamide leads to the formation of the more stable rearrangement product, namely, the trifluoromethyl selenide (equation 127).

2,4-Dinitrobenzeneselenenyl methoxide and ethoxide react with ammonia gas in chloroform to give the related selenenamides (equation 122).[329]

(127)

$$CF_3SeCl + 2C_6H_5NH_2 \longrightarrow C_6H_5\overset{+}{N}H_3Cl^- +$$

Several aromatic amines and benzylamine, when treated with a solution of selenium monobromide in carbon disulfide, gave colored amorphous solids which hydrolyzed slowly in water.[341] It is likely that a Se-N bond was formed, but the structures of the products have not been ascertained.

4. Selenocyanates

One of the most common means by which selenium is introduced into an organic molecule is through the use of selenocyanate ion. Like thiocyanate, selenocyanate is considered to be a "pseudohalide" because of the resemblance of its chemical behavior to that of halide ions.

Aliphatic selenocyanates, as a class, are stable liquids, especially when pure, and boil considerably higher than the corresponding halides. (Compare, for example, CH_3SeCN, b.p. 158°,[342] with CH_3I, b.p. 41–43°.) Some aryl selenocyanates are colorless, crystalline solids; others are high-boiling oils.

The introduction of a selenocyanate group into an organic molecule is effected mainly by potassium selenocyanate. Selenocyanogen, also an important reagent for selenocyanate preparation, is highly reactive and is generally used without isolation (cf. Chapter III).

PREPARATION

a. Displacement of Halide by Selenocyanate Ion. The reaction of potassium selenocyanate (usually in ethanol or benzene) with a primary, secondary, or tertiary alkyl halide rapidly affords the corresponding organic selenocyanate (equation 128). In this manner simple alkyl,[226,342–346]

(128) $$RX + KSeCN \rightarrow RSeCN + KX$$

haloalkyl,[228,347] olefinic,[347–349] amidoalkyl,[229] and phthalimidoalkylselenocyanates[227,228,347] have been prepared. The halide displacement reaction has also been used to make diselenocyanato compounds,[230–234,350–352] as well as α-keto,[235,353] benzyl,[22,210,245,354] and benzhydrylselenocyanates.[349] In addition, selenocyanatocarboxylic acids[246,247,353,355,356] and esters[204,351,357] have been made from the requisite halides. Glycerol tri(11-selenocyanatoundecanoate) (**31**) was derived from glycerol tri(11-bromoundecanoate).[358]

$$\underset{\textbf{31}}{\overset{\displaystyle O}{\underset{\displaystyle |}{\overset{\displaystyle ||}{CH_2—O—C—(CH_2)_{10}SeCN}}}}$$

$$\overset{\displaystyle O}{\overset{\displaystyle ||}{CH—O—C—(CH_2)_{10}SeCN}}$$

$$\overset{\displaystyle O}{\overset{\displaystyle ||}{CH_2—O—C—(CH_2)_{10}SeCN}}$$

31

Steroids of the 4,6-pregnadien-3-one type have been synthesized with a selenocyanatomethyl moiety in the 6-position.[359]

Nucleophilic displacement reactions of halide from 4-bromoquinoline-1-oxide[360] and 2-[361] and 3-iodopyrroles[124] by selenocyanate ion have also been reported.

When 1-bromocyclobutane-1-carboxylic acid was treated with potassium selenocyanate by Fredga et al.,[361a] ring contraction took place by a Demjanov rearrangement to give 1-selenocyanatomethylcyclopropane-1-carboxylic acid (equation 128a).

(128a)

$$\underset{Br}{\overset{COOH}{\triangleright\!\!\triangleleft}} + KSeCN \longrightarrow \underset{CH_2SeCN}{\overset{COOH}{\triangleright\!\!\triangleleft}} + Br^-$$

Trityl selenocyanate, believed to be the product of the reaction between trityl chloride and potassium selenocyanate,[362] was shown by Tarantelli and Pecile,[363] on the basis of its i.r. spectrum and its ability to form selenoureas with amines, to be, in actuality, trityl isoselenocyanate.

b. Displacement of Groups Other Than Halide by Selenocyanate. Cyclohexane-1,1-dimethylbenzenesulfonate, on heating at 140° with potassium selenocyanate in diethylene glycol, gave the corresponding diselenocyanate (equation 129).[248] Similarly, tosyl groups have been displaced to yield such

(129)

$$\overset{CH_2OSO_2C_6H_5}{\underset{CH_2OSO_2C_6H_5}{\bigcirc\!\!<}} + 2KSeCN \longrightarrow \overset{CH_2SeCN}{\underset{CH_2SeCN}{\bigcirc\!\!<}}$$

compounds as butyl selenocyanate,[225] ethyl α-selenocyanatopropionate[364] and 5-deoxy-1,2-O-isopropylidine-5-selenocyanate-α-D-xylofuranose.[365]

Cupric selenocyanate, obtained by the reaction of cupric bromide with potassium selenocyanate in acetonitrile, has been used to displace the bromomercuri group in **32** to give selenocyanato-ferrocene (equation 130).[223]

$$(130) \quad \underset{\textbf{32}}{\text{Fc—HgBr}} \quad + \text{ Cu(SeCN)}_2 \xrightarrow{\text{CH}_3\text{CN}} \text{Fc—SeCN}$$

c. **Displacement on Selenium.** In 1914, Lesser and Schoeller[366] reported that *o*-(chloroseleno)benzoyl chloride reacts with silver cyanide to generate the related selenocyanate (equation 131). This reaction has subsequently

$$(131) \qquad \underset{\text{SeCl}}{\text{COCl}} + \text{AgCN} \xrightarrow{70°} \underset{\text{SeCN}}{\text{COCl}}$$

been used to make other aromatic selenocyanates[367] and has been modified through the use of potassium cyanide.[238,336] Trifluoromethyl seleno-cyanate[54,287,368] and other polyhaloalkyl selenocyanates[285,288] have been made by the latter technique.

Rheinboldt and Perrier[369] have shown that thiocyanate can similarly be displaced by cyanide in aromatic selenenyl thiocyanates to give selenocyanates (equation 132). In a related reaction, Foss[370] reported that *o*-nitrobenzene-

$$(132) \qquad \text{ArSeSCN} + \text{CN}^- \rightarrow \text{ArSeCN} + \text{SCN}^-$$

selenenyl sulfinates react with aqueous cyanide to form the corresponding selenocyanates (equation 133).

$$(133) \qquad \underset{\text{NO}_2}{\text{SeSO}_2\text{R}} + \text{CN}^- \longrightarrow \underset{\text{NO}_2}{\text{SeCN}} + \text{RSO}_2^-$$

Aromatic selenenyl selenocyanates, when heated above their melting points, were observed to extrude selenium, giving the corresponding seleno-cyanate (equation 134).[250]

$$(134) \qquad \text{RSeSeCN} \xrightarrow[\Delta]{\geq \text{m.p.}} \text{RSeCN} + \text{Se}$$

d. **Via Diazonium Salts.** The preferred method for the preparation of aromatic selenocyanates is by a Sandmeyer-type reaction in which seleno-cyanate ion is added to a diazotized arylamine (equation 135). This method,

$$(135) \quad \text{C}_6\text{H}_5\text{NH}_2 + \text{HNO}_2 \longrightarrow \text{C}_6\text{H}_5\text{N}_2{}^+ \xrightarrow{\text{SeCN}^-} \text{C}_6\text{H}_5\text{SeCN} + \text{N}_2$$

first reported by Bauer[127] in 1913, has been used to make phenyl selenocyanates substituted by halo,[127,220,336] nitro,[118,127,134,220,336] methyl,[129,220] carboxy,[237] amido,[131] alkoxy,[118,220] phenoxy,[371] acetyl,[284] and benzoyl[328] groups. In addition, biphenyl[239,241] and naphthyl selenocyanates[238,372–375] have also been synthesized utilizing the diazonium salt procedure.

e.   Aromatic Selenocyanation.   Thiocyanation of aromatic amines such as aniline and 1- and 2-naphthylamines by means of thiocyanogen was reported in 1926 by Kaufmann and Oehring.[376] At the same time, Challenger *et al.*[218] described the reaction of aniline with cyanogen triselenide in which *p*-selenocyanation took place (equation 136). This was also achieved by Chao

$$(136) \quad \underset{\text{SeCN}}{\bigcirc} \text{NH}_2 + \text{Se}_3(\text{CN})_2 \longrightarrow \bigcirc \text{NH}_2 + \text{HSeCN} + \text{Se}$$

and Lyons[377] through the use of cupric selenocyanate, generated *in situ* (equation 137). The reaction, which is suggested to proceed *via* a seleno-

$$(137) \quad \bigcirc \text{NH}_2 + \text{Cu(OAc)}_2 + 2\text{NaSeCN} \longrightarrow \underset{\text{SeCN}}{\bigcirc} \text{NH}_2$$

cyanogen intermediate, is said to be of limited applicability. Recently, Nefedov[223] reported success in the selenocyanation of azulene by the use of cupric selenocyanate (equation 138). The selenocyanation of 2,6-*tert*-

$$(138) \quad \bigotimes \xrightarrow{\text{Cu(ScCN)}_2} \bigotimes^{\text{SeCN}}$$

50%

butylphenol in the 4-position was accomplished by Mueller *et al.*[125] at 0°, using selenocyanogen, a relatively unstable compound. The latter was made by the action of bromine on potassium selenocyanate in a manner analogous to the preparation of thiocyanogen.[376] Agenäs *et al.* have performed selenocyanation reactions by allowing various aromatic amines to react with selenocyanogen at −60° and in this manner have produced 3-selenocyanato-indoles (equation 139),[242,243] 5-selenocyanatoindoline,[244] 2-methyl- and

$$(139) \quad \text{[indole structure]} + (\text{SeCN})_2 \longrightarrow \text{[3-selenocyanato indole structure]} + \text{HSeCN}$$

2,5-dimethylselenocyanatopyrroles,[378] and 4-selenocyanatoanilines.[379]

The *p*-selenocyanation of dimethyl- and diethylanilines was achieved in low yield by electrolyzing them in the presence of an aqueous ethanolic solution of potassium selenocyanate.[380]

The organometallics, triphenyl bismuthine[218] and diphenyl- and dimethylmercury,[381] have been reported to react with cyanogen triselenide to give phenyl selenocyanate and methyl selenocyanate, respectively.

f. **Miscellaneous Methods.** *β*-Selenocyanatopropionic acid has been prepared by the facile ring opening of *β*-propiolactone with potassium selenocyanate (equation 140).[62] The latter reagent can also open sultones to give

$$(140) \quad \text{[β-propiolactone structure]} + \text{KSeCN} \longrightarrow \text{NCSeCH}_2\text{CH}_2\text{COOH}$$

selenocyanatoalkanesulfonates.[66]

Selenocyanic acid adds to the activated acetylenes, propynal and butynone, to give 3-selenocyanatoacrolein and 3-selenocyanatobutenone, respectively.[382]

Dimethyl and diethyl diselenides were reported to be cleaved by mercuric cyanide to give methyl and ethyl selenocyanates and mercury-containing byproducts of uncertain structure.[37]

Benzylselenourea, on refluxing with an ethanolic solution of 2,4,5-trinitrotoluene, gave a product identified as 2,4-dinitro-5-selenocyanatotoluene.[383]

REACTIONS

a. **Hydrolysis.** Acid- and base-catalyzed hydrolysis of selenocyanates constitutes an important route to diselenides (*cf.* Section IV.B).

b. **Displacement.** The attack of cyanide-^{14}C on the Se atom of benzyl and 2- and 4-nitrobenzyl selenocyanates causes exchange to occur (equation 141).[348] A limited but measurable exchange takes place between seleno-

$$(141) \quad \text{RSeCN} + {}^{14}\text{CN}^- \rightleftharpoons \text{RSe}^{14}\text{CN} + \text{CN}^-$$

cyanate-^{14}C and the above-mentioned selenocyanates according to proposed scheme 142.[384,385]

(142)   (a) $RSeCN + Se^{14}CN^- \rightleftharpoons RSeSe^{14}CN + CN^-$

      (b) $CN^- + Se^{14}CN^- \rightleftharpoons [NCSe^{14}CN]^{2-} \rightleftharpoons NCSe^- + {}^{14}CN^-$

Aryl selenocyanates react vigorously with aryl thiols and selenols with the elimination of cyanide to form selenenyl sulfides (equation 143)[252] and diselenides, respectively.

(143) $ArSeCN + Ar'SH \rightarrow ArSeSAr' + HCN$

Another displacement on selenium which has been reported is the reaction of silver acetate or other silver or copper salts in methanol with o-nitrophenyl selenocyanate in the presence of catalytic quantities of pyridine (equation 118)[331] to give the corresponding selenenyl methoxide.

c. Halogenation.   The light-catalyzed chlorination or bromination of aliphatic or aromatic selenocyanates, when performed with cooled chloroform solutions of the halogen, leads to selenenyl chlorides[367] or selenenyl bromides,[284,328,336,386,387] respectively, if the reactants are in an approximately equimolar ratio (equation 144). However, if excess chlorine or bromine is

(144) $RSeCN + X_2 \longrightarrow RSeX \xrightarrow{X_2} RSeX_3$
$$+$$
$$XCN$$

used, a further reaction takes place to yield the selenium trichloride[228,239,336] or selenium tribromide,[131,239,283,336,347,367] respectively (equation 144).

Cyclization of 2-selenocyanatobiphenyl with excess bromine in chloroform gives dibenzoselenophene dibromide.[388] (See Chapter XIB for related reactions.)

Whereas bis(2,4-dinitrophenyl) disulfide reacts with sulfuryl chloride in the presence of pyridine to give the corresponding sulfenyl chloride, it was found that chlorinolysis of the analogous diselenide or selenocyanate gave 2,4-dinitrophenylselenium trichloride rather than the anticipated selenenyl chloride.[389]

d. Oxidation.   Chlorine-water oxidation of chloroalkyl selenocyanates gives the related seleninic acid hydrochloride (equation 145).[228,347]

(145) $Cl(CH_2)_xSeCN + Cl_2 \xrightarrow{H_2O} Cl(CH_2)_xSeO_2H \cdot HCl$
        $(x = 2, 3)$

Nitric acid, in concentrations ranging from dilute to fuming, oxidizes selenocyanates to the corresponding seleninic acids,[132,234,351,371–373,375,390,391] which are usually obtained as their hydronitrate salts. Ring nitration of aromatic selenocyanates can occur concomitantly with oxidation[375,391] or with little or no oxidation,[354,390-392] depending on the reaction conditions.

Proskauer[350] reported in 1874 that concentrated nitric acid oxidized 1,2-ethylenediselenocyanate to the corresponding diselenonic acid.

Oxidation of aromatic selenocyanates to seleninic acids has also been accomplished in glacial acetic acid by a 40% peracetic acid solution[109] and by 5% potassium permanganate.[279]

e. Reduction. Selenocyanates have been reduced with zinc dust in acetic, hydrochloric, or sulfuric acid media. In some instances the intermediate, synthetically useful zinc selenide salt has been isolated (equation 146);[24,125,127]

$$(146) \quad 2RSeCN + Zn \xrightarrow{H^+} RSeZnSeR \xrightarrow{H^+} 2RSeH \xrightarrow{[O]} (RSe-)_2$$

this can be freed from zinc by further treatment with concentrated hydrochloric acid.[24,125] One can also obtain selenols[123,124] or diselenides (formed by the air oxidation of the selenols)[361] without an intermediate isolation step.

The reduction of *o*-nitrophenyl selenocyanate with zinc in alkaline media has been reported to give *o*-aminoselenophenol.[165] In addition, the conversion of selenocyanates to selenols has also been effected by sodium in methanol[365] and sodium hydrosulfite.[113]

f. Cyclization. 3-Selenocyanatoacroleins cyclize in liquid ammonia to give isoselenazoles (equation 147).[382] Appropriately *o*-substituted phenyl

(147)

selenocyanates have been utilized as precursors of other heterocyclic systems, for example, benzisoselenazolone (**33**),[393] selenanaphthenes (**34**),[394] and 2-chloro-1,3-benzoselenazin-4-one (**35**).[395]

33   34   35

The heterocyclic compounds, ψ-selenophthalimidine[245] and 1,2-diselenolane-3,5-dicarboxylic acid monoethyl ester,[204] have been synthesized by the acid hydrolysis of selenocyanate precursors.

g. Other Reactions. Raney nickel was found to deselenize a number of aromatic selenocyanates in yields ranging from 14 to 84%.[312] The action of aryllithium compounds or aryl Grignard reagents on aryl selenocyanates leads to unsymmetrical diaryl selenides.[396]

The isomerization of selenocyanates to isoselenocyanates is an equilibrium process which is not suitable for the preparation of the latter type of compound[346,349] (*cf.* Chapter VIIIB).

MISCELLANEOUS*

The electric moments of some *p*-substituted phenyl thiocyanates and phenyl selenocyanates have been measured and compared.[397] The results suggest that the selenocyanate group donates electrons to a smaller extent than does the thiocyanate group when located *para* to an electron-withdrawing group.

Conditions were established for the separation by gas chromatography of a number of organoselenium compounds, including ethyl selenocyanate.[398]

The remarkable light sensitivity of a number of *o*- and *p*-nitrophenyl selenocyanates has been reported.[374]

Aromatic selenocyanates, thiocyanates, halides, and cyanides are isomorphous, forming binary systems that give a continuous series of mixed crystals.[362,399]

Long-chain selenocyanates have been studied for their ability to promote the condensation of steam.[357,400]

5. Selenenyl Halides

Three types of organic selenium halides are known: monohalides ($RSeX$), dihalides (R_2SeX_2), and trihalides ($RSeX_3$). Compounds of the type $RSeX_2$ are unknown, as they would require the existence of trivalent selenium.

The monohalides (selenenyl halides) are generally highly reactive solids which range in appearance from colorless to yellow in the case of chlorides and yellow to intense red in the case of bromides. No successful syntheses of selenenyl fluorides or iodides have been reported.

PREPARATION

a. From Diselenides. The most common synthetic method for the preparation of selenenyl halides is the addition of halogen to a diselenide. The quantity of the halogen must be limited to prevent selenium trihalide formation. Chlorinolysis is usually performed in carbon tetrachloride solution at low temperatures (equation 148). In this manner, simple

$$(148) \qquad\qquad (RSe—)_2 + Cl_2 \xrightarrow{\;CCl_4\;} 2RSeCl$$

---

* A discussion of spectral data pertaining to selenocyanates will be found in Chapter XV.

aliphatic[303,401] fluoroalkyl[49,54,286,287,290,402] and aromatic[325,337] selenenyl chlorides have been prepared. Carbon diselenide reacts exothermically with chlorine in carbon tetrachloride to give trichloromethaneselenenyl chloride (Cl_3CSeCl).[285,403]

Lesser and Schoeller[366] found that the action of thionyl chloride on bis(o-carboxyphenyl) diselenide (diselenosalicylic acid) (**36**) cleaves the Se-Se bond to give the selenenyl chloride **37** (equation 149). o-Nitro-

(149)

$$\left(\begin{array}{c}\text{COOH}\\\text{Se—}\end{array}\right)_2 + SOCl_2 \longrightarrow 2\begin{array}{c}\text{COCl}\\\text{SeCl}\end{array}$$

36 **37**

phenylselenenyl chloride has also been made in this manner.[336] It should be noted that the S-S bond of dithiosalicylic acid was cleaved by chlorine to yield the corresponding sulfenyl chloride.[404]

Behaghel and Seibert[336] and Pitombo[335] have used sulfuryl chloride to effect the cleavage of diselenides in their preparation of aromatic selenenyl chlorides (equation 150). Lawson and Kharasch[389] observed that the

(150) $$(RSe—)_2 + SO_2Cl_2 \rightarrow 2RSeCl + SO_2$$

selenium trichloride was obtained in good yield if the reaction was performed in the presence of pyridine.

In a similar manner, the addition of a stoichiometric quantity of bromine in a solvent such as carbon tetrachloride to diselenides gives selenenyl bromides (equation 151).[54,238,288,306,333,386,387,402,405] When a chloroform

(151) $$(RSe—)_2 + Br_2 \rightarrow 2RSeBr$$

solution of bis(2-nitrophenyl) diselenide was heated under reflux with iodine, the desired selenenyl iodide was not formed.[325]

A mixture of perfluoromethylselenium trichloride (**38**) and bis(perfluoromethyl) diselenide (**39**) gave perfluoromethylselenenyl chloride (**40**) (equation 152).[54]

(152) $$CF_3SeCl_3 + (CF_3Se—)_2 \rightarrow 3CF_3SeCl$$

38 **39** **40**

Chlorine, bromine, and hydrogen bromide in acetic acid convert aromatic triselenides into the corresponding selenenyl halides[323] (*cf.* Section IV.B.2). Hydrogen bromide treatment of bis(o-nitrophenylselenenyl) sulfide (**41**) led to the formation of o-nitrophenylselenenyl bromide (equation 153).[325]

(153)

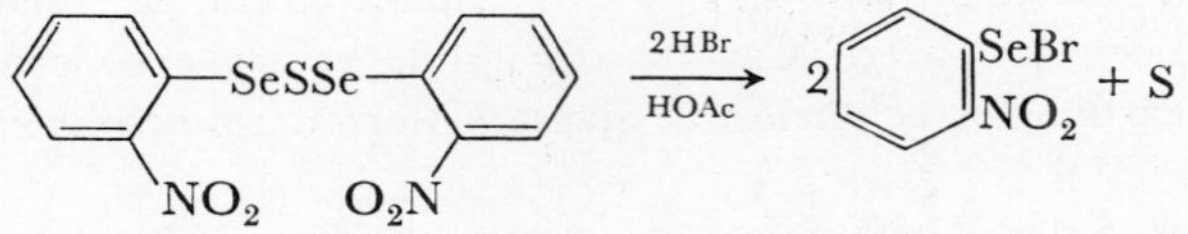

$$\text{(structure 41: } o\text{-nitrophenyl-SeSSe-}o\text{-nitrophenyl)} \xrightarrow[\text{HOAc}]{2\,H\,Br} 2\;\text{(}o\text{-nitrophenyl-SeBr)} + \text{S}$$

41

b. From Selenocyanates. Aromatic selenocyanates have been converted to the corresponding selenenyl chlorides[367] and bromides[283,328,336,386,387] by the action of chlorine and bromine, respectively. Selenenyl thiocyanates, on treatment with bromine, gave selenenyl bromides (equation 154).[369] In

$$(154) \qquad\qquad ArSeSCN + Br_2 \rightarrow ArSeBr + BrSCN$$

contrast, selenenyl selenocyanates and sulfenyl selenocyanates gave rise only to diselenides and disulfides, respectively.

c. From Selenides and Their Derivatives. Dihalide derivatives of aryl alkyl selenides have been found to dissociate upon heating at their melting points to give arylselenenyl halides and alkyl halides (equation 155).[406,407]

$$(155) \qquad\qquad \overset{Br}{\underset{+}{C_6H_5Se}}C_2H_5\;Br^- \xrightarrow[\Delta]{130°} C_6H_5SeBr + C_2H_5Br$$

p-Tolylselenoglycolic acid dibromide decomposed thermally to give p-tolyl-selenenyl bromide (equation 156).[406] Arylselenoethylenes are cleaved by

(156)

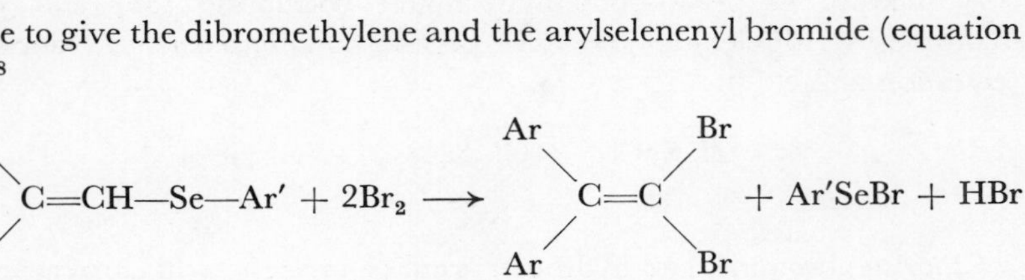

$$H_3C-\text{(C}_6H_4\text{)}-\underset{+}{\overset{Br}{Se}}-CH_2COOH \;\overset{Br^-}{\underset{\Delta}{\longrightarrow}}$$

$$H_3C-\text{(C}_6H_4\text{)}-SeBr + BrCH_2COOH$$

bromine to give the dibromethylene and the arylselenenyl bromide (equation 157).[408]

$$(157) \qquad \underset{Ar}{\overset{Ar}{\diagdown}}C{=}CH{-}Se{-}Ar' + 2Br_2 \longrightarrow \underset{Ar}{\overset{Ar}{\diagdown}}C{=}C\underset{Br}{\overset{Br}{\diagup}} + Ar'SeBr + HBr$$

Bis(trifluoromethyl) selenide is unreactive to chlorine in the dark at temperatures up to 100°, but exposure of the mixture to u.v. radiation results in the quantitative formation of trifluoromethylselenenyl chloride

(**42**, red liquid, b.p. 32°) and trifluoromethylselenium trichloride (**43**, m.p. 118°) (equation 158).[402]

(158) $$F_3CSeCF_3 + Cl_2 \xrightarrow{h\nu} F_3CSeCl + F_3CSeCl_3$$
$$\hspace{6.5cm} \textbf{42} \hspace{2.2cm} \textbf{43}$$

Renson *et al.*[367,409] were able to convert the selenide, **44**, to the corresponding selenenyl chloride, **37**, with thionyl chloride in the presence of catalytic amounts of DMF (equation 159). The rupture of the Se–C bond of aryl alkyl

(159)

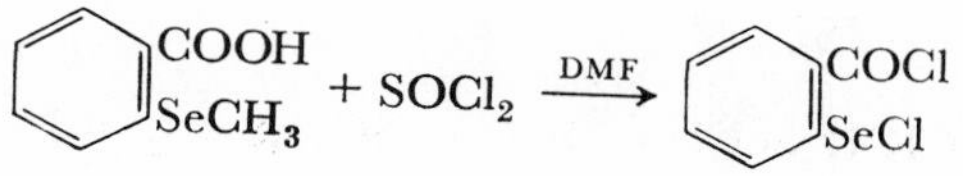

selenides in this manner is a phenomenon not observed with sulfur analogs.[404]

d. From Selenium Trihalides.

These compounds, upon the application of heat or under reduced pressure, dissociate readily to give halogen and the selenenyl halide (equation 160).[131,239,283,337,389,410] A suspension of 4-

(160) $$RSeX_3 \rightleftharpoons RSeX + X_2$$

biphenylselenium trichloride or tribromide in chloroform containing acetone was also reported[239] to give the corresponding selenenyl chloride or bromide, respectively. The reaction illustrates the ability of selenium trihalides to serve as halogenating agents—in this case, of acetone.

e. From a Selenenic Acid.

When *o*-benzoylphenylselenenic acid (**45**) was heated with 48% hydrobromic acid, the selenenyl bromide, **46**, was formed (equation 161).[328]

(161)

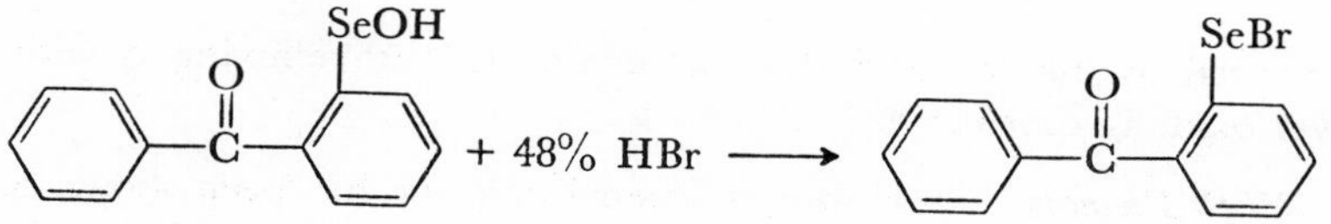

f. From a Mercury Derivative.

Bisheptafluoropropylselenenyl mercury, when treated with excess chlorine or bromine in a sealed tube at room temperature, gave the heptafluoropropylselenenyl chloride or bromide, respectively.[286]

g. From Selenium Monochloride.

When acetylene was passed into a benzene solution of selenium monochloride at 2–4°, precipitation of selenium occurred and there slowly formed a crystalline product identified as 2-chloroetheneselenenyl chloride (equation 161a).[410a]

(161a) $HC \equiv CH + Se_2Cl_2 \rightarrow Se + ClCH = CHSeCl$

REACTIONS

a. With Halogen. Selenenyl halides combine with 1 mole of halogen to form organic selenium trihalides (*cf.* Section IV.D.6). Iodine monochloride reacts with chloromethylselenenyl chloride to give chloromethylselenium trichloride (equation 162).[303]

(162) $ClCH_2SeCl + 2ICl \rightarrow ClCH_2SeCl_3 + I_2$

b. With Water. Aromatic selenenyl halides are rapidly hydrolyzed by water and are reported to give either the selenenic acid,[326] a mixture of the corresponding seleninic acid and the diselenide,[283,326,410] or only the diselenide,[326] depending on the stabilizing ability of the other substituents. A mechanism has been proposed by Behaghel and Seibert[336] (scheme 163)

(163)   (*a*) $3ArSeBr + 3H_2O \rightarrow 3ArSeOH + 3HBr$

       (*b*) $2ArSeOH \rightarrow ArSeO_2H + ArSeH$

       (*c*) $ArSeH + ArSeOH \rightarrow (ArSe-)_2 + H_2O$

$$3ArSeBr + 2H_2O \rightarrow ArSeO_2H + (ArSe-)_2 + 3HBr$$

which accounts for the formation of the above-mentioned products. Trifluoromethylselenenyl chloride, on hydrolysis, gave the related seleninic acid and diselenide (equation 164).[54] Anthraquinone-1-selenenyl bromide was

(164) $3CF_3SeCl + 2H_2O \rightarrow CF_3SeO_2H + (CF_3Se-)_2 + 3HCl$

converted to the corresponding selenenic acid by refluxing it with moist silver oxide in dioxane.[327]

c. With Phenols. 2-Naphthylselenenyl chloride has been shown to react with phenol at its *p*-position, resorcinol in the 6-position, and 2-naphthol in the 1-position to give the corresponding selenides.[335]

d. With Other Nucleophiles. The reaction of selenenyl halides with amines is discussed in Section IV.C.3. Triethylamine (or triphenylphosphine) in chloroform was noted by Petragnani and de Moura Campos[411] to convert phenylselenenyl bromide to diphenyl diselenide.

Nucleophilic displacement on the Se atom of selenenyl halides can yield a variety of interesting functional groups, as shown in Table IV-3.

Table IV-3   Reaction of various nucleophiles with selenenyl halides

$$RSeX + N \rightarrow RSeN + X$$

NUCLEOPHILE (N)	PRODUCT (RSeN)	REF.
CN^-	RSeCN	54, 287, 288, 336, 343, 366–368
OCN^-	RSeNCO	368
SCN^-	RSeSCN	335, 368–370
$SeCN^-$	RSeSeCN	368
$SSO_3{}^{2-}$	$RSeSSO_3{}^-$	329, 370
$SPO(OR')_2{}^-$	$RSeSPO(OR')_2$	370
$R'S_2O_2{}^-$	$RSeS_2O_2R'$	370
$R'SO_2{}^-$	$RSeSO_2R'$	370
H_2S	RSeSSeR	325
$R'SH$	RSeSR'	251, 367, 387
$R'SeH$	RSeSeR'	191
OAc^-	RSeOAc	326, 332, 333, 338
$OCH_3{}^-$	$RSeOCH_3$	327, 330

e.   **With Grignard Reagents.**   A selenide is formed when a selenenyl halide reacts with a Grignard reagent (equation 165).[336,367,412]

$$(165) \qquad C_6H_5SeBr + C_2H_5MgBr \rightarrow C_6H_5SeC_2H_5 + MgBr_2$$

f.   **With Ketones.**   Pitombo[335] and Yarovenko *et al.*[287] have found that selenenyl chlorides react readily at the α-position of ketones such as acetone and acetophenone to give selenides (equation 166). However, Rheinboldt

$$(166) \qquad RSeCl + CH_3\overset{\text{O}}{\underset{\|}{C}}CH_3 \rightarrow RSeCH_2{-}\overset{\text{O}}{\underset{\|}{C}}{-}CH_3 + HCl$$

and Perrier,[413] working with *o*-nitrophenylselenenyl bromide, found that the reaction with dry acetone gave only a small amount of condensation product, yielding mainly the diselenide and bromoacetone (equation 167). *o*-Nitro-

(167)

phenylselenenyl thiocyanate, however, was found to react almost quantitatively with acetone to give the selenide (equation 168).[413]

(168)

$$\text{(168)} \quad \underset{\text{NO}_2}{\overset{\text{SeSCN}}{\bigcirc}} + CH_3CCH_3 \longrightarrow \underset{\text{NO}_2}{\overset{\text{SeCH}_2-C-CH_3}{\bigcirc}}$$

g. With Oxidizing Agents.   Ozone[296,401] or nitrogen dioxide[401] treatment of selenenyl chlorides results in the formation of seleninyl chlorides (equation 169). The latter compounds react with water to give seleninic acids, which in

$$\text{(169)} \qquad RSeCl + O_3 \xrightarrow{\text{(or NO}_2)} RSeOCl$$

some instances are obtained as their hydrochloride salts. The seleninyl chlorides also react with alkoxides to yield seleninic esters (*cf.* Section IV.D.3).

h. With Heavy Metals and Their Compounds.   Phenylselenenyl bromide, when heated with zinc dust in ether or benzene, was reported to be converted to diphenyl diselenide[283] (equation 170). Treatment of trichloro-[285] and

$$\text{(170)} \qquad 2C_6H_5SeBr + Zn \rightarrow (C_6H_5Se{-})_2 + ZnBr_2$$

trifluoromethylselenenyl chlorides[49] with tin and hydrochloride acid also yielded the corresponding diselenides.

Trifluoromethylselenenyl bromide, on shaking with mercury or mercuric oxide, gave bis(trifluoromethyl) diselenide;[54] however, trifluoromethylselenenyl chloride, when treated with mercury, yielded the mercury derivative, $(CF_3Se{-})_2Hg$.

Selenenyl bromides were reported to react with mercury diaryls to give selenides (equation 171).[306] This reaction is related to the one between

$$\text{(171)} \qquad RSeBr + Ar_2Hg \rightarrow RSeAr + ArHgBr$$

selenium tetrabromide and mercury diaryls discovered earlier by Leicester (equation 172).[414] 2-Naphthylselenenyl chloride reacted with phenyl-

$$\text{(172)} \qquad SeBr_4 + 3R_2Hg \rightarrow R_2Se + RBr + 3RHgBr$$

mercuric chloride in dioxane to give phenyl 2-naphthyl selenide (equation 173).[405]

$$\text{(173)} \quad \underset{}{\overset{\text{SeBr}}{\bigcirc\bigcirc}} + C_6H_5HgCl \longrightarrow \underset{}{\overset{\text{SeC}_6H_5}{\bigcirc\bigcirc}}$$

i. With Unsaturated Compounds.   Yarovenko *et al.* reported that trichloromethylselenenyl chloride[285] (and difluoromethylselenenyl bromide[288]) add to ethylene, as indicated in equation 174. Aromatic selenenyl halides have

$$\text{(174)} \qquad Cl_3CSeCl + CH_2{=}CH_2 \rightarrow Cl_3CSeCH_2CH_2Cl$$

been shown to add to alkynes to give 1-arylseleno-2-haloethylenes (equation 175).[153,154,415]

$$(175) \qquad ArSeX + RC{\equiv}CR' \rightarrow ArSe\overset{\overset{\displaystyle R}{|}}{C}{=}\overset{\overset{\displaystyle R'}{|}}{C}{-}X$$

j.   Cyclization Reactions.   Phenylselenenyl halides possessing a carbonyl-activated functional group in the *ortho* position have been cyclized in the presence of ammonia, aniline, sodium sulfide, and sodium selenide (scheme 176),[416] as well as pyridine (reaction 177),[407] to give the heterocycles shown below.

(176)

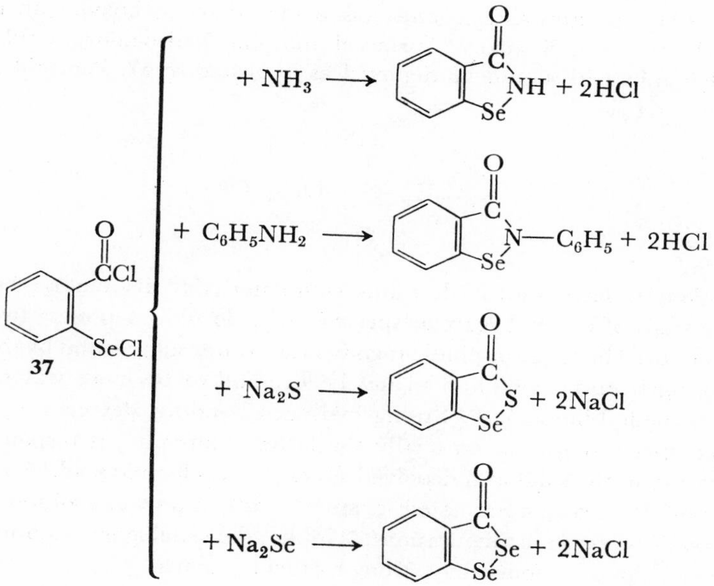

(177)

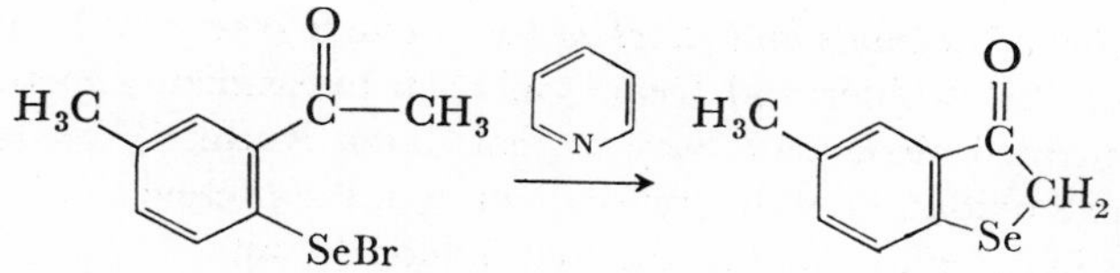

k.   Disproportionation.   Bridger and Pittman[303] found that heating chloromethylselenenyl chloride caused it to disproportionate to selenium tetrachloride, bis(chloromethyl) selenide, and bis(chloromethyl) diselenide (scheme 178). Foster[410] proposed that a similar disproportionation took place on pyrolysis of phenylselenenyl halides.

(178) $3ClCH_2SeCl \rightarrow ClCH_2SeCl_3 + (ClCH_2Se—)_2$

$ClCH_2SeCl_3 + ClCH_2SeCl \rightarrow (ClCH_2)_2Se + SeCl_4$

D. SELENINIC ACIDS AND DERIVATIVES

1. Seleninic Acids

Seleninic acids, which may be considered to be derivatives of selenious acid, $HOSeO_2H$, are amphoteric and less acidic than carboxylic acids.[216,417] As early as 1869, Rathke[418] surmised that the hydrochloric acid salt of ethylseleninic acid should be depicted as in structure **47**. Paetzold and his

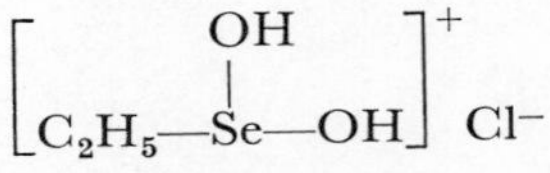

$$\left[\begin{array}{c} OH \\ | \\ C_2H_5—Se—OH \end{array} \right]^+ Cl^-$$

47

co-workers[419] have shown that this compound and its methyl homolog, on the basis of i.r. and Raman spectroscopy, do indeed possess two single Se-O bonds. The lower aliphatic members are water soluble and hygroscopic. Surprisingly, ethylseleninic acid has been found to be more water soluble than its methyl homolog.[420] Strong hydrogen bonding, detected by i.r. and Raman spectroscopy, in especially the latter compound, is responsible for association in the solid and dissolved states. Ethylseleninic acid,[420] as well as its anionic[421] and its protonated forms,[419] exists in aqueous solution in two rotational isomeric configurations. Unlike the analogous sulfinic acids, seleninic acids are moderately strong oxidizing agents.

PREPARATION

a. Oxidation of Selenols and Diselenides. The first reported seleninic acid was prepared by Wöhler and Dean[13] in 1856 by oxidizing methaneselenol with concentrated nitric acid. Several years later, Rathke[418] recognized that the nitric acid oxidation of diethyl diselenide in the presence of hydrochloric acid gave ethylseleninic acid as the hydrochloride salt.

The nitric acid oxidation of aliphatic[14] and aromatic selenols,[100,275] as well as aliphatic[54,197,199,209–212,216,417,422–424] and aromatic diselenides[86,131,276,383,425–427] and diselenacycloalkanes,[231,233] constitutes a convenient route to seleninic acid hydronitrates. Oxidation of 1,2-diselenacyclohexane gives 1,4-butanediseleninic acid dihydronitrate (equation 179),

$$(179) \qquad \underset{\text{Se}}{\overset{\text{Se}}{\bigcirc}} \longrightarrow \begin{array}{l} CH_2\!\!-\!\!CH_2SeO_2H\cdot HNO_3 \\[4pt] \;\;| \\[4pt] CH_2\!\!-\!\!CH_2SeO_2H\cdot HNO_3 \end{array}$$

which decomposes at 136° with explosive violence and the liberation of selenium.[231] In an attempt to prepare 1,2-ethanediseleninic acid, the compound 1,2,5,6-tetraselenacyclooctane was oxidized with nitric acid; however, the main product formed was ethanediseleninic acid anhydride.[234] 1,2-Ethanediseleninic acid hydronitrate, obtained by an alternative route, exhibited a marked propensity to form the corresponding anhydride when dissolved in water. A seleninic acid hydronitrate can be transformed into the free acid by the addition of an equivalent of base.[424]

The oxidation of p-phenylene diselenide polymer with 70% nitric acid yielded 1,4-benzenediseleninic acid in high yield.[43]

The oxidative cleavage of diselenides with hydrogen peroxide is another means by which seleninic acids have been synthesized in the aliphatic[246,247,258] and the aromatic[109,258,276,428,429] series. The action of 3% hydrogen peroxide on selenoketals of the type $R_2C(SeC_2H_5)_2$, followed by the addition of hydrochloric acid, gave ethylseleninic acid hydrochloride.[14] The compound 2-aminoethaneseleninic acid (**48**, selenohypotaurine) was produced by the peroxide oxidation of bis(2-aminoethyl) diselenide (**49**, selenocystamine)[41] (equation 180) and resisted further oxidation to the selenonic acid. In

$$(180) \quad (H_2NCH_2CH_2Se\!-\!)_2 \xrightarrow{\text{H}_2\text{O}_2} 2H_2NCH_2CH_2SeO_2H \xrightarrow{\text{H}_2\text{O}_2}$$
$$\quad\quad\quad \textbf{49} \qquad\qquad\qquad\qquad \textbf{48}$$
$$2H_2NCH_2CH_2SeO_3H$$

contrast, it has been demonstrated that bis(2-aminoethyl) disulfide dihydrochloride can be oxidized with hydrogen peroxide to the sulfonic acid stage to give taurine.[41,430]

Bromine water, which also transformed **49** into **48**,[228] was observed by Fredga,[430a] using polarographic methods, to convert diselenides to seleninic acids instantaneously.

Bis(p-chlorophenyl) diselenide has been oxidized to the corresponding seleninic acid with alkaline permanganate.[431]

b.   Oxidation of Selenocyanates.

Nitric acid oxidation of aliphatic[234,351,424] and aromatic[132,276,371,373,375,391,432] selenocyanates generally affords seleninic acids in good yields. Rebane[351] has prepared α,ω-alkylenediseleninic acids from the corresponding diselenocyanates by the use of fuming nitric acid. Concentrated hydrogen peroxide was inert towards p-nitrophenyl selenocyanate, but 40% peracetic acid in glacial acetic acid gave p-nitrophenylseleninic acid in 45% yield.[109]

Oxidation of 2-chloroethyl selenocyanate with chlorine-water gave 2-chloroethylseleninic acid as its hydrochloric acid salt.[347] This, on treatment with ammonium hydroxide, gave a low yield of selenohypotaurine (**48**).[228]

2-Naphthyl selenocyanate has been oxidized with permanganate in glacial acetic acid to the seleninic acid.[279,290]

c. Oxidation of Selenides.
Morgan and Porritt[102] claimed that the peroxide oxidation of substituted arylselenoglycolic acids gave arylselenoxyglycolic acids. However, Porritt[103] later showed that the compounds obtained were, in reality, seleninic acids (scheme 181).

$$(181) \quad C_6H_5SeCH_2COOH \xrightarrow[\text{H}_2\text{O}_2]{\text{H}_2\text{O}_2} \begin{array}{c} O \\ \parallel \\ C_6H_5-Se-CH_2COOH \end{array}$$
$$\xrightarrow{\text{H}_2\text{O}_2} C_6H_5SeO_2H$$

d. From Grignard Reagents.
Ethylseleninic acid was made by the reaction of ethylmagnesium bromide with selenium dioxide, followed by acid hydrolysis of the complex (equation 182).[433] Phenylseleninic acid, however,

$$(182) \quad C_2H_5MgBr + SeO_2 \longrightarrow C_2H_5SeO_2MgBr \xrightarrow[\text{H}_2\text{O}]{\text{H}^+} C_2H_5SeO_2H$$

could not be synthesized by this approach.[434]

e. Hydrolysis of Organoselenium Halides.
On hydrolysis, trifluoromethylseleninoyl chloride gives trifluoromethylseleninic acid (equation 183), while methyl- and ethylseleninoyl chlorides yield the corresponding seleninic acids as their hydrochloride salts.[401] Trifluoromethylseleninic acid is a strong acid and hence does not form a hydrochloric acid salt.

$$(183) \quad CF_3SeOCl + H_2O \rightarrow CF_3SeO_2H + HCl$$

Trifluoromethylselenenyl chloride, on hydrolysis, gives a mixture of the corresponding seleninic acid (**50**) and diselenide (**51**) (equation 184).[54] Similarly, hydrolysis of phenylselenenyl bromide yields phenylseleninic

$$(184) \quad 3CF_3SeCl + 2H_2O \rightarrow \underset{50}{CF_3SeO_2H} + \underset{51}{(CF_3Se-)_2} + 3HCl$$

acid and diphenyl diselenide.[283] The mechanism proposed for this reaction (scheme 185) requires the disproportionation of the intermediate selenenic acid.

$$(185) \quad \begin{array}{ll} (a) & C_6H_5SeBr + H_2O \rightarrow C_6H_5SeOH + HBr \\ (b) & 2C_6H_5SeOH \rightarrow C_6H_5SeO_2H + C_6H_5SeH \\ (c) & C_6H_5SeH + C_6H_5SeOH \rightarrow (C_6H_5Se-)_2 + H_2O \end{array}$$

Dialkyl selenide dibromides on treatment with moist silver oxide give seleninic acids in the cases of the diethyl and di-n-propyl compounds. The methyl congener, however, appears to yield the selenoxide under identical conditions.[37]

Organic selenium trihalides have been used to generate seleninic acids by alkaline hydrolysis. Phenyl-[99,283] (equation 186), biphenyl-,[239] benzoyl-

$$(186) \qquad C_6H_5SeBr_3 + 2H_2O \xrightarrow{\text{OH}^-} C_6H_5SeO_2H + 3HBr$$

phenyl-,[328] nitro- and chlorophenyl-,[336] p-tolyl-,[336] and o-ethylphenyl-seleninic[239] acids have been prepared in this manner. Backer and Winter[435] have transformed alkylselenium trichlorides and tribromides into seleninic acids by shaking them with a suspension of silver hydroxide in water. Similarly, 1-phthalimidoethane-2-selenium trichloride was converted into the seleninic acid by agitating it with a silver hydroxide suspension.[228]

Behaghel and Hofmann[239] have oxidized o-ethylphenylselenium tribromide with nitric acid to the corresponding seleninic acid.

f.   Other Methods.   Methyl- and ethylseleninic acids have been made by the reaction of the corresponding seleninic acid anhydrides with water.[436] Seleninic acids have also been prepared by the reduction of selenonic acids (*cf.* Section IV.E).

REACTIONS

a.   With Acids.   As indicated previously, seleninic acids, because of their amphoteric character, form salts with acids or bases. A series of methyl- and ethylseleninic acid salts derived from strong acids such as benzenesulfonic and methionic (methane disulfonic) has been prepared.[417]

b.   With Bases.   The stability of various seleninic acids in boiling 5% aqueous sodium hydroxide was found to vary considerably.[302] Benzyl-seleninic acid gave dibenzyl diselenide under these conditions, whereas n-propylseleninic acid was very stable. Seleninoacetic (**52**) and β-selenino-propionic (**53**) acids divested themselves quantitatively of selenium.

$$HO_2SeCH_2COOH \qquad HO_2SeCH_2CH_2COOH$$

$$\text{52} \qquad\qquad\qquad \text{53}$$

c.   Thermolysis.   It was suggested by Bird and Challenger[37] that methane-seleninic acid decomposes on gentle warming according to equation 187.

$$(187) \qquad 2CH_3SeO_2H \xrightarrow{\Delta} CH_3SeH + CH_2O + SeO_2 + H_2O$$

Backer and van Dam[417] also reported on the complex thermal decomposition of this seleninic acid. 1-Naphthaleneseleninic acid gave at 105–110° (15–16 hours) an unstable seleninic anhydride which on further heating led to the formation of 1-naphthalene selenoxide and selenium dioxide (equation 188).[279]

$$(188) \quad 2 \underset{\text{SeO}_2\text{H}}{\text{[naphthalene]}} \longrightarrow \underset{\overset{\|}{O}}{\text{[naphthalene]}-\text{Se}-\text{[naphthalene]}} + \text{SeO}_2 + \text{H}_2\text{O}$$

d. Oxidation. Some seleninic acids have been oxidized to selenonic acids (*cf.* Section IV.E). However, many instances have been recorded in which attempts to oxidize a seleninic acid failed to give the desired selenonic acid.

e. Reduction. The reduction of seleninic acids proceeds far more readily than their oxidation. Benzeneseleninic acid was converted with zinc and hydrochloric acid to diphenyl diselenide.[426] It was demonstrated that aromatic seleninic acids could be reduced to selenenic acids by the use of ethyl and *tert*-butyl mercaptans, thiophenol (equation 189), *p*-thiocresol,

$$(189) \quad \text{ArSeO}_2\text{H} + 2\text{C}_6\text{H}_5\text{SH} \rightarrow \text{ArSeOH} + (\text{C}_6\text{H}_5\text{S}-)_2 + \text{H}_2\text{O}$$

and selenophenol.[278] By increasing the quantity of ethanethiol relative to seleninic acid it was possible to reduce 1-anthraquinonylseleninic acid to the diselenide (equation 190) and even to the selenol stage (equation 191).[278]

$$(190) \quad 2\text{ArSeO}_2\text{H} + 6\text{RSH} \rightarrow (\text{ArSe}-)_2 + 3(\text{RS}-)_2 + 4\text{H}_2\text{O}$$

$$(191) \quad \text{ArSeO}_2\text{H} + 4\text{RSH} \rightarrow \text{ArSeH} + 2(\text{RS}-)_2 + 2\text{H}_2\text{O}$$

Aromatic seleninic acids were also found to be reduced by hypophosphorous acid at pH < 1 in high yield to the selenenic acid stage (equation 192), and with additional hypophosphorous acid, to the diselenide (equation

$$(192) \quad \text{ArSeO}_2\text{H} + 3\text{H}_3\text{PO}_2 \rightarrow \text{ArSeOH} + \text{H}_3\text{PO}_3$$

193).[277] Diminished yields were obtained at higher pH's, possibly due to further reduction of the diselenides to selenols.[10,45,46]

$$(193) \quad 2\text{ArSeO}_2\text{H} + 3\text{H}_3\text{PO}_2 \rightarrow (\text{ArSe}-)_2 + 3\text{H}_3\text{PO}_3 + \text{H}_2\text{O}$$

Phosphorous acid can also reduce aromatic seleninic acids to selenenic acids (equation 194) and diselenides (equation 195), but the yields are low.[277]

$$(194) \quad \text{ArSeO}_2\text{H} + \text{H}_3\text{PO}_3 \rightarrow \text{H}_3\text{PO}_4 + \text{ArSeOH}$$

$$(195) \quad 2\text{ArSeO}_2\text{H} + 3\text{H}_3\text{PO}_3 \rightarrow (\text{ArSe}-)_2 + 3\text{H}_3\text{PO}_4 + \text{H}_2\text{O}$$

The use of stoichiometric quantities of cold aqueous hydrazine hydrate, chloride, or sulfate to reduce seleninic acids to selenenic acids gave the latter in practically quantitative yield and with virtually no interfering by-products (equation 196).[276,328] When the reaction mixtures were heated or

$$(196) \quad 2RSeO_2H + N_2H_4 \cdot H_2SO_4 \rightarrow 2RSeOH + N_2 + H_2O + H_2SO_4$$

when excess hydrazine was used, reduction to the diselenide stage occurred (equation 197).[276] Semicarbazide and hydroxylamine hydrochlorides, as

$$(197) \quad 4ArSeO_2H + 3N_2H_4 \rightarrow (ArSe-)_2 + 3N_2 + 8H_2O$$

well as phenylhydrazine, acting as reducing agents, gave mixtures of selenenic acids and diselenides.[276] Phenylseleninic acid was reduced with hydrazine sulfate to selenophenol, which was isolated as the cadmium salt.[276]

1-[279] and 2-Naphthaleneseleninic acids[280] were reduced by the action of glucose in aqueous alcoholic alkali solution to the diselenide. Sodium borohydride was used to convert 1,4-benzenediseleninic acid to the corresponding diselenol.[43]

Methyl- and ethylseleninic acids were assayed iodometrically on the basis of their ability to oxidize hydriodic acid (equation 198).[421] This reaction,

$$(198) \quad 2RSeO_2^- + 8H^+ + 6I^- \rightarrow (RSe-)_2 + 3I_2 + 4H_2O$$

studied kinetically with a series of benzeneseleninic acids, was found to be first order in seleninic acid, iodide, and hydrogen ion.[436a] 1-Anthraquinoneseleninic acid has been shown to even be capable of oxidizing ethanol and hydroquinone to acetaldehyde and quinone, respectively.[326] The oxidizing capacity of phenylseleninic acid toward methyl ketones, desoxybenzoin, cyclohexanone, and benzylamine has also been described.[434]

f. Conversion to Selenium Halides. Ethylseleninic acid hydronitrate, on treatment with concentrated hydrobromic acid, gave ethylselenium tribromide (equation 199).[14] Phenylselenium tribromide and trichloride were

$$(199) \quad C_2H_5SeO_2H \cdot HNO_3 + 3HBr \rightarrow C_2H_5SeBr_3 + HNO_3 + 2H_2O$$

prepared similarly by the treatment of phenylseleninic acid with concentrated hydrobromic[99,101,283] and hydrochloric acids,[99] respectively. Phosphorus pentachloride converted phenylseleninic acid to diphenyl selenide dichloride (equation 200), whereas thionyl chloride transformed *p*-nitrophenylseleninic

$$(200) \qquad C_6H_5SeO_2H \xrightarrow{PCl_5} C_6H_5\underset{\underset{Cl}{|}}{\overset{\overset{Cl}{|}}{Se}}C_6H_5$$

acid into the seleninyl chloride (equation 201).[432] Other attempts at the

(201)

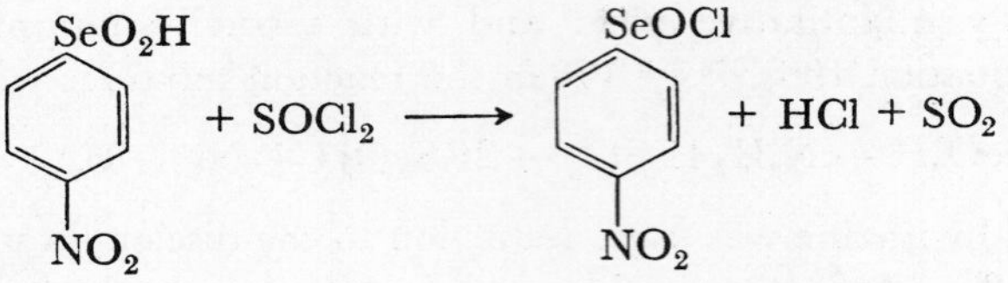

$$\underset{NO_2}{\overset{SeO_2H}{\bigcirc}} + SOCl_2 \longrightarrow \underset{NO_2}{\overset{SeOCl}{\bigcirc}} + HCl + SO_2$$

latter transformation, however, have been unsuccessful.[296]

Phenylseleninic acid was reported to react with aqueous hydrochloric, hydrobromic, or hydriodic acid (presumably concentrated) to form sparingly soluble dihalides, but to combine with 40 % hydrofluoric acid to give a water-soluble compound.[437]

g. With Thiosulfate. o-Nitrobenzeneseleninic acid has been found to react rapidly and quantitatively with thiosulfate ion to give o-nitrobenzene-selenenyl thiosulfate (equation 202).[438] The reaction, which may be con-

(202)

$$\underset{NO_2}{\overset{SeO_2H}{\bigcirc}} + 3S_2O_3^{2-} + 3H^+ \longrightarrow \underset{NO_2}{\overset{SeS_2O_3}{\bigcirc}} + S_4O_6^{2-} + 2H_2O$$

sidered to be the basis for an assay of the seleninic acid, is performed by adding excess thiosulfate and may be followed by an iodometric titration of the unreacted thiosulfate ion.

h. Cyclization Reactions. Seleninic acids have been utilized in several cyclization reactions. γ-Halosubstituted seleninic acids (as their sodium salts) have been made to ring-close to give selenacyclobutanes (equations 203, 204).[435] The seleninic acid (**54**) could be cyclized to 2-carboxyphenox-

(203)

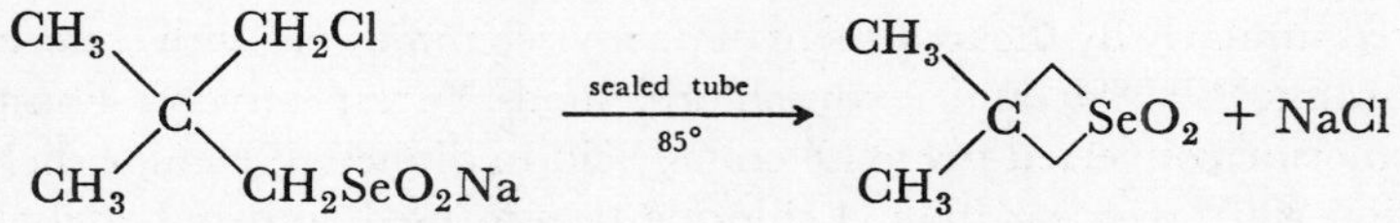

$$\underset{CH_3}{\overset{CH_3}{>}}C\overset{CH_2Cl}{\underset{CH_2SeO_2Na}{<}} \xrightarrow[85°]{\text{sealed tube}} \underset{CH_3}{\overset{CH_3}{>}}C\overset{\diagup}{\underset{\diagdown}{}}SeO_2 + NaCl$$

(204)

$$\bigcirc\overset{CH_2Br}{\underset{CH_2SeO_2Na}{<}} \xrightarrow[90°]{\text{sealed tube}} \bigcirc\overset{\diagup}{\underset{\diagdown}{}}SeO_2 + NaBr$$

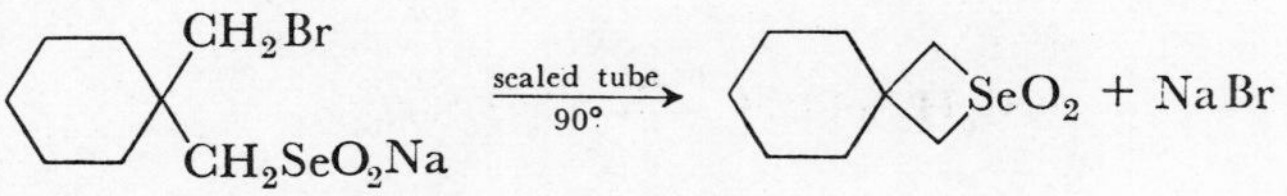

selenine-10-oxide in the presence of 85% H_2SO_4 at $40°$ (equation 205).[371]

(205)

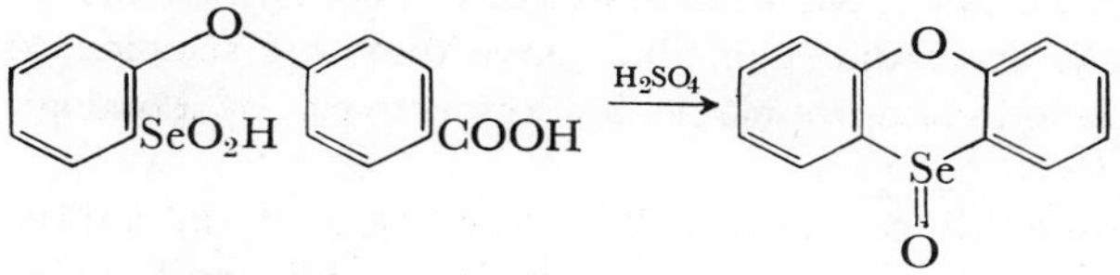

54

i. Deselenization. o-Biphenylylseleninic acid underwent deselenization with Raney nickel in benzene containing $10–20\%$ ethanol to give a 62% yield of biphenyl (equation 206).[312]

(206)

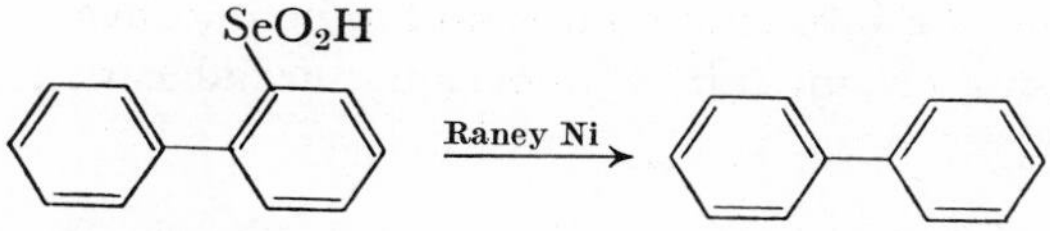

j. Formation of Derivatives. Benzylisothiuronium salts of aromatic[279,280] and aliphatic[471] seleninic acids have been formed (equation 207); however,

(207) $C_6H_5CH_2S{-}\overset{\overset{+NH_2}{\|}}{C}{-}NH_2 \cdot Cl^- + RSeO_2H \longrightarrow$

$C_6H_5CH_2S{-}\overset{\overset{+NH_2}{\|}}{C}{-}NH_2 \cdot RSeO_2^- + HCl$

those derived from the latter are reported not to crystallize sufficiently well to serve as derivatives for identification purposes.

MISCELLANEOUS*

The conversion of seleninic acids to their anhydrides, esters, and amides is discussed in Sections IV.D.2, IV.D.3, and IV.D.4, respectively.

The determination of the dissociation constants of a large number of substituted benzeneseleninic acids[109,428,439] revealed that they are weak acids comparable in strength to acetic acid.

Since the discovery by Vorozhtsov and Kozlov[440] that sulfur-substituted α-nitronaphthalenes are frequently light sensitive, Kozlov et al.[372-375] have compared numerous nitronaphthaleneseleninic acids with their analogous sulfinic acids for this property. In some instances pairs of analogous compounds were found to be of approximately equal light sensitivity, whereas in others the selenium analog was considerably more light sensitive.

The formation of monomolecular films by long-chain alkylseleninic acids on water has been studied by Rebane.[441]

* A discussion of the spectral properties of seleninic acids will be found in Chapter XV.

The sodium salts of methyl-, ethyl- and isopropylseleninic acids (or their hydrogen nitrate salts), when added to media containing the mold *Scopulariopsis brevicaulis* or certain penicillia, gave dimethyl selenide, methyl ethyl selenide, and methyl isopropyl selenide, respectively, by a biological methylation process.[442]

Alimarin and his co-workers[443–446] have studied the utility of a large number of aromatic and aliphatic seleninic acids (and their ammonium salts) as selective precipitants for a variety of metallic cations. Benzeneseleninic acid has been examined for its ability to aid in the separation of protactinium from other elements.[447] Aromatic seleninic acids form blue crystalline copper salts.[100,426]

The detergency of a lubricating oil is said to be improved by the addition of calcium, barium, or zinc salts of dodecane-, hexadecane-, or octadecaneseleninic acid.[448]

2. Seleninic Anhydrides

The anhydrides of seleninic acids are colorless, hygroscopic solids which, except for certain cyclic species, react readily with water. In contrast, the so-called sulfinic anhydrides are stable in water since they are, in actuality, disulfide trioxides (sulfinyl sulfones).[449]

PREPARATION

a. From Seleninic Acids. Seleninic anhydrides can be synthesized by heating a seleninic acid at atmospheric pressure or *in vacuo* with the elimination of 1 equivalent of water. Phenyl,[86,296] 2-carboxyphenyl,[450] and α-naphthylseleninic anhydrides[279] have been prepared in this manner (equation 208). The last-named anhydride is unstable and rearranges on further

$$(208) \qquad 2C_6H_5SeO_2H \xrightarrow[30\ hr]{130°} \begin{array}{c} C_6H_5Se{=}O \\ \diagdown \\ O \ + \ H_2O \\ \diagup \\ C_6H_5Se{=}O \end{array}$$

heating to the selenoxide with elimination of SeO_2. 1,2-Ethanediseleninic acid spontaneously dehydrates as it is formed to give the more stable 1,2-ethanediseleninic acid anhydride.[234]

b. From Diselenides. Diselenides, dissolved in carbon tetrachloride, nitromethane, or chloroform at temperatures ranging from $-10°$ to $-50°$, were found to react with 3 moles of ozone to a well-defined end point.[296] It has been suggested that the seleninic anhydrides, which are prepared in good yield by this method, are formed *via* three plausible but, as yet, un-detected intermediates indicated in scheme 209. Seleninic anhydrides are

$$(209) \quad RSeSeR \longrightarrow R\overset{\displaystyle O}{\overset{\|}{Se}}SeR \longrightarrow RSeOSeR \longrightarrow$$

$$R\overset{\displaystyle O}{\overset{\|}{Se}}\!-\!OSeR \longrightarrow R\overset{\displaystyle O}{\overset{\|}{Se}}\!-\!O\!-\!\overset{\displaystyle O}{\overset{\|}{Se}}R$$

also afforded on oxidation of diselenides with *tert*-butyl hydroperoxide under anhydrous conditions.[295,296] By these oxidative procedures it was similarly possible to form unsymmetrical seleninic anhydrides. It is of interest that true sulfinic anhydrides have not been observed in the oxidation of disulfides.[451]

c. From Selenenyl and Seleninyl Chlorides. The oxidation of trifluoro-methylselenenyl chloride, for example, using an argon stream saturated with nitrogen dioxide, gave the corresponding anhydride **55** (equation 201).[401,436]

$$(210) \qquad\qquad F_3CSeCl \xrightarrow[-15°]{NO_2} (F_3CSeO)_2O$$
$$\mathbf{55}$$

Benzeneseleninyl chloride reacted with selenophenol under anhydrous conditions in the presence of a base (*e.g.*, pyridine) to give diphenyl diselenide and benzeneseleninic anhydride (equation 211).[296]

$$(211) \quad 3C_6H_5SeH + 3C_6H_5SeOCl \longrightarrow$$

$$2(C_6H_5Se\!-\!)_2 + C_6H_5\overset{\displaystyle O}{\overset{\|}{Se}}\!-\!O\!-\!\overset{\displaystyle O}{\overset{\|}{Se}}C_6H_5 + 3HCl$$

d. From a Selenium-Containing Heterocycle. Oxidation of 2-α-pyridyl-selenonaphthene with hydrogen peroxide gave 2,2′-dicarboxyphenylseleninic anhydride (equation 212).[281]

$$(212)$$

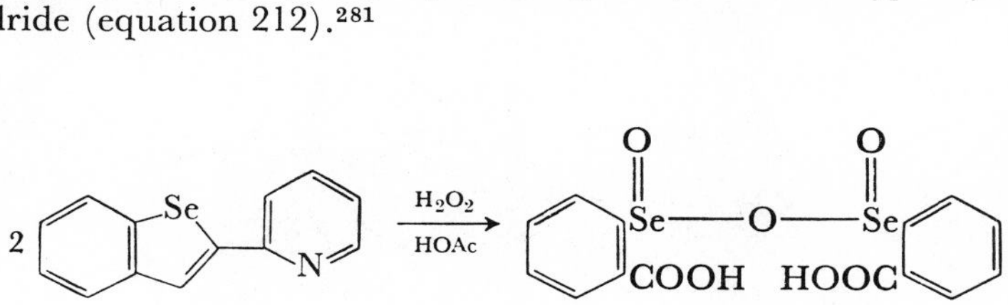

REACTIONS

a. Hydrolysis and Alcoholysis. As indicated earlier, seleninic anhydrides combine readily with water to form seleninic acids. With anhydrous alcohols they give the corresponding seleninic acids and esters (equation 213).[296,436]

(213) $\underset{\text{RSe}}{\overset{O}{\underset{\|}{}}}\!\!-O-\!\!\underset{\text{SeR}}{\overset{O}{\underset{\|}{}}} + C_2H_5OH \longrightarrow \underset{\text{RSe}}{\overset{O}{\underset{\|}{}}}\!\!-OH + \underset{\text{RSeOC}_2H_5}{\overset{O}{\underset{\|}{}}}$

1,2-Ethanediseleninic acid anhydride opens in aqueous base to give the 1,2-ethanediseleninate ion.[234]

b. **Reduction.** 2,2'-Dicarboxyphenylseleninic anhydride could be reduced to 2,2'-dicarboxydiphenyl diselenide by sodium metabisulfite.[281]

3. Seleninic Esters

Esters of seleninic acids are marked by their very high sensitivity to water. Hydrolysis of these esters leads to seleninic acid formation.

PREPARATION

a. From Salts of Seleninic Acids. The reaction of the sodium salt of β-naphthaleneseleninic acid (**56**) with methyl chlorocarbonate was reported to give the methyl seleninic ester **57** (equation 214).[132] This esterification

(214)

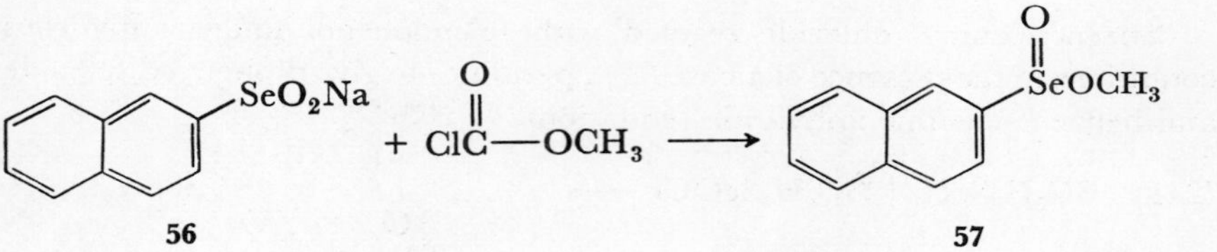

method was also used to prepare the methyl ester of p-biphenylseleninic acid;[109] however, it is claimed that **56** combines with methyl iodide in refluxing methanol to give the methylselenone **58** (equation 215).[132]

(215) $\textbf{56} + CH_3I \longrightarrow$

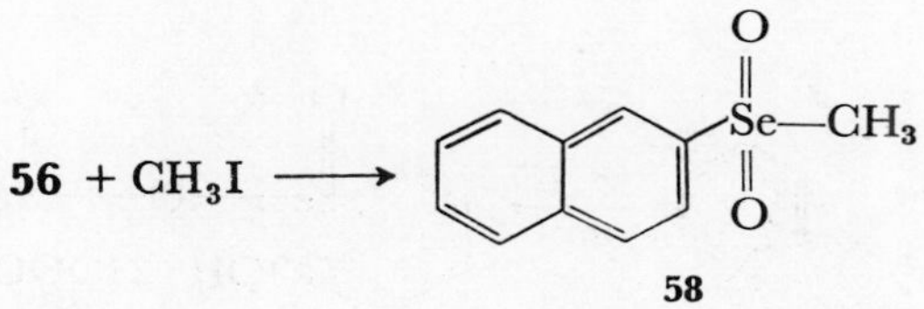

Silver salts of seleninic acids, obtained from the corresponding ammonium salts, react in the dry state with a large excess of methyl or ethyl iodide (which also serves as the reaction medium) to give the methyl or ethyl seleninate, respectively, as a distillable oil (scheme 216).[422] The esterification

(216) $RSeO_2H + NH_3 \longrightarrow RSeO_2NH_4 \xrightarrow{\text{AgNO}_3}$

$$RSeO_2Ag \xrightarrow{\text{CH}_3\text{I}} RSeO_2CH_3$$

temperature must not be permitted to rise above room temperature.

b. From Seleninic Anhydrides.   Seleninic anhydrides react with alcohols to give a mixture of the corresponding seleninic ester and acid.[296,436]

c. From Seleninyl Chlorides.   The action of sodium methoxide in methanol on methyl- or ethylseleninyl chloride produces the corresponding methyl ester (equation 217).[401] Ethyl benzeneseleninate was prepared by Ayrey

(217) $RSeOCl + NaOCH_3 \rightarrow RSeO_2CH_3 + NaCl$

et al.[296] by the reaction of benzeneseleninyl chloride with sodium ethoxide in ethanol.

d. From a Selenenic Ester.   Ozonization of ethyl 2,4-dinitrobenzene-selenenate gave the corresponding seleninic ester **59** (equation 218).[296]

(218)

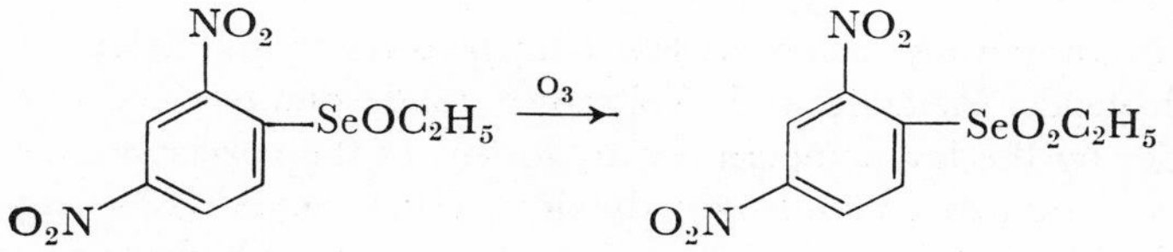

59

e. From Selenium Dioxide and a Diene.   The adduct derived from SeO_2 and 2,3-dimethylbutadiene has been shown by Mock and McCausland[452] to possess a cyclic seleninic ester structure (**60**). A symmetric cyclic selenone structure proposed previously by Backer and Strating[453] was held untenable

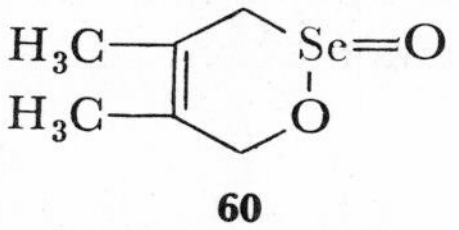

60

on the basis of the n.m.r. spectrum of the compound.

4. Seleninamides

The only known method for the synthesis of this class of compounds involves the reaction of a *para*-substituted phenylseleninic acid with thionyl chloride, followed by treatment of the resultant seleninyl chloride with concentrated ammonium hydroxide[454] or ammonia in dioxane[432] (equation 219).

(219)

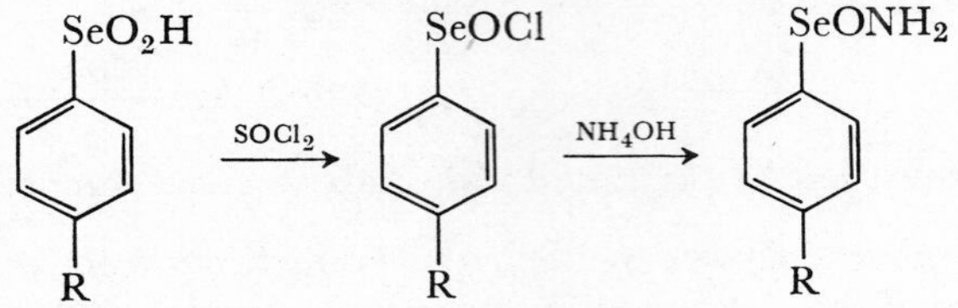

5. Seleninyl Halides

The first member of this class of compounds was prepared by Pirisi and Serreli,[432] who treated p-nitrophenylseleninic acid with thionyl chloride to give p-nitrophenylseleninyl chloride as yellow crystals (cf. equation 219). Woodbridge,[455] however, reported failure to convert seleninic acids to seleninyl chlorides in this manner.

Ayrey et al.[296] arrived at these compounds by the ozonization of aryl-selenenyl chlorides (equation 220). The latter react with exactly 1 mole of

$$(220) \qquad RSeCl + O_3 \rightarrow RSeOCl + O_2$$

ozone in anhydrous solvent at low temperatures to give seleninyl chlorides in high yields. Paetzold and Wolfram[401] made methyl- and ethylseleninyl chlorides by the low-temperature treatment of the corresponding selenenyl chlorides with ozone or nitrogen dioxide. Trifluoromethylselenenyl chloride reacted with nitrogen dioxide, however, to give the anhydride **61** as the final product (scheme 221). Hurd and Fancher[410a] prepared 2-chloroethene-

$$(221) \qquad CF_3SeCl + NO_2 \rightarrow CF_3SeOCl + NO$$

$$2CF_3SeOCl + 3NO_2 \rightarrow (CF_2SeO)_2O + 2NO_2Cl + NO$$
$$\textbf{61}$$

seleninyl chloride in 90 % yield by bubbling acetylene into a benzene solution of selenium oxychloride (equation 221a).

$$(221a) \qquad HC\equiv CH + SeOCl_2 \longrightarrow ClCH=CH-\overset{\overset{\displaystyle O}{\|}}{Se}-Cl$$

REACTIONS

a. With Water, Ammonia, or Alkoxides. Seleninyl chlorides are hydrolyzed readily to give seleninic acids, usually as hydrochloride salts (equation 222).[296,401] Ammonia in aqueous or dioxane solution converts seleninyl

$$(222) \qquad RSeOCl + H_2O \rightarrow RSeO_2H \cdot HCl$$

halides to seleninamides (*cf.* equation 219).[432,454] Sodium ethoxide or methoxide reacts with seleninyl chlorides to give seleninic esters (*cf.* equation 217).[296,401]

b. With Selenols. Selenophenol, when added to a cooled chloroform solution of benzeneseleninyl chloride in the presence of a base such as triethylamine, pyridine, or anhydrous sodium carbonate, gave diphenyl diselenide and benzeneseleninic anhydride (equation 233).[296]

$$(223) \quad 3C_6H_5SeOCl + 3C_6H_5SeH \rightarrow 2(C_6H_5Se-)_2 + (C_6H_5SeO-)_2O$$

6. Selenium Trihalides

Organic selenium trihalides are generally crystalline solids of low stability which behave as if they are in equilibrium with the corresponding selenenyl halides and halogen (equation 224). Whereas selenium trichlorides are

$$(224) \qquad\qquad RSeX_3 \rightleftharpoons RSeX + X_2$$

colorless, aromatic selenium tribromides are highly colored, ranging from orange to deep red. The chemical properties of organosulfur trichlorides, described in a review by Douglass,[456] and those of their selenium isosteres are very similar.

The aliphatic selenium trichlorides rank in order of stability as follows: $CH_3 > C_2H_5 > $ i-C_3H_7.[457] The simplest of the series, methylselenium trichloride, has been found to exist in methylene chloride mainly as a dimeric species (**62**) and has been obtained in a methylene chloride-soluble form

62

(6 g/100 ml at 25°) and in a less soluble form (0.4 g/100 ml at 25°).[458] The selenium analyses, i.r. spectra, and thermal behavior of the two variants are identical.

Organoselenium trihalides can act as Lewis bases. Wynne and Pearson[457] have reported that methyl-, ethyl-, and phenylselenium trichlorides form adducts with antimony pentachloride ($RSeCl_3 \cdot SbCl_5$) that are moisture-sensitive solids. Complexes (1:1) are also formed by methyl-[459] and phenyl-[460] selenium trihalides with tetramethylthiourea (tmtu), as exemplified in equation 225. The trapping in this manner of selenium dichloride, a

$$(225) \qquad CH_3SeCl_3 + tmtu \rightarrow SeCl_2 \cdot (tmtu) + CH_3Cl$$

compound ordinarily accessible only at high temperature, suggests that it may be an important intermediate in the decomposition of alkylselenium trichlorides.[459]

PREPARATION

a.   From Seleninic Acids.   The reaction of ethylseleninic acid hydronitrate with hydrobromic acid did not give the anticipated hydrobromide salt of the seleninic acid but yielded instead ethylselenium tribromide (equation 226) as a yellow crystalline solid.[14] Treatment of phenylseleninic acid

$$(226)\quad C_2H_5SeO_2H \cdot HNO_3 + 3HBr \rightarrow C_2H_5SeBr_3 + 2H_2O + HNO_3$$

with aqueous HBr or HCl produced the corresponding phenylselenium trihalide in good yield;[99,283] however, use of 40% HF failed to give phenylselenium trifluoride.[437]

b.   From Selenenyl Halides.   Another method for the synthesis of organic selenium trihalides is the reaction of a selenenyl halide with excess halogen.[99,283] For example, one can form phenylselenium trichloride by chlorine treatment of phenylselenenyl chloride (equation 227), phenylselenium tribromide (equation 228), or phenylselenenyl bromide (equation 229).[303]

$$(227)\qquad\qquad C_6H_5SeCl + Cl_2 \rightarrow C_6H_5SeCl_3$$

$$(228)\qquad 2C_6H_5SeBr_3 + 3Cl_2 \rightarrow 2C_6H_5SeCl_3 + 3Br_2$$

$$(229)\qquad 2C_6H_5SeBr + 2Cl_2 \rightarrow 2C_6H_5SeCl_3 + Br_2$$

The last compound, on treatment with bromine in carbon tetrachloride, gives phenylselenium tribromide.[283] Attempts to prepare phenylselenium triiodide led to a very unstable material,[99] while the action of iodine monochloride on a selenenyl chloride gave the corresponding selenium trichloride (equation 230).[303]

$$(230)\qquad\qquad ClCH_2SeCl + 2ICl \rightarrow ClCH_2SeCl_3 + I_2$$

c.   From Selenocyanates.   Direct oxidative halogenation of other divalent selenium-containing functional groups also produces organoselenium trihalides. Thus, the action of bromine on aliphatic[347] and aromatic[131,239,283,328,336] selenocyanates gives the corresponding selenium tribromide (equation 231).

$$(231)\qquad\qquad RSeCN + 2Br_2 \rightarrow RSeBr_3 + BrCN$$

Chlorine or sulfuryl chloride[228,239,336] can convert selenocyanates to selenium trichlorides in a reaction that is sometimes run in the presence of pyridine.[389]

d. From Diselenides. Diselenides, on treatment with bromine,[239,326] chlorine (equation 232),[54,239,336,337,458] iodine monochloride,[303] or sulfuryl

(232) $$(RSe—)_2 + 3Cl_2 \rightarrow 2RSeCl_3$$

chloride (equation 233),[336,389,460a] can give rise to organoselenium trihalides.

(233) $$(RSe—)_2 + 3SO_2Cl_2 \rightarrow 2RSeCl_3 + 3SO_2$$

Lawson and Kharasch[389] have noted that the chlorinolysis of bis(2,4-dinitrophenyl) disulfide with sulfuryl chloride in the presence of pyridine gave the corresponding sulfenyl chloride but that identical treatment of the analogous diselenide led to the formation of 2,4-dinitrophenylselenium trichloride.

e. From Selenides. Selenacyclobutanes, on reaction with sufficient chlorine or bromine, give selenium trihalides (equations 234, 235).[435] Bis(trifluoro-

(234)

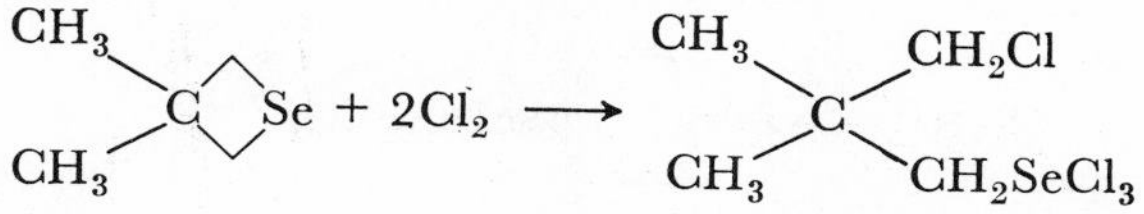

(235)

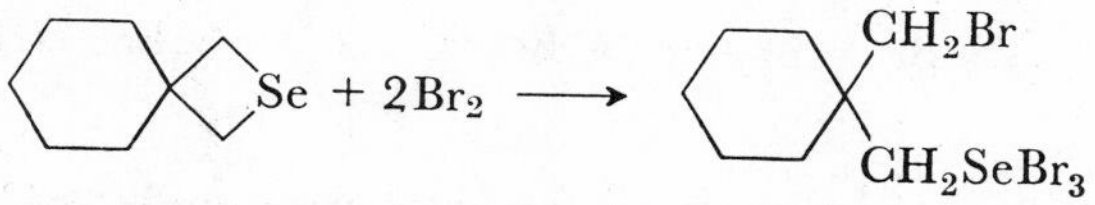

methyl) selenide (**63**) combined with chlorine in a sealed tube to yield trifluoromethylselenium trichloride (**64**).[54] Heptafluoropropylselenium trichloride is unstable, for only the selenenyl chloride is obtained under the

$$F_3CSeCF_3 \qquad F_3CSeCl_3$$

63   **64**

reaction conditions mentioned above.[286]

f. From Selenium Tetrachloride. Selenium tetrachloride reacts with esters of salicylic acid in an electrophilic aromatic substitution reaction to introduce the selenium trichloride moiety *para* to the hydroxyl group (equation 236).[461]

(236)

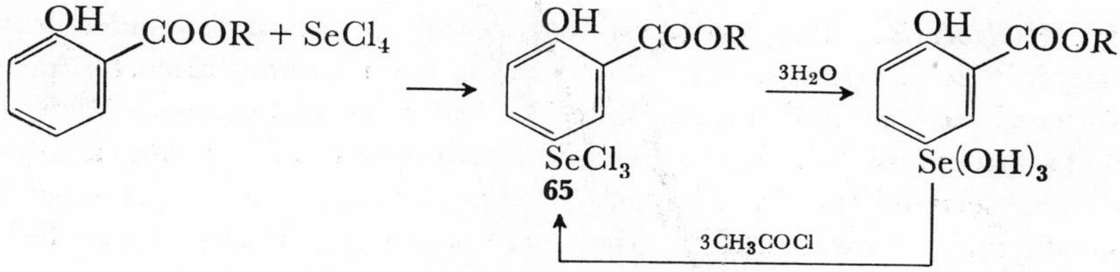

The resultant organoselenium trichloride (**65**) reacts slowly with moisture in the air to give a hydrated form of the corresponding seleninic acid. Based on the formulation of the product as a trihydroxide, an attempt was made to prepare a triacetate using acetyl chloride; instead, the selenium trichloride **65** was regenerated.[461]

REACTIONS

a. Loss of Halogen.   One of the most prominent characteristics of organic selenium trihalides is their propensity to lose halogen and thus revert to the corresponding selenenyl halide (*cf.* equation 224). This is accelerated by subjecting these compounds to reduced pressure,[131,238,336,389] by heating them near the melting point,[239] or by treating them with acetone.[337]

When phenylselenium tribromide was heated between 110° and 115°, little free bromine was evolved; instead, hydrobromic acid was evolved because of halogenation in the *p*-position of the phenyl ring (scheme 237).[410]

(237)

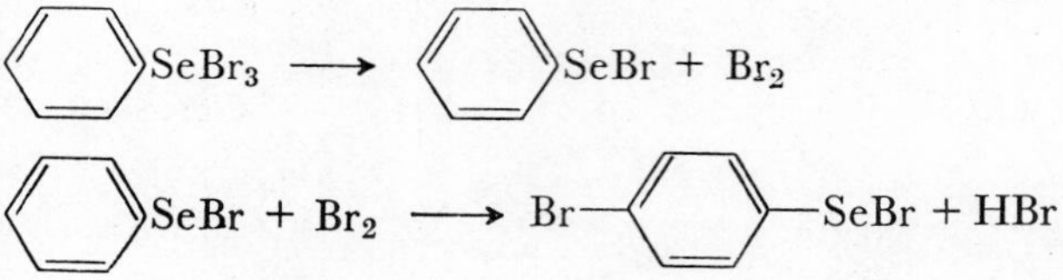

$$\langle\!\!\!\bigcirc\!\!\!\rangle\!\!-SeBr_3 \longrightarrow \langle\!\!\!\bigcirc\!\!\!\rangle\!\!-SeBr + Br_2$$

$$\langle\!\!\!\bigcirc\!\!\!\rangle\!\!-SeBr + Br_2 \longrightarrow Br\!\!-\!\!\langle\!\!\!\bigcirc\!\!\!\rangle\!\!-SeBr + HBr$$

When the *p*-position of an aromatic selenium tribromide was already occupied, as in compound **66**, nuclear halogenation was averted, resulting in bromine evolution (equation 238).[410] The selenium tribromide, rather than

(238)

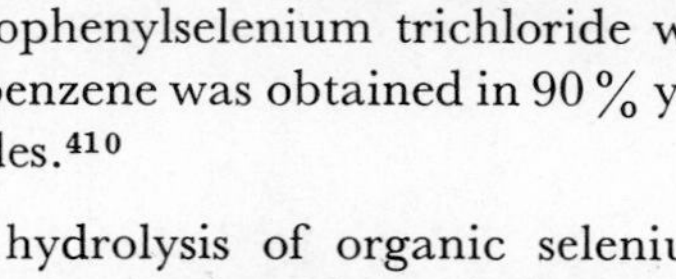

$$\underset{Br}{\overset{SeBr_3}{\bigcirc}} \longrightarrow \underset{Br}{\overset{SeBr}{\bigcirc}} + Br_2$$

66

bromine, may actually be the halogenating agent. Different behavior was observed when *p*-chlorophenylselenium trichloride was heated above 160°. In this case, *p*-dichlorobenzene was obtained in 90 % yield, as well as selenium mono- and tetrachlorides.[410]

b. Hydrolysis.   The hydrolysis of organic selenium trihalides leads to seleninic acids (*cf.* equation 186) and has been accomplished by means of water alone,[54,336] aqueous alkali,[99,239,283,328] or silver hydroxide.[228,435]

Treatment of 2-biphenylylselenium tribromide with strong alkali gave dibenzoselenophene in low yield.[239] The latter could be prepared more satisfactorily, however, from the corresponding selenenyl bromide[462] (*cf.* Chapter XIB).

c. Reduction. Phenylselenium tribromide has been reduced by zinc dust in ether or benzene to diphenyl diselenide (equation 239).[283]

$$(239) \qquad 2C_6H_5SeBr_3 + 3Zn \rightarrow (C_6H_5Se—)_2 + 3ZnBr_2$$

d. Other Reactions. Many of the reactions of 2,4-dinitrophenylselenium tribromide (**67**) are what might be expected of the corresponding selenenyl

67

bromide. This point is illustrated by equations 240–243.[389]

$$(240) \qquad \mathbf{67} + \text{(cyclohexene)} \longrightarrow \text{(cyclohexane with SeAr and Br)} + Br_2$$

$$(241) \qquad \mathbf{67} + C_2H_5OH \longrightarrow ArSeOC_2H_5 + HBr + Br_2$$

$$(242) \qquad \mathbf{67} + CH_3COCH_3 \longrightarrow ArSeCH_2\overset{O}{\overset{\|}{—C}}—CH_3 + HBr + Br_2$$

$$(243) \qquad \mathbf{67} + C_6H_5NH_2 \longrightarrow C_6H_5NHSeAr + HBr + Br_2$$

E. SELENONIC ACIDS

The chemistry of selenonic acids has been comparatively unexplored, probably because of their high reactivity and low stability. Selenonic acids, which may be thought of as derivatives of selenic acid, $HOSeO_2OH$, are very acidic and are strong oxidizing agents. They also tend to be hygroscopic. Some selenonic acids, as well as their potassium salts, are unstable to the point of being explosive when heated. Only recently has the first known selenonic acid been reinvestigated, and the structure previously accepted has now been proven to be incorrect (*vide infra*).

PREPARATION

a. From Selenic Acid or Selenium Trioxide. In 1909, Doughty[86] heated benzene with selenic acid for 100 hours and obtained, *via* the barium salt, what he thought was benzeneselenonic acid ("$C_6H_5SeO_3H$," m.p. 142°). Later, he repeated this procedure with toluene,[463] and subsequently Anschütz *et al.* performed the selenation in acetic anhydride with *o*-, *p*-,[464] and *m*-xylenes.[465] The selenation reaction was extended by the use of highly reactive selenium trioxide in liquid sulfur dioxide. This solution was added dropwise at −50° to benzene to give what Schmidt *et al.*[466,467] also believed to be benzeneselenonic acid.

Paetzold and Lienig,[468] in a more recent study of Doughty's benzene-selenonic acid, found by the Bunsen method that it had only 50 % hexavalent selenium and proposed, therefore, that it actually was the benzeneselenonic salt of benzeneseleninic acid (**68**). Benzeneselenonic acid (m.p. 64°), a

$$\left[C_6H_5Se \overset{\displaystyle OH}{\underset{\displaystyle OH}{\diagup}} \right]^+ \quad [C_6H_5SeO_3]^-$$

68

strongly hygroscopic substance, was isolated by ion-exchange chromatography. The work of Paetzold and Lienig was confirmed by Dostal et al.,[469,469a] who isolated from the reaction of 3 % SeO_3 in liquid SO_2 (or 20 % SeO_3 in H_2SeO_4) with benzene an 85 % yield of benzeneselenonic acid and a small quantity of diphenyl selenone.

Fluorene has been selenated with $H_2{}^{75}SeO_4$ in acetic acid–acetic anhydride solution to give a compound whose structure was reported to be **69**.[470]

69

In view of Paetzold and Lienig's findings, the structures of many compounds described as selenonic acids in the literature before their report are now in doubt and may very well be incorrect.

b. By the Oxidation of Diselenides. Aromatic diselenides have been cleaved with strong oxidizing agents such as fuming nitric acid,[454] potassium permanganate,[450] and chlorine water[426] to give what were claimed to be selenonic acids. Benzeneselenonic acid, prepared by Stoecker and Krafft[426] in 1906 by this method, using chlorine water, was described as being highly hygroscopic. Although the melting point was not given, it is probable that this was the first actual synthesis of the compound.

c. By the Oxidation of Seleninic Acids. Seleninic acids are the most frequently used precursors for the synthesis of selenonic acids. Free aliphatic selenonic acids decompose rapidly and have not been isolated in the pure state. Their stable potassium salts, however, have been prepared by oxidizing methyl-, ethyl-, and isopropylseleninic acids (or their hydronitrate salts) with potassium permanganate in neutral solution (equation 244).[37] Rebane[471]

$$(244) \quad 3RSeO_2H + 2KMnO_4 + KOH \rightarrow 3RSeO_3K + MnO_2 + H_2O$$

has prepared octyl- and decylselenonic acids as their potassium and *S*-benzylthiopseudourea salts. Aromatic seleninic acids have been reported to give the corresponding selenonic acids when oxidized with permanganate solution.[131,258,275,450,472]

In contrast to the great stability of 2-aminoethanesulfonic acid (taurine), its selenium analog has had but a fleeting existence. Failure to obtain a stable product was encountered in attempting to oxidize 2-aminoethaneseleninic acid (**48**) with alkaline permanganate,[228] nitric acid,[228] or 30% hydrogen peroxide or by the reaction of sodium selenite with either 2-bromoethylamine hydrobromide or ethylenimine.[41] The desired selenonic acid **70** was finally isolated in poor yield by the oxidation of **48** with chlorine water at 50–60° (equation 245), but it decomposed in several hours to give selenium and **48**.[228]

$$(245) \qquad 48 + Cl_2 \xrightarrow{H_2O} H_2NCH_2CH_2SeO_3H$$
$$70$$

d. Other Methods.   1,2-Ethanediselenonic acid and vinylselenonic acid, as their potassium salts, have been synthesized by the permanganate oxidation of 1,2-ethanediseleninic anhydride.[473] Aromatic selenonic acids, substituted by triazine moieties, have been prepared by Friedheim.[474,475]

REACTIONS

Being strong oxidizing agents, selenonic acids can be readily reduced. Hydrogen chloride treatment of selenonic acids gave the corresponding seleninic acid,[86,450,463] while hydriodic acid reduced **70** to bis(2-aminoethyl) diselenide (equation 246).[228]

$$(246) \qquad 2(70) + 10HI \rightarrow (H_2NCH_2CH_2Se—)_2 + 5I_2 + 6H_2O$$

Reducing agents such as hydrogen sulfide, sulfur dioxide, and zinc and hydrochloric acid were reported to take benzeneselenonic acid to selenophenol.[86] Tin and concentrated hydrochloric acid reduced *m*-carboxyphenylselenonic acid to the corresponding selenol, which was isolated as the diselenide after air oxidation.[258]

Potassium ethylselenonate is gradually decomposed to acetaldehyde by warm aqueous potassium permanganate solution; however, the methyl homolog is resistant to such treatment.[37]

Methyl esters of *p*-chloro- and *p*-bromobenzeneselenonic acids have been prepared by treating them with ethereal diazomethane.[472]

An unsuccessful attempt to make the acid chloride of benzeneselenonic acid by the action of phosphorus pentachloride was described as being violent and accompanied by flashes of light.[86]

MISCELLANEOUS*

o- and p-Nitrobenzeneselenonic acids have been found to be remarkably sensitive to sunlight.[374] The reactions of the ammonium salts of phenyl-selenonic acids with numerous cations have been studied[464,465] and compared with the behavior of aromatic seleninic and telluronic acids.[443,444]

F. SELENIUM-SULFUR COMPOUNDS

1. Simple Selenenyl Sulfides

A compound bearing the general structure RSeSR′ will be called here a selenenyl sulfide, rather than a sulfenyl selenide, on the grounds that a modified *sulfide* should be given precedence over a modified *selenide*. The name thiaselane has also been proposed for this type of compound (*cf.* Chapter I).

Like unsymmetrical disulfides, acyclic selenenyl sulfides can readily disproportionate, especially under alkaline conditions,[387] to the corresponding diselenides and disulfides (equation 247), thus frequently presenting

$$(247) \qquad 2RSeSR' \rightleftharpoons (RSe{-})_2 + (R'S{-})_2$$

difficult isolation and purification problems. Cyclic selenenyl sulfides tend to be more stable.

An unusual member of this group, a compound in which a Se atom is part of a "no-bond" resonance system, is depicted in **71**.[476,477] It is interesting that the Se-S bond length in this compound is shorter than the S–S bond length, a difference which may reflect the predominance of structure **71b** over structure **71a**. Such a difference in bond polarization could not be

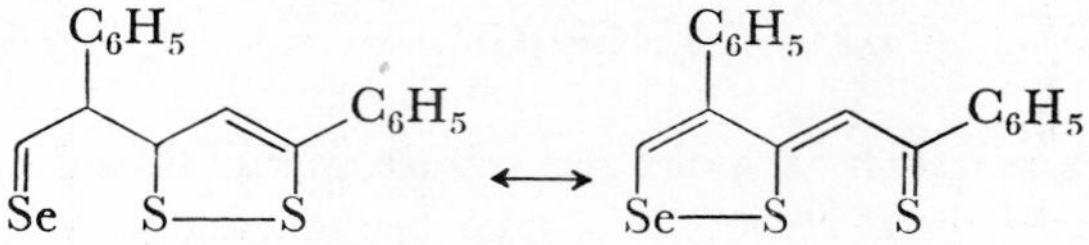

71a   **71b**

detected when the dipole moments of 1,2-dithiolane, 1,2-diselenolane, and 1-thia-2-selenolane derivatives were compared.[478]

PREPARATION

a. From Selenocyanates. Nakasaki[252,479] found that the reaction between an aryl selenocyanate and an aryl thiol proceeds vigorously with the evolution of hydrogen cyanide to give a selenenyl sulfide (equation 248). While the

$$(248) \qquad ArSeCN + Ar'SH \rightarrow ArSeSAr' + HCN$$

* A discussion of the spectral properties of selenonic acids will be found in Chapter XV.

corresponding reaction of an aryl thiocyanate with an aryl selenol showed evidence of hydrogen cyanide evolution, no major product could be isolated.[252]

b. From Selenenyl and Sulfenyl Thiocyanates. Thiophenol reacts with aromatic selenenyl thiocyanates to give selenenyl sulfides (equation 249);[369]

(249) $C_6H_5SH + ArSeSCN \rightarrow C_6H_5SSeAr + HSCN$

however, when a selenenyl selenocyanate is the precursor, the reaction takes a different course, as indicated in equation 250.[323]

(250) $2RSH + 2ArSeSeCN \rightarrow ArSe_3Ar + (RS—)_2 + Se + 2HCN$

Sulfenyl thiocyanates (equation 251),[251,369,480] as well as sulfenyl seleno-

(251) $RSeH + R'SSCN \rightarrow RSeSR' + HSCN$

cyanates,[251] behave in the expected manner with selenols to give selenenyl sulfides. Recently, Roy et al.[480] synthesized the selenenyl sulfide N-carbobenzoxy-Se-(naphthyl-2-thio)-L-selenocysteine diphenylmethyl ester (**72a**) from 2-naphthylsulfenyl thiocyanate and the selenide **72b** (equation 251a).

(251a)

$$\overset{\displaystyle Se^-}{\underset{\displaystyle 72b}{\overset{|}{C_6H_5CH_2OCONHCHCOOCH(C_6H_5)_2}}} + \quad \text{(naphthyl)—SSCN} \longrightarrow$$

$$\overset{\displaystyle S}{\underset{72a}{\overset{\displaystyle |}{\underset{\displaystyle C_6H_5CH_2OCONHCHCOOCH(C_6H_5)_2}{\overset{\displaystyle (naphthyl)}{\overset{\displaystyle |}{Se}}}}}}$$

c. From Selenenyl and Sulfenyl Halides. Aromatic selenenyl bromides react with thiols to give selenenyl sulfides (equation 252).[251,387,481] Grignard

(252) $ArSeBr + RS^- \rightarrow ArSeSR + Br^-$

intermediates of type **72c** may be substituted for thiols (equation 253).[251]

(253) $C_6H_5SeBr + \text{(naphthyl)SMgBr} \quad (72c) \longrightarrow \text{(naphthyl)SSeC}_6H_5$

Sulfenyl halides yield selenenyl sulfides in analogous reactions (equations 254, 255).[251] Sisler and Kotia[27a] have recently demonstrated that both

$$(254) \qquad C_6H_5SBr + C_6H_5SeH \rightarrow C_6H_5SSeC_6H_5$$

$$(255) \qquad C_6H_5SBr + C_6H_5SeMgBr \rightarrow C_6H_5SSeC_6H_5$$

chloramine and dimethylchloramine can assist in the oxidative coupling of selenophenol with thiophenol to give diphenylselenenyl sulfide in *ca.* 85 % yield. It was proposed that the reaction may proceed *via* the formation of an intermediate sulfenyl or selenenyl chloride, which condenses with the selenol or thiol present, as in equation 254.

d. From an Alkyl Selenosulfate or Thiosulfate. Just as sodium 2-amino-ethylthiosulfate reacts in methanol with a thiolate to give an unsymmetrical disulfide,[482] sodium 2-aminoethylselenosulfate, on combination with sodium *n*-decylmercaptide, yields 1-amino-3-selena-4-thiatetradecane (**73**) (equation 256).[41] Selenide displacement on an organic thiosulfate similarly gives a selenenyl sulfide.[480]

$$(256) \quad H_2NCH_2CH_2SeSO_3^- + n\text{-}C_{10}H_{21}S^- \rightarrow$$

$$H_2NCH_2CH_2SeSC_{10}H_{21} + SO_3{}^{2-}$$
$$\textbf{73}$$

e. From a Disulfide and a Diselenide. When bis(pentafluorophenyl) di-sulfide was heated with bis(pentafluorophenyl) diselenide, the selenenyl sulfide **74** was obtained and was separated from the starting materials by t.l.c. (equation 257).[308]

$$(257) \qquad (C_6F_5S{-})_2 + (C_6F_5Se{-})_2 \xrightarrow[\Delta]{} 2C_6F_5SSeC_6F_5$$
$$\textbf{74}$$

f. By Intramolecular Oxidation. After the protective benzyl groups of compound **75** were removed to give an intermediate thiol-selenol, oxidation of the latter yielded 1-thia-2-selenacyclopentane-4-carboxylic acid (**76a**) (equation 258).[68]

$$(258)$$

    75   76a

g. Incorporation of Selenium into a Thiacyclobutane Ring. When 2-thiaspiro[3.5]nonane (**76b**) was heated with elemental selenium in diethyl-eneglycol containing a trace of potassium cyanide, the cyclobutane ring expanded to include the Se atom, giving 2-thia-3-selenaspiro[4.5]decane (**77**) (equation 259).[483] This product has an orange-red color, said to be

(259)

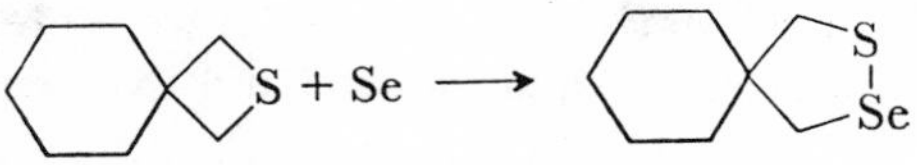

76b   **77**

characteristic of thiaselenacyclopentanes. The attempt to accommodate sulfur in 2-selenaspiro[3.5]nonane in an analogous manner resulted only in polymer formation.

REACTIONS

a. With Bromine.   The selenenyl sulfide **78** was reported to react with bromine to give benzeneselenium tribromide and *o*-nitrophenylsulfenyl bromide (equation 260).[251]

(260)

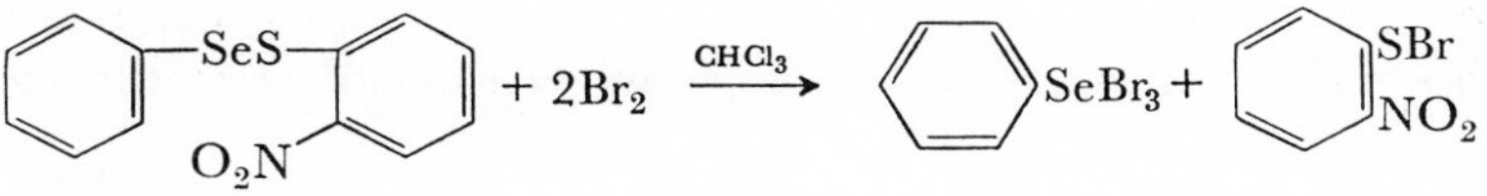

78

b. With Mercury.   When bis(pentafluorophenyl)selenenyl sulfide (**74**) was heated with mercury in a sealed tube, the Se-S bond was cleaved and mercury was incorporated into the molecule to give **79** (equation 261).[308]

$$(261) \qquad 74 + Hg \xrightarrow{\text{sealed tube}} C_6H_5SeHgSC_6F_5$$
$$79$$

c. With Bis(phenylsulfonyl)methane.   Several aromatic selenenyl sulfides have been treated with bis(phenylsulfonyl)methane anion (**80**) to give compounds of the type shown in structure **81** with displacement of the selenophenolate ion (equation 262).[484]

$$(262) \quad ArSeSAr' + (C_6H_5SO_2)_2CH^- \rightarrow (C_6H_5SO_2)_2CHSAr + ArSe^-$$
$$\qquad\qquad 80 \qquad\qquad\qquad\qquad 81$$

2. Complex Selenenyl Sulfides

In this section will be discussed selenenyl sulfides containing additional chalcogen atoms as part of the functional group. Since this is a loosely defined class of compounds, each type will be dealt with individually.

a.   Thiodiselenides and Selenodisulfides. In 1929, Levi and Baroni[254] reported the synthesis of the first triselenide by the reaction of selenium oxychloride and ethaneselenol (*cf.* equation 103). The corresponding diethyl thiodiselenide **82** and selenodisulfide **83** were prepared in an analogous manner, as indicated in equations 263 and 264, respectively. Diphenyl

(263) $2C_2H_5SeH + SOCl_2 \rightarrow C_2H_5SeSSeC_2H_5$
82

(264) $2C_2H_5SH + SeOCl_2 \rightarrow C_2H_5SSeSC_2H_5$
83

thiodiselenide and selenodisulfide were made similarly.[485] By the action of $SSeCl_2$ and $SeSCl_2$ on ethanethiol and ethaneselenol, a series of tetrachalcogens was said to be formed in 7–15% yields (*e.g.*, equation 265).[486]

(265) $2C_2H_5SH + SSeCl_2 \rightarrow C_2H_5SSSeSC_2H_5 + 2HCl$

Rheinboldt and Giesbrecht[325] made bis(*o*-nitrophenylselenenyl) sulfide (**84**) by the action of sulfide ion on a variety of selenenyl derivatives, as shown in equations 266–269. Compound **84** is described[325] as being thermo-

(266)

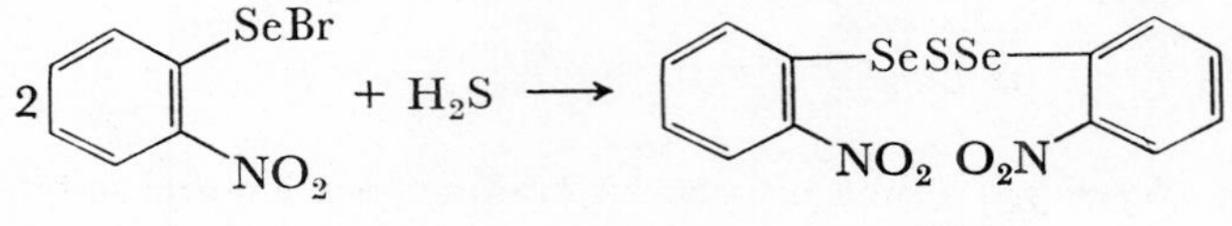

84

(267) $2ArSeSCN + H_2S \rightarrow$ **84**

(268) $2ArSeSeCN + H_2S \rightarrow$ **84**

(269) $2ArSeBr + HgS$ (or PbS, Ag_2S, Th_2S) $\rightarrow$ **84**

stable and unaffected by boiling water. Its reactions, for the most part, mimic those of an aromatic diselenide. The actions of chlorine and bromine give the corresponding selenenyl halides, while iodine is unreactive. Acids cause cleavage of the Se-S bond, and heating **84** with aqueous ethanolic potassium hydroxide gives bis(*o*-nitrophenyl) diselenide and *o*-nitrophenylseleninic acid (equation 270).[325] Oxidation of **84** with hydrogen peroxide in acetic acid gives the corresponding seleninic acid, whereas permanganate

(270)

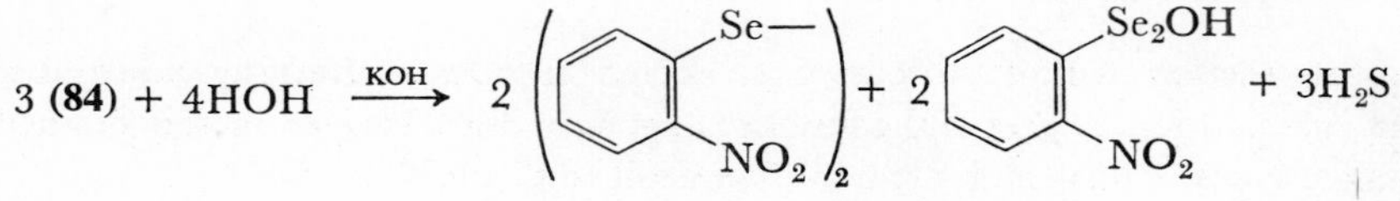

in acetone yields the diselenide. Mercury, copper, Raney nickel, and ammoniacal silver nitrate all cause desulfurization of **84**, leading to the formation of the corresponding diselenide.

The reaction of the mercaptan, 2-hydroxy-3-bromo-5-methylthiophenol (**85**), with selenium dioxide has been reported to give the selenodisulfide **86** (equation 271).[487] In 1941 Painter[488] suggested that the related reaction

(271)

$$2 \left(\begin{array}{c} \text{Br} \quad \text{OH} \\ \\ \text{CH}_3 \end{array} \text{SH} \right) + \text{SeO}_2 \longrightarrow \left(\begin{array}{c} \text{Br} \quad \text{OH} \\ \\ \text{CH}_3 \end{array} \text{S} - \right)_2 \text{Se}$$

85   86

of selenite ion with cysteine produced selenium dicysteine (**87**) according to equation 272. Stekol,[489] however, claimed to have isolated selenium tetra-cysteine (**88**) from this reaction; later Klug and Petersen[490] explained that

(272) $4\text{HOOC}-\underset{\underset{\text{NH}_2}{|}}{\text{CH}}-\text{CH}_2\text{SH} + \text{SeO}_2 \longrightarrow$

$(\text{HOOC}-\underset{\underset{\text{NH}_2}{|}}{\text{CH}}-\text{CH}_2\text{S}-)_2\text{Se} + (\text{HOOC}-\underset{\underset{\text{NH}_2}{|}}{\text{CH}}-\text{CH}_2\text{S}-)_2 + 2\text{H}_2\text{O}$

87

$$\left(\text{HOOC}-\underset{\underset{\text{NH}_2}{|}}{\text{CH}}-\text{CH}_2\text{S}- \right)_4 \text{Se}$$

88

Stekol's product was a mixture of cysteine and selenium dicysteine (**87**). Tsen and Tappel[491] noted that selenite ion was a better catalyst than cupric ion for the oxidation of glutathione and suggested that selenium diglutathione was an active intermediate in this transformation. Ganther[492] confirmed the stoichiometry proposed by Painter by following the enhanced absorption in the 260–380 nm region due to the combination of selenious acid with the sulfhydryl-containing compounds cysteine, 2-mercaptoethanol, gluta-thione, and coenzyme A. Spots corresponding to the selenodisulfide and the disulfide were observed by t.l.c. Selenium dicysteine itself was later isolated by means of column chromatography on copper-chelex.[492] Reduced ribo-nuclease A also appears to react with selenious acid to form a selenodisulfide at pH 2.[493]

Several dithiols, on reaction with selenious acid, have been reported to give selenium tetrasulfides of the type shown in structure **89**.[494]

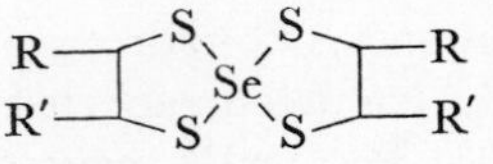

89

MISCELLANEOUS

An improved silver-plating process utilizing seleninylbis(thioacetic acid), $OSe(SCH_2CO_2H)_2$, whose synthesis is not described, has been patented.[495] The selenium compound was used in solution with cyanide and DMSO.

b. Diselenium Disulfinates. These compounds, first synthesized and so named by Foss,[496] were prepared from sodium benzene- (or toluene-) sulfinate and selenium monochloride suspended in benzene (equation 273). If the

$$(273)\quad 2C_6H_5SO_2Na + Se_2Cl_2 \xrightarrow[C_6H_6]{O^\circ} C_6H_5\overset{O}{\underset{O}{\overset{\|}{\underset{\|}{S}}}}SeSe\overset{O}{\underset{O}{\overset{\|}{\underset{\|}{S}}}}C_6H_5 + 2NaCl$$

reaction was performed in dry ether, a triselenium disulfinate, **90**, and a selenium disulfinate, **91**, were formed, probably by disproportionation of the diaryl diselenium disulfinate in a manner similar to that occurring with the analogous sulfur compounds (equation 274).[497,498]

$$(274)\quad 2C_6H_5\overset{O}{\underset{O}{\overset{\|}{\underset{\|}{S}}}}SeSe\overset{O}{\underset{O}{\overset{\|}{\underset{\|}{S}}}}C_6H_5 \longrightarrow \underset{\mathbf{90}}{Se_3(SO_2C_6H_5)_2} + \underset{\mathbf{91}}{Se(SO_2C_6H_5)_2}$$

Selenium disulfinates can also be made by the reaction of selenium tetrachloride with aromatic sodium sulfinates (equation 275).[499,500] Use of

$$(275)\quad 3C_6H_5SO_2Na + SeCl_4 \rightarrow Se(SO_2C_6H_5)_2 + C_6H_5SO_2Cl + 3NaCl$$

selenium oxychloride also can lead to a selenium disulfinate (equation 276),[500] which can react with a nucleophile, such as thiosulfate, resulting

$$(276)\quad 3RSO_2Na + SeOCl_2 \rightarrow Se(SO_2R)_2 + RSO_3Na + 2NaCl$$

in the displacement of the sulfinate group (equation 277).[500]

$$(277)\quad Se(SO_2R)_2 + 2S_2O_3{}^{2-} \rightarrow Se(S_2O_3)_2{}^{2-} + 2RSO_3{}^-$$

Selenium monochloride reacts with sodium methyl thiosulfonate in dry ether to give selenium di(methylthiosulfonate) (equation 278).[501] The

(278) $\quad 2CH_3SO_2SNa + Se_2Cl_2 \rightarrow Se(S_2O_2CH_3)_2 + Se + 2NaCl$

reaction also takes place with benzene- and p-toluenethiosulfonates.[499,502]

3. Selenosulfates, Thioselenates, and Selenenyl Thiosulfates

a. Selenosulfates. Selenosulfate ion, which is probably more nucleophilic towards carbon than thiosulfate, is a convenient agent by which selenium may be introduced into an organic molecule. It is prepared by adding excess elemental selenium to a hot concentrated solution of potassium sulfite, followed by removal of the unreacted selenium (cf. Chapter III for details).

PREPARATION

Organic selenosulfates, which have, on occasion, been termed seleno Bunte salts,* can be prepared by the reaction of potassium selenosulfate with an alkyl chloride or bromide in water or in aqueous ethanol (equation 279).

(279) $\quad\quad\quad RX + SeSO_3{}^{2-} \rightarrow RSeSO_3{}^- + X^-$

Various benzyl halides,[198,265,266,503] β-bromoethers,[267] simple alkyl bromides,[190] 1,3-dibromopropane,[504] ethyl γ-bromobutyrate,[505] ethyl 6,8-dichlorooctanoate,[264] and 3,3-bis(chloromethyl)oxetane[43] have been treated with selenosulfate ion in this manner, but in most instances the resultant organic selenosulfate was converted *in situ* to the corresponding diselenide. Nucleophilic displacement by $SeSO_3{}^{2-}$ has also been performed on 9-chloroacridines.[506]

Selenosulfates containing an amine function have been prepared by the reaction of aminoethyl halide hydrohalide with selenosulfate ion (equation 280).[40] An alternative preparation of 2-aminoethaneselenosulfuric acid

(280) $\quad R^1R^2R^3\overset{+}{N}CH_2CH_2X\cdot X^- + K_2SeSO_3 \rightarrow$

$$R^1R^2R^3\overset{+}{N}CH_2CH_2SeSO_3{}^- + 2KX$$

was developed in which ethylenimine is cleaved by selenosulfate ion in weakly acidic solution (equation 281).[41]

---

* Bunte salts are organic thiosulfates and are named after their discoverer, Hans Bunte [*Ber.*, **7**, 646 (1874)].

(281)

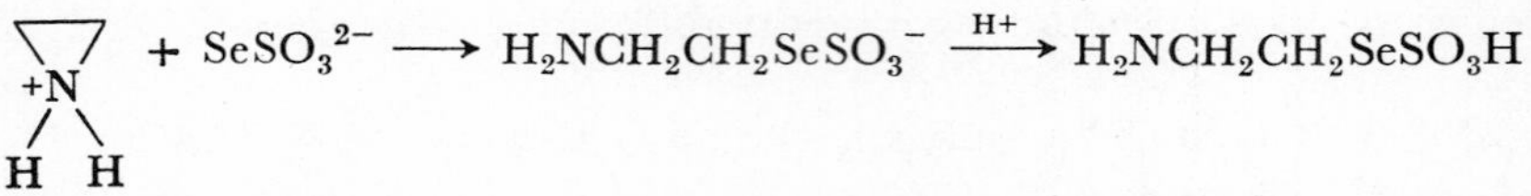

$$\text{(aziridinium)} + SeSO_3{}^{2-} \longrightarrow H_2NCH_2CH_2SeSO_3{}^- \xrightarrow{H+} H_2NCH_2CH_2SeSO_3H$$

Selenosulfate ion has also been reported to ring-open γ-butyrolactone (equation 282) in the first of a series of reactions leading to the synthesis of γ-selenobutyrolactone.[10]

(282)

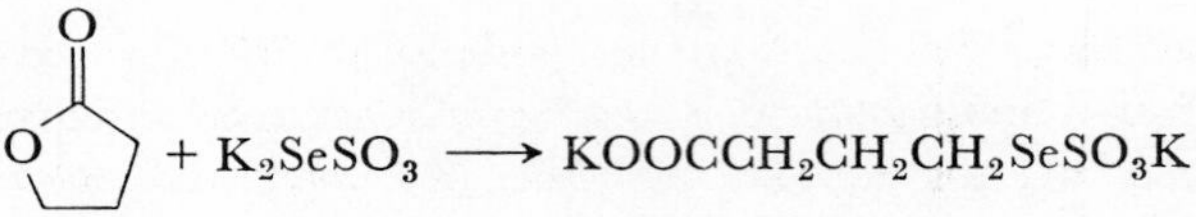

$$\text{(γ-butyrolactone)} + K_2SeSO_3 \longrightarrow KOOCCH_2CH_2CH_2SeSO_3K$$

REACTIONS

The reactions of organic selenosulfates parallel, for the most part, those of organic thiosulfates but reflect the presence of the weaker Se-S bond. Organic selenosulfates are readily converted to their corresponding diselenides by hydrolysis (*cf.* Section IV.B). It is of interest that heating an aqueous solution of 2-aminoethaneselenosulfuric acid under reflux for 62 hours gave a quantitative yield of bis(2-aminoethyl) diselenide hydrosulfate. However, on similar treatment of 2-aminoethanethiosulfuric acid, $>90\%$ of the starting material was recovered unchanged.[41]

Cyanide ion attacks the Se atom of sodium 2-aminoethaneselenosulfate to give 2-aminoethyl selenocyanate; the latter cyclizes rapidly to 2-amino-2-selenazoline (equation 283).[41] This reaction occurs analogously with

(283) $CN^- + SeCH_2CH_2NH_2 \longrightarrow SO_3{}^{2-} + [H_2NCH_2CH_2SeCN]$
                    $|$
                  $SO_3{}^-$

$$\underset{NH_2}{\overset{N\diagdown Se}{\bigtriangleup}}$$

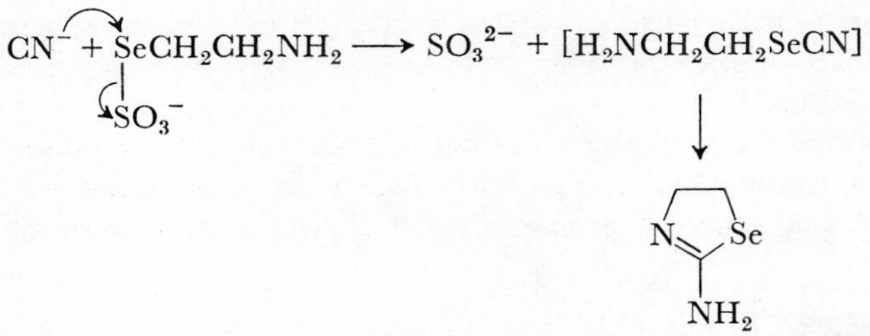

aminoethylthiosulfates.[507] The reaction of sodium 2-aminoethylseleno-sulfate with sodium decylmercaptide gives the selenenyl sulfide, 1-amino-3-selena-4-thiatetradecane (*cf.* equation 256).[41]

MISCELLANEOUS

Inasmuch as 2-aminoethanethiosulfuric acid is a good radiation-protective agent in mammals, its selenium analog, 2-aminoethaneselenosulfuric acid, was tested for the same property but found to be inactive.[508] The polarographic behavior of the two above-mentioned aminoethyl isosteres has been compared.[509]

b. Thioselenates. Thioselenic acid, which is less stable than thiosulfuric acid, has been prepared at $-78°$ by the reaction of hydrogen sulfide with selenium trioxide (equation 284) and by the reaction of chloroselenic acid with

$$(284) \qquad SeO_3 + H_2S \rightarrow HSSeO_3H$$

hydrogen sulfide (equation 285).[510] It has been reported[511] that sodium thioselenate may be synthesized by dissolving elemental sulfur in sodium selenite

$$(285) \qquad HO_3SeCl + H_2S \xrightarrow[-78°]{Et_2O} HSSeO_3H + HCl$$

contained in a 1 N sodium hydroxide solution (equation 286). After refluxing

$$(286) \qquad Na_2SeO_3 + S \xrightarrow{NaOH} Na_2SSeO_3$$

in water for 50–60 minutes, the thioselenate was reported to isomerize to selenosulfate.

PREPARATION

Organic thioselenates have been made by the reaction of selenium trioxide in ether solution with thiols at $-78°$ (equation 287).[466,467,510] Butylthioselenic acid, $C_4H_9SSeO_3H$, was obtained at $-40°$ as an oily monoetherate

$$(287) \qquad RSH + SeO_3 \xrightarrow[-78°]{Et_2O} RSSeO_3H$$

which decomposed at room temperature. Greater stability was shown, however, by benzylthioselenic acid, especially after conversion to its sodium salt.[510] Selenium trioxide also formed a "Lewis-adduct" with diethyl sulfide (equation 288).[510]

$$(288) \qquad (C_2H_5)_2S + SeO_3 \xrightarrow[-78°]{Et_2O} \begin{matrix} C_2H_5 \\ \diagdown \\ S-SeO_3 \\ \diagup \\ C_2H_5 \end{matrix}$$

Butylthioselenic acid, mentioned above, was also prepared in about 50% yield by the action of chloroselenic acid on butylmercaptan (equation 289).[510]

$$(289) \qquad C_4H_9SH + HO_3SeCl \xrightarrow[CF_2Cl_2]{-78°} C_4H_9SSeO_3H + HCl$$

Highly reactive selenium trioxide, as its pyridine and dioxane complexes, has been extensively studied.[466,467,512-514] While the pyridine-sulfur trioxide adduct has been used in the Baumgarten[515] synthesis of Bunte salts (equation 290), the pyridine-selenium trioxide complex does not appear to have been

$$(290) \qquad C_5H_5N \cdot SO_3 + C_6H_5SH \rightarrow C_6H_5SSO_3^- \cdot C_5H_6N^+$$

successfully applied to the formation of thioselenates.

c. Selenenyl Thiosulfates. Foss[370] has reported that *o*-nitrobenzeneselenenyl thiocyanate, di-*O*-alkyl monothiophosphates, and thiosulfonates react with aqueous sodium thiosulfate to give the corresponding selenenyl thiosulfate **92**. This compound, which may also be made readily by the reaction of *o*-nitro-benzeneseleninic acid with thiosulfate,[438] undergoes a displacement reaction with sodium cyclopentamethylene dithiocarbamate, as shown in equation 291.[438]

(291)

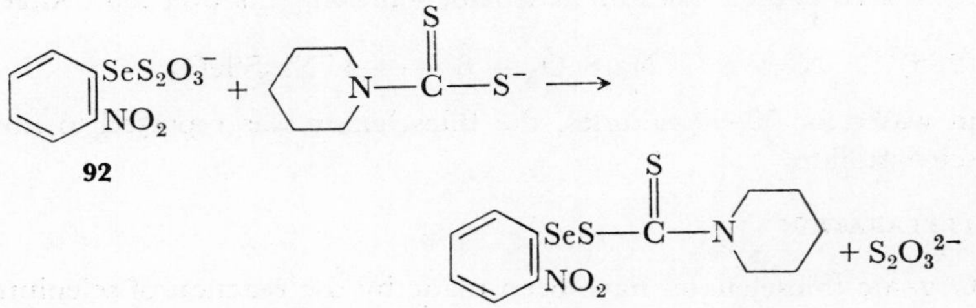

92

4. Selenenyl Thiocyanates and Related Compounds

The first selenenyl thiocyanates, RSeSCN, were reported by Foss in 1947, and shortly thereafter Rheinboldt and his co-workers began a study of this group of compounds, as well as selenenyl selenocyanates, RSeSeCN, and the isomeric sulfenyl selenocyanates, RSSeCN. In most reactions of these compounds, the thiocyanate (or selenocyanate) moiety serves as a readily displaceable leaving group.

a. Selenenyl Thiocyanates

PREPARATION

Foss[370] found that *o*-nitrobenzeneselenenyl bromide reacts with potassium or sodium thiocyanate in ethyl acetate-methanol solution to give *o*-nitro-benzeneselenenyl thiocyanate (equation 292). This method was used also

(292)

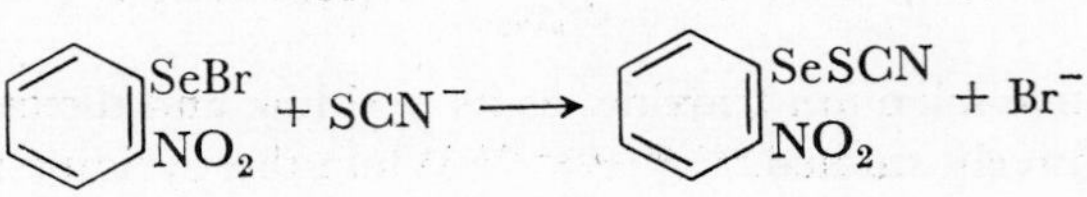

for the preparation of several other aromatic selenenyl thiocyanates,[369,516] including 1-anthraquinoneselenenyl thiocyanate.[517] Lead thiocyanate was used in the preparation of 2-naphthylselenenyl thiocyanate.[335] Trifluoromethylselenenyl thiocyanate, $F_3CSeSCN$, prepared from trifluoromethylselenenyl chloride and silver thiocyanate, is reported to be unstable above $10°$.[368] In contrast, its isomer, trifluoromethylsulfenyl selenocyanate, $F_3CSSeCN$, is stable up to $300°$.[518]

REACTIONS

Thiocyanate is considered to be a pseudohalide; therefore, it is not unexpected that selenenyl thiocyanates (or selenocyanates) behave in a manner similar to selenenyl halides. Actually, selenenyl thiocyanates are somewhat more reactive than selenenyl halides and are good electrophilic agents, as demonstrated by their reactions with various anilines (cf., for example, equation 296). Water at $100°$ converts selenenyl thiocyanates to diselenides[369] but is without effect at room temperature.[250,369] Some of the reactions of aromatic selenenyl thiocyanates are summarized in equations 293–300.

REF.

(293) $3ArSeSCN + 4KOH \longrightarrow$

$(ArSe—)_2 + ArSeO_2K + 3KSCN + 2H_2O$ 369

(294) $ArSeSCN + Br_2 \xrightarrow{CHCl_3} ArSeBr + BrSCN$ 369

(295) $ArSeSCN + KCN \longrightarrow ArSeCN$ 369

(296) $ArSeSCN +$ $\longrightarrow ArSe$ 338, 369, 519

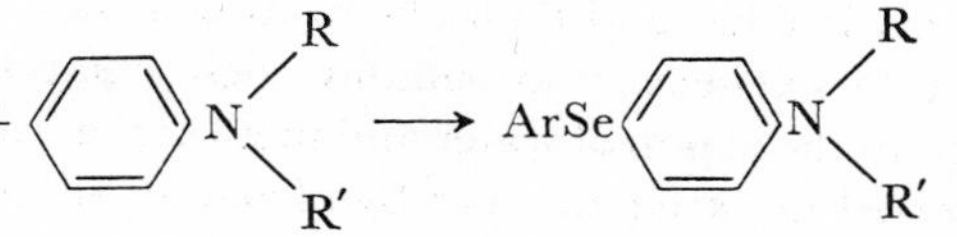

(297) $ArSeSCN + RSH \longrightarrow ArSeSR$ 369

(298) $ArSeSCN + H_2S \xrightarrow{C_6H_6} ArSeSSeAr$ 325

(299)   520

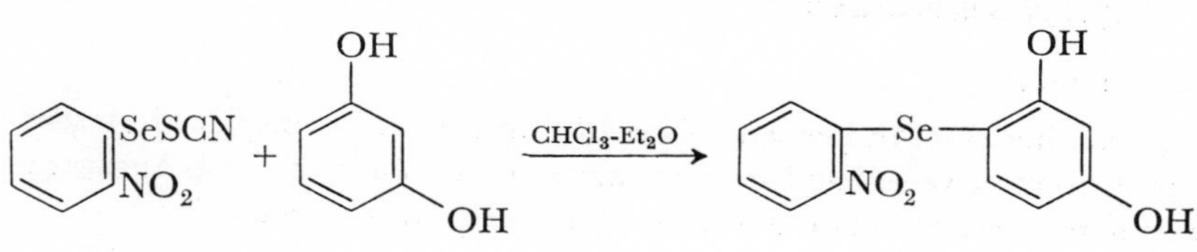

(300)

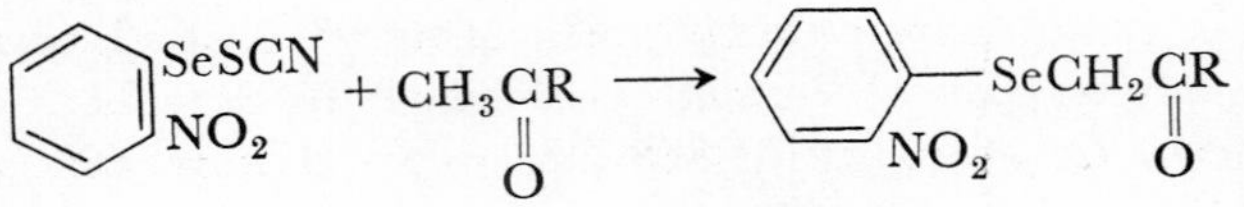

413, 521

The reactions of aromatic selenenyl thiocyanates have been reviewed by Perrier.[522]

b. Selenenyl Selenocyanates

PREPARATION

This class of compounds was first prepared by Rheinboldt and Giesbrecht[250] by the reaction of an aromatic selenenyl bromide with dry potassium selenocyanate in benzene (reaction 301). These workers later obtained improved yields by a modified procedure.[323]

(301) $ArSeBr + KSeCN \rightarrow ArSeSeCN + KBr$

Trifluoromethylselenenyl chloride reacted *in vacuo* with potassium seleno-cyanate at $-23°$ to give trifluoromethylselenenyl selenocyanate, $F_3CSeSeCN$, an orange-yellow liquid which is unstable above $-5°$.[368]

REACTIONS

Aromatic selenenyl selenocyanates melt without decomposition; however, at higher temperatures they extrude selenium to give the corresponding selenocyanates (*cf.* equation 134).[250] Selenenyl selenocyanates, as indicated earlier, may be considered to be unsymmetrical diselenides, and their chemical behavior reflects the presence of this moiety. Like selenenyl thiocyanates, these compounds are compara-tively insensitive to the action of water and methanol at room temperature.[250] They are converted in excellent yield by alcoholic ammonia or bromine to the corresponding diselenides.[250] Aromatic selenenyl selenocyanates react quantitatively with mercaptans to give triselenides according to equation 104.[323] With hydrogen sulfide the reaction takes a different course to give a bisselenenyl sulfide, as shown earlier in equation 268.[325]

c. Sulfenyl Selenocyanates

PREPARATION

The bromine atom in benzenesulfenyl bromides can be displaced by seleno-cyanate ion to give sulfenyl selenocyanates (equation 302).[523] 1-Anthraquin-onesulfenyl selenocyanate has been prepared in this manner.[517]

(302) 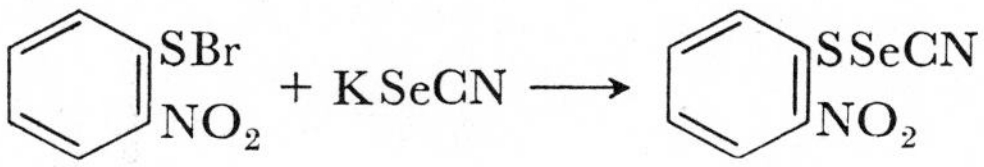   + KSeCN ⟶

The reaction of trichloromethanesulfenyl chloride with potassium seleno-cyanate in benzene proceeds *via* the unstable sulfenyl selenocyanate, $Cl_3CSSeCN$, which decomposes to give bis(trichloromethyl) disulfide.[524] Trifluoromethylsulfenyl selenocyanate (**93**), made according to reaction 303, is reported to be stable up to *ca*. 300°.[518]

(303) $F_3CSCl + AgSeCN \xrightarrow[\text{tube}]{\text{sealed}} F_3CSSeCN + AgCl$

93

REACTIONS

Aromatic sulfenyl selenocyanates do not have definite melting points but undergo thermal decomposition to the corresponding disulfide, cyanogen, and cyanogen triselenide.[523] Whereas sulfenyl thiocyanates are resistant to water,[525] *o*-nitrophenylsulfenyl selenocyanate (**94**) reacts with it in the cold to give the corresponding organic thiocyanate and a small quantity of the disulfide.[523] Treatment with alcoholic ammonia or bromine in chloroform also results in disulfide formation; however, dry ammonia in benzene affords the sulfenamide (equation 304).[523] The sulfenyl selenocyanate (**94**) reacts

(304) 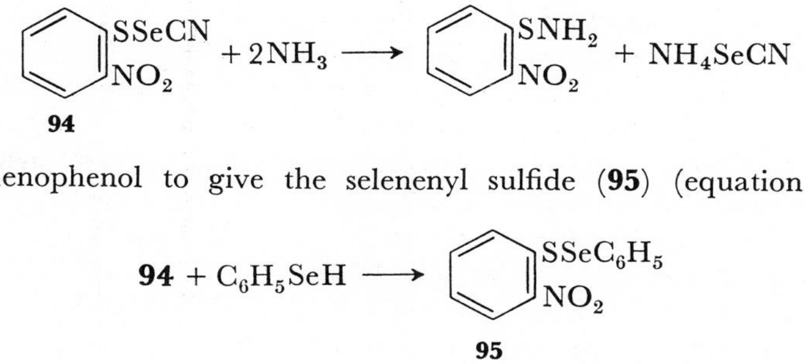   + 2NH₃ ⟶ + NH_4SeCN

94

with selenophenol to give the selenenyl sulfide (**95**) (equation 305).[251]

(305) 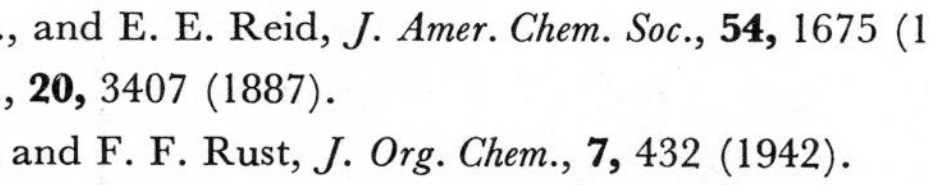   **94** + C_6H_5SeH ⟶

95

REFERENCES

1. L. M. Ellis, Jr., and E. E. Reid, *J. Amer. Chem. Soc.*, **54**, 1675 (1932).

2. P. Klason, *Ber.*, **20**, 3407 (1887).

3. W. E. Vaughn and F. F. Rust, *J. Org. Chem.*, **7**, 432 (1942).

4. J. W. Kimball and E. E. Reid, *J. Amer. Chem. Soc.*, **38**, 2757 (1916).

5. G. E. Coates, *J. Chem. Soc.*, 2839 (1953).

6. J.-P. Mila and J.-F. Labarre, *C. R. Acad. Sci., Paris, Ser. C*, **264**, 1157 (1967).

7. A. Baroni, *Atti Accad. Naz. Lincei Mem., Classe Sci. Fis., Mat. Nat.*, **27**, 238 (1938); *Chem. Abstr.*, **33**, 163 (1939).

8. R. C. Brasted, "Comprehensive Inorganic Chemistry," Vol. 8, D. Van Nostrand, Princeton, N. J., 1961, p. 37.

9. H. Tanaka, H. Sakurai, and A. Yokoyama, *Chem. Pharm. Bull.* (Tokyo), **18**, 1015 (1970).

10. W. H. H. Günther, *J. Org. Chem.*, **31**, 1202 (1966).

11. C. Siemens, *Ann. Chem. Pharm.*, **61**, 360 (1847).

12. C. A. Joy, *Justus Liebigs Ann. Chem.*, **86**, 35 (1853).

13. F. Wöhler and J. Dean, *Ann. Chem. Pharm.*, **97**, 1 (1856).

14. E. H. Shaw, Jr., and E. E. Reid, *J. Amer. Chem. Soc.*, **48**, 520 (1926).

15. A. Baroni, *Atti Accad. Naz. Lincei Mem.*, **12**, 234 (1930); *Chem. Abstr.*, **25**, 2687 (1931).

16. L. Tschugaeff, *Ber.*, **42**, 49 (1909).

17. J. von Braun, W. Teuffert, and K. Weissbach, *Justus Liebigs Ann. Chem.*, **472**, 121 (1929).

18. D. W. Goheen and W. R. Vaughn, *J. Org. Chem.*, **20**, 1019 (1955).

19. D. S. Margolis and R. W. Pittman, *J. Chem. Soc.*, 799 (1957).

20. A. Baroni, *Atti Accad. Naz. Lincei Mem., Classe Sci., Fis., Mat. Nat.*, **26**, 460 (1937).

21. G. Schroeter, C. Seidler, M. Sulzbacher, and R. Kanitz, *Ber.*, **65**, 432 (1932).

22. A. Drory, *Ber.*, **24**, 2563 (1891).

23. W. T. Stewart, Brit. Pat. 629,638 (1949); *Chem. Abstr.*, **44**, 3514h (1950).

24. W. T. Stewart, U.S. Pat. 2,510,765 (1950); *Chem. Abstr.*, **44**, 7864h (1950).

25. H. Wuyts, *Bull. Soc. Chim. Fr.*, [4], **5**, 405 (1909).

26. A. Mailhe and M. Murat, *Bull. Soc. Chim. Fr.*, [4], **7**, 288 (1910).

27. E. P. Painter, *J. Amer. Chem. Soc.*, **69**, 229 (1947).

27a. H. H. Sisler and N. K. Kotia, *J. Org. Chem.*, **36**, 1700 (1971).

28. H. Rheinboldt, F. Mott, and E. Motzkus, *J. Prakt. Chem.*, [2], **134**, 257 (1932).

29. I. Lalezari and N. Sharghi, *Spectrochim. Acta*, **23A**, 1948 (1967).

30. E. E. Reid, "Organic Chemistry of Bivalent Sulfur," Vol. 1, Chemical Publishing Co., New York, N.Y., 1958, p. 32.

31. R. E. Dunbar and E. P. Painter, *J. Amer. Chem. Soc.*, **69**, 1834 (1947).

32. G. Wagner and P. Nuhn, *Z. Chem.*, **3**, 64 (1963).

33. A. Segaloff and R. B. Gabbard, *Steroids*, **5**, 219 (1965); U.S. Pat. 3,372,173.

34. S.-H. Chu and H. G. Mautner, *J. Org. Chem.*, **27**, 2899 (1962).

35. L. V. Pavlova and F. Yu. Rachinskii, *Zh. Obshch. Khim.*, **35**, 493 (1965).

36. D. L. Klayman and P. T. McIntyre, *J. Org. Chem.*, **33**, 884 (1968).

37. M. L. Bird and F. Challenger, *J. Chem. Soc.*, 570 (1942).

38. L. Brandsma and H. E. Wijers, *Rec. Trav. Chim. Pays-Bas*, **82**, 68 (1963).

39. E. E. Aynsley, N. N. Greenwood, and J. B. Leach, *Chem. Ind.* (London), 379 (1966).

39a. G. Zdansky, *Ark. Kemi*, **19**, 559 (1962).

40. W. H. H. Günther and H. G. Mautner, *J. Med. Chem.*, **7**, 229 (1964).

41. D. L. Klayman, *J. Org. Chem.*, **30**, 2454 (1965).

42. J. L. Kurz and J. C. Harris, *J. Org. Chem.*, **35**, 3086 (1970).

43. W. H. H. Günther and M. N. Salzman, *Ann. N.Y. Acad. Sci.*, **192**, 25 (1972)

44. S.-H. Chu, W. H. H. Günther, and H. G. Mautner, *Biochem. Prep.*, **10**, 153 (1963).

45. W. H. H. Günther and H. G. Mautner, *J. Med. Chem.*, **8**, 845 (1965).
46. H. Schmidt and H. D. Block, *Chem. Ber.*, **103**, 3348 (1970).
47. W. H. H. Günther, *J. Org. Chem.*, **32**, 3931 (1967).
48. C. Draguet and M. Renson, *Bull. Soc. Chim. Belges*, **75**, 243 (1966).
49. N. N. Yarovenko, M. A. Raksha, and V. N. Shemanina, *Zh. Obshch. Khim.*, **30**, 4069 (1960).
50. G. Natta, *Giorn. Chim. Ind. Appl.*, **8**, 367 (1926).
51. L. I. Zakharkin and V. V. Gavrilenko, *Izv. Akad. Nauk. SSSR, Otd. Khim. Nauk*, 1391 (1960).
52. S. Landa, O. Weisser, and J. Mostecký, *Collection Czech. Chem. Commun.*, **24**, 2197 (1959).
53. M. Scheithauer and R. Mayer, *Z. Chem.*, **6**, 375 (1966).
54. J. W. Dale, H. J. Eméléus, and R. N. Haszeldine, *J. Chem. Soc.*, 2939 (1958).
55. H. G. Mautner, S.-H. Chu, and W. H. H. Günther, *J. Amer. Chem. Soc.*, **85**, 3458 (1963).
56. M. Renson and C. Draguet, *Bull. Soc. Chim. Belges*, **71**, 260 (1962).
57. S.-H. Chu and H. G. Mautner, *J. Med. Chem.*, **11**, 446 (1968).
58. L. A. Khazemova and V. M. Al'bitskaya, *Zh. Org. Khim.*, **5**, 1926 (1969).
59. L. A. Khazemova and V. M. Al'bitskaya, *Zh. Org. Khim.*, **6**, 935 (1970).
60. W. H. H. Günther and H. G. Mautner, *J. Amer. Chem. Soc.*, **87**, 2708 (1965).
61. H. Plieninger, *Chem. Ber.*, **83**, 265 (1950).
62. L.-B. Agenäs and B. Persson, *Acta Chem. Scand.*, **21**, 837 (1967).
63. L.-B. Agenäs, *Ark. Kemi*, **24**, 415 (1965).
64. L.-B. Agenäs, *Ark. Kemi*, **24**, 573 (1965).
65. L.-B. Agenäs and B. Persson, *Acta Chem. Scand.*, **21**, 835 (1967).
66. W. Cluass, Ger. Offen. 1,933,712 (1971); *Chem. Abstr.*, **74**, 63914 (1971).
67. G. Zdansky, *Ark. Kemi*, **21**, 211 (1963).
68. G. Bergson and A. Biezais, *Ark. Kemi*, **18**, 143 (1961).
69. L.-B. Agenäs, *Ark. Kemi*, **23**, 463 (1965).
70. J. Gosselck and E. Wolters, *Z. Naturforsch., Part b*, **17**, 131 (1962).
71. G. Shtacher and T. Sadeh, *Proc. Israel J. Chem.*, 10p (1969).
72. L. R. Williams and A. Ravve, *J. Amer. Chem. Soc.*, **70**, 1244 (1948).
73. L. R. Williams and A. Ravve, *J. Amer. Chem. Soc.*, **70**, 3524 (1948).
74. A. Yokoyama, H. Sakurai, and H. Tanaka, *Chem. Pharm. Bull.* (Tokyo), **18**, 1021 (1970).
74a. A. Yokoyama, H. Sakurai, and H. Tanaka, *Chem. Pharm. Bull.* (Tokyo), **19**, 1089 (1971).
74b. H. Sakurai, A. Yokoyama, and H. Tanaka, *Chem. Pharm. Bull.* (Tokyo), **19**, 1270 (1971).
75. I. A. Vostokov and V. T. Bychkov, *Zh. Obshch. Khim.*, **40**, 319 (1970).
76. J. F. Nelson, *Iowa State Coll. J. Sci.*, **12**, 145 (1937).
77. G. E. Coates and R. G. Hayter, *J. Chem. Soc.*, 2519 (1953).
78. H. Köpf, B. Block, and M. Schmidt, *Z. Naturforsch., Part b*, **22**, 1077 (1967).
79. N. S. Vyazankin, I. A. Vostokov, and V. T. Bychkov, *Zh. Obshch. Khim.*, **38**, 2485 (1968).

80. M. Schmidt and H. D. Block, *Z. Anorg. Allg. Chem.*, **377**, 305 (1970).

81. M. Schmidt and H. D. Block, *J. Organometal. Chem.*, **25**, 17 (1970).

82. H. H. Anderson, *J. Chem. Eng. Data*, **9**, 272 (1964).

83. W. T. Stewart and J. O. Clayton, U.S. Pat. 2,543,734; *Chem. Abstr.*, **45**, 4033 (1951).

84. B. Bennett, U.S. Pat. 2,787,609; *Chem. Abstr.*, **51**, 10112c (1957).

85. F. Krafft and R. E. Lyons, *Ber.*, **27**, 1761 (1894).

86. H. W. Doughty, *Amer. Chem. J.*, **41**, 326 (1909).

87. F. Taboury, *Bull. Soc. Chim. Fr.*, [3], **35**, 668 (1906).

88. M. J. Copley, C. S. Marvel, and E. Ginsberg, *J. Amer. Chem. Soc.*, **61**, 3161 (1939).

89. G. Guanti, C. Dell'Erba, and D. Spinelli, *Gazz. Chim. Ital.*, **100**, 184 (1970).

90. A. J. Parker, *Acta Chem. Scand.*, **16**, 855 (1962).

91. F. Taboury, *Bull. Soc. Chim. Fr.*, [3], **29**, 761 (1903).

92. H. Wuyts and G. Cosyns, *Bull. Soc. Chim. Fr.*, [3], **29**, 689 (1903).

93. F. Taboury, *Bull. Soc. Chim. Fr.*, [3], **31**, 646 (1904).

94. F. Taboury, *Bull. Soc. Chim. Fr.*, [3], **33**, 836 (1905).

95. F. Taboury, *C. R. Acad. Sci., Paris*, **138**, 982 (1904).

96. F. Taboury, *Bull. Soc. Chim. Fr.*, [3], **35**, 457 (1906).

97. F. Taboury, *Ann. Chim. Phys.*, [8], **15**, 5 (1908).

98. D. G. Foster and S. F. Brown, *J. Amer. Chem. Soc.*, **50**, 1182 (1928).

99. D. G. Foster, *J. Amer. Chem. Soc.*, **55**, 822 (1933).

100. D. G. Foster, *J. Amer. Chem. Soc.*, **61**, 2972 (1939).

101. D. G. Foster, "Organic Syntheses," Coll. Vol. 3, John Wiley, New York, N.Y., 1955, p. 771.

102. G. T. Morgan and W. H. Porritt, *J. Chem. Soc.*, **127**, 1755 (1925).

103. W. H. Porritt, *J. Chem. Soc.*, 27 (1927).

104. B. A. Arbuzov, L. M. Kataeva, E. G. Kataev, and A. V. Il'yasov, *Izv. Akad. Nauk SSSR, Otd. Khim. Nauk*, 360 (1962).

105. S. Keimatsu, K. Yokota, and I. Satoda, *J. Pharm. Soc. Japan*, **53**, 994 (1933).

106. T. Naito, S. Ina, S. Inoue, H. Hanai, and S. Katajiri, *Nagoya Shiritsu Daigaku Yakugakubu Kiyô*, **5**, 43 (1957); *Chem. Abstr.*, **52**, 6248e (1958).

107. N. Sharghi and I. Lalezari, *Spectrochim. Acta*, **20**, 237 (1964).

108. B. Sjöberg and S. Herdevall, *Acta Chem. Scand.*, **12**, 1347 (1958).

109. E. S. Gould and J. D. McCullough, *J. Amer. Chem. Soc.*, **73**, 1109 (1951).

110. E. Niwa, H. Aoki, H. Tanaka, K. Munakata, and M. Namiki, *Chem. Ber.*, **99**, 3215 (1966).

111. L. Christiaens and M. Renson, *Bull. Soc. Chim. Belges*, **79**, 133 (1970).

112. F. Dallacker and H. Zegers, *Justus Liebigs Ann. Chem.*, **689**, 156 (1965).

113. S. Keimatsu and I. Satoda, *J. Pharm. Soc. Japan*, **56**, 600 (1936); *Chem. Abstr.*, **33**, 154[8] (1939).

114. D. G. Foster, *J. Amer. Chem. Soc.*, **63**, 1361 (1941).

115. M. T. Bogert and C. N. Andersen, *Proc. Natl. Acad. Sci.*, **11**, 217 (1925).

116. M. T. Bogert and A. Stull, *J. Amer. Chem. Soc.*, **49**, 2011 (1927).

117. G. B. L. Smith, J. P. Miale, and C. W. Mason, *J. Amer. Chem. Soc.*, **55**, 3759 (1933).

118. E. Ochiai, J. Haginiwa, and K. Komatsu, *J. Pharm. Soc. Japan.*, **70**, 372 (1950).

119. R. Lesser and R. Weiss, *Ber.*, **45**, 1835 (1912).

120. J. W. Baker, G. F. C. Barrett, and W. T. Tweed, *J. Chem. Soc.*, 2831 (1952).

121. S. Keimatsu, K. Yokota, and I. Satoda, *J. Pharm. Soc. Japan.*, **52**, 531 (1932).

122. S. Keimatsu and I. Satoda, *J. Pharm. Soc. Japan.*, **55**, 58 (1935).

123. J. Develotte, *Ann. Chim.* (Paris), [12], **5**, 215 (1950).

124. L. Chierici, *Farmaco, Ed. Sci.*, **7**, 618 (1952).

125. E. Mueller, H. B. Stegmann, and K. Scheffler, *Justus Liebigs Ann. Chem.*, **657**, 5 (1962).

126. G. Wagner and P. Nuhn, *Arch. Pharm.* (Weinheim), **296**, 374 (1963).

127. H. Bauer, *Ber.*, **46**, 92 (1913).

128. Ger. Pat. 264,940; *Chem. Abstr.*, **8**, 1351 (1914).

129. S. Keimatsu, I. Satoda, and T. Kobayashi, *J. Pharm. Soc. Japan.*, **57**, 190 (1937).

130. S. Umezawa, *Bull. Chem. Soc. Japan.*, **14**, 363 (1939).

131. P. L. N. Rao, *J. Ind. Chem. Soc.*, **18**, 1 (1941).

132. J. Loevenich, H. Fremdling, and M. Föhr, *Ber.*, **62**, 2856 (1929).

133. S. Keimatsu and K. Yokota, *J. Pharm. Soc. Japan.*, **50**, 531 (1930).

134. S. Keimatsu and I. Satoda, *J. Pharm. Soc. Japan.*, **55**, 233 (1935).

135. J. Matti, *Bull. Soc. Chim. Fr.*, [5], **7**, 617 (1940); *Chem. Abstr.*, **36**, 2850 (1942).

136. E. Sekido, Q. Fernando, and H. Freiser, *Anal. Chem.*, **36**, 1768 (1964).

137. A. I. Wuertz, D. P. Graham, and M. A. Perkins, U.S. Pat. 1,991,646; *Chem. Abstr.*, **29**, 2177 (1935).

138. H. G. Mautner, S.-H. Chu, and C. M. Lee, *J. Org. Chem.*, **27**, 3671 (1962).

139. G. Wagner and G. Valz, *Pharmazie*, **22**, 548 (1967).

140. C. Chabrié, *C. R. Acad. Sci., Paris*, **109**, 182 (1889).

141. C. Chabrié, *Ann. Chim. Phys.*, [4], **20**, 202 (1890).

142. F. Krafft and A. Kaschau, *Ber.*, **29**, 428 (1896).

143. W. E. Bradt and J. F. Green, *J. Org. Chem.*, **1**, 540 (1937).

144. S. Keimatsu, K. Yokota, and S. Suzuki, *J. Pharm. Soc. Japan.*, **52**, 961 (1932).

145. S. Keimatsu, I. Satoda, and T. Chigono, *J. Pharm. Soc. Japan.*, **56**, 139 (1936).

146. N. M. Cullinane, A. G. Rees, and C. A. J. Plummer, *J. Chem. Soc.*, 151 (1939).

147. H. Gilman and L. F. Cason, *J. Amer. Chem. Soc.*, **73**, 1074 (1951).

148. J. M. Danze and M. Renson, *Bull. Soc. Chim. Belges*, **75**, 169 (1966).

149. E. G. Kataev and F. G. Gabdrakhmanov, *Zh. Obshch. Khim.*, **37**, 772 (1967).

150. F. G. Gabdrakhmanov, Yu. Yu. Samitov, and E. G. Kataev, *Zh. Obshch. Khim.*, **37**, 761 (1967).

151. F. G. Gabdrakhmanov, *Sb. Aspir. Rabot, Kazan Gos. Univ., Khim., Geol.*, 85 (1967); *Chem. Abstr.*, **70**, 96328 (1969).

152. E. G. Kataev and V. N. Petrov, *Zh. Obshch. Khim.*, **32**, 3699 (1962).

153. L. Chierici and F. Montanari, *Boll. Sci. Fac. Chim. Ind. Bologna*, **14**, 78 (1956); *Chem. Abstr.*, **51**, 5721i (1957).

154. L. Chierici and F. Montanari, *Gazz Chim. Ital.*, **86**, 1269 (1956).

155. L. M. Kataeva, E. G. Kataev, Z. S. Titova, and N. A. Aleksandrova, *Zh. Strukt. Khim.*, **7**, 715 (1966); *Chem. Abstr.*, **66**, 115,196 (1967).

156. I. N. Azerbaev, A. B. Azmanova, L. A. Tsoi, and V. S. Bazalitskaya, *Izv. Akad. Nauk Kazan SSR, Ser. Khim.*, **20,** 53 (1970); *Chem. Abstr.*, **73,** 76788 (1970).

157. I. N. Azerbaev, A. V. Azmanova, L. A. Tsoi, and M. K. Dzhamaletdinova, *Izv. Akad. Nauk Kazan SSR, Ser. Khim.*, **20,** 75 (1970); *Chem. Abstr.*, **74,** 12766 (1971).

158. G. I. Zaitseva and V. M. Al'bitskaya, *Zh. Org. Khim.*, **4,** 2057 (1968).

159. A. Ruwet, D. Janne, and M. Renson, *Bull. Soc. Chim. Belges*, **79,** 87 (1970).

160. K. Šindelař, J. Metyšova, E. Svátek, and M. Protiva, *Collection Czech. Chem. Commun.*, **34,** 2122 (1969).

161. F. Bossert, *Angew. Chem. (Intern. Ed.)*, **4,** 879 (1965).

162. A. Ruwet and M. Renson, *Bull. Soc. Chim. Belges*, **75,** 260 (1966).

163. C. Hansch and C. F. Geiger, *J. Org. Chem.*, **24,** 1025 (1959).

164. L. M. Clark, *J. Chem. Soc.*, 2805 (1927).

165. C. Courtot and J. Develotte, *C. R. Acad. Sci., Paris*, **221,** 101 (1945).

166. C. Courtot and J. Develotte, *C. R. Acad. Sci., Paris*, **223,** 64 (1946).

167. L. K. Mushkalo and D. I. Sheiko, *Ukr. Khim. Zh.*, **30,** 384 (1964).

168. L. K. Mushkalo, D. I. Sheiko, and E. I. Lanovaya, *Ukr. Khim Zh.*, **30,** 502 (1964).

169. L. K. Mushkalo and D. I. Sheiko, *Zh. Obschch. Khim.*, **33,** 157 (1963).

170. V. A. Chuyguk and L. S. Borodulya, *Ukr. Khim. Zh.*, **35,** 1178 (1969).

171. M. O. Lozinskii and P. S. Pel'kis, *Zh. Org. Khim.*, **1,** 1793 (1965).

172. I. E. Pollack and G. F. Grillot, *J. Org. Chem.*, **31,** 3514 (1966).

173. I. E. Pollack and G. F. Grillot, *J. Org. Chem.*, **32,** 3101 (1967).

174. I. E. Pollack, *Dissertation Abstr. B*, **27,** 4324 (1967).

175. H.-D. Jakubke, *Z. Chem.*, **3,** 65 (1963).

176. H.-D. Jakubke, *Chem. Ber.*, **97,** 2816 (1964).

177. H.-D. Jakubke, *Justus Liebigs Ann. Chem.*, **682,** 244 (1965).

178. T. Wieland, *Angew. Chem.*, **66,** 507 (1954).

179. D. Spinelli and C. Dell'Erba, *Ann. Chim.* (Rome), **51,** 45 (1961).

180. H. J. Backer and J. B. G. Hurenkamp, *Rec. Trav. Chim. Pays-Bas*, **61,** 802 (1942).

181. M. Shamma, N. C. Deno, and J. F. Remar, *Tetrahedron Lett.*, 1375 (1966).

182. V. Šimánek and A. Klásek, *Tetrahedron Lett.*, 3039 (1969).

183. W. A. Herbst, U.S. Pat. 2,422,484 (1947); *Chem. Abstr.*, **42,** 217h (1948).

184. I. I. Lapkin and N. N. Pavlova, *Zh. Org. Khim.*, **4,** 803 (1968).

185. I. I. Lapkin and N. N. Pavlova, USSR Pat. 239,947; *Chem. Abstr.*, **71,** 49575 (1969).

186. I. I. Lapkin, N. N. Pavlova, and G. S. Pavlov, *Zh. Org. Khim.*, **6,** 71 (1970).

187. I. I. Lapkin, N. N. Pavlova, and V. I. Proshutinskii, *Zh. Org. Khim*, **7,** 473 (1971).

188. N. Petragnani and G. Schill, *Chem. Ber.*, **103,** 2271 (1970).

189. A. Schönberg, S. Nickel, and D. Cernick, *Ber.*, **65,** 289 (1932).

190. G. G. Stoner and R. W. Williams, *J. Amer. Chem. Soc.*, **70,** 1113 (1948).

191. H. Rheinboldt and E. Giesbrecht, *Chem. Ber.*, **85,** 357 (1952).

192. T. W. Campbell, H. G. Walker, and G. M. Coppinger, *Chem. Rev.*, **50,** 279 (1952).

193. J. E. Kuder and M. A. Lardon, *Ann. N.Y. Acad. Sci.*, **192,** 147 (1972).

194. C. S. Evans, C. J. Asher, and C. M. Johnson, *Australian J. Biol. Sci.*, **21,** 13 (1968).

195. S. C. Abrahams, *Quart. Rev.*, **10,** 407 (1956).

196. W. H. Green and A. B. Harvey, *J. Chem. Phys.*, **49,** 3586 (1968).

197. G. Speroni and G. Mannelli, *Gazz. Chim. Ital.*, **64,** 506 (1934).

198. G. Speroni and G. Mannelli, *Gazz. Chim. Ital.*, **70,** 246 (1940); *Chem. Abstr.*, **35,** 2869 (1941).

199. R. Poggi and G. Speroni, *Gazz. Chim. Ital.*, **64,** 501 (1934).

200. G. Speroni and B. Simi, *Atti X° Congr. Intern. Chim.*, **3,** 358 (1939).

201. Yu. N. Shlyk, G. M. Bogolyubov, and A. A. Petrov, *Zh. Obshch. Khim.*, **38,** 1199 (1968).

202. G. M. Bogolyubov, Yu. N. Shlyk, and A. A. Petrov, *Zh. Obshch. Khim.*, **39,** 1804 (1969).

203. L.-B. Agenäs, *Acta Chem. Scand.*, **16,** 1809 (1962).

204. G. Bergson, *Ark. Kemi*, **19,** 195 (1962).

205. A. Fredga, *Acta Chem. Scand.*, **17,** S51 (1963).

206. L.-B. Agenäs, *Ark. Kemi*, **23,** 145 (1964).

207. E. Rebane, *Ark. Kemi*, **25,** 363 (1966).

208. W. H. H. Günther, *J. Org. Chem.*, **32,** 3929 (1967).

208a. D. L. Klayman and T. S. Griffin, *J. Amer. Chem. Soc.*, in press (1973).

209. C. L. Jackson, *Ber.*, **7,** 1277 (1874).

210. C. L. Jackson, *Justus Liebigs Ann. Chem.*, **179,** 1 (1875).

211. H. J. Backer and W. van Dam, *Rec. Trav. Chim. Pays-Bas*, **48,** 1287 (1929).

212. H. J. Backer and W. van Dam, *Rec. Trav. Chim. Pays-Bas*, **49,** 482 (1930).

213. A. Fredga and G. Bendz, *Svensk. Kem. Tidskr.*, **54,** 119 (1942); *Chem. Abstr.*, **38,** 2317[9] (1944).

213a. A. Fredga, *Acta Chem. Scand.*, **52,** 1896 (1971).

214. H. J. Backer and H. J. Winter, *Rec. Trav. Chim. Pays-Bas*, **56,** 691 (1937).

215. F. Wrede, *Ber.*, **52,** 1756 (1919).

216. H. J. Backer and W. van Dam, *Rec. Trav. Chim. Pays-Bas*, **49,** 479 (1930).

217. J. W. Baker and W. G. Moffitt, *J. Chem. Soc.*, 1722 (1930).

218. F. Challenger, A. T. Peters, and J. Halevy, *J. Chem. Soc.*, 1648 (1926).

219. H. Rheinboldt, "Houben-Weyl Methoden der Organischen Chemie," Vol. IX, E. Müller, Ed., Georg Thieme Verlag, Stuttgart, 1955, p. 949.

220. O. Behaghel and M. Rollmann, *J. Prakt. Chem.*, [2], **123,** 336 (1929).

221. O. Behaghel and M. Rollmann, *Ber.*, **62,** 2696 (1929).

222. S. Keimatsu and I. Satoda, *J. Pharm. Soc. Japan.*, **56,** 703 (1936); *Chem. Abstr.*, **31,** 2589 (1937).

223. V. A. Nefedov, *Zh. Obshch. Khim.*, **38,** 2191 (1968).

224. L.-B. Agenäs, Dissertation, University of Uppsala, 1969.

225. J. Kenyon, H. Phillips, and V. P. Pittman, *J. Chem. Soc.*, 1072 (1935).

226. S. Morimoto, *J. Chem. Soc. Japan.*, **75,** 557 (1954); *Chem. Abstr.*, **51,** 11234c (1957).

227. V. Coblentz, *Ber.*, **24,** 2131 (1891).

228. L. Pichat, M. Herbert, and M. Thiers, *Tetrahedron*, **12,** 1 (1961).

229. W. H. H. Günther and H. G. Mautner, *J. Amer. Chem. Soc.*, **82,** 2762 (1960).

230. L. Hagelberg, *Ber.*, **23,** 1083 (1890).

231. G. T. Morgan and F. H. Burstall, *J. Chem. Soc.*, 1096 (1929).

232. G. T. Morgan and F. H. Burstall, *J. Chem. Soc.*, 2197 (1929).

233. G. T. Morgan and F. H. Burstall, *J. Chem. Soc.*, 173 (1931).

234. R. Paetzold and D. Lienig, *Z. Anorg. Allg. Chem.*, **335,** 289 (1965).

235. H. Frerichs, *Arch. Pharm.*, **241,** 177 (1903).

236. E. Fromm and K. Martin, *Justus Liebigs Ann. Chem.*, **401,** 177 (1913).

237. W. R. Gaythwaite, J. Kenyon, and H. Phillips, *J. Chem. Soc.*, 2280 (1928).

238. W. Jenny, *Helv. Chim. Acta*, **35,** 1429 (1952).

239. O. Behaghel and K. Hofmann, *Ber.*, **72,** 582 (1939).

240. G. E. Wiseman and E. S. Gould, *J. Amer. Chem. Soc.*, **77,** 1061 (1955).

241. A. Cerniani and R. Passerini, *Boll. Sci. Fac. Chim. Ind. Bologna*, **14,** 107 (1956); *Chem. Abstr.*, **53,** 286i (1959).

242. L.-B. Agenäs, *Acta Chem. Scand.*, **17,** 268 (1963).

243. L.-B. Agenäs, *Ark. Kemi*, **23,** 155 (1964).

244. L.-B. Agenäs, *Ark. Kemi*, **31,** 159 (1969).

244a. A. Biezais-Zirnis and A. Fredga, *Acta Chem. Scand.*, **25,** 1171 (1971).

245. M. Renson and R. Collienne, *Bull. Soc. Chim. Belges*, **73,** 491 (1964).

246. A. Fredga, *J. Prakt. Chem.*, [2], **123,** 129 (1929).

247. A. Fredga, *Svensk. Kem. Tidskr.*, **42,** 66 (1930).

248. G. Bergson and A. Biezais, *Ark. Kemi*, **22,** 475 (1964).

249. K. W. Rosenmund and H. Harms, *Ber.*, **53,** 2226 (1920).

250. H. Rheinboldt and E. Giesbrecht, *J. Amer. Chem. Soc.*, **72,** 866 (1950).

251. H. Rheinboldt and E. Giesbrecht, *Justus Liebigs Ann. Chem.*, **568,** 198 (1950).

252. M. Nakasaki, *J. Inst. Polytech. Osaka City Univ., Ser. C*, **4,** 100 (1953).

253. F. von Konek and O. Schleifer, *Ber.*, **51,** 844 (1918).

254. G. R. Levi and A. Baroni, *Atti Accad. Naz. Lincei Mem.*, [6], **9,** 1019 (1929).

255. G. Bergson, *Ark. Kemi*, **10,** 127 (1956).

256. H. A. Silverwood and M. Orchin, *J. Org. Chem.*, **27,** 3401 (1962).

257. A. Schoeller, *Ber.*, **52,** 1517 (1919).

258. A. Fredga and C. Evertsdotter, *Acta Chem. Scand.*, **13,** 1042 (1959).

259. W. Strecker and A. Willing, *Ber.*, **48,** 196 (1915).

260. A. Pieroni and G. Balduzzi, *Gazz. Chim. Ital.*, II, **45,** 106 (1915); *Chem. Abstr.*, **10,** 1514 (1916).

261. J. Loevenich and K. Sipmann, *J. Prakt. Chem.*, [2], **124,** 127 (1930).

262. S. C. Cohen, M. L. N. Reddy, and A. G. Massey, *J. Organometal. Chem.*, **11,** 563 (1968).

263. D. L. Klayman, D. Kenny, R. B. Silverman, J. E. Tomaszewski, and R. J. Shine, *J. Org. Chem.*, **36,** 3681 (1971).

264. Y. Deguchi, *J. Pharm. Soc. Japan.*, **80,** 931 (1960).

265. T. S. Price and L. M. Jones, *J. Chem. Soc.*, **95,** 1729 (1909).

266. D. F. Twiss, *J. Chem. Soc.*, **105,** 36 (1914).

267. H. P. Ward and I. L. O'Donnell, *J. Amer. Chem. Soc.*, **67,** 883 (1945).

268. R. E. Lyons and W. E. Bradt, *Ber.*, **60,** 60 (1927).

269. K. Florey and A. R. Restivo, *J. Org. Chem.*, **22**, 406 (1957).

270. K. G. Florey, U.S. Pats. 2,875,196 and 2,917,507.

271. J. S. Baran, *J. Amer. Chem. Soc.*, **80**, 1687 (1958).

272. P. Rona, *J. Chem. Soc.*, 3629 (1962).

273. J. F. K. Wilshire, *Australian J. Chem.*, **20**, 359 (1967).

274. H. J. Emeléus and M. J. Dunn, *J. Inorg. Nucl. Chem.*, **27**, 752 (1965).

275. F. L. Pyman, *J. Chem. Soc.*, **115**, 169 (1919).

276. H. Rheinboldt and E. Giesbrecht, *Chem. Ber.*, **88**, 666 (1955).

277. H. Rheinboldt and E. Giesbrecht, *Chem. Ber.*, **88**, 1974 (1955).

278. H. Rheinboldt and E. Giesbrecht, *Chem. Ber.*, **88**, 1037 (1955).

279. V. V. Kozlov and V. M. Pronyakova, *Zh. Org. Khim.*, **1**, 493 (1965).

280. V. V. Kozlov and V. M. Pronyakova, *Zh. Vses. Khim. Obshchestva*, **12**, 471 (1967); *Chem. Abstr.*, **68**, 59,344 (1968).

281. C. A. Buehler, J. O. Harris, and W. F. Arendale, *J. Amer. Chem. Soc.*, **72**, 4953 (1950).

282. D. L. Klayman and J. W. Lown, *J. Org. Chem.*, **31**, 3396 (1966).

283. O. Behaghel and H. Seibert, *Ber.*, **65**, 812 (1932).

284. J. Gosselck and E. Wolters, *Chem. Ber.*, **95**, 1237 (1962).

285. N. N. Yarovenko, G. B. Gazieva, V. N. Shemanina, and N. A. Fedorova, *Zh. Obshch. Khim.*, **29**, 940 (1959).

286. H. J. Emeléus and N. Welcman, *J. Chem. Soc.*, 1268 (1963).

287. N. N. Yarovenko, V. N. Shemanina, and G. B. Gazieva, *Zh. Obshch. Khim.*, **29**, 942 (1959).

288. N. N. Yarovenko and M. A. Raksha, *Zh. Obshch. Khim.*, **30**, 4064 (1960).

289. R. G. R. Bacon and S. G. Pande, *J. Chem. Soc.*, *C*, 1967 (1970).

290. N. Welcman and H. Regev, *J. Chem. Soc.*, 7511 (1965).

291. W. McFarlane, *J. Chem. Soc.*, *A*, 670 (1969).

292. G. Speroni and G. Mannelli, *Gazz. Chim. Ital.*, **70**, 472 (1940).

293. R. Walter, D. H. Schlesinger, and I. L. Schwartz, *Anal. Biochem.*, **27**, 231 (1969).

294. R. C. Dickson and A. L. Tappel, *Arch. Biochem. Biophys.*, **130**, 547 (1969).

295. D. T. Woodbridge, *J. Chem. Soc.*, *Phys. Org.*, 50 (1966).

296. G. Ayrey, D. Barnard, and D. T. Woodbridge, *J. Chem. Soc.*, 2089 (1962).

297. K. A. Caldwell and A. L. Tappel, *Arch. Biochem. Biophys.*, **127**, 259 (1968).

298. K. A. Caldwell and A. L. Tappel, *Biochemistry*, **3**, 1643 (1964).

299. G. Bergson, *Acta Chem. Scand.*, **15**, 1611 (1961).

300. G. Frerichs and E. Wildt, *Justus Liebigs Ann. Chem.*, **360**, 105 (1908).

301. A. Fredga, *Ark. Kemi*, **11B**, [46], 1 (1934).

302. E. P. Painter, K. W. Franke, and R. A. Gortner, *J. Org. Chem.*, **5**, 579 (1940).

303. H. J. Bridger and R. W. Pittman, *J. Chem. Soc.*, 1371 (1950).

304. A. Schönberg, A. Stephenson, H. Kaltschmidt, E. Petersen, and H. Schulten, *Ber.*, **66**, 237 (1933).

305. N. Petragnani and V. G. Toscano, *Chem. Ber.*, **103**, 1652 (1970).

306. T. W. Campbell and J. D. McCullough, *J. Amer. Chem. Soc.*, **67**, 1965 (1945).

307. M. Pallotti and A. Tundo, *Boll. Sci. Fac. Chim. Ind. Bologna*, **17,** 71 (1959).

308. E. Kostiner, M. L. N. Reddy, D. S. Urch, and A. G. Massey, *J. Organometal. Chem.*, **15,** 383 (1968).

309. S. Morimoto, *J. Chem. Soc. Japan.*, **71,** 166 (1950); *Chem. Abstr.*, **45,** 6594 (1951).

310. H. Hauptmann and W. F. Walter, *J. Amer. Chem. Soc.*, **77,** 4929 (1955).

311. H. Hauptmann, *Selecta Chim.*, **16,** 41 (1957).

312. G. E. Wiseman and E. S. Gould, *J. Amer. Chem. Soc.*, **76,** 1706 (1954).

313. A. Fredga, *Ark. Kemi*, **11B,** [44], 1 (1934).

314. W. E. Bradt and J. F. Green, *J. Org. Chem.*, **1,** 540 (1937).

315. E. S. Gould and A. Amendola, *J. Amer. Chem. Soc.*, **77,** 2103 (1955).

315a. Y. Okamoto and T. Yano, *J. Organometal. Chem.*, **29,** 99 (1971).

316. N. Welcman and I. Rot, *J. Chem. Soc.*, 7515 (1965).

317. N. Welcman and P. Rozenbuch, *Israel J. Chem., Proc.*, **5,** 10p (1967).

318. E. W. Tillay, E. D. Schermer, and W. H. Bradley, *Inorg. Chem.*, **7,** 1925 (1968).

319. G. H. Denison, Jr., and P. C. Condit, U.S. Pat. 2,528,346; *Chem. Abstr.*, **45,** 1339 (1951).

320. R. L. Dueltgen, M. N. Lugasch, and S. L. Cosgrove, *Sci. Lubrication*, **13,** 21 (1961).

321. R. J. Pollet and J. F. Willems, Ger. Offen. 2,010,555; *Chem. Abstr.*, **74,** 18009 (1971).

322. B. Holmberg, *Justus Liebigs Ann. Chem.*, **359,** 81 (1908).

323. H. Rheinboldt and E. Giesbrecht, *Chem. Ber.*, **88,** 1 (1955).

324. H. A. Silverwood, *Dissertation Abstr.*, **23,** 1519 (1962).

325. H. Rheinboldt and E. Giesbrecht, *Justus Liebigs Ann. Chem.*, **574,** 227 (1951).

326. O. Behaghel and W. Müller, *Ber.*, **68,** 1540 (1935).

327. W. Jenny, *Helv. Chim. Acta*, **35,** 1591 (1952).

328. H. Rheinboldt and E. Giesbrecht, *Chem. Ber.*, **89,** 631 (1956).

329. O. Foss and S. R. Svendsen, *Acta Chem. Scand.*, **8,** 1352 (1954).

330. W. S. Cook and R. A. Donia, *J. Amer. Chem. Soc.*, **73,** 2275 (1951).

331. G. Hölzle and W. Jenny, *Helv. Chim. Acta*, **41,** 331 (1958).

332. W. Jenny, *Helv. Chim. Acta*, **35,** 845 (1952).

333. W. Jenny, *Helv. Chim. Acta*, **41,** 317 (1958).

334. W. Jenny, *Helv. Chim. Acta*, **36,** 1278 (1953).

334a. G. Hölzle and W. Jenny, *Helv. Chim. Acta*, **41,** 593 (1958).

335. L. R. M. Pitombo, *Chem. Ber.*, **92,** 745 (1959).

336. O. Behaghel and H. Seibert, *Ber.*, **66,** 708 (1933).

337. O. Behaghel and W. Müller, *Ber.*, **67,** 105 (1934).

338. H. Rheinboldt and M. Perrier, *Bull. Soc. Chim. Fr.*, 251 (1956).

339. L. M. Yagupol'skii and V. G. Voloshchuk, *Zh. Obshch. Khim.*, **36,** 165 (1966).

340. L. M. Yagupol'skii and V. G. Voloshchuk, *Ukr. Khim. Zh.*, **36,** 66 (1970).

341. S. Prasad and B. L. Khandelwal, *J. Ind. Chem. Soc.*, **37,** 645 (1960).

342. H. Stolte, *Ber.*, **19,** 1577 (1886).

343. H. L. Wheeler and H. F. Merriam, *J. Amer. Chem. Soc.*, **23,** 299 (1901).

344. W. E. Weaver and W. M. Whaley, *J. Amer. Chem. Soc.*, **68,** 2115 (1946).

345. L. C. F. Blackman and M. J. S. Dewar, *J. Chem. Soc.*, 162 (1957).

346. W. J. Franklin and R. L. Werner, *Tetrahedron Lett.*, 3003 (1965).

347. H. Brintzinger, K. Pfannstiel, and H. Vogel, *Z. Anorg. Chem.*, **256,** 75 (1948).

348. R. Perrot and R. Berger, *C. R. Acad. Sci., Paris,* **235,** 185 (1952).

349. T. Tarantelli and D. Leonesi, *Ann. Chim.* (Rome), **53,** 1113 (1963).

350. B. Proskauer, *Ber.,* **7,** 1281 (1874).

351. E. Rebane, *Acta Chem. Scand.,* **21,** 652 (1967).

352. D. C. Goodall, *J. Inorg. Nucl. Chem.,* **30,** 1267 (1968).

353. G. Hofmann, *Justus Liebigs Ann. Chem.,* **250,** 294 (1889).

354. C. L. Jackson, *Ber.,* **8,** 321 (1875).

355. A. Fredga, *J. Prakt. Chem.,* [2], **121,** 56 (1929).

356. A. Fredga, *Ber.,* **71,** 286 (1938).

357. M. J. S. Dewar, H. Hampson, and L. C. F. Blackman, Brit. Pat. 839,351; *Chem. Abstr.,* **55,** 7296a (1961).

358. L. C. F. Blackman and M. J. S. Dewar, *J. Chem. Soc.,* 165 (1957).

359. Neth. Appl. 6,604,166; *Chem. Abstr.,* **66,** 85956 (1967).

360. Y. Suzuki, *Yakugaku Zasshi,* **81,** 1151 (1961).

361. L. Chierici, *Farmaco, Ed. Sci.,* **8,** 156 (1953).

361a. A. Fredga, L.-B. Agenäs, and E. Jonsson, *Acta Chem. Scand.,* **24,** 3751 (1970).

362. H. Rheinboldt and H. V. de Campos, *J. Amer. Chem. Soc.,* **72,** 2784 (1950).

363. T. Tarantelli and C. Pecile, *Ann. Chim.* (Rome), **52,** 75 (1962).

364. W. Gerrard, J. Kenyon, and H. Phillips, *J. Chem. Soc.,* 153 (1937).

365. T. van Es and R. L. Whistler, *Tetrahedron,* **23,** 2849 (1967).

366. R. Lesser and A. Schoeller, *Ber.,* **47,** 2505 (1914).

367. M. Renson and J.-L. Piette, *Bull. Soc. Chim. Belges,* **73,** 507 (1964).

368. N. Welcman and M. Wulf, *Israel J. Chem.,* **6,** 37 (1968).

369. H. Rheinboldt and M. Perrier, *Bull. Soc. Chim. Fr.,* 245 (1950).

370. O. Foss, *J. Amer. Chem. Soc.,* **69,** 2236 (1947).

371. M. C. Thompson and E. E. Turner, *J. Chem. Soc.,* 29 (1938).

372. V. V. Kozlov, *Zh. Obshch. Khim.,* **26,** 1755 (1956).

373. V. V. Kozlov, *Zh. Obshch. Khim.,* **27,** 3011 (1957).

374. V. V. Kozlov, *Acta Chim. Acad. Sci. Hung.,* **12,** 189 (1957).

375. V. V. Kozlov and S. E. Suvorova, *Zh. Obshch. Khim.,* **31,** 3034 (1961).

376. H. P. Kaufmann and W. Oehring, *Ber.,* **59,** 187 (1926).

377. T. H. Chao and R. E. Lyons, *Proc. Indiana Acad. Sci.,* **46,** 105 (1937).

378. L.-B. Agenäs and B. Lindgren, *Ark. Kemi,* **28,** 145 (1967).

379. L.-B. Agenäs, *Ark. Kemi,* **30,** 417 (1969).

380. N. N. Mel'nikov and E. M. Cherkasova, *J. Gen. Chem.* (*USSR*), **16,** 1025 (1946); *Chem. Abstr.,* **41,** 2697c (1947).

381. E. E. Aynsley, N. N. Greenwood, and M. J. Sprague, *J. Chem. Soc.,* 2395 (1965).

382. F. Wille, A. Ascherl, G. Kaupp, and L. Capeller, *Angew. Chem.,* **74,** 753 (1962).

383. M. Giua and R. Bianco, *Gazz. Chim. Ital.,* **89,** 693 (1959).

384. G. Schiavon, *Ric. Sci. Rend., Ser. A*, **32**, 69 (1962).

385. G. Schiavon, *Ric. Sci.*, [IIA], **1**, 346 (1961).

386. G. Hölzle and W. Jenny, *Helv. Chim. Acta*, **41**, 356 (1958).

387. G. Bergson and G. Nordström, *Ark. Kemi*, **17**, 569 (1961).

388. L. Chierici and R. Passerini, *J. Chem. Soc.*, 3249 (1954).

389. D. D. Lawson and N. Kharasch, *J. Org. Chem.*, **24**, 857 (1959).

390. F. Challenger and A. T. Peters, *J. Chem. Soc.*, 1368 (1928).

391. A. Cerniani and R. Passerini, *Ann. Chim.* (Rome), **47**, 58 (1957).

392. F. Challenger and D. I. James, *J. Chem. Soc.*, 1609 (1936).

393. G. Wagner and P. Nuhn, *Pharm. Zentralhalle*, **104**, 328 (1965).

394. E. Giesbrecht and I. Mori, *Anais Acad. Brasil. Cienc.*, **30**, 521 (1958).

395. G. Simchen, *Angew. Chem.* (*Intern. Ed.*), **7**, 464 (1968).

396. B. Greenberg, E. S. Gould, and W. Burlant, *J. Amer. Chem. Soc.*, **78**, 4028 (1956).

397. T. W. Campbell and M. T. Rogers, *J. Amer. Chem. Soc.*, **70**, 1029 (1948).

398. C. S. Evans and C. M. Johnson, *J. Chromatog.*, **21**, 202 (1966).

399. H. Rheinboldt and M. Perrier, *Quimica*, 118, 130 (1951); *Chem. Abstr.*, **46**, 7554b, 7554f (1952).

400. V. P. Isachenko and A. S. Bogorodskii, *Teploenergetika*, **16**, 79 (1969); *Chem. Abstr.*, **70**, 69594 (1969).

401. R. Paetzold and E. Wolfram, *Z. Anorg. Allg. Chem.*, **353**, 167 (1967).

402. H. J. Eméléus and R. N. Haszeldine, *Science*, **117**, 311 (1953).

403. D. J. G. Ives, R. W. Pittman, and W. Wardlaw, *J. Chem. Soc.*, 1080 (1947).

404. E. W. McClelland and A. J. Gait, *J. Chem. Soc.*, 921 (1926).

405. G. Vicentini, E. Giesbrecht, and L. R. M. Pitombo, *Chem. Ber.*, **92**, 40 (1959).

406. O. J. K. Edwards, W. R. Gaythwaite, J. Kenyon, and H. Phillips, *J. Chem. Soc.*, 2293 (1928).

407. J. Gosselck, *Chem. Ber.*, **91**, 2345 (1958).

408. G. Hölzle and W. Jenny, *Helv. Chim. Acta*, **41**, 712 (1958).

409. A. Ruwet and M. Renson, *Bull. Soc. Chim. Belges*, **75**, 157 (1966).

410. D. G. Foster, *Rec. Trav. Chim. Pays-Bas*, **53**, 405 (1934).

410a. C. D. Hurd and O. Fancher, *Intern. J. Sulfur Chem.*, *A*, **1**, 18 (1971).

411. N. Petragnani and M. de Moura Campos, *Tetrahedron*, **21**, 13 (1965).

412. L. M. Yagupol'skii and V. G. Voloshchuk, *Zh. Obshch. Khim.*, **37**, 1543 (1967).

413. H. Rheinboldt and M. Perrier, *Bull. Soc. Chim. Fr.*, 759 (1950).

414. H. M. Leicester, *J. Amer. Chem. Soc.*, **60**, 619 (1938).

415. L. M. Kataeva, N. S. Podkovyrina, T. G. Mannafov, and E. G. Kataev, *Zh. Strukt. Khim.*, **10**, 1124 (1969); *Chem. Abstr.*, **72**, 66229 (1970).

416. R. Lesser and R. Weiss, *Ber.*, **57**, 1077 (1924).

417. H. J. Backer and W. van Dam, *Rec. Trav. Chim. Pays-Bas*, **54**, 531 (1935).

418. B. Rathke, *Ann. Chem. Pharm.*, **152**, 181 (1869).

419. R. Paetzold, H.-D. Schumann, and A. Simon, *Z. Anorg. Allg. Chem.*, **305**, 98 (1960).

420. R. Paetzold, H.-D. Schumann, and A. Simon, *Z. Anorg. Allg. Chem.*, **305**, 88 (1960).

421. R. Paetzold, H.-D. Schumann, and A. Simon, *Z. Anorg. Allg. Chem.*, **305**, 78 (1960).

422. R. Paetzold and E. Rönsch, *Z. Anorg. Allg. Chem.*, **338**, 195 (1965).

423. L.-B. Agenäs, *Acta Chem. Scand.*, **19**, 764 (1965).

424. E. Rebane, *Ark. Kemi*, **26**, 345 (1966).

425. F. Krafft and R. E. Lyons, *Ber.*, **29**, 424 (1896).

426. M. Stoecker and F. Krafft, *Ber.*, **39**, 2197 (1906).

427. F. L. Pyman, *Proc. Chem. Soc.*, **30**, 302 (1914).

428. J. D. McCullough and E. S. Gould, *J. Amer. Chem. Soc.*, **71**, 674 (1949).

429. D. De Filippo and F. Momicchioli, *Tetrahedron*, **25**, 5733 (1969).

430. A. Schöberl, *Z. Physiol. Chem.*, **216**, 193 (1933).

430a. A. Fredga, *Uppsala Univ. Årsskrift*, **5**, 134 (1935); quoted in ref. 299.

431. G. T. Morgan and J. C. Elliott, *Proc. Chem. Soc.*, **30**, 248 (1914).

432. R. Pirisi and G. Serreli, *Rend. Seminario Fac. Sci. Univ. Cagliari*, **31**, 45 (1961); *Chem. Abstr.*, **58**, 467e (1963).

433. D. D. Karve, *J. Ind. Chem. Soc.*, **2**, 141 (1925).

434. J. W. White, Dissertation, University of Arizona, 1967; *University Microfilms*, Ann Arbor, Mich., Order 67-11,365.

435. H. J. Backer and H. J. Winter, *Rec. Trav. Chim. Pays-Bas*, **56**, 492 (1937).

436. R. Paetzold, S. Borek, and E. Wolfram, *Z. Anorg. Allg. Chem.*, **353**, 53 (1967).

436a. F. Ferranti and D. De Filippo, *J. Chem. Soc.*, B, 1925 (1971).

437. H. J. Eméleus and H. G. Heal, *J. Chem. Soc.*, 1126 (1946).

438. O. Foss, *J. Amer. Chem. Soc.*, **70**, 421 (1948).

439. D. De Filippo, E. Moretti, and C. Preti, *Ann. Chim.* (Rome), **58**, 603 (1968).

440. N. N. Vorozhtsov and V. V. Kozlov, *Org. Chem. Ind.* (USSR), **4**, 399 (1937).

441. E. Rebane, *Ark. Kemi*, **29**, 503 (1968).

442. M. L. Bird and F. Challenger, *J. Chem. Soc.*, 574 (1942).

443. I. P. Alimarin and V. S. Sotnikov, *Dokl. Akad. Nauk SSSR*, **113**, 105 (1957).

444. I. P. Alimarin and V. S. Sotnikov, *Chem. Anal.*, **2**, 222 (1957).

445. V. S. Sotnikov and I. P. Alimarin, *Talanta*, **8**, 588 (1961).

446. I. P. Alimarin and N. V. Shakhova, *Zh. Anal. Khim.*, **16**, 412 (1961).

447. B. F. Myasoedov, E. S. Pal'shin, and N. P. Molochnikova, *Zh. Anal. Khim.*, **23**, 66 (1968).

448. J. O. Clayton and D. H. Etzler, U.S. Pat. 2,516,616; *Chem. Abstr.*, **44**, 10312e (1950).

449. H. Bredereck, A. Wagner, H. Beck, and R.-J. Klein, *Chem. Ber.*, **93**, 2736 (1960).

450. R. Lesser and R. Weiss, *Ber.*, **46**, 2640 (1913).

451. P. Allen and J. W. Brook, *J. Org. Chem.*, **27**, 1019 (1962).

452. W. L. Mock and J. H. McCausland, *Tetrahedron Lett.*, 391 (1968).

453. H. J. Backer and J. Strating, *Rec. Trav. Chim. Pays-Bas*, **53**, 1113 (1934).

454. C. K. Banks and C. S. Hamilton, *J. Amer. Chem. Soc.*, **62**, 1859 (1940).

455. D. T. Woodbridge, M.Sc. Thesis, London, 1959; quoted in ref. 296.

456. I. B. Douglass, "Organic Sulfur Compounds," Vol. 1, N. Kharasch, Ed., Pergamon Press, New York, N.Y., 1961, p. 358.

457. K. J. Wynne and P. S. Pearson, *Inorg. Chem.*, **10,** 1868 (1971).

458. K. J. Wynne and J. W. George, *J. Amer. Chem. Soc.*, **91,** 1649 (1969).

459. K. J. Wynne and P. S. Pearson, *Chem. Commun.*, 293 (1971).

460. K. J. Wynne and P. S. Pearson, *Inorg. Chem.*, **11,** 1196 (1972).

460a. R. Paetzold and D. Knaust, *Z. Chem.*, **10,** 269 (1970).

461. R. E. Nelson, E. F. Degering, and J. A. Bilderbeck, *J. Amer. Chem. Soc.*, **60,** 1239 (1938).

462. J. D. McCullough, T. W. Campbell, and E. S. Gould, *J. Amer. Chem. Soc.*, **72,** 5753 (1950).

463. H. W. Doughty and F. R. Elder, *8th Intern. Congr. Appl. Chem.*, **6,** 93 (1912).

464. R. Anschütz, J. Kallen, and K. Riepenkröger, *Ber.*, **52,** 1860 (1919).

465. R. Anschütz and F. Teutenberg, *Ber.*, **57,** 1018 (1924).

466. M. Schmidt, P. Bornmann, and I. Wilhelm, *Angew. Chem. (Intern. Ed.)*, **2,** 691 (1963).

467. M. Schmidt and I. Wilhelm, *Ber.*, **97,** 872 (1964).

468. R. Paetzold and D. Lienig, *Z. Chem.*, **4,** 186 (1964).

469. K. Dostál, P. Mošna, and Z. Žák, *Z. Chem.*, **6,** 153 (1966).

469a. K. Dostál, Z. Žák, and M. Černik, *Chem. Ber.*, **104,** 2004 (1971).

470. K. C. Agrawal and F. E. Ray, *Intern. J. Cancer*, **2,** 257 (1967).

471. E. Rebane, *Acta Chem. Scand.*, **21,** 657 (1967).

472. E. Rebane, *Acta Chem. Scand.*, **23,** 1817 (1969).

473. D. Lienig, Dissertation, University of Dresden, 1965; quoted in R. Paetzold, *Fortschr. Chem. Forsch.*, **5,** 590 (1966).

474. E. A. H. Friedheim, U.S. Pat. 2,391,452; *Chem. Abstr.*, **40,** 4482[5] (1946).

475. E. A. H. Friedheim, U.S. Pat. 2,415,555; *Chem. Abstr.*, **41,** 3133i (1947).

476. J. H. van den Hende and E. Klingsberg, *J. Amer. Chem. Soc.*, **88,** 5045 (1966).

477. E. Klingsberg, U.S. Pat. 3,407,213.

478. M. H. Krackov, G. Bergson, A. Biezais, and H. G. Mautner, *J. Amer. Chem. Soc.*, **88,** 1759 (1966).

479. M. Nakasaki, *J. Chem. Soc. Japan*, **75,** 338 (1954).

480. J. Roy, I. L. Schwartz, and R. Walter, *J. Org. Chem.*, **35,** 2840 (1970).

481. G. Bergson and S. Wold, *Ark. Kemi*, **19,** 215 (1962).

482. D. L. Klayman, J. D. White, and T. R. Sweeney, *J. Org. Chem.*, **29,** 3737 (1964).

483. A. Biezais and G. Bergson, *Acta Chem. Scand.*, **18,** 815 (1964).

484. I. Degani and A. Tundo, *Ann. Chim.* (Rome), **50,** 140 (1960).

485. A. Baroni, *Atti Accad. Naz. Lincei Mem.*, [6], **11,** 579 (1930); *Chem. Abstr.*, **24,** 4771 (1930).

486. A. Baroni, *Atti Accad. Naz. Lincei Mem., Classe Sci. Fis., Mat. Nat.*, **26,** 456 (1937).

487. T. Bersin and W. Logemann, *Justus Liebigs Ann. Chem.*, **505,** 1 (1933).

488. E. P. Painter, *Chem. Rev.*, **28,** 179 (1941).

489. J. A. Stekol, *J. Amer. Chem. Soc.*, **64,** 1742 (1942).

490. H. L. Klug and D. F. Petersen, *Proc. S. Dakota Acad. Sci.*, **28,** 87 (1949).

491. C. C. Tsen and A. L. Tappel, *J. Biol. Chem.*, **233,** 1230 (1958).

492. H. E. Ganther, *Biochemistry*, **7,** 2898 (1968).

493. H. E. Ganther and C. Corcoran, *Biochemistry*, **8,** 2557 (1969).

494. E. A. Friedheim, U.S. Pat. 3,544,593; *Chem. Abstr.*, **74,** 87991 (1971).

495. Fr. Pat. 1,533,268; *Chem. Abstr.*, **71,** 35525 (1969).

496. O. Foss, *Acta Chem. Scand.*, **4,** 1499 (1950).

497. R. Otto and J. Troeger, *Ber.*, **24,** 1125 (1891).

498. J. Troeger and V. Hornung, *J. Prakt. Chem.*, [2], **60,** 113 (1899).

499. O. Foss, *Acta Chem. Scand.*, **5,** 967 (1951).

500. O. Foss, *Acta Chem. Scand.*, **6,** 508 (1952).

501. O. Foss, *Acta Chem. Scand.*, **5,** 115 (1951).

502. O. Foss, *Acta Chem. Scand.*, **6,** 521 (1952).

503. D. F. Twiss, *J. Chem. Soc.*, **105,** 1672 (1914).

504. A. Geens, G. Swallens, and M. Anteunis, *Chem. Commun.*, 439 (1969).

505. R. P. Spencer and K. R. Brody, *J. Nucl. Med.*, **9,** 621 (1968).

506. K. Gleu and R. Schaarschmidt, *Ber.*, **72,** 1246 (1939).

507. D. L. Klayman and G. W. A. Milne, *J. Org. Chem.*, **31,** 2349 (1966).

508. D. L. Klayman, M. M. Grenan, and D. P. Jacobus, *J. Med. Chem.*, **12,** 510 (1969).

509. W. Stricks and R. G. Mueller, *Anal. Chem.*, **36,** 40 (1964).

510. M. Schmidt and I. Wilhelm, *Z. Anorg. Allg. Chem.*, **330,** 324 (1964).

511. A. Kozhakova, E. A. Buketov, M. I. Bakeev, and A. K. Shokanov, *Zh. Neorgan. Khim.*, **11,** 1782 (1966); *Chem. Abstr.*, **65,** 16468g (1966).

512. K. Dostál and J. Krejčí, *Z. Anorg. Allg. Chem.*, **296,** 29 (1958).

513. A. Ruzicka and K. Dostál, *Z. Chem.*, **7,** 394 (1967).

514. M. Schmidt and I. Wilhelm, *Chem. Ber.*, **97,** 876 (1964).

515. P. Baumgarten, *Ber.*, **63,** 1330 (1930).

516. H. Rheinboldt and M. Perrier, *Bull. Soc. Chim. Fr.*, 484 (1953).

517. A. M. Giesbrecht, *Anais Acad. Brasil. Cienc.*, **30,** 155 (1958).

518. H. J. Emeléus and A. Haas, *J. Chem. Soc.*, 1272 (1963).

519. H. Rheinboldt and M. Perrier, *Bull. Soc. Chim. Fr.*, 445 (1955).

520. H. Rheinboldt and M. Perrier, *Bull. Soc. Chim. Fr.*, 264 (1953).

521. H. Rheinboldt and M. Perrier, *Bull. Soc. Chim. Fr.*, 379 (1953).

522. M. Perrier, *Selecta Chim.*, 81 (1957).

523. H. Rheinboldt and E. Giesbrecht, *J. Amer. Chem. Soc.*, **71,** 1740 (1949).

524. A. Haas and D. Y. Oh, *Chem. Ber.*, **98,** 3353 (1965).

525. H. Lecher and K. Simon, *Ber.*, **54,** 632 (1921).

V Selenides and their derivatives

LARS-BÖRGE AGENÄS

University of Uppsala
Institute of Chemistry
Uppsala, Sweden

A. INTRODUCTION

The first synthesis of an organoselenide was reported by Siemens[1] in the middle of the nineteenth century. In an attempt to prepare ethaneselenol, he evidently isolated a product containing large amounts of the corresponding diethyl selenide. Many different types of organoselenium compounds have since been described. Those containing C–Se–C bonds have been subjected to detailed studies from synthetic as well as physicochemical points of view. For investigations published up to 1950, the reviews by Bradt,[2,3] Campbell et al.,[4] and Rheinboldt[5] have been found to cover the literature excellently.

B. SYMMETRICAL SELENIDES

1. Dialkyl Selenides

Low-molecular-weight dialkyl selenides are evil-smelling, volatile liquids, very similar in nature to the analogous dialkyl sulfides. The lowest homolog, dimethyl selenide (**1**), b.p. 58.2°, seems to be very difficult to handle because of its volatility and odor. The higher homologs are colorless, crystalline compounds similar to long-chain alkanes. The reactivity of this group of compounds is closely connected with the Se atom present in the chain. In contrast to alkanes and dialkyl sulfides, the analogous dialkyl selenides show some lability, arising from the fact that C–Se bonds are less stable than C–C and C–S bonds. Thus, occasionally, dialkyl selenides are found to split off elemental selenium, for example, on exposure to bright sunlight. Although the preparation of dialkyl selenides is frequently carried out by methods analogous to those used for dialkyl sulfides, some special methods unknown in sulfur chemistry have also been used.

a. Synthesis. Low-molecular-weight dialkyl selenides have been investigated by numerous workers in recent years. For the preparation of these compounds, the synthetic procedure by Bird and Challenger,[6] as described for dimethyl selenide (**1**) in scheme 1, has often been employed. A sodium

$$\text{(1)} \quad HO-CH_2-SO-O^{\ominus} + Se + 3OH^{\ominus} \longrightarrow H-CHO + Se^{2-} + SO_3^{2-} + 2H_2O$$

$$2CH_3I + Se^{2-} \longrightarrow CH_3-Se-CH_3 + 2I^{\ominus}$$

1

selenide solution was prepared by heating powdered gray selenium with an excess of Rongalite (sodium formaldehyde sulfoxylate) and a large excess of sodium hydroxide in water. The reducing agent Rongalite[7] will produce selenide ions, which must be protected against oxidation to diselenide ions. On reaction with an appropriate alkyl halide the reagent gives very good yields of dialkyl selenide. Products so obtained are given in Table V-1.

Employing the same method for the reaction between 1-bromocyclobutane-1-carboxylic acid and sodium selenide, Fredga *et al.*[30] found a rearrangement reaction to occur. This gave rise to a complex reaction mixture, the main component of which was identified as di(1-carboxy-1-cyclopropylmethyl) selenide.

A different method for the preparation of a sodium selenide solution was used by Denison and Condit.[31–35] Elemental selenium and sodium were allowed to react in liquid ammonia. After evaporation of the solvent, the resulting solid was dissolved in ethanol and allowed to react with alkyl halides.

Table V-1   *Dialkyl selenides prepared from aqueous sodium selenide and alkyl halides*

COMPOUND	YIELD (%)	REF.
Dimethyl selenide (**1**)	88	6–22
Dimethyl selenide-d_6 (**2**)	. . .	23
^{14}C-Dimethyl selenide (**3**)	81	16
Diethyl selenide (**4**)	60–80	6, 8, 11–14, 16–18
Dipropyl selenide (**5**)	54	6, 11, 12, 14, 16, 17
Dibutyl selenide (**6**)	. . .	10, 12, 13, 16, 18
Dipentyl selenide (**7**)	. . .	12
Dibenzyl selenide (**8**)	90	24–29

By a similar method Brandsma and Wijers[36] allowed elemental selenium and sodium to react in liquid ammonia. When this reaction was complete, an alkyl halide or dialkyl sulfate was added directly, and after evaporation of the ammonia very good yields (80–90%) of dialkyl selenides were obtained.[37,38]

Organometallic compounds, such as Grignard reagents, have frequently been employed as intermediates in the preparation of selenols (*e.g.*, benzyl-selenol).[39] Considerable amounts of by-products are often obtained, usually of the selenide type. Zakharin and Gavrilenko[40] have used trialkylaluminum for the preparation of organoselenium compounds, as shown in scheme 2.

$$(2)\quad R_3Al + Se \xrightarrow{\Delta} R_2Al{-}SeR \xrightarrow{H_2O} 2RH + RSeH + Al(OH)_3$$

$$R_3Al + 2Se \xrightarrow[(2) H_2O]{(1)\ \Delta} RSeH + R_2Se + R_2Se_2 + RH$$

Trialkylaluminum and powdered selenium reacted in the absence of solvent at about 80°. After the addition of water or dilute acid, an organoselenium compound was isolated. The nature of the product was found to be very dependent on the molar ratios of elemental selenium to trialkylaluminum. At a ratio of 1:1, the selenol was the main product; at 2:1, the dialkyl selenide dominated; while at 3:1, the main product was the dialkyl di-selenide. In this way 26% of diethyl selenide and 18% of 2,2'-dimethyl-dipropyl selenide, based on the elemental selenium used, were isolated. Similar results were also obtained for the analogous sulfur compounds.

A standard method for the preparation of alkaneselenols, starting with the reaction between an alkyl halide and selenourea and a following hydrolysis, was employed by Wagner and Nuhn[41] in the preparation of selenium-containing carbohydrates. These authors allowed 2,3,4,6-tetraacetyl-α-D-glucopyranosyl bromide to react with selenourea to give a 60–70% yield of 2,3,4,6-tetraacetyl-β-D-glucopyranosylisoselenouronium bromide (**9**). Alka-line hydrolysis of **9** to the corresponding selenol in the presence of another equivalent of the halide produced an 80% yield of di(2,3,4,6-tetraacetyl-β-D-glucopyranosyl) selenide (10) (scheme 3). Hydrolysis of 10 by sodium methanolate gave a good yield of di(β-D-glucopyranosyl) selenide.

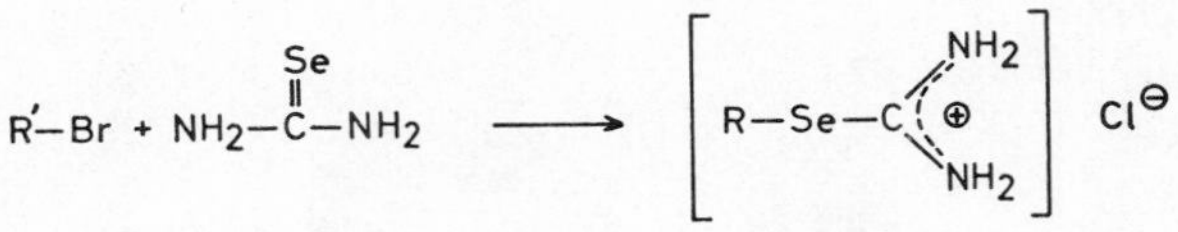

$$R'-Br + NH_2-\underset{\underset{Se}{\|}}{C}-NH_2 \longrightarrow \left[R-Se-C{\overset{\nearrow NH_2}{\underset{\searrow NH_2}{\oplus}}}\right] Cl^{\ominus}$$

9

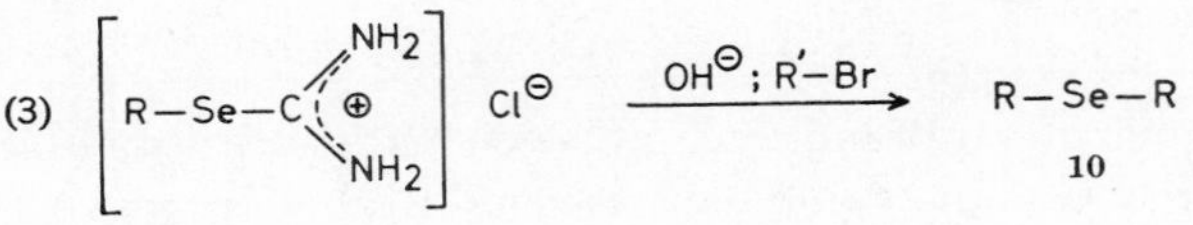

$$(3) \quad \left[R-Se-C{\overset{\nearrow NH_2}{\underset{\searrow NH_2}{\oplus}}}\right] Cl^{\ominus} \xrightarrow{OH^{\ominus};\ R'-Br} R-Se-R$$

10

$$R = $$

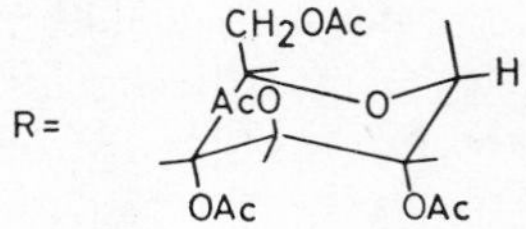

The addition of hydrogen selenide to alkenes has only rarely been employed for the preparation of symmetrical dialkyl selenides. Landa *et al.*[42] allowed ethylene to react with hydrogen selenide at 120 atm and 350° in the presence of molybdenum sulfide as a catalyst; a 65% yield of 4 was obtained, together with a 14% yield of ethaneselenol as well as high-boiling by-products (equation 4).

$$(4) \quad 2CH_2{=}CH_2 + H_2Se \xrightarrow[350°]{120\ atm} (CH_3CH_2)_2Se$$

In an experiment by Kemppainen and Compere,[43] hydrogen selenide and tropylium bromide in water were combined to give a 95% yield of dicycloheptatrienyl selenide. Gosselck and Wolters[44] allowed ethyl β-phenylpropiolate to react with hydrogen selenide in the presence of piperidine as a catalyst. An exothermic reaction gave a good yield of diethyl 2,2'-selenodicinnamate, which was hydrolyzed to 2,2'-selenodicinnamic acid.

By use of an electrochemical process, Kaabak *et al.*[45] allowed a mixture of acrylonitrile and elemental selenium to react in an aqueous sodium sulfate solution. In this way a 23% yield of 2,2'-dicyanodiethyl selenide was obtained, which was hydrolyzed to 3,3'-selenodipropionic acid.

The occasional formation of di(3-indolylmethyl) selenide, reported by Agenäs[46,47] when reacting gramine with benzylselenol at 110–135°, may best be explained by assuming that hydrogen selenide or 3-indolylmethane-selenol is formed as an intermediate in a radical process.

Fluoro-substituted dialkyl selenides have been extensively studied. The first method of preparation, published by Haszeldine[48] for hexafluoro-dimethyl selenide (**11**) and later described in detail by Dale *et al.*,[49] is given in scheme 5. Trifluoroiodomethane and elemental selenium were heated in a

$$(5) \quad 2CF_3I \;+\; Se \quad \xrightarrow{285^\circ} \quad CF_3-Se-CF_3 \;+\; I_2$$

11

Carius tube to 260–285°; a radical reaction occurred, giving under optimal conditions a 45–50% yield of **11**, elemental iodine, and, as a by-product, a 10–15% yield of the analogous diselenide.[48–50]

Compound **11** has been prepared from silver trifluoroacetate,[51] as shown in scheme 6. A similar result was obtained by the use of trifluoroacetic

$$(6) \quad 2CF_3COOAg \;+\; 2Se \quad \xrightarrow{280^\circ} \quad CF_3-Se-CF_3 \;+\; 2CO_2 \;+\; Ag_2Se$$

11

anhydride.[51] Starting with mercury trifluoroacetate in a related experiment, Yarovenko *et al.*[52] obtained only the corresponding diselenide. By the method indicated in scheme 6, dipentafluoroethyl selenide[53,54] and diheptafluoro-propyl selenide[55] were prepared in low yields, accompanied by equal amounts of the corresponding diselenides. Good yields of di(1*H*,1*H*,7*H*-dodecafluoroheptyl) selenide[56] and di(1*H*,1*H*,11*H*-eicosafluoroundecyl) sel-enide[56] were obtained by the route given in scheme 5.

Brintzinger *et al.*[57] allowed a mixture of selenium tetrachloride and elemental selenium to react with chloro-substituted alkenes in the presence of anhydrous aluminum trichloride as a catalyst, as shown for 1,2-dichloro-ethylene in scheme 7. This procedure was also applied to the synthesis of

$$(7) \quad \begin{aligned} SeCl_4 \;+\; Se \quad &\rightleftharpoons \quad 2SeCl_2 \\ 2CHCl{=}CHCl \;+\; SeCl_2 \quad &\longrightarrow \quad (CHCl_2{-}CHCl)_2Se \end{aligned}$$

12

bis(2,2-dichloroethyl), bis(1,2-dichlorovinyl), and bis(2-chlorovinyl) sele-nides.

The reaction between selenium tetrachloride and alkenes gave the corresponding chloro-substituted dialkyl selenide dichloride (scheme 8).[57,58]

$$(8) \quad \begin{aligned} SeO_2 \;+\; 4HCl \quad &\longrightarrow \quad SeCl_4 \;+\; 2H_2O \\ 2CH{\equiv}CH \;+\; SeCl_4 \quad &\longrightarrow \quad (CHCl{=}CH)_2SeCl_2 \quad \longrightarrow \quad (CHCl_2{-}CHCl)_2Se \end{aligned}$$

12

Selenium dioxide was dissolved in concentrated hydrochloric acid, after which acetylene was added to the solution to yield 2,2'-dichlorodivinyl selenide dichloride, which rearranged to **12** upon standing. No rearrangement can occur when starting with alkenes. The dialkyl selenide dichloride obtained in these cases was reduced with potassium bisulfite to give the corresponding selenide. Numerous analogous preparations have been reported,[58-60] including applications to tellurium species.[58]

In a recent mechanistic study of the reactions given in scheme 7 and applied to propylene, Lautenschlaeger[61] showed the product to be a mixture of 2,2'-dichlorodipropyl selenide and the isomers, 1,1'-dichloro-2,2'-dipropyl and 1,2'-dichloro-1',2-dipropyl selenides. The mixture was, however, not separated into its pure components.

The reaction between ethylene and selenium dioxide in acetic acid solution at 100–125° at a pressure of about 3 atm (scheme 9) has been studied by

$$
(9) \quad CH_2{=}CH_2 + CH_3COOH + SeO_2 \longrightarrow CH_3COOCH_2CH_2OCOCH_3 +
$$

$$
+ (CH_3COOCH_2CH_2)_2Se + (CH_3COOCH_2CH_2Se)_2
$$

Olson[62] from a mechanistic point of view. A maximum 40 % yield of mono-selenide was obtained in the presence of sodium acetate, with only a minor amount of the diselenide formed as a by-product.

Similarly, Takaoka and Toyama[63,64] combined equimolar amounts of alkenes and selenium dioxide in acetic acid solution for 5 hours at 115° with a stream of air bubbling through the liquid. The product, after chromatographic purification, was found to contain approximately equal amounts of selenide and the analogous selenoxide. In this way octene was converted into 2,2'-diacetoxydioctyl selenide.[63] When methyl 10-undecenoate and selenium dioxide in the molar ratio 2:1 were similarly treated, oxidation of the C-C double bond was also observed. After hydrolysis and re-esterification of the crude product, a chromatographic purification gave 52 % of 2,2'-dihydroxy-10,10'-diacetoxydecyl selenide.[64]

The reaction between aldehydes and hydrogen chloride at a low temperature, known to give α,α'-dichlorodialkyl ethers, has been applied by Brandsma and Arens[65] to the synthesis of the corresponding sulfides and selenides, as indicated in scheme 10. Paraldehyde, hydrogen selenide, and

$$
2\,CH_3{-}CH{=}O + H_2Se + 2\,HCl \xrightarrow{-10°-\,0°} \underset{\textbf{13}}{CH_3{-}\overset{Cl}{\underset{|}{CH}}{-}Se{-}\overset{Cl}{\underset{|}{CH}}{-}CH_3} + 2\,H_2O
$$

$$
(10) \quad \underset{\textbf{13}}{CH_3{-}\overset{Cl}{\underset{|}{CH}}{-}Se{-}\overset{Cl}{\underset{|}{CH}}{-}CH_3} \xrightarrow[145-150°]{base} CH_2{=}CH{-}Se{-}CH{=}CH_2 + 2\,HCl
$$

hydrogen chloride reacted in a nitrogen atmosphere at $-10°$ to give a 40–50% yield of crude 1,1'-dichlorodiethyl selenide (**13**); 1,1'-dichlorodipropyl selenide (**14**) was obtained analogously from propionaldehyde. Heating **13** with diethylaniline to 145–150° gave a 40% yield of divinyl selenide, whereas this treatment of **14** produced a 20% yield of dipropenyl selenide.[65,66]

In a recent reinvestigation of earlier reported selenium derivatives of acetylacetone,[67-69] Dewar *et al.*[70] allowed copper acetylacetonate to react with selenium tetrachloride to give a product with the proposed structure **15**. The use of n.m.r. and i.r. for the structure determination did not provide a definite answer to the question of whether the enol or the keto structure was predominant.

15

b. Reactions. Like ethers, dialkyl selenides act as Lewis bases and form complexes with suitable agents. Thus, Allkins and Hendra[20] have prepared complexes with dimethyl selenide and palladium(II) and platinum(II) halides by the reaction with aqueous solutions of various potassium tetra-halopalladite and tetrahaloplatinate salts. Complexes of the type

$$MX_2[(CH_3)_2Se]_2$$

were isolated, the *cis* and *trans* forms of which were separated by fractional crystallization. Palladium has frequently been found to form "bridged" complexes, and structures of the type

$$[PdX_2Se(CH_3)_2]_2$$

where X = Cl or Br, were obtained. These complexes have also been studied by i.r. spectroscopy.[15,17]

Similarly, Le Calve and Lascombe[19] found dimethyl selenide to form a complex with boron trifluoride, both at room temperature and at $-180°$. The corresponding BCl_3 and BBr_3 complexes have been prepared by Wynne and George[71] in the crystalline state. The adducts decompose slowly in contact with the atmosphere.

Dialkyl selenides, like the corresponding sulfides, also react with alkyl halides to form trialkyl selenonium compounds (*cf.* Chapter VI).

When subjected to oxidizing agents, dialkyl selenides form the corresponding selenoxides (*cf.* Section V.E) or occasionally selenones (*cf.* Section

V.F). Dialkyl selenides react readily with halogens to form dialkyl selenide dihalides, as described in Section V.G.

Dialkyl selenides are usually stable towards reducing agents and under conditions of hydrolysis. The alkylseleno group constitutes a poor leaving group when dialkyl selenides are treated with nucleophilic reagents. Polar groups, such as α-carboxyl functions, may, however, lead to elimination reactions, thus forming an alkene and an alkaneselenol.

Painter et al.[24] found dialkyl selenides and the analogous sulfides to be very stable to alkaline solutions. Similarly, Dale et al.[49] reported that hexafluoro-dimethyl selenide was unusually stable to aqueous bases, as well as to concentrated nitric acid at 80–90°. Heating the compound with 20 % alcoholic potassium hydroxide to 100°, however, led to decomposition. The action of liquid hydrogen chloride on dimethyl selenide and the analogous sulfide and telluride was studied by Peach,[76] who found these compounds to act as solvobases. A comparison of the conductivities indicated a similar ease of protonation. Heating hexafluorodimethyl selenide with chlorine in a Carius tube at moderate temperature showed that most of the compound was recovered unchanged after a long period of heating; only traces of tri-fluoromethylselenium trichloride could be observed. The higher homologs, perfluorodiethyl selenide[53] and perfluorodipropyl selenide,[61] were found to be similarly stable towards alkali and chlorine. Finally, Brandsma and Schuijl[66] found that dipropyl selenide reacted with lithium in liquid ammonia to cleave one of the C–Se bonds and form lithium propaneselenolate.

The stability of dialkyl selenides to heating was demonstrated by the gas chromatographic investigations of Evans and Johnson[14] and Beneš and Procházková.[16] At low column temperatures a good separation of the mixtures studied could be obtained. At a column temperature of 150° or higher, additional peaks, assigned to the formation of unsymmetrical dialkyl selenides by a thermal decomposition of the original symmetrical compounds, were observed.

c. *Structure and Physical Properties.* The properties of dialkyl selenides have been studied by various methods. Cecchini and Giesbrecht[25] determined, through the phase diagram of the binary system dibenzyl sulfide–dibenzyl selenide, that these compounds were isodimorphous.

An electron diffraction investigation of dimethyl selenide by Goldish et al.[72] showed the bond length C–Se to be 1.977 ± 0.012 Å and the bond angles C–Se–C and Se–C–H to be $98 \pm 10°$ and $110.5 \pm 3.5°$, respectively. Bowen[73] similarly studied hexafluorodimethyl selenide (**11**) and found the bond length C–Se to be 1.958 ± 0.022 Å and the bond angle C–Se–C to be $104.4 \pm 5°$.

The dipole moments of dialkyl selenides have been determined by Chierici et al.,[74] who reported the value for **1** to be 1.32 D and for **4** to be 1.52 D.

In an investigation by West *et al.*,[75] the hydrogen bonding of phenol to analogous ethers, sulfides, and selenides was studied and the base properties were found to decrease in the order $O > S > Se$.

d. *Spectroscopy.* A discussion of the spectral properties of selenides will be found in Chapter XV.

The e.s.r. spectra of dialkyl selenides have been recorded[77,78] and compared with those for the analogous sulfides. It was found that irradiation of dialkyl selenides with u.v. light causes one C-Se bond to break, leading to the formation of alkyl radicals.

2. Diaryl Selenides

Diaryl selenides, in comparison to the dialkyl analogs, are more easily handled because of their higher molecular weight and the greater stability of their C–Se bonds. The simplest compound of this type, diphenyl selenide, is a high-boiling colorless liquid, similar in character to the analogous sulfide. Diaryl selenides are readily obtained in a pure state; their reactivity is very similar to that of dialkyl selenides and will be briefly discussed in Section V.B. Methods for the preparation of diaryl selenides are generally different from those employed for dialkyl selenides.

a. *Synthesis.* A modification of the Sandmeyer reaction is the most frequently employed method of preparation for diaryl selenides, as described by Leicester,[79] and is exemplified in scheme 12 by diphenyl selenide. A

$$(12)\quad 2\ \underset{}{C_6H_5}-\overset{\oplus}{N}\equiv N\ +\ Se^{2-}\ \longrightarrow\ C_6H_5-Se-C_6H_5\ +\ 2\,N_2$$

16

sodium or potassium selenide solution was prepared by the Rongalite method shown in scheme 1. Two equivalents of the selenide solution were allowed to react with a cooled benzenediazonium chloride solution. A large amount of elemental selenium was recovered. For large batches, very good yields of diaryl selenide based on the diazonium salt could be obtained. This method of preparation has been used for a number of substituted diaryl selenides.[10,79–91] For the preparation of 2,2'-dinaphthyl selenide Bergson[91] added sodium acetate as a buffer to the aryldiazonium salt solution. This resulted in only minor decomposition of the potassium selenide; equivalent amounts of selenide and diazonium salt may consequently be used, giving a more economical and convenient overall procedure.

The formation of diphenyl selenide has also been reported by Bock *et al.*[92] from a similar reaction between benzenediazonium bromide and sodium benzeneselenolate.

Nesmeyanov *et al.*[93] have reported the preparation of diphenyl selenide in 33% yield by the addition of a solution of benzenediazonium tetrafluoroborate, followed by powdered zinc, to an acetone solution of selenium tetrachloride at 0°. The reaction was found to give low yields of substituted diphenylselenides.

The method of Krafft and Lyons[94] for the preparation of diaryl selenides by heating a mixture of a diaryl sulfone with elemental selenium has also been used recently for the synthesis of **16** (equation 13).[31, 95]

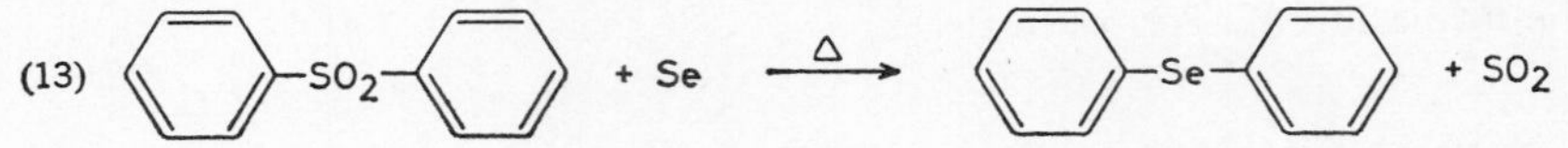

(13)   ⟨phenyl⟩—SO$_2$—⟨phenyl⟩ + Se $\xrightarrow{\Delta}$ ⟨phenyl⟩—Se—⟨phenyl⟩ + SO$_2$

                                                   16

Burlant[96] has prepared **16** by the reaction of phenyl selenocyanate in anhydrous ether with phenyllithium. After hydrolysis of the reaction mixture with water a 44% yield of **16** was obtained (equation 14). This method has

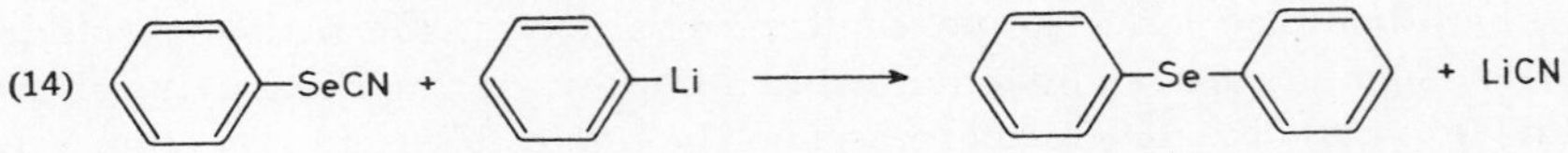

(14)   ⟨phenyl⟩—SeCN + ⟨phenyl⟩—Li $\longrightarrow$ ⟨phenyl⟩—Se—⟨phenyl⟩ + LiCN

                                                   16

been found especially valuable for the preparation of unsymmetrical diaryl selenides (*cf.* Section V.C).

The conversion of diphenyl diselenide has occasionally been used to prepare the monoselenide analog, **16**. Thus, Hauptmann and Walter[97] treated the diselenide with Raney nickel in benzene at 80° for 15 hours to give an 89% yield of **16**. Using xylene as the solvent and performing the reaction at 140–180°, they obtained a mixture of large amounts of biphenyl and **16**, a result which indicates the radical character of the reaction. Similarly, phenyl selenobenzoate, when treated with Raney nickel in benzene at 80°, gave a 54% yield of **16**, and 2,2′-dimethyldiphenyl diselenide in xylene at 140° produced a 67% yield of the corresponding monoselenide.

In 1896 Krafft and Kaschau[98] showed that selenium tetrachloride reacts with benzene in the presence of anhydrous aluminum trichloride to give a moderate yield of **16**. This method has more recently been employed by Chierici and Passerini.[86, 99] Alquist and Nelson[100] have shown that aromatic ethers are substituted by the action of selenium oxychloride in anhydrous ether, forming 4,4′-dialkoxydiaryl selenide dichlorides. The selenide dichloride may be reduced by powdered zinc to give the corresponding selenide (scheme 15). A number of other diaryl selenides have been prepared in excellent yield[80, 83, 84, 100–102] by this technique.

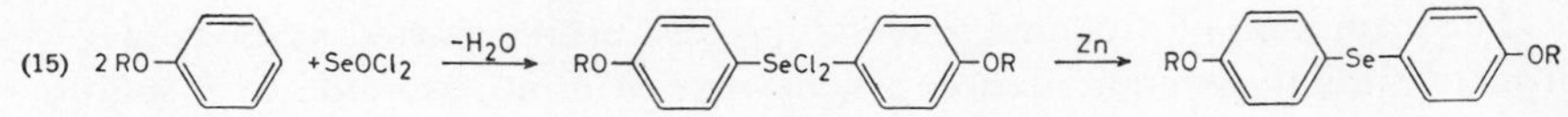

(15)   2 RO—⟨phenyl⟩ + SeOCl$_2$ $\xrightarrow{-H_2O}$ RO—⟨phenyl⟩—SeCl$_2$—⟨phenyl⟩—OR $\xrightarrow{Zn}$ RO—⟨phenyl⟩—Se—⟨phenyl⟩—OR

Starting with phenols, selenium dioxide, and hydrochloric acid, Funk and Weiss[58] prepared substituted diaryl selenides. Thus, treatment of β-naphthol gave a 70% yield of product, for which the structure 2,2'-dihydroxydinaphthyl selenide[58] was suggested. The analogous procedure gave products tentatively assigned the following structures: 5,5'-dichloro-2,2'-dihydroxydiphenyl selenide,[59] 5,5'-dichloro-4,4'-dimethyl-2,2'-dihydroxydiphenyl selenide,[59] and 3,3'-dicarboxy-4,4'-dihydroxydiphenyl selenide.[59]

Treatment of anisole with selenium dioxide under anhydrous conditions was found by Boyd et al.[88] to result in no reaction. In the presence of 1 mole of water, however, 2.2 moles of selenium dioxide and 2.5 moles of anisole gave a 20% yield of 4,4'-dimethoxydiphenyl selenide.[82,88] Other aromatic ethers gave diaryl selenides by this procedure. Although the structures of the products obtained have not been fully established, the following have been suggested: 3,3',4,4'-tetramethoxydiphenyl selenide, 2,2',4,4'-tetramethoxydiphenyl selenide, 5,5'-dibromo-2,2'-dimethoxydiphenyl selenide, 4,4'-dimethoxydinaphthyl selenide, and 2,2'-dimethoxydinaphthyl selenide.[88]

Witkop[103] used selenium dioxide to oxidize indole and obtained a byproduct which has recently been studied by Wilshire.[104] The latter has isolated a 23% yield of 3,3'-diindolyl selenide from this reaction. The analogous treatment of 1-methyl- and 2-methylindole was, however, reported to give mainly the corresponding triselenides. In similar experiments, Bergman[105] isolated 3,3'-diindolyl selenide and 1,1'-dimethyl-3,3'-diindolyl selenide.

Burdon et al.[106] found that dipentafluorophenylmercury and elemental selenium, heated in a sealed tube to 240° for 7 days, gave a 78% yield of dipentafluorophenyl selenide (**17**, scheme 16). The reaction is applicable

(16)

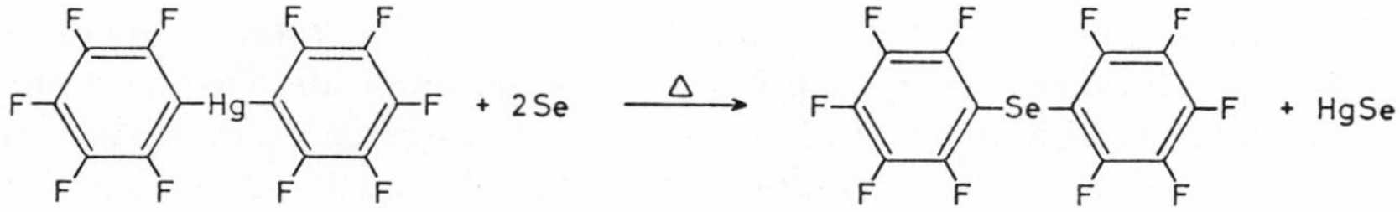

17

also to elemental sulfur and has been employed by Cohen et al.[107] for the preparation of analogous sulfides, selenides, and tellurides, sometimes in very good yields. Pentafluorophenylmercuric chloride and selenium heated in a sealed tube at 230° produced a 10% yield of **17**; pentafluoroiodobenzene and selenium, a 60% yield of **17**. The reaction between pentafluorobromobenzene and butyllithium in anhydrous ether gave pentafluorophenyllithium, which was allowed to react with selenium tetrachloride at −78° for a 20% yield of **17**. The use of ether-hexane as the solvent raised the yield to 60%.

Nitro-substituted halobenzenes are sufficiently reactive to participate in nucleophilic displacement reactions and can be employed conveniently for the preparation of diaryl selenides. In this way, Giua and Bianco[108] combined 2,4,6-trinitrochlorobenzene and potassium selenocyanate in absolute ethanol to form bis(2,4,6-trinitrophenyl) selenide. The initially formed aryl seleno-cyanate was probably attacked by another trinitrochlorobenzene molecule to give the monoselenide. Similarly, Litvinenko and Cheshko[109] and other workers have prepared 4,4'-dinitrodiphenyl selenide (**18**) (equation 17).[86,109–111]

(17) $2 NO_2$—⟨benzene⟩—Cl $+ Se^{2-}$ ⟶ NO_2—⟨benzene⟩—Se—⟨benzene⟩—NO_2 $+ 2Cl^-$

18

b. Reactions. The reactivity of diaryl selenides is very similar to that of dialkyl selenides, provided the greater stability of the aromatic analogs is taken into consideration. Thus, reaction with halogens results in diaryl selenide dihalides (*cf.* Section V.G), and oxidation under various conditions gives the corresponding selenoxides (Section V.E) or selenones (Section V.F).

Like dialkyl selenides, the aromatic analogs are relatively stable towards reduction and hydrolysis, so long as their Lewis base properties are not employed. The action of Raney nickel (180°, 8 hours, N_2, no solvent) on diaryl selenides was studied by Hauptmann and Walter,[97] who found diphenyl selenide and the 2,2'-dimethyldiphenyl selenide to give biphenyl and 2,2'-dimethylbiphenyl, respectively. Similarly, 4,4'-diethoxydiphenyl selenide, on heating with Raney nickel under reflux for 5 hours in benzene-ethanol, gave phenetole.[112] Similar treatment with nickel of 3,3'-diindolyl selenide in dioxane gave a mixture of indole and 3,3'-biindolyl;[105] 1,1'-dimethyl-3,3'-diindolyl selenide was also found to split off selenium.[105] Illumination of 3,3'-diindolyl selenide for 48 hours at room temperature in benzene solution resulted in a photochemical decomposition to give a mixture of indole, 3,3'-biindolyl, and unchanged starting material. This result is very similar to the fragmentation of 3,3'-diindolyl selenide under electron bombardment (*cf.* Chapter XVG).

Bartolotti and Cerniani[111] studied the stability of 4,4'-dinitrodiphenyl selenide towards alkaline reagents and observed that the C-Se bonds were stable to reagents such as hydrazine hydrate, piperidine, and aniline in re-fluxing ethanolic solutions. In a further experiment with hydrazine hydrate, the nitro groups attached to the aromatic rings were partially reduced to amino groups without effect on the selenide moiety.

c. Structure and Physical Properties. The crystal structure of 4,4'-dimethyldiphenyl selenide has been studied by Blackmore and Abrahams.[113]

The C-Se bond length was determined to be 1.92 Å; the C–Se–C bond angle, 106°.

The dipole moment for diphenyl selenide was calculated by Chierici et al.[8] to be 1.50 D, and a similar calculation for 4,4'-dichlorodiphenyl selenide gave the value 0.80 ± 0.1 D. Rogers and Campbell[85] determined the dipole moments for 4,4'-dimethyldiphenyl selenide and for 4,4'-dichlorodiphenyl selenide to be 1.81 D and 0.77 D, respectively. It was suggested that the C–Se–C bond angle in these compounds was 115°. Le Fèvre and Saxby[114] gave the value 1.37 D for diphenyl selenide and found the phenyl groups rotated 48° out of the C–Se–C plane.

Additional dipole moments of symmetrically substituted diphenyl selenides have been determined by Krishnamurthy and Soundararajan,[82] who found the values 1.00 D (2,2'-dimethyl); 1.65 D (3,3'-dimethyl); 1.74 D (4,4'-dimethyl), and 2.35 D (4,4'-dimethoxy).

Spectral properties of diaryl selenides are discussed in Chapter XV.

C. UNSYMMETRICAL SELENIDES

1. Dialkyl Selenides

The general properties of unsymmetrical dialkyl selenides are similar to those of the symmetrical analogs, the main differences between the classes being the methods of preparation. Certain substituents on one of the alkyl groups may have a great influence on the stability of the dialkyl selenide, resulting in a preferred path of decomposition upon attack by various reagents. Difficulties in the attempted preparation of certain dialkyl selenides may arise from similar causes.

a. Synthesis. The most common method of preparation used for the synthesis of unsymmetrical dialkyl selenides is a modified Williamson synthesis.[5] A suitable dialkyl diselenide is reduced to form alkaneselenolate ions, for which various reducing agents may be employed. Günther and Mautner[115] used sodium borohydride; Bergson and Delin,[116] Rongalite. The sodium alkaneselenolate thus obtained then reacts with an alkyl halide (scheme 18). A large number of selenides have been prepared[30,115–122] in this manner.

(18) $R-Se-Se-R \xrightarrow{\text{redn.}} R-Se^{\ominus} \xrightarrow{R'-X} R-Se-R' + X^-$

Mock and McCausland[123] reduced a cyclic seleninic ester by sodium borohydride and reacted the resulting mixture with benzyl chloride, giving benzyl 4-hydroxy-2,3-dimethyl-2-butenyl selenide. Further treatment with acetic anhydride yielded benzyl 4-acetoxy-2,3-dimethyl-2-butenyl selenide.

Suitable alkaneselenols (*e.g.*, benzylselenol) may also be dissolved in ethanolic sodium ethanolate, followed by alkylation of the resulting sodium alkaneselenolate. In this way Bergson[124] prepared a 92 % yield of 8-benzyl-seleno-6-hydroxyoctanoic acid.

Alkaneselenolates may also be prepared from various organometallic compounds (*e.g.*, Grignard reagents). Bergson and Delin[116] thus prepared an ethereal solution of ethylselenium magnesium bromide from ethyl-magnesium bromide and diethyl diselenide. Adding chloroacetone and heating the mixture under reflux gave a 19 % yield of ethylselenoacetone. Boiko *et al.*[125] prepared methylselenoacetone in a 34 % yield and methyl phenacyl selenide[126] in a 56 % yield by this route.

Brandsma *et al.*[127] reacted alkynes in liquid ammonia with sodium and selenium to form the corresponding sodium alkyneselenolates, which, on addition of an alkyl halide, gave unsymmetrical dialkyl selenides (scheme 19).[125,127−130]

$$(19) \quad R-C\equiv CNa + Se \xrightarrow{NH_3 (l)} R-C\equiv C-SeNa \xrightarrow{R'-X} R-C\equiv C-Se-R'$$

For the preparation of selenium-containing carbohydrates (*cf.* scheme 3) Wagner and Nuhn[41] allowed 2,3,4,6-tetraacetyl-β-D-glucopyranosyliso-selenouronium bromide (**9**) to react with 2,3,4,6-tetraacetyl-α-D-galacto-pyranosyl bromide in acetone solution in the presence of potassium hydroxide to obtain a 75 % yield of 2,3,4,6-tetraacetyl-β-D-glucopyranosyl 2,3,4,6-tetraacetyl-β-D-galactopyranosyl selenide. 2,3,4,6-Tetraacetyl-β-D-glucopyranosyl 2,3,4-triacetyl-β-D-xylopyranosyl selenide (**19**) was also prepared and on hydrolysis by sodium methanolate gave β-D-glucopyran-osyl β-D-xylopyranosyl selenide in an 85 % yield. When **9** and ethyl iodide in acetone were treated with potassium hydroxide, a 40 % yield of ethyl 2,3,4,6-tetraacetyl-β-D-glucopyranosyl selenide was obtained.

Alkaneselenols have also been reported to react readily with cyclic ethers such as ethylene oxide to yield alkyl hydroxyalkyl selenides (equation 20).

$$(20) \quad R-SeH + R'-\overset{O}{\overset{\diagup\diagdown}{CH-CH_2}} \xrightarrow{base} R-Se-CH_2-\overset{OH}{\underset{|}{CH}}-R'$$

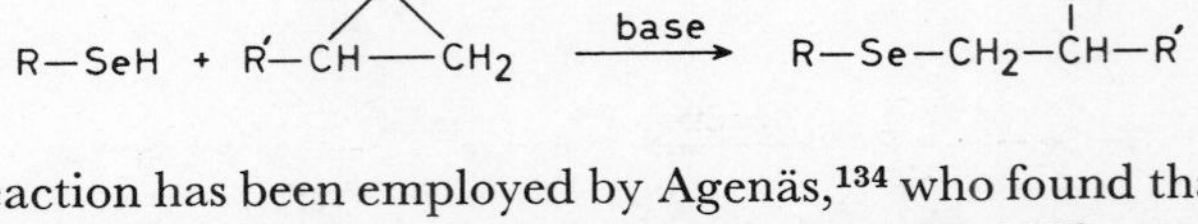

A similar reaction has been employed by Agenäs,[134] who found that sodium benzylselenolate reacted with lactones in DMF at 120–130° to give good yields of the corresponding benzylseleno-substituted carboxylic acids. The general reaction (scheme 21), here illustrated for 4-benzylselenobutyric acid (**20**), has been applied to a number of species.[134−137]

(21) [structure: lactone] + [structure: phenyl-CH₂—SeNa] $\xrightarrow[\text{(2) H}^+]{\text{(1) DMF}}$ [structure] —CH₂SeCH₂CH₂CH₂COOH

20

For the preparation of symmetrical dialkyl selenides, addition reactions of various selenium reagents to alkenes have often been employed. In some cases these reactions have been reported to give unsymmetrical dialkyl selenides. Thus, Yarovenko et al.[138,139] allowed trichloromethylselenenyl chloride to react with ethylene to form a quantitative yield of trichloromethyl 2-chloroethyl selenide. Similarly, difluoromethylselenenyl bromide and ethylene gave a nearly quantitative yield of difluoromethyl 2-bromoethyl selenide.[140]

A number of workers have reported the addition of alkaneselenols to alkenes and alkynes, usually in the presence of a catalyst such as piperidine, triethylamine, or trimethyl benzyl ammonium hydroxide (Triton B). Yields in the 65–85 % range of unsymmetrical dialkyl selenides are obtained from this reaction (equation 22), provided a polar group (carbonyl, carboxyl,

(22) $R—SeH + R'—CH=CH_2 \xrightarrow{\text{base}} R—Se—CH_2—CH_2—R'$

ether, thioether) is present in the alkene.[44,141-148]

The preparation of unsymmetrical dialkyl selenides has been achieved by starting with an organoselenium compound which was altered in the non-selenium part of the molecule. Thus, Gronowitz and Frejd,[149] when treating 2,5-dimethyl-3-iodoselenophene with ethyllithium in ether at −70°, found that the selenophene ring was split and a 75 % yield of 2-ethylseleno-2-hexen-4-yne was obtained. Mayer et al.[150] reported that the Clemmensen reduction of selenobenzoic acid Se-methyl ester gave a 23 % yield of methyl benzyl selenide in addition to benzoic acid and methaneselenol. Brintzinger et al.[57] reported the elimination of hydrogen chloride from di(1,2,2-trichloroethyl) selenide and the consequent formation of 1,2,2-trichloroethyl 1,2-dichlorovinyl selenide. Brandsma and Arens[151] allowed divinyl selenide to react with bromine at room temperature. After adding diethylaniline and heating the mixture to 100°, a 20 % yield of 2-bromodivinyl selenide was isolated. When the latter compound was treated with lithium amide in liquid ammonia, followed by alkylation with methyl iodide, a 50 % yield of vinyl propynyl selenide was obtained. Similarly, the action of lithium amide on 2-bromodivinyl selenide, followed by decomposition of the intermediate lithium salt by water, gave a 40 % yield of vinylselenoacetylene.

Yarovenko et al.[139] reported that trifluoromethylselenenyl chloride will substitute acetone to form trifluoromethylselenoacetone.

b.   Reactions.   The chemical properties of unsymmetrical dialkyl selenides are very similar to those observed for the symmetrical analogs. Attack on the Se atom of certain dialkyl selenides has been employed for the preparation of selenols and dialkyl diselenides. As known from the work by Patterson and du Vigneaud,[264] benzylthio-substituted carboxylic acids undergo reductive cleavage upon treatment with sodium in liquid ammonia. Benzylseleno-substituted analogs react similarly, forming toluene and a sodium alkane-selenolate.[124,134,136,141–144,146] The presence of the carboxyl group in these structures appears to enhance the yields of product. If the two alkyl groups attached to selenium are similar in character, the reaction is complex and has little synthetic interest. In such cases the two C-Se bonds will show almost identical reactivity towards reduction, and the products either will not contain selenium or will be a difficultly separable mixture of diselenides. This behavior has also been observed in mass spectrometric studies (*cf.* Chapter XVG).

2.   Diaryl Selenides

a.   Synthesis.   The general method for the synthesis of symmetrical diaryl selenides from an aryldiazonium salt has only occasionally been used for the preparation of unsymmetrical analogs. Thus, Chierici and Passerini[152] reported the preparation of 3-nitrodiphenyl selenide by the reaction between 3-nitrobenzenediazonium salt and sodium benzeneselenolate with sodium acetate as a buffer.

Unsymmetrical diaryl sulfones, like their symmetrical analogs, can be heated with elemental selenium to form diaryl selenides, according to reaction 13. Chierici and Passerini[86] obtained 2-methyldiphenyl selenide, 4-methyldiphenyl selenide, 4-aminodiphenyl selenide, and 4-acetamido-diphenyl selenide using this technique.

Two methods for the preparation of unsymmetrical diaryl selenides, introduced by Campbell and McCullough,[153] make use of the reaction of arylmagnesium halides with diaryl diselenides (reaction 23) and the reaction

$$(23)\quad (Ar-Se)_2 + Ar'-MgBr \longrightarrow Ar-Se-Ar' + Ar-SeMgBr$$

of diarylmercury with arylselenenyl halides (reaction 24). Both methods are

$$(24)\quad Ar_2Hg + Ar'-SeBr \longrightarrow Ar-Se-Ar' + Ar-HgBr$$

reported to give good yields. Alternatively to the Grignard reagent in reaction 23, aryllithium can be used with similar results.[154] Many compounds have been prepared by these methods in yields of 80–100 %.[153-155] A modification of these reactions was employed by Caccia Bava and Chierici,[156] who reacted 2,4-dimethyl-3-pyrrylmagnesium bromide with phenylselenenyl bromide in ether to form 2,4-dimethyl-3-phenyl-selenopyrrole. This method was also applied to the synthesis of 3-methyldiphenyl selenide.[86]

The reaction of aryl selenocyanates with an aryllithium (equation 14), introduced by Burlant,[96] has proved to be a valuable method for the preparation of unsymmetrical diaryl selenides.[96,157]

Nitrobenzeneselenenyl thiocyanates, prepared by the reaction between the corresponding selenenyl bromide and potassium thiocyanate, were found by Rheinboldt and Perrier[158] to react with arylamines and phenols in anhydrous benzene solution. 2-Nitrobenzeneselenenyl thiocyanate and N,N-dimethylaniline gave 97% of crude 2-nitro-4'-(dimethylamino)diphenyl selenide (**21**) (equation 25). Many other selenides have also been

(25)

21

prepared by use of this reaction.[86,110,158-162]

In a similar fashion, Chierici and Passerini allowed arylseleninic acids to react with aromatic amines at 115–120° for 6 hours, yielding 4-methyl-4'-aminodiphenyl selenide,[86] 4-(dimethylamino)diphenyl selenide,[86] 4-aminodiphenyl selenide,[109] 4-nitro-4'-aminodiphenyl selenide,[109] and related compounds.[86,110]

Šindelář et al.[163] reacted sodium benzeneselenolate with 2-iodobenzoic acid in water in the presence of powdered copper to obtain, after 7 hours of reflux, a 74% yield of 2-carboxydiphenyl selenide (**22**) (equation 26).

(26)

22

Reduction of **22** with lithium aluminum hydride gave 2-phenylselenobenzyl alcohol,[164] which was converted by thionyl chloride to 2-phenylselenobenzyl chloride.[164] The latter was further treated with sodium cyanide to yield 2-phenylselenobenzyl cyanide,[164] which was hydrolyzed to give 2-phenylselenophenylacetic acid.[164] A number of other compounds have been prepared in an analogous manner.[163,164]

The modified Williamson synthesis, generally applicable to the preparation of unsymmetrical dialkyl selenides, has frequently been employed for the synthesis of nitro-substituted diaryl selenides. Chierici and Passerini[86] reported the preparation of 4-nitrodiphenyl selenide (**23**) from 4-nitro-bromobenzene and benzeneselenolate in dilute ethanol by heating at reflux for 3 hours (equation 27). This method has also been used for the preparation of numerous related selenides.[86,110,111,165]

(27) NO_2—⟨benzene⟩—Br + ⟨benzene⟩—SeNa $\longrightarrow$ NO_2—⟨benzene⟩—Se—⟨benzene⟩ + NaBr

23

The method of equation 27 was applied by Chierici[166,167] to the synthesis of heterocyclic diaryl selenides in the pyrrole series and has also been employed for the preparation of pyridine[168] and thiophene[169] derivatives of this type.

In a similar method, employed by Rheinboldt and Giesbrecht[170] for the synthesis of nitrodiaryl selenides, an ethanolic solution of phenyl seleno-cyanate and 4-nitrochlorobenzene reacted with potassium hydroxide to give a 75 % yield of 4-nitrodiphenyl selenide. Modena[171] applied this method to the preparation of numerous analogs and species derived by a sequence of reactions not affecting the selenium moiety.

A number of amino-substituted diphenyl selenides have been prepared by reducing the corresponding nitro compounds.[86,109,152,168] Chierici and Passerini[152] investigated the use of various reducing agents for this purpose.

b.   Reactions.   As mentioned above, unsymmetrical diaryl selenides are very similar in nature to their symmetrical analogs. The stability of these compounds is well demonstrated, for example, by the reduction of nitro-diaryl selenides to the corresponding aminodiaryl selenides.[86,109,152,168] Bartolotti and Cerniani[111] investigated the stability of nitrodiaryl selenides to organic bases. When 2,4-dinitrodiphenyl selenide and similar compounds were heated to reflux with hydrazine hydrate, scission of one C-Se bond occurred and 2,4-dinitrophenylhydrazine was formed. However, 4-nitro-diphenyl selenide (**23**) was unaffected. Similarly, dinitrodiaryl selenides, when treated with piperidine, gave 2,4-dinitrophenylpiperidine derivatives, while **23** remained unchanged.

In a comparative investigation of analogous ethers, sulfides and selenides, Müller et al.[154] recorded the e.s.r. spectrum of 3,5-di-*tert*-butyl-4-hydroxy-diphenyl selenide in order to study the proposed selenium radical formation. No indication was found for Se atom participation in the formation of radicals. This observation may be compared to the mass spectrometric studies of diaryl selenides (*cf.* Chapter XVG).

Litvinenko and Cheshko[109] have investigated the substituent effect from the Se atom in 4-aminodiphenyl selenide and 4-amino-4'-nitrodiphenyl selenide by means of the reaction with picryl chloride to give alkylamino derivatives. Kinetic experiments showed the substituent effects of S and Se atoms to be very similar. The reaction rate was lower than that observed for the corresponding ethers, but similar to that found for analogous biphenyl derivatives.

3. Alkyl Aryl Selenides

The properties of alkyl aryl selenides fall between those characteristic for dialkyl and those for diaryl selenides. Molecular modifications of these species have had various purposes; for example, by the introduction of one aromatic group, a low-molecular-weight aliphatic selenide is made easier to handle. On the other hand, the introduction of an alkyl group makes it possible to study a single arylseleno group by spectroscopic methods.

a. Synthesis. The modified Williamson synthesis is the most frequently employed method for the preparation of alkyl aryl selenides (equation 28);

$$(28) \quad Ar-Se^{\ominus} + R-X \longrightarrow Ar-Se-R + X^{\ominus}$$

this procedure was developed by Foster and Brown.[172] Equivalent amounts of an areneselenol and sodium hydroxide are jointly dissolved in water. After dilution with ethanol, an alkyl halide is added and the product is obtained after heating at reflux for 15 min. By this simple and easily handled procedure a very large number of compounds have been prepared in excellent yields.[8,99,131,154,171,173-216]

In a similar approach, aryl selenocyanates were cleaved with base in the presence of an alkyl halide or dialkyl sulfate to give good yields of alkyl selenides.[8,86,176,186,199,204,217] Treatment with base alone was found to give diaryl diselenides. An intermediate areneselenolate was formed by reduction of the diselenide with sodium metal in methanol, to which an alkyl halide was added to obtain the desired alkyl aryl selenide.

For the preparation of selenium-containing carbohydrates, Wagner and Nuhn[218] employed glucosylisoselenouronium bromides, which were treated successively with base and nitroaryl halides to give glucosyl aryl selenides (equation 29). A number of related compounds have also been obtained[41,218,219] in this manner.

The reaction between areneselenols and cyclic ethers (equation 20) has been employed for the preparation of alkyl aryl selenides. The selenol is usually dissolved in an alkaline solution, and the cyclic ether is added with cooling to give a rapid formation of the alkyl aryl selenide.[132,220,221] In the presence of alkali the reaction proceeds to give a single product, whereas in neutral solution a mixture of isomers is obtained.

In a reaction similar to that reported by Agenäs[137] and described in equation 21, Šindelář *et al.*[222] combined benzeneselenol and phthalide in ethanolic solution in the presence of sodium ethanolate to provide an 89% yield of 2-phenylselenomethylbenzoic acid.

The addition reaction of areneselenols to alkenes or alkynes has frequently been employed for the preparation of alkyl aryl selenides. Two alternative conditions have been used: the reaction is performed either in the presence of a catalytic amount of base or in a neutral solution. In the first case (equation 30), the components are mixed or dissolved in ethanol, and sodium

(30) $\langle\;\rangle$—SeH + CH$_2$=CH—COOH $\xrightarrow{\text{base}}$ $\langle\;\rangle$—Se—CH$_2$—CH$_2$—COOH

24

ethanolate or another base such as piperidine is added.[44,187,201,223–226] In the second case, the components are mixed, possibly with the addition of ethanol as a solvent, and the product is formed by an exothermic reaction.[201,205,223,225,227–230]

Areneselenenyl halides and alkynes have been shown to undergo an addition reaction leading to halogenated alkenyl aryl selenides (equation 31).[201,202,231–233] It was found that the solvent employed has a dominating

(31) $\langle\;\rangle$—SeCl + CH≡CH $\longrightarrow$ $\langle\;\rangle$—Se—CH=CH—Cl

25

influence on the path of the reaction. Thus, with ethyl acetate a *cis* addition predominates, whereas acetic acid favors the *trans* addition. The synthesis was performed by dissolving the areneselenenyl halide in the appropriate solvent, followed by gradual addition of the alkyne. After absorption of the calculated amount, the mixture was kept at room temperature until the reaction was complete.

In a related process, Jenny[234] reacted 1-anthraquinoneselenenyl acetate with cyclohexene in acetic acid; a 69% yield of 2-acetoxycyclohexyl 1-anthraquinonyl selenide (26) was obtained (equation 32). Several species have been prepared by following the same procedure.[234–237]

Rheinboldt and Perrier[238] reacted 4-methyl-2-nitrophenylselenenyl bromide and acetophenone at room temperature in an inert atmosphere, to

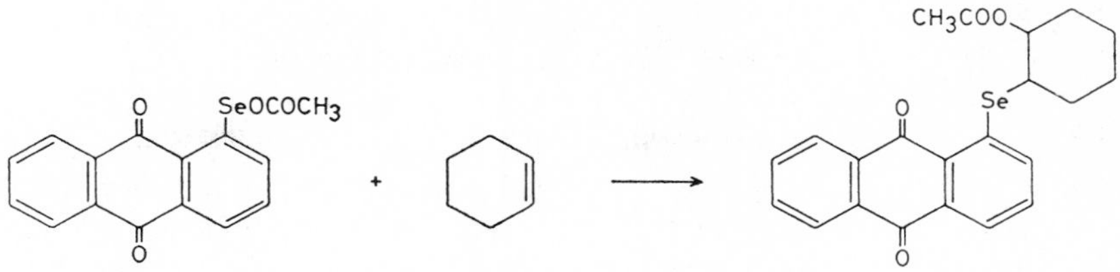

26

obtain an 82% yield of phenacyl 4-methyl-2-nitrophenyl selenide (**27**) (equation 33). In other cases, acetic acid or pyridine was used as the solvent.[235,238,239]

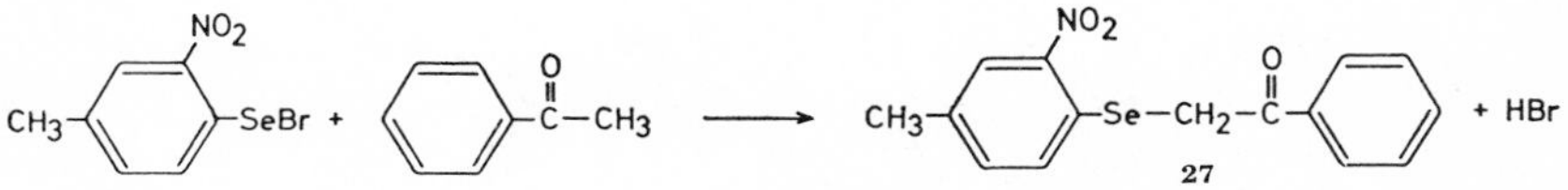

27

It was found, in some instances, that areneselenenyl halides did not substitute the aliphatic compounds, but rather led to a quantitative formation of diaryl diselenides. Rheinboldt and Perrier[240] then made the observation that in these cases areneselenenyl thiocyanates give good yields of the desired alkyl aryl selenide without any formation of diaryl diselenide. The gross reaction proceeds analogously to that given in equation 33, but details of the differing reaction mechanisms for the halide and the thiocyanate case have not been investigated. The selenenyl thiocyanate method was employed extensively by Rheinboldt and his co-workers.[87,160,238,240]

Areneselenenyl acetates have also been employed for substitution reactions similar to the one in equation 33. Hölzle and Jenny[234–237,241] allowed aryl-substituted ethenes to react with areneselenenyl acetates in refluxing acetic acid, and thus obtained a variety of substituted aryl alkyl selenides. In some instances pyridine was used as the solvent.

Alkaneselenenyl halides have also been employed for the preparation of alkyl aryl selenides through substitution reactions on aromatic amines and phenols.[242–244] Yagupol'skii and Voloshchuk,[242] for instance, reacted aniline with trifluoromethaneselenenyl chloride in ether at −20° and obtained trifluoromethyl 4-aminophenyl selenide in a 77% yield.

Grignard reagents have frequently been used as intermediates in the synthesis of alkyl aryl selenides. Usually, arylselenomagnesium halides are treated with alkyl halides in the presence of sodium hydroxide. This procedure represents another example of a modified Williamson synthesis. The more direct procedure shown in equation 34 has been employed by

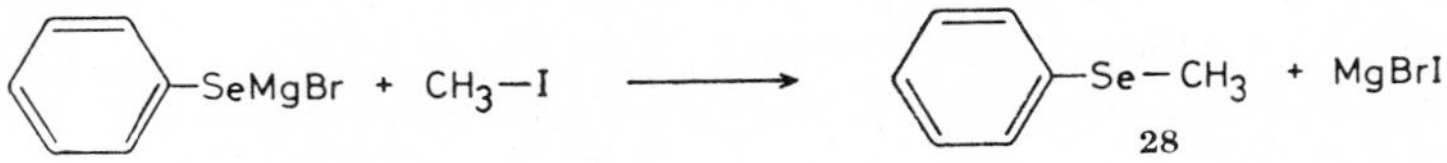

28

Supniewski *et al.*[245] Methyl 4-biphenylyl selenide[246] and methyl 2,4,6-trimethylphenyl selenide[8,184] were prepared analogously.

Unsymmetrical diaryl selenides have been obtained by the reaction between an arylselenomagnesium halide and a diaryl diselenide, as given in equation 23. This method was employed for the preparation of butyl phenyl selenide[153] and benzyl phenyl selenide.[153] The alternative reaction between diarylmercury and alkaneselenenyl bromide, following the procedure indicated in equation 24, was also found applicable to these latter compounds.[153] Yagupol'skii and Voloshchuk[243,247] have employed arylmagnesium halides for the preparation of alkyl aryl selenides from alkaneselenenyl halides (equation 35). The reaction was performed in anhydrous ether

$$\text{(35)} \quad \boxed{}\text{—MgBr} \; + \; \text{CF}_3\text{SeCl} \quad \longrightarrow \quad \boxed{}\text{—Se—CF}_3 \; + \; \text{MgClBr}$$
$$\underset{\textbf{29}}{}$$

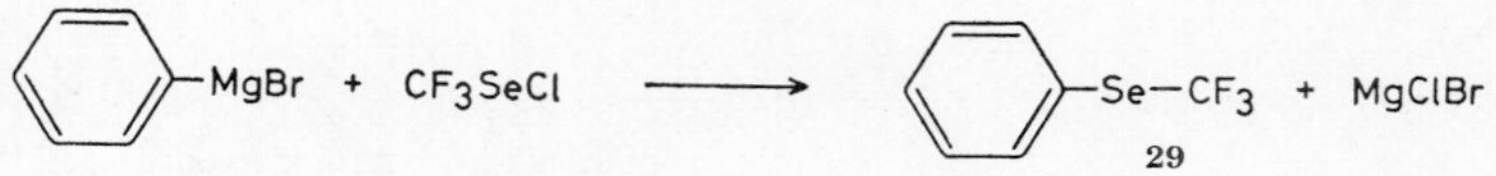

at —50°, and trifluoromethyl phenyl selenide (**29**)[247] was obtained in a 50 % yield. In this way, trifluoromethyl 4-methylphenyl selenide[247] and trifluoromethyl 4-methoxyphenyl selenide[243] were also obtained.

Aryllithium compounds have also been employed for the preparation of alkyl aryl selenides by their reaction with elemental selenium to form an intermediate lithium areneselenolate. The latter reacted with alkyl halides as shown in scheme 36. In this way Christiaens and Renson[248] prepared

$$\text{(36)} \quad \underset{\text{Li}}{\overset{\text{CH(OC}_2\text{H}_5)_2}{\boxed{}}} \; \xrightarrow{\text{Se}} \; \underset{\text{SeLi}}{\overset{\text{CH(OC}_2\text{H}_5)_2}{\boxed{}}} \; \xrightarrow[\text{(2) HCl}]{\text{(1) CH}_3\text{I}} \; \underset{\underset{\textbf{30}}{\text{Se—CH}_3}}{\overset{\text{CH=O}}{\boxed{}}}$$

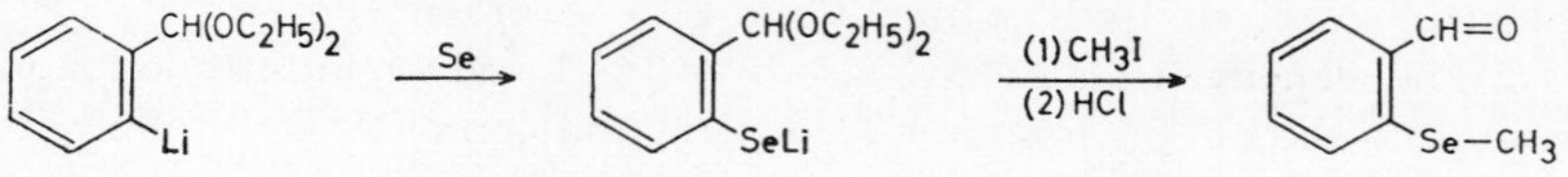

methyl 2-formylphenyl selenide (**30**) in 34 % yield. A number of other compounds, including some which are heterocyclic, have been obtained by this route.[248–254]

When Seebach and Peleties[255] allowed bis(phenylseleno)methane to react with butyllithium in the presence of benzophenone, butyl phenyl selenide and 2,2-diphenyl-2-hydroxyethyl phenyl selenide were isolated. This reaction probably occurs by the initial attack of butyllithium on one Se atom, thus forming a butyl phenyl selenide molecule and phenylselenomethyllithium. The latter, in a succeeding step, reacts with benzophenone.

Burlant[96] has reported the reaction between methyllithium and phenyl selenocyanate, by which an 83 % yield of methyl phenyl selenide (**28**) was obtained. The reaction is analogous to that given in equation 14.

Renson and Piette[256] have employed a dialkylcadmium compound for the preparation of alkylseleno-substituted aryl ketones, as shown in scheme 37.

(37) 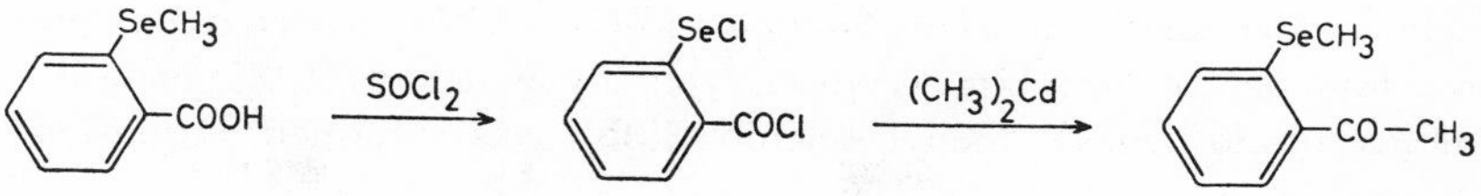 ⟶ SOCl₂ ⟶ (CH₃)₂Cd ⟶ 31

By this method and other very similar ones, a number of homologous alkylseleno-substituted aryl ketones were obtained.[196,198,256] When **31** was treated with an excess of the dimethylcadmium, the organometallic compound added to the carbonyl group, forming the tertiary alcohol, methyl 2-(2′-hydroxy-2′-propyl)phenyl selenide (**32**). At a higher reaction temperature **32** was found to eliminate water to form methyl 2-(propen-2′-yl)-phenyl selenide (**33**) (scheme 38). This procedure has been applied to a variety of similar species.[196,208]

(38) 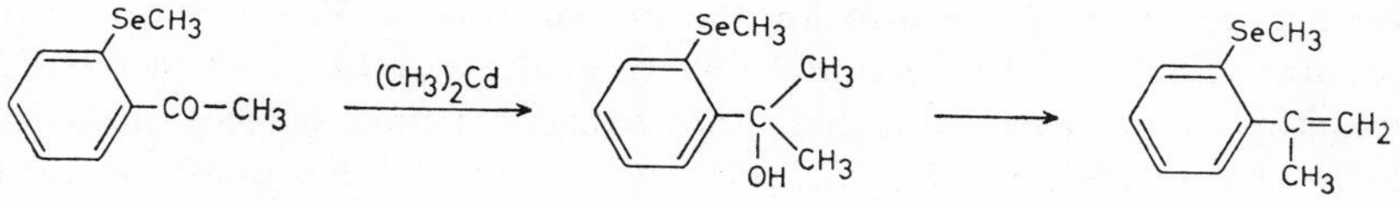   31   (CH₃)₂Cd ⟶   32   ⟶   33

Pallotti and Tundo[257] found that symmetrical diaryl diselenides react with bis(benzenesulfonyl)methane in methanol on the addition of sodium (scheme 39).

(39)

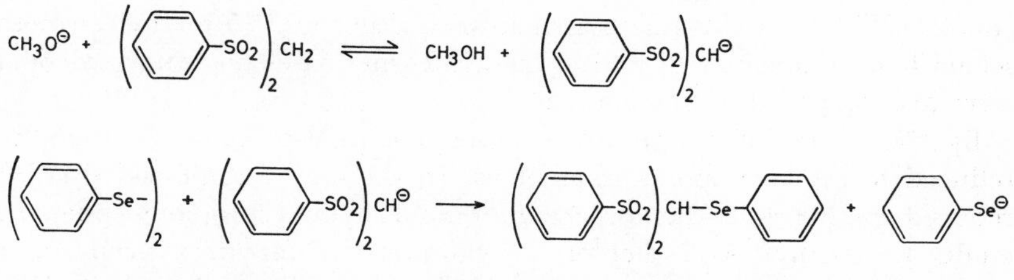

In an aminoalkylation reaction of the Mannich type, Pollak and Grillot[258] combined benzeneselenol with formaldehyde and N-methylaniline and obtained an 85% yield of N-methyl-N-phenylaminomethyl phenyl selenide (**34**) (equation 40). A 94% yield of N,N-dibenzylaminomethyl phenyl selenide[258] was also obtained. Furthermore, the related reaction of benzeneselenol with formaldehyde and hydrochloric acid was reported to give chloromethyl phenyl selenide.[264]

By an alternative process, the reaction between benzeneselenol, formaldehyde, and N-methylaniline was performed in the presence of hydrochloric

(40)   ⟨ ⟩–SeH +   ⟨ ⟩–NH–CH₃ + CH₂=O ⟶   ⟨ ⟩–Se–CH₂–N⟨ ⟩ + H₂O

34

acid. Under these conditions Pollack and Grillot[258] observed a condensation reaction of the hydroxyalkylation type, by which a 76% yield of 4-(N-methylamino)benzyl phenyl selenide (35) was obtained (equation 41).

(41) [structure] —SeH + [structure] —NH—CH₃ + CH₂=O $\xrightarrow{HCl}$ CH₃NH—[structure]—CH₂—Se—[structure] + H₂O

                                                        35

4-Aminobenzyl phenyl selenide (95%)[258] and 4-(N,N-dimethylamino)-benzyl phenyl selenide (36) (7%)[258] were also generated by this method. The low yield of 36 was raised considerably by the use of N-methyl-N-phenyl-4-(N',N'-dimethylamino)benzylamine and benzeneselenol as the starting materials. In the presence of hydrochloric acid a 76% yield of 36 was obtained by essentially the procedure depicted in equation 41.[258,259] A further improved procedure for the preparation of 36 raised the yield of product to 95%; this method involved the reaction between 4-(N,N-dimethylamino)benzyl alcohol and benzeneselenol in the presence of hydrochloric acid, which went, presumably, via a Williamson-type mechanism.

Alkyl aryl selenides have also been employed as starting materials for the preparation of other alkyl aryl selenides without influencing the C–Se–C moiety. Thus, 2-(methylseleno)benzoic acid when reduced by lithium aluminum hydride gave the corresponding 2-methylselenobenzyl alcohol.[196,198] 4-Methyl-2-(methylseleno)benzyl alcohol,[198] 5-methyl-2-(methylseleno)benzyl alcohol,[198] and 6-methyl-2-(methylseleno)benzyl alcohol[198] were also prepared in this manner.

By the action of aluminum isopropoxide in 2-propanol, ketones were reduced to the corresponding alcohols. In this way Supniewski et al.[180,245] reduced methyl 4-(2'-dichloroacetylamino-3'-hydroxypropionyl)phenyl selenide to methyl 4-(2'-dichloroacetylamino-1',3'-dihydroxypropyl)phenyl selenide. Methyl 4-(2'-acetylamino-1',3'-dihydroxypropyl)phenyl selenide was obtained by this method in 80% yield.

Nitro-substituted alkyl aryl selenides have been reduced to form the corresponding amines. Tin and hydrochloric acid reduced methyl 2-nitrophenyl selenide to methyl 2-aminophenyl selenide.[86] Raney nickel and hydrogen were also used for its preparation,[86] as well as for 2-aminobenzyl phenyl selenide.[190] Raney nickel and hydrogen were also employed as reducing agents for the synthesis of aminophenylseleno-substituted carbohydrates. Thus, 2,3,4,6-tetraacetyl-β-D-glucopyranosyl 2-aminophenyl selenide[41] was obtained and subsequently hydrolyzed to give β-D-glucopyranosyl 2-aminophenyl selenide.[41] 2,3,4,6-Tetraacetyl-β-D-glucopyranosyl 4-aminophenyl selenide[219] and β-D-glucopyranosyl 4-aminophenyl selenide[219] were similarly obtained.

Various oxidizing agents have been employed to change the structural components of alkyl aryl selenide molecules. By the use of iodine in pyridine, Sukiasyan *et al.*[253] have oxidized methyl 5-acetyl-2-thienyl selenide to give methyl 5-carboxy-2-thienyl selenide. The oxidation of methyl 5-formyl-2-thienyl selenide by means of silver oxide was reported to give the latter in 87% yield.

Nitric acid has frequently been used for the oxidation of organoselenium compounds to prepare selenoxides or seleninic acids. Hannig and Ziebandt[179] have reported the treatment of ethyl 4-propionylphenyl selenide (**39**) with dilute nitric acid at 80°, followed by a reductive step to yield 4-(ethylseleno)benzoic acid (**40**) (scheme 42). The same method was also

(42) 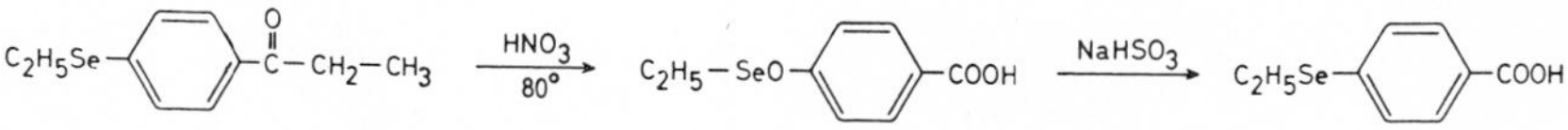 C_2H_5Se—⟨⟩—$\overset{\overset{O}{\|}}{C}$—$CH_2$—$CH_3$ $\xrightarrow[80°]{HNO_3}$ C_2H_5—SeO—⟨⟩—COOH $\xrightarrow{NaHSO_3}$ C_2H_5Se—⟨⟩—COOH

       39   40

employed for the preparation of higher homologs of **40** in yields of 45–50%.[179]

The oxidation of selenium-containing ketones to α-diketones by selenium dioxide in acetic anhydride was reported by Christiaens and Renson[248] and applied to the conversion of methyl 2-phenylacetylphenyl selenide to 2-(methylseleno)benzil.[248]

A halogen exchange reaction has been reported by Yagupol'skii and Kondratenko.[260] When trifluoromethyl phenyl selenide was treated with boron trichloride in a sealed tube at 100° for 3 hours, a 65% yield of trichloromethyl phenyl selenide was isolated. Similarly, tribromomethyl phenyl selenide was obtained in a 51% yield with boron tribromide.

Nucleophilic substitution reactions at aliphatic carbons have been utilized to effect changes of structure within alkyl aryl selenide molecules. Ruwet and Renson[196] reacted 2-(methylseleno)benzoic acid with dichloromethyl butyl ether and zinc chloride under reflux to obtain a 90% yield of 2-(methylseleno)benzoyl chloride. The latter reacted with ammonia to form the corresponding benzamide, which on further heating with phosphorus pentoxide gave 2-(methylseleno)benzonitrile.[198]

In a similar reaction (scheme 43), 2-(methylseleno)benzyl alcohol (**41**) was converted to 2-(methylseleno)benzyl chloride (**42**).[198,261] The latter, without isolation, reacted with cyanide ions to give 2-(methylseleno)benzyl cyanide (**43**).[261] Alkaline hydrolysis of the latter yielded 2-(methylseleno)phenylacetic acid (**45**).[261] Compound **42** was heated to 150° with dimethyl sulfoxide and sodium carbonate to give 2-(methylseleno)benzaldehyde (**44**). The synthesis of the latter was also performed by the Sommelet reaction with hexamethylenetetramine in chloroform or by reaction with 2-nitropropane and sodium ethanolate. The reactions in scheme 43 were applied to several analogous seleno-substituted aldehydes.[261]

(43)

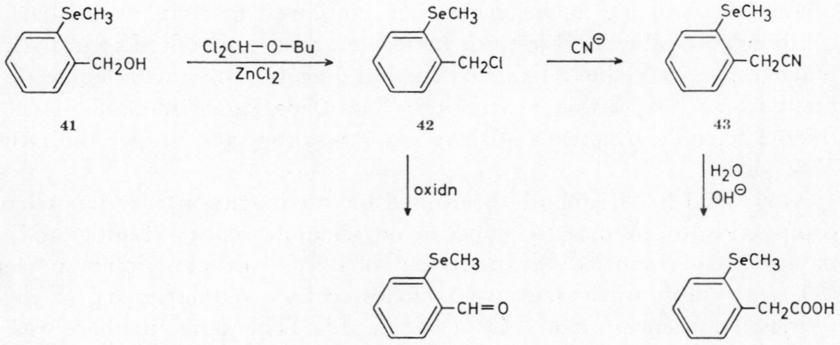

The introduction of sulfur substituents into aryl alkyl selenides was reported by Hannig and Ziebandt.[262] Methyl 4-bromoacetylphenyl selenide,[262] on reaction with thiocyanate ions in ethanol, gave an 80–85% yield of methyl 4-thiocyanatoacetylphenyl selenide.[262] The higher homologs, ethyl and propyl 4-thiocyanatoacetylphenyl selenide, were also prepared.[262] Moreover, similar reactions were carried out using thiourea to give iso-thiuronium salts.[262]

Methyl 4-bromoacetylphenyl selenide and the related ethyl and propyl analogs reacted with hexamethylenimine (azepin) in ether to give 80–85% yields of 4-alkylselenophenacylazepin hydrochlorides.[262] The adduct formed in the Delépine reaction of methyl 4-bromoacetylphenyl selenide with hexamethylenetetramine in chloroform was treated with hydrochloric acid to give 75–80% yields of methyl 4-aminoacetylphenyl selenide hydrochloride.[180,245,246,262]

A related procedure was employed by Hannig,[263] who reacted methyl 4-(3'-chloropropionyl)phenyl selenide with piperidine in ethanol in the presence of potassium acetate to obtain methyl 4-(3'-piperidinopropionyl)-phenyl selenide hydrochloride.[263]

2-(Methylseleno)benzoyl chloride (46) has been employed in a malonic ester synthesis for the preparation of the corresponding ketone, 2-(methyl-seleno)acetophenone (47) (scheme 44).[198] By the same procedure 4-methyl-2-(methylseleno)acetophenone,[198] 5-methyl-2-(methylseleno)acetophenone,[198] 6-methyl-2-(methylseleno)acetophenone,[198] and 2-(methylseleno)phenylace-tone[264] were also obtained.

The preparation of esters derived from carboxyl-substituted alkyl aryl selenides has been described. Ruwet and Renson[198] allowed 2-(methyl-seleno)benzoyl chloride (46) to react with ethanol in toluene, thereby obtaining ethyl 2-(methylseleno)benzoate. Baker et al.[197] used the silver salts of

(44)

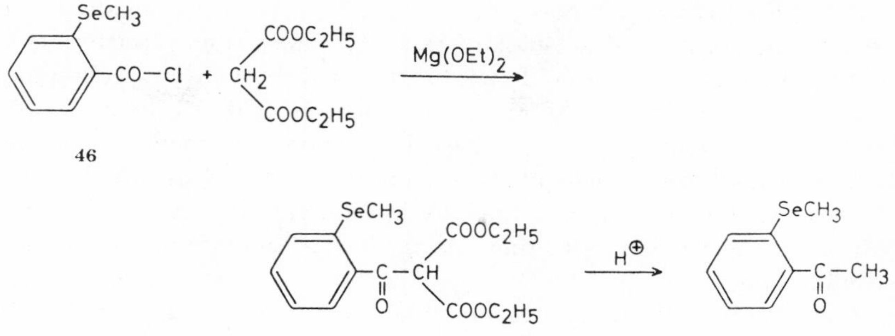

46

47

carboxylic acids, which were heated to reflux with an alkyl halide in alcohol. In this way a number of esters were obtained from comparatively complex alkyl halides with alkylseleno-substituted benzoic acids.[179,197,262]

Methyl 4-(methylseleno)benzoate also reacted with hydrazine hydrate to give 4-(methylseleno)benzoyl hydrazide.[197] In a succeeding reaction the hydrazide was treated with benzenesulfonyl chloride to give 4-(methylseleno)benzoyl benzenesulfonyl hydrazide,[197] which upon hydrolysis gave 4-(methylseleno)benzaldehyde.[197]

The preparation of methyl 4-bromoacetylphenyl selenide (**48**) by treating 4-(methylseleno)acetophenone with bromine in chloroform was reported by Supniewski et al.[180,245] No observable reaction of bromine occurred with the selenium moiety. Hannig and Ziebandt[262] used a somewhat different procedure for the preparation of **48**, in which the action of sulfuryl chloride in benzene on methyl 4-acetylphenyl selenide generated the dihalide, which was then treated with bromine in chloroform. After reduction of the intermediate with sodium hydrogen sulfite, a 78–85% yield of **48** was obtained. Ethyl 4-bromacetylphenyl selenide and propyl 4-bromoacetylphenyl selenide were similarly prepared.[262]

Chierici and Montanari[201,202] reported an addition of bromine to cis- and trans-3-(phenylseleno)propenoic acids in acetic acid solution at 0°. The adduct was treated with water to give a 50–73% yield of a mixture of cis- and trans-2-bromovinyl phenyl selenides. When this mixture was treated with ethanolic potassium hydroxide, a 71% yield of vinyl phenyl selenide was obtained.

The action of Grignard reagents of nitriles, known to produce ketones, has been used for the synthesis of ketones containing a selenide group. Thus, Christiaens and Renson[248] reacted 2-(methylseleno)benzonitrile with benzylmagnesium chloride in ether to obtain a 60% yield of benzyl 2-(methylseleno)phenyl ketone.

Butyllithium, on the other hand, reacts with alkyl aryl selenides at the Se atom, as demonstrated by Goldfarb et al.[252] for alkyl thienyl selenides. Thus, when methyl 2-thienyl selenide was treated with butyllithium followed by a subsequent reaction with carbon dioxide, an 84 % yield of 2-thienylcarboxylic acid was obtained. An interesting observation was made when methyl 5-methylthio-2-thienyl selenide was similarly treated with butyllithium. In this case a 68 % yield of 5-methylthio-2-thienylcarboxylic acid was obtained,[253] clearly demonstrating the difference in stability of sulfide and selenide bonds.

The action of strong bases on chloromethyl compounds is known to produce carbenes. Thus, the effect of potassium *tert*-butoxide on chloromethyl phenyl selenide (**49**) in *tert*-butyl alcohol has been shown by Schöllkopf and Küppers[265] to produce selenium-containing carbenes which, on reaction with isobutene, gave a 70 % yield of 2,2-dimethylcyclopropyl phenyl selenide (**50**) (scheme 45). In the presence of *cis*-2-butene and

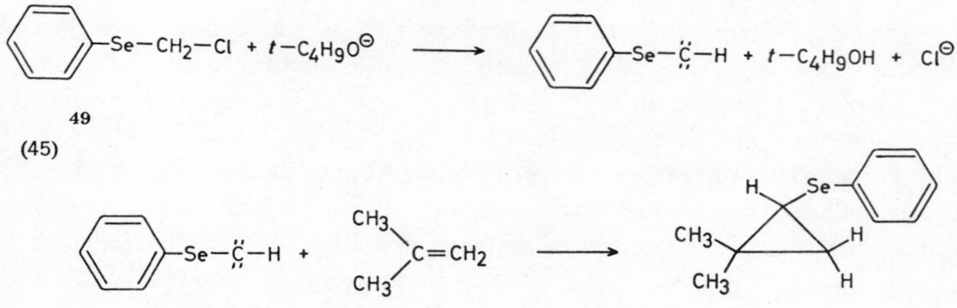

trans-2-butene the three isomeric products, **51**, **52**, and **53**, were obtained.

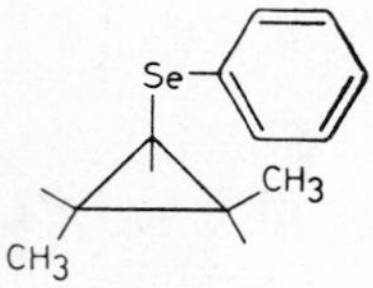

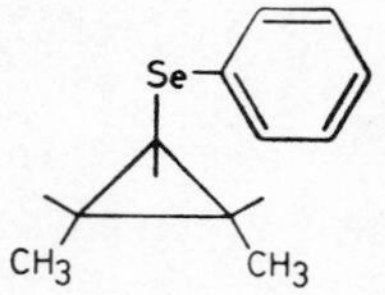

Phenylselenocarbene was found to react with cyclohexene to give a 70 % yield of a 2:1 mixture of *exo-* and *endo*-7-phenylselenonorcarane whereas 1,3-cyclohexadiene gave a 2:1 mixture of *exo-* and *endo*-7-phenylseleno-2-norcarenes.

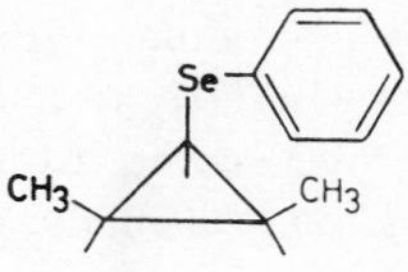

Condensation reactions of the aldol type have also been performed with substituted alkyl aryl selenides. Supniewski et al.[180,245] allowed 4-(methylseleno)-α-dichloroacetylaminoacetophenone to react with formaldehyde in the presence of sodium hydrogen carbonate to form 2-(4′-methylseleno-benzoyl)-2-dichloroacetylaminopropanol. A 70% yield of 2-(4′-methyl-selenophenyl-4″-benzoyl)-2-acetylaminopropanol[246] was obtained in this manner.

When a mixture of 2-(methylseleno)acetophenone and benzaldehyde, in acetic acid containing piperidine as a catalyst, was heated to reflux, the condensation product methyl 2-cinnamoylphenyl selenide[187] was obtained. Analogous reactions gave a 51% yield of methyl 2-cinnamoyl-4-methylphenyl selenide and a 43% yield of methyl 2-(4′-hydroxycinnamoyl)-4-methylphenyl selenide.[187]

Many condensation products, derived from methylseleno-substituted aldehydes and ketones by reactions of the Reformatsky, Perkin, and Knoevenagel types, have been reported.[266,267] Further condensation reactions, in which 2-methylselenobenzyl cyanide reacted with ethyl formate in the presence of sodium ethanolate as a catalyst, have been described by Christiaens and Renson.[261] In this way 2-(2′-methylselenophenyl)-2-cyano-acetaldehyde was obtained. Hannig and Ziebandt[268] condensed 4-(methyl-seleno)acetophenone (**54**) with ethyl acetate in ether with the addition of sodium metal to the solution to give 4-(methylseleno)benzoylacetone (**55**) (equation 47). The series 4-methyl- to 4-(heptylseleno)benzoylacetone was prepared by the same procedure.[268]

By use of a Mannich reaction, **54** and related species were condensed with formaldehyde and azepine in dioxane containing a small amount of hydrochloric acid as a catalyst. After the addition of acetic anhydride and heating to reflux, methyl 4-(3′-azepinopropionyl)phenyl selenide hydrochloride and analogs were obtained.[178]

A number of condensation reactions leading to heterocyclic compounds have been explored. Thus, Hannig[177] reacted methyl 4-(3′-chloropropionyl)-phenyl selenide with phenylhydrazine in ethanol to obtain methyl 4-(1′-phenyl-2′-pyrazolin-3′-yl)phenyl selenide. Likewise, when **55** was treated with hydrazine hydrate or with other suitable compounds containing two amino groups, various pyrazolines, isoxazolines, pyrimidines, and diazepines were obtained.[177,268]

Other cyclization reactions have been performed by heating methyl 4-thiocyanatoacetylphenyl selenide in acetic acid and sulfuric acid to give methyl 4-(2′-hydroxy-4′-thiazolyl)phenyl selenide.[262] When 4-(methyl-seleno)phenacylisothiuronium bromide was heated with ethanolic potassium hydroxide, a good yield of methyl 4-(2'-amino-4'-thiazolyl)phenyl selenide[262] was obtained.

Methyl 4-dibromoacetylphenyl selenide,[262] prepared by bromination of the corresponding 4-(methylseleno)acetophenone, was treated with dilute potassium hydroxide. A Favorskii rearrangement occurred, giving 4-(methylseleno)-phenylhydroxyacetic acid. The same procedure led to 4-(ethylseleno)phenyl-hydroxyacetic acid[262] and 4-(propylseleno)phenylhydroxyacetic acid.[262]

Various aromatic substitution reactions have also been performed with alkyl aryl selenides as the starting materials. Thus, the treatment of methyl 2-thienyl selenide in concentrated hydrochloric acid with a bromide-bromate solution gave a 45% yield of methyl 5-bromo-2-thienyl selenide.[252]

Alkyl aryl selenides have frequently been subjected to the Friedel-Crafts acylation reaction. When methyl phenyl selenide reacted with acetyl chloride and anhydrous aluminum trichloride in chloroform at −5°, a high yield of 4-(methylseleno)acetophenone[180,245] was obtained. Numerous other related species have also been prepared.[177–180,186,245,253]

A few examples of Vilsmeier formylations of alkyl aryl selenides have also been reported. When methyl 3,4-methylenedioxyphenyl selenide was treated with dimethylformamide and phosphoryl chloride, 3,4-methylene-dioxy-6-(methylseleno)benzaldehyde was obtained.[269] In the same way 5-methylseleno-2-thiophenealdehyde[253] and 5-methyl-2-methylseleno-3-thio-phenealdehyde[253] were synthesized.

Trifluoromethyl aminoaryl selenides have been diazotized, and the intermediate diazonium salts undergo various reactions. Thus, trifluoromethyl 4-aminophenyl selenide (56), on diazotization and treatment with hypophosphorous acid, gave a 35% yield of the deaminated product, trifluoromethyl phenyl selenide.[242,243] Trifluoromethyl 3-fluorophenyl selenide[244] and trifluoromethyl 3-carboxyphenyl selenide[244] were obtained in a like manner. Furthermore, the diazonium salt produced from 56 was allowed to react with other reagents: the Schiemann reaction with tetrafluoroborate ions gave trifluoromethyl 4-fluorophenyl selenide;[242,243] the Sandmeyer reaction with cuprous bromide in hydrobromic acid gave trifluoromethyl 4-bromophenyl selenide;[243] the modified Sandmeyer reaction with iodide ion gave trifluoromethyl 4-iodophenyl selenide.[242] Similarly, the Sandmeyer reaction with cyanide ion produced trifluoromethyl 4-cyanophenyl selenide,[244] which was hydrolyzed to the corresponding benzoic acid.

b. *Reactions.*   An outstanding characteristic of the selenide group appears to be its nonreactivity towards numerous reagents capable of affecting other parts of the molecule; this is illustrated by the synthetic procedures in the latter part of Section V.C.3a. The heat stability of the C–Se–C bonds in selenides is often remarkably high. When, for example, *cis*-3-(phenylseleno)-propenoic aid was heated to 260°, the selenide group was found to be unaffected, but an isomerization to *trans*-3-(phenylseleno)propenoic acid was observed.[201,202] Although the action of acids does not, in general, influence the selenide group, strong bases are likely to cause scission of the C–Se bond. Interesting exceptions have, however, been observed. The action of potassium *tert*-butoxide in DMSO on 2-propenyl phenyl selenide was reported to give only the isomeric 1-propenyl phenyl selenide.[191,270,271] The participation of the Se atom in this rearrangement reaction, however, was not proved. Similar rearrangements have also been observed for other alkenyl and alkynyl groups,[194,195] as well as for certain compounds with substituents in the aromatic ring of the alkyl aryl selenides. On the other hand, Gilman and Cason[227] observed that γ-ketoselenides, when treated with 1 M sodium hydroxide in ethanol, readily decompose, forming the corresponding selenol and an unsaturated ketone. Numerous examples of condensation reactions have similarly shown the C–Se bonds to be stable also to moderately strong bases.

Alkyl aryl selenides are normally stable towards reducing agents. The action of sodium and other alkali metals in liquid ammonia or amines, however, causes scission of one C–Se bond, predominantly on the alkyl side of the molecule.

Oxidizing agents give alkyl aryl selenoxides in a reaction which will be discussed further in Section V.E. Parts of the molecule other than the selenide group are not usually oxidized without affecting the conversion of the compound to the analogous selenoxide.

The action of halogen on alkyl aryl selenides will, with a few exceptions, give the corresponding selenide dihalides. Similarly, sulfuryl chloride has been found to generate the corresponding selenide dichloride, in an attempt to prepare acid chlorides from carboxylic acids containing a selenide group. The method of Ruwet and Renson,[196] which employs dichloromethyl butyl ether and zinc chloride for this purpose, eliminates the risk of formation of selenide dihalides. On the other hand, Supniewski *et al.*[180,245,246] reported that bromine in chloroform reacted with selenium-containing ketones without influencing the selenide group. This result is contrary to that of Chierici and Montanari,[201,202] who added bromine to the C–C double bond of aryl-selenoalkenoic acids without being able to avoid the formation of a selenide dibromide group. Occasionally, bromine has been found to cause scission of the alkyl-Se bond of alkyl aryl selenides containing nitro or methoxy substituents in the aromatic group, thus forming an areneselenenyl bromide.[241]

Organometallic compounds, important intermediates for the preparation of numerous organoselenium compounds, react with alkyl aryl selenides in various ways. Grignard reagents combine with the cyano group of alkyl-selenobenzonitriles without influencing the selenide group.[261,268] However, alkyl- and aryllithium often split the C–Se bond, as was shown by Seebach and Peleties[255] for methyl phenyl selenide. In that case, the alkyl-Se bond was split by butyllithium, whereas Goldfarb et al.[252] found the aryl-Se bond of alkyl thienyl selenides to be cleaved.

D. COMPOUNDS CONTAINING TWO OR MORE ISOLATED SELENIDE GROUPS

Organoselenium compounds containing two or more isolated selenide groups have chemical and physical properties very similar to those of dialkyl selenides or other monoselenides. Although the preparation of such compounds has often been achieved by methods well known for simple monoselenides, a few special procedures have also been employed for this class.

1. Synthesis

Aynsley et al.,[272] in a modification of the Williamson ether synthesis, allowed dimethyl diselenide to react with sodium metal in liquid ammonia. The sodium methaneselenolate thus formed reacted with 1,2-dibromopropane to give a good yield of 2,5-diselenahexane (**57**) (scheme 48).[272,273] The

$$(CH_3Se)_2 + 2\,Na \xrightarrow{NH_3\,(l)} 2\,CH_3SeNa$$

(48)

$$2\,CH_3SeNa + Br-CH_2CH_2-Br \longrightarrow CH_3SeCH_2CH_2SeCH_3$$

57

same procedure was also employed for the preparation of 2,6-diselenaheptane,[272,273] 2,7-dimethyl-3,6-diselenaoctane,[274,275] 5,16-diethyl-7,14-diselenaeicosane,[276] and 1,10-dicyclopentyl-2,9-diselenadecane.[276] The reaction of sodium benzeneselenolate with diiodomethane in ethanol under reflux, analogous to the preparation of the corresponding sulfur compound,[277] gave a 95% yield of bis(phenylseleno)methane (**60**).[255] In a similar way, 1,4-dichloro-2-butyne led to the synthesis of 1,4-bis(phenylseleno)-2-butyne.[195]

For the preparation of acetylene derivatives containing two isolated selenide groups, elemental selenium was combined with sodium acetylide in liquid ammonia. After the addition of ethyl iodide, more selenium, and sodium amide, a 35–38% yield of 3,6-diselena-4-octyne (**58**) was obtained (scheme 49).[128,278] Kataeva et al.[279] have prepared 2,5-diselena-3-hexene

$$CH\equiv C-Na + Se \xrightarrow{NH_3\,(l)} CH\equiv C-SeNa$$

(49) $CH\equiv C-SeNa + 2\,C_2H_5I + Se + NaNH_2 \xrightarrow{NH_3\,(l)}$

$$C_2H_5Se-C\equiv C-SeC_2H_5 + 2\,NaI + NH_3$$

58

(**59**) by allowing dimethyl diselenide and acetylene to react in the presence of ethanolate or other bases, such as piperidine (equation 50). A number of related species were obtained by the same procedure.[279]

$$(50) \quad (CH_3Se)_2 \; + \; CH{\equiv}CH \; \xrightarrow{\text{base}} \; CH_3-Se-CH{=}CH-Se-CH_3$$
$$59$$

For the preparation of tri(phenylseleno)methane, Seebach and Peleties[255] adapted a method previously described for the sulfur analog,[280] that is, benzeneselenol and trimethyl orthoformate were heated in chloroform in the presence of hydrogen chloride. Organometallic compounds have also been employed for the preparation of this type of organoselenium compound. Cinquini *et al.*[281] reacted phenylselenomagnesium bromide with 1,4-dichloromethylbenzene to form 1,4-bis(phenylselenomethyl)benzene, whereas other compounds were prepared by the use of organolithium compounds.[255]

When lithium diisobutylamide reacted with bis(phenylseleno)methane (**60**), a lithium derivative was obtained which reacted with diphenyl ketone as indicated in scheme 51 to give 1,1-diphenyl-2,2-di(phenylseleno)ethanol

$$(51) \quad \left(\!\!\left\langle\bigcirc\right\rangle\!\!-Se\right)_{\!2}\!CH_2 \; \xrightarrow{(i\text{-}C_4H_9)_2NLi} \; \left(\!\!\left\langle\bigcirc\right\rangle\!\!-Se\right)_{\!2}\!CHLi \; \xrightarrow{Ph_2CO} \; \left(\!\!\left\langle\bigcirc\right\rangle\!\!-Se\right)_{\!2}\!CH{-}\overset{OH}{\underset{}{C}}\!\!\left(\!\!-\!\!\left\langle\bigcirc\right\rangle\right)_{\!2}$$
$$60 \qquad\qquad\qquad\qquad\qquad\qquad\qquad\qquad\qquad 61$$

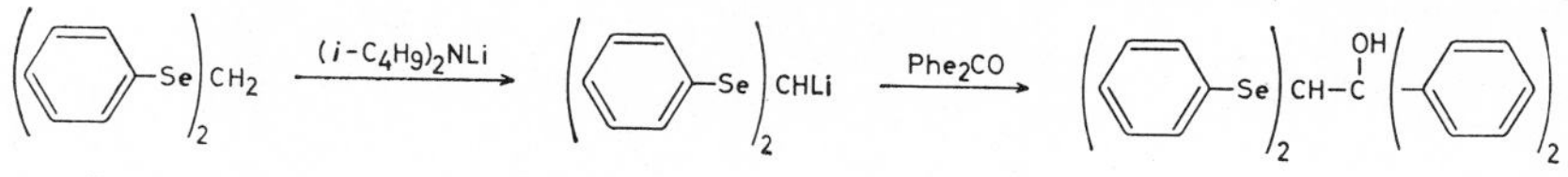

(**61**). 1,1,1-Tri(phenylseleno)ethane, 1-phenyl-2,2,2-tri(phenylseleno)-ethanol, tetra(phenylseleno)methane, and 1,1-di(phenylseleno)-2,2-di-(phenylthio)cyclopropane were obtained in a similar way.

Welcman *et al.*[54,55] reacted perfluoroalkylselenylmercury derivatives with diiodomethane for 48 hours at 110–120° in a Carius tube to give bis(trifluoromethylseleno)methane,[55] bis(pentafluoroethylseleno)methane,[54] and bis(heptafluoropropylseleno)methane.[54]

An alternative method for the preparation of **60** has been described by Petragnani and Schill,[282] as indicated in equation 52. This procedure also gave 1,5-diphenyl-2,4-diselenapentane.[282]

$$(52) \quad CH_2N_2 \; + \; \left(\!\!\left\langle\bigcirc\right\rangle\!\!-Se-\right)_{\!2} \; \xrightarrow{h\nu} \; 60$$

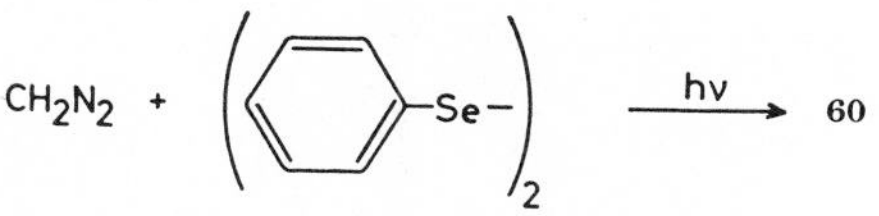

Pollak and Grillot,[258] when condensing aniline and formaldehyde with benzeneselenol (*cf.* equation 40), obtained a 42% yield of *N*,*N*-di(phenylselenomethyl)aniline. The same method also produced tri(phenylselenomethyl)amine and *N*,*N*-di(phenylselenomethyl)benzyl amine.[258]

The preparation of 1,2-diarylseleno-substituted benzene derivatives was described by Petragnani and Toscano.[283] When diphenyliodonium-2-carboxylate was heated with diphenyl diselenide, the trapping of a benzyne intermediate occurred, as shown in scheme 53, to give 1,2-bis(phenyl-

(53)

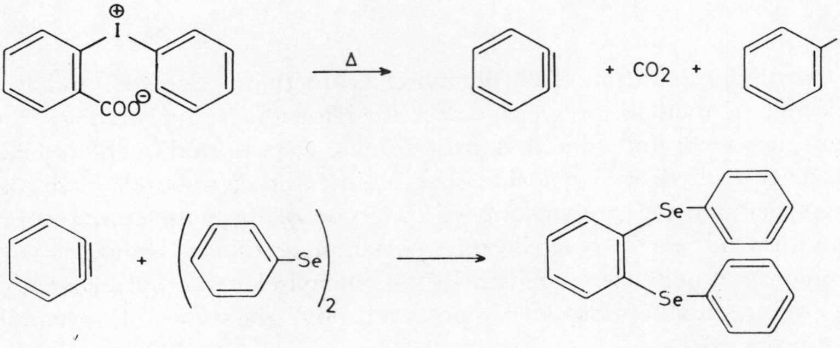

62

seleno)benzene (62), which could not be isolated. The crude material was converted into the stable tetrachloride derivative.

2. Reactions

The relatively few known organoselenium compounds containing two or more isolated selenide groups have not been very widely investigated.

The action of organolithium compounds on this type of selenide has been reported for two cases. Seebach and Peleties[255] found that bis(phenylseleno)-methane (60) reacted with butyllithium to give butyl phenyl selenide and phenylselenomethyllithium. Similarly, Brandsma[278] observed that 58, on reaction with phenyllithium, gave ethyl phenyl selenide and dilithium acetylide.

As with other selenides, the action of oxidizing agents produced the corresponding selenoxides.[281] Similarly, the corresponding selenide dihalides were formed when this type of compound reacted with halogens or other suitable reagents. When, for instance, 62 was treated with sulfuryl chloride, the corresponding tetrachloride was isolated.[283]

E. SELENOXIDES

The properties of organic selenoxides differ significantly from those of the selenides since the Se atom is present in a tetravalent oxidation state. The nature of the Se–O bond has been extensively investigated by many chemists, and the results have been reviewed by Paetzold.[284] From the coordinate covalent character of the Se–O bond follows a certain polarity, which, in conjunction with the lone pair of electrons remaining at selenium, makes selenoxides very likely to form adducts with a variety of substances. In comparison to the analogous sulfoxides, selenoxides are stronger bases and form hydrates more easily. At the same time, the special properties of the Se–O bond cause it to be less stable than the corresponding S–O bond, thereby giving rise to certain problems in product formation and isolation.

1. Synthesis

Essentially two general reactions have been employed for the preparation of selenoxides: the oxidation of selenides and the hydrolysis of selenide dihalides.

Ayrey et al.[10] employed ozone as the oxidizing agent for the preparation of dimethyl selenoxide (**63**) (equation 54). Dimethyl selenide was dissolved

$$(54) \quad CH_3-Se-CH_3 \quad \xrightarrow[CHCl_3]{O_3} \quad CH_3-\overset{\overset{O}{\uparrow}}{Se}-CH_3$$

$$\mathbf{1} \hspace{5cm} \mathbf{63}$$

in chloroform, and at −50° an equivalent amount of ozone was added, giving a 75% yield of **63**.[10,285,286] By the use of excess ozone, the corresponding selenone was formed. The same method has also been used for the preparation in high yield of dibutyl selenoxide,[10] didodecyl selenoxide,[10] diphenyl selenoxide,[10] butyl phenyl selenoxide,[10,287] and 6β-phenylselen-inyl-5α-cholestane.[288]

Gould and McCullough[80] oxidized 4,4′-dimethoxydiphenyl selenide in dioxane with hydrogen peroxide at 80° to obtain a 63% yield of 4,4′-dimethoxydiphenyl selenoxide. The same method also generated 4,4′-diethoxydiphenyl selenoxide[80] and 2,2,2′,2′-tetrachlorodiethenyl selen-oxide.[57] Attempts to prepare selenoxides with carbohydrate structures by this method are reported to have failed.[211] Peracetic acid in acetic acid solu-tion was employed[80] for the preparation of 4,4′-dichlorodiphenyl selenoxide. The method appears to give somewhat lower yields than the hydrogen peroxide oxidation.

Paetzold et al.[37] employed dinitrogen tetroxide in ether for the preparation of dimethyl selenoxide from dimethyl selenide; the dinitrogen tetroxide adduct of the selenoxide was reported to be formed by this method.

Cinquini et al.[289] have reported the use of sodium periodate for the oxidation of selenides to selenoxides and applied this method to a variety of aromatic and benzylic species.[289] Iodobenzene dichloride was employed for the oxidation of selenides to selenoxides[289] by adding a solution of the reagent in anhydrous pyridine to a dilute pyridine solution of the selenide at −40°. The same procedure has been used also for the preparation of sulfoxides from sulfides and was found to be general for the preparation of aromatic selenoxides.[289]

Spinelli and Monaco[290] have treated diaryl selenides with chloramine-T to produce N-4-methylphenylsulfonylselenimides. Occasionally, diaryl selenoxides were isolated, and in this way 4,4′-dichlorodiphenyl selenoxide and 4-carboxybenzylseleninylbenzene were obtained.

The second general procedure for the preparation of selenoxides has been employed by, for example, Riley et al.[291] In this method a suspension of 2,2′-dichlorodipentyl selenide dichloride (**64**) in water was stirred at room temperature for 2 hours (equation 55). From the reaction mix-

$$(55) \quad (CH_3-CH_2-CH_2-\underset{\underset{Cl}{|}}{CH}-CH_2-)_2 SeCl_2 \xrightarrow{H_2O} (CH_3-CH_2-CH_2-\underset{\underset{Cl}{|}}{CH}-CH_2)_2 SeO$$

$$64 \qquad\qquad\qquad\qquad\qquad\qquad 65$$

ture a 93% yield of 2,2′-dichlorodipentyl selenoxide (**65**) was isolated. This procedure gave a number of dialkyl and diaryl selenoxides in high yields, starting with the corresponding selenide dichlorides as well as selenide dibromides.[110,247,291–295] For some of the preparations, the addition of a base such as sodium hydroxide was found to favor the reaction. In the case of 4,4′-dimethoxydiphenylselenide dichloride or dibromide, the corresponding selenoxide hydrate was isolated.[88,269]

For the preparation of low-molecular-weight dialkyl selenoxides, Paetzold et al.[37,38] hydrolyzed the corresponding selenide dibromides with silver oxide in water to obtain dimethyl, diethyl, dipropyl, and dibutyl selenoxides in good yield.

Kozlov and Pronyakova[297] have reported the oxidation of 1-selenocyanatonaphthalene by potassium permanganate in acetic acid, to give the corresponding seleninic acid. On prolonged heating, 1-naphthaleneseleninic acid yielded the labile anhydride, which decomposed to 1,1′-dinaphthyl selenoxide (97%) and selenium dioxide. 2,2′-Dinaphthyl selenoxide[298] was prepared similarly.

Brintzinger et al.[57] reacted vinyl chloride with selenium oxychloride in an exothermic reaction to form 2,2,2′,2′-tetrachlorodiethyl selenoxide.

Finally, Cinquini et al.[281] reported the preparation of 1,4-di(phenylseleninylmethyl)benzene without describing the synthetic procedure.

2. Reactions

The basicity of selenoxides, studied by potentiometric titration of *N*-2-methylseleninylethylphthalimide by Štefanac and Tomašković,[299] makes them associate with one molecule of water to form a hydrate.[88,287,296] They also react with strong acids to form adducts. Thus, Paetzold *et al.*[37] prepared a hydrochloride of dimethyl selenoxide (**63**) by dissolving the selenoxide in aqueous hydrochloric acid. The corresponding selenide dichloride also resulted as a by-product. The hydronitrate salt of **63** was similarly obtained,[37] as well as the hydronitrate of diphenyl selenoxide.[172] The hydroperchlorate salt of **63** has also been described.[37]

Dialkyl selenoxides also form coordination complexes with a number of metal salts, such as perchlorates[300,301] and chlorides.[285,286] Moreover, diphenyl selenoxide has been reported to be a suitable donor for similar complexes.[293] Usually divalent, but also tri- or tetravalent, metal ions have proved suitable as acceptors, as have metal-organic compounds, such as dimethyltin dichloride.[286] Dimethyl selenoxide has been found to coordinate with two molecules of pyridine[37] to form a stable complex.

3. Spectroscopy and Physical Properties

Selenoxides have frequently been investigated by spectroscopic methods in order to study the nature of the Se–O bond. Tanaka and Kamitani[286] and, especially, Paetzold *et al.*,[37,302,303] on studying the i.r. spectra of selenoxides, found the characteristic frequencies of the Se–O bond to occur at 735–820 cm^{-1} and sometimes also at 385–466 cm^{-1}. The u.v. spectra of trifluoroseleninylbenzene and similar compounds were investigated by Lutskii *et al.*;[304] an absorption band, assigned to the Se–O bond, was generally observed at 260–302 nm.

As a consequence of the pyramidal configuration of the Se atom in selenoxides, a chiral center is present, so that two enantiomorphs of unsymmetrical selenoxides are theoretically possible. This interesting question has been extensively discussed.[96,287,305] Recently, Jones *et al.*[288] and Cinquini *et al.*[281] have claimed the first isolation of diastereomers of selenoxides.

F. SELENONES

Only a very limited number of dialkyl selenones and diaryl selenones have thus far been prepared, and consequently their properties are not completely known. From the data in the literature, this class of compounds seems to contain fairly stable structures that can be purified and stored for a reasonable period of time.

1. Synthesis

For the preparation of dialkyl selenones, only the method by Paetzold[38] has so far been described. For the preparation of dimethyl selenone (**66**), diethyl selenone, dipropyl selenone, and dibutyl selenone, ozonization of the corresponding selenoxides was employed (equation 56). The yields of selenone

$$
(56) \quad CH_3-\overset{\overset{O}{\uparrow}}{Se}-CH_3 \quad \xrightarrow{\;O_3\;} \quad CH_3-\overset{\overset{O}{\uparrow}}{\underset{\underset{O}{\downarrow}}{Se}}-CH_3
$$

$$
\qquad\qquad 63 \qquad\qquad\qquad\qquad\qquad\qquad 66
$$

were reported to be generally very good; for example, **66** was obtained in a 75 % yield. This contradicts statements by Ayrey et al.,[10] who claimed that an excess of ozone in the oxidation of dimethyl selenide did not give any selenone. As a side reaction, one C–Se bond was severed by ozone and small amounts of methaneseleninic acid or anhydride were formed.

In the aromatic series Rheinboldt and Giesbrecht[292] oxidized diphenyl selenoxide with potassium permanganate to obtain an 81 % yield of diphenyl selenone. Yagupol'skii and Voloshchuk[243] reacted trifluoromethyl aryl selenides or selenoxides with trifluoroperacetic acid at −40°, resulting in very good yields of the corresponding selenones.

2. Reactions

In comparison with the analogous selenoxides, selenones are found to be considerably less reactive to most reagents. Because of their lower basicity,[38] there is no tendency to protonation or hydrogen-bond formation to the selenone O atoms. In the same way, no coordination complexes with metal salts, so frequently occurring with selenoxides, have been found for selenones.

The postulated oxidative effect of selenones on suitable reagents has been studied by Paetzold and Bochmann[38] for **66** and other dialkyl selenones. In concentrated hydrochloric acid, no chlorine formation occurred; rather, a splitting of the molecule to form methyl chloride and methaneseleninic acid hydrochloride was observed (equation 57).

In similar experiments with trifluoromethaneselenonylbenzene (**67**), Yagupol'skii and Voloshchuk[243] found no reaction with potassium iodide in neutral solution. In dilute sulfuric acid a very slow formation of iodine was observed, thus providing some limited evidence for the oxidative effect of

$$(57) \quad CH_3\overset{O}{\underset{O}{\overset{\uparrow}{\underset{\downarrow}{Se}}}}CH_3 \ + \ 2HCl \quad \longrightarrow \quad CH_3Cl \ + \ \left[CH_3\overset{\oplus}{\underset{OH}{\overset{OH}{Se}}} \right] Cl^{\ominus}$$

66

selenones. Diphenyl selenone and **67** in cold concentrated hydrochloric acid, however, were observed to cause the evolution of chlorine, while the corresponding sulfones were unreactive.

When **67** was dissolved in a 10% aqueous sodium carbonate, the compound decomposed and a reaction, similar to the haloform reaction, was observed (equation 58).

$$(58) \quad CF_3\overset{O}{\underset{O}{\overset{\uparrow}{\underset{\downarrow}{Se}}}}\!\!-\!\!\bigcirc \ + \ OH^{\ominus} \quad \longrightarrow \quad CHF_3 \ + \ \bigcirc\!\!-\!\!SeO_2\!\!-\!\!O^{\ominus}$$

67

3. Spectroscopy

Paetzold and Bochmann[38,302,303] have studied the i.r. and Raman spectra of dialkyl selenones. Characteristic frequencies assigned to the selenonyl group were reported to occur at 860–970 cm^{-1} and 912–1059 cm^{-1}.

G. SELENIDE DIHALIDES

A large number of selenide dihalides have been prepared, most of which are stable in neutral or acid solutions and can be stored for a reasonable time. In neutral or alkaline aqueous solutions, the selenide dihalides hydrolyze to give the corresponding selenoxides. Selenide dihalides are, therefore, valuable intermediates for the synthesis of selenoxides and selenones.

1. Synthesis

The method most frequently employed for the preparation of selenide dihalides makes use of monoselenides as starting materials. The latter react with a halogen, usually dissolved in an inert solvent, such as carbon tetrachloride, diethyl ether, chloroform, or carbon disulfide. The reactants are mixed with cooling to undergo the reaction given in equation 59 for the preparation of dimethyl selenide dichloride (**68**). By this basic method many dichlorides,[57,59,71,155,157,238,247,306] dibromides,[6,9,22,37,38,88,110,155,157,186,238] and diiodides[9,22,84,307,308] have been prepared.

$$(59) \quad CH_3-Se-CH_3 \ + \ Cl_2 \quad \xrightarrow{CCl_4} \quad CH_3-\overset{\overset{\displaystyle Cl}{\uparrow}}{\underset{\underset{\displaystyle Cl}{|}}{Se}}-CH_3$$

1   68

For the preparation of halogenated dialkyl selenide dihalides, the procedure described by Funk *et al.*[58,59] for the synthesis of 2,2'-dichlorodiethyl selenide dichloride (**69**) and similar compounds has been found to be very useful. Thus, alkenes in the gaseous state were passed into a solution of selenium dioxide in concentrated hydrochloric or hydrobromic acid (equation 60) to give 80–95 % yields of **69** and other products derived from

$$(60) \quad 2\,CH_2{=}CH_2 \ + \ SeO_2 \ + \ 4HCl \ \longrightarrow \ Cl-CH_2-CH_2-\overset{\overset{\displaystyle Cl}{\uparrow}}{\underset{\underset{\displaystyle Cl}{\downarrow}}{Se}}-CH_2-CH_2Cl$$

69

ethylene, acetylene, styrene, vinyl acetate, cyclopentene, cyclohexene, and related species.[58–60] This procedure, with suitable modifications, also gave the tellurium analog of **69**, as well as 2,2'-dibromodicyclo-hexyltelluride dibromide.[58]

In the reaction shown in equation 60, selenium oxychloride is a very probable intermediate. When acetone was treated as described for the preparation of **69**, a substitution reaction occurred in which di(acetylmethyl) selenide dichloride[58] was formed. The analogous method, using hydrobromic acid as the solvent, gave di(acetylmethyl) selenide dibromide.[60] Also, the treatment of 2-butanone and acetophenone with selenium dioxide in hydrobromic acid produced species that were presumably di(propionylmethyl) selenide dibromide[60] and diphenacyl selenide dibromide,[60] respectively.

By use of selenium oxychloride in the reaction originally described by Alquist and Nelson,[100] anisole dissolved in ether gave 4,4'-dimethoxydiphenyl selenide dichloride.[80]

Hannig and Ziebandt[262] combined 4-(methylseleno)acetophenone with sulfuryl chloride in benzene in an exothermic reaction which gave a 90–95 % yield of methyl 4-acetylphenyl selenide dichloride. Similarly prepared were ethyl 4-acetylphenyl selenide dichloride,[262] propyl 4-acetylphenyl selenide dichloride,[262] and 1,2-di(phenylseleno)benzene tetrachloride.[283]

For the preparation of diphenyl selenide dichloride,[79,81] diphenyl selenide was treated with a mixture of nitric acid and hydrochloric acid. The product precipitated immediately to give an 85–87 % yield of the dichloride. 4,4'-Dimethyldiphenyl selenide dichloride[82] was also prepared in this manner.

Selenide difluorides were prepared for the first time by Wynne and Puckett[309,310] by a method in which a selenide was treated with silver difluoride in 1,1,2-trifluoroethane (equation 61). In this way, dimethyl selenide di-

(61) $CH_3-Se-CH_3 + 2AgF_2 \longrightarrow CH_3-Se-CH_3 + 2AgF$

with F atoms above and below the Se in the product.

    1   70

fluoride (**70**) was obtained, as well as diethyl selenide difluoride, dipropyl selenide difluoride, 2,2'-dipropyl selenide difluoride, and diphenyl selenide difluoride.

A halogen-exchange reaction was employed by Funk *et al.*[58,59] for the preparation of chloro-substituted dialkyl selenide dibromides from the dichlorides. By treating, for example, **69** with hydrobromic acid, there resulted 2,2'-dichlorodiethyl selenide dibromide.[58] 2,2'-Dichlorodipropyl selenide dibromide,[59] 2,2'-dichloro-3,3'-dibromodipropyl selenide dibromide,[59] and 2,2',3,3'-tetrachloro-2,2'-dimethyldipropyl selenide dibromide[58] were similarly prepared.

2. Reactions

As mentioned above, selenide dihalides are fairly stable under neutral or acidic conditions; however, it has been noted that after several months' storage, selenide dihalides deposited elemental selenium.[6] Greenberg *et al.*[157] observed that a large number of unsymmetrical selenide dihalides containing a biphenylyl group rearranged very readily to form selenonium salts.

In a number of cases, selenide dihalides have been employed as intermediates for the preparation of other organoselenium compounds. By hydrolysis, the corresponding selenoxides may be prepared (*cf.* Section V.E). Furthermore, Paetzold and Lindner[311] have reported the reaction of diphenyl selenide dichloride, and of the dibromide and diiodide analogs, with methanolate and ethanolate ion to give diphenyl selenide dimethanolate and diphenyl selenide diethanolate, respectively.[311]

Selenide dihalides have also been treated with potassium iodide,[80] bisulfite,[58] or zinc in chloroform[100] to generate the corresponding selenides through reduction.

Dimethyl and diphenyl selenide dihalides form 1:1 adducts with boron trifluoride.[71,76]

214   L.-B. Agenäs

3. Spectroscopy and Physical Properties

The detailed investigations by McCullough *et al.*[9,80,84,155,307,308] of selenide dihalides (mainly diaryl selenide dibromides and diiodides) have resulted in good understanding of the dissociation constants of the equilibrium shown in equation 62.

$$(62) \quad R_2SeI_2 \rightleftharpoons R_2Se + I_2$$

The crystallographic properties of 4,4'-dimethyl diphenyl selenide dichloride and dibromide have been studied by Neuerberg[312] and McCullough and Marsh.[313]

The i.r. and Raman spectra of dialkyl selenide dihalides have been investigated by Hayward and Hendra[22] and compared with those for the analogous sulfur and tellurium compounds. No specific absorption could be assigned to the Se–X bonds.

REFERENCES

1. C. Siemens, *Ann. Chem. Pharm.*, **61,** 360 (1847).
2. W. E. Bradt, *Proc. Indiana Acad. Sci.*, **40,** 141 (1930).
3. W. E. Bradt, *Proc. Indiana Acad. Sci.*, **43,** 72 (1934).
4. T. W. Campbell, H. G. Walker, and G. M. Coppinger, *Chem. Rev.*, **50,** 279 (1952).
5. H. Rheinboldt, "Houben-Weyl Methoden der Organischen Chemie," Vol. IX, E. Müller, Ed., Georg Thieme Verlag, Stuttgart, 1955, pp. 972, 1005, 1020, 1030.
6. M. L. Bird and F. Challenger, *J. Chem. Soc.*, 570 (1942).
7. I. M. Kolthoff and N. Tamberg, *J. Polarog. Soc.*, 54 (1958).
8. L. Chierici, H. Lumbroso, and R. Passerini, *Bull. Soc. Chim. Fr.*, 686 (1955).
9. N. W. Tideswell and J. D. McCullough, *J. Amer. Chem. Soc.*, **79,** 1031 (1957).
10. G. Ayrey, D. Barnard, and D. T. Woodbridge, *J. Chem. Soc.*, 2089 (1962).
11. H. H. Anderson, *J. Chem. Eng. Data*, **9,** 272 (1964).
12. J.-P. Mila and J.-F. Labarre, *C. R. Acad. Sci., Paris, Ser. C*, **263,** 1481 (1966).
13. J. F. Beecher, *J. Mol. Spectry.*, **21,** 414 (1966).
14. C. S. Evans and C. M. Johnson, *J. Chromatog.*, **21,** 202 (1966).
15. J. R. Allkins and P. J. Hendra, *Spectrochim. Acta*, **22,** 2075 (1966).
16. J. Beneš and V. Procházková, *J. Chromatog.*, **29,** 239 (1967).
17. J.-P. Mila and J.-P. Laurent, *Bull. Soc. Chim. Fr.*, 2735 (1967).
18. T. Hashimoto, M. Sugita, H. Kitani, and K. Fukui, *Nippon Kagaku Zasshi*, **88,** 991 (1967).
19. J. Le Calve and J. Lascombe, *Spectrochim. Acta*, **24A,** 737 (1968).
20. J. R. Allkins and P. J. Hendra, *J. Chem. Soc., A*, 1325 (1967).
21. J. R. Allkins and P. J. Hendra, *Spectrochim. Acta*, **24A,** 1305 (1968).

22. G. C. Hayward and P. J. Hendra, *J. Chem. Soc.*, *A*, 1760 (1969).

23. J. R. Allkins and P. J. Hendra, *Spectrochim. Acta*, **23A**, 1671 (1967).

24. E. P. Painter, K. W. Franke, and R. A. Gortner, *J. Org. Chem.*, **5**, 579 (1940).

25. M. A. Cecchini and E. Giesbrecht, *J. Org. Chem.*, **21**, 1217 (1956).

26. Yu. N. Shlyk, G. M. Bogolyubov, and A. A. Petrov, *Zh. Obshch. Khim.*, **38**, 1199 (1968).

27. L.-B. Agenäs, *Acta Chem. Scand.*, **22**, 1763 (1968).

28. G. M. Bogolyubov, Yu. N. Shlyk, and A. A. Petrov, *Zh. Obshch. Khim.*, **39**, 1804 (1969).

29. L. R. M. Pitombo, *Anal. Chim. Acta*, **46**, 158 (1969).

30. A. Fredga, L.-B. Agenäs, and E. Jonsson, *Acta Chem. Scand.*, **24**, 3751 (1970).

31. G. H. Denison and P. C. Condit, *Ind. Eng. Chem.*, **41**, 944 (1949).

32. G. H. Denison and P. C. Condit, U.S. Pat. 2,398,414 (1946); *Chem. Abstr.*, **40**, 3598 (1946).

33. G. H. Denison and P. C. Condit, U.S. Pat. 2,398,415 (1946); *Chem. Abstr.*, **40**, 3598 (1946).

34. G. H. Denison and P. C. Condit, U.S. Pat. 2,398,416 (1946); *Chem. Abstr.*, **40**, 3598 (1946).

35. G. H. Denison and P. C. Condit, U.S. Pat. 2,473,510 (1949); *Chem. Abstr.*, **43**, 6820 (1949).

36. L. Brandsma and H. E. Wijers, *Rec. Trav. Chim. Pays-Bas*, **82**, 68 (1963).

37. R. Paetzold, U. Lindner, G. Bochmann, and P. Reich, *Z. Anorg. Allg. Chem.*, **352**, 295 (1967).

38. R. Paetzold and G. Bochmann, *Z. Anorg. Allg. Chem.*, **360**, 293 (1968).

39. E. P. Painter, *J. Amer. Chem. Soc.*, **69**, 229 (1947).

40. L. I. Zakharkin and V. V. Gavrilenko, *Izv. Akad. Nauk SSSR, Otd. Khim. Nauk*, 1391 (1960).

41. G. Wagner and P. Nuhn, *Arch. Pharm.* (Weinheim), **297**, 461 (1964).

42. S. Landa, O. Weisser, and J. Mostecký, *Collection Czech. Chem. Commun.*, **24**, 2197 (1959).

43. A. E. Kemppainen and E. L. Compere, Jr., *J. Chem. Eng. Data*, **11**, 588 (1966).

44. J. Gosselck and E. Wolters, *Z. Naturforsch.*, *Part B*, **17**, 131 (1962).

45. L. V. Kaabak, A. P. Tomilov, and S. L. Varshchavskii, *Zh. Vses. Khim. Obshchestva*, **9**, 700 (1964).

46. L.-B. Agenäs, *Ark. Kemi*, **31**, 31 (1969).

47. L.-B. Agenäs, *Acta Univ. Upsaliensis, Abstr. Uppsala Dissertations Sci.*, **132**, 11 (1969).

48. R. N. Haszeldine, *Angew. Chem.*, **66**, 698 (1954).

49. J. W. Dale, H. J. Emeléus, and R. N. Haszeldine, *J. Chem. Soc.*, 2939 (1958).

50. T. Birchall, R. J. Gillespie, and S. L. Vekris, *Can. J. Chem.*, **43**, 1672 (1965).

51. H. J. Emeléus and M. J. Dunn, *J. Inorg. Nucl. Chem.*, **27**, 752 (1965).

52. N. N. Yarovenko, V. N. Shemanina, and G. B. Gazieva, *Zh. Obshch. Khim.*, **29**, 942 (1959).

53. N. Welcman, *Israel J. Chem.*, **1**, 307 (1963).

54. N. Welcman and H. Regev, *J. Chem. Soc.*, 7511 (1965).

55. H. J. Emeléus and N. Welcman, *J. Chem. Soc.*, 1268 (1963).

56. P. D. Faurote and J. G. O'Rear, *J. Amer. Chem. Soc.*, **78**, 4999 (1956).

57. H. Brintzinger, K. Pfannstiel, and H. Vogel, *Z. Anorg. Allg. Chem.*, **256**, 75 (1948).

58. H. Funk and W. Weiss, *J. Prakt. Chem.*, **273**, 33 (1954).

59. H. Funk and W. Papenroth, *J. Prakt. Chem.*, **280**, 256 (1959).

60. H. Funk and W. Papenroth, *J. Prakt. Chem.*, **283**, 191 (1960).

61. F. Lautenschlaeger, *J. Org. Chem.*, **34**, 4002 (1969).

62. D. H. Olson, *Tetrahedron Lett.*, 2053 (1966).

63. K. Takaoka and Y. Toyama, *Nippon Kagaku Zasshi*, **89**, 618 (1968).

64. K. Takaoka and Y. Toyama, *Nippon Kagaku Zasshi*, **89**, 405 (1968).

65. L. Brandsma and J. F. Arens, *Rec. Trav. Chim. Pays-Bas*, **81**, 33 (1962).

66. L. Brandsma and P. J. W. Schuijl, *Rec. Trav. Chim. Pays-Bas*, **88**, 513 (1969).

67. G. T. Morgan and H. D. K. Drew, *J. Chem. Soc.*, **117**, 1456 (1920).

68. G. T. Morgan and H. D. K. Drew, *J. Chem. Soc.*, **121**, 922 (1922).

69. G. T. Morgan, H. D. K. Drew, and T. V. Barker, *J. Chem. Soc.*, **121**, 2432 (1922).

70. D. H. Dewar, J. E. Fergusson, P. R. Hentschel, C. J. Wilkins, and P. P. Williams, *J. Chem. Soc.*, 688 (1964).

71. K. J. Wynne and J. W. George, *J. Amer. Chem. Soc.*, **87**, 4750 (1965).

72. E. Goldish, K. Hedberg, R. E. Marsh, and V. Schomaker, *J. Amer. Chem. Soc.*, **77**, 2948 (1955).

73. H. J. M. Bowen, *Trans. Faraday Soc.*, **50**, 452 (1954).

74. L. Chierici, H. Lumbroso, and R. Passerini, *C. R. Acad. Sci., Paris*, **237**, 611 (1953).

75. R. West, D. L. Powell, M. K. T. Lee, and L. S. Whatley, *J. Amer. Chem. Soc.*, **86**, 3227 (1964).

76. M. E. Peach, *Can. J. Chem.*, **47**, 1675 (1969).

77. J. J. Windle, A. K. Wiersma, and A. L. Tappel, *Nature*, **203**, 404 (1964).

78. J. J. Windle, A. K. Wiersma, and A. L. Tappel, *J. Chem. Phys.*, **41**, 1996 (1964).

79. H. M. Leicester, "Organic Syntheses," Coll. Vol. 2, John Wiley, New York, N.Y., 1943, p. 238.

80. E. S. Gould and J. D. McCullough, *J. Am. Chem. Soc.*, **73**, 3196 (1951).

81. M. A. A. Beg and A. R. Shaikh, *Tetrahedron*, **22**, 653 (1966).

82. S. S. Krishnamurthy and S. Soundararajan, *J. Organometal. Chem.*, **15**, 367 (1968).

83. J. D. McCullough and B. A. Eckerson, *J. Amer. Chem. Soc.*, **67**, 707 (1945).

84. J. D. McCullough and D. Mulvey, *J. Phys. Chem.*, **64**, 264 (1960).

85. M. T. Rogers and T. W. Campbell, *J. Amer. Chem. Soc.*, **69**, 2039 (1947).

86. L. Chierici and R. Passerini, *Atti Accad. Naz. Lincei, Rend., Classe Sci., Fis., Mat., Nat.*, [8] **15**, 69 (1953).

87. H. Rheinboldt, M. Perrier, E. Giesbrecht, A. Levy, M. A. Cecchini, and H. Viera de Campos, *Univ. São Paolo, Fac. Filosof., Ciênc. Lett., Bol.* 129, *Quim.*, **3**, 29 (1951).

88. G. V. Boyd, M. Doughty, and J. Kenyon, *J. Chem. Soc.*, 2196 (1949).

89. Á. I. Kiss and B. R. Muth, *J. Chem. Phys.*, **23**, 2189 (1955).

90. V. M. Ivanova, S. A. Nemleva, E. N. Seina, E. G. Kaminskaya, S. S. Gitis, and A. Ya. Kaminskii, *Zh. Org. Khim.*, **3**, 146 (1967).

91. G. Bergson, *Ark. Kemi*, **10**, 127 (1956).

92. H. Bock, G. Rudolph, E. Baltin, and J. Kroner, *Angew. Chem.*, **77**, 473 (1965).

93. A. N. Nesmeyanov, V. N. Vinogradova, and L. G. Makarova, *Izv. Akad. Nauk SSSR, Otd. Khim. Nauk*, 1710 (1960).

94. F. Krafft and R. E. Lyons, *Ber.*, **27**, 1761 (1894).

95. Á. I. Kiss and B. R. Muth, *Acta Chim. Acad. Sci. Hung.*, **24**, 231 (1960).

96. W. J. Burlant, Dissertation, Polytechnic Institute of Brooklyn, New York, N.Y., 1955, p. 44; *University Microfilms*, Ann. Arbor, Mich., No. 12,498.

97. H. Hauptmann and W. F. Walter, *J. Amer. Chem. Soc.*, **77**, 4929 (1955).

98. F. Krafft and A. Kaschau, *Ber.*, **29**, 428 (1896).

99. L. Chierici and R. Passerini, *Atti Accad. Naz. Lincei, Rend., Classe Sci., Fis., Mat., Nat.*, [8] **14**, 99 (1953).

100. F. N. Alquist and R. E. Nelson, *J. Amer. Chem. Soc.*, **53**, 4033 (1931).

101. Š. Korček, Š. Holotík, J. Leško, and V. Veselý, *Chem. Zvesti*, **23**, 281 (1969).

102. E. Solčániová and Š. Kováč, *Chem. Zvesti*, **23**, 687 (1969).

103. B. Witkop, *Justus Liebigs Ann. Chem.*, **558**, 98 (1947).

104. J. F. K. Wilshire, *Australian J. Chem.*, **20**, 359 (1967).

105. J. Bergman, *Acta Chem. Scand.*, **22**, 1883 (1968).

106. J. Burdon, P. L. Coe, and M. Fulton, *J. Chem. Soc.*, 2094 (1965).

107. S. C. Cohen, M. L. N. Reddy, and A. G. Massey, *J. Organometal. Chem.*, **11**, 563 (1968).

108. M. Giua and R. Bianco, *Gazz. Chim. Ital.*, **89**, 693 (1959).

109. L. M. Litvinenko and R. S. Cheshko, *Zh. Obshch. Khim.*, **30**, 3682 (1960).

110. L. Bartolotti and R. Passerini, *Ric. Sci.*, **25**, 1095 (1955).

111. L. Bartolotti and A. Cerniani, *Boll. Sci. Fac. Chim. Ind. Bologna*, **14**, 33 (1956).

112. G. E. Wiseman and E. S. Gould, *J. Amer. Chem. Soc.*, **76**, 1706 (1954).

113. W. R. Blackmore and S. C. Abrahams, *Acta Cryst.*, **8**, 323 (1955).

114. R. J. W. Le Fèvre and J. D. Saxby, *J. Chem. Soc.*, B, 1064 (1966).

115. W. H. H. Günther and H. G. Mautner, *J. Med. Chem.*, **7**, 229 (1964).

116. G. Bergson and A.-L. Delin, *Ark. Kemi*, **18**, 441 (1962).

117. W. H. H. Günther and H. G. Mautner, *J. Amer. Chem. Soc.*, **87**. 2708 (1965).

118. W. H. H. Günther and H. G. Mautner, *J. Med. Chem.*, **8**, 845 (1965).

119. R. Golmohammadi, *Acta Chem. Scand.*, **20**, 563 (1966).

120. A. Fredga, *Acta Chem. Scand.*, **24**, 1117 (1970).

121. L.-B. Agenäs, *Ark. Kemi*, **23**, 145 (1964).

122. M. Renson and P. Pirson, *Bull. Soc. Chim. Belges*, **75**, 456 (1966).

123. W. L. Mock and J. H. McCausland. *Tetrahedron Lett.*, 391 (1968).

124. G. Bergson, *Ark. Kemi*, **21**, 439 (1964).

125. Yu. A. Boiko, B. S. Kupin, and A. A. Petrov, *Zh. Org. Khim.*, **5**, 1553 (1969).

126. Yu. A. Boiko, B. S. Kupin, and A. A. Petrov, *Zh. Org. Khim.*, **4**, 1355 (1968).

127. L. Brandsma, H. E. Wijers, and J. F. Arens, *Rec. Trav. Chim. Pays-Bas*, **81**, 583 (1962).

128. L. Brandsma, H. E. Wijers, and C. Jonker, *Rec. Trav. Chim. Pays-Bas*, **82**, 208 (1963).

129. A. A. Petrov, S. I. Radchenko, K. S. Mingaleva, I. G. Savich, and V. B. Lebedev, *Zh. Obshch. Khim.*, **34**, 1899 (1964).

130. S. I. Radchenko and A. A. Petrov, *Zh. Org. Khim.*, **1**, 2115 (1965).

131. W. H. H. Günther, *J. Org. Chem.*, **31**, 1202 (1966).

132. L. A. Khazemova and V. M. Al'bitskaya, *Zh. Org. Khim.*, **5**, 1926 (1969).

133. L. A. Khazemova and V. M. Al'bitskaya, *Zh. Org. Khim.*, **6**, 935 (1970).

134. L.-B. Agenäs, *Ark. Kemi*, **24**, 415 (1965).

135. L.-B. Agenäs and B. Persson, *Acta Chem. Scand.*, **21**, 837 (1967).

136. L.-B. Agenäs, *Ark. Kemi*, **24**, 573 (1965).

137. L.-B. Agenäs and B. Persson, *Acta Chem. Scand.*, **21**, 835 (1967).

138. N. N. Yarovenko, G. B. Gazieva, V. N. Shemanina, and N. A. Fedorova, *Zh. Obshch. Khim.*, **29**, 940 (1959).

139. N. N. Yarovenko, V. N. Shemanina, and G. B. Gazieva, *Zh. Obshch. Khim.*, **29**, 942 (1959).

140. N. N. Yarovenko and M. A. Raksha, *Zh. Obshch. Khim.*, **30**, 4064 (1960).

141. G. Zdansky, *Ark. Kemi*, **19**, 559 (1963).

142. G. Zdansky, *Ark. Kemi*, **21**, 211 (1963).

143. G. Zdansky, *Ark. Kemi*, **29**, 47 (1968).

144. G. Zdansky, *Ark. Kemi*, **29**, 437 (1968).

145. G. Bergson and A. Biezais, *Ark. Kemi*, **18**, 143 (1961).

146. L.-B. Agenäs, *Ark. Kemi*, **23**, 463 (1965).

147. H. C. Volgers and J. F. Arens, *Rec. Trav. Chim. Pays-Bas*, **77**, 1170 (1958).

148. N. V. Elsakov and A. A. Petrov, *Opt. i. Spektroskopiya*, **16**, 797 (1964).

149. S. Gronowitz and T. Frejd, *Acta Chem. Scand.*, **23**, 2540 (1969).

150. R. Mayer, S. Scheithauer, and D. Kunz, *Chem. Ber.*, **99**, 1393 (1966).

151. L. Brandsma and J. F. Arens, *Rec. Trav. Chim. Pays-Bas*, **81**, 539 (1962).

152. L. Chierici and R. Passerini, *Boll. Sci. Fac. Chim. Ind. Bologna*, **12**, 56 (1954).

153. T. W. Campbell and J. D. McCullough, *J. Amer. Chem. Soc.*, **67**, 1965 (1945).

154. E. Müller, H. B. Stiegmann, and K. Scheffler, *Justus Liebigs Ann. Chem.*, **657**, 5 (1962).

155. J. D. McCullough and M. K. Barsh, *J. Amer. Chem. Soc.*, **71**, 3029 (1949).

156. A. M. Caccia Bava and L. Chierici, *Boll. Chim. Farm.*, **89**, 397 (1950).

157. B. Greenberg, E. S. Gould, and W. Burlant, *J. Amer. Chem. Soc.*, **78**, 4028 (1956).

158. H. Rheinboldt and M. Perrier, *Bull. Soc. Chim. Fr.*, 245 (1950).

159. H. Rheinboldt and M. Perrier, *Bull. Soc. Chim. Fr.*, 445 (1955).

160. H. Rheinboldt and M. Perrier, *Bull. Soc. Chim. Fr.*, 484 (1953).

161. H. Rheinboldt and M. Perrier, *Bull. Soc. Chim. Fr.*, 264 (1953).

162. W. S. Cook and R. A. Donia, *J. Amer. Chem. Soc.*, **73**, 2275 (1951).

163. K. Šindelář, E. Svátek, J. Metyšová, J. Metyš, and M. Protiva, *Collection Czech. Chem. Commun.*, **34**, 3792 (1969).

164. K. Šindelář, J. Metyšová, and M. Protiva, *Collection Czech. Chem. Commun.*, **34**, 3801 (1969).

165. R. J. Knopf and T. K. Brotherton, U.S. Pat. 3,284,502 (1966); *Chem. Abstr.*, **67**, 64045 (1967).

166. L. Chierici, *Farmaco, Ed. Sci.*, **7**, 618 (1952).

167. L. Chierici, *Farmaco, Ed. Sci.*, **8**, 156 (1953).

168. L. Chierici and R. Passerini, *Ric. Sci.*, **25**, 2316 (1955).

169. G. Guanti, C. Dell'Erba, and D. Spinelli, *Gazz. Chim. Ital.*, **100**, 184 (1970).

170. H. Rheinboldt, "Houben-Weyl Methoden der Organischen Chemie," Vol. IX, E. Müller, Ed., Georg Thieme Verlag, Stuttgart, 1955, p. 1000.

171. G. Modena, *Advan. Mol. Spectry.*, **2**, 483 (1962).

172. D. G. Foster and S. F. Brown, *J. Amer. Chem. Soc.*, **50**, 1182 (1928).

173. M. Nardelli and L. Chierici, *Ann. Chim.* (Rome), **42**, 111 (1952).

174. A. Arcoria and G. Cordella, *Boll. Sedute Accad. Gioenia Sci. Nat. Catania*, [4] **3**, 314 (1956).

175. Á. I. Kiss and B. R. Muth, *Központi Fizi kutató Intözet. Közleményei* (Budapest), **7**, 147 (1959).

176. N. Marziano and R. Passerini, *Gazz. Chim. Ital.*, **94**, 1137 (1964).

177. E. Hannig, *Arch. Pharm.* (Weinheim), **296**, 441 (1963).

178. E. Hannig and H. Ziebandt, *Pharm. Zentralhalle*, **104**, 301 (1965).

179. E. Hannig and H. Ziebandt, *Pharmazie*, **22**, 626 (1967).

180. J. Supniewski, S. Misztal, and J. Krupińska, *Bull. Acad. Pol. Sci., Classe II*, **2**, 153 (1954).

181. A. Magelli and R. Passerini, *Boll. Sci. Fac. Chim. Ind. Bologna*, **14**, 52 (1956).

182. K. Bowden and E. A. Braude, *J. Chem. Soc.*, 1068 (1952).

183. L. Chierici and R. Passerini, *Boll. Sci. Fac. Chim. Ind. Bologna*, **11**, 104 (1953).

184. L. Chierici, H. Lumbroso, and R. Passerini, *Boll. Sci. Fac. Chim. Ind. Bologna*, **12**, 127 (1954).

185. A. Mangini, A. Trombetti, and C. Zauli, *J. Chem. Soc.*, B, 153 (1967).

186. J. Gosselck, *Chem. Ber.*, **91**, 2345 (1958).

187. J. Gosselck and E. Wolters, *Chem. Ber.*, **95**, 1237 (1962).

188. F. Dallacker, F.-E. Eschelbach, and H. Zegers, *Justus Liebigs Ann. Chem.*, **689**, 163 (1965).

189. A. Gasco, G. Di Modica, and E. Barni, *Ann. Chim.* (Rome), **58**, 385 (1968).

190. I. Degani, R. Fochi, and G. Spunta, *Boll. Sci. Fac. Chim. Ind. Bologna*, **23**, 165 (1965).

191. E. G. Kataev, L. M. Kataeva, and G. A. Chmutova, *Zh. Org. Khim.*, **2**, 2244 (1966).

192. E. G. Kataev and G. A. Chmutova, *Zh. Org. Khim.*, **3**, 2192 (1967).

193. E. G. Kataev, G. A. Chmutova, and E. G. Yarkova, *Zh. Org. Khim.*, **3**, 2188 (1967).

194. G. Pourcelot, *C. R. Acad. Sci., Paris*, **260**, 2847 (1965).

195. G. Pourcelot and P. Cadiot, *Bull. Soc. Chim. Fr.*, 3016 (1966).

196. A. Ruwet and M. Renson, *Bull. Soc. Chim. Belges*, **75**, 157 (1966).

197. J. W. Baker, G. F. C. Barrett, and W. T. Tweed, *J. Chem. Soc.*, 2831 (1952).

198. A. Ruwet and M. Renson, *Bull. Soc. Chim. Belges*, **78**, 571 (1969).

199. L. D. Pettit, A. Royston, C. Sharrington, and R. J. Whewell, *J. Chem. Soc.*, B, 588 (1968).

200. N. Bellinger, P. Cagniant, and M. Renson, *C. R. Acad. Sci., Paris, Ser. C*, **269**, 532 (1969).

201. L. Chierici and F. Montanari, *Boll. Sci. Fac. Chim. Ind. Bologna*, **14**, 78 (1956).

202. L. Chierici and F. Montanari, *Gazz. Chim. Ital.*, **86**, 1269 (1956).

203. N. Bellinger and P. Cagniant, *C. R. Acad. Sci., Paris, Ser. C*, **268**, 1385 (1969).

204. B. Sjöberg and S. Herdevall, *Acta Chem. Scand.*, **12**, 1347 (1958).

205. J. M. Danze and M. Renson, *Bull. Soc. Chim. Belges*, **75**, 169 (1966).

206. Á. I. Kiss and B. R. Muth, *J. Chem. Phys.*, **23**, 2187 (1955).

207. S. Patai and A. K. Muszkat, *Bull. Res. Council Israel*, **10A**, 73 (1961).

208. M. Vafai and M. Renson, *Bull. Soc. Chim. Belges*, **75**, 145 (1966).

209. M. Renson, *Bull. Soc. Chim. Belges*, **73**, 483 (1964).

210. W. A. Bonner and A. Robinson, *J. Amer. Chem. Soc.*, **72**, 354 (1950).

211. G. Wagner and G. Lehmann, *Pharm. Zentralhalle*, **100**, 160 (1961).

212. G. Wagner and P. Nuhn, *Arch. Pharm.* (Weinheim), **298**, 686 (1965).

213. G. Wagner and P. Nuhn, *Arch. Pharm.* (Weinheim), **298**, 692 (1965).

214. G. Wagner and P. Nuhn, *Arch. Pharm.* (Weinheim), **296**, 374 (1963).

215. G. Wagner and P. Nuhn, *Pharm. Zentralhalle*, **104**, 328 (1965).

216. G. Wagner and G. Valz, *Pharmazie*, **22**, 548 (1967).

217. L. Christiaens and M. Renson, *Bull. Soc. Chim. Belges*, **77**, 153 (1968).

218. G. Wagner and P. Nuhn, *Z. Chem.*, **3**, 64 (1963).

219. G. Wagner, E. Fickweiler, P. Nuhn, and H. Pischel, *Z. Chem.*, **3**, 62 (1963).

220. G. I. Zaitseva and V. M. Al'bitskaya, *Zh. Org. Khim.*, **4**, 2057 (1968).

221. H. Frenzel, P. Nuhn, and G. Wagner, *Arch. Pharm.* (Weinheim), **302**, 62 (1969).

222. K. Šindelář, J. Metyšová, E. Svátek, and M. Protiva, *Collection Czech. Chem. Commun.*, **34**, 2122 (1969).

223. E. G. Kataev and F. G. Gabdrakhmanov, *Zh. Obshch. Khim.*, **37**, 772 (1967).

224. E. G. Kataev and V. N. Petrov, *Zh. Obshch. Khim.*, **32**, 3699 (1962).

225. L. M. Kataeva, I. V. Anonimova, L. K. Yuldasheva, and E. G. Kataev, *Zh. Obshch. Khim.*, **32**, 3965 (1962).

226. S. J. Kuhn and J. S. McIntyre, U.S. Pat. 3,489,774 (1970); *Chem. Abstr.*, **72**, 100517 (1970).

227. H. Gilman and L. F. Cason, *J. Amer. Chem. Soc.*, **73**, 1074 (1951).

228. F. G. Gabdrakhmanov, Yu. Yu. Samitov, and E. G. Kataev, *Zh. Obshch. Khim.*, **37**, 761 (1967).

229. L. M. Kataeva, E. G. Kataev, Z. S. Titova, and N. A. Aleksandrova, *Zh. Strukt. Khim.*, **7**, 715 (1966).

230. A. N. Pudovik and E. I. Kashevarova, *Dokl. Akad. Nauk SSSR*, **158**, 167 (1964).

231. L. M. Kataeva, E. G. Kataev, and T. G. Mannafov, *Zh. Strukt. Khim.*, **7**, 226 (1966).

232. L. M. Kataeva, E. G. Kataev, and T. G. Mannafov, *Zh. Strukt. Khim.*, **10**, 830 (1969).

233. L. M. Kataeva, N. S. Podkovyrina, T. G. Mannafov, and E. G. Kataev, *Zh. Strukt. Khim.*, **10**, 1124 (1969).

234. W. Jenny, *Helv. Chim. Acta*, **36**, 1278 (1953).

235. G. Hölzle and W. Jenny, *Helv. Chim. Acta*, **41**, 593 (1958).

236. W. Jenny, Swiss Pat. 322,623 (1957); *Chem. Abstr.*, **52**, 3351 (1958).

237. Brit. Pat. 788,725 (1958), Ger. Pat. 1,015,447 (1957), French Pat. 1,103,317 (1955), and U.S. Pat. 2,807,631 (1957).

238. H. Rheinboldt and M. Perrier, *Bull. Soc. Chim. Fr.*, 379 (1953).

239. M. de Moura Campos and N. Petragnani, *Chem. Ber.*, **93**, 317 (1960).

240. H. Rheinboldt and M. Perrier, *Bull. Soc. Chim. Fr.*, 759 (1950).

241. G. Hölzle and W. Jenny, *Helv. Chim. Acta*, **41**, 712 (1958).

242. L. M. Yagupol'skii and V. G. Voloshchuk, *Zh. Obshch. Khim.*, **36**, 165 (1966).

243. L. M. Yagupol'skii and V. G. Voloshchuk, *Zh. Obshch. Khim.*, **38**, 2509 (1968).

244. V. G. Voloshchuk, L. M. Yagupol'skii, G. P. Syrova, and V. F. Bystrov, *Zh. Obshch. Khim.*, **37**, 118 (1967).

245. J. Supniewski, S. Misztal, and J. Krupińska, *Arch. Immunol. Terapii Doświadczalnej*, **3**, 531 (1955).

246. J. Supniewski, F. Rogóz, and J. Krupińska, *Bull. Acad. Pol. Sci., Ser. Sci. Biol.*, **9**, 231 (1961).

247. L. M. Yagupol'skii and V. G. Voloshchuk, *Zh. Obshch. Khim.*, **37**, 1543 (1967).

248. L. Christiaens and M. Renson, *Bull. Soc. Chim. Belges*, **79**, 133 (1970).

249. E. Niwa, H. Aoki, H. Tanaka, K. Munakata, and M. Namiki, *Chem. Ber.*, **99**, 3215 (1966).

250. O. S. Chizhov, B. M. Zolotarev, A. N. Sukiasian, V. P. Litvinov, and Ya. L. Goldfarb, *Org. Mass Spectrom.*, **3**, 1379 (1970).

251. Ya. L. Goldfarb, V. P. Litvinov, and A. N. Sukiasian, *Izv. Akad. Nauk SSSR, Ser. Khim.*, 2585 (1967).

252. Ya. L. Goldfarb, V. P. Litvinov, and A. N. Sukiasian, *Dokl. Akad. Nauk SSSR*, **182**, 340 (1968).

253. A. N. Sukiasyan, V. P. Litvinov, and Ya. L. Goldfarb, *Izv. Akad. Nauk SSSR, Ser. Khim.*, 1345 (1970).

254. Ya. L. Goldfarb and V. P. Litvinov, *Izv. Akad. Nauk SSSR, Ser. Khim.*, 2088 (1964).

255. D. Seebach and N. Peleties, *Angew. Chem.*, **81**, 465 (1969).

256. M. Renson and J.-L. Piette, *Bull. Soc. Chim. Belges*, **73**, 507 (1964).

257. M. Pallotti and A. Tundo, *Boll. Sci. Fac. Chim. Ind. Bologna*, **17**, 17 (1959).

258. I. E. Pollack and G. F. Grillot, *J. Org. Chem.*, **31**, 3514 (1966).

259. I. E. Pollack and G. F. Grillot, *J. Org. Chem.*, **32**, 3101 (1967).

260. L. M. Yagupol'skii and N. V. Kondratenko, *Zh. Obshch. Khim.*, **37**, 1770 (1967).

261. L. Christiaens and M. Renson, *Bull. Soc. Chim. Belges*, **79**, 235 (1970).

262. E. Hannig and H. Ziebandt, *Pharmazie*, **23**, 552 (1968).

263. E. Hannig, *Pharmazie*, **19**, 201 (1964).

264. W. J. Patterson and V. Du Vigneaud, *J. Biol. Chem.*, **111**, 393 (1935).

265. U. Schöllkopf and H. Küppers, *Tetrahedron Lett.*, 105 (1963).

266. A. Ruwet and M. Renson, *Bull. Soc. Chim. Belges*, **77**, 465 (1968).

267. A. Ruwet and M. Renson, *Bull. Soc. Chim. Belges*, **79**, 61 (1970).

268. E. Hannig and H. Ziebandt, *Pharmazie*, **23**, 688 (1968).

269. F. Dallacker and F.-E. Eschelbach, *Justus Liebigs Ann. Chem.*, **689**, 171 (1965).

270. E. G. Kataev, G. A. Shmutova, E. G. Yarkova, and T. G. Plotnikova, *Zh. Org. Khim.*, **5**, 1514 (1969).

271. E. G. Kataev, G. A. Shmutova, A. A. Musina, and E. G. Yarkova, *Dokl. Akad. Nauk SSSR*, **187**, 1308 (1969).

272. E. E. Aynsley, N. N. Greenwood, and J. B. Leach, *Chem. Ind.* (London), 379 (1966).

273. E. W. Abel and G. V. Hutson, *J. Inorg. Nucl. Chem.*, **31**, 3333 (1969).

274. N. N. Greenwood and G. Hunter, *J. Chem. Soc.*, A, 1520 (1967).

275. H. J. Whitfield, *J. Chem. Soc.*, *A*, 113 (1970).

276. D. W. Goheen and W. R. Vaughan, *J. Org. Chem.*, **20**, 1019 (1955).

277. E. J. Corey and D. Seebach, *J. Org. Chem.*, **31**, 4097 (1966).

278. L. Brandsma, *Rec. Trav. Chim. Pays-Bas*, **83**, 307 (1964).

279. L. M. Kataeva, E. G. Kataev, and D. Ya. Idiyatullina, *Zh. Strukt. Khim.*, **7**, 380 (1966).

280. D. Seebach, *Angew. Chem.*, **79**, 468 (1967).

281. M. Cinquini, S. Colonna, and D. Landini, *Boll. Sci. Fac. Chim. Ind. Bologna*, **27**, 207 (1969).

282. N. Petragnani and G. Schill, *Chem. Ber.*, **103**, 2271 (1970).

283. N. Petragnani and V. G. Toscano, *Chem. Ber.*, **103**, 1652 (1970).

284. R. Paetzold, *Z. Chem.*, **4**, 321 (1964).

285. K. A. Jensen and V. Krishnan, *Acta Chem. Scand.*, **21**, 1988 (1967).

286. T. Tanaka and T. Kamitani, *Inorg. Chim. Acta*, **2**, 175 (1968).

287. M. Oki and H. Iwamura, *Tetrahedron Lett.*, 2917 (1966).

288. D. N. Jones, D. Mundy, and R. D. Whitehouse, *Chem. Commun.*, 86 (1970).

289. M. Cinquini, S. Colonna, and R. Giovini, *Chem. Ind.* (London), 1737 (1969).

290. D. Spinelli and G. Monaco, *Boll. Sci. Fac. Chim. Ind. Bologna*, **20**, 56 (1962).

291. R. F. Riley, J. Flato, and D. Bengels, *J. Org. Chem.*, **27**, 2651 (1962).

292. H. Rheinboldt and E. Giesbrecht, *J. Amer. Chem. Soc.*, **68**, 2671 (1946).

293. R. Paetzold and P. Vordank, *Z. Anorg. Allg. Chem.*, **347**, 294 (1966).

294. E. Rebane, *Acta Chem. Scand.*, **24**, 717 (1970).

295. A. Cerniani, *Ric. Sci.*, **26**, 2100 (1956).

296. A. R. Poggi and R. Pirisi, *Rend. Seminario Fac. Sci. Univ. Cagliari*, **30**, 134 (1960).

297. V. V. Kozlov and V. M. Pronyakova, *Zh. Org. Khim.*, **1**, 493 (1965).

298. V. V. Kozlov and V. M. Pronyakova, *Zh. Vses. Khim. Obshchestva*, **12**, 471 (1967).

299. Z. Štefanac and M. Tomašković, *Bull. Sci. Conseil Acad. RSF Yougoslavie*, **10A**, 317 (1965).

300. R. Paetzold and G. Bochmann, *Z. Chem.*, **8**, 308 (1968).

301. R. Paetzold and G. Bochmann, *Z. Anorg. Allg. Chem.*, **368**, 202 (1969).

302. R. Paetzold, *Spectrochim. Acta*, **24A**, 717 (1968).

303. R. Paetzold and G. Bochmann, *Spectrochim. Acta*, **26A**, 391 (1970).

304. A. E. Lutskii, E. M. Obukhova, V. A. Granshan, S. A. Volchenok, Z. M. Kanevskaya, L. M. Yagupol'skii, and V. G. Voloshchuk, *Teor. Eksp. Khim.*, **5**, 614 (1969).

305. G. Bergson, *Acta Chem. Scand.*, **11**, 580 (1957).

306. K. J. Wynne and J. W. George, *Inorg. Chem.*, **4**, 256 (1965).

307. J. D. McCullough and I. G. Zimmermann, *J. Phys. Chem.*, **64**, 1084 (1960).

308. J. D. McCullough and B. A. Eckerson, *J. Amer. Chem. Soc.*, **73**, 2954 (1951).

309. K. J. Wynne and J. Puckett, *Chem. Commun.*, 1532 (1968).

310. K. J. Wynne, *Inorg. Chem.*, **9**, 299 (1970).

311. R. Paetzold and U. Lindner, *Z. Anorg. Allg. Chem.*, **350**, 295 (1967).

312. G. J. Neuerberg, *Anal. Chem.*, **23**, 1042 (1951).

313. J. D. McCullough and R. E. Marsh, *Acta Cryst.*, **3**, 41 (1950).

VI Selenonium compounds

ROBERT J. SHINE

Ramapo College of New Jersey
School of Theoretical and Applied Science
Mahwah, New Jersey

A. INTRODUCTION

Although the term selenonium has been used to denote compounds containing tetravalent selenium bonded to one, two, three, or four organic groups,[1] in this chapter "selenonium" has been restricted to denote selenium-containing compounds possessing three C-Se bonds. Selenonium compounds of the type R_3SeX show ionic character, the ions being the cation R_3Se^+ and the anion X^-. The ion X^- can be any one of a number of inorganic ions, for example, halide, nitrate, hydroxide, chloroplatinate, or some similar species.

Selenonium compounds of the general structure $R_3Se^+X^-$, although easily formed, have not been studied to any great extent. A few reviews have been published on the subject. The first extensive review, by Bradt,[1] appeared in 1934; a more recent one is by Rheinboldt.[2] Other reviews on organic selenium chemistry that discuss selenonium compounds to a limited extent are also available.[3-6]

B. PREPARATION OF SELENONIUM COMPOUNDS

1. Aliphatic

a. Reaction of a Dialkyl Selenide with an Alkyl Halide. The common method of preparation of aliphatic selenonium compounds is by addition of an alkyl halide to a dialkyl selenide (equation 1). This method parallels

$$(1) \qquad R_2Se + RX \rightleftharpoons R_3Se^+X^-$$
$$\mathbf{1}$$

that for the preparation of sulfonium compounds by the addition of alkyl halides to sulfides. Reaction 1 is general and proceeds smoothly with or without solvent, giving high yields of the selenonium compound **1**. For example, trimethylselenonium iodide (**1**, R = CH_3, X = I) has been prepared[7] in 91 % yield by the reaction of methyl iodide with dimethyl selenide at room temperature for 1 hour.

McFarlane[8] also prepared the same compound, trimethylselenonium iodide, using an ethereal solution of methyl iodide and dimethyl selenide, whereas Wynne[9] prepared trimethylselenonium chloride, (**1**, R = CH_3, X = Cl) by treating dimethyl selenide with methyl chloride in a sealed tube for 1 week. α-Bromoacids[10] and α-bromoesters[11] react smoothly with selenides to give selenetine (selenonium carboxylate) compounds, as illustrated by the preparation of dimethylselenetine bromide (**2**) in reaction 2.

$$(2) \quad (CH_3)_2Se + BrCH_2COOCH_2CH_3 \longrightarrow \overset{Br^-}{\underset{\underset{\mathbf{2}}{+}}{(CH_3)_2SeCH_2COOCH_2CH_3}}$$

b. Reaction of Elemental Selenium with an Alkyl Halide. Aliphatic selenonium compounds can be made also by other, less general methods. Trimethylselenonium iodide was prepared[12-14] by the reaction of elemental selenium with methyl iodide in a sealed tube for 8 hours at 220° (equation 3).

$$(3) \qquad 3CH_3I + Se \xrightarrow[\Delta]{\text{sealed tube}} (CH_3)_3Se^+I^- + I_2$$

c. Reaction of an Alkyl Selenide with Nitric Acid. Tribenzylselenonium nitrate was made by treating benzyl selenide with nitric acid.[15]

d. Reaction of Selenium Halides with an Organometallic Reagent.
Selenium tetrachloride reacts with diethylzinc in ether, giving zinc chloride and triethylselenonium tetrachlorozincate (equation 4).[16]

(4) $2SeCl_4 + 3Zn(C_2H_5)_2 \rightarrow 2ZnCl_2 + [(C_2H_5)_3Se]_2ZnCl_4$

Ethylmagnesium bromide, ethyl bromide, and selenium monobromide react to give triethylselenonium hydroselenide (equation 5).[17]

(5) $3C_2H_5MgBr + Se_2Br_2 + C_2H_5Br \xrightarrow{H_2O}$
$$(C_2H_5)_3Se^+HSe^- + 3MgBr_2 + C_2H_5OH$$

e. Anion-Exchange Reactions. Since selenonium compounds are ionic, as shown by their water solubility and electric conductivity, the anion can be exchanged by simple metathetical (ion-exchange) reactions.[7,9,18-20] For example, Hashimoto et al.[7] made a series of trialkylselenonium tetrafluoroborates, using silver tetrafluoroborate; the silver ion precipitates the iodide ion from the aliphatic selenonium iodide. Platinum chloride complexes of aliphatic selenonium compounds have also been reported,[18-20] and a boron trichloride complex of trimethylselenonium chloride has been made (equation 6).[9] The physical properties of other aliphatic selenonium compounds are listed in the chemical literature.[7,9,14,21-23]

(6) $(CH_3)_3Se^+Cl^- + BCl_3 \rightarrow (CH_3)_3Se^+BCl_4^-$

f. Reaction of a Cyclic Selenium Compound with an Alkyl Halide. If the Se atom forms part of an aliphatic ring system, an alkyl iodide will add normally to the cyclic selenide to give a selenonium compound, as shown by the reaction of selenoisochroman (**3**) with methyl iodide to yield methylselenoisochromanium iodide (**4**) (equation 7).[24] However, if the ring system

(7)

3   **4**

is strained, ring opening occurs, giving a noncyclic selenonium cation. For example, Backer and Winter[25] reported the facile ring opening of three selenacyclobutane derivatives, using methyl iodide (equation 8). Truce and Emrick,[26] in studying the selenacycloheptane system, reported that two molecules of methyl iodide added without ring opening to give the dipositive

(8)

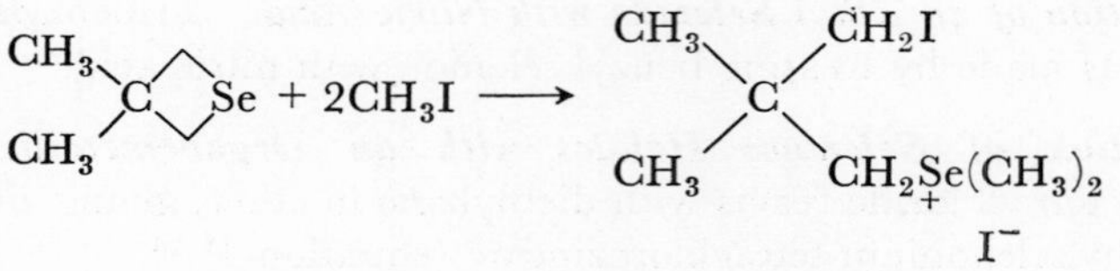

$$CH_3 \text{-}C\text{-}Se + 2CH_3I \longrightarrow$$

$$\underset{CH_3}{\overset{CH_3}{\diagdown}}C\underset{CH_2Se(CH_3)_2}{\overset{CH_2I}{\diagup}}$$

$$I^-$$

cation 2,7-dihydro-1,1-dimethyl-3,4-5,6-dibenzoselenepinium diiodide (**5**) (equation 9).

(9)

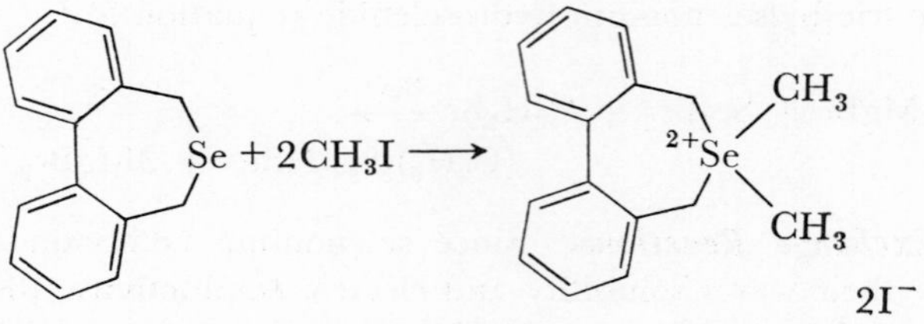

$$\text{(structure)} Se + 2CH_3I \longrightarrow \text{(structure)}^{2+}Se\underset{CH_3}{\overset{CH_3}{\diagup}}$$

$$2I^-$$

5

2. Aromatic

a. *Selenium Not Contained in an Aromatic Ring*

i. Reaction of an Alkylselenium Halide with an Aromatic Hydrocarbon. In contrast to aliphatic selenonium salts, triarylselenonium compounds of the general formula $Ar_3Se^+X^-$ (**6**) cannot be made by the addition of an aryl halide to a diaryl selenide.[27] The method that has found the widest use in the preparation of triarylselenonium compounds is the reaction of an aromatic hydrocarbon with a diaryl selenide dihalide in the presence of aluminum trichloride (equation 10). Utilizing this method, Leicester and

(10) $$Ar_2SeCl_2 + ArH \xrightarrow{\text{AlCl}_3} Ar_3Se^+Cl^-$$
6

Bergstrom[27,28] prepared triphenylselenonium chloride (**6**, $Ar = C_6H_5$, $X = Cl$).

ii. Reaction of Selenium Tetrachloride with an Aromatic Hydrocarbon. Starting with selenium tetrachloride, benzene, and anhydrous aluminum trichloride, Bradt and Green[29] obtained, among other products, triphenylselenonium chloride (equation 11), whereas Leicester and Bergstrom[27]

(11) $$SeCl_4 + 3ArH \xrightarrow{\text{AlCl}_3} Ar_3Se^+Cl^- + 3HCl$$

reported low yields because of many side reactions.

iii. Reaction of an Arylselenium Halide with an Arylmercury Chloride.
Leicester and Bergstrom[30,31] prepared triarylselenonium compounds
(isolated as the mercuric chloride double salt) by the reaction of mercury
diaryls and diarylselenium dihalides (equation 12) in a two-step reaction
shown by equations 13 and 14.

(12) $$2R_2SeCl_2 + R_2Hg \rightarrow R_3Se^+Cl^- \cdot HgCl_2 + R_2Se + RCl$$

(13) $$R_2SeCl_2 + R_2Hg \rightarrow R_2Se + RCl + RHgCl$$

(14) $$RHgCl + R_2SeCl_2 \rightarrow R_3Se^+Cl^- \cdot HgCl_2$$

iv. Reaction of Selenium Dioxide with an Aromatic Hydrocarbon. Another
useful method[13,32-34] for preparing triarylselenonium compounds is the
reaction of an aromatic hydrocarbon with selenium dioxide in the presence
of aluminum trichloride (equation 15). Hilditch and Smiles,[13,32] the first

(15) $$SeO_2 + 3ArH \xrightarrow{AlCl_3} Ar_3Se^+Cl^-$$

to utilize this method, obtained trianisylselenonium chloride (**6**, Ar =
$CH_3OC_6H_4$, X = Cl) when anisole and selenium dioxide were heated with
aluminum trichloride. Although the position of the methoxyl groups in the
product was not determined, Crowell and Bradt[33] reported later that
heating toluene, selenium dioxide, and aluminum trichloride gave tri-*p*-
tolylselenonium chloride, which was isolated as the zinc chloride addition
compound in 77% yield. Excessive heating must be avoided because of the
thermal instability of the selenonium compound. Funk[34] used hydrogen
chloride instead of aluminum chloride in the reaction of selenium dioxide
with *o*-cresol, and obtained tris(1-hydroxy-2-methylphenyl)selenonium
chloride (**7**).

7

v. Reaction of Selenium Oxychloride with an Aromatic Hydrocarbon.
Morgan and Burstall,[35] in studying the reaction of phenol with selenium
oxychloride in ether or chloroform (equation 16), obtained two isomeric

(16) $$SeOCl_2 + ArH \xrightarrow{ether} Ar_3Se^+Cl^-$$

selenonium chlorides. The less soluble isomer was characterized as tri-*p*-hydroxyphenylselenonium chloride (**6**, Ar = *p*-HOC$_6$H$_4$, X = Cl); the more soluble isomer, obtained in low yield, was not fully identified. Reaction 16 does not appear to be a general one for the preparation of selenonium compounds since the reaction of *β*-naphthol with selenium oxychloride results in selenide formation.[35]

vi. Reaction of an Alkyl Selenide with an Aryliodonium Tetrafluoroborate. Using diphenyliodonium tetrafluoroborate [(C$_6$H$_5$)$_2$IBF$_4$] as a phenylating agent, Makarova and Nesmeyanov[36,37] obtained triphenylselenonium tetrafluoroborate (**6**, Ar = C$_6$H$_5$, X = BF$_4$) from diphenyl selenide at 210° in 80% yield (equation 17).

(17) $(C_6H_5)_2Se + (C_6H_5)_2IBF_4 \rightarrow (C_6H_5)_3Se^+BF_4^- + C_6H_5I$

As with aliphatic selenonium compounds, the anion of triarylselenonium compounds can be readily exchanged.[7,14,27,29,32,35,38]

b. Selenium Contained in a Heterocyclic System.

i. Reaction of a Cyclic Selenide with an Oxidizing Agent. Aromatic heterocyclic compounds containing a positively charged Se atom are made by treatment of the cyclic selenide with a suitable oxidizing agent (equation 18). Degani *et al.*[39–41] used triphenylmethyl perchlorate as the oxidant to

(18)

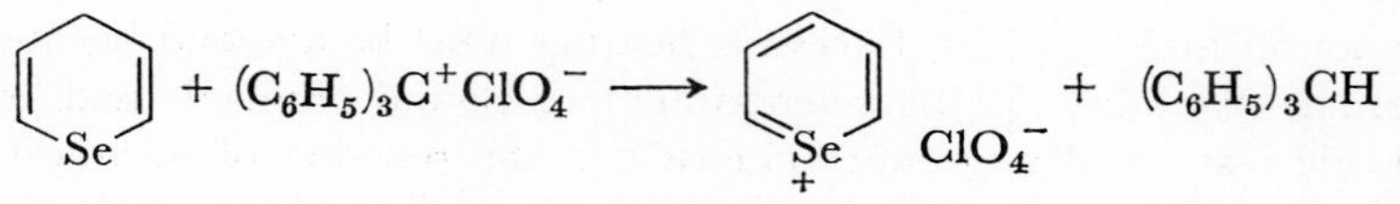

prepare selenapyrylium perchlorate (**8**) (equation 18),[39] 4-methylselenapyrylium perchlorate (**9**),[41] and 1-selenachromylium perchlorate (**10**) (equation 19).[40] Danze and Renson[42] prepared a series of 1-selenachromylium

(19)

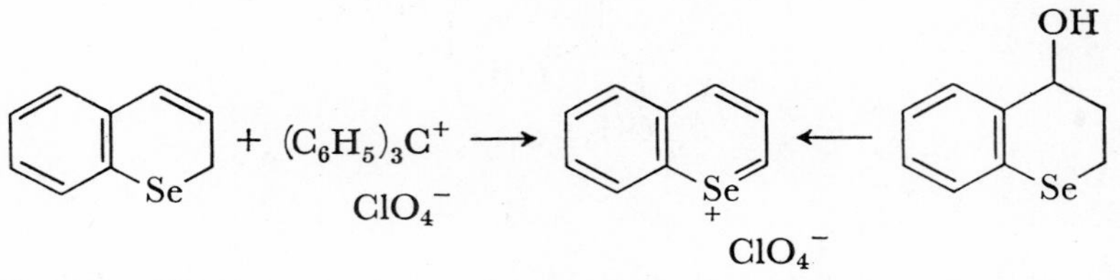

perchlorates by a similar oxidation. Thionyl chloride was used in the final step of the preparation of the dibenzo[b,d]selenapyrylium cation (**11**) (equation 20),[43] while tin chloride was employed for dehydrogenation in the

$$(20) \qquad + \text{SOCl}_2 \longrightarrow$$

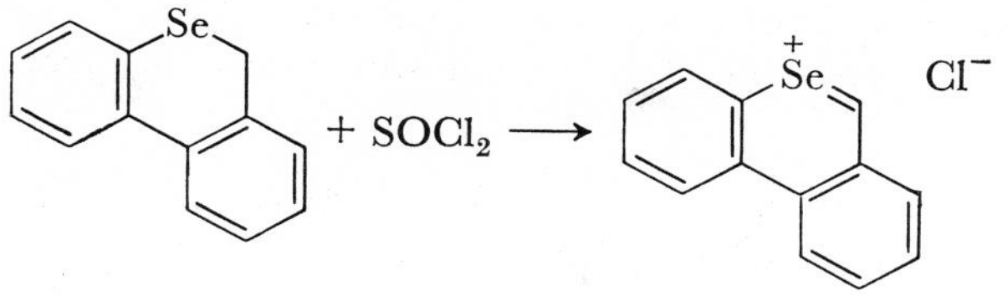

11

preparation of 1,3-diaminophenazselenonium chloride (**12**).[44]

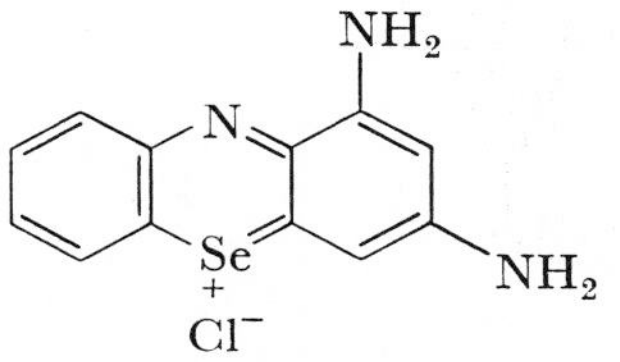

12

The resonance-stabilized diselenolylium cation, bearing methyl substituents in the 3- and 5-positions (**13**, R = CH$_3$), has been prepared by the reaction of acetonylacetone with hydrogen selenide.[45,46] Oxidation of seleno-malonamide with iodine gives the 3,5-diaminodiselenolylium cation (**13**, R = NH$_2$) in 68% yield, while ferric chloride produces the same cation (**13**, R = NH$_2$) in 76% yield.[47]

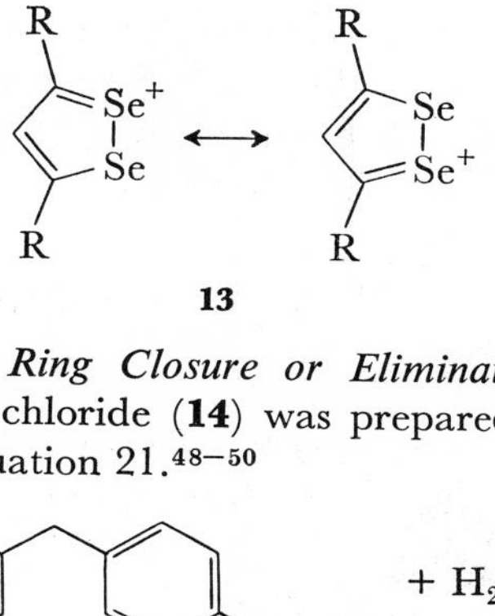

13

ii. Reactions Involving Ring Closure or Elimination. 3,6-Bis(dimethylamino)selenoxanthylium chloride (**14**) was prepared by the selenious acid ring closure shown in equation 21.[48–50]

$$(21) \qquad (\text{CH}_3)_2\text{N} \qquad \qquad \text{N(CH}_3)_2 \qquad + \text{H}_2\text{SeO}_3 \longrightarrow$$

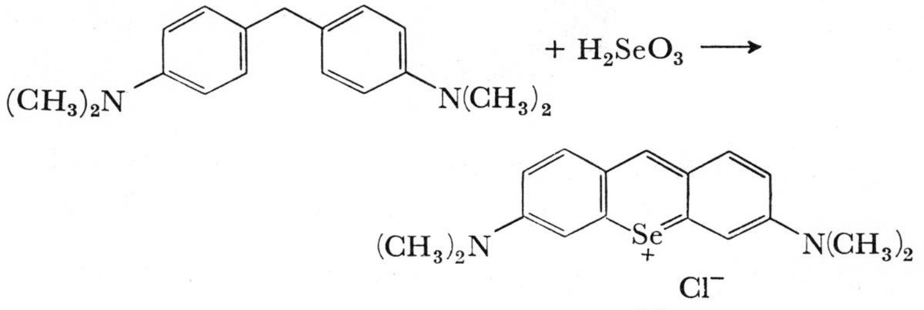

14

Selenium oxychloride reacted with acetone to give the selenapyrylium compound **15** (equation 22).[51] Selenoxanthylium perchlorate (**16**) was

(22)

$$SeOCl_2 + 2CH_3\overset{O}{\overset{\|}{C}}CH_3 \xrightarrow{HCl}$$

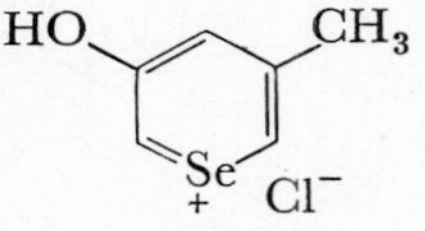

15

prepared by the reaction of selenoxanthydrol in acetic anhydride with 70% perchloric acid (equation 23).[52]

(23)

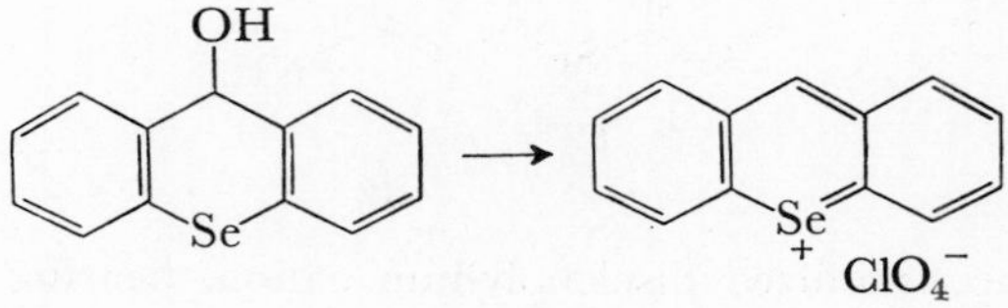

16

3.   Mixed Aliphatic-Aromatic

a.   Reaction of a Selenide with an Alkyl Sulfate or Halide.   Alkyl halides can be added with difficulty to alkyl aryl selenides or diaryl selenides to give selenonium compounds. The addition of dimethyl sulfate to methyl phenyl selenide[21,53,54] produced dimethylphenylselenonium methyl sulfate (**17**) (equation 24). Methyl iodide gave poor yields in this reaction.[21] Diphenyl-

(24) $C_6H_5SeCH_3 + (CH_3)_2SO_4 \rightarrow C_6H_5\overset{+}{Se}(CH_3)_2\ CH_3SO_4^-$

17

phenacylselenonium derivatives (**18**), isolated as the tetrafluoroborate salts, were prepared[23] by the addition of substituted phenacyl bromides to diphenyl selenide in 1,2-dichloroethane (equation 25). Alkyl halides also added

(25) $Ar\overset{O}{\overset{\|}{C}}CH_2Br + (C_6H_5)_2Se \longrightarrow Ar\overset{O}{\overset{\|}{C}}CH_2\overset{+}{Se}(C_6H_5)_2\ BF_4^-$

18

quantitatively[55] to o-acetyl and o-formylselenoanisole to give selenonium compounds.

α-Halocarboxylic acids add to aryl alkyl selenides[56-58] to form selenetine-type compounds (**19**), as exemplified in equation 26. However, heating such

$$(26) \qquad C_6H_5SeCH_3 + BrCH_2COOH \longrightarrow \overset{+}{C_6H_5\underset{\underset{CH_3}{|}}{Se}}CH_2COOH \quad Br^-$$

19

compounds at 100° for 24 hours can result in displacement of a methyl group, giving product **20**, shown in equation 27, in 66% yield.[58]

(27)

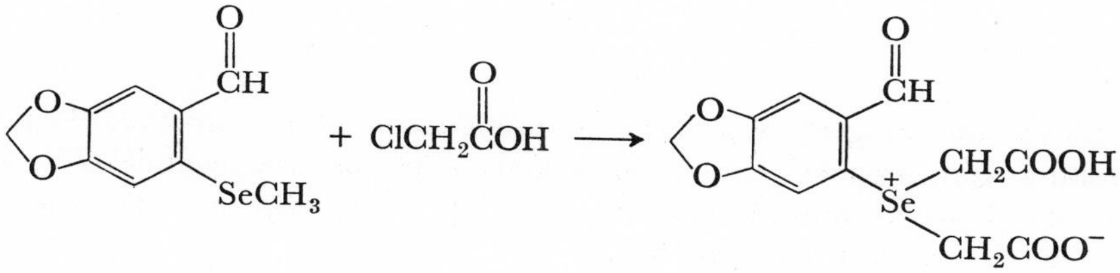

20

The alkylation of methyl phenylethynyl selenide with triethyloxonium tetrafluoroborate without solvent produced the methylethyl(phenylethynyl)-selenonium cation (**21**), isolated as the picrate salt (equation 28).[59]

$$(28) \qquad C_6H_5C{\equiv}CSeCH_3 + (C_2H_5)_3O^+BF_4^- \longrightarrow C_6H_5C{\equiv}C\overset{+}{\underset{\underset{CH_3}{|}}{Se}}C_2H_5$$

21

b. Reaction of a Selenide with Benzaldehyde.   The addition of benzaldehyde to 2-acetyl-4-methylselenoanisole (**22**) in the presence of hydrogen chloride gave the 1,6-dimethylselenoflavanonium ion (**23**) in 73% yield, as shown in equation 29.[60]

(29)

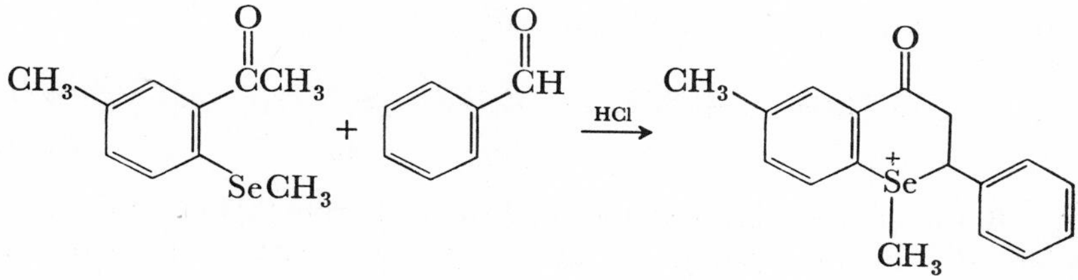

22   **23**

4. Miscellaneous

Removal of a hydrogen ion from the α-position of a selenonium compound should result, theoretically, in ylid formation. Treatment of fluorenyl-9-dimethylselenonium bromide (**24**) with aqueous alkali resulted in a black precipitate that is reported to have the ylid structure **25** (equation 30).[3,61]

(30)

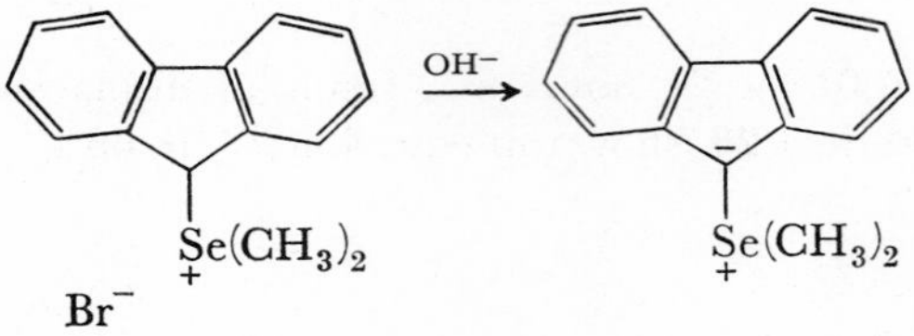

                    24   **25**

Lloyd and Singer[62] reported the preparation of an air-stable, light-unstable ylid, diphenylselenonium tetraphenylcyclopentadienylide (**26**), by heating diazotetraphenylcyclopentadiene (**27**) with diphenyl selenide at 140° for 10 minutes (equation 31).

(31)

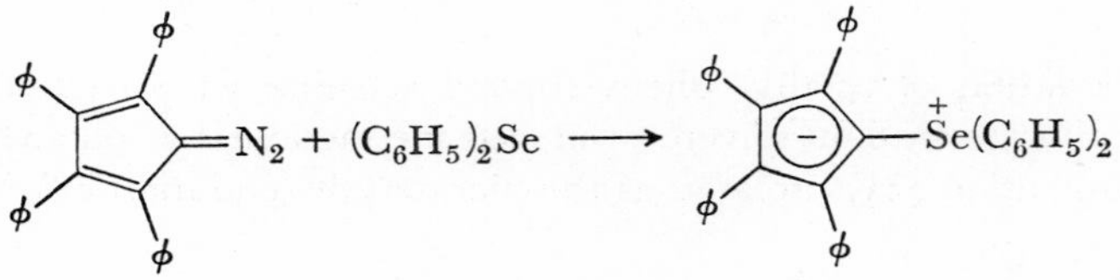

          27   **26**

C. REACTIONS OF SELENONIUM COMPOUNDS

1. Aliphatic

a. Stability.   Aliphatic selenonium compounds have been shown to form new trialkylselenonium cations as transformation products on reaction with alkyl halides. For example, Jackson[19] reported the isolation of trimethyl-selenonium iodide as one of the products of the reaction of dibenzyl diselenide with an excess of methyl iodide (equation 32).

$$(32) \quad (C_6H_5CH_2Se-)_2 + 5CH_3I \rightarrow \overset{+}{C_6H_5CH_2Se}(CH_3)_2 + (CH_3)_3Se^+I^-$$
$$I_3^- \qquad + C_6H_5CH_2I$$

Von Braun *et al.*[63] observed the formation of trimethylselenonium iodide when decyl methyl selenide was allowed to stand with methyl iodide for a few days (equation 33).

$$(33) \qquad C_{10}H_{21}SeCH_3 + CH_3I \rightarrow (CH_3)_3Se^+I^-$$

Heating triethylselenonium iodide at 80–126° results in its decomposition,[20] giving diethyl selenide and ethyl iodide (equation 34). Refluxing an aqueous

$$(34) \qquad (C_2H_5)_3Se^+I^- \rightarrow (C_2H_5)_2Se + CH_3CH_2I$$

solution of fluorenyl-9-dimethylselenonium bromide (**24**) for 30 minutes results in the formation of dimethyl selenide and 9-fluorenyl alcohol.[61]

b. Optical Resolution. Attempts to resolve asymmetric selenonium compounds into optically active isomers have met with some success. Biilmann and Jensen[10] prepared the selenetine system **28**, which contains an

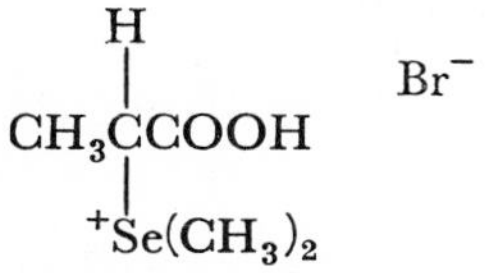

$$\begin{array}{c} \text{H} \\ | \\ \text{CH}_3\text{CCOOH} \qquad \text{Br}^- \\ | \\ {}^+\text{Se(CH}_3)_2 \end{array}$$

28

asymmetric C atom attached to the Se atom. Holliman and Mann[24,64] resolved the 2-*p*-chlorophenacylselenoisochromanium ion **29** by use of

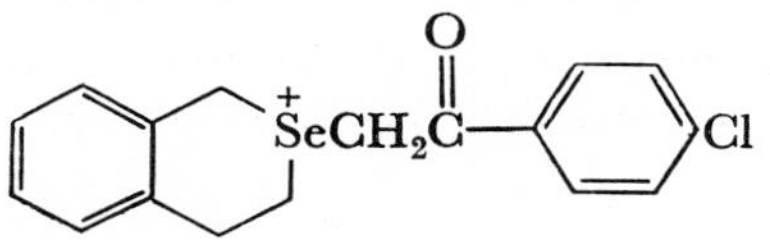

29

d-bromocamphorsulfonate. The picrate salt of this resolved selenonium ion **29**, which has an $[M_D]$ of $-533°$, is reported to have high optical stability. The sulfur analog also has high optical stability, whereas the telluronium compound does not.[24]

2. Aromatic

a. Selenium Not Contained in an Aromatic Ring.

i. Stability. Like aliphatic selenonium salts, aromatic selenonium compounds undergo thermal decomposition. Leicester and Bergstrom[27,30] observed that triphenylselenonium halides gave diphenyl selenides and aryl halides when heated (equation 35). However, the thermal decomposition

$$(35) \qquad Ar_3Se^+X^- \xrightarrow{\Delta} Ar_2Se + ArX$$

of triarylselenonium compounds, unlike that of aliphatic selenonium ions, is an irreversible reaction.[27] Crowell and Bradt,[33] in the course of their preparation of tri-*p*-tolylselenonium chloride, obtained small amounts of *p*-tolyl selenide and *p*-chlorotoluene because of thermal decomposition.

Heating diphenyl *p*-biphenylselenonium chloride to 130° gave chlorobenzene and phenyl-*p*-biphenyl selenide.[65]

ii. Halogenation. The action of halogen on aromatic selenonium compounds has resulted in ring halogenation[35] and the formation of salts containing trihalide ions.[7] Bromination of tri-*p*-hydroxyphenylselenonium chloride (**6**, Ar = *p*-HOC$_6$H$_4$, X = Cl) resulted in the formation of a hexabromide derivative which was not fully characterized.[35]

iii. Reaction with Base. An amphoteric oxide **30** precipitated when tri-*p*-hydroxyphenylselenonium chloride was dissolved in an alkaline solution and acidified with acetic acid.[35]

$$[(HOC_6H_4)_3Se]_2O$$
30

iv. Reaction with an Organometallic Reagent. Phenyllithium reacts with triphenylselenonium bromide (**6**, Ar = C$_6$H$_5$, X = Br) in absolute ether to give a 23% yield of biphenyl and a 31% yield of diphenyl selenide isolated as the dichloride.[66] On the basis of the observation that triphenyltelluronium bromide forms tetraphenyltellurium with phenyllithium, Wittig and Fritz[66] proposed the possible intermediacy of unstable tetraphenylselenium (**31**), which decomposes to the products indicated in equation 36. This postulate

$$(36) \quad (C_6H_5)_3Se^+Br^- + C_6H_5Li \rightarrow [(C_6H_5)_4Se] \rightarrow$$

$$\textbf{31} \quad C_6H_5\!\!-\!\!C_6H_5 + (C_6H_5)_2Se$$

is further substantiated by the subsequent work of Hellwinkel and Fahrbach[66a] These workers prepared the hypervalent selenium compound bis(2,2′-biphenylylene)selenium, which underwent ring opening on treatment with water and potassium iodide to give 2,2′-biphenylylene-2-biphenylyl-selenonium iodide.

b. Selenium Contained in Heterocyclic System.

i. Oxidation and Reduction. In a study of the oxidation of various oxygen, sulfur, and selenium heteroaromatic cations, Degani and Fochi[67,68] treated 1-selenachromylium perchlorate and selenaxanthylium perchlorate with manganese dioxide. Oxidation of the 1-selenachromylium cation resulted in ring contraction, giving a 92% yield of 2-benzoselenophenecarboxaldehyde (**32**) (equation 37), whereas oxidation of the selenaxanthylium cation gave

(37)

10   **32**

an 88 % yield of selenoxanthone (**33**) (equation 38). Nealey and Driscoll[50]

(38) 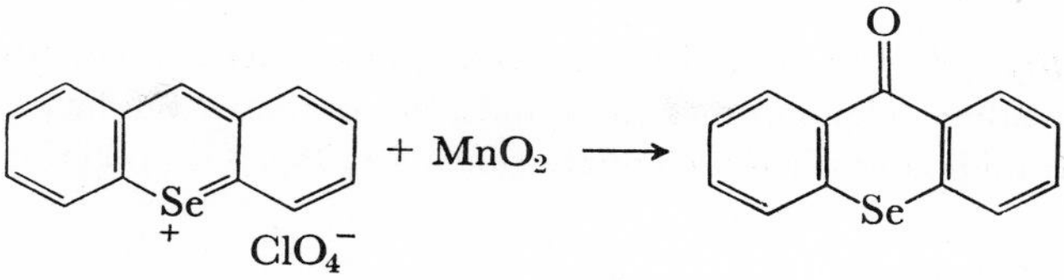 [structure of **16**] + MnO$_2$ $\longrightarrow$ [structure of **33**]

16   **33**

studied the oxidation and reduction reactions of the 3,6-bis(dimethylamino)-selenaxanthylium cation (**34**) and its sulfur analog. The selenium compound, when treated with alkaline permanganate, gave 3,6-bis(dimethylamino)sel-enoxanthone (**35**) in 21 % yield, analogously to Degani and Fochi's results with the same ring system. When the 3,6-bis(dimethylamino)selenaxanthyl-ium cation (**34**) was reduced with lithium aluminum hydride in THF, 3,6-bis(dimethylamino)selenoxanthene (**36**) was obtained in 83 % yield heme 39).

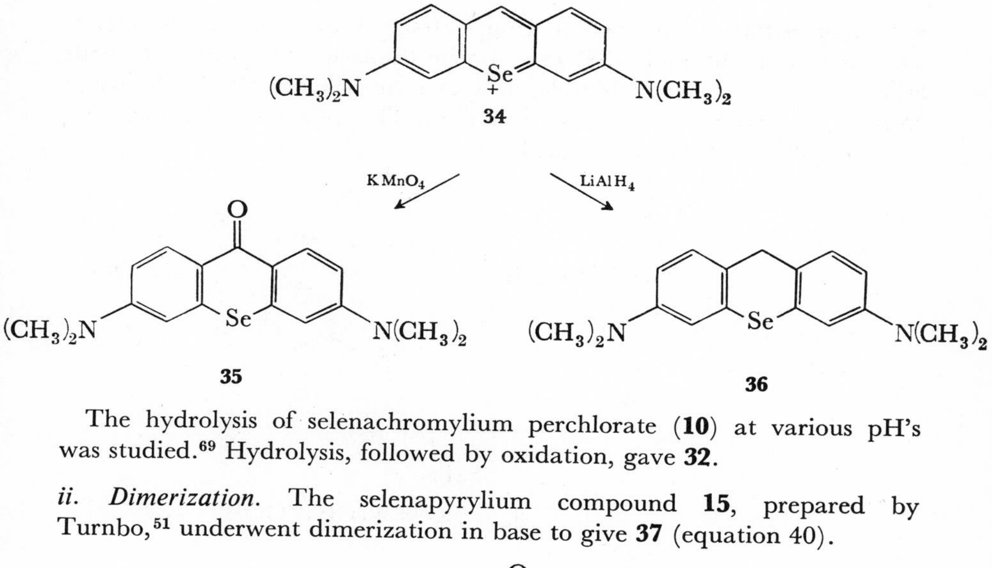

[structure of **34**]

34

K MnO$_4$ /   \ LiAlH$_4$

[structure of **35**]   [structure of **36**]

35   **36**

The hydrolysis of selenachromylium perchlorate (**10**) at various pH's was studied.[69] Hydrolysis, followed by oxidation, gave **32**.

ii.   Dimerization. The selenapyrylium compound **15**, prepared by Turnbo,[51] underwent dimerization in base to give **37** (equation 40).

(40) 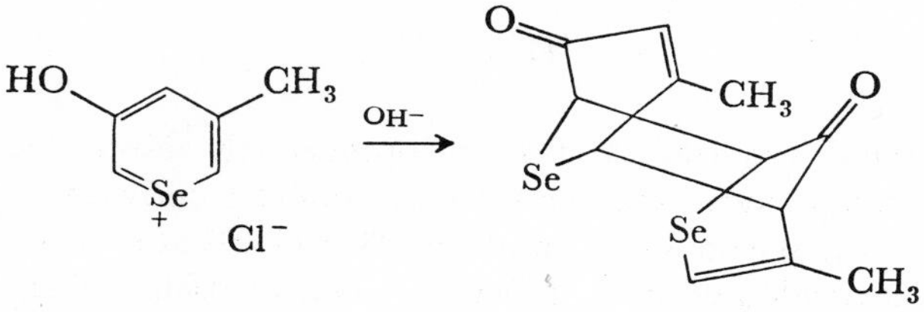 [structure of **15**] $\xrightarrow{\text{OH}^-}$ [structure of **37**]

15   **37**

3. Mixed Aliphatic-Aromatic

a. Stability. It has been shown[57] that phenylmethylselenetine bromide (**19**), like other selenonium compounds, decomposes on heating to give methyl bromide and phenylselenoglycolic acid (**38**). This reaction (equation

(41)

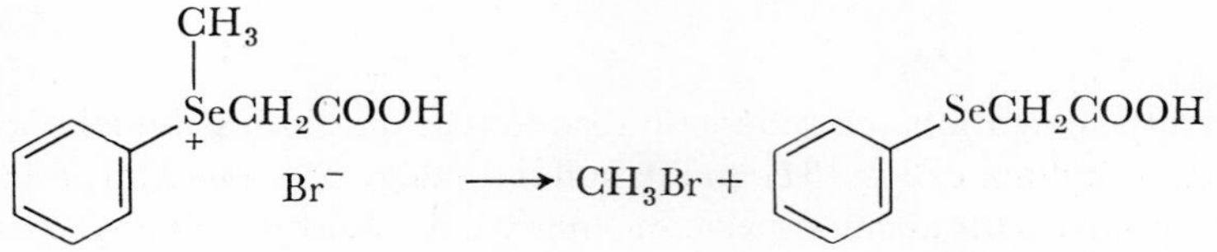

41), like the decomposition of triarylselenonium compounds, is irreversible. *p*-Tolylmethylselenetine bromide (**39**) undergoes a similar thermal decomposition.

b. Halogenation. The methyl group attached to selenium in *p*-tolylmethylselenetine bromide (**39**) can also be displaced[57] by treatment with bromine in carbon tetrachloride as shown in scheme 42. With excess bromine, *p*-tolylmethylselenetine bromide (**39**) gives *p*-tolylselenium tribromide (**40**) and bromoacetic acid.

(42)

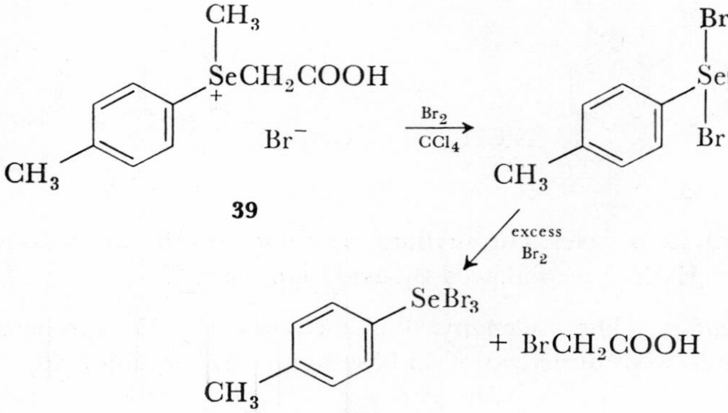

c. Cyclization. Cyclization of some appropriately substituted β-carbonyl selenonium compounds gave benzo[b]selenophene derivatives (**42**).[55,58,70] For example, the selenonium compound **41** cyclized after 4 hours of reflux in acetic anhydride-pyridine, giving the benzo[b]selenophene derivative

42 (equation 43). The same cyclization reaction occurred when **41** was

(43)

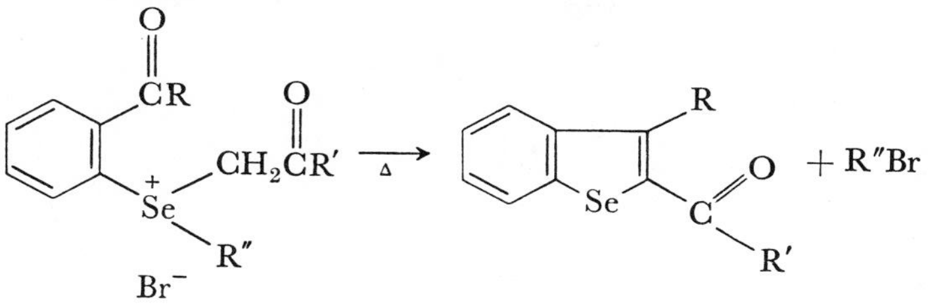

41   **42**

heated with potassium hydrogen sulfate until a reaction started and alkyl halide was liberated. Christiaens and Renson[55] prepared eight benzo[b]-selenophenes by the above methods.

Another cyclization method[58] involved the reaction shown in equation 44, performed at 210° with copper bronze in quinoline until carbon dioxide evolution ceased.

(44)

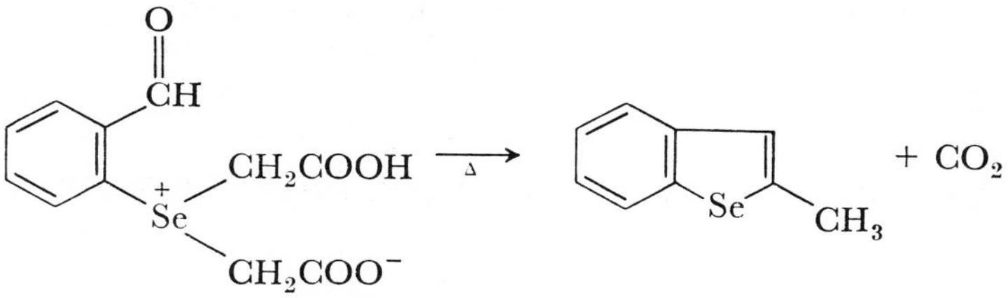

d. Electrophilic Aromatic Substitution. Selenonium groups attached to an aromatic ring have been used in a study of the directive influence of a positive group in electrophilic aromatic substitution. Baker and Moffitt[21] reported almost exclusive *meta* nitration of the dimethylphenylselenonium cation, whereas the dimethylbenzylselenonium cation gave a lower percentage of the *meta* nitro isomer. These workers also concluded that *meta* nitration increases in going from selenium to sulfur. Gilow and his coworkers,[53,54] in studying the same reaction, reported that nitration of dimethylphenylselenonium methyl sulfate (**17**) with concentrated nitric acid in concentrated sulfuric acid gave 2.6% ortho, 91.3% *meta*, and 6.1% *para* substitution. Analysis of the percentages of *ortho*, *meta*, and *para* substitution was accomplished by dealkylation of the nitrated selenonium compounds (**43**) by sodium methoxide at −70° (scheme 45), followed by gas chromatographic analysis of the methyl nitrophenyl selenides (**44**). Treatment of the selenonium compounds with sodium methoxide at room temperature resulted in the formation of nitroanisole derivatives **45** (scheme 45). Gilow[54]

(45)

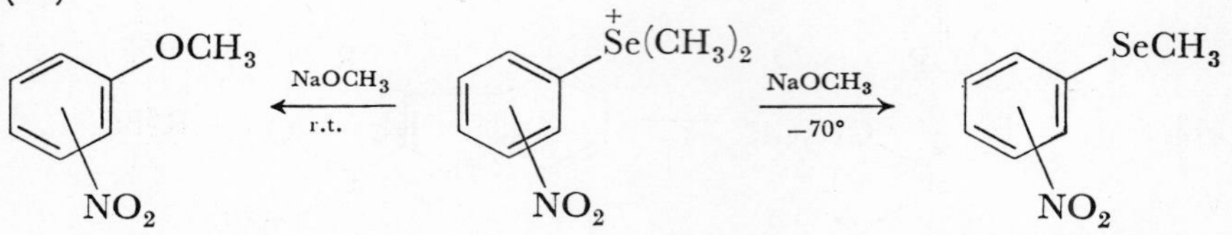

also studied the chlorination and bromination of dimethylphenylselenonium methyl sulfate (**17**). The *meta/para* ratios for halogenation and nitration were similar; however, the *ortho/para* ratio is higher for halogenation than for nitration. The findings of Gilow *et al.*[53,54] concerning electrophilic aromatic substitution are thoroughly discussed.

e. Optical Resolution.   Optically active selenetine compounds have been resolved with silver *d*-bromocamphorsulfonate. By this method Pope and Neville[56] resolved the phenylmethylselenetine ion (**19**) and noted that racemization of the optically pure enantiomers occurred during formation of the mercuric iodide complexes. Later, Balfe and Phillips[71] confirmed these results.

D. PROPERTIES AND USES OF SELENONIUM COMPOUNDS

1. Aliphatic

Trimethylselenonium iodide was used in a deuterium exchange study to show *d*-orbital resonance.[12] The rate of deuterium exchange decreases in going from the sulfur to the tellurium compound. The contribution of *d*-orbital resonance is reportedly the same for all three compounds, and the rate decrease was due to increased bond length. In a related study, McDaniel[72] stated that the stability of trimethylselenonium iodide can be attributed to hyperconjugation involving the low-lying vacant orbitals of selenium (equation 46).

(46)
$$CH_3\overset{+}{\underset{\underset{CH_3}{|}}{Se}}CH_3 \leftrightarrow \overset{H^+}{HC}\underset{\underset{H}{|} \underset{CH_3}{|}}{=}SeCH_3$$

The n.m.r. spectrum of trimethylselenonium iodide has been studied in detail[8,9,73,74] (*cf.* also Chapter XVD).

The i.r. spectrum of trimethylselenonium chloride[9] supports the ionic formulation of this compound. No bands occur in the 300–400 cm^{-1} region, where covalent Se–Cl bonds absorb.

The large Se–I bond length found in an X-ray study further supports the ionic representation of aliphatic selenonium compounds. Hope[75] determined that crystals of trimethylselenonium iodide are orthorhombic with the structure being built up of pairs of selenonium and iodide ions. The C–Se bond length is 1.962 Å, and the Se–I bond length is 3.776 Å.

In biological studies, the trimethylselenonium ion was identified as the normal excretory product of selenite fed to rats.[76,77] Subsequent work[78] has shown that the trimethylselenonium cation is a major excretory product from selenate, selenocystine, and selenomethionine metabolism. This selenonium cation accounts for about 20–50% of the selenium excreted in the urine.

2. Aromatic

a. Selenium Not Contained in an Aromatic Ring. Aromatic selenonium compounds are also ionic in character, as has been shown by conductivity measurements of tri-*p*-tolylselenonium hydroxide, a good electrical conductor in solution.[30] Triphenylselenonium chloride in acetonitrile has a molar conductivity of 127 ohm^{-1} cm^2, while diphenylselenium dichloride has a value of 6.7 ohm^{-1} cm^2.[79] The maximum Se–Cl distance in triphenylselenonium chloride has been reported[79] as no less than 3.40 Å, showing the bond to be ionic.

Emeléus and Heal,[14] who studied the stabilities of various organic sulfur, selenium, and tellurium "onium" compounds, found that stability increased as the Group VI element was varied from sulfur to tellurium.

An X-ray study was made[80] of triphenylselenonium chloride. The minimum observed Se–Cl bond length was 3.60 Å, the large value showing the bond to be ionic. The water solubility of the compound also supports an ionic structure.

The i.r. and u.v. spectra of triphenylselenonium chloride, as well as of diphenyl selenide and diphenylselenium dichloride, have been recorded.[81] All three compounds showed bands at 1600–1300 cm^{-1}, 1175–1000 cm^{-1}, and 750–660 cm^{-1}. The assignment of the 1600–1300 cm^{-1} band is due to C–C skeletal vibrations; 1175–1000 cm^{-1}, to in-plane hydrogen deformation; and 750–660 cm^{-1}, to the conjugated phenyl ring. The selenonium ion gives three bands in the 750–660 cm^{-1} region, suggesting that this ion is unsymmetrical. It may have an unsymmetrical pyramidal structure in which the three phenyl groups are differently oriented. The u.v. spectrum shows fine structure in the 250–270 nm region.

The analytical use of the triphenylselenonium cation has been reviewed by Bowd et al.[82] Potratz and Rosen[83] found the triphenylselenonium ion to be a good, specific, and sensitive analytical reagent for the determination of bismuth and cobalt. Subsequent workers[84,85] studied the composition of the reaction product of triphenylselenonium chloride with Bi^{3+} in the presence of iodide ions. The confirmed composition of triphenylselenonium tetraiodobismuthate, which precipitates from solution, agreed well with polarographic results.[86,87] The triphenylselenonium cation could itself be determined by amperometric titration with sodium tetraphenyl borate.[88] The slightly soluble salt that forms is a 1:1 complex of the selenonium and tetraphenyl borate ions. Ziegler and his co-workers[89-91] found that the triphenylselenonium cation precipitates dichromate, perchlorate, and molybdate ions. As little as 5 γ of perchlorate can be detected in the presence of chlorate. Extraction into organic solvents and colorimetric methods were used by Ziegler, triphenylselenonium dichromate being measured at 362 or 445 nm.[89] Perrhenates and molybdates can also be determined quantitatively by use of the triphenylselenonium cation.[91] Mercuric ion can be separated from iron, aluminum, cobalt, nickel, manganese, and copper ions by extraction with triphenylselenonium ions. The mercury can then be determined as mercuric sulfide.[90] Recently, Ziegler and Ziegeler[92] described an analytical method in which bismuth can be determined gravimetrically by the use of the triphenylselenonium cation.

Selenonium salts were found to be effective in the homogeneous liquid-phase oxidation of cumene and α-pinene.[93] In a study of charge-transfer compounds,[94] the triphenylselenonium cation formed a 2:1 charge-transfer complex with 7,7,8,8-tetracyanoquinodimethane.

b.   Selenium Contained in a Heterocyclic System.   The u.v. spectra of various heterocyclic selenonium compounds have been recorded. The 3,5-dimethyl-1,2-diselenolylium cation (**13**, R = CH_3) in 1 M hydrochloric acid shows a λ_{max} at 300 nm (log ε = 3.75) and 320 nm (log ε = 3.81).[45] The u.v. spectra of selenapyrylium perchlorate (**8**)[39] and some of its methyl derivatives[95] have been reported. The unsubstituted derivative, selenapyrylium perchlorate (**8**), shows a λ_{max} at 267 nm (log ε = 3.90) and 298 nm (log ε = 3.53). The u.v. spectrum of 1-selenachromylium perchlorate (**10**)[40] a has λ_{max} at 259 nm (log ε = 4.53), 348 nm (log ε = 3.74), and 408 nm (log ε = 3.69).

The i.r. spectrum of the 3,5-dimethyl-1,2-diselenolylium cation (**13**, R = CH_3) has been recorded and discussed in detail.[45,96] The similarity between the i.r. spectrum of this selenium compound and that of its sulfur analog, the 3,5-dimethyl-1,2-dithiolylium cation, suggests that the two ions possess the same molecular geometry.

The n.m.r. spectra of the pyrylium, thiapyrylium, and selenapyrylium ions have been studied.[97]

The stability of various heterocyclic selenonium ions was measured by equilibrium techniques.[52,98,99] It was found that sulfur-containing heterocycles were more stable than the corresponding oxygen and selenium compounds. In the selenium series, the dibenzoselenapyrylium ion was more stable than the benzo derivative, which, in turn, was more stable than the unsubstituted selenapyrylium ion.

REFERENCES

1. W. E. Bradt, *Proc. Indiana Acad. Sci.*, **43,** 72 (1934).

2. H. Rheinboldt, "Houben-Weyl Methoden der Organischen Chemie," Vol. IX, E. Müller, Ed., Georg Thieme Verlag, Stuttgart, 1955, p. 1034.

3. T. W. Campbell, H. G. Walker, and G. M. Coppinger, *Chem. Rev.*, **50,** 279 (1952).

4. J. Gosselck, *Angew. Chem. (Intern. Ed.)*, **2,** 660 (1963).

5. E. P. Painter, *Chem. Rev.*, **28,** 179 (1940).

6. M. Shinagawa and H. Matsuo, *Kagaku no Ryôiki*, **10,** 111 (1956).

7. T. Hashimoto, M. Sugita, H. Kitano, and K. Fukui, *Nippon Kagaku Zasshi*, **88,** 991 (1967).

8. W. McFarlane, *Mol. Phys.*, **12,** 243 (1967).

9. K. J. Wynne and J. W. George, *J. Amer. Chem. Soc.*, **91,** 1649 (1969).

10. E. Biilmann and K. A. Jensen, *Bull. Soc. Chim. Fr.*, [5] **3,** 2310 (1936).

11. M. L. Bird and F. Challenger, *J. Chem. Soc.*, **161,** 570 (1942).

12. W. von E. Doering and A. K. Hoffmann, *J. Amer. Chem. Soc.*, **77,** 521 (1955).

13. T. P. Hilditch and S. Smiles, *J. Chem. Soc.*, **93,** 1384 (1908).

14. H. J. Emeléus and H. G. Heal, *J. Chem. Soc.*, 1126 (1946).

15. E. Fromm and K. Martin, *Justus Liebigs Ann. Chem.*, **401,** 177 (1913).

16. B. Rathke, *Justus Liebigs Ann. Chem.*, **152,** 210 (1869).

17. A. Pieroni and C. Coli, *Gazz. Chim. Ital.*, II, **44,** 349 (1914).

18. C. L. Jackson, *Ber.*, **7,** 1277 (1874).

19. C. L. Jackson, *Justus Liebigs Ann. Chem.*, **179,** 1 (1875).

20. L. von Pieverling, *Justus Liebigs Ann. Chem.*, **185,** 331 (1877).

21. J. W. Baker and W. G. Moffitt, *J. Chem. Soc.*, **137,** 1722 (1930).

22. C. T. Bahner, P. P. Neblett, Jr., and H. A. Rutter, Jr., *J. Amer. Chem. Soc.*, **74,** 3453 (1952).

23. T. Hashimoto, H. Kitano, and K. Fukui, *Nippon Kagaku Zasshi*, **89,** 83 (1968).

24. F. G. Holliman and F. G. Mann, *J. Chem. Soc.*, 37 (1945).

25. H. J. Backer and H. J. Winter, *Rec. Trav. Chim. Pays-Bas*, **56,** 492 (1937).

26. W. E. Truce and D. D. Emrick, *J. Amer. Chem. Soc.*, **78,** 6130 (1956).

27. H. M. Leicester and F. W. Bergstrom, *J. Amer. Chem. Soc.*, **51,** 3587 (1929).

28. H. M. Leicester, *Organic Syn.*, **18,** 30 (1938).

29. W. E. Bradt and J. F. Green, *J. Org. Chem.*, **1,** 540 (1937).

30. H. M. Leicester and F. W. Bergstrom, *J. Amer. Chem. Soc.*, **53,** 4428 (1931).

31. H. M. Leicester, *J. Amer. Chem. Soc.*, **57,** 1901 (1935).

32. S. Smiles and T. P. Hilditch, *Proc. Chem. Soc.*, **23,** 12 (1907).

33. J. H. Crowell and W. E. Bradt, *J. Amer. Chem. Soc.*, **55,** 1500 (1933).

34. H. Funk and W. Papenroth, *J. Prakt. Chem.*, **11,** 191 (1960).

35. G. T. Morgan and F. H. Burstall, *J. Chem. Soc.*, **133,** 3260 (1928).

36. L. G. Makarova and A. N. Nesmeyanov, *Bull. Acad. Sci. USSR, Div. Chem. Sci.*, 617 (1945).

37. A. N. Nesmeyanov and L. G. Makarova, *Uch. Zap. Mosk. Gosudarst. Univ. im. M.V. Lomonosova, No. 132, Org. Khim.*, **7,** 109 (1950).

38. E. Amberger and E. Gut, *Chem. Ber.*, **101,** 1200 (1968).

39. I. Degani, R. Fochi, and C. Vincenzi, *Gazz. Chim. Ital.*, **94,** 203 (1964).

40. I. Degani, R. Fochi, and C. Vincenzi, *Gazz. Chim. Ital.*, **94,** 451 (1964).

41. I. Degani and C. Vincenzi, *Boll. Sci. Fac. Chim. Ind. Bologna*, **25,** 51 (1967).

42. J. M. Danze and M. Renson, *Bull. Soc. Chim. Belges*, **75,** 169 (1966).

43. I. Degani, R. Fochi, and G. Spunta, *Boll. Sci. Fac. Chim. Ind. Bologna*, **23,** 165 (1965).

44. H. Bauer, *Ber.*, **47,** 1873 (1914).

45. G. A. Heath, R. L. Martin, and I. M. Stewart, *Australian J. Chem.*, **22,** 83 (1969).

46. G. A. Heath, I. M. Stewart, and R. L. Martin, *Inorg. Nucl. Chem. Lett.*, **5,** 169 (1969).

47. K. A. Jensen and U. Henriksen, *Acta Chem. Scand.*, **21,** 1991 (1967).

48. M. Battegay and G. Hugel, *Bull. Soc. Chim. Fr.*, **27,** 557 (1920).

49. M. Battegay and G. Hugel, *Bull. Soc. Chim. Fr.*, **33,** 1103 (1923).

50. R. H. Nealey and J. S. Driscoll, *J. Heterocycl. Chem.*, **3,** 228 (1966).

51. R. G. Turnbo, Ph.D. Thesis, University of Texas, 1965.

52. I. Degani, R. Fochi, and C. Vincenzi, *Boll. Sci. Fac. Chim. Ind. Bologna*, **23,** 21 (1965).

53. H. M. Gilow and G. L. Walker, *J. Org. Chem.*, **32,** 2580 (1967).

54. H. M. Gilow, R. B. Camp, Jr., and E. C. Clifton, *J. Org. Chem.*, **33,** 230 (1968).

55. L. Christiaens and M. Renson, *Bull. Soc. Chim. Belges*, **77,** 153 (1968).

56. W. J. Pope and A. Neville, *J. Chem. Soc.*, **81,** 1552 (1902).

57. O. J. K. Edwards, W. R. Gaythwaite, J. Kenyon, and H. Phillips, *J. Chem. Soc.*, **133,** 2293 (1928).

58. F. Dallacker, E. Kaiser, and R. Uddrich, *Justus Liebigs Ann. Chem.*, **689,** 179 (1965).

59. J. Gosselck, L. Beress, H. Schenk, and G. Schmidt, *Angew. Chem.* (*Intern. Ed.*), **4,** 1080 (1965).

60. J. Gosselck, *Chem. Ber.*, **91,** 2345 (1958).

61. E. D. Hughes and K. I. Kuriyan, *J. Chem. Soc.*, **147,** 1609 (1935).

62. D. Lloyd and M. I. C. Singer, *Chem. Commun.*, 390 (1967).

63. J. von Braun, W. Teuffert, and K. Weissbach, *Justus Liebigs Ann. Chem.*, **472,** 121 (1929).

64. F. G. Mann and F. G. Holliman, *Nature*, **152,** 749 (1943).

65. O. Behaghel and K. Hofmann, *Ber.*, **72B,** 697 (1939).

66. G. Wittig and H. Fritz, *Justus Liebigs Ann. Chem.*, **577,** 39 (1952).

66a. D. Hellwinkel and G. Fahrbach, *Justus Liebigs Ann. Chem.*, **715,** 68 (1968).

67. I. Degani and R. Fochi, *Ann. Chim.* (Rome), **58,** 251 (1968).

68. I. Degani, *Corsi Semin. Chim.*, 177 (1968).

69. I. Degani, R. Fochi, and G. Spunta, *Boll. Sci. Fac. Chim. Ind. Bologna*, **23,** 151 (1965).

70. L. Christiaens and M. Renson, *Bull. Soc. Chim. Belges*, **79,** 133 (1970).

71. M. P. Balfe and H. Phillips, *J. Chem. Soc.*, **143,** 127 (1933).

72. D. H. McDaniel, *Science*, **125,** 545 (1957).

73. W. McFarlane and J. A. Nash, *Chem. Commun.*, 524 (1969).

74. C. J. Jameson, *J. Amer. Chem. Soc.*, **91,** 6232 (1969).

75. H. Hope, *Acta Cryst.*, **20,** 610 (1966).

76. I. S. Palmer, D. D. Fischer, and A. W. Halverson, *Biochim. Biophys. Acta*, **177,** 336 (1969).

77. J. L. Byard, *Arch. Biochem. Biophys.*, **130,** 556 (1969).

78. I. S. Palmer, R. P. Gunsalus, A. W. Halverson, and O. E. Olson, *Biochim. Biophys. Acta*, **208,** 260 (1970).

79. D. A. Couch, P. S. Elmes, J. E. Ferguson, M. L. Greenfield, and C. J. Wilkins, *J. Chem. Soc., A*, 1813 (1967).

80. J. D. McCullough and R. E. Marsh, *J. Amer. Chem. Soc.*, **72,** 4556 (1950).

81. M. A. A. Beg and A. R. Shaikh, *Tetrahedron*, **22,** 653 (1966).

82. A. J. Bowd, D. T. Burns, and A. G. Fogg, *Talanta*, **16,** 719 (1969).

83. H. A. Potratz and J. M. Rosen, *Anal. Chem.*, **21,** 1276 (1948).

84. M. Shinagawa, H. Matsuo, and M. Yoshida, *Japan Analyst*, **3,** 139 (1955).

85. M. Shinagawa and H. Matsuo, *Japan Analyst*, **4,** 211 (1955).

86. M. Shinagawa, H. Matsuo, and S. Isshiki, *Japan Analyst*, **3,** 199 (1954).

87. H. Matsuo, *J. Sci. Hiroshima Univ., Ser. A*, **22,** 281 (1958).

88. H. Nezu, *Bunseki Kagaku*, **10,** 575 (1961).

89. M. Ziegler and K. D. Pohl, *Z. Anal. Chem.*, **204,** 413 (1964).

90. M. Ziegler and N. P. Riedel, *Z. Anal. Chem.*, **212,** 291 (1965).

91. M. Ziegler and M. Gindl, *Naturwissenschaften*, **54,** 19 (1967).

92. M. Ziegler and L. Ziegeler, *Talanta*, **17,** 641 (1970).

93. K. Fukui, K. Ohkubo, and T. Yamabe, *Bull. Chem. Soc. Japan*, **42,** 312 (1969).

94. D. S. Acker and D. C. Blomstrom, U.S. Pat. 3,162,641 (1964); *Chem. Abstr.*, **63,** 549a (1965).

95. I. Degani and C. Vincenzi, *Boll. Sci. Fac. Chim. Ind. Bologna*, **23,** 249 (1965).

96. O. Siimann and J. Fresco, *Inorg. Chem.*, **9,** 294 (1970).

97. I. Degani, F. Taddei, and C. Vincenzi, *Boll. Sci. Fac. Chim. Ind. Bologna*, **25,** 61 (1967).

98. I. Degani, R. Fochi, and G. Spunta, *Boll. Sci. Fac. Chim. Ind. Bologna*, **23,** 243 (1965).

99. I. Degani, R. Fochi, and G. Spunta, *Boll. Sci. Fac. Chim. Ind. Bologna*, **26,** 3 (1968).

VII Selenium analogs of aldehydes and ketones

RICHARD B. SILVERMAN

Harvard University
Department of Chemistry
Cambridge, Massachusetts

A. INTRODUCTION

Selenoaldehydes and selenoketones, represented by the general formulas **1** and **2**, respectively, are classes of compounds that have received relatively

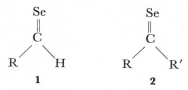

little attention in the chemical literature. The paucity of authentic selenium analogs of aldehydes arises from the fact that attempts to synthesize the lower members of this series have resulted in the formation of diselenides and polymers. However, when many of these compounds were initially reported, they were assumed to be monomeric despite the fact that solubility difficulties were encountered and, therefore, molecular weight determinations could not be made. In many instances inexplicable physical properties were attributed to the existence of various modifications of the compounds. It is now doubtful that the lower members of the selenoketone series exist. Although these examples were originally described as dimeric selenoketones, these compounds, in addition to others, were found later to be nonketonic in composition. It appears that, in general, resonance or ligand stabilization is required for a selenoaldehyde or selenoketone to survive.

The most common method of synthesis for these compounds is the reaction of an aldehyde or ketone with hydrogen selenide or one of its salts. In rare instances, phosphorus pentaselenide is used in their preparation.

B. SELENOALDEHYDES

1. Aliphatic Selenoaldehydes

a. Selenoformaldehyde. The first member of this series was reported[1] in 1915 as the product of the reaction of aqueous formaldehyde in concentrated hydrochloric acid with hydrogen selenide. Although the compound was earlier suspected[2] of being polymeric in nature, it was not until 1950 that its molecular weight was determined[3] to be that of a trimer, and a cyclic structure (3) was proposed. An X-ray analysis[4] confirmed this conjecture

3

in 1965. Trimeric selenoformaldehyde is discussed further in Chapters XIA and XIV.

b. Selenoacetaldehyde. Selenoacetaldehyde (5) was described[1,2] as the product obtained when hydrogen selenide was bubbled into a solution of acetaldehyde in saturated alcoholic hydrogen chloride. Depending on the crystallization solvent used, the semi-solid product was reported to form crystalline modifications having different melting points, namely, 139° (ethanol), 123–124° (acetone), and 117° (chloroform). More recently, however, the reaction of anhydrous hydrogen chloride and hydrogen selenide at 0 to −10° with neat acetaldehyde was shown[4a] to yield bis(1-chloroethyl) selenide.

Another method of preparation[5] for the compound was the reaction of acetaldehyde with bromomagnesium selenol (**4**) (equation 1). In 1967,

$$(1) \quad CH_3CHO + BrMgSeH \longrightarrow CH_3\overset{\displaystyle SeH}{\underset{\displaystyle OMgBr}{\overset{\diagup}{\underset{\diagdown}{CH}}}} \quad \xrightarrow{H_2O}$$

$$\mathbf{4}$$

$$CH_3\overset{\displaystyle SeH}{\underset{\displaystyle OH}{\overset{\diagup}{\underset{\diagdown}{CH}}}} \quad \longrightarrow \quad CH_3\overset{\displaystyle Se}{\underset{\mathbf{5}}{\overset{\|}{CH}}}$$

however, the structure of this so-called selenoaldehyde was determined[6] by X-ray crystallography to be the trimer, 2,4,6-trimethyl-1,3,5-triselenane (*cf.* Chapter XIA).

c. **Selenoisovaleraldehyde.**   In 1871 the preparation of selenovaleraldehyde was reported.[7] However, from the boiling point of the starting material used, it appears that this reagent was in actuality isovaleraldehyde (**6**), and therefore the compound whose synthesis was attempted was really selenoisovaleraldehyde (**7**) (equation 2). Inasmuch as this reaction has not

$$(2) \quad \overset{\displaystyle CH_3}{\underset{\displaystyle CH_3}{\overset{\diagdown}{\underset{\diagup}{CH}}}}CH_2CHO + H_2Se \longrightarrow \overset{\displaystyle CH_3}{\underset{\displaystyle CH_3}{\overset{\diagdown}{\underset{\diagup}{CH}}}}CH_2\overset{\displaystyle Se}{\overset{\|}{CH}}$$

$$\mathbf{6} \qquad\qquad\qquad\qquad\qquad \mathbf{7}$$

been successful with other aliphatic aldehydes for the preparation of their corresponding selenoaldehydes, the authenticity of this selenoaldehyde should be questioned. Those working with **7** have frequently noted palpitation of the heart and breathing difficulty.[2,8]

d. **11-Phenylselenoundecapentaenal.**   This selenoaldehyde (**9**) was synthesized[9,10] by treating 11-phenylundecapentaenal (**8**) with hydrogen selenide (equation 3). The selenium was removed by copper, metal carbonates, various reducing agents, and certain amines (*e.g.*, piperidine) to give

$$(3) \quad C_6H_5CH{=}CHCH{=}CHCH{=}CHCH{=}CHCH{=}CHCHO + H_2Se \longrightarrow$$

$$\mathbf{8}$$

$$C_6H_5CH{=}CHCH{=}CHCH{=}CHCH{=}CHCH{=}CH\overset{\displaystyle Se}{\overset{\|}{CH}}$$

$$\mathbf{9}$$

1,22-diphenyldocosaundecaene (**10**) (equation 4). This reaction was developed because

(4)   **9** $\longrightarrow$

$$C_6H_5CH{=}CHCH{=}CHCH{=}CHCH{=}CHCH{=}CHCH$$
$$C_6H_5CH{=}CHCH{=}CHCH{=}CHCH{=}CHCH{=}CHCH{\overset{\|}{C}}H$$

10

the preparation of the series of diphenyl polyenes could not be extended beyond eight conjugated double bonds by means of the existing method (*i.e.*, lead oxide-catalyzed condensation of two phenyl-polyunsaturated aldehydes to a dicarboxylic acid). With the newer procedure, 1,30-diphenyl-tricontapentadecaene could be obtained.

2. Aromatic Selenoaldehydes

a. Selenobenzaldehyde. This member of the aromatic subclass (**11**) has been investigated by several workers, no two of whom have completely agreed as to its properties. It was first reported[11] in 1875 as yellow needles (m.p. 70°) from the reaction of benzal chloride with ethanolic potassium selenide (equation 5). Although no molecular weight determination was

(5) $$C_6H_5CHCl_2 + K_2Se \xrightarrow[\Delta]{C_2H_5OH} C_6H_5\overset{\overset{\displaystyle Se}{\|}}{C}H + 2KCl$$

11

made, it was assumed to be monomeric, since it resembled its sulfur analog, thiobenzaldehyde. Later the product obtained when hydrogen selenide was passed into an ethanolic hydrogen chloride solution of benzaldehyde (equation 6) was described[1] as existing in three modifications: α, monomeric

(6) $$C_6H_5CHO + H_2Se \xrightarrow[\text{conc. HCl}]{C_2H_5OH} C_6H_5\overset{\overset{\displaystyle Se}{\|}}{C}H + H_2O$$

yellow columns (m.p. 83–84°); β, trimeric lustrous golden needles (m.p. 205°) having the composition $(C_6H_5CHSe)_3 \cdot C_6H_6$ (from benzene); and γ, pale yellow needles (m.p. 166°). However, when this work was repeated[12] in 1932 only two products were obtained: a dimer (m.p. 92–93°) and a trimer with the same properties as the β-modification, but with a melting point of 189–193° (dec.). In light of the findings of Margolis and Pittman,[23] the suspected dimer may actually be dibenzyl diselenide (*cf.* equation 14).

In 1928 the synthesis of selenobenzaldehyde was also attempted[5] by the use of BrMgSeH, and the product was described as possessing the properties of the so-called β-modification (equation 7).

(7) 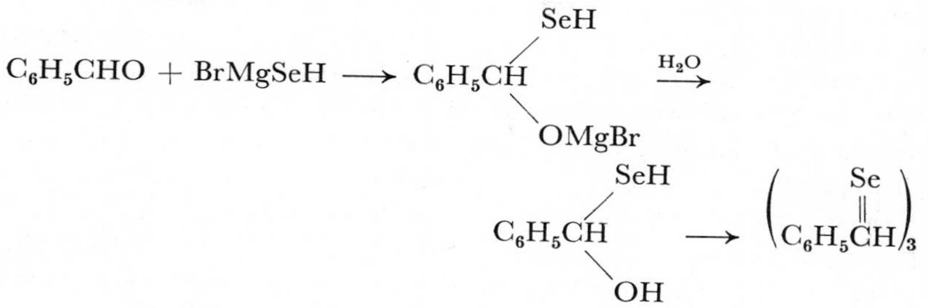 $C_6H_5CHO + BrMgSeH \longrightarrow C_6H_5\overset{\displaystyle SeH}{\underset{\displaystyle OMgBr}{CH}} \xrightarrow{H_2O}$

$C_6H_5\overset{\displaystyle SeH}{\underset{\displaystyle OH}{CH}} \longrightarrow \left(C_6H_5\overset{\displaystyle \overset{Se}{\parallel}}{CH} \right)_3$

The β-form was reported[1] to react with copper at 300° to yield *trans*-stilbene (**12**) (equation 8).

(8) 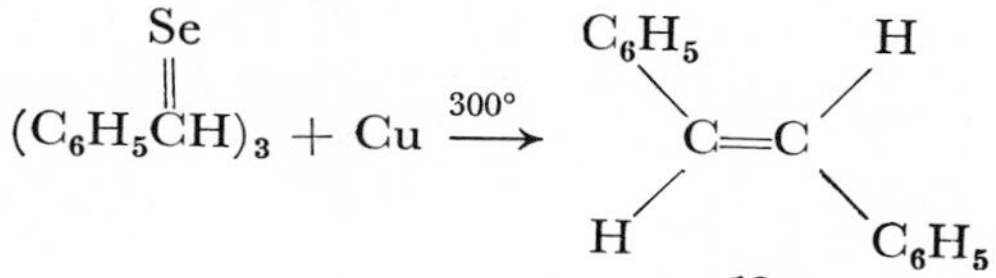 $(C_6H_5\overset{\displaystyle \overset{Se}{\parallel}}{CH})_3 + Cu \xrightarrow{300°}$

12

b. **3-Selenoformylindole.** 3-Selenoformylindole (**13**) has been proposed[13] as an intermediate in the synthesis of di(3-indolylmethyl) diselenide (**14**). Although **13** was not isolated, there are analogous cases in the sulfur series,[14,15] which led the authors to postulate scheme 9.

13

(9)

14

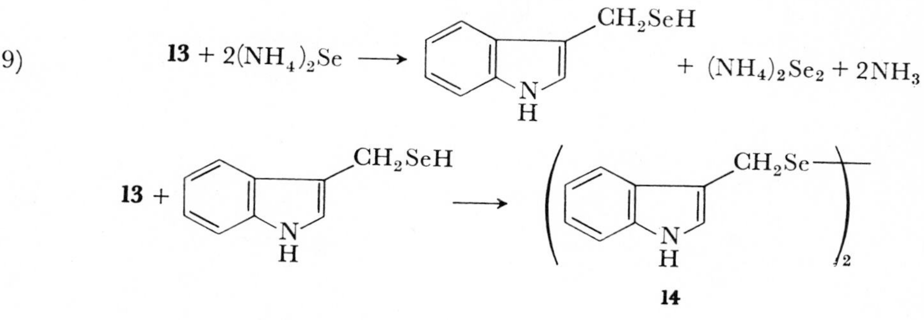

3. Thiaselenolodithioles

Compounds of this basic structure were classified[16] as having "no-bond resonance," a term used[17] to describe the sulfur analog, thiothiophthene. This term refers to the exceptionally long bond length relative to the normal S–S and S–Se distances, and the condition is represented by two extreme resonance structures, **15a** and **15b**. The "no-bond resonance" is responsible for the aromatic character of the two rings, in agreement with the properties of the compound. The diphenyl analog (**15**), 2,4-diphenyl[1,2]thiaselenolo-[5,1-b][1,2]dithiole (also named 3,5-epidithio-2,5-diphenyl-2,4-pentadiene selenal) was first reported[16] in 1966 and was prepared by an elegant sequence of reactions (scheme 10). Compound **15** was later described[18] as an effective

(10)

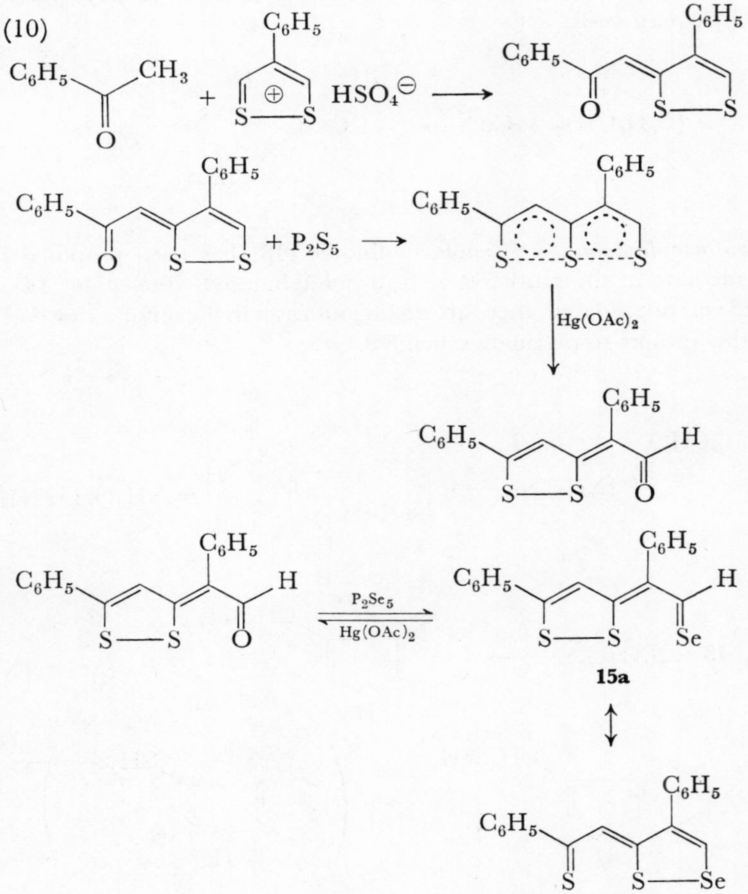

miticidal agent. The use of phosphorus pentaselenide to introduce the Se atom into the molecule constitutes one of only a few known reactions with P_2Se_5 which does not lead to a phosphorus-containing organoselenium compound.

Compound 15 was prepared[19] in 1968 by a different route (scheme 11).

(11)

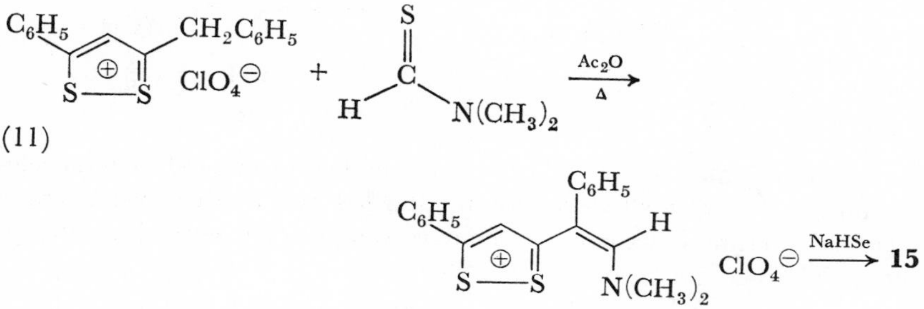

16

The reaction of Vilsmeier salts (e.g., 16) with monoacidic nucleophiles other than sodium hydrogen selenide was utilized[19] as a general route to isosteres of thiothiophthenes.

It was shown[16] by n.m.r. analysis and by X-ray crystallography that, although there are two theoretical resonance structures (i.e., 15a and 15b), 15b contributes more to the true electronic representation than does 15a. Also, selenium is accommodated in the thiothiophthene system without distortion.

A comparison[16] of the visible and u.v. spectra of 17, where X = O, S,

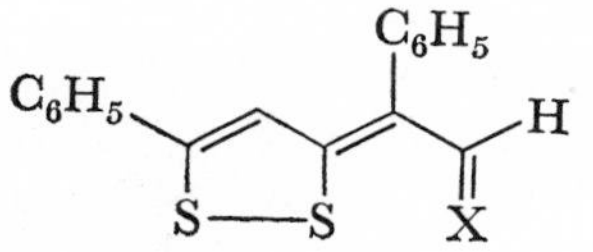

17

or Se, revealed that by replacing oxygen with sulfur a shift in the visible absorption to a longer wavelength with a weakening of intensity was displayed. The u.v. absorption was shifted to a shorter wavelength, and the intensity was greatly strengthened. When selenium was substituted for sulfur, the pattern was continued in the visible region. In the u.v., however, the absorption was intermediate in position and intensity between the compounds when X = O and X = S.

4. Selenoaldehydes and Selenoketones as Dye Intermediates

Merocyanine dyes can be obtained[20] from the selenium analog of 2-acyl-methylene heterocycles (e.g., 18), which were prepared from the carbonyl

compound by treatment with phosphorus pentaselenide (see, *e.g.*, equation
12). Alkylation of the selenium-containing compounds and condensation of

(12)

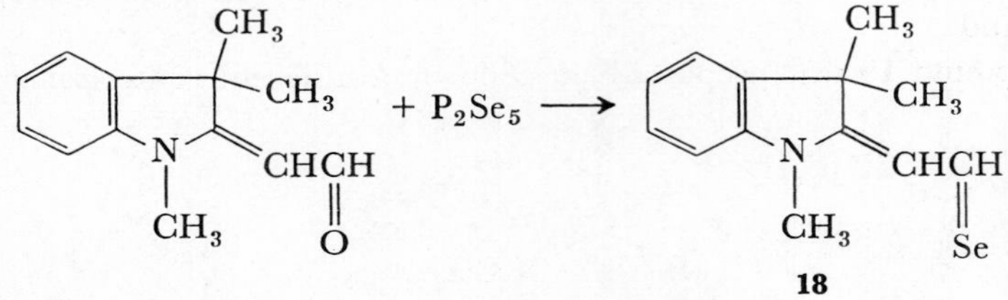

the resultant quaternary salts with suitable five-membered heterocycles
containing an active methylene group afforded the desired merocyanine
(scheme 13). Trimethinecarbocyanine dyes were synthesized[21] similarly (*cf.*
Chapter XIE).

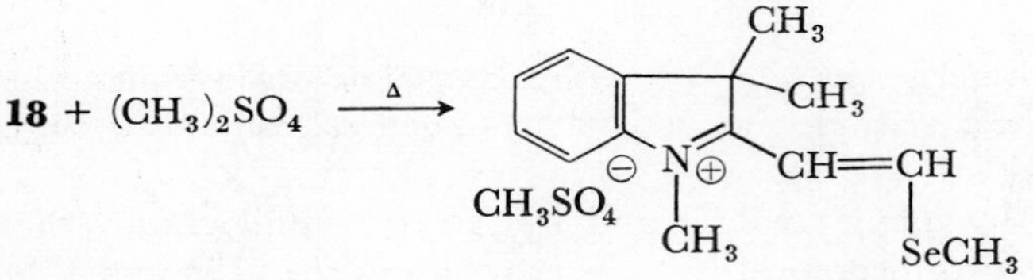

$18 + (CH_3)_2SO_4 \xrightarrow{\Delta}$

(13)

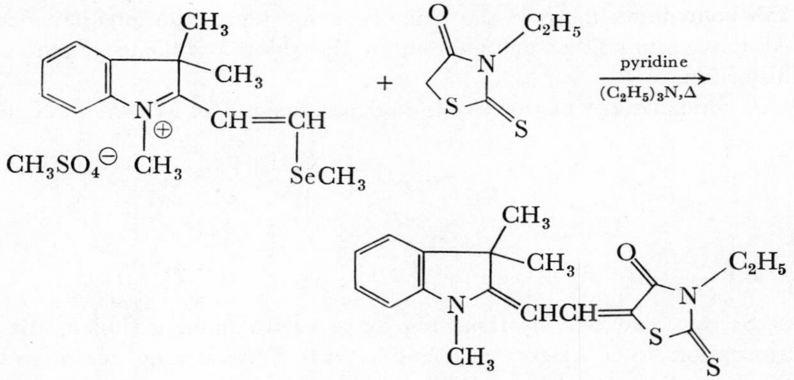

C. SELENOKETONES

1. Aliphatic and Aromatic Selenoketones

Dimers of selenoacetone, methyl ethyl selenoketone, selenoacetophenone, and
selenobenzophenone were originally proposed[22] in 1927 as the products
from the reaction of the corresponding ketone with hydrogen selenide in
concentrated hydrochloric acid. Thirty years later, however, it was proved[23]

that the products were actually bis(dialkylmethyl) diselenides (**19**), obtained from a reduction involving hydrogen selenide (equation 14). The diselenide

$$(14) \qquad 2R\overset{\overset{\displaystyle O}{\|}}{C}R' + 3H_2Se \longrightarrow (RR'HCSe-)_2 + 2H_2O + Se$$
$$\textbf{19}$$

proposal was substantiated by melting point and mixture melting point determinations with authentic diselenides. As further proof of the diselenide structure, what was thought to be selenobenzophenone was heated at 200° and 1,1,2,2-tetraphenylethane (**20**) was obtained[23] quantitatively (equation 15). Since the pyrolysis of a selenoaldehyde[1,9,10] or a cyclic selenoketone[24]

$$(15) \qquad [(C_6H_5)_2HCSe-]_2 \xrightarrow{200°} (C_6H_5)_2CHCH(C_6H_5)_2$$
$$\textbf{20}$$

has been shown to yield the corresponding unsaturated dimer with the elimination of elemental selenium, then, in the pyrolysis of selenobenzophenone, the expected product would be tetraphenylethylene $(C_6H_5)_2C=C(C_6H_5)_2$, not tetraphenylethane.

It was suggested[23] that the initial step of the reaction shown in equation 14 is the same for both aldehydes and ketones, namely, formation of hydroselenohydrins (**21**) (equation 16). In the case of simple aldehydes, three

$$(16) \qquad R\overset{\overset{\displaystyle O}{\|}}{C}R' + H_2Se \longrightarrow HO-\overset{\overset{\displaystyle R}{|}}{\underset{\underset{\displaystyle R'}{|}}{C}}-SeH$$
$$\textbf{21}$$

molecules condense, yielding the cyclic triselenide. This would explain why there is no precipitation of elemental selenium with simple aldehydes. With ketones, however, the possibility remains that the elements of water are eliminated from one molecule, giving rise to a monomeric selenoketone (**22**) which, in turn, would be subject to reduction by hydrogen selenide according to equation 17.

$$(17) \qquad 2RR'CSe + H_2Se \rightarrow (RR'HCSe-)_2 + Se$$
$$\textbf{22}$$

2. Selenoacetone Derivatives

Although the parent compound, selenoacetone, has never been isolated, some compounds related to it have been synthesized.

a.   *Diselenoacetylacetonate[Ni(II)]*.   Bis(4-selenopent-3-ene-2-selenoxo)-nickel(II) (**23**) was recently prepared[25,26] by the reaction of an acidified solution of acetylacetone with hydrogen selenide in the presence of Ni(II) ions (equation 18). The similarity of its i.r. spectrum to that of the sulfur

$$(18) \quad CH_3\overset{\overset{O}{\|}}{C}CH_2\overset{\overset{O}{\|}}{C}CH_3 + H_2Se \xrightarrow{Ni(II)} (C_5H_7Se_2)_2Ni$$
$$\textbf{23}$$

analog,[27] bis(4-thiopent-3-ene-2-thioxo)nickel(II), indicates comparable bonding and stereochemistry. Since the sulfur analog was shown[28] to have a square planar structure, the expected conformation of the selenium complex can be depicted as **24**.

24

b.   *1,2-Diselenolene complexes*.   Bis(trifluoromethyl)-1,2-diselenetene (**25**) forms bis and tris complexes, **26**, with transition metals[29,30] (equation 19).

$$n = 2,3$$

25   **26**

For sulfur analogs there is evidence[31] that in the neutral species the ligand loses its dithiolate character and becomes more dithioketonic, as represented by resonance structures **27a** and **27b**.

27a   **27b**

In general, no appreciable changes in structure or properties occur from the interchange of selenium for sulfur in corresponding metal complexes.[30] It may be inferred, therefore, that the selenium complexes have a reasonable amount of diselenoketonic character.

c. *Diselenodiacetylacetone and selenoketonic analogs of thiothiophthenes.*
In 1957 it was reported[32] incorrectly that the products obtained when
2,6-dimethyl-4-seleno-γ-pyrone (**28**) (whose preparation and further
reactions are discussed in Section VII.C) was treated with sodium selenide
were bis(selenoacetyl)acetone (**29**) and 3,7-dimethyl-1,2-diselenacyclo-
hepta-3,6-dien-5-one (**30**), the product of the air oxidation of **29**
(equation 20). However, it was later found,[33] as suggested earlier,[34] that the

(20)

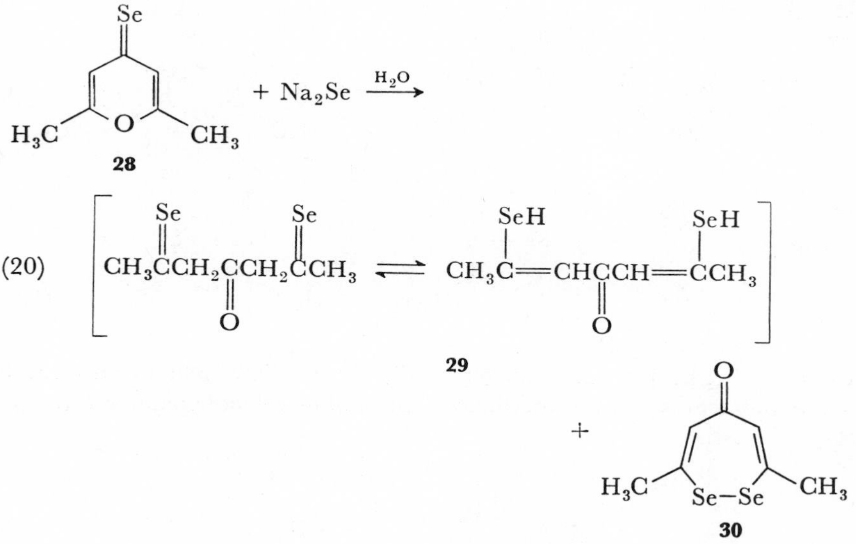

$$CH_3\overset{\underset{\parallel}{Se}}{C}CH_2\overset{\underset{\parallel}{O}}{C}CH_2\overset{\underset{\parallel}{Se}}{C}CH_3 \rightleftharpoons CH_3\overset{\underset{|}{SeH}}{C}=CHC\overset{\underset{\parallel}{O}}{C}H=\overset{\underset{|}{SeH}}{C}CH_3$$

29

compound originally suspected of being **29** was actually diselenoacetylacetone
(**31**), and the oxidation product was 4,6-epidiselenohepta-3,5-dien-2-one
(**32**, equation 21). It was further shown[33] that substituting a S atom for a Se

28 + Na$_2$Se $\xrightarrow{\text{H}_2\text{O}}$

(21)

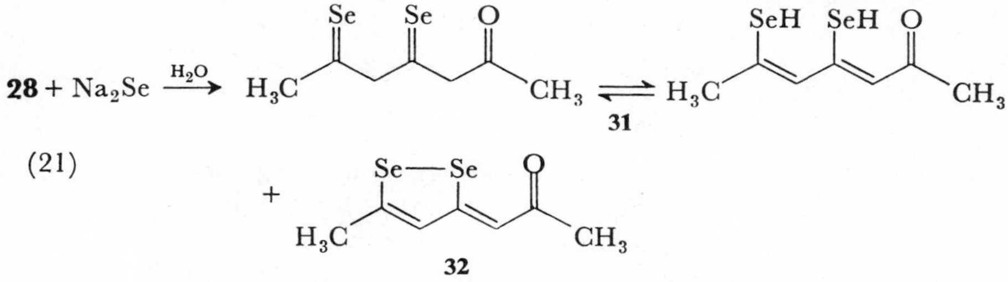

atom did not substantially alter the conformation or electronic distribution
of the molecule.

The starting material (*i.e.*, **28**) could be regenerated quantitatively[32] by heating **31** in methanolic hydrogen chloride or by fusion.

Treatment of **32** with phosphorus pentasulfide in refluxing benzene, which was originally thought[32] to give the cyclic diselenide thione, was later inferred[34] to yield 2,5-dimethyldiselenothiophthene (**34**). This reaction was subsequently found[35] to produce a mixture of 2,5-dimethylselenothiophthene (**33**, 88%) and **34** (12%), as in equation 22. The main product of

$$32 + P_2S_5 \xrightarrow[\Delta]{C_6H_6}$$

(22)

reaction 22 (*i.e.*, **33**) had been reported[36] the previous year in an electric dipole moment study as meribicycloepi-2-thio-4-selenoheptadien-6-thione (**35**).

An i.r. study was made[35] of selenium-, sulfur-, and oxygen-containing thiophthene derivatives (**36** and **37**, where X, Y, Z = Se or S). Results

showed aromatic character in the non-oxygen-containing compounds (**36**), whose electronic configuration shifted towards a highly conjugated system when one Se or S atom was substituted by oxygen (**37**).

Selenoaldehydic analogs of thiothiophthenes were discussed in Section VII.B.

d. Selenomalonodiarylamides. In 1930 it was claimed[37] that a malonodiarylamide (**38**) reacted with selenium tetrachloride to yield the corresponding dimeric selenomalonodiarylamide (**39**), the malonodichloroarylamide, hydrogen chloride, and elemental selenium (equation 23). However, no concrete evidence was given for any of the products.

(23) $5CH_2(CONHR)_2 + 4SeCl_4 \rightarrow 2[(RHNCO)_2C=Se]_2$

　　　　38　　　　　　　　　　　　　　**39**

$$+ 3CH_2(CONHRCl)_2 + 10HCl + 2Se$$

3. Cyclic Selenoketones

a. γ-Selenolutidones. When lutidone (**40**) was chlorinated by phosphorus pentachloride, alkylated, and then heated with potassium hydrogen selenide, γ-selenolutidones (**41**) were obtained[38] (scheme 24). Although termed

(24)

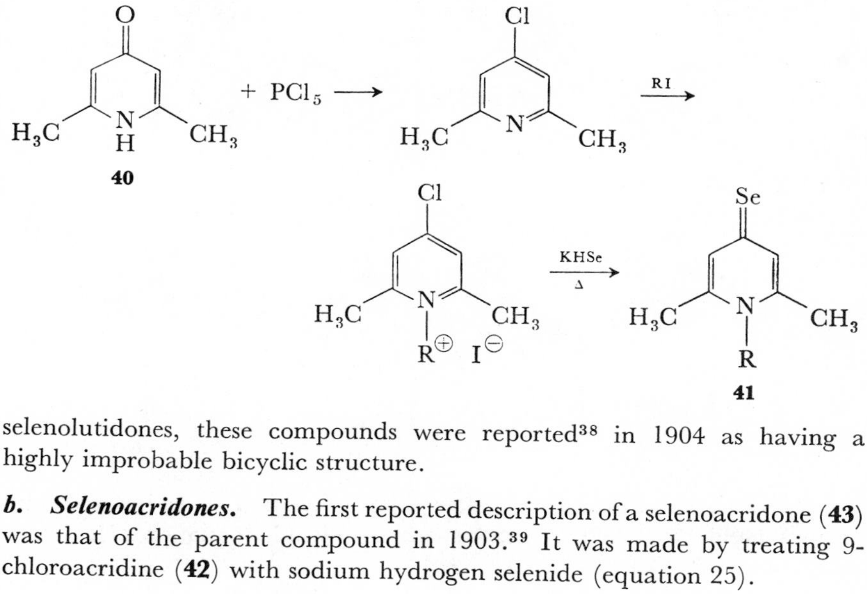

selenolutidones, these compounds were reported[38] in 1904 as having a highly improbable bicyclic structure.

b. Selenoacridones. The first reported description of a selenoacridone (**43**) was that of the parent compound in 1903.[39] It was made by treating 9-chloroacridine (**42**) with sodium hydrogen selenide (equation 25).

(25)

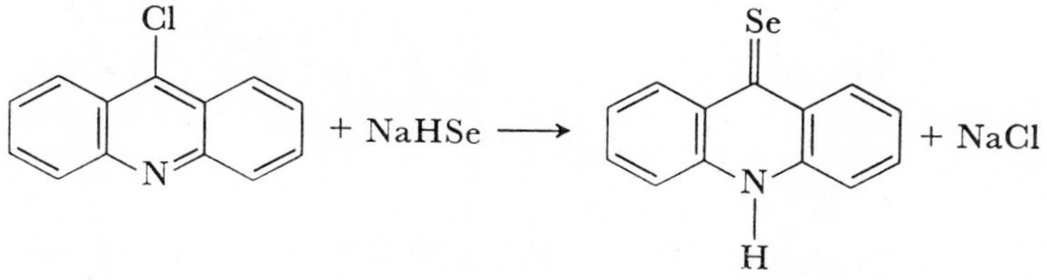

A similar route was employed[40] in 1939 to prepare the *N*-methyl, *N*-ethyl, and *N*-phenyl derivatives (**45**). However, instead of introducing the selenium as sodium hydrogen selenide, potassium selenosulfate was used. The seleno-sulfate salt (**44**) thus formed was hydrolyzed to the selenol, which is the enol tautomer of the corresponding selenoketone (scheme 26).

(26)

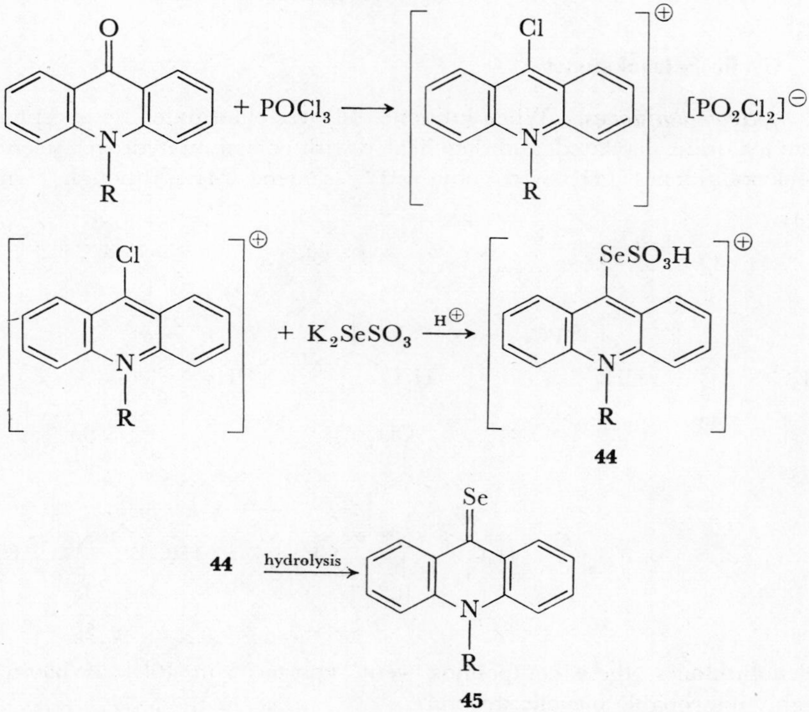

Selenoacridone was shown[39] to undergo alkylation when treated with various halides to yield the corresponding selenide (equation 27).

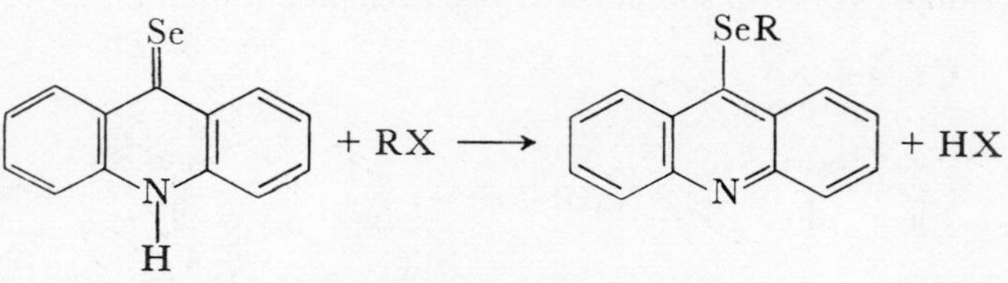

c. 4-Seleno-γ-pyrones. The synthesis of 4-seleno-γ-pyrones (**46**) in *ca.* 50% yield has been reported[24,32] (equation 28, where Y = O or S and

(28)

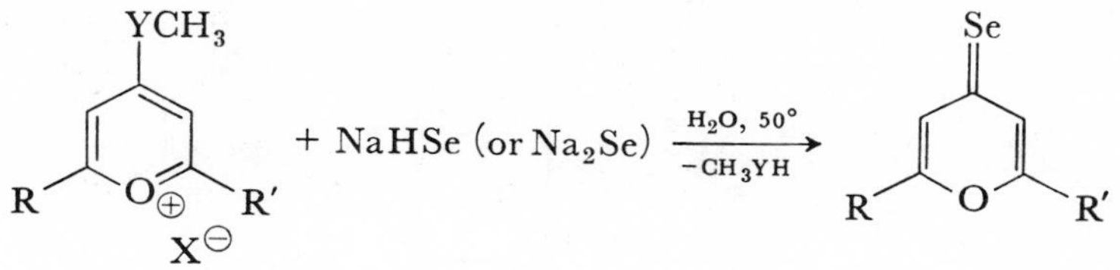

$X^- = ClO_4^-$, I^-, or $CH_3SO_4^-$). Derivatives of 4-seleno-1-thia-γ-pyrones (**48**) have also been prepared.[24,32] When 2 equivalents of sodium hydrogen sulfide were allowed to react with **46**, 4-thio-1-thia-γ-pyrones (**47**) were obtained. These were methylated, and the methylthio group displaced by selenide ion, producing the corresponding **48** (scheme 29).

(29) 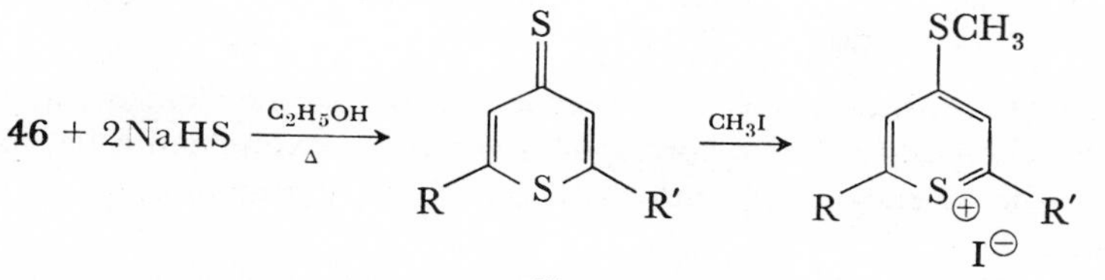 **46** + 2 NaHS $\xrightarrow[\Delta]{C_2H_5OH}$

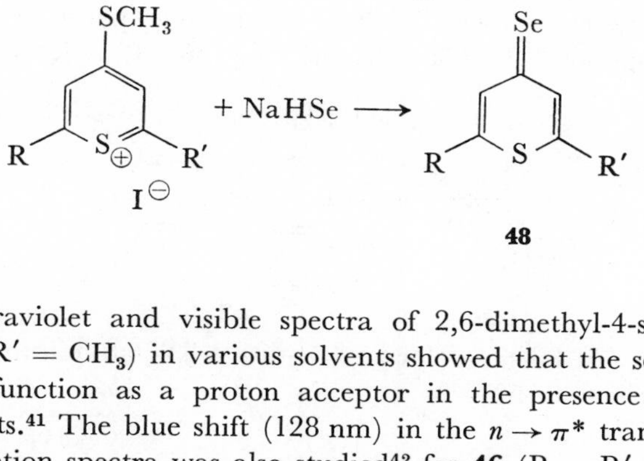

Ultraviolet and visible spectra of 2,6-dimethyl-4-seleno-γ-pyrone (**46**, $R = R' = CH_3$) in various solvents showed that the selenocarbonyl group may function as a proton acceptor in the presence of proton-donating solvents.[41] The blue shift (128 nm) in the $n \rightarrow \pi^*$ transition in the visible absorption spectra was also studied[42] for **46** ($R = R' = CH_3$), and it was found that Kundt's law[43] (there is an increase in the absorption band as the refractive index rises) was obeyed. The variation of the absorption maximum as a function of dielectric constant for **46** ($R = R' = CH_3$) was very similar to that for the corresponding 4-thio-γ-pyrone and **47** ($R = R' = CH_3$).[42]

The 4-thio-γ-pyrones (**49**) can be prepared[24,32] from the corresponding **46** by treating the selenopyrone methiodide with sodium hydrogen sulfide (equation 30).

(30) 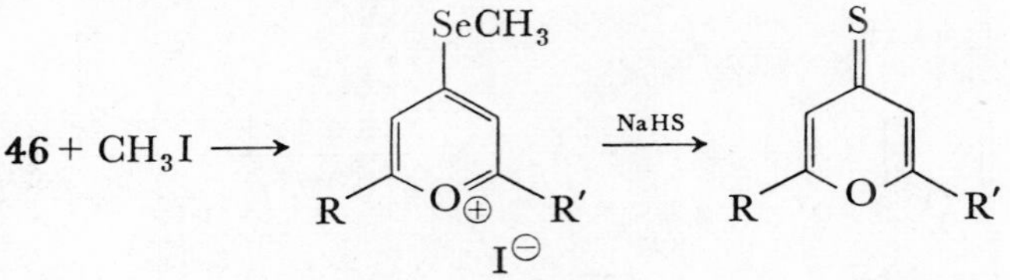   **46** + $CH_3I \longrightarrow$ [structure with SeCH$_3$ group, R and R' substituents, O$^{\oplus}$, I$^{\ominus}$] $\xrightarrow{\text{NaHS}}$ [structure with S, R, O, R']

49

2,6-Dimethyl-4-thio-γ-pyrone (**51**) was also prepared[32] from the analogous selenium compound (**50**) by heating with phosphorus pentasulfide in boiling benzene (equation 31).

(31) 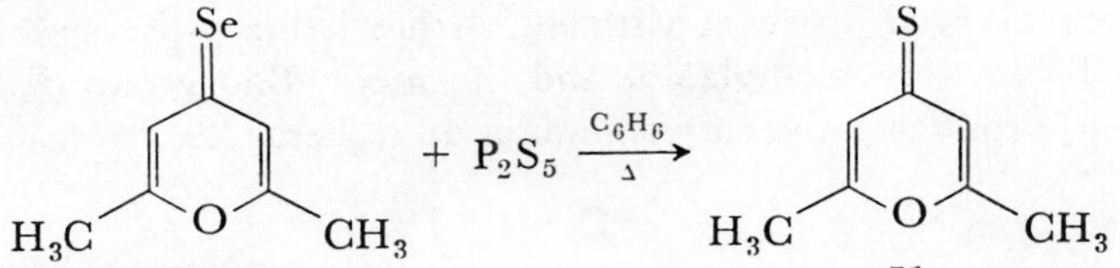   [structure **50** with Se, H_3C, O, CH_3] + $P_2S_5 \xrightarrow[\Delta]{C_6H_6}$ [structure **51** with S, H_3C, O, CH_3]

50   **51**

Water was sufficiently basic to regenerate the pyrone from the analogous selenium compound.[24] The corresponding pyrone oxime could be made by heating the selenoketone with hydroxylamine hydrochloride in boiling methanol.[24]

When 2,6-diphenyl-4-seleno-γ-pyrone was pyrolyzed at 250° in a hydrogen atmosphere, the carbene dimer, 2,2',6,6'-tetraphenyldipyrylene (**52**), was obtained[24] (equation 32). Likewise, heating **48** (R = R' = CH_3)

(32) 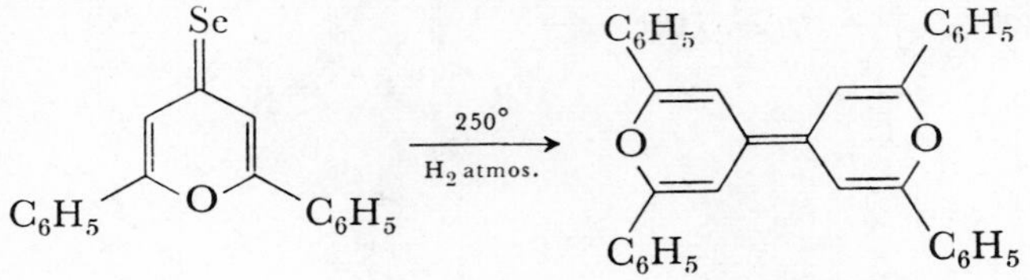   [structure with Se, C_6H_5, O, C_6H_5] $\xrightarrow[\text{H}_2 \text{ atmos.}]{250°}$ [structure **52** with C_6H_5 groups, O, O]

52

in boiling ligroin (70–90°) yielded 1,1'-dithia-2,2',6,6'-tetramethyl-dipyrylene (**53**) (equation 33).[24]

(33) 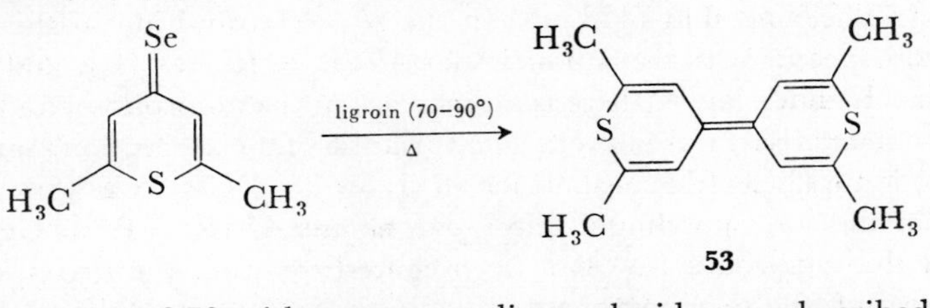   [structure with Se, H_3C, S, CH_3] $\xrightarrow[\Delta]{\text{ligroin (70–90°)}}$ [structure **53** with H_3C, CH_3, S, S, H_3C, CH_3]

53

The reaction of **50** with aqueous sodium selenide was described earlier (*cf*. equation 20).[24]

REFERENCES

1. L. Vanino and A. Schinner, *J. Prakt. Chem.*, **91**, 116 (1915).
2. W. E. Bradt and M. Van Valkenburgh, *Proc. Indiana Acad. Sci.*, **39**, 165 (1929).
3. H. J. Bridger and R. W. Pittman, *J. Chem. Soc.*, 1371 (1950).
4. L. Mortillaro, L. Credali, M. Mammi, and G. Valle, *J. Chem. Soc.*, 807 (1965).
4a. L. Brandsma and J. F. Arens, *Rec. Trav. Chim. Pays-Bas*, **81**, 33 (1962).
5. Q. Mingoia, *Gazz. Chim. Ital.*, **58**, 667 (1928).
6. L. Credali, M. Russo, L. Mortillaro, C. De Checchi, G. Valle, and M. Mammi, *J. Chem. Soc.*, B, 117 (1967).
7. A. Schröder, *Ber.*, **4**, 400 (1871).
8. W. E. Bradt, *J. Chem. Educ.*, **12**, 363 (1935).
9. R. Kuhn, *Angew. Chem.*, **50**, 703 (1937).
10. R. Kuhn, *J. Chem. Soc.*, 605 (1938).
11. G. Cole, *Ber.*, **8**, 1165 (1875).
12. L. Szperl and W. Wiorogórski, *Rocz. Chem.*, **12**, 270 (1932).
13. L.-B. Agenäs, *Ark. Kemi*, **31**, 31 (1969).
14. L. Nutting, R. M. Silverstein, and C. M. Himel, U.S. Pat. 3,033,875 (1962); *Chem. Abstr.*, **57**, 12438e (1962).
15. F. Asinger and K. Halcour, *Monatsh. Chem.*, **94**, 1029 (1963).
16. J. H. van den Hende and E. Klingsberg, *J. Amer. Chem. Soc.*, **88**, 5045 (1966).
17. S. Bezzi, M. Mammi, and C. Garbuglio, *Nature*, **182**, 247 (1958).
18. E. Klingsberg, U.S. Pat. 3,407,213 (1968); *Chem. Abstr.*, **70**, 28906u (1969).
19. J. G. Dingwall, S. McKenzie, and D. H. Reid, *J. Chem. Soc.*, C, 2543 (1968).
20. Agfa Akt.-Ges. für Photofabrikation, Ger. Pat. 910,199 (1954); *Chem. Abstr.*, **53**, 936h (1959).
21. Agfa Akt.-Ges. für Photofabrikation, Ger. Pat. 913,215 (1954); *Chem. Abstr.*, **53**, 937i (1959).
22. R. E. Lyons and W. E. Bradt, *Ber.*, **60**, 824 (1927).
23. D. S. Margolis and R. W. Pittman, *J. Chem. Soc.*, 799 (1957).
24. G. Traverso, *Ann. Chim.* (Rome), **47**, 1244 (1957).
25. C. G. Barraclough, R. L. Martin, and I. M. Stewart, *Australian J. Chem.*, **22**, 891 (1969).
26. G. A. Heath, I. M. Stewart, and R. L. Martin, *Inorg. Nucl. Chem. Lett.*, **5**, 169 (1969).
27. R. L. Martin and I. M. Stewart, *Nature*, **210**, 522 (1966).
28. R. Beckett and B. F. Hoskins, *Chem. Commun.*, 909 (1967).
29. A. Davison and E. T. Shawl, *Chem. Commun.*, 670 (1967).
30. A. Davison and E. T. Shawl, *Inorg. Chem.*, **9**, 1820 (1970).
31. J. A. McCleverty, *Progr. Inorg. Chem.*, **10**, 49 (1968).
32. G. Traverso, *Ann. Chim.* (Rome), **47**, 3 (1957).
33. S. Bezzi, *Gazz. Chim. Ital.*, **92**, 859 (1962).
34. S. Bezzi, C. Garbuglio, M. Mammi, and G. Traverso, *Gazz. Chim. Ital.*, **88**, 1226 (1958).
35. S. Pietra, C. Garbuglio, and M. Mammi, *Gazz. Chim. Ital.*, **94**, 48 (1964).

36. M. Sanesi, G. Traverso, and M. Lazzarone, *Ann. Chim.* (Rome), **53,** 548 (1963).
37. K. G. Naik and R. K. Trivedi, *J. Indian Chem. Soc.*, **7,** 239 (1930).
38. A. Michaelis and A. Hölken, *Justus Liebigs Ann. Chem.*, **331,** 245 (1904).
39. A. Edinger and J. C. Ritsema, *J. Prakt. Chem.*, **68,** 72 (1903).
40. K. Gleu and R. Schaarschmidt, *Ber.*, **72,** 1246 (1939).
41. M. Rolla and P. Franzosini, *Gazz. Chim. Ital.*, **88,** 837 (1958).
42. P. Franzosini, *Gazz. Chim. Ital.*, **88,** 1109 (1958).
43. A. Kundt, *Ann. Phys. Chem.*, 615 (1874).

VIII Selenium analogs of carboxylic acids

A. SELENOCARBOXYLIC ACIDS AND ESTERS

K. A. JENSEN

University of Copenhagen
Chemistry Laboratory II
The H. C. Ørsted Institute
Copenhagen, Denmark

1. Monoselenocarboxylic Acids

Whereas aliphatic thiocarboxylic acids can easily be prepared from acid chlorides and hydrogen sulfide in pyridine,[1,2] it has only been possible to isolate diacyl selenides from the corresponding reaction with hydrogen selenide.[3] The lower aliphatic diacyl selenides are distillable liquids, but only dipropionyl selenide has been subjected to a close examination. It reacted with sodium methoxide to form the selenopropionate ion, which could be trapped by its reaction with benzyl bromide to form *Se*-benzyl selenopropionate. With aniline, dipropionyl selenide formed 2 moles of propionanilide, indicating that selenopropionic acid, like thio and dithio acids (*cf.* the review by Janssen[4]), acts as a strong acylating agent.

According to Mingoia[5] and Lewis,[6] selenobenzoic acid is a solid (m.p. 133°). The compound that they reported, however, is not the acid but is in fact dibenzoyl diselenide, which is formed easily by access of air to the true acid. Selenobenzoic acid could be isolated[3] as an unstable pink oil that slowly eliminated hydrogen selenide at room temperature, being thereby transformed into dibenzoyl selenide (**1**, equation 1). This product in turn

(1) $2PhCOSeH \rightleftharpoons (PhCO)_2Se + H_2Se$

                                     1

undergoes a redox reaction with selenobenzoic acid by which dibenzoyl diselenide (**2**) and benzylidene bis(selenobenzoate) (**3**) are formed (equation 2).

(2)

    $(PhCO)_2Se + 3PhCOSeH \rightarrow (PhCOSe)_2 + PhCH(SeCOPh)_2 + H_2O$

                                  2   **3**

Reaction 1 is reversible and is suppressed when selenobenzoic acid is kept in a closed ampoule. The second reaction is catalyzed by pyridine. Benzaldehyde probably cannot be an intermediate in the formation of the benzylidene compound (**3**) because benzaldehyde does not react in this way with selenobenzoic acid. All three compounds, **1–3**, had been isolated earlier by Szperl and Wiorogórsky[7] from the reaction of hydrogen selenide with benzoyl chloride in the presence of aluminum chloride.

The i.r. and ^1H n.m.r. spectra of selenobenzoic acid show that it consists exclusively or predominantly of the selenol tautomer, $C_6H_5CO-SeH$. The i.r. spectrum shows carbonyl and Se–H frequencies but no hydroxyl bands. Likewise the position of the Se–H proton signal ($\tau = 6.15$) indicates that exchange of the proton between oxygen and selenium does not take place to any appreciable extent. This result is in agreement with the view, likewise based on spectroscopic evidence,[8,9] that thiocarboxylic acids have the thiol structure RCO–SH.

Selenobenzoic acid is soluble in aqueous alkali with the formation of a yellow color. The anion is readily *Se*-alkylated, as would be expected because it is probably a very strong nucleophile (*cf.* the quantitative data[10] on the relative nucleophilicities of PhS⁻ and PhSe⁻).

2. Esters of Monoselenocarboxylic Acids

Two types of esters can be derived from a monoselenocarboxylic acid:

(*a*) R—C(=O)—SeR′   *Se*-alkyl (or -aryl) monoselenoates

(*b*) R—C(=Se)—OR′   *O*-alkyl (or -aryl) monoselenoates

Compounds of type (a) have been obtained almost exclusively by treatment of metal (Na, Mg, Pb, etc.) selenolates or selenols (dissolved in pyridine) with acyl halides (equation 3).

(3) $$R—COCl + R'—Se^- \rightarrow R—C(=O)—SeR' + Cl^-$$

The second general method for synthesizing compounds of type (a), the reaction of metal selenocarboxylates with alkyl or aralkyl halides, has been tried only recently (equation 4).[3] It is advantageous only in so far as it circumvents the use of selenols.

(4) $$R—COSe^- + R'X \rightarrow R—C(=O)SeR' + X^-$$

Some compounds of this type have been known from the beginning of this century.[11] More recently Renson and Draguet[12] and Renson and Prette[13] prepared several monoselenocarboxylic esters from butyl, phenyl, o-tolyl, and o-methoxyphenyl selenols by reaction with the acid chlorides of acetic, dimethylacrylic, sorbic, and benzoic acids, as well as various substituted benzoic acids by the method depicted in equation 3. The yields ranged from 45 to 83%. The butyl ester, which was the first aliphatic selenol ester to be prepared, is a colorless liquid. The aromatic derivatives are as a rule colorless solids, which can be purified by recrystallization from, for example, petroleum ether.

Günther and Mautner[14] prepared the Se-(2-dimethylamino)ethyl ester of selenobenzoic acid from benzoyl chloride and the sodium salt of 2-(dimethylamino)ethaneselenol (obtained by reduction of the diselenide with sodium borohydride) (equation 5).

(5)

$$(CH_3)_2NCH_2CH_2Se^- + C_6H_5COCl \rightarrow (CH_3)_2NCH_2CH_2SeCOC_6H_5 + Cl^-$$

Reduction of 4,4'-diselenodibutyric acid followed by thermal cyclization gave the internal ester seleno-γ-butyrolactone (tetrahydroselenophen-2-one).[15] The highest yield (88%) was obtained when the reduction was carried out with hypophosphorous acid (phosphinic acid).

An alternative method for the preparation of esters of selenocarboxylic acids is the addition of water (catalyzed by $HgSO_4$) to (alkylseleno)acetylenes. Boiko et al.[16] have prepared Se-methyl 2-phenylselenoacetate by this method from methyl phenylethynyl selenide (equation 6). Although these authors

(6) $$C_6H_5C\equiv CSeCH_3 + H_2O \xrightarrow{HgSO_4} C_6H_5CH_2COSeCH_3$$

have prepared several acetylene derivatives of this type, this is apparently the only case in which the addition of water has been investigated.

Only three O-substituted monoselenoates of type (b) have been described in the literature. They were obtained by reaction of imidoesters with hydrogen selenide, a method analogous to the usual procedure for the preparation of O-alkyl thiocarboxylates (equation 7).

(7)
$$ R—C \overset{\displaystyle NH}{\underset{\displaystyle OR'}{}} + H_2Se \longrightarrow R—C \overset{\displaystyle Se}{\underset{\displaystyle OR'}{}} + NH_3 $$

Collard-Charon and Renson[17] prepared O-methyl selenobenzoate and O-methyl seleno-p-toluate in this manner, although in very low yields (5.6 and 3%, respectively). The same authors tried to prepare an aliphatic ester of this type, namely, O-methyl selenoacetate, but the i.r. spectrum of the product showed that the desired compound was not obtained. The yield of O-methyl selenobenzoate has been increased to 36% by treatment of a solution of the imidoester hydrochloride in pyridine with hydrogen selenide.[18]

Renson and Collienne[19] have obtained an internal ester of this type, isobenzofuran-1-selone, in good yield (62%) according to reaction 8.

(8)

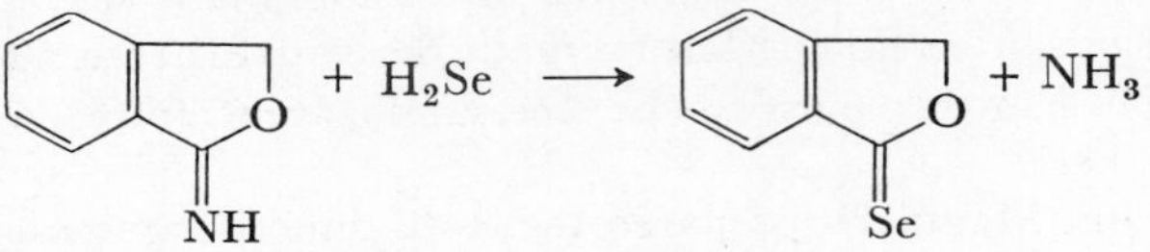

3. Esters of Selenothiocarboxylic Acids

Free selenothiocarboxylic acids or salts of these acids have not yet been prepared, but both possible types of esters, (a) and (b), are known:

(a) R—C(=S)—SeR' Se-alkyl (or -aryl) selenothioates

(b) R—C(=Se)—SR' S-alkyl (or -aryl) selenothioates

Collard-Charon and Renson[17] tried without success to prepare S-butyl selenothioacetate from S-butyl thioimidoacetate and hydrogen selenide. Two products were obtained which may have been formed as secondary products from the desired selenothioacetate, namely, selenoacetamide and di[1-(butylthio)ethyl] selenide, $[CH_3CH(SC_4H_9)]_2Se$.

A cyclic ester of type (a) has been prepared from the corresponding imine and hydrogen sulfide (equation 9).[19]

(9)

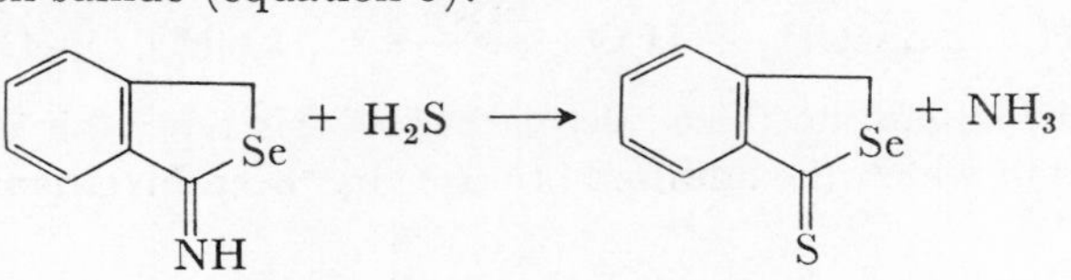

Schuijl *et al.*[20] claimed to have prepared an ester of type (*b*), C_3H_7—$C(=Se)SC_2H_5$, from $C_2H_5C\equiv CLi$, selenium, and ethanethiol. The product was stated to be impure; also, since no analytical or spectral data were given, it is doubtful whether the postulated product was in fact obtained.

Esters of both types can, however, be obtained from thio- or seleno-piperidides[21] by an extension of a method used by Jensen and Pedersen[22] for the preparation of several esters of dithiocarboxylic acids (equations 10 and 11).

(10)

(11)

The experiments were carried out with $R'X$ = methyl iodide or bromo-acetic acid; that is, methyl esters and carboxymethyl esters were prepared. The latter have the advantage of being solid compounds that can be purified by recrystallization. However, they could not be obtained in all cases.

Although both types of esters, (*a*) and (*b*), were successfully obtained by this method, the yields were low because of a competing reaction in which the SR' or SeR' group is displaced instead of the piperidine group (*cf.* the method of Klayman and Shine[23] for the preparation of selenoureas). In fact, the selenopiperidides, which were first prepared by the reaction of piperidides with phosphorus pentaselenide, could be prepared more conveniently from the thiopiperidide methiodides and hydrogen selenide.

S-Methyl selenothioates, $R—C(=Se)SCH_3$, with R = methyl, *tert*-butyl, or phenyl, and *S*-carboxymethyl selenothioates with R = methyl, ethyl, benzyl, phenyl, *m*-chlorophenyl, *p*-chlorophenyl, or *p*-methoxyphenyl, were prepared by this method. The esters are beautifully colored, the aromatic substances dark green and the aliphatic ones purple or violet. The crystalline carboxymethyl esters can be stored at $-20°$ without decomposition when they are protected from light and oxygen.

Se-Alkyl selenothioates are more stable that the isomeric *S*-alkyl selenothioates. Nevertheless, the preparation of such esters by this method met with great difficulties because of the predominating competing reaction that leads to thiopiperidides by exchange of SeR' with S. Only *Se*-methyl and *Se*-carboxymethyl selenothiobenzoate and *Se*-methyl selenothioacetate were prepared. Attempts to prepare *Se*-carboxymethyl selenothioacetate resulted in the formation of thioacetopiperidide.

Probably Se-alkyl selenothiobenzoates could generally be prepared more conveniently from thiobenzoyl chloride and selenols. So far, however, this possibility has been tried in only one case, the preparation of Se-(2-dimethylaminoethyl) selenothiobenzoate by Chu and Mautner.[24]

4. Esters of Diselenocarboxylic Acids

Several attempts have been made in the laboratory of the author to prepare diselenocarboxylic acids or their salts by reaction of carbon diselenide with Grignard reagents or other organometallic compounds. The Grignard reactions yielded only elemental selenium, even when carried out at −80°. Dialkylzincs yielded amorphous yellowish products that had approximately the composition of diselenoates.[25] However, the products were insoluble in all solvents with which they did not react, so they could not be purified by recrystallization. Also, it has not been possible to prepare either the free acids or any derivatives from these zinc compounds. Probably they contain polymeric anions of the following type:

$$\left[\begin{array}{c} Se^{\ominus} \\ | \\ -C-Se- \\ | \\ R \end{array} \right]_n$$

It was found, however, that esters of diselenocarboxylic acids could in principle be prepared analogously to selenothiocarboxylates, that is, by treatment of Se-alkyl selenoacylpiperidinium salts with hydrogen selenide (equation 12).[21]

$$(12) \quad R-\underset{\underset{SeR'}{|}}{C}=\overset{+}{N}\bigcirc + H_2Se \longrightarrow R-\underset{\underset{SeR'}{|}}{C}=Se + H_2\overset{+}{N}\bigcirc$$

This preparation met with the same difficulty encountered in the case of the selenothiocarboxylates: the competing reaction resulting in breaking of the C-Se bond instead of the C-N double bond predominates, so that selenopiperidides rather than diselenocarboxylic esters are formed.

Methyl diselenobenzoate and carboxymethyl diselenobenzoate were obtained in fair yields as dark green oils (the latter compound ought to be crystalline, but all efforts to induce it to crystallize failed).

Methyl diselenoacetate (purple) was obtained only in solution. The ester is volatile with ether and could not be separated from it, but N,Ndimethylselenoacetamide was obtained on addition of dimethylamine. The presence of methyl diselenoacetate was also proved by i.r. and visible spectra.

Several other *Se*-methyl or *Se*-carboxymethyl selenoacylpiperidinium salts yielded purple or green solutions on treatment with hydrogen selenide in methanol, but the pure diselenolates could not be obtained.

Some special derivatives of diselenoformic acid are the trialkyl phosphonio-diselenoformates, $R_3P^+CSeSe^-$, formed from trialkylphosphines and carbon diselenide,[26] and the reaction product of carbon diselenide and cyanide ion.[27] The primary product of the latter reaction is undoubtedly the cyanodiselenoformate ion, $NC-CSeSe^-$. However, the only well-defined substance that has so far been isolated from the solution is a tetraphenyl-arsonium salt of the anion, $NC-CSe_3^-$. It is concluded from X-ray evidence[27] that this anion has the cyclic structure, **4**.

4

Attempts to prepare salts of bis(hydroseleno)maleonitrile, analogous to the famous dimercaptomaleonitrile, from carbon diselenide and cyanides have been unsuccessful (Prokop *et al.*,[27] Bähr and Schleitzer,[28] and unpublished results from the author's laboratory).

Triselenoorthoformates have been prepared in low yields from carbon tetrachloride and sodium selenolates.[29]

5. Selenocarbonic Acids and Their Esters

A comprehensive review of chalcogenocarbonic acids and their salts and derivatives has been published by Dräger and Gattow.[30] Although the free acids are very unstable, they can as a rule be isolated at low temperatures. Triselenocarbonic acid, H_2CSe_3, is, however, not as well characterized as carbonic and trithiocarbonic acids, both of which have been isolated in the crystalline state. The triseleno acid (or its ether adduct) is formed as a red oil by treatment of a suspension of barium triselenocarbonate in ether containing a little water with hydrogen chloride.[31] It decomposes rapidly with evolution of hydrogen selenide. Selenodithiocarbonic acid, H_2CSeS_2, has been isolated as a bluish red oil by addition of hydrochloric acid to a suspension of barium selenodithiocarbonate in water.[30]

The formation of dimethyl and diethyl triselenocarbonate from barium triselenocarbonate and the alkyl halides has been claimed by Gattow and Dräger.[31] However, the methyl ester was reported to be insoluble in the common solvents, which is very improbable. Pure dialkyl triselenocarbonates were first prepared and characterized by Henriksen.[32] They are formed in a somewhat obscure reaction between carbon diselenide, alkyl halides, and a base in dimethyl sulfoxide. They are red, distillable liquids which are miscible with most organic solvents. Interestingly, aminolysis of these esters can proceed by two different routes.[33] In the first, diselenocarbamates, isoselenocyanates, and selenoureas are formed *via* a nucleophilic displacement by the amine on carbon (equation 13). In the second, selenoformamides, together with diselenides, are formed through initial attack of a selenolate ion on selenium (equation 14).

$$(13) \quad (RSe)_2CSe \xrightarrow{R'NH_2} R'NHCSe_2R \longrightarrow$$
$$R'NCSe \xrightarrow{R'NH_2} (R'NH)_2CSe$$

$$(14) \quad (RSe)_2CSe \xrightarrow{RSe^-,H^+} (RSeCHSe) + RSe-SeR \xrightarrow{R'NH_2} R'NHCHSe$$

Alkali-metal O-alkyl diselenocarbonates ("diselenoxanthates") can be prepared analogously to O-alkyl dithiocarbonates (xanthates) but are less stable.[34-37] Sodium O-ethyl selenothiocarbonate has been prepared similarly from carbon selenide sulfide.[38] Solid alkali-metal O-alkyl diselenocarbonates can be alkylated in good yields to O,Se-dialkyl diselenocarbonates by treatment with alkyl iodides in toluene.[39] O-Alkyl Se-carboxymethyl diselenocarbonates can be prepared by alkylation with bromoacetate ion in aqueous solution.[40] The carboxymethyl esters have the advantage of being solid compounds that can be purified by recrystallization. They have been used for the preparation of O-alkyl monoselenocarbazates[37] and for the study of the i.r.[40] and electronic absorption spectra of the C=Se group.[41] Analogous S-alkyl Se-carboxymethyl diselenothiocarbonates were obtained in low yields by a similar procedure (equation 15).[40] Attempts to prepare

$$(15) \quad RSH + CSe_2 \xrightarrow{KOH} RS-C-Se^- \xrightarrow{BrCH_2CO_2^-}$$
$$\underset{\displaystyle \| }{}$$
$$Se$$

$$RS-C-Se-CH_2CO_2^- \xrightarrow{H^+} RS-C-Se-CH_2CO_2H$$
$$\underset{Se}{\|} \qquad\qquad\qquad \underset{Se}{\|}$$

the corresponding triselenocarbonates from a selenolate and carbon diselenide, however, were unsuccessful.

The alkali-metal O-alkyl diselenocarbonates are oxidized by iodine[36] or persulfate[41] to bis(alkoxyselenocarbonyl) diselenides, $(Se-CSeOR)_2$, and with cyanogen bromide they form the corresponding monoselenides,[36] $Se(CSeOR)_2$. Mingoia[42] has possibly prepared an analogous oxygen compound, $Se(COOEt)_2$, from ethyl chloroformate and $MgBr(SeH)$. By the same reaction he also obtained a solid with a melting point of 122–123°, which was alleged to be the acid EtO–COSeH; considering the lability of the corresponding sulfur acid, however, this claim can hardly be valid.

Salts of cyanimidodiselenocarbonic acid, NC–N=C(SeH)$_2$, are formed by the reaction of cyanamide with carbon diselenide and a strong base.[43] They have been transformed into *Se,Se'*-dialkyl derivatives and heterocyclic compounds (diselenolanes and diselenanes), which have been studied in connection with the closely related compounds obtained from the reaction of carbon diselenide with active methylene compounds.

REFERENCES

1. S. Sunner, *Svensk Kem. Tidskr.*, **62,** 204 (1950).

2. A. Fredga and H. Bauer, *Ark. Kemi*, **2,** 113 (1950).

3. K. A. Jensen, L. Bøje, and L. Henriksen, *Acta Chem. Scand.*, **26,** 1465 (1972).

4. M. J. Janssen, "Chemistry of Carboxylic Acids and Esters," S. Patai, Ed., Interscience Publishers, New York, N.Y., 1969, Chap. 15.

5. Q. Mingoia, *Gazz. Chim. Ital.*, **56,** 835 (1926).

6. D. T. Lewis; *J. Chem. Soc.*, 831 (1940).

7. L. Szperl and W. Wiorogórski, *Rocz. Chem.*, **12,** 71 (1932).

8. N. Sheppard, *Trans. Faraday Soc.*, **45,** 693 (1949).

9. W. W. Crouch, *J. Amer. Chem. Soc.*, **74,** 2926 (1952).

10. R. G. Pearson, H. Sobel, and J. Songstad, *J. Amer. Chem. Soc.*, **90,** 319 (1968).

11. See H. Rheinboldt, "Houben-Weyl Methoden der Organischen Chemie," 4th ed., Vol. IX, Georg Thieme Verlag, Stuttgart, 1955, p. 1205.

12. M. Renson and C. Draguet, *Bull. Soc. Chim. Belges*, **71,** 260 (1962).

13. M. Renson and J. L. Prette, *Bull. Soc. Chim. Belges*, **73,** 507 (1967).

14. W. H. H. Günther and H. G. Mautner, *J. Med. Chem.*, **7,** 229 (1964).

15. W. H. H. Günther, *J. Org. Chem.*, **31,** 1202 (1966).

16. Yu. A. Boiko, B. S. Kupin, and A. A. Petrov, *Zh. Org. Khim.*, **4,** 1355 (1968).

17. C. Collard-Charon and M. Renson, *Bull. Soc. Chim. Belges*, **71,** 563 (1962).

18. R. Mayer, S. Scheithauer, and D. Kunz, *Chem. Ber.*, **99,** 1393 (1966).

19. M. Renson and R. Collienne, *Bull. Soc. Chim. Belges*, **73,** 491 (1964).

20. P. J. W. Schuijl, L. Brandsma, and J. F. Arens, *Rec. Trav. Chim. Pays-Bas,* **85,** 889 (1966).

21. K. A. Jensen, H. Mygind, and P. H. Nielsen, *Acta Chem. Scand.*, to be published.

22. K. A. Jensen and C. Pedersen, *Acta Chem. Scand.*, **15,** 1087 (1961).

23. D. L. Klayman and R. J. Shine, *J. Org. Chem.*, **34,** 3549 (1969).

24. S.-H. Chu and H. G. Mautner, *J. Med. Chem.*, **11,** 446 (1968).

25. K. A. Jensen, J. B. Carlsen, A. Holm, and P. H. Nielsen, *Acta Chem. Scand.*, **17,** 550 (1963).

26. K. A. Jensen and P. H. Nielsen, *Acta Chem. Scand.*, **17,** 549 (1963).

27. P. Prokop, B. Lorenz, and E. Hoyer, *Z. Chem.*, **9,** 156 (1969).

28. G. Bähr and G. Schleitzer, *Chem. Ber.*, **88,** 1171 (1955); **90,** 438 (1957).

29. A. Baroni, *Atti Accad. Ital.*, *Rend. Classe Sci. Fis.*, *Mat. Nat.*, [7], **1,** 543 (1940).

30. M. Dräger and G. Gattow, *Angew. Chem.*, **80,** 954 (1968).

31. G. Gattow and M. Dräger, *Z. Anorg. Allg. Chem.*, **348,** 229 (1966).

32. L. Henriksen, *Acta Chem. Scand.*, **21,** 1981 (1967).

33. L. Henriksen and E. S. S. Kristiansen, *Ann. N.Y. Acad. Sci.*, **192,** 101 (1972).

34. B. Rathke, *Justus Liebigs Ann. Chem.*, **152,** 206 (1969).

35. H. G. Grimm and H. Metzger, *Ber.*, **69,** 1356, 1362 (1936).

36. A. Rosenbaum, H. Kirchberg, and E. Leibnitz, *J. Prakt. Chem.*, [4], **19,** 1 (1963).

37. K. A. Jensen, P. A. A. Frederiksen, and L. Henriksen, *Acta Chem. Scand.*, **24,** 2061 (1970).

38. A. Stock and E. Wilfroth, *Ber.*, **47,** 144, 152 (1914).

39. A. Rosenbaum, *J. Prakt. Chem.* [4], **37,** 200 (1968).

40. K. A. Jensen and U. Anthoni, *Acta Chem. Scand.*, **24,** 2055 (1970).

41. M. L. Shankaranarayana, *Acta Chem. Scand.*, **24,** 2065 (1970).

42. Q. Mingoia, *Gazz. Chim. Ital.*, **58,** 670, 671 (1928).

43. K. A. Jensen and L. Henriksen, *Acta Chem. Scand.*, **24,** 3213 (1970).

VIII Selenium analogs of carboxylic acids

B. NITROGEN DERIVATIVES OF SELENOCARBOXYLIC ACIDS

ROBERT J. SHINE

Ramapo College of New Jersey
School of Theoretical and Applied Science
Mahwah, New Jersey

1. Introduction

This part of Chapter VIII discusses the chemistry of selenium analogs of carboxylic acids containing nitrogen and some chemically related compounds that are not strictly derivatives of carboxylic acids. The classes of compounds which are included are arranged in the order of increasing number of Se atoms and N atoms in their functional groups. Each section discusses, in the following sequence, the method of preparation, the chemical reactions, and the properties and uses of the particular class of compound.

This group of selenium compounds has not been extensively surveyed in the past; however, there are a few reviews that mention some or all of these classes.[1-5] The most recent one was written by Jensen.[6]

2. Isoselenocyanates

a. Preparation.

i. Reaction of an Isonitrile with Selenium. Isoselenocyanates (**1**) are generally unstable liquids or low-melting solids that are best used as chemical intermediates.[7] They are most conveniently prepared by the addition of elemental selenium to an alkyl or aryl isocyanide (equation 1). Jensen and

$$(1) \qquad RN{\equiv}C + Se \rightarrow RN{=}C{=}Se$$
$$\mathbf{1}$$

Frederiksen,[8] the first to employ this method in their preparation of phenyl isoselenocyanate (**1**, R = C_6H_5), performed the reaction in refluxing chloroform for 2 hours, obtaining a 20% yield of the isoselenocyanate. These reaction conditions were employed essentially unchanged by subsequent workers.[9–12] Collard-Charon and Renson[7] used a 1:1 mole ratio of isocyanide to selenium, the unreacted selenium being removed by filtration at the end of 48 hours. Bulka and Ahlers[13] and others[10,11] reported the use of chloroform or petroleum ether as the solvent, obtaining yields of 80–90% for a series of aryl isoselenocyanates.

ii. Reaction of an Alkyl or Acyl Halide with Selenocyanate Ion. The selenocyanate ion can displace a reactive halogen from an alkyl or acyl halide to give an alkyl or an acyl isoselenocyanate (**1**) (equation 2).

$$(2) \qquad RX + SeCN^- \rightarrow RN{=}C{=}Se + X^-$$
$$\mathbf{1}$$

Douglass[14] treated an acetone solution of potassium selenocyanate with benzoyl chloride to obtain benzoyl isoselenocyanate (**1**, R = C_6H_5CO), which was immediately treated with amines to give selenoureas. Collard-Charon and Renson[7] also used this reaction. Pedersen[10] prepared triphenylmethyl isoselenocyanate and diphenylmethyl isoselenocyanate from the corresponding halides by selenocyanate displacement in acetone solution, while Kirsanova and Derkach[15] made an isoselenocyanate of an N-substituted iminocarboxylic acid (equation 3).

$$(3) \qquad \underset{\begin{array}{c} \| \\ C_6H_5C-Cl \end{array}}{\overset{N-C_6H_5}{}} + NaSeCN \longrightarrow \underset{\begin{array}{c} \| \\ C_6H_5C-N{=}C{=}Se \end{array}}{\overset{N-C_6H_5}{}} + NaCl$$

This displacement reaction is hardly as unequivocal as it might appear because of the fact that selenocyanate is a bidentate ion. Franklin and Werner[11] reported that treatment of silver selenocyanate with methyl iodide gave only methyl selenocyanate. The isomerization of selenocyanates to isoselenocyanates is discussed later in this section.

Some organometallic isoselenocyanates have been prepared by displacement (equation 2). Silyl isoselenocyanates have been made by Ebsworth and Mays,[16] Thayer,[17] and Bürger and Goetze;[18] organophosphorus isoselenocyanates, by Gabrio and Barnikow.[19]

iii.   Reaction of a Dichloroisocyanate with Selenide Ion. Treatment of phenyl dichloroisocyanate (**2**) with excess sodium selenide gives phenyl isoselenocyanate (**1**, R = C_6H_5) (equation 4).[7,8,20]

(4) $C_6H_5N{=}CCl_2 + Na_2Se \rightarrow C_6H_5N{=}C{=}Se + 2NaCl$
            2

iv.   Reaction of an Isocyanate with Phosphorus Pentaselenide. Collard-Charon and Renson[7] reported that the reaction of excess phosphorus pentaselenide with phenyl isocyanate gave phenyl isoselenocyanate (equation 5).

(5) $5C_6H_5N{=}C{=}O + P_2Se_5 \rightarrow 5C_6H_5N{=}C{=}Se + P_2O_5$

v.   Isomerization of Selenocyanates. Selenocyanates do not isomerize quantitatively[21] to isoselenocyanates, but rather reach a state of equilibrium (equation 6) whose position depends strongly on the organic moiety. Franklin

(6) $RSeCN \rightleftharpoons RNCSe$

and Werner[11] reported that isomerization of selenocyanates does not afford a general preparative method for the formation of isoselenocyanates.

The possibility that organic selenocyanates may have either the normal structure (RSeCN) or the iso structure (RNCSe) was recognized in 1889 by Hofmann.[22] He postulated that the isoselenocyanate should lose selenium readily. Tarantelli *et al.*[21,23] studied the isomerization of organic selenocyanates and devised methods of determining which isomer prevailed. Isoselenocyanates and selenocyanates differ in their i.r. spectra;* the isoselenocyanates giving a broad, intense band at 2098 cm^{-1} with a shoulder at 2011 cm^{-1},[23] whereas selenocyanates show a very sharp peak at 2140 cm^{-1}.[10] The u.v. spectra also exhibit slight differences that can be used for identification.[21] Beg[24] used i.r. data to show that a series of aryl selenocyanates possessed the normal structure, as they all had sharp peaks at 2160 cm^{-1}. Many other workers have also used i.r. evidence in structural elucidation.[12,18,25–27]

---

* The spectroscopic properties of organic isoselenocyanates are discussed further in Chapter XV.

Fava[28] has reviewed the subject of isomerization of organic selenocyanates and has concluded that, where equilibrium exists, the mixture consists of approximately equal amounts of the normal and iso structures. This is in contrast to organic thiocyanates, in which the iso form predominates.[28]

b. Reactions.

i. Reaction of Isoselenocyanates with Amines. The addition of amines or hydrazines to isoselenocyanates is often used to demonstrate that one has the isoselenocyanate rather than the selenocyanate structure. Ammonia or amines add to organic isoselenocyanates to give selenoureas in high yield (*cf.* Section B.4).

ii. Reaction of Isoselenocyanates with Hydrazines. Hydrazine and its derivatives add to isoselenocyanates to give selenosemicarbazides (*cf.* Section B.5).

iii. Reduction of Isoselenocyanates. Alkyl isoselenocyanates can be reduced[11] with zinc in hydrochloric acid to give amines (equation 7).

$$(7) \qquad RNCSe + Zn/HCl \rightarrow RNH_2$$

Amines are also the product when lithium aluminum hydride is used as the reducing agent (equation 8).[21,23]

$$(8) \qquad RNCSe + LiAlH_4 \rightarrow RNH_2$$

iv. Displacement of Selenium from Isoselenocyanates. Alkyl isoselenocyanates react immediately with silver nitrate,[11] precipitating black silver selenide. Although isoselenocyanates are only very slowly hydrolyzed by water, decomposition by aqueous acids rapidly gives a precipitate of red elemental selenium.[11] The C-Se bond in organosilicon isoselenocyanates (**3**) is weak and susceptible to hydrolysis by water (equation 9).[17] The Se

$$(9) \qquad 2R_3SiNCSe + H_2O \rightarrow (R_3Si)_2O + 2HCN + 2Se$$
$$\mathbf{3}$$

atom of triphenylmethyl isoselenocyanate can be removed quantitatively by heating in dry acetonitrile to give triphenylacetonitrile as the product (equation 10).[23]

$$(10) \qquad (C_6H_5)_3CNCSe \rightarrow (C_6H_5)_3CCN + Se$$

3. Selenoamides

a. Preparation.

i. Reaction of Nitriles with Hydrogen Selenide. Selenoamides (**4**) are reported[29] to be unstable solids that discolor rapidly in light and are slowly converted to resinous materials. The stability increases from primary selenoamides to secondary and tertiary selenoamides. Preparation of primary selenoamides is accomplished most easily by adding hydrogen selenide to a nitrile (equation 11).[30,31] For example, addition of hydrogen selenide

$$(11) \qquad RC{\equiv}N + H_2Se \longrightarrow \overset{\overset{\displaystyle Se}{\|}}{R-C-NH_2}$$

4

to an ammoniacal solution of benzonitrile in the absence of air gave selenobenzamide (**4**, R = C_6H_5).[32] Kindler,[33] on treatment of an ethanolic solution containing sodium ethoxide and phenylacetonitrile with hydrogen selenide, obtained phenylselenoacetamide (**4**, R = $C_6H_5CH_2$). Starting with acetonitrile and using similar conditions, he secured a 17% yield of selenoacetamide. Jensen and Nielsen[34] obtained diselenomalonamide in 36% yield by adding hydrogen selenide to malononitrile. This route has also been used by Collard-Charon and Renson[29] and by Asinger et al.[35]

ii. Reaction of Amides with Phosphorus Pentaselenide. Secondary and tertiary selenoamides (**5**) can be prepared by slowly adding powdered phosphorus pentaselenide to a refluxing solution of the corresponding amide in dry benzene (equation 12).[29,36] Jensen and Nielsen[34] used this method to

$$(12) \qquad 5R^1\overset{\overset{\displaystyle O}{\|}}{C}NR^2R^3 + P_2Se_5 \longrightarrow 5R^1\overset{\overset{\displaystyle Se}{\|}}{C}NR^2R^3$$

5

prepare N,N-dimethylselenobenzamide (**5**, R^1 = C_6H_5, R^2 = R^3 = CH_3) from N,N-dimethylbenzamide in 4.7% yield. In their preparation of some selenolactams, Hallam and Jones[37] modified the above procedure by using red phosphorus and gray selenium instead of phosphorus pentaselenide. Low yields of selenolactams were obtained.

iii. Nucleophilic Displacement by Selenide Ion. Michaelis et al.[38,39] were the first to report the displacement of halogens from nitrogen-containing heterocycles by hydroselenide ion. For example, N-methyl-α-selenopyridone was prepared by the displacement of iodide from N-methyl-α-iodopyridinium iodide. Mautner et al.[40] reported a similar displacement in their preparation of α-selenopyridone in refluxing ethylene glycol monoethyl ether. 6-Chloropurines were converted to 6-selenopurines by displacement with sodium hydrogen selenide.[41] (See Chapter XID for further details.)

Yarovenko and Raksha[42] reported that F atoms attached to the α-position of fluorinated amines are very labile and are easily removed by hydrogen selenide. The resultant product is a fluorinated selenoamide, obtained in about 35 % yield, as shown in reaction 13.

$$(13) \quad CHF_2CF_2N(CH_2CH_3)_2 + H_2Se \longrightarrow CHF_2\overset{\overset{\displaystyle Se}{\|}}{C}N(CH_2CH_3)_2 + 2HF$$

Recently, Hartmann[42a] synthesized N,N-dialkyl-substituted selenoamides by the hydroselenide displacement of the methylthio group of the corresponding S-methylthioamides (equation 13a). This method is analogous to one developed earlier for the synthesis of selenoureas from the related

$$(13a) \quad C_6H_5-\overset{\overset{\displaystyle CH_3S}{|}}{C}=\overset{+}{N}\!\!\bigcirc\!\!O + HSe^- \longrightarrow C_6H_5-\overset{\overset{\displaystyle Se}{\|}}{C}-N\!\!\bigcirc\!\!O + CH_3SH$$

S-methylthiopseudoureas (*vide infra*).

b. Reactions.

i. Hydrolysis. The acid hydrolysis of 2-amino-2-methylselenopropion-amide (**6**) gave 2-amino-2-methylpropionic acid, ammonium ion, and hydrogen selenide (equation 14).[35]

$$(14) \quad CH_3\overset{\overset{\displaystyle CH_3}{|}}{\underset{\underset{\displaystyle NH_2}{|}}{C}}-\overset{\overset{\displaystyle }{}}{\underset{\underset{\displaystyle Se}{\|}}{C}}-NH_2 + H^+ + 2H_2O \longrightarrow$$

6

$$CH_3\overset{\overset{\displaystyle CH_3}{|}}{\underset{\underset{\displaystyle NH_2}{|}}{C}}-COOH + NH_4^+ + H_2Se$$

ii. Cyclization. Hofmann[22] reported the cyclization of a selenoamide and an α-chloroketone. Selenobenzamide and α-chloroacetone condensed smoothly, giving 2-phenyl-4-methylselenazole (**7**) (reaction 15). Brooker

$$(15) \quad C_6H_5\overset{\overset{\displaystyle Se}{\|}}{C}NH_2 + CH_3\overset{\overset{\displaystyle O}{\|}}{C}CH_2Cl \longrightarrow$$

7

et al.[43] obtained 2,4-dimethylselenazole by the same reaction, starting with selenoacetamide. Asinger *et al.*[35] reported the condensation of α-aminoselenoamides with ketones, resulting in high yields of 4-selenoxo-imidazolidines (**8**) (equation 16). The reaction of selenoacetamide with

(16)

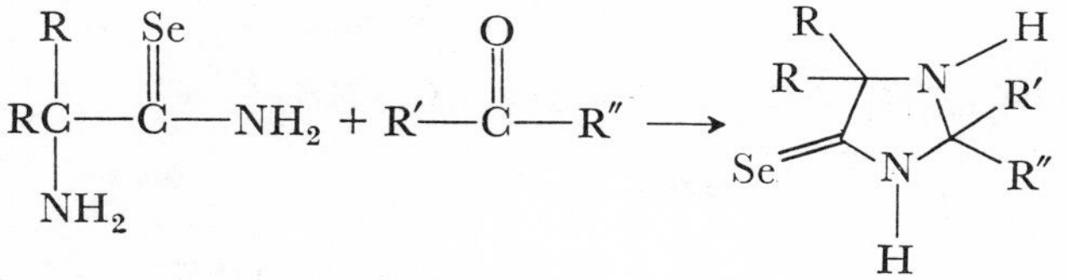

$$RC\overset{R}{\underset{NH_2}{|}}\overset{Se}{\overset{||}{C}}-NH_2 + R'-\overset{O}{\overset{||}{C}}-R'' \longrightarrow$$

8

2-bromoethylamine hydrobromide results in the formation of 2-methyl-selenazoline hydrobromide (**9**) (equation 17).[44]

(17)

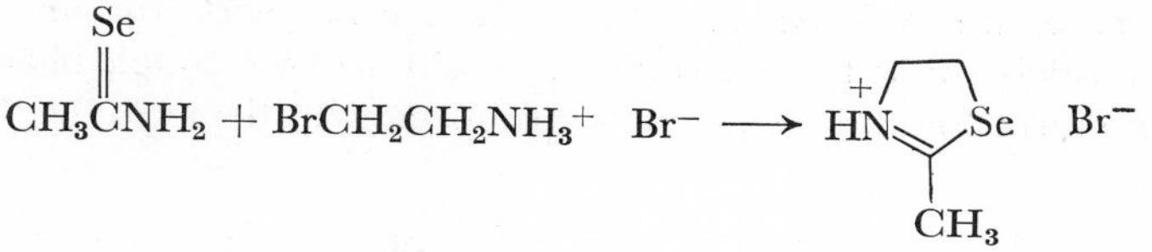

$$CH_3\overset{Se}{\overset{||}{C}}NH_2 + BrCH_2CH_2NH_3^+ \ Br^- \longrightarrow$$

9

iii. Oxidation. Becker and Meyer[31] reported that oxidation of seleno-amides with iodine results in selenium deposition. The organic product was assigned structure **10** (reaction 18). Iodine oxidation of selenomalonamide[45]

(18)

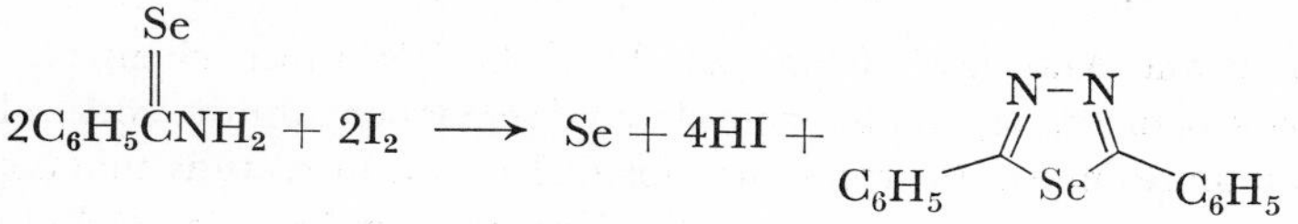

$$2C_6H_5\overset{Se}{\overset{||}{C}}NH_2 + 2I_2 \longrightarrow Se + 4HI +$$

10

results in ring formation, giving a diselenolylium compound (**11**) (equation 19) (*cf.* Chapter VI).

(19)

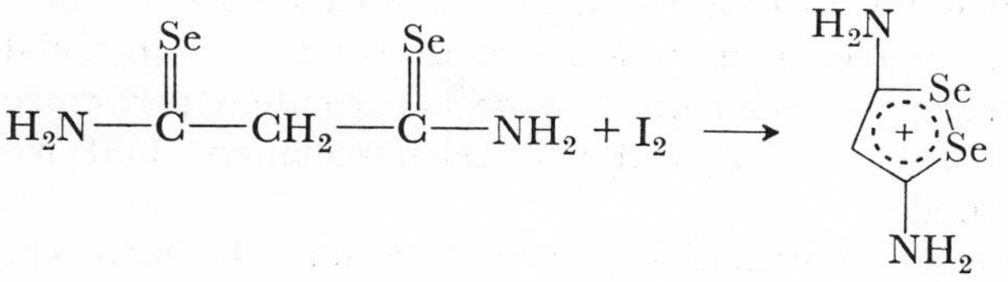

$$H_2N-\overset{Se}{\overset{||}{C}}-CH_2-\overset{Se}{\overset{||}{C}}-NH_2 + I_2 \longrightarrow$$

11

iv. Alkylation. The alkylation of 2-hydroselenopyridine with methyl iodide results in the formation of 2-(methylseleno)pyridine.[40] Jensen and Nielsen[34] also reported the addition of alkyl halides to selenoamides, giving selenoamidium salts.

v. Reaction of Selenobenzamide with an N-Isothiocyanatoamine. Anthoni *et al.*[46] found that the reaction of selenobenzamide with *N*-isothiocyanato-diisopropylamine (**12**) under oxidizing conditions results in the formation of bis(*N*,*N*-diisopropylthiocarbazoyl) diselenide (**13**), as shown in reaction 20.

$$(20) \quad 2[(CH_3)_2CH]_2NN\!=\!C\!=\!S + 2C_6H_5\overset{\overset{\displaystyle Se}{\|}}{C}\!-\!NH_2 \longrightarrow$$
$$\mathbf{12}$$

$$\{[(CH_3)_2CH]_2NNH\overset{\overset{\displaystyle S}{\|}}{C}Se\!-\!\}_2 + C_6H_5C\!\equiv\!N$$
$$\mathbf{13}$$

c. Properties and Uses*

i. Nature of the Selenoamide Bond. In recent years, the amide group has been widely studied to determine the position of the double bond. Resonance structures for the selenoamide group can be depicted by **I** and **II**.

Hampson and Mathias,[47] from a study of the ^{14}N n.m.r. chemical shifts of the series benzamide, thiobenzamide, and selenobenzamide, concluded that **II** is the more important resonance form. In an n.m.r. study that compared the coalescence temperature of the methyl signals for *N,N*-dimethyl-benzamide, *N,N*-dimethylthiobenzamide, and *N,N*-dimethylselenobenz-amide, Schwenker and Rosswag[48] concluded that hindered rotation about the C-N bond increased as the heteroatom was changed from oxygen to sulfur to selenium. Electric moments of the same series of compounds gave results which agreed with conclusions drawn from n.m.r. data.[49] Jensen and Sandström,[50] in their study of barriers to rotation, performed a complete line shape analysis of *N,N*-dimethylselenoacetamide and *N,N*-dimethyl-selenobenzamide.

The increased contribution of resonance form **II** has been explained[51] as due to a decrease in orbital overlap between the carbon $2p$ and the selenium $4p$ orbitals. Hückel molecular orbital calculations were made[51] and compared with the experimental values for C-N bond length and dipole moments.

A review by Hallam and Jones[52] presents the spectroscopic and related evidence for the exact nature of the $-CX\!-\!N\!\!\diagdown$ group in cyclic and acyclic oxy-, thio-, and selenoamides.

---

* The spectroscopic properties of selenoamides are discussed in Chapter XV.

ii. Use of Selenoamides as Photographic Sensitizers. Organic compounds that readily deposit selenium have found use as sensitizers in photographic products.[53,54] Included are selenoamides, isoselenocyanates, and selenoureas.

4. Selenoureas

Selenoureas (**14**) are easily prepared crystalline solids and, in general, are reasonably stable to air and light in the pure state.[55] The barrier to hindered rotation in 1,1-dimethyl-2-selenourea has recently been determined.[55a]

$$R^1 \diagdown \quad \overset{\displaystyle Se}{\underset{\displaystyle \|}{}} \quad \diagup R^3$$

$$N-C-N$$

$$R^2 \diagup \qquad\qquad \diagdown R^4$$

14

a. Preparation

i. Addition of Hydrogen Selenide to Cyanamides or Carbodiimides. Selenourea (**14**, $R^1 = R^2 = R^3 = R^4 = H$) itself can be conveniently prepared by the addition of hydrogen selenide to a solution of cyanamide (equation 21). Verneuil,[56] the first to use this method, employed a 2% ethereal solution of cyanamide.

(21) $$H_2N-C \equiv N + H_2Se \longrightarrow H_2N-\overset{\displaystyle Se}{\underset{\displaystyle \|}{C}}-NH_2$$

The selenourea was recrystallized from water, giving a colorless solid. Dunbar and Painter[57] obtained a 22% yield of selenourea by this method, using hydrogen chloride as a catalyst. Hope[58] employed a 50% aqueous solution of cyanamide and obtained a good yield of selenourea after a 1-day reaction time. Badiello and Breccia[59] prepared selenourea by the cyanamide plus hydrogen selenide route and then proceeded to make the radioactive ^{75}Se-selenourea by irradiation of the inactive material in a BEPO reactor. The radioactive ^{75}Se-selenourea is less stable than the non-radioactive form. Selenourea has also been prepared by other workers[60,61] using this method.

Use of substituted cyanamides results in the formation of either mono-substituted selenoureas or 1,1-disubstituted selenoureas. Thus, the addition of hydrogen selenide to phenylcyanamide gives 1-phenyl-2-selenourea (**14**, $R^1 = R^2 = R^3 = H$, $R^4 = C_6H_5$),[62] while 1-methyl-1-naphthylcyanamide in ethanol produces a 93% yield of 1-methyl-1-(1-naphthyl)-2-selenourea.[63] This route has also been used successfully to prepare 1-allyl-2-selenourea,[64] 1-ethyl-2-selenourea,[65] 1-guanidino-2-selenourea,[66] *p*-ethoxyphenyl-2-selenourea,[67] and 1,1-dimethyl-2-selenourea.[68]

Carbodiimides and hydrogen selenide give 1,3-disubstituted selenoureas.[69-71]

*ii. **Reaction of Isoselenocyanates with Amines.*** The reaction of isoselenocyanates with amines (*cf.* Section B.2) proceeds smoothly, giving high yields of the selenourea. The reaction can be used to prepare alkyl, aryl, and acyl selenoureas that are mono-, di-, or trisubstituted. Stolte[20] was the first to use this method in his preparation of 1,3-diphenyl-2-selenourea[8] from phenyl isoselenocyanate and aniline (equation 22). The preparation of

$$(22) \quad C_6H_5N{=}C{=}Se + C_6H_5NH_2 \longrightarrow C_6H_5NH{-}\overset{\overset{\displaystyle Se}{\|}}{C}{-}NHC_6H_5$$

1-phenyl-3-methyl-2-selenourea from phenyl isoselenocyanate and methylamine has been reported.[71] Lipp *et al.*[71a] prepared four selenoureas by the reaction of isocyanides, elemental selenium, and amines in absolute alcohol (equation 23). A series of three substituted 1-butyl-2-selenoureas was pre-

$$(23) \quad R{-}N{\equiv}C + Se + NH_2R' \longrightarrow RNH{-}\overset{\overset{\displaystyle Se}{\|}}{C}{-}NHR'$$

pared[72] by the reaction of butyl isoselenocyanate with ammonia or amines. Bulka *et al.*[73-75] found that the reaction of aryl isoselenocyanates with aryl amines is very vigorous in the absence of solvent but proceeds smoothly in organic solvents, giving excellent yields of aryl selenoureas. The reaction of isoselenocyanates with amines was found[21,23,76] to be a useful way to demonstrate the iso structure of an isoselenocyanate (*cf.* Section B.2).

A few organophosphorus selenoureas were prepared[19] from phosphorus-substituted isoselenocyanates. Isoselenocyanates of *N*-substituted iminocarboxylic acids react with aromatic amines to give the iminosubstituted selenourea (**15**) as shown by reaction 24.[15]

$$(24) \quad C_6H_5{-}\overset{\overset{\displaystyle NC_6H_5}{\|}}{C}{-}N{=}C{=}Se + C_6H_5NH_2 \longrightarrow C_6H_5{-}\overset{\overset{\displaystyle NC_6H_5}{\|}}{C}{-}NH{-}\underset{\underset{\displaystyle Se}{\|}}{C}{-}NHC_6H_5$$

$$\mathbf{15}$$

Acyl-substituted selenoureas can be prepared by treating acyl isoselenocyanates with ammonia or amines.[14,73,77] 1-Benzoyl-2-selenourea was prepared by this method[14] in 27% yield. The benzoyl group was then hydrolyzed by 5–10% aqueous sodium hydroxide, resulting in a good yield of selenourea.[73]

Amines also react with the selenocyanate ion to form selenoureas. For example, an aqueous or alcoholic solution of benzylamine hydrochloride and potassium selenocyanate gave 1-benzyl-2-selenourea.[78-80]

iii. Reaction of Carbon Diselenide with Amines. Grimm and Metzger[81] observed in 1936 that the reaction of carbon diselenide with two equivalents of aniline resulted in the formation of 1,3-diphenyl-2-selenourea (equation 25). Warner[82] developed experimental conditions that minimized the

$$\text{(25)} \quad CSe_2 + 2C_6H_5NH_2 \longrightarrow C_6H_5NH-\overset{\overset{\textstyle Se}{\|}}{C}-NHC_6H_5 + H_2Se$$

problems associated with carbon diselenide. Using primary aliphatic amines, he obtained good yields of 1,3-disubstituted selenoureas. Although it appears that this reaction would be useful primarily in the preparation of symmetrical selenoureas, Warner was nevertheless successful in obtaining an unsymmetrical compound of this type, namely, 1,1-dimethyl-3-benzyl-2-selenourea. Rosenbaum *et al.*[83,84] obtained 1,3-diisopropyl-2-selenourea from the reaction of isopropylamine and carbon diselenide. It is possible that a diselenocarbamate is an intermediate in this reaction. Rosenbaum[85] reported that the oxidation of the piperidine or isopropyl derivatives of diselenocarbamates gave the corresponding selenourea.

In a somewhat related reaction, Henriksen[86] obtained 1,3-diphenyl-2-selenourea by treating esters of triselenocarbonic acid with aniline (equation 26).

$$\text{(26)} \quad Se{=}C(SeR)_2 + 2C_6H_5NH_2 \longrightarrow C_6H_5NH-\overset{\overset{\textstyle Se}{\|}}{C}-NHC_6H_5 + 2RSeH$$

iv. Displacement by Hydroselenide Ion. The displacement of the methyl-thio group from a thiopseudourea by hydroselenide ion has been shown by Klayman and Shine[55,87,87a] to be a general method for the preparation of selenoureas (equation 27). It is particularly well suited for the preparation of tetrasubstituted selenoureas,[87] a class of selenoureas difficult to prepare by other methods. For example, 1,1,3,3-tetramethyl-2-selenourea was prepared in 70% yield. Yields are usually 60–70% for all types of nitrogen substitution, and the attached groups may be alkyl or aryl.[55] The optimum pH of the reaction medium was found to be about 9. At lower pH (*ca.* pH 5), however, a selenothiocarbamate ester is obtained as the product with elimination of an amino group (*vide infra*). 2-Thio-4-selenobiuret could also be prepared by the method depicted in equation 27.[55]

Analogous displacement reactions were reported later by Ramakrishna and Irving[88,89] in their preparation of 1,5-diphenylselenocarbazide (**16**) (equation 28). These authors also reported the displacement of a methylthio

$$
\text{(28)} \quad \underset{\displaystyle \overset{\textstyle NO_2}{|}}{C_6H_5N{=}N{-}C{=}N{-}NHC_6H_5} + H_2Se \longrightarrow
$$

$$
\underset{\textstyle \mathbf{16}}{C_6H_5NHNH{-}\overset{\textstyle Se}{\overset{\|}{C}}{-}NHNHC_6H_5}
$$

group as shown in equation 29.[88] The displacement of a methylthio group

$$
\text{(29)}
$$

$$
\underset{}{C_6H_5NHN{=}\overset{\textstyle SCH_3}{\overset{|}{C}}{-}N{=}NC_6H_5} + H_2Se \longrightarrow C_6H_5NHN{=}\overset{\textstyle SeH}{\overset{|}{C}}{-}N{=}NC_6H_5
$$

by hydroselenide ion, developed by Klayman and Shine, has been extended to the synthesis of 6-selenopurines.[41]

v. Conversion of a Carbon-Oxygen Double Bond to a Carbon-Selenium Double Bond. Jensen *et al.*[90] converted 1,1,3,3-tetramethylurea to 1,1,3,3-tetramethyl-2-selenourea by treatment with phosphorus pentaselenide. After a 72-hour reflux time in benzene, a 5.6% yield of the selenourea was obtained. Giudicelli *et al.*[71] reported the conversion of *N,N*-dimethylformamide into 1-methyl-2-selenourea by treatment with phosphorus oxychloride, selenium, and ammonia.

b. Reactions

i. Alkylation of Selenourea and its Derivatives. Selenourea, like thiourea, is readily alkylated by alkyl sulfates or alkyl halides. Dimethyl sulfate gives a high yield of 2-methyl-2-selenopseudourea sulfate (**17**) (equation 30).[57] This reaction has been used also with many other selenoureas.[64,65,72] Yields

$$
\text{(30)} \quad 2H_2N{-}\overset{\textstyle Se}{\overset{\|}{C}}{-}NH_2 + (CH_3)_2SO_4 \longrightarrow \underset{\textstyle \mathbf{17}}{(H_2N{-}\overset{\textstyle SeCH_3}{\overset{|}{C}}{=}NH)_2{\cdot}H_2SO_4}
$$

are generally in the 60–90% range,[91] and the products are reportedly more stable than the selenourea precursor.[64] That the methylation occurs at the Se atom and not the N atom was demonstrated by Warner and Page.[92]

α,ω-Dibromoalkanes react with 2 equivalents of selenourea at reflux in ethanol for 30 minutes, giving bis(selenopseudourea) salts in high yield.[91]

The formation of selenopseudourea derivatives has found some limited application in carbohydrate and steroid research. Wagner and Nuhn[93,94] converted tetraacetyl-1-β-D-glucosyl bromide to tetraacetyl-1-β-D-glucosyl diselenide *via* hydrolysis of the intermediate selenopseudourea hydrobromide. Segaloff and Gabbard[95,96] and Hiscock *et al.*[97] prepared 3-β-selenosteroids by the hydrolysis of the 3-β-selenopseudourea tosylates.

Selenopseudourea derivatives may also be prepared by electrophilic addition of selenourea to a conjugated C-C multiple bond (equation 31).

$$
(31) \quad CH_2{=}CHCOOH + H_2N{-}\overset{\displaystyle \overset{Se}{\|}}{C}{-}NH_2 + HCl \longrightarrow
$$

$$
\begin{array}{c}
SeCH_2CH_2COOH \\
| \\
H_2N{-}\overset{+}{C}{=}NH_2 \\
Cl^-
\end{array}
$$

Kataev *et al.*[98–100] have shown that selenourea adds to acrylic acid, acrylonitrile, or propriolic acid in methanol solution saturated with hydrogen chloride, giving high yields of the selenopseudourea hydrochloride. However, when one forms a selenopseudourea from a compound containing a functional group, such as carboxyl, there is the possibility of further reaction resulting in cyclization of the selenopseudourea (*vide infra*).[100]

An example of another class of selenopseudoureas that can undergo further intramolecular reaction is *Se*-(2-ammonioethyl)selenouronium dibromide (**18**).[101,102] When this substance is treated with aqueous base, it undergoes a transguanylation reaction to give 2-guanidinoethaneselenol, whereas heating in water alone results in cyclization, yielding 2-aminoselenazoline hydrobromide (equation 32).[101] Thus, the nature of the product formed depends on the pH of the solution.[102]

(32)

$$
\begin{array}{ccccc}
HNCH_2CH_2SeH & & SeCH_2CH_2NH_3^+ & & \\
| & \overset{OH^{\ominus}}{\longleftarrow} & | & \longrightarrow & \\
H_2N{-}C{=}NH & & H_2N{-}\overset{+}{C}{=}NH_2 \quad 2Br^- & &
\end{array}
$$

18

In a related reaction, ethyl 2-amino-6-selenopurine-9-carboxylate was prepared by treating ethyl 2-amino-6-chloropurine-9-carboxylate with selenourea in ethanol.[103,104]

ii. Cyclization Reactions Involving Selenourea and Its Derivatives. In 1889, Hofmann[22] reported that selenourea reacts with α-chloroketones, giving substituted selenazole compounds (**19**) (equation 33). Using this

$$
\text{(33)} \quad \underset{\text{H}_2\text{N}}{\overset{\overset{\displaystyle \text{Se}}{\|}}{}} \text{—} \overset{\text{Se}}{\underset{}{\text{C}}} \text{—NH}_2 + \text{ClCH}_2\overset{\overset{\displaystyle \text{O}}{\|}}{\text{C}}\text{—R} \longrightarrow
$$

19

method, Backer[60] prepared a series of 2-aminoselenazoles, while Knott[105] made a few selenazoles containing carboxylic acid groups in the 5-position. Extensive use of this cyclization reaction has been made by other workers.[70,106–108]

King and Hlavacek[61] employed the ketone plus iodine in place of the α-haloketone used in the reaction above. Presumably, an α-iodoketone is formed *in situ*.

2-Aminobenzoselenazole derivatives (**20**) have been prepared[67,109] by treating 1-phenyl-2-selenourea derivatives with bromine (equation 34).

$$
\text{(34)} \quad \text{C}_6\text{H}_5\text{NH}\text{—}\overset{\overset{\displaystyle \text{Se}}{\|}}{\text{C}}\text{—NH}_2 + \text{Br}_2 \longrightarrow
$$

20

2-Aminonaphtho[1,2]selenazoles have been prepared similarly.[63,109]

2-Amino-4-selenazolidinone derivatives (**21**) have been prepared by the reaction of selenourea with α-haloacids[22,70,106] or α-haloesters[110,111] (equation 35). Whether **21** exists as the amino or imino form[112,113] has generated

$$
\text{(35)} \quad \text{H}_2\text{N}\text{—}\overset{\overset{\displaystyle \text{Se}}{\|}}{\text{C}}\text{—NH}_2 + \text{R}\text{—}\overset{\overset{\displaystyle \text{X}}{|}}{\text{CH}}\text{COOH} \longrightarrow
$$

21

some controversy.[71,110,111] Giudicelli *et al.*[71] state that the position of the double bond in **21** depends on the substituent attached to the exocyclic nitrogen.

2-Selenohydantoins[114] and 2-selenobarbiturates[115] have also been prepared from selenourea.

The condensation of 2-ethyl-2-selenopseudourea with β-trichloromethylpropiolactone in ethanol gives a 65% yield of 2-imino-4-oxo-6-trichloromethylperhydro-1,3-selenazine (**22**) (equation 36).[116,117]

$$(36) \quad \underset{\underset{\text{H}_2\text{N}-\text{C}=\text{NH}}{|}}{\overset{\text{SeCH}_2\text{CH}_3}{|}} \; + \; \underset{\text{O}}{\overset{\text{Cl}_3\text{C}}{\diagdown}}\!\!\!\!\!\!\!\!\overset{}{\underset{\text{O}}{\square}}\!\!=\!\!\text{O} \quad \longrightarrow \quad$$

22

iii. Oxidation of Selenourea Compounds. Selenourea, in analogy to thiourea, can be oxidized by hydrogen peroxide,[118] ferricyanide,[119] or *p*-benzoquinone[99] to the α,α'-diselenobisformamidinium cation (**23**) (equation 37). Preisler and Scortia[119] studied the redox potential of this readily

$$(37) \quad 2\text{H}_2\text{N}-\overset{\overset{\text{Se}}{\|}}{\text{C}}-\text{NH}_2 + \text{H}_2\text{O}_2 \xrightarrow[\text{C}_2\text{H}_5\text{OH}]{\text{HCl}} \; \text{H}_2\text{N}-\overset{\overset{+}{\overset{\text{NH}_2}{\|}}}{\text{C}}-\text{Se}-\text{Se}-\overset{\overset{+}{\overset{\text{NH}_2}{\|}}}{\text{C}}-\text{NH}_2$$
$$2\text{Cl}^-$$
$$\textbf{23}$$

reversible system in acid solution and reported its value as being $E^0 = +0.240$ V. Infrared and crystallographic data for **23** have also been reported.[118,120]

α,α'-Diselenobisformamidine has also been formed when selenourea complexes of certain metal ions have been made. For example, the reaction of copper(II) chloride with selenourea results in the formation of the copper(I) chloride-selenourea complex and α,α'-diselenobisformamidine dihydrochloride.[121] Silver chloride-selenourea complexes have been reported to decompose, giving **23**.[118]

The oxidation of 1,5-diphenylselenocarbazide (**16**) by air gives 1,5-diphenylselenocarbazone (**24**), which can be oxidized further to **25** (equation 38).[88,89]

$$(38) \quad 2\text{C}_6\text{H}_5\text{NHNH}-\overset{\overset{\text{Se}}{\|}}{\text{C}}-\text{NHNHC}_6\text{H}_5 \xrightarrow{[\text{O}]} 2\text{C}_6\text{H}_5\text{NHNH}\overset{\overset{\text{Se}}{\|}}{\text{C}}-\text{N}=\text{NC}_6\text{H}_5$$
$$\textbf{16} \qquad\qquad [\text{O}] \qquad\qquad \textbf{24}$$

$$\left(\begin{array}{c} \text{C}_6\text{H}_5\text{N}=\text{N}-\overset{}{\text{C}}-\text{Se}- \\ \overset{\|}{} \\ \text{C}_6\text{H}_5\text{NHN} \end{array} \right)_2$$
$$\textbf{25}$$

The imino-substituted selenourea (**15**) can be oxidized by chlorine or sulfuryl chloride as shown in equation 39.[15]

$$
\text{(39)} \quad
\underset{\underset{\displaystyle \text{15}}{\overset{\displaystyle \|}{\underset{\displaystyle \text{Se}}{\|}}}{\overset{\displaystyle \text{NC}_6\text{H}_5}{\text{C}_6\text{H}_5\text{C}}}\!\!-\!\!\text{NH}\!\!-\!\!\text{C}\!\!-\!\!\text{NHC}_6\text{H}_5 + \text{Cl}_2 \longrightarrow
\underset{\overset{\displaystyle |}{\displaystyle \text{SeCl}}}{\overset{\displaystyle \text{NC}_6\text{H}_5}{\overset{\displaystyle \|}{\text{C}_6\text{H}_5\text{C}}}}\!\!-\!\!\text{N}\!\!=\!\!\text{C}\!\!-\!\!\text{Cl}
$$

Oxidation of selenourea by sodium hypochlorite in cold aqueous hydrochloric acid results in the formation of triselenourea dichloride hydrate, $[\text{SeC}(\text{NH}_2)_2]_3\text{Cl}_2\cdot\text{H}_2\text{O}$,[121a] the structure of which has been studied by X-ray crystallography.[122]

iv. Miscellaneous Reactions of Selenourea Compounds.   The decomposition of selenourea and its derivatives by the silver ion and the lead ion has been reported. With silver nitrate many selenoureas give a black precipitate of silver selenide.[71a] The organic product from the treatment of 1,3-diisopropyl-2-selenourea with silver nitrate is 1,3-diisopropylurea.[84] Alkaline lead acetate also removes selenium from selenoureas, giving lead selenide.[65] This reaction has been found to be a convenient means of depositing a photoconductive layer of lead selenide on a surface.[123] The fate of the selenourea molecule in this reaction has been reported by Lundin and Kitaev.[124]

Spica[79] has shown that some selenoureas decompose in hydrochloric acid. For example, the products obtained from 1,1-dibenzyl-2-selenourea and hydrochloric acid are dibenzylamine hydrochloride, hydrogen cyanide, and selenium.

Selenourea, like urea and thiourea, forms inclusion compounds with a number of hydrocarbons,[125] as well as coordination complexes with many metal ions (*cf.* Chapter XVI).

c. Properties and Uses*

i. Analytical Applications of Selenourea Compounds.   Selenourea forms insoluble complexes with copper, bismuth, mercury, silver, tin, antimony, arsenic, palladium, and gold ions.[126] All these complexes are soluble in excess reagent, whereas the precipitate with lead is not. Other ions, such as mercurous, molybdate, and vanadate, are reduced by selenourea. Analytical use of this complex formation has been made by Pilipenko and Sereda[127–129] in their simultaneous colorimetric determination of 2.60 γ/ml of ruthenium and osmium.

---

* The biological and spectroscopic properties of selenoureas are discussed in Chapters XIIID and XV, respectively.

Acyl selenoureas have also found analytical applications.[130] 1-Acetyl-3-phenyl-2-selenourea[131] and 1-benzoyl-3-phenyl-2-selenourea[132] can be used for the colorimetric determination of palladium. Photometric determination of osmium[133] and palladium[77] is also possible using 1-benzoyl-3-carbo-ethoxymethyl-2-selenourea as the complexing agent.

ii. Miscellaneous Uses of Selenourea Compounds. Selenourea has served as the selenium source in the preparation of fine films of lead selenide that are uniform in appearance and electrical resistance.[134,135] The mechanism and kinetics of this film formation have been studied in detail.[124,136]

Organic compounds that readily deposit selenium have found application as photographic sensitizers. Selenoureas are one class of organic selenium compounds that have been used for this objective.[53,137,138] Selenourea compounds have also been employed in photographic bleach-fixing solutions.[139] The use of selenourea in the separation of organic hydrocarbons, based on the formation of crystallized complexes, has been reported.[140,141] Its application as a vulcanization accelerator has been suggested.[142]

5. Selenosemicarbazides and Derivatives

a. Preparation

i. Reaction of Isoselenocyanates with Hydrazines. Selenosemicarbazides (**26**) are generally stable, crystalline solids that are easily formed, in high yield, by the reaction of an isoselenocyanate with a hydrazine (equation 40).[13,71a,143,144] Jensen and Frederiksen[8] reported the formation of the first

$$(40) \quad R^1N{=}C{=}Se + R^2R^3N{-}NR^4H \longrightarrow R^1NH{-}\overset{\overset{\displaystyle Se}{\displaystyle \|}}{C}{-}NR^4NR^3R^2$$

$$\textbf{26}$$

selenosemicarbazide, 4-phenylselenosemicarbazide (**26**, $R^1 = C_6H_5$, $R^2 = R^3 = R^4 = H$), by the treatment of phenyl isoselenocyanate with hydrazine hydrate. Later, Huls and Renson[9] prepared 4-ethylselenosemicarbazide by this method. Collard-Charon *et al.*[143] treated an ethanol solution of the appropriate isoselenocyanate with 1 equivalent of hydrazine in ethanol to give the same 4-substituted selenosemicarbazide in about 50 % yield. Other 4-substituted selenosemicarbazide compounds have been prepared by Pedersen[10] and by Bulka *et al.*[13,145]

Substitution at other than the 4-position is possible if one starts with a substituted hydrazine. Phenyl isoselenocyanate and phenylhydrazine gave a 65% yield of 1,4-diphenylselenosemicarbazide, whereas butyl isoseleno-cyanate reacted with methylhydrazine to give 2-methyl-4-butylselenosemi-carbazide in 30% yield.[143] The method of determining the position of the substituents in the final selenosemicarbazide has been reported.[143,144] In general, if a monoaryl hydrazine is used, the product is 1,4-substituted, whereas monoalkyl hydrazines give 2,4-substitution.[143] Eleven of the 17 possible methyl-substituted selenosemicarbazides have been prepared by Jensen *et al.*[144] The 4-unsubstituted derivatives were prepared by removing a *tert*-butyl protective group from the 4-position by refluxing in concentrated hydrochloric acid.

Attempts to prepare selenosemicarbazide derivatives by rearrangement of hydrazine selenocyanate (equation 41) were not successful unless an

$$(41) \qquad \text{H}_2\text{NNH}_2\cdot\text{HCNSe} \;\;\overset{}{-\!\!/\!\!\!\rightarrow}\;\; \text{H}_2\text{NNH}\overset{\displaystyle\text{Se}}{\underset{\displaystyle\|}{\text{--C--}}}\text{NH}_2$$

aldehyde or ketone was added.[146–148] The products then obtained were selenosemicarbazone derivatives (*vide infra*).

The preparation of selenosemicarbazide compounds can also be accomplished by treating isocyanides with elemental selenium and hydrazines (equation 42).[71a]

$$(42) \qquad \text{R--N}{\equiv}\text{C} + \text{Se} + \text{H}_2\text{NNHR}' \longrightarrow \text{RNH}\overset{\displaystyle\text{Se}}{\underset{\displaystyle\|}{\text{--C--}}}\text{NHNHR}'$$

ii. Reaction of Hydrogen Selenide with a Diazonium Cyanide or Cyano-hydrazine. Aromatic diazonium cyanides (**27**) react with hydrogen selenide to give selenosemicarbazide compounds in low yield (equation 43).[149]

$$(43) \qquad \underset{\textbf{27}}{\text{ArN}{=}\text{N}^+\text{C}{\equiv}\text{N}^-} + 2\text{H}_2\text{Se} \longrightarrow \text{ArNHNH}\overset{\displaystyle\text{Se}}{\underset{\displaystyle\|}{\text{--C--}}}\text{NH}_2 + \text{Se}$$

Hydrogen selenide also adds to cyanohydrazines (**28**), giving selenosemi-carbazides (equation 44).[149,150]

$$(44) \qquad \underset{\textbf{28}}{\underset{\displaystyle\overset{\displaystyle|}{\text{C}{\equiv}\text{N}}}{\text{C}_6\text{H}_5\text{N--NH}_2}} + \text{H}_2\text{Se} \longrightarrow \text{H}_2\text{N}\overset{\displaystyle\text{Se}}{\underset{\displaystyle\|}{\text{--C--}}}\underset{\displaystyle\overset{\displaystyle|}{\text{C}_6\text{H}_5}}{\text{N--NH}_2}$$

b. Reactions

i. Selenosemicarbazone Formation. In analogy with semicarbazide, selenosemicarbazides react with aldehydes and ketones to give selenosemicarbazone derivatives (**29**) (equation 45).[9,13,143,145,150,151] The experimental conditions reported by Bednarz[152] include treatment of the selenosemicarbazide with the carbonyl compound in absolute ethanol containing a trace of acetic acid. Selenocarbonohydrazine can condense with two equivalents of an aldehyde or ketone in dilute sulfuric acid solution.[153]

$$(45) \quad RNH{-}\overset{\overset{Se}{\|}}{C}{-}NRNH_2 + R'{-}\overset{\overset{O}{\|}}{C}{-}R'' \longrightarrow RNH{-}\overset{\overset{Se}{\|}}{C}{-}NRN{=}C\overset{\displaystyle R'}{\underset{\displaystyle R''}{\big\langle}}$$

$$\textbf{29}$$

Another route to selenosemicarbazone derivatives is by an exchange reaction with acetone selenosemicarbazone (**30**) (equation 46).[9] This

$$(46) \quad H_2N{-}\overset{\overset{Se}{\|}}{C}{-}NHN{=}C\overset{\displaystyle CH_3}{\underset{\displaystyle CH_3}{\big\langle}} + R{-}\overset{\overset{O}{\|}}{C}{-}H \longrightarrow$$

$$\textbf{30}$$

$$H_2N{-}\overset{\overset{Se}{\|}}{C}{-}NHN{=}C\overset{\displaystyle R}{\underset{\displaystyle H}{\big\langle}} + CH_3\overset{\overset{O}{\|}}{C}CH_3$$

method is successful only when aldehydes are used in the displacement reaction.[9,154,155]

Another method for preparing selenosemicarbazones is the isomerization of hydrazine selenocyanate in the presence of an aldehyde or ketone (equation 47).[146–148]

$$(47) \quad H_2NNH_2{\cdot}HSeCN + CH_3\overset{\overset{O}{\|}}{C}CH_3 \longrightarrow H_2N{-}\overset{\overset{Se}{\|}}{C}{-}NHN{=}C\overset{\displaystyle CH_3}{\underset{\displaystyle CH_3}{\big\langle}}$$

ii. Cyclization. α-Diketones can condense with selenosemicarbazide compounds to yield a substituted 1,2,4-triazine, as shown by equation

48.[152] Reaction of a selenosemicarbazide with an α-halo ketone results in the

(48)

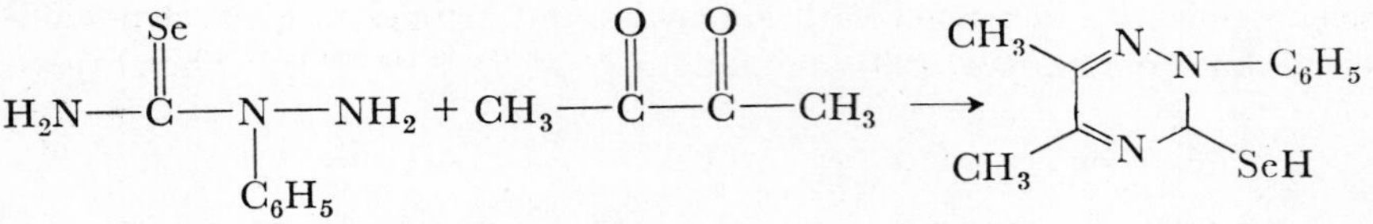

$$H_2N-\underset{\underset{C_6H_5}{|}}{\overset{\overset{Se}{\|}}{C}}-N-NH_2 + CH_3-\overset{\overset{O}{\|}}{C}-\overset{\overset{O}{\|}}{C}-CH_3 \longrightarrow$$

2-amino-1,3,4-selenodiazine ring system (31), as shown by reaction 49.[156]

(49)

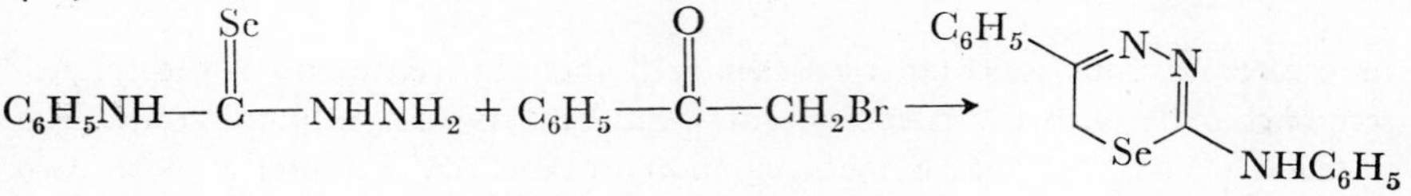

$$C_6H_5NH-\overset{\overset{Se}{\|}}{C}-NHNH_2 + C_6H_5-\overset{\overset{O}{\|}}{C}-CH_2Br \longrightarrow$$

31

With β-keto esters, selenosemicarbazide compounds give the intermediate selenosemicarbazone, which can cyclize to yield a 1-selenocarbamoyl-pyrazol-5-one compound (32) (equation 50).[151]

(50)

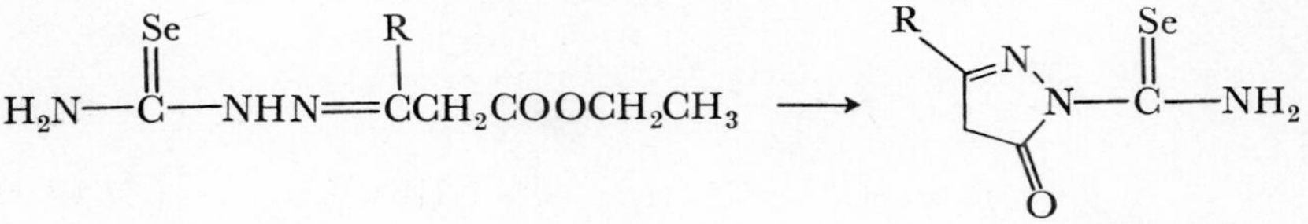

$$H_2N-\overset{\overset{Se}{\|}}{C}-NHN=\overset{\overset{R}{|}}{C}CH_2COOCH_2CH_3 \longrightarrow$$

32

Finally, the reaction of a selenosemicarbazone with an α-halo acid is reported[113] to give a 2-iminoselenazolidin-4-one (33) (equation 51).

(51)

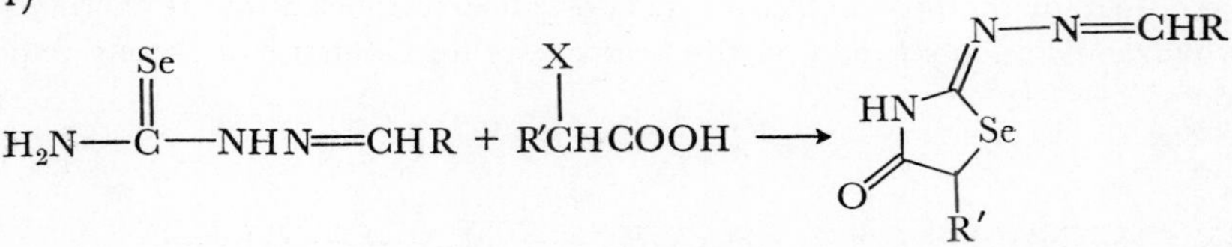

$$H_2N-\overset{\overset{Se}{\|}}{C}-NHN=CHR + R'\overset{\overset{X}{|}}{C}HCOOH \longrightarrow$$

33

The condensation of 1,1,4-trimethylselenocarbonohydrazide (34) with carbon disulfide in the presence of pyridine results in the formation of 4-dimethylamino-1-methyl-3-thio-1,2,4-triazol-5-thione (35) (equation 52) by an unusual selenium-sulfur interchange reaction.[157]

$$(52) \quad (CH_3)_2NNH—\overset{\overset{\displaystyle Se}{\|}}{C}—\underset{\underset{\displaystyle CH_3}{|}}{N}NH_2 + CS_2 \longrightarrow$$

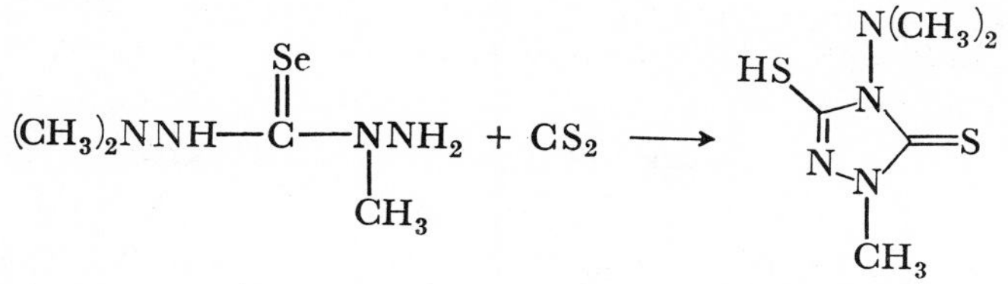

34   35

iii. Isoselenosemicarbazide Formation. Treatment of a selenosemicarbazide[143] or a selenosemicarbazone[155] with an alkyl halide or alkyl sulfate can result in isoselenosemicarbazide formation (equation 53).

$$(53) \quad C_4H_9NH—\overset{\overset{\displaystyle Se}{\|}}{C}—\underset{\underset{\displaystyle CH_3}{|}}{N}N(CH_3)_2 + (CH_3)_2SO_4 \longrightarrow$$

$$C_4H_9\overset{H}{\underset{+}{N}}{=}\overset{\overset{\displaystyle SeCH_3}{|}}{C}—\underset{\underset{\displaystyle CH_3}{|}}{N}N(CH_3)_2 \; CH_3SO_4^-$$

iv. Miscellaneous Reactions of Selenosemicarbazide. Bednarz[152] treated 2-phenylselenosemicarbazide with several aryl isothiocyanates and obtained 1-(arylaminothioformyl)-2-phenylselenosemicarbazide derivatives (**36**) (equation 54). In a similar reaction, Bulka *et al.*[145] treated 4-*p*-tolyl-

$$(54) \quad H_2N—\overset{\overset{\displaystyle Se}{\|}}{C}—\underset{\underset{\displaystyle C_6H_5}{|}}{N}NH_2 + ArN{=}C{=}S \longrightarrow ArNH—\overset{\overset{\displaystyle S}{\|}}{C}—\underset{\underset{\displaystyle C_6H_5}{|}}{N}HN—\overset{\overset{\displaystyle Se}{\|}}{C}—NH_2$$

36

selenosemicarbazide with *p*-tolyl isoselenocyanate and obtained **37** (equation 55). Selenocarbonohydrazine reacts with thiocyanic acid to

$$(55) \quad p\text{-}CH_3C_6H_4NH—\overset{\overset{\displaystyle Se}{\|}}{C}—NHNH_2 + p\text{-}CH_3C_6H_4N{=}C{=}Se \longrightarrow$$

$$(p\text{-}CH_3C_6H_4NH\overset{\overset{\displaystyle Se}{\|}}{C}NH—)_2$$

37

give 1-thiocarbamoylselenocarbonohydrazide (equation 56).[158,159] Treatment of this product with potassium thiocyanate gives 1,5-dithiocarbamoylseleno-carbonohydrazide.[159]

$$(56) \qquad \underset{\displaystyle \overset{\displaystyle \text{Se}}{\|}}{\text{H}_2\text{NNH}-\text{C}-\text{NHNH}_2} + \text{HSCN} \longrightarrow \underset{\displaystyle \overset{\displaystyle \text{S}}{\|}}{\text{H}_2\text{N}-\text{C}-\text{NHNH}}-\underset{\displaystyle \overset{\displaystyle \text{Se}}{\|}}{\text{C}-\text{NHNH}_2}$$

A method[160] has been reported for the rapid quantitative analysis of selenium in selenosemicarbazones by its precipitation as silver selenide. Selenosemicarbazide derivatives form complexes with many metal ions (*cf*. Chapter XVI).

c. Properties and Uses.* Dipole moment data have been given by Mautner and Kumler.[150] The acid-base properties of salicylaldehyde selenosemicarbazone have been determined.[161]

6. Selenocarbamates

a. Preparation

i. Diselenocarbamates. Diselenocarbamates, prepared by the reaction of carbon diselenide with amines (equation 57), are stable as salts (**38**)

$$(57) \qquad \text{CSe}_2 + 2\text{R}_2\text{NH} \longrightarrow \text{R}_2\text{N}-\text{C} \overset{\displaystyle \text{Se}}{\underset{\displaystyle \text{Se}^-}{\big\langle}} \quad \text{R}_2\text{NH}_2{}^+$$

38

but diselenocarbamic acids are unstable. The parent compound, ammonium diselenocarbamate, was prepared by Grimm and Metzger[81] from ammonia and carbon diselenide in alcohol. One disadvantage of this preparation is the tendency of carbon diselenide to polymerize. Barnard and Woodbridge[162] minimized this problem by using a dilute solution of carbon diselenide in dioxane. By this method, they prepared a series of *N,N*-dialkyl diselenocarbamates, isolated as sodium salts, from secondary amines. *N,N*-Diaryl diselenocarbamates can be prepared starting with secondary aryl amines.[163] It was reported later, however, that primary amines did not give the desired alkyl diselenocarbamate;[83] instead, selenourea formation

---

* The biological and spectroscopic properties of selenosemicarbazides are discussed in Chapters XIIID and XV, respectively.

was observed. Indeed, Warner[82] postulated that diselenocarbamates were intermediates in his selenourea preparation from carbon diselenide and primary amines. This preparation of diselenocarbamates has been utilized by other workers.[164–167]

Dimethyldiselenocarbazic acid (**39**), prepared from dimethylhydrazine and carbon diselenide,[168] was determined to be zwitterionic by i.r. spectroscopy.[168,169] Hydrazine reacted with carbon diselenide to give the un-

$$(58) \qquad (CH_3)_2NNH_2 + CSe_2 \longrightarrow (CH_3)_2\overset{+}{N}HNHC\overset{\displaystyle\nearrow^{Se}}{\searrow_{Se^-}}$$

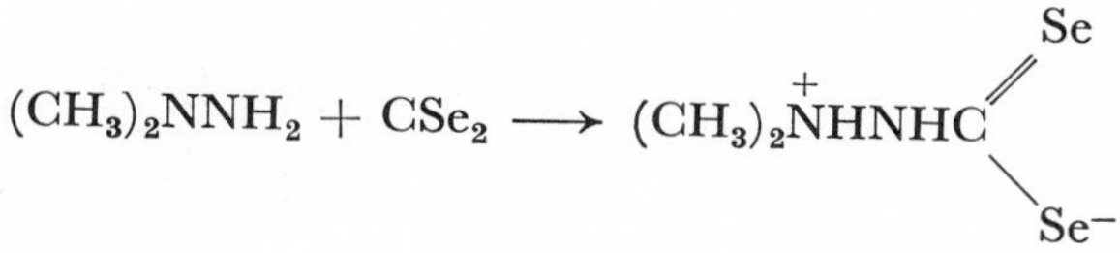

39

substituted diselenocarbazic acid, isolated as the hydrazinium salt.[170] Treatment of this salt with DCl gave the completely deuterated inner salt, **40**.

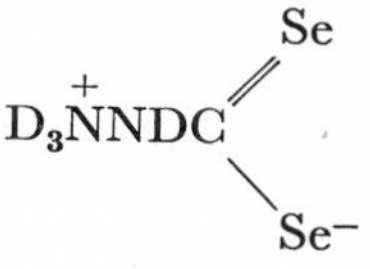

$$D_3\overset{+}{N}NDC\overset{\displaystyle\nearrow^{Se}}{\searrow_{Se^-}}$$

40

The hydrazinium salt of 2-methyldiselenocarbazic acid reacted with acetone, resulting in the formation of a cyclized product, 3,5,5-trimethyl-1,3,4-selenadiazolidine-2-selone. The preparation of diselenocarbazic acid hydrazine salt in essentially quantitative yield was also reported by Feher *et al.*[153,172] However, when these authors[153] allowed hydrazine and carbon diselenide to react in the presence of air for 1 day, the product was selenocarbonohydrazide.

ii. Selenocarbamates Containing Oxygen or Sulfur. Selenocarbamates and selenothiocarbamates were first prepared only recently. Their functional groups can also be part of various heterocyclic systems (*cf.* Chapter XI).[82,109,173–175]

The synthesis of selenothiocarbamic esters is possible by displacement of the amino group of thiopseudourea derivatives by hydroselenide ion at pH 5–6 (equation 59).[55] In more alkaline solution, the selenourea is the major product.

$$(59) \quad \begin{array}{c} \text{CH}_3 \\ R^1 \quad | \quad R^3 \\ \diagdown \quad S \quad \diagup \\ N—C{=}N \\ \diagup \quad + \quad \diagdown \\ R^2 \qquad R^4 \end{array} \xrightarrow[\text{pH 5-6}]{\text{HSe}^-} \left[\begin{array}{c} \text{CH}_3 \\ R^1 \quad | \quad R^3 \\ \diagdown \quad S \quad \diagup \\ N—C{\rightleftharpoons}N \\ \diagup \quad | \quad \diagdown \\ R^2 \quad Se \quad R^4 \\ | \\ H \end{array} \right] \longrightarrow$$

$$\begin{array}{cc} R^1 \quad Se & R^3 \\ \diagdown \quad || & \diagup \\ N—C—SCH_3 + HN \\ \diagup & \diagdown \\ R^2 & R^4 \end{array}$$

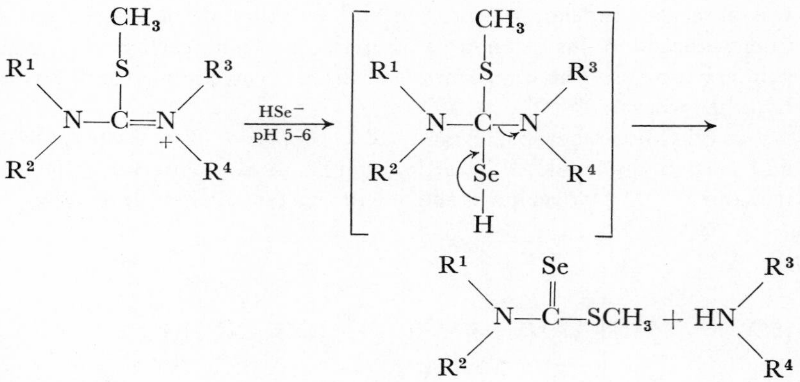

Selenocarbamates have been made by a displacement reaction of seleno-xanthogenacetic acid by ammonia[176,177] (equation 60) or by the addition of hydrogen selenide to alkyl cyanates (equation 61).[178,179]

$$(60) \quad CH_3CH_2O—\overset{\overset{\text{Se}}{||}}{C}—SeCH_2COOH + NH_3 \longrightarrow CH_3CH_2O—\overset{\overset{\text{Se}}{||}}{C}—NH_2$$

$$(61) \quad RO—C{\equiv}N + H_2Se \longrightarrow RO—C\diagup^{\overset{\text{Se}}{}}_{\diagdown NH_2}$$

Diisopropylselenothiocarbazic acid (**41**) was made by the addition of hydrogen selenide to N-isothiocyanatodiisopropylamine (equation 62).[180]

$$(62)$$

$$[(CH_3)_2CH]_2N—N{=}C{=}S + H_2Se \longrightarrow [(CH_3)_2CH]_2\overset{+}{N}HNH—C\diagup^{\overset{\text{S}}{}}_{\diagdown Se^-}$$

41

A series of alkoxyselenocarbonylhydrazine compounds (**42**) has been made by the reaction of hydrazine derivatives with [(alkoxyselenocar-bonyl)seleno]acetic acid derivatives (reaction 63).[177]

$$(63) \quad RO—\overset{\overset{\text{Se}}{||}}{C}—SeCH_2COOH + R_2'NNHR'' \longrightarrow RO—\overset{}{C}—NR''NR_2' \\ \qquad\qquad\qquad\qquad\qquad\qquad\qquad\qquad\qquad || \\ \qquad\qquad\qquad\qquad\qquad\qquad\qquad\qquad\quad Se$$

42

b. Reactions

i. Oxidation.

Barnard and Woodbridge[162] reported that dialkyl diseleno-carbamates can be oxidized by hydrogen peroxide, potassium ferricyanide, or oxygen to give a mixture of bis(N,N-dialkylselenocarbamoyl) mono-selenide (**43**) and bis(N,N-dialkylselenocarbamoyl) triselenide (**44**) (equation 64). This behavior has also been noted by other workers.[83,181] Oxidation

$$
(64) \quad R_2N-C\underset{Se^-}{\overset{Se}{\diagup}} + [O] \longrightarrow
$$

$$
\underset{\textbf{43}}{R_2N-\overset{Se}{\overset{\|}{C}}-Se-\overset{Se}{\overset{\|}{C}}-NR_2} + \underset{\textbf{44}}{R_2N-\overset{Se}{\overset{\|}{C}}-SeSeSe-\overset{Se}{\overset{\|}{C}}-NR_2}
$$

of O-ethyl selenocarbamate with mercuric oxide gave ethyl isocyanate, hydrogen selenide, and water,[176] while air oxidation of diisopropyl-selenothiocarbazic acid (**41**) produced bis(N,N-diisopropylthiocarbazoyl) diselenide.[46,180] Oxidation of 1,1-dimethylhydrazinium 3,3-dimethyl diselenocarbazate with iodine resulted in the formation of bis(3,3-dimethyl-selenocarbazoyl) triselenide.[169]

ii. Selenium Alkylation.

Sodium diethyl diselenocarbamate reacts with α-chloroacetic acid, giving the carboxymethyl ester (**45**) in high yield (equation 65).[167]

$$
(65) \quad (CH_3CH_2)_2N-C\underset{Se^-Na^+}{\overset{Se}{\diagup}} + ClCH_2COOH \longrightarrow
$$

$$
\underset{\textbf{45}}{(CH_3CH_2)_2N-C\underset{SeCH_2COOH}{\overset{Se}{\diagup}}}
$$

c. Properties and Uses.

Sodium dimethyl diselenocarbamate was investigated as a potential analytical reagent for metal ions inasmuch as many metal ions form water-insoluble dimethyl diselenocarbamate complexes (cf. Chapter XVI).[181–183]

REFERENCES

1. D. Ruffilli, *Studi Sassaresi*, **26,** 98 (1948).
2. T. W. Campbell, H. G. Walker, and G. M. Coppinger, *Chem. Rev.*, **50,** 279 (1952).
3. M. Dräger and G. Gattow, *Angew. Chem.*, **80,** 954 (1968).
4. M. O. Lozinskii and P. S. Pel'kis, *Usp. Khim.*, **37,** 840 (1968); *Russ. Chem. Rev.*, **37,** 363 (1968).
5. H. Rheinboldt, "Houben-Weyl Methoden der Organischen Chemie," Vol. IX, E. Müller, Ed., Georg Thieme Verlag, Stuttgart, 1955.
6. K. A. Jensen, *Quart. Rept. Sulfur Chem.*, **5,** 45 (1970).
7. C. Collard-Charon and M. Renson, *Bull. Soc. Chim. Belges*, **71,** 531 (1962).
8. K. A. Jensen and E. Frederiksen, *Z. Anorg. Allg. Chem.*, **230,** 31 (1936).
9. R. Huls and M. Renson, *Bull. Soc. Chim. Belges*, **65,** 684 (1956).
10. C. T. Pedersen, *Acta Chem. Scand.*, **17,** 1459 (1963).
11. W. J. Franklin and R. L. Werner, *Tetrahedron Lett.*, 3003 (1965).
12. E. Bulka, K.-D. Ahlers, and E. Tuček, *Chem. Ber.*, **100,** 1367 (1967).
13. E. Bulka and K.-D. Ahlers, *Z. Chem.*, **3,** 348 (1963).
14. I. B. Douglass, *J. Amer. Chem. Soc.*, **59,** 740 (1937).
15. N. A. Kirsanova and G. I. Derkach, *Ukr. Khim. Zh.*, **36,** 372 (1970); *Chem. Abstr.*, **73,** 45417q (1970).
16. E. A. V. Ebsworth and M. J. Mays, *J. Chem. Soc.*, 3893 (1963).
17. J. S. Thayer, *J. Organometal. Chem.*, **9,** P30 (1967).
18. H. Bürger and U. Goetze, *J. Organometal. Chem.*, **10,** 380 (1967).
19. T. Gabrio and G. Barnikow, *Z. Chem.*, **9,** 183 (1969).
20. H. Stolte, *Ber.*, **19,** 2350 (1886).
21. Y. Tarantelli and D. Leonesi, *Ann. Chim.* (Rome), **53,** 1113 (1963).
22. G. Hofmann, *Justus Liebigs Ann. Chem.*, **250,** 294 (1889).
23. T. Tarantelli and C. Pecile, *Ann. Chim.* (Rome), **52,** 75 (1962).
24. M. A. A. Beg and S. K. Hasan, *Pakistan J. Sci. Ind. Res.*, **7,** 220 (1964).
25. M. O. Lozinskii and P. S. Pel'kis, *Zh. Org. Khim.*, **1,** 1415 (1965).
26. E. E. Aynsley, N. N. Greenwood, G. Hunter, and M. J. Sprague, *J. Chem. Soc.*, *A*, 1344 (1966).
27. J. S. Thayer and R. West, *Advan. Organometal. Chem.*, **5,** 169 (1967).
28. A. Fava, "Organic Sulfur Compounds," Vol. 2, N. Kharasch and C. Y. Meyers, Eds., Pergamon Press, New York, N.Y., 1966, p. 73.
29. C. Collard-Charon and M. Renson, *Bull. Soc. Chim. Belges*, **72,** 304 (1963).
30. F. von Dechend, *Ber.*, **7,** 1273 (1874).
31. W. Becker and J. Meyer, *Ber.*, **37,** 2550 (1904).
32. P. L. Compagnon and M. Miocque, *Ann. Chim.* (Paris), **5,** 11 (1970).
33. K. Kindler, *Justus Liebigs Ann. Chem.*, **431,** 187 (1923).
34. K. A. Jensen and P. H. Nielsen, *Acta Chem. Scand.*, **20,** 597 (1966).
35. F. Asinger, H. Berding, and H. Offermanns, *Monatsh. Chem.*, **99,** 2072 (1968).

36. Belg. Pat. 444,545 (1942); *Chem. Abstr.*, **39**, 528 (1945).
37. H. E. Hallam and C. M. Jones, *J. Chem. Soc.*, *A*, 1033 (1969).
38. A. Michaelis and M. Stein, *Justus Liebigs Ann. Chem.*, **320**, 32 (1902).
39. A. Michaelis and A. Hölken, *Justus Liebigs Ann. Chem.*, **331**, 245 (1904).
40. H. G. Mautner, S.-H. Chu, and C. M. Lee, *J. Org. Chem.*, **27**, 3671 (1962).
41. F. Bergmann and M. Rashi, *Israel J. Chem.*, **7**, 63 (1969).
42. N. N. Yarovenko and M. A. Raksha, *J. Gen. Chem. USSR*, **29**, 2125 (1959).
42a. H. Hartmann, *Z. Chem.*, **11**, 60 (1971).
43. L. G. S. Brooker, G. H. Keyes, and F. L. White, *J. Amer. Chem. Soc.*, **57**, 2492 (1935).
44. F. L. White, Brit. Pat. 392,410 (1933); *Chem. Abstr.*, **27**, 5550 (1933). U.S. Pat. 1,957,870 (1934); *Chem. Abstr.*, **28**, 4074 (1934).
45. K. A. Jensen and U. Henriksen, *Acta Chem. Scand.*, **21**, 1991 (1967).
46. U. Anthoni, C. Larsen, and P. H. Nielsen, *Acta Chem. Scand.*, **21**, 2580 (1967).
47. P. Hampson and A. Mathias, *Mol. Phys.*, **13**, 361 (1967).
48. G. Schwenker and H. Rosswag, *Tetrahedron Lett.*, 4237 (1967).
49. H. Lumbroso, C. Pigenet, H. Rosswag, and G. Schwenker, *C.R. Acad. Sci., Paris, Ser. C.*, **266**, 1479 (1968).
50. K. A. Jensen and J. Sandström, *Acta Chem. Scand.*, **23**, 1911 (1969).
51. A. Azman, M. Drofenik, D. Hadzi, and B. Lukman, *J. Mol. Struct.*, **1**, 181 (1968).
52. H. E. Hallam and C. M. Jones, *J. Mol. Struct.*, **5**, 1 (1970).
53. Neth. Pat. Appl. 6,501,562 (1965); *Chem. Abstr.*, **63**, 17369c (1965).
54. French Pat. 1,499,808 (1967); *Chem. Abstr.*, **69**, 82317v (1968).
55. D. L. Klayman and R. J. Shine, *J. Org. Chem.*, **34**, 3549 (1969).
55a. L. W. Reeves, R. C. Shaddick, and K. N. Shaw, *J. Phys. Chem.*, **75**, 3372 (1971).
56. A. Verneuil, *Bull. Soc. Chim. Fr.*, **41**, 599 (1884).
57. R. E. Dunbar and E. P. Painter, *J. Amer. Chem. Soc.*, **69**, 1833 (1947).
58. H. Hope, *Acta Chem. Scand.*, **18**, 1800 (1964).
59. R. Badiello and A. Breccia, *Ric. Sci.*, **36**, 335 (1966).
60. H. J. Backer, *Rec. Trav. Chim. Pays-Bas*, **62**, 580 (1943).
61. C. King and R. J. Hlavacek, *J. Amer. Chem. Soc.*, **73**, 1864 (1951).
62. H. Stolte, *Ber.*, **19**, 1577 (1886).
63. H. W. J. Cressman, Brit. Pat. 610,566 (1948); *Chem. Abstr.*, **43**, 7848e (1949).
64. Ger. Pat. 305,262 (1918); *Chem. Abstr.*, **13**, 325 (1919). Ger. Pat. 205,263 (1918); *Chem. Abstr.*, **13**, 325 (1919).
65. H. Schmidt, *Ber.*, **54B**, 2067 (1921).
66. W. Geisel, Ger. Pat. 607,382 (1934); *Chem. Abstr.*, **29**, 1836 (1935).
67. J. Haginiwa, *J. Pharm. Soc. Japan*, **69**, 565 (1949).
68. F. Bennett and R. Zingaro, "Organic Syntheses," Coll. Vol. 4, N. Rabjohn, Ed., John Wiley, New York, N.Y., 1963, p. 359.
69. F. Zetzsche and H. Pinske, *Ber.*, **74B**, 1022 (1941).
70. R. A. Zingaro, F. C. Bennett, Jr., and G. W. Hammar, *J. Org. Chem.*, **18**, 292 (1953).
71. J. F. Giudicelli, J. Menin, and H. Najer, *Bull. Soc. Chim. Fr.*, 1099 (1968).
71a. M. Lipp, F. Dallacker, and I. Meier zu Köchen, *Monatsh. Chem.*, **90**, 41 (1959).

72. C. Collard-Charon and M. Renson, *Bull. Soc. Chim. Belges*, **72**, 149 (1963).

73. E. Bulka and K.-D. Ahlers, *Z. Chem.*, **3**, 387 (1963).

74. E. Bulka, K.-D. Ahlers, and E. Tuček, *Chem. Ber.*, **100**, 1459 (1967).

75. E. Bulka, D. Ehlers, and E. Storm, *Z. Chem.*, **10**, 403 (1970).

76. T. Tarantelli and C. Furlani, *J. Chem. Soc.*, *A*, 1717 (1968).

77. D. P. Bendito, J. Casillas, and F. Pino, *Inform. Quim. Anal.* (Madrid), **20**, 69 (1966).

78. H. Schiff, *Ber.*, **10**, 888 (1877).

79. P. Spica, *Gazz. Chim. Ital.*, **7**, 90 (1877).

80. M. Giua and R. Bianco, *Gazz. Chim. Ital.*, **89**, 693 (1959).

81. H. G. Grimm and H. Metzger, *Ber.*, **69**, 1356 (1936).

82. J. S. Warner, *J. Org. Chem.*, **28**, 1642 (1963).

83. A. Rosenbaum, H. Kirchberg, and E. Leibnitz, *J. Prakt. Chem.*, **19**, 1 (1963).

84. A. Rosenbaum, *J. Prakt. Chem.*, **37**, 200 (1968).

85. A. Rosenbaum, *Freiberger Forschungsh.*, *A*, **328**, 59 (1964).

86. L. Henriksen, *Acta Chem. Scand.*, **21**, 1981 (1967).

87. D. L. Klayman and R. J. Shine, *Chem. Commun.*, 372 (1968).

87a. D. L. Klayman and R. J. Shine, U.S. Pat. 3,597,444 (1971).

88. R. S. Ramakrishna and H. M. N. H. Irving, *Anal. Chim. Acta*, **48**, 251 (1969).

89. R. S. Ramakrishna and H. M. N. H. Irving, *Chem. Ind.* (London), 325 (1969).

90. K. A. Jensen, G. Felbert, and B. Kagi, *Acta Chem. Scand.*, **20**, 281 (1966).

91. A. Senning and O. N. Sørensen, *Acta Chem. Scand.*, **20**, 1445 (1966).

92. J. S. Warner and T. F. Page, Jr., *J. Org. Chem.*, **31**, 606 (1966).

93. G. Wagner and P. Nuhn, *Z. Chem.*, **3**, 64 (1963).

94. G. Wagner and P. Nuhn, *Arch. Pharm.*, **297**, 461 (1964).

95. A. Segaloff and R. B. Gabbard, U.S. Pat. 3,372,173 (1968); *Chem. Abstr.*, **69**, 77632q (1968).

96. A. Segaloff and R. B. Gabbard, *Steroids*, **5**, 219 (1965).

97. S. M. Hiscock, D. A. Swann, and J. H. Turnbull, *Chem. Commun.*, 1310 (1970).

98. E. G. Kataev and L. K. Barinova, *Dokl. Akad. Nauk SSSR*, **141**, 1373 (1961); *Proc. Acad. Sci. USSR*, **141**, 1293 (1961).

99. E. G. Kataev and L. K. Konovalova, *Zh. Org. Khim.*, **3**, 949 (1967); *J. Org. Chem. USSR*, **3**, 912 (1967).

100. E. G. Kataev, L. K. Konovalova, and E. G. Yarkova, *Zh. Org. Khim.*, **5**, 621 (1969).

101. S.-H. Chu and H. G. Mautner, *J. Org. Chem.*, **27**, 2899 (1962).

102. L. V. Pavlova and F. Yu. Rachinskii, *Zh. Obshch. Khim.*, **35**, 493 (1965).

103. H. G. Mautner, *J. Amer. Chem. Soc.*, **78**, 5292 (1956).

104. E. Dyer and C. E. Minnier, *J. Med. Chem.*, **11**, 1232 (1968).

105. E. B. Knott, *J. Chem. Soc.*, 628 (1945).

106. A. N. Roy and P. C. Guha, *J. Ind. Chem. Soc.*, **22**, 82 (1945).

107. E. Bulka and K.-D. Ahlers, *Z. Chem.*, **3**, 388 (1963).

108. E. Bulka, K.-D. Ahlers, H. Ewert, W. Haase, P. Meier, and W. Pilz, *Chem. Ber.*, **101**, 1910 (1968).

109. C. Hasan and R. F. Hunter, *J. Chem. Soc.*, 1762 (1935).

110. J. F. Giudicelli, J. Menin, and H. Najer, *C.R. Acad. Sci., Paris, Ser. C,* **262,** 285 (1966).

111. J. Menin, J. F. Giudicelli, and H. Najer, *C.R. Acad. Sci., Paris, Ser. C,* **262,** 788 (1966).

112. A. M. Comrie, D. Dingwall, and J. B. Stenlake, *J. Chem. Soc.,* 5713 (1963).

113. A. M. Comrie, D. Dingwall, and J. B. Stenlake, *J. Pharm. Pharmacol.,* **16,** 268 (1964).

114. P. Bergmann, F. Pragst, and H. Paul, *Arch. Pharm.,* **299,** 499 (1966).

115. H. G. Mautner and E. M. Clayton, *J. Amer. Chem. Soc.,* **81,** 6270 (1959).

116. F. I. Luknitskii, D. O. Taube, and B. A. Vovsi, *Dokl. Akad. Nauk SSSR,* **184,** 355 (1969).

117. F. I. Luknitskii, D. O. Taube, and B. A. Vovsi, *Zh. Org. Khim.,* **5,** 1844 (1969); *J. Org. Chem. USSR,* **5,** 1790 (1969).

118. A. Chiesi, G. Grossoni, M. Nardelli, and M. E. Vidoni, *Chem. Commun.,* 404 (1969).

119. P. W. Preisler and T. N. Scortia, *J. Amer. Chem. Soc.,* **80,** 2309 (1958).

120. A. Chiesi Villa, M. Nardelli, and M. E. Vidoni Tani, *Acta Cryst.,* **26B,** 1504 (1970).

121. V. L. Varand, V. M. Shul'man, and E. V. Khlystunova, *Izv. Akad. Nauk SSSR, Ser. Khim.,* 450 (1970).

121a. A. Verneuil, *Ann. Chim. Phys.,* **9,** 289 (1886).

122. S. Hauge, D. Opedal, and J. Aarskog, *Acta Chem. Scand.,* **24,** 1107 (1970).

123. H. E. Spencer and J. V. Morgan, U.S. Pat. 3,121,023 (1964); *Chem. Abstr.,* **60,** 15298h (1964).

124. A. B. Lundin and G. A. Kitaev, *Izv. Akad. Nauk SSSR, Neorgan. Materialy,* **1,** 2107 (1965); *Chem. Abstr.,* **64,** 9400g (1966).

125. H. Van Bekkum, J. D. Remijnse, and B. M. Wepster, *Chem. Commun.,* 67 (1969).

126. A. T. Pilipenko and I. P. Sereda, *Zh. Anal. Khim.,* **13,** 3 (1958).

127. A. T. Pilipenko and I. P. Sereda, *Zh. Anal. Khim.,* **16,** 73 (1961).

128. A. T. Pilipenko and I. P. Sereda, *Zh. Neorg. Khim.,* **6,** 413 (1961).

129. A. T. Pilipenko and I. P. Sereda, *Vopr. Analiza Blagorodn. Metal.,* 64 (1963); *Chem. Abstr.,* **61,** 10039a (1964).

130. J. M. Balcells Garcia de la Torre, Thesis, Laguna University, 1958; *Selenium Tellurium Abstr.,* **0,** 1311 (1959).

131. F. Pino Perez, F. Burriel, and L. M. Conejero, *Inform. Quim. Anal.* (Madrid). **13,** 38 (1959).

132. F. Pino Perez, F. Burriel, and J. M. Balcells, *Anales Real Soc. Españ. Fis. Quim., Ser. B,* **55,** 579 (1959).

133. D. P. Bendito and F. Pino, *Inform. Quim. Anal.* (Madrid), **21,** 9 (1967).

134. R. A. Zingaro and D. O. Skovlin, *J. Electrochem. Soc.,* **111,** 42 (1964).

135. T. H. Johnson, U.S. Pat. 3,178,312 (1965); *Chem. Abstr.,* **63,** 2548g (1965).

136. A. B. Ludin and G. A. Kitaev, *Izv. Akad. Nauk SSSR, Neorgan. Materialy,* **1,** 2102 (1965); *Chem. Abstr.,* **64,** 9393d (1966).

137. P. A. McVeigh, Swiss. Pat. 476,325 (1969); *Chem. Abstr.,* **72,** 138324b (1970).

138. J. S. Dunn, U.S. Pat. 3,297,446 (1967); *Chem. Abstr.,* **66,** 80798s (1967).

139. D. Alcock and E. Piccotti, Ger. Pat. 1,937,727 (1970); *Chem. Abstr.,* **73,** 9430e (1970).

140. Dutch Pat. 73,715 (1953); *Chem. Abstr.,* **48,** 8532c (1954).

141. G. H. Reman, Dutch Pat. 75,372 (1954); *Chem. Abstr.,* **49,** 6590c (1955).

142. W. Scott, U.S. Pat. 2,140,272 (1938); *Chem. Abstr.,* **33,** 2765 (1939).

143. C. Collard-Charon, R. Huls, and M. Renson, *Bull. Soc. Chim. Belges,* **71,** 541 (1962).

144. K. A. Jensen, G. Felbert, C. T. Pedersen, and U. Svanholm, *Acta Chem. Scand.*, **20**, 278 (1966).

145. E. Bulka, K.-D. Ahlers, and E. Tuček, *Chem. Ber.*, **100**, 1373 (1967).

146. R. Huls and M. Renson, *Bull. Soc. Chim. Belges*, **65**, 511 (1956).

147. R. Huls and M. Renson, *Bull. Soc. Chim. Belges*, **66**, 265 (1957).

148. Belgium Pat. 544,031 (1956); *Chem. Abstr.*, **54**, 12068h (1960).

149. C. Collard-Charon, R. Huls, and M. Renson, *Bull. Soc. Chim. Belges*, **71**, 554 (1962).

150. H. G. Mautner and W. D. Kumler, *J. Amer. Chem. Soc.*, **78**, 97 (1956).

151. E. Bulka, W. Dietz, H.-G. Patzwaldt, and H. Beyer, *Chem. Ber.*, **96**, 1996 (1963).

152. K. Bednarz, *Dissertationes Pharm.*, **9**, 249 (1957).

153. F. Feher, D. Hirschfeld, and K. H. Linke, *Z. Naturforsch.*, *Part B*, **17**, 625 (1962).

154. K. Bednarz, *Dissertationes Pharm.*, **10**, 93 (1958).

155. B. A. Gingras, T. Suprunchuk, and C. H. Bayley, *Can. J. Chem.*, **43**, 1650 (1965).

156. E. Bulka and K.-D. Ahlers, *Z. Chem.*, **3**, 349 (1963).

157. U. Anthoni, C. Larsen, and P. H. Nielsen, *Acta Chem. Scand.*, **23**, 537 (1969).

158. K. H. Linke and D. Skupin, *Z. Naturforsch.*, *Part B*, **25**, 1 (1970).

159. K. H. Linke and D. Skupin, *Z. Naturforsch.*, *Part B*, **25**, 3 (1970).

160. R. Huls and M. Renson, *Bull. Soc. Chim. Belges*, **65**, 696 (1956).

161. K. Volka and Z. Holzbacher, *Collection Czech. Chem. Commun.*, **34**, 3160 (1969).

162. D. Barnard and D. T. Woodbridge, *J. Chem. Soc.*, 2922 (1961).

163. G. M. Larin, P. M. Solozhenkin, N. I. Kopitsya, and H. Kirspuu, *Dokl. Akad. Nauk Tadzh. SSR*, **12**, 32 (1969); *Chem. Abstr.*, **72**, 84787j (1970).

164. K. A. Jensen and V. Krishnan, *Acta Chem. Scand.*, **21**, 2904 (1967).

165. B. Lorenz and E. Hoyer, *Z. Chem.*, **8**, 230 (1968).

166. G. M. Larin, P. M. Solozhenkin, N. I. Kopitsya, and H. Kirspuu, *Izv Akad. Nauk SSSR, Ser. Khim.*, 968 (1969).

167. M. L. Shankaranarayana, *Acta Chem. Scand.*, **24**, 351 (1970).

168. U. Anthoni, *Acta Chem. Scand.*, **20**, 2742 (1966).

169. U. Anthoni, B. M. Dahl, C. Larsen, and P. H. Nielsen, *Acta Chem. Scand.*, **24**, 959 (1970).

170. U. Anthoni, B. M. Dahl, C. Larsen, and P. H. Nielsen, *Acta Chem. Scand.*, **23**, 1061 (1969).

171. B. M. Dahl and P. H. Nielsen, *Acta Chem. Scand.*, **24**, 1468 (1970).

172. F. Feher, D. Hirschfeld, and K. H. Linke, *Z. Naturforsch.*, *Part B*, **17**, 625 (1962).

173. R. K. Blake, Belgium Pat. 670,823 (1966); *Selenium Tellurium Abstr.*, **7**, 5384 (1966).

174. L. G. S. Brooker and R. H. Sprague, U.S. Pat. 2,332,433 (1943).

175. O. Gorgon and M. Simek, Czech. Pat. 127,919 (1968); *Chem. Abstr.*, **70**, 96799m (1969).

176. K. A. Jensen and A. Holm, *Acta Chem. Scand.*, **18**, 2417 (1964).

177. K. A. Jensen. P. A. A. Frederiksen, and L. Henriksen, *Acta Chem. Scand.*, **24**, 2061 (1970).

178. G. Frerichs and E. Wildt, *Justus Liebigs Ann. Chem.*, **360**, 118 (1908).

179. K. A. Jensen, M. Due, A. Holm, and C. Wentrup, *Acta Chem. Scand.*, **20**, 2091 (1966).

180. U. Anthoni, C. Larsen, and P. H. Nielsen, *Acta Chem. Scand.*, **21,** 2571 (1967).

181. H. Kirspuu and A. I. Busev, *Zh. Anal. Khim.*, **23,** 354 (1968); *J. Anal. Chem. USSR*, **23,** 291 (1968).

182. A. I. Busev and H. Kirspuu, *Tartu Riikliku Ulikooli Toim.*, **219,** 215 (1968); *Chem. Abstr.*, **71,** 56283c (1969).

183. A. I. Busev, H. Kirspuu, H. Kook, and T. Tuisk, *Tartu Riikliku Ulikooli Toim,* **235,** 153 (1969); *Chem. Abstr.*, **74,** 19031k (1971).

IX Compounds containing a selenium-oxygen-carbon moiety

ROLAND PAETZOLD and **MANFRED REICHENBÄCHER**

Friedrich Schiller University
Jena
German Democratic Republic

A. INTRODUCTION

Of the many possible compounds possessing a selenium-oxygen-carbon moiety (Table IX-1), the selenium tetraalkoxides, the dialkyl selenites, and

Table IX-1   Possible classes of Se—O—C compounds

(a) Se(II):	$Se(OR)_2$
(b) Se(IV):	$Se(OR)_4$, $OSe(OR)_2$
(c) Se(VI):	$Se(OR)_6$, $OSe(OR)_4$, $O_2Se(OR)_2$

the dialkyl selenates and their derivatives have been most thoroughly studied, from both the preparative and the spectroscopic viewpoints. Selenium dialkoxides, selenium hexaalkoxides, and tetraalkyl selenates have not, as yet, been described. Only a single compound, dimethyl diselenate, is known to contain more than one Se atom combined with the selenium-oxygen-carbon moiety (cf. Section IX.D.1).

This chapter summarizes the results of investigations of the known types of compounds. An attempt has been made to consider all the published literature for a critical assessment of the field.

B. SELENIUM TETRAALKOXIDES AND DERIVATIVES

1. Selenium Tetraalkoxides

Only a few members of this class of selenium compounds or their sulfur or tellurium analogs are known. Selenium tetraalkoxides, as esters of the hypothetical orthoselenious acid, have been described only in the cases of selenium tetramethoxide (1),[1] tetraethoxide,[1] tetrapropoxide,[1] tetrabut-

$$
\begin{array}{c}
OCH_3 \\
| \quad OCH_3 \\
\llap{/\!\!/\!\!/\!\!/} Se \\
| \quad OCH_3 \\
OCH_3
\end{array}
$$

1

oxide,[1] the bicyclic ester derived from ethylene glycol,[1] the dioxalate[2] and the tetrakis[N,N-bis(trifluoromethyl)hydroxylaminato] selenium.[3]

Selenium tetramethoxide (1) was formed by the reaction of selenium tetrachloride with 4 equivalents of sodium methoxide at −70° (equation 1). The ethyl, propyl, and butyl derivatives have been prepared by trans-

(1) $SeCl_4 + 4NaOCH_3 \rightarrow 1 + 4NaCl$

esterification of the tetramethoxide. Selenium tetrachloride also reacted with propyl and butyl alcohols in the presence of triethylamine in THF solution (equation 2).

(2) $SeCl_4 + 4C_3H_7OH + 4(C_2H_5)_3N \rightarrow Se(OC_3H_7)_4 + 4[(C_2H_5)_3NH]Cl$

The preparation of selenium tetraethoxide according to equation 3 is not

$$(3)\quad NaH[Se(OC_2H_5)_6] + HCl \rightarrow Se(OC_2H_5)_4 + NaCl + 2C_2H_5OH$$

reproducible, since it has not been possible to prepare the starting salt according to the original reference[4] or by any other route.

Methods analogous to the above procedures have been used to prepare tellurium tetraalkoxides,[4-8] whereas, of the sulfur analogs, only perfluoropinacol orthosulfite[9] and sulfur tetraphenoxide[10] have been shown to exist.

The bicyclic ester of ethylene glycol (2) is prepared by a reaction analogous

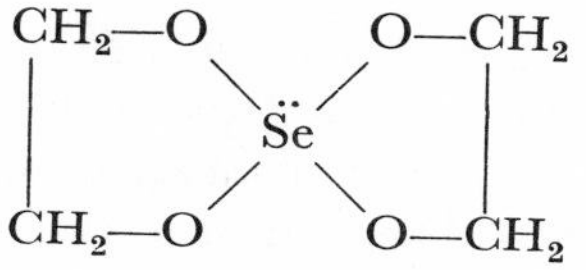

2

to that shown in equation 2 or by the transesterification of the selenium tetraalkoxides. The crystalline compound has a monomeric structure.

Selenium dioxalate (3) is obtained as a THF or a dioxane adduct by the reaction of selenium tetrachloride with silver oxalate in THF or dioxane solution (equation 4). The THF adduct, when dissolved in THF solution,

$$(4)\quad SeCl_4 + 2Ag_2C_2O_4 \xrightarrow[\text{dioxane}]{\text{THF}} \quad$$

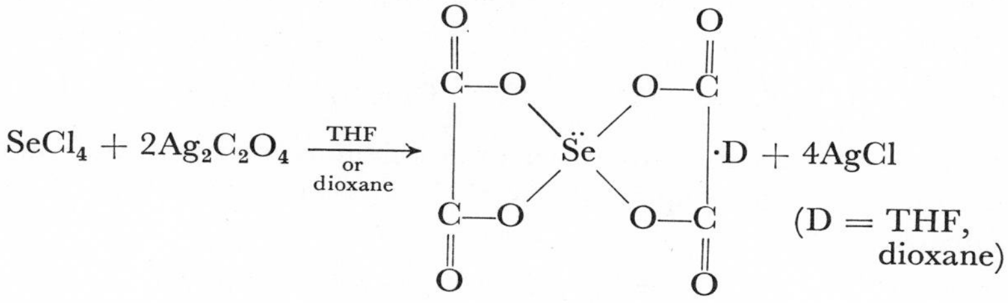

(D = THF, dioxane)

3

yields, on addition of pyridine, the pyridine adduct. Tetrakis[N,N-bis(trifluoromethyl)hydroxylaminato] selenium (4) is prepared by reaction of bis(trifluoromethyl) nitroxide with elemental selenium (equation 5).[3]

$$(5)\quad Se + 4ON(CF_3)_2 \rightarrow Se[ON(CF_3)_2]_4$$

4

Tetraalkoxides are colorless liquids that are soluble in organic solvents and can be separated by distillation from the dialkyl selenites, the main byproducts of their preparation. The tetraalkoxides, although stable below 0°, tend to decompose slowly at room temperature according to equation 6.

$$(6)\quad Se(OR)_4 \rightarrow O{=}Se(OR)_2 + R_2O$$

Selenium tetramethoxide reacts with sodium or potassium methoxide, forming pentamethoxyselenate(IV) (equation 7).[11] Selenium tetrafluoride

$$(7) \qquad Se(OCH_3)_4 \xrightleftharpoons[+HCl, -CH_3OH, -NaCl]{+NaOCH_3} Na[Se(OCH_3)_5]$$

reacts analogously with sodium fluoride.[12] In methanolic solution sodium pentamethoxyselenate(IV) is not stable. The anion dissociates quantitatively according to equation 8.

$$(8) \qquad Se(OCH_3)_5^- \rightarrow Se(OCH_3)_4 + OCH_3^-$$

The Raman and i.r. spectra of selenium tetramethoxide indicate a tetrahedral C_{2v} symmetry of the SeO_4 skeleton with some distortion towards a trigonal bipyramid, thus demonstrating a steric effect of the free $4s^2$ electron pair.[11] The n.m.r. spectrum of the tetramethoxide shows only one signal at room temperature, indicating an equivalence of all methyl protons.[11] This may be explained by postulation of a rapid intramolecular exchange of the methoxide functions. Thus, the tetramethoxide closely resembles the tetrafluoride, in regard to both structure and exchange properties.[13–15]

2. Monosubstituted Selenium Trialkoxides

Reaction of phenylselenium tribromide with sodium methoxide yields phenylselenium trimethoxide (equation 9),[16] a nondistillable liquid. Chloroselenium

$$(9) \qquad C_6H_5SeBr_3 + 3NaOCH_3 \rightarrow C_6H_5Se(OCH_3)_3 + 3NaBr$$

trimethoxide can be synthesized from selenium tetramethoxide and selenium tetrachloride (equation 10).[17] The bromo derivative is prepared similarly.[17]

$$(10) \qquad 3Se(OCH_3)_4 + SeCl_4 \rightarrow 4(CH_3O)_3SeCl$$

Selenium trimethoxide nitrate is obtained by the reaction of chloroselenium trimethoxide with silver nitrate (equation 11).[17] Fluoroselenium trimethoxide is prepared analogously, utilizing silver fluoride.[17]

$$(11) \qquad ClSe(OCH_3)_3 + AgNO_3 \rightarrow (CH_3O)_3SeONO_2 + AgCl$$

Selenium trimethoxide nitrate and chloro- and bromoselenium trimethoxide are crystalline and very hygroscopic, whereas the fluoro compound is a nondistillable liquid. All have limited stability at $< -10°$, but at room temperature decomposition takes place slowly, liberating the methyl halides or the methyl nitrate (equation 12).

$$(12) \qquad (CH_3O)_3SeX \rightarrow (CH_3O)_2Se{=}O + CH_3X$$
$$(X = F, Cl, Br, NO_3)$$

Aluminum trichloride reacts with chloroselenium trimethoxide, forming compound **5**.

$$(CH_3O)_3Se^+AlCl_4^-$$

5

The mean values of the wave numbers of the Se–O stretching bands constitute a measure of the strength of the Se–O bond. Table IX-2 indicates

Table IX-2   Mean values of the wave numbers of the Se—O stretching vibrations

GROUP	COMPOUND	ν_{Se-O} (cm^{-1})
I	Na[Se(OCH$_3$)$_5$]	505
II	Se(OCH$_3$)$_4$	538
	(CH$_3$O)$_3$SeF	551[a]
	(CH$_3$O)$_3$SeC$_6$H$_5$	531
III	(CH$_3$O)$_3$SeBr	591
	(CH$_3$O)$_3$SeCl	596
IV	(CH$_3$O)$_3$SeAlCl$_4$	632

[a] This mean value includes vibrational contributions of ν_{Se-F}.

that the mean values of the wave numbers increases from Group I to Group IV, corresponding to an increasing Se-O bond strength. This order is attributable to an increase in the effective nuclear charge of the central atom, resulting from the decrease in the number of ligands from 5 (Group I) to 4 (Group II) to 3 (Groups III and IV). The bond polarity of the neighboring Se-X bond increases in the same direction, resulting in saltlike compounds of Group III and especially Group IV (*cf.* Section IX.E).

The conclusion that the Se-F and Se-C bonds in fluoro- and phenylselenium trimethoxide (in Group II) are covalent, which is reached through the interpretation of their i.r. spectra, is in accord with the physical properties of these compounds.

3. Disubstituted Selenium Dialkoxides (X$_2$Se(OR)$_2$)

The known compounds within this classification are diphenylselenium dimethoxide[18] and diethoxide,[18] dimethylselenium dimethoxide,[18] and dichloroselenium dimethoxide.[19]

Diphenylselenium dimethoxide (**6**) is prepared from diphenylselenium dichloride or dibromide and sodium methoxide (equation 13). Diphenyl-

(13) $(C_6H_5)_2SeCl_2 + 2NaOCH_3 \rightarrow (C_6H_5)_2Se(OCH_3)_2 + 2NaCl$

6

selenium diethoxide and dimethylselenium dimethoxide are synthesized in the same manner.

Dichloroselenium dimethoxide is prepared by reaction of selenium tetramethoxide with 1 mole of selenium tetrachloride (equation 14), or by

$$(14) \qquad Se(OCH_3)_4 + SeCl_4 \rightarrow 2Cl_2Se(OCH_3)_2$$

halogenation of dimethyl selenite with thionyl chloride (equation 15).

$$(15) \qquad (CH_3O)_2SeO + SOCl_2 \rightarrow Cl_2Se(OCH_3)_2 + SO_2$$

Mehrotra and Mathur[20] reported that dialkyl selenites yield the corresponding dichloroselenium dialkoxides upon reaction with chlorine gas in carbon tetrachloride solution. However, in the case of dimethyl selenite, the reaction could not be confirmed.[19]

Diphenylselenium dimethoxide and diethoxide hydrolyze to diphenyl selenoxide in moist air (equation 16). Dimethylselenium dimethoxide

$$(16) \qquad (C_6H_5)_2Se(OCH_3)_2 \xrightarrow{H_2O} (C_6H_5)_2Se{=}O + 2CH_3OH$$

decomposes >0° according to equation 17. Dichloroselenium dimethoxide

$$(17) \qquad (CH_3)_2Se(OCH_3)_2 \rightarrow (CH_3)_2Se + CH_2O + CH_3OH$$

decomposes at room temperature, yielding methyl chloroselenite and methyl chloride (equation 18).

$$(18) \qquad Cl_2Se(OCH_3)_2 \rightarrow CH_3OSeOCl + CH_3Cl$$

The considerable diminution of the Se–O–C stretching vibrations of **6** and of dimethylselenium dimethoxide, in comparison with that of dialkyl selenite, supposes a bond weakening. This is in accordance with a nearly trigonal-bipyramidal structure, with the alkoxide groups in the axial and the alkyl or aryl groups and the free $4s^2$ electron pair of selenium in the equatorial positions.

The known compounds of the $XYSe(OR)_2$ type are limited to the phenylselenium dimethoxide chloride, bromide, and nitrate. The chloro compound is prepared by reaction of phenylselenium trimethoxide with phenylselenium trichloride (equation 19).[16] The bromo compound is similarly prepared.[16]

$$(19) \qquad 2C_6H_5Se(OCH_3)_3 + C_6H_5SeCl_3 \rightarrow 3C_6H_5Se(OCH_3)_2Cl$$

From either halo compound, the nitrate may be obtained by reaction with silver nitrate.

4. Trisubstituted Selenium Alkoxides
(X_3SeOR, X_2YSeOR, XY_2SeOR)

Only a few examples of this class are known, all of which have been described recently. Trichloroselenium methoxide[19] is synthesized by reaction of selenium tetramethoxide with 3 moles of selenium tetrachloride (equation 20). The compound decomposes at room temperature according to equation 21.

$$(20) \qquad Se(OCH_3)_4 + 3SeCl_4 \rightarrow 4Cl_3SeOCH_3$$

$$(21) \qquad Cl_3SeOCH_3 \rightarrow Cl_2SeO + CH_3Cl$$

Dichlorophenylselenium methoxide may be prepared according to equation 22, as may also the dibromo compound. Chloro- and bromo-

$$(22) \qquad C_6H_5Se(OCH_3)_3 + 2C_6H_5SeCl_3 \rightarrow 3C_6H_5Se(OCH_3)Cl_2$$

diphenylselenium methoxide are prepared according to equation 23.

$$(23) \qquad (C_6H_5)_2Se(OCH_3)_2 + (C_6H_5)_2SeCl_2 \rightarrow 2(C_6H_5)_2Se(OCH_3)Cl$$

C. SELENIOUS ACID ESTERS AND DERIVATIVES

1. Dialkyl Selenites

The lower dialkyl selenites are distillable liquids, fuming in damp air, whereas the higher ones are solids at room temperature. Esters of selenious acid can be synthesized in various ways.

a. From Selenium Dioxide or Selenious Acid. Dialkyl selenites have been prepared by reaction of selenium dioxide with alcohols (equation 24),[21,22] with methanol and diazomethane (equation 25),[23] with ethanol and ethyl

$$(24) \qquad SeO_2 + 2ROH \rightarrow (RO)_2SeO + H_2O$$

$$(25) \qquad SeO_2 + CH_3OH + CH_2N_2 \rightarrow (CH_3O)_2SeO + N_2$$

diazoacetate (equation 26),[23] with dialkyl sulfites (equation 27),[24] or with

$$(26) \quad SeO_2 + C_2H_5OH + N_2CHCO_2C_2H_5 \rightarrow$$
$$C_2H_5O\text{--}SeO\text{--}OCH_2CO_2C_2H_5 + N_2$$

selenium tetraalkoxides (equation 28).[11]

$$(27) \qquad (RO)_2SO + SeO_2 \rightarrow (RO)_2SeO + SO_2$$

$$(28) \qquad SeO_2 + Se(OCH_3)_4 \rightarrow 2(CH_3O)_2SeO$$

Dialkyl selenites have also been obtained as products of the reaction of selenious acid with alcohols (equation 29)[25,26] or with diazoalkanes (equation 30).[26,27]

$$(29) \qquad (HO)_2SeO + 2ROH \rightarrow (RO)_2SeO + 2H_2O$$

$$(30) \qquad (HO)_2SeO + 2RN_2 \xrightarrow{\text{ether}} (RO)_2SeO + 2N_2$$

Schemes 24 and 29 should be considered the methods of choice. Investigations utilizing Raman spectroscopy have revealed some interesting aspects of these two reactions.[26,28] When selenium dioxide is dissolved in an alcohol (equation 24), it is completely converted to an alkylselenious acid by degradation of its polymeric chain structure. Alkylselenious acids are also formed upon dissolving selenious acid in alcohols; the reactants and products exist in the equilibrium described by equation 31. The alkylselenious acid formed in both reactions is also in equilibrium with dialkyl selenite (equation 32).

$$(31) \qquad (HO)_2SeO + ROH \leftrightharpoons ROSeOOH + H_2O$$

$$(32) \qquad ROSeOOH + ROH \leftrightharpoons (RO)_2SeO + H_2O$$

Therefore removal of water from the reaction mixture in either reaction (24 or 29) results in a displacement of equilibrium in the direction of the dialkyl selenite. Dimethyl selenite is prepared by azeotropic distillation of a mixture of selenium dioxide or selenious acid (equations 24 or 29), methanol, and chloroform.

The diols, 1,2-dihydroxyethane, 1,3-dihydroxypropane, and 1,4-dihydroxybutane, react with both selenium dioxide and selenious acid, yielding the corresponding cyclic esters (equation 33).[29] In contrast to the

$$(33) \qquad SeO_2 + HOCH_2CH_2OH \longrightarrow \begin{array}{c} CH_2O \\ | \\ CH_2O \end{array}\!\!\!\Big\rangle SeO + H_2O$$

alkylselenious acids,[26,28] which can be isolated, no halfesters have been obtained. The entropy effect apparently results in a rapid conversion to the cyclic selenite (equation 34).[7] For the same reason, the halfesters cannot be

$$(34) \qquad HOCH_2CH_2OSeOOH \longrightarrow \begin{array}{c} CH_2O \\ | \\ CH_2O \end{array}\!\!\!\Big\rangle SeO + H_2O$$

obtained by partial hydrolysis of the cyclic esters, again in contrast to the dialkyl selenites.

b. From Selenium Oxychloride. Selenium oxychloride reacts with sodium alkoxides (equation 35)[30,31] or with 2-chloroethanol (equation 36),[32]

(35) $$SeOCl_2 + 2NaOR \rightarrow (RO)_2SeO + 2NaCl$$

yielding dialkyl selenites.

(36) $$SeOCl_2 + 2HOCH_2CH_2Cl \rightarrow (ClCH_2CH_2O)_2SeO + 2HCl$$

Carboxylates of selenious acid have also been prepared: the diacetate from selenium oxychloride and sodium acetate (equation 37),[33] and the oxalate

(37) $$SeOCl_2 + 2CH_3COONa \rightarrow (CH_3COO)_2SeO + 2NaCl$$

from selenium oxychloride and silver oxalate (equation 38).[33]

(38) $$SeOCl_2 + Ag_2C_2O_4 \longrightarrow \quad \begin{matrix} OCO \\ | \quad \diagdown \\ \quad \quad SeO + 2AgCl \\ | \quad \diagup \\ OCO \end{matrix}$$

The dibenzoate of selenious acid has been made from a mixed melt of selenium dioxide and benzoic anhydride (equation 39).[34]

(39) $$SeO_2 + (C_6H_5CO)_2O \rightarrow (C_6H_5COO)_2SeO$$

c. By Other Methods. Silver selenite reacts with alkyl iodides to yield dialkyl selenites (equation 40).[35] Alkylselenious acids react with diazomethane, giving methyl alkyl selenites (equation 41).[26] Other esters are

(40) $$Ag_2SeO_3 + 2RI \rightarrow (RO)_2SeO + 2AgI$$

(41) $$ROSeOOH + CH_2N_2 \xrightarrow{ether} ROSeOOCH_3 + N_2$$

best obtained by transesterification of dimethyl or diethyl selenite (equation 42).[22,26]

(42) $$(RO)_2SeO + 2R'OH \rightarrow (R'O)_2SeO + 2ROH$$

Mehrotra and Mathur[25] concluded that transesterification of diethyl selenite with secondary and tertiary alcohols results in the formation of mixed esters (equation 43). This opinion is not well founded, however,

(43) $$(C_2H_5O)_2SeO + ROH \rightarrow C_2H_5OSeOOR + C_2H_5OH$$

since the characterization of the final products was performed only by selenium analysis, so that mixed esters could not be excluded, but neither were they demonstrated.

When diethyl selenite reacts with ammonia in ether solution at $-50°$, the diamide of selenious acid, a thermally unstable compound, is obtained (equation 44).[36] The bis(dimethylamide) is prepared analogously.[36]

$$(44) \qquad (C_2H_5O)_2SeO + 2NH_3 \rightarrow (NH_2)_2SeO + 2C_2H_5OH$$

Dialkyl selenites decompose at high temperature, usually by the redox reaction described in equation 45. Thus, alcohols can be converted into

$$(45) \qquad (RCH_2O)_2SeO \rightarrow RCHO + SeO_2 + H_2O$$

aldehydes.[21,22]

The Raman and i.r. spectra of dimethyl selenite, diethyl selenite, methyl ethyl selenite, and of the cyclic ester of ethylene glycol have been studied,[7,29,37] and it is concluded that the SeO_3 skeleton is trigonal-pyramidal. Dialkyl selenites hydrolyze more rapidly than the corresponding dialkyl sulfites.[38] This is the result of the greater polarity of the Se-O bond and the lesser steric hindrance of the Se atom, thus facilitating nucleophilic attack.

The difference in behavior of the dialkyl selenites and the dialkyl sulfites towards nucleophilic attack by alkoxide ions is also intelligible on the same basis. For example, complex salts crystallize out of equimolar alcoholic solutions of dialkyl selenite and alkali alkoxides (equation 46).[39] In contrast,

$$(46) \qquad (CH_3O)_2SeO + NaOCH_3 \rightarrow Na^+[OSe(OCH_3)_3]^-$$

alkoxide ions are alkylated by dialkyl sulfite (equation 47).[40] Obviously, the nucleophilic attack of the alkoxide ion takes place at the Se atom in the

$$(47) \qquad (CH_3O)_2SO + NaOCH_3 \rightarrow CH_3OSO_2Na + CH_3OCH_3$$

case of dialkyl selenite, and at the C atom with dialkyl sulfite.

Dialkyl sulfites can readily be converted to the thermodynamically more stable isomeric sulfonic esters (equation 48).[41,42] Dialkyl selenites, however,

$$(48) \qquad (CH_3O)_2SO \rightarrow CH_3SO_2OCH_3$$

do not undergo this reaction. For further discussions of this topic the reader is referred to the review by Paetzold.[43]

2. Alkylselenious Acids and Derivatives

a. *Halogen-Substituted Alkylselenious Esters (ROSeOHal).* Until 1950, the only known compound within this class was ethylchloroselenite (**7**),[32,35] but now various fluoro-, chloro-, and bromoselenious esters have been

$$C_2H_5O\!-\!\overset{\displaystyle O}{\overset{\displaystyle \|}{Se}}\!-\!Cl$$

7

described. The best, and most generally applicable, preparative procedure is to mix equimolar quantities of selenium oxyhalides and dialkyl selenites (equation 49). Methyl and ethyl fluoroselenites,[44] and methyl,[44,45] ethyl,[44,45]

$$(49) \qquad (CH_3O)_2SeO + OSeF_2 \rightarrow 2CH_3OSeOF$$

and n-butyl chloroselenites[45] were prepared in this way. Chloroselenious esters can also be prepared by reaction of selenium oxychloride with excess alcohol (equation 50), wherein, in contrast to thionyl chloride, only one

$$(50) \qquad SeOCl_2 + CH_3CH_2OH \rightarrow CH_3CH_2OSeOCl + HCl$$

halogen atom is exchanged.[32,35] According to Mehrotra and Mathur,[20] methyl, propyl, isopropyl, butyl, *sec*-butyl, and isoamyl chloroselenite should be obtainable by this technique. It is our experience, however, that the methyl ester cannot be obtained satisfactorily in this way because of the poor conversion to the ester product and the inability to separate the product from selenium oxychloride by distillation, due to the near identity of their boiling points.[44]

Methyl chloroselenite and methyl bromoselenite can be prepared by reaction of the appropriate selenium oxyhalide with 1 mole of sodium methoxide (equation 51).[45] Selenium oxychloride reacts with ethylene and

$$(51) \qquad SeOBr_2 + NaOCH_3 \rightarrow CH_3OSeOBr + NaBr$$

propylene oxides to yield C-substituted chloroselenious esters (equation 52).[45] Ethyl fluoroselenite can be prepared from the corresponding chloro

$$(52) \qquad SeOCl_2 + CH_3CH{-}CH_2 \longrightarrow CH_3CHClCH_2OSeOCl$$
$$\underset{O}{\diagdown \diagup}$$

compound by an exchange reaction with potassium fluoride.[44] Methyl chloroselenite can be quantitatively transesterified with excess ethanol (equation 53), whereas the reverse reaction takes place to only a slight

$$(53) \qquad CH_3OSeOCl + C_2H_5OH \rightarrow C_2H_5OSeOCl + CH_3OH$$

extent.[44]

The halogenoselenious esters are all readily soluble in organic solvents and are hydrolyzed rapidly (equation 54). Although they can be used in the

$$(54) \qquad C_2H_5OSeOCl + H_2O \rightarrow HCl + H_2SeO_3 + C_2H_5OH$$

preparation of mixed esters and other derivatives of selenious acid, little investigation has been conducted in this area. Thus, methyl ethyl selenite is obtainable by reaction of equimolar quantities of ethyl chloroselenite, methanol, and pyridine (equation 55);[26] however, the simplest procedure

(55) $C_2H_5OSeOCl + CH_3OH + C_5H_5N$
$$\rightarrow C_2H_5OSeOOCH_3 + C_5H_5NH^+Cl^-$$

would be to mix equimolar quantities of dimethyl and diethyl selenites.

The Raman and i.r. spectra of the methyl and ethyl esters of fluoro- and chloroselenious acids have been studied extensively,[46] and results indicate that two rotational isomers of these compounds in the liquid state exist in a temperature-dependent equilibrium.

b. Alkylselenious Acids (ROSeOOH) and Their Salts. Methyl- and ethylselenious acids were obtained as early as 1890 by evaporating alcoholic selenium dioxide solutions to dryness; however, there was little agreement concerning the structure of the compounds.[47–49] Interpretation of Raman spectra[26] leads to the conclusion that the alkylselenious acids present in alcoholic solutions of selenium dioxide or selenious acid are in equilibria with dialkyl selenites (equation 32). The higher alkylselenious acids (C_{10}–C_{18}) have been prepared by melting the alcohol with an excess of selenium dioxide (equation 56).[22] Methyl- and ethylselenious acids hydrolyze spon-

(56) $SeO_2 + C_{14}H_{29}OH \rightarrow C_{14}H_{29}OSeOOH$

taneously in water. Raman spectroscopy indicates that alkylselenious acid molecules form strong hydrogen bonds in the crystalline state;[28] furthermore, in liquid ethylselenious acid, two rotational isomers in equilibrium are found, originating from hindered rotation of the methyl group about the C–O bond.[28]

Sodium or potassium salts of alkylselenious acids are products of the reaction of alkali alkoxides with (a) alkylselenious acids in dioxane, (b) alcoholic solutions of selenium dioxide, or (c) alcoholic solutions of dialkyl selenites and water (in 1:1 molar ratio).[50] These preparative methods are based on the acid-base reaction illustrated in equation 57. The alkylselenious

(57) $C_2H_5OSeOOH + NaOC_2H_5 \rightarrow C_2H_5OSeO_2^-Na^+ + C_2H_5OH$

acid is formed by degradation of the polymeric chain structure of selenium dioxide during the reaction with alcohol in the case of method (b) and according to equation 32 in the case of method (c). Reaction (c) does not take place in the absence of water, because dialkyl selenites, in contrast to dialkyl sulfites, are unable to alkylate alkoxide ions. Alkali alkyl selenites are also obtained from alcoholic solutions of hydrogen selenites (**8**) or di-selenites (**9**) by precipitation with dioxane.[50] Hydrogen selenites in alcohol

$$Cat^+\ HOSeO_2^-   \qquad 2Cat^+\ Se_2O_5^{2-}$$

8   **9**

are in equilibrium with the corresponding alkylselenious acid salt (equation 58). In the case of diselenites dissolved in alcohol, the equilibria depicted

$$(58) \qquad HOSeO_2^-Na^+ + C_2H_5OH \leftrightharpoons C_2H_5OSeO_2^-Na^+ + H_2O$$

in equations 58 and 59 exist. The corresponding reactions are unknown in

$$(59) \quad O_2SeOSeO_2^{2-} \, 2Na^+ + C_2H_5OH \leftrightharpoons$$

$$C_2H_5OSeO_2^-Na^+ + HOSeO_2^-Na^+$$

sulfur chemistry because of the structure of the hydrogen sulfite and disulfite ions (*cf.* Paetzold[43]). Raman and i.r. spectra prove the existence of the alkyl selenite anions in both salts and their alcoholic solutions.[51] These salts exist with C_s symmetry in a trigonal-pyramidal selenium-oxygen configuration. Alkyl selenite anions hydrolyze rapidly to selenite anions (equation 60)

$$(60) \qquad CH_3OSeO_2^-K^+ + H_2O \rightarrow HOSeO_2^-K^+ + CH_3OH$$

and are easily re-esterified at room temperature.[50]

c.   Ester Amides of Selenious Acid.   The only known amide of an alkyl-selenious acid is methylselenious acid dimethylamide, which exists only at room temperature in equilibrium with dimethyl selenite and the bis(di-methylamide) of selenious acid (equation 61). The ester amide may, in

$$(61) \qquad 2CH_3OSeON(CH_3)_2 \leftrightharpoons (CH_3O)_2SeO + [(CH_3)_2N]_2SeO$$

addition, be prepared as indicated in equations 62 and 63.[52]

$$(62) \quad CH_3OSeOCl + 2(CH_3)_2NH \rightarrow$$

$$CH_3OSeON(CH_3)_2 + [(CH_3)_2NH_2]^+Cl^-$$

$$(63) \quad [(CH_3)_2N]_2SeO + CH_3OH \rightarrow CH_3OSeON(CH_3)_2 + (CH_3)_2NH$$

D. SELENIC ACID ESTERS AND DERIVATIVES

1. Dialkyl Selenates

Dialkyl selenates are distillable without decomposition at low pressure. They hydrolyze more slowly than the dialkyl selenites and are also much less soluble in water, dissolving only to the extent of hydrolysis.

Esters of selenic acid have been synthesized by the following methods.

a.   From Silver Selenate.   Alkyl iodides react with silver selenate to yield dialkyl selenates (equation 64).[53-56]

$$(64) \qquad Ag_2SeO_4 + 2RI \rightarrow (RO)_2SeO_2 + 2AgI$$

b. From Selenic Acid. Dimethyl selenate is prepared by the reaction of selenic acid with diazomethane (equation 65).[55] The compound may also be

(65) $(HO)_2SeO_2 + 2CH_2N_2 \rightarrow (CH_3O)_2SeO_2 + 2N_2$

prepared by the reaction of selenic acid with methanol (scheme 66).[57]

(66) $(HO)_2SeO_2 + CH_3OH \leftrightharpoons CH_3OSeO_2OH + H_2O$

$2CH_3OSeO_2OH \leftrightharpoons (CH_3O)_2SeO_2 + (HO)_2SeO_2$

Raman spectroscopy yields detailed information concerning the mechanism of this reaction, which proves to be more complex than illustrated. Reactions 67–69 occur as the ratio of alcohol is increased. In contrast to the

(67) $2(HO)_2SeO_2 + ROH \leftrightharpoons ROSeO_2OH + HOSeO_3^- + H_3O^+$

(68) $ROSeO_2OH + ROH \leftrightharpoons ROSeO_3^- + ROH_2^+$

(69) $HOSeO_3^- + ROH \leftrightharpoons ROSeO_3^- + H_2O$

situation with alkylselenious acid (equation 32), esterification of alkylselenic acid does not occur, because the great acidic strength of alkylselenic acid results in the predominance of the acid-base reaction described in equation 68. Dialkyl selenate is thus formed exclusively by the equilibria of the second equation 66.

c. From Selenium Trioxide. Selenium trioxide, consisting of tetrameric molecules, reacts with dimethyl ether to yield dimethyl selenate (equation 70).[57] The reaction of selenium trioxide with alcohols is an additional route

(70) $(SeO_3)_4 + 4CH_3OCH_3 \rightarrow 4(CH_3O)_2SeO_2$

for the preparation of dialkyl selenates (equation 71).[57] Raman spectroscopy reveals that the reaction of selenium trioxide with an alcohol gives the

(71) $(SeO_3)_4 + 4ROH \rightarrow 4ROSeO_2OH$

$2ROSeO_2OH \rightleftharpoons (RO)_2SeO_2 + (HO)_2SeO_2$

alkylselenic acid in quantitative yield.[59] As above, the alkylselenic acid is stabilized by the equilibrium with dialkyl selenate and selenic acid.[59]

d. By Transesterification. Dimethyl or diethyl selenate can be transesterified to yield higher esters (equations 72 and 73).[57,58] Such transesterifications

(72) $(CH_3O)_2SeO_2 + 2ROH \longrightarrow (RO)_2SeO_2 + 2CH_3OH$

(73) $(C_2H_5O)_2SeO_2 + ROH \xrightarrow{C_6H_6} C_2H_5OSeO_2OR + C_2H_5OH$

do not proceed without complications, in contrast to the case of the dialkyl selenites (equation 42), for alkyl selenates will alkylate alcohols to some extent (equation 74).[59] It seems to us that the transesterification described

$$(74) \qquad (RO)_2SeO_2 + 2R'OH \rightarrow R'OR + ROSeO_3^- + R'OH_2^+$$

by Mehrotra and Mathur (equation 73)[58] is not secured.

Kinetic investigation of the hydrolysis of dimethyl selenate reveals that the first methoxide group hydrolyzes rapidly, but the second one slowly.[60] It is interesting that the Se–O bond is broken during both steps, as compared with dimethyl sulfate, wherein the C–O bond is broken. This is apparently a consequence of the greater strength[43] and the lesser polarity[43] of the S–O bond.[60] The hydrolysis of dialkyl selenates proceeds equally well when acid- or base-catalyzed.[60]

Similar behavior is seen towards ammonia. Dimethyl selenate reacts in liquid ammonia to yield diaminoselenate, by splitting of the Se–O bond (equation 75).[61] Methylamine and dimethylamine react similarly.[61] Under

$$(75) \qquad (CH_3O)_2SeO_2 + 2NH_3 \rightarrow SeO_2(NH_2)_2 + 2CH_3OH$$

the same conditions, the C–O bond is split in dimethyl sulfate.[62]

Acetate ions react with dialkyl selenates as indicated in equation 76.[56] This procedure may be applied to the preparation of salts of alkylselenic

$$(76) \qquad CH_3CO_2^- + (RO)_2SeO_2 \xrightarrow{ROH} ROSeO_3^- + CH_3COOR$$

acids, but a different one is usually utilized for the preparation of analogous salts of alkylsulfuric acids (equation 77).[63] This reaction is not applicable to

$$(77) \qquad (RO)_2SO_2 + OR^- \xrightarrow{ROH} ROSO_3^- + ROR$$

selenium chemistry, inasmuch as alkyl selenate anions will alkylate alkoxide ions (equation 78), whereas this behavior is not observed with alkyl sulfate

$$(78) \qquad ROSeO_3^- + OR^- \xrightarrow{ROH} SeO_4^{-2} + ROR$$

anions.[56]

Dimethyl selenate forms adducts with pyridine [$(CH_3O)_2SeO_2 \cdot Py$] and trimethylamine [$(CH_3O)_2SeO_2 \cdot (CH_3)_3N$]. These adducts rearrange readily, yielding the isomeric salts

$$[CH_3OSeO_3^- PyCH_3^+] \qquad and \qquad [CH_3OSeO_3^-(CH_3)_4N^+],$$

respectively.[64]

Selenium trioxide reacts with dimethyl selenate (equation 79) to form dimethyl diselenate, which has explosive tendencies.[65]

$$(79) \qquad 4(CH_3O)_2SeO_2 + (SeO_3)_4 \rightarrow 4(CH_3O)_2Se_2O_5$$

2. Alkyl Selenate Derivatives

a. Fluoroselenic Esters.   Methyl and ethyl fluoroselenate are obtainable by reaction of selenyl fluoride with the appropriate dialkyl selenate (equation 80).[66] These esters are liquids and are unreactive towards glass. Unlike the

(80) $$SeO_2F_2 + (CH_3O)_2SeO_2 \rightarrow 2CH_3OSeO_2F$$

analogous fluorosulfuric esters, they exhibit considerable association in the liquid state. This is in accord with the generalization that selenium compounds always exhibit stronger autoassociation than do the corresponding sulfur compounds, for the larger Se atom and the more polar selenium-substituent bonds result in a lesser steric and electronic degree of saturation of the Se atom.

Hydrolysis of fluoroselenic esters results, in the first step, in the cleavage of the Se–F bond. Alcoholysis gives rise to dialkyl selenates (equation 81).

(81) $$CH_3OSeO_2F + CH_3OH \rightarrow (CH_3O)_2SeO_2 + HF$$

Raman and i.r. spectra indicate that ethyl fluoroselenate exhibits two rotational isomers in the liquid state.

b. Alkylselenic Acids and Their Salts.   Methyl- and ethylselenic acids, the only members of the class which have been studied in detail, are prepared by reaction of the dialkyl selenate with 1 mole of anhydrous selenic acid (equation 82). They are also formed in highly concentrated solutions

(82) $$(C_2H_5O)_2SeO_2 + H_2SeO_4 \rightarrow 2C_2H_5OSeO_2OH$$

of selenic acid in alcohol. These acids exist only in the equilibrium described by the second part of scheme 66.

Salts of alkylselenic acids have been prepared according to equation 76, as well as by potassium acetate precipitation of the alkyl selenate ions formed according to the scheme of equations 67–69.[56] These two methods are the most suitable for preparative purposes.

Additional procedures for the preparation of salts of alkylselenic acids are the methanolysis of amidoselenates (equation 83),[67] the esterification of

(83) $$NH_2SeO_3^-NH_4^+ + CH_3OH \rightarrow CH_3OSeO_3^-NH_4^+ + NH_3$$

hydrogen selenate ions (equation 84),[56,68] the alcoholysis of diselenate ions

(84) $$HOSeO_3^-Cat^+ + ROH \rightarrow ROSeO_3^-Cat^+ + H_2O$$

(scheme 85),[56,68] and the alcoholysis of adducts of selenium trioxide with

(85) $$2Cat^+O_3SeOSeO_3^{2-} + ROH \rightarrow ROSeO_3^-Cat^+ + HOSeO_3^-Cat^+$$
$$HOSeO_3^-Cat^+ + ROH \rightarrow ROSeO_3^-Cat^+ + H_2O$$

pyridine, quinoline, and 8-hydroxyquinoline (equation 86).[69]

(86) $\quad SeO_3 \cdot C_5H_5N + ROH \rightarrow ROSeO_3^- C_5H_5NH^+$

$\qquad SeO_3 \cdot C_9H_7N + ROH \rightarrow ROSeO_3^- C_9H_7NH^+$

$\qquad SeO_3 \cdot C_9H_7ON + ROH \rightarrow ROSeO_3^- C_9H_7ONH^+$

From Raman and i.r. spectroscopy of alkyl selenate ions and the corresponding calculated Se–O valence force constants, it is concluded that Se–O bonds have smaller angles, and thus greater polarities, than the S–O bonds in the analogous alkyl sulfate ions.[68] As is the case for dialkyl selenites and dialkyl selenates, nucleophilic attack on the selenium of alkyl selenate ions is easier than on the sulfur of the corresponding sulfur compounds. Thus, the hydrolysis of alkyl selenate ions proceeds much more rapidly than that of alkyl sulfate ions.[56,60,67] Transesterifications exhibit similar behavior.

Alkali alkyl selenates dissociate upon heating according to equation 87.[56]

(87) $\qquad 2CH_3OSeO_3^- Na^+ \rightarrow CH_3OCH_3 + Na_2Se_2O_7$

Ethyl selenate ions exhibit rotational isomerism in ethanolic solution.[68]

E. SPECTROSCOPY OF THE SELENIUM—OXYGEN—CARBON MOIETY*

The directional character of both bonding orbitals of the O atom suggests an angular Se–O–C arrangement. The value of this angle, however, has not yet been determined experimentally; it might well vary from 113° to 120°, depending on the compound. The angularity of the bond is inferred experimentally by the existence of two rotational isomers of the ethyl selenate ion in ethanolic solution.[68] This could be the result of steric hindrance of free rotation of the methyl group about the C–O bond.

Compounds which exhibit rotational isomerism are ethylselenious acid,[59] ethyl fluoroselenite,[46] ethyl chloroselenite,[46] alkaneseleninic acid ethyl esters,[70] the ethyl selenate anion,[68] and ethyl fluoroselenate.[66] Rotational isomerism occurs only in the liquid state, dissolved or molten, and is associated with two temperature-dependent Se–O stretching bands, observable in both Raman and i.r. spectra. Only one isomer can be found in the crystalline state, which is associated with the stretching band of the higher wave number. For example, the ethyl selenate anion in solution absorbs at 617 cm^{-1} and 676 cm^{-1}, and in the crystal at 679 cm^{-1}. The exact structures of the rotational isomers are not known.

The Se–O stretching band of the Se–O–C group appears in the range of 440–700 cm^{-1}, its position depending on the strength of the Se–O bond and

* For further discussion of the spectroscopic properties of Se–O–C compounds, see Chapter XV.

the coupling within the molecule. Although the vibrational coupling influences the frequency of the Se–O single bond stretching more than that of the Se–O double bond, the position of this stretching band (or the mean value in the case of several bands) can be taken as a qualitative measure of the strength, and hence the electron density, of the Se–O single bond. The frequency increases with an increasing effective nuclear charge on the Se atom, which is in turn dependent on the degree of oxidation, the electronegativity of the ligands, and the total charge of the molecule. Examples are given in Table IX-3 to further demonstrate these relations. An increase in

Table IX-3   *The position of the Se—O stretching band of the Se—O—C group*

X	CH_3OSeOX, cm^{-1} (Ref.)	CH_3OSeO_2X, cm^{-1} (Ref.)
CH_3	567 (70)	. . .
OCH_3	576[a] (37)	645[a] (65)
Cl	584 (46)	. . .
OH	590 (28)	648 (59)
F	615 (46)	656, 665[b] (66)

[a] Mean value.

[b] Both bands include vibrational contributions of ν_{Se-O} and ν_{Se-F}.

ν_{Se-O} is observed from left to right (increase in degree of oxidation) and downward (increase in electronegativity of the ligand). The influence of the total charge can be discerned by the transition from $CH_3OSeO_2^-$ (540 cm^{-1})[51] to CH_3OSeO_2H (590 cm^{-1})[28] and from $CH_3OSeO_3^-$ (622 cm^{-1})[68] to CH_3OSeO_2OH (648 cm^{-1}).[59]

The n.m.r. spectra of methyl fluoroselenate,[66] ethyl fluoroselenate,[66] selenium tetramethoxide,[11] diphenylselenium dimethoxide,[16] and phenylselenium trimethoxide[16] have been studied. Rotational isomers are not distinguished by n.m.r. spectroscopy, since their lifetimes are shorter than 10^{-4} second.

REFERENCES

1. R. Paetzold and M. Reichenbächer, *Z. Chem.*, **8,** 307 (1970).
2. M. Reichenbächer and R. Paetzold, *Z. Chem.*, in press (1972).
3. H. G. Ang, *J. Inorg. Nucl. Chem.*, **31,** 877 (1969).
4. H. Meerwein and T. Bersin, *Justus Liebigs Ann. Chem.*, **476,** 149 (1929).

5. P. Dupuy, *C.R. Acad. Sci., Paris*, **240**, 2238 (1955).

6. R. G. Mehrotra and S. N. Mathur, *J. Ind. Chem. Soc.*, **42**, 1 (1965).

7. A. Simon and G. Heintz, *Chem. Ber.*, **95**, 2333 (1962).

8. R. G. Mehrotra and S. N. Mathur, *J. Ind. Chem. Soc.*, **42**, 749 (1965).

9. M. Allan, A. F. Janzen and C. J. Willis, *Can. J. Chem.*, **46**, 3671 (1968).

10. J. I. Darragh and D. W. A. Sharp, *Angew. Chem.*, **82**, 45 (1970).

11. M. Reichenbächer and R. Paetzold, *Z. Anorg. Allg. Chem.*, in press (1973).

12. E. E. Aynsley, R. D. Peacock, and P. L. Robinson, *J. Chem. Soc.*, 1231 (1952).

13. J. A. Rolff, L. A. Woodward, and D. A. Long, *Trans. Faraday Soc.*, **49**, 1388 (1953).

14. H. J. M. Bowen, *Nature*, **172**, 171 (1953).

15. E. L. Muetterties and W. D. Phillips, *J. Amer. Chem. Soc.*, **81**, 1084 (1959).

16. V. Horn and R. Paetzold, *Z. Anorg. Allg. Chem.*, in press (1973).

17. M. Reichenbächer and R. Paetzold, *Z. Anorg. Allg. Chem.*, in press (1973).

18. R. Paetzold and U. Lindner, *Z. Anorg. Allg. Chem.*, **350**, 295 (1966).

19. M. Reichenbächer, G. Hopf, and R. Paetzold, *Z. Anorg. Allg. Chem.*, in press (1973).

20. R. G. Mehrotra and S. N. Mathur, *Ind. J. Chem.*, **5**, 375 (1967).

21. N. N. Melnikov and M. S. Rockickaja, *Zh. Obshch. Khim.*, **7**, 1532 (1937).

22. H. P. Kaufmann and D. Spannuth, *Chem. Ber.*, **91**, 2127 (1958).

23. G. Hesse and S. Majmudar, *Chem. Ber.*, **93**, 1129 (1960).

24. R. Paetzold, G. Hopf, and K. Aurich, *Z. Chem.*, in press (1972).

25. R. G. Mehrotra and S. N. Mathur, *J. Ind. Chem. Soc.*, **41**, 111 (1964).

26. A. Simon and R. Paetzold, *Z. Anorg. Allg. Chem.*, **303**, 53 (1960).

27. F. C. Palazzo and F. Maggiocomo, *Atti Accad. Naz. Lincei, Rend., Classe Sci., Fis., Mat., Nat.*, [**5**], **17 I**, 438 (1908); *Gazz. Chim. Ital.*, **38 II**, 122 (1908).

28. A. Simon and R. Paetzold, *Z. Anorg. Allg. Chem.*, **303**, 72 (1960).

29. A. Simon and G. Heintz, *Naturwissenschaften*, **47**, 468 (1960).

30. W. R. Orndroff, *Amer. Chem. J.*, **9**, 462 (1887).

31. W. Strecker and W. Daniel, *Justus Liebigs Ann. Chem.*, **462**, 186 (1928).

32. H. G. Cook, J. D. Ilett, B. C. Saunders and G. J. Stacey, *J. Chem. Soc.*, 3125 (1950).

33. R. Paetzold, *Z. Chem.*, **6**, 72 (1966).

34. F. Nerdel and J. Kleinwächter, *Naturwissenschaften*, **42**, 577 (1955).

35. A. Michaelis and B. Landmann, *Justus Liebigs Ann. Chem.*, **241**, 150 (1887).

36. G. Hopf and R. Paetzold, *Z. Chem.*, in press (1972).

37. A. Simon and R. Paetzold, *Z. Anorg. Allg. Chem.*, **303**, 79 (1960).

38. C. A. Bunton and B. N. Hendy, *J. Chem. Soc.*, 3137 (1963).

39. R. Paetzold and K. Aurich, *Z. Anorg. Allg. Chem.*, **348**, 94 (1966).

40. W. Voss and E. Blanke, *Justus Liebigs Ann. Chem.*, **485**, 258 (1931).

41. H. F. van Woerden, *Chem. Rev.*, **63**, 557 (1963).

42. A. Simon, R. Paetzold and H. Kriegsmann, *Chem. Ber.*, **89**, 883 (1956).

43. R. Paetzold, *Z. Chem.*, **4**, 321 (1964).

44. R. Paetzold and K. Aurich, *Z. Anorg. Allg. Chem.*, **317**, 149 (1962).

45. N. N. Yarovenko, M. A. Raksha, and G. B. Gaziewa, *Zh. Obshch. Khim.*, **31**, 4006 (1961).

46. R. Paetzold and K. Aurich, *Z. Anorg. Allg. Chem.*, **317,** 156 (1962).

47. O. Hinsberg, *Justus Liebigs Ann. Chem.*, **260,** 42 (1890).

48. S. Astin, L. de V. Moulds, and H. L. Riley, *J. Chem. Soc.*, 901 (1935).

49. H. Rheinboldt, "Houben-Weyl Methoden der Organischen Chemie," Vol. IX, E. Müller, Ed., Georg Thieme Verlag, Stuttgart, 1955, p. 1121.

50. R. Paetzold and E. Rönsch, *Z. Anorg. Allg. Chem.*, **314,** 91 (1962).

51. R. Paetzold and E. Rönsch, *Z. Anorg. Allg. Chem.*, **315,** 64 (1962).

52. R. Paetzold and E. Rönsch, *Z. Anorg. Allg. Chem.*, **338,** 22 (1965).

53. J. Meyer and W. Wagner, *Ber.*, **55,** 1216 (1922).

54. W. Strecker and W. Daniel, *Justus Liebigs Ann. Chem.*, **462,** 192 (1928).

55. J. Meyer and W. Hinke, *Z. Anorg. Chem.*, **204,** 29 (1932).

56. R. Paetzold, *Z. Anorg. Allg. Chem.*, **323,** 97 (1963).

57. J. Krejoi, L. Zborilova, and I. Horsak, *Collection Czech. Chem. Commun.*, **32,** 3468 (1967).

58. R. G. Mehrotra and S. N. Mathur, *J. Ind. Chem. Soc.*, **43,** 448 (1966).

59. R. Paetzold, Habilitationsschrift, Technical University, Dresden, 1963.

60. C. A. Bunton and B. N. Hendy, *J. Chem. Soc.*, 3130 (1963).

61. K. Dostál and L. Zborilova, *Collection Czech. Chem. Commun.*, **32,** 2809 (1967).

62. F. Ephraim and M. Gurewitsch, *Ber.*, **43,** 138 (1910).

63. J. Houben, "Die Methoden der Organischen Chemie," 3rd ed., Vol. III, Georg Thieme Verlag, Stuttgart, 1930, p. 150.

64. R. Kurze and R. Paetzold, *Z. Anorg. Allg. Chem.*, **387,** 361 (1972).

65. R. Paetzold and H. Amoulong, *Z. Chem.*, **6,** 29 (1966).

66. R. Paetzold, R. Kurze, and G. Engelbrecht, *Z. Anorg. Allg. Chem.*, **353,** 62 (1967).

67. K. Dostál and L. Zborilova, *Z. Anorg. Allg. Chem.*, **316,** 335 (1962).

68. R. Paetzold, *Z. Anorg. Allg. Chem.*, **325,** 47 (1963).

69. B. Blanka and J. Touzin, *Collection Czech. Chem. Commun.*, **32,** 3284 (1967).

70. E. Rönsch, Dissertation, Technical University, Dresden, 1964.

X Organoselenophosphorus compounds

JAN MICHALSKI AND ANNA MARKOWSKA

Institute of Organic Chemistry
Technical University of Łódź (Politechnika)
and
Institute of Organic Chemistry
Polish Academy of Sciences,
Łódź, Poland

INTRODUCTION

In this chapter an attempt will be made to survey all aspects of the chemistry of organic compounds containing selenium linked to phosphorus. The literature has been surveyed to the end of 1969, but many references from 1970 and some from 1971 are also included. No attempt has been made to discuss comprehensively topics like metal complexes, spectroscopic methods, and biological activity.

The most notable feature of the chemistry of phosphorus is its multiple valence. Stable compounds are known in which the P atom carries three, four, five, and six substituents. Almost all known organoselenophosphorus compounds belong to the class $P(IV)$—$Se(II)$. The Se atom can be bonded to phosphorus by a single σ bond or by a double bond of the p_π-d_π type. Although there is a lack of physical and particularly thermochemical data for P-Se bonds, important conclusions can be drawn about bond lengths and angles from the results of X-ray crystallography. Some representative values of these parameters are recorded in Table X-1. In selenophosphoryl compounds, $\geqslant$PSe, as in other types of tetracovalent $P(IV)$ phosphorus compounds, four substituents are arranged tetrahedrally about a central P atom that is essentially sp^3 hybridized. Angles Se=P—C are larger than 113°, and other bond angles at phosphorus are less than 109°. The P-Se bond length of 2.24–2.27 Å is close to the sum of the covalent radii of phosphorus and selenium corrected for bond polarity,[1] and the P=Se length, 2.09–2.13 Å, is in good agreement with the expected value of 2.07 Å, based on covalent double-bond radii for the two atoms.[2]

In $P(IV)$ phosphoryl and thiophosphoryl compounds the fourth bond is derived by coordination of the nonbonding $3s$ electrons of $P(III)$ phosphorus with an electron acceptor, resulting in a formal charge at the P atom. When the fourth substituent is neutral, phosphonium-type compounds result, for example, $(R_3\overset{+}{P}\text{–}SeR')Y^-$. If, however, one of the substituents bears a

Table X-1   Geometry of some selected organoselenophosphorus compounds

COMPOUND	BOND	BOND LENGTH (Å)	BOND ANGLES DEGREES		REF.
Et_3PSe	P=Se	1.963	Se=P—C	114	3
			C—P—C	106	
$MeOP(Se)Ph_2$	P=Se	2.09	Se=P—C	113	4
$(o\text{-}MeC_6H_4)_3PSe$	P=Se	2.13	Se=P—C	112.5	5
			C—P—C	107	
$[Et_2P(Se)]_2Se$	P=Se	2.095	Se=P—C	116	2
	P—Se(P)	2.24	Se—P—C	108	
$[Et_2P(S)Se]_2Te$	P—Se(Te)	2.26			6
$[Et_2P(S)Se]_2$	P—Se(Se)	2.275			7
P_4Se_5	P=Se	2.123[a]			8
	P^{IV}—Se	2.24[b]			
	P^{III}—Se	2.26–2.30[b]			
P_4Se_3	P^{III}—Se	2.24[b]			9

[a] Exocyclic Se atom.

[b] Selenium atom within the cages.

negative charge, as in the case of selenophosphoryl compounds, $\gg\!\!\overset{+}{P}\!-\!\overset{-}{Se}$, stabilization by back-donation of a lone pair of electrons from the negatively charged substituent is to be expected. Whereas $R_3\overset{+}{N}\!-\!\overset{-}{O}$ is a "true" semi-polar bond, because nitrogen cannot form more than four covalent bonds, the P-X bonds (where X = O, S, Se) can be considered as essentially double because the 3d orbitals of phosphorus are available for π bonding. The d_π–p_π bonding in selenophosphoryl compounds arises from the donation of nonbonding 4p electrons of the Se atom into vacant 3d orbitals of the P atom. Such a double bond should be appreciably shorter than a single bond, and this is consistent with the crystallographic data.

The nomenclature of the organic compounds of phosphorus is somewhat confusing even to a specialist in the field. The nomenclature rules accepted in English-speaking countries can be found in *Chem. Eng. News*, **30**, 4515 (1952) and in *J. Chem. Soc.*, 5122 (1952). To avoid the difficulties arising from different systems of nomenclature, structural formulae are given for all the compounds mentioned. All names in this chapter are based on the rules mentioned above with the exception of esters of phosphorous acid, $(HO)_2P(O)H$, which are referred to as phosphites. Thus $(EtO)_2P(O)H$ and $(EtO)_2P(Se)H$ are called diethyl phosphite and diethyl selenophosphite, respectively. A group R attached to phosphorus is used to represent a substituent with a direct P-C bond.*

For readers who would like to have deeper insight into phosphorus chemistry, and especially into analogies with organic phosphorus compounds containing sulfur, two sources of information are recommended: "The Organic Chemistry of Phosphorus" by A. J. Kirby and S. G. Warren, Elsevier Publishing Company, Amsterdam, 1967, which deals with the reactions of organic compounds of phosphorus; and "Organophosphorus Compounds" by K. Sasse, Vols. XII/1 and XII/2 of "Houben-Weyl Methoden der Organischen Chemie," edited by E. Müller, Georg Thieme Verlag, Stuttgart, 1963 and 1964. These latter volumes also describe methods of preparation. Current progress in phosphorus chemistry has been reviewed annually since 1970 in "Organophosphorus Chemistry," The Chemical Society, London. Several excellent reviews have appeared in "Topics in Phosphorus Chemistry," including structural problems of phosphorus chemistry and spectroscopy.

Preparative work with selenophosphorus compounds calls for special precautions. In addition to the obvious need for good fume-hood facilities and a device to wash reaction vessels under the hood, oxidizing agents like nitric acid and hypochlorites are recommended. Calcium hypochlorite can be used to neutralize the smell of selenophosphorus compounds in water running from the sinks. Most operations should be performed in a moisture- and oxygen-free atmosphere. Organoselenophosphorus compounds should not be subjected to prolonged heating. Therefore, careful attention to technique is required during distillation of high-boiling viscous liquids.

It has been known for a long time that numerous organothiophosphorus compounds are powerful inhibitors of cholinesterase.[10] Certain organo-thiophosphorus compounds are effective and relatively safe pesticides currently produced on a large industrial scale. Very little is known, however, about the toxicity and biological activity of the corresponding seleno compounds.

The organoselenophosphorus compounds described by Åkerfeldt and Fagerlind[11] (cf. equations 34 and 35) are among the most toxic phosphorus

---

* Editors' Note. This chapter uses the affixes selenolo and selenono to designate replacement of oxygen by selenium in P–O–R and P=O groups, respectively. Application of these terms is restricted to cases where ambiguity in the name might result from the presence of more than one chalcogen (O, S, Se, Te) atom in different bonding relationships to phosphorus. This usage is based on the 1952 Nomenclature Agreement between the Chemical Society [J. Chem. Soc., 5122 (1952)] and the American Chemical Society [Chem. Eng. News, 30, 4515 (1952)]. It is realized that this agreement was not internally consistent, having assigned the selenono prefix to both the P=Se case and to denote a selenonic acid (–SeO₃H). This latter definition has now also been adopted into IUPAC Rule C-701-1 (cf. Chapter I) without resolution of the conflicting phosphorus nomenclature.

compounds known, having LD_{50} values ranging from 0.02 to 0.06 mg/kg when injected subcutanously into mice. They are more toxic than the corresponding sulfur analogs. Their toxicity is due to an ability to inhibit cholinesterase, as shown in tests with human erythrocyte enzyme.

Several selenono- and selenolophosphorus esters are claimed in the patent literature to be effective pesticidal agents, including so-called systemic insecticides.[12,13,14]

A. ORGANIC MONOSELENOACIDS OF PHOSPHORUS AND THEIR DERIVATIVES

1. Free Acids, Salts, and Complexes

Organic monoselenoacids of phosphorus can be divided into three main types, which differ in the number of P-C bonds and in the number of alkoxy (aroxy) groups:

$(RO)_2P(Se)OH$ *O,O*-Diorganyl *O*-hydrogen phosphoroselenoates
1   *O,O*-Diorganyl phosphoroselenoic acids
$(RO)R'P(Se)OH$ *O*-Organyl *O*-hydrogen organylphosphonoselenoates
2   *O*-Organyl organylphosphonoselenoic acids
$R_2P(Se)OH$ *O*-Hydrogen diorganylphosphinoselenoates
3   Diorganylphosphinoselenoic acids

(R = alkyl or aryl)

The free acids have been described only recently,[11,15] although their salts were prepared and characterized in Pishchimuka's early classic investigations, published in 1911.[16]

a. Preparation.   Methods of preparation of monoselenoacids are based on nucleophilic dealkylation of esters, $>P(Se)OR$; hydrolysis of acid chlorides, $>P(Se)Cl$; and addition of selenium to metal salts of dialkyl phosphites $(RO)_2POM$, and their structural analogs. Pishchimuka[16] obtained the salt of *O,O*-diethyl phosphoroselenoic acid from the thermal decomposition at 95° of an *O,O,O*-triethyl phosphoroselenoate mercuric iodide adduct *in vacuo* (equation 1). At atmospheric pressure and at 78° decomposition

$$(1) \qquad (EtO)_3PSe \cdot HgI_2 \rightarrow (EtO)_2P(Se)O^-HgI^+ + EtI$$

resulted in the *O,O,Se*-triethyl phosphoroselenoate mercuric iodide adduct. Dealkylation of the same compound by alkyl mercaptide was also described by Pishchimuka (equation 2).[16] Dealkylation of *O,O,O*-trimethyl phosphoro-

$$(2) \qquad EtS^-Na^+ + (EtO)_3PSe \rightarrow Et_2S + (EtO)_2P(Se)O^-Na^+$$

selenoate by trimethylamine gives tetramethylammonium O,O-dimethyl phosphoroselenoate (equation 3).[17]

(3) $\qquad$ $(MeO)_3PSe + NMe_3 \rightarrow (MeO)_2P(Se)O^-Me_4N^+$

Diethylphosphinoselenoic acid has been obtained in high yield by alkaline hydrolysis of diethyl phosphinochloridoselenoate (equation 4).[15]

(4) $\qquad$ $R_2P(Se)Cl \xrightarrow{\text{NaOH}} R_2P(Se)ONa \xrightarrow{\text{HCl}} R_2P(Se)OH$

A similar method of preparation of the O-ethyl ethylphosphonoselenoic acid (**2**, R = R′ = Et) starts from ethylphosphonous dichloride (ethyl dichlorophosphine) (equation 5).[11] The intermediate O-ethyl ethyl phospho-

(5) $\quad$ $EtPCl_2 \xrightarrow{\text{Se}} EtP(Se)Cl_2 \xrightarrow{\text{EtOH}}$

$\qquad\qquad\qquad\qquad [(EtO)EtP(Se)Cl] \xrightarrow{\text{HOH}} (EtO)EtP(Se)OH$

nochloridoselenoate was not isolated.

A general method for the preparation of sodium and potassium salts of O,O-dialkyl phosphoroselenoic acids (**1**) has been described by Foss.[18] The synthesis is relatively simple and employs readily available dialkyl phosphites (**4**). Selenium is added to the sodium or potassium salts of these phosphites (which show the unsaturated character of tricovalent phosphorus derivatives), converting them to salts of O,O-dialkyl phosphoroselenoic acids (**1**) in almost quantitative yields (equation 6). This method has

(6)

$\qquad\qquad$ **4** $\qquad$ (M = Na, K)

recently been extended to the synthesis of O-alkyl alkylphosphonoselenoic acids (**2**).[11,15]

Dialkylphosphinoselenoic acids (**3**) are best synthesized by addition of selenium to a magnesium salt prepared *in situ* from a dialkyl phosphite (**4**) (equation 7).[15]

(7) $\qquad$ $(RO)_2P(O)H \xrightarrow{\text{3R′MgBr}} R_2'POMgBr \xrightarrow[(2)\text{HCl}]{(1)\text{Se}} R_2'P(Se)OH$

$\qquad\qquad\qquad$ **4** $\qquad\qquad\qquad\qquad\qquad$ **3**

The free acids are prepared from their salts by acidification with a mineral acid; they can then be extracted and are usually converted at once into more stable derivatives.

b. Chemical and Physical Properties. All known monoselenoacids are viscous liquids. O,O-Diorganylphosphoroselenoic acids (**1**) are thermally unstable and decompose if distillation is attempted. Some O-organyl organylphosphonoselenoic (**2**) and diorganylphosphinoselenoic (**3**) acids are stable enough to be distilled *in vacuo*. Monoselenoacids tend to decompose on storage, especially in the presence of water. In the acid-catalyzed hydrolysis of these acids (equation 8) selenium is exchanged for oxygen, suggesting the participation of a quinquecovalent intermediate.[19-21]

(8)
$$\begin{matrix} RO \\ \diagdown \\ \diagup \quad P \diagup \quad Se \\ RO \quad \diagdown OH \end{matrix} + HOH \xrightarrow{H^+} \begin{matrix} RO \quad SeH \\ \diagdown \diagup \\ P—OH \\ \diagup \diagdown \\ RO \quad OH \end{matrix} \longrightarrow \begin{matrix} RO \quad O \\ \diagdown \diagup \\ P \\ \diagup \diagdown \\ RO \quad OH \end{matrix} + H_2Se$$

The tautomerism of monoselenoacids has been investigated by analysis of Hammett constants.[22] The equilibrium was found to be shifted towards the selenono form (equation 9), just as phosphorus monothioacids exist predominantly in the thionoform.[23,24]

(9)
$$\begin{matrix} \diagdown \quad Se \\ \diagup P \diagup \\ \diagup \diagdown \\ OH \end{matrix} \rightleftharpoons \begin{matrix} \diagdown \quad SeH \\ \diagup P \diagup \\ \diagup \diagdown \\ O \end{matrix}$$

Monoselenoacids of phosphorus are medium to strong acids; for example, the pK_a value for **3** (R = Et) is 4.67 in 80% ethanol.[22] Although structure and acid strength are difficult to correlate because of the limited experimental data available, phosphinic acids, $R_2P(Se)XH$ (X = Se, S, or O), may be ranked in the following order of decreasing strength:

$$R_2P(Se)SeH > R_2P(Se)SH > R_2P(Se)OH$$

The mesomeric anion of the acid **1** is an ambident nucleophile which reacts at either the Se or the O atom. Alkylation with alkyl halides leads to formation of a Se-C bond (equation 10a), but in phosphorylation a P-O bond is formed (equation 10b).

In general terms, the ambident character of the monoselenoacid anion is very much like that of the monothioacid anion, but the nucleophilicity of selenium towards sp^3 carbon is, as expected, higher than that of sulfur in analogous systems.[25] Data so far available indicate that alkylation of the

(10a)

$$\begin{array}{ccc} RO & & Se \\ & \diagdown \; \diagup \\ & P\; - \\ & \diagup \; \diagdown \\ RO & & O \end{array} \quad \xrightarrow{\;R'X\;} \quad \begin{array}{ccc} RO & & SeR' \\ & \diagdown \; \diagup \\ & P \\ & \diagup \; \diagdown \\ RO & & O \end{array}$$

(10b)

5

$$\xrightarrow{\;(RO)_2P(O)Cl\;} \quad (RO)_2P-O-P(OR)_2$$
$$\qquad\qquad\qquad\qquad \overset{\|}{Se} \qquad \overset{\|}{O}$$

6

monoselenoacid anion leads selectively to O,O,Se-trialkyl phosphoroselenoates (**5**) and that phosphorylation gives tetraalkyl monoselenopyrophosphates (**6**), but detailed mechanistic studies are required. Alkylation of the corresponding monothioacid anions leads in some cases to a mixture of isomeric thiolo- and thionoesters.[26-28] O-Acylation of a monoselenoacid anion may be considered as an intermediate stage in its reaction with ketene (see equation 61).

Dimethyl sulfoxide oxidizes all types of monoselenoacids of phosphorus (equation 11).[29] The reaction is presumably initiated by proton transfer

(11) $R_2P(Se)OH + Me_2SO \rightarrow R_2P(O)OH + Me_2S + Se$

since salts of these acids are oxidized only under drastic conditions.[29,30]

Oxidation of sodium and potassium O,O-dialkyl phosphoroselenoic acid (**1**) salts with iodine[18] and sulfuryl chloride[31] gives bis(dialkoxyphosphinyl) diselenides (**19**) (see equations 62, 63). Monoselenoacids $ROP(Se)(OH)_2$ and $RP(Se)(OH)_2$ and their salts have not been described because of difficulties in removing alkoxy groups from **12** (see equation 20) and **13** without exchange of selenium for oxygen.

The instability shown by free monoselenoacids is not so marked in their salts and complexes. Sodium salts are more soluble in organic solvents than potassium salts, and salts of selenoacids are more soluble than those of thioacids, with the exception of methyl derivatives.[18] These differences in solubility match the degrees of covalence expected. The sodium and potassium salts of **1** are quantitatively oxidized by iodine in sodium hydrogen carbonate buffers (equation 12).[18] This method provides an accurate and

(12) $(RO)_2PSeO^- + 3I_2 + 8OH^- \rightarrow (RO)_2POO^- +$
$$SeO_3{}^{2-} + 6I^- + 4H_2O$$

self-consistent analytical procedure. Larionov and Il'ina[32] have studied both the sodium salt and the Co^{2+} complex of **1** (R = Et), and their interpretation of i.r. spectra indicates that the ligand is attached to oxygen in the former and to selenium in the latter. Claims for the existence of two isomeric forms of selenoacid salts, $>P(O)Se^- \, ^+NMe_4$ and $>P(Se)O^- \, ^+NMe_4$, are erroneous.[17] Complexes of diorganylselenophosphinic acid with Co^{2+}, Zn^{2+}, and Cd^{2+} have been studied by Kuchen and Hertel[33] and are described in their excellent review. They are complexes of lesser stability within the series $Et_2P(X)YM$ (M = metal; X = S, Se; Y = O, S, Se). Attempts to prepare Ni^{2+} and Bi^{3+} complexes failed. Zn^{2+} and Cd^{2+} complexes exhibit concentration-dependent association in benzene and chloroform because of the presence of ligand bridges.

2. Acid Halides: $>P(Se)Cl$, $-P(Se)Cl_2$, $-P(Se)Br_2$

a. Preparation. Diphenyl phosphorochloridoselenoate (**7a**, R = Ph) was described in 1916 by Strecker and Grossmann,[34] who prepared it by addition of elemental selenium to O,O-diphenyl phosphorochloridite (di-

(13a)
$$R_2P\!\!-\!\!Cl \xrightarrow[140°]{Se} R_2P(Se)Cl$$
$$\mathbf{7}$$

(13b)
$$(RO)_2P\!\!-\!\!Cl \xrightarrow[200°]{Se} (RO)_2P(Se)Cl$$
$$\mathbf{7a}$$

phenyl chlorophosphite). This method has been extended to the preparation of **7** (R = Et, R = Ph)[15,36] and **7a** (R = Et).[35]

It must be emphasized that phosphorochloridites, $(RO)_2P\!\!-\!\!Cl$, like most tricovalent phosphorus compounds, are "biphilic reagents," able to donate electrons to form σ bonds and to accept electrons by back-donation into the vacant $3d$ orbitals of phosphorus. Here the chlorophosphite acts as a nucleophile, but it can exhibit pronounced electrophilic character in other reactions.

The addition of selenium to alkylphosphinous dichloride (alkyl dichlorophosphine) was described by Gryszkiewicz-Trochimowski *et al.*[37] and by Ivin and Shelyakova,[38] and the method (equation 14) was later confirmed by other

(14)
$$RPCl_2 \xrightarrow[AlCl_3]{Se} RP(Se)Cl_2$$
$$\mathbf{8}$$

authors.[11,39] The presence of a catalyst is essential. Addition of red selenium to methylphosphinous dibromide proceeds without catalyst, however, and in high yield (equation 15).[40]

(15)
$$MePBr_2 \xrightarrow{Se} MeP(Se)Br_2$$

Ivin and Shelyakova[38] developed a method of preparation of **8** (R = Et) based on the reaction of elemental selenium with the complex formed by ethyl chloride, phosphorus trichloride, and aluminum chloride (equation 16).

$$(16) \qquad [RPCl_3]^+[AlCl_4]^- \xrightarrow{\text{Se}} RP(Se)Cl_2$$
$$\textbf{8}$$

Pudovik and Ishmayeva[41] obtained the dichloride (**9**) by action of hydrogen selenide on the addition product of phosphorus pentachloride and butadiene (equation 17).

$$(17) \quad CH_2{=}CH{-}CH{=}CH_2 + 2PCl_5 \longrightarrow$$
$$(Cl{-}CH_2CH{=}CH{-}CH_2{-}\overset{+}{P}Cl_3)\overset{-}{P}Cl_6 \xrightarrow{2H_2Se}$$
$$Cl{-}CH_2CH{=}CH{-}CH_2P(Se)Cl_2 + Cl_3PSe + 4HCl$$
$$\textbf{9}$$

Chlorination of dialkyl selenophosphites (**10**) by sulfuryl chloride leads to formation of O,O-dialkyl phosphorochloridoselenoates.[19,20] The method is

$$(18) \qquad \begin{array}{cc} RO & Se \\ & \diagdown \quad \diagup\diagup \\ & P \\ & \diagup \quad \diagdown \\ RO & H \end{array} \xrightarrow{SO_2Cl_2} \begin{array}{cc} RO & Se \\ & \diagdown \quad \diagup\diagup \\ & P \\ & \diagup \quad \diagdown \\ RO & Cl \end{array} + HCl + SO_2$$
$$\qquad\qquad \textbf{10} \qquad\qquad\qquad \textbf{7a}$$

very convenient provided that dialkyl selenophosphites (*cf.* equation 114) are available. The main problem in all these preparative methods is the thermal instability of the acid chlorides, $>P(Se)Cl$ and $-P(Se)Cl_2$.

b.   Chemical Properties.   Chlorides of type **7a** are typical examples of selenophosphorylating agents which readily undergo nucleophilic substitution at the P atom (equation 19). Examples are given in this chapter of

$$(19) \qquad Nu:^- \begin{array}{cc} RO & Se \\ & \diagdown \quad \diagup\diagup \\ & P \\ & \diagup \quad \diagdown \\ RO & Cl \end{array} \longrightarrow \begin{array}{cc} RO & Se \\ & \diagdown \quad \diagup\diagup \\ & P \\ & \diagup \quad \diagdown \\ RO & Nu \end{array} + Cl^-$$
$$\qquad\qquad\qquad \textbf{7a}$$

hydrolysis, leading to monoselenoacids (equations 4 and 5); alcoholysis, yielding esters, $>P(Se)OR$ (equation 21); and aminolysis, yielding amides, $>P(Se)NR_2$ (equation 105). Many examples involve compounds containing P-N bonds, for example, $R_2N(R'O)P(Se)Cl$ and $(R_2N)_2P(Se)Cl$ (equations 103a, 103b, 104). On the basis of the behavior of thio and oxo analogs,[42,43] it can be assumed that nucleophilic displacement at the selenophosphoryl center is of $S_N2(P)$ type with a trigonal bipyramidic transition state or intermediate and colinear attacking and leaving groups.

3. Selenonoesters: >P(Se)OR

a. Preparation. The most general method of preparation of seleno-noesters, based on the addition of elemental selenium to trialkyl phosphites (**11**) and their structural analogs (equation 20), has been described by

(20)

$$\begin{array}{ccc}
RO & & RO \quad Se \\
\diagdown & \xrightarrow{\text{Se}} & \diagdown \diagup\diagup \\
P{-}OR & & P \\
\diagup & & \diagup \diagdown \\
RO & & RO \quad OR \\
\mathbf{11} & & \mathbf{12}
\end{array}$$

Pishchimuka,[16] and several investigators have applied it to the preparation of *O,O,O*-trialkyl and *O,O,O*-triaryl phosphoroselenoates (**12**),[44-48] *O,O*-dialkyl phosphonoselenoates (**13**)[49,50] and *O*-alkyl phosphinoselenoates (**14**).[4]

$$(RO)_2P(Se)R' \qquad (RO)P(Se)R'_2$$
$$\mathbf{13} \qquad\qquad \mathbf{14}$$

Cyclic selenophosphates have been prepared by adding elemental selenium to cyclic phosphites, but yields in these reactions[51,52] are low in comparison with those obtained with acyclic systems. The preparative procedure can be simplified by adding selenium to a crude P(III) ester prepared *in situ*.[37] More recently, acyclic[41] (equation 21) and cyclic[39] selenonoesters (**15**) and (**16**) (equations 22 and 23) have been prepared by nucleophilic displacement at the P atom.

Three methods specific for the preparation of esters of *O*-organyl organyl-phosphonoselenoic acids are based on dialkyl selenophosphites (**10**) as starting materials. These are additions to α,β-unsaturated compounds, additions to a carbonyl group, and condensations of sodium dialkyl seleno-phosphite with alkyl halides as described in Section X.F (see equations 118–120).

b. Structure, Physical and Chemical Properties. The structure of **14** (R = Me, R' = Ph) has been examined by X-ray crystallography.[4] The four substituents at phosphorus are arranged tetrahedrally. The ^{1}H n.m.r. spectrum of **15** (R = R' = Me) indicates the magnetic nonequi-valence of methyl groups,[39,53] and of **15** (R = Me, R' = H) indicates *cis-trans* isomerism with three isomers possible.[39] Most selenonoesters are high-boiling liquids with a very unpleasant odor, insoluble or partly soluble in water but soluble in organic solvents.

Little is known about the chemical properties of selenonoesters. The classic work of Pishchimuka is still an important source of information.[16] The chemistry of alkyl selenonoesters should be determined largely by their

(21) $Cl-CH_2-CH=CH-CH_2P(Se)Cl_2 + 2MeOH \xrightarrow{2Et_3N}$

      9

$$Cl-CH_2CH=CHCH_2P(Se)(OMe)_2 + 2Et_3N \cdot HCl$$

(22)

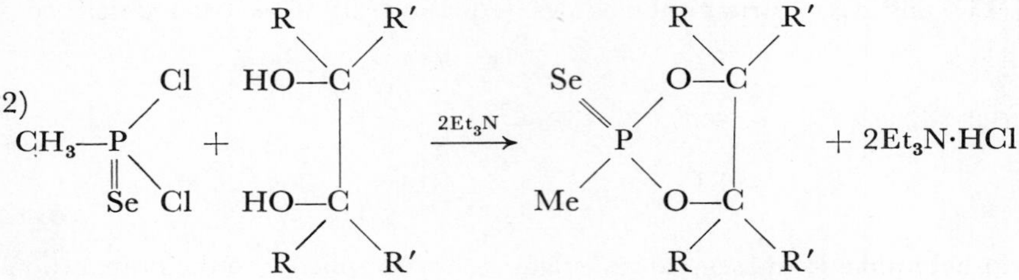

$$(R = Me; R' = Me \text{ or } H)$$

(23)

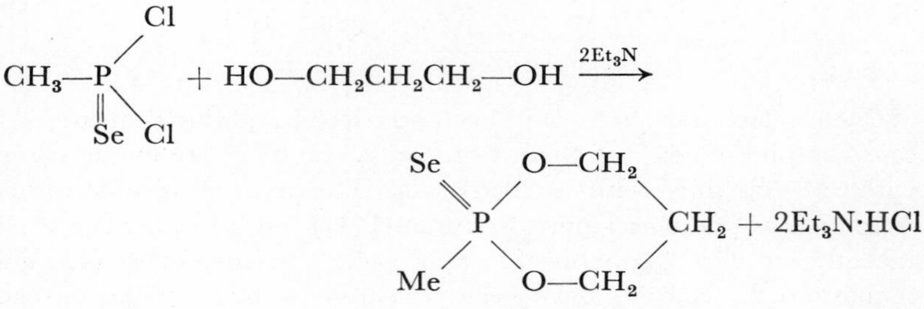

alkylating properties. Although little experimental work has been carried out in this field, many analogies can be drawn from the chemistry of alkyl thiophosphates, which is described in an excellent review by Hilgetag and Teichmann.[54] Some dealkylations have already been mentioned in connection with the preparation of monoselenoacids salts (see equations 2 and 3). Nucleophilic reagents like the mercaptide ion (polarizable, soft)[55] should cause an S_N2 reaction at the α-C atom of the ester group (equation 24).

(24)

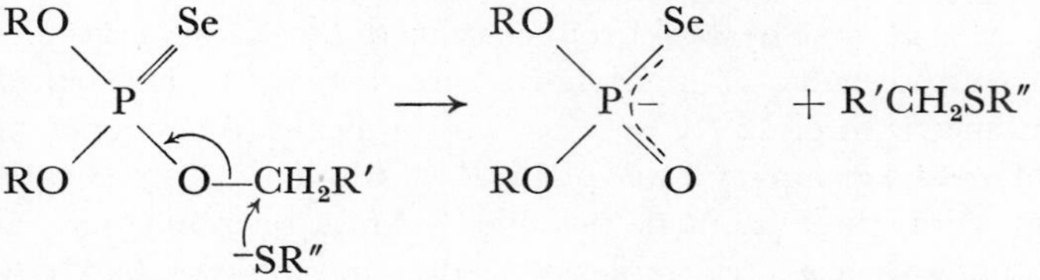

Alkoxide anions (strong base, hard),[55] on the other hand, should attack the P atom either in an $S_N2(P)$ reaction or through a pentacoordinate intermediate (equation 25). This follows from a comparison of the nucleophilicity constants for tetrahedral carbon and tetrahedral phosphorus.[56]

(25)

$$R'O^- \overset{\displaystyle RO \quad \; Se}{\underset{\displaystyle RO \quad \; OR}{\diagup\hspace{-0.5em}P}} \longrightarrow \overset{\displaystyle RO \quad \; Se}{\underset{\displaystyle R'O \quad \; OR}{\diagup\hspace{-0.5em}P}} + RO^-$$

The Se atom in selenonoesters is nucleophilic, and alkylation results in formation of phosphonium complexes that may or may not undergo further transformation, depending on the nature of the anion and the substituents at phosphorus. Formation of stable complexes with an anion of low nucleophilicity has been described by Schmidpeter and Brecht[36] (equation 26).

(26) $\quad MeOP(Se)Ph_2 + Me_3OSbCl_6^- \longrightarrow \left[\underset{\displaystyle SeMe}{\overset{\displaystyle +}{MeOPPh_2}} \right] SbCl_6^- + Me_2O$

A similar complex with **12** (R = Me) was found to be very unstable.[36]

Complexes with boron trifluoride (equation 27)[57] and mercuric diiodide (equation 28)[16] have also been thermally isomerized into the corresponding

(27) $\qquad (MeO)_3PSe \cdot BF_3 \xrightarrow{140°} (MeO)_2P(O)(SeMe) \cdot BF_3$

(28) $\qquad (EtO)_3PSe \cdot HgI_2 \xrightarrow{78°} (EtO)_2P(O)(SeEt) \cdot HgI_2$

complexes of selenoloesters (**5**). The isomerization of the HgI_2-selenonoester complex was interpreted by Pishchimuka[16] as dealkylation followed by subsequent alkylation with ethyl iodide (equation 29). The complex with HgI_2

(29) $\quad HgI_2 \cdot SeP(OEt)_3 \rightarrow [(EtO)_2P(Se)O^-HgI^+ + EtI] \rightarrow$
$$(EtO)_2P(O)SeEt \cdot HgI_2$$

may itself act as the alkylating agent.[58] A thermal Pishchimuka isomerization can also occur without complex formation, since the selenonoester PhP(Se)(OMe)$_2$, prepared by addition of elemental selenium to the tricovalent phosphorus ester, has been found to be contaminated with the selenoloisomer PhP(O)(SeMe)(OMe).[59]

More recent examples include methyl esters, (PhO)$_2$P(Se)OMe (equation 30) and (PhO)(Et$_2$N)P(Se)OMe. Isomerization occurs at 150–180° and probably involves an intermolecular reaction of the Pishchimuka type (see equation 36).[60]

For thermal isomerization of selenonoesters containing an allyl group a cyclic mechanism (equation 31) should also be considered.[61] In addition, isomerization catalyzed by tetraethylammonium iodide has been described (equation 32).[62]

(30)

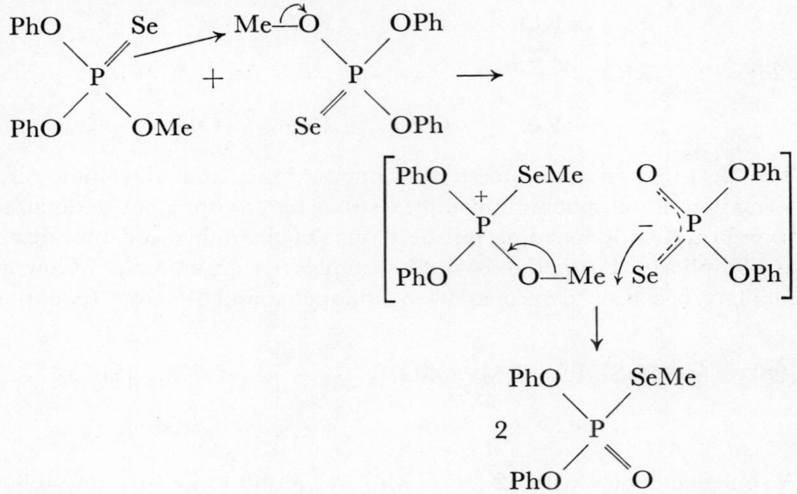

(31)

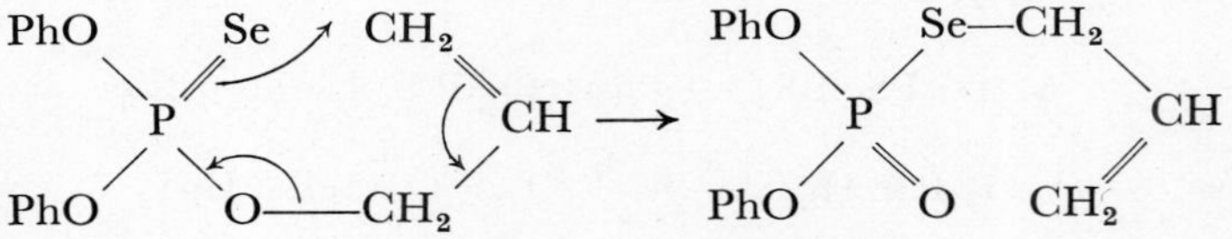

(32)

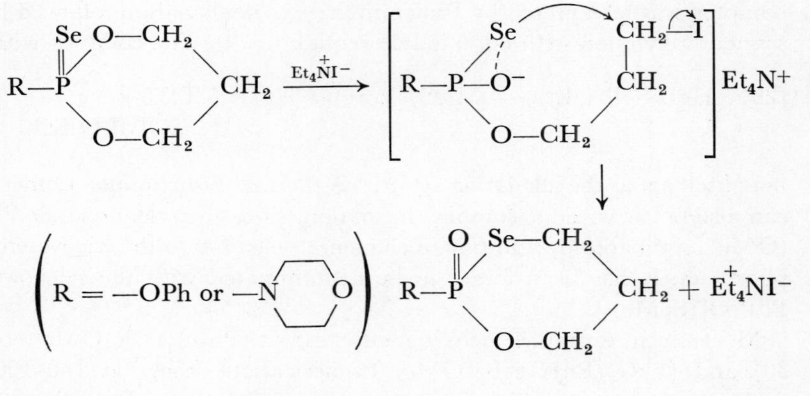

$$\left(R = \text{—OPh or } \text{—N} \bigcirc O \right)$$

(33)

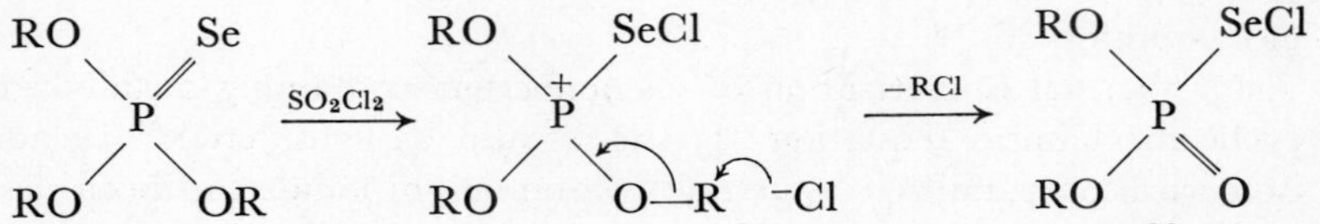

12

20

Soft electrophiles like halogens and halogenating agents attack the Se atom in the selenonoester (**12**), yielding selenenyl chlorides (**20**) (equation 33).[63,64]

The possibility of oxidation of selenonophosphates to phosphates has not yet been explored, although the corresponding oxidation of thiono compounds is well known.

4. Selenoloesters: $>P(O)SeR$

a. Preparation. The ambident nature of the anion $>P(Se)O^-$ has already been discussed (see equations 10a and 10b), and its use in selenoloester preparation has been indicated. More recent examples (equations 34, 35) illustrate the wide application of this method.[11] The reaction with aziridine

$$(34) \quad Et(EtO)P(Se)O^- Na^+ + ClCH_2CH_2NEt_2 \xrightarrow{-NaCl}$$
$$Et(EtO)P(O)SeCH_2CH_2NEt_2$$

$$(35) \quad (EtO)_2P(Se)O^- Na^+ + H_2C\underset{\underset{H}{N}}{\overset{}{\diagdown\diagup}}CH_2 \longrightarrow$$

$$(EtO)_2P(O)SeCH_2CH_2NHNa \xrightarrow{H_2O} (EtO)_2P(O)SeCH_2CH_2NH_2 + NaOH$$

takes place at room temperature and is presumably assisted by ring strain energy. Alkylation of selenonoesters, well known for thionophosphates as the Pishchimuka reaction,[67] should be a convenient method of preparing selenoloesters (**5**) (equation 36). Very little information is presently available

(36)

about this interesting reaction. One recent example, however, is illustrated in equation 37.[36]

$$(37) \quad (MeO)_3P{=}Se \xrightarrow{(MeO)_2SO_2} (MeO)_2P(O)SeMe$$

Synthetic methods based on thermal and catalytic isomerizations of selenonoesters have been already described (see equations 30–32).

A wide group of reactions leading to selenoloesters involves the interaction of phosphorus nucleophiles of the trialkyl phosphite (**11**) (*cf.* equation 20) and dialkyl phosphite (**4**) (*cf.* equation 6) types with selenium electrophiles R–Se–Y, where Y is a suitable leaving group (Y = Cl, Br, CN, SeR) (equations 38–40). The mechanistic scheme of reactions with trialkyl

$$(38) \qquad R{-}SeY + {>}P{-}OR \longrightarrow RSe{-}P(O){<} + R{-}Y$$

$$(39) \qquad R{-}SeY + {>}P(O)M \longrightarrow RSe{-}P(O){<} + M^+Y^-$$

$$(40) \qquad R{-}SeY + {>}P(O)H \longrightarrow RSe{-}P(O){<} + HY$$

phosphites involving formation of an intermediate phosphonium complex follows the pattern of the Arbuzov reaction (equation 41). The most inter-

(41)

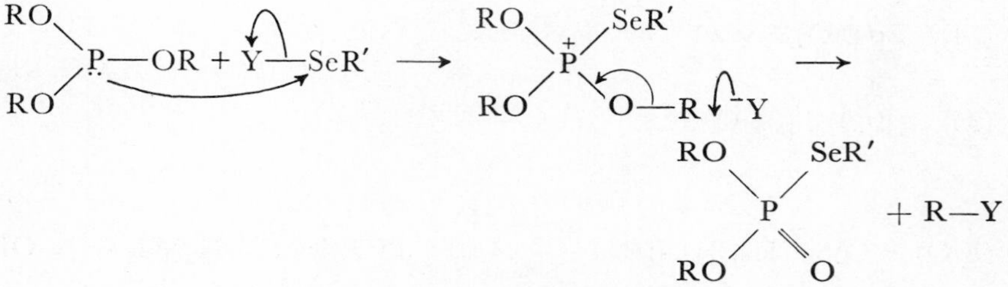

esting examples are those in which Y = Cl, Br (Kataev *et al.*[68,69] and Petragnani *et al.*)[70] and Y = CN, SeR (Michalski and Wieczorkowski).[31,71]

Much higher temperatures are required in the cases of selenocyanates, RSeCN, and diselenides, (RSe–)$_2$, than are needed when selenenyl halides, RSeX, are used. Yields are usually good. These reactions can be classified as nucleophilic substitution at selenium. Attack of the phosphorus nucleophile on positive halogen is also possible (equation 42).[72]

(42)

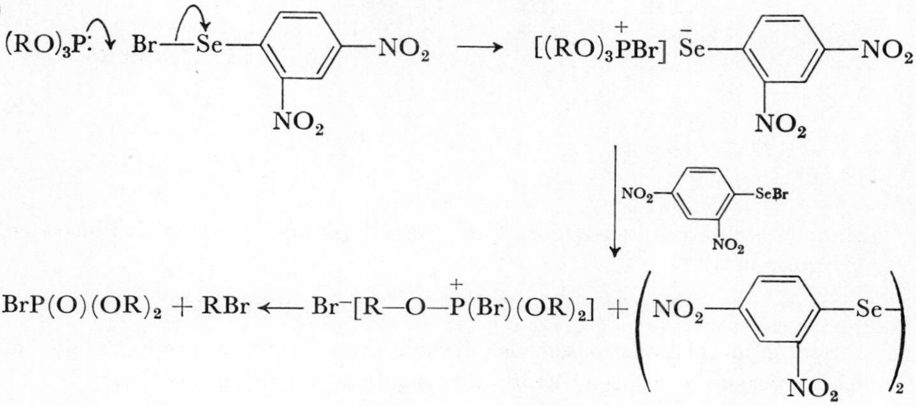

A free-radical mechanism is less likely to occur, although no work has been reported to rule out such a possibility. Dialkyl phosphorochloridites (dialkyl chlorophosphites) can also act as nucleophiles and have been employed to synthesize selenoloesters (**5**) containing three different substituents at the P atom (equation 43).[73]

$$(43) \quad PhSeCl + (EtO)_2PCl \longrightarrow [(EtO)_2\overset{+}{P}(SePh)Cl]Cl^- \longrightarrow$$

$$(EtO)(PhSe)P(O)Cl \xrightarrow{ROH} (EtO)(PhSe)P(O)OR$$

A Michaelis-Becker type of reaction (see equation 118 and relevant text) has been reported between sodium dialkyl phosphites and alkyl or aryl selenocyanates (equation 44).[74-76] Selenium analogs of organophosphorus

$$(44) \quad (RO)_2PONa + NCSeR' \rightarrow (RO)_2P(O)SeR' + NaCN$$

pesticides have been prepared in this way. Mercury derivatives of dialkyl phosphites obtained by the Arbuzov reaction react with selenenyl halides (equation 45).[70] The mercury derivatives are less reactive than sodium

$$(45) \quad (RO)_3P \xrightarrow[-RCl]{HgCl_2} (RO)_2P(O)HgCl \xrightarrow[dioxane]{PhSeBr}$$

$$(RO)_2P(O)SePh + HgBrCl$$

dialkyl phosphites because the nucleophiles in the two cases have different structures—tricovalent in the former, and tetracovalent in the latter. Free dialkyl phosphites react readily with organic selenenyl halides (equation 46).

$$(46) \quad ArSeCl + [(RO)_2P(O)H \rightleftharpoons (RO)_2POH] \rightarrow (RO)_2P(O)SeAr + HCl$$

The reaction of dialkyl phosphites with organic selenocyanates (equation 47) is of special interest[76] because the transition state appears to be stabilized

by hydrogen bonding. In both reactions the tautomeric form with P(III) (see equation 46) is probably the reactive species.[77] Disubstituted phosphines and their metal derivatives can be used in reactions with R–SeY type reagents (equations 48–50).[70] The intermediate *Se*-phenyl diphenyl phosphinoselenoite is very readily oxidized (*cf.* equations 49, 50).[78]

Two routes based on nucleophilic substitution at tetracovalent phosphorus and tricovalent phosphorus have been described (equations 51, 52).[70,79] A reaction similar to equation 51 between phenyl phosphonic dichloride and phenylselenomagnesium bromide gives a product (presumably the acid

(48) $2 PhSeBr + Ph_2PH \longrightarrow [Ph_2P(Br_2)SePh] + \frac{1}{2}(PhSe-)_2$

$$\downarrow H_2O$$

$$Ph_2P(O)SePh$$
17

(49) $PhSeBr + Ph_2PMgCl$

$$[PhSePPh_2] \xrightarrow{[O]} Ph_2P(O)SePh$$

(50) $PhSe-SePh + Ph_2PMgCl$

(51) $PhSeMgBr + Ph_2P(O)Cl \longrightarrow Ph_2P(O)SePh + MgClBr$

$$\uparrow [O]$$

(52) $PhSeMgBr + Ph_2PCl \longrightarrow [Ph_2PSePh] + MgClBr$

chloride), which can be hydrolyzed to *Se*-phenyl hydrogen phenylphosphono-
selenoate (equation 53).[79] The extension of these reactions to phosphorus

(53) $PhSeMgBr + PhP(O)Cl_2 \longrightarrow$ (product) $\xrightarrow{HOH}$
$$PhP(O)(SePh)(OH)$$

acid chlorides containing alkoxy and aroxy groups, $(RO)_2P(Se)Cl$ and
$R(RO)P(Se)Cl$, should be possible. Excess of the RSe^- nucleophile must
be avoided, however, to prevent dealkylation of the products.

Two other types of reactions were described by Petragnani *et al.*,[70] and
the schemes shown in equations 54, 55 were proposed.[70] Phosphinoselenoates,

(54) $PhSe-SePh + 2Ph_2PCl \longrightarrow Ph_2P(Cl_2)SePh \xrightarrow{H_2O} Ph_2P(O)SePh$

(55) $Ph_2P(O)-P(O)Ph_2 + PhSeBr \longrightarrow Ph_2P-O-SePh + Ph_2P(O)Br$

$$\downarrow$$

$$Ph_2P(O)SePh$$

$R_2P(O)SeR'$, containing alkyl groups attached to the P atom have not been
described. A special group of selenolophosphates and phosphonates can be
obtained by addition of the corresponding oxophosphoraneselenenyl chlorides (see equations 33 and 65) to ethylenic hydrocarbons (equation 56).[63,64]

(56)

The addition product is presumably in the *trans* form.

b. Physical and Chemical Properties. Selenoloesters are high-boiling liquids with the exception of compounds like **17**, which are crystalline. Their physical properties are similar to those of selenonoisomers.

Practically no published information concerning the chemical properties of selenoloesters is available. They should act as phosphorylating agents, as there is reason to suppose that –SeR is a relatively good leaving group. Vulnerability to attack by halogenating agents can also be expected. The formation of a 1:1 complex with BF_3 and HgI_2 has been reported.[16,40] The comparative reactivity of $>P(O)SeR$ and $>P(O)SR$ compounds is an open field for research.

5. Selenopyrophosphates and Mixed Anhydrides

a. Tetraalkyl Monoselenopyrophosphates. Tetraalkyl monoselenopyrophosphates (**6**; see equation 57) have been prepared by phosphorylation of sodium dialkyl selenophosphates[80–84] (see equation 10b). Structure **6** was confirmed by n.m.r. and i.r. spectra.[83,85] The isomeric symmetric monoselenopyrophosphates (**18**) are unknown. If an analogy with thiopyrophosphates is justified, symmetric monoselenopyrophosphates (**18**) should be unstable and should readily isomerize to **6**. Symmetric tetraalkyl monothiopyrophosphates have been obtained only recently, and some of them isomerize rapidly at ambient temperature.[86] One reaction has been reported, however, in which intermediate formation of **18** is almost certain (equation 57).[31]

(57)

$$(RO)_2P\!\!\underset{\underset{O}{\|}}{\overset{\overbrace{\,:P(OR)_3}}{{-}Se{-}Se{-}P(OR)_2}}\;\longrightarrow\; \left[(RO)_2P\overset{Se}{\underset{\underset{O}{\|}}{\Big\langle}}\;\;R\!-\!O\!-\!\overset{+}{\underset{\underset{OR\;O}{\|}}{P}}\!-\!Se\!-\!P(OR)_2\right]$$

19

$$\underset{\overset{\|}{Se}}{(RO)_2P}\underset{\overset{\|}{O}}{-O-P(OR)_2}\;\longleftarrow\;\underset{\overset{\|}{O}}{(RO)_2P}\underset{\overset{\|}{O}}{-Se-P(OR)_2}\;+\;\underset{\overset{\|}{O}}{RSeP(OR)_2}$$

6   **18**   **5**

(R = Et)

b. Tetraalkyl Diselenopyrophosphates. Tetraethyl diselenopyrophosphate has been prepared by addition of selenium to tetraethyl pyrophosphite (equation 58).[80,87] The intermediate diethylphosphorous-diethylseleno-

(58) $\;(EtO)_2P\!-\!O\!-\!P(OEt)_2 \xrightarrow{\;Se\;} (EtO)_2P\!-\!O\!-\!P(OEt)_2 \xrightarrow{\;Se\;}$
$$\underset{\overset{\|}{Se}}{}$$
$$(EtO)_2P\!-\!O\!-\!P(OEt)_2$$
$$\underset{\overset{\|}{Se}}{}\;\;\underset{\overset{\|}{Se}}{}$$

phosphoric anhydride has been isolated.[80]

c. *Tetraalkyl Selenothiopyrophosphates.* Thiophosphorylation of sodium diethyl phosphoroselenoate leads to tetraethyl P¹-thio-P²-selenopyrophosphate (equation 59).[80] For similar anhydrides, see Section X.B.4 and

(59)

$$
\begin{array}{c}
\underset{EtO}{\overset{EtO}{>}}\!\!P\!\!\underset{O}{\overset{Se}{<}}{}^{-} \; + \; \underset{S}{\overset{Cl}{>}}\!\!P\!\!\underset{OEt}{\overset{OEt}{<}} \xrightarrow{40\%} \underset{EtO}{\overset{EtO}{>}}\!\!P\!\!\underset{Se}{\overset{||}{}}\!\!-O-\!\!P\!\!\underset{S}{\overset{||}{\underset{OEt}{}}}\!\!\overset{OEt}{<} \; + \; Cl^{-}
\end{array}
$$

equation 86.

d. *Anhydrides with Sulfonic and Carboxylic Acids.* It has been claimed that mixed anhydrides of dialkyl selenophosphoric acids and sulfonic acids can be obtained (equation 60).[80,81] This is a somewhat unexpected result

(60) $(EtO)_2PSeO^- Na^+ + ClSO_2R \rightarrow (EtO)_2P(Se)OSO_2CH_3 + NaCl$

because an anhydride with such a good leaving group should react with the selenoacid salt to give tetraethyl diselenopyrophosphate. Similar reactions with sulfur analogs lead invariably to tetraalkyl dithiopyrophosphates, $(RO)_2P(S)OP(S)(OR)_2$.

Monoselenophosphoric-acetic acid mixed anhydrides have been obtained in almost quantitative yield by the action of ketene on selenoacids (equation 61).[88] The reaction is probably initiated by a proton transfer.

(61) $R_2P(Se)OH + CH_2{=}C{=}O \xrightarrow{slow}$

$$[CH_3{-}\overset{+}{C}{=}O] R_2P(Se)O^- \xrightarrow{fast} R_2P(Se)OC(O)CH_3$$

6. Bis(dialkoxyphosphinyl) Diselenides: $[(RO)_2P(O)Se{-}]_2$

a. *Preparation.* The bis(dialkoxyphosphinyl) diselenides (**19**) were prepared by Foss[18] by oxidation of the corresponding selenoacid sodium or

(62) $2(RO)_2P(Se)O^- + I_3^- \rightleftharpoons [(RO)_2P(O)Se{-}]_2 + 3I^-$

19

potassium salts. The equilibria are established in neutral and in acidic aqueous solutions. The dialkyl phosphoroselenoates are more readily oxidized by iodine than the corresponding phosphorothioates, the latter being indifferent to iodine in the presence of iodide. Diselenides (**19**) can also be obtained by action of sulfuryl chloride on sodium or potassium diethyl phosphoroselenoate (equation 63).[31]

(63) $2(RO)_2P(Se)ONa + SO_2Cl_2 \rightarrow [(RO)_2P(O)Se{-}]_2 + NaCl + SO_2$

b. Physical and Chemical Properties. On the basis of their oxidation-reduction potentials diselenides of type **19** can be considered as pseudohalogens (e.g., $E_0^{Se} = 0.37$ V for R = Me; $E_0^{Se} = 0.31$ V for R = i-Pr).[18] They are capable of oxidizing iodide, thiosulfate (to tetrathionate), and thiocarbonyl salts (to the corresponding disulfides). The redox potentials of **19** are 0.19 V less negative than those of the corresponding disulfides, and in consequence the almost quantitative transformation shown in equation 64

$$(64) \quad [(RO)_2P(O)S-]_2 + 2(RO)P(Se)O^- \rightarrow [(RO)_2P(O)Se-]_2 + 2(RO)P(S)O^-$$

has been demonstrated.[18] Further details and a comparison of **19** with bis(dialkoxyphosphinyl) disulfides can be found in the elegant work of Foss.[18]

7. Oxophosphoraneselenenyl Chlorides: $(RO)_2P(O)SeCl$

If the diselenides (**19**) can be considered as pseudohalogens, the oxophosphoraneselenenyl chlorides (**20**) may be described as halogeno-pseudohalogens.[63,64]

a. Preparation. Selenenyl chlorides (**20**) are prepared by three methods: chlorination of selenoesters (equation 65), of selenoacids (equation 66), and of diselenides (equation 67). Reactions of sulfuryl chloride with **1** and **19** are

$$(65) \quad (RO)_3P{=}Se + SO_2Cl_2 \rightarrow (RO)_2P(O)SeCl + RCl + SO_2$$
$$ \mathbf{12} \mathbf{20}$$

$$(66) \quad (RO)_2P(Se)OH + SO_2Cl_2 \rightarrow (RO)_2P(O)SeCl + HCl + SO_2$$
$$ \mathbf{1}$$

$$(67) \quad [(RO)_2P(O)Se-]_2 + SO_2Cl_2 \rightarrow 2(RO)_2P(O)SeCl + SO_2$$
$$ \mathbf{19}$$

closely related since condensation of **1** with **20** leads to **19** (equation 68).

$$(68) \quad (RO)_2P(Se)OH + (RO)_2P(O)SeCl \rightarrow [(RO)_2P(O)Se-]_2 + HCl$$
$$ \mathbf{1} \mathbf{20} \mathbf{19}$$

b. Chemical Properties. Oxophosphoraneselenenyl chlorides (**20**) and their structural analogs, $R(RO)P(O)SeCl$, oxidize iodide and add to ethylenic hydrocarbons (see equation 56).[63,64] Oxophosphoraneselenenyl chlorides (**20**) and the corresponding sulfenyl chlorides, $(RO)_2P(O)SCl$, differ in their behavior towards phosphorus nucleophiles.[89] For example, the latter react with trialkyl phosphites and dialkyl phosphites to give thiopyrophosphates, $(RO)_2P(S)-O-P(O)(OR)_2$, but the former do not give selenopyrophosphates under the same conditions.

B. ORGANIC SELENOTHIOACIDS OF PHOSPHORUS AND THEIR DERIVATIVES

1. Free Acids, Salts, and Complexes

Investigation of the chemistry of thioselenoacids of phosphorus has started only recently. Major contributions have come from Kataev, Kuchen, Michalski, Zemlyanskii, and their colleagues.

a. Preparation. Five approaches to selenothioacid preparation have been made. Two involve addition reactions, that is, of selenium to dialkyl thiophosphites and of sulfur to dialkyl selenophosphites (**10**) (equations 69a,

$$(69a) \quad (RO)_2 P\overset{S}{\underset{H}{\diagup}} + Se \xrightarrow[EtOH]{C_6H_{11}NH_2} \quad \overset{RO}{\underset{RO}{\diagdown}} P\overset{Se}{\underset{S}{\diagup}}^- \quad C_6H_{11}\overset{+}{N}H_3$$

$$(69b) \quad (RO)_2 P\overset{Se}{\underset{H}{\diagup}} + S \xrightarrow[EtOH]{C_6H_{11}NH_2}$$

10

69b).[90] The anion $>PSeS^-$ has mesomeric character, and therefore the same cyclohexylamine salt is formed in both reactions.

Reaction 70 has been used in the synthesis of an asymmetric *O*-ethyl ethylphosphonoselenothioic acid with four different ligands attached to

$$(70) \quad \overset{EtO}{\underset{Et}{\diagdown}} P\overset{S}{\underset{H}{\diagup}} \xrightarrow{EtONa} \quad \overset{EtO}{\underset{Et}{\diagdown}} P-SNa \xrightarrow[EtOH]{Se}$$

$$\overset{EtO}{\underset{Et}{\diagdown}} P\overset{Se}{\underset{S}{\diagup}}^- \quad Na^+ \xrightarrow{HCl} Et(EtO)P(Se)SH$$

phosphorus.[90] The acid was resolved through diastereomeric quinine salts into optical isomers, $[\alpha]_D^{20} -6.7°$ and $[\alpha]_D^{20} +4.9°$, and was the first selenophosphorus compound to be resolved successfully.

A third method involves nucleophilic substitution at a selenophosphoryl center, $>P(Se)X$ (equation 71).

(71) $\qquad$ $R_2P(Se)Cl + 2SH^- \rightarrow R_2P(Se)S^- + Cl^- + H_2S$

$\qquad$ (R = PhO;[91,92] R = Et[93,94])

Fourth, nucleophilic substitution at a thiophosphoryl center, $>P(S)X$, may be used (equation 72).[93,94]

(72) $\qquad$ $Et_2P(S)Cl + 2NaSeH \xrightarrow{0°} Et_2P(Se)SNa + H_2Se + NaCl$

Finally, scission of tetraethyldiphosphine disulfide by a mixture of sodium selenide and selenium[93,94] is possible (equation 73). This method is rather

(73) $\qquad$ $Et_2P(S)—P(S)Et_2 + Na_2Se + Se \xrightarrow{200°} 2Et_2P(Se)SNa$

drastic and probably suitable only for systems without alkoxy groups attached to phosphorus.

b. Chemical and Physical Properties. The problem of tautomeric equilibria of phosphorus selenothioacids (equation 74) is of great interest,

(74)

$$\underset{SH}{\overset{Se}{P}} \rightleftharpoons \underset{S}{\overset{SeH}{P}}$$

but no experimentally justified conclusions can be presented. The lack of data for the P-Se double-bond energy does not allow any of the quantitative predictions of tautomeric equilibrium which have been made for monothio-acids of phosphorus.[95] The two nucleophilic centers in the ambident anion of selenothioacids differ only slightly in electronegativity on Pauling's scale. Monoselenoacid anions have been found to undergo Se-alkylation only. Although S-alkylation is the rule for monothioacids anions, in some cases concomitant O-alkylation has been observed.[26–28] The ratio of S- and Se-alkylation in the case of the selenothioacid anion depends on the type of alkylating agent used,[96,97] that is, its hardness.

(75a) $\qquad$ $\xrightarrow{EtI}$ $(EtO)_2P(S)SeEt$ $\qquad$ (100%)

(75b) $\qquad$ $\xrightarrow{ClCH_2OCH_3}$ $(EtO)_2P(S)SeCH_2OCH_3$ $\qquad$ (60%)
$\qquad\qquad\qquad\qquad\qquad\qquad\qquad\qquad$ $(EtO)_2P(Se)SCH_2OCH_3$ $\qquad$ (40%)

$$\underset{EtO}{\overset{EtO}{\underset{}{}}}\underset{S}{\overset{Se}{P}}^-$$

(75c) $\qquad$ $\xrightarrow{Et_3\overset{+}{O}\;BF_4^-}$ $(EtO)_2P(S)SeEt$ $\qquad$ (20%)
$\qquad\qquad\qquad\qquad\qquad\qquad\qquad\qquad$ $(EtO)_2P(Se)SEt$ $\qquad$ (80%)

The distribution of products shown in equations 75a–75c was estimated by the method of Purdy and Truter[98] and is only approximate ($\pm 10\%$). The influences of substitution at phosphorus and of the reaction medium merit investigation. *Se*-Alkylation has been observed in several other cases: sodium diphenyl phosphoroselenothioate, $(PhO)_2P(Se)SNa$;[91] sodium diethyl phosphinoselenothioate, $Et_2P(Se)SNa$;[93] and sodium *O*-ethyl ethylphosphonoselenothioate, $Et(EtO)P(Se)SNa$ (equation 76).[96]

(76)
$$\underset{/}{\overset{\backslash}{P}}\!\!\!\overset{Se}{\underset{S}{\diagup\!\!\!\diagdown}} - \;+\; RX \longrightarrow \underset{/}{\overset{\backslash}{P}}\!\!\!\overset{SeR}{\underset{S}{\diagup\!\!\!\diagdown}} \;+\; X^-$$

The diversity of reactions of phosphorus selenothioacid salts with acylating and phosphorylating reagents is best considered in terms of the types of structure formed, and these will be discussed in Section X.B.4. *O*-Ethyl ethylphosphonoselenothioic acid is moderately stable at room temperature, but when it is distilled *in vacuo* there is partial decomposition to compound **25** (see equation 88b).[96, 99] An aqueous solution of diethylphosphinoselenothioic acid, $Et_2P(Se)SH$, prepared by ion exchange from its sodium salt, can be kept unchanged in the dark for 24 hours. Its pK_a, determined potentiometrically in 80% aqueous 2-propanol at 20°, is 2.29.[100] An attempt to isolate this acid by acidification of its sodium salt led to an oil that decomposed rapidly into a crystalline product of presumed structure $Et_2P(S)-Se_3-P(S)Et_2$.[94]

Formation of complexes was observed for all types of selenothioacids mentioned. Acid complexes of $(PhO)_2P(Se)SH$ have been prepared, and diamagnetic Ni^{2+} and Pd^{2+} complexes have been studied in some detail.[101] Acid $Et_2P(Se)SH$ complexes are described in a review article.[33] Complexes of the type $[Et_2P(Se)S]_nM$ are known (M = Zn^{2+}, Cd^{2+}, Pd^{2+}, Pb^{2+}, Tl^+, Bi^{3+}, Rh^{3+}), and all are readily soluble in organic solvents. They are distinctly less stable than the corresponding dithio complexes but more stable than complexes of diethylphosphinoselenoic acid (**3**, R = Et).

2. Selenonothioesters: $>P(Se)SR$

A phosphoroorganic ester containing both sulfur and selenium was first described in 1907 by the pioneer in the organic chemistry of phosphorus, Michaelis.[102] This compound was *S,S,S*-triphenyl phosphorotrithioselenoate, $(PhS)_3PSe$.

The general method of preparation of selenonothioesters (**21**) is illustrated in equation 77.[91, 96] Cyclic esters can be similarly synthesized (equation 78).[39, 103]

(77) $\qquad (RO)_2P(Se)Cl + NaSR' \rightarrow (RO)_2P(Se)SR' + NaCl$

$$(78) \quad \underset{CH_3}{\overset{Se}{\underset{|}{\overset{\|}{P}}}}\overset{Cl}{\underset{Cl}{<}} + \underset{HS—CH_2}{\overset{HS—CH_2}{<}}\overset{CH_2}{\underset{CH_2}{>}} \xrightarrow{2R_3N}$$

$$\underset{CH_3}{\overset{Se}{\underset{|}{\overset{\|}{P}}}}\overset{S—CH_2}{\underset{S—CH_2}{<}}\overset{}{\underset{}{>}}CH_2 + 2R_3N \cdot HCl$$

Five other specific methods are known. Addition of selenium to a suitable P(III) thio derivative has so far been employed only for cyclic esters (equation 79).[39,103] S-Alkylation of the anion $>$P(Se)S$^-$ is limited to

$$(79) \quad CH_3—P\overset{Cl}{\underset{Cl}{<}} + \underset{HS—CH_2}{\overset{HS—CH_2}{|}} \xrightarrow{2R_3N}$$

$$CH_3—P\overset{S—CH_2}{\underset{S—CH_2}{<}}\overset{}{\underset{}{|}} \xrightarrow{Se} CH_3—\overset{Se}{\overset{\|}{P}}\overset{S—CH_2}{\underset{S—CH_2}{<}}\overset{}{\underset{}{|}}$$

certain alkylating agents, such as triethyl oxonium fluoroborate (equation 75c).[96] O-Ethyl S-(2-chlorocyclohexyl) ethylphosphonoselenoate has been prepared by addition of a selenophosphoranesulfenyl chloride (**29**) (see equation 91) to cyclohexene (equation 80).[96,97]

$$(80) \quad (EtO)(Et)P(Se)SCl + \underset{}{\bighexagon} \longrightarrow (EtO)(Et)P(Se)S—\underset{Cl}{\bighexagon}$$

$$\mathbf{29}$$

Phosphoroselenonothioates of type **21** can be obtained by addition of O,O-diethyl phosphoroselenothioic acid to olefins or to α,β-unsaturated compounds (equations 81a–81c).[104] The structures of these phosphoro-

$$(81a) \quad \xrightarrow[\text{without catalyst}]{CH_2=CHCN} (EtO)_2P(Se)SCH(CH_3)CN$$

$$(81b) \quad (EtO)_2\overset{Se}{\overset{\|}{P}}—SH \xrightarrow[\text{MeONa}]{CH_2=CHCN} (EtO)_2P(Se)SCH_2CH_2CN$$

$$(81c) \quad \xrightarrow[\text{without catalyst}]{CH_3CH=CHCH_2CH_3}$$

$$(EtO)_2P(Se)SCH(CH_3)CH_2CH_2CH_3$$

selenonothioates have been deduced by oxidative chlorinolysis (equation 82).

$$\text{(82)} \quad \underset{\underset{\displaystyle CH_3}{|}}{(EtO)_2\overset{\overset{\displaystyle Se}{\|}}{P}SCH(CH_2)_2CH_3} \xrightarrow[H_2O]{Cl_2} \underset{\underset{\displaystyle CH_3}{|}}{ClSO_2CHCH_2CH_2CH_3} \xrightarrow{PhNH_2}$$

$$\underset{\underset{\displaystyle CH_3}{|}}{PhNHSO_2CHCH_2CH_2CH_3}$$

The uncatalyzed addition to acrylonitrile and to 2-pentene seems to involve a free-radical mechanism, but the base-catalyzed heterolytic addition to acrylonitrile is an example of the Michael reaction. It is most surprising that in these reactions S-C bonds are formed almost exclusively. The relationship of these results to the alkylation reactions already described needs to be clarified.

3. Selenolothioesters: >P(S)SeR and —P(O)(SR)SeR'

In addition to the alkylations of phosphorus selenothioacids already discussed (equations 75a–c), two methods of preparation of selenothionoesters (22), based on the condensation of dialkyl thiophosphites with organic seleno-cyanides[105] and selenenyl chlorides,[73] have been described (equations 83a, 83b). For comparison see equations 44 and 46. Recently Se-methyl

$$\text{(83)} \quad (RO)_2\overset{\overset{\displaystyle S}{\displaystyle\diagup\diagup}}{\underset{\displaystyle\diagdown}{P}}H \quad \xrightarrow[\text{(b) } ClSeR']{\text{(a) } NCSeR'} \quad (RO)_2\overset{\overset{\displaystyle S}{\|}}{P}\text{—SeR'} + HCN\ (HCl)$$

$$\mathbf{22}$$

dimethylphosphinoselenothioate has been prepared by addition of elemental sulfur to Se-methyl dimethylphosphinoselenoite (equation 84).[106,107]

$$\text{(84)} \qquad\qquad MeSePMe_2 \xrightarrow{S} MeSeP(S)Me_2$$

O-Ethyl S-n-butyl Se-phenyl phosphate is the only known representative of its kind (equation 85).[73]

$$\text{(85)} \quad (EtO)_2PCl + ClSePh \xrightarrow{-EtCl} \underset{\text{not isolated}}{[(EtO)(PhSe)P(O)Cl]} \xrightarrow{n\text{-BuSH}}$$

$$(EtO)(PhSe)P(O)(n\text{-BuS})$$

4. Tetraalkyl Selenodithiopyrophosphates and Structural Analogs

One method of preparation of the selenothiopyrophosphate **24** has already been mentioned (see equation 59). It can also be obtained by phosphorylation of sodium diethyl phosphoroselenothioate, and this fact strongly implies an intermediate selenothiopyrophosphate (**23**) and isomerization to **24** (equation 86). The isomerization of **23** is similar to that undergone by

(86)
$$(EtO)_2P \overset{Se}{\underset{S}{\lessgtr}} - \xrightarrow{(EtO)_2P(O)Cl} (EtO)_2\underset{\underset{O}{\|}}{P} - S - \underset{\underset{Se}{\|}}{P}(OEt)_2 \longrightarrow$$

23

$$(EtO)_2\underset{\underset{Se}{\|}}{P} - O - \underset{\underset{S}{\|}}{P}(OEt)_2$$

24

symmetric monothiopyrophosphates and monoselenopyrophosphates (**18**) (see equation 57), but contrasts with the behavior of dithiopyrophosphates, $>P(S)SP(O)<$.[108] The latter isomerize only under drastic thermal conditions. The structure of **24** has been confirmed by its i.r. spectrum and by nucleophilic degradation (equation 87).[96] Similar degradation of the isomer

(87)
$$\underset{EtO}{\overset{EtO}{\diagdown}}\underset{\underset{Se}{\|}}{P} - O - \underset{\underset{S}{\|}}{P}\overset{OEt}{\diagup}_{OEt} \xrightarrow{Nu^-} \underset{EtO}{\overset{EtO}{\diagdown}}\underset{\underset{S}{\|}}{P} - Nu + {}^-O - \underset{\underset{Se}{\|}}{P}\overset{OEt}{\diagup}_{OEt}$$

24

(**23**) would lead to different products: $(EtO)_2P(O)Nu$ and $(EtO)_2P(Se)S^-$.

The selenodithiopyrophosphonate **25** has been obtained by thiophosphorylation of the *O*-ethyl ethylphosphonoselenothioic acid sodium salt (equation 88a) or by thermal decomposition of the free acid (equation 88b).

(88a)
$$Et(EtO)P(Se)SNa \xrightarrow{Et(EtO)P(S)Cl}$$

$$Et(EtO)\underset{\underset{Se}{\|}}{P} - S - \underset{\underset{S}{\|}}{P}(OEt)Et$$

25

(88b)
$$2Et(EtO)P(Se)SH \xrightarrow[-H_2Se]{\Delta}$$

Tetraethyl selenodithiopyrophosphate (**27**) has been prepared by deselenization of **26** (see equation 95) with triphenylphosphine (equation 89).[85] The

(89) $(EtO)_2P$—Se—S—$P(OEt)_2$ $\xrightarrow{Ph_3P}$ $(EtO)_2P$—S—$P(OEt)_2$ + Ph_3PSe

$\quad\quad\quad\;\;\overset{\|}{S}\quad\quad\quad\overset{\|}{Se}\quad\quad\quad\quad\quad\quad\quad\quad\overset{\|}{Se}\quad\;\overset{\|}{S}$

$\quad\quad\quad\quad\quad\mathbf{26}\quad\quad\quad\quad\quad\quad\quad\quad\quad\quad\quad\quad\quad\mathbf{27}$

asymmetric structure of **27** is evident from its proton n.m.r. spectrum[85] and is additionally confirmed by comparison of its i.r. spectrum with the spectra of model selenophosphorus compounds and other reported P–Se and P–S–P absorptions.[109–111]

5. Mixed Anhydrides of Selenothioacids and Carboxylic and Carbonic Acids

Mixed anhydrides of selenothioacids and carboxylic[96,112] (**28**) (equation 90)

(90) $Et(EtO)P(Se)SNa + ClCPh \longrightarrow Et(EtO)P$—S—C—Ph + NaCl

$\quad\quad\quad\quad\quad\quad\quad\quad\overset{\|}{O}\quad\quad\quad\quad\quad\quad\quad\overset{\|}{Se}\quad\overset{\|}{O}$

$\quad\quad\quad\quad\quad\quad\quad\quad\quad\quad\quad\quad\quad\quad\quad\quad\quad\mathbf{28}$

and carbonic acids[112] (**30**) (equation 92) have been described. The structure of **28** is consistent with its conversion to **29** by chlorinating agents (equation 91) and subsequent addition of **29** to cyclohexene (see equation 80).[96]

(91) $Et(EtO)P$—S—C—Ph + $SO_2Cl_2 \longrightarrow$

$\quad\quad\quad\;\overset{\|}{Se}\quad\;\overset{\|}{O}$

$\quad\quad\quad\quad\quad\quad\quad\quad\quad Et(EtO)P$—S—Cl + Cl—C—Ph + SO_2 + HCl

$\quad\quad\quad\quad\quad\quad\quad\quad\quad\quad\quad\quad\;\overset{\|}{Se}\quad\quad\quad\quad\;\overset{\|}{O}$

$\quad\quad\quad\quad\quad\quad\quad\quad\quad\quad\quad\quad\quad\mathbf{29}$

(92) $(EtO)_2PSeSK + ClC$—OEt $\longrightarrow (EtO)_2P$—Se—C—OEt

$\quad\quad\quad\quad\quad\quad\quad\quad\;\;\overset{\|}{O}\quad\quad\quad\quad\quad\quad\quad\;\overset{\|}{S}\quad\;\overset{\|}{O}$

$\quad\quad\quad\quad\quad\quad\quad\quad\quad\quad\quad\quad\quad\quad\quad\quad\quad\mathbf{30}$

Formation of both selenono, $>P(Se)S–C(O)–$, and thiono, $>P(S)Se–C(O)–$, isomers has been claimed.[112]

6. Diselenides: $>P(S)Se–SeP(S)<$, and Related Compounds

Careful oxidation of sodium diethylphosphinoselenothioate with iodine in aqueous potassium iodide solution gives the diselenide (**31**) (equation 93).[94]

(93) $2Et_2PSeSNa + KI_3 \longrightarrow Et_2P$—Se—Se—$PEt_2$ + KI + 2NaI

$\quad\quad\quad\quad\quad\quad\quad\quad\quad\quad\quad\;\;\overset{\|}{S}\quad\quad\quad\;\overset{\|}{S}$

$\quad\quad\quad\quad\quad\quad\quad\quad\quad\quad\quad\quad\quad\mathbf{31}$

The description of **31** as a diselenide follows from X-ray analysis.[7] Oxidation of **34** gives a mixture of **35** and **36** (equation 94).[85] Evidence to support a

(94)

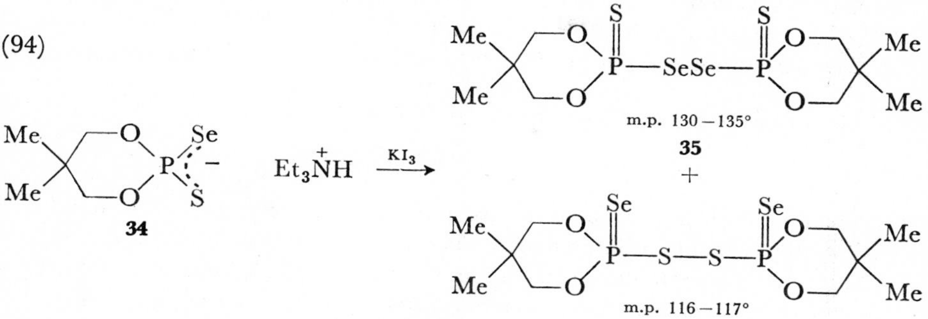

m.p. 130–135°

35

+

m.p. 116–117°

36

34

diselenide structure for **35** and a disulfide structure for **36** was obtained from the integration of their mass spectra.[85]

There is also some evidence for the formation of an –Se–S– link (**26**) in a similar oxidation (equation 95).[85] The proton n.m.r. spectrum of **26** is con-

(95) $2(EtO)_2PSeSK + NaI_3 \longrightarrow (EtO)_2P-Se-S-P(OEt)_2$

                                                                                    ‖   ‖
                                                                                    S   Se

26

sistent with an asymmetric structure for the compound. Other products are formed in addition to **26**. The structure $[Et_2P(S)]_2Se_3$ has been ascribed to a product formed by decomposition of the acid Et_2PSeSH.[94]

7. Selenophosphoranesulfenyl Chlorides: >P(Se)SCl

The properties and preparation of (ethoxy)ethylselenophosphoranesulfenyl chloride (**29**) are described elsewhere (equations 80 and 91, respectively).

C. ORGANIC DISELENOACIDS OF PHOSPHORUS AND THEIR DERIVATIVES

1. Free Acids, Salts, and Complexes

Three synthetic routes leading to organic diselenoacids are available. Reaction of phosphorus pentaselenide with an alcohol, followed by the addition of alcoholic potassium hydroxide, has been studied by Zingaro *et al.* (equation 96).[113] The reaction between P_2Se_5 and ethanol was reported

(96) $4ROH + P_2Se_5 \xrightarrow[-H_2Se]{} 2[(RO)_2PSeSeH] \xrightarrow{KOH} 2(RO)_2PSeSeK$

                                                        37

over a century ago by Carius and Bogen.[114] The products, $EtOP(O)(SeEt)_2$ and $EtO(EtSe)P(Se)SeH$, described then differ from those observed by Zingaro. Reaction 96 is limited to preparation of diselenoacid salts of type **37**, since the acids themselves are unstable.

Two other routes, equation 97[94,100] and equation 98,[85,97] are of general

$$(97) \qquad Et_2P(Se)Cl + 2NaSeH \rightarrow Et_2P(Se)SeNa + H_2Se + NaCl$$

$$(98) \qquad (RO)_2PSeNa \overset{Se}{\rightarrow} (RO)_2P(Se)SeNa$$

application and can be used to synthesize acids containing C-P bonds. Free dialkyl selenophosphites do not add selenium unless an amine is present.

Free diethylphosphinodiselenoic acid has not been isolated, but an aqueous solution of it has been prepared from its sodium salt, using a strongly acidic ion-exchange resin.[100] The pK_a of the acid, determined potentiometrically in 80% aqueous 2-propanol at 20°, is 2.18. Potassium salts of **37** are unstable, and special precautions should be taken when storing them. The diselenoacids form stable complexes[33,99,113,115,116] with the following metal ions: Tl^+, Zn^{2+}, Cd^{2+}, Sn^{2+}, Ni^{2+}, Pd^{2+}, Pb^{2+}, Rh^{3+}, Cr^{3+}, As^{3+}, Sb^{3+}, Ir^{3+}, Bi^{3+}. They are soluble in organic solvents and decompose in aqueous solutions.

2.   Esters of Diselenoacids and Triselenoacids of Phosphorus

Alkylation of sodium diethylphosphinodiselenoate with alkyl bromides leads to Se-alkyl diethylphosphinodiselenoates (equation 99).[94,100]

$$(99) \qquad Et_2P(Se)SeNa + RBr \rightarrow Et_2P(Se)SeR + NaBr$$

Se,Se,Se-Triphenyl phosphorotriselenoite has been postulated as an intermediate in the reaction of selenophenol with phosphorus trichloride.[70] It is readily oxidized to $(PhSe)_3PO$ (equation 100).

$$(100) \qquad (PhSe)_3P \overset{[O]}{\rightarrow} (PhSe)_3PO$$

3.   Selenophosphinyl Selenides: $>P(Se)-Se_n-P(Se)<$

Zingaro et al.[113] attempted the preparation of the acid **37** by reaction of alcohols with P_2Se_5 but isolated selenides **38** ($n = 2$) and **38** ($n = 3$) instead. These workers suggested two possible routes to **38** ($n = 2$) and

$$(RO)_2\underset{\underset{Se}{\parallel}}{P}-Se_n-\underset{\underset{Se}{\parallel}}{P}(OR)_2$$

38

38 $(n = 3)$. One involved atmospheric oxidation of the acid **37**, and the other required reaction of **37** with phosphorus pentaselenide in the absence of air, to form lower phosphorus selenides and elemental selenium as coproducts. Selenide **38** $(n = 2)$ is distinctly more stable than **38** $(n = 3)$.

Attempts to isolate the free acid $Et_2PSeSeH$ led to formation of an oil which decomposed rapidly, giving selenides **39** $(n = 1)$ and **39** $(n = 3)$.[100]

$$Et_2P\!\!-\!\!Se_n\!\!-\!\!PEt_2$$
$$\underset{Se}{\|} \qquad \underset{Se}{\|}$$
$$\mathbf{39}$$

It is possible that the diselenide (**39**, $n = 2$) is formed first but then disproportionates to the monoselenide (**39**, $n = 1$) and the triselenide (**39**, $n = 3$).

D. PHOSPHONIUM SALTS AND PENTACOVALENT SELENOPHOSPHORUS DERIVATIVES

Several types of phosphonium salts, $(\!\geqslant\!\overset{+}{P}\!\!-\!\!SeR)Y^-$, have been prepared by alkylation of the corresponding selenophosphoryl compounds, $\geqslant\!PSe$ (see equations 26, 108, 126). The stability of these complexes depends on the nature of the anion and the type of ligands at the P atom. Complexes with anions of very low nucleophilicity, like $SbCl_6^-$, or with no alkoxy group attached to phosphorus are relatively stable. When an alkoxy group is present, the phosphonium salts are transformed into $>\!P(O)SeR$ derivatives by nucleophilic dealkylation. In many of the reactions described in this chapter, unstable phosphonium-type intermediates are formed (see equations 36, 37, 41, 43, 44, 46, 57, 85, 130). In order to avoid repetition, the phosphonium salts are discussed in connection with the appropriate reactions of corresponding selenophosphoryl compounds.

There are no authentic data about the formation of pentacovalent phosphorus compounds with a Se atom attached to phosphorus. Such intermediates can be postulated, however, in some chemical reactions (*e.g.*, equation 8).

E. ORGANIC SELENOPHOSPHORUS COMPOUNDS CONTAINING A P–N LINKAGE

1. Selenophosphorus Acid Amides

The majority of phosphoro-organic compounds containing selenium and a P–N bond have been prepared by adding elemental selenium to tricovalent phosphorus compounds. When tricovalent phosphorus is attached to an amido group, its ability to add selenium is enhanced (equations 101, 102).[117-124]

(101) $(RO_2)P{-}NR'_2 \xrightarrow{Se} (RO)_2P(Se)NR'_2$

(102) $RO(R'S)P{-}NR''_2 \xrightarrow{Se} RO(R'S)P(Se)NR''_2$

Another method of preparation is based on nucleophilic exchange at a selenophosphoryl center (equations 103a, 103b, 104). The amide group may already be attached to the P atom[117,121,125] or may be introduced in the

(103a)
$$
\underset{R'_2N}{\overset{RO}{\diagdown}} P \underset{Cl}{\overset{Se}{\diagup}} \xrightarrow{R''OH} \underset{R'_2N}{\overset{RO}{\diagdown}} P \underset{OR''}{\overset{Se}{\diagup}}
$$

(103b)
$$
\underset{R'_2N}{\overset{RO}{\diagdown}} P \underset{Cl}{\overset{Se}{\diagup}} \xrightarrow{R''SH} \underset{R'_2N}{\overset{RO}{\diagdown}} P \underset{SR''}{\overset{Se}{\diagup}}
$$

(104)
$$
(R_2N)_2P \overset{Se}{\underset{Cl}{}} \xrightarrow{R'OH} (R_2N)_2P \overset{Se}{\underset{OR'}{}}
$$

course of substitution (equation 105).[34,36,37,121,122] In the former case a tertiary amine must be present.

(105) $(RO)_2P(Se)Cl + 2HNR'_2 \rightarrow (RO)_2P(Se)NR'_2 + R'_2NH \cdot HCl$

Selenophosphoric triamides and alkylammonium N,N'-bis(alkyl) diamido-diselenophosphates have been prepared by the reaction of phosphorus pentaselenide with primary amines (equation 106, 107).[126] The yield of

(106) $9RNH_2 + P_2Se_5 \rightarrow 2(RNH)_3P{=}Se + 3\overset{+}{R}NH_3 \, HSe^-$

(107) $7RNH_2 + P_2Se_5 \rightarrow 2(RNH)_2P(Se)Se^- \, \overset{+}{R}NH_3 + \overset{+}{R}NH_3 \, HSe^-$

amide in this reaction is low, but no systematic attempt has been made to maximize it. In contrast with the reaction of phosphorus pentaselenide with alcohols (*cf.* Section X.B.3), no bridged Se_2 compounds (*e.g.*, **38**, $n = 2$) or Se_3 compounds (*e.g.*, **38**, $n = 3$) are formed. This is probably due to the fast conversion of the free acid to the stable ammonium salt.

It must be emphasized that the selenophosphoryl group, $\geqslant$P(Se), in these amides should be even more readily attacked by alkylating agents than the same group in selenonoesters.[67] Amides of selenophosphoric and seleno-phosphonic acids react readily with methyl iodide, giving the corresponding phosphonium salts (equation 108).[123]

(108) $\qquad (R_2N)_3PSe + MeI \rightarrow [(R_2N)_3\overset{+}{P}SeMe]I^-$

Complexes with the hexachlorostibinate anion have been prepared from phosphonoamidoselenoate, $PhP(Se)(NMe_2)_2$; phosphinoamidoselenoate, $Ph_2P(Se)(NMe_2)$; and trimethyloxonium hexachlorostibinate (see also equation 26).[36]

$$[PhP(NMe_2)_2SeMe]^+ SbCl_6^- \qquad [Ph_2P(NMe_2)SeMe]^+ SbCl_6^-$$

It is probable that, if an alkoxy group is attached to the P atom and an anion of higher nucleophilicity is present, the phosphonium complex will decompose according to the scheme of the Pishchimuka reaction (equations 36, 109).

(109) $\qquad [(R_2N)_2\overset{+}{P}(OR')SeR'']Y^- \rightarrow (R_2N)_2P(O)SeR'' + R'Y$

2. Acid Chlorides: $R_2N\overset{|}{P}(Se)Cl$

Acid chlorides containing an amido group (amidoselenoic chlorides) have been prepared by adding elemental selenium to the corresponding P(III) amidochloro derivatives (equations 110–112).[117,121,125,127] The use of

(110) $\qquad RO(R'_2N)P\!-\!Cl \xrightarrow[130-160°]{Se} RO(R'_2N)P(Se)Cl$

(111) $\qquad (R_2N)_2P\!-\!Cl \xrightarrow[120-130°]{Se} (R_2N)_2P(Se)Cl$

(112) $\qquad R(R'_2N)P\!-\!Cl \xrightarrow[120-130°]{Se} R(R'_2N)P(Se)Cl$

amidoselenoic chlorides as selenophosphorylating agents has already been mentioned in connection with the preparation of selenophosphorus acid amides (see equations 103a, 103b, 104).

3. Dialkyl Phosphoroisothiocyanatoselenoates: $(RO)_2P(Se)NCS$

The unstable diethyl phosphoroisothiocyanatoselenoate has been obtained by thiocyanogenation of diethyl selenophosphite (**10**, R = Et) and char-acterized by formation of its thioureido derivative (equation 113).[20]

(113) $(EtO)_2P(Se)H \xrightarrow{(SCN)_2} (EtO)_2\overset{\parallel}{\underset{Se}{P}}\!-\!NCS \xrightarrow{R_2NH}$

$$(EtO)_2\overset{\parallel}{\underset{Se}{P}}\!-\!NH\!-\!\overset{\parallel}{\underset{S}{C}}\!-\!NR_2$$

F. DIALKYL SELENOPHOSPHITES: $(RO)_2P(Se)H$

Dialkyl selenophosphites (**10**) have been obtained by the action of hydrogen selenide on dialkyl phosphorochloridites (dialkyl chlorophosphites) in the presence of a tertiary amine (equation 114).[19,20] Selenophosphites (**10**) are

(114) $(RO)_2PCl + H_2Se + R'_3N \longrightarrow (RO)_2P(Se)H + R'_3N\cdot HCl$
                                                                                10

relatively stable and exhibit an interesting spectrum of chemical activity. Their reactions with sulfuryl chloride (equation 18) and thiocyanogen (equation 113) are described elsewhere. Compounds **10** are seleno analogs of dialkyl phosphites, $(RO)_2P(O)H$ (**4**), and show very much the same chemical activity.

Tautomerism similar to that generally accepted[77] for **4** can also be envisaged for **10**, with the equilibrium shifted considerably towards the selenono form (equation 115).[20] The position of the resonance in the ^{31}P

(115) $(RO)_2\overset{\parallel}{\underset{Se}{P}}\!-\!H \rightleftharpoons (RO)_2PSeH$

n.m.r. spectrum of **10** (-71.0 ppm from H_3PO_4) and the very large splitting observed (J_{PH} 630 Hz) show conclusively that a proton is directly attached to phosphorus.[20]

Dialkyl selenophosphites (**10**) react quantitatively with sodium hydroxide solution in aqueous alcohol (equation 116).[20] This reaction can be used as a

(116) $(RO)_2P(Se)H + NaOH \rightarrow NaO(RO)P(Se)H + ROH$

comparatively accurate analytical method.

Selenophosphites (**10**) react more slowly than both dialkyl phosphites (**4**) and thiophosphites, $(RO)_2P(S)H$, with sodium and potassium, probably because of the low solubility of the alkali-metal derivatives formed (equation 117). The sodium derivative of **10** shows properties typical of tricovalent

(117) $(RO)_2P(Se)H \xrightarrow{Na} (RO)_2PSeNa$

phosphorus compounds and readily adds sulfur (to form a selenothioacid salt; see equation 69) and selenium (to form a diselenoacid salt; equation 98). Free selenophosphites (**10**) do not add elemental sulfur or selenium unless an amine is present. The classical Michaelis-Becker reaction for synthesis of compounds containing a C–P link can be applied to sodium dialkyl seleno-phosphites provided that suitable reaction medium (*e.g.*, 1,2-dimethoxy-ethane) is used (equation 118).

(118) $n\text{-BuCl} + (n\text{-BuO})_2\text{PSeNa} \rightarrow n\text{-BuP(Se)(O-}n\text{-Bu})_2$

In solvents like benzene, cyclohexane, and ether the yield is very low. The addition of **10** to a carbonyl group and to an α,β-unsaturated compound is catalyzed by sodium ethoxide and results in the formation of a C-P bond (equations 119, 120). Both reactions are examples of nucleophilic addition

(119) $(EtO)_2P(Se)H + O_2N\!-\!\!\langle\ \rangle\!-\!C\overset{\displaystyle O}{\underset{\displaystyle H}{\diagup}}\ \xrightarrow{\ \text{EtONa}\ }$

$$(EtO)_2P\!\!-\!\!CH\!-\!\!\langle\ \rangle\!-\!NO_2$$
$$\underset{\displaystyle Se\ \ OH}{\overset{\displaystyle \|\quad\ |}{}}$$

(120) $(EtO)_2P(Se)H + CH_2\!\!=\!\!CHCN \xrightarrow{\ \text{EtONa}\ } (EtO)_2P(Se)CH_2CH_2CN$

to a polarized double bond. These are only a few of the many possible applications of dialkyl selenophosphites to the synthesis of organic seleno-phosphorus compounds that await investigation.

G. SECONDARY PHOSPHINE SELENIDES: $R_2P(Se)H$

Secondary phosphine selenides, formed in the reaction of elemental selenium with secondary phosphines (equation 121), seem to be less stable and also

(121) $R_2PH + Se \rightarrow R_2P(Se)H$

less reactive than dialkyl selenophosphites and the corresponding sulfides and oxides.[128] The selenono P=Se structure has been confirmed by [31]P n.m.r. spectra.

H. TERTIARY PHOSPHINE SELENIDES: R_3PSe

1. Preparation

Trimethyl- and triethylphosphine selenides appear to have been described first by Cahours and Hofmann[129] as early as 1857. Tertiary phosphine selenides (**39a**) have been prepared by addition of selenium to the corresponding phosphines,[130–163] either by direct fusion or by reaction in an inert solvent (equation 122).

(122) $RR'R''P + Se \rightarrow RR'R''PSe$
                                        39a

A racemic asymmetric triarylphosphine selenide containing an amino group in the aromatic ring has been prepared by Davies and Mann (equation 123),[134] but resolution was not described. Optically active methyl-

(123)

$$ArPCl_2 \xrightarrow{Ph_2Hg} ArPhPCl \xrightarrow{Ar'MgX} ArAr'PhP \xrightarrow{Se} ArAr'PhP(Se)$$

$$[Ar = p\text{-}BrC_6H_4\text{—}; Ar' = p\text{-}(Me_2N)C_6H_4\text{—}]$$

phenylpropylphosphine selenide has been prepared by stereoselective addition of elemental selenium to the corresponding optically active phosphine and was oxidized to the phosphine oxide with either an inversion or a retention mechanism.[164]

An interesting transformation of a tertiary phosphine by potassium selenocyanate in acetonitrile has been described (equation 124).[165] This is a

(124) $Ph_3P + KSeCN \rightarrow Ph_3PSe + KCN$
                                        40

simple and efficient method of preparing aromatic tertiary phosphine selenides. The reaction is specific for selenium. Analogous reactions with cyanates and thiocyanates yield neither tertiary phosphine oxides nor sulfides. The application of this method to trialkyl phosphines has not been investigated. It has been reported recently that triphenylphosphine selenide (**40**) can be obtained by rearrangement of Se-phenyl diphenylphosphinoselenoite (**46**) (see equation 129).

Another interesting reaction, in which the formation of a cyclic tertiary phosphine selenide (**42**) is accompanied by aromatization, has been described by Campbell and Cookson.[65,66] Dehydrogenation of the cyclic phosphine oxide (**41**) with a mixture of elemental selenium and dihydrogen potassium phosphate (2:1) at 270–370° gave **42** in 25% yield (equation 125).

(125)

41   **42**

The same transformation with selenium alone was much less effective, giving only 4 % of **42**.

Two other cyclic tertiary phosphine selenides containing a five-membered ring have also been described.[66,166]

2. Structure and Chemical Properties

The structures of Et_3PSe and of **43** have been studied by X-ray crystallography.[3] By means of [1]H n.m.r. studies of the selenide **43** two distinct

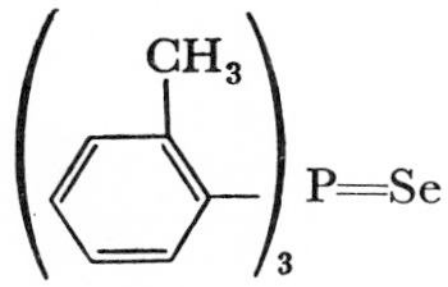

$$\left(\underset{\substack{}}{\overset{CH_3}{\bigcirc}}\right)_3 P{=}Se$$

43

methyl environments in the ratio $2:1$ ($\delta = 2.32$ and 2.37) at 20° have been observed. The sulfur analog shows only one environment at this temperature but resolves into two at 5°. The methyl signals of the oxide and the phosphine show only one environment down to −60°. All the methyl groups in the corresponding phosphine and in its oxide occupy similar positions, but in the selenide **43** one methyl group lies behind the P atom.[5]

Alkylation of triphenylphosphine selenide (**40**) leading to formation of phosphonium hexafluoroantimonate has been described (equation 126).[36]

(126) $Ph_3P{=}Se \xrightarrow{Me_3\overset{+}{O}Sb\bar{C}l_6} [Ph_3\overset{+}{P}SeMe]\bar{S}bCl_6 + Me_2O$

Triphenylphosphine selenide (**40**) readily adds boron trichloride and tribromide to form crystalline adducts in quantitative yield.[57]

$$Ph_3P{=}Se{\cdot}BCl_3 \qquad Ph_3P{=}Se{\cdot}BBr_3$$

Information about the oxidative conversions of tertiary phosphine selenides into the corresponding oxides is scarce. Conversion of the selenide **42** into the oxide **44** was not achieved using hydrogen peroxide as oxidizing agent, and the methylation process shown in scheme 127 had to be adopted.[66]

Triphenylphosphine selenide (**40**) can be oxidized quantitatively to the triphenylphosphine oxide by the sodium salt of N-chloro-p-toluenesulfonamide (Chloramine-T); 8 equivalents of the oxidant are consumed. The procedure is suggested as being general for all selenophosphorus compounds.[167]

(127)

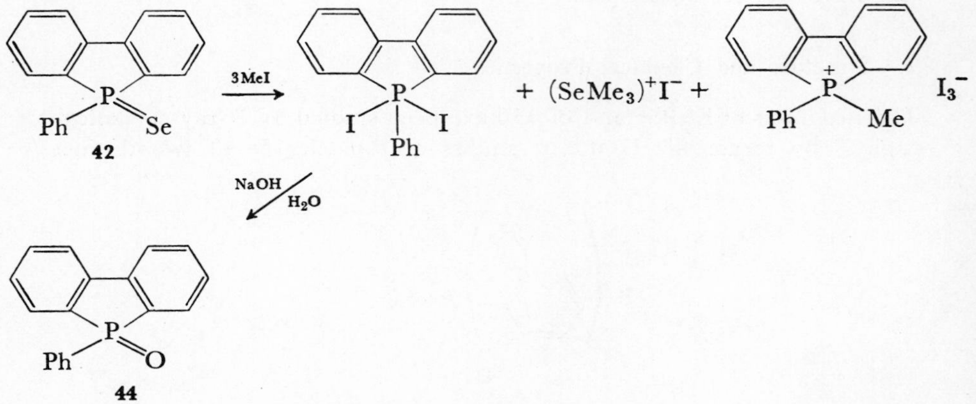

I. POLYPHOSPHINE SELENIDES

The following selenides derived from polyphosphines are known:

$$Ph_2P(S)CH_2(Se)PPh_2,^{168} \qquad Ph_2P(Se)CH_2CH_2(Se)PPh_2,^{165}$$

$$[Ph_2P(Se)]_3C_3N_3,^{169} \quad and \quad [Ph_2P(Se)CH_2]_4C.^{170}$$

Pentaphenylcyclopentaphosphine reacts with red selenium to give the following compound:[171]

$$(Ph{-}PSe)_n$$

45

The same compound was obtained by Maier by action of red selenium on phenylphosphine[172] and was given a structure (**45**, $n = 3$) analogous to that of the sulfur-containing compound $(Ph–PS)_3$.[173] In contrast to the sulfur compound, however, the selenium compound (**45**) does not react further with elemental selenium. Reaction of selenium with aliphatic cyclophosphines has not been described.[174]

J. TRICOVALENT SELENOPHOSPHORUS COMPOUNDS: >P–SeR

Se-Phenyl diphenylphosphinoselenoite has been prepared from diphenylphosphinous chloride (diphenylchlorophosphine) and selenophenol in the presence of triethylamine (equation 128).[78]

(128) $Ph_2P{-}Cl + HSePh + Et_3N \rightarrow Ph_2P{-}SePh + Et_3N \cdot HCl$

The phosphinoselenoite **46** isomerizes readily to triphenylphosphine selenide. The fact that the selenium compound **46** undergoes an Arbuzov-type rearrangement (equation 129) much more readily than its sulfur analog

$$(129) \qquad Ph_2P—SePh \rightarrow Ph_3P=Se$$
$$\textbf{46}$$

is explained by consideration of the relative strengths of the organosulfur and organoselenium bonds.[78] The rearrangement seems to be subject to catalytic acceleration in a manner which is not clear, because the condensation of diphenylphosphinous chloride, Ph_2PCl, with sodium seleno-phenolate gives the triphenylphosphine selenide as the main product. The selenium compound **46** was not isolated even when this condensation was carried out at $-10°$. Phosphinoselenoite **46** undergoes very rapid oxidation to phosphinoselenoate **17** (equations 49, 50, 52).[78] Se-Methyl dimethyl-phosphinoselenoite has recently been prepared by the action of dimethyl diselenide on tetramethyldiphosphine (equation 130).[106,107]

$$(130) \qquad \begin{matrix} MeSe—SeMe \\ Me_2P—PMe_2 \end{matrix} \longrightarrow \left[\begin{matrix} MeSe\overset{+}{P}Me_2 \\ Me_2P\overset{-}{}\,^-SeMe \end{matrix} \right] \longrightarrow 2MeSePMe_2$$

A preparation of Se-trifluoromethyl bis(trifluoromethyl)phosphinoselenoite has been achieved according to equation 131 by heating the reactants in a

$$(131) \qquad 2(CF_3)_2P—I + Hg(SeCF_3)_2 \rightarrow 2(CF_3)_2PSeCF_3 + HgI_2$$

sealed glass tube. After separation from HgI_2, the product was isolated by vacuum distillation.[175] Se,Se,Se-Triphenyl phosphorotriselenoite has been prepared by action of selenophenol on phosphorus trichloride (equation 132).[70] It is readily oxidized to $(PhSe)_3P(O)$ (see equation 100).

$$(132) \qquad 3PhSeH + PCl_3 \rightarrow (PhSe)_3P$$

Another type of tricovalent selenophosphorus compound which can be considered as a mixed anhydride has been prepared from phosphorus trichloride and potassium alkyl diselenocarbonates (equation 133).[176,177]

$$(133) \qquad 3ROC(Se)SK + PCl_3 \rightarrow [ROC(Se)Se]_3P + 3KCl$$

$$(R = Et, i\text{-}Pr, s\text{-}Bu)$$

K. SPECTROSCOPIC METHODS

1. Nuclear Magnetic Resonance Spectra

Of all the physical methods available for the study of organophosphorus compounds, n.m.r. spectroscopy has attracted by far the most interest. Specific information about the bonding and environment of the P atom can be obtained directly from ^{31}P spectra and indirectly from the spectra of other atoms coupled to phosphorus. Problems posed by the low sensitivity of high-resolution ^{31}P resonance can be overcome by the application of new techniques such as double resonance.[178,179] Studies utilizing ^{1}H and ^{19}F resonance are also of great importance in determining the structure of organophosphorus compounds; other nuclei resonances, including [77]Se,[106] have been employed only occasionally. It can be predicted that [13]C n.m.r. and the use of paramagnetic ions as shift reagents in [1]H n.m.r.[180] will be very useful in this field.

[31]P and [1]H n.m.r. are fully complementary techniques that constitute the key to further study of the chemistry of organoselenophosphorus compounds. In addition n.m.r. spectroscopy offers a means of checking the purity of many organoselenophosphorus compounds already been described in the literature. This is particularly necessary for high-boiling liquids, the purity of which cannot usually be determined in any other way.

Organoselenophosphorus compounds and organothiophosphorus compounds give n.m.r. spectra that are similar in many ways; for example, chemical shifts, relative peak areas, and coupling constants are comparable. Organoselenophosphorus compounds obey orientation rules that allow the type of bonding at the P atom to be assessed from the ^{31}P chemical shift. Large negative shifts are attributable to tricovalent phosphorus derivatives. Tetracovalent phosphoryl compounds ($\gg$PO) generally appear in the range from -25 to $+50$ ppm, but tetracovalent thiophosphoryl, $\gg$PS, and selenophosphoryl, $\gg$PSe, compounds are found at lower fields in the range from -100 to 0 ppm and sometimes slightly higher. For pentacovalent and hexacovalent phosphorus compounds, in which phosphorus uses its d-orbitals to form σ bonds, positive shifts are observed. The only selenophosphorus compounds which have been systematically studied are tetracovalent ones.[36,59,85,181–183]

Recent results based on a relatively large number of selenophosphorus compounds have shown that the chemical shift (δ ^{31}P) of phosphorus chalcogenides ($\gg$P$=$Y, $\gg$P–YMe; Y = O, S, Se) increases steadily as phenyl groups are replaced by methoxy or dimethylamino groups, and that the oxides are more deshielded than the sulfides or selenides.[36] A special situation arises for $>$P(Se)S, $>$P(S)Se, and $>$P(S)S groupings. They all possess similar ^{31}P shifts which are not resolvable.[85] Extensive study of organophosphorus compounds and observations of selenophosphorus compounds indicate that the J_{PXCH} coupling constants vary markedly with the valence of phosphorus and also depend on the nature of the other groups attached to this element. A relationship of the Karplus type links these J values with the dihedral angle.[85]

Selected data from the n.m.r. spectra of organoselenophosphorus compounds are collected in Table X-2. All the ^{31}P and ^{1}H chemical shifts are

Table X-2   Nuclear magnetic resonance spectra of organoselenophosphorus compounds

COMPOUND	$\delta^{31}P$ (ppm)[a]	$\delta^{1}H$ (ppm)[b]	J (Hz)	REF.
MeP(Se)Cl$_2$	16		PCH = 13.5	183
MeP(Se)Br$_2$			PCH = 12.2	40, 183
(MeO)$_3$PSe	−78.4	O—Me 3.70 / 3.50	POCH = 14.2 / 14.0	36, 57, 184
(EtO)$_3$PSe	−71, −72.1			184, 187
(i-PrO)$_3$PSe	−67.9			184
(n-BuO)$_3$PSe	−73.0			184
(Me$_3$CO)$_3$PSe	−31.0			184
(Me$_3$CCH$_2$O)$_3$PSe	−72.6			184
(PhO)$_3$PSe	−58.0			184
(EtO)(Me$_3$CO)$_2$PSe	−45.2			184
(MeO)$_2$PhPSe	−98.6,[c] 14.7[d]	O—Me 3.72 / 3.60	POCH = 14.2 / 14.0	36, 59
(MeO)Ph$_2$PSe	−88.0,[c] 24.0[d,e]	O—Me 3.55 / 3.60[f]	POCH = 14.4 / 14.3	36, 59

Table X-2 (continued)

COMPOUND	$\delta\,^{31}P$ (ppm)[a]	$\delta\,^{1}H$ (ppm)[b]	J (Hz)	REF.
(structure: Se=P—Me, 1,3,2-dioxaphospholane)		P—Me 1.10[g]	PCH = 13.5	39
(structure: Se=P—Me, dimethyl dioxaphospholane)		P—Me 1.68, 1.72, 1.65[g] C—Me 0.73 (multipl.)	PCH = 13.5	39
(structure: Se=P—Me, tetramethyl dioxaphospholane)		P—Me 1.25 (doubl.) C—Me 0.65, 0.30 (singl.)	PCH = 13.5	39
(structure: Se=P—Me, 1,3,2-dioxaphosphorinane)		P—Me 1.27 (doubl.)[g] C—CH₂—C 0.73 (multipl.) CH₂—C—CH₂ 3.50 (multipl.)	PCH = 14.5	39

Compound	Chemical shift (^{31}P / ^{13}C)	^{1}H data	Coupling constants	Ref.
(MeO)$_2$P(O)SeMe		Me—O 3.77 Me—Se 2.27	POCH = 13.5 PSeCH = 13.0	57
(MeO)PhP(O)SeMe		Me—O 3.71 Me—Se 1.93	POCH = 12.1 PSeCH = 13.1	59
(EtO)$_2$P(Se)OP(O)(OEt)$_2$	α 14.7 β −55.8 −56.1			85, 184
(MeS)$_3$PSe	−81.9			184
(EtS)$_3$PSe	−75.8			184
(n-PrS)$_3$PSe	−75.7			184
(n-BuS)$_3$PSe	−75.1			184
		P—Me 2.57 C—CH$_2$—C 2.28 (multipl.) CH$_2$—C—CH$_2$ 3.44 (multipl.)	PCH = 13.0	39
Me$_2$P(S)SeMe			^{13}CH(Se) = +145.0 ^{13}CH(P) = +131.2 ^{13}C^{31}P = +50 ± 4 ^{31}PCH = −12.7 ^{31}P^{77}Se = −341	107

Table X-2 (continued)

COMPOUND	$\delta^{31}P$ (ppm)[a]	δ^1H (ppm)[b]	J (Hz)	REF.
[general structure: 5,5-dimethyl-1,3,2-dioxaphosphorinane ring with X=P, then P—Y—P linked to a second ring with Z=P, O's and CMe₂ groups]				
$X = Z = S;\ Y = Se\!-\!Se$		$CH_2\begin{Bmatrix}\text{ax. }4.25\\\text{eq. }4.01\end{Bmatrix}\ CMe_2\begin{Bmatrix}\text{ax. }1.16\\\text{eq. }0.88\end{Bmatrix}$		85
$X = Z = Se;\ Y = S\!-\!S$		$CH_2\begin{Bmatrix}\text{ax. }4.35\\\text{eq. }4.02\end{Bmatrix}\ CMe_2\begin{Bmatrix}\text{ax. }1.29\\\text{eq. }0.98\end{Bmatrix}$		85
$(EtO)_2P(X)YP(Z)(OEt)_2$				
$X = Y = S;\ Z = Se$	-73.6	$CH_2\ \ 4.26$ $Me\ \ 1.39$	POCH = 10.8 POCH = 7.1	85
$X = S;\ Y = Se\!-\!S;\ Z = Se$	-73.0	$CH_2\ \ 4.26$ $Me\ \ 1.40$	POCH = 10.4 POCH = 7.1	85
$(EtO)_2P(Se)SeK$		$Me\!-\!C\ \ 1.15$ $C\!-\!CH_2\ \ 3.91$	POCH = 9.5	113
$(n\text{-PrO})_2P(Se)SeK$		$Me\!-\!C\!-\ \ 0.92$ $Me\!-\!CH_2\!-\ \ 3.94$ $C\!-\!CH_2\!-\!C\ \ 1.62$	POCH = 9.1	113
$[(MeO)_2PhPSeMe]^+SbCl_6{}^-$	-84.6	$Me\!-\!Se\ \ 2.39$ $Me\!-\!O\ \ 4.28$	PSeCH = 16.0 POCH = 13.5	36

Compound	Shift	Assignment		Coupling		Ref.
$[(MeO)Ph_2PSeMe]^+SbCl_6^-$	-87.8	Me—Se	2.34	PSeCH = 12.9		36
		Me—O	4.12	POCH = 14.5		
$[Ph_3PSeMe]^+SbCl_6^-$	-35.8^c	Me—Se	2.35	PSeCH = 12.6		36
$[(Me_2N)_3PSeMe]^+SbCl_6^-$	-63.5^h	Me—Se	2.42	PSeCH = 14.0		36
		Me₂—N	2.92	PNCH = 11.5		
$[(Me_2N)_3PSeMe]^+I^-$		Me₂—N	2.86	PNCH = 11.4		123
		Me—Se	2.23	PSeCH = 12.3		
$[(Me_2N)Ph_2PSeMe]^+SbCl_6^-$	-65.7^c	Me—Se	2.55	PSeCH = 12.2		36
		Me₂—N	3.02	PNCH = 13.2		
$(Me_2N)_3PSe$	$-83.5, -83.2^c$	Me₂—N	2.54	PNCH = 11.6		36, 184
$(Me_2N)_2PhPSe$	-84.6^c	Me₂—N	2.50	PNCH = 12.5		36
$(Me_2N)Ph_2PSe$	-72.0^c	Me₂—N	2.39	PNCH = 15.0		36
$(EtO)_2P(Se)H$	-71.0			PH = 630		184
$(i\text{-}Bu)_2P(Se)H$	$-3.3, +14.0^g$			PH = 420		128

Table X-2 (continued)

COMPOUND	$\delta^{31}P$ (ppm)[a]	$\delta^{1}H$ (ppm)[b]	J (Hz)	REF.
$Ph_2P(Se)H$	$-15.1, +3.5$[g]		$PH = 450$	128
Et_3PSe	-45.8			188
$(n\text{-}Pr)_3PSe$	-72.6			184
$(n\text{-}Bu)_3PSe$	-36.9			184
Ph_3PSe	-35.0[c]			36, 185
$(o\text{-}MeC_6H_4)_3PSe$		Me 2.32, 2.37		5
Me_2PSeMe			$^{13}C\text{—}H(P) = +129$[c] $^{13}C\text{—}H(Se) = +140$ $^{13}C\text{—}^{31}P = -25 \pm 2$ $^{31}P\text{—}^{77}Se = +205$	107

a 85% $H_3PO_4 = 0$ ppm if not otherwise stated.
b TMS $= 0$ ppm.
c In CH_2Cl_2.
d Relative to P_4O_6.
e In $CHCl_3$.
f In $CDCl_3$.
g In benzene.
h In DMSO.

relative to 85% phosphoric acid and tetramethylsilane, respectively. In cases where phosphorus trioxide, P_4O_6 (-112.5 ± 1 ppm referred to 85% H_3PO_4), has been used, ^{31}P shifts are given without recalculation.

Several reviews have been published on the various aspects of n.m.r. spectra of organophosphorus compounds.[178,181,182,184-186]

2. Other Methods

The electronic spectra of some organoselenophosphorus compounds (mainly salts and metal complexes) have been investigated.[32,68,101,113,189] The only reported application of mass spectrometry was mentioned in Section X.B.6. X-ray crystallographic analyses are discussed in the introduction, and characteristic parameters are given in Table X-1. Infrared spectroscopy of organoselenophosphorus compounds is discussed in Chapter XVA, Section 8b.

REFERENCES

1. V. Shomaker and D. P. Stevenson, *J. Amer. Chem. Soc.*, **63**, 37 (1941).

2. S. Husebye and G. Helland-Madsen, *Acta Chem. Scand.*, **23**, 1398 (1969).

3. M. van Meerssche and A. Léonard, *Acta Cryst.*, **12**, 1053 (1959).

4. G. Lepicard, D. de Saint-Giniez-Liebig, A. Laurant, and C. Rerat, *Acta Cryst.*, **25B**, 617 (1969).

5. R. A. Shaw, M. Woods, T. S. Cameron, and B. Dahlen, *Chem. Ind.* (London), 151 (1971).

6. S. Husebye, *Acta Chem. Scand.*, **23**, 1389 (1969).

7. S. Husebye, *Acta Chem. Scand.*, **20**, 51 (1966).

8. G. J. Penney and G. M. Sheldrick, *J. Chem. Soc.*, *A*, 1100 (1971).

9. E. Keulen and A. Vos, *Acta Cryst.*, **12**, 323 (1959).

10. E. Heilbronn-Wilkström, *Svensk. Kem. Tidskr.*, **75**, 508 (1965).

11. S. Åkerfeldt and L. Fagerlind, *J. Med. Chem.*, **10**, 115 (1967).

12. G. Schrader, German Pat. 830,262 (1952); *Chem. Abstr.*, **52**, 8447b (1958).

13. T. B. David and J. W. Apple, *J. Econ. Entomol.*, **44**, 528 (1951).

14. G. Schrader, Brit. Pat. 691,267 (1953); *Chem. Abstr.*, **48**, 7047f (1954). U.S. Pat. 2,680,132 (1954); *Chem. Abstr.*, **48**, 9402a (1954).

15. A. Markowska and J. Michalski, *Rocz. Chem.*, **34**, 1675 (1960).

16. P. Pishchimuka, *J. Prakt. Chem.*, **84**, 746 (1911); *Zh. Fiz. Khim.*, **44**, 1406 (1912).

17. P. Chabrier and N. T. Thuong, *C.R. Acad. Sci.*, Paris, **258**, 3738 (1964).

18. O. Foss, *Acta Chem. Scand.*, **1**, 8 (1947).

19. J. Michalski and C. Krawiecki, *Rocz. Chem.*, **31**, 715 (1957).

20. C. Krawiecki, J. Michalski, R. A. Y. Jones, and A. R. Katritzky, *Rocz. Chem.*, **43**, 869 (1969).

21. "Organophosphorus Chemistry," Vol. 1, The Chemical Society, London, 1970, p. 150.

22. A. Markowska, *Bull. Acad. Pol. Sci.*, *Ser. Sci. Chim.*, **13**, 149 (1965).

23. M. I. Kabachnik, T. A. Mastryukova, A. E. Shipov, and T. A. Melentyeva, *Tetrahedron*, **9**, 10 (1960).

24. T. A. Mastryukova and M. I. Kabachnik, *Usp. Khim.*, 1751 (1969); *Russ. Chem. Rev.*, **38**, 795 (1969).

25. V. Šimánek and A. Klásek, *Tetrahedron Lett.*, 3039 (1969).

26. T. A. Mastryukova, A. E. Shipov, V. V. Abalyaeva, E. M. Popov, and M. I. Kabachnik, *Dokl. Akad. Nauk SSSR*, **158**, 1373 (1964).

27. G. Hilgetag and H. Teichmann, *Z. Chem.*, **5**, 179 (1965).

28. J. Michalski and J. Wieczorkowski, *Bull. Acad. Pol. Sci.*, *Classe III*, **5**, 917 (1957).

29. M. Mikołajczyk, *Chem. Ind.* (London), 2059 (1966).

30. M. Mikołajczyk and M. Para, *Chem. Commun.*, 1192 (1969).

31. J. Michalski and J. Wieczorkowski, *J. Chem. Soc.*, 885 (1960).

32. S. V. Larionov and L. A. Il'ina, *Zh. Obshch. Khim.*, **39**, 1587 (1969).

33. W. Kuchen and H. Hertel, *Angew. Chem.*, **81**, 127 (1969), *Angew. Chem.* (*Intern. Ed.*), **8**, 89 (1969).

34. W. Strecker and C. Grossmann, *Chem. Ber.*, **49**, 63 (1961).

35. N. I. Zemlyanskii, N. M. Chernaya, V. V. Turkevich, and V. I. Krasnoshchek, *Zh. Obshch. Khim.*, **36**, 1240 (1966).

36. A. Schmidpeter and H. Brecht, *Z. Naturforsch.*, *Part B*, **24**, 179 (1969).

37. E. Gryszkiewicz-Trochimowski, J. Quinchon, and O. Gryszkiewicz-Trochimowski, *Bull. Soc. Chim. Fr.*, 1794 (1960).

38. S. Z. Ivin and I. D. Shelyakova, *Zh. Obshch. Khim.*, **31**, 4052 (1961).

39. M. Wieber and H. U. Werther, *Monatsh. Chem.*, **99**, 1153 (1968).

40. L. Maier, *Helv. Chim. Acta*, **46**, 2667 (1963).

41. A. N. Pudovik and E. A. Ishmayeva, *Zh. Obshch. Khim.*, **35**, 358 (1965).

42. A. J. Kirby and S. G. Warren, "The Organic Chemistry of Phosphorus," Elsevier Publishing Co., Amsterdam, 1967, Chap. 10.

43. J. Michalski, M. Mikołajczyk, B. Młotkowska, and J. Omelańczuk, *Tetrahedron*, **25**, 1743 (1969).

44. G. R. Waitkins, U.S. Pat. 2,506,049 (1950); *Chem. Abstr.*, **44**, 6618f (1950).

45. R. E. Heiks and F. C. Croxton, *Ind. Eng. Chem.*, **43**, 876 (1951).

46. V. Tichy, *Chem. Zvesti*, **9**, 3 (1955).

47. V. Mark and J. R. Van Wazer, *J. Org. Chem.*, **29**, 1006 (1964).

48. I. A. Nuretdinov and N. P. Grechkin, *Izv. Akad. Nauk SSSR*, *Ser. Khim.*, 2831 (1968).

49. A. I. Razumov, O. A. Mukhacheva, and Sim-Do-Khen, *Izv. Akad. Nauk SSSR*, *Ser. Khim.*, 894 (1952).

50. A. I. Razumov and O. A. Mukhacheva, *Zh. Obshch. Khim.*, **26**, 2463 (1956).

51. A. E. Arbuzov and N. A. Razumov, *Izv. Akad. Nauk SSSR*, *Ser. Khim.*, 187 (1956).

52. W.-H. Chang, *J. Org. Chem.*, **29**, 3711 (1964).

53. B. Fontal and H. Goldwhite, *Tetrahedron*, **22**, 3275 (1966).

54. G. Hilgetag and H. Teichmann, *Angew. Chem.*, **77**, 1001 (1965); *Angew. Chem.* (*Intern. Ed.*), **4**, 914 (1965).

55. R. G. Pearson, *J. Amer. Chem. Soc.*, **85**, 3533 (1963); R. G. Pearson, "Survey of Progress in Chemistry," Vol. V, A. F. Scott, Ed., Academic Press, New York, N.Y., 1969, p. 1.

56. J. O. Edwards and R. G. Pearson. *J. Amer. Chem. Soc.*, **84**, 16 (1962); R. F. Hudson, *Chimia*, **16**, 173 (1962).

57. L. Elegant, J. F. Gal, and M. Azzaro, *Bull. Soc. Chim. Fr.*, 4273 (1969).

58. H. Teichmann and G. Hilgetag, *Chem. Ber.*, **98**, 856 (1965); G. Hilgetag, H. Teichmann, and M. Krüger, *Chem. Ber.*, **98**, 864 (1965).

59. G. Mavel, R. Mankowski-Favelier, and T. N. Tanh, *J. Chim. Phys.*, **64**, 1692 (1966).

60. I. A. Nuretdinov, N. A. Buina, N. P. Grechkin, and S. G. Salikhov, *Zh. Obshch. Khim.*, **39**, 930 (1969).

61. I. A. Nuretdinov, N. A. Buina, N. P. Grechkin, and E. I. Loginova, *Izv. Akad. Nauk SSSR, Ser. Khim.*, 131 (1971).

62. H. P. Nguyen, N. T. Thuong, and P. Chabrier, *C.R. Acad. Sci., Paris, Ser. C.* **272**, 1145 (1971).

63. J. Michalski and A. Markowska, *Dokl. Akad. Nauk SSSR*, **136**, 108 (1961).

64. A. Markowska, *Bull. Acad. Pol. Sci., Ser. Sci. Chim.*, **15**, 153 (1967).

65. I. G. M. Campbell, R. C. Cookson, and M. B. Hocking, *Chem. Ind.* (London), 359 (1962).

66. I. G. M. Campbell, R. C. Cookson, M. B. Hocking, and A. N. Hughes, *J. Chem. Soc.*, 2184 (1965).

67. A. J. Burn and J. J. Cadogan, *J. Chem. Soc.*, 5532 (1961).

68. E. G. Kataev and T. G. Mannafov, *Zh. Obshch. Khim.*, **36**, 254 (1966).

69. E. G. Kataev, T. G. Mannafov, and G. I. Kostina, *Zh. Obshch. Khim.*, **37**, 2059 (1967).

70. N. Petragnani, V. G. Toscano, and M. de Moura Campos, *Chem. Ber.*, **101**, 3070 (1968).

71. J. Michalski and J. Wieczorkowski, *Rocz. Chem.*, **33**, 105 (1959).

72. N. Petragnani and M. de Moura Campos, *Tetrahedron*, **21**, 13 (1965).

73. E. G. Kataev, T. G. Mannafov, and G. I. Kostina, *Zh. Obshch. Khim.*, **38**, 363 (1968).

74. G. Schrader and W. Lorenz, Ger. Pat. 824,046; *Chem. Zentr.*, 6375 (1955).

75. G. Schrader, Ger. Pat. 830,262 (1952); *Chem. Abstr.*, **52**, P8447b (1958).

76. Brit. Pat. 691,267 (1953); *Chem. Zentr.*, 3471 (1955).

77. S. G. Warren, *Angew. Chem.*, **80**, 649 (1968); *Angew. Chem. (Intern. Ed.)*, **7**, 606 (1968).

78. R. A. N. McLean, *Inorg. Nucl. Chem. Lett.*, **5**, 745 (1969).

79. N. Petragnani and M. de Moura Campos, *Chem. Ind.* (London), 1076 (1965).

80. G. Schrader, German Pat. 818,046 (1951).

81. C. Lutter and E. Cauer, French Pat. 998,726 (1952); *Chem. Zentr.*, 4198 (1955).

82. A. E. Arbuzov, *Tr. Kazansk. Filiala Akad. Nauk SSSR, Ser. Khim.*, No. 2, 7 (21) (1956); *Chem. Abstr.*, **51**, 1036g (1957).

83. D. G. Coe, B. J. Perry, and R. K. Brown, *J. Chem. Soc.*, 3604 (1957).

84. J. Michalski and J. Wieczorkowski, *Rocz. Chem.*, **28**, 233 (1954).

85. A. R. Katritzky, M. R. Nesbit, J. Michalski, Z. Tulimowski, and A. Zwierzak, *J. Chem. Soc., B*, 140 (1970).

86. J. Michalski, M. Mikołajczyk, and B. Młotkowska, *Chem. Ber.*, **102**, 90 (1969).

87. A. E. Arbuzov, B. A. Arbuzov, P. I. Alimov, K. V. Nikonorov, N. I. Rizpolozhenskii,

and A. N. Fedorova, *Tr. Kazansk. Filiala Akad. Nauk SSSR, Ser. Khim.*, 21 (1956); *Chem. Abstr.*, **51,** 10363b (1957).

88. M. Mikołajczyk, J. Omelańczuk, and J. Michalski, *Bull. Acad. Pol. Sci., Ser. Sci. Chim.*, **17,** 155 (1969).

89. J. Michalski and A. Skowrońska, *J. Chem. Soc., C*, 703 (1969).

90. C. Krawiecki, J. Michalski, and Z. Tulimowski, *Chem. Ind.* (London), 34 (1965).

91. N. I. Zemlyanskii, N. M. Chernaya, and V. V. Turkevich, *Dokl. Akad. Nauk SSSR*, **163,** 1397 (1965).

92. N. I. Zemlyanskii and N. M. Chernaya, *Ukr. Khim. Zh.*, **33,** 182 (1967).

93. W. Kuchen and B. Knop, *Angew. Chem.*, **76,** 496 (1964); *Angew. Chem. (Intern. Ed.)*, **3,** 507 (1964).

94. W. Kuchen and B. Knop, *Chem. Ber.*, **99,** 1663 (1966).

95. R. F. Hudson, "Structure and Mechanisms in Organophosphorus Chemistry," Academic Press, New York, N.Y., 1965, p. 122.

96. J. Michalski and Z. Tulimowski, *Bull. Acad. Pol. Sci., Ser. Sci. Chim.*, **14,** 217 (1966).

97. Z. Tulimowski, Ph.D. Thesis, Łódź, 1965.

98. S. J. Purdy and E. V. Truter, *Chem. Ind.* (London), 506 (1962); *Analyst* (London), **87,** 802 (1962).

99. S. E. Livingstone, *Quart. Rev.*, **19,** 386 (1965).

100. W. Kuchen and B. Knop, *Angew. Chem.*, **77,** 259 (1965); *Angew. Chem. (Intern. Ed.)*, **4,** 244 (1965).

101. L. A. Il'ina, N. I. Zemlyanskii, S. V. Larionov, and N. M. Chernaya, *Izv. Akad. Nauk SSSR, Ser. Khim.*, 198 (1969).

102. A. Michaelis and G. L. Linke, *Ber.*, **40,** 3419 (1907).

103. M. Wieber, J. Otto, and M. Schmidt, *Angew. Chem.*, **76,** 648 (1964); *Angew. Chem. (Intern. Ed.)*, **3,** 586 (1964).

104. J. Michalski and Z. Tulimowski, *Bull. Acad. Pol. Sci., Ser. Sci. Chim.*, **14,** 303 (1966).

105. W. Lorenz and G. Schrader, German Pat. 1,074,035 (1960); *Chem. Zentr.*, 15545 (1960).

106. Yu. N. Shlyk, G. M. Bogolyubov, and A. A. Petrov, *Zh. Obshch. Khim.*, **38,** 193 (1968).

107. W. McFarlane and J. A. Nash, *Chem. Commun.*, 913 (1969).

108. J. Michalski and J. Wasiak, *J. Chem. Soc.*, 5056 (1962); L. Almasi and L. Paskucz, *Monatsh. Chem.*, **99,** 187 (1968).

109. L. C. Thomas and R. A. Chittenden, *Chem. Ind.* (London), 1913 (1961).

110. C. N. R. Rao, "Chemical Applications of Infrared Spectroscopy," Academic Press, New York, N.Y., 1963, p. 291.

111. D. E. C. Corbridge, "Topics in Phosphorus Chemistry," M. Grayson and E. J. Griffith, Eds., Vol. 6, Interscience Publishers, New York, N.Y., 1969, p. 235.

112. N. I. Zemlyanskii, N. M. Chernaya, and V. V. Turkevich, *Zh. Obshch. Khim.*, **37,** 495 (1967).

113. M. V. Kudchadker, R. A. Zingaro, and K. J. Irgolic, *Can. J. Chem.*, **46,** 1415 (1968).

114. L. Carius and W. Bogen, *Justus Liebigs Ann. Chem.*, **124,** 57 (1862).

115. C. K. Jørgensen, *Mol. Phys.*, **5,** 485 (1962).

116. V. Krishnan and R. A. Zingaro, *Inorg. Chem.*, **8,** 2337 (1969).

117. L. K. Nikonorova, N. P. Grechkin, and I. A. Nuretdinov, *Izv. Akad. Nauk SSSR, Ser. Khim.*, 464 (1969).

118. V. S. Abramov and Z. S. Druzhina, *Zh. Obshch. Khim.*, **36**, 923 (1966).

119. I. A. Nuretdinov, N. A. Buina, and N. P. Grechkin, *Zh. Obshch. Khim.*, **37**, 959 (1967).

120. K. Sasse, "Houben-Weyl Methoden der Organischen Chemie," Vol. XII/2, E. Müller, Ed., Georg Thieme Verlag, Stuttgart, 1964, p. 841.

121. I. A. Nuretdinov, L. K. Nikonorova, and N. P. Grechkin, *Zh. Obshch. Khim.*, **39**, 2265 (1969).

122. N. A. Buina, I. A. Nuretdinov, and N. P. Grechkin, *Izv. Akad. Nauk SSSR, Ser. Khim.*, 1606 (1967).

123. I. A. Nuretdinov, N. A. Buina, and N. P. Grechkin, *Izv. Akad. Nauk SSSR, Ser. Khim.*, 169 (1969).

124. N. I. Rizpolozhenskii, V. D. Akamsin, and T. M. Dosova, *Izv. Akad. Nauk SSSR, Ser. Khim.*, 622 (1970).

125. I. A. Nuretdinov, N. A. Buina, N. P. Grechkin, and E. I. Loginova, *Izv. Akad. Nauk SSSR, Ser. Khim.*, 708 (1970).

126. R. G. Melton and R. A. Zingaro, *Can. J. Chem.*, **46**, 1425 (1968).

127. I. A. Nuretdinov, N. P. Grechkin, N. A. Buina, and L. K. Nikonorova, *Izv. Akad. Nauk SSSR, Ser. Khim.*, 1535 (1969).

128. L. Maier, *Helv. Chim. Acta*, **49**, 1000 (1966).

129. A. Cahours and A. W. Hofmann, *Justus Liebigs Ann. Chem.*, **104**, 1 (1857).

130. R. A. Zingaro and R. E. McGlothlin, *J. Chem. Eng. Data*, **8**, 226 (1963).

131. A. Michaelis, *Justus Liebigs Ann. Chem.*, **315**, 43 (1901).

132. R. R. Renshaw and F. K. Bell, *J. Amer. Chem. Soc.*, **43**, 916 (1921).

133. A. W. Hoffmann, *Phil. Trans.*, **150**, 409 (1860).

134. W. C. Davies and F. G. Mann, *J. Chem. Soc.*, 276 (1944).

135. A. H. Cowley and J. L. Mills, *J. Amer. Chem. Soc.*, **91**, 2915 (1969).

136. E. Bannister and F. A. Cotton, *J. Chem. Soc.*, 1959 (1960).

137. C. F. Baranauckas, R. D. Carlson, E. E. Harris, and R. J. Lisanke, U.S. Dept. Com., Office Tech. Serv., AD 263,891 (1961); *Chem. Abstr.*, **58**, 3089e (1963).

138. M. Becke-Goehring and H. Thielemann, *Z. Anorg. Allg. Chem.*, **308**, 33 (1961).

139. A. M. Brodie, G. A. Rodley, and C. J. Wilkins, *J. Chem. Soc.*, A, 2927 (1969).

140. E. F. Bugerenko, E. A. Chernyshev, and A. D. Petrov, *Izv. Akad. Nauk SSSR, Ser. Khim.*, 286 (1965); *Chem. Abstr.*, **62**, 14721h (1965).

141. B. R. Condray, R. A. Zingaro, and M. V. Kudchadker, *Inorg. Chem. Acta*, **2**, 309 (1968).

142. J. R. Durig, J. S. DiYorio, and D. W. Wertz, *J. Mol. Spectr.* **28,**, 444 (1968).

143. L. S. D. Glasser, L. Ingram, M. G. King, and G. P. McQuillan, *J. Chem. Soc.*, A, 2501 (1969).

144. S. O. Grim, A. W. Yankowsky, S. A. Bruno, W. J. Bailey, E. F. Davidoff, and T. J. Marks, *J. Chem. Eng. Data*, **15,** 497 (1970).

145. H. Hartmann and A. Meixner, *Naturwissenschaften*, **50,** 403 (1963).

146. K. A. Jensen and P. H. Nielsen, *Acta Chem. Scand.*, **17,** 1875 (1963).

147. M. G. King and G. P. McQuillan, *J. Chem. Soc.*, A, 898 (1967).

148. R. A. N. McLean, *Inorg. Nucl. Chem. Lett.*, **5,** 745 (1969).

149. L. Maier, *Chem. Ber.*, **94**, 3043 (1961).

150. A. Michaelis, *Justus Liebigs Ann. Chem.*, **181**, 265 (1876).

151. A. Michaelis and H. Köhler, *Ber.*, **9**, 519, 1053 (1876).

152. A. Michaelis and H. von Soden, *Justus Liebigs Ann. Chem.*, **229**, 295 (1885).

153. N. Müller, P. C. Lauterbur, and J. Goldenson, *J. Amer. Chem. Soc.*, **78**, 3557 (1956).

154. P. Nicpon and D. W. Meek, *Chem. Commun.*, 398 (1966).

155. P. Nicpon and D. W. Meek, *Inorg. Syn.*, **10**, 157 (1967).

156. E. Niwa, H. Aoki, H. Tanaka, and K. Munakata, *Chem. Ber.*, **99**, 712 (1966).

157. A. Schmidpeter, private communication.

158. C. Screttas and A. F. Isbell, *J. Org. Chem.*, **27**, 2573 (1962).

159. G. P. Sollot and W. R. Peterson, *J. Organometal. Chem.*, **4**, 491 (1965).

160. M. van Meerssche and A. Léonard, *Bull. Soc. Chim. Belges*, **69**, 45 (1960); *Chem. Abstr.*, **54**, 16982e (1960).

161. R. A. Zingaro, *Inorg. Chem.*, **2**, 192 (1963).

162. R. A. Zingaro, R. E. McGlothlin, and E. A. Meyers, *J. Phys. Chem.*, **66**, 2579 (1962).

163. R. A. Zingaro and E. A. Meyers, *Inorg. Chem.*, **1**, 771 (1962).

164. W. Stec, A. Okruszek, and J. Michalski, *Angew. Chem.*, **83**, 491 (1971); *Angew. Chem. (Intern. Ed.)*, **10**, 494 (1971).

165. P. Nicpon and D. W. Meek, *Inorg. Chem.*, **5**, 1297 (1966).

166. E. H. Braye, W. Hübel, and I. Caplier, *J. Amer. Chem. Soc.*, **83**, 4406 (1961).

167. R. A. Shaw, Abstracts of International Symposium on Organic Sulphur Compounds, Venice, June 1970.

168. A. J. Carty and R. K. Harris, *Chem. Commun.*, 234 (1967).

169. W. Hewertson, R. A. Shaw, and B. C. Smith, *J. Chem. Soc.*, 1020 (1964).

170. J. Ellermann and D. Schirmacher, *Chem. Ber.*, **100**, 2220 (1967).

171. H. L. Krauss and H. Jung, *Z. Naturforsch., Part B*, **15**, 545 (1960).

172. L. Maier, *Helv. Chim. Acta*, **48**, 1190 (1965).

173. L. Maier, *Helv. Chim. Acta*, **49**, 1119 (1966).

174. L. Maier, "Fortschritte der Chemischen Forschung," Vol. 8, Springer-Verlag, Berlin, 1967, p. 1.

175. H. J. Eméléus, K. J. Packer, and N. Welcman, *J. Chem. Soc.*, 2529 (1962).

176. A. Rosenbaum, H. Kirchberg, and E. Leibnitz, *J. Prakt. Chem.*, [4], **19**, 1 (1963).

177. A. Rosenbaum, *J. Prakt. Chem.*, [4], **37**, 200 (1968).

178. R. A. Dwek, R. E. Richards, and D. Taylor, "Annual Review of NMR Spectroscopy," Vol. 2, E. F. Mooney, Ed., Academic Press, London, 1969, p. 293.

179. J. A. Potenza, E. H. Poindexter, P. J. Caplan, and R. A. Dwek, *J. Amer. Chem. Soc.*, **91**, 4356 (1969).

180. J. K. M. Sanders and D. H. Williams, *Chem. Commun.*, 422 (1970).

181. R. A. Y. Jones and A. R. Katritzky, *Angew. Chem.*, **74**, 60 (1962); *Angew. Chem. (Intern. Ed.)*, **1**, 32 (1962).

182. G. Mavel, "Progress in Nuclear Magnetic Resonance Spectroscopy," Vol. 1, J. W. Emsley, J. Feeney, and L. H. Sutcliffe, Eds., Pergamon Press, Oxford, 1966, p. 251.

183. G. Martin and G. Mavel, *C.R. Acad. Sci., Paris*, **253**, 2523 (1961).

184. V. Mark, C. H. Dungan, M. M. Crutchfield, and J. R. Van Wazer, "Topics in Phosphorus Chemistry," Vol. 5, M. Grayson and E. J. Griffith, Eds., Interscience Publishers, New York, N.Y., 1967, pp. 1, 277.

185. E. Fluck, "Die Kernmagnetische Resonanz und ihre Anwendung in der anorganichen Chemie," Springer-Verlag, Heidelberg, 1968.

186. J. Nakayama, *J. Synth. Org. Chem. Japan*, **28**, 177 (1970).

187. R. A. Y. Jones, A. R. Katritzky, and J. Michalski, *Proc. Chem. Soc.*, 321 (1959).

188. N. Muller, P. C. Lauterbur, and J. Goldenson, *J. Amer. Chem. Soc.*, **78**, 3557 (1956).

189. A. I. Busev and M.-T. Huang, *Zh. Neorg. Khim.*, **7**, 88 (1962).

XI Heterocyclic selenium compounds

A. SELENACYCLOALKANES

LUIGI MORTILLARO and MARIO RUSSO

Montecatini Edison Research Center
Milan, Italy

1. Introduction

This chapter concerns saturated rings containing carbon, selenium, and other heteroatoms (with the exception of nitrogen and R–O–Se-compounds). In addition (although not strictly classified as selenacycloalkanes), selena-heterocycles without carbon atoms in the ring are discussed.

Selenacycloalkanes are generally very stable liquid or solid compounds and are soluble in common organic solvents. Their odor is often unpleasant and pungent, sometimes because of the presence of impurities.

As is true of all cyclic compounds, the stability of selenacycloalkanes increases with increasing ring size. In fact, seleniranes (three-membered rings) have never been isolated; they have only been observed as transient intermediates in the reaction between selenium and olefins. Selenetane, a four-membered ring, is a labile compound that can be stored only under cool, dark conditions. Its 3,3-disubstituted derivatives are much more stable, but the action of iodomethane, chlorine, or bromine on them causes ring opening. In contrast, rings having more than five members are very stable. Another series of unstable selenacycloalkanes consists of unsubstituted cyclic 1,2-diselenides. These are known to exist only in solution because of their marked tendency to polymerize. Substituted derivatives of cyclic 1,2-diselenides, however, are much more stable.

The physical characteristics of selenacycloalkanes and thiacycloalkanes are very similar, and often these analogs are isomorphous. Pairs of optical antipodes of selenacycloalkanes and the corresponding sulfur compounds of opposite sign give quasi racemates. The similarity of sulfur and selenium is even more evident from the study of the "mixed" heterocycles, the 1,2-thiaselenacycloalkanes. In fact, it was ascertained that charge separation of the S–Se bond in 1-thia-2-selenolane-4-carboxylic acid and in 2-thia-3-selenaspiro[4.5]decane does not take place to an extent detectable by dipole moment measurements.

2. Selenacycloalkanes with One Selenium Atom in the Ring

a. *Three-Membered Rings*

Seleniranes.   The simplest member of this class is selenirane, also named episelenide (**1**). As already mentioned, no seleniranes have been isolated,

$$\text{H}_2\text{C} \overset{\displaystyle \text{Se}}{\underset{\textstyle }{\diagup \diagdown}} \text{CH}_2$$

1

but some have been observed as transient compounds during the reaction of Se atoms with olefins. In flash-photolyzed CSe_2-olefin mixtures, transient spectra were observed and tentatively assigned to episelenides.[1] Tyerman et al.[2] and Strausz and Gunning[3] developed a mass spectrometric technique for the time-resolved detection of transient intermediates in flash-photolyzed systems, by which the spectral assignments were confirmed.

The following olefins were examined for their reaction with CSe_2: ethylene, propylene, allene, 1,3-butadiene, vinyl fluoride, di-, tri-, and tetrafluoroethylene, 2-fluoropropene, pentafluoro-1-butene, 2-(trifluoromethyl)-1-propene, vinyl chloride, cis- and trans-2-dichloroethylene, and vinyltrifluorosilane. Most seleniranes were found to be inherently unstable and to undergo decomposition, with half-lives ranging from about 20 msec to over 5 sec.[3] Allene appears to be the only olefin giving products with a higher molecular weight than that of the corresponding selenirane. The allene and butadiene episelenides are more stable than the episelenides of the simple olefins, such as ethylene. Their decomposition probably is bimolecular (equation 1).

$$(1) \qquad \begin{array}{c} H_2C \\ \\ \\ H_2C \end{array} \!\!\! \diagdown\!\!\diagup \!\! Se \; + \; Se \!\! \diagdown\!\!\diagup \!\!\! \begin{array}{c} CH_2 \\ \\ \\ CH_2 \end{array} \longrightarrow Se_2 + 2CH_2\!\!=\!\!CH_2$$

b. Four-Membered Rings

Selenetane. Morgan and Burstall[4] and Backer and Winter[5] prepared selenetane (**2**) in low yield by causing 1,3-dibromopropane to react with an alkali selenide (equation 2). Selenetane is a colorless oil having an

$$(2) \qquad H_2C \begin{array}{c} \diagup CH_2Br \\ \\ \diagdown CH_2Br \end{array} + Se^{2-} \xrightarrow{C_2H_5OH} H_2C \begin{array}{c} \diagup CH_2 \diagdown \\ \qquad\qquad Se \\ \diagdown CH_2 \diagup \end{array} + 2Br^-$$
$$\textbf{2}$$

exceedingly powerful and penetrating smell. The vapor has an irritant effect on the mucous membrane of the nose. Selenetane is stable when stored under cool, dark conditions but polymerizes easily either on distillation or by the action of mineral acids.

Harvey et al.[6] studied the vibrorotational spectrum of selenetane in order to identify the normal mode of vibration corresponding to ring inversion. A potential barrier corresponding to 373 cm^{-1} was found, in fairly good agreement with that revealed by microwave investigations and similar to that determined for thietane,[7] where microwave investigation has shown that the barrier has a value of 274 cm^{-1}. By n.m.r. investigations, Moniz[8] confirmed the nonplanarity of the ring of selenetane and the occurrence of rapid inversion.

Compound **2** reacts with iodomethane to open the ring and with an alcoholic iodine solution to give 1,1-diiodoselenetane and an amorphous violet polymer. When **2** is treated with alcoholic mercuric chloride, the complex separates as white crystals, which decompose on heating into mercuric selenide and 1,3-dichloropropane and regenerate **2** on treatment with sodium hydroxide.[4]

Like the corresponding sulfur derivatives, the 3,3-disubstituted selenetanes are considerably more stable than selenetane itself. Backer and Winter[5] prepared 3,3-dimethylselenetane (**3**) by reaction of 1,3-dibromo-2,2-dimethylpropane with potassium selenide (equation 3). Compound **3** is a

$$
(3) \qquad
\begin{array}{c}
H_3C \qquad CH_2Br \\
\diagdown \ \diagup \\
C \\
\diagup \ \diagdown \\
H_3C \qquad CH_2Br
\end{array}
+ K_2Se \xrightarrow[N_2]{C_2H_5OH}
\begin{array}{c}
H_3C \qquad CH_2 \\
\diagdown \ \diagup \ \diagdown \\
C \qquad \qquad Se \\
\diagup \ \diagdown \ \diagup \\
H_3C \qquad CH_2 \\
\mathbf{3}
\end{array}
+ 2KBr
$$

colorless liquid. At 760 mm it boils at 139–140° and is transformed into a viscous oil (polymerization or decomposition).

Backer and Winter[5] also prepared 3,3-dimethylselenetane derivatives, such as 1,1-diiodo and addition compounds with mercuric halides. The reaction between iodomethane and **3** results in ring opening and yields 3-iodo-2,2-dimethylpropanedimethylselenonium iodide (**4**) (equation 4).

$$
(4) \qquad \mathbf{3} + 2CH_3I \xrightarrow{C_2H_5OH}
\begin{array}{c}
H_3C \qquad CH_2Se^+(CH_3)_2I^- \\
\diagdown \ \diagup \\
C \\
\diagup \ \diagdown \\
H_3C \qquad CH_2I \\
\mathbf{4}
\end{array}
$$

The action of chlorine and bromine causes ring opening and the addition of four halogen atoms; hydrolysis of these selenium trihalides with silver hydroxide yields seleninic acids (scheme 5).

$$
(5) \qquad \mathbf{3} + 2X_2 \longrightarrow
\begin{array}{c}
H_3C \qquad CH_2X \\
\diagdown \ \diagup \\
C \\
\diagup \ \diagdown \\
H_3C \qquad CH_2SeX_3
\end{array}
\xrightarrow{AgOH}
$$

$$
\begin{array}{c}
H_3C \qquad CH_2X \\
\diagdown \ \diagup \\
C \\
\diagup \ \diagdown \\
H_3C \qquad CH_2SeO_2H
\end{array}
$$

Compound **3** was oxidized by hydrogen peroxide to give the 1,1-dioxide (**5**), which may be also obtained by cyclization of sodium 3-chloro-2,2-dimethylpropane-1-seleninate (scheme 6).[5]

(6)   **3**

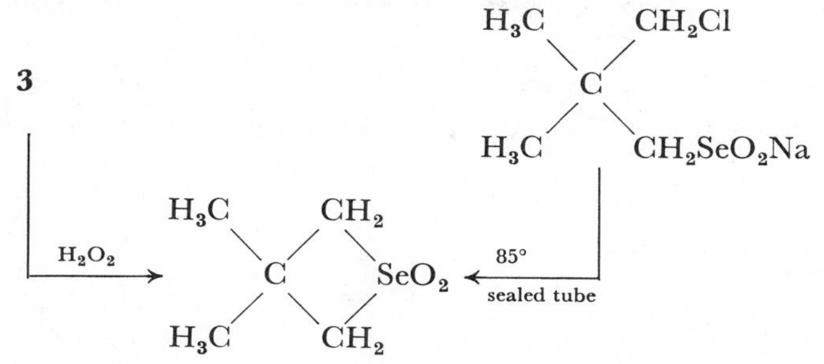

$$H_3C \diagdown \diagup CH_2Cl$$
$$C$$
$$H_3C \diagup \diagdown CH_2SeO_2Na$$

$$\xrightarrow{H_2O_2} \quad H_3C \diagdown \quad CH_2 \diagdown$$
$$C \qquad SeO_2 \xleftarrow[\text{sealed tube}]{85°}$$
$$H_3C \diagup \quad CH_2 \diagup$$
5

In 1935, Fredga[9] prepared 2,2-bis(methylenecarboxylic)-1-oxa-2-selenetane-4-one (**6**) by reaction of equivalent amounts of bis(methylenecarboxylic) selenide with monochloroacetic acid (equation 7). According to

(7) $O{=}C \diagdown \diagup CH_2Cl \qquad + Se \diagup CH_2{-}COOH \qquad \longrightarrow$
$$\diagdown OH \qquad \qquad \diagdown CH_2{-}COOH$$

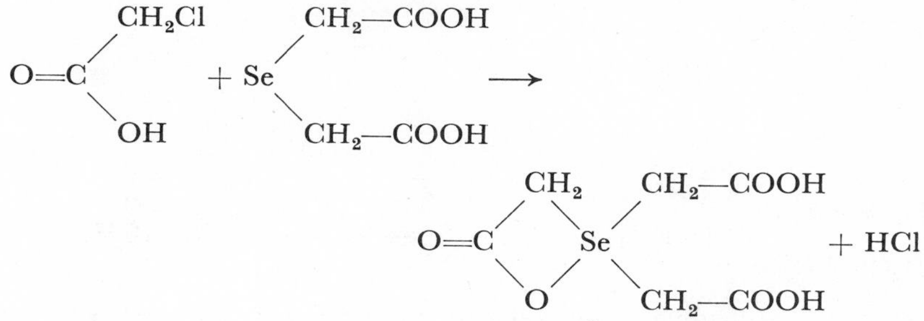

$$CH_2 \quad CH_2{-}COOH$$
$$O{=}C \diagdown \diagup \diagup$$
$$Se \qquad \qquad + HCl$$
$$\diagdown O \diagup \diagdown CH_2{-}COOH$$
6

Fredga, it was impossible to establish whether **6** has a monomeric or a polymeric form. He suggested that in nondissociating solvents **6** is in the cyclic form, whereas in dissociating media a zwitterionic linear structure (**7**) may appear.

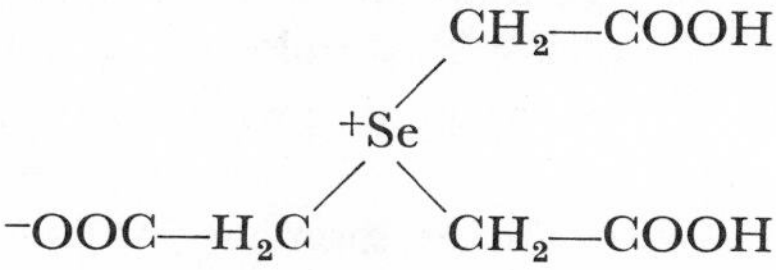

$$CH_2{-}COOH$$
$$\diagup$$
$$^+Se$$
$$\diagup \diagdown$$
$$^-OOC{-}H_2C \qquad CH_2{-}COOH$$
7

c. Five-Membered Rings

Selenolane. This selenacycloalkane (**8**), also known as tetrahydroselenophene, was prepared for the first time by Morgan and Burstall[10] by reaction of 1,4-dibromobutane with sodium selenide. Two other methods for the preparation of selenolane are also available. Yur'ev[11] prepared **8** by passing oxolane over Al_2O_3 at 400° in a H_2Se stream. The procedure of McCullough and Lefohn[12] consists in allowing 1,4-dibromobutane to react with a mixture of selenium powder, sodium formaldehyde sulfoxylate (Rongalite), NaOH, and water.

Selenolane is also formed by heating a low-molecular-weight poly-[(diseleno)tetramethylene] (regarded as 1,2-diselenane by Morgan and

(8)

Burstall[10]) (scheme 8). Compound **8** was purified by conversion to the 1,1-dibromo derivative, and reduction back to the selenolane. Selenolane is a liquid that, in the pure liquid state, is practically colorless but has a very pungent and somewhat unpleasant smell.

De Vries Robles[13] measured the dipole moment of **8** in benzene, and in a subsequent paper[14] he proposed a probable structure for this molecule. According to De Vries Robles,[14] the molecule is not completely planar, since atoms 2 and 5 are out of the plane formed by atoms 1, 3, and 4, and on the opposite sides.

Strom et al.[15] compared the dipole moment values in CCl_4 of selenolane and other five-membered heterocycles with those in benzene. They found that, since π-electrons of benzene should act to neutralize the dipole, the dipole moments of oxolane and thiolane are, as anticipated, greater in CCl_4, whereas selenolane exhibits the opposite effect.

Milazzo[16] studied the u.v. spectrum of selenolane in vapor form, at room temperature, in the range 2475–1690 Å. He found a spectrum with twenty most pronounced absorption maxima, with minima at 1849 and 1942 Å, and at least four electronic transitions (2475–2300, 2285–2215, 2163–2084, and 2080–1690 Å).

1,1-Dichloro-, 1,1-dibromo-, and 1,1-diiodoselenolane were prepared by reaction of the respective halogen with selenolane in carbon tetrachloride.[10] Hope and McCullough[17] found by X-ray analysis that 1,1-diiodoselenolane crystals display orthorhombic symmetry. Pedersen and Hope[18] determined by n.m.r. that 1,1-diiodoselenolane is a puckered ring in which adjacent methylene groups are in an approximately staggered conformation, giving rise to two enantiomorphic molecular conformations. This agrees with the calculation of De Vries Robles[14] for the selenolane molecule. McCullough and Brunner[19] performed spectrophotometric studies of the dissociation of the 1,1-diiodoselenolane complex, at various wavelengths and temperatures in carbon tetrachloride. The small dissociation-constant values found by the authors indicate that the iodine complex of selenolane is a very stable compound (more stable than the sulfur analog).

1,1-Dibromoselenolane reacts with bromine in chloroform solution to give a perbromo-1,1-dibromoselenolane, $C_4H_8SeBr_2 \cdot Br_5$. This compound is unstable and decomposes slowly into the 1,1-dibromo derivative.[10] By treatment of 1,1-dibromoselenolane with silver oxide, 1,1-dihydroxyselenolane was obtained as highly hygroscopic, colorless crystals. Addition of aqueous hydrobromic acid regenerated the original 1,1-dibromo derivative.

A methiodide derivative of **8** was obtained from selenolane and iodomethane.[10] According to Morgan and Burstall,[10] by the reaction of selenolane and 1,4-dibromobutane **9** or **10** may be obtained (scheme 9). Finally, selenolane reacts with alcoholic mercuric chloride to give a mercurichloride complex.

(9)

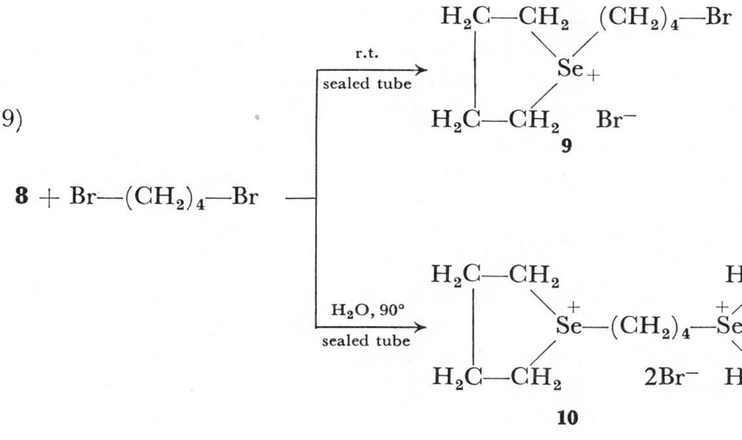

Brewer et al.[20] prepared a coordination compound of gallium tetrachloro-gallate and selenolane by allowing vapor of selenolane to diffuse into a benzene solution of gallium dichloride in a closed device. The resulting complex was formulated as $GaSel_4^+GaCl_4^-$, where "Sel" represents a selenolane monodentate ligand.

Substituted Selenolanes. Suginome and Umezawa[21] prepared several halogen-substituted selenolanes such as 2,2,3,4,5,5-hexachloroselenolane, obtained from selenophene and excess chlorine. Oxidation of this compound with nitric acid gave the corresponding 1-oxide. When chlorine was passed through a solution of selenophene in carbon disulfide at $-15°$, addition took place, giving 2,3,4,5-tetrachloroselenolane (**11**). Oxidation of this compound with nitric acid yielded oxalic acid. These authors[21] heated **11** to 95° until evolution of HCl ceased; after distillation of 2,5-dichloroselenophene the residue gave 2,2,5,5-tetrachloroselenolane (**12**) (equation 10). This com-

(10)

pound is more stable than its isomer, **11**. On oxidation with nitric acid, the 1-oxide derivative was obtained.

Bromination of selenophene in carbon disulfide gave 1,1,2,2,5,5-hexabromoselenolane. On treatment with water, 2,2,5,5-tetrabromoselenolane was obtained in 9% yield, as well as an ether-insoluble product, 2,2,5,5-tetrabromoselenolane-1-oxide. The latter was also prepared by the nitric acid oxidation of 1,1,2,2,5,5-hexabromoselenolane. By reaction of excess chlorine with a solution of 2,5-dibromoselenophene in carbon disulfide, 2,5-dibromo-2,3,4,5-tetrachloroselenolane was obtained.

Funk and Papenroth[22] prepared 1,1-dibromo-3,4-dichloroselenolane by reaction of 3,4-dichloroselenophene with concentrated aqueous hydrobromic acid.

Krespan and Langkammerer[23,24] prepared several fluoroselenolanes. Thus, 2,2,3,3,4,4,5,5-octafluoroselenolane (**13**) was obtained in 10% yield by heating a mixture of selenium, tetrafluoroethylene, and iodine (catalyst) at 250° under autogenous pressure for 7 hours (equation 11). Krespan[25,26] found that modification of reaction 11 by the addition of an unsaturated third component led to a general preparative method for partially fluorinated selenolanes. Thus, from selenium, propylene, tetrafluoroethylene, and iodine,

(11)

$$\begin{matrix} CF_2 \\ \| \\ CF_2 \end{matrix} + Se + \begin{matrix} CF_2 \\ \| \\ CF_2 \end{matrix} \longrightarrow \begin{matrix} F_2C \!-\!-\!-\! CF_2 \\ | \quad\quad | \\ F_2C \quad\quad CF_2 \\ \diagdown \quad \diagup \\ Se \end{matrix}$$

13

there was obtained 2,2,3,3-tetrafluoro-5-methylselenolane in 8 % yield. In a similar manner,[26] the following fluoroselenolanes were obtained: 2,2,3,3-tetrafluoro-5-n-butylselenolane, 2,2,3-trifluoro-3-bromo-5-cyclohexylselenolane, 2,2,3,3-tetrafluoro-5-iodomethylselenolane, 2,2,3,3-tetrafluoro-5-phenyl-5-cyanoselenolane, and 2,2,3,3-tetrafluoro-4-methyl-5-methoxyselenolane.

Yur'ev *et al.*[27] prepared 2,3,4,5-tetradeuterioselenolane in 85 % yield by heating 2,3,4,5-tetraiodoselenophene with anhydrous zinc filings, deuterium oxide, and anhydrous acetic anhydride. Duffield *et al.*[28] obtained 2,2,5,5-tetradeuterioselenolane (97 % d_4 species) by preparative vapor-phase chromatography of the reaction product from 1,4-dibromobutane-1,1,4,4-d_4 and sodium selenide. These authors measured the mass spectra of selenolane and other five-membered heterocyclic analogs and determined from deuterium-labeling studies the different modes of fragmentation upon electron impact.

Fredga[29] prepared selenolane-2,5-dicarboxylic acid (**14**) by reaction of α,α'-dibromoadipic acid and potassium diselenide (equation 12) or mono-

(12)

$$\begin{matrix} COOH \\ | \\ H_2C\!-\!CHBr \\ | \\ H_2C\!-\!CHBr \\ | \\ COOH \end{matrix} + K_2Se_2 \xrightarrow{\text{acetone}} \begin{matrix} COOH \\ | \\ H_2C\!-\!CH \\ | \quad\quad \diagdown \\ \quad\quad\quad Se + 2KBr + Se \\ | \quad\quad \diagup \\ H_2C\!-\!CH \\ | \\ COOH \end{matrix}$$

14

selenide. Compound **14** exists in *cis* and *trans* forms. *meso*-α,α'-Dibromoadipic acid gives the *cis* (*meso*) acid; *rac*-α,α'-dibromoadipic acid, the *trans* (*rac*) acid.

Fredga[30] also separated the (+) and (−) antipodes of the *trans* isomer, using brucine and quinine salts. Melting point curves for binary systems (+)-**14** and (−)-**14**, for (−)-**14** and the (+)-sulfur analog of **14**, and for the (+)-sulfur analog and (−)-sulfur analog of **14** were reported by Fredga.[31] All three curves show the same general form, thus demonstrating the formation of a quasi racemate between (−)-**14** and its (+)-sulfur analog.

The formation of quasi racemates from **14** and its sulfur analog antipode systems was supported by Rosenberg and Schotte,[32] who ascertained by i.r. spectroscopy that (+)-**14** and its (−)-sulfur analog also form a quasi-racemic compound. Fredga[33] obtained (−)-selenolane-2,5-dicarboxylic acid-1-oxide by oxidation of (−)-**14** with iodine.

1-Thia-2-selenolane. Thiaselenacycloalkane (**15**) is unknown; however, its derivatives, 1-thia-2-selenolane-4-carboxylic acid (**16**) and 1-thia-2-seleno-

$$
\begin{array}{cc}
S\!\!-\!\!-\!\!-\!\!Se \\
| \qquad | \\
H_2C \qquad CH_2 \\
\diagdown \quad \diagup \\
CH_2 \\
\mathbf{15}
\end{array}
$$

lane-4-aminomethyl hydrochloride (**17**), have been synthesized. Bergson and Biezais[34] prepared **16** by debenzylation of 1-benzylthio-3-benzylseleno-propane-2-carboxylic acid, according to reaction 13. Compound **16** shows

(13)

$$
\begin{array}{ccc}
C_6H_5\!-\!CH_2\!-\!Se\!-\!CH_2 & & Se\!-\!CH_2 \\
\diagdown & & | \qquad \diagdown \\
CH\!-\!COOH \xrightarrow[\text{(2) }O_2 + FeCl_3]{\text{(1) Na in liq. }NH_3} & & CH\!-\!COOH \\
\diagup & & | \qquad \diagup \\
C_6H_5\!-\!CH_2\!-\!S\!-\!CH_2 & & S\!-\!CH_2 \\
& & \mathbf{16}
\end{array}
$$

a pronounced tendency to polymerize (see Chapter XIV).

Bergson[35] studied the rate of oxidation of **16** by ammonium persulfate in aqueous ethanol spectrophotometrically and observed that it followed a path similar to that of 1,2-diselenolane-4-carboxylic acid, the rate for **16** being slower than that for the diselenolane but faster than that for 5-(1,2-dithiolan-3-yl)pentanoic acid (α-lipoic acid). Bergson postulated that in oxidation the primary oxidation product is the thioseleninate (equation 14).

(14) $\mathbf{16} \xrightarrow[25°]{(NH_4)_2S_2O_8}$

$$
\begin{array}{c}
HO_2SeCH_2 \\
\diagdown \\
CH\!-\!COOH \\
\diagup \\
HSCH_2
\end{array}
$$

Krackov *et al.*[36] measured the dipole moment and the acid dissociation constant of **16**. From their study on isologous 1,2-dithiolane, 1,2-diselenolane, and 1-thia-2-selenolane derivatives, it was concluded that no significant polarization of the S-Se bond could be detected, and that there is no significant interaction between Se–S and carboxyl groups by hydrogen bonding.

Bergson[37] synthesized **17** by debenzylation of 1-benzylthio-3-benzylseleno-2-aminomethylpropane hydrochloride, followed by oxidation (equation 15).

(15)

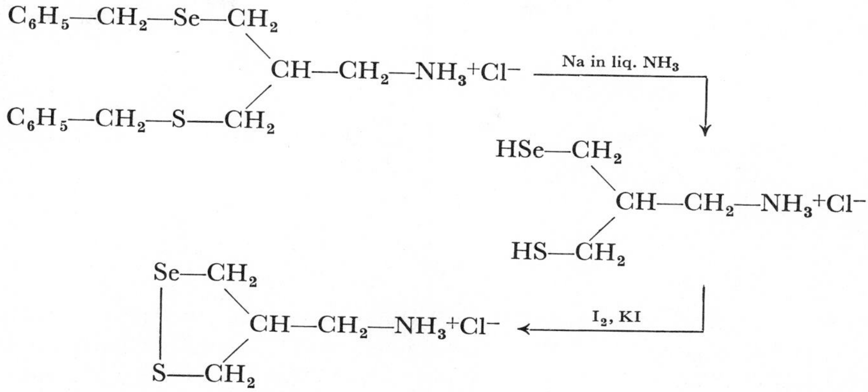

$$C_6H_5—CH_2—Se—CH_2$$
$$\qquad\qquad\qquad CH—CH_2—NH_3^+Cl^- \quad \xrightarrow{\text{Na in liq. NH}_3}$$
$$C_6H_5—CH_2—S—CH_2$$

$$HSe—CH_2$$
$$\qquad\qquad CH—CH_2—NH_3^+Cl^-$$
$$HS—CH_2$$

$$Se—CH_2$$
$$\mid \qquad\quad\; CH—CH_2—NH_3^+Cl^- \quad \xleftarrow{\text{I}_2,\text{ KI}}$$
$$S—CH_2$$
17

Compound **17** is a yellow-orange crystalline substance that polymerizes on heating at about 120°. Attempts to obtain the free amine failed because of its tendency towards polymerization.

The u.v. spectrum of **17** shows a characteristic peak at 389 nm, which is very close to that observed for **16** (396 nm), thus confirming the cyclic structure.

1-Selena-2-silolanes.   In 1968 Dubac and Mazerolles[38] prepared 2,2-di-*n*-butyl-1-selena-2-silolane (**18**) by a ring-expansion reaction starting from 1,1-di-*n*-butylsiletane and selenium (equation 16).

(16)

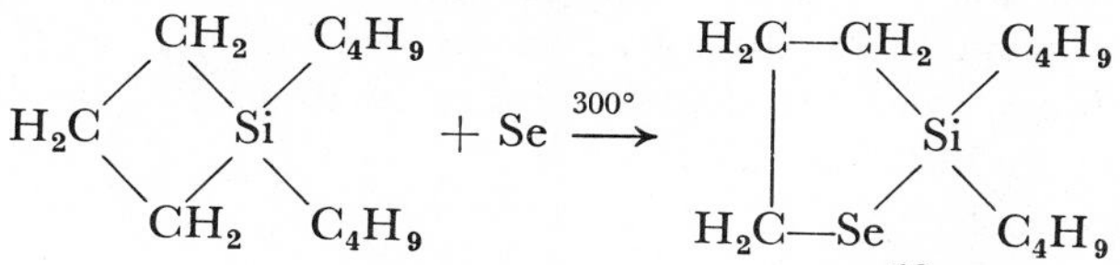

1-Selena-2-germolanes.   A ring-expansion reaction similar to that used to prepare **18** was reported by Mazerolles *et al.*[39] for the preparation of 2,2-di-*n*-butyl-1-selena-2-germolane (**19**) on heating 1,1-di-*n*-butylgermetane and selenium (equation 17).

(17)

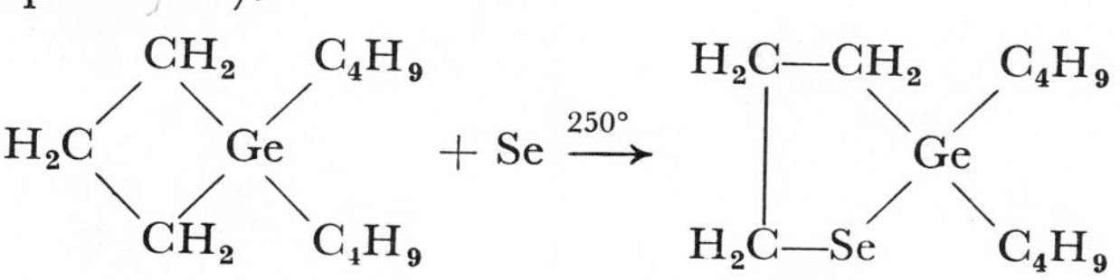

d. Six-Membered Rings

Selenane.   This compound (**20**) was prepared for the first time by Morgan and Burstall[40] by the reaction of 1,5-dibromopentane with sodium selenide in ethanol. These authors also obtained **20** on heating poly[(diseleno)-

$$
\begin{array}{c}
\text{Se} \\
\diagup \quad \diagdown \\
\text{H}_2\text{C} \qquad \text{CH}_2 \\
| \qquad\qquad | \\
\text{H}_2\text{C} \qquad \text{CH}_2 \\
\diagdown \quad \diagup \\
\text{CH}_2 \\
\mathbf{20}
\end{array}
$$

pentamethylene] (regarded as 1,2-diselenepane by Morgan and Burstall). Yur'ev[11] prepared **20** by passing oxane over Al_2O_3 at 400° in a stream of H_2Se. McCullough and Lefohn[12] also prepared **20** by allowing 1,5-dibromopentane to react with a mixture of selenium powder, Rongalite, NaOH, and water. All these methods are the same as those followed for the preparation of selenolane. Compound **20** is a practically colorless liquid possessing a pungent and rather unpleasant smell. By the same kinds of reaction reported for the synthesis of selenolane complexes[10] (see Section XI.A.2.c), Morgan and Burstall[40] prepared numerous selenane complexes.

McCullough and Brunner[19] performed spectrophotometric studies of the dissociation of the 1,1-diiodoselenane complex in carbon tetrachloride. The small dissociation-constant values found by these authors show that the iodine complex of selenane (like the selenolane complex) is a very stable compound and more stable than the sulfur analog.

Morgan and Burstall[41] obtained 2-methylselenane (**21**) on heating poly-(selenohexamethylene), poly[(diseleno)hexamethylene], or 1,8-diselenacyclotetradecane (equation 18).

(18) $[-\text{Se}-(\text{CH}_2)_6-\text{Se}-]_n \xrightarrow[-\text{Se}]{250°}$

$$
\begin{array}{c}
\text{Se} \\
\diagup \quad \diagdown \\
\text{H}_2\text{C} \qquad \text{CH}-\text{CH}_3 \\
| \qquad\qquad | \\
\text{H}_2\text{C} \qquad \text{CH}_2 \\
\diagdown \quad \diagup \\
\text{CH}_2 \\
\mathbf{21}
\end{array}
\xrightarrow{220°}
\begin{array}{c}
\left[-(\text{CH}_2)_6-\text{Se}-\right]_n \\[4pt]
\left[
\begin{array}{c}
(\text{CH}_2)_6 \\
\diagup \quad\quad \diagdown \\
\text{Se} \qquad\qquad \text{Se} \\
\diagdown \quad\quad \diagup \\
(\text{CH}_2)_6
\end{array}
\right]
\end{array}
$$

Fredga and Styrman[42] prepared selenane-2,6-dicarboxylic acid (**22**) by reaction of α,α'-dibromopimelic acid and sodium selenide (from Rongalite, NaOH, and selenium) (equation 19). Like selenolane-2,5-dicarboxylic acid

(19)

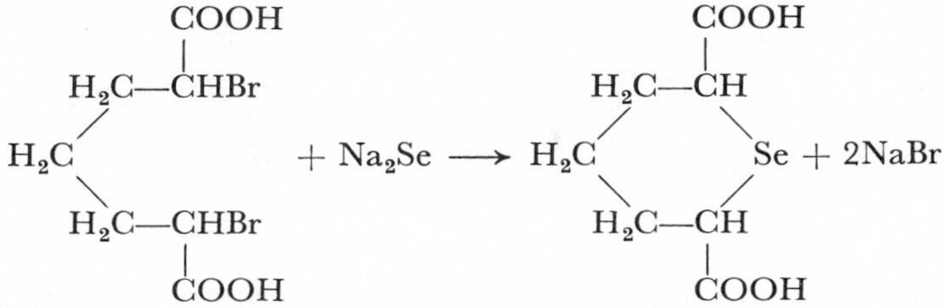

$$\begin{array}{ccc}
\text{COOH} & & \text{COOH} \\
| & & | \\
\text{H}_2\text{C—CHBr} & & \text{H}_2\text{C—CH} \\
\diagup \qquad \diagdown & & \diagup \qquad \diagdown \\
\text{H}_2\text{C} \qquad\quad + \text{Na}_2\text{Se} \longrightarrow \text{H}_2\text{C} & & \text{Se} + 2\text{NaBr} \\
\diagdown \qquad \diagup & & \diagdown \qquad \diagup \\
\text{H}_2\text{C—CHBr} & & \text{H}_2\text{C—CH} \\
| & & | \\
\text{COOH} & & \text{COOH} \\
& & \mathbf{22}
\end{array}$$

(**14**), compound **22** exists in *cis* and *trans* forms. The resolution of *trans*-**22** into its optical antipodes was performed with quinine and brucine salts.

In 1968 Hänsel and Haller[43] prepared 2,6-dimethylselenane-4-one-3,5-dicarboxylic acid diethyl ester (**23**) by condensation of acetaldehyde, acetonedicarboxylic acid diethyl ester, and H_2Se in the presence of 1,4-diazabicyclo[2.2.2]octane (equation 20). As shown by n.m.r., the two

(20)

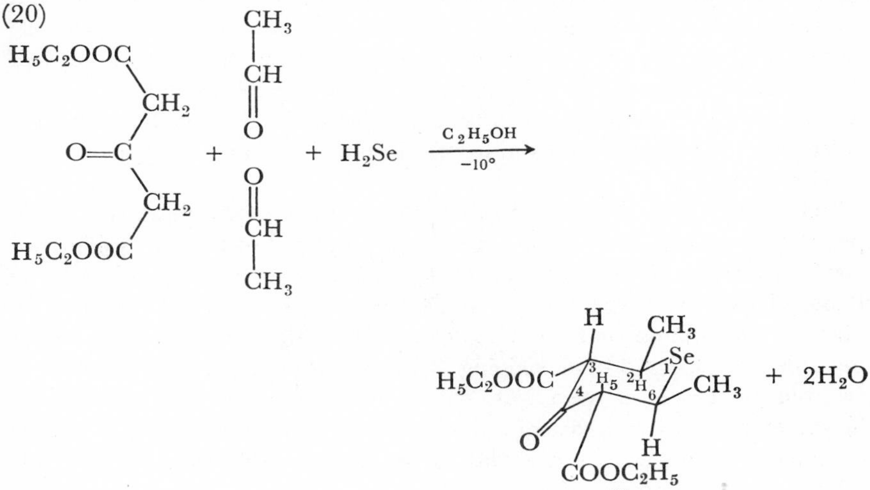

methyl groups are in *cis* configuration to each other, and the two carbethoxy groups also are in the *cis* configuration to each other. The chair form is found to be the preferred conformation with equatorial substituents on the C(2), C(3), C(5), and C(6) atoms.

1-Oxa-4-selenane. 1-Oxa-4-selenane (**24**) was first prepared from sodium selenide and β,β'-dichloro- or β,β'-diiododiethyl ether by Gibson and

Johnson.[44,45] McCullough and Lefohn[12] obtained **24** from essentially the same reaction, preparing sodium selenide *in situ* from Rongalite, selenium, and NaOH (equation 21). Compound **24** is a colorless liquid with a pene-

$$(21) \quad \begin{array}{c} H_2C\!-\!CH_2X \\ / \\ O \\ \backslash \\ H_2C\!-\!CH_2X \end{array} + Na_2Se \xrightarrow[\;H_2\;]{H_2O,\,100°} \begin{array}{c} H_2C\!-\!CH_2 \\ / \quad \backslash \\ O \qquad Se \\ \backslash \quad / \\ H_2C\!-\!CH_2 \end{array} + 2NaX$$

24

trating but not unpleasant smell. The physical constants of **24** were also reported in other papers.[46,47]

1-Oxa-4-selenane forms several complexes[45] with halogens to give 4,4-dichloro, 4,4-dibromo, and 4,4-diiodo complexes, as well as the methiodide and the mercurichloride derivatives. When ammonia was passed through a benzene solution of the 4,4-dibromo complex of **24,** a compound was obtained that had composition $C_4H_8OBr_2Se\cdot2NH_3$, was colorless and insoluble in organic solvents, and decomposed on treatment with water. Similar compounds were formed between 4,4-dibromo-1-oxa-4-selenane and pyridine, piperidine, and aniline. Treatment of **24** with nitric acid under the conditions leading to the formation of a sulfoxide from a sulfide yielded, instead of the selenoxide, a hydroxynitrate with explosive violence. The hydroxynitrate decomposed at 140–141° and detonated on heating in a sealed tube. It was also prepared from cold nitric acid and 4,4-dihydroxy-1-oxa-4-selenane.

McCullough and Zimmermann[48] performed thermodynamic studies by spectrophotometric analysis of the iodine complexes of the five sulfur and selenium analogs of 1,4-dioxane, *i.e.*, 1,4-dithiane, 1,4-diselenane, 1-oxa-4-thiane, 1-oxa-4-selenane, and 1-thia-4-selenane, in carbon tetrachloride solution. They found that the complexing tendencies towards iodine are in the order $Se > S > O$, in keeping with the order of the availability of electrons on the respective atoms. Where the iodine molecule has a choice, the attachment will be at selenium rather than at sulfur or oxygen. Thermodynamic constants for the dissociation of iodine complexes at 25° in CCl_4 solution were reported.

The crystal and molecular structures of the diiodine and of the iodine monochloride complex of **24** were studied by Maddox and McCullough[49] and by Knobler and McCullough,[50] respectively. The iodine monochloride complex was obtained by mixing equimolar quantities of the reactants in ethylene chloride. It is interesting to note that the Se-I bond in the iodine monochloride complex is the shortest of its kind reported so far, and (presumably as a result) the I-Cl bond is lengthened to 2.73 Å from its uncomplexed value of 2.32 Å. This situation and the fact that the chlorine forms no other bonds than the weak one to iodine suggest that there is a considerable ionic character in the I-Cl bond.

1-Oxa-4-selena-2,6-sila- and germacyclohexanes.   Wieber *et al.*[51,52] prepared several silicon- and germanium-containing selenium heterocycles. 2,2,6,6-Tetramethyl-1-oxa-4-selena-2,6-disilacyclohexane (**25**) was prepared by Schmidt and Wieber[51] by causing 1,3-bis(chloromethyl)tetramethyldisiloxane to react with sodium selenide, according to reaction 22. Compound

(22)

25 is insoluble in water and is not highly sensitive to hydrolysis. It reacts with iodomethane to give the selenonium salt (**26**) (equation 23).

(23)   **25** + CH$_3$I $\longrightarrow$

In a similar manner Wieber and Schwarzmann[52] prepared 2,2,6,6-tetramethyl-1-oxa-4-selena-2-sila-6-germacyclohexane (**27**) and 2,2,6,6-tetramethyl-1-oxa-4-selena-2,6-digermacyclohexane (**28**). The chemical properties of these compounds are similar to those of **25**.

1,3-Dioxa-5-selenane.   The only known derivative of this selenacycloalkane is 2,4,6-trimethyl-1,3-dioxa-5-selenane (**29**), also known as monoseleno-paraldehyde. This compound was prepared in 1967 by Credali *et al.*[53] by

$$
\begin{array}{c}
\text{Se} \\
\diagup \quad \diagdown \\
\text{H}_3\text{C—HC} \qquad \text{CH—CH}_3 \\
| \qquad\qquad | \\
\text{O} \qquad\quad \text{O} \\
\diagdown \quad\quad \diagup \\
\text{CH} \\
| \\
\text{CH}_3 \\
\mathbf{29}
\end{array}
$$

the reaction of hydrogen selenide with acetaldehyde in an aqueous solution, at 0–5°, in the presence of hydrochloric acid. Compound **29** has a characteristic biting odor.[54]

1-Thia-4-selenane.   1-Thia-4-selenane (**30**) was prepared by Gibson and Johnson[55] in 1933 by a method similar to that used for the synthesis of 1-oxa-4-selenane.[45] In this case, the reaction of β,β'-dichlorodiethyl sulfide on aqueous sodium selenide (equation 24) gave **30** in low yield. Gibson and

$$
(24) \quad
\begin{array}{c}
\text{H}_2\text{C—CH}_2\text{Cl} \\
\diagup \\
\text{S} \\
\diagdown \\
\text{H}_2\text{C—CH}_2\text{Cl}
\end{array}
+ \text{Na}_2\text{Se} \xrightarrow{\text{H}_2\text{O}}
\begin{array}{c}
\text{H}_2\text{C—CH}_2 \\
\diagup \qquad \diagdown \\
\text{S} \qquad\qquad \text{Se} \\
\diagdown \qquad \diagup \\
\text{H}_2\text{C—CH}_2 \\
\mathbf{30}
\end{array}
+ 2\text{NaCl}
$$

Johnson[55] attempted unsuccessfully to obtain **30** by the interaction of β,β'-dichlorodiethyl sulfide and aluminum selenide or of β,β'-dichloro-diethyl selenide and sodium sulfide.

According to McCullough and Radlick,[56] the poor yield from the Gibson and Johnson method was due to the inability of Se^{2-} to compete favorably with water in the reaction with the cyclic sulfonium ion. McCullough and Radlick prepared **30** in 44% yield according to reaction 24, using absolute ethanol as the reaction medium instead of water. McCullough and Lefohn[12] prepared **30** in *ca.* 3% yield according to reaction 24, using sodium selenide obtained *in situ* from Rongalite, selenium, and NaOH. Compound **30** is isomorphous with 1,4-dithiane and 1,4-diselenane.[56]

McCullough[57] prepared several halogen and interhalogen addition compounds of **30**. The structure of the diiodine complex of **30** ($C_4H_8SSe2I_2$) was reported by Hope and McCullough,[58] who found this complex to be iso-morphous with $C_4H_8S_22I_2$.

McCullough and Zimmermann,[48] in an attempt to determine the dissociation constants of the $C_4H_8SSe2I_2$ complex in CCl_4 solution according to
equations 25 and 26, found that, even when iodine is in 40-fold excess with

$$(25) \qquad C_4H_8SSe2I_2 \rightleftharpoons C_4H_8SSeI_2 + I_2$$

$$(26) \qquad C_4H_8SSeI_2 \rightleftharpoons C_4H_8SSe + I_2$$

respect to 1-thia-4-selenane, equation 25 is essentially complete. The single
equilibrium (26) based on a 1:1 complex also takes place. The thermodynamic
constants for dissociation of the iodine complex at 25° were compared by
McCullough and Zimmermann with those of analogous heterocycles
(*cf.* 1-oxa-4-selenane).

1-Thia-4-selena-2,6-disila- and digermacyclohexanes.   Wieber and Schwarzmann[52] prepared 2,2,6,6-tetramethyl-1-thia-4-selena-2,6-disilacyclohexane
(**31**) and 2,2,6,6-tetramethyl-1-thia-4-selena-2,6-digermacyclohexane (**32**)
by the route followed to synthesize **25** (equations 27 and 28). Compounds

(27)

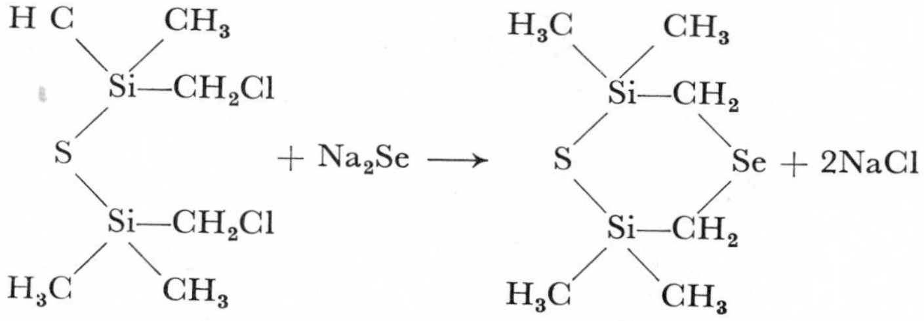

31

(28)

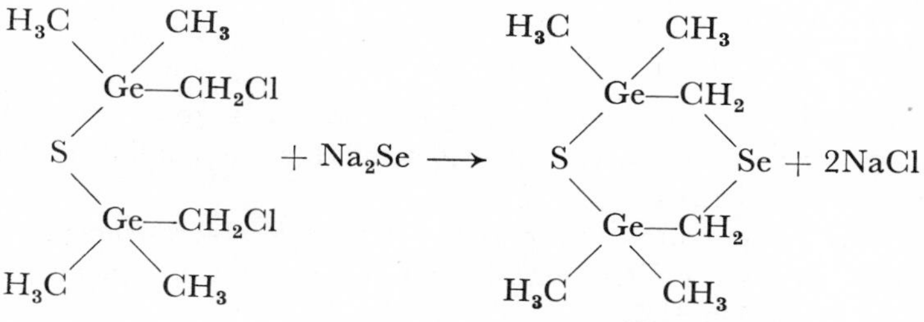

32

31 and **32** are highly sensitive to hydrolysis and react with iodomethane to
give selenonium salts.

1-Selena-2,6-digermacyclohexanes.   In 1969 Mazerolles *et al.*[59] prepared
2,2,6,6-tetraethyl-1-selena-2,6-digermacyclohexane (**33**) by a ring-expansion
reaction from 1,1,2,2-tetraethyl-1,2-digermolane and selenium (equation
29).

(29)

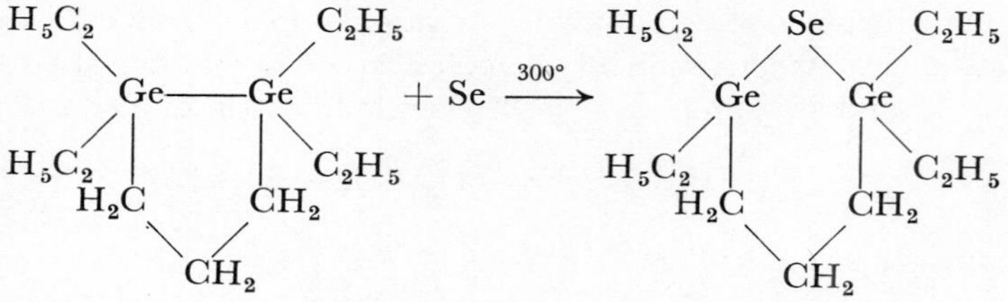

$$H_5C_2,\ C_2H_5,\ Ge——Ge,\ H_5C_2,\ C_2H_5,\ H_2C,\ CH_2,\ CH_2 \quad +\ Se\ \xrightarrow{300°}$$

$$H_5C_2,\ Se,\ C_2H_5,\ Ge,\ Ge,\ H_5C_2,\ C_2H_5,\ H_2C,\ CH_2,\ CH_2$$

33

e. Seven-Membered Rings

Selenepane. Morgan and Burstall[41] in 1931 prepared selenepane (**34**) in low yield, together with 1,8-diselenacyclotetradecane (**35**) and a low-molecular-weight poly(selenohexamethylene), by the reaction of 1,6-dibromohexane with sodium selenide (scheme 30). Compound **34** forms

(30)

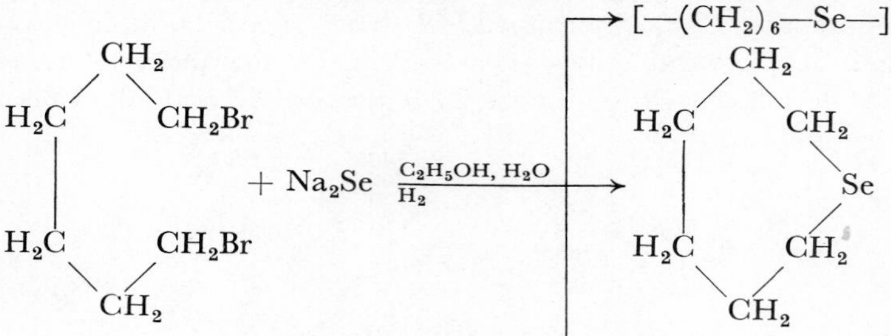

$$H_2C,\ CH_2,\ CH_2Br,\ H_2C,\ CH_2Br,\ CH_2 \quad +\ Na_2Se\ \xrightarrow[H_2]{C_2H_5OH,\ H_2O}$$

$$\rightarrow [—(CH_2)_6—Se—]_n$$

$$H_2C,\ CH_2,\ CH_2,\ Se,\ H_2C,\ CH_2,\ CH_2$$

34

$$(CH_2)_6,\ Se,\ Se,\ (CH_2)_6$$

35

diiodo, dibromo, and dichloro complexes.[41] With mercuric chloride and with iodomethane, the mercurichloride and the methiodide derivatives, respectively, were formed.

Funk and Papenroth[22] prepared 1,1,3,6-tetrachloroselenepane (**36**) by reaction of 1,5-hexadiene with selenium dioxide and aqueous concentrated HCl (equation 31).

(31)

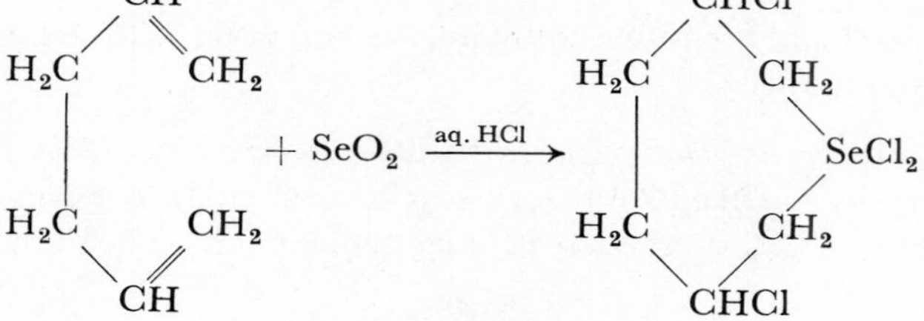

$$CH,\ H_2C,\ CH_2,\ H_2C,\ CH_2,\ CH \quad +\ SeO_2\ \xrightarrow{aq.\ HCl}$$

$$CHCl,\ H_2C,\ CH_2,\ SeCl_2,\ H_2C,\ CH_2,\ CHCl$$

36

3. Selenacycloalkanes with Two Selenium Atoms in the Ring

a. Four-Membered Rings

1,3-Diselenetane. 2,2,4,4-Tetraacetyl-1,3-diselenetane (**37**) and 2,4-di-methyl-1,3-diselenetane-2,4-diselenol diacetate (**38**) (see equation 35) are the only diselenetanes known so far. Compound **37** was prepared by Morgan *et al.*[60, 61] by reaction of selenium tetrachloride with acetylacetone or copper acetylacetone (equation 32); these workers assigned an incorrect structure

$$2SeCl_4 + 4CH_3-CO-CH_2-CO-CH_3$$

(32)

to this compound. In 1964 Dewar *et al.*[62] ascertained by n.m.r. and optical spectra that the compound obtained by Morgan and his co-workers has the diselenetane structure (**37**).

Compound **37** was converted quantitatively by hydriodic acid to 3,3′-diselenodi(4-hydroxypent-3-en-2-one) (diselenium bisacetylacetone) (**39**).[61]

Compound **37** reacts with hydrogen cyanide to give 2,4-pentanedione-3-selenocyanate (**40**) (equation 33) and with α-naphthyl mercaptan to give

$$(33) \qquad 37 \xrightarrow{\text{HCN}} CH_3-CO-\underset{\underset{\textbf{40}}{SeCN}}{C}=\overset{\overset{OH}{|}}{C}-CH_3$$

3-(α-naphthylthioseleno)-2,4-pentanedione (**41**), α,α'-dinaphthyl disulfide, and **39** (equation 34). Compound **37** is not affected by iodomethane,

$$(34) \qquad 37 \xrightarrow{\alpha-C_{10}H_7SH} \begin{array}{l} \rightarrow \textbf{39} \\ \\ \rightarrow CH_3-CO-\underset{\underset{\textbf{41}}{SeS-C_{10}H_7}}{C}=\overset{\overset{OH}{|}}{C}-CH_3 + (\alpha-C_{10}H_7S)_2 \end{array}$$

iodoethane, or acetone. Hydrogen chloride in dry ether dissolves **37** without reaction, whereas hydrochloric acid decomposes it into selenium and chloroacetylacetone. Compound **37** is decomposed by ammonia in a sealed tube to give selenium, acetamide, and hydrocyanic acid. It is reduced by potassium bisulfite to acetylacetone and potassium selenodithionate.[63]

2,4-Dimethyl-1,3-diselenetane-2,4-diselenol diacetate (**38**) was prepared by Olsson and Almqvist[64] in 1969, together with 1,3,5,7-tetramethyl-2,4,6,8,9,10-hexaselenatricyclo[3.3.1.1[3,7]]decane, essentially by the route followed for the preparation of the corresponding sulfur compound. In this case acetyl chloride and liquid H_2Se reacted at low temperature in the presence of anhydrous $AlCl_3$ in about 12 % yield (equation 35).

$$(35) \qquad CH_3COCl + \text{liq. } H_2Se \xrightarrow{AlCl_3}$$

38

b. Five-Membered Rings

1,2-Diselenolane.   According to Hagelberg[65] and to Morgan and Burstall,[4] hydrolysis of 1,3-propanediselenocyanate yields a yellow powder, which was regarded as 1,2-diselenolane (**42**) (m.p. 59°) by Morgan and Burstall and as a dimer (m.p. 54.5°) by Hagelberg. Morgan and Burstall obtained the

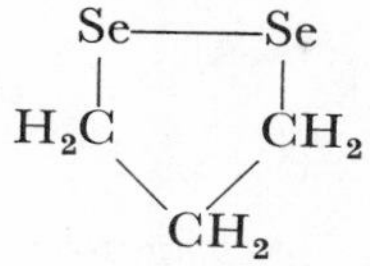

$$Se\text{------}Se$$
$$H_2C \qquad CH_2$$
$$CH_2$$

42

same product also by depolymerization on heating of poly(selenotrimethylene). Bergson and Claeson[66,67] found that this powder does not dissolve unless it is heated to 60°. They suggested that the powder probably is a dimer or a polymer of **42** which reversibly depolymerizes to 1,2-diselenolane on heating at about 60°. According to Bergson, 1,2-diselenolane has never been prepared in the pure form but has been shown to exist in solution (*cf.* Chapter XIV). Ultraviolet absorption measurements of the solution[66–69] show the characteristic peak of the 1,2-diselenolane ring at about 440 nm.

On treating a chloroform solution of **42** with iodine Morgan and Burstall[4] obtained a red precipitate of 1,3-propanediselenenyl diiodide, while oxidation of **42** with nitric acid gave 1,3-propanedisenininic acid dihydronitrate.

Backer and Winter[70] prepared 4,4-dimethyl-1,2-diselenolane (**43**) in 80% yield by reaction of 2,2-dimethyl-1,3-propanediselenocyanate with sodium ethoxide. Oxidation of **43** with nitric acid gave 2,2-dimethyl-1,3-propanediseleninic acid dihydronitrate (**44**) (scheme 36).

(36)

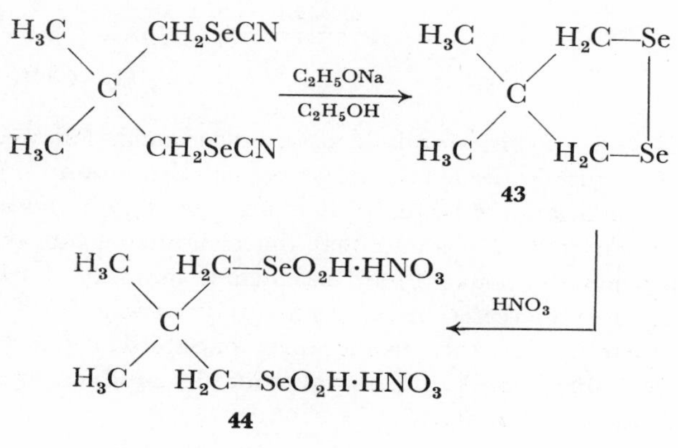

$$H_3C\diagdown \diagup CH_2SeCN$$
$$C$$
$$H_3C\diagup \diagdown CH_2SeCN$$
$$\xrightarrow[C_2H_5OH]{C_2H_5ONa}$$
$$H_3C\diagdown \diagup H_2C\text{---}Se$$
$$C \qquad |$$
$$H_3C\diagup \diagdown H_2C\text{---}Se$$

43

$$H_3C\diagdown \diagup H_2C\text{---}SeO_2H\cdot HNO_3$$
$$C$$
$$H_3C\diagup \diagdown H_2C\text{---}SeO_2H\cdot HNO_3$$

$$\xleftarrow{HNO_3}$$

44

Backer and Winter[70] also prepared 4-methyl-4-phenyl-1,2-diselenolane (**45**) by the reaction of 1,3-dibromo-2-methyl-2-phenylpropane with potassium selenide (1 % yield) or diselenide (2 % yield) (equation 37). Oxidation

(37)

$$
\begin{array}{ccc}
H_3C \quad CH_2Br & & H_3C \quad H_2C{-}Se \\
\diagdown \diagup & \xrightarrow{\text{K}_2\text{Se or K}_2\text{Se}_2} & \diagdown \diagup \quad | \\
C & & C \\
\diagup \diagdown & & \diagup \diagdown \quad | \\
H_5C_6 \quad CH_2Br & & H_5C_6 \quad H_2C{-}Se
\end{array}
$$

45

of **45** with nitric acid gave the corresponding diseleninic acid dihydronitrate, from which the free acid could be obtained.

1,2-Diselenolane-4-carboxylic acid (**46**) was prepared by Bergson[67] by two methods. The first was the reaction of β,β'-dibromoisobutyric acid sodium salt and sodium diselenide (equation 38); the second involved the

(38)

$$
\begin{array}{ccc}
BrCH_2 & & Se{-}CH_2 \\
\diagdown & \xrightarrow[\text{H}_2\text{O}]{\text{Na}_2\text{Se}_2} & | \qquad \diagdown \\
CH{-}COOH & & CH{-}COOH \\
\diagup & & | \qquad \diagup \\
BrCH_2 & & Se{-}CH_2
\end{array}
$$

46

(39)

$$
\begin{array}{c}
C_6H_5{-}CH_2{-}Se{-}CH_2 \\
\diagdown \\
\quad CH{-}COOH \\
\diagup \\
C_6H_5{-}CH_2{-}Se{-}CH_2
\end{array}
$$

$\xrightarrow{\text{Na in liq. NH}_3}$

$$
\begin{array}{c}
HSe{-}CH_2 \\
\diagdown \\
\quad CH{-}COOH \\
\diagup \\
HSe{-}CH_2
\end{array}
$$

$\uparrow \; O_2$

debenzylation of β,β'-bis(benzylseleno)isobutyric acid, followed by oxidation of the resulting diselenolisobutyric acid (equation 39). Bergson[35] followed the oxidation of **46** with ammonium persulfate in aqueous ethanol spectrophotometrically. He found that the characteristic peak of the 1,2-diselenolane ring at about 440 nm reached a minimum after about 16 minutes and then increased to approximately two-thirds of the original value; this corresponds to the oxidation of one-third of the **46** to the diseleninic acid (equation 40). The initial oxidation product, presumably the monoxide, could not be isolated.

Krackov *et al.*[36] measured the dipole moment and the acid dissociation constant of **46**. In analogy with what was found for the isologous 1,2-dithiolane and 1-thia-2-selenolane derivatives, they concluded that there is no significant interaction between Se–Se and carboxylic groups.

(40)

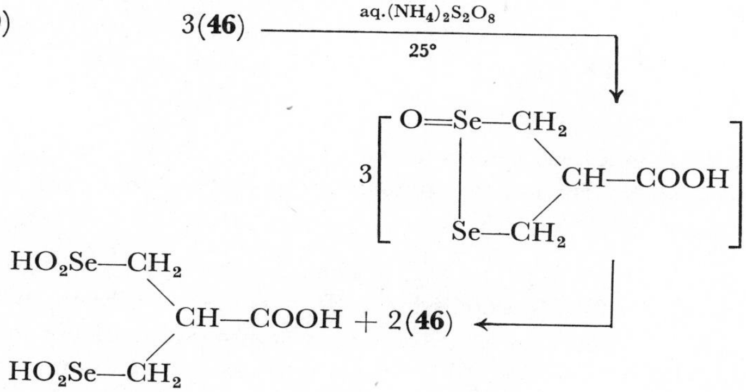

$$3(\mathbf{46}) \xrightarrow[\text{25°}]{\text{aq.(NH}_4)_2\text{S}_2\text{O}_8}$$

$$3 \left[\begin{array}{c} O=Se-CH_2 \\ | \qquad\qquad\diagdown \\ | \qquad\qquad\qquad CH-COOH \\ | \qquad\qquad\diagup \\ Se-CH_2 \end{array} \right]$$

$$\begin{array}{c} HO_2Se-CH_2 \\ \diagdown \\ \qquad CH-COOH + 2(\mathbf{46}) \longleftarrow \\ \diagup \\ HO_2Se-CH_2 \end{array}$$

4,4-Bis(hydroxymethyl)-1,2-diselenolane (**47**) was prepared by Bergson by treatment of 2,2-bis(iodomethyl)propane-1,3-diol[67,71] or the analogous dichloro compound[67] with sodium diselenide (equation 41). The dichloro

(41)

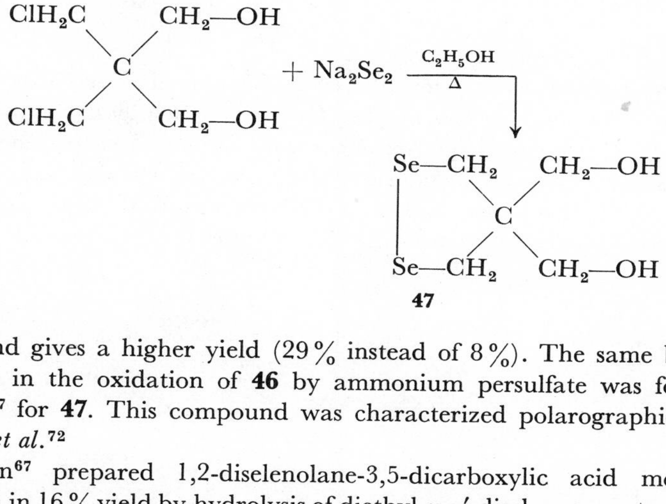

$$\begin{array}{cc} ClH_2C & CH_2-OH \\ \diagdown \diagup \\ C \\ \diagup \diagdown \\ ClH_2C & CH_2-OH \end{array} + Na_2Se_2 \xrightarrow[\Delta]{C_2H_5OH}$$

$$\begin{array}{cc} Se-CH_2 & CH_2-OH \\ | \qquad\diagdown \diagup \\ | \qquad\qquad C \\ | \qquad\diagup \diagdown \\ Se-CH_2 & CH_2-OH \end{array}$$

47

compound gives a higher yield (29% instead of 8%). The same behavior observed in the oxidation of **46** by ammonium persulfate was found by Bergson[67] for **47**. This compound was characterized polarographically by Nygard *et al.*[72]

Bergson[67] prepared 1,2-diselenolane-3,5-dicarboxylic acid monoethyl ester (**48**) in 16% yield by hydrolysis of diethyl α,α′-diselenocyanatoglutarate (equation 42). The structure of **48** was confirmed by i.r. (C=O stretching frequencies at 5.76 and 5.86 μ, attributed to one ester group and one carboxylic group, respectively) and u.v. spectra (presence of the characteristic 1,2-diselenolane peak at about 440 nm).

$$(42) \quad \begin{array}{c} \text{COOC}_2\text{H}_5 \\ | \\ \text{NCSe—CH} \\ \diagdown \\ \text{CH}_2 \\ \diagup \\ \text{NCSe—CH} \\ | \\ \text{COOC}_2\text{H}_5 \end{array} \xrightarrow{\text{C}_2\text{H}_5\text{OH, 2 } M \text{ H}_2\text{SO}_4} \begin{array}{c} \text{COOC}_2\text{H}_5 \\ | \\ \text{Se—CH} \\ \diagdown \\ \text{CH}_2 \\ \diagup \\ \text{Se—CH} \\ | \\ \text{COOH} \\ \textbf{48} \end{array}$$

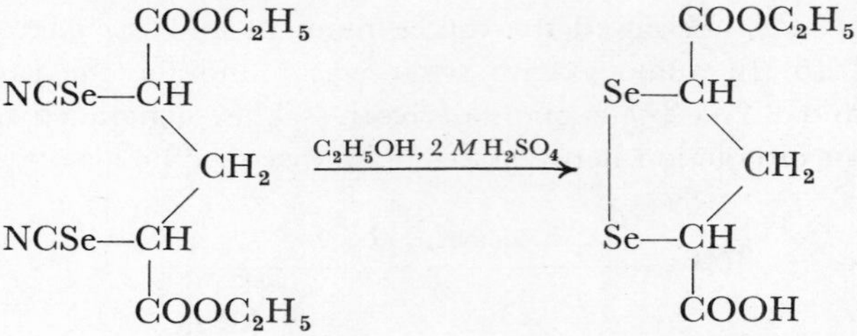

Bergson[68] studied the electronic and steric effects on the u.v. and i.r. spectra of several cyclic selenium compounds and concluded that diselenides behave similarly to disulfides.

1,3-Diselenolane. The compound 1,3-diselenolane is unknown, but its 2-selenoxo derivative (**49**) was prepared in 1967 by Henriksen[73] by treatment of 1,2-dibromoethane and carbon diselenide in a suspension of potassium carbonate in DMSO containing 10% of water (equation 43). Compound

$$(43) \quad \begin{array}{c} \text{CH}_2\text{Br} \\ | \\ \text{CH}_2\text{Br} \end{array} + \text{CSe}_2 \xrightarrow{\text{K}_2\text{CO}_3 \text{ in DMSO/H}_2\text{O}} \begin{array}{c} \text{H}_2\text{C—Se} \\ | \qquad\quad \diagdown \\ \qquad\qquad \text{C}=\text{Se} \\ | \qquad\quad \diagup \\ \text{H}_2\text{C—Se} \\ \textbf{49} \end{array}$$

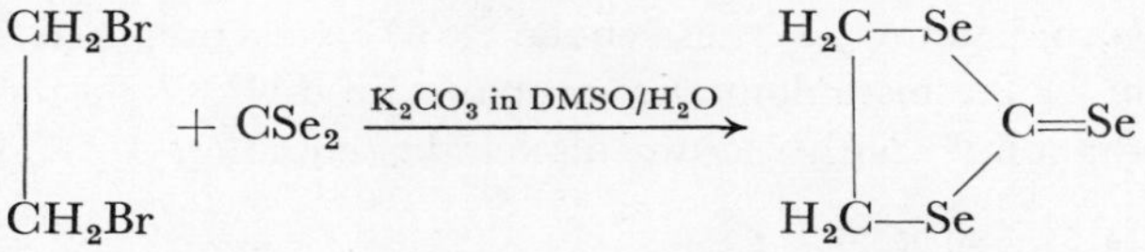

49 may be regarded as an ester of triselenocarbonic acid (ethylenetriselenocarbonate). The presence of a C=Se group was shown by its ability to form 1,3-diphenyl-2-selenourea on treatment with aniline. The presence of the 1,3-diselenolane ring was demonstrated by treating **49** with iodomethane, followed by malononitrile and triethylamine dissolved in pyridine; 1,3-diselenolane-2-ylidene-malononitrile (**50**) was formed (scheme 44).

$$(44) \quad \textbf{49} \xrightarrow{\text{CH}_3\text{I}} \begin{array}{c} \text{H}_2\text{C—Se} \\ | \qquad\quad \diagdown \\ \qquad\qquad \text{C}=\text{Se}^+\text{—CH}_3\text{I}^- \\ | \qquad\quad \diagup \\ \text{H}_2\text{C—Se} \end{array} \xrightarrow{\text{CH}_2(\text{CN})_2} \begin{array}{c} \text{H}_2\text{C—Se} \qquad \text{CN} \\ | \qquad\quad \diagdown \qquad \diagup \\ \qquad\qquad \text{C}=\text{C} \\ | \qquad\quad \diagup \qquad \diagdown \\ \text{H}_2\text{C—Se} \qquad \text{CN} \\ \textbf{50} \end{array}$$

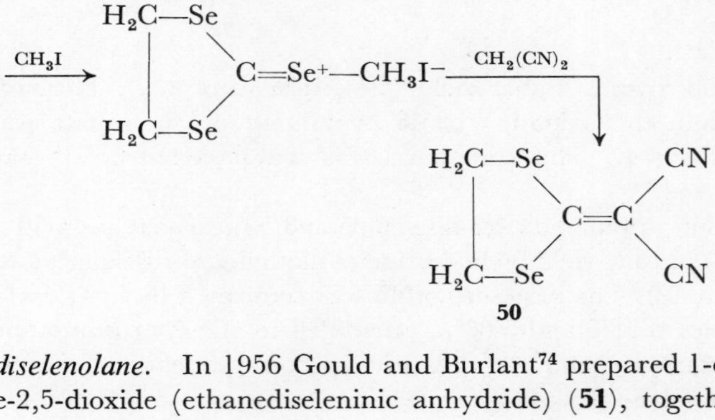

1-Oxa-2,5-diselenolane. In 1956 Gould and Burlant[74] prepared 1-oxa-2,5-diselenolane-2,5-dioxide (ethanediseleninic anhydride) (**51**), together with

1,4-diselenane-1,4-dioxide, by the oxidation of 1,4-diselenane with per-acetic acid. Paetzold and Lienig[75] prepared **51** either by the oxidation of 1,2-ethanediselenocyanate or 1,2,5,6-tetraselenocane (**52**) with nitric acid (scheme 45). According to Paetzold and Lienig,[75] **51** is transformed in nitric

(45)

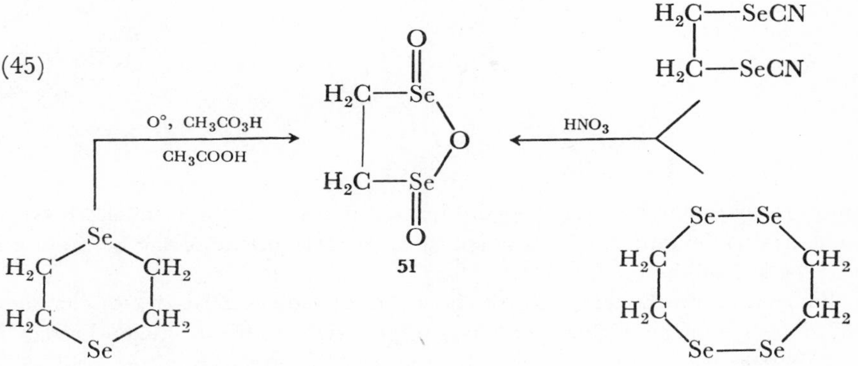

$$O°, \ CH_3CO_3H \quad / \quad CH_3COOH$$

$$HNO_3$$

H₂C—SeCN / H₂C—SeCN

51

52

acid solution into the diseleninic acid hydronitrate (**53**), whereas in alkaline solution the diseleninate (**54**) is obtained (equation 46). Gould and Post[76]

(46)

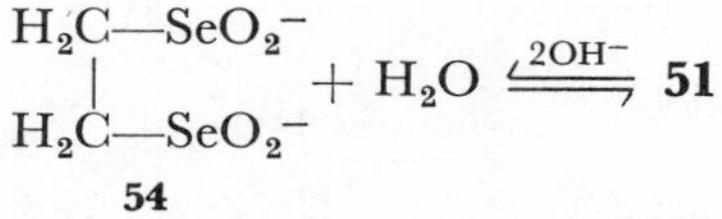

$$\begin{array}{c} H_2C\!-\!SeO_2^- \\ | \\ H_2C\!-\!SeO_2^- \end{array} + H_2O \ \underset{2OH^-}{\overset{}{\rightleftharpoons}} \ 51$$

54

$$51 \ \xrightarrow{\text{aq. HNO}_3} \ \left[\begin{array}{c} OH \\ H_2C\!-\!Se^+ \\ OH \\ OH \\ H_2C\!-\!Se^+ \\ OH \end{array} \right]^{2+} 2(NO_3)^-$$

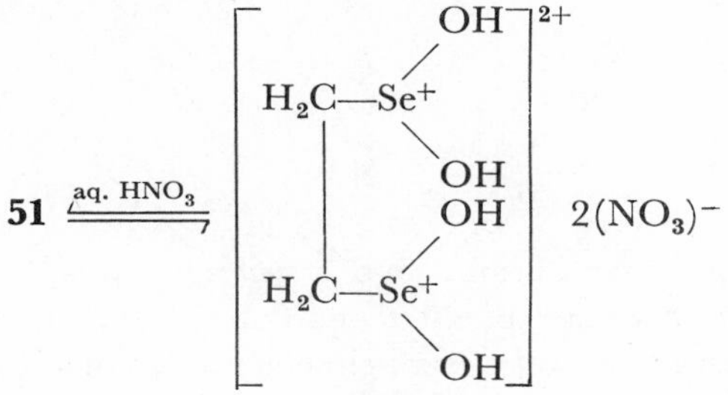

53

have determined the crystal structure of **51**.

c. Six-Membered Rings

1,2-Diselenane. By air oxidation of 1,4-butanediselenocyanate in an alkaline medium Morgan and Burstall[10] obtained a yellow powder (m.p. 41–42°), regarded as 1,2-diselenane (**55**). According to Bergson[68] and to Brown *et al.*[77] the product obtained by Morgan and Burstall was not

1,2-diselenane, but rather a low-molecular-weight polymeric form of this monomer (*cf.* Chapter XIV). Hence **55** is not known in the free state; however, it was obtained by Bergson[68] and by Brown *et al.*[77] in chloroform

$$
\begin{array}{c}
\text{Se} \\
\diagup \quad \diagdown \\
\text{Se} \qquad \text{CH}_2 \\
| \qquad\quad | \\
\text{H}_2\text{C} \qquad \text{CH}_2 \\
\diagdown \quad \diagup \\
\text{CH}_2 \\
\textbf{55}
\end{array}
$$

solution by depolymerization of Morgan and Burstall's product. It shows a peak at 364–365 nm in the u.v. spectrum, characteristic of the Se-Se bond in 1,2-diselenane.

The only known derivative of the 1,2-diselenane class is 1,2-diselenane-3,6-dicarboxylic acid (**56**), which was prepared for the first time by Fredga[78] in the *trans* (*rac*) form by gently heating *rac*-diselenocyanoadipic acid (**57**) with dilute sulfuric acid (equation 47).[79] Pure (−) and (+) forms of **56**

(47)

$$
\begin{array}{ccc}
\begin{array}{c}
\text{COOH} \\
| \\
\text{CH} \\
\diagup \quad \diagdown \\
\text{H}_2\text{C} \qquad \text{SeCN} \\
| \qquad\qquad | \\
\text{H}_2\text{C} \qquad \text{SeCN} \\
\diagdown \quad \diagup \\
\text{CH} \\
| \\
\text{COOH} \\
\textbf{57}
\end{array}
&
\xrightarrow{\text{dil. H}_2\text{SO}_4}
&
\begin{array}{c}
\text{COOH} \\
| \\
\text{CH} \\
\diagup \quad \diagdown \\
\text{H}_2\text{C} \qquad \text{Se} \\
| \qquad\qquad | \\
\text{H}_2\text{C} \qquad \text{Se} \\
\diagdown \quad \diagup \\
\text{CH} \\
| \\
\text{COOH} \\
\textbf{56}
\end{array}
\end{array}
$$

were obtained *via* their quinine and strychnine salts, respectively. In a subsequent paper Fredga[80] pointed out that 1,2-dithiane-3,6-dicarboxylic acid and **56** resemble each other very closely. Fredga described the formation of a quasi racemate from (−)-**56** with its (+)-sulfur analog, and a quasi racemate from (+)-**56** with its (−)-sulfur analog. The formation of quasi racemates in **56** and its sulfur analog antipode systems was supported by i.r. absorption investigations by Rosenberg and Schotte.[32] Equivalent amounts of (+)-**56** and its (+)-sulfur analog give mixed crystals. Fredga[81] also prepared *cis*- (*meso*)-**56** by boiling the potassium salt of *meso*-**57** with dilute sulfuric acid. This compound is obtained in an unstable form that changes to the stable one on heating or on standing.

It is interesting to note that an attempted preparation of **56** by the direct introduction of the diselenide moiety failed. In fact, when α,α'-dibromo-adipic acid was treated with potassium diselenide (equation 12), the reaction led to the five-membered cyclic monoselenide.

Fredga,[82] on treating **56** with mercuric chloride, obtained **58** as a heavy white precipitate, which on acidification reverted to **56** (equation 48).

(48) 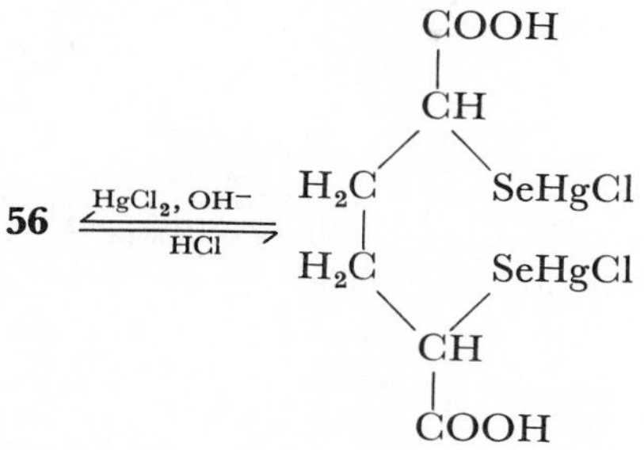 $56 \; \underset{\text{HCl}}{\overset{\text{HgCl}_2, \text{OH}^-}{\rightleftharpoons}}$

$$
\begin{array}{c}
\text{COOH} \\
| \\
\text{CH} \\
\diagup \quad \diagdown \\
\text{H}_2\text{C} \qquad \text{SeHgCl} \\
| \\
\text{H}_2\text{C} \qquad \text{SeHgCl} \\
\diagdown \quad \diagup \\
\text{CH} \\
| \\
\text{COOH}
\end{array}
$$

58

Djerassi et al.[83,84] have studied the optical rotatory dispersion curves of $(+)$-1,2-dithiane-3,6-dicarboxylic acid and $(+)$-**56**, in which the hetero-atoms are adjacent to the asymmetric center, and noted strong Cotton effects related to the u.v. absorptions at 340 and 270 nm. They also point out that the vicinal action from the carboxyl group may be important with respect to the anisotropy of the disulfide and diselenide.

Foss and Schotte,[85] in an X-ray crystallographic study of cyclic 1,2-disulfides and diselenides, reported the crystal data of $(+)$-**56** and of rac-**56**. The crystal and molecular structures of the latter compound were reported later by Foss et al.[86] Husebye[87] performed an X-ray crystallographic study of the quasi racemate of $(+)$-**56** and its $(-)$-sulfur analog.

1,3-Diselenane. Recently Geens et al.[88] prepared 1,3-diselenane (**59**) from propane-1,3-diselenol dipotassium salt, either by heating it in strongly acidic aqueous formaldehyde or by stirring it with dichloromethane in DMF (scheme 49). The former method produces the higher yield; the latter gives **59** together with some polymeric product.

(49)

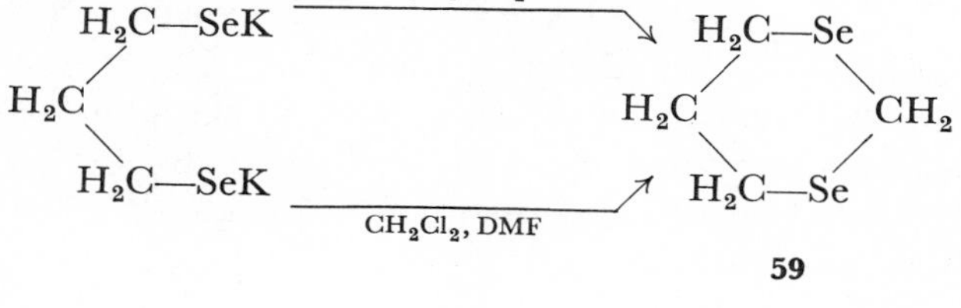

$$
\begin{array}{c}
\text{H}_2\text{C}-\text{SeK} \\
\diagup \\
\text{H}_2\text{C} \\
\diagdown \\
\text{H}_2\text{C}-\text{SeK}
\end{array}
\quad
\begin{array}{c}
\xrightarrow{\text{H}^+, \text{ aq. CH}_2\text{O}} \\
\\
\xrightarrow[\text{CH}_2\text{Cl}_2, \text{ DMF}]{}
\end{array}
\quad
\begin{array}{c}
\text{H}_2\text{C}-\text{Se} \\
\diagup \qquad \diagdown \\
\text{H}_2\text{C} \qquad \text{CH}_2 \\
\diagdown \qquad \diagup \\
\text{H}_2\text{C}-\text{Se}
\end{array}
$$

59

Henriksen[73] prepared 1,3-diselenane-2-selone (**60**) by a route similar to that followed for the preparation of its five-membered homolog (**49**) (equation 50). Compound **60**, also named trimethylenetriselenocarbonate, is

$$(50) \quad H_2C \Big\langle {}^{CH_2Br}_{CH_2Br} + CSe_2 \xrightarrow{K_2CO_3 \text{ in DMSO/H}_2O} H_2C \Big\langle {}^{H_2C-Se}_{H_2C-Se} \Big\rangle C{=}Se$$

60

somewhat less stable than its lower homolog.

1,4-Diselenane. The compound 1,4-diselenane (**61**) was prepared for the first time in 1951 by Gould and McCullough.[89] Using anhydrous acetone, they obtained **61** in 0.3% yield from the reaction of β,β'-dichlorodiethyl selenide with powdered lithium selenide (equation 51). Compound **61** was

$$(51) \quad Se \Big\langle {}^{H_2C-CH_2Cl}_{H_2C-CH_2Cl} + Li_2Se \xrightarrow{N_2, \, CH_3COCH_3} Se \Big\langle {}^{H_2C-CH_2}_{H_2C-CH_2} \Big\rangle Se + 2LiCl$$

61

not obtained when 1,2-dibromoethane or β,β'-dichlorodiethyl selenide was treated with solutions of sodium or lithium selenide in aqueous alcohol.

McCullough and Tideswell[90] obtained a 2% yield of **61** from the reaction of aluminum selenide with 1,2-dibromoethane (equation 52). By controlling

$$(52) \qquad BrCH_2CH_2Br + Al_2Se_3 \rightarrow \textbf{61}$$

the reaction conditions, that is, gradually raising the temperature, Gould and Burlant[74] increased the yield to 10%.

Marsh and McCullough[91] determined the crystal structure of **61** by X-ray diffraction. 1,4-Diselenane readily reacts with halogens to form 1,1,4,4-tetrahalo derivatives.[90] The crystal structure of 1,1,4,4-tetraiodo-1,4-diselenane was studied by McCullough et al.[92] and refined by Chao and McCullough.[93] This compound has been investigated by Hendra and Sadasivan[94] in the i.r. and far-i.r. regions and compared with 1,4-dithiane and 1,4-dioxane. The data suggest that the strength of the halogen-hetero-atom bonds follows the order O < S < Se, in agreement with the previous conclusions of McCullough and Zimmermann.[48]

Amendola et al.[95] determined by X-ray analysis the crystal structure of 1,1,4,4-tetrachloro-1,4-diselenane. It is interesting to note that iodine forms a molecular complex with 1,4-diselenane rather than a true iodide, whereas chlorine forms a true tetrachloride.

Krespan,[23,24] while preparing perfluoroselenolane (**13**) by heating a mixture of selenium, tetrafluoroethylene, and iodine, obtained a minor product, 2,2,3,3,5,5,6,6-octafluoro-1,4-diselenane (**62**). This perfluoro-1,4-diselenane, in its n.m.r. spectrum, had only one peak in a region compatible

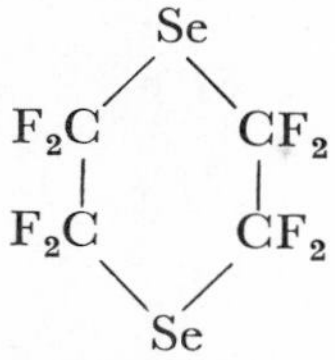

$$Se$$
$$F_2C \diagup \quad \diagdown CF_2$$
$$F_2C \quad \quad CF_2$$
$$\diagdown Se \diagup$$

62

with CF_2–Se. Both **13** and **62** are thermally less stable than the corresponding sulfides.

Careful oxidation by Gould and Burlant[74] of 1,4-diselenane with peracetic acid at 0° yielded a compound that was rapidly reduced with sulfurous acid or hydroxylamine to 1,4-diselenane. The possibility that the product was the 1,1-dioxide was ruled out by its reaction with aqueous hydrogen halides to form 1,1,4,4-tetrahalo derivatives; thus, the compound was the 1,4-dioxide (**63**) previously reported by Gould and Post.[76] During the oxidation, some degradation of the ring occurred and gave *trans*-**51** as a minor product (see 1-oxa-2,5-diselenolane) (equation 53). Compound **63** decom-

(53) 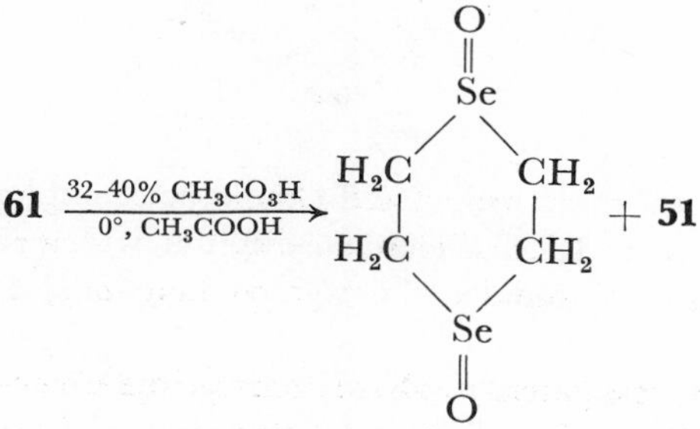   **61** $\xrightarrow[\text{0°, CH}_3\text{COOH}]{32-40\% \text{ CH}_3\text{CO}_3\text{H}}$

$$O$$
$$\parallel$$
$$Se$$
$$H_2C \diagup \quad \diagdown CH_2$$
$$H_2C \quad \quad CH_2 \quad + \textbf{51}$$
$$\diagdown Se \diagup$$
$$\parallel$$
$$O$$

63

poses rapidly if heated above 80°, darkens slowly if allowed to stand in moist air for more than 2 days, and is extremely sensitive to bases.

Oxidation of 1,4-diselenane with concentrated nitric acid at 0° gives the bishydroxynitrate (**64**), which on treatment with HCl or HBr is converted to the tetrahalo derivative.[74] The formation of the methiodide (**65**) has also been described (scheme 54).

1,4-Diselenane has marked donor characteristics and forms several complexes with inorganic salts. According to Hendra and Sadasivan,[96] who

(54)

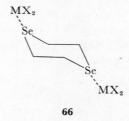

65   64

studied the i.r. spectra of these complexes, they fall into two groups: those having the general formula $(MX_2)_2C_4H_8Se_2$ and those having the formula $MX_n(C_4H_8Se_2)$, where M = metal and X = halogen. Hendra and Sadasivan suggest a bridging structure (**66**) for the complexes of the first group.

66

Several other structures are proposed for the other complexes. It is suggested that the complexes of 1,4-diselenane exhibit a wider range of structures than has as yet been found for the related ligands, 1,4-dioxane and 1,4-dithiane.

Bjorvatten[97] investigated by X-ray analysis the crystal structure of the 2:1 addition compound, iodoform-1,4-diselenane, obtained from a solution of reactants in carbon disulfide. Hayward and Hendra[98] studied the i.r. and Raman spectra in the solid state of this iodoform-diselenane addition compound. Holmesland and Rømming[99] studied the crystal structure of the 1:1 addition compound of diiodoacetylene with 1,4-diselenane. Dahl and Hassel[100] investigated by X-ray analysis the crystal structure of the tetra-iodoethylene-1,4-diselenane 1:1 addition compound. The structure is similar to that found for the diiodoacetylene-1,4-diselenane complex, with endless chains of alternating acceptor and donor molecules and with a nearly linear arrangement of donor atom, iodine, and carbon.

According to Hassel,[101] who reported the heteroatom-iodine distances of several complexes of oxa- and thiaselenacycloalkanes with iodine-containing molecules, it appears that the nature of the atom responsible for the donor properties is of much greater importance than the chemical constitution of the donor molecule. The structure of the acceptor molecule, if derived from a saturated aliphatic, an olefinic, or an acetylenic hydrocarbon, also appears to be of secondary importance for the donor atom-iodine distance.

1-Oxa-3,5-diselenane.   Russo *et al.*[102] obtained 1-oxa-3,5-diselenane (**67**), together with a linear copolymer $(-CH_2O-CH_2Se-)_n$, by the reaction of α,α'-dichlorodimethyl ether with sodium selenide in anhydrous methanol or in an aqueous methanol medium (equation 55). Compound **67** polymerizes

$$(55)$$

in bulk (100°) with cationic catalysts to give a linear copolymer $[-CH_2O-(CH_2Se)_2-]_n$. Mammi and Valle[103] found that **67** has the chair conformation.

Mortillaro and Russo[104] prepared 2,4,6-triethyl-1-oxa-3,5-diselenane (**68**) by reaction of aqueous propionaldehyde with hydrogen selenide at 0–5° in the presence of hydrochloric acid (equation 56).

$$(56)$$

d.   Seven-Membered Rings.   Morgan and Burstall,[40] by treating 1,5-pentanediselenocyanate with alcoholic KOH, obtained a yellow gum which, on the basis of elemental analysis, was regarded as 1,2-diselenepane (**69**). It is reasonable to assume, in analogy with what has been found for five- and six-membered 1,2-diselenacycloalkanes, that **69** does not exist in the free state and that Morgan and Burstall's product is a low-molecular-weight polymer. This product is discussed further in Chapter XIV.

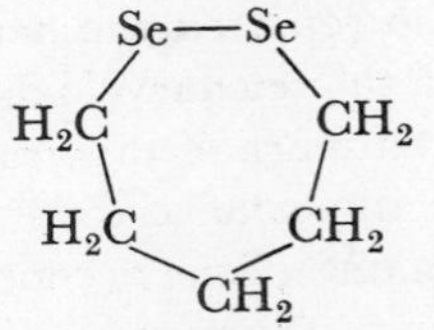

$$\text{Se—Se}$$
$$\text{H}_2\text{C} \quad\quad \text{CH}_2$$
$$\text{H}_2\text{C} \quad\quad \text{CH}_2$$
$$\text{CH}_2$$

69

e. Eight-Membered Rings. 1,2-Diselenocane is unknown, whereas its linear polymeric form has been prepared by Morgan and Burstall.[41] On heating poly(selenotrimethylene), low-molecular-weight poly[(diseleno)-trimethylene] was obtained (cf. Chapter XIV). This forms, according to Morgan and Burstall,[4] via the intermediate 1,5-diselenocane (**70**).

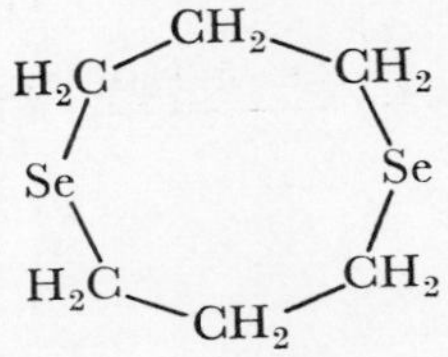

$$\text{CH}_2 \quad \text{CH}_2$$
$$\text{H}_2\text{C} \quad\quad\quad \text{CH}_2$$
$$\text{Se} \quad\quad\quad\quad \text{Se}$$
$$\text{H}_2\text{C} \quad\quad\quad \text{CH}_2$$
$$\text{CH}_2$$

70

f. Fourteen-Membered Rings. During the preparation of selenepane by the reaction of 1,6-dibromohexane with sodium selenide, Morgan and Burstall[41] obtained a yellow gummy mixture of at least two polymeric materials. The first was isolated and identified by the investigators as a dimer of selenepane, 1,8-diselenacyclotetradecane, to which structure **71** was assigned. Compound **71** decomposed above 220° into 2-methylselenane and combined

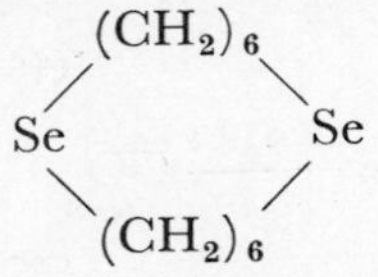

$$\text{(CH}_2)_6$$
$$\text{Se} \quad\quad\quad \text{Se}$$
$$\text{(CH}_2)_6$$

71

with bromine in carbon tetrachloride to give the dibromo derivative.

4. Selenacycloalkanes with Three Selenium Atoms in the Ring

a. Six-Membered Rings

Triselenanes. The first known selenacycloalkane of this class, 1,3,5-tri-selenane (**72**), was prepared by Vanino and Schinner[105] in 1915 by the reaction of hydrogen selenide with formaldehyde in the presence of hydrochloric acid. These authors, on the basis of the elemental analysis, mistakenly called the product "selenoformaldehyde" (m.p. 215°). This compound was also described by Pearson et al.[106] In 1950 Bridger and Pittman[107] prepared

$$
\begin{array}{ccc}
 & \text{Se—CH}_2 & \\
 \diagup & & \diagdown \\
 \text{H}_2\text{C} & & \text{Se} \\
 \diagdown & & \diagup \\
 & \text{Se—CH}_2 &
\end{array}
$$

72

Vanino's "selenoformaldehyde" (m.p. 210°) and found by a molecular weight determination that this product is a trimer, $(CH_2Se)_3$; on the basis of its chemical behavior, they advanced the hypothesis of a cyclic structure. Bridger and Pittman also prepared this trimer in 70% yield by hydrolyzing bis(chloromethyl) selenide with aqueous bicarbonate (equation 57).

(57) $(ClCH_2)_2Se \xrightarrow{\text{aq. NaHCO}_3} 2HCl + CH_2O + HOCH_2SeH$
$$\downarrow$$
72

In 1965 Mortillaro et al.[108] investigated Vanino's experiment by X-ray crystallography and found the "selenoformaldehyde" to be isomorphic with 1,3,5-trithiane, further demonstrating its cyclic structure. Mortillaro et al.[109] ascertained that the reaction of hydrogen selenide with aqueous formaldehyde in the presence of concentrated hydrochloric acid actually gives 1,3,5-triselenane (yield 28%, m.p. 226–228° in sealed capillary), together with the linear polymer of selenoformaldehyde poly(selenomethylene) (**73**), (yield 29%, m.p. 174–178°) (equation 58). The mechanism of

(58) $CH_2O + H_2Se \xrightarrow[\text{H}_2\text{O}]{\text{conc. HCl}} \mathbf{72} + (\text{—CH}_2Se\text{—})_n$
73

this reaction, suggested by Russo and Credali,[54] is reported in Chapter XIV.

Compound **72** was also obtained by Russo et al.[110] (yield 30%), together with **73** and 1,3,5,7-tetraselenocane (**74**), by reaction of dichloromethane with sodium selenide (equation 59). By using essentially the same reaction but with ethanol as the reaction medium, Schmidt et al.[111] obtained 1,3,5-triselenane in 45% yield.

In 1934 Rice and Glasebrook[112] claimed that they had prepared poly-(selenomethylene) when investigating methylene reactions (from diazomethane decomposition) on selenium mirror. The same conclusion was drawn by Pearson et al.[106] while investigating methylene reactions (from ketene photolysis) on selenium mirror. Williams and Dunbar[113] found in

1968 that methylene radicals, obtained by decomposition of diazomethane, when passed over selenium mirror actually give 1,3,5-triselenane.

(59) CH_2Cl_2 + Na_2Se $\xrightarrow[N_2]{CH_3OH-H_2O}$

$$72 + \begin{array}{c} \text{Se} \\ H_2C \diagup \quad \diagdown CH_2 \\ \mid \qquad\qquad \mid \\ \text{Se} \qquad\qquad \text{Se} \\ \mid \qquad\qquad \mid \\ H_2C \diagdown \quad \diagup CH_2 \\ \text{Se} \end{array} + 73$$

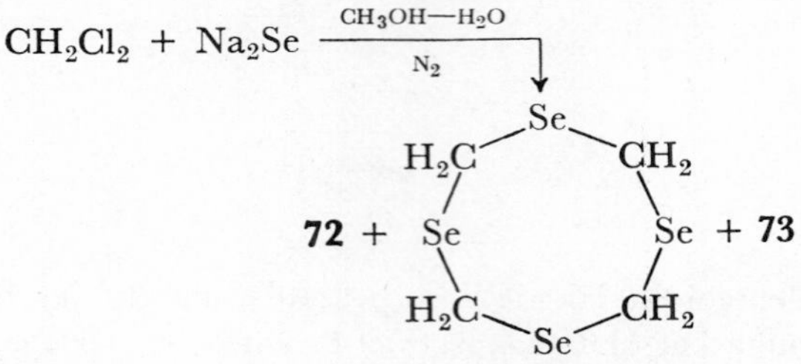

$$74$$

Mammi et al.[114] determined the crystal structure of **72** by X-ray analysis.

Compound **72** reacts with dry chlorine[107] in cold anhydrous carbon tetrachloride, giving bis(chloromethyl) selenide, chloromethylselenyl chloride, and bis(chloromethyl) diselenide (equation 60). The products from the

(60)   **72** $\xrightarrow{Cl_2, CCl_4}$ $(ClCH_2)_2Se$ + $ClCH_2SeCl$ + $ClCH_2SeSeCH_2Cl$

reaction of 1,3,5-triselenane with selenium monochloride depend on the relative proportions of the reactants. Chloromethylselenenyl chloride is obtained when 0.05 mole of **72** and 0.14 mole of Se_2Cl_2 are used, while a mixture of bis(chloromethyl) selenide and bis(chloromethyl) diselenide is formed when 0.2 mole of **72** and 0.3 mole of Se_2Cl_2 react (scheme 61).

(61)   **72** $\xrightarrow{Se_2Cl_2, CCl_4}$
$\longrightarrow ClCH_2SeCl$

$\longrightarrow ClCH_2SeCH_2Cl$ + $ClCH_2SeSeCH_2Cl$

Mortillaro et al.[109] found that 1,3,5-triselenane polymerizes in bulk to hexagonal poly(selenomethylene). Carazzolo and Mammi[115] found that by solid-state radiation polymerization of **72** orthorhombic poly(selenomethylene) was obtained (cf. Chapter XIV). The crystalline structure of 1,3,5-triselenane and of its oxygen and sulfur analogs were compared by Carazzolo and Mammi.[116] All three molecules are in the chair conformation.

Credali et al.[53] reported the preparation and preliminary X-ray analysis of 2,4,6-trimethyl-1,3,5-triselenane (**75**), also known as triselenoparaldehyde. This selenacycloalkane was obtained in low yield by reaction of hydrogen selenide with an ethanolic acetaldehyde solution saturated with HCl (equation 62). This reaction had been previously reported by Vanino and Schinner,[105] who obtained some crystalline fractions having somewhat

(62) 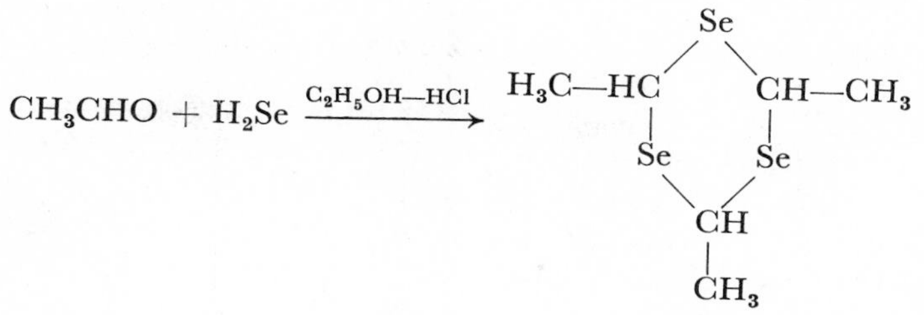 $CH_3CHO + H_2Se$ $\xrightarrow{C_2H_5OH—HCl}$

$$
\begin{array}{c}
\text{Se} \\
\text{H}_3\text{C—HC} \diagup \quad \diagdown \text{CH—CH}_3 \\
| \qquad \qquad | \\
\text{Se} \qquad \text{Se} \\
\diagdown \quad \diagup \\
\text{CH} \\
| \\
\text{CH}_3
\end{array}
$$

75

different melting points and described as different modifications of "seleno-acetaldehyde." X-ray analysis[53] shows that the molecules are in the chair conformation.

5.   Selenacycloalkanes with Four Selenium Atoms in the Ring

a.   Eight-Membered Rings

1,2,5,6-Tetraselenocane. Hagelberg[65] in 1890 was the first to prepare 1,2,5,6-tetraselenocane (**52**) by the reaction of 1,2-ethanediselenocyanate with alcoholic KOH (equation 63). In a similar way **52** was obtained by

(63) 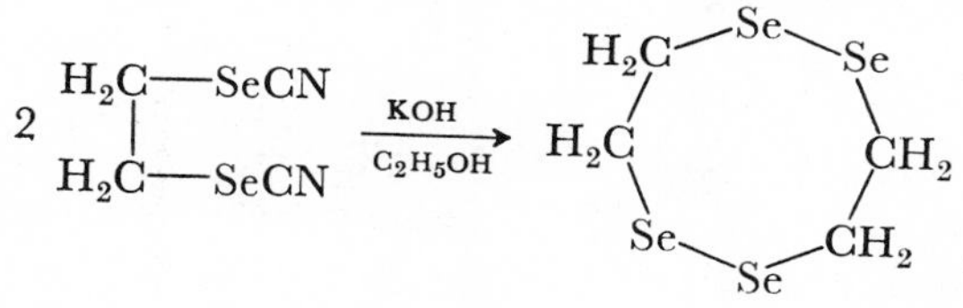 $2 \begin{array}{c} H_2C—SeCN \\ | \\ H_2C—SeCN \end{array}$ $\xrightarrow[C_2H_5OH]{KOH}$

$$
\begin{array}{c}
\text{H}_2\text{C} \diagup^{\text{Se}} \diagdown \text{Se} \\
\text{H}_2\text{C} \qquad \qquad \text{CH}_2 \\
\diagdown_{\text{Se}} \diagdown_{\text{Se}} \diagup \text{CH}_2
\end{array}
$$

52

Paetzold and Lienig,[75] using methanol as the reaction medium. By the reaction of **52** with nitric acid, 1-oxa-2,5-diselenolane-2,5-dioxide (**51**) was obtained.

1,3,5,7-Tetraselenocane. This selenacycloalkane (**74**) was obtained by Russo *et al.*[110] (yield 4%), together with hexagonal poly(selenomethylene) (**73**) and 1,3,5-triselenane (**72**), by the reaction of dichloromethane with sodium selenide. Compound **74** polymerizes in bulk to orthorhombic poly-(selenomethylene) (*cf.* Chapter XIV).

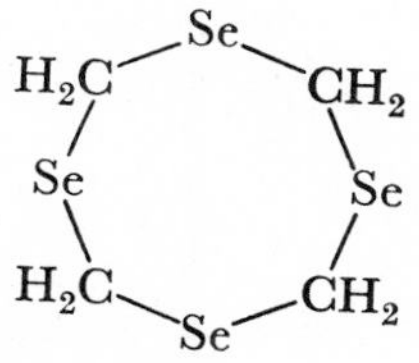

$$
\begin{array}{c}
\text{H}_2\text{C} \diagup^{\text{Se}} \diagdown \text{CH}_2 \\
\text{Se} \qquad \qquad \text{Se} \\
\text{H}_2\text{C} \diagdown_{\text{Se}} \diagup \text{CH}_2
\end{array}
$$

74

6. A Selenacycloalkane with Five Selenium Atoms in the Ring

Schmidt *et al.*[111] obtained a precipitate of 1,3,5-triselenane by the reaction of dichloromethane with sodium selenide in ethanol. The filtrate was evaporated, and the yellow oil obtained was allowed to stand at room temperature, under vacuum, for 3 days. A crystalline solid then separated, which on the basis of elemental analysis and molecular weight determination was reported to be 1,3,5,7,9-pentaselenecane (**76**).

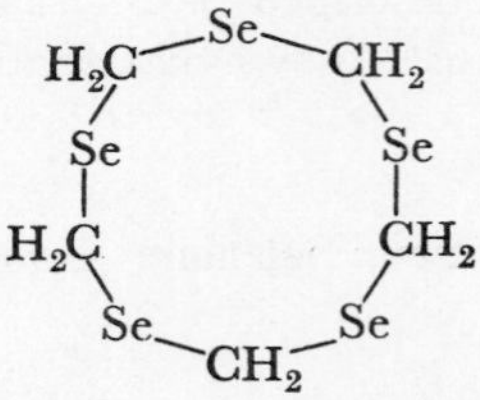

$$H_2C \overset{Se}{\diagup} CH_2$$

76

7. Spiro Selenacycloalkanes

a. With a Selenetane Ring

2,6-Diselenaspiro[3.3]heptane. In 1937 Backer and Winter[5] prepared 2,6-diselenaspiro[3.3]heptane (**77**) in about 94% yield by reaction of tetra(bromomethyl)methane with potassium selenide (equation 64). Compound **77** is isomorphous with the corresponding sulfur derivative and is very

(64)

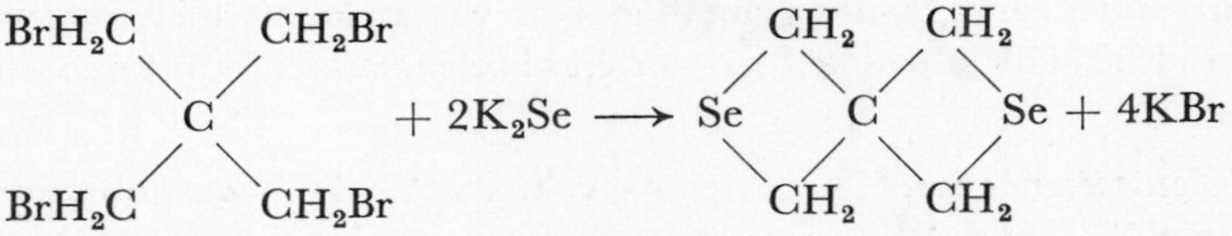

77

sensitive to oxidation. It forms addition compounds with mercuric chloride and with iodine and ring-opens on treatment with iodomethane to give the selenonium compound, **78** (equation 65).[5]

(65)

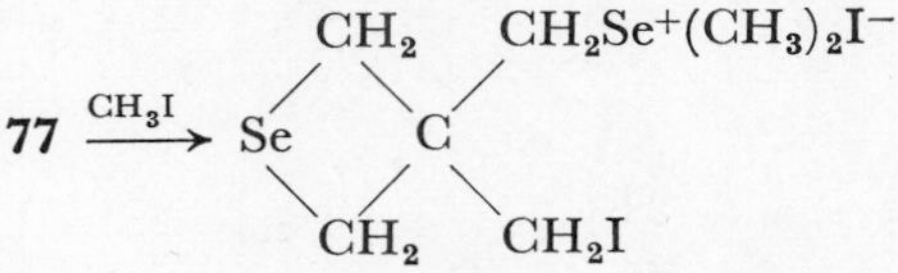

$$77 \xrightarrow{CH_3I}$$

78

2-Selenaspiro[3.5]nonane. By the general route followed to prepare **77**, Backer and Winter[5] obtained 2-selenaspiro[3.5]nonane (**79**) by the reaction of 1,1-bis(bromomethyl)cyclohexane with potassium selenide (equation 66).

$$(66) \quad \begin{array}{c} H_2C-CH_2 \quad CH_2Br \\ H_2C \qquad C \\ H_2C-CH_2 \quad CH_2Br \end{array} + K_2Se \xrightarrow[N_2]{C_2H_5OH}$$

$$\begin{array}{c} H_2C-CH_2 \quad CH_2 \\ H_2C \qquad C \qquad Se + 2KBr \\ H_2C-CH_2 \quad CH_2 \\ \textbf{79} \end{array}$$

Compound **79** forms addition compounds with mercuric chloride, mercuric bromide, and iodine. At room temperature a molecule of **79** takes up four atoms of bromine, or two and then four atoms of chlorine, with cleavage of the ring to give a selenium trihalide (**80**), which may be hydrolyzed by silver hydroxide to the corresponding seleninic acid (**81**) (equation 67).

$$(67) \quad \textbf{79} \xrightarrow[CCl_4]{X_2} \begin{array}{c} H_2C-CH_2 \quad CH_2X \\ H_2C \qquad C \\ H_2C-CH_2 \quad CH_2SeX_3 \\ \textbf{80} \end{array} \xrightarrow[AgOH]{C_2H_5OH}$$

$$\begin{array}{c} H_2C-CH_2 \quad CH_2X \\ H_2C \qquad C \\ H_2C-CH_2 \quad CH_2SeO_2H \\ \textbf{81} \end{array}$$

The reaction of **79** with iodomethane is similar to that reported for **77**.

Backer and Winter[5] also prepared 2-selenaspiro[3.5]nonane-2,2-dioxide (**82**) by heating the seleninate, **83**, in a sealed tube (equation 68).

$$(68) \quad \begin{array}{c} H_2C-CH_2 \quad CH_2Br \\ H_2C \qquad C \\ H_2C-CH_2 \quad CH_2SeO_2Na \\ \textbf{83} \end{array} \xrightarrow[\Delta]{C_2H_5OH}$$

$$\begin{array}{c} H_2C-CH_2 \quad CH_2 \\ H_2C \qquad C \qquad SeO_2 \\ H_2C-CH_2 \quad CH_2 \\ \textbf{82} \end{array}$$

b. With a Thiaselenolane Ring.   2-Thia-3-selenaspiro[4.5]decane (**84**) was prepared by Biezais and Bergson[117] by a ring-expansion reaction on heating 2-thiaspiro[3.5]nonane (**85**) with selenium to 180–190° (16 hours) in diethylene glycol containing a trace of potassium cyanide (equation 69). An

(69)

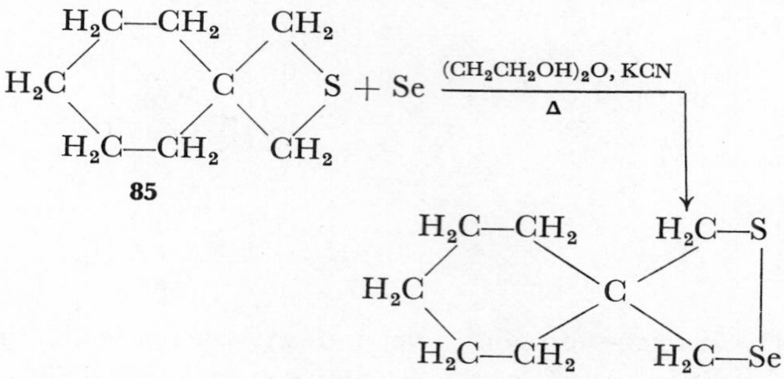

85

84

analogous attempt to treat 2-selenaspiro[3.5]nonane with sulfur led only to polymer formation. The u.v. spectrum of **84** is very similar to that reported for 1-thia-2-selenolane-4-carboxylic acid (**16**).[34]

Krackov *et al.*[36] measured the dipole moments of **84** and its analogs possessing a dithiolane and a diselenolane ring. The three moments were found to be almost identical, indicating a lack of polarization of the S-Se bond in the thiaselenolane ring.

c. With a Diselenolane Ring.   Bergson and Biezais[118] prepared 2,3-diselenaspiro[4.5]decane (**86**) by heating cyclohexane-1,1-dimethylbenzenesulfonate (**87**) with potassium selenocyanate (equation 70). The reaction

(70)

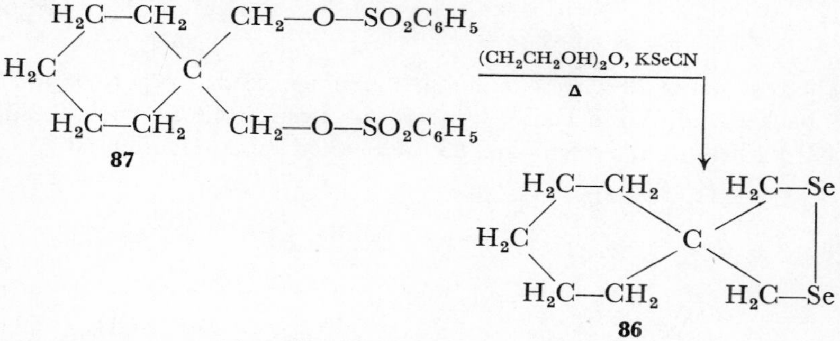

87

86

leads directly to the cyclic diselenide without isolation of the intermediate diselenocyanate. Bergson and Biezais found that starting with 1,1-bis(bromomethyl)cyclohexane lowers the yield of **86**.

Krackov *et al.*[36] have measured the dipole moment of **86** in benzene.

Günther and Salzman[118a] prepared 2-oxa-6,7-diselenaspiro[3.4]octane (**88**) in 58% yield from 3,3-bis(chloromethyl)oxetane by reaction with potassium selenosulfate in water, followed by alkaline hydrolysis of the intermediate bisselenosulfate (**89**) (equation 71).

The highly crystalline **88** has an $\varepsilon^{max}_{453\,nm}$ of 142 (ethanol), and n.m.r. and i.r. spectra characteristic for the diselenolane and the oxetane moieties. The species was remarkably resistant to polymerization through the di-selenide or the oxetane group, but readily formed an elastomeric composite with elemental selenium (*cf.* also Section XI.A.9).

8. Bridged Selenacycloalkanes

Selenatricyclo[3.3.1.1³,⁷]decanes. Olsson and Almqvist[64] prepared 1,3,5,7-tetramethyl-2,4,6,8-tetraselenatricyclo[3.3.1.1³,⁷]decane (**90**) and 1,3,5, 7-tetramethyl-2,4,6,8,9,10-hexaselenatricyclo[3.3.1.1³,⁷]decane (**91**), also (72)

named 1,3,5,7-tetramethyl-2,4,6,8-tetraselenaadamantane and 1,3,5,7-tetra-
methyl-2,4,6,8,9,10-hexaselenaadamantane, respectively, by a route similar
to that followed for the preparation of the sulfur analogs.

In the preparation of **90**, H_2Se was bubbled into a solution of 2,4-pentane-
dione in acetic acid, in the presence of $ZnCl_2$, to give the selenaadamantane
in 70% yield, together with the oxa and 2,4-dioxa analogs (equation 72).

Compound **91** was obtained as a by-product from the preparation of
2,4-dimethyl-1,3-diselenetane-2,4-diselenol diacetate (**38**). After crystallization, pure **91** was recovered in about 7% yield.[64]

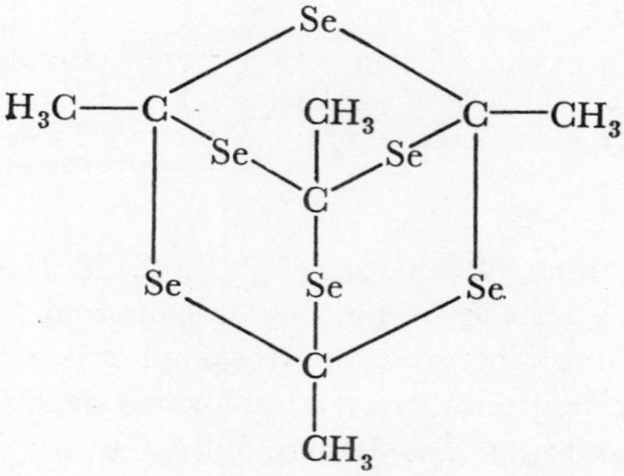

91

9. Condensed Selenacycloalkanes

A well-defined crystalline cyclic dimer (mol.wt. 530) of α,α'-diseleno-*m*-
xylene was obtained by Günther and Salzman[118a] from α,α'-dibromo-*m*-
xylene *via* the intermediate bisselenosulfate (**92**) (equation 73). The dimer
(**94**) was found to convert to the polymer (**93**) at temperatures above 145°
and could be recovered in crystalline form by slow equilibration in 1,1,2,2-
tetrachloroethane (TCE) solution at 80–100°. This species and related
ones[118a] have the remarkable property of forming single-phase, nonstoichiometric "alloys" with elemental selenium. The reaction with the element
occurs at 220–260° and is believed to involve intercalation of the organic

(73)

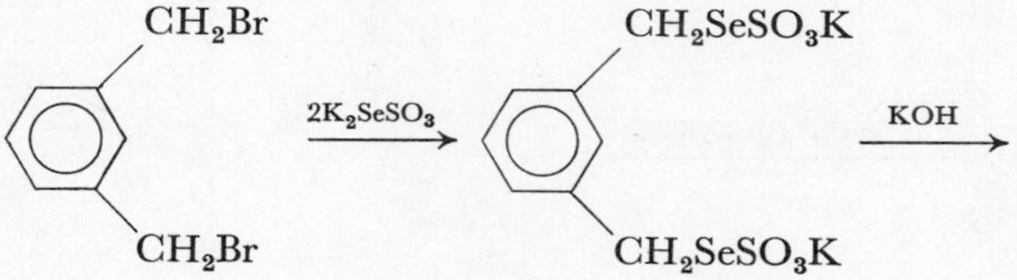

92

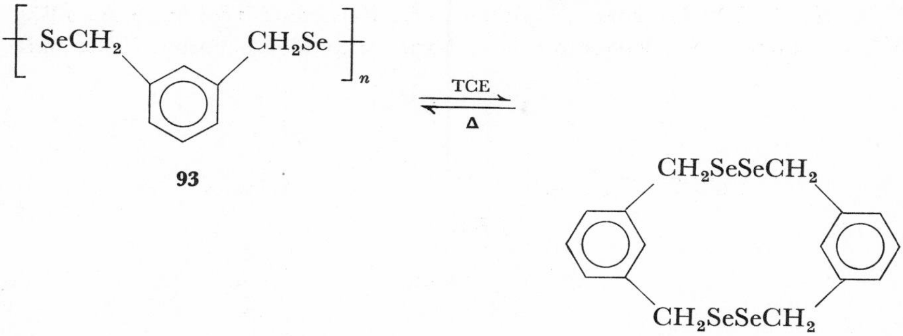

$$\left[SeCH_2 \underset{\text{(m-phenylene)}}{\bigcirc} CH_2Se \right]_n \underset{\Delta}{\overset{TCE}{\rightleftharpoons}}$$

93

$$\underset{CH_2SeSeCH_2}{\bigcirc \quad \bigcirc}$$

94

moiety into selenium polymer chains through Se—Se bond interchange reactions. The mixed polymers are odorless and stable at room temperature.

10. Heterocycles without Carbon Atoms in the Ring

a. Five-Membered Rings. Mikhailov and Shchegoleva[119] prepared 3,5-dibutyl-1,2,4-triselena-3,5-diborolane (**95**) by heating tributylboron and selenium (equation 74). Compound **95** was also obtained by Schmidt *et al.*[120] on heating selenium with dibutylboron iodide (equation 75).

(74) $[CH_3—(CH_2)_3—]_3B + Se \xrightarrow{220-250°}$

(75)

$[CH_3—(CH_2)_3—]_2BI + Se \xrightarrow{160°}$

$$H_3C—(CH_2)_3—B \underset{Se}{\overset{Se—Se}{\diagdown}} B—(CH_2)_3—CH_3$$

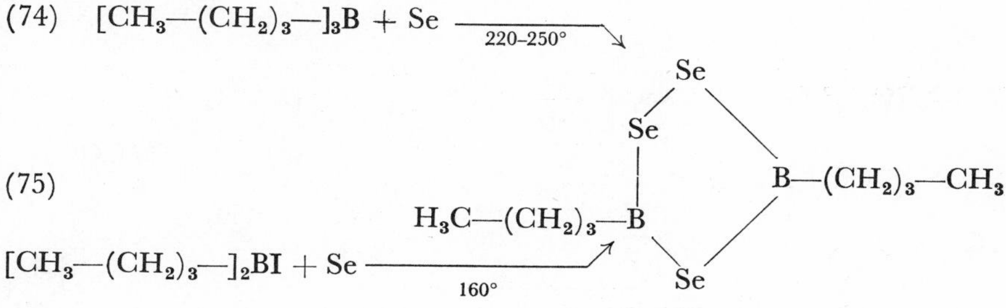

95

By a similar route, 3,5-diisobutyl-1,2,4-triselena-3,5-diborolane (**96**) was obtained[119] (equation 76). Treatment of **95** with aqueous THF gave an immediate precipitate of selenium and butylboronic acid.

(76)

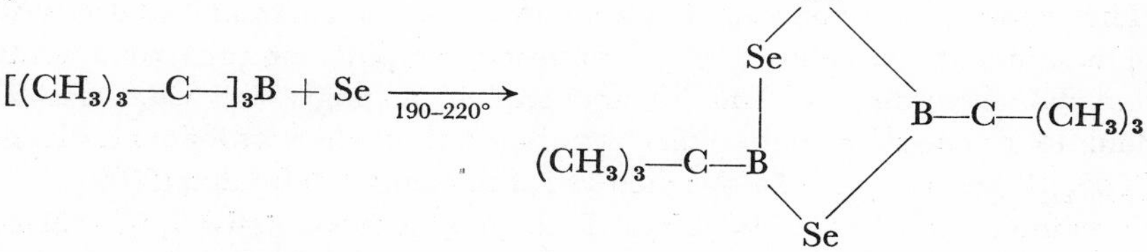

$[(CH_3)_3—C—]_3B + Se \xrightarrow{190-220°}$

$$(CH_3)_3—C—B \underset{Se}{\overset{Se—Se}{\diagdown}} B—C—(CH_3)_3$$

96

Schmidt *et al.*[120] prepared 3,5-diiodo-1,2,4-triselena-3,5-diborolane (**97**) by the reaction of red selenium with boron triiodide (equation 77). Com-

(77) 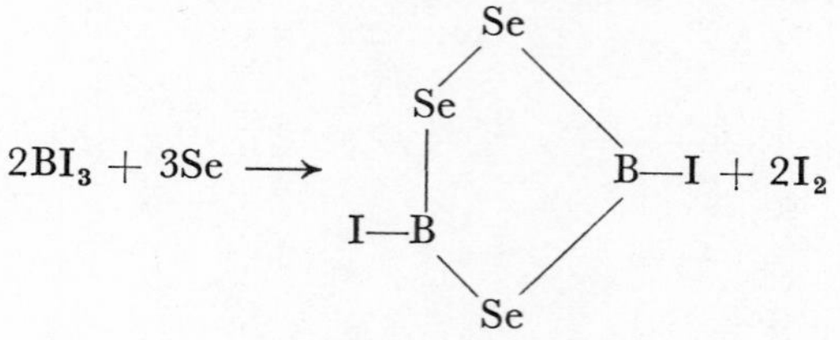 $2BI_3 + 3Se \longrightarrow$

97

pound **97** is very sensitive to moisture and to oxygen, and very soluble in aprotic solvents.

In contrast to the corresponding sulfur analog, **97** did not form a six-membered ring with excess BI_3 and is decomposed into selenium and I_2 by dimethyl sulfide.

b.   Six-Membered Rings.   By the reaction of dimethyldichlorosilane with sodium selenide (equation 78) Schmidt and Ruf[121] obtained an oil which,

(78) 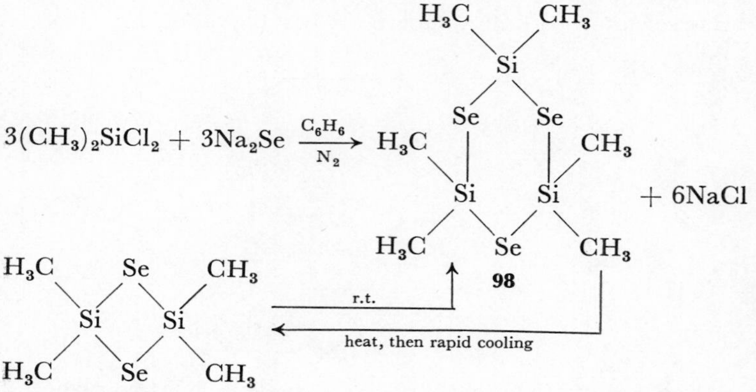 $3(CH_3)_2SiCl_2 + 3Na_2Se \xrightarrow[N_2]{C_6H_6}$

$+ 6NaCl$

98

99

on distillation at 2 mm Hg, gave a white, crystalline sublimate and 2,2,4,4,6,6-hexamethyl-1,3,5-triselena-2,4,6-trisilacyclohexane (**98**) as a yellow oil. These workers were unable to determine the molecular weight of the crystalline sublimate, because it was transformed rapidly at room temperature into **98**. According to Schmidt and Ruf, in analogy with what has been found for the sulfur analog, the crystalline sublimate is probably a dimer of $(CH_3)_2SiSe$, that is, 2,2,4,4-dimethyl-1,3-diselena-2,4-disiletane (**99**).

Schmidt and Ruf[122] obtained 2,2,4,4,6,6-hexamethyl-1,3,5-triselena-2,4,6-trigermacyclohexane (**100**) by the reaction of dimethylgermanium dichloride with sodium selenide (equation 79). Compound **100** is unaffected by air.

(79) $3(CH_3)_2GeCl_2 + 3Na_2Se \xrightarrow{C_6H_6}$

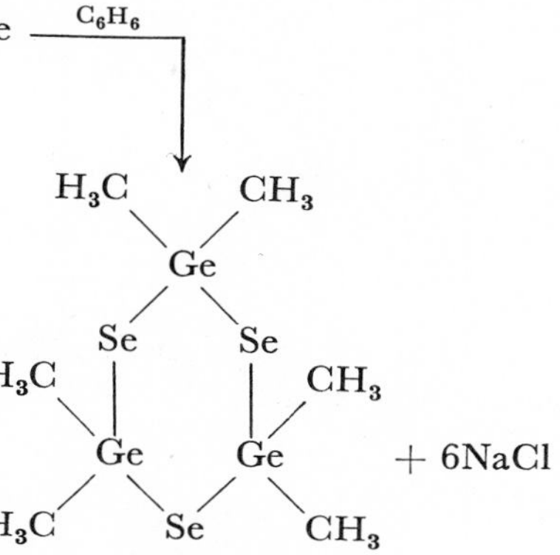

$$
\begin{array}{c}
H_3C \quad \quad CH_3 \\
Ge \\
Se \quad \quad Se \\
H_3C \quad \quad \quad \quad CH_3 \\
Ge \quad \quad Ge \quad \quad + 6NaCl \\
H_3C \quad Se \quad CH_3
\end{array}
$$
100

The same workers also synthesized 2,2,4,4,6,6-hexamethyl-1,3,5-tri-selena-2,4,6-tristannacyclohexane (**101**) by the reaction of dimethyltin dichloride with sodium selenide (equation 80).[123]

(80) $3(CH_3)_2SnCl_2 + 3Na_2Se \xrightarrow{C_6H_6}$

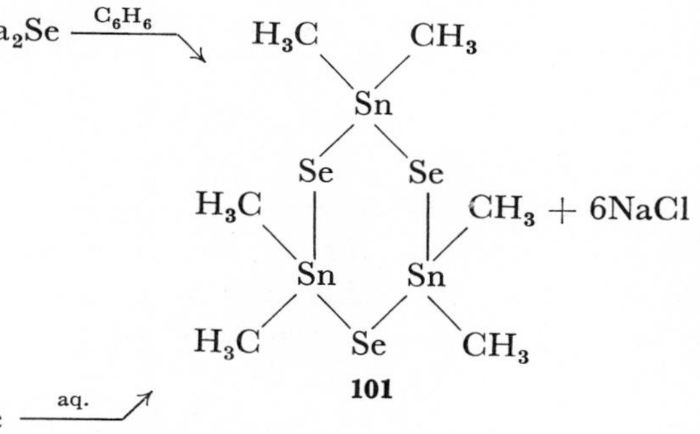

$$
\begin{array}{c}
H_3C \quad \quad CH_3 \\
Sn \\
Se \quad \quad Se \\
H_3C \quad \quad \quad \quad CH_3 \quad + 6NaCl \\
Sn \quad \quad Sn \\
H_3C \quad Se \quad CH_3
\end{array}
$$
101

(81) $(CH_3)_2SnBr_2 + H_2Se \xrightarrow{aq.}$

Compound **101** is unaffected by air and boiling water but decomposes in u.v. light.

Kriegsmann and Hoffmann[124] prepared **101** by passing hydrogen selenide through an aqueous solution of dimethyltin dibromide (equation 81). In a similar way Kriegsmann et al.[125] prepared 2,2,4,4,6,6-hexaethyl-1,3,5-triselena-2,4,6-tristannacyclohexane (**102**). The same workers[125] studied the vibrational spectra of **101** and **102** in CCl_4 solution, together with those of the corresponding sulfur-containing cyclic compounds. On the basis of the number of bands observed, the boat conformation seemed the most probable one.

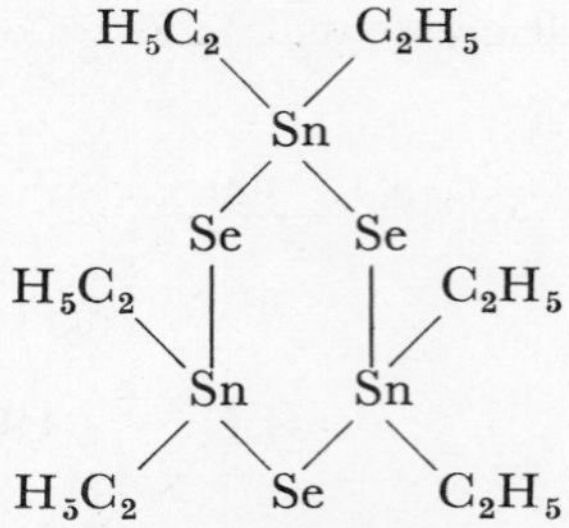

$$
\begin{array}{c}
H_5C_2 \qquad C_2H_5 \\
\diagdown \diagup \\
Sn \\
\diagup \diagdown \\
Se \qquad Se \\
H_5C_2 \quad | \qquad | \quad C_2H_5 \\
\diagdown | \qquad | \diagup \\
Sn \qquad Sn \\
\diagup \qquad \diagdown \\
H_5C_2 \qquad Se \qquad C_2H_5 \\
\mathbf{102}
\end{array}
$$

Schmidt and Schumann[126] obtained 2,2,4,4,6,6-hexabutyl-1,3,5-triselena-2,4,6-tristannacyclohexane (**103**) by stirring tetrabutyltin and selenium (equation 82).

(82) $3[CH_3-(CH_2)_3-]_4Sn + 3Se \xrightarrow{\Delta}$

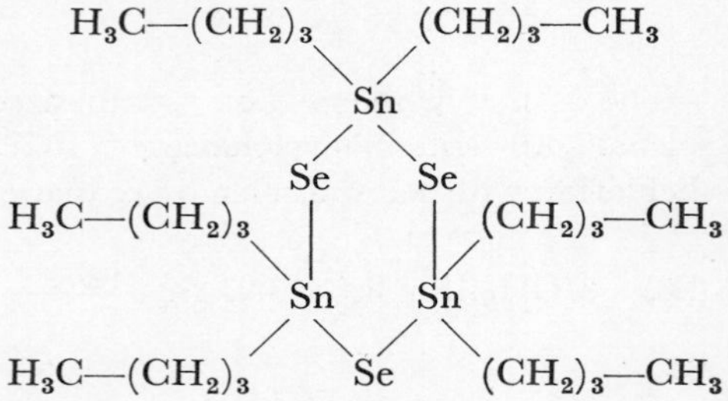

$$
\begin{array}{c}
H_3C-(CH_2)_3 \qquad (CH_2)_3-CH_3 \\
\diagdown \diagup \\
Sn \\
\diagup \diagdown \\
Se \qquad Se \\
H_3C-(CH_2)_3 \quad | \qquad | \quad (CH_2)_3-CH_3 \\
\diagdown | \qquad | \diagup \\
Sn \qquad Sn \\
\diagup \qquad \diagdown \\
H_3C-(CH_2)_3 \qquad Se \qquad (CH_2)_3-CH_3 \\
\mathbf{103}
\end{array}
$$

Schumann et al.[127] prepared 2,2,4,4,6,6-hexaphenyl-1,3,5-triselena-2,4,6-tristannacyclohexane (**104**) by the reaction of lithium triphenyltin selenide and diphenyltin dichloride (equation 83). Compound **104** is unaffected by water but decomposes slowly (with separation of selenium) on boiling with dilute acids.

(83) $6(C_6H_5)_3SnSeLi + 3(C_6H_5)_2SnCl_2 \longrightarrow$

$$
3(C_6H_5)_3Sn-Se-Sn(C_6H_5)_3 +
\begin{array}{c}
H_5C_6 \qquad C_6H_5 \\
\diagdown \diagup \\
Sn \\
\diagup \diagdown \\
Se \qquad Se \\
H_5C_6 \quad | \qquad | \quad C_6H_5 \\
\diagdown | \qquad | \diagup \\
Sn \qquad Sn \\
\diagup \qquad \diagdown \\
H_5C_6 \qquad Se \qquad C_6H_5 \\
\mathbf{104}
\end{array}
+ 6LiCl
$$

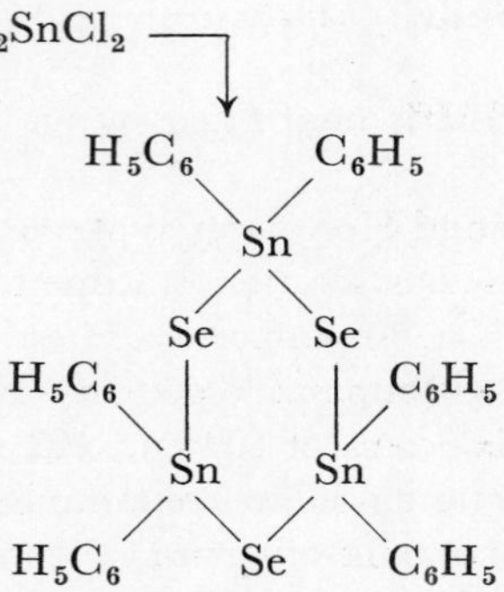

REFERENCES

1. A. B. Callear and W. J. R. Tyerman, *Proc. Chem. Soc.*, 296 (1964); *Trans. Faraday Soc.*, **61,** 2395 (1965); **62,** 371 (1966).
2. W. J. R. Tyerman, W. B. O'Callaghan, P. Kebarle, O. P. Strausz, and H. E. Gunning, *J. Amer. Chem. Soc.*, **88,** 4277 (1966).
3. O. P. Strausz and H. E. Gunning, "The Chemistry of Sulfides," A. V. Tobolsky, Ed., Interscience Publishers, New York, N.Y., 1968, p. 39.
4. G. T. Morgan and F. H. Burstall, *J. Chem. Soc.*, 1497 (1930).
5. H. J. Backer and H. J. Winter, *Rec. Trav. Chim. Pays-Bas*, **56,** 492 (1937).
6. A. B. Harvey, J. R. Durig, and A. C. Morrissey, *J. Chem. Phys.*, **47,** 4864 (1967).
7. D. O. Harris, W. H. Harrington, A. C. Luntz, and W. D. Gwinn, *J. Chem. Phys.*, **44,** 3467 (1966).
8. W. B. Moniz, *J. Phys. Chem.*, **73,** 1124 (1969).
9. A. Fredga, *Uppsala Univ. Arsskrift*, No. 5, 79 (1935).
10. G. T. Morgan and F. H. Burstall, *J. Chem. Soc.*, 1096 (1929).
11. Yu. K. Yur'ev, *J. Gen. Chem. USSR*, **16,** 851 (1946).
12. J. D. McCullough and A. Lefohn, *Inorg. Chem.*, **5,** 150 (1966).
13. H. De Vries Robles, *Rec. Trav. Chim. Pays-Bas*, **58,** 111 (1939).
14. H. De Vries Robles, *Rec. Trav. Chim. Pays-Bas*, **59,** 184 (1940).
15. E. T. Strom, B. S. Snowden, Jr., H. C. Custard, D. E. Woenner, and J. R. Norton, *J. Org. Chem.*, **33,** 2555 (1968).
16. G. Milazzo, *Rend. Ist. Super. Sanità*, **22,** 479 (1959).
17. H. Hope and J. D. McCullough, *Acta Cryst.*, **17,** 712 (1964).
18. B. Pedersen and H. Hope, *Acta Cryst.*, **19,** 473 (1965).
19. J. D. McCullough and A. Brunner, *Inorg. Chem.*, **6,** 1251 (1967).
20. F. M. Brewer, J. R. Chadwick, and G. Garton, *J. Inorg. Nucl. Chem.*, **23,** 45 (1961).
21. H. Suginome and S. Umezawa, *Bull. Chem. Soc. Japan*, **11,** 157 (1936).
22. H. Funk and W. Papenroth, *J. Prakt. Chem.*, **8,** 256 (1959).
23. U.S. Pat. 2,931,803 (1960); *Chem. Abstr.*, **55,** 12436h (1961).
24. C. G. Krespan and C. M. Langkammerer, *J. Org. Chem.*, **27,** 3584 (1962).
25. C. G. Krespan, *J. Org. Chem.*, **27,** 3588 (1962).
26. U.S. Pat. 3,149,124 (1964); *Chem. Abstr.*, **62,** 7728e (1965).
27. Yu. K. Yur'ev, N. N. Magdesieva, and L. Ya. Petrova, *Khim. Geterotsikl. Soedin*, 910 (1966); *Chem. Abstr.*, **67,** 11384n (1967).
28. A. M. Duffield, H. Budzikiewicz, and C. Djerassi, *J. Amer. Chem. Soc.*, **87,** 2920 (1965).
29. A. Fredga, *J. Prakt. Chem.*, **127,** 103 (1930).
30. A. Fredga, *J. Prakt. Chem.*, **130,** 180 (1931).
31. A. Fredga, *J. Prakt. Chem.*, **150,** 124 (1938).
32. A. Rosenberg and L. Schotte, *Ark. Kemi*, **8,** 143 (1955).
33. A. Fredga, *Uppsala Univ. Arsskrift*, No. 5, 69 (1935).
34. G. Bergson and A. Biezais, *Ark. Kemi*, **18,** 143 (1961).

35. G. Bergson, *Acta Chem. Scand.*, **15**, 1611 (1961).

36. M. H. Krackov, G. Bergson, A. Biezais, and H. G. Mautner, *J. Amer. Chem. Soc.*, **88**, 1759 (1966).

37. G. Bergson, *Ark. Kemi*, **19**, 75 (1962).

38. J. Dubac and P. Mazerolles, *C. R. Acad. Sci., Paris, Ser. C*, **267**, 411 (1968).

39. P. Mazerolles, J. Dubac, and M. Lesbre, *J. Organometal. Chem.*, **12**, 143 (1968).

40. G. T. Morgan and F. H. Burstall, *J. Chem. Soc.*, 2197 (1929).

41. G. T. Morgan and F. H. Burstall, *J. Chem. Soc.*, 173 (1931).

42. A. Fredga and K. Styrman, *Ark. Kemi*, **14**, 461 (1959).

43. W. Hänsel and R. Haller, *Naturwissenschaften*, **55**, 83 (1968).

44. C. S. Gibson and J. D. A. Johnson, *Chem. Ind.* (London), **49**, 896 (1930).

45. C. S. Gibson and J. D. A. Johnson, *J. Chem. Soc.*, 266 (1931).

46. J. D. A. Johnson, *J. Chem. Soc.*, 1530 (1933).

47. J. Böeseken, F. Tellegen, and P. Cohen Henriquez, *Rec. Trav. Chim. Pays-Bas*, **54**, 733 (1935).

48. J. D. McCullough and I. C. Zimmermann, *J. Phys. Chem.*, **65**, 888 (1961).

49. H. Maddox and J. D. McCullough, *Inorg. Chem.*, **5**, 522 (1966).

50. C. Knobler and J. D. McCullough, *Inorg. Chem.*, **7**, 365 (1968).

51. M. Schmidt and M. Wieber, *Chem. Ber.*, **194**, 1426 (1961).

52. M. Wieber and G. Schwarzmann, *Monatsh. Chem.*, **100**, 74 (1969).

53. L. Credali, M. Russo, L. Mortillaro, C. De Checchi, G. Valle, and M. Mammi, *J. Chem. Soc., B*, 117 (1967).

54. M. Russo and L. Credali, *J. Makromol. Sci.* (*Chem.*), **1A**, 387 (1967).

55. C. S. Gibson and J. D. A. Johnson, *J. Chem. Soc.*, 1529 (1933).

56. J. D. McCullough and P. Radlick, *Inorg. Chem.*, **3**, 924 (1964).

57. J. D. McCullough, *Inorg. Chem.*, **3**, 1425 (1964).

58. H. Hope and J. D. McCullough, *Acta Cryst.*, **15**, 806 (1962).

59. P. Mazerolles, M. Lesbre, and M. Joanny, *J. Organometal. Chem.*, **16**, 227 (1969).

60. G. T. Morgan and H. D. K. Drew, *J. Chem. Soc.*, 1456 (1920).

61. G. T. Morgan, H. D. K. Drew, and T. V. Barker, *J. Chem. Soc.*, 2432 (1922).

62. D. H. Dewar, J. E. Fergusson, P. R. Hentschel, C. J. Wilkins, and P. P. Williams, *J. Chem. Soc.*, 688 (1964).

63. G. T. Morgan and J. D. M. Smith, *J. Chem. Soc.*, 1066 (1921).

64. K. Olsson and S. O. Almqvist, *Acta Chem. Scand.*, **23**, 3271 (1969).

65. L. Hagelberg, *Ber.*, **23**, 1083 (1890).

66. G. Bergson and G. Claeson, *Acta Chem. Scand.*, **11**, 911 (1957).

67. G. Bergson, *Ark. Kemi*, **19**, 195 (1962).

68. G. Bergson, *Ark. Kemi*, **13**, 11 (1958).

69. G. Bergson, G. Claeson, and L. Schotte, *Acta Chem. Scand.*, **16**, 1159 (1962).

70. H. J. Backer and H. J. Winter, *Rec. Trav. Chim. Pays-Bas*, **56**, 691 (1937).

71. G. Bergson, *Acta Chem. Scand.*, **12**, 582 (1958).

72. B. Nygard, E. Johanson, and J. Olofsson, *J. Electroanal. Chem.*, **12**, 564 (1966).

73. L. Henriksen, *Acta Chem. Scand.*, **21**, 1981 (1967).

74. E. S. Gould and W. Burlant, *J. Amer. Chem. Soc.*, **78**, 5825 (1956).

75. R. Paetzold and D. Lienig, *Z. Anorg. Allg. Chem* , **335**, 289 (1965).

76. E. S. Gould and B. Post, *J. Amer. Chem. Soc.*, **78**, 5161 (1956).

77. J. R. Brown, G. P. Gillman, and M. H. George, *J. Polymer Sci.*, *Part A-1*, **5**, 903 (1967).

78. A. Fredga, *Ark. Kemi, Mineral. Geol.*, **11B**, No. 15, 6 pp. (1933).

79. A. Fredga, *Uppsala Univ. Arsskrift*, No. 5, 124 (1935).

80. A. Fredga, *Ark. Kemi, Mineral. Geol.*, **12A**, No. 27, 28 pp. (1938).

81. A. Fredga, *Uppsala Univ. Arsskrift*, No. 5, 128 (1935).

82. A. Fredga, *Ber.*, **71B**, 286 (1938).

83. C. Djerassi, A. Fredga, and B. Sjöberg, *Acta Chem. Scand.*, **15**, 417 (1961).

84. C. Djerassi, H. Wolf, and E. Bunnenberg, *J. Amer. Chem. Soc.*, **84**, 4552 (1962).

85. O. Foss and L. Schotte, *Acta Chem. Scand.*, **11**, 1424 (1957).

86. O. Foss, K. Johnsen, and T. Reistad. *Acta Chem. Scand.*, **18**, 2345 (1964).

87. S. Husebye, *Acta Chem. Scand.*, **15**, 1215 (1961).

88. A. Geens, G. Swallens, and M. Anteunis, *Chem. Commun.*, 439 (1969).

89. E. S. Gould and J. D. McCullough, *J. Amer. Chem. Soc.*, **73**, 1105 (1951).

90. J. D. McCullough and N. W. Tideswell, *J. Amer. Chem. Soc.*, **76**, 3091 (1954).

91. R. E. Marsh and J. D. McCullough, *J. Amer. Chem. Soc.*, **73**, 1106 (1951).

92. J. D. McCullough, G. Y. Chao, and D. E. Zuccaro, *Acta Cryst.*, **12**, 815 (1959).

93. G. Y. Chao and J. D. McCullough, *Acta Cryst.*, **14**, 940 (1961).

94. P. J. Hendra and N. Sadasivan, *Spectrochim. Acta*, **21**, 1127 (1965).

95. A. Amendola, E. S. Gould, and B. Post, *Inorg. Chem.*, **3**, 1199 (1964).

96. P. J. Hendra and N. Sadasivan, *J. Chem. Soc.*, 2063 (1965).

97. T. Bjorvatten, *Acta Chem. Scand.*, **17**, 2292 (1963).

98. G. C. Hayward and P. J. Hendra, *Spectrochim. Acta*, **23A**, 1937 (1967).

99. O. Holmesland and C. Rømming, *Acta Chem. Scand.*, **20**, 2601 (1966).

100. T. Dahl and O. Hassel, *Acta Chem. Scand.*, **19**, 2000 (1965).

101. O. Hassel, *Acta Chem. Scand.*, **19**, 2259 (1965).

102. M. Russo, L. Mortillaro, L. Credali, and C. De Checchi, *J. Polymer Sci.*, *Part B*, **4,** 167 (1966).

103. M. Mammi and G. Valle, private communication.

104. L. Mortillaro and M. Russo, unpublished results.

105. L. Vanino and A. Schinner, *J. Prakt. Chem.*, **91**, 116 (1915).

106. T. G. Pearson, R. H. Purcell, and G. S. Saigh, *J. Chem. Soc.*, 409 (1938).

107. H. J. Bridger and R. W. Pittman, *J. Chem. Soc.*, 1371 (1950).

108. L. Mortillaro, L. Credali, M. Mammi, and G. Valle, *J. Chem. Soc.*, 807 (1965).

109. L. Mortillaro, L. Credali, M. Russo, and C. De Checchi, *J. Polymer Sci.*, *Part B*, **3,** 581 (1965).

110. M. Russo, L. Mortillaro, L. Credali, and C. De Checchi, *J. Polymer Sci.*, *Part A-1*, **4,** 248 (1966).

111. M. Schmidt, K. Blaettner, P. Kochendörfer, and H. Ruf, *Z. Naturforsch.*, *Part B*, **21,** 622 (1966).

112. F. O. Rice and A. L. Glasebrook, *J. Amer. Chem. Soc.*, **56**, 2381 (1934).

113. F. D. Williams and F. X. Dunbar, *Chem. Commun.*, 459 (1968).

114. M. Mammi, G. Carazzolo, G. Valle, and A. Del Pra, *Z. Kristr.*, **127**, 401 (1968).

115. G. Carazzolo and M. Mammi, *J. Polymer Sci., Part C*, **16**, 1521 (1967).

117. A. Biezais and G. Bergson, *Acta Chem. Scand.*, **18**, 815 (1964).

118. G. Bergson and A. Biezais, *Ark. Kemi*, **22**, 475 (1964).

118a. W. H. H. Günther and M. N. Salzman, *Ann. N.Y. Acad. Sci.*, **192**, 25 (1972).

119. B. M. Mikhailov and T. A. Shchegoleva, *Izv. Akad. Nauk. SSSR, Otd. Khim. Nauk*, 356 (1959); *Chem. Abstr.*, **53**, 20041i (1959).

120. M. Schmidt, W. Siebert, and E. Gast, *Z. Naturforsch., Part B*, **22**, 557 (1967).

121. M. Schmidt and H. Ruf, *Z. Anorg. Allg. Chem.*, **321**, 270 (1963).

122. M. Schmidt and H. Ruf, *J. Inorg. Nucl. Chem.*, **25**, 557 (1963).

123. M. Schmidt and H. Ruf, *Chem. Ber.*, **96**, 784 (1963).

124. H. Kriegsmann and H. Hoffmann, *Z. Chem.*, **3**, 268 (1963).

125. H. Kriegsmann, H. Hoffmann, and H. Geissler, *Z. Anorg. Allg. Chem.*, **359**, 58 (1968).

126. M. Schmidt and H. Schumann, *Chem. Ber.*, **96**, 780 (1963).

127. H. Schumann, K. F. Thom, and M. Schmidt, *J. Organometal. Chem.*, **2**, 361 (1964).

General References

A. Weissberger, "The Chemistry of Heterocyclic Compounds," Vol. XIX (Part 2), A. Weissberger, Ed., John Wiley, New York, N.Y., 1964, p. 716.

H. D. Hartough, "The Chemistry of Heterocyclic Compounds," Vol. III, A. Weissberger, Ed., John Wiley, New York, N.Y., 1952, p. 490.

D. S. Breslow and H. Skolnik, "The Chemistry of Heterocyclic Compounds," Vol. XXI (Part 2), A. Weissberger, Ed., John Wiley, New York, N.Y., 1964, p. 1272.

K. W. Bagnall, "Topics in Inorganic and General Chemistry," Monograph 7, P. L. Robinson, Ed., Elsevier, Amsterdam, 1966, p. 180.

H. Rheinboldt, "Houben-Weyl Methoden der Organischen Chemie," Vol. IX, E. Müller, Ed., Georg Thieme Verlag, Stuttgart, 1955, p. 917.

XI Heterocyclic selenium compounds

B. SELENOPHENES AND RELATED COMPOUNDS

NATALIA N. MAGDESIEVA and **NICOLAI S. ZEFIROV**

Moscow State University
Faculty of Chemistry
Moscow, U.S.S.R.

1. Introduction

The first compound of the selenophene series to be prepared was 2,5-dimethylselenophene, which Paal[1] obtained in 1885 by heating acetonyl-acetone in a sealed tube with phosphorus pentaselenide, followed by treatment of the product with dilute sodium hydroxide solution (equation 1).

(1)

In 1923 Bogert and Herrera[2] described a synthesis of 2,4-diarylselenophenes from acetophenone anils by fusion with selenium. Foa[3] in 1909 reported the preparation of the parent compound, but from the physical properties he attributed to it, one may assume that he had not actually obtained selenophene. The latter was synthesized initially by Mazza and Solazzo[4] and later by Briscoe et al.[5,6] and Suginome and Umezawa,[7-10] all of whom made it by the reaction of acetylene with selenium (equation 2).

(2)
$$2HC{\equiv}CH + Se \longrightarrow$$

Although selenophene and certain of its homologs have been known for about 40 years, selenophene chemistry has advanced very slowly. Only a few simple electrophilic substitution reactions of the nucleus were known[8-10] when Yur'ev and his co-workers started their investigations. Probably the development was hindered because there were no convenient procedures to prepare selenophene, its homologs, and its derivatives.

The chemistry of selenophene, a five-membered selenium-containing heterocyclic compound, has been studied intensively over the last 10–15 years. One of the objectives of this chapter is to compare selenophene with its isologs, thiophene and furan, and to clarify the theoretically challenging question of the specific effect of selenium as a heteroatom.

Selenophene chemistry was reviewed previously by Harthough,[11] Yur'ev et al.,[11a] and Magdesieva.[12]

2. Molecular Structure and Spectral Properties

a. Electronic Structure. Selenophene is a heteroaromatic system whose aromatic π-electron system is formed by the interaction of the π-electrons of two C-C double bonds with the lone pair of electrons of the Se heteroatom. The conjugation of the selenium pair with π-electrons of the adjacent atoms

is, to a certain degree, a criterion for comparing the aromaticity of selenophene with that of other heterocycles. In selenophene six electrons are in
the field of five nuclei. The sixfold axial symmetry characteristic of a usual
aromatic system is replaced by C_{2v} symmetry.[13] Therefore, a uniform
distribution of electron density is impossible in selenophene. This was confirmed by extended Hückel MO calculations[14] of the π-electron structure of
selenophene using the following parameters:

Hpp (eV) = Se 9.75, C 11.18; ζ(au) = Se 2.257, C 1.5679.

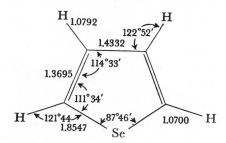

b. Molecular Geometry. In the past there was a question as to the
planarity of the selenophene molecule.[15-19] Also, the electric moment was
in doubt (0.41 or 0.77 D).[20,21] Therefore, Pozdeev et al.[13] undertook a
microwave spectral investigation of selenophene in order to determine the

geometry and the electric moment of this molecule. The structure of the
selenophene molecule was determined by the Kraitchman-Costain[22,23]
method. The findings of Pozdeev and his co-workers[13] pertaining to the
geometry of selenophene are compared with those obtained for thiophene
and furan in Table XIB-1. It should be mentioned that the selenophene
ring was the first example found in which the C–Se–C angle is less than 90°.
Recently, 1,3,4-selenadiazole was determined to have a C–Se–C angle
of 81.8°.[26]

Brown[27] analyzed the microwave spectra of selenophene for two isotopic forms of the molecule ($C_4H_4{}^{80}Se$ and $C_4H_4{}^{78}Se$) and gave an approximate structure. He also determined its electric moment (0.398 D).

Nardelli and Fava[28] and, later, Mario et al.[29] used X-ray analysis to
determine the bond lengths and angles in selenophene-2-carboxylic acid.
These workers believe that the Se atom in the latter compound is
about 0.06 Å above the plane of the ring. Tsukerman et al.[30-32] measured

Table XIB-1 Geometry of selenophene, thiophene, and furan as determined by the microwave method

	BOND LENGTH (Å)		
BOND	Selenophene[13]	Thiophene[24]	Furan[25]
Heteroatom—C(2)	1.8547	1.7140	1.3621
C(2)—C(3)	1.3695	1.3696	1.3609
C(3)—C(4)	1.4332	1.4232	1.4309
C(2)—H(2)	1.0700	1.0776	1.0750
C(3)—H(3)	1.0792	1.0805	1.0768
ATOM GROUP	ANGLE		
Heteroatom—C(2)—C(3)	111°34′	111°28′	110°41′
C(5)—heteroatom—C(2)	87°46′	92°10′	106°33′
C(2)—C(3)—C(4)	114°33′	112°27′	106° 3′
Heteroatom—C(2)—H(2)	121°44′	119°51′	115°55′
C(4)—C(3)—H(3)	122°52′	124°16′	127°57′

the electric moments of selenophene and its derivatives in dilute benzene solutions (the Debye method) at 25° and, in addition, calculated the moments by a vector additive scheme. They also determined the electric moments of selenophene α- and β-unsaturated ketones.[31,32]

c. Nuclear Magnetic Resonance Spectra. In the n.m.r. spectrum of selenophene the α-protons appear at δ 7.70 ppm and the β-protons at δ 7.12 ppm.[16] The differences between the chemical shifts at the α- and β-protons in furan,[33] thiophene,[34] and selenophene[33,35] are 1.05, 0.12, and 0.57 ppm, respectively, suggesting that thiophene resembles benzene most closely. The n.m.r. spectra of selenophene and its derivatives have also been studied by Read and his co-workers[36] and Tsukerman and his co-workers[37] (*cf.* Chapter XVD for additional details).

d. Infrared and Raman Spectra. Gerding *et al.*,[25] who studied the i.r. and Raman spectra of selenophene, concluded that the molecule was not planar. Later, however, several workers disproved that idea.[13] Chumayevskii *et al.*[38] found that selenophene shows ν(C–H) bands at 2960 and 3050 cm^{-1}, while the ring ν(C–C) band was reported to be at 1582 cm^{-1}.

The Raman spectra of some unsaturated and cyclopropyl derivatives of selenophene have been studied.[39,40] It was found that the intensities of certain bands depend on the length of the conjugated systems, as compared to the influence of an *n*-propyl side chain. The presence of a cyclopropyl group

also increases the intensity of some bands. This phenomenon of band intensification occurs also with phenylcyclopropanes.[41,42]

A full interpretation of the vibrational spectra of selenophene has been
given by Trombetti and Zauli[18] and Alexanyan and his co-workers.[43,44]
Assignment of the principal vibrations permitted the calculation of the force
field of thiophene and selenophene with respect to the symmetry coordinates.[45] Kimel'fel'd et al.[46] studied the substituent effect on the vibrational
spectra of furan, thiophene, and selenophene.

e. *Ultraviolet Spectra.* The u.v. spectra of selenophene and its various
methyl and dimethyl homologs were studied by Chumayevskii et al.[38]
Later, Treshchova and her colleagues[40] reported the spectra of furans,
thiophenes, and selenophenes possessing n-propyl, propenyl, and cyclopropyl substituents in the 2-position. They found that these compounds
exhibit a conjugation effect as shown by the bathochromic shift of the
absorption maxima, the shifts increasing in the order O < S < Se. Various
chemists[47–53] have studied the u.v. spectra of selenophene and a number of
its derivatives with the ring mono- and disubstituted by electron-withdrawing groups. The u.v. spectra suggest that the aromaticity of the system
decreases in the series: benzene > selenophene > thiophene > furan.

Trombetti and Zauli[18] studied the u.v. spectrum of selenophene and
showed that this compound is characterized by absorption at 37,000–
47,500 cm^{-1}, caused by two superimposed transitions: the $\pi \to \pi^*$ transition
and the one characteristic of other aromatic systems. Calculations of the
electronic spectra of selenophene were made by Fabian et al.[19]

f. *Conjugation of the Selenophene Ring with Functional Groups.* A study
of the effect displayed by various heterocyclic radicals (2-furyl, 2-thienyl,
2-selenophene-yl*) attached to 3,5-substituted pyrazols on the basicity of the
latter revealed[54] that in all cases in which there is an aromatic system at
position 5(3) of the pyrazole nucleus the basicity of the pyrazole is markedly
reduced.

The electronic effect of five-membered heterocyclic groups, thienyl and
selenophene-yl, can be judged by the ability of these aromatic systems to
enolize the neighboring carbonyl group in β-diketones that contain thienyl,
phenyl, or selenophene-yl substituents.[55–57] This ability was found to increase
in the series: 3-selenophene-yl < phenyl < 2-selenophene-yl < 2-thienyl.
An assessment of the electronic properties of these heterocycles by the
strengths of the corresponding monocarboxylic acids and on the basis of their
effect on the dissociation constant of β-diketones[58–61] gives the following
series: phenyl < 2-selenophene-yl < 2-thienyl.

* The designation for derivatives of selenophene recommended by *Chemical Abstracts*
is "selenophene-yl."

In the static state the comparative effect of five-membered heterocycles was found by studying the i.r. and u.v. spectra of the isomeric heterocyclic analogs of chalcones (**1** and **2**).[62,63] The 2-selenophene-yl moiety was

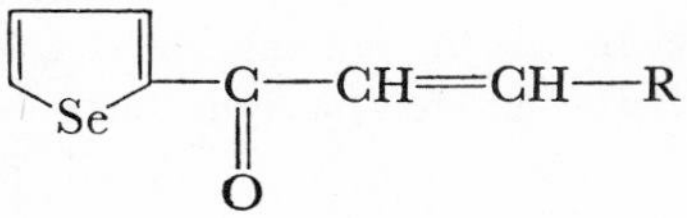

found to have a stronger bathochromic and electron-donating effect in the systems studied than the 2-furyl and 2-thienyl moieties in similarly structured compounds. That five-membered heterocycles are electron donors in the ground state is confirmed by comparing the electric moments of chalcones of the above structure. The electric moment was found to be a function of carbonyl group polarization, which increases with the electron-withdrawing capacity of the substituents.[31,32]

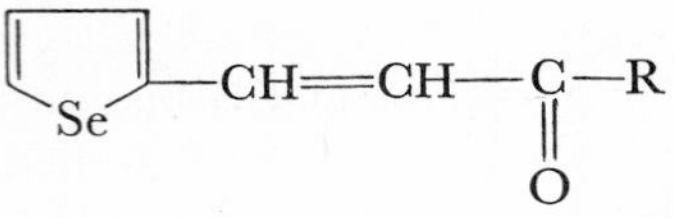

Protolytic equilibrium constants were measured[64,65] spectrophotometrically for the isomeric chalcones (**1** and **2**) dissolved in sulfuric acid (monohydrate) mixed with glacial acetic acid. In the heterocyclic analogs of chalcones the five-membered heterocycles display an electron-withdrawing effect that decreases in the series: 2-furyl > 2-selenophene-yl > 2-thienyl.

Further investigations have provided insight into the physicochemical properties of condensed selenophenes.[19,66-78]

3. The Synthesis of Selenophenes

a. Simple Selenophenes

i. From Acetylenes. As mentioned earlier, heating acetylene with selenium gives selenophene (equation 2).[4-10] Perveev and his co-workers[79] could obtain alkyl-, vinyl-, and hydroxyalkylselenophenes by the reaction of hydrogen selenide with acetylenic, vinylacetylenic, or hydroxyacetylenic epoxides (equation 3).

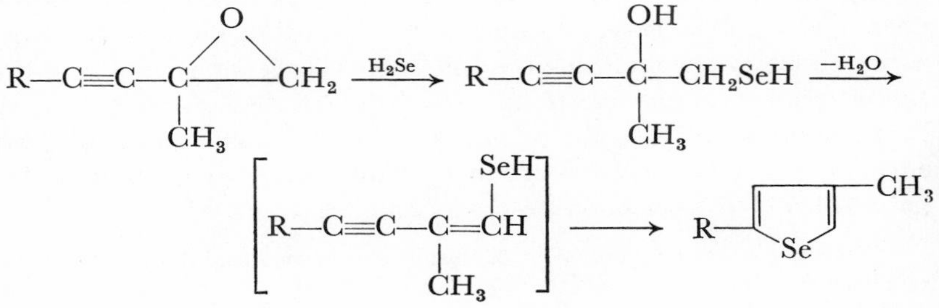

2,5-Disubstituted selenophenes were obtained by the addition of hydrogen selenide to a diyne system, catalyzed by silver or cuprous ions (equation 4).[80] Another method of synthesis of 2,5-disubstituted selenophenes is based

(4) 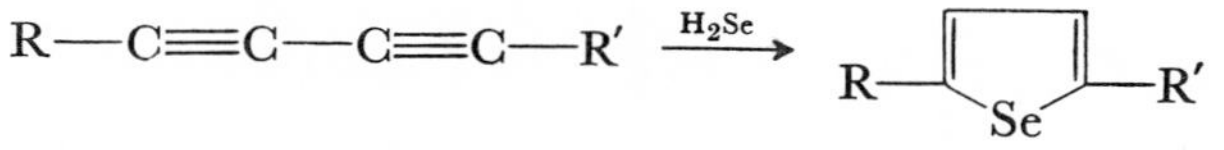 $\quad$ R—C≡C—C≡C—R' $\xrightarrow{H_2Se}$ R—[selenophene ring]—R'

on the condensation of ketoacetylenic esters with hydrogen selenide (equation 5).[81]

(5)

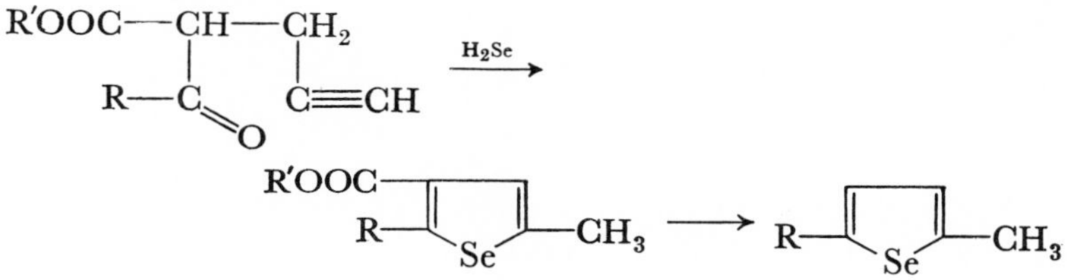

$$R'OOC—CH—CH_2$$
$$R—C \underset{O}{\diagdown} \quad C≡CH \xrightarrow{H_2Se}$$

R—[selenophene ring]—CH₃ → R—[selenophene ring]—CH₃

ii. *From Olefins.* Yur'ev and Khmel'nitskii[82] developed a synthesis of selenophene and its homologs from the reaction of paraffins, olefins, or conjugated dienes with selenium dioxide in the presence of chromic oxide on alumina at 450–500°. Arbuzov and Kataev[83,84] obtained selenophene and its methyl homologs from conjugated dienes and elemental selenium at 380–420°. Yur'ev and Magdesieva[85] found a very convenient procedure to obtain selenophene and its homologs in the reaction of butylenes with elemental selenium at 580–600°.

Tetrachloroselenophene is formed from the reaction of selenium with an equimolar amount of hexachlorobutadiene at 250° (equation 6).[86]

(6) 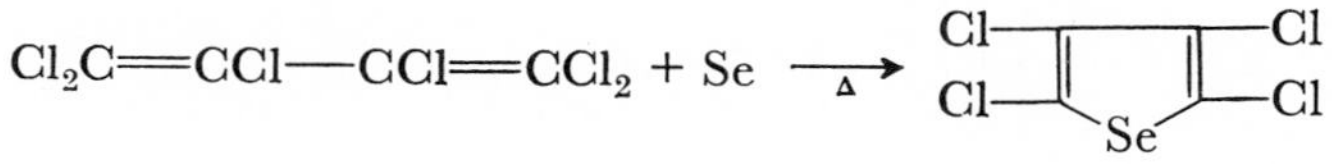 $\quad$ Cl_2C=CCl—CCl=CCl_2 + Se $\xrightarrow{\Delta}$ [tetrachloroselenophene ring]

Monoaryl selenophenes (**3**) can be prepared by interaction of a styrene derivative (**4**) with elemental selenium (equation 7).[87] Fusion of tetra-

(7)

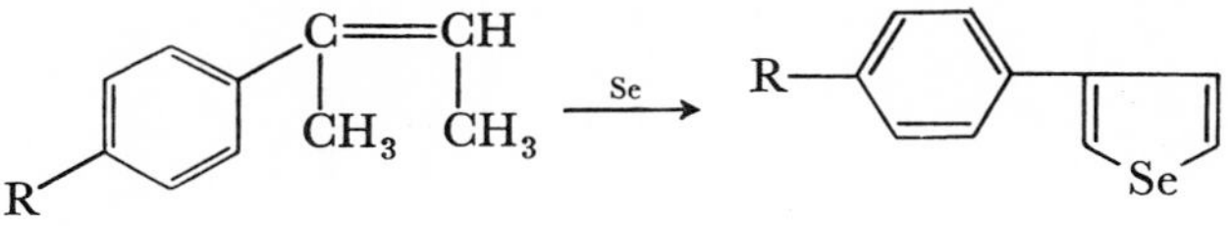

[structure 4] $\xrightarrow{Se}$ [structure 3]

4 $\qquad\qquad$ **3**

phenylcyclopentadienone with selenium gives tetraphenylselenophene (equation 8).[88]

(8)

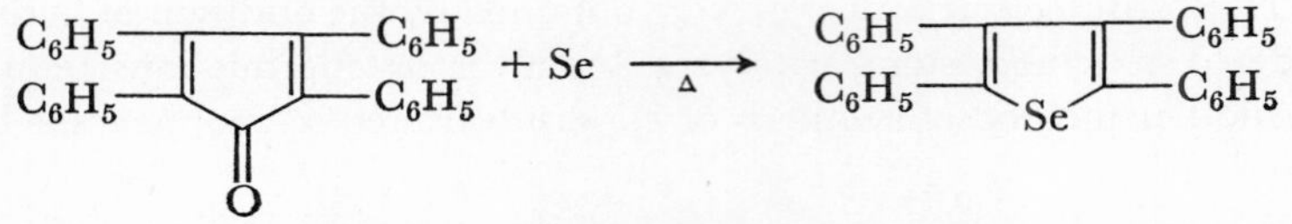

$$C_6H_5 \diagdown \diagup C_6H_5 \diagup C_6H_5 + Se \xrightarrow{\Delta} C_6H_5 \diagdown \diagup C_6H_5$$

iii. From Benzil and Selenodiacetic Ester. 3,4-Diaryl selenophenes are made by condensing selenodiacetic ester (**5**) with benzil (equation 9), followed by hydrolysis and decarboxylation of the products.[89]

(9)

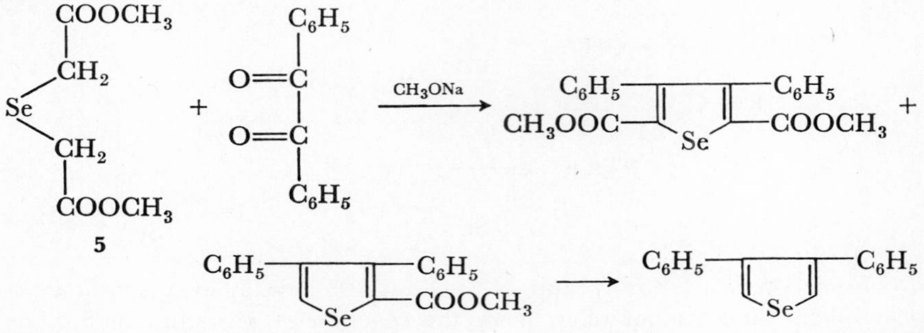

iv. Ring Closure of 1,4-Difunctional Butadienes. Tetraphenylselenophene (**6**) was obtained from 1,4-dilithiotetraphenylbutadiene (**7**) and Se$_2$Br$_2$, as well as from 1,4-diiodotetraphenylbutadiene (**8**) and lithium selenide (scheme 10).[90]

(10)

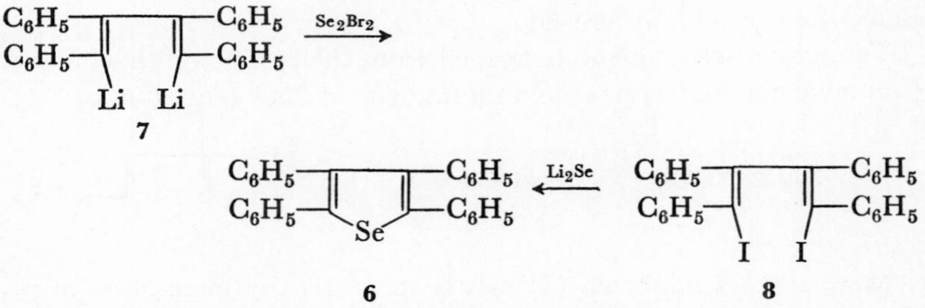

v. From Furan. Yur'ev[91] showed that furan can be converted to seleno-phene with hydrogen selenide in the presence of magnesium oxide at 450° (equation 11).

(11)

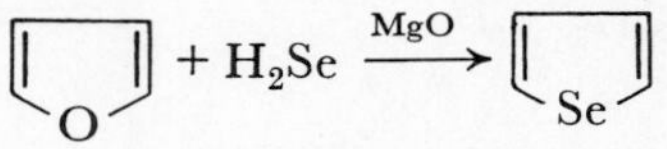

$$\text{furan} + H_2Se \xrightarrow{MgO} \text{selenophene}$$

vi. From Anils. 2,4-Diaryl selenophenes are made by fusing anils of aliphatic aromatic ketones with elemental selenium at 240° (equation 12).[2,92,93]

(12)

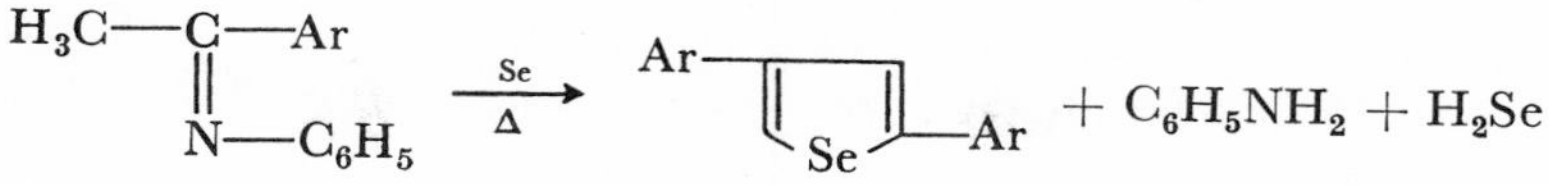

$$\text{(structure: } H_3C-\underset{\underset{C_6H_5}{\|}}{C}-Ar \xrightarrow[\Delta]{Se} \text{ Ar-[selenophene]-Ar} + C_6H_5NH_2 + H_2Se\text{)}$$

vii. Transformation of Functional Groups. Yur'ev *et al.*[94] obtained dimethyl-, trimethyl-, and tetramethylselenophene by the Wolff-Kishner reduction of the appropriate formyl derivatives of selenophene. 2-Vinyl-selenophene (**9**) has been synthesized in three ways: by decarboxylation of β-(selenophene-2-yl)acrylic acid,[95,96] by the Wittig reaction,[97] and by dehydration of methyl(selenophene-2-yl)carbinol (equation 13).[95] It was

(13)

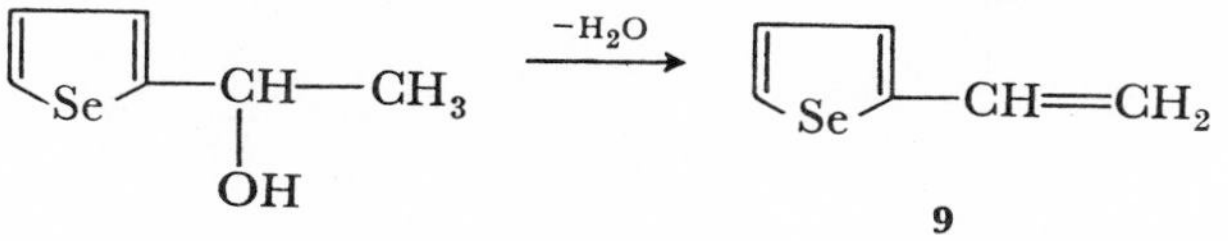

$$\text{[selenophene]}-\underset{\underset{OH}{|}}{C}H-CH_3 \xrightarrow{-H_2O} \text{[selenophene]}-CH{=}CH_2$$

9

found that the most convenient method is a vapor-phase dehydration of methyl(selenophene-2-yl)carbinol in the presence of alumina at 200°.

2-Cyclopropyl- and 2-(2-methylcyclopropyl)selenophene were obtained by the interaction of the respective carbonyl compounds with hydrazine hydrate, followed by the Kishner decomposition of the pyrazoline.[96]

b. Deuteriated Selenophenes. Yur'ev *et al.*[98] prepared deuteriated seleno-phenes from halogenated selenophenes by halogen-deuterium replacement (equation 14). Thus, 2-deuterio-, 2,5-dideuterio-, and tetradeuterioseleno-

(14)

$$\text{[selenophene]}-I \xrightarrow{Zn;\ (CH_3CO)_2O;\ D_2O} \text{[selenophene]}-D$$

phene, and 3-methyl- and 5-methyl-2-deuterioselenophene, were obtained. 3-Deuterioselenophene was obtained from selenophene-3-yllithium, which was then hydrolyzed with deuteriated acetic acid at −70° (equation 15).

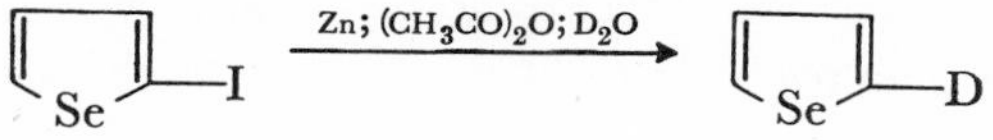

(15)

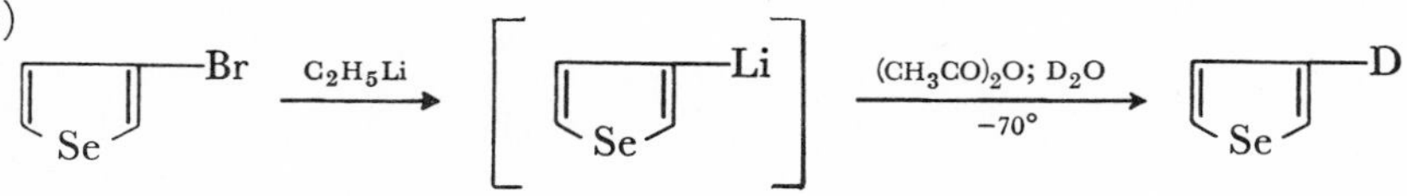

$$\text{[selenophene]}-Br \xrightarrow{C_2H_5Li} \left[\text{[selenophene]}-Li\right] \xrightarrow[-70°]{(CH_3CO)_2O;\ D_2O} \text{[selenophene]}-D$$

The lithium derivative was obtained from 3-bromoselenophene and ethyl-lithium at −70°.

c. Condensed Selenophenes. If acetylene is passed over elemental selenium at 250–300°, benzo[b]selenophene (**10**) is formed in addition to selenophene.[4]

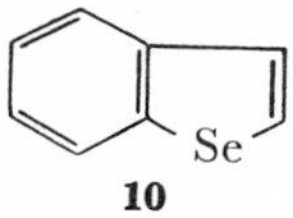

10

Careful investigation of this reaction showed that the high-boiling fraction contained **10** and the three isomeric selenophthenes (**11a, 11b, 11c**).[99,100]

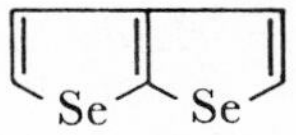

11a 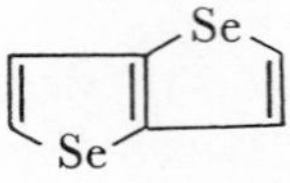   **11b** 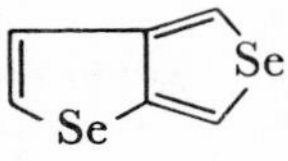   **11c**

Recently it was determined that phenylacetylene reacts with selenium tetrabromide to give a mixture of 3-bromo- (**12**) and 2,3-dibromobenzo[b]-selenophene (**13**).[158a]

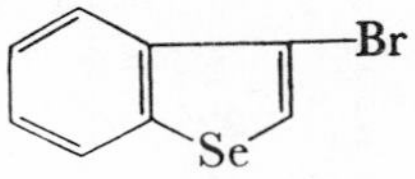

12 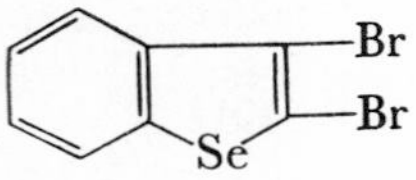   **13**

Benzo[b]selenophene is also formed by passing styrene and selenium dioxide over a Cr_2O_3-Al_2O_3 catalyst[101] or by dehydrogenation of o-ethyl-phenylselenol;[102] however, yields are low for these pyrolytic processes. Preparative methods for the synthesis of benzoselenophene are generally based on cyclization reactions of o-substituted benzenes such as o-selenolcinnamic acid.[99] An important route involves the reduction of 3-keto-2,3-dihydro-benzoselenophene (**14**), which is derived from 2-(o-carboxyphenylseleno)-acetic acid (scheme 16).[103–111] (It should be pointed out that **14** exists in the keto[107–109] rather than the hydroxy form, as had been assumed earlier.[103–106])

(16)

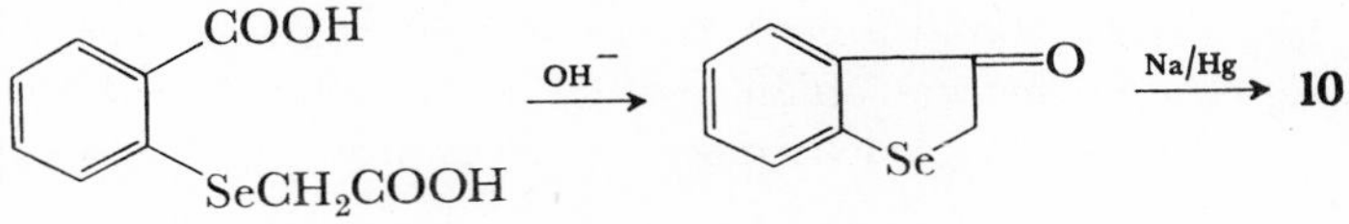

14

It has also been possible to utilize a Diekmann-type condensation[113a] as well as crotonic-type condensations of aromatic aldehydes and ketones[111–113] for the preparation of condensed selenophenes. The latter method was used in the synthesis of thieno[2,3-b]-[114] and thieno[3,2-b]selenophene (scheme 17).[115] Another method is based on the cyclization of o-selenophene-yl-substituted acetophenones.[112–117]

(17)

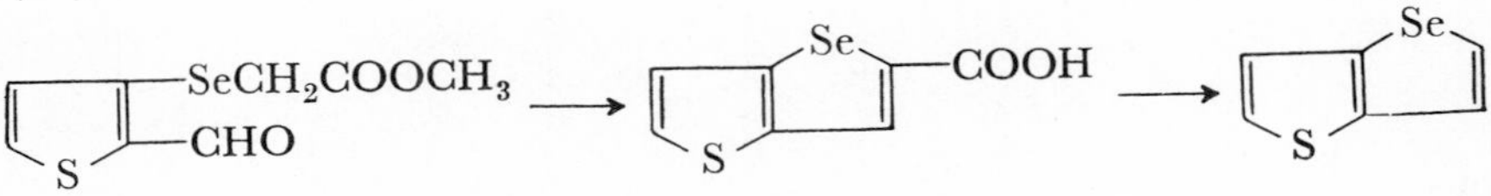

It is interesting that **14** reacts readily with aryl hydrazines (Fischer reaction), giving selenopheneindoles (equation 18).[118,119] Compound **14** also

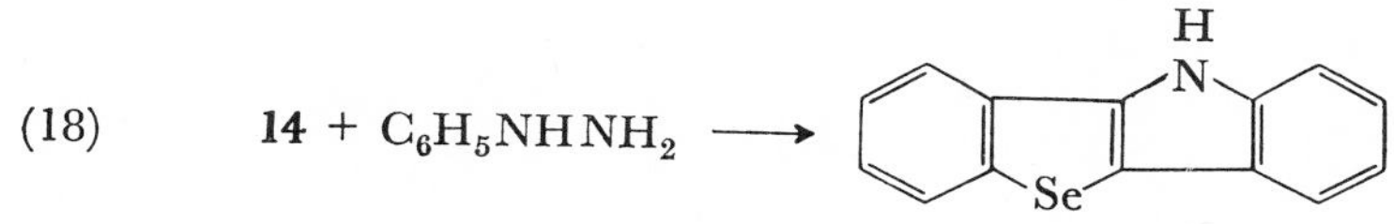

(18) **14** + $C_6H_5NHNH_2$ ⟶

undergoes condensation with 1-naphthylamine and paraformaldehyde to give benzo[h][1]benzoselenopheno[2,1-b]quinoline (**15**) in poor yield (equation 19).[118]

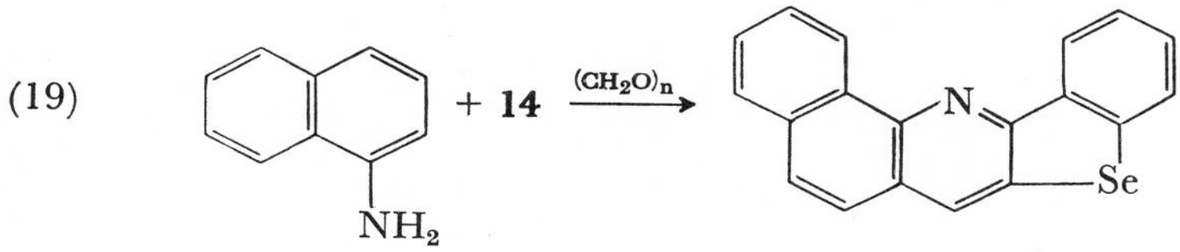

(19)

15

One of the well-known methods for the synthesis of benzothiophenes consists of the intramolecular alkylation or acylation of a benzene ring to form a thiophene nucleus.[120] This technique has also been used for the synthesis of benzoselenophene derivatives (equation 20).[111]

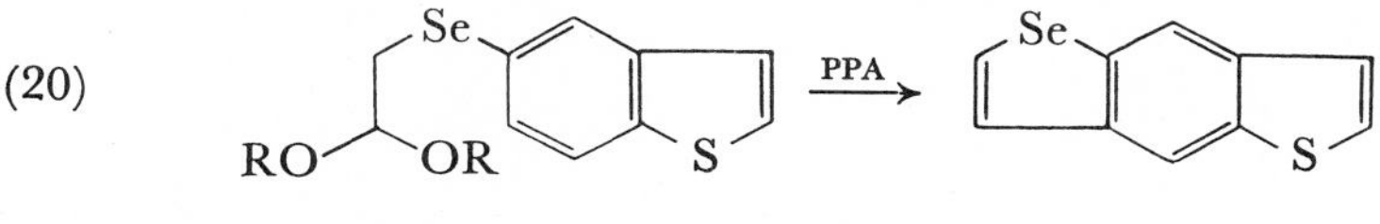

(20)

Somewhat related compounds have been synthesized by the condensation of thioglycolic acid with 2-formyl-3-bromobenzo[b]selenophene (equation 21).[120a] Benzo[b]selenophenes that are substituted in the 2-position by

(21)

+ $HSCH_2COOH$ ⟶

$-(CH_2)_3COOH$ and $-S(CH_2)_2COOH$ moieties have been cyclized by the use of stannic chloride in carbon disulfide to give **15a**[120b] and **15b**,[120c] respectively. The condensation to 7-membered rings was accomplished through the cyclization of side chains possessing an additional methylene group. Kirsch

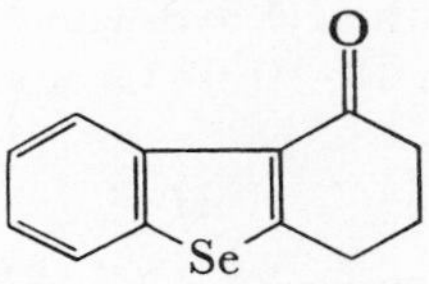

15a 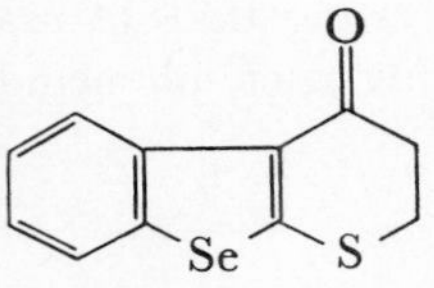   **15b**

and   Gagniant[120d]   prepared   [1]benzoselenopheno[2,3-b]benzothiophene

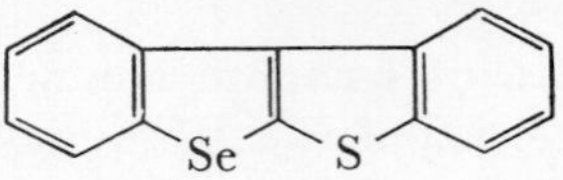

15c

(**15c**) by the cyclodehydration of 2-(2-oxocyclohexylthio)benzo[b]seleno-
phene.

Quinones of the benzoselenophene series were synthesized by the selenium
dioxide oxidation of carvone (equation 22).[121] An extensive series of benzo-

(22)

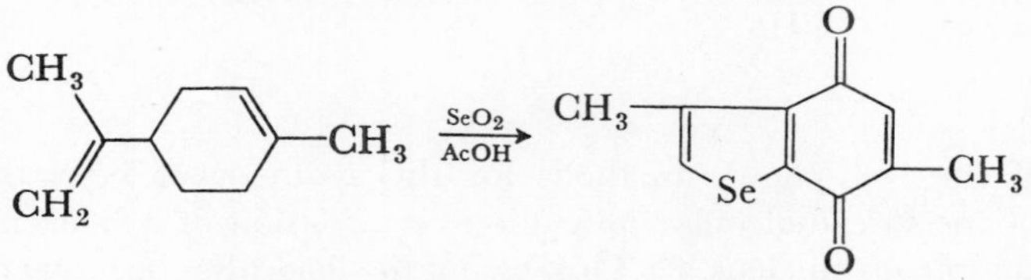

[2,3-b]benzoselenophenes was reported by Patai et al.[122–124] to result from
the reaction of selenium dioxide or selenium oxychloride with 1,1-diaryl
ethylenes (equation 23). In 1971 these workers reported that 1,4-di-α-

(23)

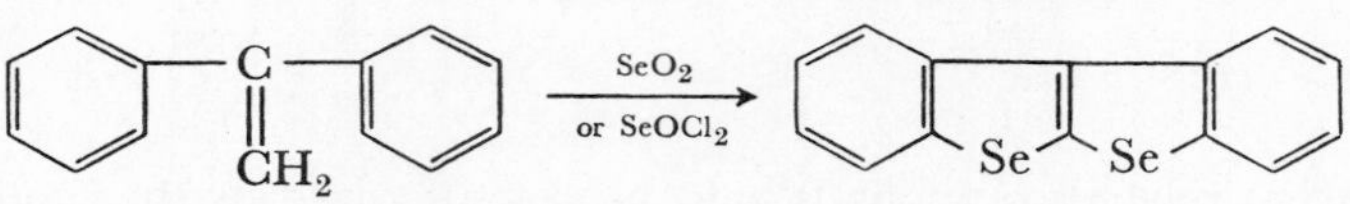

styrylbenzene reacts with selenium oxychloride to give a compound whose
structure is either **15d** or **15e**.[124a] The synthesis of 2,2′-diselenophene-yl and

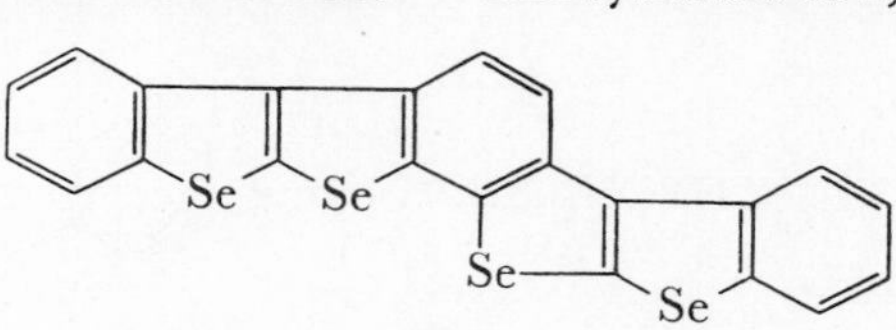

15d

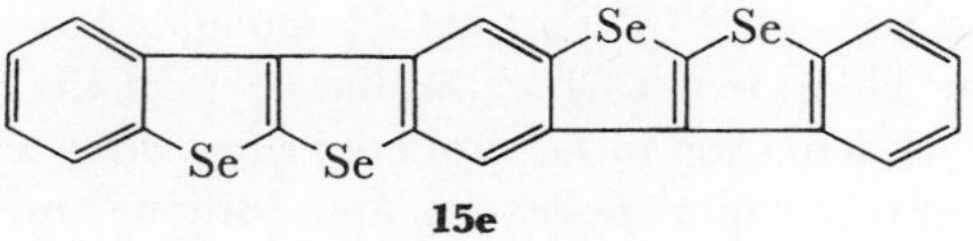

15e

its derivatives can be accomplished by the Ullmann reaction (equation 24).[125,126]

(24)

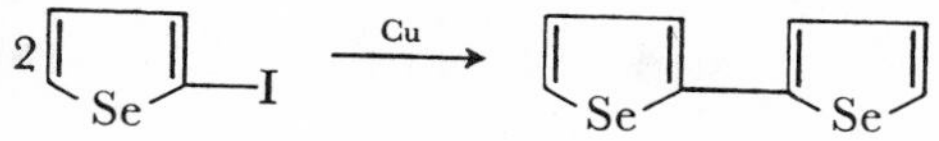

Several methods for the synthesis of dibenzoselenophenes are based on cyclization reactions of diphenyl selenides. For example, cyclization of o-aminodiphenyl selenide *via* a diazonium intermediate (Pschorr synthesis) gave dibenzoselenophene (**16**) in low yield (equation 25).[127] Compound **16**

(25)

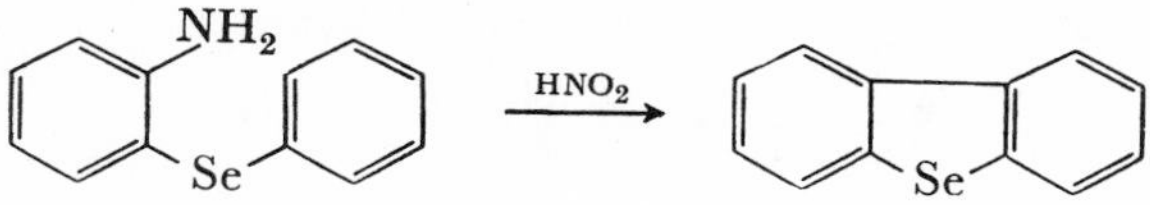

was also obtained by cyclodehydration of diphenyl selenoxide (**17**)[128] or by heating selenanthrene (**18**) with copper bronze (scheme 26).[127] A further

(26)

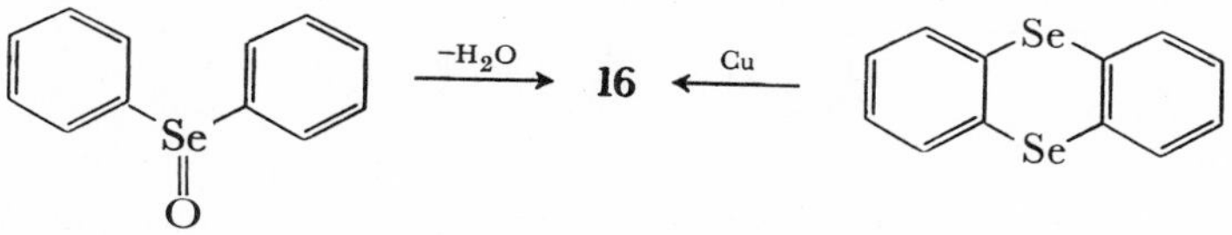

synthesis of **16** is based on the cyclization of biphenyls having –SeX or –SeX$_3$ groups in the o-position. Treatment of o-biphenylselenium trichloride (**19**) with potassium hydroxide gave **16**.[129] It is of interest that heating **19** alone produced a mixture of brominated **16**, *i.e.*, **20** and dibenzoselenophene dibromide (**21**) (equation 27). Compound **21** is readily obtained by the

(27)

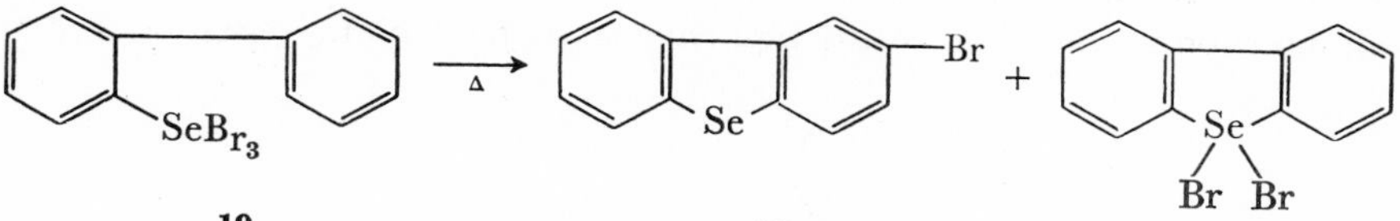

reaction of bromine with o-selenocyanatobiphenyl (**22**) (equation 28).[130] Analogously, cyclization of phenyl o-biphenylyl selenides (as dichlorides)

(28)

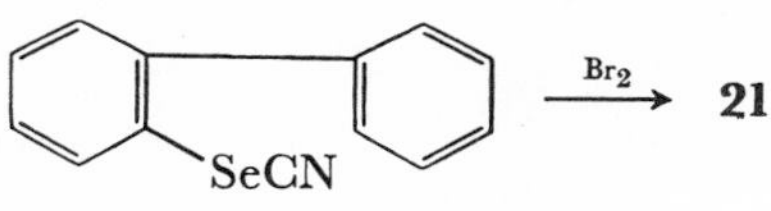

(29)

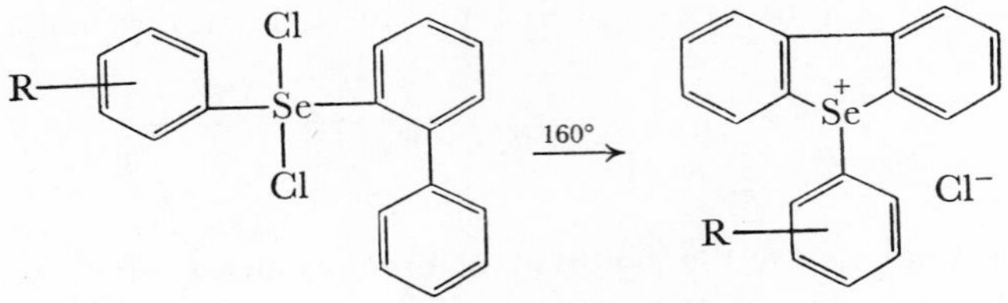

gives 5-phenyldibenzselenonium chlorides (equation 29).[131] The most convenient method for the synthesis of dibenzoselenophenes is based on the bromination of bis(o-biphenylyl) diselenide (equation 30).[132,133]

(30)

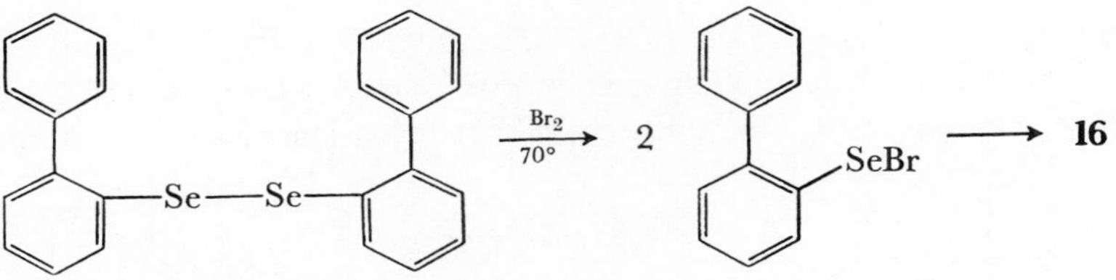

Functional derivatives of dibenzoselenophene are readily obtained by using substitution reactions and by transformation of existing substituents.[134-142] Condensed polycyclic derivatives of selenophene have been made by intramolecular acylation, reduction of keto groups, and dehydrogenation.[143] Octafluorodibenzoselenophene was obtained by heating 2,2'-diiodooctafluorobiphenyl with selenium at 375°.[144,145] It is interesting that heating 2,2'-biphenylenemercury tetramer (**23**) with selenium yields **16**, the tosylimine (**24**) of which gives the selenospirane, **25**, after treatment with o,o'-dilithiumbiphenyl (scheme 31).[146] Dibenzoselenophene was obtained as an insertion product by heating biphenylene with selenium (equation 32).[147]

(31)

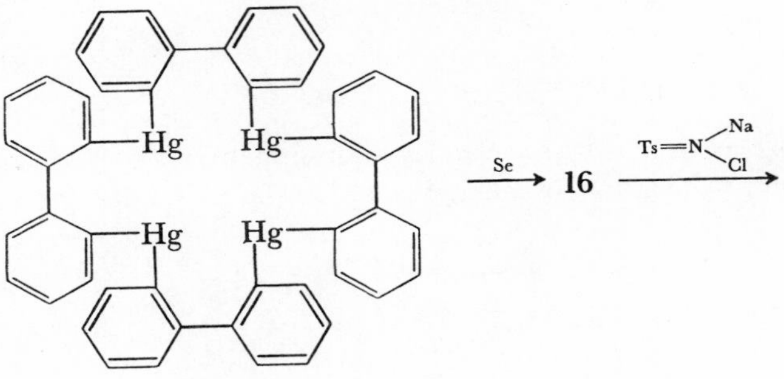

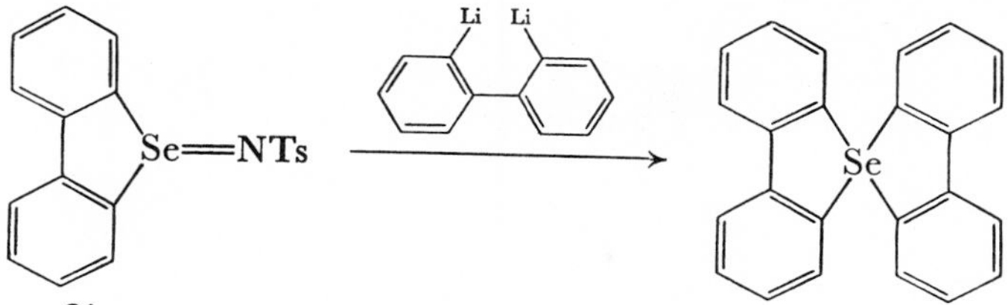

24

25

(32)

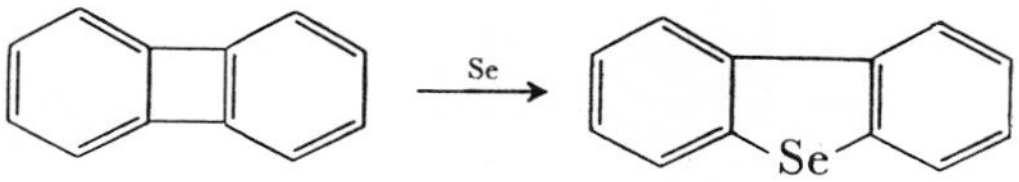

4. Reactions of Selenophenes

a. Electrophilic Substitution.   Electrophilic substitution is easier on selenophene than on benzene. Selenophene is normally substituted first at the
α-position because the transition state, **26a**, seems to be more stable than
26b, which would lead to β-substitution. There are no clear-cut orientation

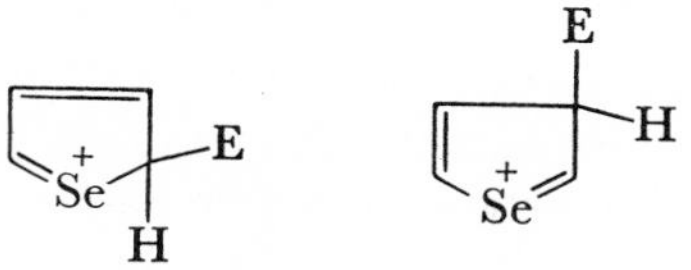

26a   26b

rules in electrophilic substitution at the selenophene nucleus. However,
systematic studies of electrophilic substitution have established that, regardless of what substituent is in the α-position of the nucleus, electrophilic
substitution occurs chiefly at the free α-position.

i. Nitration.   Nitration with fuming nitric acid in acetic anhydride gives
85 % of 2-nitro- and 15 % of 3-nitroselenophene.[148] 2-Nitro- and 3-nitro-
selenophenes obtained by the decarboxylation of 5-nitro- and 4-nitroseleno-
phene-2-carboxylic acids,[149] respectively, were the reference compounds
which proved that Suginome and Umezawa[7] had nitrated selenophene to
obtain a mixture of 2-nitro- (70 %) and 3-nitroselenophene (30 %) rather
than the pure 2-isomer.

2,2′-Diselenophene-yl is nitrated to obtain the 5,5′-dinitro derivative.[125-126]

When selenophene-2-carboxaldehyde is nitrated, the nitro group enters
the 4- and 5-positions.[52,150,150a] When nitrated, 2-acetylselenophene gives
products similar to those obtained from selenophene-2-carboxaldehyde,
that is, 4-nitro-(50 %) and 5-nitro-2-acetylselenophene (8 %), together with
2,4-dinitroselenophene (41 %).[149]

5-Nitro-2-acetylselenophene is best obtained by the reaction of 5-nitro-selenophene-2-carboxylic acid chloride with ethoxymagnesium malonic ester, followed by hydrolysis and decarboxylation of the product (scheme 33).[151]

(33)

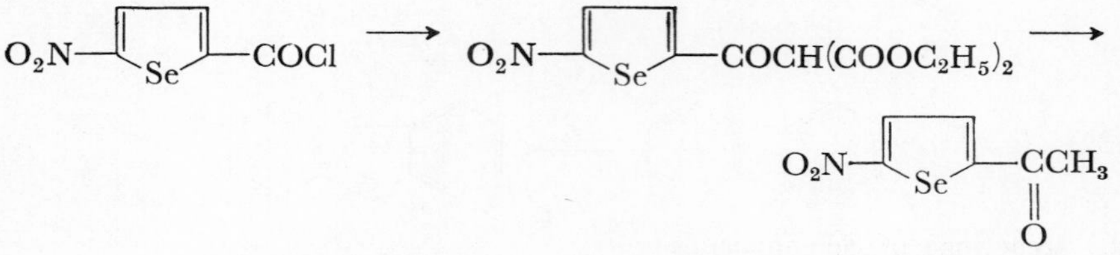

O_2N—⟨Se⟩—COCl $\longrightarrow$ O_2N—⟨Se⟩—COCH(COOC$_2$H$_5$)$_2$ $\longrightarrow$

O_2N—⟨Se⟩—CCH$_3$
‖
O

ii. Sulfonation. Yur'ev and Sadovaya[152] showed that, when selenophene was sulfonated by sulfuric acid by Umezawa,[8] or by pyridine-sulfur tri-oxide,[152,153] the sulfo group entered the 2-position. When selenophene-2-carboxaldehyde or its diacetate was sulfonated with dioxane-sulfur trioxide, 5-sulfoselenophene-2-carboxaldehyde was formed. The 2-carboxylic acid with oleum gives 5-sulfoselenophene-2-carboxylic acid containing an ad-mixture (∼20 %) of the 4-sulfo isomer. The sulfo group is readily replaced by a nitro group through the action of fuming nitric acid.[154]

iii. Halogenation. Bromo and chloro derivatives of selenophene are obtained by direct halogenation. Iodo derivatives are obtained either through exchange of HgX by iodine or through iodination with iodine in the presence of yellow mercuric oxide.

When chlorinated at 50–60°, selenophene gives a mixture of 2-chloro- and 2,5-dichloroselenophene. Excess chlorine results in addition and sub-stitution, leading to hexachloroselenolane. Chlorination in carbon disulfide at —15° produces tetrachloroselenolane.[7] Chlorination with sulfuryl chloride yields only monochloroselenophene.[155]

Bromination in carbon disulfide at —20° converts selenophene into 2-bromoselenophene, while excess bromine leads to 2,5-dibromo- and 2,3,5-tribromoselenophene.[7] Chierici *et al.*[156] synthesized several isomeric halo-genated mononitro- and dinitroselenophenes. Bromination of 2,4-diphenyl-selenophene leads to the formation of 3,5-dibromo-2,4-diphenylselenophene; an excess of bromine leads to the tribromo derivative, while the tetrabromo-derivative is formed in the reaction of 2,4-diphenylselenophene with bro-mine under irradiation without solvent.[93] 3-Bromoselenophene is readily obtained from 2,3,5-tribromoselenophene by debromination with zinc in acetic acid.[15]

Iodine in the presence of yellow mercuric oxide converts selenophene and

3-methyl-, 2,4-dimethyl-, and 3,4-dimethylselenophene into the respective α-monoiodo derivatives.[158]

Selenophene-2-carboxaldehyde and its diacetate are halogenated at the free α-position of the selenophene ring.[155]

The bromination of benzo[b]selenophene (**10**) gave a complex mixture of products, from which were isolated the bromo derivatives **27a**, **27b**, **27c**, and **27d**.[158a]

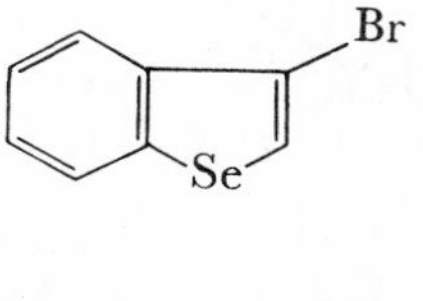

27a 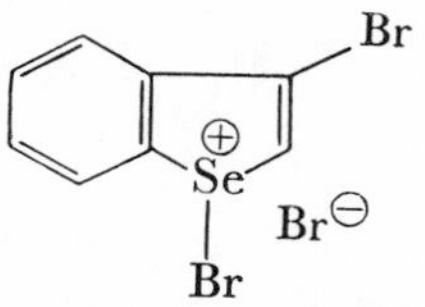   **27b**

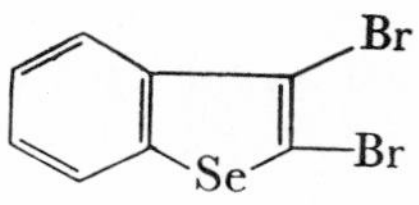

27c 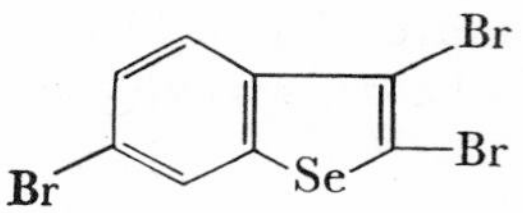   **27d**

iv. Mercuration. The selenophene nucleus is readily mercurated by mercuric salts, the HgX group entering the 2-position.[10,92,98,159,160] 2,5-(Diacetoxymercury)selenophene is obtained in the reaction of selenophene with mercuric acetate in dilute acetic acid.[161] In mercurated selenophene derivatives the group HgX is readily exchanged for various halogens[98,161] and for the cyano group,[92] but in the presence of sodium iodide in acetone coupling occurs (equation 34).[92]

(34)

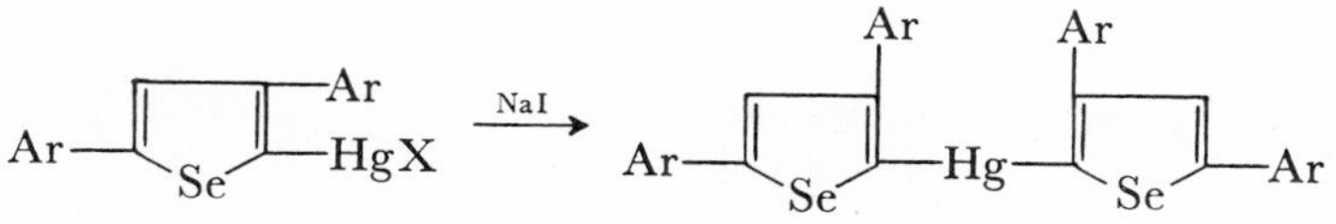

v. Chloro-, Amino-, and Acylaminomethylation. Chloromethylation of selenophene gives 2-(chloromethyl)selenophene, together with small amounts of the 2,5-bis(chloromethyl) derivative (equation 35).[162] Methylselenophenes

(35)

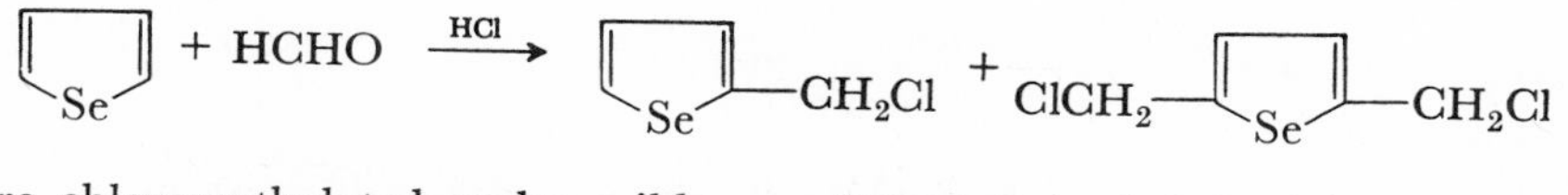

are chloromethylated under milder conditions, 3-methylselenophene being chloromethylated more readily than the 2-methyl compound.[162]

Acyl selenophenes are chloromethylated at the free α-position. A second chloromethyl group can be introduced *ortho* to the first and *meta* to the acyl group.[163] 5-Methyl-2-acetylselenophene is chloromethylated at the 4-position.

Selenophene and its homologs are dialkylaminomethylated by dimethylamine hydrochloride and paraformaldehyde in anhydrous alcohol (Mannich reaction) (equation 36).[164]

(36)

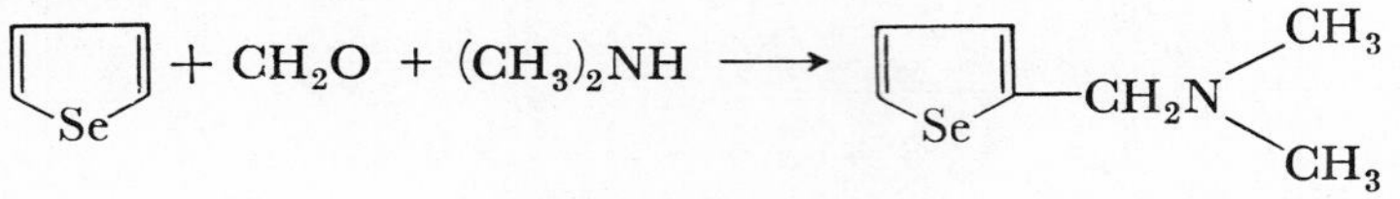

vi. Acylation. The selenophene ring can be acylated under Friedel-Crafts conditions.[8,92,165,166] Organic silicoanhydrides (tetraacyloxysilanes) may be used in the presence of stannic chloride.[167–169] When acylated by silico-anhydrides of dibasic organic acids in the presence of anhydrous stannic chloride, selenophene is converted into the respective keto acid (equation 37).[170]

(37)

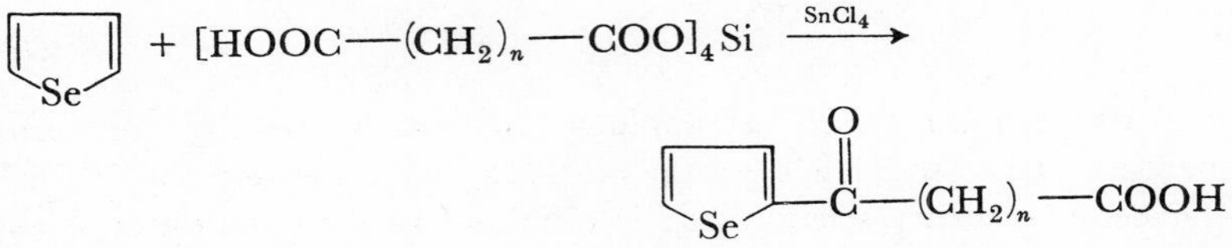

No diacylated selenophene can be obtained by direct acylation; however, 2,5-diacyl-3,4-hydroxyselenophenes were obtained from 1,3,4,6-tetraketones and selenium dioxide in dioxane (equation 38).[171]

(38)

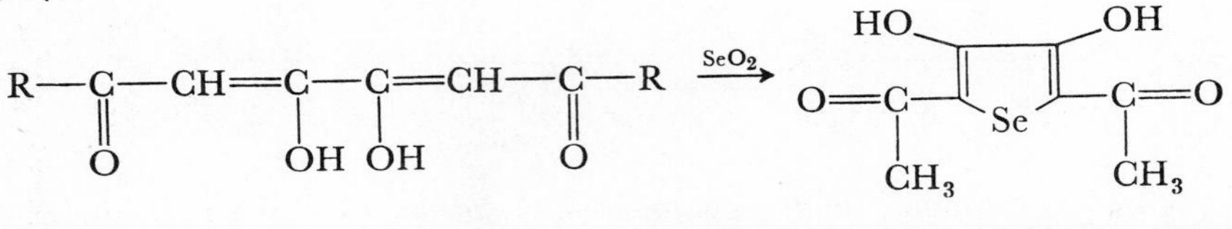

3-Acetylselenophene, which cannot be made by the Friedel-Crafts method, is synthesized from the corresponding acid chloride (equation 39).[157] Acylation of 2,2'-diselenophene-yl gives the 5,5'-diacetyl derivative.[125]

(39)

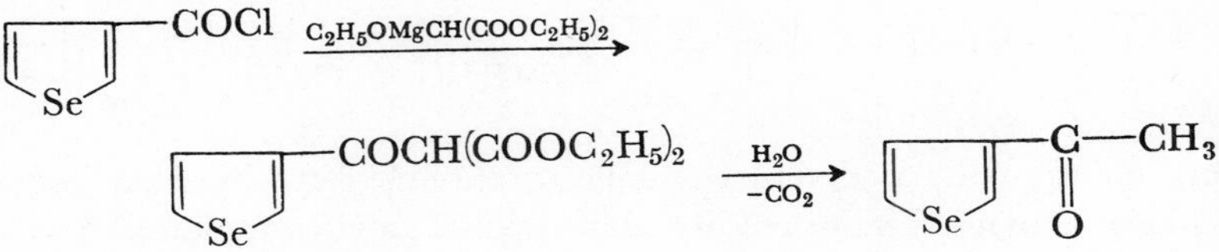

It has been found that acetylation and benzoylation of benzo[b]seleno-phene (**10**) by the Friedel-Crafts method gives 85 % of the 2-acylated and 15 % of the 3-acylated isomer.[100a] Thus, **10** behaves more like benzofuran than benzothiophene.

vii. Formylation. Selenophene is easily formylated by DMF in the presence of phosphorus oxychloride, the formyl group entering the α-position.[172] Methylselenophenes are formylated under milder conditions than unsubstituted selenophene.[173] 2-Methylselenophene gives the 5-formyl-[173] and 3-methylselenophene gives the 2-formyl derivative.[94]

5-Halogenated selenophene-2-carboxaldehydes can be obtained by the reaction of *N*-halosuccinimides with selenophene-2-carboxaldehydes[155] or by formylation of the 2-haloselenophenes.[50,155]

Selenophene-3-carboxaldehyde has been obtained by the Sommelet reaction of 3-bromomethylselenophene, by the reaction of DMF with 3-lithioselenophene[174] and by the reduction of selenophene-3-carbonitrile.[55]

2,5-Dimethyl-3-formylselenophene was obtained by the Sommelet reaction of 2,5-dimethyl-3-chloromethylselenophene. 2,5-Diformylselenophene has been prepared in two ways: by the reaction of DMF with 3-lithio-5-formyl-selenophene and by the hydrolysis of the nitrone resulting from the reaction of *p*-nitrosodimethylaniline with the pyridinium salt obtained from 2,5-bis(chloromethyl)selenophene.[175] Dell'Erba[126] found that formylation of 2,2'-diselenophene-yl involves electrophilic attack at the 5- and 5'-positions.

2-Bromobenzo[b]selenophene has been formylated under Friedel-Crafts conditions in the 3-position by the method of *Rieche et al.*[175a] through the use of butyl dichloromethyl ether and $TiCl_4$ in either carbon disulfide or methylene chloride.[100a]

viii. Isotopic Exchange of Deuterioselenophenes with Acids. Isotopic exchange, a type of electrophilic aromatic substitution reaction, has been studied in various selenophenes by *Shatenstein et al.*[176]

b. Other Nuclear Substitution Reactions

i. Metallation. When selenophene or its homologs are treated with butyllithium,[157] metallation occurs at the α-position.[157] Metallation can also be effected by the reaction of 2-iodoselenophene with phenyllithium[158] or magnesium.[177,178] 3-Metallated selenophenes, prepared from the correspond-ing 3-bromo or 3-iodo compounds, have been used to make 3-selenophene-yl carbinols, carboxaldehydes, and carboxylic acids.[157,179]

When treated with butyllithium,[175] the acetal of selenophene-2-car-boxaldehyde is metallated at the free α-position. It has been shown that selenophene is metallated in much the same manner as thiophene.[180,181] It is surprising, however, that treatment of 2,5-dimethyl-3-iodoselenophene gives 3-methyl-3-selena-4-octene-6-yne (**28**) as the main product (equation 40).[182]

(40)

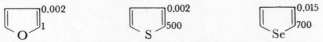

28

ii. Base-Catalyzed Hydrogen Exchange. Deuteriated selenophenes have been subjected to isotopic exchange in DMSO containing lithium butoxide catalyst or potassium *tert*-butoxide. The reaction seems to proceed by a protophilic mechanism[176,183] in which the exchange of deuterium takes place 50,000 times faster at an α- than at a β-position of selenophene.

5. The Reactivity of Selenophene

a. Substitution at the Nucleus. The results from kinetic data on isotopic exchange in deuteriated selenophenes, thiophenes, and furans may be summarized as follows.

1. When treated with lithium *tert*-butoxide in DMSO, selenophene[176,183] exchanges α-deuterium atoms approximately 1.5 times faster than thiophene[184,185] and 700 times faster than furan. The formulas below summarize the hydrogen activities of selenophene, thiophene, and furan in protophilic isotopic exchange.

<table>
<tr><td>0.002</td><td>0.002</td><td>0.015</td></tr>
<tr><td>O · 1</td><td>S · 500</td><td>Se · 700</td></tr>
</table>

2. In an acidic medium, α-deuterium atoms in selenophene[176] are exchanged approximately 6–10 times faster than those in thiophene.[185]

3. A methyl group, which displays electron-donor properties (+I + M), facilitates the acid- and hinders the alkali-catalyzed exchange in deuteriated methylselenophenes.[185] Partial rate factors (*f*) obtained from deuterium exchange in 2-deuteriated 3- and 5-methylselenophenes and thiophenes have been compared with the *f* values of deuterium exchange in *o*-, *m*-, and *p*-deuteriated toluenes.[186,187] The comparison shows that the acid- and base-catalyzed exchange of deuterium is similarly affected by methyl substituents in the selenophene, thiophene, and benzene series. Most interesting is the comparison of the reactivities of selenophene and of other aromatic systems in electrophilic substitutions. It was found that 2-bromo-, 2-acetyl-, 2-carbethoxy-, and 2-nitroselenophenes[160] are mercurated 1.5–3 times faster than

the corresponding α-derivatives of thiophene.[138] Linda and Marino[189] measured the relative reactivities of selenophene and thiophene in five electrophilic substitutions. The maximal value of $K_{selenophene}/K_{thiophene} = 47.5$ was found for bromination in acetic acid; the minimal one is 1.9 for acetylation with acetic anhydride and $SnCl_4$. In all the reactions examined, selenophene exhibits a reactivity intermediate between the values for furan and thiophene. The same order of reactivity was found for protodemercuration of 2-HgX-heterocycles [the relative rates are 2.45 (furan), 2.27 (selenophene), 1 (thiophene)].[190]

Selenophene compounds are also more reactive in nucleophilic substitutions.[191,194] Thus, 3-nitro- and 5-nitro-2-bromoselenophenes[191] react with piperidine faster than do the corresponding thiophenes.[193] The effect of electron-accepting substituents on the nucleophilic substitution of bromine located at the 2-position of 3,5-disubstituted selenophenes has been studied, as also has the applicability of the Hammett equation to this piperidine-debromination.[194] Both selenophenol and thiophenol displace the bromo group in 3-bromo-2-nitroselenophene more readily than in the corresponding thiophene.[194a]

Thus, the kinetic results show that selenophene is more susceptible to electrophilic and nucleophilic substitution reactions than is thiophene.

b. Solvolysis of Chloromethyl Derivatives. On the basis of solvolysis kinetics, the reactivity of chlorine in chloromethylated selenophenes has been quantitatively compared with that in the derivatives of thiophene and furan.[195] The compounds are solvolyzed some 10^4 times faster than benzyl chloride in methanol. The solvolysis proceeds by the unimolecular mechanism shown in equation 41.

The Hammett equation using $\rho+$ parameters correlated the effect of substituents at the 5-position on the solvolysis rate ($\rho = 6.42$).[195,196] The Hammett relation was found applicable for 4- and 5-substituted selenophene-2-carboxylic acids. It was observed that ρ values measured for pK_a have the following order: ρ_{furan} (1.394) $>$ $\rho_{selenophene}$ (1.229) $>$ $\rho_{thiophene}$ (1.076) $>$ $\rho_{benzene}$ (1.00).[61] Application of the Hammett equation was found for nucleophilic substitution in 2-bromo-3-nitro-substituted selenophenes.[194]

c. Other Properties and Reactions. The aromaticity of a five-membered heterocyclic compound may be estimated from its reactivity in the Diels-Alder reaction.[197] Spectrophotometry shows that furan, thiophene, and selenophene resemble benzene in that maleic anhydride forms 1:1 complexes with them. Heating selenophene and maleic anhydride gives selenium and the diene **29**, which yields a further adduct, **30**, with another molecule of maleic anhydride (scheme 42).

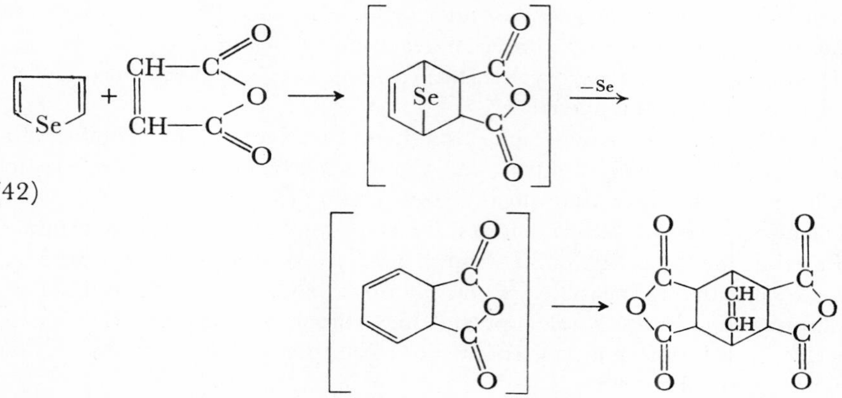

(42)

29   30

Polarographic reduction of 2-nitro compounds showed that 2-nitrofuran was the easiest to reduce. 2-Nitrothiophene and 2-nitroselenophene required an additional 20–30 mV, and nitrobenzene 40 mV more.[198,199]

6. Selenophene Functional Derivatives

a. Chloromethyl Derivatives. A detailed study of selenophene chloromethyl derivatives has revealed that the Cl atom is very reactive and readily enters into many nucleophilic substitutions.[200–205]

b. Carbonyl Derivatives. Aldehydes in the selenophene series behave as typical aromatic aldehydes. They undergo the usual aldehyde reactions, such as the Gantsch,[174,206,207] Darzens,[208] Perkin,[172] Claisen,[208] and crotonic-type[150,155,173,208] condensations, as well as reduction.[209,210]

The reactivity of the acyl group in the ketones of the selenophene series also has been studied in numerous reactions.[174,206–212] Bromination of 2-acetylselenophene with dioxane dibromide and reduction of the bromo-ketone with LiAlH$_4$ in a basic solution gave the epoxide, **31** (scheme 43).[211]

α-Dicarbonyl compounds of the selenophene series are obtained by the oxidation of 2-acetylselenophene with selenium dioxide.[213] Selenophene-2-yl glyoxal,[214] the simplest α-diketone of the selenophene series, was obtained by oxidation of 2-acetylselenophene with selenium dioxide, and by hydrolysis

(43)

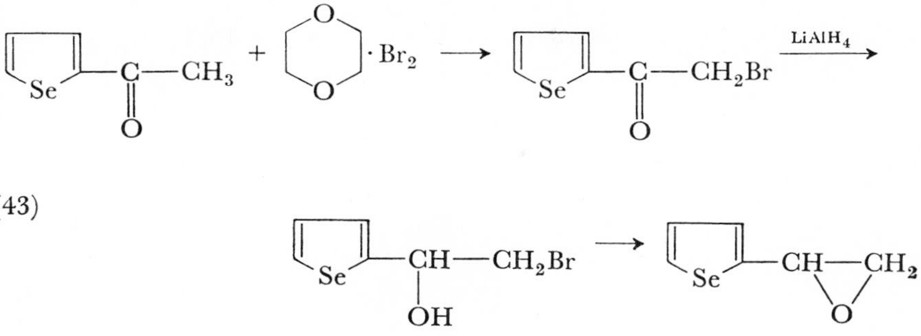

31

of the product resulting from nitrosation of that ketone. Most β-diketones of the selenophene series are obtained by the general Claisen method, the condensation of 2-acetyl- or 3-acetylselenophene with aliphatic, aromatic, and heterocyclic esters in the presence of sodamide.[215-217]

The structures of β-diketones of the selenophene series were found by physical methods[222] (i.r., u.v., n.m.r. spectra) and by chemical means (reaction with hydroxyamine).[55-57,71-73,218-227] β-Diketoselenophenes containing aroyl or heteroaroyl groups exist completely as the *cis*-enols.[222] 2,5-Diketones with hydroxy groups in the 3- and 4-positions were obtained from aliphatic tetraketones by treatment with selenium tetrachloride or selenium dioxide (equation 44).[171,228]

(44)

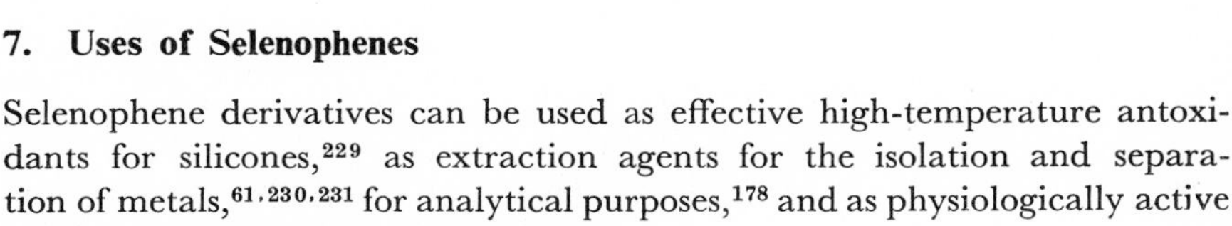

7. Uses of Selenophenes

Selenophene derivatives can be used as effective high-temperature antoxidants for silicones,[229] as extraction agents for the isolation and separation of metals,[61,230,231] for analytical purposes,[178] and as physiologically active compounds (*cf.* Chapter XIIID).[233]

REFERENCES

1. C. Paal, *Ber.* **18,** 2255 (1885).
2. M. T. Bogert and P. P. Herrera, *J. Amer. Chem. Soc.*, **45,** 238 (1923).

3. L. Foa, *Gazz. Chim. Ital.*, **39**, 527 (1909).

4. F. P. Mazza and L. Solazzo, *Rend. Accad. Sci. Napoli*, **33**, [3], 236 (1927); *Chem. Abstr.*, **23**, 2417 (1929).

5. H. Briscoe and J. Peel, *J. Chem. Soc.*, **119**, 1741 (1928).

6. H. Briscoe, J. Peel, and P. Robinson, *J. Chem. Soc.*, **119**, 2628 (1928).

7. H. Suginome and S. Umezawa, *Bull. Chem. Soc. Japan*, **11**, 157 (1936).

8. S. Umezawa, *Bull. Chem. Soc. Japan*, **11**, 775 (1936).

9. S. Umezawa, *Bull. Chem. Soc. Japan*, **12**, 4 (1937).

10. S. Umezawa, *Bull. Chem. Soc. Japan*, **14**, 155 (1939).

11. H. D. Harthough, "Thiophene and Its Derivatives," Interscience Publishers, New York, N.Y., 1952, p. 479.

11a. Yu. K. Yur'ev, N. N. Mezentsova, and N. K. Sadovaya, *Vestn. Mosk. Univ.*, No. 4, 201 (1957).

12. N. N. Magdesieva, "Advances in Heterocyclic Chemistry," Vol. 12, A. R. Katritzky and A. J. Boulton, Eds., Academic Press, New York, N.Y., 1970, p. 1.

13. N. M. Pozdeev, O. B. Akulinin, A. A. Shapkin, and N. N. Magdesieva, *Dokl. Akad. Nauk SSSR*, **185**, 384 (1969).

14. G. A. Shchembelov, Moscow State University, private communication.

15. H. Gerding, G. Milazzo, and H. H. K. Rossmark, *Rec. Trav. Chim. Pays-Bas*, **72**, 957 (1953).

16. M. L. Heffernan and A. A. Humffray, *Mol. Phys.* **7**, 527 (1964).

17. M. Nardelli, G. Fava, and G. Giraldi, *Acta Cryst.*, **15**, 737 (1962).

18. A. Trombetti and C. Zauli, *J. Chem. Soc., A*, 1106 (1967).

19. J. Fabian, A. Mehlhorn, and R. Zagradnic, *Theoret. Chim. Acta*, **12**, 247 (1968).

20. H. Robles, *Rec. Trav. Chim. Pays-Bas*, **58**, 111 (1939).

21. B. Tamamusi, H. Akiyama, and S. Umezawa, *Bull. Chem. Soc. Japan*, **14**, 310 (1939).

22. J. Kraitchman, *Amer. J. Phys.*, **21**, 17 (1953).

23. C. Costain, *J. Chem. Phys.*, **29**, 864 (1958).

24. B. Bak, D. Christensen, L. Hansen-Nygaard, and J. Rastrup-Andersen, *J. Mol. Spectry.*, **7**, 58 (1961).

25. B. Bak, D. Christensen, W. B. Dixon, L. Hansen-Nygaard, J. Rastrup-Andersen, and M. Schottlander, *J. Mol. Spectry.*, **9**, 124 (1962).

26. R. V. Kendall and R. A. Olofson, *J. Org. Chem.*, **35**, 806 (1970).

27. R. D. Brown, F. R. Burden, and P. D. Godfrey, *J. Mol. Spectry.*, **25**, 415 (1968).

28. M. Nardelli and G. Fava, *Gazz. Chim. Ital.*, **88**, 229 (1958).

29. N. Mario, F. Giovanna, and G. Giulia, *Acta Cryst.* **15**, 737 (1962).

30. S. V. Tsukerman, V. D. Orlov, and V. F. Lavrushin, *Zh. Strukt. Khim.*, **10**, 263 (1969).

31. S. V. Tsukerman, V. D. Orlov, L. N. Tkhien, and V. F. Lavrushin, *Khim. Geterotsikl. Soedin*, 974 (1969).

32. S. V. Tsukerman, V. D. Orlov, Yu. N. Surov, and V. F. Lavrushin, *Zh. Strukt. Khim.*, **9**, 67 (1968).

33. T. Isole, *Bull. Chem. Res. Inst. Non-aqueous Solutions, Tohoku Univ.*, **9**, 115 (1960).

34. R. Hoffman and S. Gronowitz, *Ark. Kemi*, **15**, 45 (1959).

35. N. N. Magdesieva, Doctoral Dissertation, Moscow State University, 1967.

36. J. M. Read Jr., C. T. Mathis, and J. H. Goldstein, *Spectrochim. Acta*, **21,** 85 (1965).

37. S. V. Tsukerman, A. I. Yatsenko, V. D. Orlov, and V. F. Lavrushin, *Teor. Eksp. Khim.*, in press.

38. N. A. Chumayevskii, V. M. Tatevskii, and Yu. K. Yur'ev, *Opt. Spektrosk.*, **6,** 47 (1959).

39. Yu. K. Yur'ev, N. N. Magdesieva, and Ye. G. Treshchova, *Vest. Mosk. Univ.*, No. 1, 60 (1962).

40. Ye. G. Treshchova, D. Eckhardt, and Yu. K. Yur'ev, *Zh. Fiz. Khim.*, **38,** 295 (1964).

41. Ye. G. Treshchova, R. Ya. Levina, Yu. S. Shavarov, and K. S. Shanazarov, *Vestn. Mosk. Univ.*, No. 5, 145 (1957).

42. F. Ya. Perveev and D. N. Glebovskii, *Vestn. Leningrad Univ.*, No. 22, 135 (1956).

43. V. T. Alexanyan, Ya. M. Kimel'fel'd, N. N. Magdesieva, and Yu. K. Yur'ev, *Opt. Spectrosk.*, **22,** 216 (1967).

44. V. T. Alexanyan, Ya. M. Kimel'fel'd, and N. N. Magdesieva, *Opt. Spektrosk.*, Coll. Papers, **III**, 168, 178 (1967).

45. V. T. Alexanyan, Ya. M. Kimel'fel'd, and N. N. Magdesieva, *Zh. Strukt. Khim.*, **9,** 633 (1968).

46. Ya. M. Kimel'fel'd, V. T. Alexanyan, N. N. Magdesieva, and Yu. K. Yur'ev, *Zh. Strukt. Khim.*, **7,** 42 (1966).

47. G. Milazzo and E. Meischer, *Gazz. Chim. Ital.*, **83,** 782, 787 (1953).

48. G. Milazzo and L. Paoloni, *Gazz. Chim. Ital.*, **82,** 576 (1952).

49. L. Chierici and G. Pappalardo, *Gazz. Chim. Ital.*, **88,** 453 (1958).

50. L. Chierici and G. Pappalardo, *Gazz. Chim. Ital.*, **89,** 560 (1959).

51. L. Chierici and G. Pappalardo, *Gazz. Chim. Ital.*, **89,** 1900 (1959).

52. A. Belotti and L. Chierici, *Gazz. Chim. Ital.*, **90,** 1125 (1960).

53. L. Chierici and G. Pappalardo, *Gazz. Chim. Ital.*, **90,** 69 (1960).

54. I. I. Grandberg and A. N. Kost, *Zh. Obshch. Khim.*, **32,** 3025 (1962).

55. Yu. K. Yur'ev, N. N. Magdesieva, and A. T. Monakhova, *Khim. Geterotsikl. Soedin.*, 649 (1968).

56. Yu. K. Yur'ev. N. N. Magdesieva, and V. V. Titov, *Zh. Obshch. Khim.*, **34,** 1078 (1964).

57. Yu. K. Yur'ev, N. N. Magdesieva, and V. V. Titov, *Zh. Obshch. Khim.*, **33,** 2158 (1963).

58. I. P. Yefinov, O. D. Lagunova, N. N. Magdesieva, V. V. Titov, Yu. K. Yur'ev, and V. M. Peshkova, *Vestn. Mosk. Univ.*, No. 5, 49 (1963).

59. I. P. Yefimov, V. V. Titov, N. N. Magdesieva, A. T. Monakhova, Yu. K. Yur'ev, and V. M. Peshkova, *Vestn. Mosk. Univ.*, No. 2, 90 (1966).

60. A. N. Zozulya, N. N. Mezentsova, V. M. Peshkova, and Yu. K. Yur'ev, *Zh. Anal. Khim.*, **14,** 17 (1959).

61. D. Spinelli, G. Guanti, and C. Dell'Erba, *Ric. Sci.*, **38,** 1048 (1968).

62. S. V. Tsukerman, V. D. Orlov, Yu. S. Rozum, and V. F. Lavrushin, *Khim. Geterotsikl. Soedin.*, 625 (1969).

63. S. V. Tsukerman, V. D. Orlov, V. F. Lavrushin, and Yu. K. Yur'ev, *Khim. Geterotsikl. Soedin.*, 913 (1966); 67 (1969).

64. S. V. Tsukerman, L. A. Katulya, Yu. N. Surov, and V. F. Lavrushin, *Dokl. Akad. Nauk. SSSR*, **164,** 354 (1965).

65. S. V. Tsukerman, L. A. Katulya, and Yu. N. Surov, *Khim. Geterotsikl. Soedin.*, 36 (1966); 204 (1968).

66. N. P. Buu-Hoi, M. Mangane, M. Renson, and L. Christiaens, *J. Chem. Soc., B*, 971 (1969).

67. J. M. Bonnier and P. Jardon, *C. R. Acad. Sci., Paris, Ser. C.*, **266,** 62 (1968).

68. J. M. Bonnier and P. Jardon, *Bull. Soc. Chim. Fr.*, 4787 (1968).

69. E. Sawicki and H. Johnson, *Microchem. J.*, **8,** 85 (1964).

70. R. Passerini and G. Purrello, *Ann. Chim.* (Rome), **48,** 738 (1958).

71. A. Cerniani and R. Passerini, *Boll. Sci. Fac. Chim. Ind. Bologna*, **12,** 65 (1957); *Chem. Abstr.* **49,** 721 (1965).

72. R. Gerdil and E. A. C. Lucken, *J. Amer. Chem. Soc.*, **87,** 213 (1965).

73. R. Gerdil and E. A. C. Lucken, *J. Amer. Chem. Soc.*, **88,** 733 (1966).

74. E. S. Gould and J. D. McCullough, *J. Amer. Chem. Soc.*, **73,** 3196 (1951).

75. J. D. McCullough and D. Mulvey, *J. Phys. Chem.*, **64,** 264 (1960).

76. A. I. Kiss, *Acta Univ. Szegediensis, Acta Phys. Chem.*, **5,** 45 (1959), *Chem. Abstr.*, **55,** 13037 (1961).

77. A. I. Kiss, *Acta Chim. Acad. Sci. Hung.*, **11,** 57 (1957).

78. M. R. Padhye and J. C. Patel, *J. Sci. Ind. Res.*, **15B,** 171 (1956); *Chem. Abstr.*, **51,** 854 (1957).

79. F. Ya. Perveev, N. I. Kudryashova, and D. N. Glebovskii, *Zh. Obshch. Khim.*, **26,** 3331 (1956).

80. R. F. Curtis, S. N. Hasnain, and J. A. Taylor, *Chem. Commun.*, 365 (1968).

81. K. Schulte, J. Reisch, and D. Bergenthal, *Chem. Ber.*, **101,** 1540 (1968).

82. Yu. K. Yur'ev and L. I. Khmel'nitskii, *Dokl. Akad. Nauk SSSR*, **94,** 265 (1954).

83. B. A. Arbuzov and E. G. Kataev, *Dokl. Akad. Nauk SSSR*, **96,** 983 (1954).

84. B. A. Arbuzov and E. G. Kataev, *Dokl. Akad. Nauk SSSR*, **65,** 493 (1949).

85. Yu. K. Yur'ev and N. N. Magdesieva, *Bull. Inst. Pure Reagents* (Moscow), 6 (1962).

86. W. Mack, *Angew. Chem.* **77,** 260 (1965).

87. Yu. K. Yur'ev, N. N. Mezentsova, T. A. Melent'eva, and Ye. G. Treshchova, *Zh. Obshch. Khim.*, **27,** 2260 (1957).

88. W. Dilthey, German Pat. 628,954 (1936); *Chem. Abstr.*, **30,** 6009 (1936).

89. H. J. Backer and W. Stevens, *Rec. Trav. Chim. Pays-Bas.*, **59,** 423 (1940).

90. E. H. Braye, W. Hübel, and I. Caplier, *J. Amer. Chem. Soc.*, **83,** 4406 (1961).

91. Yu. K. Yur'ev, *Zh. Obshch. Khim.*, **16,** 851 (1946).

92. M. T. Bogert and C. N. Andersen, *J. Amer. Chem. Soc.*, **48,** 223 (1926).

93. P. Demerzeman, N. P. Buu-Hoi, R. Royer, and A. Cheutin, *J. Chem. Soc.*, 2720 (1954).

94. Yu. K. Yur'ev, N. K. Sadovaya, and M. A. Gal'bershtam, *Zh. Obshch. Khim.* **28,** 620, 622 (1958); **29,** 1917 (1959).

95. Yu. K. Yur'ev, N. N. Mezentsova, and V. Ye. Vas'kovskii, *Zh. Obshch. Khim.*, **28,** 3262 (1958).

96. Yu. K. Yur'ev, N. N. Mezentsova, and V. Ye. Vas'kovskii, *Zh. Obshch. Khim.*, **30,** 1628 (1960).

97. Yu. K. Yur'ev, N. N. Mezentsova, and B. I. Keda, *Zh. Obshch. Khim.*, **32,** 1820 (1962).

98. Yu. K. Yur'ev, N. N. Magdesieva, and L. Ya. Petrova, *Khim. Geterotsikl. Soedin.*, 910 (1966).

99. S. Umezawa, *Bull. Chem. Soc. Japan.*, **14,** 363 (1939).

100. B. Tamamusi, H. Akiyama, and S. Umezawa, *Bull. Chem. Soc. Japan*, **14,** 318 (1939).

100a. T. Q. Minh, L. Christiaens, F. Mantovani, P. Faller, and M. Renson, *Bull. Soc. Chim. Fr.*, in press.

101. Yu. K. Yur'ev, N. N. Mezentsova, T. A. Melent'eva, and Ye. G. Treshchova, *Zh. Obshch. Khim.*, **27,** 2260 (1957).

102. C. Hansch and C. F. Geiger, *J. Org. Chem.*, **24,** 1025 (1959).

103. R. Lesser and R. Weiss, *Ber.*, **45,** 1835 (1912).

104. R. Lesser and R. Weiss, *Ber.*, **46,** 2640 (1913).

105. R. Lesser and A. Schoeller, *Ber.*, **47,** 2294 (1914).

106. G. Komppa and G. A. Nyman, *J. Prakt. Chem.*, **139,** 229 (1934).

107. G. Grandolini, A. Ricci, N. P. Buu-Hoi, and F. Perin, *J. Heterocycl. Chem.*, **5,** 133 (1968).

108. A. I. Kiss and B. R. Muth, *Acta Chim. Acad. Sci. Hung.*, **11,** 57 (1957).

109. A. I. Kiss and B. R. Muth, *Magy. Tud. Acad. Kozp. Fiz. Kut. Int. Kozlemen.* **3,** 213 (1955); *Chem. Abstr.*, **53,** 3876 (1959).

110. B. R. Muth and A. I. Kiss, *J. Org. Chem.*, **21,** 576 (1956).

111. M. Vafai and M. Renson, *Bull. Soc. Chim. Belges*, **75,** 145 (1966).

112. L. Christiaens and M. Renson, *Bull. Soc. Chim. Belges*, **77,** 153 (1968).

113. L. Christiaens and M. Renson, *Bull. Soc. Chim. Belges*, **79,** 133 (1970).

113a. T. Q. Minh, L. Christiaens, and M. Renson, *Tetrahedron*, in press.

114. A. Bugge, *Acta Chem. Scand.*, **23,** 1823 (1969).

115. Ya. L. Gol'dfarb, V. P. Litvinov, and S. A. Ozolin, *Izv. Akad. Nauk SSSR, Ser. Khim.*, 1419 (1968).

116. E. Giesbrecht and J. Mori, *Anais. Acad. Brasil Cienc.*, **30,** 521 (1958).

117. J. Gosselck, *Chem. Ber.*, **91,** 2345 (1958).

118. N. P. Buu-Hoi, G. Saint-Ruf, A. Martani, A. Ricci, and D. Balucani, *J. Chem. Soc.*, C, 609 (1968).

119. N. P. Buu-Hoi and Hien-Do-Phuoc, *Biochem. Pharmacol.*, **17,** 1227 (1968).

120. O. Dann and M. Kokorudz, *Chem. Ber.*, **91,** 172 (1958).

120a. T. Q. Minh, L. Christiaens, P. Thibaut, and M. Renson, *Tetrahedron*, in press.

120b. P. Cagniant and G. Kirsch, *C. R. Acad. Sci., Ser. C*, **272,** 1978 (1971).

120c. P. Cagniant, G. Kirsch, and M. Renson, *C. R. Acad. Sci., Ser. C*, **272,** 1369 (1971).

120d. G. Kirsch and P. Cagniant, *C. R. Acad. Sci., Ser. C*, **273,** 902 (1971).

121. J. Schmitt and J. Seilert, *Justus Liebigs Ann. Chem.*, **562,** 15 (1949).

122. S. Patai and K. A. Muszkat, *Bull. Res. Council Israel*, **10A,** 73 (1961).

123. S. Patai, M. Sokolovsky, and A. Friedlander, *Proc. Chem. Soc.*, 181 (1960).

124. S. Patai, K. A. Muszkat, and M. Sokolovsky, *J. Chem. Soc.*, 734 (1962).

124a. S. Patai, D. Elmaleh, and Z. Rappoport, *J. Chem. Soc. C*, 2637 (1971).

125. L. Chierici, C. Dell'Erba, A. Guareschi, and D. Spinelli, *Ric. Sci., A*, **8,** 1537 (1965), *Chem. Abstr.*, **65,** 2203 (1966).

126. C. Dell'Erba, *Corsi Semin. Chim.*, 175 (1968).

127. N. M. Cullinane, A. G. Rees, and C. A. J. Plummer, *J. Chem. Soc.*, 151, (1939).

128. C. Courtot and A. Moytamedi, *C. R. Acad. Sci., Paris*, **199,** 531 (1934).

129. O. Behaghel and K. Hofmann, *Ber.*, **72,** 582 (1939).

130. L. Chierici and R. Passerini, *J. Chem. Soc.*, 3249 (1954).

131. B. Greenberg, E. S. Gould, and W. Burlant, *J. Amer. Chem. Soc.*, **78,** 4028 (1956).

132. J. D. McCullough, T. W. Campbell, and E. S. Gould, *J. Amer. Chem. Soc.*, **72,** 5753 (1950).

133. G. E. Wiseman and E. S. Gould, *J. Amer. Chem. Soc.*, **77,** 1061 (1955).

134. N. Marziano, *Ric. Sci.*, **30,** 743 (1960).

135. E. Sawicki and F. E. Ray, *J. Amer. Chem. Soc.*, **74,** 4120 (1952).

136. E. Sawicki and F. E. Ray, *J. Org. Chem.*, **18,** 946 (1953).

137. E. Sawicki, *J. Org. Chem.*, **19,** 1163 (1954).

138. E. Sawicki, *J. Amer. Chem. Soc.*, **77,** 957 (1955).

139. G. E. Wiseman, *Dissertation Abstr.*, **17,** 244 (1957).

140. G. E. Wiseman and E. S. Gould, *J. Amer. Chem. Soc.*, **77,** 1061 (1955).

141. A. Magelli and R. Passerini, *Boll. Sedute Accad. Gioenia Sci. Nat. Catania*, [4], **3,** 185 (1956); *Chem. Abstr.*, **52,** 10049f (1958).

142. N. P. Buu-Hoi and N. Hoan, *J. Org. Chem.*, **17,** 643 (1952).

143. N. P. Buu-Hoi and N. Hoan, *J. Chem. Soc.*, 3745 (1952).

144. S. C. Cohen, M. L. N. Reddy, and A. G. Massey, *Chem. Commun.*, 451 (1967).

145. S. C. Cohen, M. L. N. Reddy, and A. G. Massey, *J. Organometal. Chem.*, **11,** 563 (1968).

146. D. Hellwinkel and G. Fahrbach, *Justus Liebigs Ann. Chem.*, **715,** 68 (1968).

147. J. Gaidis, *J. Org. Chem.* **35,** 2811 (1970).

148. Yu. K. Yur'ev, Ye. L. Zaitseva, and G. G. Rozantsev, *Zh. Obshch. Khim.*, **30,** 2207 (1960).

149. Yu. K. Yur'ev and Ye. L. Zaitseva, *Zh. Obshch. Khim.*, **30,** 859 (1960).

150. Yu. K. Yur'ev and Ye. L. Zaitseva, *Zh. Obshch. Khim.*, **28,** 2164 (1958).

151. Yu. K. Yur'ev and Ye. L. Zaitseva, *Zh. Obshch. Khim.*, **29,** 3664 (1959).

152. Yu. K. Yur'ev and N. K. Sadovaya, *Zh. Obshch. Khim.*, **34,** 1803 (1964).

153. E. G. Kataev and A. Ye. Zimkin, *Uch. Zap. Kazansk. Gos. Univ.*, **117,** 174 (1957).

154. Yu. K. Yur'ev and N. K. Sadovaya, *Zh. Obshch. Khim.*, **34,** 2190 (1964).

155. Yu. K. Yur'ev, N. N. Magdesieva, and A. T. Monakhova, *Zh. Obshch. Khim.*, **30,** 2726 (1960).

156. L. Chierici, C. Dell'Erba, and D. Spinelli, *Ann. Chim.* (Rome), **55,** 1069 (1965).

157. Yu. K. Yur'ev, N. K. Sadovaya, and Ye. A. Grekova, *Zh. Obshch. Khim.*, **34,** 847 (1964).

158. Yu. K. Yur'ev and N. K. Sadovaya, *Zh. Obshch. Khim.*, **26,** 3154 (1956).

158a. T. Q. Minh, L. Christiaens, and M. Renson, private communication.

159. Yu. K. Yur'ev, N. K. Sadovaya, and M. A. Gal'bershtam, *Zh. Obshch. Khim.*, **29,** 1970 (1959).

160. Yu. K. Yur'ev, M. A. Gal'bershtam, and I. I. Kondror, *Khim. Geterotsikl. Soedin*, 897 (1966).

161. H. Briscoe and J. Peel, *J. Chem. Soc.*, **120,** 2589 (1929).

162. Yu. K. Yur'ev, N. K. Sadovaya, and M. A. Gal'bershtam, *Zh. Obshch. Khim.*, **32,** 259 (1962).

163. Yu. K. Yur'ev, N. K. Sadovaya, and Ye. N. Lubimova, *Zh. Obshch. Khim.*, **30,** 2732 (1960).

164. Yu. K. Yur'ev, N. K. Sadovaya, and A. B. Ibragimova, *Zh. Obshch. Khim.*, **29,** 3647 (1959).

165. N. P. Buu-Hoi, P. Demerzeman, and R. Royer, *C. R. Acad. Sci., Paris,* **237,** 397 (1953).

166. E. G. Kataev and M. V. Palkina, *Uch. Zap. Kazansk. Gos. Univ.*, **113,** 115 (1953).

167. Yu. K. Yur'ev and G. B. Yelyakov, *Dokl. Akad. Nauk SSSR*, **102,** 763 (1955).

168. Yu. K. Yur'ev and N. K. Sadovaya, *Zh. Obshch. Khim.*, **26,** 930 (1956).

169. Yu. K. Yur'ev, N. K. Sadovaya, and V. V. Titov, *Zh. Obshch. Khim.*, **28,** 3036 (1958).

170. Yu. K. Yur'ev, G. B. Yelyakov, and Z. V. Belyakova, *Zh. Obshch. Khim.*, **26,** 2353 (1956).

171. K. Balenovic, D. Cerar, and L. Filipovic, *J. Org. Chem.*, **19,** 1556, (1954).

172. Yu. K. Yur'ev and N. N. Magdesieva, *Zh. Obshch. Khim.*, **27,** 179 (1957).

173. Yu. K. Yur'ev, N. N. Mezentsova, and V. Ye. Vas'kovskii, *Zh. Obshch. Khim.*, **27,** 3155 (1957).

174. C. Paulmier and P. Pastour, *C. R. Acad Sci., Paris, Ser. C,* **265,** 926 (1967).

175. C. Paulmier and P. Pastour, *Bull. Soc. Chim. Fr.*, 4021 (1966).

175a. A. Rieche, H. Gross, and E. Höft, *Chem. Ber.*, **93,** 88 (1960).

176. A. I. Shatenstein, N. N. Magdevieva, I. O. Shapiro, Yu. I. Ranneva, and A. I. Serebryanskaya, *Teor. Eksp. Khim.*, **3,** 343 (1967).

177. Yu. K. Yur'ev and N. K. Sadovaya, *Zh. Obshch. Khim.*, **28,** 2162 (1958).

178. A. N. Nesmeyanov, V. A. Sazonova, and V. N. Drozd, *Izv. Akad. Nauk SSSR, Otd. Khim. Nauk*, 1389 (1957).

179. C. Paulmier, J. Morel, P. Pastour, and D. Semard, *Bull. Soc. Chim. Fr.*, 2511 (1969).

180. S. Gronowitz, *Ark. Kemi*, **7,** 361 (1954).

181. S. Gronowitz and K. Halvarson. *Ark. Kemi*, **8,** 343 (1955).

182. S. Gronowitz and T. Frejd, *Acta Chem. Scand.*, **23,** 2540 (1969).

183. A. I. Shatenstein, I. O. Shapiro, Yu. I. Ranneva, N. N. Magdesieva, and Yu. K. Yur'ev, *Reaktsionnaya Sposobnost. Org. Soed. Tartu*, **9,** 236 (1964).

184. A. I. Shatenstein, I. O. Shapiro, F. S. Yakushin, Y. G. Isaeva, and Yu. I. Ranneva, *Kinet. Katal.*, **5,** 752 (1964).

185. A. I. Shatenstein, A. G. Kamrad, I. O. Shapiro, Yu. I. Ranneva, and Ye. N. Zvyagintseva, *Dokl. Akad. Nauk SSSR*, **168,** 364 (1966).

186. W. Lauer and G. Stedman, *J. Amer. Chem. Soc.*, **80,** 6349 (1958).

187. Ye. N. Yarygina, P. A. Alikhanov, Ye. A. Izraelevich, P. N. Manochkina, and A. I. Shatenstein, *Zh. Fiz. Khim.*, **34,** 587 (1960).

188. R. Matoyama, S. N. Nishimura, E. Iomoto, and J. Murakani, *Nippon Kagaku Zasshi*, **78,** 962 (1957).

189. P. Linda and G. Marino, *J. Chem. Soc., B*, 43 (1970).

190. R. D. Brown, A. S. Buchanan, and A. A. Humffray, *Australian J. Chem.*, **18,** 1513 (1965).

191. L. Chierici, C. Dell'Erba, A. Guareschi, and D. Spinelli, *Ann. Chim.* (Rome), **57**, 632 (1967).

192. D. Spinelli, G. Guanti, and C. Dell'Erba, *Boll. Sci. Fac. Chim. Ind. Bologna*, **25**, 71 (1967).

193. C. Dell'Erba and D. Spinelli, *Tetrahedron*, **21**, 1061 (1965).

194. C. Dell'Erba, A. Guareschi, and D. Spinelli, *J. Heterocycl. Chem.*, **4**, 438 (1967).

194a. G. Guanti, C. Dell'Erba, and G. Garbarino, *J. Heterocycl. Chem.*, **7**, 1425 (1970).

195. Yu. K. Yur'ev, M. A. Gal'bershtam, and A. F. Prokov'eva, *Izv. Vyssh. Uchebn, Zaved.*, *Khim. Teckhnol.*, **7**, 419, 598 (1964).

196. Yu. K. Yur'ev, M. A. Gal'bershtam, and A. F. Prokov'eva, *Izv. Vyssh. Uchebn. Zaved.*, *Khim. Tecknol.*, **8**, 421 (1965).

197. B. A. Arbuzov and B. V. Konovalov, *Izv. Akad. Nauk SSSR*, 2130 (1959).

198. Ya. P. Stradin', S. A. Hiller, and Yu. K. Yur'ev, *Dokl. Akad. Nauk SSSR*, **129**, 816 (1959).

199. Ya. P. Stradin', S. A. Hiller, Yu. K. Yur'ev, and Ye. L. Zaitseva, *Izv. Akad. Nauk Latv. SSR*, No. 3, 85 (1960).

200. Yu. K. Yur'ev, M. A. Gal'bershtam, and N. K. Sadovaya, *Zh. Obshch. Khim.*, **32**, 1301 (1962).

201. Yu. K. Yur'ev, M. A. Gal'bershtam, and G. G. Rozantsev, *Zh. Obshch. Khim.*, **32**, 3562 (1962).

202. Yu. K. Yur'ev, and M. A. Gal'bershtam, *Zh. Obshch. Khim.*, **32**, 3922, (1962).

203. Yu. K. Yur'ev and M. A. Gal'bershtam, *Zh. Obshch. Khim.*, **32**, 3249 (1962).

204. Yu. K. Yur'ev and M. A. Gal'bershtam, *Zh. Obshch. Khim.*, **33**, 462 (1963).

205. K. Yu. Novitskii, H. Groesle, and Yu. K. Yur'ev, *Khim. Geterotsikl. Soedin.*, 829, 832 (1966).

206. Yu. K. Yur'ev, N. N. Mezentsova, and T. A. Balashova, *Zh. Obshch. Khim.*, **27**, 2536 (1957).

207. Yu. K. Yur'ev and N. N. Mezentsova, *Zh. Obshch. Khim.*, **28**, 3041 (1958).

208. Yu. K. Yur'ev, N. N. Mezentsova, and V. Ye. Vas'kovskii, *Zh. Obshch. Khim.*, **29**, 3240 (1959).

209. C. Paulmier, J. Bourguignon, J. Morel, and P. Pastour, *C. R. Acad. Sci., Paris, Ser. C*, **270**, 494 (1970).

210. Yu. K. Yur'ev and N. K. Sadovaya, *Zh. Obshch. Khim.*, **31**, 3535 (1961).

211. N. N. Magdesieva and T. A. Balashova, *Khim. Geterotsikl. Soedin.*, 716 (1970).

212. Yu. K. Yur'ev, Ye. L. Zaitseva, and A. N. Nikiforova, *Zh. Obshch. Khim.*, **30**, 2209 (1960).

213. Yu. K. Yur'ev, N. N. Mezentsova and E. A. Kashutina, *Zh. Obshch. Khim.*, **29**, 2597 (1959).

214. Yu. K. Yur'ev, N. N. Magdesieva and A. T. Monakhova, *Zh. Obshch. Khim.*, **35**, 68 (1965).

215. Yu. K. Yur'ev and N. N. Mezentsova, *Zh. Obshch. Khim.*, **31**, 1449 (1961).

216. Yu. K. Yur'ev, N. N. Magdesieva, and V. V. Titov, *Zh. Obshch. Khim.*, **32**, 3252 (1962).

217. Yu. K. Yur'ev, N. N. Magdesieva, and A. T. Monakhova, *Zh. Org. Khim.*, **1**, 1094 (1965).

218. Yu. K. Yur'ev, N. N. Magdesieva, and V. V. Titov, *Zh. Obshch. Khim.*, **33,** 1156 (1965).
219. Yu. K. Yur'ev, N. N. Magdesieva, and V. V. Titov, *Zh. Obshch. Khim.*, **33,** 2577 (1963).
220. Yu. K. Yur'ev, N. N. Magdesieva, and T. Lesiak, *Khim. Geterotsikl. Soedin.*, 902 (1966).
221. Yu. K. Yur'ev, N. N. Magdesieva, and V. V. Titov, *Zh. Org. Khim.*, **1,** 163 (1965).
222. N. N. Magdesieva, V. V. Titov, V. F. Bystrov, V. P. Lezina, and Yu. K. Yur'ev, *Zh. Strukt, Khim.*, **6,** 402 (1965).
223. Yu. K. Yur'ev, N. N. Mezentsova, and M. B. Saporovskaya, *Zh. Obshch. Khim.*, **32,** 1444 (1962).
224. Yu. K. Yur'ev, N. N. Magdesieva, V. V. Titov, and V. P. Brysova, *Zh. Obshch. Khim.*, **33,** 3517 (1963).
225. S. V. Tsukerman, V. D. Orlov, V. F. Lavrushin, and Yu. K. Yur'ev, *Zh. Org. Khim.*, **1,** 650 (1965).
226. S. V. Tsukerman, V. D. Orlov, V. P. Izvekov, and V. F. Lavrushin, *Khim. Geterotsikl. Soedin.*, 34 (1966).
227. S. V. Tsukerman. V. D. Orlov, V. F. Lavrushin, and Yu. K. Yur'ev, *Khim. Geterotsikl, Soedin.*, 918 (1966).
228. K. Balenovic, A. Jeljac, B. Gospert, and Z. Stefanac, *Monatsh. Chem.*, **98,** 1344 (1967).
229. R. I. Kobzova, Ye. M. Oparina, N. K. Levkina, N. N. Magdesieva, and Yu. K. Yur'ev, *Zh. Prikl. Khim.*, **39,** 1638 (1966).
230. N. V. Mel'chakova, N. N. Magdesieva, Yu. K. Yur'ev, and V. M. Peshkova, *Vestn. Mosk. Univ.*, 82 (1966).
231. V. M. Peshkova, I. P. Yefimov, and N. N. Magdesieva, *Zh. Anal. Khim.*, **21,** 499 (1966).
232. A. N. Kudrin, L. F. Chernysheva, M. A. Gal'bershtam, and Yu. K. Yur'ev, *Farmakol. Toksikol*, **6,** 692 (1964).

XI Heterocyclic selenium compounds

C. SELENIUM-CONTAINING NITROGEN HETEROCYCLES

EHRENFRIED BULKA

Ernst-Moritz-Arndt-University
Department of Chemistry
Greifswald, German Democratic Republic

1. Introduction

Selenium-containing nitrogen heterocyclic compounds are less well known than the corresponding oxygen- or sulfur-containing species, and the results of synthetic studies are usually widely scattered in the literature. This is mainly due to the fact that frequently the essential selenium-containing starting materials are less stable and less readily prepared than the analogous sulfur intermediates. Some of the selenium-organic intermediates are, furthermore, highly toxic. It is readily seen, therefore, why some classes of selenium-containing nitrogen heterocycles are represented by only a single example and why information on reactivity and other properties is frequently lacking.

The following discussion is an attempt to survey the known species and classes. Since most systematic work has been done on five-membered ring systems, these represent the major portion of the chapter. Six- and seven-membered ring systems are summarized at the beginning.

2. Six- and Seven-Membered Ring Systems with Nitrogen and Selenium

a. 1,3-Selenazines. The first 1,3-selenazine compound, namely, 2-imino-2,3,5,6-tetrahydro-1,3-selenazine (**1**), was obtained by Baringer[1] from 1-bromo-3-aminopropane and potassium selenocyanate in boiling aqueous solution (equation 1). Additional tetrahydro-1,3-selenazine derivatives (1,3-selenazanes) were recently prepared by Luknitskii *et al.*[2]

The reaction of β-(trichloromethyl)-β-propiolactone with *Se*-ethyl-selenopseudourea in ethanol according to equation 2 yielded 2-imino-4-

oxo-6-(trichloromethyl)-2,3,5,6-tetrahydro-1,3-selenazine (**2**). Through acid hydrolysis it was possible to convert the imino compound (**2**) into 2,4-dioxo-6-(trichloromethyl)-2,3,5,6-tetrahydro-1,3-selenazine (**3**). By the use of β-(trichloromethyl)-β-ethanesultone in an analogous reaction, 3-imino-5-(trichloromethyl)-2*H*,6*H*-1,4,2-thiaselenazine-1,1-dioxide (**4**) was obtained.[3]

4

Simchen[4a,b] has recently described a derivative of benzo-1,3-selenazine. On cyclization of *o*-(selenocyanato)benzoyl chloride with hydrogen chloride at 60°, he obtained 2-chloro-1,3-benzoselenazin-4-one (**5**).

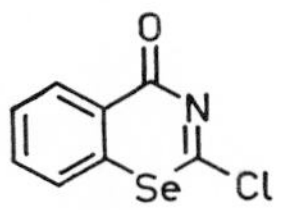

5

b. 1,4-Selenazines. Asinger *et al.*[5] found that the combined action of ethylenimine and elemental selenium on 3-pentanone yielded 2-methyl-3-ethyl-5,6-dihydro-4*H*-1,4-selenazine (**6**) (equation 3). The reaction products

(3)

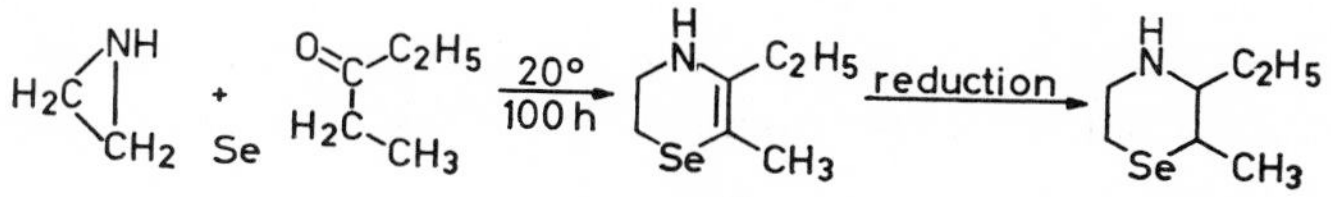

originally expected (*i.e.*, 3-selenazolines) could be prepared neither by direct synthesis from ketones, elemental selenium, and ammonia nor by attempted reconversion of the corresponding hydroselenoketones. The reduction of **6** with concentrated formic acid yielded 2-methyl-3-ethyl-2,3,5,6-tetrahydro-1,4-selenazine (2-methyl-3-ethylselenomorpholine, **7**) (equation 3). Behaghel and Rollmann[6] reported the synthesis of a benzo derivative of 1,4-selenazine (**8**). On reduction of *o*-nitrophenylselenoacetic acid with iron filings in acetic acid solution, the intermediate *o*-amino compound was observed to dehydrate spontaneously (equation 4) and could be isolated only as 3,4-dihydro-2*H*-1,4-benzoselenazin-3-one (**8**).

(4)

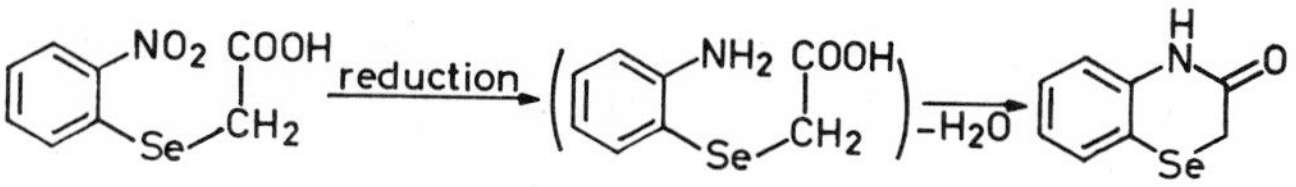

Kiprianov and Khilya[7] prepared the same compound by heating the zinc salt of *o*-aminophenylselenol under reflux with chloroacetic acid. The reaction of **8** with phosphorus pentasulfide produced the corresponding 3-thione (**9**). This, in turn, was condensed with bromoacetone to give 1-methyl-thiazolo[2,3-c][1,4]benzoselenazine (**10**). The latter compound was further used for the synthesis of cyanine dyes of the general type **11** (equation 5).

(5)

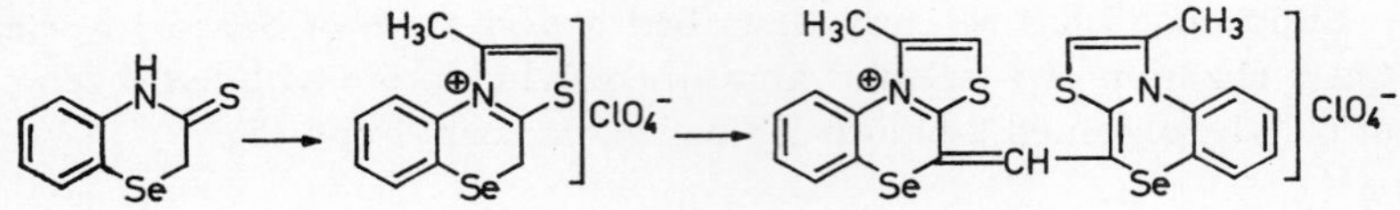

In a similar fashion the condensation of zinc bis(o-aminophenyl) selenide with acetylenedicarboxylic acid proceeded to form 3,4-dihydro-3-oxo-2H-1,4-benzoselenazine-$\Delta^{2,a}$-acetic acid (12).[8]

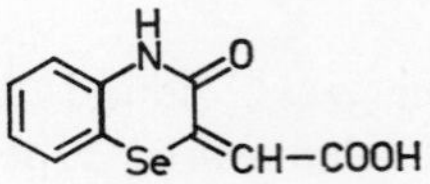

Phenoselenazine (13) and analogous derivatives[9a] are readily formed, even in the cold, by reaction of selenium dichloride with suitable diaryl-amines (equation 6) in an inert solvent such as benzene. The vigorous reaction proceeds with rapid evolution of hydrogen chloride.

(6)

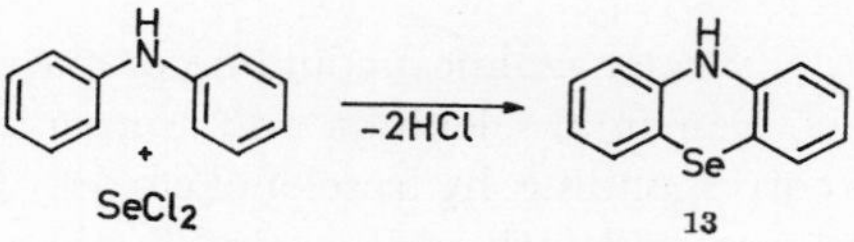

Several phenoselenazine analogs of phenothiazine drugs were prepared, and their pharmacological properties investigated. They were found to be central nervous system depressants.[9b,c,d] The acetylation of phenoselenazines by means of the Friedel-Crafts reaction was studied by Cordella and Sparatore.[9e] Simov and Khristova[9f] have described the alkylation of pheno-selenazine in liquid ammonia with various alkyl halides.

c. **Selenadiazines.** By reaction of 1,8-diaminonaphthalene with selenium dioxide in dilute acetic acid, Sachs[10] succeeded in preparing 1H,3H-naphtho[1,8-cd][1,2,6]selenadiazine (14).

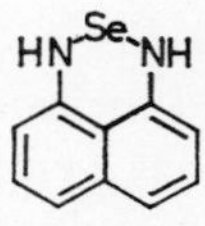

Starting with 4-arylselenosemicarbazides, Bulka and Ahlers[11] prepared derivatives of 2-amino-6H-1,3,4-selenadiazines (15) by condensation with α-haloketones in an acidic medium according to equation 7. The 2-aryl-amino-5-phenyl-6H-1,3,4-selenadiazines are remarkably stable towards

(7)

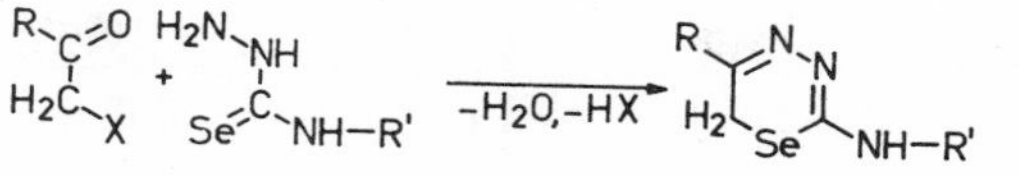

hydrogen ions and survive several hours of refluxing in concentrated hydrochloric acid. A structure proof is based on their reaction with p-nitrobenzaldehyde.

By analogy with the corresponding 2-amino-5-phenyl-1,3,4-thiadiazines,[12] a ring-opening reaction with elimination of selenium yields 4-arylsemicarbazones of 4-nitrochalcone.[11,13] In the case of 2-p-chloranilino-5-phenyl-6H-1,3,4-selenadiazine (**16**) the intermediate product, **17**, can be isolated and then converted into the final product in a definitive fashion (equation 8).[11]

(8)

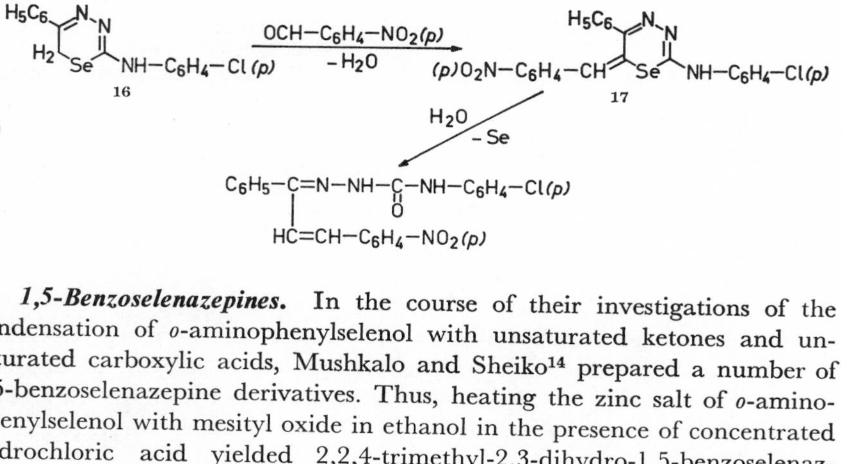

d. 1,5-Benzoselenazepines. In the course of their investigations of the condensation of o-aminophenylselenol with unsaturated ketones and unsaturated carboxylic acids, Mushkalo and Sheiko[14] prepared a number of 1,5-benzoselenazepine derivatives. Thus, heating the zinc salt of o-aminophenylselenol with mesityl oxide in ethanol in the presence of concentrated hydrochloric acid yielded 2,2,4-trimethyl-2,3-dihydro-1,5-benzoselenazepine (**18**).[14] With 3-methylhept-3-en-5-one the 2,4-diethyl-2-methyl-1,5-benzoselenazepine (**19**) was formed,[15] and crotonic acid yielded 2,3-dihydro-2-methyl-1,5-benzoselenazepin-4(5H)-one (**20**).[8]

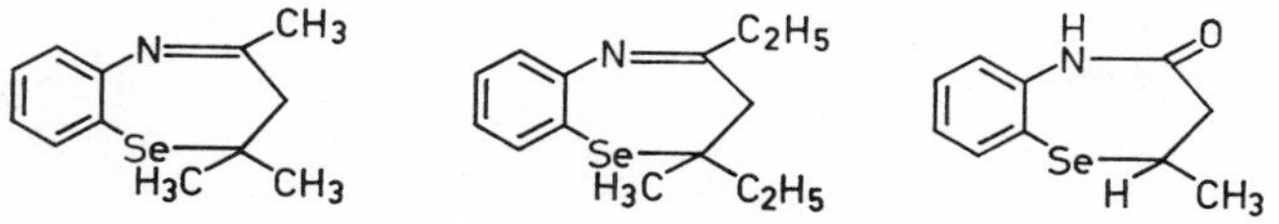

3. Five-Membered Rings with Nitrogen and Selenium

Brief descriptions of a number of five-membered heterocyclic compounds that contain selenium and nitrogen were given by Bambas,[16] who reviewed the literature that had been published until about 1950.

a. 1,2,3-Selenadiazoles. Derivatives of all four possible selenadiazole isomers are known.

1,2,3-Benzoselenadiazole (**21**), as well as its 4- and 6-methyl derivatives, were prepared by Keimatsu and Satoda[17a] by diazotization of *o*-amino-phenylselenol and its methyl derivatives, respectively (equation 9). The

(9)

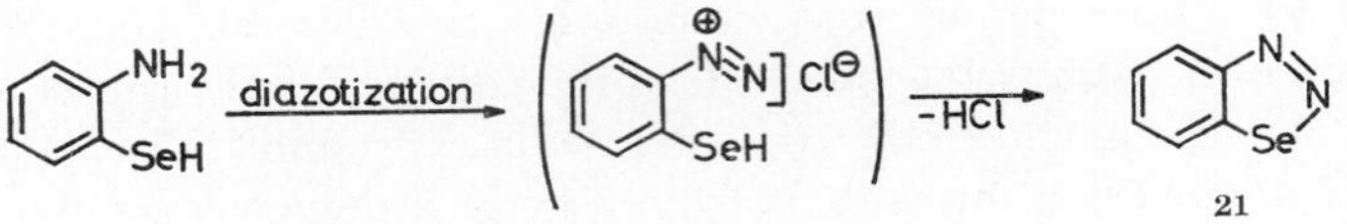

1,2,3-selenadiazoles without fused benzene rings were unknown until 1969, when Lalezari *et al.*[17b,c] described the synthesis of some 1,2,3-selenadiazoles by selenium dioxide oxidation of aryl ketone semicarbazones.

b. 1,2,4-Selenadiazoles. Becker and Meyer[18] have described some derivatives of 1,2,4-selenadiazoles. By oxidation of selenobenzamide with iodine they obtained 3,5-diphenyl-1,2,4-selenadiazole (**22**) (equation 10). In an analogous fashion 3,5-di-*p*-tolyl-1,2,4-selenadiazole was obtained.

(10)

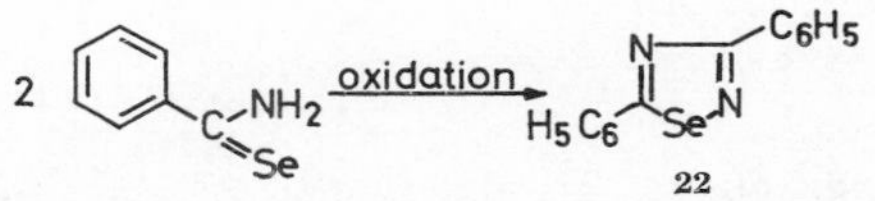

Some 5-amino-1,2,4-selenadiazoles (**23**) were prepared by Goerdeler *et al.*[19] in analogy to the corresponding thiadiazoles. Thus, reaction of potassium selenocyanate with *N*-halides of benzamidine or acetamidine yielded 5-amino-3-phenyl-1,2,4-selenadiazole or the 3-methyl derivative, respectively (equation 11).[19]

(11)

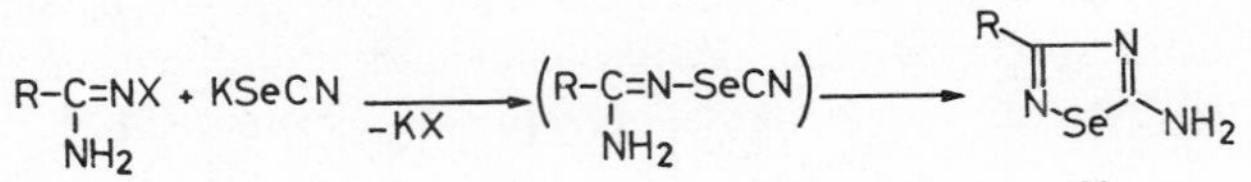

c. 1,2,5-Selenadiazoles. Aromatic condensed systems, such as 2,1,3-benzo-selenadiazole, represent the best-known examples of the 1,2,5-selenadiazole ring system. 2,1,3-Benzoselenadiazole (**24**) was prepared in 1889 by Hins-berg[20] as one of the first examples of selenium-containing nitrogen hetero-cyclic ring systems. It was then called "piaselenole" and later, occasionally, "piazselenole." Hinsberg's synthesis of **24** and of some simple derivatives by condensation of *o*-phenylenediamine and selenium dioxide (equation 12)

(12)

in ethanol or dilute acids still represents the major synthetic route to this class, since condensations occur quantitatively in the cold or on slight warming. The smooth reaction and the reproducibility of this condensation are highly suitable for analytical purposes in which the product may be determined by absorption or fluorescence spectrophotometry. The process allows detection of traces of *o*-phenylenediamine in the presence of the *m*- and *p*-isomers.[21] On the other hand, numerous procedures have been generated to determine trace amounts of selenium by the same reaction.[22,23] In place of *o*-phenylenediamine it is possible to use many other diamines, such as 2,3-diaminonaphthalene.[24] For additional details see Chapter XVII.

Major contributions to the chemistry of 2,1,3-benzoselenadiazoles have been made by the groups of Sawicki,[25-32] Efros,[33-38] and Pesin.[39-50] Since there are essentially no limits to the choice of *o*-diamino compound, it is readily seen why numerous more highly condensed ring systems with 1,2,5-selenadiazole components have been prepared. Some examples are derivatives of naphtho[1,2-c][1,2,5]selenadiazole[20] (naphthopiaselenole, **25**), phenanthro[9,10-c][1,2,5]selenadiazole (**26**),[51] indeno[1,2-f][2,1,3]benzoselenadiazole (indeno[1,2-f]piaselenole, **27**),[29] 5*H*-[1,2,5]selenadiazolo-

[3,4-b]carbazole (5*H*-indolo[2,3-f]piaselenole, **28**),[26] 6*H*-[1,2,5]selenadiazolo[3,4-c]carbazole (6*H*-indolo[3,2-e]piaselenole, **29**),[26] and [1]benzothieno[2,3-e][2,1,3]benzoselenadiazole (thianaphtheno[2,3-e]piaselenole, **30**).[25] Further examples include [1]benzoselenopheno[2,3-e][2,1,3]benzoselenadiazole (selenanaphtheno[2,3-e]piaselenole, **31**),[27] [1]benzoselenopheno[3,2-e][2,1,3]benzoselenadiazole (selenanaphtheno[3,2-e]piaselenole,

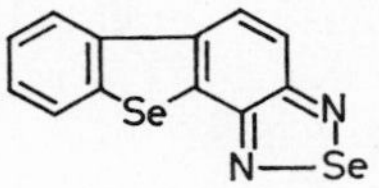

31 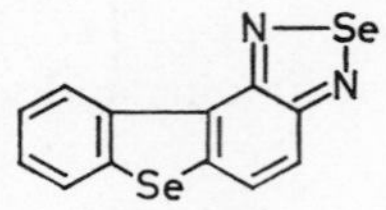   32 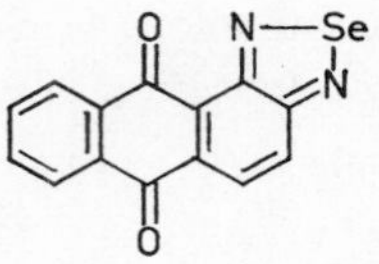   33

32),[28] and anthra[1,2-c][1,2,5]selenadiazol-6,11-dione (**33**),[52a,b] as well as 1,2,5-selenadiazolo[3,4-b]pyridine (**34**),[53a,b] 1,2,5-selenadiazolo[3,4-c]-pyridine (**35**),[53a,b] 1,2,5-selenadiazolo[3,4-d]pyrimidine (8-selena-purine, **36**),[32,39,54] and thiazolo[4,5-g][2,1,3]benzoselenadiazole (**37**).[40-42,55]

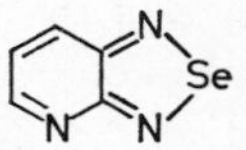

34 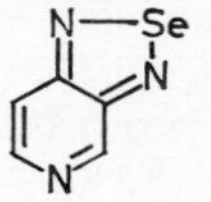   35 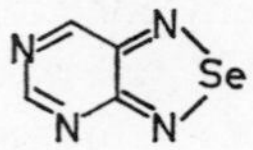   36 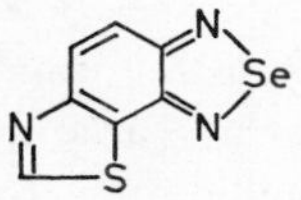   37

Reviews on the structure and reactivity of 2,1,3-benzoselenadiazoles have appeared in the literature.[30,33,36,50] The chemical properties of these compounds are very similar to those of the analogous benzothiadiazoles. In both systems substitution reactions on the attached rings are possible, and aminations,[43,56] chloromethylations,[44,46,47] nitrations,[45,49] and halogenations[45,49,57a] have been reported. On halogenation of 2,1,3-benzoselenadiazoles the monosubstitution occurs in the 4-position, followed by formation of the 4,7-disubstituted product. The halogen is subject to nucleophilic displacement reactions.[58] Sulfonation occurs also in the 4-position and is followed by 4,5-disubstitution when forcing conditions are used.[33,57a] Alkylation of 2,1,3-benzoselenadiazole yields alkyl-2,1,3-benzoselenadiazolium salts.[59a,b,c] Reactions with aromatic aldehydes of suitably activated methyl derivatives, such as 5-methyl-4-nitro-2,1,3-benzoselenadiazole, generate 5-styryl derivatives.[31]

Although the selenadiazole ring remains intact in all the above reactions, strong reducing conditions such as zinc and hydrochloric acid or, better, stannous chloride and hydrochloric acid lead to ring scission with production of the corresponding substituted o-phenylenediamines (equation 13).[48,57a]

(13)

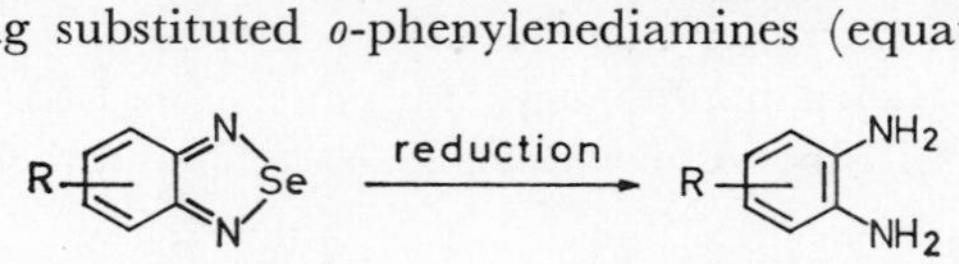

The process has some practical application for the transient protection of o-diamines during chemical reactions.[57a,b] N-Substituted o-phenylenedia-mines are generated by a similar scission through the action of phenyl-lithium on 2,1,3-benzoselenadiazoles (equation 14).[60] The ring-opening

(14)   R—⟨benzoselenadiazole⟩ $\xrightarrow{(C_6H_5Li)}$ R—⟨benzene-NH$_2$, NH—C$_6$H$_5$⟩ + C$_6$H$_5$—Se—C$_6$H$_5$

reaction of 1,2,5-selenazolo[3,4-a]anthraquinone with boiling aqueous sodium hydroxide yielded the corresponding substituted 1,2-diamino-anthraquinones.[61]

Spectroscopic measurements on new synthetic products in the u.v.-visible region have been performed by Sawicki and Carr,[25-28,30,31] and some further data are given by Efros and Todres-Selektor[34] and Brizzi et al.[56] Recently, some detailed spectroscopic investigations have also been undertaken, and some of the data obtained relate to isosteric series of 2,1,3-benzoxa-, benzothia-, and benzoselenadiazoles. Hollas and Wright[62a] reported the vibrational analysis of the near-u.v. spectrum of 2,1,3-benzoselenadiazole. Dal Monte and Sandri[62b] have discussed the u.v. spectra of some amino derivatives of this series and their pK values. Bird and Cheeseman[63] undertook a detailed investigation of 2,1,3-benzoselenadiazole i.r. spectra. Vibrational spectra of 2,1,3-benzothia- and benzoselenadiazoles have been reported by Korobkov et al.[64]

Brown and Bladon[53a] measured and evaluated the i.r., u.v., and n.m.r. spectra of 1,2,5-selenadiazolopyridines. The same authors also compared the n.m.r. spectra of 2,1,3-benzoxa, benzothia-, and benzoselenadiazoles.[65] Similar work was treated in papers by Fedin and Todres,[66] Tobiason and Goldstein,[67] and Braier et al.,[68a] as well as Kamiya et al.[68b] The spectroscopic data agree well with an o-quinoid structure of 2,1,3-benzoselena-diazole (24).[37] This assignment was further confirmed by crystallographic analysis.[69] Polarographic studies on benzoxa-, benzothia- and benzoselena-diazoles in DMF[70] and in aqueous solution[71a] were carried out by Zhdanov et al. Polarographic data were reported also for nitro derivatives of 2,1,3-benzoselenadiazole.[71b] Buu-Hoi et al.[72] reported further on the electron-impact fragmentation of compounds containing condensed 1,2,5-selena-diazole rings.

It is remarkable that 1,2,5-selenadiazoles without other condensed rings have been described only recently.[73] Shealy et al.[74,75] utilized for such syntheses a specific ring scission of the condensed system that left the selena-diazole system intact, in contrast to the other methods mentioned above. Aqueous potassium hydroxide or alcoholic ammonia was used to degrade the pyrimidine ring in 1,2,5-selenadiazolo[3,4-d]pyrimidin-7(6H)ones,[32] yielding derivatives of 4-amino-1,2,5-selenadiazole-3-carboxylic acid (38). These compounds were tested for cytotoxic and antimicrobial effects, and a low level of activity was noted.[74] Further investigations into the antimicro-bial effects of 1,2,5-selenadiazoles were undertaken by Hunt and Pittillo.[76]

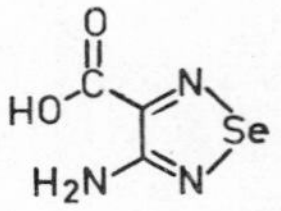

38

The synthesis of other 1,2,5-selenadiazoles was reported almost simultaneously from two other sources. Weinstock et al.[77] and Bertini[78] succeeded in preparing 1,2,5-selenadiazole (39) by reaction of ethylenediamine in DMF with selenium monochloride and selenium dioxide (equation 15), respectively. Starting from other vicinal diamines or from dioximes,

(15)

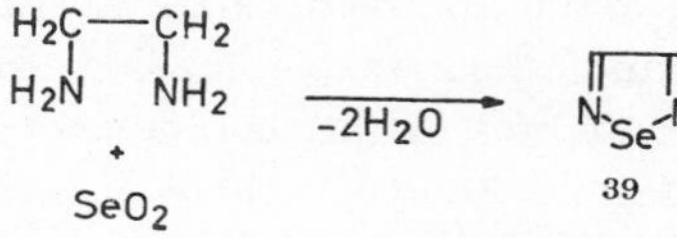

Bertini[78,79] prepared in the same manner a number of 1,2,5-selenadiazoles substituted in the 3- and 4-positions. Bertini et al. have also undertaken n.m.r. investigations of 1,2,5-selenadiazole and its 3-methyl derivative,[80] as well as i.r. and Raman spectral measurements.[81a]

Blackman et al.[81b] have examined the microwave spectra of the [80]Se and [78]Se isotopic species of 1,2,5-selenadiazole and have determined the dipole moment and the quadrupole coupling constants.

d. 1,3,4-Selenadiazoles.

Stollé and Gutmann[82] were the first to prepare 1,3,4-selenadiazole derivatives. By heating N,N'-dibenzoylhydrazine with phosphorus pentaselenide they obtained 2,5-diphenyl-1,3,4-selenadiazole (40) in about 10 % yield, with the major ring-closure product represented by the 1,3,4-oxadiazole compound, 41 (equation 16). A 2,5-dimethyl derivative was obtained by an analogous route.

(16)

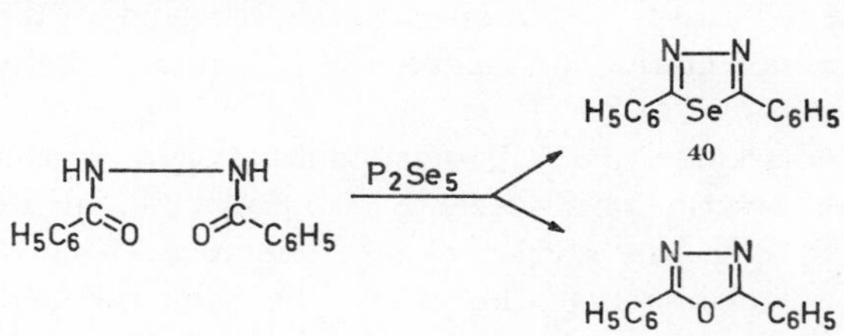

More recently, a number of 2-aryl(alkyl)amino-1,3,4-selenadiazoles (42) were obtained by a procedure analogous to that used to prepare 1,3,4-thiadiazoles from thiosemicarbazides.[16] Heating 1-acyl-4-aryl(alkyl)selenosemicarbazides with acetyl chloride in glacial acetic acid led to the cyclization depicted in equation 17.[83]

(17)

$$\underset{42}{\underset{\text{R}^{C}\overset{}{\sim}\text{O} \quad \text{Se}^{C}\text{NH}-\text{R}'}{\overset{\text{HN}\text{---}\text{NH}}{}}} \quad \xrightarrow{-\text{H}_2\text{O}} \quad \underset{\text{R}^{} \text{Se}^{} \text{NH}-\text{R}'}{\overset{\text{N}\text{---}\text{N}}{}}$$

In the same manner a number of 2-amino-1,3,4-selenadiazoles (**44**) were prepared from 1,4-diacylselenosemicarbazides *via* 2-benzamido-1,3,4-selenadiazoles (**43**) and subsequent hydrolysis (equation 18).[83] The 2-amino-1,3,4-

(18)

$$\underset{\text{R}^{C}\overset{}{\sim}\text{O} \quad \text{Se}^{C}\text{NH}-\text{COC}_6\text{H}_5}{\overset{\text{HN}\text{---}\text{NH}}{}} \xrightarrow{-\text{H}_2\text{O}} \underset{43}{\underset{\text{R} \quad \text{Se} \quad \text{NH}-\text{COC}_6\text{H}_5}{\overset{\text{N}\text{---}\text{N}}{}}} \xrightarrow{\text{hydrolysis}} \underset{44}{\underset{\text{R} \quad \text{Se} \quad \text{NH}_2}{\overset{\text{N}\text{---}\text{N}}{}}}$$

selenadiazoles (**44**) exhibit the typical behavior of aromatic amines; that is, they may be diazotized and coupled to suitable aromatic nuclei,[83] and they can be condensed with aromatic aldehydes to yield azomethines.[84]

The synthesis of the unsubstituted 1,3,4-selenadiazole nucleus (**45**) was reported in 1970 by Kendall and Olofson.[85a] This compound was obtained in 25% yield by reaction of N,N-dimethylformamidazine with hydrogen selenide in the presence of a precise amount of pyridine (equation 19).

(19)

$$(\text{CH}_3)_2\text{N}-\text{CH}=\text{N}-\text{N}=\text{CH}-\text{N}(\text{CH}_3)_2 \xrightarrow{\text{H}_2\text{Se}} \underset{45}{\underset{\text{H} \quad \text{Se} \quad \text{H}}{\overset{\text{N}\text{---}\text{N}}{}}}$$

N,N-Dimethylselenoformamide resulted, in approximately the same yield, as a second product of this reaction. The authors reported detailed spectroscopic data and physical properties for **45**, and comparisons were drawn with 1,3,4-thia- and oxadiazole, as well as 1,2,5-selenadiazole. Data pertaining to the microwave spectrum, structure, and dipole moment of 1,3,4-selenadiazole were given also by Levine *et al.*[85b]

2-Imino-3-aryl-Δ⁴-1,3,4-selenadiazolines (**46**) were obtained from α-halohydrazones by reaction with potassium selenocyanate (equation 20).[86]

(20)

$$\underset{\text{KSe}}{\underset{\text{R}^{C}\text{X}}{\overset{\text{N}\text{---}\text{NH}-\text{R}'}{}}} + \underset{}{\text{C}\equiv\text{N}} \xrightarrow{-\text{KX}} \underset{46}{\underset{\text{R} \quad \text{Se} \quad \text{NH}}{\overset{\text{N}\text{---}\text{N}\text{-}\text{R}'}{}}}$$

The reaction of **46** with sodium nitrite yielded 3,5-disubstituted 2-nitrosamino-Δ⁴-1,3,4-selenadiazolines. The latter lose nitrogen on heating in xylene to give 2-oxo compounds.[86]

e. 2,1,3-Benzoxaselenazoles. On treatment of cyclohexane-1,3-dioximes with selenium dioxide King and Felton[87a] obtained 7-nitroso-Δ³,⁷-dihydrobenz-2,1,3-oxaselenazoles (**47**) (equation 21). Additional oxaselenazoles

(21)

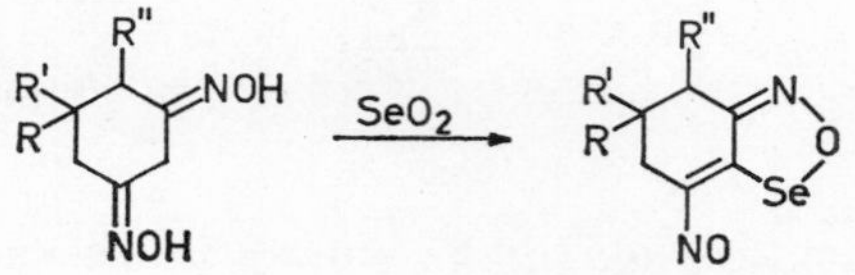

$$\text{(structure 47)}$$

were described by Takeda *et al.*[87b,c,d] The products showed bacteriostatic effects *in vitro* and *in vivo*, especially on *Mycobacterium tuberculosis*, and a therapeutic effect on Ehrlich carcinoma of mice.

f. 1,2-Isoselenazoles. 1,2-Benzisoselenazole derivatives, specifically 2-aryl(alkyl)-1,2-benzisoselenazol-3(2)ones (**48**), were prepared by Lesser and Weiss[88] by the reaction of 2-chlorocarbonylphenylselenenyl chloride with ammonia or primary amines (equation 22). These authors also reported the

(22)

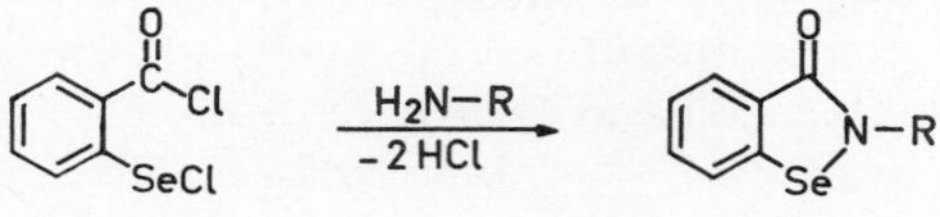

$$\xrightarrow[\text{- 2 HCl}]{\text{H}_2\text{N—R}}$$

48

synthesis of the selenium analog of saccharin, namely, 1,2-benzisoselenazolin-3-one-1,1-dioxide (**49**).[89] Although the chemical nature of **49** barely

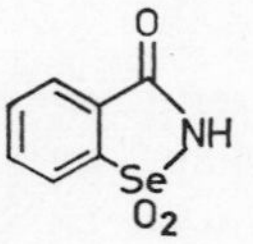

49

differs from that of the analogous sulfur compound, the physiological effect is completely different, since the typically sweet taste of saccharin is absent.

The isoselenazole ring system (**50**) and its 3-methyl derivative have recently been obtained by ring closure of 3-cyanoselenoacrolein and the corresponding butenone in liquid ammonia (equation 23).[90]

(23)

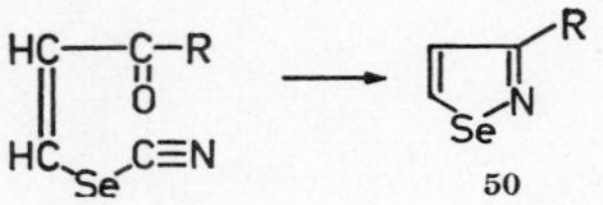

$$\xrightarrow{}$$

50

g. 1,3-Selenazoles

i. Selenazoles. Numerous derivatives of 1,3-selenazole (**51**) have been described in the literature, since this class of selenium-containing nitrogen heterocycles has been known for a long time. A review of their chemistry by

Bulka[91] covers the literature until 1963. With few exceptions, the general method of synthesis for 1,3-selenazoles is a modification of the Hantzsch thiazole synthesis. In place of thioamides, derivatives of selenocarboxamides are reacted with α-halocarbonyl compounds to give the corresponding selenazoles smoothly in good yield (equation 24). With a variety of seleno-carboxamides a series of 2-alkyl(aryl)selenazoles was thus prepared.[92–94]

(24)

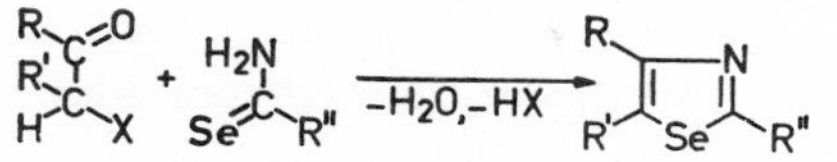

51

   To avoid difficulties in the preparation of labile aliphatic selenoamides, Haginiwa[95] attempted to use these in *statu nascendi* by passing hydrogen selenide into a mixture of acetonitrile and chloroacetone. In an extensive series of experiments he determined the most favorable conditions for this reaction and found that the components should be heated for 10 hours on a boiling water bath in the presence of anhydrous zinc chloride or phosphorus oxychloride. The condensation was further enhanced by the use of fresh aluminum chloride, but tarry side products resulted in a difficult work-up in this case. Although the average yield in the above reaction was only 10%, the process also allowed the preparation of the first selenazole which was unsubstituted in the 2-position. For the preparation of 4-methylselenazole (**52**),[96] hydrogen selenide and hydrogen cyanide were jointly passed into chloroacetone (equation 25).

(25)

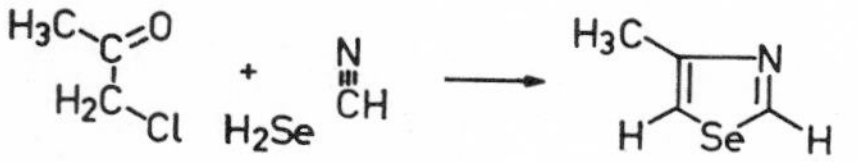

52

   Alkylselenazoles are alkaline liquids with an odor reminiscent of the corresponding thiazole and pyridine derivatives. Their characterization is possible *via* crystalline picrate or methiodide derivatives.

   When selenourea was used as the selenoamide component in the Hantzsch synthesis, 2-aminoselenazoles (**53**) resulted.[92, 96–103] Because of the large variety of available α-halocarbonyl compounds and the fact that selenourea is more readily accessible than selenocarboxamides, numerous 2-amino-selenazoles have been prepared. In these cases it is not always necessary to utilize a haloketone in the reaction; the ketone and iodine are frequently sufficient.[104]

   2-Aminoselenazoles are crystalline and, with the exception of the un-substituted species, are generally stable and practically odorless substances. They are more strongly basic than 2-aryl(alkyl)selenazoles.

2-Benzamidoselenazoles (**55**) may be prepared by using 1-benzoyl-2-seleno-urea (**54**) as depicted in scheme 26.[105] This procedure may be extended to

(26)

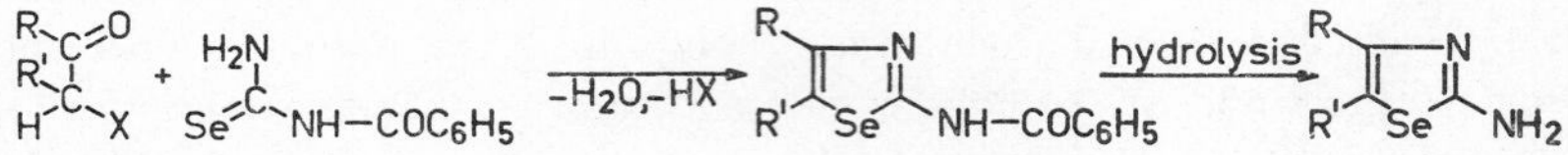

other acylselenoureas. 1-Benzoyl-2-selenourea is easily obtained, according to the method of Douglass,[106] by reaction of potassium selenocyanate with benzoyl chloride in acetone and subsequent addition of ammonia. Hydrolysis of the benzamides **55** with phosphoric or sulfuric acid then yielded 2-amino-selenazoles (**53**) identical to those obtained directly from selenourea.[105] The chief advantage of the latter method is the avoidance of the unpleasant and dangerous handling of hydrogen selenide in the preparation of selenourea from cyanamide.

The condensation of monosubstituted alkyl or arylselenoureas[107] and α-haloketones gave the analogous 2-alkylamino-[108,109] or 2-arylamino-selenazoles (**56**).[110] In a corresponding fashion 2-dialkylaminoselenazoles

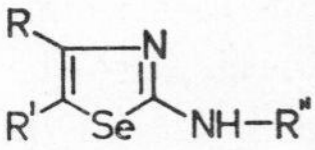

(**57**) were obtained from 1,1-dimethyl(diethyl)selenoureas.[108,109,111]

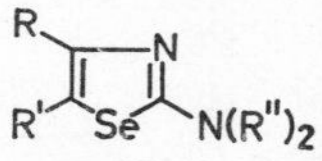

There is little difference in the properties of 2-alkyl(aryl)aminoselenazoles and those of the simple 2-amino derivatives, except that the aryl-substituted species are generally less basic and their salts are subject to hydrolytic cleavage. When selenosemicarbazones were used as the selenoamide component in the Hantzsch synthesis, 2-selenazolylhydrazones (**58**) were obtained[112,113] according to equation 27.

(27)

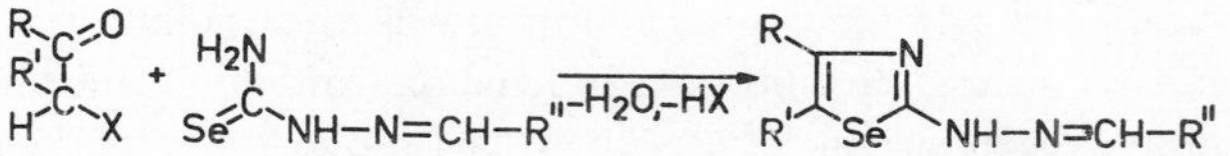

Arylidenehydrazones of type **58** show the typical behavior of this compound class and may be coupled with diazonium salts to yield formazanes containing a selenazole moiety.[114] Products such as 1,3-diaryl-5-(2-selenazolyl)formazanes (**59**) are crystalline, highly colored compounds which may

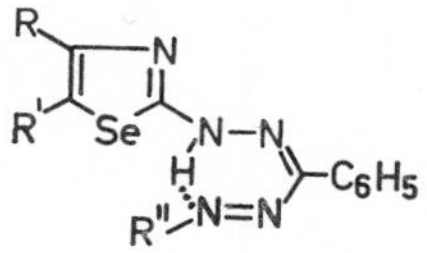

59

be dehydrogenated to the corresponding tetrazolium salts. 2-Selenazolyl-hydrazines (**61**) are obtained through hydrolysis of the corresponding 2-selenazolylhydrazones (**60**) (equation 28).[112] Compounds of type **61** show the

(28)

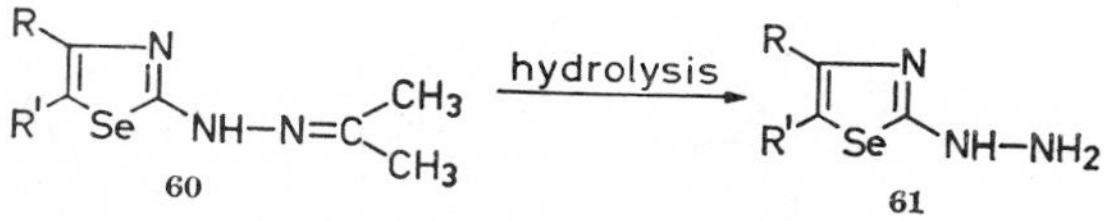

characteristic properties of substituted hydrazines. They reduce Fehling's solution and ammoniacal silver nitrate solution, and react with carbonyl compounds to yield hydrazones of type **60**. By condensation of **61** with β-ketoesters a number of 1-(2-selenazolyl)-3-alkyl-5-pyrazolones were prepared.[115] The latter may also be obtained by first cyclizing selenosemi-carbazones of β-ketoesters to 1-selenocarbamoyl-3-alkyl-5-pyrazolones (**63**) and then using the products in a Hantzsch synthesis according to equation 29.

(29)

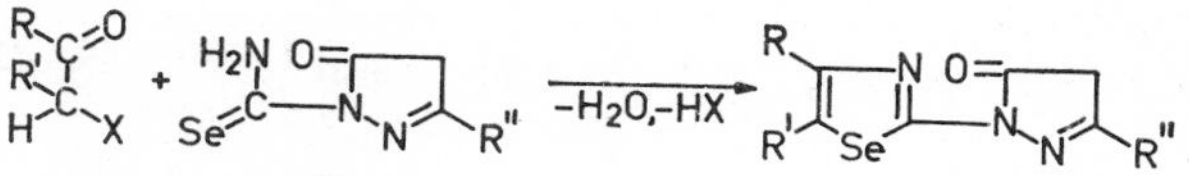

By reaction of pyrazolones of type **62** with p-nitrosodialkylanilines a series of azomethines was prepared and tested for suitability as magenta dyes in color photography.[115] The patent literature further contains a variety of miscellaneous selenazole derivatives.[116–118]

Little information is available concerning the reactivity of the 2-position of selenazoles, since only a single species is known without a substituent in this position. The amination of 4-methylselenazole (**52**) was attempted as a test of the susceptibility of this class to nucleophilic substitution.[119] However, the experiments were unsuccessful and resulted in ring scission. Attempts to produce other selenazoles without 2-substitution by diazotization of 2-aminoselenazoles and a special Sandmeyer reaction were also unsuccessful[97] and led to complete destruction of the selenazole ring system.[120]

In contrast, the 5-position of selenazoles is fairly readily accessible to electrophilic substitution. Nitration of 2-alkylselenazoles unsubstituted in the 5-position (**51**) with nitric-sulfuric acid under relatively mild conditions gave 5-nitro derivatives.[119] Attempts at direct nitration of 2-aminoselenazoles

(53) resulted in their complete decomposition,[120] but the 2-acetamido derivatives nitrated smoothly in the 5-position.[120]

When 2-arylaminoselenazoles (56) were treated with sodium nitrite in glacial acetic acid at room temperature, there were instantaneously formed the corresponding nitroso compounds.[121] According to recent investigations,[122] these are 2-arylamino-5-nitrososelenazoles, which exist in a tautomeric equilibrium, as indicated in structures 64a and 64b, since their properties differ significantly from those of analogous aromatic nitroso compounds.

64a   64b

If the mobile H atom is absent, as in 2-diethylaminoselenazoles (57), 5-nitroso compounds of type 65 are formed; these have properties comparable to those of p-nitrosodialkylanilines. N-Nitrosamino derivatives[121]

65

are formed only if the 5-position is blocked, as in 2-arylamino-4,5-diphenyl-selenazoles. Sulfonations are known to occur only with 2-alkylselenazoles,[119] where heating with fuming sulfuric acid produced the corresponding sulfonic acid.

The halogenation of selenazoles is more complex than the substitution reactions mentioned up to this point. It is possible to convert 2,4-dimethyl-selenazole on treatment with bromine *via* an unstable intermediate into the hydrobromide of 2,4-dimethyl-5-bromoselenazole.[119] Direct bromination of 2-aminoselenazoles[120,123] yields 2-amino-5-bromoselenazoles, stable only as salts. Attempts to generate the free bases always led to complete decomposition. By contrast, it was found[120] that 2-acylamino-5-bromoselenazoles are comparatively stable. Diazonium salts couple readily with selenazoles to give the corresponding 5-azo compounds.[108,121] This reaction may occur in preference to the formation of formazanes when 2-arylidenehydrazono-selenazoles with an available 5-position are used in such syntheses.[114]

Hydrazones of 2-selenazolylhydrazines (58) may be converted into quinoid dyes of the general type 2,2'-dioxo-$\Delta^{3,3'}$-biselenazol-5,5'-inylidenebis-(hydrazone) (66) by oxidation with ferric chloride and hydrogen peroxide.[113]

66

Compounds **56** and **58** react readily with *p*-nitrosodialkylanilines to give deeply colored azomethines of the general type represented by 5-(*p*-dialkylaminophenylimino)-2-selenazolone hydrazones and arylimines (**67**).[121,124]

$$(p)(R'')_2N-C_6H_4-N \overset{R}{\underset{Se}{\bigvee}} \overset{N}{\underset{}{\bigvee}} N-R'$$

67

The reaction of 2-diethylamino-4-methylselenazole with benzaldehyde in the presence of anhydrous zinc chloride leads to the corresponding carbinol.[108] With formaldehyde, compounds related to 5-[(2-amino-5-selenazolyl)methylene]selenazoline (**68**) are formed.[125] On the other hand, a methyl

$$R'-HN \overset{N}{\underset{Se}{\bigvee}} \overset{R}{\underset{}{\bigvee}} -CH \overset{R}{\underset{Se}{\bigvee}} \overset{N}{\underset{}{\bigvee}} N-R'$$

68

group in 2-position of the selenazole ring is sufficiently reactive to give the corresponding styryl compound with benzaldehyde.[119] Alkylation of selenazoles yields corresponding selenazolium salts.[94]

Ochiai[126] has published a comparative study on pyridines, thiazoles, and selenazoles. From purely theoretical considerations it was deduced that selenazole exhibits aromatic character, but that the degree of aromaticity is least in selenazole and increases from thiazole to pyridine. At the same time the more strongly metallic character of selenium should weaken the C-Se bond and, consequently, increase the tendency towards ring-opening reactions. The basicity of the three classes of heterocycles was considered to be similar, the azole nitrogen showing properties similar to those of the pyridine nitrogen. Of the two C atoms in the positions *ortho* to the nitrogen, that is, C(2) and C(4), only the 2-position should remain sufficiently activated by interaction with the Se (or S) atom. This activation would be anticipated to be smaller than in pyridine and lowest in the selenium analog. The 4- and 5-positions in the azoles were postulated to be more strongly susceptible to electrophilic substitution than the 3- and 5-positions in pyridine, this being especially apparent at the 5-position in selenazole. The C(2) and C(5) in selenazoles and thiazoles thus correspond to the *p*-position in benzene, and C(2) and C(4) are similar to the *m*-position. A substituent with a strongly positive mesomeric effect would activate the 5-position even more, whereas C(4) would remain unaffected.

A substituent with an active hydrogen at C(2) may give rise to a tautomeric system, as shown in structures **69a** and **69b**, further decreasing the aromaticity by electron withdrawal from the π-electron system. This could especially affect the stability of selenazole and increase its ring-opening tendencies.

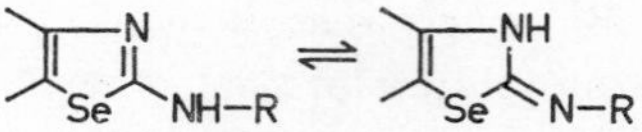

69a   ⇌   69b

These theoretical considerations were found to be in general agreement with experimental work by Haginiwa[108,119,120] and also fit the known data discussed earlier in this section. A few specific supporting examples may be cited. Thus, the low activity of the 2-position towards nucleophilic attack is shown by the failure of 4-methylselenazole (**52**) to react with sodamide. Furthermore, the reaction of 2,4-dimethylselenazole with an excess of benzaldehyde yields exclusively a monostyryl compound. On oxidation of 2,4-dimethylselenazole with selenium dioxide, 4-methylselenazole (**52**) can be isolated. This indicates that only the methyl group in the 2-position is sufficiently activated for oxidation to the 2-carboxylic acid, which, as is known for the corresponding thiazole-2-carboxylic acid, decarboxylates readily to give **52** (equation 30).

(30)

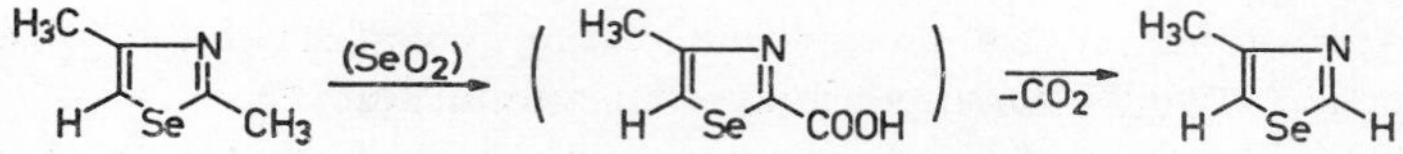

The nitration of 4-methyl- and of 2,4-dimethylselenazole occurs more rapidly than is the case in the corresponding thiazoles. Sulfonation and bromination of 2,4-dimethylselenazole occur readily, but the products are relatively unstable and decompose with formation of elemental selenium. This shows both that the activation of the 5-position is higher in selenazole than in thiazole and that the ring lability has increased in going from sulfur to selenium. The activity of the 5-position, however, does not approach that of benzene or toluene towards electrophilic substitution, as demonstrated by the failure of the methyl compounds to undergo the Gattermann aldehyde synthesis or Friedel-Crafts reactions.

Substituents with a positive mesomeric effect in the 2-position indeed increase the reactivity of the 5-position towards electrophilic substitution, as shown by the examples of 2-amino, 2-acylamino, and 2-hydrazino compounds. With a free amino group, a greater tendency towards ring opening is observed at the same time. From 2-acetamido-4-methylselenazole, the 5-nitro derivative could be isolated; and from the 5-bromo compound, even the free base. By reaction with mercuric acetate the 5-mercuriacetate is formed; this may further be converted into the chloride by reaction with sodium chloride. On treatment with potassium iodide, reduction occurs and the starting material is reformed with loss of mercury. Scheme 31 recapitulates the various types of reaction discussed above.

(31)

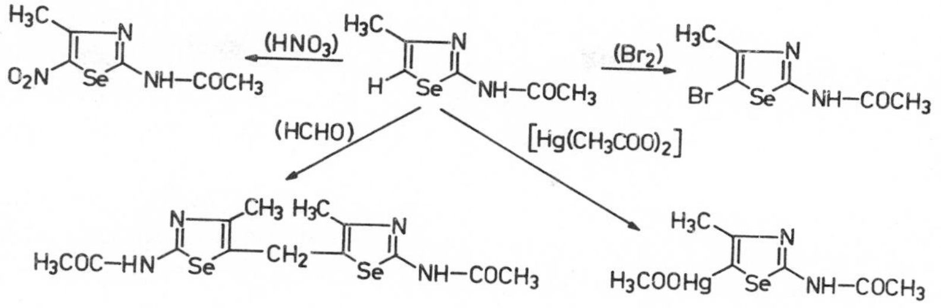

A certain degree of stabilization is conferred on the ring by 4-substituents. This is especially true when the behavior of 4-aryl-substituted selenazoles is considered.

In summary, comparison of the chemical properties of thiazole and selenazole gives the following picture. A methyl group in position 2 is equally reactive in the two classes. Of the two positions *ortho* to nitrogen, only the 2-position is activated. The 5-position is most reactive towards electrophilic reagents and has a *para* relationship to the 2-position. On the other hand, thiazole and selenazole differ markedly by a decreased susceptibility of the selenium system to nucleophilic attack in the 2-position, an increased tendency of the 5-position to electrophilic substitution, and a clear tendency towards easier ring cleavage. Difficulties are encountered in attempted 2-aminoselenazole diazotization reactions.

The preceding discussion indicates a steady decrease in the aromaticity of the three heterocyclic systems from pyridine to thiazole to selenazole. A comparison of the boiling points (Table XIC-1) shows that the thiazole compounds are generally more volatile than the pyridines, while the boiling

Table XIC-1   Comparison of the boiling points of related N, S, and Se heterocycles

COMPOUND	BOILING POINT ($^{\circ}$c)	COMPOUND	BOILING POINT ($^{\circ}$c)	COMPOUND	BOILING POINT ($^{\circ}$c)
Pyridine	114.8	Thiazole	116.8		
2-Picoline	129	2-Methylthiazole	128	2-Methylselenazole	149
3-Picoline	143	4-Methylthiazole	130	4-Methylselenazole	152–153
2,4-Lutidine	157	2,4-Dimethylthiazole	144	2,4-Dimethylselenazole	162–164

points of the selenazole derivatives are considerably higher than those of the other two.

A comparison of the u.v. spectra for sulfur and selenium analogs shows that there is little difference in the general shape of the curves, except for a small bathochromic shift in the latter system. Infrared spectroscopic studies and comparisons with imidazoles have been undertaken by Bassignana et al.[127]

Quaternary salts of selenazoles have been used for the preparation of photographically useful cyanine dyes.[93,94,117,118,128,129] A number of selenazole derivatives such as the selenium analog of vitamin B_1 have been tested for physiological activity.[130] A number of sulfonamides may be prepared from 2-aminoselenazoles by the usual methods.[99,131-134] Recently some potential radioprotective selenazole derivatives were reported.[135]

ii. Selenazolines and Selenazolidines. The first derivatives with a defined selenazoline structure were reported in 1892 by Michels.[136] In the course of his investigations on sulfur and selenium derivatives of ethylamine, he prepared 2-methyl-2-selenazoline (**70**) by reaction of bis(2-aminoethyl) diselenide with acetic anhydride to give the diacetyl derivative, which underwent ring closure with phosphorus pentachloride (equation 32).

(32)

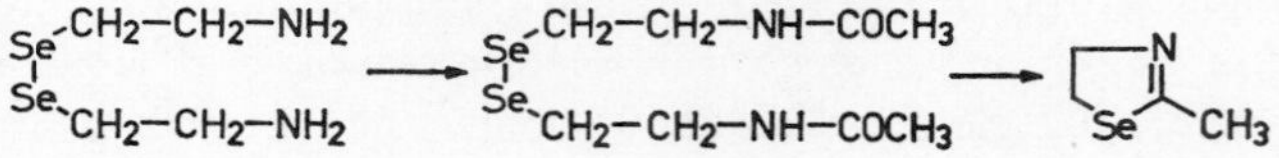

Compound **70** was also reported from the reaction of 2-bromoethylamine hydrobromide with selenoacetamide.[137] Van Dormael[138] developed another method for the preparation of **70** by heating acetamidoethanol with phosphorus pentaselenide (equation 33).

(33)

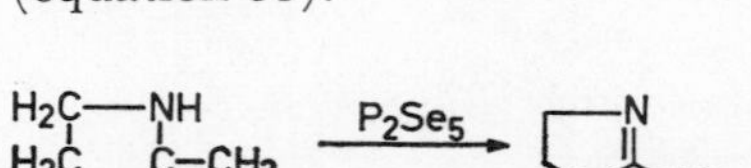

Other 2-alkyl(aryl)-substituted compounds were obtained in the same manner by use of suitably substituted N-acylethanolamines.[139,140] Starting from 2-bromoethylamine hydrobromide and selenourea, Chu and Mautner[141] prepared the 2-aminoethylselenopseudourea salt, which was cyclized to 2-amino-2-selenazoline (**71**) by heating in aqueous solution (equation 34).

(34)

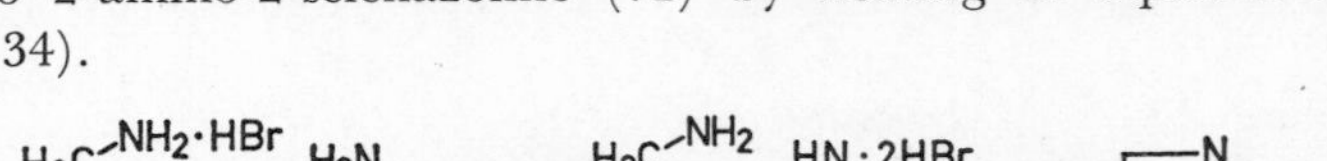

Compound **71** and its 5-methyl derivative had been prepared earlier by Baringer[1] from 2-bromoethylamine hydrobromide and 1-amino-2-bromo-propane, respectively, by reaction with potassium selenocyanate. He arbitrarily assigned an imino structure to his products. Klayman[142] obtained **71** by reaction of cyanide ion with the sodium salt of 2-aminoethaneseleno-sulfuric acid (equation 35).

(35)

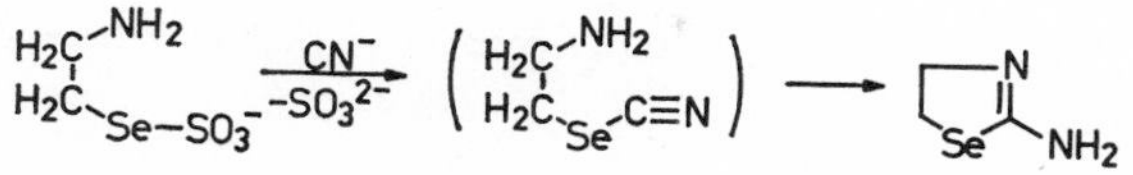

Compound **71** has been ring-opened in aqueous solution by treatment with alkali[142a] and hydrogen sulfide[142b]. Air-oxidation of the resulting selenols to corresponding diselenides gave 1,1'-(diselenodiethylene)bis(2-urea) and 1,1'-(diselenodiethylene)bis(2-thiourea) (**71a**), respectively.

$$\text{(H}_2\text{NCNHCH}_2\text{CH}_2\text{Se--)}_2$$
$$\overset{\overset{\text{S}}{\parallel}}{}$$

71a

Addition of phenyl isothiocyanate to **71** yielded 1-phenyl-3-(2-selen-azolin-2-yl)-2-thiourea (**72**). This was converted to 2-phenylimino-3-phenyl-4-thioxoselenazolo[3,2-a]tetrahydro-s-triazine (**73**)[143] on heating with an excess of phenyl isothiocyanate in acetonitrile (equation 36).

(36)

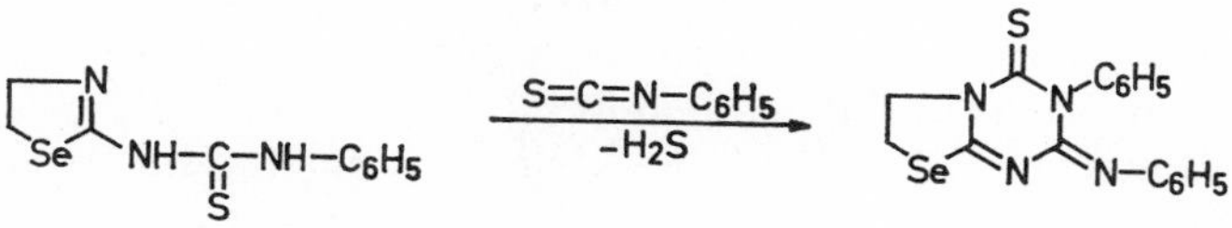

The crystal structure of the antiradiation drug, 2-amino-1,3-selenazol-2-inium bromide, was determined by X-ray diffraction with the help of a Patterson synthesis.[143a]

4-Oxoselenazolines with a disubstituted 2-amino function (**74**) were obtained by reaction of α-halocarboxylic acids with 1,1-dimethyl-, 1,1-diethyl-, and 1-methyl-1-phenyl-2-selenourea (equation 37).[109,144-147]

(37)

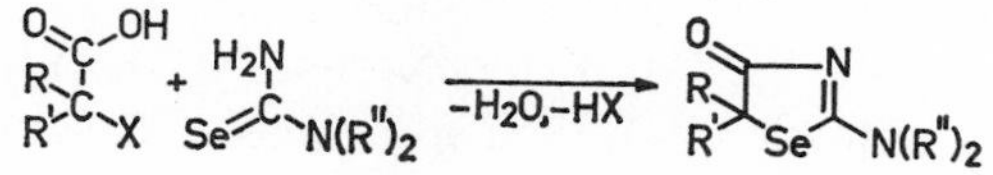

Although compounds of type **74** have a fixed 2-selenazoline structure, any species containing at least one H atom on the amino group offers the possibility for tautomerism, as shown in structures **75a** and **75b**. This problem has recently been investigated by means of spectroscopic measurements. Comrie *et al.*[147] prepared for this purpose a number of compounds of type **75**

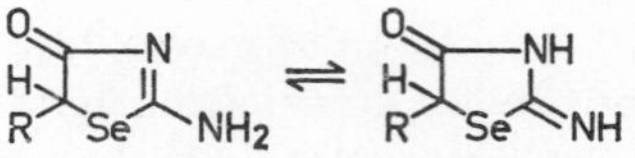

$$75a \rightleftharpoons 75b$$

by condensation of selenoureas with α-halocarboxylic acids. The parent compound of this class had been obtained by Hofmann[92] in 1889 through chloroacetic acid and selenourea. By using 1,3-diethyl-2-selenourea, a number of derivatives (76) with a fixed 2-imino structure were also prepared.

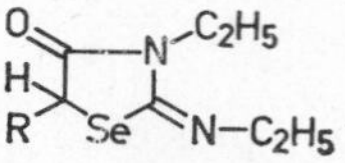

$$76$$

From a comparison of the u.v. spectra of the three types, 74, 75, and 76, Comrie et al.[147] deduced that in the equilibrium mixture the imino form of 75 predominates. This result does not agree, however, with the findings obtained for the analogous oxygen and sulfur compounds.

Since the spectra were measured in the presence of an equivalent amount of acid and thus represent only the cations and not the neutral molecules, the matter was studied further by Najer et al.[144–146] An additional class of compounds with the general structure 77 was prepared. The spectroscopic data indicated that the amino form, 75a, predominates in the case of alkyl-substituted compounds, while the imino form, 75b, is in excess only with aryl substitution. Thus the results fit the general scheme. New spectroscopic evidence indicated that the amino form, 77a, predominates when R represents a H atom or an alkyl residue, while the imino form, 77b, predominates when R stands for an aryl group. Since all measurements were made in alcoholic solution, the authors restrict their conclusions to this solvent.

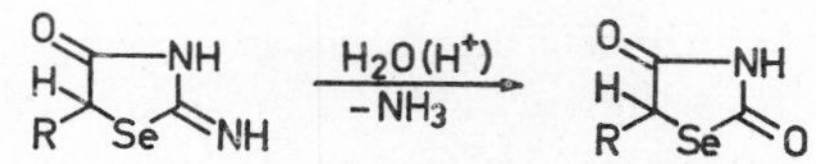

$$77a \rightleftharpoons 77b$$

(X = O, S, Se)

Generally, the reactivity of the compounds indicates predominance of the tautomer 77b, since this explains the ready hydrolysis of these compounds to 2-oxo derivatives (equation 38).[148] For example, Hofmann[92] noted the hydrolysis of 2-imino-4-oxoselenazolidine (75) to 78. This type of hydrolysis

(38)

$$75 \xrightarrow[-NH_3]{H_2O(H^+)} 78$$

always occurs when the possibility for tautomerism to the imino form exists. The reaction fails with a fixed amino structure, as in the case of **74**.

A number of 2-alkylidene(arylidene)-2,4-dioxoselenazolidines (**79**) were

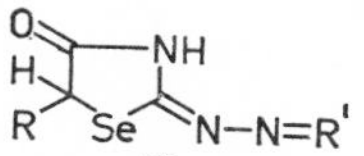

prepared by the reaction of α-halocarboxylic acids with selenosemi-carbazones derived from acetone or benzaldehyde.[148]

2-Thio-4-oxoselenazolidines (**80**), substituted in position 3, have been described in the patent literature.[149,150] They serve as intermediates for pharmaceutical products and photographic sensitizing dyes. The compounds were obtained by reaction of isothiocyanates with 2-hydroselenoacetic acid in the presence of a strong acid and with exclusion of atmospheric oxygen (equation 39).

(39)

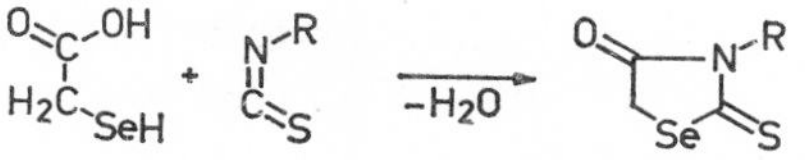

A number of selenazolidines (**81**) were prepared by Draguet and Renson,[151] starting with an aziridine, *via* two synthetic pathways. The first involved reaction of an aziridine with an aldehyde or ketone and treatment of the intermediate product with hydrogen selenide (equation 40). Selenazolidines

(40)

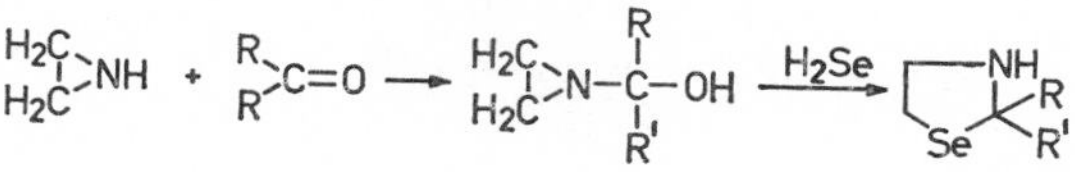

were also obtained when an aziridine was first treated with hydrogen selenide and the product was condensed with a carbonyl compound (equation 41).

(41)

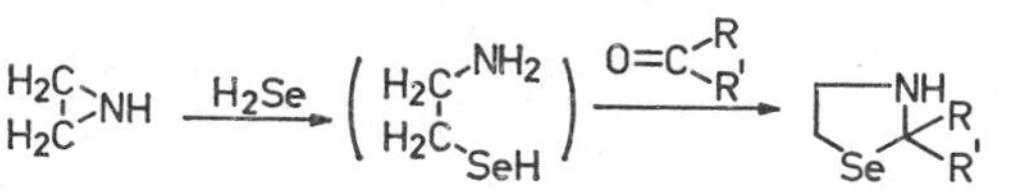

In a similar fashion 2-methyl- and 2,2-dimethylaziridines were converted to 2,4-disubstituted selenazolidines (**82**) (equation 40), whereas *N*-substituted aziridines were converted to 2,3-disubstituted selenazolidines (**83**) according to equation 41.

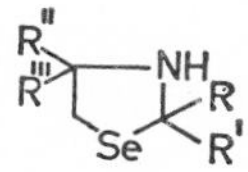

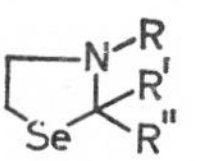

On treatment of 2,2-dimethylselenazolidine (**84**) with reactive aromatic aldehydes, an exchange of the acetone moiety occurred to form **85** according to equation 42.[151]

(42)

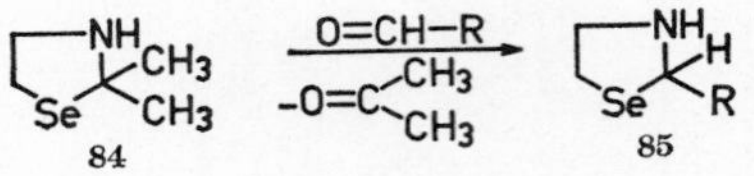

Benzoylation of selenazolidines **81** in alkaline solution cleaved the selenazolidine ring and led to *N,Se*-dibenzoylselenoalkylamines (**86**). The selena-

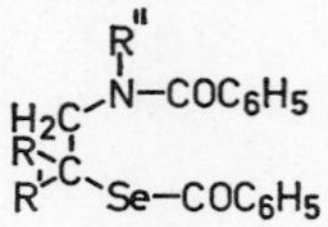

zolidines were acylated at the N atom without ring cleavage when the reaction was carried out in benzene in the presence of pyridine. The hydrolysis and subsequent oxidation in air of the benzoylated selenazolidines led to *N,N'*-dibenzoylselenocystamine (**87**).

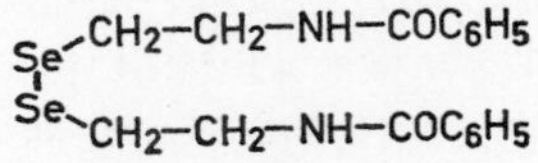

Methylation of 2-phenylselenazolidine (**88**) yielded benzalamino-*Se*-methylselenocysteamine (**89**) (equation 43).[151]

(43)

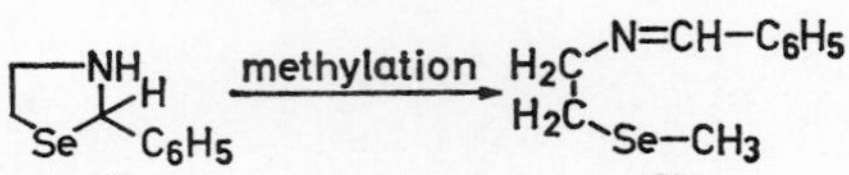

iii. Benzoselenazoles. The first example of this compound class was described in 1913 by Bauer.[152] By reaction of zinc bis(*o*-aminophenyl-selenolate) with benzoyl chloride he obtained 2-phenylbenzoselenazole (**90**) (equation 44). The same method, with only minor variations, has remained

(44)

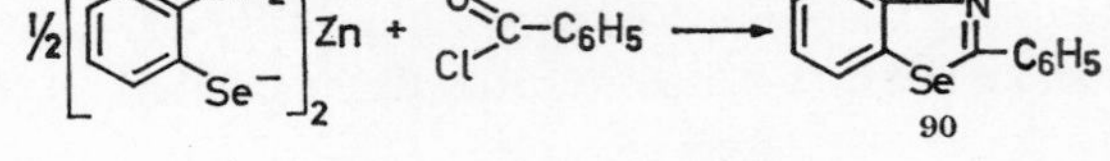

the most important pathway for the preparation of benzoselenazoles. Starting from aliphatic,[153–157] aromatic,[156,158,159] and heterocyclic[159] acid chlorides, various workers have obtained numerous 2-substituted benzoselenazoles by this procedure. Compound **90** was also prepared by heating benzanilide with selenium.[160]

The unsubstituted ring system (**91**) was obtained by synthesis with formic acid.[154,156,161] In a corresponding manner 1,2-di[benzoselenazolyl-(2)]-alkanes (**92**) were generated[156] from dichlorides of dicarboxylic acids.

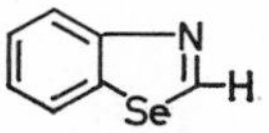

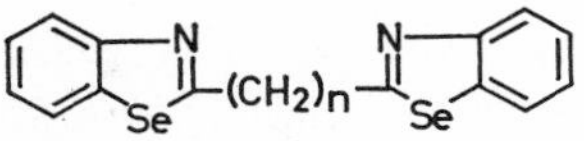

Cyclic anhydrides of dicarboxylic acids, however, react with one equivalent of zinc bis(o-aminophenylselenolate) to give 2-benzoselenazolylalkane(arene)-carboxylic acids (**93, 94, 95**).[162]

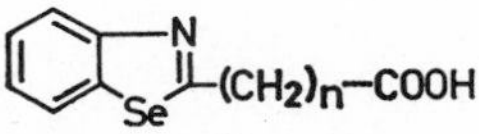

Hasan and Hunter[163] obtained 2-aminobenzoselenazole (**96**) by the oxidation of 1-phenyl-2-selenourea with bromine (equation 45). Starting from

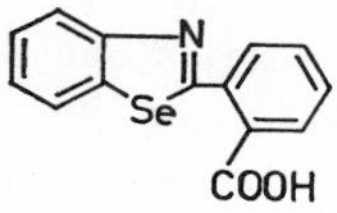

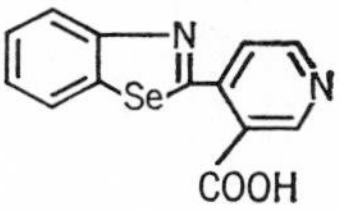

(45)

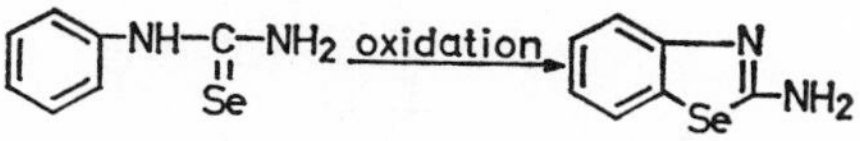

other substituted selenoureas, these authors[163] and Haginiwa[164] prepared a variety of 2-aminobenzoselenazoles with substituents on the benzene ring or on the amino function.

Another method for the preparation of 2-aminobenzoselenazoles utilizes the cyclization of o-aminoaryl selenocyanates (equation 46).[165]

(46)

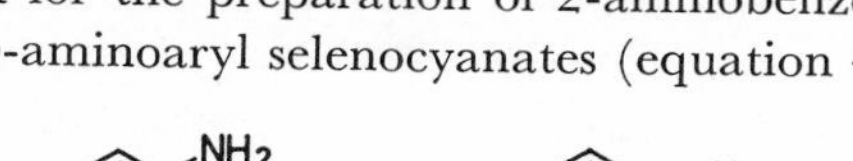

Routes leading to 2-hydroxybenzoselenazole (**97**) involve either combination of zinc (o-aminophenylselenolate) with phosgene or a reaction sequence in which bis(o-aminophenyl) diselenide and ethyl chloroformate yield bis(o-urethanophenyl) diselenide, followed by a cyclization reduction (equation 47).[163] In an analogous fashion, 2-mercaptobenzoselenazole (**98**) was obtained

(47)

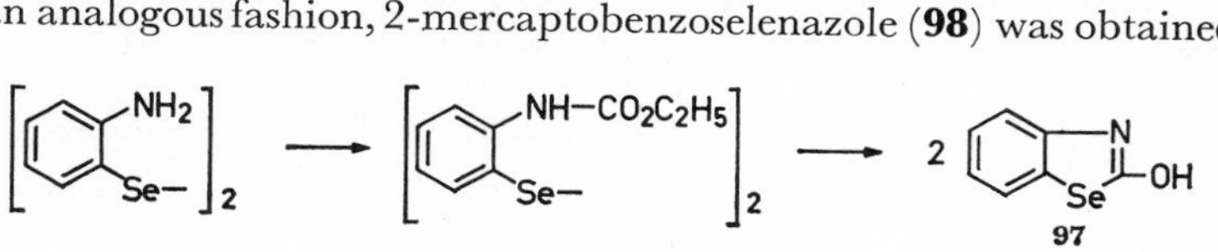

by reaction with thiophosgene (equation 48) or by reaction of bis(o-amino-phenyl) diselenide with carbon disulfide.[163,166]

(48)

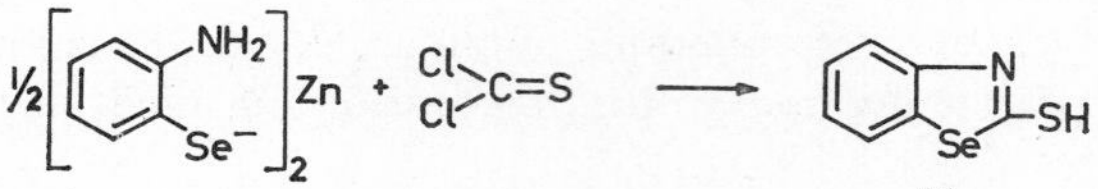

93

Nuclear magnetic resonance investigations of benzoselenazoles and comparisons with the benzoxazole and benzothiazole derivatives were carried out by Di Modica et al.[167,168] and by Tobiason and Goldstein.[169] The 2-alkyl(aryl)benzoselenazoles are weak bases that form hydrolytically unstable mineral acid salts. Reaction with alkyl halides yields benzoselenazolium salts. The 2-position of benzoselenazole (91) appears to be very unreactive; thus, attempted amination with sodium amide failed.[154,156] The action of benzoyl chloride in the presence of potassium cyanide results in ring scission,[154] as is the case for the corresponding benzothiazole.

A methyl group in the 2-position of benzoselenazole is activated and reacts readily with aromatic aldehydes to yield the corresponding styrene derivatives.[154,156,170,171] The case for benzaldehyde is illustrated in equation 49. Product 100 can also be obtained, however, by the general method

(49)

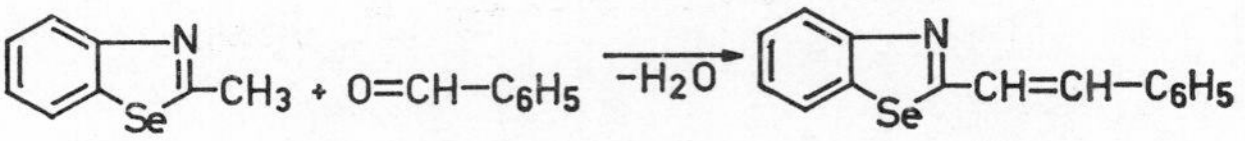

99   100

(equation 44), which involves treatment of the zinc salt of o-aminophenyl-selenol with cinnamoyl chloride.[156]

Benzoselenazole-2-aldehyde (103) was prepared by Hamer[172] by the reaction of 2-methylbenzoselenazole (99) with chloral to 2-(3,3,3-trichloro-2-hydroxypropyl)benzoselenazole (101). Saponification of the latter with sodium ethoxide to β-2-benzoselenazolylacrylic acid (102), followed by oxidation with potassium permanganate in alkaline solution, gave the desired compound. A by-product of the last step of scheme 50 is the carboxylic acid 104.

(50)

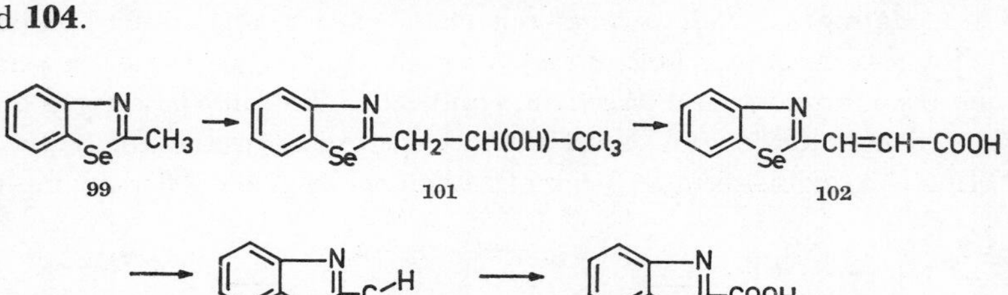

99   101   102

103   104

The aldehyde **103** was also obtained[173] by oxidation of **99** with selenium dioxide in xylene at 90°. Condensation with aromatic amines gave the corresponding anils (**105**),[172] and the benzoin condensation yielded benzoselenazoin (**106**).[174] The latter compound (**106**) can be oxidized to benzoselenazil with oxygen in boiling xylene or with bromine in chloroform.

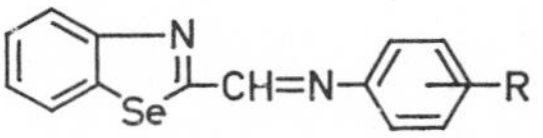

105

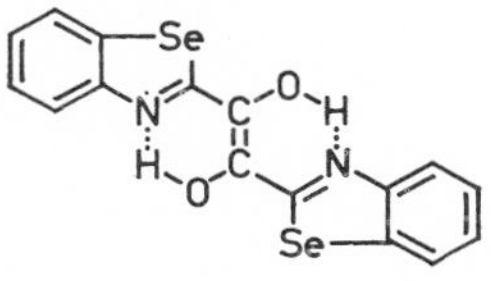

106

Oxidation with atmospheric oxygen in alkaline solution or with hydrogen peroxide led to benzoselenazole-2-carboxylic acid (**104**), which, in turn, decarboxylates to benzoselenazole (**91**) upon melting.[174]

Under mild conditions the nitration of benzoselenazoles gave 6-nitro derivatives.[154,158] The nitration of 2-phenylbenzoselenazole, under more vigorous conditions, yielded a dinitro compound that is probably the 4,6-substituted species.[158] The reduction of 6-nitro-2-phenylbenzoselenazole (**107**) gave the amino derivative **108**, which condensed readily with benzaldehyde to the corresponding benzylidene anil **109** (equation 51). Attempted halogenation of benzoselenazoles did not yield any well-defined products.[156]

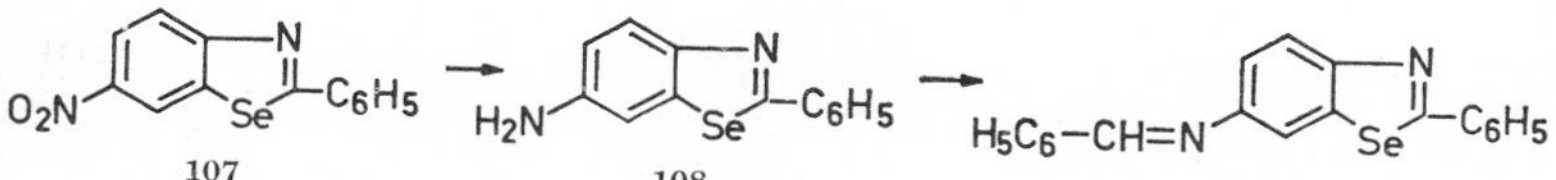

The amino group in position 2 shows behavior typical of an aromatic amino group, giving, for instance, normal diazotization products.[163] The diazonium salt, when subjected to the Sandmeyer reaction, yielded the 2-chloro derivative, which, in turn, gave **97** on hydrolysis,[163] **98** by reaction with sodium hydrogen sulfide, and 2-selenylbenzoselenazole (**110**) by reaction with sodium hydrogen selenide.[175]

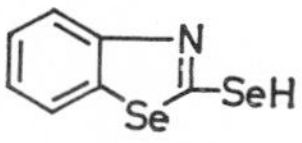

110

Upon methylation of **96**, 2-imino-3-methyl-2,3-dihydrobenzoselenazole (**111**) was formed. The isomeric 2-methylamino compound (**112**) was obtained according to equation 45 from 1-methyl-3-phenyl-2-selenourea.[163]

The alkali salts of 2-mercaptobenzoselenazole (**98**) on treatment with

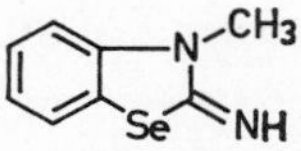

111

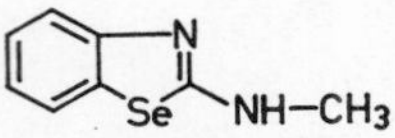

112

alkyl halides readily gave the corresponding thioethers.[176] Some 2-alkyl-thio-6-nitrobenzoselenazoles (**113**) showed *in vitro* tuberculostatic activity.[176]

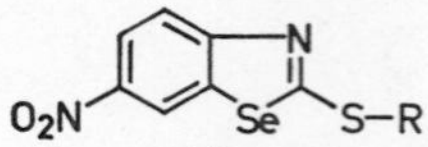

113

The patent literature contains descriptions of 2-mercapto- and 2-alkyl-(aryl)thiobenzoselenazoles used as vulcanization accelerators.[177] In the course of investigations on reactive heterocyclic 2-hydrazones, Hünig and Balli[178] prepared *N*-methylbenzoselenazolone-2-hydrazone (**114**). The latter was made from 2-mercaptobenzoselenazole (**98**), *via* the methylthio deriv-ative, which was converted to the quaternary onium salt, and subjected to hydrazinolysis. The hydrazone **114** was isolated as its *p*-toluenesulfonate salt.[178]

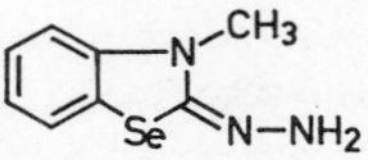

114

The Se–H moiety in 2-selenylbenzoselenazoles (**110**) undergoes reactions similar to those of the mercapto group. Thus oxidation of **110** yielded bis-(benzoselenazolyl) diselenide (**115**). Reaction of the sodium salt with aryl halides gave 2-arylselenobenzoselenazolines (**116**), whereas treatment with formaldehyde and an aromatic amine yielded the corresponding 2-(aryl-aminomethyleneseleno)benzoselenazoline (**117**).[175] These and similar selenyl derivatives were also described as vulcanization accelerators.[179]

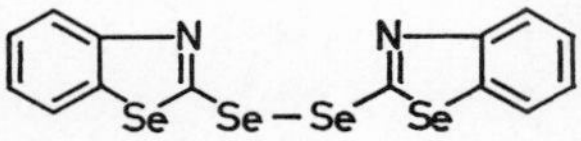

115

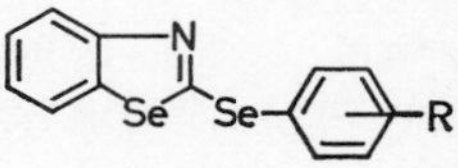

116

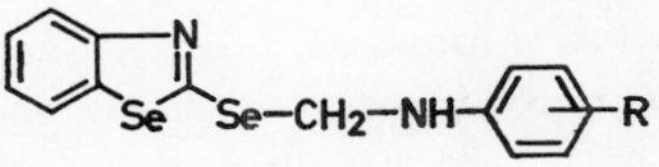

117

iv. Benzoselenazolines. Of considerable practical importance is the above-mentioned ability of benzoselenazoles to form quaternary salts. These are used in numerous ways in the synthesis of cyanine dyes. Furthermore, it is easy to convert benzoselenazolium salts into derivatives of benzoselenazoline, which are also used in dye syntheses. The considerable patent literature in this field is covered in Chapter XIE, and only a few examples will be cited here.

By coupling 2,3-dihydrobenzoselenazolium salts with diazotized amines, Wahl[180] generated bisazo dyes of the 2,3-dihydrobenzoselenazole type (**118.**)

Taber *et al.*[181] succeeded in generating the corresponding fulvene (**119**) by treating the β-acetanilidovinyl derivative of 3-ethyl-2-methylbenzoselenazolium iodide with tetraphenylcyclopentadiene.

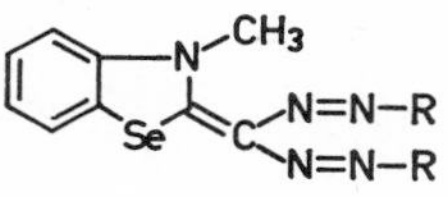

118

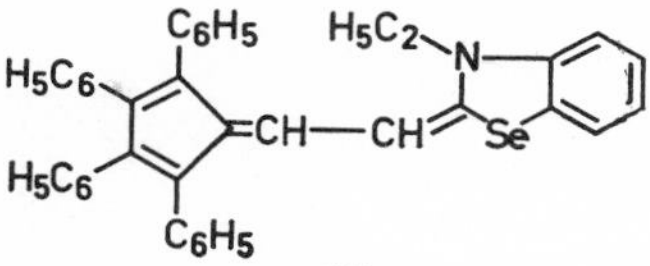

119

Steric hindrance in quaternary benzoselenazolium salts was described by Kiprianov and Shrubovich.[170] 2-Acylmethylene-3-ethylbenzoselenazolines (**120**) are formed by treatment of 3-ethyl-2-methylbenzoselenazolium salts with various acid halides in dry pyridine.[182]

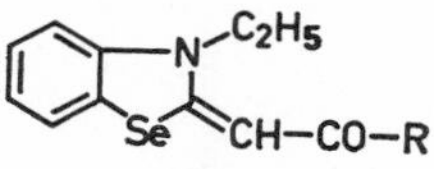

120

The action of a cold, saturated aqueous solution of potassium cyanide on a saturated solution of a 3-methylbenzoselenazolium salt generated 2-cyanobenzoselenazoline (**121**).[183] Brunken and Poppe[184] described dimethinemero-

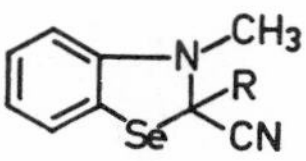

121

cyanines of type **122**, which contain a benzoselenazoline ring system.

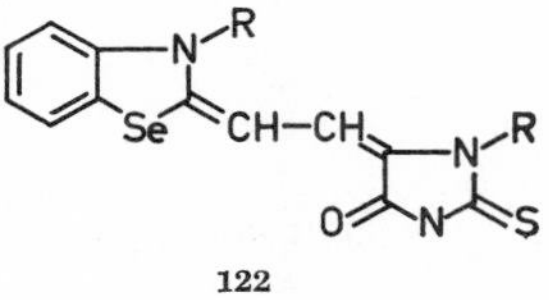

122

Pianka and Hall[185] noted that some selenacarbocyanines exhibited noticeable fungicidal properties.

Benzoselenazole analogs containing a hydrogenated benzene ring, as, for instance, the 2-amino-4,5,6,7-tetrahydrobenzoselenazole (**123**), were

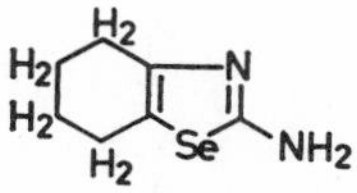

123

obtained by the general selenazole synthesis (*cf.* equation 24), starting from α-chlorocyclohexanone and selenourea.[98,104] 2-Alkyl(aryl)hexahydro-benzoselenazoles (**124**) were synthesized by treatment of *o*-chlorocyclohexylamine with selenocarboxamides.[186]

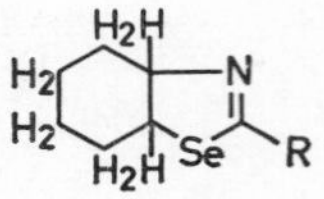

124

v. Selenazoles with Various Condensed Rings. Derivatives of two isomer series are known for naphthoselenazoles. Knott[100,187,188] synthesized 2-acetamido derivatives of naphtho[1,2-d]selenazole (β-naphthoselenazole, **126**) by cyclization of 2-amino-4-arylselenazole-5-acetic acid (**125**) under reflux with acetic anhydride (equation 52). These compounds were referred

(52)

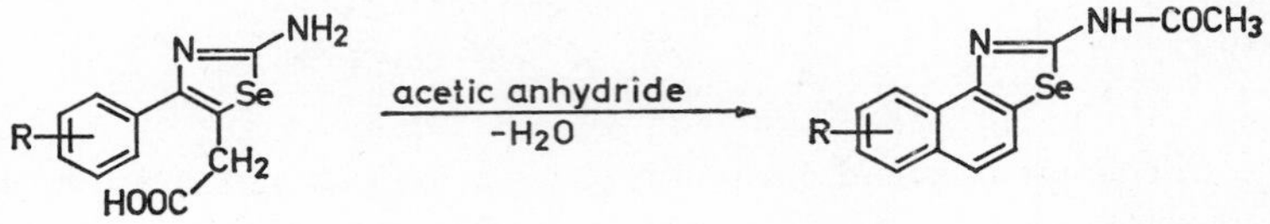

125   126

to in the literature as naphtha-1':2':4:5-selenazoles.

2-Amino derivatives of naphtho[2,1-d]selenazole, 1-amino-α-naphthoselenazole (**127**), were prepared by Hasan and Hunter[163] by heating 1-bromo-2-naphthylamine with potassium selenocyanate for 12 hours (equation 53).

(53)

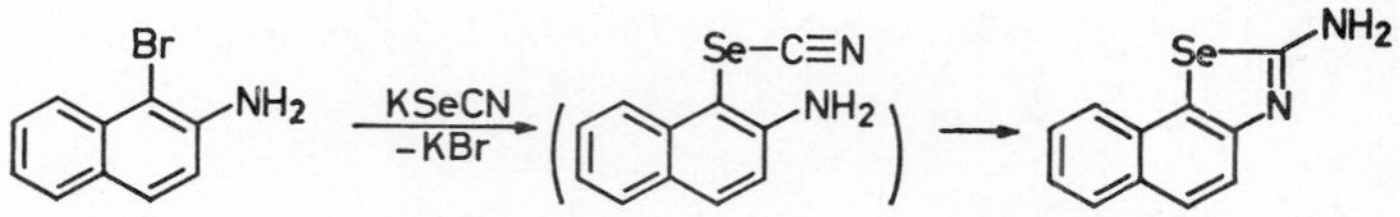

127

By heating the benzalamino compound **109** with sulfur or with selenium, Bogert and Hopkins[189] obtained 2,6-diphenylselenazolo[5,4-f]benzothiazole (**128**) or 2,6-diphenylbenzo[1,2-d:4,5-d']bisselenazole (**129**), respectively.

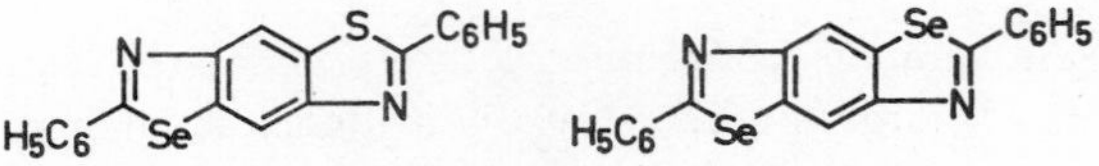

128   129

Treibs and Poppe[190] described 2'-methylbenzoselenazolo[6',5':2,3]-1-azulene (**130**). This was prepared by reduction of 2-methyl-6-nitrobenzoselenazole to the 6-amino derivative, diazotization, and reduction to the hydrazino compound. Further treatment with cyclohexanone and dehydrogenation of the intermediate hexahydro derivative with chloranil in boiling toluene gave **130** as the final product. 2-Methylbenzoselenazole reacts with

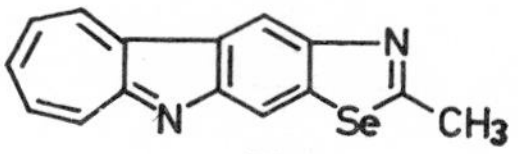

130

2 moles of dimethyl acetylenedicarboxylate in THF solution over a 12-hour period at 0° to give tetramethyl 9,10-dihydrobenzoselenazolo[3,2-a]-azepine-7,8,9,10-tetracarboxylate (**131**).[191]

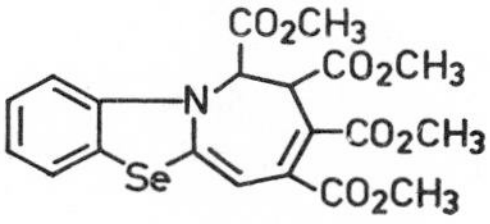

131

Quaternary salts of types **132** and **133** were obtained by cyclization of 2-benzoselenazolylalkyl(aryl)carbinols with phosphorus tribromide.[193] The carbinols are obtained by reduction of the esters of carboxylic acids **93–95**.[192]

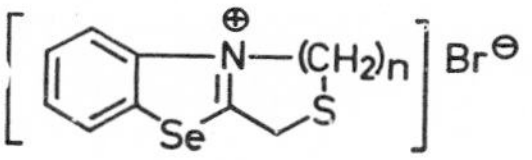

132

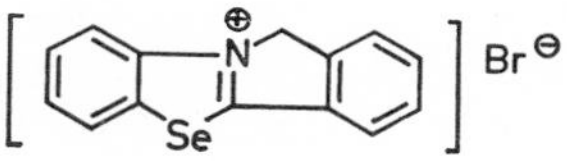

133

Treatment of these quaternary salts with alkali yields compounds such as the pyrido[2,3-c]benzoselenazolo[3,2-a]pyrrole (**134**).[193]

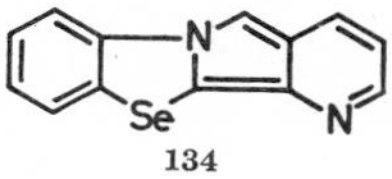

134

REFERENCES

1. W. Baringer, *Ber.*, **23**, 1003 (1890).
2. F. I. Luknitskii, D. O. Taube, and B. A. Vovsi, *Zh. Org. Khim.*, **5**, 1844 (1969); *Chem. Abstr.*, **72**, 21671 (1970).
3. F. I. Luknitskii, D. O. Taube, and B. A. Vovsi, *Dokl. Akad. Nauk SSSR*, **184**, 355 (1969); *Chem. Abstr.*, **70**, 96772 (1969).
4a. G. Simchen, *Angew. Chem. (Intern. Ed.)*, **7**, 464 (1968); *Angew. Chem.*, **80**, 484 (1968).
4b. G. Simchen and J. Wenzelburger, *Chem. Ber.*, **103**, 413 (1970).
5. F. Asinger, H. Berding, and H. Offermanns, *Monatsh. Chem.*, **99**, 2084 (1968).
6. O. Behaghel and M. Rollmann, *J. Prakt. Chem.*, [2], **123**, 341 (1929).

7. A. I. Kiprianov and V. P. Khilya, *Zh. Org. Khim.*, **3**, 1097 (1967); *Chem. Abstr.*, **69**, 37081c (1968).

8. L. K. Mushkalo and D. I. Sheiko, *Ukr. Khim. Zh.*, **30**, 384 (1964); *Chem. Abstr.*, **61**, 4359e (1964).

9a. W. Cornelius, *J. Prakt. Chem.*, [2], **88**, 395 (1913).

9b. B. D. Podolesov and V. B. Jordanovska, *Croat. Chem. Acta*, **42**, 61 (1970); *Chem. Abstr.*, **73**, 3895p (1970).

9c. G. Cordella and F. Sparatore, *Farmaco, Ed. Sci.*, **20**, 446 (1965).

9d. U.S. Pat. 3,043,839 (1962); *Chem. Abstr.*, **57**, 16636b (1962).

9e. G. Cordella and F. Sparatore, *Ann. Chim.* (Rome), **56**, 182 (1966).

9f. D. Simov and N. Khristova, *Godishnik Sofistiya Univ., Khim. Fak.*, **59**, 237 (1964–1965) (Pub. 1966); *Chem. Abstr.*, **67**, 108642h (1967).

10. F. Sachs, *Justus Liebigs Ann. Chem.*, **365**, 142 (1909).

11. E. Bulka and K.-D. Ahlers, *Z. Chem.*, **3**, 349 (1963).

12. E. Bulka, K.-D. Ahlers, and H. Beyer, *Chem. Ber.*, **94**, 1122 (1961).

13. E. Bulka and K.-G. Thiemann, unpublished work.

14. L. K. Mushkalo and D. I. Sheiko, *Zh. Obshch. Khim.*, **33**, 157 (1963); *Chem. Abstr.*, **59**, 631e (1963).

15. L. K. Mushkalo, D. I. Sheiko, and E. I. Lanovaya, *Ukr. Khim. Zh.*, **30**, 502 (1964); *Chem. Abstr.*, **61**, 5654c (1964).

16. L. L. Bambas, Five-Membered Nitrogen Compounds with Nitrogen and Sulfur or Nitrogen, Sulfur, or Oxygen (Except Thiazole), Interscience Publishers, New York N.Y., 1952, p. 215.

17a. S. Keimatsu and I. Satoda, *Yakugaku Zasshi*, **55**, 233 (1935); *Chem. Abstr.*, **31**, 6661[9] (1937).

17b. I. Lalezari, A. Shafiee, and M. Yalpani, *Tetrahedron Lett.*, 5105 (1969).

17c. I. Lalezari, A. Shafiee, and M. Yalpani, *Angew. Chem.*, **82**, 484 (1970).

18. W. Becker and J. Meyer, *Ber.*, **37**, 2553 (1904).

19. J. Goerdeler, D. Gross, and H. Klinke, *Chem. Ber.*, **96**, 1289 (1963).

20. O. Hinsberg, *Ber.*, **22**, 862, 2895 (1889).

21. L. Barcza, *Acta Chim. Acad. Sci. Hung.*, **41**, 91 (1964); *Chem. Abstr.*, **62**, 4621f (1965).

22. H. Aritoshi, M. Kiniwa, and K. Toei, *Talanta*, **5**, 112 (1960); *Chem. Abstr.*, **55**, 3298b (1961).

23. L. J. Throop, *Anal. Chem.*, **32**, 1807 (1960).

24. P. F. Lott, P. Cukor, G. Moriber, and J. Solga, *Anal. Chem.*, **35**, 1159 (1963).

25. E. Sawicki, *J. Org. Chem.*, **18**, 1492 (1953).

26. E. Sawicki, *J. Amer. Chem. Soc.*, **76**, 665 (1954).

27. E. Sawicki, *J. Org. Chem.*, **19**, 1163 (1954).

28. E. Sawicki, *J. Amer. Chem. Soc.*, **77**, 957 (1955).

29. E. Sawicki, B. Chastain, and H. Bryant, *J. Org. Chem.*, **21**, 754 (1956).

30. E. Sawicki and A. Carr, *J. Org. Chem.*, **22**, 503, 507 (1957).

31. E. Sawicki and A. Carr, *J. Org. Chem.*, **23**, 610 (1958).

32. A. Carr, E. Sawicki, and F. E. Ray, *J. Org. Chem.*, **23**, 1940 (1958).

33. L. S. Efros and Z. V. Todres-Selektor, *Zh. Obshch. Khim.*, **27**, 983 (1957); *Chem. Abstr.*, **52**, 2845g (1958).

34. L. S. Efros and Z. V. Todres-Selektor, *Zh. Obshch. Khim.*, **27**, 3127 (1957); *Chem. Abstr.*, **52**, 9089c (1958).

35. L. S. Efros and A. V. Eltsov, *Zh. Obshch. Khim.*, **28**, 2172 (1958); *Chem. Abstr.*, **53**, 2207b (1959).

36. L. S. Efros, *Usp. Khim.*, **29**, 162 (1960); *Chem. Abstr.*, **54**, 14091d (1960).

37. V. A. Pozdyshev, Z. V. Todres-Selektor, and L. S. Efros, *Zh. Obshch. Khim.*, **30**, 2551 (1960); *Chem. Abstr.*, **55**, 12040c (1961).

38. L. S. Efros, R. P. Polyakova, and M. G. Argitti, *Zh. Obshch. Khim.*, **32**, 516 (1962); *Chem. Abstr.*, **58**, 1450e (1963).

39. V. G. Pesin, A. M. Khaletskii, and L. V. Zolotova-Zolotukhina, *Zh. Obshch. Khim.*, **31**, 3000 (1961); *Chem. Abstr.*, **57**, 4651g (1962).

40. V. G. Pesin, A. M. Khaletskii, and L. V. Zolotova-Zolotukhina, *Zh. Obshch. Khim.*, **33**, 1101 (1963); *Chem. Abstr.*, **59**, 13968f (1963).

41. V. G. Pesin, A. M. Khaletskii, and L. V. Zolotova-Zolotukhina, *Zh. Obshch. Khim.*, **33**, 3716 (1963); *Chem. Abstr.*, **60**, 8021b (1964).

42. V. G. Pesin, L. V. Zolotova-Zolotukhina, and A. M. Khaletskii, *Zh. Obshch. Khim.*, **34**, 255 (1964); *Chem. Abstr.*, **60**, 10672g (1964).

43. V. G. Pesin, A. M. Khaletskii, and V. A. Sergeev, *Zh. Obshch. Khim.*, **34**, 261 (1964); *Chem. Abstr.*, **60**, 10673b (1964).

44. V. G. Pesin, S. A. Dyachenko, and A. M. Khaletskii, *Zh. Obshch. Khim.*, **34**, 1258 (1964); *Chem. Abstr.*, **61**, 1851h (1964).

45. V. G. Pesin and R. S. Muravnik, *Latv. PSR Zinat. Akad. Vestis*, 725 (1964); *Chem. Abstr.*, **63**, 4279c (1965).

46. V. G. Pesin and S. A. Dyachenko, *Khim. Geterotsikl. Soedin.*, 533 (1966); *Chem. Abstr.*, **66**, 28711 (1967).

47. V. G. Pesin and S. A. Dyachenko, *Khim. Geterotsikl. Soedin.*, 100 (1967); *Chem. Abstr.*, **67**, 82163a (1967).

48. V. G. Pesin and S. A. Dyachenko, *Khim. Geterotsikl. Soedin.*, 950 (1967); *Chem. Abstr.*, **68**, 87533b (1968).

49. V. G. Pesin, V. A. Sergeev, and A. G. Nesterova, *Khim. Geterotsikl. Soedin.*, 95 (1968); *Chem. Abstr.*, **69**, 96603 (1968).

50. V. G. Pesin, *Khim. Geterotsikl. Soedin.*, 235 (1969); *Chem. Abstr.*, **71**, 22076 (1969).

51. N. P. Buu-Hoi, *J. Chem. Soc.*, 2884 (1949).

52a. M. V. Gorelik, *Khim. Geterotsikl. Soedin.*, 541 (1967); *Chem. Abstr.*, **68**, 21730g (1968).

52b. M. V. Gorelik, S. B. Lantsman, and T. P. Kononova, *Khim. Geterotsikl. Soedin.*, 808 (1968); *Chem. Abstr.*, **71**, 22083 (1969).

53a. N. M. D. Brown and P. Bladon, *Tetrahedron*, **24**, 6577 (1968).

53b. G. H. Harts, K. B. DeRoos, and C. A. Salemink, *Rec. Trav. Chim. Pays-Bas*, **89**, 5 (1970).

54. F. Sachs and G. Meyerheim, *Ber.*, **41**, 3957 (1908).

55. J. T. Ralph and C. E. Marks, *Tetrahedron*, **22**, 2487 (1966).

56. C. Brizzi, D. Dal Monte, and E. Sandri, *Ann. Chim.* (Rome), **54**, 476 (1964).

57a. C. W. Bird, G. W. H. Cheeseman, and A. A. Sarsfield, *J. Chem. Soc.*, 4767 (1963).

57b. J. A. Elvidge, G. T. Newbold, A. Percival, and I. R. Senciall, *J. Chem. Soc.*, 5119 (1965).

58. D. Dal Monte and E. Sandri, *Ann. Chim.* (Rome), **53,** 1697 (1963).

59a. A. J. Nunn and J. T. Ralph. *J. Chem. Soc.,* 6769 (1965).

59b. A. J. Nunn and J. T. Ralph, *J. Chem. Soc., C,* 1568 (1966).

59c. G. A. Jaffari, A. J. Nunn, and J. T. Ralph, *J. Chem. Soc., C,* 2060 (1970).

60. E. S. Lane and C. Williams, *J. Chem. Soc.,* 1468 (1955).

61. M. V. Gorelik and V. V. Puchkova, *Zh. Org. Khim.,* **5,** 376 (1969); *Chem. Abstr.,* **70,** 114872 (1969).

62a. J. M. Hollas and R. A. Wright, *Spectrochim. Acta,* **25A,** 1211 (1969).

62b. D. Dal Monte and E. Sandri, *Boll. Sci. Fac. Chim. Ind. Bologna,* **22,** 33 (1964); *Chem. Abstr.,* **62,** 1542f (1965).

63. C. W. Bird and G. W. H. Cheeseman, *Tetrahedron,* **20,** 1701 (1964).

64. V. S. Korobkov, A. V. Sechkarev, L. P. Zubanova, and N. I. Dvorovenko, *Izv. Vyssh. Ucheb. Zaved., Fiz.,* **11,** 158 (1968); *Chem. Abstr.,* **69,** 47806g (1968).

65. N. M. D. Brown and P. Bladon, *Spectrochim. Acta,* **24A,** 1869 (1968).

66. E. I. Fedin and Z. V. Todres, *Khim. Geterosikl. Soedin.,* 416 (1968); *Chem. Abstr.,* **69,** 111910 (1968).

67. F. L. Tobiason and J. H. Goldstein, *Spectrochim. Acta,* **25A,** 1027 (1969).

68a. L. Braier, P. V. Petrovskii, Z. V. Todres, and E. I. Fedin, *Khim. Geterotsikl. Soedin.,* 62 (1969); *Chem. Abstr.,* **70,** 114423 (1969).

68b. M. Kamiya, S. Katayama, and Y. Akahori, *Chem. Pharm. Bull.* (Tokyo), **17,** 1821 (1969); *Chem. Abstr.,* **71,** 118170 (1969).

69. V. Luzzati, *Acta Cryst.* **4,** 193 (1950).

70. S. I. Zhdanov, V. Sh. Tsveniashvili, and Z. V. Todres, *J. Polarog. Soc.,* **13,** 100 (1967).

71a. V. Sh. Tsveniashvili, S. I. Zhdanov, and Z. V. Todres, *Fresenius' Z. Anal. Chem.,* **224,** 389 (1967).

71b. Z. V. Todres, V. Sh. Tsveniashvili, S. I. Zhdanov, and D. N. Kursanov, *Dokl. Akad. Nauk SSSR,* **181,** 906 (1968); *Chem. Abstr.,* **69,** 105759g (1968).

72. N. P. Buu-Hoi, P. Jaquignon, O. Roussel-Perin, F. Perin, and M. Mangane, *J. Heterocycl. Chem.,* **4,** 415 (1967).

73. D. Shew, *Dissertation Abstr.,* **20,** 1593 (1959); *Chem. Abstr.,* **54,** 4548a (1960).

74. Y. F. Shealy, J. D. Clayton, G. J. Dixon, E. A. Dulmadge, R. F. Pittillo, and D. E. Hunt, *Biochem. Pharmacol.,* **15,** 1610 (1966).

75. Y. F. Shealy and J. D. Clayton, *J. Heterocycl. Chem.,* **4,** 96 (1967).

76. D. E. Hunt and R. F. Pittillo, *Antimicrobial Agents Chemother.,* 551 (1966); *Chem. Abstr.,* **67,** 89502 (1967).

77. L. M. Weinstock, P. Davis, D. M. Mulvey, and J. C. Schaeffer, *Angew. Chem.* (*Intern. Ed.*), **6,** 364 (1967); *Angew. Chem.,* **79,** 315 (1967).

78. V. Bertini, *Angew. Chem.* (*Intern. Ed.*), **6,** 563 (1967); *Angew. Chem.,* **79,** 530 (1967).

79. V. Bertini, *Gazz. Chim. Ital.,* **97,** 1870 (1967).

80. P. Bucci, V. Bertini, G. Ceccarelli, and A. De Munno, *Chem. Phys. Lett.,* **1,** 473 (1967).

81a. E. Benedetti and V. Bertini, *Spectrochim. Acta,* **24A,** 57 (1968).

81b. G. L. Blackman, R. D. Brown, F. R. Burden, and J. E. Kent, *Chem. Phys. Lett.,* **1,** 379 (1967).

82. R. Stollé and L. Gutmann, *J. Prakt. Chem.,* [2], **69,** 509 (1904).

83. E. Bulka and D. Ehlers, unpublished work.

84. E. Bulka and H. Storm, unpublished work.

85a. R. V. Kendall and R. A. Olofson, *J. Org. Chem.*, **35,** 806 (1970).

85b. D. M. Levine, W. D. Krugh, and L. P. Gold, *J. Mol. Spectry.*, **30,** 459 (1969).

86. R. Fusco and C. Musante, *Gazz. Chim. Ital.*, **68,** 665 (1938).

87a. F. E. King and D. G. I. Felton, *J. Chem. Soc.*, 275 (1949).

87b. K. Takeda, R. Kitagawa, T. Konaka, R. Noguchi, H. Nishimura, N. Shimaoka, N. Kimura, K. Kaida, K. Sugiyama, H. Ikeda, and T. Maki, *Ann. Rept. Shionogi Res. Lab.*, **2,** 12 (1952); *Chem. Abstr.*, **51,** 15703c (1957).

87c. K. Takeda, H. Nishimura, N. Shimaoka, R. Noguchi, and K. Nakajima, *Ann. Rept. Shionogi Res. Lab.*, **5,** 1 (1955); *Chem. Abstr.*, **51,** 4559f (1957).

87d. Jap. Pat. 4967 (1958); *Chem. Abstr.*, **53,** 13179g (1959).

88. R. Lesser and R. Weiss, *Ber.*, **57,** 1077 (1924).

89. R. Lesser and R. Weiss, *Ber.*, **45,** 1835 (1912).

90. F. Wille, A. Ascherl, G. Kaupp, and C. Capeller, *Angew. Chem.*, **74,** 753 (1962).

91. E. Bulka, "Advances in Heterocyclic Chemistry," Vol. II, A. R. Katritzky, Ed., Academic Press, New York. N.Y. 1963, p. 343.

92. G. Hofmann, *Justus Liebigs Ann. Chem.*, **250,** 294 (1889).

93. Brit. Pat. 405,028 (1934); *Chem. Zentr.*, **II,** 1250 (1934).

94. L. G. S. Brooker, G. H. Keyes, and F. L. White, *J. Amer. Chem. Soc.*, **57,** 2492 (1935).

95. J. Haginiwa, *Yakugaku Zasshi*, **68,** 191 (1948); *Chem. Abstr.*, **47,** 8074f (1953).

96. J. Metzger and P. Bailly, *C.R. Acad. Sci., Paris*, **237,** 906 (1953).

97. J. Metzger and P. Bailly, *Bull. Soc. Chim. Fr.*, 1249 (1955).

98. H. J. Backer and H. Bos, *Rec. Trav. Chim. Pays-Bas*, **62,** 580 (1943).

99. A. N. Roy and P. C. Guha, *J. Indian Chem. Soc.*, **22,** 82 (1945).

100. E. B. Knott, *J. Chem. Soc.*, 628 (1945).

101. U.S. Pat. 2,423,709 (1947); *Chem. Abstr.*, **41,** 6582i (1947).

102. Brit. Pat. 593,024 (1947); *Chem. Abstr.*, **43,** 8401d (1949).

103. W. Davies, J. A. Maclaren, and L. R. Wilkinson, *J. Chem. Soc.*, 3491 (1950).

104. L. C. King and R. J. Hlavacek, *J. Amer. Chem. Soc.*, **73,** 1864 (1951).

105. E. Bulka, A. Ehlers, D. Ehlers, and H.-E. Nürnberg, unpublished work.

106. I. B. Douglass, *J. Amer. Chem. Soc.*, **59,** 740 (1937).

107. E. Bulka, K.-D. Ahlers, and E. Tuček, *Chem. Ber.*, **100,** 1459 (1967).

108. J. Haginiwa, *Yakugaku Zasshi*, **69,** 566 (1949); *Chem. Abstr.*, **44,** 4465d (1950).

109. R. A. Zingaro, F. C. Bennett, Jr., and G. W. Hammar, *J. Org. Chem.*, **18,** 292 (1953).

110. E. Bulka, K.-D. Ahlers, H. Ewert, W. Haase, P. Meier, and W. Pilz, *Chem. Ber.*, **101,** 1910 (1968).

111. E. Bulka and F. Riemer, unpublished work.

112. E. Bulka, H.-G. Patzwaldt, F.-K. Peper, and H. Beyer, *Chem. Ber.*, **94,** 1127 (1961).

113. E. Bulka, M. Mörner, and H. Beyer, *Chem. Ber.*, **94,** 2763 (1961).

114. E. Bulka, G. Rodekirch, and H. Beyer, *Chem. Ber.*, **95,** 658 (1962).

115. E. Bulka, W. Dietz, H.-G. Patzwaldt, and H. Beyer, *Chem. Ber.*, **96,** 1996 (1963).

116. U.S. Pat. 2,481,674 (1949); *Chem. Abstr.*, **44,** 7173b (1950).

117. U.S. Pat. 2,500,142 (1950); *Chem. Abstr.*, **44,** 7174g (1950).

118. Brit. Pat. 645,901 (1950); *Chem. Abstr.*, **45,** 2985i (1951).

119. J. Haginiwa, *Yakugaku Zasshi*, **68,** 195 (1948); *Chem. Abstr.*, **47,** 8074h (1953).

120. J. Haginiwa, *Yakugaku Zasshi*, **68,** 197 (1948); *Chem. Abstr.*, **47,** 8075b (1953).

121. E. Bulka and K.-D. Ahlers, *Z. Chem.*, **3,** 388 (1963).

122. E. Bulka and R. Thiemann-Ortmann, unpublished work.

123. U.S. Pat. 2,457,078 (1948); *Chem. Abstr.*, **43,** 3042b (1949).

124. E. Bulka, H.-G. Patzwaldt, F.-K. Peper, and H. Beyer, *Chem. Ber.*, **94,** 2759 (1961).

125. Belg. Pat. 707,384 (1968); *Chem. Abstr.*, **71,** 17539 (1969).

126. E. Ochiai, *Yakugaku Zasshi*, **69,** 59 (1949); *Chem. Abstr.*, **44,** 3494a (1950).

127. P. Bassignana, C. Cogrossi, M. Gandino, and P. Merli, *Spectrochim. Acta*, **21,** 605 (1965).

128. Can. Pat. 441,072 (1947); *Chem. Abstr.*, **41,** 5039a (1947).

129. U.S. Pat. 2,450,400 (1948); *Chem. Abstr.*, **43,** 514c (1949).

130. F. Schultz, *Hoppe-Seylers Z. Physiol. Chem.*, **265,** 113 (1940).

131. H. J. Backer and J. De Jonge, *Rec. Trav. Chim. Pays-Bas*, **60,** 495 (1941).

132. K. A. Jensen and K. Schmith, *Dansk. Tids. Farm.*, **15,** 197 (1941); *Chem. Abstr.*, **38,** 1483[1] (1944).

133. P. C. Guha and A. N. Roy, *Current Sci.*, **12,** 150 (1943); *Chem. Abstr.*, **37,** 6653[6] (1943).

134. A. R. Frisk, *Acta Med. Scand.*, *Suppl.*, **142,** 1 (1943); *Chem. Abstr.*, **38,** 4692[6] (1944).

135. W. Brucker and E. Bulka, *Studia Biophys.*, **1,** 253 (1966); *Chem. Abstr.*, **69,** 93475 (1968).

136. W. Michels, *Ber.*, **25,** 3048 (1892).

137. Brit. Pat. 392,410 (1932); *Chem. Zentr.*, **II,** 1934 (1933).

138. A. van Dormael, *Bull. Soc. Chim. Belges*, **57,** 341 (1948).

139. Y. Mizuhara, *Bull. Chem. Soc. Japan*, **25,** 263 (1952); *Chem. Abstr.*, **48,** 3966c (1954).

140. Y. Mizuhara, *Proc. Fujihara Mem. Fac. Eng. Keio Univ.*, **5,** No. 18, 17 (70) (1952); *Chem. Abstr.*, **49,** 5441e (1955).

141. S.-H. Chu and H. G. Mautner, *J. Org. Chem.*, **27,** 2899 (1962).

142. D. L. Klayman, *J. Org. Chem.*, **30,** 2454 (1965).

142a. L. V. Pavlova and F. Yu. Rachinskii, *Zh. Obshch. Khim.*, **35,** 493 (1965).

142b. D. L. Klayman and P. T. McIntyre, *J. Org. Chem.*, **33,** 884 (1968).

143. D. L. Klayman and G. W. A. Milne, *Tetrahedron*, **25,** 191 (1969).

143a. L. Karle, *Angew. Chem. (Intern. Ed.)*, **7,** 812 (1968); *Angew. Chem.*, **80,** 793 (1968).

144. J.-F. Giudicelli, J. Menin, and H. Najer, *C.R. Acad. Sci., Paris, Ser. C*, **262,** 285 (1966).

145. J. Menin, J.-F. Giudicelli, and H. Najer, *C.R. Acad. Sci., Paris, Ser. C*, **262,** 778 (1966).

146. J.-F. Giudicelli, J. Menin, and H. Najer, *Bull. Soc. Chim. Fr.*, 1099 (1968).

147. A. M. Comrie, D. Dingwall, and J. B. Stenlake, *J. Chem. Soc.*, 5713 (1963).

148. A. M. Comrie, D. Dingwall, and J. B. Stenlake, *J. Pharm. Pharmacol.*, **16,** 268 (1964).

149. U.S. Pat. 2,228,156 (1941); *Chem. Abstr.*, **35,** 2680[9] (1941).

150. Ger. Pat. 700,555 (1940); *Chem. Abstr.*, **37,** 5735[6] (1943).

151. C. Draguet and M. Renson, *Bull. Soc. Chim. Belges*, **75,** 243 (1966).

152. H. Bauer, *Ber.*, **46,** 92 (1913).

153. L. M. Clark, *J. Chem. Soc.*, 2313 (1928).

154. E. Ochiai and J. Haginiwa, *Yakugaku Zasshi*, **67**, 138 (1947); *Chem. Abstr.*, **45**, 9539c (1951).

155. E. Ochiai, J. Haginiwa, and K. Komatsu, *Yakugaku Zasshi*, **70**, 372 (1950); *Chem. Abstr.*, **45**, 2475 (1951).

156. J. Develotte, *Ann. Chim.* (Paris), **5**, 215 (1950).

157. H. C. Barany and M. Pianka, *J. Chem. Soc.*, 2217 (1953).

158. M. T. Bogert and Y. Chen, *J. Amer. Chem. Soc.*, **44**, 2352 (1922).

159. M. T. Bogert and A. Stull, *J. Amer. Chem. Soc.*, **49**, 2011 (1927).

160. E. Fromm and K. Martin, *Justus Liebigs Ann. Chem.*, **401**, 177 (1913).

161. L. M. Clark, *J. Chem. Soc.*, 2805 (1927).

162. F. S. Babichev and V. S. Likhitskaya, *Khim. Geterotsikl. Soedin.*, 508 (1969); *Chem. Abstr.*, **72**, 3435 (1970).

163. C. Hasan and R. F. Hunter, *J. Chem. Soc.*, 1762 (1935).

164. J. Haginiwa, *Yakugaku Zasshi*, **69**, 565 (1949); *Chem. Abstr.*, **44**, 4465f (1950).

165. Ger. Pat. 852,392 (1952); *Chem. Abstr.*, **52**, 10206c (1958).

166. Czech. Pat. 127,919 (1968); *Chem. Abstr.*, **70**, 96799 (1969).

167. G. Di Modica, E. Barni, and A. Gasco, *J. Heterocycl. Chem.*, **2**, 457 (1965).

168. E. Barni, G. Di Modica, and A. Gasco, *J. Heterocycl. Chem.*, **4**, 139 (1967).

169. F. L. Tobiason and J. H. Goldstein, *Spectrochim. Acta*, **23A**, 1385 (1967).

170. A. I. Kiprianov and V. A. Shrubovich, *Zh. Obshch. Khim.*, **26**, 2891 (1956); *Chem. Abstr.*, **51**, 8072a (1957).

171. B. P. Lugovkin, *Khim. Geterotsikl. Soedin.*, 571 (1966); *Chem. Abstr.*, **66**, 18685 (1967).

172. F. M. Hamer, *J. Chem. Soc.*, 3197 (1952).

173. M. Seyhan, *Chem. Ber.*, **86**, 888 (1953).

174. T. Ukai and S. Kanahara, *Yakugaku Zasshi*, **75**, 31 (1955); *Chem. Abstr.*, **50**, 961h (1956).

175. Brit. Pat. 489,202 (1938); *Chem. Abstr.*, **33**, 435[6] (1939).

176. V. Ettel, M. Semonský, J. Kuňák, and A. Černy, *Chem. Listy*, **46**, 756 (1952); *Chem. Abstr.*, **47**, 12358e (1953).

177. Brit. Pat. 494,602 (1938); *Chem. Abstr.*, **33**, 2764[7] (1939).

178. S. Hünig and H. Balli, *Justus Liebigs Ann. Chem.*, **609**, 160 (1957).

179. U.S. Pat. 2,259,353 (1941); *Chem. Abstr.*, **36**, 929[9] (1942).

180. H. Wahl, *Bull. Soc. Chim. Fr.*, 251 (1954).

181. D. Taber, N. Picus, E. I. Becker, and P. E. Spoerri, *J. Amer. Chem. Soc.*, **77**, 1010 (1955).

182. Ger. Pat. 670,505 (1939); *Chem. Abstr.*, **33**, 6613[2] (1939).

183. J. Bourson, *Bull. Soc. Chim. Fr.*, 4028 (1966).

184. J. Brunken and E.-J. Poppe, *Veröffentl. Wiss. Photo-Lab. Wolfen*, **10**, 101 (1965); *Chem. Abstr.*, **65**, 9994h (1966).

185. M. Pianka and J. C. Hall, *J. Sci. Food Agr.*, **8**, 432 (1957); *Chem. Abstr.*, **52**, 4611f (1958).

186. Brit. Pat. 502,109 (1939); *Chem. Abstr.*, **33**, 6344[1] (1939).

187. Brit. Pat. 593,025 (1947); *Chem. Abstr.*, **44**, 666c (1950).

188. U.S. Pat. 2,476,669 (1949); *Chem. Abstr.*, **45,** 2027h (1951).

189. M. T. Bogert and H. H. Hopkins, *J. Amer. Chem. Soc.*, **46,** 1912 (1924).

190. W. Treibs and E.-J. Poppe, *J. Prakt. Chem.*, [4] **13**, 330 (1961).

191. R. M. Acheson, M. W. Foxton, and G. R. Miller, *J. Chem. Soc.*, 3200 (1965).

192. V. S. Likhitskaya and F. S. Babichev, *Khim. Geterotsikl. Soedin.*, 164 (1970); *Chem. Abstr.*, **72,** 121449 (1970).

193. V. S. Likhitskaya and F. S. Babichev, *Ukr. Khim. Zh.*, **35,** 746 (1969); *Chem. Abstr.*, **72,** 12654 (1970).

XI Heterocyclic selenium compounds

D. HETEROCYCLES WITH SELENIUM SUBSTITUTED ON THE RING

HENRY G. MAUTNER

Department of Biochemistry and Pharmacology
Tufts University School of Medicine
Boston, Massachusetts

1. Introduction

With increasing interest in the synthesis and study of selenium analogs of biologically important sulfur compounds, coupled with the fact that many thio-substituted heterocycles exert medicinally important actions, it is not surprising that considerable attention has been centered on the synthesis of seleno-substituted heterocycles. This chapter will concentrate on selenium compounds related to thio-substituted heterocycles of biological or medicinal importance. No attempt will be made, however, to cover this material in encyclopedic fashion.

2. Methods of Synthesis

Most organic chemists approach the synthesis of selenium compounds with the feeling that, if one can make a sulfur compound, its selenium analog can be made by the same method. This approach will work on occasion, but it is certainly not infallible. Since the stabilities and dissociation constants of sulfur and selenium compounds tend to be different, it may be found that sulfur compounds survive reaction conditions under which selenium compounds decompose completely. In rare instances, the opposite situation may be seen. Furthermore, the basicities, nucleophilicities, and leaving tendencies of analogous thio and seleno reagents may differ widely.

In synthesizing seleno-substituted heterocyclic compounds two basic approaches may be used: (1) replacement of a substituent of the desired compound by a selenium-containing substituent; and (2) introduction of a selenium-containing fragment into the skeleton of the desired compound.

a. Conversion of Oxo to Seleno Compounds. Phosphorus pentasulfide has long been widely used for the conversion of amides to thioamides or the conversion of oxo to thio heterocyclic compounds. However, even in synthesizing thio compounds by the use of this reagent considerable difficulty may be encountered on many occasions. These problems are greatly augmented in working with phosphorus pentaselenide, possibly, as already discussed in Chapter III, because this poorly characterized reagent is often valueless unless freshly prepared. On the other hand, van den Hende and Klingsberg[1] have carried out such a conversion with phosphorus pentaselenide, while Hallam and Jones[2] converted lactams to selenolactams by the use of phosphorus pentaselenide generated *in situ* by heating a mixture of lactams, red phosphorus, and selenium in refluxing xylene. It should be noted that the latter synthesis produced selenolactams in rather low yield. A thorough review of the preparation and physical properties of phosphorus pentaselenide has appeared.[3]

b. Conversion of Halogeno to Seleno Compounds. This is probably the most widely used method for the synthesis of heterocyclic and other selenium compounds. A thorough survey of the nucleophilic seleno anions used to displace various substituents from the carbon being attacked is provided in Chapter III.

It should be noted that frequently milder reaction conditions have to be used for the synthesis of seleno heterocyclic compounds than for the synthesis of their thio analogs. For instance, although 6-chloropurines can be readily converted to the corresponding thio derivatives by treatment with aqueous potassium hydrogen sulfide, the conversion of chloro- to seleno-purines cannot be carried out in water without complete decomposition of the product and extensive precipitation of colloidal selenium. It is also useful, in the case of the seleno compounds, to protect the reaction mixtures from direct light.

Sodium hydrogen selenide, generated *in situ* by bubbling hydrogen selenide into an ethanolic solution of sodium ethoxide, has been used for the synthesis of 6-selenopurine (equation 1),[4] 2,4-diselenouracil,[4] 2,4-diselenothymine,[5]

(1)

and 6-selenoguanine[5] from the corresponding chloropyrimidines and chloro-purines. Similarly, sodium hydrogen selenide was used to convert 6-chloro-9-D-ribofuranosylpurine to the corresponding 6-seleno compound.[6] The same method was applied recently for the preparation of 6-selenoguano-sine[7,8] from the 2-amino-6-chloropurine riboside.

Other nucleophiles, such as selenourea,[4] potassium selenosulfate or seleno-cyanate, discussed in Chapter III, can be used for these syntheses; however, this author has found sodium hydrogen selenide to be a conveniently prepared reagent likely to give high yields of products.

c. **Conversion of Alkylthio to Seleno Compounds.**   The first conversion of an alkylthio to a seleno compound appears to have been the formation of substituted selenoureas from the corresponding S-methylthiopseudoureas, using sodium hydrogen selenide generated in ethanol.[9] This reaction has been applied to the synthesis of 6-selenopurines. 6-Alkylthiopurines, on treatment with sodium hydrogen selenide in sodium bicarbonate solution, give on refluxing the corresponding 6-selenopurines in good yield (equation 2).[10]

(2)

However, this reaction seems to be limited to purines in which the reactivity of the 6-alkylthio grouping has been enhanced by *N*-alkylation in the 3-position.[11,12] In contrast to the chloro compounds, reactions of alkylthio compounds with hydroselenide ion can be carried out in aqueous solution. Since, in many cases, thioheterocyclic compounds are more acces-sible than chloro heterocyclic compounds, this reaction deserves further exploration. (*Cf.* Chapter VIIIB for further discussion of this type of reaction.)

d.   *Conversion of Imino to Seleno Compounds.* Seleno-substituted phthalides have been prepared conveniently by the synthesis shown in equation 3.

(3)

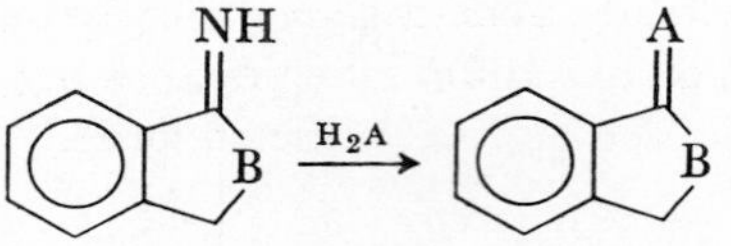

A = O, S, Se; B = O, S, Se

Of the nine possible chalcogenic analogs of phthalide eight have now been synthesized.[13,14] Only the selenocarbonylselenol ester proved to be too unstable to be isolated.[14]

e.   *Introduction of Selenourea into Heterocyclic Compounds.* If suitable halogeno-, alkylthio-, or other-substituted heterocyclic compounds to be modified by reaction with a selenium-containing nucleophile are not available, then selenium-containing fragments can be introduced directly into the rings to be formed. 2-Selenouracil,[4] 2-selenothymine,[4] and a series of 2-selenobarbiturates[15] have been synthesized by this approach (equation 4).

(4)

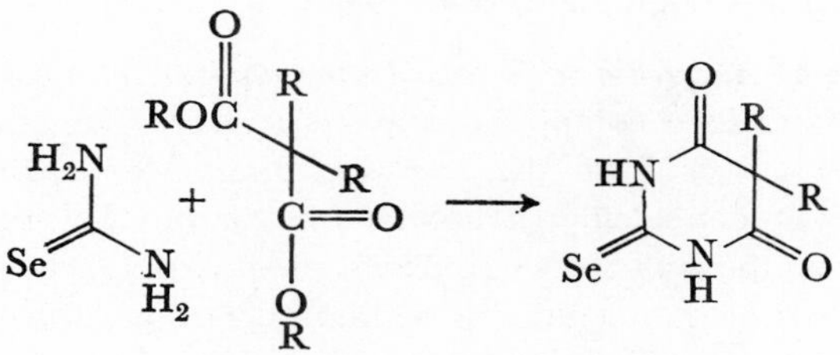

These reactions should be carried out in solvents other than water, and the reaction mixtures should be protected from direct light. Even with these precautions, however, precipitation of traces of colloidal selenium may be a problem. The preparation of selenourea is discussed in Chapters III, and VIIIB. It is desirable that this reagent be prepared just before its use.

3.   Properties of Seleno-Substituted Heterocycles with Respect to Their Oxo and Thio Analogs

Selenosubstituted heterocyclic compounds are of interest not only because of their occasionally high biological activities[4-6,8,16] but also because of the light they can throw on the physical and chemical properties of their oxygen and sulfur isologs. Such information is important since the basis for the biologically useful actions of the related sulfur compounds is often not understood. Examples of such agents are the cyclic thioamides with antithyroid activity,[17] the 2-thiobarbiturate hypnotics and anesthetics,[18] and the 6-thiopurines with antitumor activity[19] and the ability to block the immune response.[19a]

Since, in many cases, thio-substituted heterocyclic compounds are biologically active whereas their oxo analogs are not, and since the differences in activity must, presumably, be ascribed to differences in size, conformation, electron distribution, polarizability, hydrogen bonding ability, or hydrophobic bonding ability, measurements of these factors have been carried out on oxygen and sulfur isologs in an attempt to elucidate their mechanisms of action. By comparing these parameters, as well as biological activities for isologous oxygen, sulfur, *and* selenium compounds, one is at least in the position of basing one's correlation on three rather than on two points.

Sulfur and selenium isologs also have the advantage of being isosteric both in the crystal[20,21] and in solution,[22] with the result that their abilities to fit receptor sites should be very similar and differences in biological action should be attributable to other factors.

For these reasons it may be hoped that comparative studies of the spectra, dipole moments, crystal structures, binding abilities, and kinetics of isologous oxygen, sulfur, and selenium compounds may lead to new insights into the factors involved in their biological actions.

The i.r. and u.v. spectra of heterocyclic selenium compounds are discussed in Chapters XVA and XVB, respectively. The considerable bathochromic shifts noted in the u.v. spectra of heterocyclic amides as their oxygen atom is replaced progressively by sulfur and selenium (and in a recent example by tellurium[23]) have been ascribed to the progressive importance of charge-separated forms of the following type:

$$\text{R}\overset{+}{-}\text{NH}\!=\!\underset{|}{\text{C}}\!-\!\text{B}^{-} \qquad (\text{B} = \text{O, S, Se})$$

on descending the periodic table.[4,24,25] As in the case of heterocyclic amides and thioamides such as 2-pyridone and 2-pyridthione,[26] it proved possible to show that 2-pyridselenone[27] existed almost exclusively in the lactam rather than in the lactim form. As would be expected, the spectrum of 2-pyridselenone is very similar to that of N-methyl-2-selenopyridone and very different from that of 2-methylselenopyridine (Table XID-1).

In view of the increasing charge separation on passing from amides to thioamides to selenoamides, one would expect the acidity of the –NH– group to increase in parallel fashion. This is indeed the case. However, while

Table XID-1   *Ultraviolet spectra of 2-selenopyridines*[a]

COMPOUND	MOLECULE IN NEUTRAL FORM (H_2O): λ_{max} (nm)	MOLECULE IN CATIONIC FORM (70% H_2SO_4): λ_{max} (nm)
2-Selenopyridine	227, 285, 358	242, 308
N-Methyl-2-selenopyridine	232, 287, 357	238, 306
2-Methylselenopyridine	245, 254, 295	262, 328

[a] Data from Mautner *et al.*[27]

increasing charge separation leads to increasing ease of dissociation of

$$\overset{\displaystyle B^-}{\underset{\displaystyle}{|}}$$
$-C\!\!=\!\!\overset{+}{N}H-$ protons, it does not lead to an increasing ability of the heteroatoms sulfur and selenium to accept protons. 2-Pyridone is a stronger base
in 70% sulfuric acid than either its thio or seleno isolog (Table XID-2).

Table XID-2   *Dissociation constants of 2-substituted pyridines*[a]

COMPOUND	B	pK_a	$pK_a{}'$	ASSOCIATION CONSTANT (BENZENE)
	O	11.62	0.75	2700
	S	9.97	−1.07	520
	Se	9.36	−1.00	(oxidized)

[a] Data from Krackov *et al.*[28]

Additional proof for the increase in charge separation noted in passing
from amides to thioamides to selenoamides, whether acyclic or cyclic, may
be obtained from dipole moment measurements (Table XID-3). It can be

Table XID-3   *Dipole moments of 2-substituted N-methylpyridines*[a]

COMPOUND	B	DIPOLE MOMENT (DIOXANE) (debyes)
	O	4.07
	S	5.49
	Se	5.92

[a] Data from Krackov *et al.*[28]

seen that the increases in moment induced by replacing the oxygen of the amide group by sulfur and by selenium are considerably larger than would be expected from the fact that sulfur and selenium have larger atomic radii than oxygen.[24,28]

Evidence for the increase in double-bond character of the C–N bond in the urea, thiourea, and selenourea groupings of 2-selenouracil has been presented by Tsernoglou[29] on the basis of X-ray diffraction measurements.

The relatively low abilities of the S and Se atoms in heterocyclic thio- and selenocarboxamides to accept protons are matched by relatively low abilities to accept hydrogen bonds. A quantitative study of the relative abilities of 2-pyridone and its thio and seleno analogs to form hydrogen-bonded dimers has been carried out. Association constants of the dimers formed could be calculated from the dependence of the dielectric constant on concentration and from the dependence of the molecular weight on concentration. Since with increasing concentration the molecular weight of the dimer was approached asymptotically, the possibility of stacking interactions could be excluded.[28] It was found that the association constant of 2-pyridone was

(B = O, S, Se)

lower than that of 2-pyridthione. Although oxidizability interfered with the precise determination of an association constant for 2-pyridselenone, its value was similar to that of its sulfur analog.

The observation that the sulfur in thiolactams has a lower ability to accept hydrogen bonds than the oxygen in lactams was confirmed by comparing the distances of N–H$\cdots$S interaction with the distances of N–H$\cdots$O interactions in related compounds.[30] It is interesting to note that in saturated heterocyclic lactams and thiolactams, in which charge separation of the type discussed plays a less major role, thiolactams have a higher degree of association than their corresponding oxygen compounds, a finding based on i.r. measurements,[31] even though here too the N–H$\cdots$S bonds are weaker than the corresponding N–H$\cdots$O bonds.

Crystallographic evidence for the formation of N–H$\cdots$Se hydrogen bonds is discussed in Chapter XVH. As expected, the energy barriers for rotation about the C–N bond increase in passing from carbamoyl to thiocarbamoyl to selenocarbamoyl compounds[32] as the double-bond character of that bond increases.

The finding that the increase in the acidity of the –NH group in passing from amides to thioamides to selenoamides is not accompanied by an increase in the ability of sulfur or selenium to accept either a proton or a hydrogen bond has been explained in terms of the generalization that "hard acids" interact preferentially with "hard bases" and that "soft acids" interact preferentially with "soft bases."[33,34] The highly polarized S and Se atoms in cyclic or acyclic thioamides and selenoamides act as classical "soft bases." As expected, such S and Se atoms have relatively low abilities to accept protons or hydrogen bonds, but have high abilities to chelate "soft metals," accompanied by anomalously high abilities to be dissolved in "soft" nonpolar solvents. This factor is responsible for the observation that, even though heterocyclic thio and seleno compounds are ionized to a greater degree than their oxygen isologs at physiological pH, they are more lipid soluble and better able to cross the blood-brain barrier.[15] Similarly, heterocyclic thiono esters and selenocarbonyl esters are considerably more lipid soluble than analogous esters.

In synthesizing and studying thio- and seleno-substituted heterocyclic compounds it is important to consider the positional differences in their chemical and biological behavior. For instance, whereas 2-thiouracil and 2-selenouracil are colorless, the corresponding 4-thio- and 4-selenouracil analogs are yellow and orange, respectively. In accordance with this, it has been noted that replacement of oxygen by sulfur or by selenium induced much greater bathochromic shifts in the u.v. spectra of 2,4-disubstituted pyrimidines (or the corresponding 2,6-disubstituted purines) when it took place in the 4-position of the pyrimidines (6-position of the purines) than when it occurred in the 2-position of these compounds.[4,34a] 2,4-Dithiouracil, on reacting with ammonia, yields 2-thiocytosine exclusively;[35] analogously, 2,4-diselenouracil yields 2-selenocytosine.[5] Positional differences are seen in the relative ease of oxidation of 2 thiouracil and of 4 thiouracil; these have been correlated to the observation that the 2- but not the 4-thio isomer has antithyroid activity.[36] 6-Thioguanine and 6-mercaptopurine, but not the corresponding 2-thio analogs, exert antineoplastic action,[37] while 4-thiopteridines and -pyrimidines and 6-thiopurines, but not the 2-substituted isomers, bind strongly to dihydrofolic acid reductase.[38]

It was proposed that thioamide and selenoamide groups are chromophoric only when they are conjugated to a double bond ("amidic")

$$(-NH-\overset{\overset{\textstyle B}{\|}}{C}-CH=CH-)$$

but not when they are part of the $-NH-\overset{\overset{\textstyle B}{\|}}{C}-NH-$ (B = S, Se) grouping ("ureidic").

Since 2-thiouridine and 4-thiouridine are important components of tRNA,[40,41] and since formation of disulfide bonds by such residues might be involved in the regulation of the conformation of the polymer,[42] studies of the conformations and the detailed structures of these sulfur compounds and their selenium analogs are of considerable interest.

An investigation of the crystal structure of 2,4-dithiouracil[20] showed the C–S bond in the 4-position (1.684 Å) to be considerably longer than the C–S bond in the 2-position (1.645 Å), indicating the predominance in this compound of the resonance form:

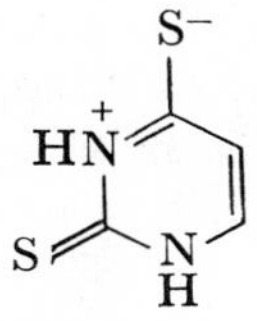

Similarly, the lengths of the S $\cdots$ H bonds were also found to be position dependent.[20] Differences were also observed in the C–S bond lengths of the riboside 2,4-dithiouridine.[43] Unfortunately, a crystallographic investigation of 2,4-diselenouracil[44] could not be carried out with enough precision to permit observation of analogous differences in the lengths of the C–Se bonds.

It may be hoped that in this system, as in other ones, comparative studies of the physical and biological properties of thio and seleno analogs will throw light on the functions of the sulfur compounds.

REFERENCES

1. J. H. van den Hende and E. Klingsberg, *J. Amer. Chem. Soc.*, **88,** 5045 (1966).

2. H. E. Hallam and C. M. Jones, *J. Chem. Soc., A*, 1033 (1969).

3. M. V. Kudchadker, R. A. Zingaro, and K. J. Irgolic, *Can. J. Chem.*, **46,** 1415 (1968).

4. H. G. Mautner, *J. Amer. Chem. Soc.*, **78,** 5292 (1956).

5. H. G. Mautner, S.-H. Chu, J. J. Jaffe, and A. C. Sartorelli, *J. Med. Chem.*, **6,** 36 (1963).

6. J. J. Jaffe and H. G. Mautner, *Cancer Res.*, **20,** 381 (1960).

7. L. B. Townsend and G. H. Milne, *J. Heterocycl. Chem.*, **7,** 753 (1970).

8. S.-H. Chu, *J. Med. Chem.*, **14,** 254 (1971).

9. D. L. Klayman and R. J. Shine, *J. Org. Chem.*, **34,** 3549 (1969).

10. F. Bergmann and M. Rashi, *Israel J. Chem.*, **7,** 63 (1969).

11. J. W. Jones and R. K. Robins, *J. Amer. Chem. Soc.*, **84,** 1914 (1962).

12. Z. Neiman and F. Bergmann, *Israel J. Chem.*, **3,** 85 (1965).

13. M. Renson and R. Collienne, *Bull. Soc. Chim. Belges*, **73,** 491 (1964).

14. I. Wallmark, M. H. Krackov, S.-H. Chu, and H. G. Mautner, *J. Amer. Chem. Soc.*, **92,** 4447 (1970).

15. H. G. Mautner and E. M. Clayton, *J. Amer. Chem. Soc.*, **81**, 6270 (1959).

16. H. G. Mautner and J. J. Jaffe, *Cancer Res.*, **18**, 295 (1958).

17. E. B. Astwood, "The Pharmacological Basis of Therapeutics," 4th ed., L. S. Goodman and A. Gilman, Eds., Macmillan, New York, N.Y., 1970, p. 1466.

18. H. G. Mautner and H. C. Clemson, "Medicinal Chemistry," 3rd ed., A. Burger, Ed., John Wiley, New York, N.Y., 1970, p. 1371.

19. G. B. Elion, S. Callahan, R. W. Rundles, and G. H. Hitchings, *Advan. Chemother.*, **2**, 91 (1965).

19a. J. E. Murray, B. S. Barnes, and J. Atkinson, *Transplant.*, **5**, 752 (1967).

20. E. Shefter and H. G. Mautner, *J. Amer. Chem. Soc.*, **89**, 1249 (1967).

21. E. Shefter and H. G. Mautner, *Proc. Natl. Acad. Sci.*, **63**, 1253 (1969).

22. R. J. Cushley and H. G. Mautner, *Tetrahedron*, **26**, 2151 (1970).

23. M. Z. Girshovich and A. V. El'tsov, *Zh. Obshch. Khim.*, **39**, 941 (1969).

24. H. G. Mautner and W. D. Kumler, *J. Amer. Chem. Soc.*, **78**, 97 (1956).

25. S. F. Mason, *J. Chem. Soc.*, 5010 (1957).

26. A. Albert and G. B. Barlin, *J. Chem. Soc.*, 2384 (1959).

27. H. G. Mautner, S.-H. Chu, and C. M. Lee, *J. Org. Chem.*, **27**, 3671 (1962).

28. M. H. Krackov, C. M. Lee, and H. G. Mautner, *J. Amer. Chem. Soc.*, **87**, 892 (1965).

29. D. Tsernoglou, Ph.D. Thesis, Yale University, 1966.

30. J. Donohue, *J. Mol. Biol.*, **45**, 231 (1969).

31. N. Kulevsky and P. M. Froehlich, *J. Amer. Chem. Soc.*, **89**, 4389 (1967).

32. U. Svanholm, *Ann. N.Y. Acad. Sci.*, **192**, 124 (1972).

33. R. G. Pearson, *J. Amer. Chem. Soc.*, **85**, 3533 (1963).

34. R. G. Pearson, "Survey of Progress in Chemistry," Vol. 5, A. F. Scott, Ed., Academic Press, New York, N.Y., 1969, p. 1.

34a. G. B. Elion, *J. Amer. Chem. Soc.*, **68**, 2137 (1946).

35. G. H. Hitchings, G. B. Elion, E. A. Falco, and P. B. Russell, *J. Biol. Chem.*, **177**, 357 (1949).

36. W. H. Miller, R. O. Roblin, and E. B. Astwood, *J. Amer. Chem. Soc.*, **67**, 2201 (1945).

37. R. K. Robins, *J. Med. Chem.*, **7**, 186 (1964).

38. J. R. Bertino and H. G. Mautner, unpublished data.

39. H. G. Mautner and G. Bergson, *Acta Chem. Scand.*, **17**, 1694 (1963).

40. J. A. Carbon, L. Hung, and D. S. Jones, *Proc. Natl. Acad. Sci.*, **53**, 979 (1965).

41. M. N. Lipsett, *J. Biol. Chem.*, **240**, 3975 (1965).

42. M. N. Lipsett, *Cold Spring Harbor Symp. Quant. Biol.*, **31**, 449 (1966).

43. P. Faerber, M. Saenger, K. H. Scheit, and D. Suck, *F.E.B.S. Letters*, **10**, 41 (1970).

44. E. Shefter, M. N. G. James, and H. G. Mautner, *J. Pharm. Sci.*, **55**, 643 (1966).

XI   Heterocyclic selenium compounds

E. SELENIUM-CONTAINING DYES

LESLIE G. S. BROOKER,* JOHN A. FORD, JR., and
EARL J. VAN LARE

Kodak Research Laboratories
Kodak Park
Rochester, New York

1. Introduction

For convenience, selenium-containing dyes will be divided in this chapter
into cyanine and related types and those that do not belong to the cyanine
series. The cyanine dyes are the more numerous and therefore will be dealt

* Deceased on December 22, 1971.

with first. Their greatest use in technology has been as photographic sensitiz-ing dyes, and it may be helpful at the outset to give an outline of the function-ing of dyes in this way.

The silver halide emulsions used in photography, of themselves, absorb only ultraviolet, violet, and blue light. It is, therefore, only light of these colors that affects them, thus rendering them developable. In 1873 H. W. Vogel of Berlin made the chance discovery that, when certain dyes were added to photographic emulsions, the silver halide in some cases became sensitive to light that the added dye itself absorbed (photographic sensitization by dyes). In this process it is necessary that the dyes be adsorbed at the surface of the silver halide particles, or "grains," which are dispersed throughout the emulsion and are also able to transfer the energy they themselves absorb to the silver halide particles, rendering these developable in the ordinary way. However, Vogel and many later experimenters found that by no means all dyes fulfilled these qualifications. Dyes of the most diverse classes were tested; but as time has passed, with but few exceptions (notably erythrosin), the best results have been obtained with dyes of the cyanine group, of which many subclasses may be distinguished.[1] Of the cyanine dyes, thousands have been prepared and tested; of the large number made, however, only a very small percentage have been selected as good enough to become production items in photographic technology—although this must be true also of dyes in general, used for purposes other than photography.

2. Cyanine Dyes

a. *Synthesis.* As a group, the cyanines are complex ammonium salts; in a simple example the molecule comprises two nitrogen-containing hetero rings linked in such a way that a single cationic charge is shared between two N atoms as in formula **1**, which depicts a quinoline derivative and was actually the first cyanine type to be prepared. Using normal valences, formula **1** may be written in two equally probable ways, **a** and **b**. Mills

1a   1b

and Wishart[2] clearly had this in mind when they referred to "virtual tautomerism" in the cyanine series, but Bury[3] gave greater precision to the concept in 1935 when he described **1** as a "resonance hybrid" represented by **a** ↔ **b**.[3] The resonance concept has since been successfully employed in treating a number of aspects of these dyes.[4]

In the cyanine series, continuing progress showed that dyes such as **1** could be obtained with a greater length of chain between the nuclei than one =CH– unit. For the special resonance hybrid character of the dyes to be maintained, however, the bridge between the nuclei has to be increased in length by two –CH= groups (–CH=CH–, or vinylene) for each increase in chain length.

Cyanines in which the two hetero rings are linked by one –CH= group are called monomethine, or "simple," cyanines. Those with =CH–CH=CH– between the rings are called carbocyanines, those with the bridge

$$=CH(-CH=CH)_2-$$

are termed dicarbocyanines, and so on, and in general the formula for a cyanine bridge may be written as $=CH(-CH=CH)_n-$, where $n = 0, 1, 2, 3, 4$, etc. The length of the bridge in any particular dye will be determined by the synthetic methods available.

The first selenium-containing cyanines to be described were two carbocyanines prepared by Clark[5] from 2-methylbenzoselenazole (**2**), it having been previously established by Mills[6] that similar carbocyanines prepared from 2-methylbenzothiazole showed promise as photographic sensitizing dyes. The 2-methylbenzoselenazole itself was prepared by Clark by the action of acetyl chloride on the zinc salt of o-aminoselenophenol,[5] and in Clark's dyes the base was first quaternated by the addition to the nitrogen of, for example, alkyl iodides, as in **2a** (equation 1).

(1)

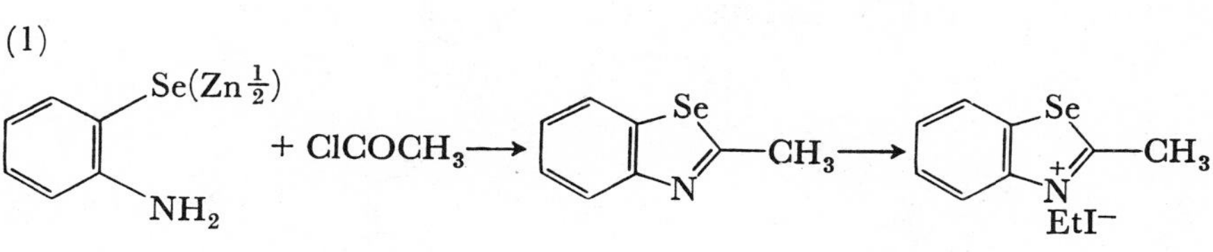

2 **2a**

The 2-methylbenzothiazole which Mills used (the structure is similar to **2**, but with sulfur in place of selenium) had originally been prepared by von Hofmann[7] in a classical series of comparisons of bases of the benzothiazole series with those of the quinoline series.

After the preparation by Clark[5] of the carbocyanine **3** ($n = 1$) the vinylogous series of selenacyanines **3** ($n = 0, 1, 2, 3$) became available by 1936,

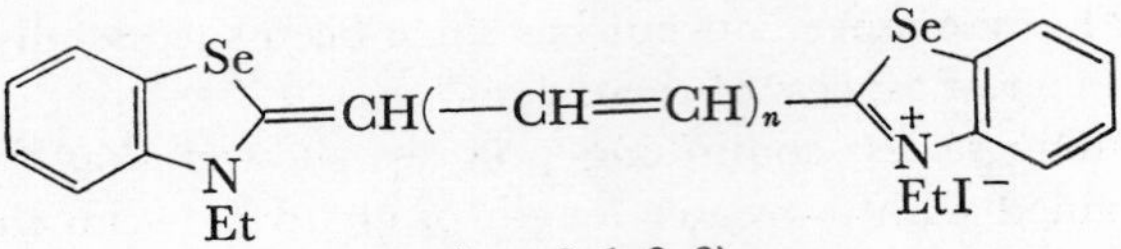

3 $(n = 0, 1, 2, 3)$

when their absorption spectra were published.[8] Monomethine selena-cyanines were prepared and patented by a method devised by I. G. Farben.[9] but only correctly interpreted by Fisher and Hamer.[10] Two molecular equivalents of 2-methylbenzoselenazole quaternary salt (preferably the alkyl chlorides) undergo condensation with amyl nitrite with loss of the elements of water, amyl alcohol, and hydrogen cyanide according to equation 2.

(2)

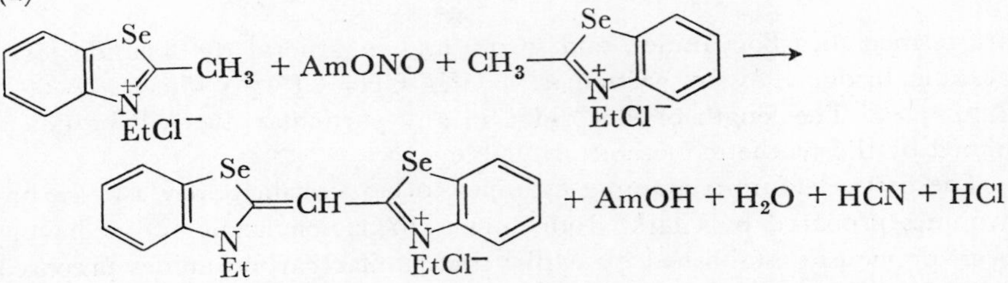

The next higher vinylog **3** $(n = 1)$, had been prepared by Clark[5] by a method originated by König[11] and perfected by Hamer.[12] This reaction (equation 3) comprises the condensation of 2 equivalents of 2-methylbenzo-

(3)

selenazole quaternary salt with ethyl orthoformate. For this reaction anhydrous pyridine was the solvent of choice.[12]

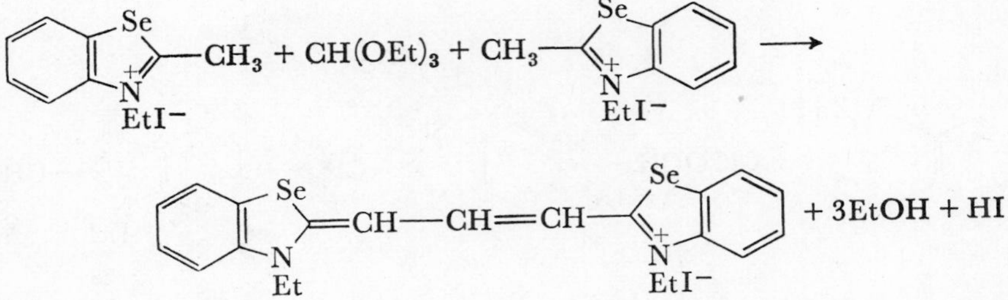

Cyanines with the next longer length of conjugated chain between the nuclei, called dicarbocyanines, were first obtained by König[11] using tri-methoxypropene, $CH_3O-CH=CH-CH(OMe)_2$, which may be regarded as the vinylog of methyl orthoformate. For **3** $(n = 2)$ the synthesis was carried out by Fisher and Hamer,[8] who published the absorption spectrum of the compound. The reaction is similar to that shown in equation 3, except that use of trimethoxypropene gives a dye with the bridge $=CH(-CH=CH)_2-$ between the hetero rings.

The series with a bridge $=CH(-CH=CH)_3-$ between the hetero rings belongs to the tricarbocyanines, but despite their longer bridge these dyes were actually easier to prepare than the corresponding dicarbocyanines. The starting point is pyridine, the ring of which is cleaved as shown in equation 4. The product of this reaction, glutaconaldehyde dianilide hydro-

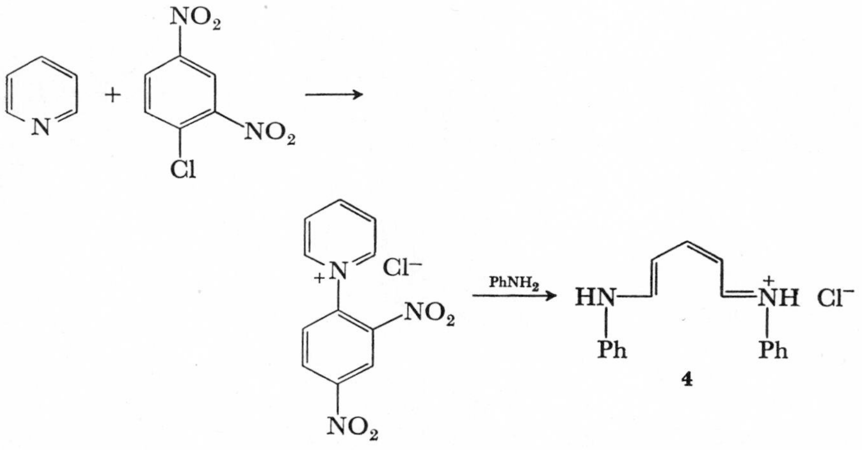

(4)

chloride (**4**), which is itself a dye and may be represented as a resonance hybrid (though only one extreme structure is shown), is then allowed to react with 2 equivalents of 2-methylbenzoselenazole quaternary salt to give selenatricarbocyanine (equation 5).[13]

(5)

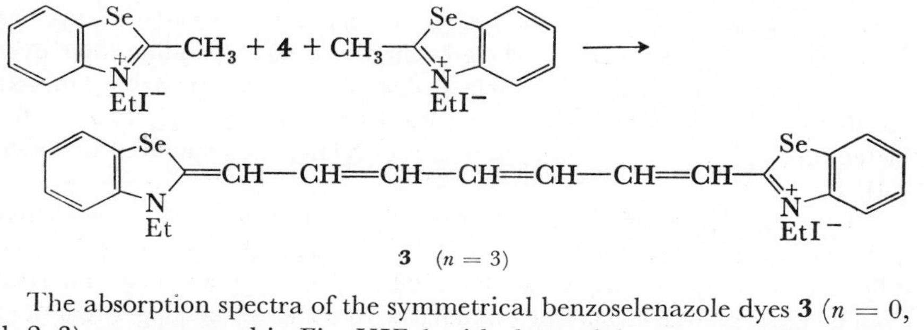

3 $(n = 3)$

The absorption spectra of the symmetrical benzoselenazole dyes **3** ($n = 0$, 1, 2, 3) are compared in Fig. XIE-1 with those of the corresponding benzo-thiazole derivatives. The spectra for the thia and selena dyes with the same

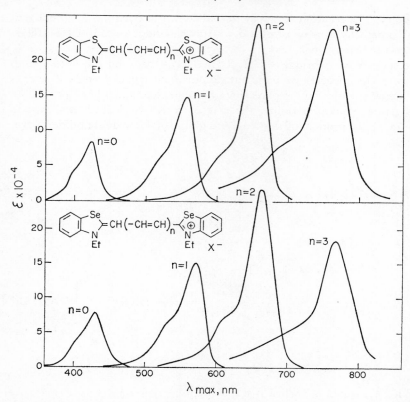

FIG. XIE-1   Absorption spectra of vinylogous thia and selenacyanines, determined in MeOH in the Eastman Kodak Research Laboratories, Rochester, New York, were obtained independently of data previously published.[8]

value of n are very similar to each other, though the selena dye always has λ_{max} at somewhat longer wavelength than the corresponding thia dye. The greatest difference in curve shape is shown by the two tricarbocyanines; the thia dye has ε_{max} 24.7×10^4, and the selena dye has ε_{max} 18.2×10^4. Values of ε_{max} and λ_{max} for the dyes in Fig. XIE-1 are compared in Table XIE-1.

The selenacyanines described thus far have comprised two benzoselenazole nuclei linked together, but a number of methods exist whereby two different heterocyclic nuclei can be combined to give an unsymmetrical cyanine dye. Such dyes may be obtained with a shorter or longer length of bridge between the dissimilar nuclei.

Of several available methods, probably the most generally applicable for

Table XIE-1   Values of ε_{max} vs. λ_{max} for dyes in Fig. XIE-1

	THIA		SELENA	
n	λ_{max} (nm)	ε_{max}	λ_{max} (nm)	ε_{max}
0	423	8.2×10^4	430	7.7×10^4
1	557.5	14.7×10^4	570	15.0×10^4
2	656	25.4×10^4	661	25.7×10^4
3	760	24.7×10^4	765	18.2×10^4

giving unsymmetrical monomethine (or simple) cyanines is that in which a methyl quaternary salt (*e.g.*, 2-methylbenzoselenazole ethiodide), after first losing the elements of the acid related to the anion to give a so-called methylene base, is subjected to electrophilic attack by a substance such as 2-phenylmercaptoquinoline ethiodide in a basic medium (equation 6).[14]

(6)

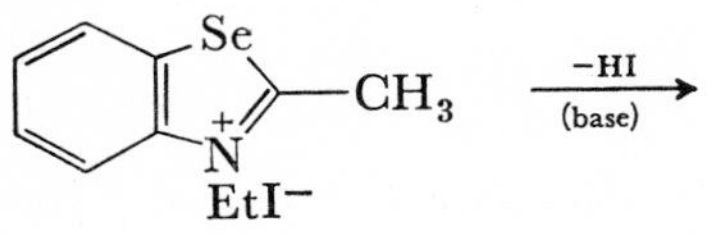

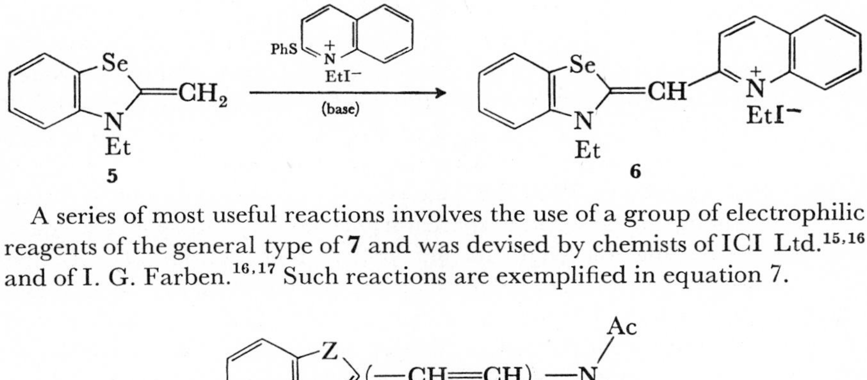

A series of most useful reactions involves the use of a group of electrophilic reagents of the general type of **7** and was devised by chemists of ICI Ltd.[15,16] and of I. G. Farben.[16,17] Such reactions are exemplified in equation 7.

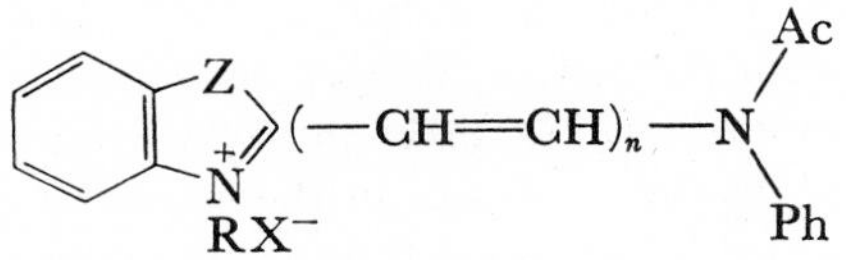

$(Z = O, S, Se, \overset{Me_2}{C}, —CH{=}CH—; R = \text{alkyl, aryl}; n = 1, 2, 3)$

7

(7)

[Structure: benzoselenazole ring with Se and N-Et, I$^-$]—CH$_3$ + Ac/Ph>N—CH=CH—[benzoxazole ring with O and N-Et, I$^-$] **8** →

2a

[Structure: benzoselenazole ring with Se and N-Et]=CH—CH=CH—[benzoxazole ring with O and N-Et, I$^-$] **9** + Ac/Ph>NH + HI

The reagents **7** were, in turn, prepared by the condensation of heterocyclic quaternary salts containing a reactive methyl group with 1 equivalent of substances such as diphenylformamidine and its vinylogs in, for example, acetic anhydride (equation 8).

(8)   [Structure: benzo ring with Z and N—R, X$^-$]—CH$_3$ + HN(Ph)—CH(=CH—CH)$_n$—NH(Ph) ⟶ **7**

b. Absorption Spectra of Symmetrical and Unsymmetrical Cyanines.

The most characteristic and valuable feature of the absorption spectra of the symmetrical vinylogous cyanine series, which includes the two series of dyes shown in Fig. XIE-1, is that, for each vinylene group added to lengthen the bridge between the hetero rings, the absorption maximum is shifted towards longer wavelengths by approximately 100 nm.[8] Thus, whereas **3**, $n = 0$, is a yellow dye with λ_{max} 423 nm, the dye with $n = 1$ is crimson with λ_{max} 570 nm, that with $n = 2$ is blue with λ_{max} 661 nm, and that with $n = 3$ is green with λ_{max} 765 nm. With the same two benzoselenazole rings, it is thus possible to change the absorption spectrum enormously, and hence the color, merely by changing the length of the "polymethine" bridge between the hetero rings. All four of these selena dyes are photographic sensitizers. The region of sensitivity conferred by each dye corresponds approximately to the region over which the particular dye absorbs in a solvent such as methanol, but usually shifted somewhat towards longer wavelengths. A mixture of the four dyes **3**, $n = 0, 1, 2$, and 3, could, in theory, serve to confer sensitivity over the whole of the visible spectrum and into the near i.r., but the sensitizing effects of the individual dyes are frequently depressed when they are used in mixtures of this sort.

A second characteristic feature, the "deviation" effect, concerns the absorption of unsymmetrical dyes, of which **9** is an example. This dye is the structural cross between two symmetrical ("parent") carbocyanines **3** ($n = 1$), the one containing two benzoselenazole rings and the other two benzoxazole rings. If the two different rings differ sufficiently in basicity, the absorption maximum of the unsymmetrical dye is found to lie at a shorter wavelength than the mean of the values of λ_{max} of the two "parent" symmetrical dyes, the difference between this mean value and the λ_{max} actually found being termed the "deviation." In **9** the two hetero rings do not differ sufficiently in basicity for the deviation in this particular dye to be appreciable, but in **10**, where benzoselenazole is linked to much less basic pyrrole, the deviation is considerable (33 nm). In **10**, incidentally, the two rings are

10

linked by a two-methine bridge, but the shortest chain connecting the two N atoms comprises an uneven number of C atoms, and the dye contains the amidinium ion skeleton (**11**) characteristic of all cyanine dyes. The nature of

$$>N-(\overset{|}{C}=\overset{|}{C}-)_n\overset{\oplus}{\overset{|}{C}}=N< \longleftrightarrow >\overset{\oplus}{N}=(\overset{|}{C}-\overset{|}{C}=)_n\overset{|}{C}-N<$$

11

the deviation has been described fully,[18] and it has been shown to afford a simple way of arranging the nuclei used in cyanine dyes in the order of basicity.

Figure XIE-2 gives the deviations of six dyes. The deviation is the greater, the greater the difference in basicity of the nuclei combined in each dye. Each of the six different nuclei shown (there is one at the left of each dye) is combined with the very feebly basic 2,5-dimethyl-1-phenylpyrrole ring. The dye showing the small deviation of 14 nm (that from N-phenyldimethyl-indolenine), therefore, contains the least basic of the six variable nuclei, and 1,3-dimethylbenzimidazole, the dye from which has a deviation of 92 nm, is the most basic. Benzoselenazole is here shown to be somewhat less basic than benzothiazole.

c. Chain-Substituted Carbocyanines.

In theory, cyanines with system **12** could be substituted anywhere along the conjugated bridge linking the two

12

FIG. XIE-2 Deviation of six unsymmetrical cyanines. Values of λ_{max} in MeOH are as

follows: • = unsymmetrical cyanine; △ = $\text{Me}\!\!\underset{\overset{|}{\text{Ph}}}{\underset{\text{N}}{\big[}}\!\!\text{Me}$ $\text{Me}\!\!\underset{\overset{|}{\text{Ph}}}{\underset{\overset{+}{\text{N}}}{\big]}}\!\!\text{Me}$; ○ = second sym-

metrical cyanine; I = average of △ and ○. Deviations are shown by ⌣, the amount being given in nanometers.

hetero rings. In actual fact trimethinecyanines such as **13**, carrying a substituent at the center of the bridge, are of considerable technical significance. Such benzoselenazole dyes were prepared by Brooker and White,[19] using a number of *ortho* esters of the type $G–C(OR)_3$ in place of the ethyl orthoformate employed by König.[11] Unsymmetrical chain-substituted carbocyanines

(R = alkyl; G = Me, Et, Ph, *etc.*)

13

were obtained by the use of reactive pseudo ketones of type **14**, obtained from acid chlorides, GCOCl, as shown in equation 9.[20] By condensation with a methyl quaternary salt (*e.g.*, 2-methyl-β-naphthothiazole metho-*p*-toluene-sulfonate) in acetic anhydride, unsymmetrical chain-substituted dyes such as **15** could be made.[21]

(9)

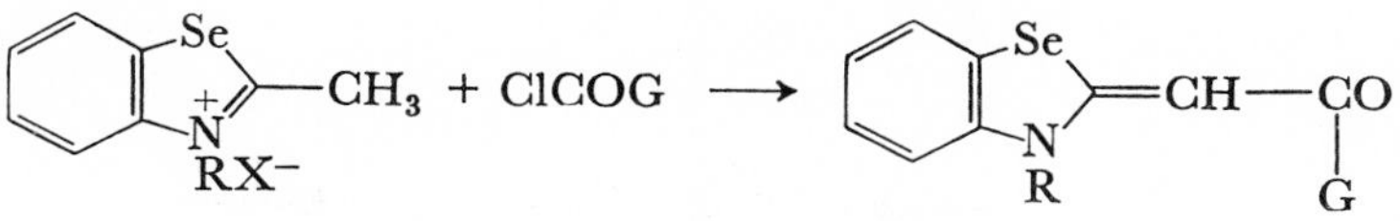

$$14$$

A number of cyanines, especially certain chain-substituted carbocyanines, are valuable in practice in that they tend to give particularly sharp absorption and sensitizing maxima, associated with an aggregated state in which

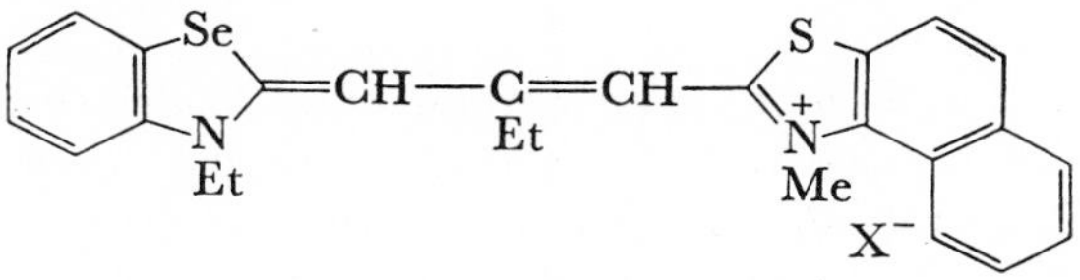

15

the dye molecules, attracted by intermolecular forces, become stacked up at the surface of the silver halide grains. At the same time such "reversible polymers" (like irreversible polymers) tend to be nondiffusing, a necessary property of sensitizers for monopack color processes, and it was the discovery of dyes of this kind that made possible the launching of the Kodachrome process in 1935.[22] When the scale projections of dyes which tend to form such "J-aggregates" are examined, it is found that their molecules are of a "compact" type, that is, they are held in a fairly rigid position. On the one hand, for the maximum resonance stabilization to be attained in the cyanine molecule, the two heterocyclic nuclei will seek to lie in the same plane.[23] If, however, the heterocyclic rings are not very bulky, as in 3,3'-dimethyloxacarbocyanine, the cation of which is shown in the scale projection in 16,[24] the nuclei in a loose molecule of this kind are free to assume a variety of positions, as indicated by the arrows, which show that the heterocyclic rings have considerable freedom of movement without disturbing the planarity of the molecule.

If the hetero oxygen atoms in 16 are replaced by sulfur as in 17, and especially by even bulkier selenium, and the chain is substituted by, for example,

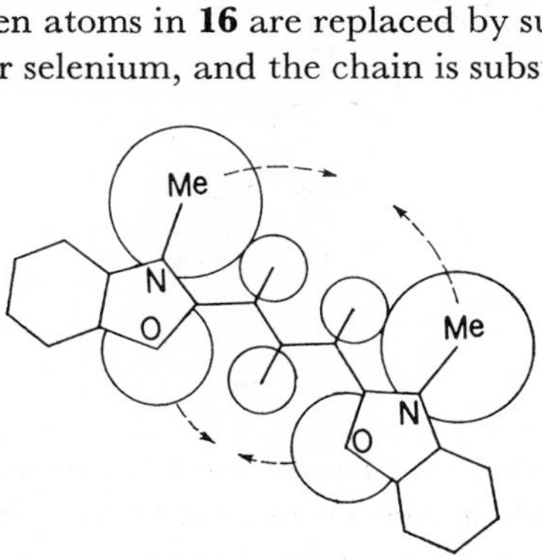

16

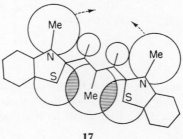

17

methyl groups, a projection using normal bond angles shows that the S (or Se) atoms will crowd the *meso*-methyl substituent (indicated by cross-hatching), forcing the >N–Me groups towards each other, as indicated by the arrows. The molecule can still preserve its planarity, at the expense of some distortion of bond angles, but will be "compact" rather than loose, as **16** is. Other things being equal, a compact molecule such as **17** will give rise to fewer position isomers than a loose molecule, so that at a given instant it is easier for a compact than for a loose molecule to encounter other molecules of nearly identical spatial arrangement. Since molecular aggregation probably takes place through the orderly lining up of (preferably) planar molecules of the same configuration, it follows that aggregation (and hence nondiffusibility of the dyes) should be favored by compactness.

d.   Other Selenium-Containing Nuclei Used in Cyanine Dyes.   So far the emphasis has been on cyanines derived from benzoselenazole, but dyes containing nuclear-substituted benzoselenazole rings have also been investigated, for example, 5-methyl (**18**)[25] and 5-methoxy (**19**).[26] Dyes from the two isomeric naphthoselenazole rings **20**[27] and **21**[28] have also been described,

as well as dyes containing the selenadiazole nuclei **22** and **23**.[29] 2-Methyl-selenazole itself (**24**) and various 4-substituted derivatives were described by Brooker and White,[30] and cyanines from these bases by Brooker.[31]

22 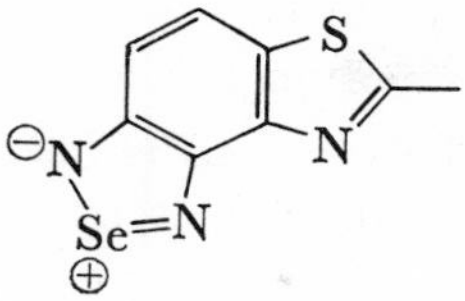   **23** 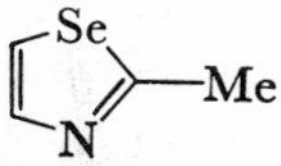   **24**

e. **Merocyanines.** These dyes comprise a large group characterized by the amidic resonance system $>N-(\overset{|}{C}=\overset{|}{C}-)_n\overset{|}{C}=O \leftrightarrow >\overset{\oplus}{N}=(\overset{|}{C}-\overset{|}{C}=)_n\overset{|}{C}-\overset{\ominus}{O}$.[32] In these dyes the nitrogen and carbonyl "ends" are used in rings as in structure **25** (only the uncharged formula is shown). Dyes of this type have been

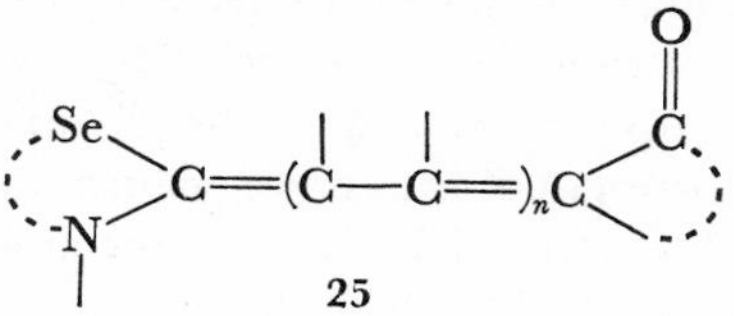

25

described in the patent literature.[33] Furthermore, selenium may be found in the carbonyl-containing ring, as in **26**.[34] Selenium was successfully introduced

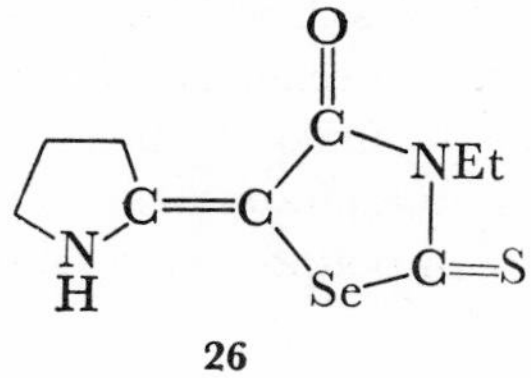

26

into merocyanines such as **27** in the two-stage synthesis shown in scheme 10 to give the selena dye **28**.[35]

(10)

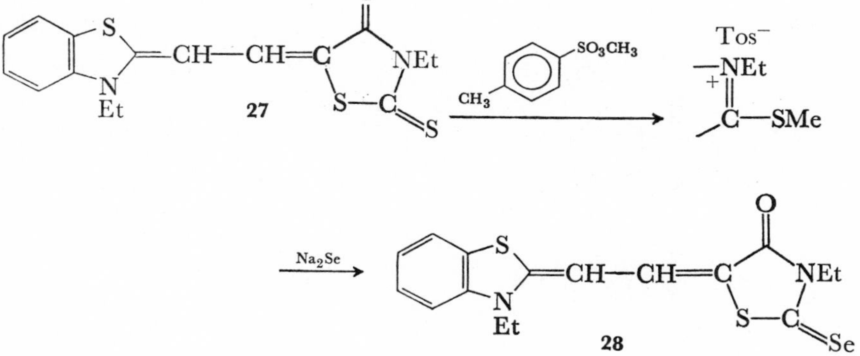

27

28

3. Noncyanine Selenium-containing Dyes

A relatively small body of literature pertains to these dyes. None is apparently of commercial importance as applied to textiles.

a. Selenazine Dyes.   Phenoselenazine (**29**) is formed by heating selenium monochloride with diphenylamine (scheme 11).[36] The reaction of picryl

(11)   Se$_2$Cl$_2$ +

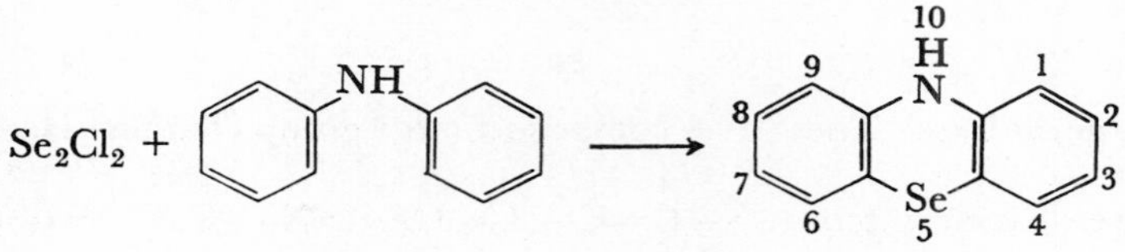

29

chloride with *o*-aminoselenophenol or its tin double salt and cyclization of the resulting 2,4,6-trinitroanilinoselenophenol with alcoholic potassium hydroxide gave 1,3-dinitrophenoselenazine (**30**). Nitration of the latter yielded the 1,3,6-trinitro compound **31**. The di- and trinitro compounds were reduced to the corresponding amines, which were oxidized by ferric chloride to the 1,3-diamino- (**32**) or 1,3,6-triaminophenoselenazinium chloride (**33**) (scheme 12).[37] The 1,3-diamino dye **32** is green-brown in water; the 1,3,6-triamino dye **33** is violet, but it dyes cotton mordanted with tannic acid a bluer shade than does its sulfur analog.[37]

Iminonaphthoselenazine dyes have been made by refluxing selenium with α-naphthylamine in the presence of a 1:1 mercury-acid amide catalyst, although an early attempt to make **29** from selenium and diphenylamine failed.[36] The synthesis of iminonaphthoselenazine dyes proceeds *via* the formation of the hydrazo compound, a *p*-semidine rearrangement, reaction

(12)

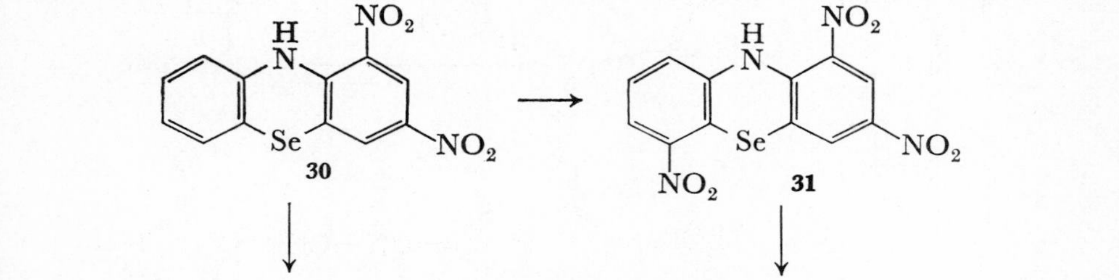

30

31

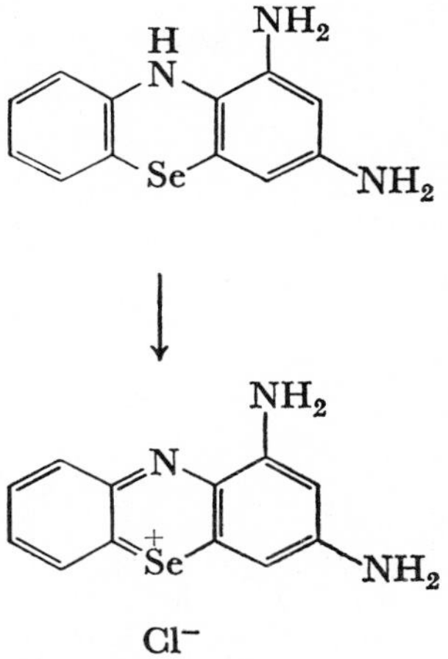

of the resulting 4-aminodiarylamine with selenium to form the selenazine, and then oxidation to the dye with atmospheric oxygen (scheme 13).[38] The reaction of amines with **29**, followed by ferric chloride oxidation, gave 3,6-diaminophenoselenazinium dyes (scheme 14).[39]

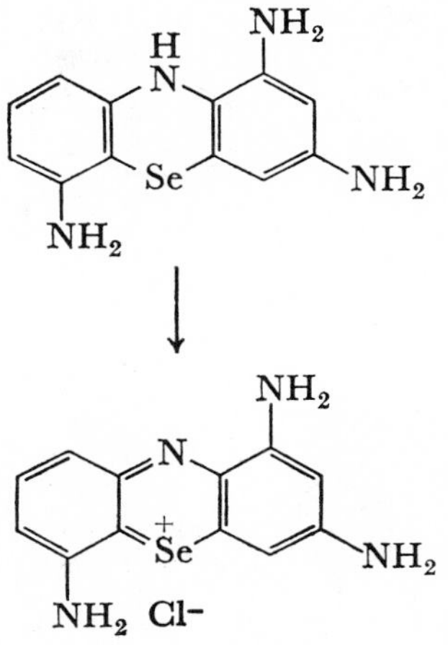

Early efforts to make the selenium analog of methylene blue, 3,6-bis-(dimethylamino)phenoselenazinium halide (**34**), by the reaction of N,N-dimethyl-p-phenylenediamine with hydrogen selenide gave an unstable blue dye.[36,40,41] The poor stability of this product caused Karrer[39] to conclude that it was not true **34**. He later made selenomethylene blue bromide by bromination of **29** and reaction of the resulting perbromide with dimethylamine (scheme 15).[42] Karrer's product proved similar in color, stability, and toxicity to methylene blue.

3-Anilinophenoselenazinium chloride (**35**) and 3-(4-arsonoanilino)-phenoselenazinium inner salt (**36**) are dark green in acid, the latter being

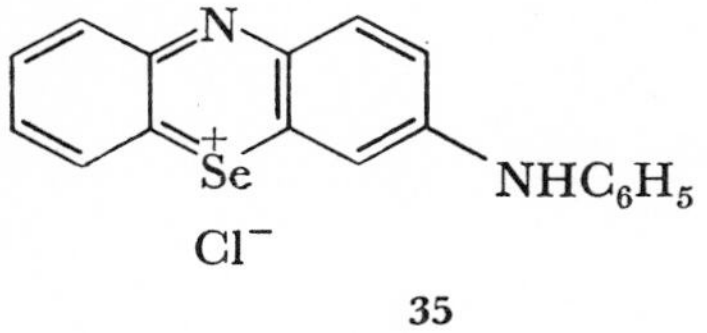

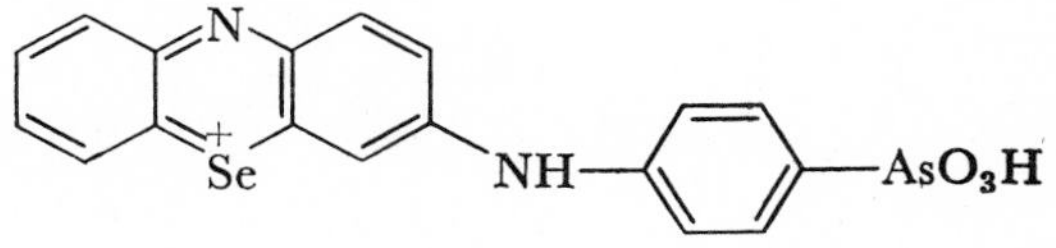

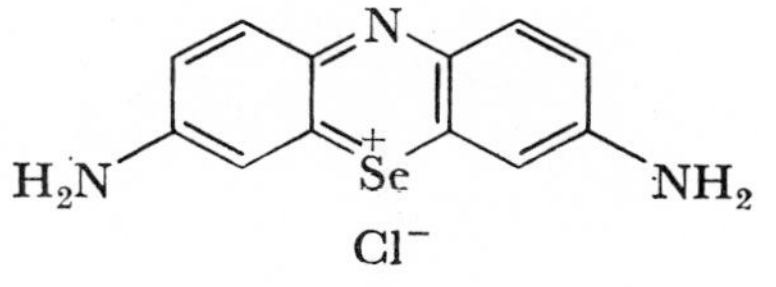

(13)

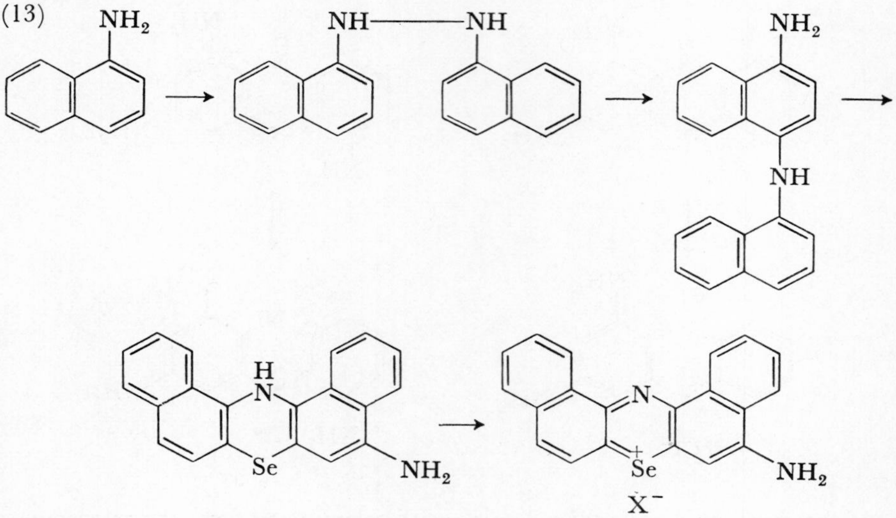

(14)

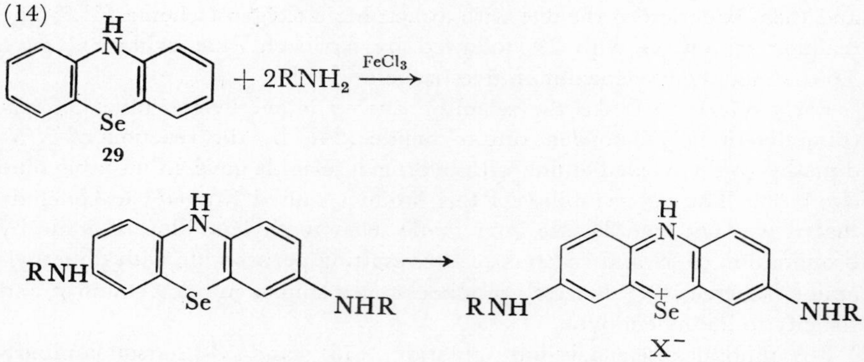

29

(15)

29

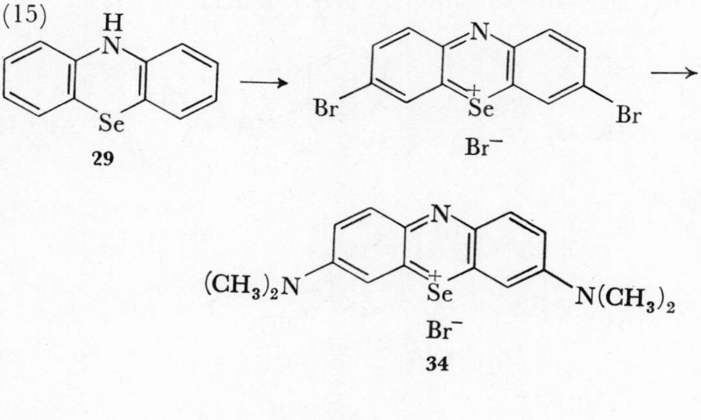

34

violet-red in base.[39]

The selenium analog of Lauth's violet, 3,6-diaminophenoselenazinium chloride (37), is deep blue.[36,39]

b. Selenoxanthenes. Tetrazotization of 4′,4″-methylenebis(3-aminoaceta-nilide) (38), followed by treatment with potassium selenocyanate, gave impure bisselenocyanate (39), which was converted by strong acid to 3,6-diaminoselenoxanthylium chloride (40) (scheme 16). It behaves chemically

(16)

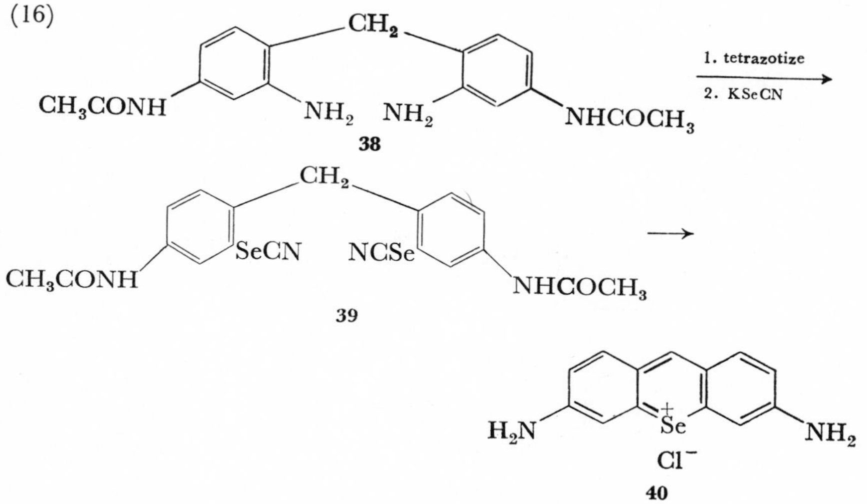

like the thio analog, but in solution and on fabric it looks bluer and is not fluorescent. Its poor wash stability on cotton is improved by a tannic acid mordant.[43]

The N,N,N′,N′-tetramethyl compound 41 was made by the reaction of 4,4′-methylenebis(N,N-dimethylaniline) with sodium selenite and fuming sulfuric acid and isolated as the zinc double salt (scheme 17). It is red in strong acid and blue in weak acid or base.[44]

(17)

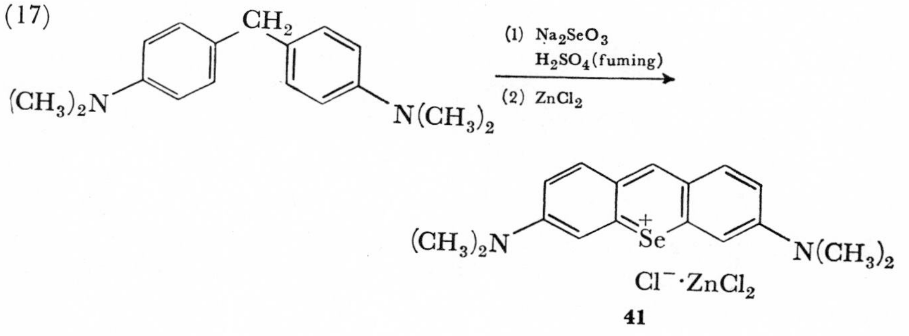

c.   Selenoindigos.   Two starting materials for these dyes are 3-hydroxy-benzo[b]selenophene (3-hydroxyselenonaphthene) (**42**) and 2*H*,3*H*-benzo-[b]selenophene-2,3-dione (selenonaphthenequinone) (**43**). The first is made by the addition to an alkali-metal selenide of diazotized anthranilic acid, reduction of the resulting 2,2′-diselenodibenzoic acid with zinc dust to seleno-salicylic acid, alkylation on selenium with potassium chloroacetate, and cyclization in acetic anhydride of the 2-carboxyphenylselenoacetic acid (scheme 18).[45] The quinone **43** is made in 70% yield by a careful and

(18)

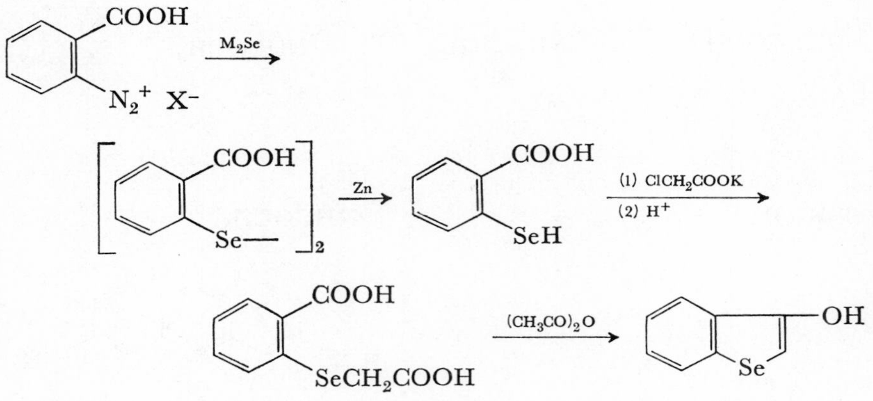

laborious acid hydrolysis of the anil from **42** and *N*,*N*-dimethyl-*p*-nitroso-aniline (scheme 19).[46]

(19)

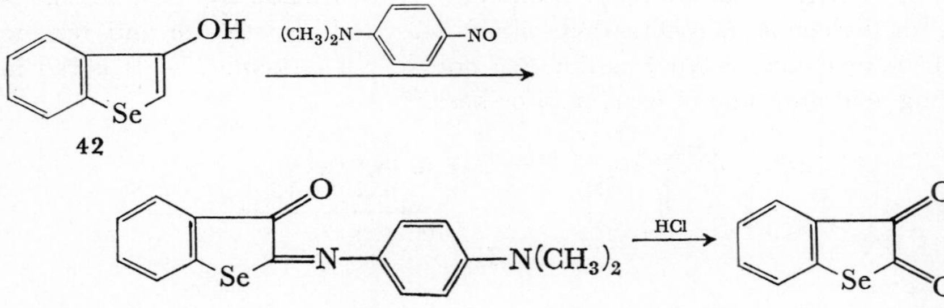

Selenoindigo (**44**) can be made by ferricyanide oxidation of **42**[45] or by room-temperature bromination of **42** and heating the resulting 2-bromo-3-hydroxyselenonaphthene (scheme 20).[46]

(20)

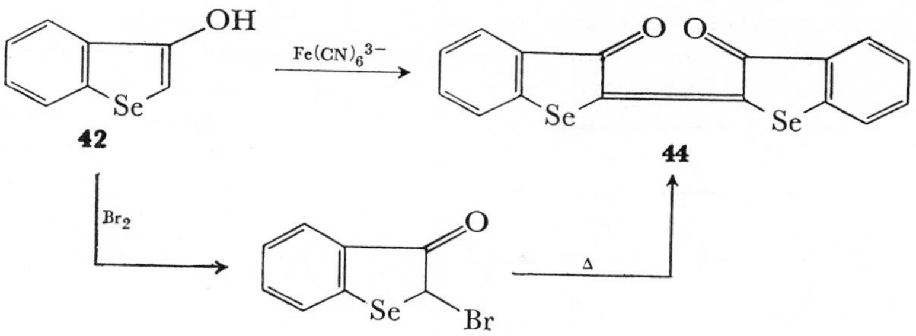

Other 2-selenonaphtheneindigoid dyes were made by the piperidine-catalyzed reaction of **42** with isatin in alcohol to give 2-selenonaphthene-3'-indoleindigo (**45**) with 5-bromoisatin to give 2-selenonaphthene-3'-(5-bromoindole)indigo (**46**), with 5-methylisatin to give 2-selenonaphthene-3'-(5-methylindole)indigo (**47**), and with acenaphthenequinone to give 2-acenaphthene-2'-selenonaphtheneindigo (**48**).[45] Heating **42** and **43** in acetic acid containing fused potassium acetate yielded 2,3'-bisseleno-naphtheneindigo (**49**), and similar treatment of **42** with 2-bromo-3-hydroxy-benzo[b]thiophene produced 2-selenonaphthene-2'-thionaphtheneindigo (**50**).[46] The reaction of **42** with isatin chloride gave 2-indole-2'-seleno-naphtheneindigo (**51**), while condensation with 2H,3H-benzo[b]thiophene-2,3-dione yielded 2-selenonaphthene-3'-thionaphtheneindigo (**52**).[47]

The 3-selenonaphtheneindigos are made by treatment of **43** with hetero-cycles containing active methylene groups. Thus, the reaction with indoxyl gave 2-indole-3'-selenonaphtheneindigo (**53**),[46] while reaction with 3-hydroxybenzo[b]thiophene afforded 2-thionaphthene-3'-selenonaphthene-indigo (**54**).[46]

The selenoindigos are similar in chemical properties to the corresponding thioindigos but appear deeper in color.[45] Recent work[46a] on 4,4,4',4'-tetra-methyl-3,3'-dioxo-2,2'-bisselenolanylidene has shown that it represents the basic chromophore in selenoindigo. The insertion of a vinylene group between the rings of a selenoindigo in contrast to the behavior of the seleno-cyanines (cf. Section 2b) causes a hypsochromic shift in the absorption spectrum of the resulting dye.[46a] Table XIE-2 gives the colors of the seleno-indigos in sulfuric acid solution and those of the corresponding sulfonic acids, all of which are water soluble.[45—47]

d. Azo Dyes. Diazonium salts couple with **42** to give 2-arylazo dyes (**55**) (scheme 21). The phenylazo compound is yellow in acetic acid and red in either sodium hydroxide or concentrated sulfuric acid, while the corre-sponding sulfonic acid is yellow.[46] The p-nitrophenylazo dye is dark red in concentrated sulfuric acid and violet with red-violet fluorescence in base.[47]

Table XIE-2   Colors of selenoindigos (A=B)

	STRUCTURE		COLOR IN	COLOR OF
	A	B	H$_2$SO$_4$	SULFONIC ACID
44			Deep green	Red-violet
45	,,		Olive-green	
46	,,		Blue-green	Intense violet
47	,,		Green	Violet
48	,,		Blue-green	
49	,,		Green	
50	,,		Blue-green	Carmine-red
51	,,		Intense blue	Red-violet

Table XIE-2   (*continued*)

STRUCTURE		COLOR IN H_2SO_4	COLOR OF SULFONIC ACID
A	B		
52		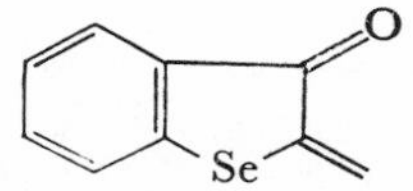 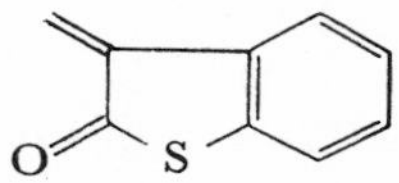 Intense green	Violet-red
53		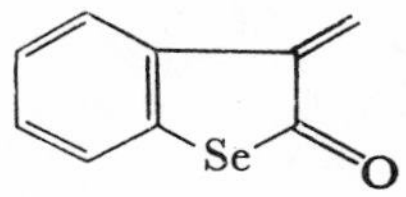 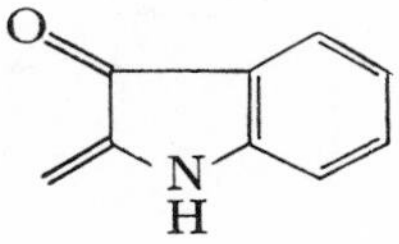 Deep green	Red
54 ,,	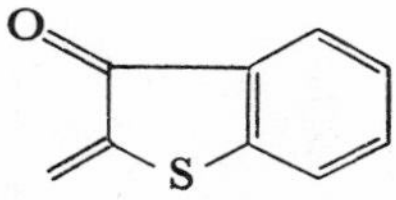	Green	

A study of the visible spectra of some 4′-substituted 4-dimethylamino-azobenzenes gave σ_p values of 0.38 and 0.63 for the trifluoromethylseleno (CF_3Se-) and trifluoromethylselenenyl [$CF_3Se(=O)-$] substituents, respectively. The absorption maxima for the former dye are 447 nm in ethanol and 505 nm in 2:1 ethanol-hydrogen chloride. Comparable values for the latter dye are 448 nm in ethanol and 510 nm in 2:1 ethanol-hydrogen chloride.[48]

(21)

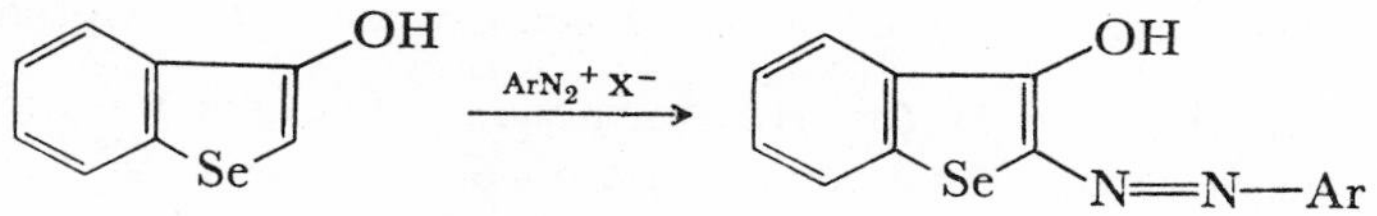

e. Vat Dyes. The reaction of 1-halo-2-aminoanthraquinone with an alkali-metal selenide gives 1-seleno-2-aminoanthraquinone. Treatment of this compound or its diselenide with an aldehyde, aldehyde liberator, or acyl halide yields a 6,11-dihydroanthra[2,1-d]selenazole-6,11-dione (**56**) (scheme 22).[49] Heating 2-halo-5-benzamidoanthraquinone and selenium in kerosene or tetralin with an acid-absorbing agent gives 2,2′-selenobis-(5-benzamidoanthraquinone) (**57**), which dyes cotton yellow to brown shades from a vat.[50] Treating benzanthrone with selenium dioxide in sulfuric acid, followed by alkali fusion, produces a blue vat dye.[51] When cobalt phthalo-cyanine is treated with aluminum chloride in dimethylformamide or in

(22)

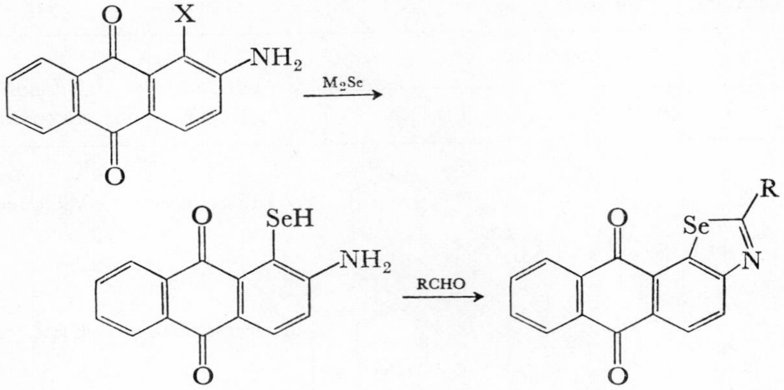

56

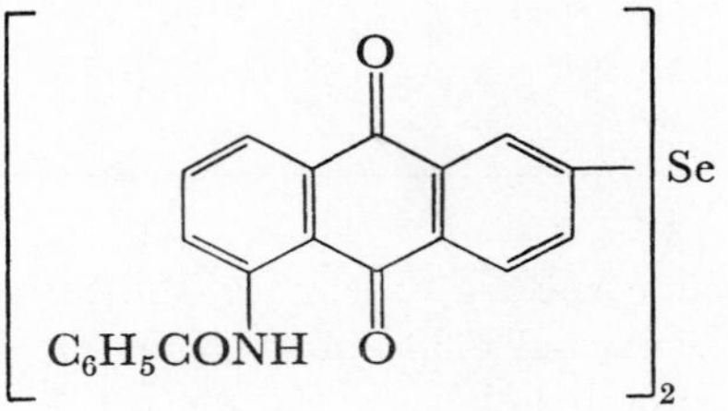

57

melts of aluminum chloride-sodium chloride, aluminum chloride-potassium chloride, or aluminum chloride-sodium acetate, and then treated with powdered selenium, a fast blue or blue-green dye for cotton is produced.[52]

f. Miscellaneous Dyes of Uncertain Structure.

The selenium analog of Sulfur Black T, made by boiling 2,4-dinitrophenol with selenium and alkaline sodium selenide, dyes cotton fast walnut-brown.[53] The reaction of ethyl *p*-aminobenzoate with red selenium in the presence of a mercury-acetamide catalyst afforded small amounts of three water-soluble selenium-containing dyes.[54] Alkali-metal selenocyanates or selenides reacted with phthaleins also gave selenium-containing dyes.[55]

REFERENCES

1. F. M. Hamer, "The Cyanine Dyes and Related Compounds," Interscience Publishers, New York, N.Y., 1964.
2. W. H. Mills and R. S. Wishart, *J. Chem. Soc.*, **117**, 579 (1920).
3. C. R. Bury, *J. Amer. Chem. Soc.*, **57**, 2115 (1935).

4. For example, L. G. S. Brooker, "Resonance and Organic Chemistry" (Frontiers in Chemistry, Vol. III), Interscience Publishers, New York, N.Y., 1945.

5. L. M. Clark, *J. Chem. Soc.*, 2313 (1928).

6. W. H. Mills, *J. Chem. Soc.*, **121,** 455 (1922).

7. A. W. von Hofmann, *Ber.*, **20,** 2257 (1887).

8. N. I. Fisher and F. M. Hamer, *Proc. Roy. Soc., Ser. A*, **154,** 703 (1936).

9. I. G. Farbenind. A.-G., Brit. Pat. 380,702 (Jan. 10, 1931).

10. N. I. Fisher and F. M. Hamer, *J. Chem. Soc.*, 2502 (1930).

11. W. König, *Ber.*, **55,** 3293 (1922).

12. F. M. Hamer, *J. Chem. Soc.*, 2796 (1927).

13. N. I. Fisher and F. M. Hamer, *J. Chem. Soc.*, 189 (1933).

14. L. G. S. Brooker, G. H. Keyes, and F. L. White, *J. Amer. Chem. Soc.*, **57,** 2492 (1935).

15. I.C.I. Ltd., H. A. Piggott, and E. H. Rodd, Brit. Pats. 344,409 (Nov. 4, 1929) and 354,898 (May 13, 1930).

16. I.C.I. Ltd., H. A. Piggott, and E. H. Rodd, Brit. Pat. 355,693 (May 31, 1930); I. G. Farbenind. A.-G., Brit. Pats. 434,234 and 434,235.

17. I. G. Farbenind. A.-G. and O. Wahl, Ger. Pat. 499,967 (June 29, 1928); I. G. Farbenind. A.-G., Brit. Pat. 438,484 (May 18, 1933).

18. L. G. S. Brooker, "Some Recent Developments in the Chemistry of Photographic Sensitizing Dyes," XIVth International Congress of Pure and Applied Chemistry, Experientia Supplementum II, Birkhäuser Verlag, Basel and Stuttgart, 1955, p. 229.

19. L. G. S. Brooker and F. L. White, *J. Amer. Chem. Soc.*, **57,** 547, 2480 (1935).

20. L. G. S. Brooker and F. L. White, U.S. Pat. 2,112,139 (1938).

21. L. G. S. Brooker and F. L. White, U.S. Pat. 2,112,140 (1938).

22. C. E. K. Mees, "From Dry Plates to Ektachrome Film," Ziff-Davis Publishing Co., New York, N.Y., 1961, pp. 215, 217.

23. L. Pauling, "Nature of the Chemical Bond," 2nd ed., Cornell University Press, Ithaca, N.Y., 1940.

24. L. G. S. Brooker, F. L. White, D. W. Heseltine, G. H. Keyes, S. G. Dent, Jr., and E. J. Van Lare, *J. Phot. Sci.*, **1,** 174 (1953).

25. E. I. duPont de Nemours and Co., Brit. Pat. 566,684 (1945).

26. H. C. Barany and M. Pianka, *J. Chem. Soc.*, 2217 (1953).

27. E. I. duPont de Nemours and Co., U.S. Pat. 2,330,791 (1943).

28. Eastman Kodak Co., U.S. Pat. 2,475,949 (1949).

29. S. G. Fridman, *J. Gen. Chem. USSR*, **31,** 1013 (1961).

30. L. G. S. Brooker and F. L. White, U.S. Pat. 2,005,411 (1935).

31. L. G. S. Brooker, U.S. Pat. 2,189,599 (1940).

32. L. G. S. Brooker, G. H. Keyes, R. H. Sprague, R. H. Van Dyke, E. Van Lare, G. Van Zandt, and F. L. White, *J. Amer. Chem. Soc.*, **73,** 5326 (1951).

33. For example, L. G. S. Brooker, U.S. Pat. 2,211,732 (1940).

34. O. Riester and F. Bauer, Ger. Pat. 936,071 (1955).

35. L. G. S. Brooker and R. H. Sprague, U.S. Pat. 2,332,433 (1943).

36. W. Cornelius, *J. Prakt. Chem.*, **88,** 395 (1913).

37. H. Bauer, *Ber.*, **47,** 1873 (1914).

38. P. S. Pishchimuka, *J. Gen. Chem. USSR*, **10,** 305 (1940); *Chem. Abstr.*, **34,** 7915 (1940).

39. P. Karrer, *Ber.*, **49,** 597 (1916).

40. A. von Wassermann a⸣ 1 E. von Wassermann, Ger. Pat. 261,793; *Chem. Abstr.*, **7,** 3239 (1913).

41. W. Fraenkel, Ger. Pat. 280,713; *Chem. Abstr.*, **9,** 1847 (1915).

42. P. Karrer, *Ber.*, **51,** 190 (1918).

43. P. Ehrlich and H. Bauer, *Ber.*, **48,** 502 (1915).

44. M. Battegay and G. Hugel, *Bull. Soc. Chim. Fr.*, **27,** 557 (1920).

45. R. Lesser and R. Weiss, *Ber.*, **45,** 1835 (1912).

46. R. Lesser and A. Schoeller, *Ber.*, **47,** 2292 (1914).

46a. L. Fitjer and W. Lüttke, *Chem. Ber.*, **105,** 919 (1972).

47. R. Lesser and R. Weiss, *Ber.*, **46,** 2640 (1913).

48. L. M. Yagupol'skii and V. G. Voloshchuk, *Ukr. Khim. Zh.*, **36,** 1 (1970); *Chem. Abstr.*, **73,** 16251 (1970).

49. E. I. duPont de Nemours and Co., French Pat. 761,570; *Chem. Abstr.*, **28,** 4245 (1934).

50. M. A. Perkins, U.S. Pat. 1,973,773; *Chem. Abstr.*, **28,** 7032 (1934).

51. R. F. Thomson, D. B. Anderson, and S. Thornley, Brit. Pat. 367,462; *Chem. Abstr.*, **27,** 2043 (1933): U.S. Pat. 1,972,960; *Chem. Abstr.*, **28,** 7032 (1934).

52. French Pat. 1,522,539; *Chem. Abstr.*, **71,** 82644 (1969).

53. O. I. Lee, *Polytechnic Eng.*, **12,** 81 (1912); *Chem. Abstr.*, **7,** 2309 (1913).

54. L. F. Shpeier, *Tr. Khar'kovsk. Sel'skokhoz. Inst.*, **35,** 138 (1961); *Chem. Abstr.*, **59,** 3799 (1963).

55. A. von Wassermann and E. von Wassermann, Ger. Pat. 261,556; *Chem. Abstr.*, **7,** 2422 (1913): Ger. Pat. 286,950; *Chem. Abstr.*, **10,** 1711 (1916): Ger. Pat. 287,020; *Chem. Abstr.*, **10,** 1935 (1916).

XI Heterocyclic selenium compounds

F. MISCELLANEOUS SELENIUM-CONTAINING HETEROCYCLES

RICHARD B. SILVERMAN

Harvard University
Department of Chemistry
Cambridge, Massachusetts

1. Introduction

Within the scope of this chapter will be found only those selenium-containing heterocycles that cannot be categorized into any of the other portions of Chapter XI. These compounds have been organized in an ascending sequential arrangement in which precedence is given to the number of rings in the molecule. Within each of these ring groups, the sizes of the rings are systematized, and then for particular ring sizes the number of Se atoms in the rings is ordered. For the sake of fluidity, minor inconsistencies in this arrangement occur. Also, some compounds are mentioned in more than one part of the text if they are by-products of other reactions.

2. Heterocycles Consisting of One Ring

a. *Three-Membered Ring.* The theoretically smallest members of the miscellaneous heterocyclic group of compounds, selenirenes (also termed "acetylene selenides") (1), are unknown. In 1962, though, the existence of these compounds was hypothesized.[1] Because of the similarity of the electronic structure of selenium to that of carbenes, it was suggested that the logical mode of synthesis would be the combination of acetylenes with selenium (equation 1). However, in a model reaction[1] using germanium diiodide

(1)
$$R—C{\equiv}C—R + Se \ {-\!\!\!/\!\!\rightarrow} \quad \underset{R}{\overset{Se}{\diagup\!\!\diagdown}} R$$

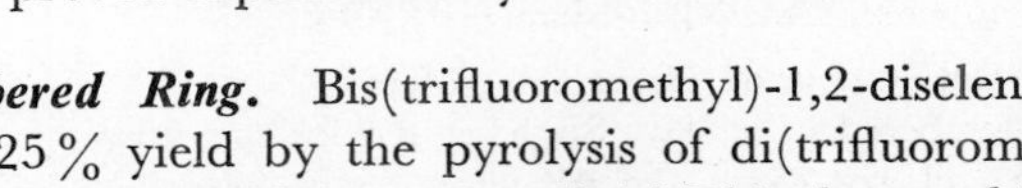

1

instead of selenium (both have the same electronic configuration), this postulation was proved experimentally to be invalid.[2,3]

b. *Four-Membered Ring.* Bis(trifluoromethyl)-1,2-diselenetene (2) was prepared in a 25% yield by the pyrolysis of di(trifluoromethyl)acetylene with selenium vapor at 700° (equation 2).[4,5] This deep red liquid forms bis and tris complexes (3) with transition metals as shown in equation 3.

(2)
$$CF_3—C{\equiv}C—CF_3 + Se \ \xrightarrow{700°} \ \underset{F_3C}{\overset{F_3C}{>}}{=}{\overset{Se}{\underset{Se}{|}}}$$

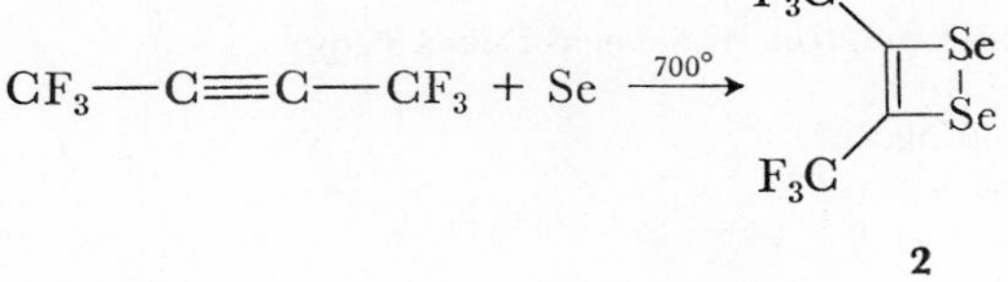

2

(3) $2 + M(CO)_x \longrightarrow$

$$\left[\begin{array}{c} F_3C \\ \\ F_3C \end{array} \begin{array}{c} Se \\ M/n \\ Se \end{array}\right]^z$$

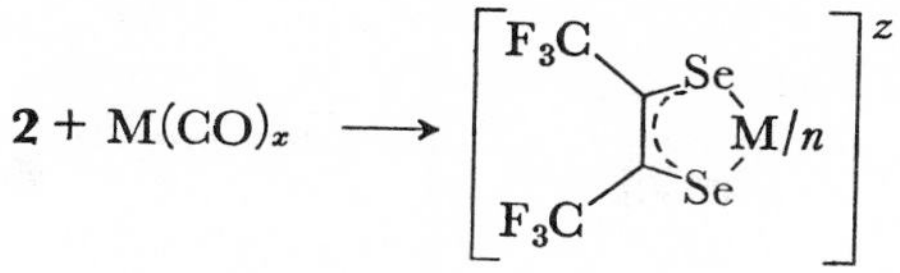

$n=2,3$

3

c. Five-Membered Rings Containing One Selenium Atom

i. 2,5-Disubstituted 3-Selenolenes and 2,5-Disubstituted 2,3-Dihydro-selenophenes. The acid hydrolysis of 2-substituted 5-*tert*-butoxyseleno-phenes (**4**) potentially gives rise to three tautomeric products, **5a**, **5b**, and **5c** (equation 4). An n.m.r. study of the product has revealed that, at equilib-

(4)   R—[Se]—OC(CH₃)₃ $\xrightarrow[\Delta]{H^\oplus}$ R—[Se]=O $\rightleftharpoons$

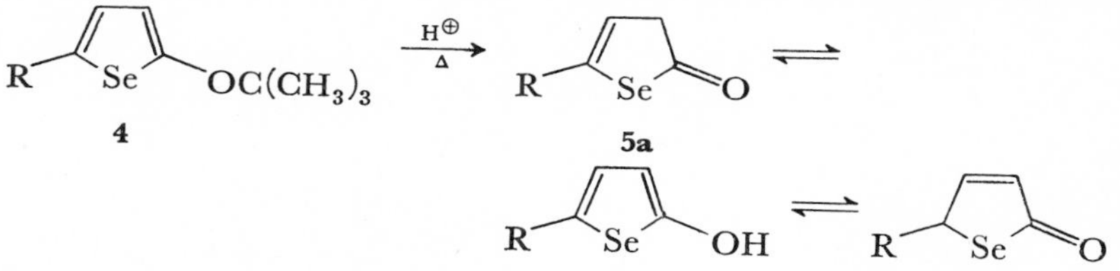

4   **5a**

R—[Se]—OH $\rightleftharpoons$ R—[Se]=O

5b   **5c**

rium, the parent compound (**5**, R = H) exists completely as 2-oxo-3-seleno-lene (**5c**, R = H).[6] Other derivatives are composed mostly of **5c** with some 5-substituted 2,3-dihydroselenophen-2-one (**5a**), but none of tautomer **5b**. These results correspond with those of a similar investigation of the sulfur analog.[7]

The selenium homolog of maleic anhydride, that is, 2,5-dioxo-3-selenolene (**6**), was prepared according to equation 5.[6]

(5)   (CH₃)₃CO—[Se]—OC(CH₃)₃ (with Br) $\xrightarrow{H^\oplus}$ O=[Se]=O

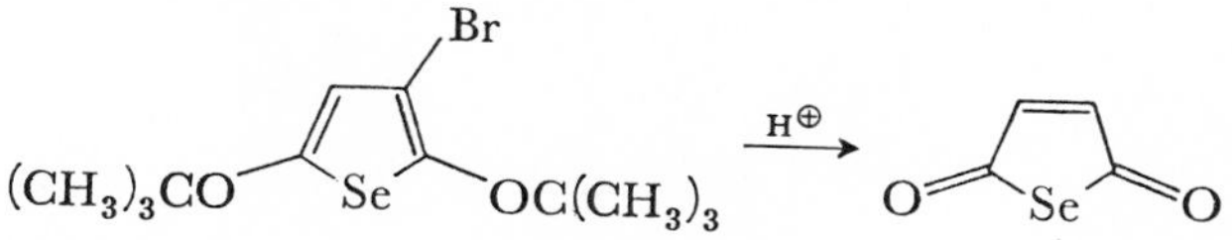

6

ii. Selenium Analogs of Thiothiophthenes

Thiaselenolodithioles. The class of compounds known as thiaselenolo-dithioles (**7**) can be considered as both heterocycles (**7a**) and selenoaldehydes (**7b**). The preparation and characterization of this group of compounds are

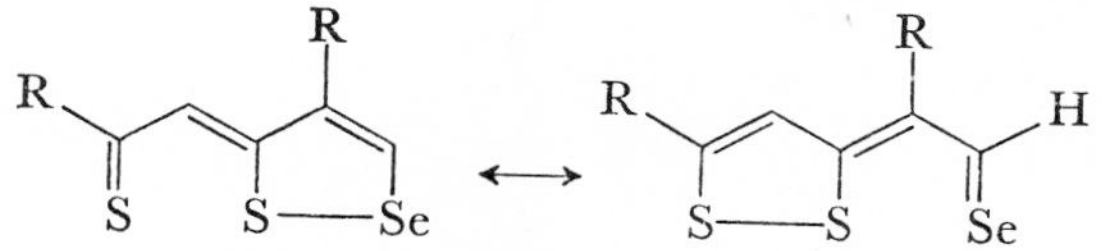

7a $\longleftrightarrow$ **7b**

described in Section VII.B. Nevertheless, it should be emphasized here that with the aid of n.m.r. analysis and X-ray crystallography **7a** was shown to contribute more to the electronic structure than **7b**.[8]

Other selenothiophthenes. It was reported incorrectly[9] that the product of the reaction of 2,6-dimethyl-4-seleno-γ-pyrone and sodium selenide, followed by treatment with phosphorus pentasulfide, was 3,7-dimethyl-1,2-diselena-cyclohepta-3,6-dien-5-thione. The actual product was later described,[10] as suggested earlier,[11] as being 2,5-dimethyldiselenothiophthene (**9**). This reaction was subsequently determined to yield a mixture of 2,5-dimethyl-selenothiophthene (**8**, 88%) and **9** (12%).[12]

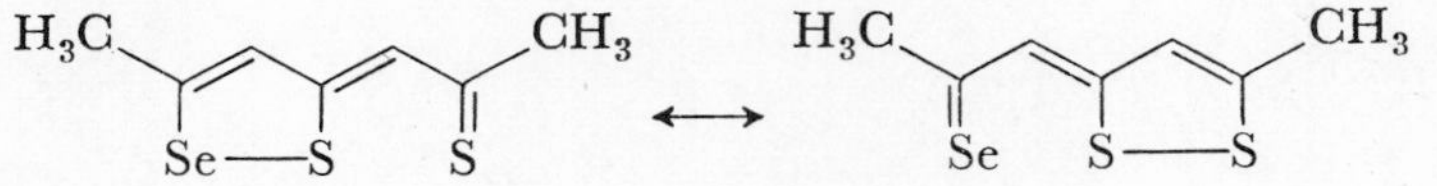

8

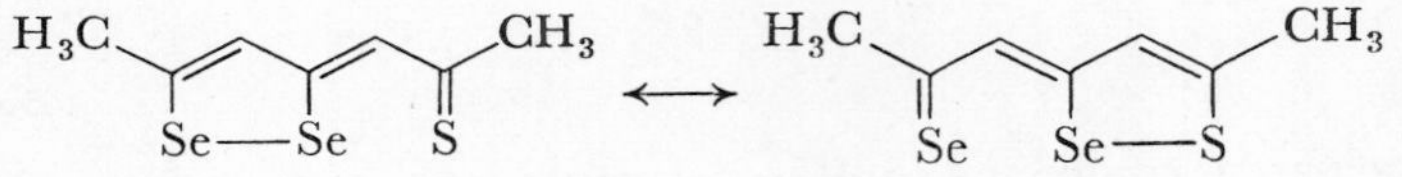

9

An investigation of the i.r. spectra of both of these compounds, as well as of a number of other selenium-, sulfur-, and oxygen-containing thiophthene derivatives, was made.[12] The structure of **9** had been reported in an electric dipole moment study the previous year as meribicycloepi-2-thio-4-selenoheptadien-6-thione (**10**).[13]

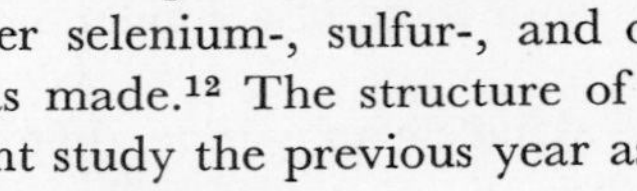

10

iii. Thiaselenole Heterofulvalenes and Related Compounds. Compounds with the general structure **11**, where X, Y, Z may be N or C and R, R', R"

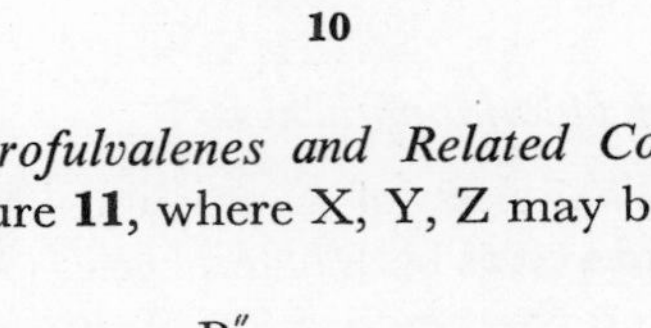

11

may be H, C_6H_5, or absent were prepared by the method exemplified in scheme 6.[14] The perchlorate salt used as the starting material for the above

(6)

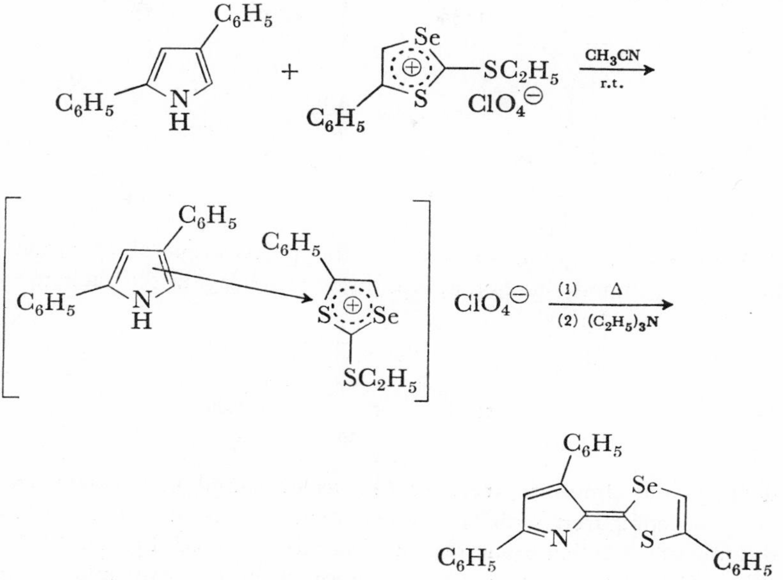

reaction was prepared from the corresponding thione by reaction with $(C_2H_5)_3OBF_4$ in perchloric acid. This thione and its corresponding ketone have been made by the route shown in scheme 6a.[14a]

(6a) $C_6H_5C\equiv CNa + Se + CS_2 \xrightarrow{CH_3CN}$

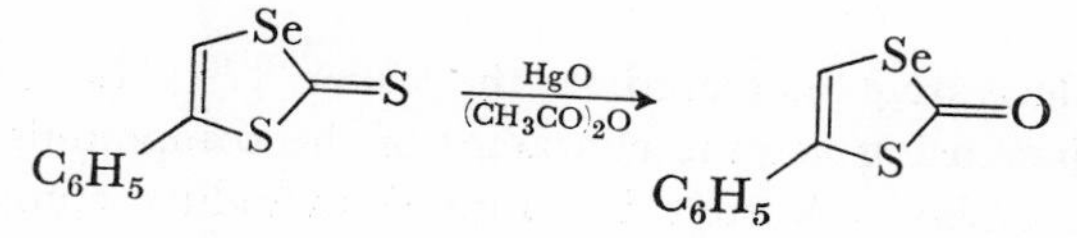

d. Five-Membered Rings Containing Two Selenium Atoms

i. 3-Benzylidene-4-phenyl-2H-1,3-diselenole. The acid dimerization of the sodium salt of 2-phenylethynylselenol (**12**) afforded the diselenole (**13**) in a 14% yield (equation 7).[15]

(7)

$$C_6H_5C\equiv CNa + Se \xrightarrow{\Delta} C_6H_5C\equiv CSeNa \xrightarrow{H\oplus}$$

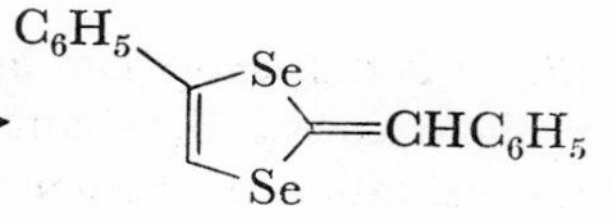

12   13

ii.   Epidiselenohepta-2,4-dien-6-one   Derivatives. The synthesis of 2,4-epidiselenohepta-2,4-dien-6-one (**14**) was unwittingly reported[9] in 1957 as shown in equation 8 (see Section VII.C for details).

(8)

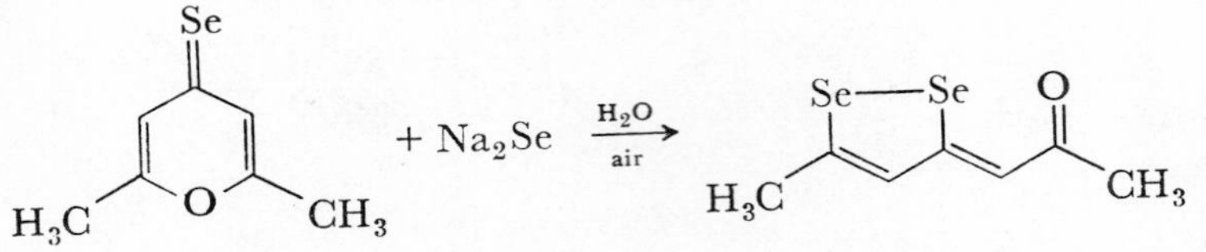

14

Also unknowingly described was the preparation of the mono sulfur analog, 2,5-dimethyldiselenothiophthene (**15**)[9] (again see Section VII.C for

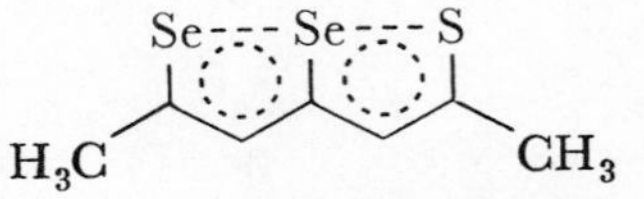

15

details). The dipole moment of **14** was measured in benzene at 20° and 50° and compared with the dipole moments of the sulfur analogs, epidithiohepta-2,4-dien-6-one (**16**) and meribicycloepidithiohepta-2,4-diene-6-thione (**17**).[16] The tendency for formation of the 2,4-dinitrophenylhydrazone

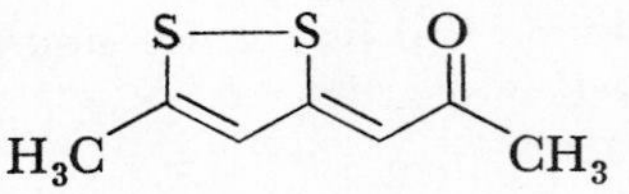

16

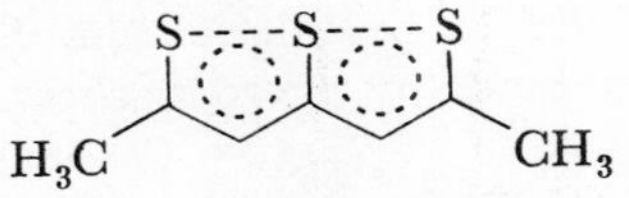

17

derivative increased in the order **17** < **16** < **14**, which agreed with the increasing ketonic character of the compounds on the basis of their dipole moments. A later i.r. study[12] including compounds **14**, **15**, **16**, and **17** showed aromatic character in the non-oxygen-containing compounds (*e.g.*, **15** and **17**), whose electronic configuration shifted towards a highly conjugated system when one Se or S atom was substituted by oxygen (*e.g.*, **14** and **16**). This result, therefore, confirmed the findings of the dipole moment study.[16] Compound **14** was also included in an electric dipole moment investigation.[13]

iii.   1,2-Diselenolylium Salts. The first reported derivative of this heterocyclic class was 3,5-diamino-1,2-diselenolylium iodide (**18**), the product of the oxidation of diselenomalonamide with iodine (equation 9).[17] Also prepared was the tetrachloroferrate(II) salt of **18** in a 76% yield.[17]

Initial attempts to prepare the diselenoacetylacetonate cobalt(II) complex under the same conditions that were used to synthesize the nickel complex

(9)

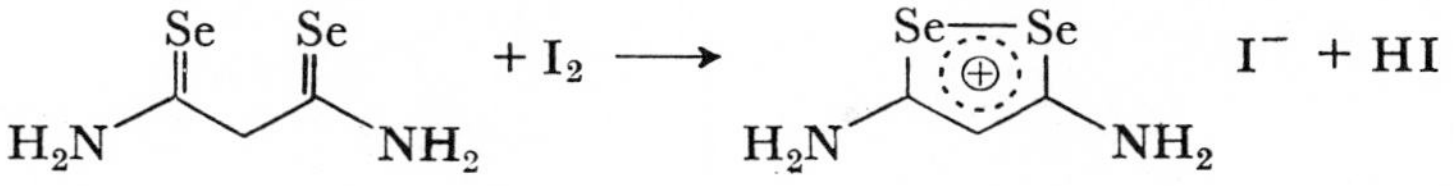

18

(*cf.* Section VII.C) produced 3,5-dimethyl-1,2-diselenolylium tetra-chlorocobaltate(II) (**19**) (equation 10).[18,19] An X-ray study of the structure

(10)

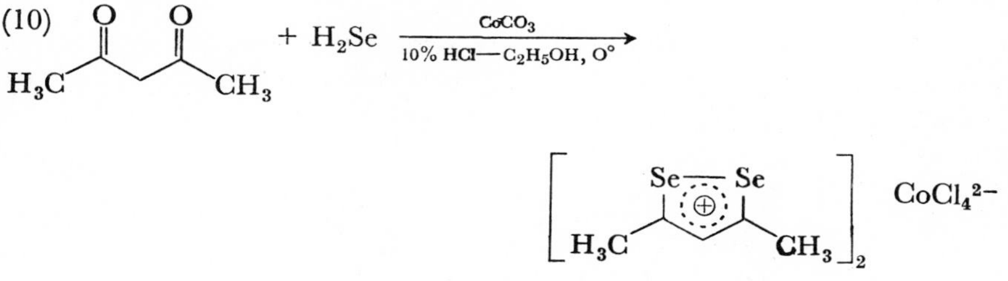

19

of 3,5-dimethyl-1,2-diselenolylium tetrachloroferrate(II) revealed the close similarity of this complex to its sulfur analog.[20]

e. Six-Membered Rings Containing One Selenium Atom

i. Selenopyrans. Selenopyrans have been made only in low yields. The parent compound (**20**, R = H), an unstable liquid, was synthesized in a 16% yield as shown in scheme 11.[21]

(11)

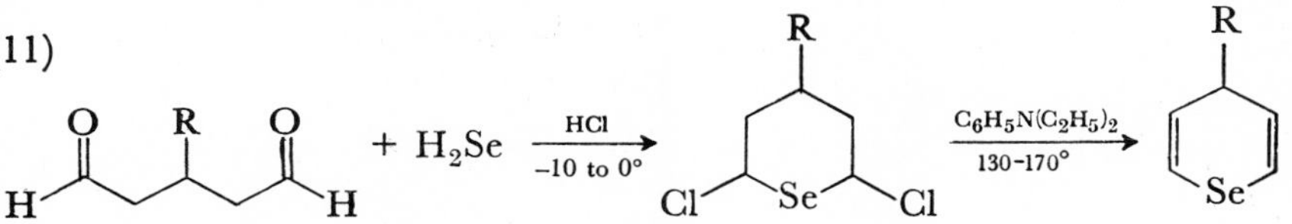

20

The first derivative of the family, that is, 4-methyl-1-seleno-γ-pyran (**20**, R = CH$_3$), was prepared later by the same route used to synthesize the parent compound.[22] Chemical shifts and coupling constants of the proton n.m.r. spectra for selenopyran and its oxygen and sulfur analogs were correlated.[23] A proton n.m.r. study of the methyl derivatives of selenopyran and its oxygen and sulfur analogs also was conducted.[24]

ii. Selenopyrylium Salts. Selenopyrylium (also called seleninium) salts (**21**, X = Se) have been made[21,22,24,25] (*cf.* Chapter VI.B.2).

Also studied were the u.v. and visible spectra,[26,27] the n.m.r. spectra,[23,24] and the stabilities[28] of these compounds. It was also postulated that for the series of cations **21** (where R = H; X = O, S, and Se) the degree of aromaticity increased in the order O < S ⩽ Se.[25]

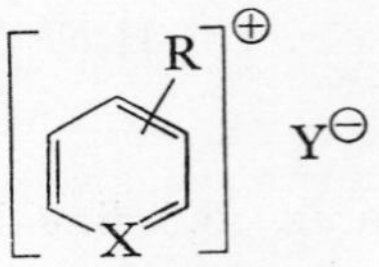

21

iii.   Selenopyrones.   Substituted selenopyrones (**23**) were reported as the products from the photolysis of di-iron carbonyl complexes having the formula $Fe_2(CO)_7(RC_2R)_2$ (**22**) with potassium polyselenide (equation 12).[29]

(12)

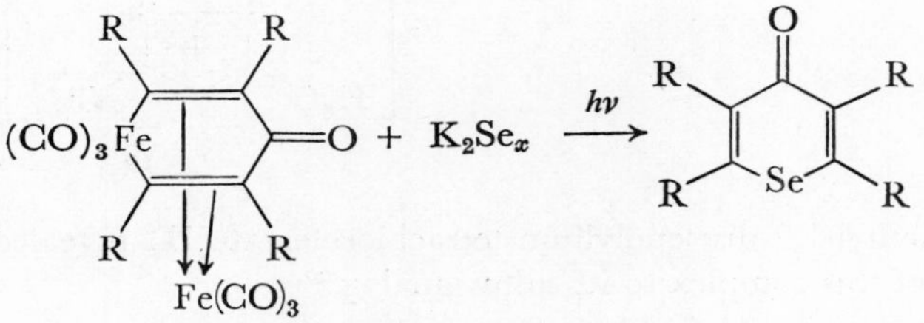

22

iv.   1-Oxa-2-selena-4-cyclohexene-2-oxides.   In 1934 the reaction of substituted butadienes with selenious acid at room temperature was studied, and the products isolated were described as substituted selenacyclopent-3-ene-1,1-dioxides (**24**).[30] However, in 1968 the structure of these products was reinvestigated, and it was concluded that the previously proposed selenones were actually cyclic selenenic esters (**25**) (equation 13).[31]

(13)

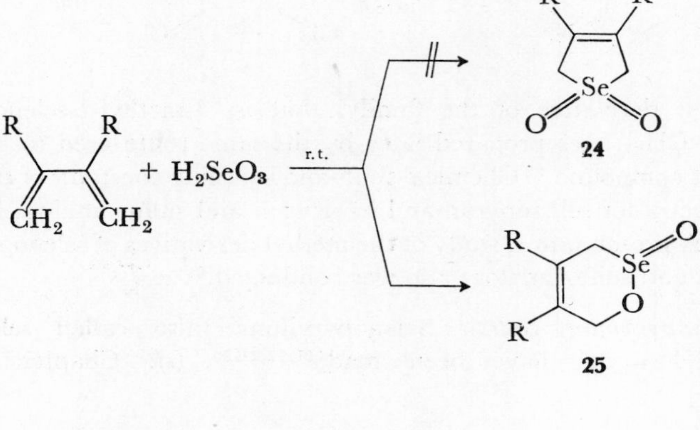

The analogous reactions with sulfur dioxide also were reported to yield the corresponding sulfones.[32] Unlike the selenium analogs, however, these products were confirmed in a recent examination of the reaction mechanism.[33]

f. Six-Membered Rings Containing Two Selenium Atoms. Diselenins have been obtained by only two methods, both of which gave very low yields. The reaction of 2,3-diiodohexafluoro-2-butene with selenium at 180° under autogenous pressure was described as a route to the preparation of 2,5-diphenyl-1,4-diselenacyclohexa-2,5-diene (**26**) in an 8% yield (equation 14).[34] The recovery of an appreciable amount of the diselenin after exposure

(14) 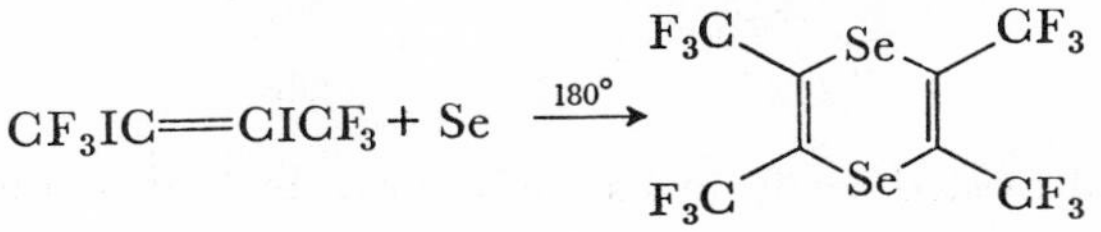 $CF_3IC = CICF_3 + Se \xrightarrow{180°}$

26

to a temperature at which the *p*-dithiin ring contracts showed that the *p*-diselenin ring was relatively stable towards loss of selenium to form the selenophene.

In a related reaction the product of the basic thermolysis of phenylacetylene with selenium which was obtained in a 10% yield was characterized as 2,5-diphenyl-1,4-diselenacyclohexa-2,5-diene (**27**).[35] The structure of the *p*-diselenin was proved by dipole moment determination and i.r. and n.m.r. analyses to be in a boat configuration (equation 15).

(15) 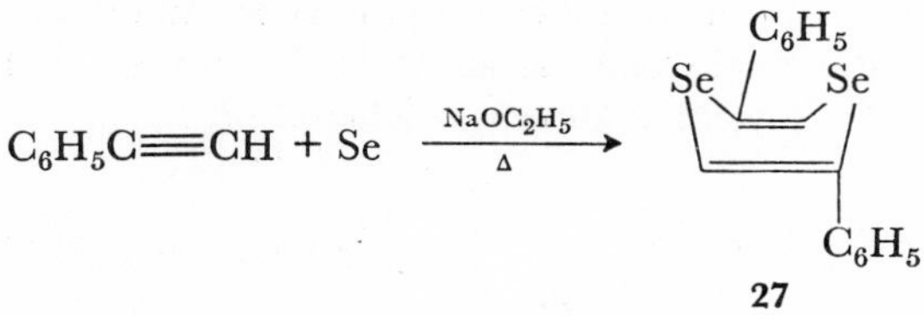 $C_6H_5C \equiv CH + Se \xrightarrow[\Delta]{NaOC_2H_5}$

27

g. Seven-Membered Rings. The seven-membered cyclic diselenide, 3,7-dimethyl-1,2-diselenacyclohepta-3,6-dien-5-one, was incorrectly reported as the minor product of the reaction of 2,6-dimethyl-4-seleno-γ-pyrone with sodium selenide.[9] This reaction and the reported preparation[9] of the thione derivative are discussed fully in Section VII.C.

3. Heterocycles Consisting of Two Fused Rings

a. One Five-Membered, One Six-Membered Ring, One Selenium Atom

i. 2,3-Dihydroselenanaphthen-3-one and Related Compounds. These compounds (**28** shows the parent compound) have been studied relatively extensively. Although some workers have chosen to name them as derivatives of 3-hydroxyselenanaphthene (**28b**), a study of the absorption spectrum of the parent compound revealed that in neutral and acidic solvents this compound exists in the keto form (**28a**).[36,37] In basic solution the enol salt, whose absorption spectrum is entirely different from that of the free enol, is formed. An n.m.r. investigation in 1968 confirmed this report.[38] Three years earlier, however, an n.m.r. study of the sulfur analog proved that

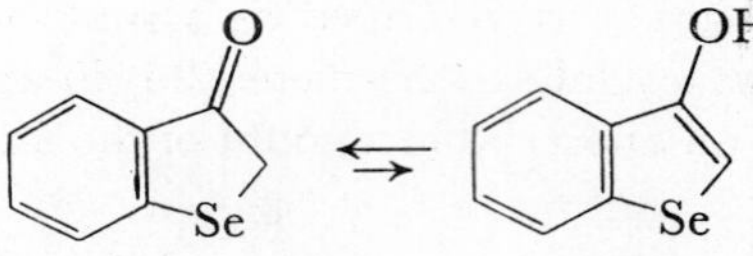

28a   **28b**

this compound existed solely in its enolic form in both neutral and acidic media.[39]

The parent compound **28** was synthesized in a 90–95% yield from 2-carboxyphenylselenoacetic acid (**29**) according to equation 16.[40]

(16)

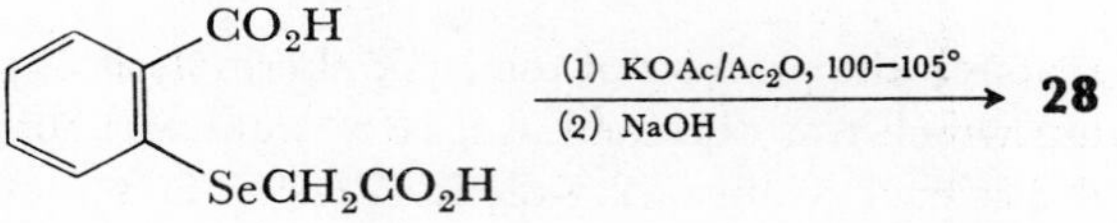

$$\xrightarrow[\text{(2) NaOH}]{\text{(1) KOAc/Ac}_2\text{O, 100–105°}} \textbf{28}$$

29

Derivatives have been synthesized with substitution on the aromatic ring[41–43] or on the methylene carbon.[44,45] The 5-methyl derivative (**30**) was made[41] in a 34% yield as shown in scheme 17.

(17)

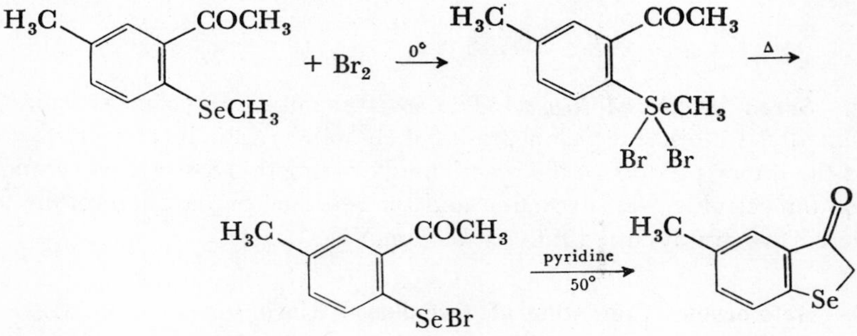

30

The preparation of the 6-chloro- and 6-methoxy-2,3-dihydroselenanaphthen-3-one (**31**, R = Cl and CH_3O, respectively) was reported as described by equation 18.[42] The 5-bromo derivative (**32**) was obtained by two routes (both in *ca.* 25 % yield), *i.e.*, the method shown in scheme 17 and the cyclization shown in equation 19.[43]

(18)

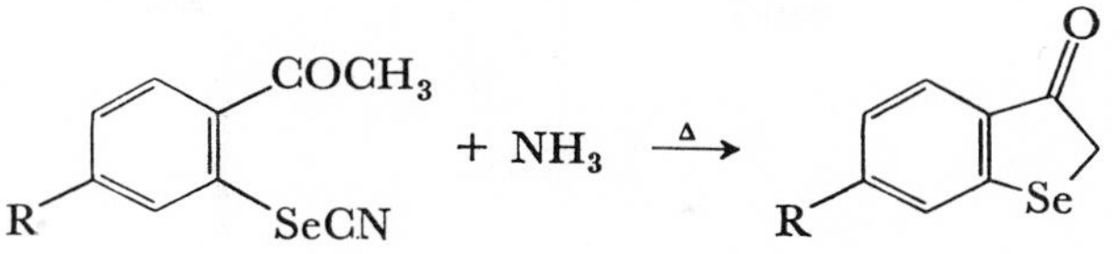

31

(19)

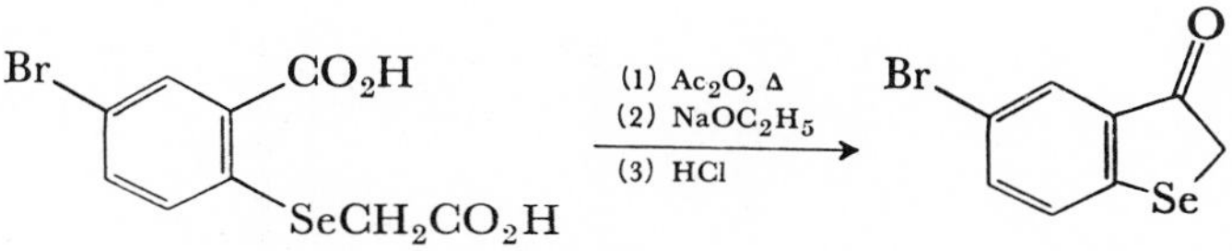

32

Numerous examples have been cited in the literature of derivatives of 2,3-dihydroselenanaphthen-3-one substituted at the methylene carbon. Included in the preparations which do not utilize the parent compound as a starting reagent is that of the 2-benzylidene derivative (**34**). This was made by the ring contraction of selenoflavanone dibromide (**33**) in boiling acetic acid (equation 20).[44]

(20)

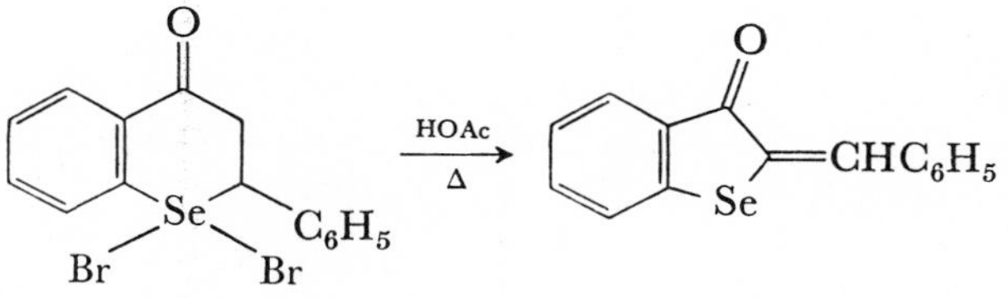

33   **34**

Selenanaphthenequinone (**35**) was prepared in a 70 % yield by the method shown in equation 21.[45]

(21)

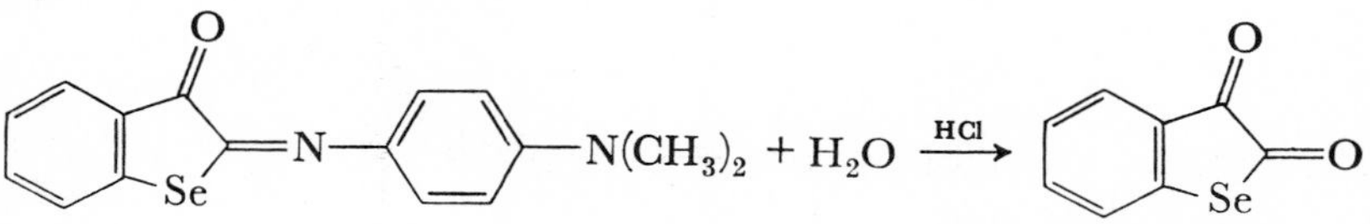

35

Most of the methylene carbon-substituted derivatives were synthesized from the parent compound **28**. Reactions at the methylene site usually involved carbonyl compounds,[40,44–46] other electrophiles[45,46] or oxidative coupling compounds.[40,41] A series of 2-benzylidene derivatives (**36**) was prepared from a reaction that possibly produced the parent compound as an unisolated intermediate (equation 22).[44]

(22)

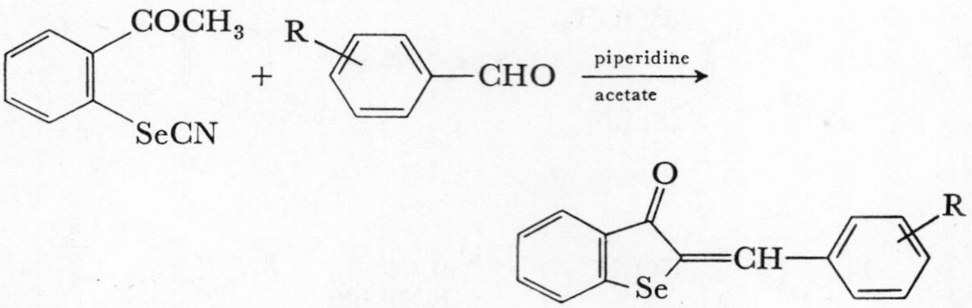

A large number of "ylidene" compounds (**37**) were obtained from the reaction of the parent substance with noncyclic aldehydes (equation 23).[46]

(23) $28 + RCHO \xrightarrow[\text{HCl, }\Delta]{\text{C}_2\text{H}_5\text{OH or HOAc}}$

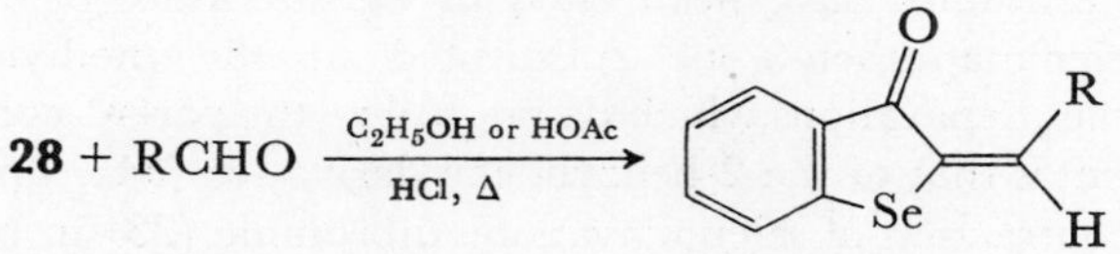

A group of indigo derivatives (**38**) was prepared by the treatment of 2,3-dihydroselenanaphthen-3-one with certain cyclic carbonyl compounds (equation 24, X may be N, S, or Se).[40,45,46] Other variations of this type of compound also were made.[40,46]

(24)   **28** +

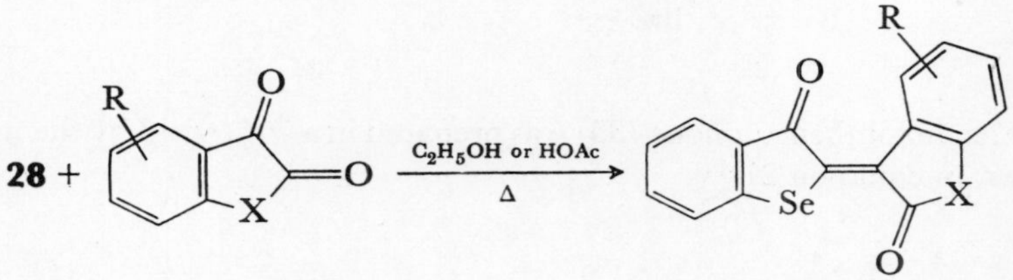

The oxidative coupling of the parent compound by heating with potassium ferricyanide in base was reported in 1912 to give only a *cis*-selenoindigo (**39**, R = H) (equation 25).[40] In 1958, however, the 5-methyl derivative was reported to yield, under these same conditions, the *trans*-selenoindigo (**40**, R = CH₃) (equation 25).[41] The initial synthesis[40] of selenoindigo actually gives a mixture of the *cis* and *trans* isomers.[47] The u.v. spectra of these isomers were recorded[37,47] without physically separating them.[47] The

(25)

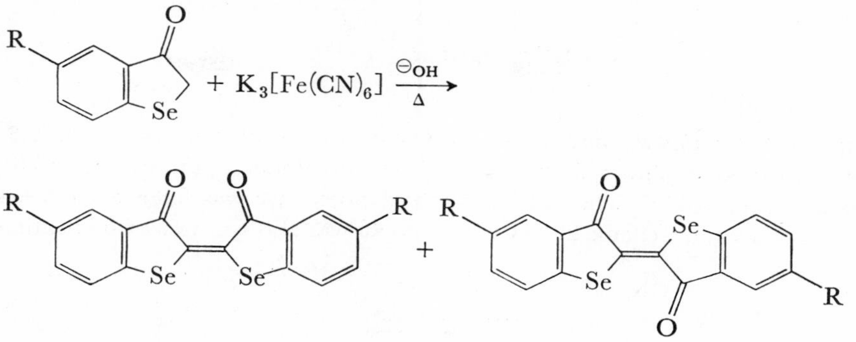

$$39 \qquad\qquad 40$$

photochemical isomerization of these compounds was studied,[47,48] and quantum yields were determined.[47]

It was noted by Gosselck[41] that the reactivity of the selenium-containing heterocycles which he studied, as compared with that of their sulfur analogs, was decreased as a result of resonance weakening by the larger Se atom.

ii. 2,3-Dihydro-2-methylbenzo[b]selenophene. When allyl phenyl selenide was heated in boiling quinoline for 1–2 hours, a Claisen rearrangement occurred and 2,3-dihydro-2-methylbenzo[b]selenophene (**41**) was obtained in a 30 % yield (equation 26).[49] It also was reported[50] that heating allyl phenyl

(26)

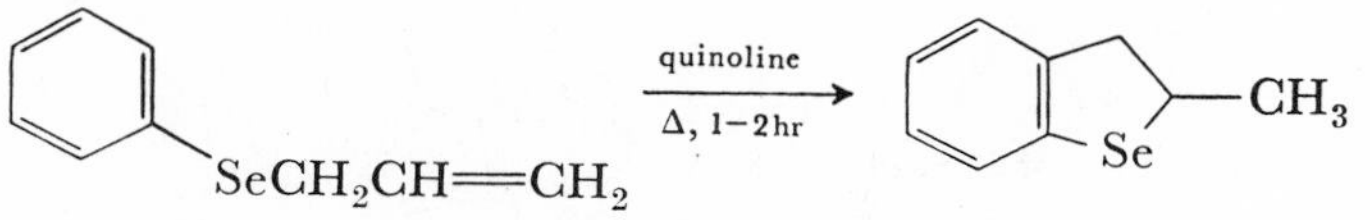

$$41$$

selenide in boiling quinoline for 3 hours produced **41** (in only a 15 % yield) plus a crystalline compound (in 25 % yield). This, it was suggested, was 1-selenochromane (**42**) (equation 27). It was expected that **42** might be

(27)

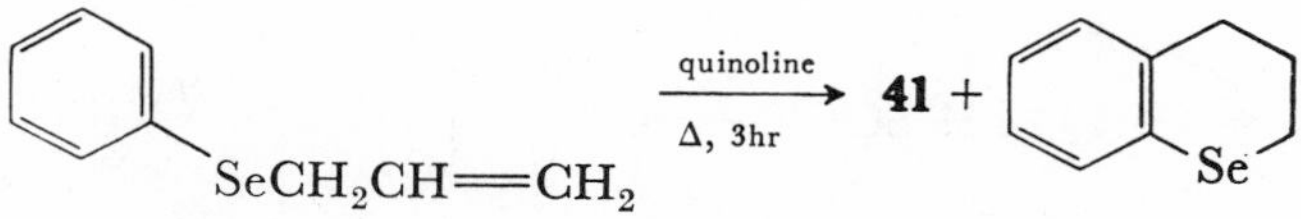

$$42$$

derived from phenyl propenyl selenide (**43**), however, when **43** was subjected to the same conditions as its isomer, no reaction occurred (equation 28).[49]

(28)

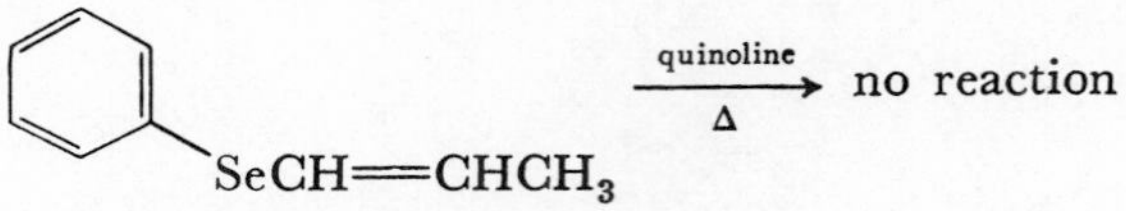

$$\xrightarrow[\Delta]{\text{quinoline}} \text{no reaction}$$

43

iii.   Selenophthalide and Isologs.   The synthesis of an isostere of 2,3-dihydroselenonaphthen-3-one, that is, 2-selenophthalide (**45**), was originally reported in 1891.[51] The almost identical procedure was followed in 1964 and is shown in scheme 29.[52] Compound **45** was also synthesized by heating

(29)

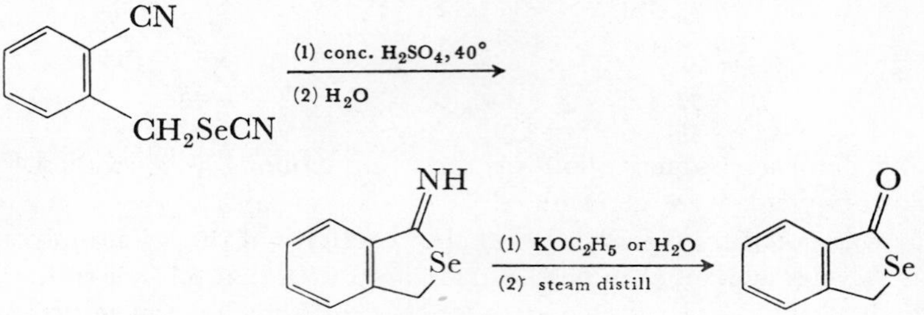

α,α-diselenodi-*o*-toluic acid with 50% aqueous hypophosphorous acid.[52a] The thione **46** was prepared by treating 2-selenophthalimide (**44**) with hydrogen sulfide (equation 30).[52]

(30)

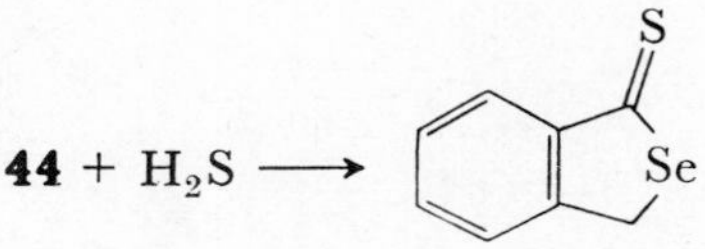

$$\textbf{44} + H_2S \longrightarrow$$

46

iv.   4,5-Benzo-3H-1,2-selenathiol-3-one.   This heterocycle (**47**), containing two heteroatoms, was synthesized in 1924 by the method shown in scheme 31.[53]

(31)

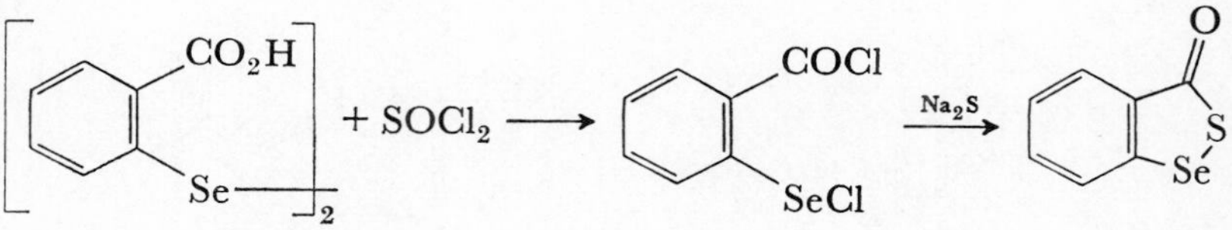

v.   4,5-Benzo-3H-1,2-thiaselenol-3-one.   This isomer of **47**, *i.e.*, **48** was prepared recently according to equation 32.[54]

(32)

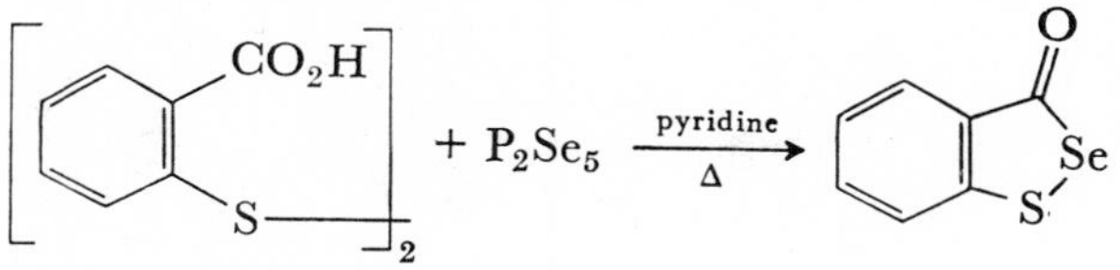

48

b. One Five-Membered, One Six-Membered Ring, Two Selenium Atoms.
Benzoylene diselenide (**49**) was made by a route analogous to that used
for the preparation of the selenosulfide (**47**) (equation 31) and is shown in
equation 33.[53]

(33)

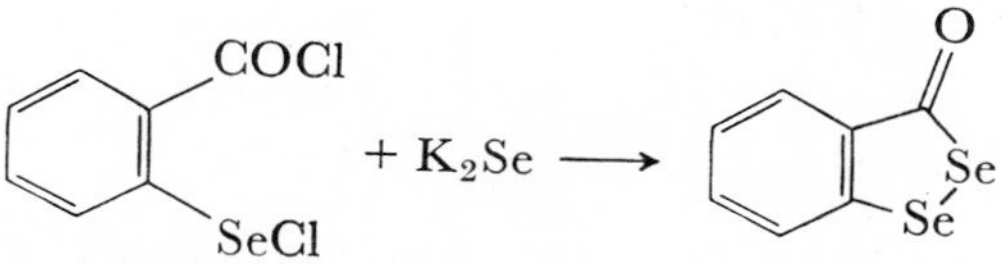

49

c. Two Six-Membered Rings

i. Selenochromane Derivatives

Selenochroman-4-ones. The parent compound (**50**) was prepared in 1958
in *ca.* 23 % yield according to equation 34.[41] In 1964 a superior synthesis

(34)

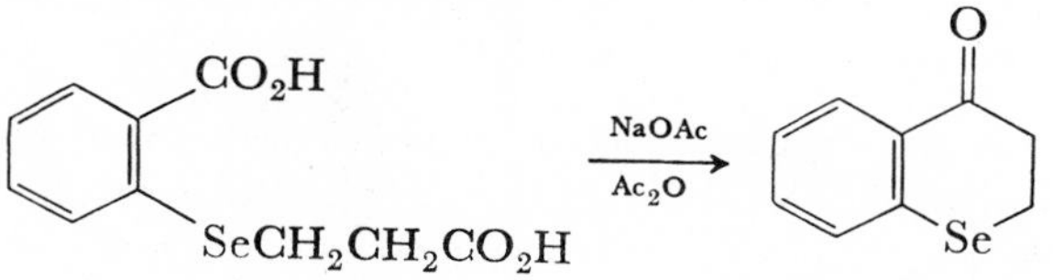

50

of **50** was reported. Using polyphosphoric acid (PPA) as the cyclizing agent
Renson[55] obtained this compound in *ca.* 82 % yield (equation 35).

(35)

A series of monomethyl-substituted selenochromanones (**51**) was compiled by combining the results of the reactions of three synthetic methods,
in which only the cyclizing reagent was varied (equation 36; R, R′, and R″
may be H or CH_3).[56] The choice of dehydrating agent for maximum yield
depended on the site of methyl substitution.

Selenoflavanones (**52**), that is, selenochromanones substituted in the 2-
position with a phenyl group, were synthesized according to equation 37

(36)

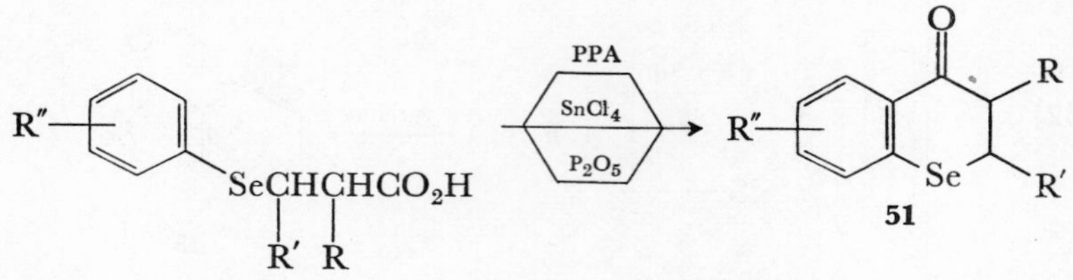

(37)

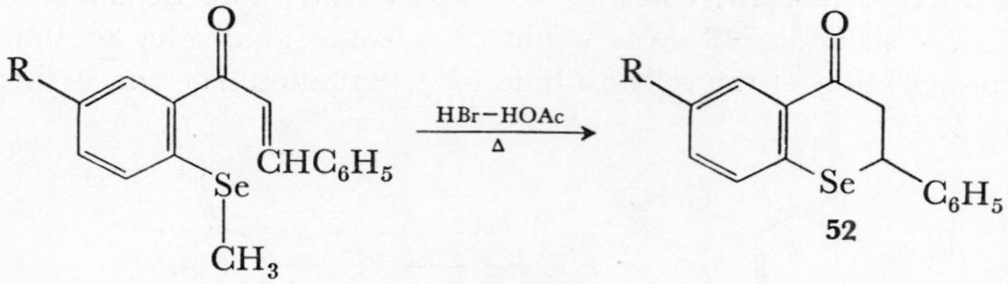

$(R = H \text{ or } CH_3)$.[44] It was noted that selenoflavanones showed an increased tendency, relative to the sulfur isologs, to contract from the heterocyclic six-membered ring to a five-membered ring.[44]

The absorption spectra of isologous chromanones were compared, and it was shown that shifts of the absorption bands of longest wavelength (heteroatomic unbound electron pair transitions) are considerably greater between the oxygen and sulfur isologs than between the sulfur and selenium compounds.[57] This indicates that the unpaired electron excitability becomes greater with decreasing electronegativity.

Selenochroman-4-ols. Selenochroman-4-one was reduced to selenochroman-4-ol by sodium borohydride[58] or lithium aluminum hydride.[55] A series of monomethyl-substituted selenochromanols was made using lithium aluminum hydride as the reducing agent.[56]

When selenochromanone was treated with phenylmagnesium bromide, 4-phenyl-selenochroman-4-ol was obtained.[55] The 4-phenyl derivatives of the monomethyl-substituted series were also prepared.[56]

1-Selenochromanes. The Clemmensen reduction of selenochroman-4-one afforded 1-selenochromane in a 40% yield.[55] Heating allyl phenyl selenide in boiling quinoline may have produced this compound in a 25% yield, as described previously (equation 27).[50]

4-Amino-1-selenochromane was synthesized by treating selenochroman-4-one with hydroxylamine and then reducing the resulting oxime with lithium aluminum hydride.[55]

Selenochrom-3-enes. Selenochrom-3-ene (**53**, $R = R' = R'' = H$) has been prepared by the elimination of water from selenochroman-4-ol. The dehydrating agents utilized were potassium bisulfate[58] (56% yield), aluminum oxide[55] (59% yield), and phosphorus pentoxide[55] (30% yield) (equation 38).

(38)

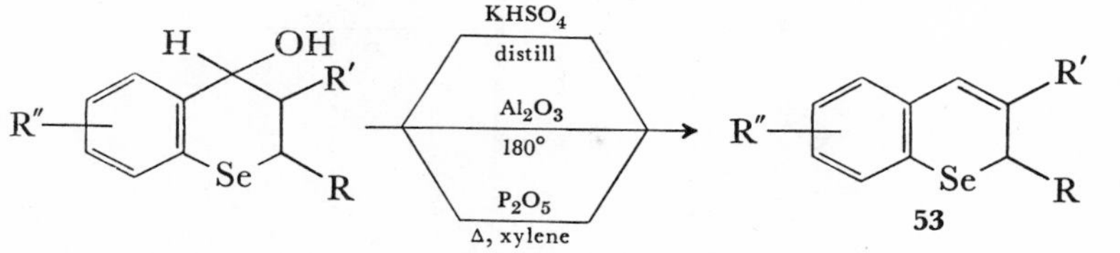

A series of monomethyl-substituted selenochrom-3-enes (**53**, R, R', R" may be H or CH$_3$) was also made using potassium bisulfate and aluminum oxide as dehydrating agents.[56] 4-Phenylselenochrom-3-ene was obtained from the reaction of the corresponding alcohol with phosphorus pentoxide.[55] The monomethyl-substituted derivatives were prepared in the same way.[56]

Also synthesized was 3-formyl-4-chloro-2H-(1)-benzoselenopyran (**54**), which was used as an intermediate in the preparation of 2-carboxy-4H-thieno[3,2-c]-(1)-benzoselenopyran (**55**) (scheme 39).[59]

(39) **50** + POCl$_3$ + (CH$_3$)$_2$NCHO $\longrightarrow$

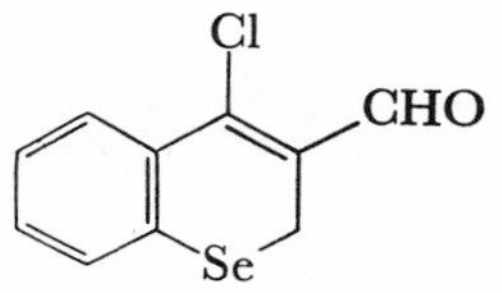

54 $\xrightarrow{\text{(1) HSCH}_2\text{CO}_2\text{H, NaHCO}_3}_{\text{(2) KOH, }\Delta}$

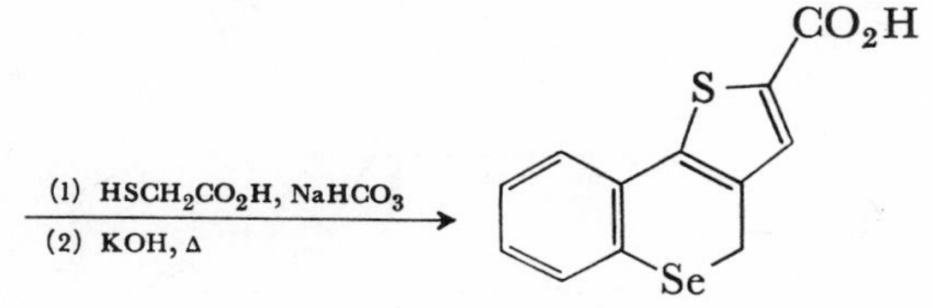

Substituted diphenyl diselenides (**56**) were obtained by treatment of substituted selenochrom-3-enes with selenium dioxide in pyridine (scheme 40).[60] When there was no substitution at the 3-position (*i.e.*, R = H), substituted 2-carboxyaldehyde derivatives of benzoselenophene (**57**) were isolated.[60]

Selenochrom-4-ones. Selenochrom-4-one (**58**) was obtained in a 20% yield by heating selenochroman-4-one (**50**) with chloranil in refluxing xylene (equation 41a),[55] or in a 67% yield by treating **50** with trityl perchlorate and then neutralizing with bicarbonate[58] (equation 41b). By means of the method

(40)

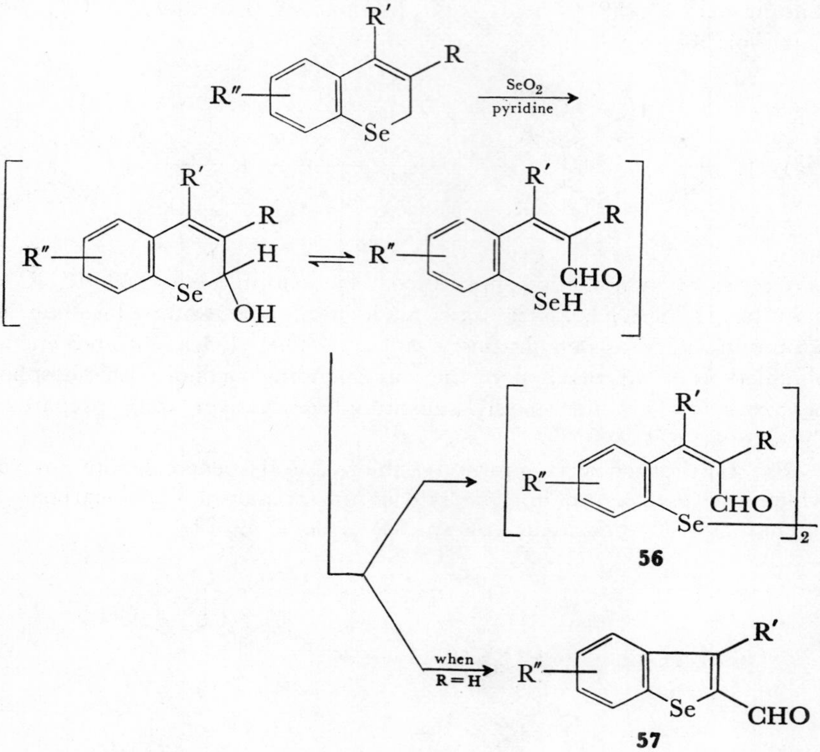

(41)   **50**

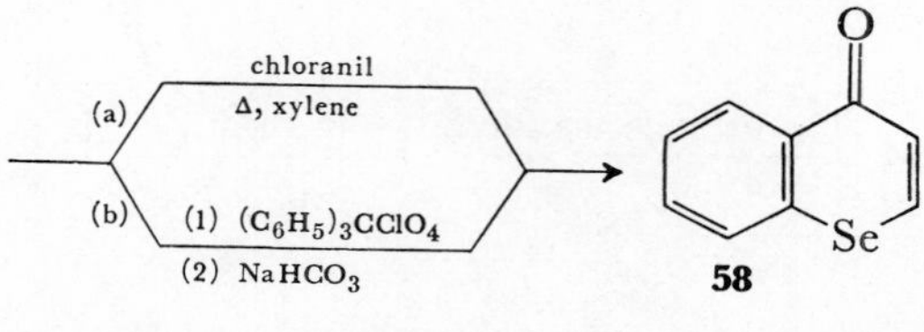

indicated by pathway 41b, a series of monomethyl-substituted selenochrom-4-ones was formed.[56]

The technique of heating a selenophenol with an acyl or aroyl acetic ester (**59**) in polyphosphoric acid as a general procedure for the preparation of selenochrom-4-ones was reported by two independent groups[61,62] (equation 42). Low yields (mostly in the 5–25 % range) were obtained for the series.[62]

Another procedure used to make selenochrom-4-ones involved the oxidation of selenochrom-3-enes with selenium dioxide in pyridine at 50° (equation 43).[60]

(42)

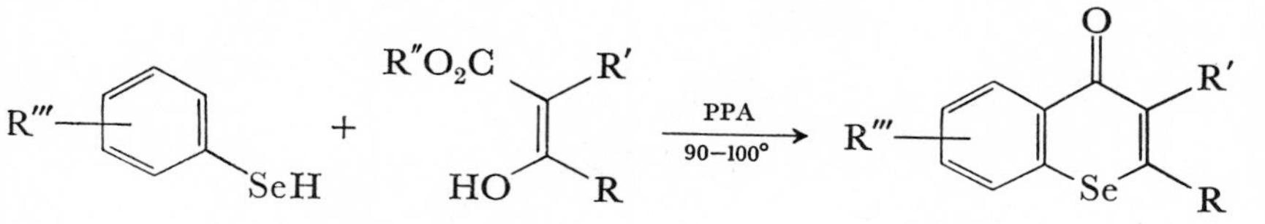

(43)

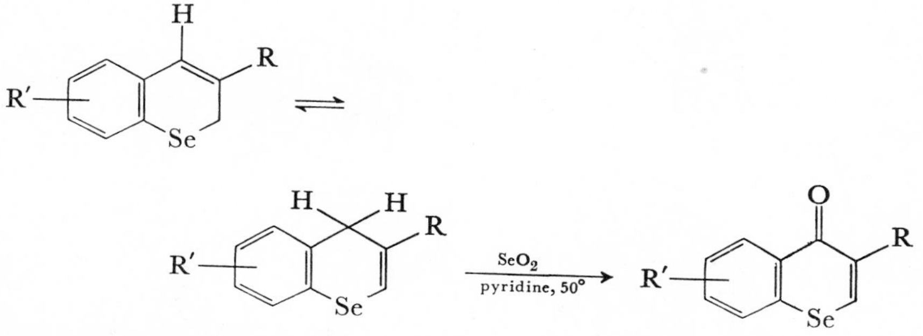

2-Methyl-1-selenochrom-4-one (**62**) was recently obtained by two related routes.[63] The first method employed the reaction of selenophenol with diketene (**60**) in the presence of *p*-toluenesulfonic acid, followed by cyclization of the intermediate **61** by polyphosphoric acid (scheme 44). Alternatively

(44)

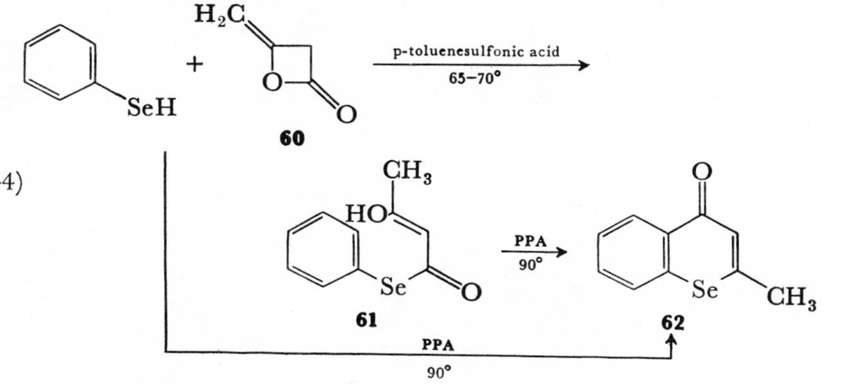

the isolation of **61** is avoided by adding the reactants directly to polyphosphoric acid. By another procedure it was found[63] that, unlike phenyl acetoacetate, which cyclizes with aluminum chloride or polyphosphoric acid in accord with the Pechmann reaction[64] to give 4-methylcoumarin, phenyl acetothioacetate and phenyl acetoselenoacetate close with these reagents to give 2-methylthio- and 2-methylselenochrom-4-one, respectively, in accord with the Simonis reaction[65] (scheme 45).

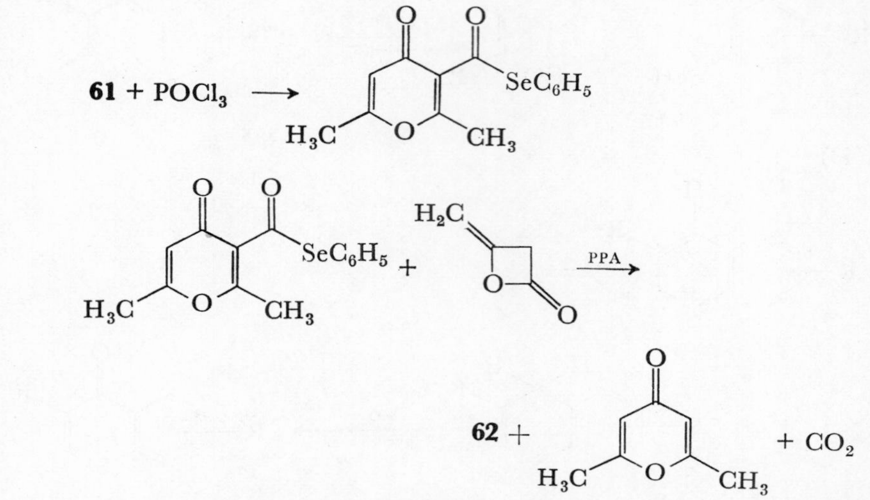

$$61 + POCl_3 \longrightarrow$$

The absorption spectra of isologous chromones were described,[57] and the same conclusions that were stated for the isologous chromanones (*cf.* p. 546) were found to be relevant also in this case. The i.r. spectra of 1-seleno-chromones, 1-selenocoumarins, and their sulfur analogs were compared.[66] The basicity (pK_b) values of selenochromone, selenoxanthone, and their oxygen and sulfur analogs also were determined.[67,68]

Selenochrom-4-ol. Selenochrom-4-ol (**63**) was made as an unpurified intermediate in the preparation of selenochromylium perchlorate (**64**) (equation 46).[58]

(46)

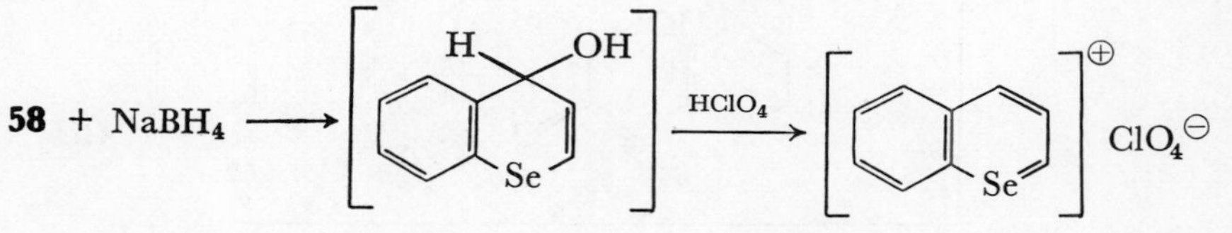

Selenochromylium salts. Selenochromylium (selenobenzopyrylium) perchlorate (**64**) and its derivatives have been made[56,58] and are described in Chapter VI. Also studied were the u.v. and visible spectra,[27,58] hydrolysis reactions at various pH values,[69] oxidation reactions,[68,69] and stabilities[28,70] of these compounds.

1-Selenocoumarins. Numerous 1-selenocoumarins (**66**) were prepared in 50–60 % yield by the cyclization of *o*-(methylseleno)cinnamic acids (**65**) with phosphorus oxychloride (equation 47) or of *o*-(methylseleno)cinnamoyl

(47)

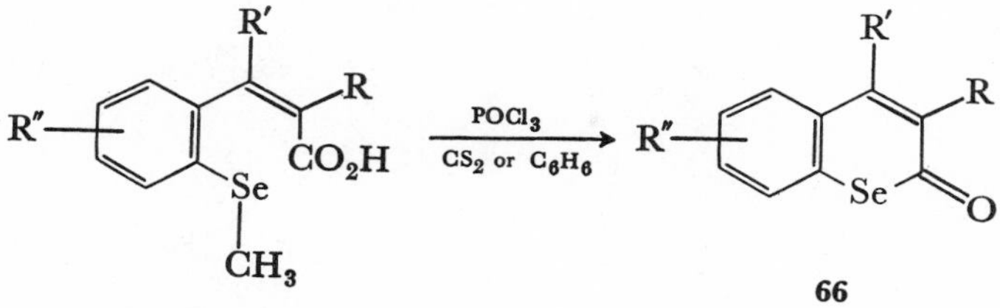

chlorides (**67**) with either phosphorus oxychloride or aluminum chloride (equation 48).[71,72]

(48)

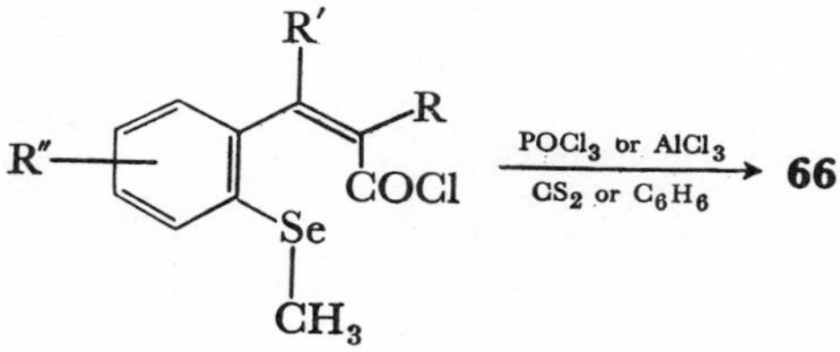

The oxidation of 1-selenochrom-3-enes with chromic acid in pyridine also was described as a route to selenocoumarins.[60,71]

Scheme 49 shows a reaction for the preparation of 4-hydroxy-1-seleno-coumarin (**68**).[73]

(49) $CH_2(COCl)_2 + 2C_6H_5SeH \xrightarrow{\Delta} CH_2(COSeC_6H_5)_2$

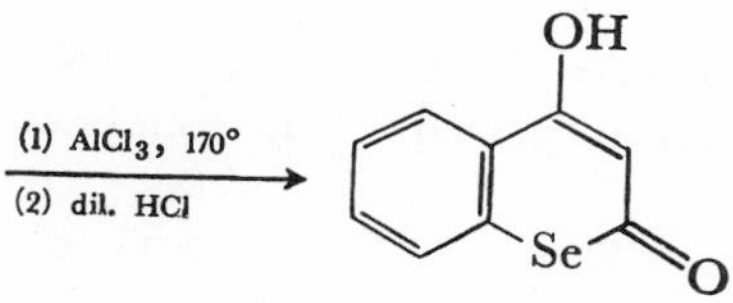

3,3′-Methylene bis(4-hydroxyselenocoumarin) (**69**) was found to be the sole product when **68** was treated with formaldehyde in water (equation 50).[73] This condensation was confirmed using ethanol as the solvent.[74] The product was found to have practically no anticoagulant activity.

The reaction of **68** with phosphorus oxychloride is shown in equation 51.[73]

(50) 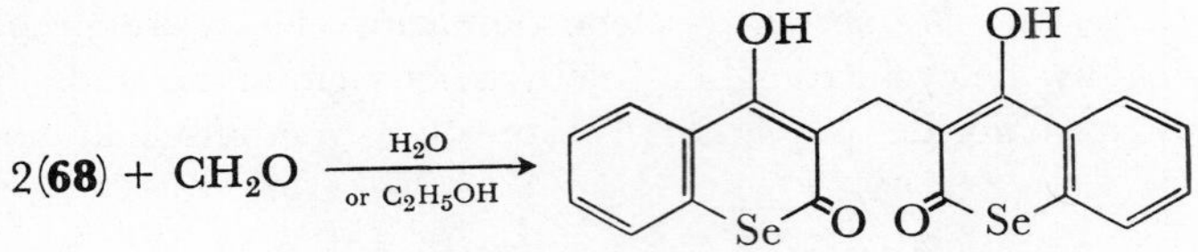 $2(68) + CH_2O \xrightarrow[\text{or } C_2H_5OH]{H_2O}$

69

(51) 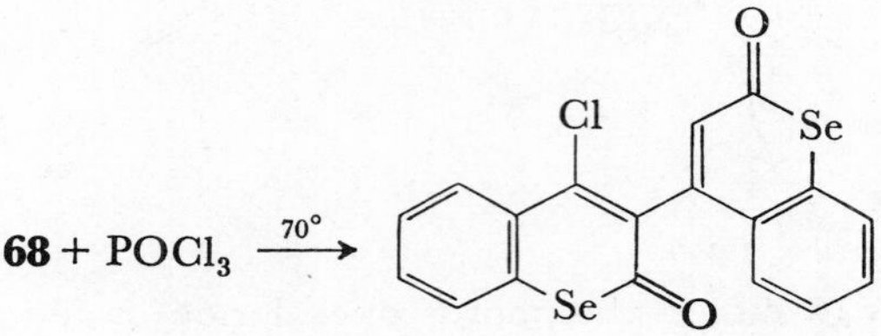 $68 + POCl_3 \xrightarrow{70°}$

The i.r. spectra of 1-selenocoumarins, 1-selenochromones, and their sulfur analogs have been compared.[66]

ii. Isoselenochromane Derivatives

Isoselenochromanone. The compound isoselenochromanone (**70**) was synthesized in low yield according to equation 52.[75]

(52)

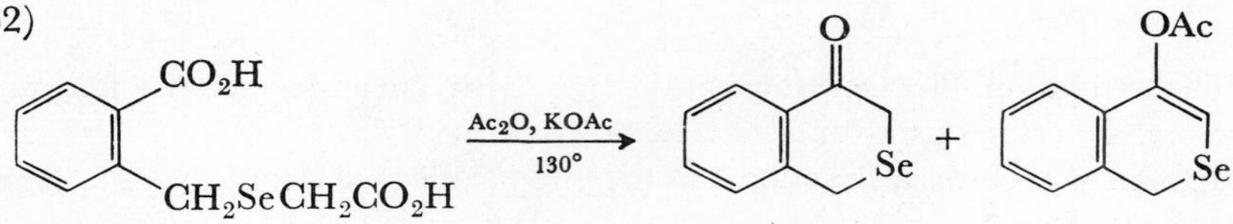

$\xrightarrow[130°]{Ac_2O, KOAc}$ +

70   **71**

Isoselenochromanol. This alcohol was obtained by the reduction of isoselenochromanone with sodium borohydride.[75]

Isoselenochromane. The synthesis of isoselenochromane (**72**) in 58% yield is presented in equation 53.[76,77] The dichloride, dibromide, and

(53) 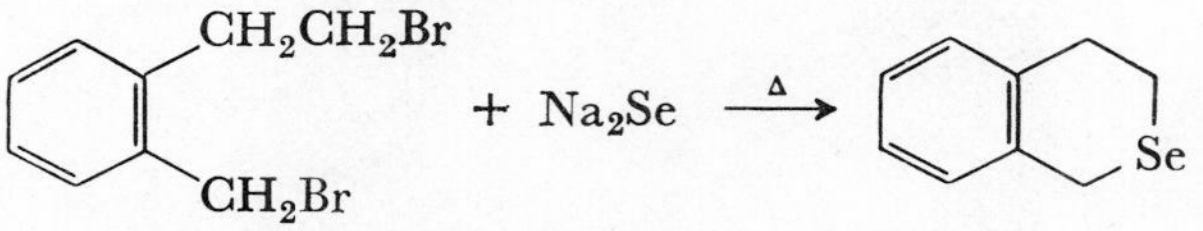 CH_2CH_2Br ... CH_2Br $+ Na_2Se \xrightarrow{\Delta}$

72

selenoxide were also made.[77]

Isoselenochromenes. The preparation of the parent compound (**73**) was accomplished by distilling isoselenochromanol from potassium bisulfate.[75]

4-Acetoxyisoselenochrom-3-ene (**71**) was shown to be the by-product in the preparation of isoselenochromanone (equation 52).[75]

Isoselenochromylium perchlorate. This salt (**74**) was obtained by treating isoselenochromene (**73**) with sulfuryl chloride and perchloric acid (equation 54).[75] Its u.v. spectrum was compared with that of the sulfur analog and of

(54)

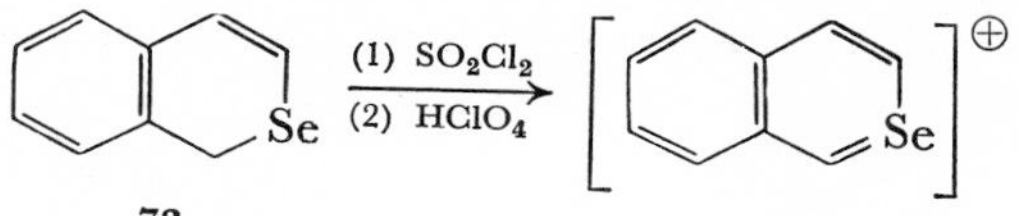

selenochromylium perchlorate.[75]

iii. A Selenium-containing Spiro Compound. The reaction of acetylacetone selenium dioxide was reported to yield the heterocycle (**75**) shown in equation 55.[78,79]

(55)

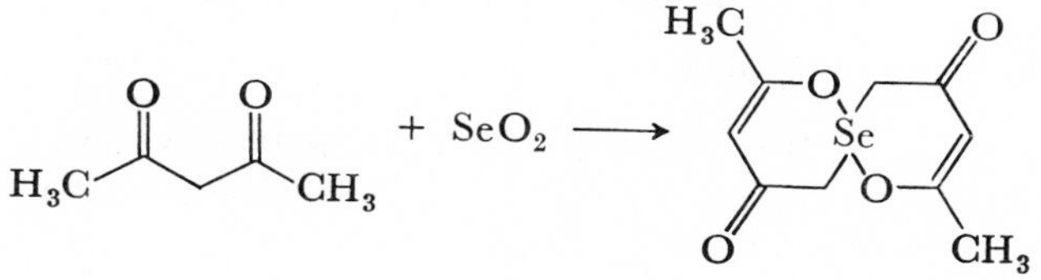

d. One Six-Membered, One Seven-Membered Ring.

i. Benzoselenepin Derivatives

Homoselenochroman-5-ones. The intramolecular condensation of 4-phenyl-selenobutyric acid (**76**, R = H) produced 2,3,4,5-tetrahydro-1-benzo-selenepin-5-one (**77**, R = H) in 14% yield (equation 56).[55] Five years

(56)

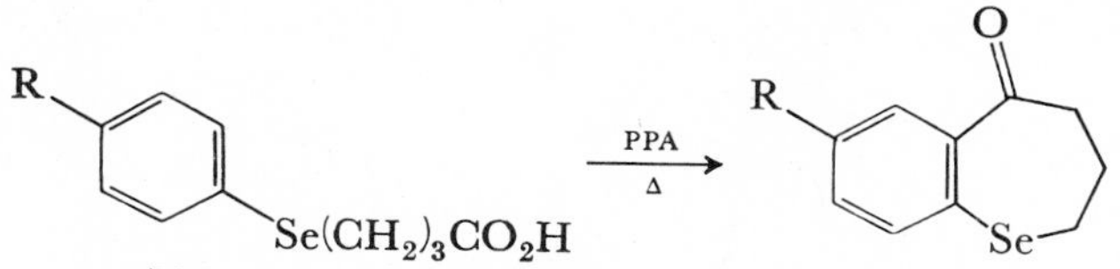

later, the same reaction was reported to give an improved yield of 75%.[80] Homoselenochroman-5-ones substituted in the 7-position were prepared by the same route used to make the parent compound (equation 56).[81]

Homoselenochroman-5-ol. The parent alcohol was made by the sodium borohydride reduction of the corresponding homoselenochroman-5-one.[80]

The preparation of substituted homoselenochroman-5-ols was also described.[81]

Homoselenochromanes.   A Wolff-Kishner reduction of the corresponding homoselenochromanone yielded 2,3,4,5-tetrahydro-1-benzoselenepins.[81]

Homoselenochrom-4-enes.   The elimination of water from homoselenochroman-5-ols with PPA was the method used to synthesize 2,3-dihydrobenzoselenepins.[81]

4. Heterocycles Consisting of Three Fused Rings

a. One Five-Membered Ring, Two Six-Membered Rings.   The preparation of 2-carboxy-4H-thieno[3,2-c]-(1)-benzoselenopyran (55) was shown in scheme 39.

b. Three Six-Membered Rings, One Selenium Atom

i. Selenoxanthene Derivatives

Selenoxanthones.   In 1914 the synthesis of the parent compound (78) was described (equation 57).[82] Fifty-five years later a similar reaction, which gave selenoxanthone in an 88 % yield, was reported.[83]

(57)

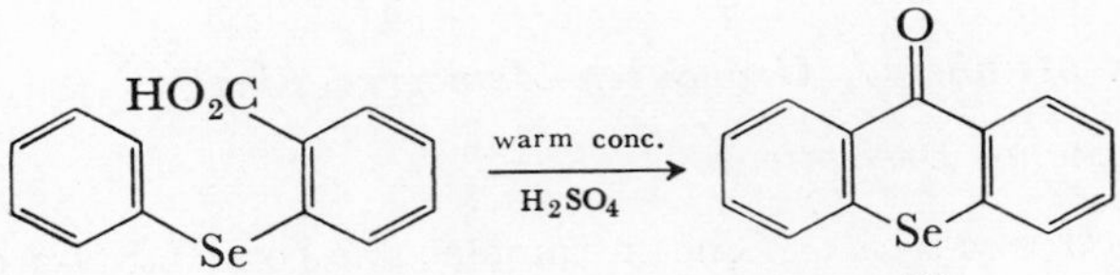

Selenoxanthone (79, R = H) also was synthesized by the procedure in equation 58.[53] Lesser and Weiss[53] were unable to differentiate between

(58)

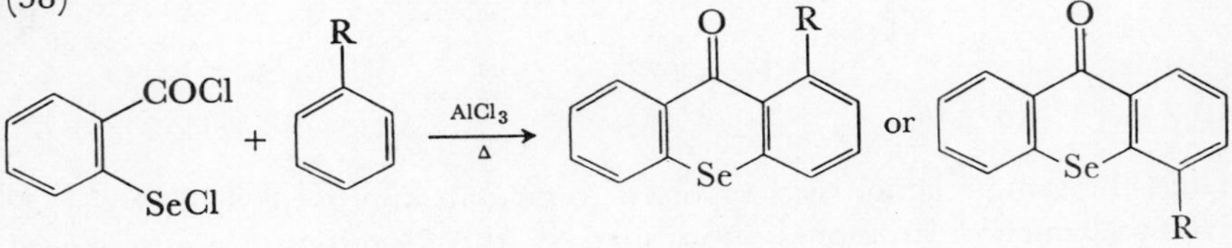

structures 79a and 79b (R = CH$_3$) when they formulated the methyl derivative.

Selenoxanthone was also made in 1968 by the oxidation of selenoxanthylium perchlorate (**80**) with manganese dioxide, the yield ($>88\%$) being dependent on the reaction solvent (equation 59).[84]

(59)

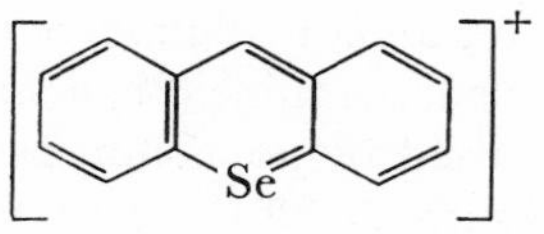

$$\left[\begin{array}{c} \text{Se} \end{array} \right]^{+} ClO_4^- + MnO_2 \xrightarrow{\Delta} \textbf{78}$$

80

The preparation of selenoxanthonecarboxylic acid (**83**) was mentioned parenthetically in 1912,[40] and later reported[46,82] in detail as shown in scheme 60. Since 2,2′-dicarboxydiphenyl selenide (**81**) was obtained in only a 5–10% yield, but the related diselenide **82** in a 50% yield, a method was developed to convert the latter compound to the selenide in an 85% yield (scheme 60).[82] The second product **84** from the ring-closure reaction was

(60)

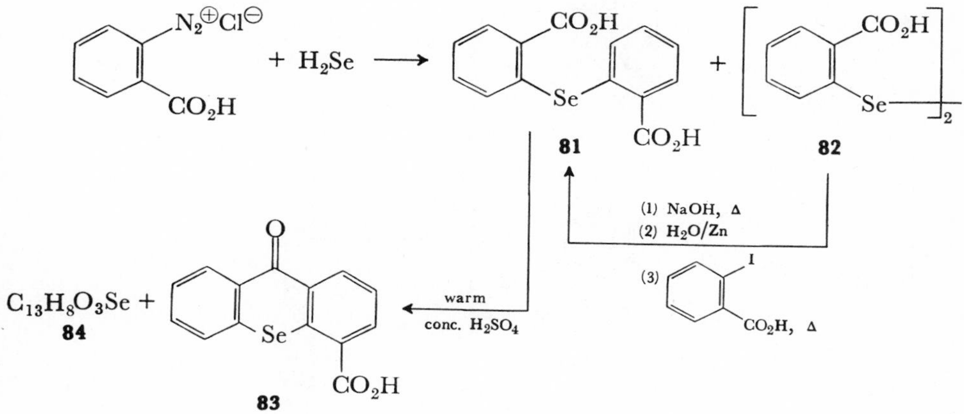

described initially as a selenone **85**,[46] but the following year as a dilactone **86**.[82]

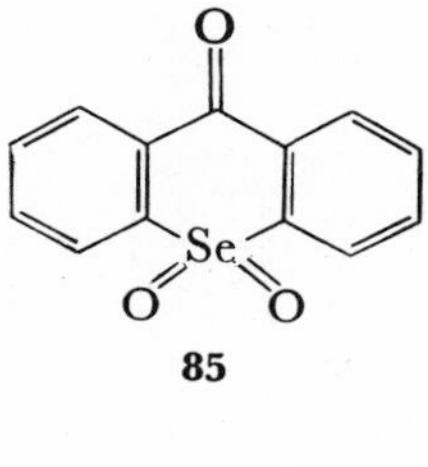

85 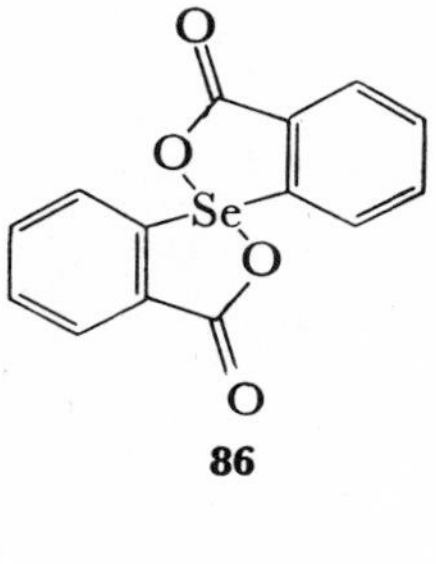   **86**

A series of esters and amides was also made from the carboxylic acid by standard methods.[82]

By heating **83** with soda-lime, the parent compound **78** was obtained.[82]

3,6-Bis(dimethylamino)selenoxanthone (**89**) was prepared in an 80% yield[85] by a procedure that utilized 3,6-tetramethyldiaminoselenopyronine (**87**)[86] and 3,6-tetramethyldiaminocyanoselenopyronine (**88**) as intermediates (equations 61 and 62). When **87** was oxidized by potassium permanganate in aqueous sodium hydroxide the selenoxanthone **89** was produced.[87]

(61)

$$87 \xrightarrow[\substack{(2)\ \text{HCl, FeCl}_3 \\ (3)\ \text{HNO}_3}]{(1)\ \text{aq. KCN, } 65°}$$

(62) **88** + NaOH $\longrightarrow$

The oxidation of the parent compound **78** with chromium trioxide produced selenoxanthone-10-oxide (**90**) (equation 63).[53]

(63)

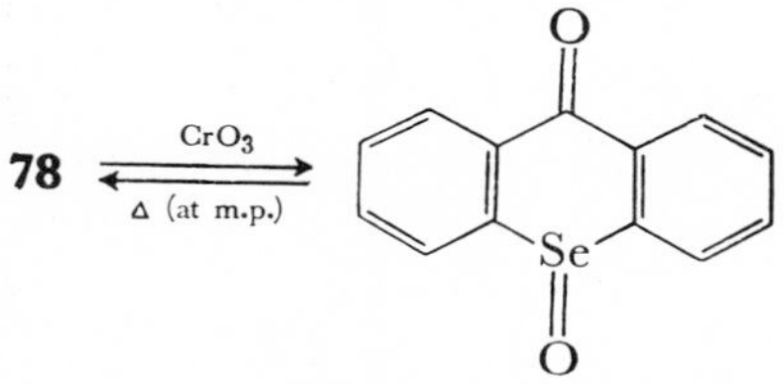

90

Selenonexanthone (**91**) was synthesized from toluene and selenic acid (scheme 64).[88]

(64)

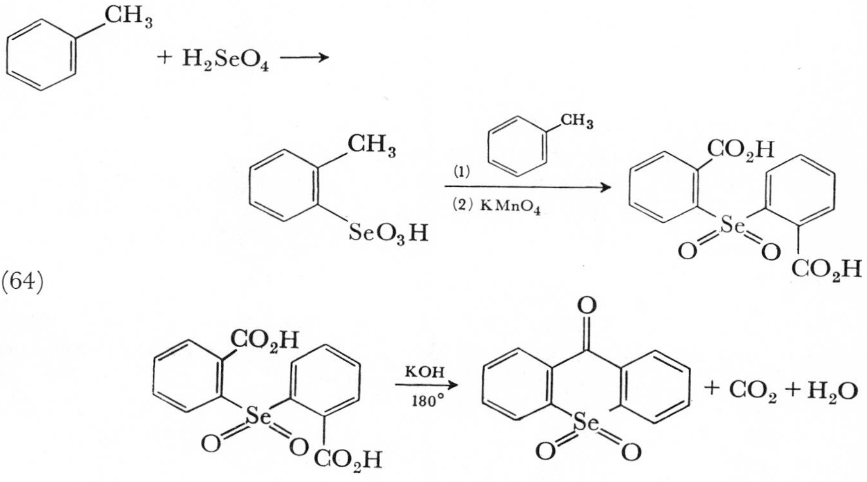

91

Electron spin resonance studies of the metal ketyls (**92** and **93**) of selenoxanthone, thioxanthone, and xanthone indicated that the covalent nature of the bond between the O atom and the metal, M, increased in the order xanthone, thioxanthone, selenoxanthone (equation 65).[89, 90]

The mass spectrum of selenoxanthone included as one of its prominent features the decarbonylation species, that is, dibenzo[b,d]selenophene ion, and the selenium extrusion species, fluorenone ion.[91] The basicity (pK_b) values of selenoxanthone, selenochromone, and their oxygen and sulfur analogs were determined.[67, 68]

(65)

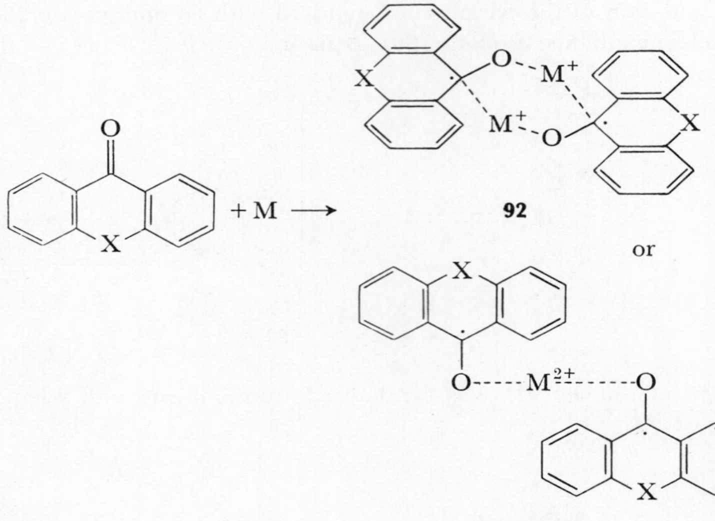

92

or

93

Selenoxanthenols. Selenoxanthen-9-ol was obtained from the reduction of the corresponding ketone with sodium amalgam.[92]

The 9-phenyl alcohol was prepared by the reaction of selenoxanthone with phenylmagnesium bromide.[92] 9-(3-Dimethylaminopropyl)selenoxanthen-9-ol was synthesized by the reaction of the corresponding Grignard reagent with selenoxanthone.[83]

Selenoxanthenes. Selenoxanthene (**94**) was prepared in a nearly quantitative yield when selenoxanthone **78** was heated with red phosphorus and fuming hydriodic acid at 160–180° (equation 66).[93]

(66) 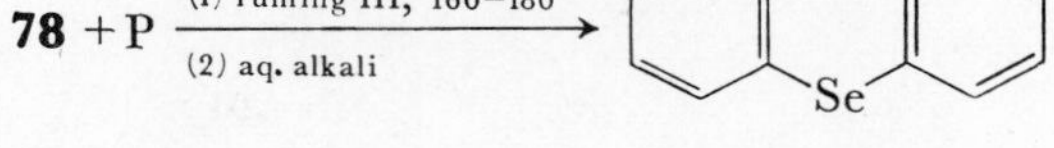   **78** + P $\xrightarrow[\text{(2) aq. alkali}]{\text{(1) fuming HI, 160–180°}}$

94

The reduction of 3,6-bis(dimethylamino)selenoxanthylium chloride-zinc chloride double salt (**87**) by lithium aluminum hydride yielded the corresponding selenoxanthene.[87] Treatment of **87** with potassium cyanide produced the 9-cyano derivative.[85]

Selenoxanthenols undergo numerous reactions to give other substituted selenoxanthenes. The parent alcohol forms a selenoxanthylium double chloride salt with either mercuric or ferric chloride in acetic and hydrochloric acids,[92] condenses in ethanol-acetic acid with ureas, thioureas, and carbamic esters,[94] and also reacts with β-diketones and ethyl acetoacetate in acetic acid.[95]

Recently, 9-chloro-9-phenylselenoxanthene was reported as the product obtained when hydrogen chloride was bubbled into a solution of the 9-phenylselenoxanthen-9-ol.[96]

9-(3-Dimethylaminopropyl)selenoxanthen-9-ol was converted into 9-(3-dimethylaminopropyl)selenoxanthene by heating with acetic and hydriodic acids at 120°.[83] When the alcohol was heated with sulfuric acid instead of acetic and hydriodic acids, the corresponding propylidene derivative was isolated.[83]

Selenoxanthylium salts. Some of these compounds were described earlier as intermediates in the preparation of various selenoxanthene derivatives (*cf.* **80**, **87**, and **88**). Other selenoxanthylium derivatives have been made[97,98] (*cf.* Chapter VI). Also studied were the oxidation reactions[68] and stabilities[28,70] of these compounds.

ii. 6H-Dibenzo[b,d]selenopyran and Dibenzo[b,d]selenopyrylium Salts. In 1965 the preparation of 6*H*-dibenzo[b,d] selenopyran (**95**) in a 20% yield by the Pschorr reaction was reported (equation 67).[99] When **95** was treated

(67)

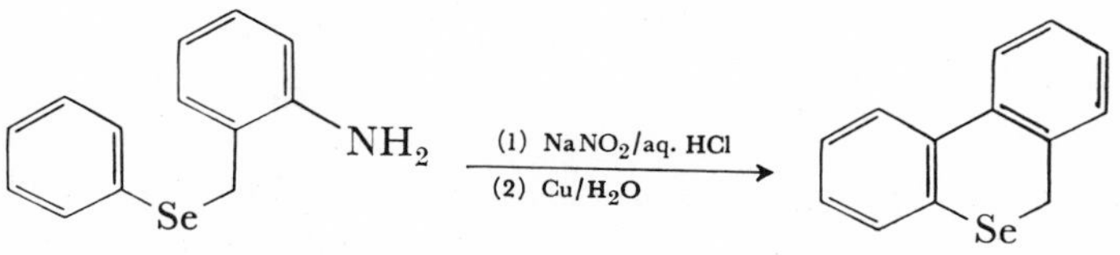

with thionyl chloride, dibenzo[b,d]selenopyrylium chloride was obtained, which, on treatment with perchloric acid, gave the perchlorate salt.[99] The stability of this cation relative to its oxygen and sulfur analog and to other heterocyclic cations was determined.[70]

iii. Naphtho[1,2-b]selenopyran Derivatives

4-Oxo-3,4-dihydro-2H-naphtho[1,2,-b]selenopyran. This ketone (**97**) was synthesized in low yields by the cyclization of β-(1-naphthylseleno)-propionic acid (**96**), using polyphosphoric acid (equation 68).[100] The yield

(68)

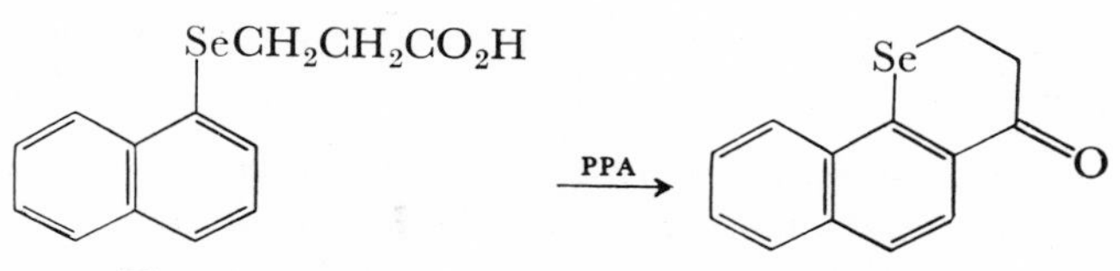

was raised to 60 % when the acid chloride derivative of **96** was treated with stannic chloride (equation 69).[100]

$$SeCH_2CH_2COCl$$

(69)

$$\xrightarrow{SnCl_4} \textbf{97}$$

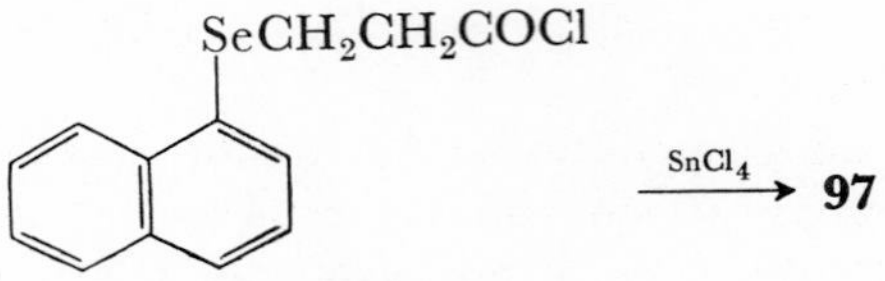

4-Hydroxy-3,4-dihydro-2H-naphtho[1,2-b]selenopyran. The alcohol **98** was made by reduction of the ketone **97** with sodium borohydride.[100]

3,4-Dihydro-2H-naphtho[1,2-b]selenopyran. A Wolff-Kishner reduction of the 4-oxo compound **97** produced this selenopyran.[100]

2H-Naphtho[1,2-b]selenopyran. Dehydration of the alcohol **98** yielded 2H-naphtho[1,2-b]selenopyran (**99**) (equation 70).[100]

(70)

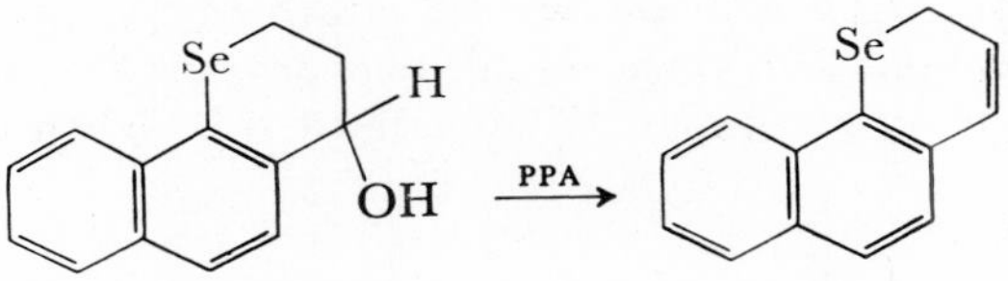

iv. Naphtho[2,1-b] Selenopyran Derivatives. 1-Oxo-2,3-dihydro-1H-naphtho[2,1-b]selenopyran (**101**) was synthesized using the same procedure as that utilized to prepare the other isomer **97**.[100] The starting material, though, was β-(2-naphthylseleno)propionic acid (**100**) (equation 71). 1-Hydroxy-2,3-dihydro-, 2,3-dihydro-, and the parent compound,

(71)

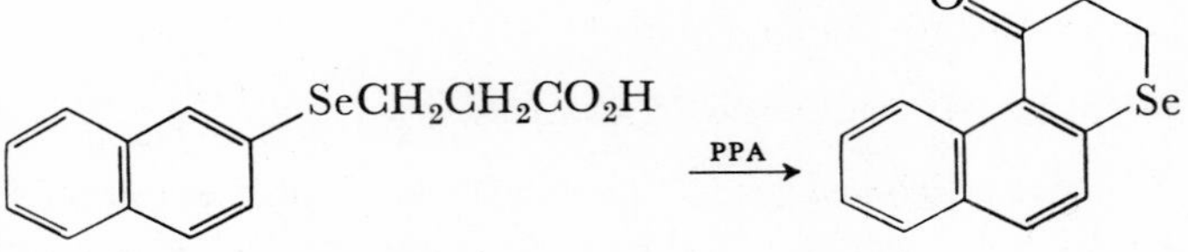

1H-naphtho[2,1-b]-selenopyran were also made by the same methods as their respective isomers (*vide supra*).[100]

v. Pyronoselenocoumarins. 4-Hydroxy-1-selenocoumarin reacted with malonyl chloride in tetrachloroethane at 135–145° to produce pyronoselenocoumarin (**102**) in a 90 % yield (equation 72).[73] Polypyronoselenocoumarin (**103**) was isolated as the by-product.

(72)

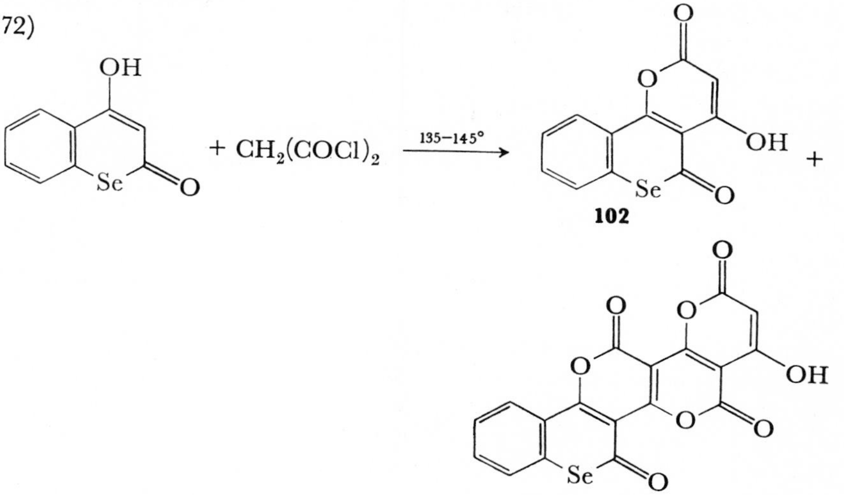

102

103

c.   Three Six-Membered Rings, One Selenium Atom Plus Another Hetero-atom

i.   Phenoxselenines.   An intricate procedure for the preparation of phenox-selenine (**105**) had to be employed because of the technical problems encountered (scheme 73).[101] The molecular complex (**104**) could not be resolved

(73)

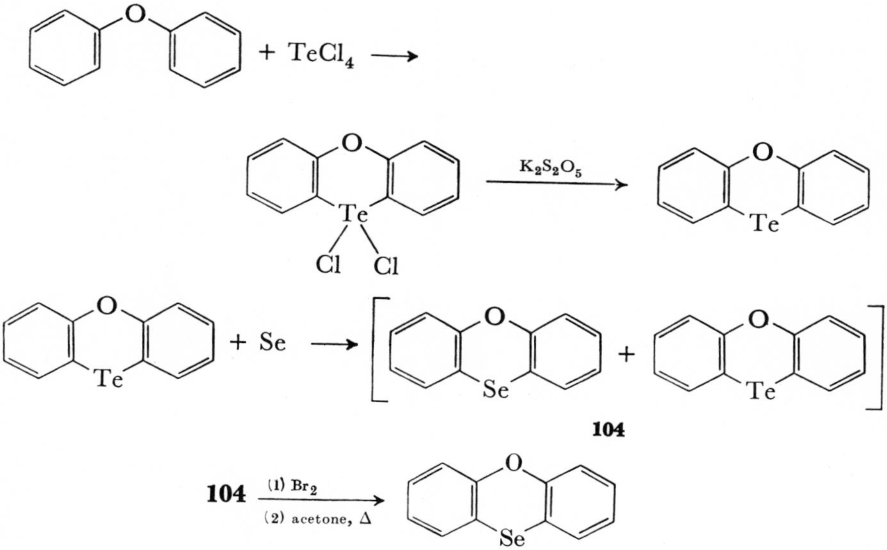

104

105

by crystallization, but was separated into its components by transformation to the mixed dibromides. The dibromide (and dichloride) of the selenium analog differs from the corresponding halide of the tellurium heterocycle in that it is a halogenating agent toward such compounds as acetone. Phenoxselenine could not be synthesized directly when selenium tetrachloride was used as the starting reagent because of the chlorinating action of the latter substance.

The selenoxide hydrate and the corresponding *Se,Se*-diacetate of **105** were also prepared.[101] Dehydration of the hydrate over phosphorus pentoxide produced the oxide, which reverted to **105** when heated above its melting point.

In contrast to phenoxthionine, phenoxselenine forms a dichloride, a dibromide, and a diacetate. Although the oxides of both can be made, that of the sulfur analog is only sparingly soluble in water and can be distilled unchanged. It does not decompose hydrogen peroxide catalytically nor is it reduced by alkali bisulfite. Phenoxselenine *Se*-oxide, however, is extremely soluble in water, quantitatively loses oxygen upon heating above its melting point, liberates oxygen from H_2O_2, and is immediately reduced by alkali bisulfite in the cold.[101]

The dibromide (or dichloride) liberates hydrogen bromide (or chloride) in the presence of sulfuric acid to form phenoxselenine disulfate (**106**), which on warming assumes the quinoid configuration (**107**) (equation 74).[101]

(74)

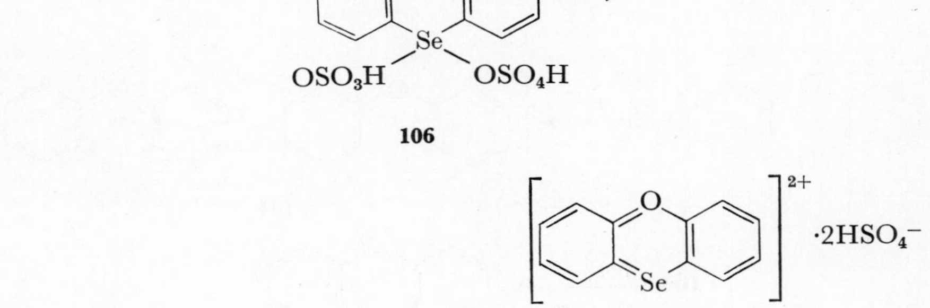

106

107

The synthesis of phenoxselenine-2-carboxylic acid (**108**) was devised to investigate the possibility of optical activity of phenoxselenines (scheme 75).[102] Before 1938, when this study was reported, it was thought that sub-

(75)

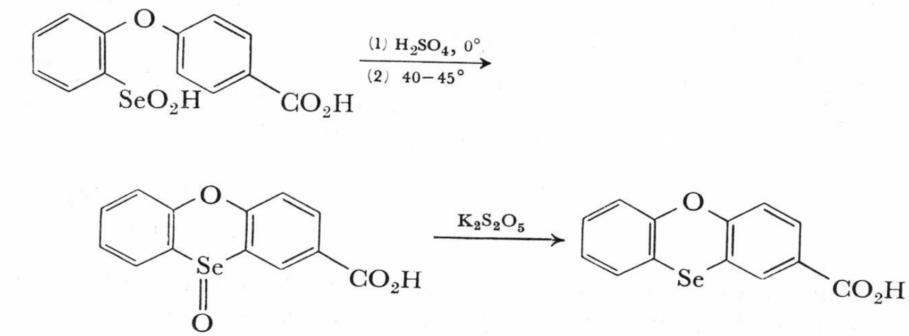

108

stituted phenoxarsines had the potential to exhibit optical activity, since the molecule was folded about the O–As axis. The belief was that the dissimilarity in size and valence angle of oxygen and arsenic stabilized this folding. However, it was found that neither phenoxselenines nor selenoxanthones could be obtained in enantiomeric forms.[102]

The original synthesis of octahydrophenoxselenin-1,9-dione-10-oxides (**110**) was reported in 1933 as shown in equation 76.[103] In this initial work

(76)

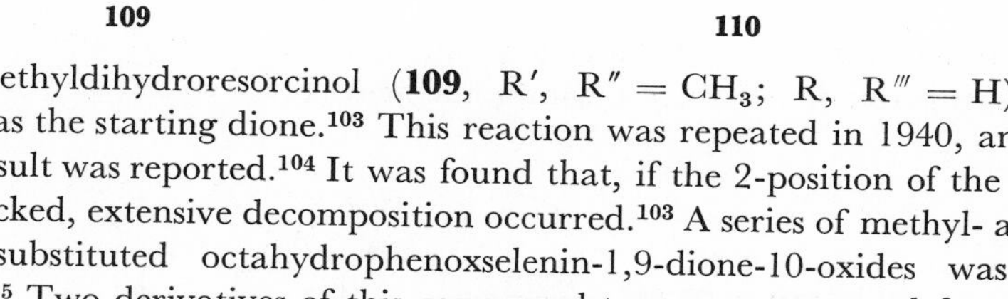

109   **110**

1,1-dimethyldihydroresorcinol (**109**, R′, R″ = CH₃; R, R‴ = H) was chosen as the starting dione.[103] This reaction was repeated in 1940, and the same result was reported.[104] It was found that, if the 2-position of the dione was blocked, extensive decomposition occurred.[103] A series of methyl- and/or phenyl-substituted octahydrophenoxselenin-1,9-dione-10-oxides was also noted.[105] Two derivatives of this compound type were prepared for testing as antitumor agents, but neither was found active.[106]

When **110** (R = R‴ = H; R′ = R″ = CH₃) was treated with bromine, 4,4′-dibromomethone (**111**) was isolated (42% yield), in addition to other oxidation products and halogen derivatives of methone (equation 77a).[103]

(77)

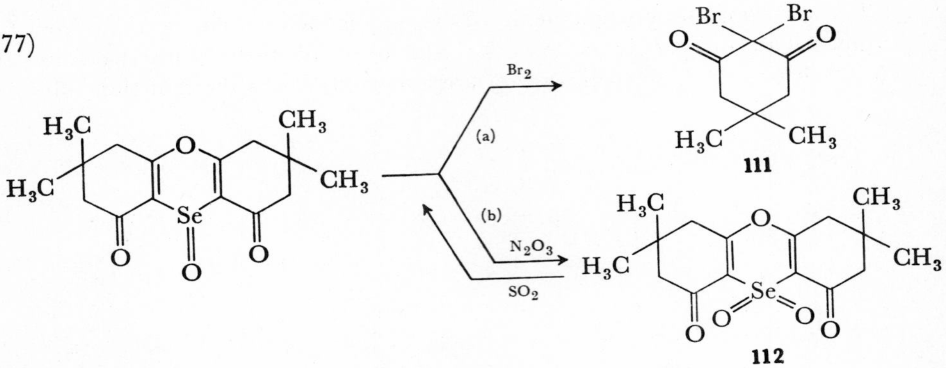

Oxidation with nitrous anhydride yielded the selenone **112**, which was converted to the starting material with sulfur dioxide (equation 77b).[103]

ii.   5,10-Episeleno-5,10-dihydroarsanthrene.   In 1970 this novel hetero-cycle **113** was prepared according to the method shown in scheme 78.[107]

(78)

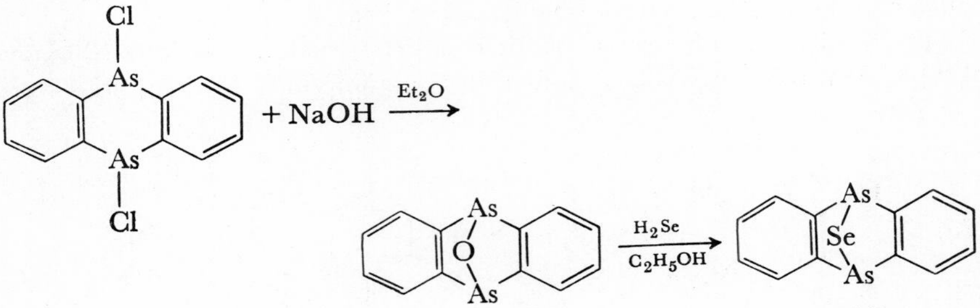

The product remained unchanged when boiled in methyl iodide, but in hydrogen peroxide it formed arsanthrenic acid (**114**) (equation 79).

(79)   **113** + H_2O_2 $\xrightarrow{\text{acetone}}$

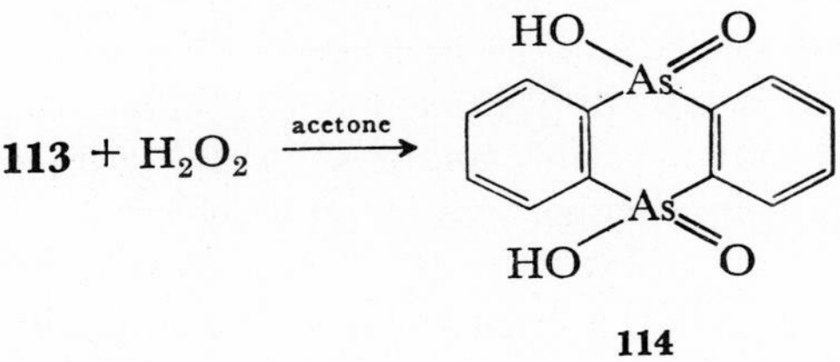

114

d. Three Six-Membered Rings, Two Selenium Atoms

i. Selenanthrenes. A low-yield method for the preparation of selenan-threne (**115**) was described in 1896 (equation 80).[108] In the late 1930's a

(80)

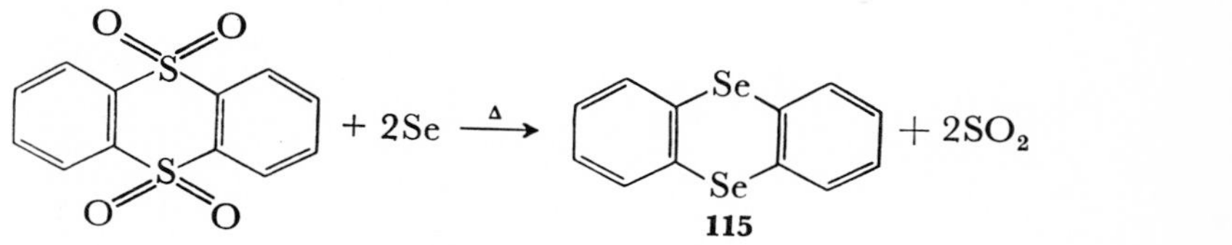

technical improvement over this original method by Cullinane *et al.*[109,110] increased the yield to 44%.

Two synthetic methods for the preparation of selenanthrene were intro-duced in 1935.[111] One procedure, which afforded the parent compound in only an 8% yield, is shown in equation 81. A second method was utilized

(81)

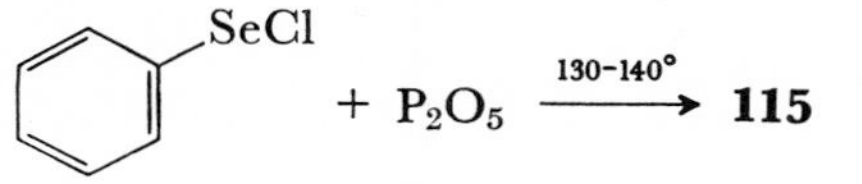

to prepare a series of methyl derivatives (**117**), which also included the parent compound.[111] These products were obtained in high yields (*e.g.*, selenanthrene in 95% yield) by heating phenylenediazoselenides (**116**) at 200° for 30–40 hours or in boiling acetic acid for 10–15 hours (scheme 82).

(82)

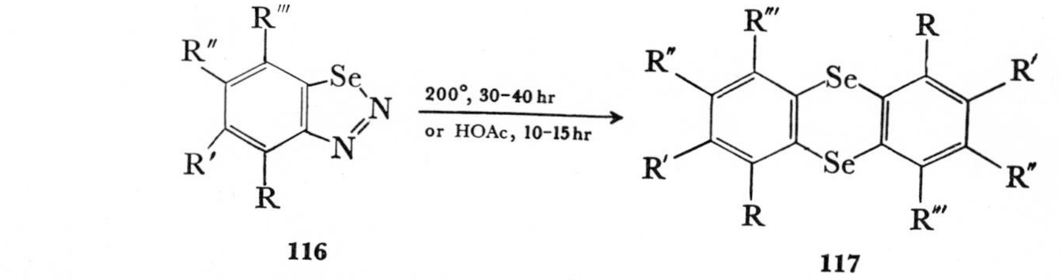

Another route to selenanthrenes was devised in 1936;[112] this produced the parent compound in a 70% yield (equation 83). In 1937 this series was

(83)

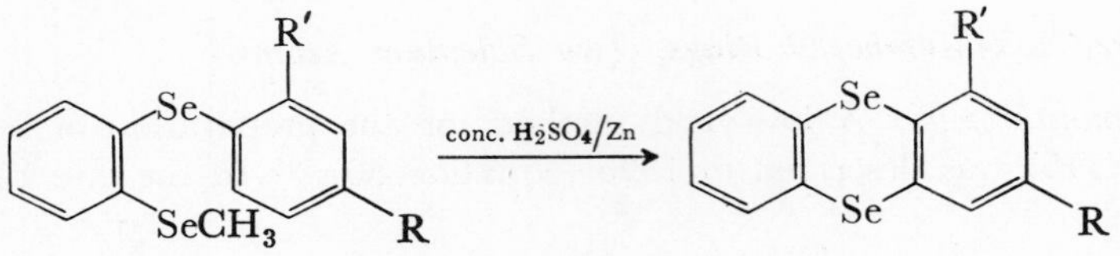

$$\xrightarrow{\text{conc. } H_2SO_4/Zn}$$

expanded to include numerous other methyl derivatives.[113] The preparation of selenanthrene in a 50% yield was also reported,[114] as shown in equation 84.

(84)

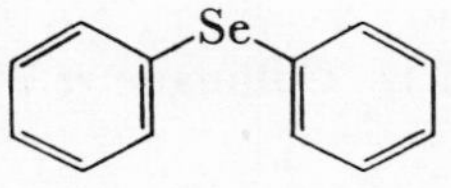

$$+ \text{ KNH}_2 \xrightarrow{\text{liq. NH}_3} \textbf{115} + C_6H_5SeK + K_2Se_x$$
$$(x \geqslant 2)$$

When tetraphenyltin was pyrolyzed with selenium at 270–310°, selenanthrene was obtained in a 9% yield (equation 85).[115]

(85) $(C_6H_5)_4Sn + Se \xrightarrow{270-310°} \textbf{115} + (C_6H_5)_2Se + H_2Se$

Heating 1,2-diiodotetrafluorobenzene with selenium at 320° for 3 days produced perfluoroselenanthrene (**118**) in an 86% yield (equation 86).[116]

(86)

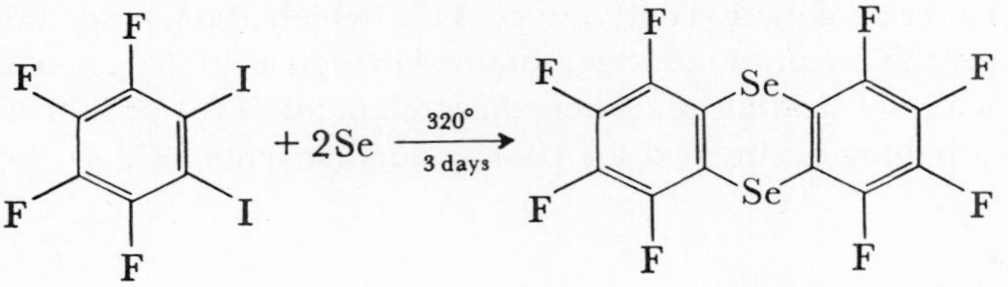

$$+ 2Se \xrightarrow[\text{3 days}]{320°}$$

118

Three additional routes to the perfluorinated compound, but in low yields (5–30%), were devised later (scheme 87).[117]

The oxidation of selenanthrene (**115**) with nitric acid gave the dioxide **119** *via* the corresponding dihydronitrate (equation 88).[108] A similar method for the preparation of the dioxide was employed later, and it was found that the starting material could be recovered by heating the dioxide at its melting point[114,118] or by reduction with sodium pyrosulfite.[118] The dioxide was made quantitatively by the hydrolysis of the tetrachloride.[119]

Although both diphenyl selenoxide and thianthrene-5-oxide yield dichlorides upon treatment with hydrochloric acid, selenanthrene 5,10-dioxide readily forms a dihydrochloride.[118] Up until 1951, when this reaction was studied, there appeared to be no other record of selenoxides reacting with hydrochloric acid in this way.

For a compound of general formula **120**, if the molecule is planar, the

(87)

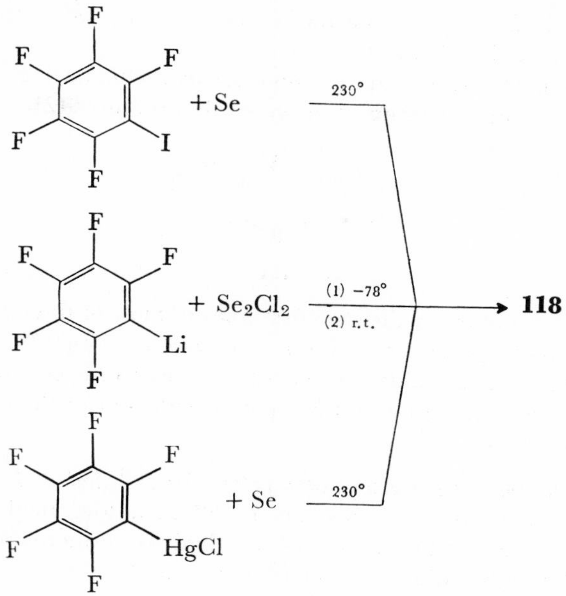

(88)

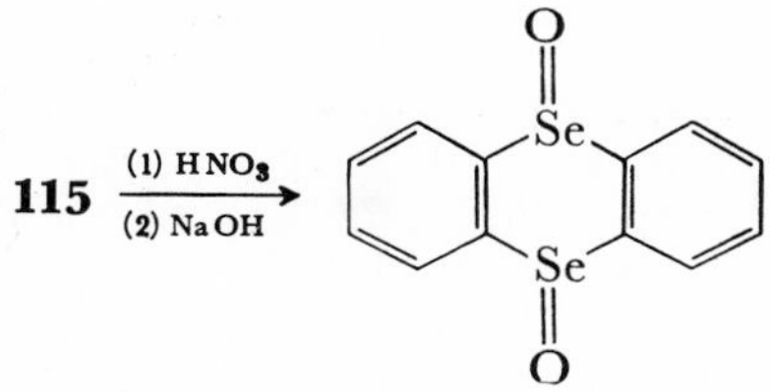

$$115 \xrightarrow[\text{(2) NaOH}]{\text{(1) HNO}_3}$$

119

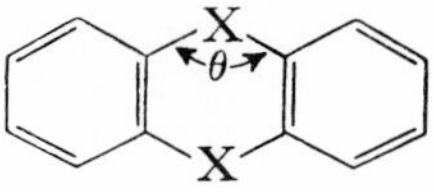

120

valence angle θ at X is 120°. If θ is 109.5° (the tetrahedral angle), however, the molecule will be folded about the axis X–X,[120] so that the angle of fold between the two planes, each containing one aromatic nucleus and both atoms X, is approximately 141°. The valence angle was found to increase in the order X = Se, S, O.[121] Dipole moments, which are a measure of folding, were determined for various compounds of this general structure.[120,121] X-ray crystallographic studies of the structure of selenanthrene also were made.[122–124] Isomorphous relationships of analogous derivatives of **120** were determined to be in harmony with the periodic relationships of oxygen, sulfur, and selenium.[125]

In a study to determine the relation between chemical constitution and u.v. absorption of compounds of general structure **121**, it was found

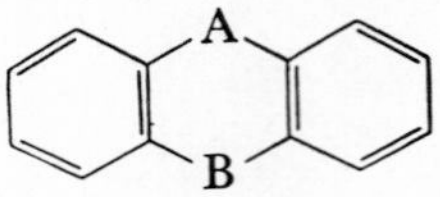

121

that the wave numbers of the strongest lines of most of these substances can be made to yield, by certain divisions, the same quotient.[126]

The electrochemical oxidation potential of selenanthrene and of phenoxselenine has been measured and compared with that of thianthrene and of phenthioxine.[127]

ii. 5,8 - Dimethyl - 10,12 - diselenatricyclo[5.3.1.1^{2,6}]dodec - 4,8 - diene - 3,11 - dione. The preparation of this dimer **123** in a low yield was reported in a doctoral thesis by the treatment of 3-hydroxy-5-methylselenopyrylium chloride (**122**) with sodium carbonate (equation 89).[25]

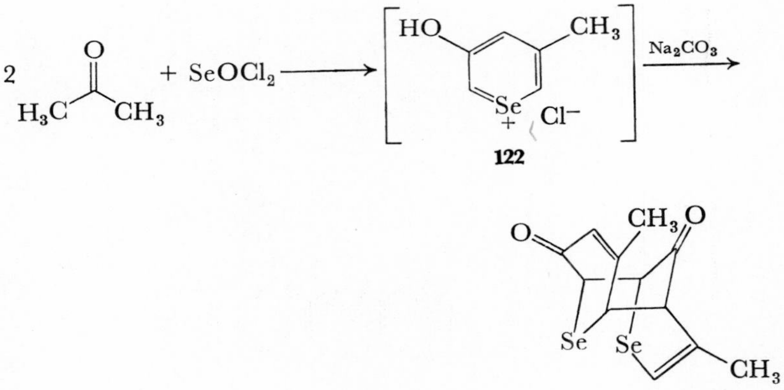

e.   Two Six-Membered Rings, One Seven-Membered Ring

i.   Dibenzoselenepins

Dibenzo[b,f]selenepins. 10,11-Dihydrodibenzo[b,f]selenepin-10-one (**124**) was synthesized in a 47% yield according to equation 90.[80] The by-product **125** was obtained in less than 1% yield. The 10-chloro compounds **126**, made by treating the corresponding alcohols with anhydrous hydrogen chloride, were used as intermediates in the preparation of 10-(4-methyl-piperazinyl) derivatives **127** (equation 91).[80]

(90)

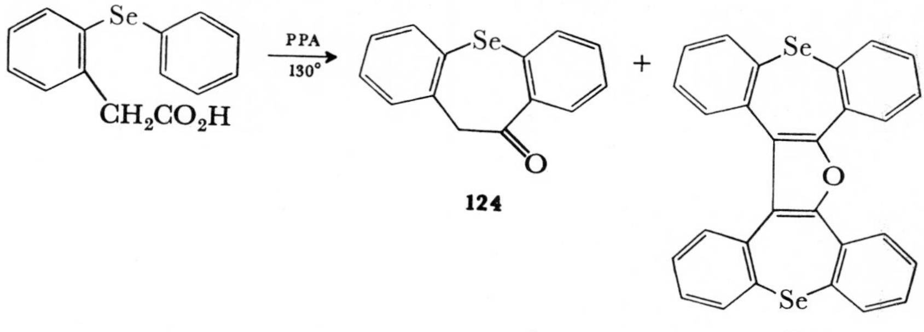

124

125

(91)

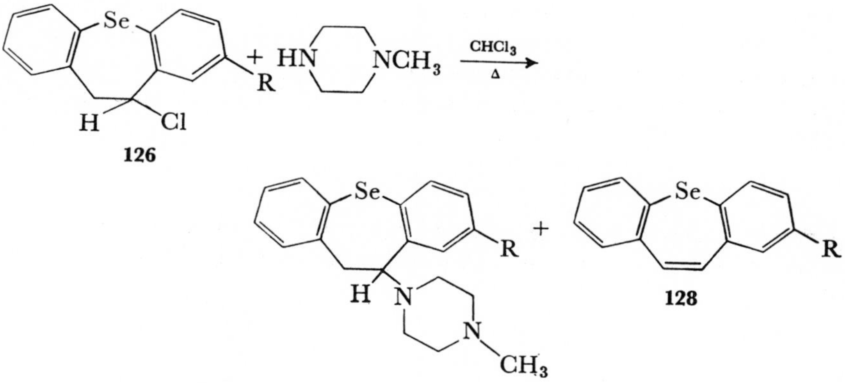

126

127   **128**

Also isolated from reaction 91 was the unsaturated compound, dibenzo-[b,f]selenepin (**128**, R = H). The reaction of **124** with 1-methylpiperazine produced 10-(4-methyl-1-piperazinyl)dibenzo[b,f]selenepin.[128]

(b) *Dibenzo[b,e]selenepins.* 6,11-Dihydrodibenzo[b,e]selenepin-11-one (**129**) was prepared according to equation 92.[129]

(92)

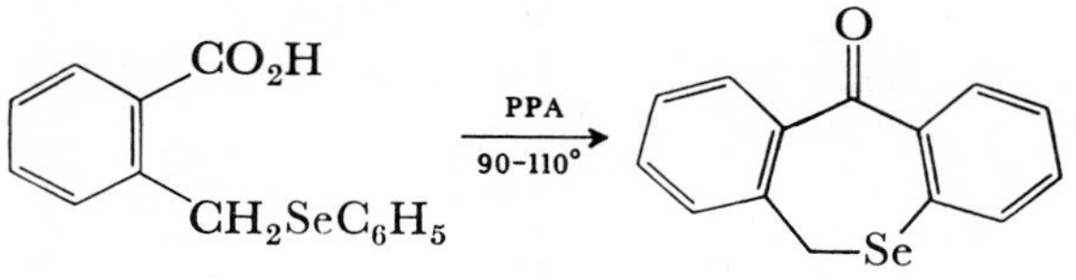

129

11-(3-Dimethylaminopropylidene)-6,11-dihydrodibenzo[b,e]selenepin was obtained by the dehydration of the corresponding alcohol.[129]

Dibenzo[c,e]selenepins. In 1956, 5,7-dihydrodibenzo[c,e]selenepin (**130**, R = H) was prepared quantitatively according to equation 93.[130] Formation

(93)

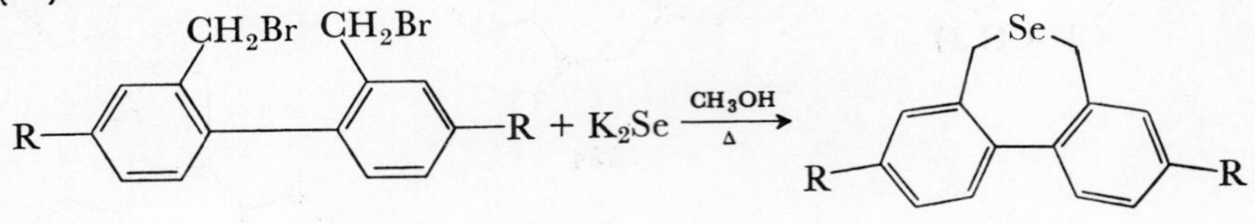

130

of the 3,9-diphenyl derivative (**130**, R = C$_6$H$_5$) was reported[131] to occur in a 55% yield by this reaction. A seven-membered ring all-carbon-atom 2,2′-bridged biphenyl is sufficiently close to coplanarity to cause racemization.[130] However, the presence of a large heteroatom (*e.g.*, Se), because of its increased covalent radius, may force the development of enantiomorphism. The u.v. absorption maxima of **130** (R = H) indicated noncoplanarity and asymmetry.[130]

ii. Naphthoselenepins

Tetrahydronaphtho[1,2-b]selenepins. 5-Oxo-2,3,4,5-tetrahydronaphtho[1,2-b]selenepin (**131**) was made as described by equation 94.[81] This ketone was

(94)

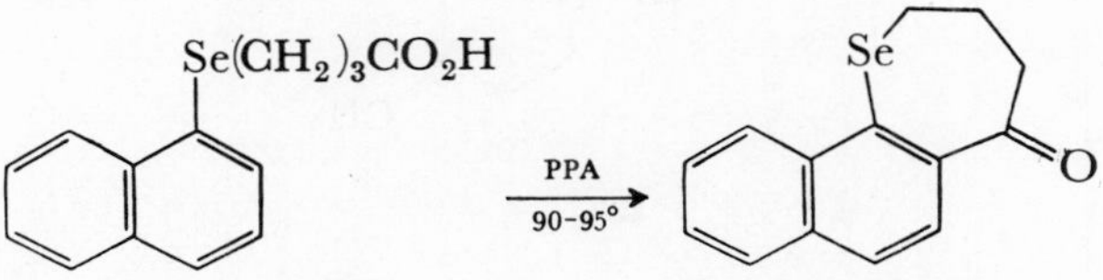

131

reduced to the parent compound, that is, 2,3,4,5-tetrahydronaphtho-[1,2-b]selenepin, *via* a Wolff-Kishner reduction.[81]

Tetrahydronaphtho[2,1-b]selenepins. The corresponding ketone **133** was prepared by the same route as its isomer **131** except that γ-(2-naphthyl-seleno)butyric acid (**132**) was used as the starting reagent (equation 95).[81]

(95)

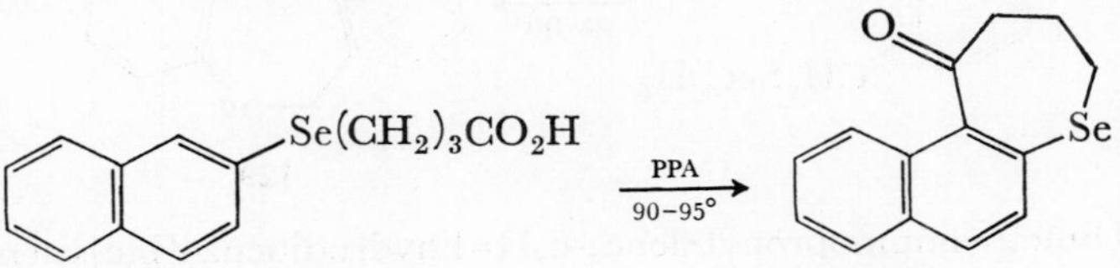

132   **133**

f. Two Six-Membered Rings, One Eight-Membered Ring

6,12-Episeleno-dibenzo[a,e]cyclooctene-5,11(6H, 12H)-dione.   The reaction of dibenzo[a,e]cyclooctene-5,11(6H,12H)-dione (**134**) with selenium dioxide in refluxing dioxane yielded three compounds as shown in equation 96.[132] The episeleno compound **135** was obtained as a minor product.

(96)

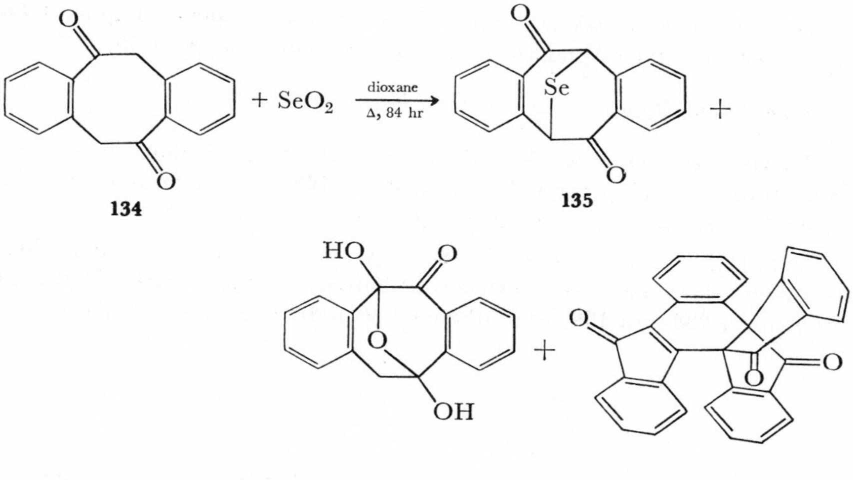

134   **135**

5. Heterocycles Consisting of Several Fused Rings

a. Four Fused Rings

i.   [1]Benzoselenino[4,3-b]indoles. 6,11-Dihydro[1]benzoselenino[4,3-b]indole (**136**) was synthesized in an 89% yield by indolization of the phenylhydrazone of selenochromanone (equation 97).[133]

(97)

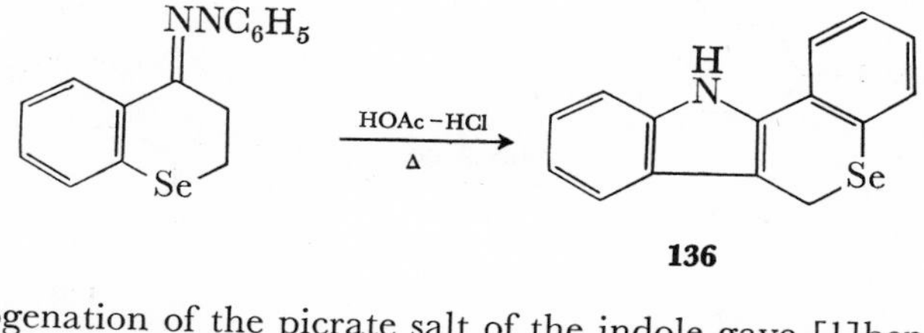

136

Dehydrogenation of the picrate salt of the indole gave [1]benzoselenino-[4,3-b]indole (**137**) in 70% yield (equation 98). This heterocycle is a

(98) 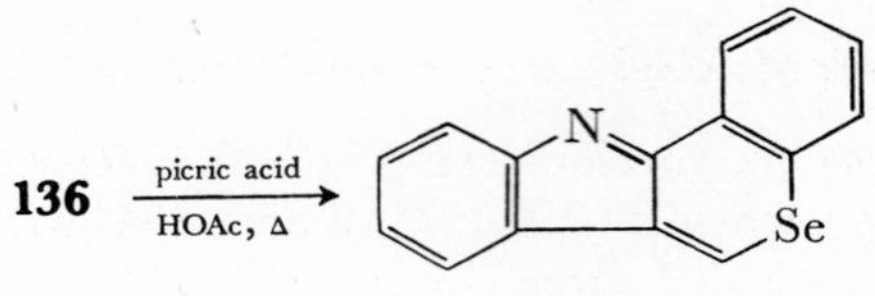   **136** $\xrightarrow[\text{HOAc, }\Delta]{\text{picric acid}}$

137

member of a class of compounds known as pseudoazulenes, since they are iso-π-electronic with dibenz[a,e]azulene.

ii. As mentioned previously, a selenium-containing spiro compound **86** was described as the secondary product (*cf.* Section 4, scheme 60).

b. Five Fused Rings

i. Polypyronoselenocoumarin. Also described earlier (*cf.* Section 4) was the preparation of polypyronoselenocoumarin (**103**), a by-product in the synthesis of pyronoselenocoumarin (equation 72).

ii. Benzo[g][1]benzoselenino[4,3-b]indoles. When the α- and β-naphthyl-hydrazones of selenochromanone were indolized, the respective 6,13-dihydro compounds, **138** and **139**, were obtained (equations 99 and 100).[133]

(99)

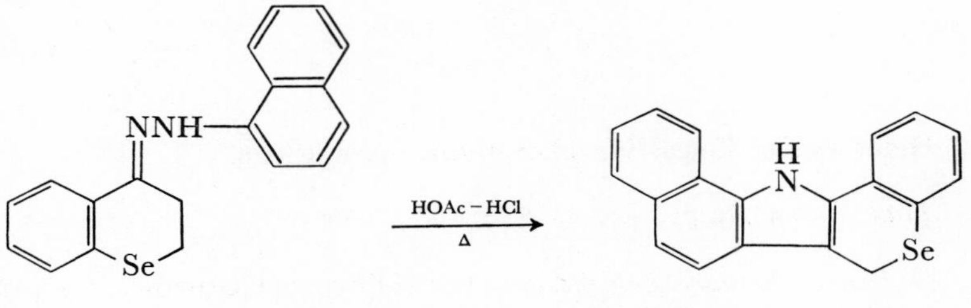

138

(100)

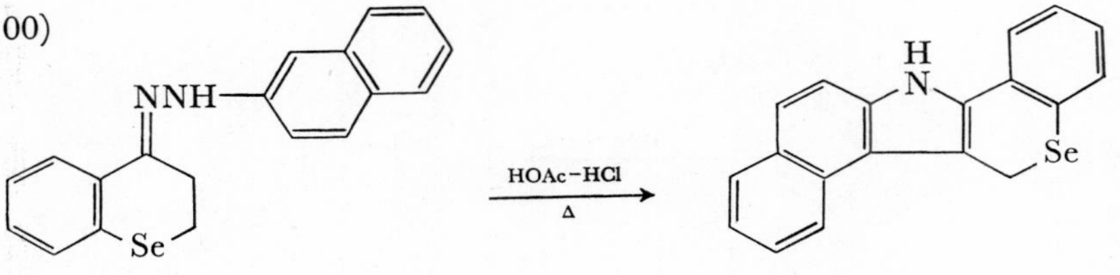

139

Dehydrogenation of the picrate salts of these indoles afforded the corresponding pseudoazulenes **140** and **141**, respectively (equations 101 and 102).

(101)   **138** $\xrightarrow[\text{HOAc, }\Delta]{\text{picric acid}}$

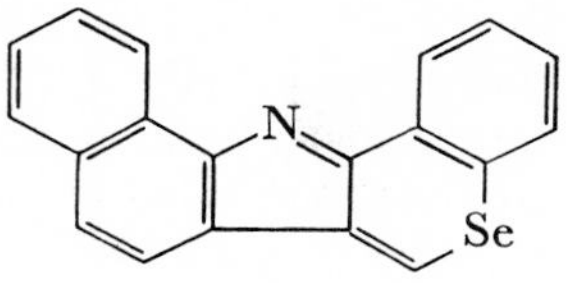

140

(102)   **139** $\xrightarrow[\text{HOAc, }\Delta]{\text{picric acid}}$

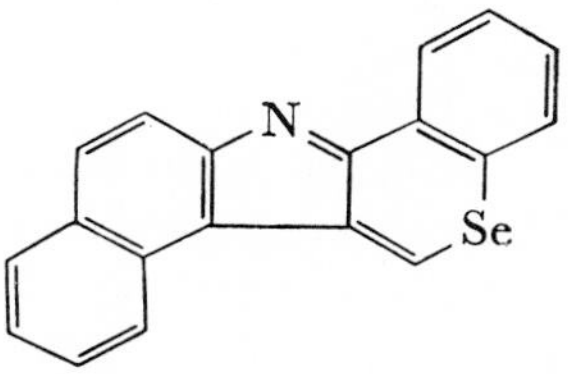

141

c. Seven Fused Rings. In the preparation of 10,11-dihydrodibenzo-[b,f]selenepin-10-one, a minor by-product **125** was obtained as previously indicated (*cf.* equation 90).

REFERENCES

1. M. E. Volpin, Yu. D. Koreshkov, V. G. Dulova, and D. N. Kursanov, *Tetrahedron*, **18**, 107 (1962).

2. F. Johnson and R. S. Gohlke, *Tetrahedron Lett.*, 1291 (1962).

3. M. E. Volpin, Yu. T. Struchkov, L. V. Vilkov, V. S. Mastryukov, V. G. Dulova, and D. N. Kursanov, *Izv. Akad. Nauk SSSR, Ser. Khim.*, 2067 (1963).

4. A. Davison and E. T. Shawl, *Chem. Commun.*, 670 (1967).

5. A. Davison and E. T. Shawl, *Inorg. Chem.*, **9**, 1820 (1970).

6. J. Morel, C. Paulmier, D. Semard, and P. Pastour, *C.R. Acad. Sci., Paris, Ser. C*, **270**, 825 (1970).

7. H. J. Jakobsen, E. H. Larsen, and S.-O. Lawesson, *Tetrahedron*, **19**, 1867 (1963).

8. J. H. van den Hende and E. Klingsberg, *J. Amer. Chem. Soc.*, **88**, 5045 (1966).

9. G. Traverso, *Ann. Chim.* (Rome), **47**, 3 (1957).

10. S. Bezzi, *Gazz. Chim. Ital.*, **92**, 859 (1962).

11. S. Bezzi, C. Garbuglio, M. Mammi, and G. Traverso, *Gazz. Chim. Ital.*, **88**, 1226 (1958).

12. S. Pietra, C. Garbuglio, and M. Mammi, *Gazz. Chim. Ital.*, **94**, 48 (1964).

13. M. Sanesi, G. Traverso, and M. Lazzarone, *Ann. Chim.* (Rome), **53**, 548 (1963).

14. R. Weiss and R. Gompper, *Tetrahedron Lett.*, 481 (1970).

14a. R. Mayer and A. K. Müller, *Z. Chem*, **4**, 384 (1964).

15. R. Mayer, B. Hunger, R. Prousa, and A. K. Müller, *J. Prakt. Chem.*, **35**, 294 (1967).

16. M. Sanesi and G. Traverso, *Chem. Ber.*, **93**, 1566 (1960).

17. K. A. Jensen and U. Henriksen, *Acta Chem. Scand.*, **21**, 1991 (1967).

18. G. A. Heath, I. M. Stewart, and R. L. Martin, *Inorg. Nucl. Chem. Lett.*, **5**, 169 (1969).

19. G. A. Heath, R. L. Martin, and I. M. Stewart, *Australian J. Chem.*, **22**, 83 (1969).

20. O. Siimann and J. Fresco, *Inorg. Chem.*, **9**, 294 (1970).

21. I. Degani, R. Fochi, and C. Vincenzi, *Gazz. Chim. Ital.*, **94**, 203 (1964).

22. I. Degani and C. Vincenzi, *Boll. Sci. Fac. Chim. Ind. Bologna*, **25**, 51 (1967).

23. I. Degani, L. Lunazzi, and F. Taddei, *Boll. Sci. Fac. Chim. Ind. Bologna*, **23**, 131 (1965).

24. I. Degani, F. Taddei, and C. Vincenzi, *Boll. Sci. Fac. Chim. Ind. Bologna*, **25**, 61 (1967).

25. R. G. Turnbo, Doctoral Dissertation, University of Texas; *University Microfilms*, Order No. 65-10772, 105 pp. (1965).

26. I. Degani and C. Vincenzi, *Boll. Sci. Fac. Chim. Ind. Bologna*, **23**, 249 (1965).

27. J. Fabian, A. Mehlhorn, and R. Zahradník, *Theoret. Chim. Acta*, **12**, 247 (1968).

28. I. Degani, R. Fochi, and C. Vincenzi, *Boll. Sci. Fac. Chim. Ind. Bologna*, **23**, 21 (1965).

29. K. W. Hubel and E. H. Braye, U.S. Pat. 3,280,017 (1966); *Chem. Abstr.*, **66**, 2462 (1967).

30. H. Backer and J. Strating, *Rec. Trav. Chim. Pays-Bas*, **53**, 1113 (1934).

31. W. L. Mock and J. H. McCausland, *Tetrahedron Lett.*, 391 (1968).

32. H. Backer and J. Strating, *Rec. Trav. Chim. Pays-Bas*, **53**, 525 (1934).

33. W. L. Mock, *J. Amer. Chem. Soc.*, **88**, 2857 (1966).

34. C. G. Krespan, *J. Amer. Chem. Soc.*, **83**, 3434 (1961).

35. E. G. Kataev, L. M. Kataeva, and Z. S. Titova, *Khim. Geterotsikl. Soedin.*, 172 (1968); *Chem. Abstr.*, **70**, 4060 (1969).

36. A. I. Kiss and B. R. Muth, *Magy. Tud. Akad. Köz. Fiz. Kut. Int. Közlemén.*, **3**, 213 (1955); *Chem. Abstr.*, **53**, 3876g (1959).

37. A. I. Kiss and B. R. Muth, *Acta Chim. Acad. Sci. Hung*, **11**, 57 (1957); *Chem. Abstr.*, **52**, 1760i (1958).

38. G. Grandolini, A. Ricci, N. P. Buu-Hoi, and F. Perin, *J. Heterocycl. Chem.*, **5**, 133 (1968).

39. N. P. Buu-Hoi, V. Bellavita, A. Ricci, and G. Grandolini, *Bull. Soc. Chim. Fr.*, 2658 (1965).

40. R. Lesser and R. Weiss, *Ber.*, **45**, 1835 (1912).

41. J. Gosselck, *Chem. Ber.*, **91**, 2345 (1958).

42. E. Giesbrecht and I. Mori, *Anales Acad. Brasil. Cienc.*, **30**, 521 (1958).

43. M. Vafai and M. Renson, *Bull. Soc. Chim. Belges*, **75**, 145 (1966).

44. J. Gosselck and E. Wolters, *Chem. Ber.*, **95**, 1237 (1962).

45. R. Lesser and A. Schoeller, *Ber.*, **47**, 2292 (1914).

46. R. Lesser and R. Weiss, *Ber.*, **46**, 2640 (1913).

47. D. L. Ross, J. Blanc, and F. J. Matticoli, *J. Amer. Chem. Soc.*, **92**, 5750 (1970).

48. R. Pummerer and G. Marondel, *Chem. Ber.*, **93**, 2834 (1960).

49. E. G. Kataev, G. A. Chmutova, A. A. Musina, and A. P. Anastas'eva, *Zh. Org. Khim.*, **3**, 597 (1967).

50. G. A. Chmutova, *Sb. Aspir. Rab. Kazan, Gos. Univ., Khim., Geol.*, 70 (1967); *Chem. Abstr.*, **70**, 87444q (1969).

51. A. Drory, *Ber.*, **24**, 2563 (1891).

52. M. Renson and R. Collienne, *Bull. Soc. Chim. Belges*, **73**, 491 (1964).

52a. W. H. H. Günther, *J. Org. Chem.*, **32**, 3929 (1967).

53. R. Lesser and R. Weiss, *Ber.*, **57B**, 1077 (1924).

54. E. Klingsberg, U.S. Pat. 3,511,853 (1970); *Chem. Abstr.* **73**, 66589 (1970).

55. M. Renson, *Bull. Soc. Chim. Belges*, **73**, 483 (1964).

56. J. M. Danze and M. Renson, *Bull. Soc. Chim. Belges*, **75**, 169 (1966).

57. J. Gosselck, *Angew. Chem. (Intern. Ed.)*, **2**, 660 (1963); ref. 30.

58. I. Degani, R. Fochi, and C. Vincenzi, *Gazz. Chim. Ital.*, **94**, 451 (1964).

59. A. Ricci, D. Balucani, C. Rossi, and A. Croisy, *Boll. Sci. Fac. Chim. Ind. Bologna*, **72**, 279 (1969).

60. A. Ruwet, J. Meessen, and M. Renson, *Bull. Soc. Chim. Belges*, **78**, 459 (1969).

61. F. Bossert, *Angew. Chem. (Intern. Ed.)*, **4**, 879 (1965).

62. A. Ruwet and M. Renson, *Bull. Soc. Chim. Belges*, **75**, 260 (1966).

63. A. Ruwet, D. Janne, and M. Renson, *Bull. Soc. Chim. Belges*, **79**, 81 (1970).

64. H. von Pechmann and C. Duisberg, *Ber.*, **16**, 2119 (1883).

65. H. Simonis and A. Elias, *Ber.*, **49**, 768 (1916).

66. A. Ruwet and M. Renson, *Bull. Soc. Chim. Belges*, **79**, 89 (1970).

67. I. Degani, R. Fochi, and G. Spunta, *Boll. Sci. Fac. Chim. Ind. Bologna*, **26**, 3 (1968).

68. I. Degani, *Corsi. Semin. Chim.*, 177 (1968); *Chem. Abstr.*, **73**, 24760 (1970).

69. I. Degani, R. Fochi, and G. Spunta, *Boll. Sci. Fac. Chim. Ind. Bologna*, **23**, 151 (1965).

70. I. Degani, R. Fochi, and G. Spunta, *Boll. Sci. Fac. Chim. Ind. Bologna*, **23**, 243 (1965).

71. A. Ruwet and M. Renson, *Bull. Soc. Chim. Belges*, **77**, 465 (1968).

72. A. Ruwet and M. Renson, *Bull. Soc. Chim. Belges*, **78**, 449 (1969).

73. E. Ziegler and E. Nölken, *Monatsh. Chem.*, **89**, 737 (1958).

74. N. P. Buu-Hoi, M. Mangane, O. Périn-Roussel, M. Renson, A. Ruwet, and M. Maréchal, *J. Heterocycl. Chem.*, **6**, 825 (1969).

75. M. Renson and P. Pirson, *Bull. Soc. Chim. Belges*, **75**, 456 (1966).

76. F. G. Mann and F. G. Holliman, *Nature*, **152**, 749 (1943).

77. F. G. Holliman and F. G. Mann, *J. Chem. Soc.*, 37 (1945).

78. P. W. Moy, U.S. Pat. 3,480,524 (1969); *Chem. Abstr.*, **72**, 38310 (1970).

79. P. W. Moy, U.S. Pat. 3,497,530 (1970); *Chem. Abstr.*, **72**, 96097 (1970).

80. K. Šindelár, J. Metyšová, and M. Protiva, *Collection Czech. Chem. Commun.*, **34**, 3801 (1969).

81. N. Bellinger, P. Cagniant, and M. Renson, *C.R. Acad. Sci., Paris, Ser. C*, **269**, 532 (1969).

82. R. Lesser and R. Weiss, *Ber.*, **47**, 2510 (1914).

83. K. Šindelár, E. Svátek, J. Metyošvá, J. Metyš, and M. Protiva, *Collection Czech. Chem. Commun.*, **34**, 3792 (1969).

84. I. Degani and R. Fochi, *Ann. Chim.* (Rome), **58**, 251 (1968).

85. M. Battegay and G. Hugel, *Bull. Soc. Chim. Fr.*, **31**, 440 (1922).

86. M. Battegay and G. Hugel, *Bull. Soc. Chim. Fr.*, **27**, 557 (1920).

87. R. H. Nealey and J. S. Driscoll, *J. Heterocycl. Chem.*, **3**, 228 (1966).

88. H. W. Doughty and F. R. Elder, *Eighth Intern. Congr. Appl. Chem.*, **6**, 93 (1912).

89. K. Maruyama, M. Yoshida, I. Tanimoto, and J. Osugi, *Rev. Phys. Chem. Japan*, **39**, 117 (1969).

90. K. Maruyama, M. Yoshida, I. Tanimoto, and J. Osugi, *Rev. Phys. Chem. Japan*, **39**, 123 (1969).

91. N. P. Buu-Hoi, M. Mangane, M. Renson, and L. Christiaens, *J. Chem. Soc., B*, 971 (1969).

92. F. François, *C.R. Acad. Sci., Paris*, **190**, 191 (1930).

93. B. R. Muth, *Chem. Ber.*, **93**, 283 (1960).

94. F. François, *C.R. Acad. Sci., Paris*, **190**, 800 (1930).

95. F. François, *C.R. Acad. Sci., Paris*, **190**, 1306 (1930).

96. K. Maruyama, M. Yoshida, and K. Murakami, *Bull. Chem. Soc. Japan*, **43**, 152 (1970).

97. P. Ehrlich and H. Bauer, *Ber.*, **48**, 502 (1915).

98. M. Battegay and G. Hugel, *Bull. Soc. Chim. Fr.*, **33**, 1103 (1923).

99. I. Degani, R. Fochi, and G. Spunta, *Boll. Sci. Fac. Chim. Ind. Bologna*, **23**, 165 (1965).

100. N. Bellinger and P. Cagniant, *C.R. Acad. Sci., Paris, Ser. C*, **268**, 1385 (1969).

101. H. D. K. Drew, *J. Chem. Soc.*, 511 (1928).

102. M. C. Thompson and E. E. Turner, *J. Chem. Soc.*, 29 (1938).

103. H. Stamm and K. Gosrau, *Ber.*, **66**, 1558 (1933).

104. W. Borsche and H. Hartmann, *Ber.*, **73B**, 839 (1940).

105. F. E. King and D. G. I. Felton, *J. Chem. Soc.*, 274 (1949).

106. K. Takeda, H. Nishemura, N. Shimaoka, R. Noguchi, and K. Nakajima, *Ann. Rept. Shionogi Res. Lab.*, **5**, 1 (1955).

107. D. W. Allen, J. C. Coppola, O. Kennard, F. G. Mann, W. D. S. Motherwell, and D. G. Watson, *J. Chem. Soc., C*, 810 (1970).

108. F. Krafft and A. Kashau, *Ber.*, **29**, 443 (1896).

109. N. M. Cullinane, N. M. E. Morgan, and C. A. J. Plummer, *Rec. Trav. Chim. Pays-Bas*, **56**, 627 (1937).

110. N. M. Cullinane, A. G. Rees, and C. A. J. Plummer, *J. Chem. Soc.*, 151 (1939).

111. S. Keimatsu and I. Satoda, *J. Pharm. Soc. Japan*, **55**, 233 (1935).

112. S. Keimatsu, I. Satoda, and T. Tigono, *J. Pharm. Soc. Japan*, **56**, 869 (1936).

113. S. Keimatsu, I. Satoda, and T. Kobayasi, *J. Pharm. Soc. Japan*, **57**, 190 (1937).

114. O. Schmitz-DuMont and B. Ross, *Angew. Chem. (Intern. Ed.)*, **6**, 1071 (1967).

115. M. Schmidt and H. Schumann, *Chem. Ber.*, **96**, 780 (1963).

116. S. C. Cohen, M. L. N. Reddy, and A. G. Massey, *Chem. Commun.*, 451 (1967).

117. S. C. Cohen, M. L. N. Reddy, and A. G. Massey, *J. Organometal. Chem.*, **11**, 563 (1968).

118. N. M. Cullinane, *J. Chem. Soc.*, 237 (1951).

119. G. Bergson, *Acta Chem. Scand.*, **11**, 580 (1957).

120. I. G. M. Campbell, C. G. LeFèvre, R. J. W. LeFèvre, and E. E. Turner, *J. Chem. Soc.*, 404 (1938).

121. K. Higasi, *Sci. Papers Inst. Phys. Chem. Res. Tokyo*, **38**, 331 (1941).

122. J. E. Crackston and R. G. Wood, *Nature*, **142**, 257 (1938).

123. R. G. Wood and J. E. Crackston, *Phil. Mag.*, **31**, 62 (1941).

124. R. G. Wood and G. Williams, *Nature*, **150**, 321 (1942).

125. N. M. Cullinane and C. A. J. Plummer, *J. Chem. Soc.*, 63 (1938).

126. J. Moir, *Trans. Roy. Soc. S. Africa*, **18, Pt. 2**, 137 (1929).

127. C. Barry, G. Cauquis, and M. Maurey, *Bull. Soc. Chim. Fr.*, 2510 (1966).

128. K. Šindelář, J. Metyšová, J. Metyš, and M. Protiva, *Naturwissenschaften*, **56,** 374 (1969).

129. K. Šindelář, J. Metyšová, E. Svátek, and M. Protiva, *Collection Czech. Chem. Commun.* **34,** 2122 (1969).

130. W. E. Truce and D. D. Emrick, *J. Amer. Chem. Soc.*, **78,** 6130 (1956).

131. R. L. Taber, G. H. Daub, F. N. Hayes, and D. G. Ott, *J. Heterocycl. Chem.*, **2,** 181 (1965).

132. P. Yates and E. G. Lewars, *Can. J. Chem.*, **48,** 796 (1970).

133. N. P. Buu-Hoi, A. Croisy, P. Jacquignon, M. Renson, and A. Ruwet, *J. Chem. Soc., C,* 1058 (1970).

XII Selenium amino acids and peptides

A. SELENOAMINO ACIDS

GÖRAN ZDANSKY

University of Uppsala
Institute of Chemistry
Uppsala, Sweden

1. Introduction

Extensive investigations in connection with a cattle ailment known as "alkali disease," which ravaged certain parts of the United States, eventually showed it to be closely connected with the selenium content of the soil. In the 1930's it was furthermore demonstrated that certain plants were capable of incorporating and accumulating selenium, and that part of the selenium so incorporated entered the protein fraction of the plants. These findings led to research on the synthesis of selenium-containing amino acids. The synthetic selenoamino acids proved to be toxic, producing the symptoms that characterize "alkali disease."

Nowadays selenoamino acids are useful tools for tracing and understanding complex biochemical processes.

2. Selenocysteine, Selenocystine, and Related Compounds

The types of compounds to be covered in this section are as follows:

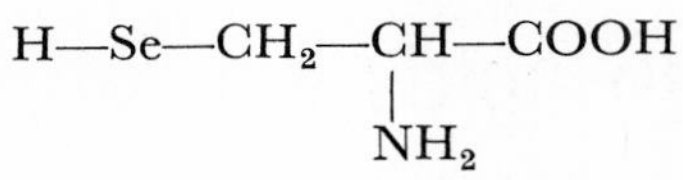

$$H—Se—CH_2—CH—COOH$$
$$|$$
$$NH_2$$

Selenocysteine

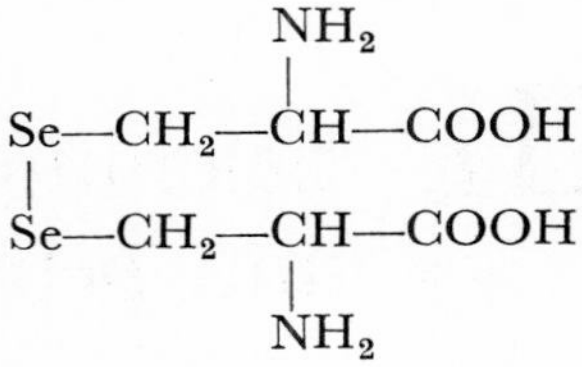

$$\begin{array}{l} NH_2 \\ | \\ Se—CH_2—CH—COOH \\ | \\ Se—CH_2—CH—COOH \\ | \\ NH_2 \end{array}$$

Selenocystine

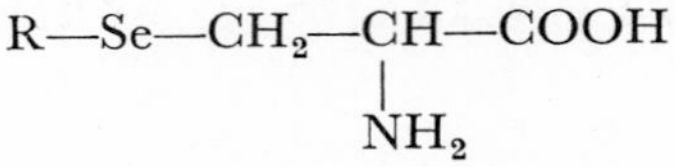

$$R—Se—CH_2—CH—COOH$$
$$|$$
$$NH_2$$

Selenocysteine derivatives

In the selenocysteine derivative structure, R denotes alkyl and aryl groups, but not other amino acid residues as in compounds such as selenolanthionine and selenocystathionine. The latter will be covered in Section XII.A.6.

a. Early Synthetic Methods. DL and *meso*-Selenocystine, the first seleno-amino acid synthesized, was made by Fredga,[1] who later[2] also synthesized L- and D-selenocystine. The method employed was a variation of the Fischer and Raske[3] synthesis of cystine (equation 1). Serine, converted to 3-chloro-

$$(1) \quad Cl—CH_2—CH—COOCH_3 \xrightarrow[\text{KOH}]{\text{K}_2\text{Se}_2} \begin{array}{l} NH_2 \\ | \\ Se—CH_2—CH—COOH \\ | \\ Se—CH_2—CH—COOH \\ | \\ NH_2 \end{array}$$
$$|$$
$$NH_3{}^+Cl^-$$

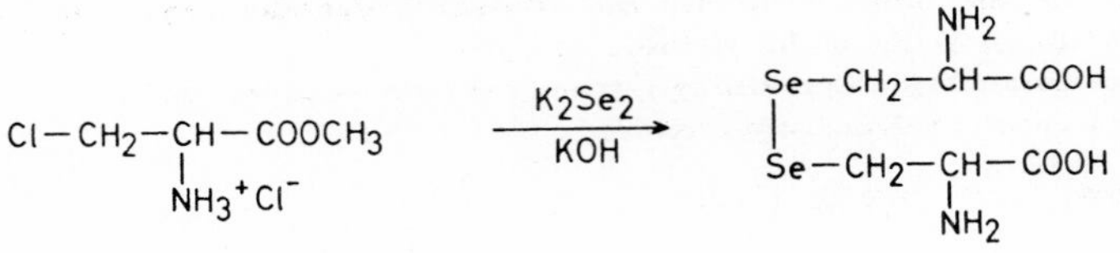

alanine methyl ester hydrochloride, and aqueous potassium diselenide were kept for 36 hours at room temperature in a closed vessel. In spite of apparent simplicity the method never became popular because of its inconsistent yield (0–30 %).

Painter[4] described two syntheses of selenocystine and an attempted synthesis, all based on the reaction of derivatives of 3-chloroalanine with sodium benzylselenolate, sodium phenyl selenolate, or sodium hydrogen selenide (scheme 2). The first two reactions were of the Williamson type, while the latter was a variation of the method used by Fredga[1,2] and likewise gave low yields (20 % at the best in the last step of the reaction series, 13–14 % overall starting from serine). The Williamson reaction-type synthesis, using either 3-chloroalanine methyl ester hydrochloride or 3-chlorobenzamido-alanine methyl ester as one component and sodium phenyl selenolate or

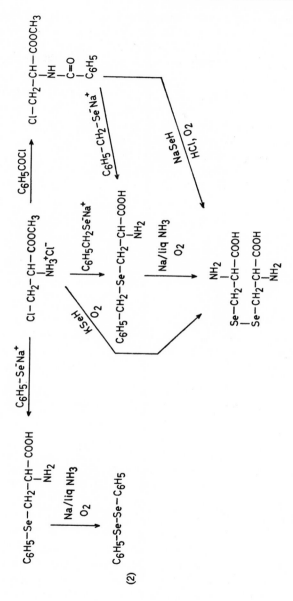

sodium benzylselenolate as the other, turned out to be more successful in part. The phenyl group in phenylselenoalanine, however, could not be removed by reduction with sodium in liquid ammonia, the whole selenium content being recovered as diphenyl diselenide. Cleavage by hydriodic acid worked, but the yields were low. The benzyl group in 3-benzylseleno-alanine was removed by reduction with sodium in liquid ammonia, and the overall yield from serine to selenocystine, using 3-chloroalanine methyl ester hydrochloride as an intermediate, was as high as 36 %. The 3-chloro-benzamidoalanine methyl ester gave a lower yield: 24 %. Painter's best method thus offered a feasible pathway to selenocystine.

Williams and Ravve[5] presented two syntheses for selenocystine (scheme 3). One synthesis was based on 3-chloroalanine and barium hydrogen selenide,

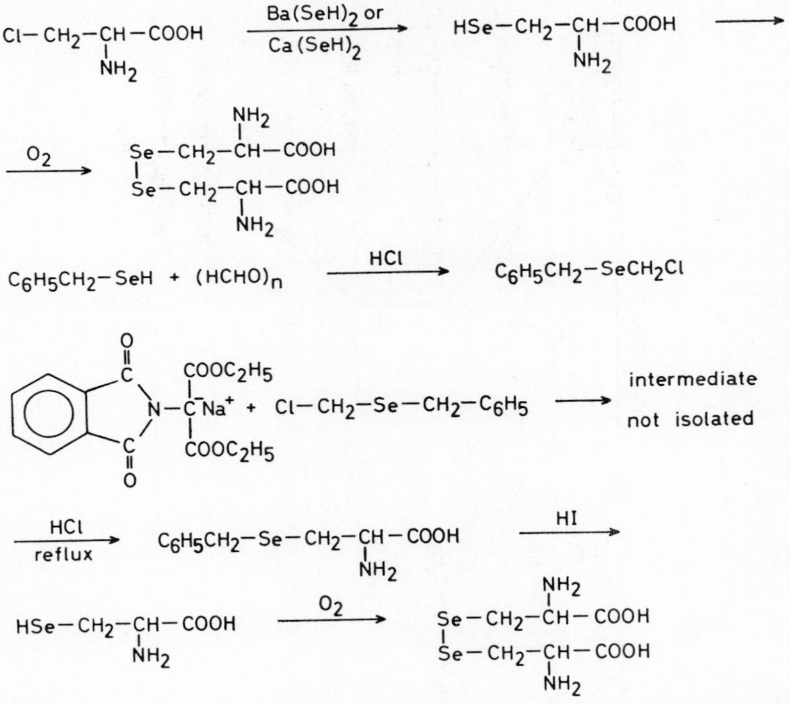

$$Cl-CH_2-\underset{\underset{NH_2}{|}}{CH}-COOH \xrightarrow[\text{Ca(SeH)}_2]{\text{Ba(SeH)}_2 \text{ or}} HSe-CH_2-\underset{\underset{NH_2}{|}}{CH}-COOH \longrightarrow$$

$$\xrightarrow{O_2} \quad \underset{|}{\overset{\overset{NH_2}{|}}{Se}}-CH_2-\underset{}{CH}-COOH$$
$$Se-CH_2-\underset{\underset{NH_2}{|}}{CH}-COOH$$

$$C_6H_5CH_2-SeH + (HCHO)_n \xrightarrow{HCl} C_6H_5CH_2-SeCH_2Cl$$

(3)

$$\text{(phthalimido)}N-\underset{\underset{COOC_2H_5}{|}}{\overset{\overset{COOC_2H_5}{|}}{C}}{}^-Na^+ + Cl-CH_2-Se-CH_2-C_6H_5 \longrightarrow \begin{array}{l}\text{intermediate}\\ \text{not isolated}\end{array}$$

$$\xrightarrow[\text{reflux}]{HCl} C_6H_5CH_2-Se-CH_2-\underset{\underset{NH_2}{|}}{CH}-COOH \xrightarrow{HI}$$

$$HSe-CH_2-\underset{\underset{NH_2}{|}}{CH}-COOH \xrightarrow{O_2} \underset{|}{\overset{\overset{NH_2}{|}}{Se}}-CH_2-CH-COOH$$
$$Se-CH_2-\underset{\underset{NH_2}{|}}{CH}-COOH$$

giving an overall yield below 11 %. The other was a modification of Wood and du Vigneaud's[6] synthesis for DL-cystine by the Gabriel-Sørensen phthal-imidomalonic ester method, using benzyl chloromethyl selenide as alkylating agent. Since the yields were low, use was made of the vesicant, benzyl chloromethyl selenide, and because the product was bound to be a mixture of DL- and *meso*-selenocystine, this method did not come into wide use.

$$
\begin{array}{c}
\text{H}_2\text{C}=\underset{\underset{\underset{\text{CH}_3}{|}}{\overset{|}{\text{C}=\text{O}}}}{\overset{|}{\underset{\text{NH}}{\text{C}}}}-\text{COOH}_3
\end{array}
\xrightarrow[\text{CH}_3\text{ONa}]{\text{C}_6\text{H}_5-\text{CH}_2\text{SeH}}
\text{C}_6\text{H}_5-\text{CH}_2-\text{Se}-\text{CH}_2-\underset{\underset{\underset{\text{CH}_3}{|}}{\overset{|}{\text{C}=\text{O}}}}{\overset{|}{\underset{\text{NH}}{\text{CH}}}}-\text{COOCH}_3
\xrightarrow[\text{HCl}]{\text{NaOH}}
$$

(4)

$$
\text{C}_6\text{H}_5-\text{CH}_2-\text{Se}-\text{CH}_2-\underset{\underset{\underset{\text{CH}_3}{|}}{\overset{|}{\text{C}=\text{O}}}}{\overset{|}{\underset{\text{NH}}{\text{CH}}}}-\text{COOH}
\xrightarrow[\text{Fe}^{3+}\,\text{O}_2]{\text{NH}_3\,\text{liq}\,\text{Na}}
\begin{array}{l}
\text{Se}-\text{CH}_2-\overset{|}{\underset{\underset{\text{CH}_3}{\underset{|}{\text{C}=\text{O}}}}{\underset{|}{\text{NH}}}}\text{CH}-\text{COOH} \\
| \\
\text{Se}-\text{CH}_2-\underset{\underset{\underset{\text{CH}_3}{|}}{\overset{|}{\text{C}=\text{O}}}}{\overset{|}{\underset{\text{NH}}{\text{CH}}}}-\text{COOH}
\end{array}
\xrightarrow{\text{HCl, H}_2\text{O}}
$$

$$
\begin{array}{l}
\text{Se}-\text{CH}_2-\overset{|}{\underset{\text{NH}_2}{\text{CH}}}-\text{COOH} \\
| \\
\text{Se}-\text{CH}_2-\underset{\underset{\text{NH}_2}{|}}{\text{CH}}-\text{COOH}
\end{array}
$$

The syntheses described above can, therefore, be classified as obsolete. Only Painter's[4] best one may perhaps be considered acceptable.

b. Currently Used Synthetic Methods. In addition to Painter's[4] best method, which has been used by Janicki *et al.*[7] and Frank[8] with minor variations, there are two newer methods for the synthesis of selenocystine. One was first reported by Zdansky,[9] who used the readily accessible acetamidoacrylic acid ester and benzylselenol, rendering DL- and *meso*-selenocystine in about 70% overall yield (scheme 4).[9] This procedure was later described in detail by Chu and his co-workers.[10] Zdansky's resolution of *Se*-benzylselenocysteine[11] will be reviewed later in this chapter.

The other method, reported by Theodoropoulos *et al.*,[12,13] utilizes nucleophilic displacement of the *O-p*-toluenesulfonate moiety of *O*-tosylated serine derivatives by the benzyl selenolate anion (scheme 5). According to a

Scheme 5

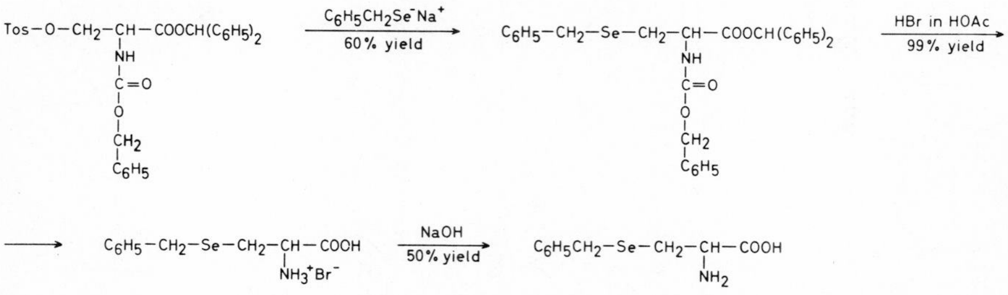

$$\text{Tos}-\text{O}-\text{CH}_2-\underset{\underset{\underset{\underset{\underset{C_6H_5}{|}}{CH_2}}{|}}{\underset{O}{|}}{\underset{|}{C=O}}}{\overset{|}{\underset{|}{NH}}}{CH}-\text{COOCH}(C_6H_5)_2 \xrightarrow[\text{60\% yield}]{C_6H_5CH_2Se^-Na^+} C_6H_5-CH_2-\text{Se}-\text{CH}_2-\underset{\underset{\underset{\underset{\underset{C_6H_5}{|}}{CH_2}}{|}}{\underset{O}{|}}{\underset{|}{C=O}}}{\overset{|}{\underset{|}{NH}}}{CH}-\text{COOCH}(C_6H_5)_2 \xrightarrow[\text{99\% yield}]{\text{HBr in HOAc}}$$

$$\longrightarrow C_6H_5-CH_2-\text{Se}-\text{CH}_2-\underset{\underset{NH_3^+Br^-}{|}}{CH}-\text{COOH} \xrightarrow[\text{50\% yield}]{\text{NaOH}} C_6H_5CH_2-\text{Se}-\text{CH}_2-\underset{\underset{NH_2}{|}}{CH}-\text{COOH}$$

later paper by Roy *et al.*,[14] sodium hydrogen selenide has been used instead of sodium benzyl selenolate.

These last-mentioned methods have been specifically worked out for peptide synthesis as regards suitable protecting and assisting groups. The yield of *Se*-benzyl-L-selenocysteine from diphenylmethyl *N*-carbobenzoxy-L-serinate is about 50% with the method of Theodoropoulos *et al.*,[12,13] outlined in scheme 5. Therefore the method of Zdansky[9,11] seems to be a more efficient one for preparing *Se*-benzylselenocysteine, *Se*-benzyl-*N*-acetylselenocysteine, or selenocystine in racemic and optically active forms in larger quantities. The methods of Theodoropoulos *et al.*[12,13] and Roy *et al.*[14] on the other hand, are more expedient and efficient for the direct small scale production of optically active selenocysteine derivatives suitable for use in peptide chemistry.

c. Stereoisomerism. Utilizing optically active serine, Fredga[2] synthesized both antipodes of selenocystine (equation 1) and recorded the fact that both forms crystallize in microscopic hexagonal plates. The molar rotation is highly dependent on the concentration of the hydrochloric acid used as solvent, but Fredga also noted that the concentration of the amino acid itself has only a minor influence. This is in agreement with the behavior of cystine.[2] The rotation is high: 1% in $0.5\ M$ HCl $[\alpha]_D^{25} = +175.5°$ $[M]_D^{25} = +573°$, which is also in agreement with cystine. In alkaline solution the rotational power dropped because of racemization; 1% in $0.2\ M$ NaOH showed $[\alpha]_D^{25} = +32.5°$, $[M]_D^{25} = +108°$, but the values diminished by some 20% in 24 hours. The acid solutions remained unaltered even after 3 days.

Frank,[8] utilizing Painter's[4] best method (scheme 2), synthesized L-Se-benzylselenocysteine from L-serine and recorded $[\alpha]_D^{23} = +35.6°$ ($C = 2$ in $1\ M$ NaOH). Frank[8] has also used D-tartaric acid in order to resolve partially racemized ethyl L-Se-benzylselenocysteinate, but he does not appear to have resolved the racemic compound itself. Frank does not state the total yield but records $[\alpha]_D^{23} = +35.5°$ ($C = 2$ in $1\ M$ NaOH). Walter and Chan[15] resolved DL-N-acetyl-Se-benzylselenocysteine, using hog acylase I, and recovered only the L-antipode in 30% yield: $[\alpha]_D^{23} = +35.2°$, $C = 2$ in $1\ M$ NaOH, m.p. 183–184°. Zdansky obtained a similar low yield using acylase; with papain, however, he has been able to resolve DL-N-benzylselenocysteine in excellent yield and to recover both antipodes with equal and higher rotations than had been recorded before (scheme 6).[11] L-Se-Benzylselenocysteine was obtained in 62% yield: $[\alpha]_D^{20} = +39.4°$, $C = 1$ in

DL—C_6H_5—CH_2—Se—CH_2—CH—COOH → $\dfrac{(CH_3\overset{O}{\overset{\|}{C}})_2O}{OH^-}$ → DL—C_6H_5—CH_2—Se—CH_2—CH—COOH
   | |
   NH_2 NH
                                                                                      |
                                                                                     C=O
                                                                                      |
                                                                                     CH_3

DL—$C_6H_5CH_2$—Se—CH_2—CH—COOH + $C_6H_5NH_2$ → $\dfrac{papain}{buffer\ pH\,5}$
   |
   NH
   |
   C=O
   |
   CH_3

(6)

L—$C_6H_5CH_2$—Se—CH_2—CH—$\overset{O}{\overset{\|}{C}}$—NH—$C_6H_5$ + D—$C_6H_5CH_2$—Se—CH_2—CH—COOH
   | |
   NH   NH
   | |
   C=O   C=O
   | |
   CH_3 CH_3

| HCl, H_2O | HCl, H_2O
| OH^- to pH 6 | OH^- to pH 6

L—$C_6H_5CH_2$—Se—CH_2—CH—COOH   D—$C_6H_5CH_2$-Se-CH_2—CH—COOH
   | |
   NH_2 NH_2

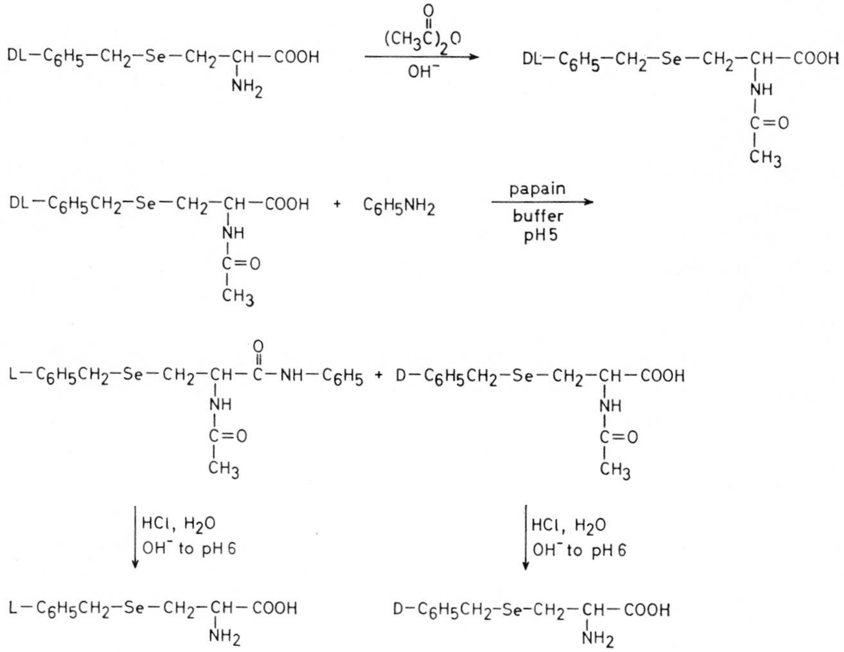

1 M NaOH; and D-Se-benzylselenocysteine in 67 % yield: $[\alpha]_D^{20} = -39.2°$, $C = 1$ in 1 M NaOH. This method thus seems to be the most expedient of those recorded.

Some doubts were expressed by Theodoropoulos et al.[13] as to the feasibility of Frank's[8] method, complaints being made that the hydrolysis of the ester group with aqueous hydrochloric acid adversely influenced the ultimate result. On the other hand, Zdansky[11] has, with consistently good results, used boiling 2 M hydrochloric acid in order to deacetylate and remove the anilide group from both D-Se-benzyl-N-acetylselenocysteine and L-Se-benzyl-N-acetylselenocysteine anilide. The latter compound, partly because of its lower solubility, admittedly requires a larger volume of hydrochloric acid and a longer reaction time in comparison with the former compound, but gives nevertheless the same degree of rotation (with opposite sign) and comparable yield: 83 % for the L-isomer and 91 % for the D-isomer. These facts indicate, as the reaction conditions are alike in both cases, that there is nothing intrinsically unsound in the hydrolysis with 2 M hydrochloric acid. On the other hand, Zdansky[16] has observed that even small amounts of alcohol, deliberately introduced into the hydrolysis mixture in order to enhance the solubility, especially of L-Se-benzyl-N-acetylselenocysteine anilide, have a marked, adverse influence on both yield and optical purity. Seen in this light, the statement by Theodoropoulos et al.[13] may be justified.

d. Properties of Selenocystine.

The solubility of optically active selenocystine in water at 25° is given as 0.00063 mole/l by Fredga.[2] Huber and Criddle[17] report the solubility at pH 7 and 25° for DL-selenocystine as 0.00235 mole/l, and for DL-cystine as 0.00038 mole/l. Huber and Criddle[17] record the pK values for the ionizations of carboxyl, selenohydryl, and amino groups of DL-selenocysteine as 2.01, 5.24, and 9.96, respectively. For DL-cysteine they report the carboxyl constant as pK 2.3 and the values of the amino and sulfhydryl group in the range of 8–10. The polarographic half-wave potentials of selenocysteine and cysteine are given as -0.212V and 0.021V, respectively.

A thorough polarographic investigation of the selenocystine-selenocysteine system and a comparison with the cystine-cysteine system have been made by Nygård,[18,19] using d.c. and a.c. polarography. Nygård[19] states that d.c. polarography revealed marked differences in behavior between selenocystine and cystine, whereas selenocysteine and cysteine were oxidized by similar mechanisms.

From comparison of their X-ray powder diffraction photographs L(+)-selenocystine and hexagonal L-cystine have been assumed to be isomorphous. The unit cell dimensions of (+)-selenocystine have been determined as $\mathbf{a} = 5.47$ Å and $\mathbf{c} = 57.91$ Å,[20a] compared with $\mathbf{a} = 5.422$ Å and $\mathbf{c} = 56.275$ Å for L-cystine.[20b]

Huber and Criddle[17] state that selenocystine is unstable in 6 M hydrochloric acid at 110°. Only 5% of the selenium compound, in contrast to 95% of the sulfur compound, survives 6 hours of treatment. Walter *et al.*[21] have confirmed Huber and Criddle's observation. It is very interesting to note that 60% of selenocystine incorporated in peptides survives under these conditions.[21]

Caldwell and Tappel[22] have studied the reactions between selenocystine or cystine and hydroperoxides. They state that the reduction of hydroperoxides is more facile and extensive with selenocystine than with cystine, and that organic hydroperoxides react more slowly than hydrogen peroxide with selenocystine. When selenocystine is treated, the reaction products are given as selenocystineseleninic acid and alanine; in the case of cystine, cystine disulfoxide, cysteinesulfinic acid, and cysteic acid are reported as reaction products.[22] It is interesting to note that, whereas cystine reacts quite rapidly with potassium cyanide to give 2-aminothiazoline-4-carboxylic acid and the potassium mercaptide of cysteine, selenocystine is unaffected.[23]

Reduction of selenocystine to selenocysteine may be accomplished electrolytically[18] or with sodium borohydride.[17,24] Reductive cleavage of the Se–Se bond was also achieved by thiols.[21] Selenocystine in acid or alkaline medium, when treated with cysteine, gives 2,7-diamino-4-thia-5-selenooctanedioic acid (**1**). Substituting 2-mercaptoethanol for cysteine gives, as

$$\underset{\underset{NH_2}{|}}{HOOC-CH}-CH_2-S-Se-CH_2-\underset{\underset{NH_2}{|}}{CH}-COOH$$

1

the reaction product, 2-amino-4-seleno-5-thia-7-hydroxyheptanoic acid (**2**),[21] whereas β-selenopropionic acid reacts with selenocystine to yield 2-

$$HO-CH_2-CH_2-S-Se-CH_2-\underset{\underset{NH_2}{|}}{CH}-COOH$$

2

amino-4,5-diselenooctanedioic acid (**3**).[21]

$$HOOC-CH_2-CH_2-Se-Se-CH_2-\underset{\underset{NH_2}{|}}{CH}-COOH$$

3

Metallic mercury in acid solution has been used by Fredga[1] to cleave the Se–Se bond. Rongalite, sodium formaldehyde sulfoxylate, in alkaline solution gives poor results, according to Zdansky.[25]

Se-Benzylselenocysteine is cleaved to selenocysteine in nearly quantitative yield by reduction with sodium in liquid ammonia,[4] but under the same conditions phenylselenocysteine gives diphenyl diselenide.

e. Selenocysteine and Derivatives. On account of its exceedingly fast uptake of oxygen selenocysteine does not appear to have been isolated in its pure state. Solutions of selenocysteine have been made by the electrolytic reduction of selenocystine by Nygård,[18] by debenzylation of *Se*-benzylselenocysteine,[23,25,27,28] and by nucleophilic displacement of *O*-tosylated serine derivatives by sodium hydrogen selenide or sodium benzyl selenolate.[14,14a]

f. Se-Alkyl and Aryl Derivatives of Selenocysteine. Fredga[1] prepared *Se*-benzylselenocysteine by first shaking selenocystine in acid solution with mercury and after neutralization treating the mixture with benzyl chloride in alcohol. Nowadays, *Se*-benzylselenocysteine is obtained as an intermediate in the synthesis of selenocystine by several routes.[4,5,9,13] Apart from being slightly light sensitive, *Se*-benzylselenocysteine has a good shelf-life and is easily debenzylated with sodium in liquid ammonia, giving a solution of selenocysteine.

Se-Methylselenocysteine has probably been synthesized by DiSomma, who, however, apparently never published the synthesis.[29,30] It was achieved by alkylating debenzylated *Se*-benzylcysteine with methyl iodide.[29–31] The amino acid is very soluble in water.[31] Spåre and Virtanen[28] have synthesized *Se*-methyl-, *Se*-propyl-, and *Se*-allylselenocysteine from debenzylated *Se*-benzylselenocysteine and the appropriate alkyl bromides and state, without giving particulars, that the amino acids were all white, crystalline compounds traveling on paper chromatograms in butanol-acetic acid exactly like their sulfur analogs. Nigam and McConnell[32] have isolated *Se*-methylselenocysteine from *Astragalus bisulcatus*.

[75]*Se*-Selenocystine has been prepared by Spencer *et al.*[32a] by irradiation for 30 hours at a flux of 1×10^{13} neutrons/cm^2/sec. The specific activity thus obtained was about 2.9 μCi/mg.

3. Selenoamino Acids of the Homocystine and the Homocysteine Type

a. Early Methods. DL-Selenohomocystine was first prepared by Painter,[34] who treated ethyl 2-benzamido-4-chlorobutyrate with either sodium hydrogen selenide (20% yield) or sodium benzyl selenolate. The second reaction gave very satisfactory yields of both *Se*-benzylselenohomocysteine (72%) and selenohomocystine (65%). However, since ethyl 2-benzamido-4-chlorobutyrate was obtained by a very tedious preparation involving a multi-step synthesis with a low overall yield, Painter searched for other methods.

Klosterman and Painter[35] then reacted 5-(2-bromoethyl)hydantoin with sodium benzyl selenolate to obtain 5-(2-benzylselenoethyl)hydantoin in about 72% yield (scheme 7). The latter compound was subsequently hydro-

(7)

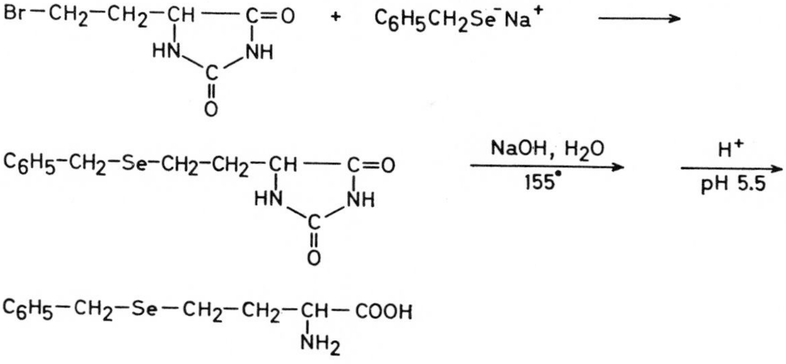

lyzed in an autoclave with aqueous NaOH to give Se-benzylselenohomocysteine in 86% yield. This was an excellent synthesis except that the overall reaction from γ-butyrolactone yielded only 17% of Se-benzylselenohomocysteine and required many steps.

Zdansky[36] used the Strecker synthesis, starting with acrolein and benzylselenol, which in a Michael addition gave 3-benzylselenopropionaldehyde. The latter was treated with hydrogen cyanide and ammonia in methanol. Subsequent acid hydrolysis produced Se-benzylselenohomocysteine in about 50% yield (scheme 8).

(8)

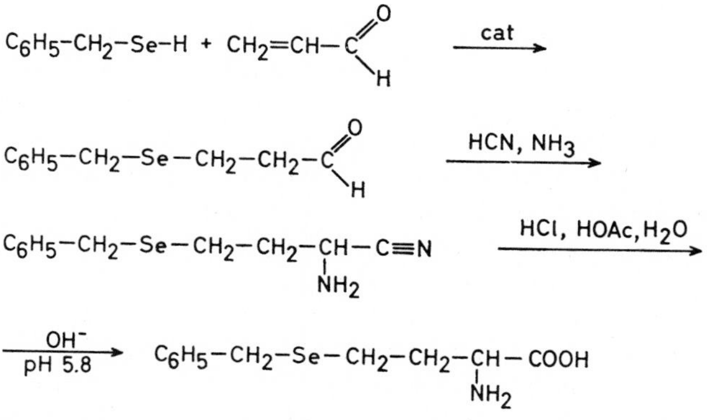

The methods mentioned so far may be classed as obsolete, being either inconvenient or disagreeable to work with.

b. Current Methods. Zdansky[37] has introduced a three-step synthesis of Se-benzylselenohomocysteine from acrolein and benzylselenol, using the

Bucherer-Bergs hydantoin synthesis (scheme 9). This method gives a 41 % overall yield and is convenient and suitable for production on a relatively

$$C_6H_5CH_2-Se-H + CH_2 = CH-C\underset{H}{\overset{O}{\diagup}} \xrightarrow{\text{cat}}$$

$$C_6H_5CH_2-Se-CH_2-CH_2-C\underset{H}{\overset{O}{\diagup}} \xrightarrow{\text{(NH}_4)_2CO_3 + NaCN}$$

(9)

$$C_6H_5CH_2-Se-CH_2-CH_2-CH\underset{HN}{\overset{\underline{}}{\big|}}\overset{}{\underset{C}{\diagdown_{NH}}}C=O \xrightarrow[108°]{OH^-,\ H_2O}$$

$$\xrightarrow[\text{pH 5.8}]{H^+} C_6H_5CH_2-Se-CH_2-CH_2-\underset{NH_2}{\underset{|}{CH}}-COOH$$

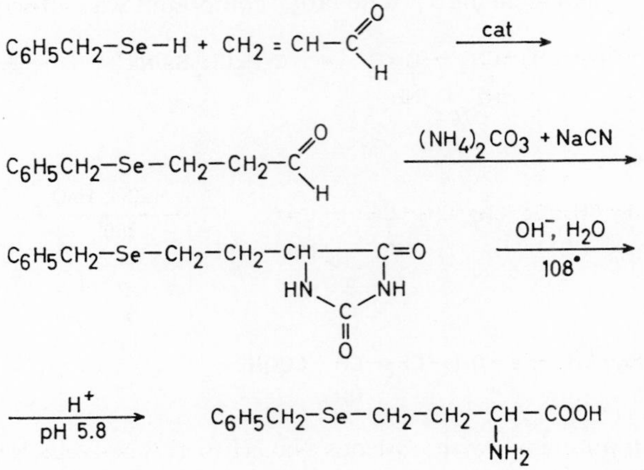

large scale.

Se-Benzylselenohomocysteine has been resolved into optical antipodes by Pan and Tarver,[38] using acylase I on *N*-acetyl-*Se*-benzyl-DL-seleno-homocysteine. They report $[\alpha]_D^{27} = +15.10°$ (L-form), $[\alpha]_D^{27} = -15.15°$ (D-form), both $C = 1$ in 1 M HCl. No yield was given.

Zdansky[37] has resolved *N*-acetyl-*Se*-benzyl-DL-selenohomocysteine, using papain, and recovered 79 % and 83 % of the L- and D-antipodes, respectively, calculated from *Se*-benzyl-DL-selenohomocysteine: $[\alpha]_D^{25} = +15.5°$ (L-form), $[\alpha]_D^{25} = -15.4°$ (D-form), both $C = 1$ in 1 M HCl.

Jakubke *et al.*[39] have synthesized *Se*-benzyl-L-selenohomocysteine by a Williamson synthesis, using sodium benzylselenolate and L-2-amino-4-bromobutyric acid hydrobromide in 79 % yield: $[\alpha]_D^{22} = +19.2°$, $C = 0.51$ in 2 M HCl.

Pande *et al.*[40] have reacted *N*-carbobenzoxy-*O*-tosyl-L-homoserine *p*-nitrobenzyl ester with sodium benzylselenolate to give *N*-carbobenzoxy-*Se*-benzyl-L-selenohomocysteine *p*-nitrobenzyl ester in 83 % yield. Selective hydrolysis of the *p*-nitrobenzyl ester group with NaOH in dioxane-water and subsequent hydrolysis of the carbobenzoxy group with HBr in glacial acetic acid gave the hydrobromide of *Se*-benzyl-L-selenohomocysteine. *Se*-Benzyl-L-selenohomocysteine as the free base was precipitated from the hydrobromide with ammonia; the overall yield was 38 % (scheme 10). This product showed $[\alpha]_D^{21} = +23.2°$, $C = 0.5$ in 2 M HCl. Pure selenohomocysteine does not appear to have been isolated because it is very easily oxidized to selenohomocystine, and it also tends to form a selenolactone. In fact, in order to obtain a smooth oxidation to selenohomocystine one

$$\text{Tos}-\text{O}-\text{CH}_2-\text{CH}_2-\text{CH}-\text{COO}-\text{CH}_2-\langle\text{NO}_2\rangle \xrightarrow{\text{C}_6\text{H}_5-\text{CH}_2\text{Se}^-} \text{C}_6\text{H}_5-\text{CH}_2-\text{Se}-\text{CH}_2-\text{CH}_2-\text{CH}-\text{COO}-\text{CH}_2-\langle\text{NO}_2\rangle$$

$$\underset{\underset{\underset{\text{C}_6\text{H}_5}{|}}{\overset{|}{\text{CH}_2}}}{\overset{|}{\underset{\text{C}=\text{O}}{\overset{|}{\text{NH}}}}} \qquad \underset{\underset{\underset{\text{C}_6\text{H}_5}{|}}{\overset{|}{\text{CH}_2}}}{\overset{|}{\underset{\text{C}=\text{O}}{\overset{|}{\text{NH}}}}}$$

(10) $\Bigg\downarrow \begin{array}{c}\text{NaOH}\\ \text{in dioxane}\end{array}$

$$\text{C}_6\text{H}_5-\text{CH}_2-\text{Se}-\text{CH}_2-\text{CH}_2-\text{CH}-\text{COOH} \xrightarrow{\begin{array}{c}\text{HBr}\\ \text{in AcOH}\end{array}} \text{C}_6\text{H}_5-\text{CH}_2-\text{Se}-\text{CH}_2-\text{CH}_2-\text{CH}-\text{COOH}$$

$$\underset{\underset{\underset{\text{C}_6\text{H}_5}{|}}{\overset{|}{\text{CH}_2}}}{\overset{|}{\underset{\text{C}=\text{O}}{\overset{|}{\text{NH}}}}} \qquad\qquad\qquad\qquad\qquad \overset{|}{\underset{\text{NH}_3^+\text{Br}^-}{}}$$

$$\Bigg\downarrow \begin{array}{c}\text{NH}_4\text{OH}\\ \text{to pH 5.5}\end{array}$$

$$\text{C}_6\text{H}_5-\text{CH}_2-\text{Se}-\text{CH}_2-\text{CH}_2-\text{CH}-\text{COOH}$$

$$\overset{|}{\underset{\text{NH}_2}{}}$$

must take care not to get the solution acidic after debenzylation before and during the oxidation stage.[41]

Acid hydrolysis of *N*-acyl esters and anilide groups must also be performed with care.[37] Hydrolysis in dilute hydrochloric acid containing some 20% ethanol in order to enhance the solubility of, for example, *N*-acetyl *Se*-benzyl-L-selenohomocysteine anilide, results in realkylation and cleavage of the bond between selenium and carbon atom 4.[37] The realkylation takes place with methionine and is used (*e.g.*, by McRorie *et al.*[42]) to prepare *S*-benzylhomocysteine from methionine. Pande *et al.*[40] noted that the specific optical rotation of *Se*-benzylselenohomocysteine in acid solution is extremely temperature dependent. They state, furthermore, that the optical rotation value rapidly diminishes in acid solution.

4. Branched Selenohomocystines and *Se*-Benzylselenohomocysteines

A series of methyl-substituted *Se*-benzylselenohomocysteines (**4**, **5**, **6**) bearing methyl groups on carbon atoms 2, 3, or 4 has been synthesized by

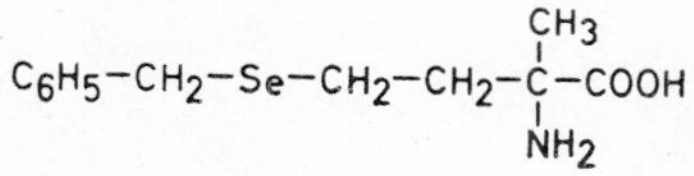

$$C_6H_5-CH_2-Se-CH_2-CH_2-\underset{\underset{NH_2}{|}}{\overset{\overset{CH_3}{|}}{C}}-COOH$$

$$C_6H_5-CH_2-Se-CH_2-\overset{\overset{CH_3}{|}}{C}H-\underset{\underset{NH_2}{|}}{C}H-COOH$$

4 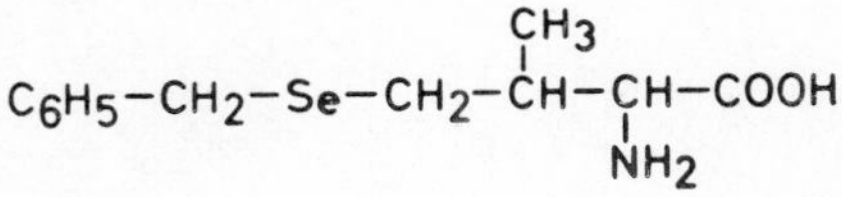   5

$$C_6H_5-CH_2-Se-\overset{\overset{CH_3}{|}}{C}H-CH_2-\underset{\underset{NH_2}{|}}{C}H-COOH$$

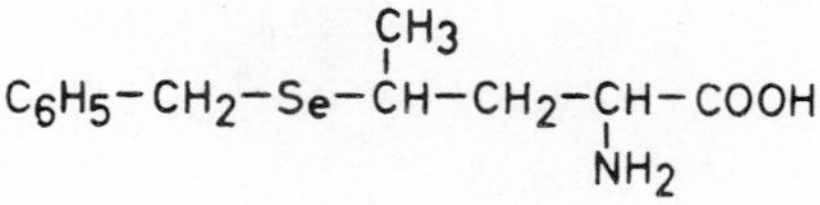

6

Zdansky.[41] Michael adducts of benzylselenol and unsaturated aldehydes or ketones served as intermediates for a Strecker synthesis with HCN and NH$_3$ under pressure, followed by acid hydrolysis (compare with scheme 8). The overall yields ranged from 24–52%. The sulfur analogs have also been prepared by means of the same technique.[41] However, attempts to synthesize 3,4-dimethyl-*Se*-benzylselenohomocysteine and 2,4,4-trimethyl-*Se*-benzyl-selenohomocysteine using the Strecker technique failed.[33]

Zdansky[33,37,43] has shown that the Bucherer-Bergs hydantoin synthesis is eminently suited for the production of selenoamino acids (scheme 9) and that it has fewer limitations than the Strecker synthesis. In addition to the monomethyl *Se*-benzylhomocysteines already mentioned, a few di- and trimethylselenohomocysteines with the methyl groups at C(2), C(3), or C(4) have been prepared in good yields using this technique.[43] All corresponding sulfur analogs have also been synthesized by means of the same technique and have proved to be strikingly similar to the selenium compounds.[33,37,43]

Attempts at the resolution by means of papain of the *N*-acetyl or *N*-formyl compounds of 2-methyl-*Se*-benzylhomocysteine and the sulfur isologs have failed.[33]

2,2'-, 3,3'-, and 4,4'-Dimethylselenohomocystines (**7, 8, 9**) and the corresponding sulfur isologs have been prepared by debenzylation with sodium in liquid ammonia, followed by oxidation with air in the presence of Fe^{3+} ions of the appropriate benzylselenohomocysteines and benzylhomocysteines, respectively.[41]

$$
\begin{array}{l}
\text{Se}-\text{CH}_2-\text{CH}_2-\overset{\overset{\text{CH}_3}{|}}{\text{C}}-\text{COOH} \\
\qquad\qquad\qquad \underset{|}{\text{NH}_2} \\
\text{Se}-\text{CH}_2-\text{CH}_2-\overset{\overset{|}{\text{NH}_2}}{\underset{|}{\text{C}}}-\text{COOH} \\
\qquad\qquad\qquad \text{CH}_3
\end{array}
$$

7

$$
\begin{array}{l}
\text{Se}-\text{CH}_2-\overset{\overset{\text{CH}_3}{|}}{\text{CH}}-\text{CH}-\text{COOH} \\
\qquad\qquad\qquad \underset{|}{\text{NH}_2} \\
\text{Se}-\text{CH}_2-\text{CH}-\overset{\overset{|}{\text{NH}_2}}{\underset{}{\text{CH}}}-\text{COOH} \\
\qquad\qquad\qquad \text{CH}_3
\end{array}
$$

8

$$
\begin{array}{l}
\text{Se}-\overset{\overset{\text{CH}_3}{|}}{\text{CH}}-\text{CH}_2-\text{CH}-\text{COOH} \\
\qquad\qquad\qquad\quad \underset{|}{\text{NH}_2} \\
\text{Se}-\text{CH}-\text{CH}_2-\overset{\overset{|}{\text{NH}_2}}{\underset{}{\text{CH}}}-\text{COOH} \\
\qquad \text{CH}_3
\end{array}
$$

9

By use of the hydantoin synthesis Zdansky[43] has also prepared some dimethyl- and trimethyl-substituted *Se*-benzylselenohomocysteines and their sulfur analogs, such as *Se*-benzyl-3,4-dimethylselenohomocysteine (**10**) and *Se*-benzyl-2,4,4-trimethylselenohomocysteine (**11**).

$$
\begin{array}{l}
\text{C}_6\text{H}_5-\text{CH}_2-\text{Se}-\overset{\overset{}{\underset{\underset{\text{CH}_3}{|}}{|}}}{\text{CH}}-\overset{\overset{}{\underset{\underset{\text{CH}_3}{|}}{|}}}{\text{CH}}-\overset{\overset{}{\underset{\underset{\text{NH}_2}{|}}{|}}}{\text{CH}}-\text{COOH}
\end{array}
$$

10

$$
\begin{array}{l}
\text{C}_6\text{H}_5-\text{CH}_2-\text{Se}-\overset{\overset{\text{CH}_3}{|}}{\underset{\underset{\text{CH}_3}{|}}{\text{C}}}-\text{CH}_2-\overset{\overset{\text{CH}_3}{|}}{\underset{\underset{\text{NH}_2}{|}}{\text{C}}}-\text{COOH}
\end{array}
$$

11

5. Selenoamino Acids of the Methionine Type

a. Selenomethionine. DL-Selenomethionine (**12**) was first prepared by Painter,[34] who reduced DL-selenohomocystine with sodium in liquid ammonia and alkylated the resulting sodium selenohomocysteinate with methyl iodide. Painter also methylated the debenzylation product of *Se*-benzyl-DL-selenohomocysteine in the same way.[34]

$$CH_3-Se-CH_2-CH_2-\underset{\underset{NH_2}{|}}{CH}-COOH$$

12

An interesting synthesis of DL-selenomethionine was published by Plieninger.[44] Starting with butyrolactone, he prepared α-aminobutyro-lactone, which was subsequently heated in toluene with sodium methyl selenide in an autoclave at 160°. In the latter reaction an addition product is first formed which can be cleaved with water. On prolonged heating at 160° this product rearranges to give sodium 2-amino-4-methyl selenobuty-rate, that is, the sodium salt of selenomethionine (scheme 11).[44]

$$(11)$$

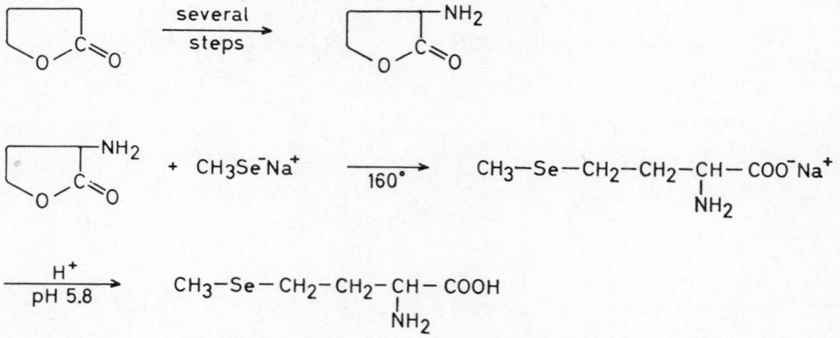

$$\xrightarrow[pH\ 5.8]{H^+} \quad CH_3-Se-CH_2-CH_2-\underset{\underset{NH_2}{|}}{CH}-COOH$$

By using sodium ethyl selenide instead of sodium methyl selenide Plieninger also prepared DL-selenoethionine (**13**).[44]

$$CH_3-CH_2-Se-CH_2-CH_2-\underset{\underset{NH_2}{|}}{CH}-COOH$$

13

Plieninger's synthesis is unique in being the sole example of a direct approach to selenomethionine. The method has, however, several severe drawbacks. The preparation of 2-aminobutyrolactone is rather difficult; furthermore, the product is not stable, being transformed rather rapidly into a piperazine. The use of the very evil-smelling methaneselenol (or ethaneselenol, in the case of selenoethionine) is also a nuisance, and these compounds do not seem to be replaceable by benzylselenol in this reaction.[45]

Currently used methods for the preparation of selenomethionine and selenoethionine start either with *Se*-benzylselenohomocysteine, prepared by one of the methods reviewed in Section XII.A.4, or with *O*-tosylated homo-serine derivatives.

DL-Selenomethionine has been prepared from Se-benzyl-DL-selenohomo-cysteine in 84% yield by Zdansky[45] by the direct methylation with methyl iodide of sodium selenohomocysteinate in liquid ammonia. The same technique has been used by Pan et al.[46] for the preparation of DL-seleno-methionine-methyl-^{14}C and DL-selenoethionine-ethyl-1-^{14}C by alkylation with methyl iodide-^{14}C and ethyl iodide-1-^{14}C, respectively. Pan and Tarver[38] have prepared L- and D-selenomethionine by the method outlined above, starting with optically active Se-benzylselenohomocysteine, and report, for L-selenomethionine, $[\alpha]_D^{27} = +17.8°$ and, for D-selenomethionine, $[\alpha]_D^{27} = -17.2°$, both $C = 1$ in 1 M HCl. They have also prepared L-selenoethionine, $[\alpha]_D^{27} = +23.6°$, $C = 2$ in 2 M HCl, from N-acetyl-DL-selenoethionine, using acylase I.

Jakubke et al.[39] have, by this method, also prepared L-selenomethionine and L-selenoethionine: $[\alpha]_D^{22} = +17.5°$, $C = 0.5$ in 2 M HCl, and $[\alpha]_D^{22} = +15.9°$, $C = 0.5$ in 2 M HCl, respectively.

L-Selenomethionine and D-selenomethionine, prepared in a similar manner, have been resolved, using papain by Zdansky,[37] $[\alpha]_D^{25} = +18.1°$, $C = 1.2$ in 1 M HCl, for the L-form, and $[\alpha]_D^{25} = -18.3°$, $C = 1$ in 1 M HCl for the D-form.

Alkylation of derivatives of O-tosylated homoserine with sodium methyl selenide or with sodium ethyl selenide has been successfully used by Pande et al.[40] for the preparation of L-selenomethionine and L-selenoethionine, respectively. The optical rotations that they give are the highest recorded in the literature: $[\alpha]_D^{22} = +21.6°$, $C = 0.5$ in 2 M HCl, for L-selenomethionine, and $[\alpha]_D^{22} = +21.5°$, $C = 0.5$ in 2 M HCl, for L-selenoethionine.

b. Methylselenomethionines. The three methylselenomethionines (**14, 15, 16**) with a methyl group on either C(2), C(3), or C(4) have been prepared

$$CH_3-Se-CH_2-CH_2-\underset{\underset{NH_2}{|}}{\overset{\overset{CH_3}{|}}{C}}-COOH$$

14

$$CH_3-Se-CH_2-\underset{\underset{NH_2}{|}}{CH}-\overset{\overset{CH_3}{|}}{CH}-COOH$$

15

$$CH_3-Se-\overset{\overset{CH_3}{|}}{CH}-CH_2-\underset{\underset{NH_2}{|}}{CH}-COOH$$

16

by Zdansky[47] from the corresponding Se-benzylselenomethylhomocyst-eines[45,47] by conventional debenzylation and methylation with methyl iodide. The sulfur analogs have also been prepared.[47]

c. Properties of Selenomethionines. Shepherd and Huber[48] have investigated some chemical and biochemical properties of selenomethionine. They report a solubility in water at 30° and pH 7 of 0.386 M and 0.108 M for L-methionine and DL-selenomethionine, respectively. Stability in 6 N HCl at 110° under anaerobic conditions has been investigated; it was revealed that after 7 hours of hydrolysis the selenomethionine had completely decomposed, but that 96% of the methionine remained. The fate of the selenomethionine during this hydrolysis is uncertain; the decomposition products elute relatively early from the amino acid analyzer column, using the buffer system of Spackman *et al.*[49]

With cyanogen bromide, selenomethionine in 0.1 M HCl solution reacts considerably faster than methionine, the end product in either case being homoserine. After 15 minutes the selenomethionine had completely reacted, whereas 24 hours was required in the case of methionine. Furthermore Shepherd and Huber[48] report that the difference in reactivity between selenomethionine and methionine toward hydrogen peroxide is lower than in the case of cyanogen bromide, but that selenomethionine is still the more reactive compound. Caldwell and Tappel[22] have not drawn any definite conclusions from their experiments on the same subject.

With methyl iodide, selenomethionine forms a methylselenonium compound which has been prepared by Virupaksha *et al.*,[50] using the method of Toennis and Kolb.[51] They give, however, no particulars concerning the product.

d. Labeled Selenomethionine and Related Amino Acids. DL-Selenomethionine-methyl-^{14}C and DL-selenoethionine-ethyl-1-^{14}C have been prepared by Pan *et al.*[46] by the alkylation of DL-selenohomocysteine with the appropriate ^{14}C-alkyl iodide.

Randomly tritiated selenomethionine has been synthesized by Bremer and Natori,[52] using the Wilzbach[53] method. ^{75}Se-Labeled selenomethionine has been prepared by McConnell and Wabnitz,[54] using the neutron activation procedure of McConnell *et al.*[55]

6. Selenoamino Acids Containing Carbon-Selenium-Carbon Bonds

Several amino acids containing thioether linkages are known to occur in nature, for example, lanthionine, cystathionine, and djenkolic acid. β-Methyllanthionine has been found in cinnamycin.

Only one amino acid containing selenoether linkages appears to have been found in nature: selenocystathionine (**17**), isolated by Kerdel-Vegas *et al.*[56]

$$HOOC-\underset{\underset{NH_2}{|}}{CH}-CH_2-Se-CH_2-CH_2-\underset{\underset{NH_2}{|}}{CH}-COOH$$

17

from the nuts of coco de mono (*Lecythis ollaria*). Several amino acids containing selenoether linkages have been synthesized, however, and will be reviewed in this section.

a. Selenolanthionine. Selenolanthionine (**18**), in an optically inactive mixture of DL- and *meso*-forms, was first synthesized by Zdansky,[27] who

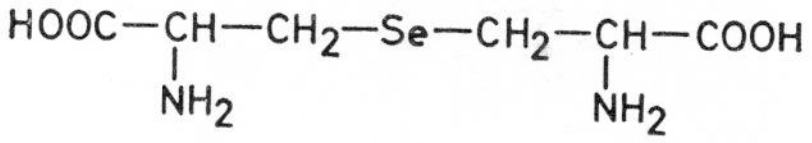

$$\underset{\underset{18}{}}{HOOC-\underset{\underset{NH_2}{|}}{CH}-CH_2-Se-CH_2-\underset{\underset{NH_2}{|}}{CH}-COOH}$$

added DL-selenocysteine to the activated double bond in ethyl N-acetyl aminoacrylate to yield the N-monoacetylated half-ester of selenolanthionine. This intermediate was not isolated in pure form but was immediately hydrolyzed with dilute hydrochloric acid to give optically inactive selenolanthionine in 78% yield (scheme 12).

(12)

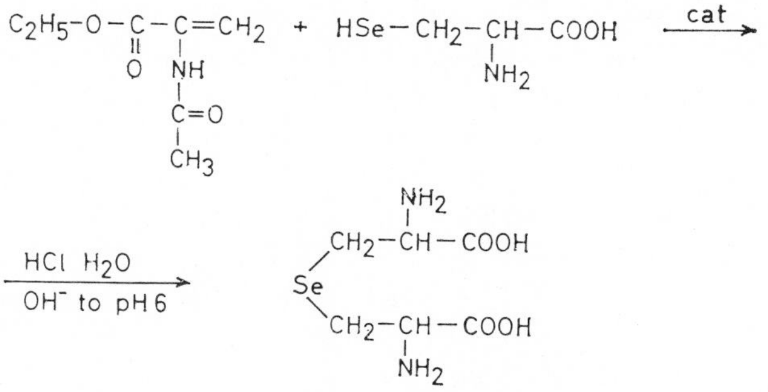

Lanthionine has proved to be difficult to resolve, either by chromatographic methods[57] or enzymatically.[58,59] Considering this fact, Zdansky[11] prepared L-, D- and *meso*-selenolanthionine from L- and D-3-chloroalanine and L- and D-selenocysteine, derived by resolution of DL-Se-benzylselenocysteine and debenzylation with sodium in liquid ammonia. L- and D-Selenolanthionine were obtained in about 80% yield, with $[\alpha]_D^{25} = +34.8°$, $C = 1$ in 1 M HCl, for the L-form, and $[\alpha]_D^{25} = -34.8°$, $C = 1$ in 1 M HCl, for the D-form. The solubility of the *meso* form seems to be considerably higher than that of either antipode or of the racemic mixture.[11]

Later, Roy et al.[14] prepared L-selenolanthionine by reacting the sodium salt of N-carbobenzoxy-L-selenocysteine diphenyl methyl ester, obtained from the reaction between N-carbobenzoxy-O-tosyl-L-serine diphenyl methyl ester and sodium hydrogen selenide, with N-carbobenzoxy-O-tosyl-L-serine diphenyl methyl ester and subsequently removing the protecting groups from the protected L-selenolanthionine: $[\alpha]_D^{21} = +34.9°$, $C = 1$ in 5 M HCl.

b. Selenocystathionine. L-Selenocystathionine has been isolated from natural sources by Kerdel-Vegas *et al.*[56] and Shrift and Virupaksha.[30]

COOH	COOH	COOH	COOH
H₂N−C−H	H−C−NH₂	H−C−NH₂	H₂N−C−H
CH₂	CH₂	CH₂	CH₂
CH₂	CH₂	CH₂	CH₂
Se	Se	Se	Se
CH₂	CH₂	CH₂	CH₂
H−C−NH₂	H₂N−CH	H−C−NH₂	H₂N−C−H
COOH	COOH	COOH	COOH
L (+)	D (−)	L (−) allo	D (+) allo

L(+)- and D(−)-Selenocystathionines and L(−)allo- and D(+)-alloselenocystathionines was first prepared by Zdansky[24] using the same technique employed for the synthesis of optically active selenolanthionine. From L-*Se*-benzylselenohomocysteine and L-3-chloroalanine, L(+)-selenocystathionine was prepared: $[\alpha]_D^{25} = +36.1°$, $C = 1$ in 1 M HCl. D-*Se*-Benzylselenohomocysteine and D-3-chloroalanine yielded D(−)-selenocystathionine: $[\alpha]_D^{25} = -36.2°$, $C = 1$ in 1 M HCl.

L(−)-Alloselenocystathionine was prepared from D-*Se*-benzylselenohomocysteine and L-3-chloroalanine: $[\alpha]_D^{25} = -3.1°$, $C = 1$ in 1 M HCl. D(+)-Alloselenocystathionine was prepared from L-*Se*-benzylselenohomocysteine and D-3-chloroalanine: $[\alpha]_D^{25} = +3.2°$, $C = 1$ in 1 M HCl.

L(+)-Selenocystathionine has also been prepared by Pande *et al.*[40] by reacting the sodium salt of *N*-carbobenzoxy-L-selenocysteine diphenylmethyl ester with *N*-carbobenzoxy-*O*-tosyl-L-homoserine *p*-nitrobenzyl ester and subsequently removing the protecting groups from the protected L-selenocystathionine.

7. Miscellaneous Selenoamino Acids

A djenkolic acid analog containing both sulfur and selenium, 2,10-diamino-4,8-diselena-6-thiaundecanedioic acid (**19**), was made by alkylating dichlorodimethyl sulfide with sodium selenocysteinate.[23] The corresponding all-sulfur compound, 2,10-diamino-4,6,8-trithiaundecanedioic acid, has been prepared by substituting L-cysteine for selenocysteine. Both amino acids resemble djenkolic acid and cystine in being poorly soluble in water.

$$HOOC-CH-CH_2-Se-CH_2-S-CH_2-Se-CH_2-CH-COOH$$
$$\quad\ \ \ \ NH_2 \qquad\qquad\qquad\qquad\qquad\qquad\qquad NH_2$$

19

Walter *et al.*[15,21] have demonstrated chromatographically the formation of 2,7-diamino-4-thia-5-selenaoctanedioic acid (**1**) by interaction of selenocystine and cysteine in dilute hydrochloric acid or sodium hydroxide. Under the same conditions 2-mercaptoethanol and selenocystine give rise to the formation of 2-amino-4-selena-5-thia-7-hydroxyheptanoic acid (**2**). β-Selenopropionic acid, on reaction with selenocystine, yields 2-amino-4,5-diselenaoctanedioic acid (**3**). These compounds seem, however, not to have been prepared *in corpore*, but were only detected chromatographically.

REFERENCES

1. A. Fredga, *Svensk Kem. Tidskr.*, **48,** 160 (1936).
2. A. Fredga, *Svensk Kem. Tidskr.*, **49,** 124 (1937).
3. E. Fischer and K. Raske, *Ber.*, **41,** 893 (1908).
4. E. P. Painter, *J. Amer. Chem. Soc.*, **69,** 229 (1947).
5. L. R. Williams and A. Ravve, *J. Amer. Chem. Soc.*, **70,** 1244 (1948).
6. J. L. Wood and V. du Vigneaud, *J. Biol. Chem.*, **131,** 267 (1939).
7. J. Janicki, J. Skopin, and B. Zagalak, *Rocz. Chem.*, **36,** 353 (1962).
8. W. Frank, *Z. Physiol. Chem.*, **339,** 202 (1964).
9. G. Zdansky, *Ark. Kemi*, **17,** 273 (1961).
10. S.-H. Chu, W. H. H. Günther, and H. G. Mautner, "Biochemical Preparations," Vol. 10, G. B. Brown, Ed., John Wiley, New York, N.Y., 1963, p. 153.
11. G. Zdansky, *Ark. Kemi*, **29,** 443 (1968).
12. D. Theodoropoulos, I. L. Schwartz, and R. Walter, *Tetrahedron Lett.*, **25,** 2411 (1967).
13. D. Theodoropoulos, I. L. Schwartz, and R. Walter, *Biochemistry*, **6,** 3927 (1967).
14. J. Roy, W. Gordon, and R. Walter, *J. Org. Chem.*, **35,** 510 (1970).
14a. R. Walter, "Peptides: Chemistry and Biochemistry," B. Weinstein, Ed., Marcel Dekker, New York, N.Y., 1970, p. 467.
15. R. Walter and W. Y. Chan, *J. Amer. Chem. Soc.*, **89,** 3892 (1967).
16. G. Zdansky, unpublished results.
17. R. E. Huber and R. S. Criddle, *Arch. Biochem. Biophys.*, **122,** 164 (1967).
18. B. Nygård, *Ark. Kemi*, **27,** 341 (1967).
19. B. Nygård, *Dissertation Acta Univ. Upsaliensis*, No. 104 (1967).
20a. Dr. K.-I. Aldén, present address: Lantbrukshögskolan, S-750 07 Uppsala 7, Sweden, private communication.
20b. B. M. Oughton and P. M. Harrison, *Acta Cryst.*, **12,** 396 (1959).
21. R. Walter, D. H. Schlesinger, and I. L. Schwartz, *Analyt. Biochem.*, **27,** 231 (1969).
22. K. A. Caldwell and A. L. Tappel, *Biochemistry*, **3,** 1643 (1964).
23. G. Zdansky, *Ark. Kemi*, **17,** 519 (1961).
24. G. Zdansky, *Ark. Kemi*, **29,** 449 (1968).
25. G. Zdansky, *Ark. Kemi*, **29,** 443 (1968).
26. W. Frank, *Z. Physiol. Chem.*, **339,** 214 (1964).
27. G. Zdansky, *Ark. Kemi*, **26,** 213 (1966).

28. C. G. Spåre and A. I. Virtanen, *Acta Chem. Scand.*, **18,** 280 (1964).

29. A. Shrift, personal communication.

30. Reference is made, for example, by A. Shrift and T. K. Virupaksha, *Biochim. Biophys. Acta*, **100,** 65 (1965) to Di Somma for a gift of sample.

31. G. Zdansky, unpublished results.

32. S. N. Nigam and W. B. McConnell, *Biochim. Biophys. Acta*, **192,** 185 (1969).

32a. R. P. Spencer, K. R. Brody, W. H. H. Günther, and H. G. Mautner, *J. Chromatog.*, **21,** 343 (1966).

33. G. Zdansky. *Dissertation Acta Univ. Upsaliensis*, No. 123 (1968).

34. E. P. Painter, *J. Amer. Chem. Soc.*, **69,** 232 (1947).

35. H. J. Klosterman and E. P. Painter, *J. Amer. Chem. Soc.*, **69,** 2009 (1947).

36. G. Zdansky, *Ark. Kemi*, **19,** 559 (1962).

37. G. Zdansky, *Ark. Kemi*, **29,** 437 (1968).

38. F. Pan and H. Tarver, *Arch. Biochem. Biophys.*, **119,** 429 (1967).

39. H.-D. Jakubke, J. Fischer, K. Jošt, and J. Rudinger, *Collection Czech. Chem. Commun.*, **33,** 3910 (1968).

40. C. S. Pande, J. Rudick, and R. Walter, *J. Org. Chem.*, **35,** 1440 (1970).

41. G. Zdansky, *Ark. Kemi*, **21,** 211 (1963).

42. R. A. McRorie, G. L. Sutherland, M. S. Lewis, A. D. Barton, M. R. Glazener, and W. Shive, *J. Amer. Chem. Soc.*, **76,** 115 (1954).

43. G. Zdansky, *Ark. Kemi*, **29,** 47 (1968).

44. H. Plieninger, *Chem. Ber.*, **83,** 265 (1950).

45. G. Zdansky, *Ark. Kemi*, **19,** 559 (1962).

46. F. Pan, Y. Natori, and H. Tarver, *Biochim. Biophys. Acta*, **93,** 521 (1964).

47. G. Zdansky, *Ark. Kemi*, **27,** 447 (1967).

48. L. Shepherd and R. E. Huber, *Can. J. Biochem.*, **47,** 877 (1969).

49. D. H. Spackman, W. H. Stein, and S. Moore, *Anal. Chem.*, **30,** 1190 (1958).

50. T. K. Virupaksha, A. Shrift, and H. Tarver, *Biochim. Biophys. Acta*, **130,** 45 (1966).

51. G. Toennis and J. J. Kolb, *J. Amer. Chem. Soc.*, **67,** 849 (1945).

52. J. Bremer and Y. Natori, *Biochim. Biophys. Acta*, **44,** 367 (1960).

53. K. E. Wilzbach, *J. Amer. Chem. Soc.*, **79,** 1013 (1957).

54. K. P. McConnell and C. H. Wabnitz, *Biochim. Biophys. Acta*, **86,** 182 (1964).

55. K. P. McConnell, H. G. Mautner, and G. W. Leddicotte, *Biochim. Biophys. Acta*, **59,** 217 (1962).

56. F. Kerdel-Vegas, F. Wagner, P. B. Russell, N. H. Grant, H. E. Alburn, D. E. Clark, and J. A. Miller, *Nature*, **205,** 1186 (1965).

57. S. Blackburn and G. R. Lee, *Analyst*, **80,** 875 (1955).

58. D. McHale, P. Mamalis, and J. Green, *J. Chem. Soc.*, 2847 (1960).

59. G. Stening, Unpublished Undergraduate Thesis, Department of Organic Chemistry, University of Uppsala, Uppsala, Sweden.

XII Selenium amino acids and peptides

B. SELENIUM-CONTAINING PEPTIDES AND RELATED AMINO ACIDS

RODERICH WALTER and J. ROY

The Mount Sinai Medical and Graduate Schools
of the City University of New York
Department of Physiology
New York, New York

1. Introduction*

Selenopeptides and selenoproteins were undiscovered territory a decade ago. The chemical basis for physical, physiochemical, and physiological studies was lacking at that time, since chemists had provided neither adequate isolation techniques for selenium-containing peptides nor convenient

---

* Abbreviations used are in accordance with IUPAC-IUB Tentative rules [*Biochem.*, **5,** 2485 (1966)]. Some of the abbreviations are the following: Z = benzyloxycarbonyl, Tos = *p*-toluenesulfonyl, Bzl = benzyl, Bzh = diphenylmethyl, Boc = *tert*-butyloxycarbonyl, β-Mpr = β-mercaptopropionyl, Np = *p*-nitrophenyl, Nps = *o*-nitrophenylsulfenyl, Su = 1-succinimidyl, DCCI = *N,N'*-dicyclohexylcarbodiimide, AcOH = acetic acid, TFA = trifluoroacetic acid.

synthetic methods for the preparation of selenoamino acids and peptides in optically pure form. With the resolution of both of these problems the foundation was laid upon which to further our knowledge of the chemical, physical, and biological properties of selenium-containing peptides. The elucidation of many biomedical phenomena—several of practical importance—may be anticipated through an interdisciplinary approach involving the close collaboration of experts from various fields in the basic sciences and the medical community.

This chapter is divided into two parts. The first deals with selective detection and isolation techniques for selenium-containing amino acids and peptides, while the second discusses general aspects concerning their synthesis. In the latter, synthetic selenium-containing analogs of the neurohypophyseal hormone oxytocin serve to illustrate the application of selenium as a probe for biological and conformational studies. Moreover, the absolute configuration of the C–S–Se–C and C–Se–Se–C bonds in selenium-containing neurohypophyseal peptides is assigned, and the approach used should be applicable to other thioselenenates and diselenides. General comments regarding the chemical behavior, reactivity, and stability of selenium-containing amino acids and peptides are interspersed throughout.

We hope to convey in this chapter the excitement currently felt by students of selenium. We also have tried to highlight past achievements and to illuminate important problems awaiting solution.

2. Detection and Separation Techniques

Major obstacles are encountered in attempting to separate a selenium-containing amino acid or peptide from its corresponding sulfur analog. The paper chromatographic and electrophoretic behaviors of selenocystine, Se-methylselenocysteine, and selenomethionine were found to be indistinguishable from those of the corresponding sulfur amino acids.[1] Analogs of oxytocin (e.g., 6-selenooxytocin) exhibited the same migration rate upon thin-layer chromatography,[2] gel filtration,[2] or paper chromatography[3] as the neurohypophyseal hormone. Seen in this light, it is readily understood why the isolation from natural sources of selenium-containing amino acids and peptides has in the past met with serious difficulties. The fact that the sulfur analog is usually present as the overwhelming component magnifies the problem of separation. Thus, Horn and Jones[4] were unable to isolate selenocystathionine free of cystathionine from leaves of *Astragalus pectinatus*, nor were early attempts to recover Se-methylselenocysteine free of S-methylcysteine successful.[5]

Fortunately, in recent years the development of methods directed at the selective detection and effective separation of selenium- and sulfur-containing amino acids and peptides has drawn increasing attention.

a. Detection. Methods, such as spot tests or colorimetric reagents, capable of selectively revealing an organoselenium moiety in the presence of its sulfur analog are highly desirable. Methods approaching the aim of preferential detection involve either colorimetric detection of selenium-containing products after oxidation on a thin-layer support and selective reduction[5a] or conversion of [76]Se by neutron activation to the γ-emitter [77m]Se. Using the latter method, Barak and Swanberg,[6] by combining paper chromatography and neutron activation analysis, quantitatively determined selenomethionine in the presence of methionine.

A more rewarding approach to date for the selective identification of selenium-containing amino acids entails their preferential modification, that is, the sulfur analog remains unchanged. The desired product is subsequently separated from the reaction mixture by conventional techniques. Two such approaches have been documented. In the first, Scala and Williams[7] exposed on paper a mixture of selenomethionine and methionine or selenocystine and cystine to hydrogen peroxide, and separated the mixtures of reaction products by paper chromatography. It was suggested[7] that the hydrogen peroxide had oxidized the selenomethionine and selenocystine *selectively* to the selenoxides; however, Shepherd and Huber[8] later showed that this oxidation method is not highly specific. Hence, peroxide treatment is not applicable for the selective identification of selenomethionine. However, in the case of selenocystine the selenoxide formed by treatment with hydrogen peroxide undergoes a rapid β-elimination of the oxidation product, resulting in the quantitative formation of dehydroalanine.[9] Therefore, the very fact of the difference in reactivity of cystine and selenocystine and their oxides renders the peroxide reaction a useful method for the identification of selenocystine.

The second approach applies the well-known conversion of methionine to homoserine by treatment with cyanogen bromide.[10] Shepherd and Huber[8] found that within the first hour selenomethionine is transformed to the homoserine lactone, while no conversion of methionine is detected.

The few methods discussed above are applicable to the detection of selenium-containing amino acids, and there is good reason to believe that they will be applicable to selenium-containing peptides as well. Several physicochemical methods, including mass spectrometry and i.r., u.v., and n.m.r. spectroscopy, have the potential to differentiate between sulfur and selenium analogs but have not yet been applied to biological problems.

b. Separation. The separation of selenium-containing amino acids and peptides from their sulfur-containing counterparts requires techniques with a resolution capability greater than that offered by paper, thin-layer, of column chromatography or gel filtration, although the separation or methionine and selenomethionine by t.l.c. has been reported.[11] Therefore, ion-exchange chromatography, countercurrent distribution, partition chromatography, and electrophoresis are the methods of choice. Some of these techniques have already been successfully applied. McConnell and Wabnitz[12] reported that on sulfonated styrene-divinylbenzene copolymer resins selenocystine and selenomethionine possess elution rates different from those of cystine and methionine. Independently, Walter and du Vigneaud,[2] while studying the biological effects of replacing the S atom by Se in the neurohypophyseal hormone oxytocin, discovered that selenocystine was separable

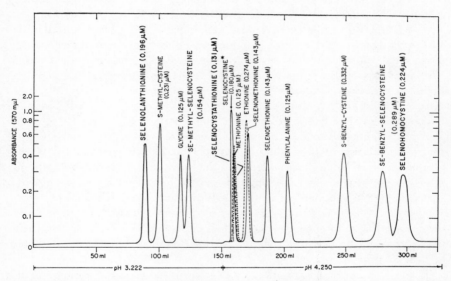

FIG. XIIB-1   Composite of the chromatographic separation of selenium-containing amino acids from their corresponding sulfur analogs. Buffers contained thiodiglycol reagent, except for the analysis of selenocystine and selenohomocystine. Glycine and phenylalanine were added as position markers.

from its corresponding sulfur analog by ion-exchange chromatography (see also Walter et al.[13]). These findings were confirmed and extended by Martin and Cummins.[14] Since then it has been found that ion-exchange chromatography generally separates selenium-containing amino acids from their corresponding sulfur analogs, as illustrated by investigations involving selenocystine, *Se*-methylselenocysteine, *Se*-benzylselenocysteine, selenohomocystine, selenomethionine, selenoethionine, selenolanthionine, and selenocystathionine.[13,15,16] These data are summarized in composite Fig. XIIB-1 and Table XIIB-1, which lists the volume of peak emergence of each selenoamino acid after the start of the chromatogram (as well as the color value and 440/570 nm ratio for each selenoamino acid-ninhydrin reaction). While it is usually advisable to perform amino acid analyses with buffers containing thiodiglycol in order to prevent oxidation of methionine,[17]

Tabel XIIB-1   Elution volumes, ratios calculated from ninhydrin absorption at 440 nm and at 570 nm, and color constants of selenium-containing amino acids[a]

AMINO ACID[b]	PEAK EMERGENCE[c]	RATIO OF ABSORBANCE AT 440/570 nm	C VALUE[d]
Selenolanthionine[e]	88.5	0.174	127.8
Se-Methylselenocysteine	121.1	0.139	123.0
$\frac{1}{2}$-Selenocystine	156.8	0.170	89.8
Selenocystathionine[f]	159.2	0.134	114.0
Selenomethionine	169.9	0.143	133.5
Selenoethionine	185.7	0.145	133.2
Se-Benzylselenocysteine	281.8	0.153	115.1
$\frac{1}{2}$-Selenohomocystine[f]	297.0	0.154	98.6

[a] Unless stated otherwise, values have been taken from Walter et al.[13] This reference also gives experimental conditions, such as column size and flow rate.

[b] At least three determinations were made for each value listed. Glycine (peak emergence 105.6 ml) and phenylalanine (peak emergence 203.1 ml) were added as position markers.

[c] Peak emergence is expressed in milliliters after start of chromatogram.

[d] Color value calculated according to $C = HW/\mu M$, where H equals height of peak and W equals width of peak at half-height.

[e] Elution volume reported by Roy et al.[46]

[f] Values taken from Schlesinger et al.[16]

difficulties are encountered in the case of diselenides. As shown by Walter et al.,[13] 2-mercaptoethanol, which is present as an impurity in thiodiglycol, gives rise to sulfhydryl-diselenide interchange; in the case of selenocystine, the thioselenenate formed was 2-amino-4-selena-5-thia-7-hydroxyheptanoic acid. Recently we found that results analogous to those with selenocystine are also obtained with selenohomocystine. In the absence of mercapto-ethanol, selenohomocystine emerges as a single peak after 297 ml of citrate buffer. In the presence of mercaptoethanol an increase of the 570 nm absorption occurs at an elution volume of 238 ml of citrate buffer, and the absorption reaches a plateau until a further increase at 297 ml of buffer, which corresponds to the elution of selenohomocystine. The initial absorption increase at 238 ml corresponds to the emergence of 2-amino-5-selena-6-thia-8-hydroxyoctanoic acid.[16]

From these studies it is apparent that, in order to avoid complications in the quantification of diselenides, thiodiglycol should be omitted from the buffers used for amino acid analysis. Unfortunately, mercaptans are not the only agents that cleave the diselenide bond, and interchange reactions are frequently encountered by components of the very mixture to be analyzed. Moreover, it is not only during amino acid analyses that interchange reactions occur, but also during gas chromatography and other separation techniques. Likewise, certain experimental conditions, including acid and base hydrolysis or the presence of certain nucleophilic reagents, will cause scission of diselenide and thioselenenate bonds. The interchange reactions reported to date are listed in Table XIIB-2.

Despite these technical difficulties amino acid analysis has become the method of choice for testing plant material for selenoamino acid and peptide content.[15,18-20] For the specific detection or quantification of selenocystine and/or selenomethionine, an accelerated amino acid analysis program, using a single buffer, constant temperature, and constant buffer flow rate, has been developed.[21] Preparative ion-exchange chromatography was success-fully employed by Walter et al.[13] for the isolation of selenium-containing amino acids, and by Nigam and McConnell[18] for the isolation of Se-methyl-selenocysteine and the dipeptide γ-glutamyl-Se-methylselenocysteine.

Countercurrent distribution as introduced by Craig et al.,[22] a highly effective purification and separation technique, was used repeatedly during synthetic studies of selenium-containing peptides, with several solvent systems found applicable.[2,3,23] In fact, deaminooxytocin, deamino-6-selenooxytocin (or deamino-1-selenooxytocin), and deaminodiselenooxytocin were all separable by countercurrent distribution.[23] These cyclic peptides (Fig. XIIB-2, e, f, g, h) are identical except that one or both S atoms are replaced by selenium. These peptides were also distinctly separable by column partition chromatography on Sephadex resin.[23] Whether these two techniques are applicable also to the separation of isologous, acyclic disele-nides and disulfides must be determined by further experiments. In this context attention should be drawn to a new form of countercurrent chroma-tography, referred to as droplet countercurrent chromatography[24]—an all-liquid separation technique having high resolving power and well suited

Table XIIB-2   Interchange reactions involving selenium

TYPE OF REACTION	REACTANTS	TYPE OF PRODUCT	PRODUCTS	REF.
—SH	(1) 2-Mercaptoethanol and selenocystine	—S—Se—	(1) 2-Amino-4-selena-5-thia-7-hydroxyheptanoic acid	13
	(2) Cysteine and selenocystine		(2) 2,7-Diamino-4-thia-5-selenaoctanedioic acid	13
	(3) 2-Mercaptoethanol and selenohomocystine		(3) 2-Amino-5-selena-6-thia-8-hydroxyoctanoic acid	16
[—SeH]	(1) β-Selenopropionic acid and selenocystine	—Se—Se—	(1) 2-Amino-4,5-diselenaoctanedioic acid	13
—SeH	(1) Selenolate of N-carbobenzoxy-L-selenocystine diphenyl methyl ester and 2-naphthylsulfenylthiocyanate	—Se—S—	(1) N-Carbobenzoxy-Se-(naphthyl-2-thio)-L-selenocysteine diphenyl methyl ester	61
	(2) Selenolate of N-carbobenzoxy-L-selenocystine diphenyl methyl ester and sodium S-benzylthiosulfate		(2) N-Carbobenzoxy-Se-(benzylthio)-L-selenocysteine diphenyl methyl ester	61

Table XIIB-2 (continued).

TYPE OF REACTION		REACTANTS	TYPE OF PRODUCT	PRODUCTS	REF.
—S—S—	—Se—Se—	(1) Cystine and selenocystine	—S—Se—	(1) Sulfoselenide (tentative)	25
—Se—Se—	—Se—Se—	(1) Selenocystine and selenohomocystine	—Se—Se—	Unsymmetrical diselenide (unidentified)	15
—Se—S—	H⁺ or OH⁻	(1) N-Carbobenzoxy-Se-(naphthyl-2-thio)-L-selenocysteine diphenyl methyl ester	—Se—Se— and —S—S—	(1) Bis(diphenylmethyl) bis(N-carbobenzoxy)-L-selenocystinate and 2-naphthyl disulfide	61
		(2) N-Carbobenzoxy-Se-(benzylthio)-L-selenocysteine diphenyl methyl ester		(2) Bis(diphenylmethyl) bis(N-carbobenzoxy)-L-selenocystinate and benzyl disulfide	61

—Se—S— or —Se—Se—	—S—S—, —Se—Se—, and —S—Se—	6 N HCl, 110°, 22 hr			
			(1) 1- and 6-Selenooxytocin	(1) Cystine, selenocystine, and 2,7-diamino-4-thia-5-selenaoctanedioic acid[a]	3
			(2) Deamino-1-selenooxytocin	(2) Cystine, 2-amino-4-thia-5-selenaoctanedioic acid,[b] and 4,5-diselenaoctanedioic acid[c]	3
			(3) Deamino-6-selenooxytocin	(3) Selenocystine, 2-amino-4-selena-5-thiaoctanedioic acid,[d] and 4,5-diselenaoctanedioic acid[c]	3
			(4) Deaminodiselenooxytocin	(4) Selenocystine and 2-amino-4,5-diselenaoctanedioic acid	13

[a–d] Compounds a, b, c, and d were previously referred to as 1,6-diamino-3-thio-4-selenohexane-1,6-dicarboxylic acid, 1-amino-3-thio-4-selenohexane-1,6-dicarboxylic acid, 3,4-diselenohexane-1,6-dicarboxylic acid, and 1-amino-3-seleno-4-thiohexane-1,6-dicarboxylic acid, respectively.[3]

$$
\begin{array}{c}
\quad\quad\quad\quad\quad\quad C_6H_4OH \quad\quad C_2H_5 \\
\quad\quad\quad\quad\quad\quad | \quad\quad\quad\quad | \\
\quad\quad Z \quad O \quad\quad CH_2 \quad O \quad\quad CH-CH_3 \\
\quad\quad | \quad\parallel \quad\quad | \quad\quad\parallel \quad\quad | \\
CH_2-CH-C-NH-CH-C-NH-CH \\
\quad | \quad\quad 1 \quad\quad\quad 2 \quad\quad\quad 3\; | \\
\quad X \quad\quad\quad\quad\quad\quad\quad\quad\quad C=O \\
\quad | \quad\quad\quad\quad\quad\quad\quad\quad\quad | \\
\quad Y \quad\quad\quad\quad O \quad\quad\quad O \quad NH \\
\quad | \quad\quad 6 \quad\parallel \quad 5 \quad\quad\parallel \; 4\; | \\
CH_2-CH-NH-C-CH-NH-C-CH-(CH_2)_2-CONH_2 \\
\quad\quad | \quad\quad\quad\quad\quad\quad | \\
\quad\quad C=O \quad\quad\quad\quad\quad CH_2 \\
\quad\quad | \quad\quad\quad\quad\quad\quad | \\
CH_2-N \quad\quad\quad O \quad CONH_2 \; O \\
\quad | \quad\quad\;\;\backslash\; 7 \quad\parallel \quad 8\; | \quad\quad\parallel \quad\quad 9 \\
\quad | \quad\quad\quad\;/ \; CH-C-NH-CH-C-NH-CH_2-CONH_2 \\
CH_2-CH_2 \quad\quad\quad\quad\quad | \\
\quad\quad\quad\quad\quad\quad\quad\quad CH_2 \\
\quad\quad\quad\quad\quad\quad\quad\quad | \\
\quad\quad\quad\quad\quad\quad\quad CH(CH_3)_2
\end{array}
$$

FIG. XIIB-2   Oxytocin, deaminooxytocin, and their selenium-containing analogs; numbers indicate the positions of the component amino acid residues. *a*. Z = NH$_2$, X–Y = S–S; *b*. Z = NH$_2$, X–Y = Se–S; *c*. Z = NH$_2$, X–Y = S–Se; *d*. Z = NH$_2$, X–Y = Se–Se; *e*. Z = H, X–Y = S–S; *f*. Z = H, X–Y = Se–S; *g*. Z = H, X–Y = S–Se; *h*. Z = H, X–Y = Se–Se.

for the isolation of minute quantities of naturally occurring selenoamino acids and peptides.

The application of gas-liquid chromatography to the separation of sulfur- and selenium-containing amino acids is still in its infancy. The only examples reported so far appear to be the separation of selenocystine and selenomethionine from their corresponding sulfur analogs on a 2 % SE-30 column.[25] In this study the amino acids were injected as their silylated derivatives. As other volatile selenium-containing amino acid derivatives become available, rapid acceptance of g.l.c. and advances in its utilization, particularly in conjunction with mass spectrometry, are anticipated.

The identification and characterization of selenium-containing amino acids present in the sequence of naturally occurring peptides and possibly proteins require the breakdown of the biopolymer into its monomeric units. Acid hydrolysis is possible in principle and has been used,[13,26-28] as has Dowex-50 resin (in the H$^+$ form) in 70 % ethanol;[29] for all practical purposes, however, acid hydrolysis is not advisable since selenocystine and selenomethionine are partially degraded[13]—a disadvantage amplified by the fact that these amino acid residues are usually present only as a minor component of the bipolymer. Therefore, enzymatic digestion is the method of choice, and it has found application through the use of pronase[20,30] and combinations of trypsin-pepsin[31] and of pepsin-pancreatin-erepsin.[32,33] In this context attention should be drawn to the potential of cathepsin,[34,35] which cleaves dipeptide units from the N-terminal of proteins, but which has not yet been applied to selenium-containing proteins.

3. Synthetic, Biological, and Conformational Studies

a. Synthesis. Synthesis has retained its key role for ultimate demonstration of the correctness of a proposed molecular structure. Synthetic peptides, containing one or more "unnatural" amino acid residues, can serve as molecular probes during studies of chemical and physical properties as well as of biological functions of their naturally occurring analogs. The possibility of securing selenium-containing peptides by direct synthesis frees the researcher from working with limited quantities and from performing cumbersome purification of selenium-containing material obtained from natural sources. Hence, synthesis places into his hands selenium-containing peptides that sometimes are essentially isosteric and isofunctional with their sulfur congeners, but more frequently display the subtle physicochemical and biological differences required of an ideal molecular probe.

In principle, two approaches exist for the synthesis of selenium-containing peptides. In the first, the selenium-containing amino acid is incorporated into the growing peptide chain at the desired loci; in the second, one or more of the amino acid residues naturally present (or specifically introduced or modified during synthesis) in the molecule are converted to yield the desired selenopeptide or protein. The latter method is illustrated by the synthesis of protected selenoglutathione (scheme 1) by Theodoropoulos *et al.*,[36,37] in which the dipeptide benzyl *N-tert*-butyloxycarbonyl-L-serinyl-glycinate (**1**) is converted to its *O*-tosyl derivative (**2**), which in turn yields the protected, selenium-containing dipeptide (**3**) upon treatment with sodium benzylselenolate. Next, **3** is elongated by the addition of α-benzyl *N*-carbobenzoxy-L-glutamate to give the protected selenoglutathione (**4**). Clearly, this method is limited in its application and may be practical only with peptides and not with proteins.

The more versatile approach to the synthesis of selenium-containing peptides consists, as mentioned above, of the incorporation of an appropriately protected selenoamino acid into the growing peptide chain. A few methods, based on well-established reactions generally suitable for the preparation of amino acids possessing a propionic acid and a butyric acid skeleton, have been applied to the synthesis of the optically active isomers of selenoamino acids. For example, the original synthesis of L-cystine by Fischer and Raske,[38] starting from *L*-α-amino-β-chloropropionic acid, was adapted for the preparation of L-selenocystine by Fredga[39] and of *Se*-benzyl-L-selenocysteine by Frank.[40] Also, *Se*-benzyl-L-selenohomocysteine (part of which was converted to L-selenomethionine) has been secured *via*

(1)

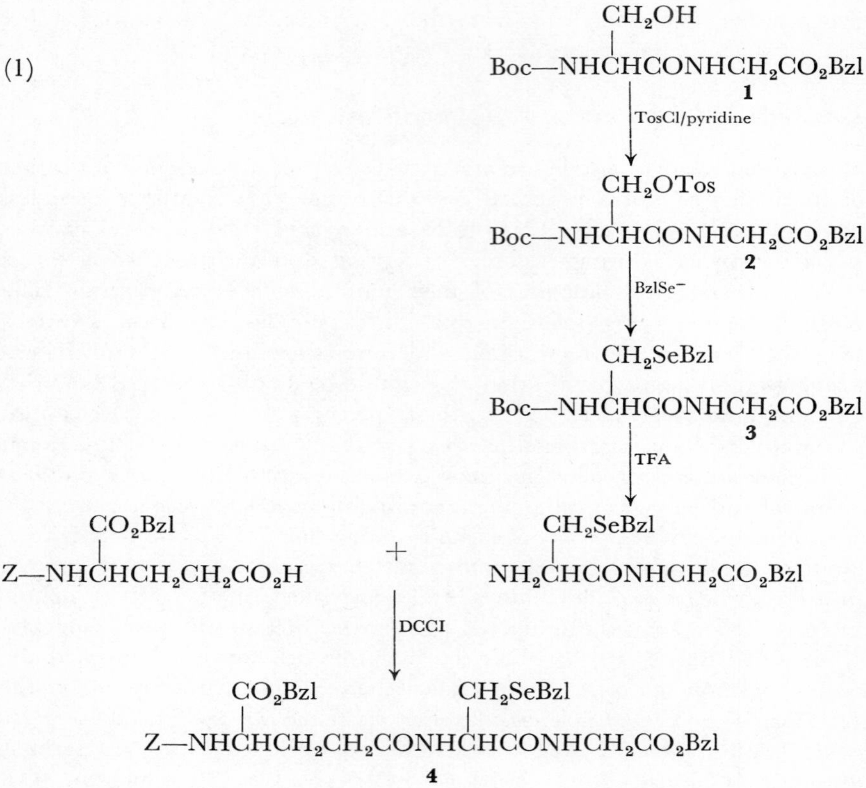

$$CH_2OH$$
$$|$$
$$Boc—NHCHCONHCH_2CO_2Bzl$$
$$\textbf{1}$$

TosCl/pyridine

$$CH_2OTos$$
$$|$$
$$Boc—NHCHCONHCH_2CO_2Bzl$$
$$\textbf{2}$$

BzlSe⁻

$$CH_2SeBzl$$
$$|$$
$$Boc—NHCHCONHCH_2CO_2Bzl$$
$$\textbf{3}$$

TFA

$$CO_2Bzl \qquad\qquad CH_2SeBzl$$
$$| \qquad\qquad\qquad\qquad |$$
$$Z—NHCHCH_2CH_2CO_2H \quad + \quad NH_2CHCONHCH_2CO_2Bzl$$

DCCI

$$CO_2Bzl \qquad\quad CH_2SeBzl$$
$$| \qquad\qquad\qquad |$$
$$Z—NHCHCH_2CH_2CONHCHCONHCH_2CO_2Bzl$$
$$\textbf{4}$$

L-α-amino-γ-bromobutyric acid.[41] Successful enzymatic resolution of a DL-selenoamino acid precursor, which can be derived from any standard pathway of synthesis, was first reported by Walter and Chan,[3] as well as by Pan and Tarver,[42] for the synthesis of Se-benzyl-L-selenocysteine and the corresponding homocysteine derivative, respectively. This general approach has since been applied by other workers for the preparation of Se-benzyl-L-selenocysteine[43] and Se-benzyl-L-selenohomocysteine.[44] Finally, a special case is the separation by fractional crystallization of N-carbobenzoxy-Se-benzyl-L-selenocysteinyl-L-prolyl-L-leucylglycinamide from the diastereo-isomeric peptide mixture, N-carbobenzoxy-Se-benzyl-DL-selenocysteinyl-L-propyl-L-leucylglycinamide.[3] The resolution of an amino acid into its optical isomers while part of a larger peptide is rare but not unique.[45]

A novel method, which is most versatile and generally applicable for the synthesis of optically active selenium-containing amino acids, is based on the displacement of the "pseudo-halogen," O-p-toluenesulfonate group of O-tosylated hydroxyamino acids by a selenium nucleophile (scheme 2).[36,37,46,47]

(2)
$$X{-}NHCHCO_2R \xrightarrow{R'Se^-} X{-}NHCHCO_2R$$

with side chains $(CH_2)_nOTos$ and $(CH_2)_nSeR'$

$(R' = \text{alkyl or } H; \, n = 1, 2)$

As can be seen from scheme 2, this reaction seems to proceed independently of the carbon number of the amino acid side chain and hence is generally applicable. In contrast to other methods described for the synthesis of selenium-containing amino acids, the tosylation of a hydroxy group and the subsequent displacement of the O-tosyl by a selenium nucleophile proceed under mild reaction conditions (no strong acid or base is required). Therefore, the amino and carboxyl groups of the amino acid can be blocked with protecting groups commonly used in peptide chemistry, even those which can be removed under exceedingly mild reaction conditions. The latter point is important since it is consistent with the trend in peptide chemistry toward ever-milder reaction conditions. Moreover, the mild reaction conditions make repeated deprotection and protection superfluous. Such repetitive steps would be required if the synthesis of a DL-selenocysteine derivative were to follow a classical route, and this DL-derivative had to be resolved and properly protected for incorporation into a growing peptide chain.

Unlike the replacement of the O-tosyl group in O-tosylserine derivatives by sulfur nucleophiles,[48] the replacement by selenium nucleophiles, which are more nucleophilic but less basic,[49] proceeds under proper reaction conditions with retention of chirality. This conclusion is based on the fact that the conversion of **5** in scheme 3 by two selenium nucleophiles (sodium

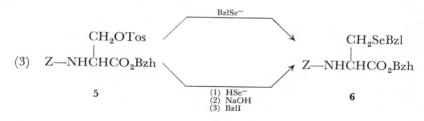

(3)

5

6

(1) HSe⁻
(2) NaOH
(3) BzlI

benzylselenolate[37] and sodium hydrogen selenide[49]) led to **6** with essentially identical properties; in addition, partial deprotection of **6** gives Se-benzyl-L-selenocysteine, which exhibits an optical rotation comparable to that recorded previously for this compound secured either by hydrolysis of methyl Se-benzyl-L-selenocysteinate[40] or by resolution of N-acetyl-Se-benzyl-DL-selenocysteine with hog acylase.[3] For a detailed description of pathways by which various selenium-containing amino acids have been prepared and for a discussion of their properties consult Chapter XIIA.

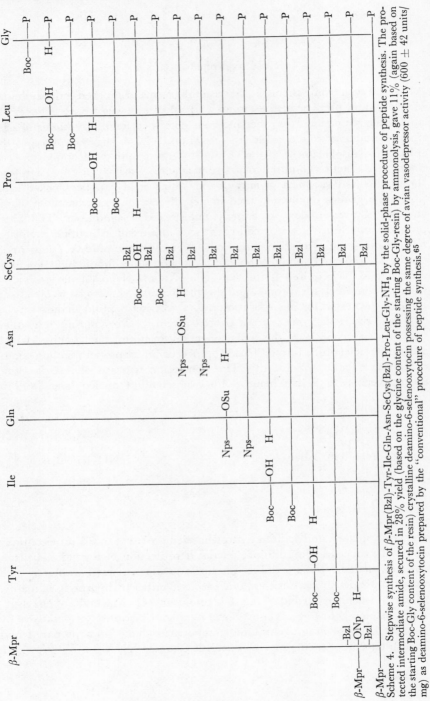

β-Mpr
Scheme 4.   Stepwise synthesis of β-Mpr(Bzl)-Tyr-Ile-Gln-Asn-SeCys(Bzl)-Pro-Leu-Gly-NH₂ by the solid-phase procedure of peptide synthesis. The protected intermediate amide, secured in 28% yield (based on the glycine content of the starting Boc-Gly-resin) by ammonolysis, gave 11% (again based on the starting Boc-Gly content of the resin) crystalline deamino-6-selenooxytocin possessing the same degree of avian vasodepressor activity (600 ± 42 units/mg) as deamino-6-selenooxytocin prepared by the "conventional" procedure of peptide synthesis.⁶⁵

Once the selenium-containing amino acid is prepared, it can be incorporated (frequently after protection of one or the other functional group) into the peptide. To date most syntheses of selenopeptides have been carried out by "classical" methods of peptide synthesis, although Walter[50] has prepared crystalline deamino-6-selenooxytocin also on a solid polymer support (scheme 4), according to the general procedure of Merrifield.[51] The preparation of some model peptides,[40] illustrating the use of dicyclohexylcarbodiimide, an activated ester or a mixed anhydride for the coupling, is given in schemes 5–7.

(5)

$$\overset{\overset{\displaystyle Bzl}{\displaystyle |}}{\text{Z--SeCys}} + \text{HCl, Gly--OEt} \xrightarrow{\text{DCCI/CHCl}_3\text{/Et}_3\text{N}} \overset{\overset{\displaystyle Bzl}{\displaystyle |}}{\text{Z--SeCys--Gly--OEt}}$$

(6) $\overset{\overset{\displaystyle Bzl}{\displaystyle |}}{\text{Z--SeCys--ONp}} + \text{HCl, Gly--Gly--OEt} \xrightarrow{\text{CHCl}_3\text{/Et}_3\text{N}}$

$$\overset{\overset{\displaystyle Bzl}{\displaystyle |}}{\text{Z--SeCys--Gly--Gly--OEt}}$$

(7) Mixed anhydride from

$$\overset{\overset{\displaystyle Bzl}{\displaystyle |}}{\text{Z--SeCys}} \text{ and ClCO}_2\text{Et} + \overset{\overset{\displaystyle Bzl}{\displaystyle |}}{\text{SeCys}} \longrightarrow$$

$$\overset{\overset{\displaystyle Bzl}{\displaystyle |}}{\text{Z--SeCys}}\overset{\overset{\displaystyle Bzl}{\displaystyle |}}{\text{--SeCys}} \xrightarrow{\text{HBr/AcOH}} \overset{\overset{\displaystyle Bzl}{\displaystyle |}}{\text{SeCys}}\overset{\overset{\displaystyle Bzl}{\displaystyle |}}{\text{--SeCys}}$$

The physiologically important peptide that has been studied most extensively from a synthetic, biological, and conformational point of view is oxytocin (Fig. XIIB-2a). This peptide hormone is the milk-ejecting principle of mammals and is released from the posterior pituitary gland.[52] One conspicuous feature of oxytocin is its disulfide bridge forming the 20-membered ring. Having ruled out a direct chemical involvement of the disulfide bridge of neurohypophyseal hormones with the receptor,[53, 54] it was felt that the contribution of this bridge to the biological properties of the hormones might be expressed through its role in determining their three-dimensional structures. With this in mind, oxytocin analogs in which one or both S atoms are replaced by selenium (Fig. XIIB-2) were prepared by Walter and Chan.[3] The general scheme of synthesis of the protected nonapeptide is illustrated for 1-selenooxytocin (scheme 8).[3] The protected key intermediates for 6-seleno-, diseleno-, deamino-1-seleno-, deamino-6-seleno-, and deaminodiselenooxytocin (the N-terminal amino group is

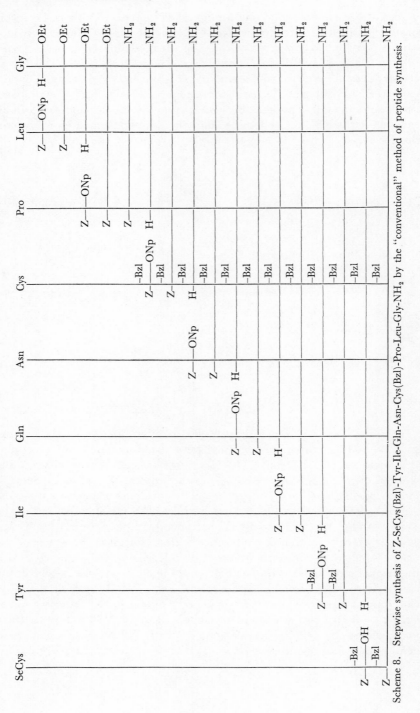

Scheme 8. Stepwise synthesis of Z-SeCys(Bzl)-Tyr-Ile-Gln-Asn-Cys(Bzl)-Pro-Leu-Gly-NH₂ by the "conventional" method of peptide synthesis.

replaced in deamino analogs by a H atom), shown below, were prepared similarly:

Z-Cys(Bzl)-Tyr-Ile-Gln-Asn-SeCys(Bzl)-Pro-Leu-Gly-NH$_2$[2,3]
Z-SeCys(Bzl)-Tyr-Ile-Gln-Asn-SeCys(Bzl)-Pro-Leu-Gly-NH$_2$[49,55]
β-SeMpr(Bzl)-Tyr-Ile-Gln-Asn-Cys(Bzl)-Pro-Leu-Gly-NH$_2$[3]
β-Mpr(Bzl)-Tyr-Ile-Gln-Asn-SeCys(Bzl)-Pro-Leu-Gly-NH$_2$[2,3]
β-SeMpr(Bzl)-Tyr-Ile-Gln-Asn-SeCys(Bzl)-Pro-Leu-Gly-NH$_2$[56]

Conversion of these intermediates to their corresponding cyclic peptides proceeds in a two-step process, that is, reduction followed by oxidation, as illustrated with 1-selenooxytocin (scheme 9).[3] Decarbobenzoxylation

(9)   Z–SeCys(Bzl)–Tyr–Ile–Gln–Asn–Cys(Bzl)–Pro–Leu–Gly–NH$_2$

(1) Na/Liquid NH$_3$
(2) Potassium ferricyanide (pH 6.8)

1–Selenooxytocin

and debenzylation are effected by reduction with sodium in liquid ammonia according to the method introduced by Sifferd and du Vigneaud.[57] Thus far, two oxidation methods have been used to convert the deprotected selenium-containing nonapeptides to the cyclic peptides: by aeration[58] and with ferricyanide,[59] both in aqueous solution at pH 6.8. Purification is effected by countercurrent distribution or partition chromatography (*cf.* Section XIIB.2.b).

b. Biological and Conformational Studies. When the three possible selenium isologs of oxytocin were compared, they were found to exhibit all the biological activities characteristic of oxytocin. In fact, they are highly active and their specific activities are of the same order of magnitude as those of oxytocin (see Table XIIB-3). Unfortunately, a detailed comparison of the differences between the biological activities of these amorphous selenium analogs and those of oxytocin is not too rewarding. As has been repeatedly noticed with disulfide-containing peptide hormones, these selenium-containing hormones become partially inactivated during the last step of isolation, lyophilization. The finding that crystalline deamino-dicarbaoxytocin, a hormone analog in which the disulfide bridge has been replaced by an ethylene bridge, retains full biological activity upon repeated lyophilization[60] implicates the S–S, S–Se, and Se–Se moieties as responsible for peptide inactivation; it is concluded that the selenium-containing peptides undergo dimerization and polymerization *via* S–Se or Se–Se interchange reactions. The degree to which this interchange reaction occurs is difficult to control during lyophilization from dilute acetic acid and depends on the nature of the compound. Unsymmetrically substituted cyclic disulfides

Table XIIB-3   Biological potencies of oxytocin, 1-selenooxytocin, 6-selenooxytocin, and diselenooxytocin[a,b]

COMPOUND	DEPRESSOR (FOWL)	OXYTOCIC (RAT)	MILK-EJECTING (RABBIT)	PRESSOR (RAT)	ANTIDIURETIC (RAT)
Oxytocin	507 ± 15	546 ± 18	410 ± 16	3.1 ± 0.1	2.7 ± 0.2
1-Selenooxy-tocin	361 ± 18	362 ± 9	351 ± 15	3.1 ± 0.2	5.7 ± 0.6
6-Selenooxy-tocin	385 ± 15	405 ± 5	398 ± 6	3.8 ± 0.2	3.4 ± 0.1
Diselenooxy-tocin	∼ 600[c]	...	...	...	...

[a] Values taken from Walter and Chan[3] unless otherwise stated.
[b] Expressed in USP units per milligram.
[c] Value taken from Walter.[49]

appear to be more resistant than the corresponding S–Se- or Se–Se-containing analogs. Moreover, the presence of a charged residue enhances polymerization by acid-base catalysis. Thus, while the "neutral" deamino-diselenooxytocin was inactivated by about 60 % during lyophilization, the "charged" diselenooxytocin was inactivated by about 85 %, as judged from avian vasodepressor assays (Fig. XIIB-3).[49] Nevertheless, the stability of aliphatic *cyclic* thioselenenates[3] is, with certain exceptions, considerably greater than that of their aliphatic *acyclic* congeners;[61] this point was a consideration when we suggested that acyclic thioselenenates (and for that matter aliphatic, acyclic diselenides) are unlikely to be involved in biological processes with a slow rate of turnover, but can be expected to function as cofactors or catalysts.[61,62] In order to minimize the polymerization of seleno-oxytocin analogs—or, generally speaking, of compounds known or suspected to contain a Se–Se or S–Se bridge—concentration to dryness during the isolation should be avoided. When diselenooxytocin and deaminodiseleno-oxytocin were tested after purification with omission of lyophilization in the avian vasodepressor assay, activities comparable to those of their amorphous sulfur analogs, oxytocin (507 U/mg)[63] and deaminooxytocin (733 U/mg)[63] (compare with data in Fig. XIIB-3), were obtained.

Most desirable with peptides of low molecular weight is, of course, the crystallization of the final product, as was achieved with deamino-1-selenooxytocin[64] and deamino-6-selenooxytocin.[65] This in turn permitted these peptides to be compared in detail for the biological activities characteristic of neurohypophyseal hormones (Table XIIB-4). What at first glance

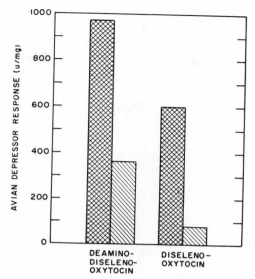

FIG. XIIB-3 The effect of lyophilization on the avian vasodepressor activity of diselenooxytocin and deaminodiselenooxytocin. The first bar for each compound represents the activity obtained with a sample ampouled immediately after countercurrent distribution and evaporation of the *organic* layer; the second bar represents residual activity of the peptides after purification by countercurrent distribution followed by lyophilization and dilution in saline. Bioassay was performed according to the procedure employed by R. A. Munsick, W. H. Sawyer, and H. B. van Dyke, *Endocrinology*, **66,** 860 (1960).

Table XIIB-4 Biological potencies of crystalline neurohypophyseal hormone analogs: deamino-1-selenooxytocin and deamino-6-selenooxytocin[a,b]

COMPOUND	DEPRESSOR (FOWL)	OXYTOCIC (RAT)	MILK-EJECTING (RABBIT)	PRESSOR (RAT)	ANTI-DIURETIC (RAT)
Deamino-1-selenooxytocin	1306 ± 46^c	1217 ± 13^c	479 ± 32^d	5.00 ± 0.18^c	29.9 ± 1.5^d
Deamino-6-selenooxytocin	586 ± 36^c	443 ± 7.0^c	282 ± 15^d	1.81 ± 0.08^c	20.5 ± 2.1^d
Ratio:	2.23	2.75	1.70	2.76	1.46

[a] Expressed in USP units per milligram.

[b] B. M. Ferrier, D. Jarvis, and V. du Vigneaud [*J. Biol. Chem.*, **240,** 4264 (1965)] report for the disulfide-containing analog, deaminooxytocin, the following values: fowl depressor, 975 ± 24; oxytocic, 803 ± 36; milk-ejecting, 541 ± 13; pressor, 1.44 ± 0.06; and antidiuretic, 19.0 ± 1.

[c] Values taken from Walter *et al.*[65]

[d] Values taken from Walter *et al.*[93]

seems to be as minimal an alteration as the substitution of a Se atom for one or the other S atom of a disulfide bridge in a 1000-molecular-weight peptide, evokes a substantial dissociation of *all* the tested biological activities between the two resulting compounds, deamino-1-selenooxytocin being always the more potent, independently of the absolute value of the specific activity.

In searching for the molecular basis of this biological nonequivalence of selenium analogs of oxytocin we embarked on the conformational analysis of neurohypophyseal peptides, starting with the disulfide region of the hormone. Generally, compounds of the type $-\ddot{X}-\ddot{Y}-$, which possess two unshared electron pairs on adjacent atoms (such as peroxides, disulfides, thioselenenates, diselenides; X and/or Y = O, S, Se), are inherently dissymmetric (no plane or center of symmetry) and are present in two molecular conformations, as predicted for peroxides in 1934 by Penney and Sutherland:[66]

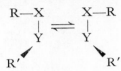

The two optical antipodes are a result of the repulsion between the unshared electron pair of $3p\pi$ and of $4p\pi$ in the cases of sulfur and selenium, respectively.[67] The dihedral angle of an aliphatic, strain-free diselenide was predicted[67] to be between 90° and 100° (this applies also to disulfides[68] and, most probably, to thioselenenates) and was confirmed experimentally.[69] When the repulsion between the lone-pair electrons of the S–S, S–Se, or Se–Se groups is at a minimum, an interchange of the electrons with the unoccupied *d*-levels of the neighboring sulfur or selenium is possible, giving rise to partial double-bond character ($p\pi$–$d\pi$ bonding). With open-chain disulfides—thus far disulfides constitute the predominant source of our information—reported values for the energy required for rotation about the S–S bond ranged from 2 to 14 kcal/mole.[70–77] For analogous S–Se and Se–Se compounds, this rotational energy is expected to be somewhat lower because of the enhanced bond length of the S–Se as well as the Se–Se group (about 2.3 Å),[78] as compared to the S–S group (about 2.05 Å).[79] Enantiomeric disulfides, thioselenenates, and diselenides (compounds in which the S–S, S–Se, or Se–Se is the sole dissymmetric center) do not exhibit any optical activity since they are present as a racemic mixture.[80] However, when the X–Y group is part of a molecule which contains additional asymmetric centers (such as a peptide), then the energy of the molecule possessing the right-handed X–Y configuration should differ from the energy of that possessing the left-handed screw sense. Hence, these diastereoisomers should not be present in equimolar amounts.

Applying these general considerations to oxytocin and its selenium-containing analogs, it is to be expected that each of these compounds possesses a preferred X–Y screw sense and, in fact, that a definite interrelationship exists between the preferred configuration of the X–Y group and the preferred conformation of the hormonal peptides.

Since oxytocin and its seleno analogs contain only three types of chromophores, that is, the S–S, S–Se, or Se–Se, on the one hand, and the tyrosine residue and peptide amides, on the other, it was readily possible, with a group of synthetic analogs (*e.g.*, substitution of Tyr in oxytocin by Ile or Ala, and of sulfur by selenium), to assign observed transitions in the circular dichroism spectra to specific groups in the peptides.[65,81] The negative peak at about 235 nm in each of the peptides studied and the positive band at around 255 nm in the case of deaminooxytocin, at 260 nm for deamino-1-selenooxytocin and deamino-6-selenooxytocin, and at 270–280 nm for deaminodiselenooxytocin were attributed to the X–Y group of these peptides (see Fig. XIIB-4). The contention that inherently dissymmetric groups are the predominant factors contributing to optical activity[82]—although dissymmetric vicinal perturbations clearly have to be considered—seems to

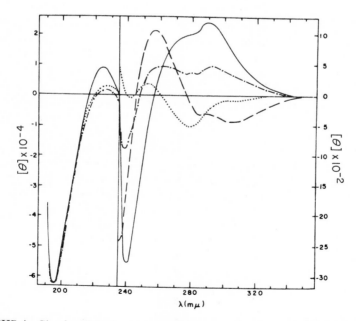

FIG. XIIB-4   Circular dichroism pattern of deaminooxytocin and its seleno analogs (in water, pH 5.5). ———— Deaminodiselenooxytocin; – – – – deamino-1-selenooxytocin; ·—·—· deamino-6-selenooxytocin; · · · · · · deaminooxytocin.

apply to neurohypophyseal peptides. The same general chromophoric pattern of a positive band in the 250–260 nm range and a negative band in the 230–240 nm range is observed upon protonation of the N-terminal amino group in oxytocin or the replacement of the tyrosine residue by isoleucine or alanine.[65,81,83]

Application to oxytocin of the rule developed by Carmack and Neubert[84] and independently by Claeson[80] for the assignment of the absolute configuration of disulfides (*i.e.*, that a negative Cotton effect associated with the longest-wavelength band of the disulfide absorption is indicative of a left-handed helical screw sense of the C–S–S–C group, while a positive CD band is associated with a right-handed screw sense) led to the conclusion that the cystine moiety in oxytocin has a preferred right-handed chirality.[65,81] Recently, we suggested that lysine-vasopressin, the mammalian antidiuretic hormone and close congener of oxytocin, likewise possesses a preferred right-handed helical screw sense.[85] The fact that the general circular dichroism patterns of oxytocin and its deamino analog, as well as those of deamino-1-selenooxytocin, deamino-6-selenooxytocin, and deaminodiselenooxytocin, are all identical with respect to the signs of the circular dichroism bands associated with the C–X–Y–C group (whereas differences in position and rotational strength would be expected) justifies an extension of the rule of disulfides[80,84] to the assignment of the absolute configuration of thioselenenates and diselenides. Thus it can be concluded that in deamino-1-selenooxytocin and deamino-6-selenooxytocin the C–S–Se–C moiety is present in a preferred right-handed screw sense, as is the case with deaminodiselenooxytocin (C–Se–Se–C).

Adding the remaining peptide structure, as recently assigned from a series of conformational studies involving 220 MHz n.m.r. spectroscopy,[86–90] we can propose a conformation of the selenium-containing analogs of oxytocin in solution as shown in Fig. XIIB-5. In principle, the backbone conformation of these compounds consists of two β-turns, one composed of the sequence -Tyr-Ile-Gln-Asn- in the ring moiety and the other of the sequence -Cys-Pro-Leu-Gly-; with analogs possessing a selenium moiety in position 6, the cysteine residue in the second β-turn is replaced by the selenocysteine residue. The sterically voluminous side chains of the residues in positions 3, 4, and 8 (Ile, Gln, and Leu) and the cyclic moiety of proline in position 7 act synergistically with the hydrogen bonds between the peptide nitrogen of asparagine and the C=O of tyrosine, on the one hand, and the peptide nitrogen of glycine with the C=O of cysteine/selenocysteine in position 6, on the other, to stabilize the two β-turns. Judging from space-filling molecular models, these peptides are tightly packed, particularly the 20-membered ring component which contains one of the β-turns. Protein energy minimization calculations confirm the general features

FIG. XIIB-5 Proposed conformation of oxytocin and its seleno analogs; X–Y stands for S–S, S–Se, and Se–Se.

derived for oxytocin by n.m.r. and c.d.[91] Two molecules of deamino-6-selenooxytocin can be fitted in such a manner as to yield unit cell dimensions compatible with a *B*2 space group in agreement with preliminary X-ray crystallographic studies of this compound.[92]

In regard to the specific contribution of the covalent linkage between residues 1 and 6 in neurohypophyseal peptides (–S–S–, –S–Se–, –Se–Se–, –S–CH$_2$, –CH$_2$–CH$_2$–, –CO–NH–, *etc.*), this linkage is responsible for the unusually great stability of the β-turn in the 20-membered ring. However, in addition to this predominant factor it also exerts a more subtle, but for the physical and biological properties very important, influence on the topography of these peptides. This influence is implemented by changing the dihedral angle of the bridge components and thereby evoking a significant alteration of the spatial relationship of practically every atom to every other atom in the molecule. Thus, any minimal change in the precise configuration of this bridge will be amplified; that is, a chemical alteration in the bridge should have a much greater overall effect on the molecule (as expressed by differences in physical, chemical, and biological properties) than a comparable alteration at some other site.

These general considerations (for a more detailed exposition see Walter *et al.*[93] and Walter[85]) are borne out and illustrated with deaminoselenooxytocin analogs. In analogy to findings with oxytocin[65, 81] and several of its disulfide-containing analogs, the dihedral angle of the –S–Se– bond is

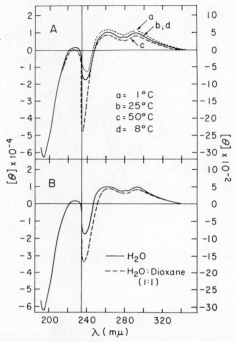

FIG. XIIB-6   *A*. The effect of temperature on the c.d. spectrum of deamino-6-seleno-oxytocin (in water, pH 5.5). *B*. The effect of 50% dioxane (by volume) on the c.d. spectrum of deamino-6-selenooxytocin.

temperature dependent (Fig. XIIB-6*A*) and solvent dependent (Fig. XIIB-6*B*), as deduced from circular dichroism studies with deamino-6-selenooxy-tocin. The difference, under a given set of experimental conditions, in the magnitude of the negative ellipticity in the 230–240 nm region (see Fig. XIIB-4) between deamino-1-selenooxytocin and deamino-6-selenooxytocin suggest that these two analogs possess different dihedral angles, and hence exist in a different conformational equilibrium, in a given environment. This might form the basis for the differences observed in the specific biological activities of these two congeners (Table XIIB-4).

REFERENCES

1. P. J. Peterson and G. W. Butler, *J. Chromatog.*, **8,** 70 (1962).
2. R. Walter and V. du Vigneaud, *J. Amer. Chem. Soc.*, **87,** 4192 (1965).
3. R. Walter and W. Y. Chan, *J. Amer. Chem. Soc.*, **89,** 3892 (1967).
4. M. J. Horn and D. B. Jones, *J. Biol. Chem.*, **139,** 649 (1941).

5. S. F. Trelease, A. A. DiSomma, and A. L. Jacobs, *Science*, **132,** 618 (1960).

5a. W. H. H. Günther and H. G. Mautner, *J. Amer. Chem. Soc.*, **87,** 2708 (1965).

6. A. J. Barak and S. C. Swanberg, *J. Chromatog.*, **31,** 282 (1967).

7. J. Scala and H. H. Williams, *J. Chromatog.*, **15,** 546 (1964).

8. L. Shepherd and R. E. Huber, *Can. J. Biochem.*, **47,** 877 (1969).

9. R. Walter and J. Roy, *J. Org. Chem.*, **36,** 2561 (1971).

10. E. Gross and B. Witkop, *J. Amer. Chem. Soc.*, **83,** 1510 (1961).

11. K. R. Millar, *J. Chromatog.*, **21,** 344 (1966).

12. K. P. McConnell and C. H. Wabnitz, *Biochim. Biophys. Acta*, **86,** 182 (1964).

13. R. Walter, D. H. Schlesinger, and I. L. Schwartz, *Anal. Biochem.*, **27,** 231 (1969).

14. J. L. Martin and L. M. Cummins, *Anal. Biochem.*, **15,** 530 (1966).

15. J. L. Martin and M. L. Gerlach, *Anal. Biochem.*, **29,** 257 (1969).

16. D. H. Schlesinger, J. Roy, and R. Walter, unpublished data.

17. S. Moore, D. H. Spackman, and W. H. Stein, *Anal. Chem.*, **30,** 1185 (1958).

18. S. N. Nigam and W. B. McConnell, *Biochim. Biophys. Acta*, **192,** 185 (1969).

19. S. N. Nigam, J.-I. Tu, and W. B. McConnell, *Phytochemistry*, **8,** 1161 (1969).

20. O. E. Olson, E. J. Novacek, E. I. Whitehead, and I. S. Palmer, *Phytochemistry*, **9,** 1181 (1970).

21. J. V. Benson, Jr. and J. A. Patterson, *Anal. Biochem.*, **29,** 130 (1969).

22. L. C. Craig, W. Hausmann, E. H. Ahrens, Jr., and E. J. Harfenist, *Anal. Chem.*, **23,** 1236 (1951).

23. W. Gordon, R. T. Havran, I. L. Schwartz, and R. Walter, *Proc. Natl. Acad. Sci. U.S.*, **60,** 1353 (1968).

24. Y. Ito and R. L. Bowman, *Science*, **167,** 281 (1970).

25. K. A. Caldwell and A. L. Tappel, *J. Chromatog.*, **32,** 635 (1968).

26. M. Blau, *Biochim. Biophys. Acta*, **49,** 389 (1961).

27. M. Hidiroglou, D. P. Heaney, and K. J. Jenkins, *Can. J. Physiol. Pharmacol.*, **46,** 229 (1968).

28. E. Hansson and M. Blau, *Biochem. Biophys. Res. Commun.*, **13,** 71 (1963).

29. T. K. Virupaksha and A. Shrift, *Biochim. Biophys. Acta*, **107,** 69 (1965).

30. G. W. Butler and P. J. Peterson, *Australian J. Biol. Sci.*, **20,** 77 (1967).

31. K. J. Jenkins and M. Hidiroglou, *Can. J. Biochem.*, **45,** 1027 (1967).

32. T. Tuve and H. H. Williams, *J. Biol. Chem.*, **236,** 597 (1961).

33. K. F. Weiss, J. C. Ayres, and A. A. Kraft, *J. Bacteriol.*, **90,** 857 (1965).

34. J. K. McDonald, B. B. Zeitman, T. J. Reilly, and S. Ellis, *J. Biol. Chem.*, **244,** 2693 (1969).

35. J. K. McDonald, P. X. Callahan, B. B. Zeitman, and S. Ellis, *J. Biol. Chem.*, **244,** 6199 (1969).

36. D. Theodoropoulos, I. L. Schwartz, and R. Walter, *Tetrahedron Lett.*, 2411 (1967).

37. D. Theodoropoulos, I. L. Schwartz, and R. Walter, *Biochemistry*, **6,** 3927 (1967).

38. E. Fischer and K. Raske, *Ber.*, **41,** 893 (1908).

39. A. Fredga, *Svensk Kem. Tidskr.*, **49,** 124 (1937).

40. W. Frank, *Z. Physiol. Chem.*, **339,** 202 (1964).

41. H.-D. Jakubke, J. Fischer, K. Jošt, and J. Rudinger, *Collection Czech. Chem. Commun.*, **33**, 3910 (1968).

42. F. Pan and H. Tarver, *Arch. Biochem. Biophys.*, **119**, 429 (1967).

43. G. Zdansky, *Ark. Kemi*, **29**, 443 (1968).

44. G. Zdansky, *Ark. Kemi*, **29**, 437 (1968).

45. J. S. Fruton, *Advan. Protein Chem.*, **5**, 4 (1949); E. Schnabel and H. Zahn, *Justus Liebigs Ann. Chem.*, **614**, 141 (1958); D. Yamashiro, D. Gillessen, and V. du Vigneaud, *J. Amer. Chem. Soc.*, **88**, 1310 (1966).

46. J. Roy, W. Gordon, I. L. Schwartz, and R. Walter, *J. Org. Chem.*, **35**, 510 (1970).

47. C. S. Pande, J. Rudick, and R. Walter, *J. Org. Chem.*, **35**, 1440 (1970).

48. I. Photaki, *J. Amer. Chem. Soc.*, **85**, 1123 (1963).

49. R. Walter, "Peptides: Chemistry and Biochemistry," B. Weinstein, Ed., Marcel Dekker, New York, N.Y., 1970, p. 467.

50. R. Walter, unpublished data.

51. R. B. Merrifield, *Advan. Enzymol.*, **32**, 221 (1969).

52. V. du Vigneaud, *Science*, **123**, 967 (1956).

53. J. Rudinger and K. Jošt, *Experientia*, **20**, 570 (1964).

54. I. L. Schwartz, H. Rasmussen, and J. Rudinger, *Proc. Natl. Acad. Sci. U.S.*, **52**, 1044 (1964).

55. W. Frank, *Z. Physiol. Chem.*, **339**, 222 (1964).

56. R. Walter and V. du Vigneaud, *J. Amer. Chem. Soc.*, **88**, 1331 (1966).

57. R. H. Sifferd and V. du Vigneaud, *J. Biol. Chem.*, **108**, 753 (1935).

58. V. du Vigneaud, C. Ressler, J. M. Swan, C. W. Roberts, and P. G. Katsoyannis, *J. Amer. Chem. Soc.*, **76**, 3115 (1954).

59. D. B. Hope, V. V. S. Murti, and V. du Vigneaud, *J. Biol. Chem.*, **237**, 1563 (1962).

60. T. Yamanaka, S. Hase, S. Sakakibara, I. L. Schwartz, B. M. Dubois, and R. Walter, *Mol. Pharmacol.*, **6**, 474 (1970).

61. J. Roy, I. L. Schwartz, and R. Walter, *J. Org. Chem.*, **35**, 2840 (1970).

62. R. Walter, I. L. Schwartz, and J. Roy, *Ann. N.Y. Acad. Sci.*, **192**, 175 (1972).

63. W. Y. Chan and V. du Vigneaud, *Endocrinology*, **71**, 977 (1962).

64. R. Walter and I. L. Schwartz, "Pharmacology of Hormonal Polypeptides and Proteins," N. Back, L. Martini, and R. Paoletti, Eds., Plenum Press, New York, N.Y., 1968, p. 101.

65. R. Walter, W. Gordon, I. L. Schwartz, F. Quadrifoglio, and D. W. Urry, "Peptides 1968," E. Bricas, Ed., North-Holland Publishing Co., Amsterdam, 1968, p. 50.

66. W. G. Penney and G. B. B. M. Sutherland, *J. Chem. Phys.*, **2**, 492 (1934).

67. L. Pauling, *Proc. Natl. Acad. Sci. U.S.*, **35**, 495 (1949).

68. O. Foss, "Organic Sulfur Compounds," Vol. I, N. Kharasch, Ed., Pergamon Press, London, 1961, p. 75.

69. S. C. Abrahams, *Quart. Rev.* (London), **10**, 407 (1956).

70. F. Fehér and R. Schulze-Rettmer, *Z. Anorg. Allg. Chem.*, **295**, 262 (1958).

71. L. Pauling, "The Nature of the Chemical Bond," 3rd ed., Cornell University Press, Ithaca, N.Y., 1960, p. 135.

72. D. W. Scott, H. L. Finke, M. E. Gross, G. B. Guthrie, and H. M. Huffman, *J. Amer. Chem. Soc.*, **72**, 2424 (1950).

73. D. W. Scott, H. L. Finke, J. P. McCullough, M. E. Gross, R. E. Pennington, and G. Waddington, *J. Amer. Chem. Soc.*, **74**, 2478 (1952).

74. G. B. Guthrie, Jr., D. W. Scott, and G. Waddington, *J. Amer. Chem. Soc.*, **76**, 1488 (1954).

75. W. N. Hubbard, D. R. Douslin, J. P. McCullough, D. W. Scott, S. S. Todd, J. F. Messerly, I. A. Hossenlopp, A. George, and G. Waddington, *J. Amer. Chem. Soc.*, **80**, 3547 (1958).

76. H. Kessler and W. Rundel, *Chem. Ber.*, **101**, 3350 (1968).

77. G. Bergson, *Ark. Kemi*, **18**, 409 (1961).

78. S. Husebye and G. Helland-Madsen, *Acta Chem. Scand.*, **23**, 1398 (1969) and citations contained therein.

79. A. F. Wells, "Structural Inorganic Chemistry," Oxford University Press, London, 1962, p. 418.

80. G. Claeson, *Acta Chem. Scand.*, **22**, 2429 (1968).

81. D. W. Urry, F. Quadrifoglio, R. Walter, and I. L. Schwartz, *Proc. Natl. Acad. Sci. U.S.*, **60**, 967 (1968).

82. K. Mislow, "Introduction to Stereochemistry," W. A. Benjamin, New York, N.Y., 1966, p. 158.

83. S. Beychok and E. Breslow, *J. Biol. Chem.*, **243**, 151 (1968).

84. M. Carmack and L. A. Neubert, *J. Amer. Chem. Soc.*, **89**, 7134 (1967).

85. R. Walter, "Proceedings of Second International Symposium on Structure-Activity Relationships of Proteins and Polypeptide Hormones," M. Margoulies and F. C. Greenwood, Eds., Excerpta Medica, Amsterdam, 1971, p. 181.

86. R. Walter and L. F. Johnson, 13th Annual Biophysical Society Meeting, *Biophys. J.*, **9**, A-159 (1969).

87. L. F. Johnson, I. L. Schwartz, and R. Walter, *Proc. Natl. Acad. Sci. U.S.*, **64**, 1269 (1969).

88. D. W. Urry, M. Ohnishi, and R. Walter, *Proc. Natl. Acad. Sci. U.S.*, **66**, 111 (1970).

89. D. W. Urry and R. Walter, *Proc. Natl. Acad. Sci. U.S.*, **68**, 956 (1971).

90. R. Walter, R. T. Havran, I. L. Schwartz, and L. F. Johnson, "Peptides 1969," E. Scoffone, Ed., North-Holland Publishing Co., Amsterdam, 1971, p. 255.

91. D. Kotelchuck, H. A. Scheraga, and R. Walter, *Proc. 3rd Amer. Peptide Symp.*, Boston, 1972, in press.

92. D. F. Koenig, C. C. Chiu, and R. Walter, "Abstracts of 8th International Congress of Crystallography, Stony Brook, N.Y., 1969," p. S187.

93. R. Walter, I. L. Schwartz, J. H. Darnell, and D. W. Urry, *Proc. Natl. Acad. Sci. U.S.*, **68**, 1355 (1971).

XIII Selenium compounds in nature and medicine

A. NUTRITIONAL IMPORTANCE OF SELENIUM

MILTON L. SCOTT

Cornell University
Department of Poultry Science
Ithaca, New York

1. Introduction

Nutritional research over the past decade has shown that selenium is an essential nutrient required for the prevention of a number of serious deficiency diseases in various species of livestock and poultry. This unexpected nutritional role of selenium emerged in 1957 with the discovery by Schwarz and Foltz[1] that the element was the unknown factor, previously found in yeast, kidney, and liver, required for prevention of dietary liver necrosis in vitamin E-deficient rats.

During the years since 1957, inorganic selenium compounds, such as sodium selenite or sodium selenate, have been shown to be intimately involved, together with vitamin E, in the prevention of a wide variety of nutritional deficiency diseases in various animals. Proof of the specific nutritional essentiality of selenium was obtained by Thompson and Scott,[2] who showed that chicks and Japanese quail receiving diets containing synthetic amino acids and other purified ingredients extremely low in selenium fail to grow and suffer high mortality unless their diet is supplemented with selenium. Addition of all known nutrients, including high levels of vitamin E, failed to prevent severe deficiency signs and death, whereas addition to the basal high-vitamin E diet of as little as 0.02 ppm of selenium (as sodium selenite) completely prevented all deficiency signs and promoted a good rate of growth in the chicks and quail.

Although selenium has now been shown to be required entirely apart from its interaction with vitamin E, the action of selenium in the prevention of deficiency diseases in animals is so intimately connected with that of vitamin E that a brief description of the vitamin E-deficiency diseases of animals is in order.

Vitamin E was discovered by Evans and Bishop[3] in 1922 as an unknown fat-soluble factor in vegetable oils required for normal reproduction in rats. Over the ensuing 45 years vitamin E has been shown to be concerned in maintaining the health of the brain, the vascular system, the erythrocytes, the skeletal muscles, the liver, the heart, the gonads, and the incisor teeth of rats, and in the prevention of yellow fat and ceroid in the adipose and other tissues. It appears to function in at least two metabolic roles: (1) as a fat-soluble antioxidant, and (2) in a more specific role interrelated with the metabolism of selenium and sulfur amino acids. Most of the vitamin E-deficiency diseases are now quite well defined. Recent research has provided information concerning the influence of polyunsaturated fatty acids, of antioxidants, of sulfur amino acids, and of selenium upon these various diseases. A summary of this information is presented in Table XIIIA-1.

Some "vitamin E-deficiency diseases" are now known to respond primarily to antioxidants; others have a more specific requirement for vitamin E; some require selenium as the primary preventive factor; while certain diseases may respond to either vitamin E or selenium, and others may require both vitamin E and selenium. The "selenium-responsive diseases" are necrotic liver degeneration in rats, mice, rabbits, and pigs; muscular dystrophies (myopathies) in lambs, calves, pigs, horses, and turkeys; infertility in ewes; "unthriftiness" in cattle and sheep; poor hair and feather development in pigs, horses, and chickens; exudative diathesis in chicks; and atrophy and fibrosis of the pancreas in chicks.

Descriptions of the selenium-deficiency diseases, and a review of the

Table XIIIA-1 Vitamin E deficiency diseases

DISEASE OR CONDITION	EXPERIMENTAL ANIMAL	TISSUE AFFECTED	INFLUENCED BY POLYUNSATURATED ACIDS	PREVENTED BY:			
				Vitamin E	Selenium	Antioxidant	Sulfur amino acid
I. Reproductive failure							
Embryonic degeneration							
Type A	Female rat, hen, turkey	Vascular system of embryo	X	X		X	
Type B	Cow, ewe			[a]			
Sterility	Male rat, guinea pig, hamster, dog, cock	Male gonads		X	X[b]		
II. Ailments of liver, blood, brain, capillaries, pancreas							
Liver necrosis	Rat, pig, beef calves	Liver		X	X		
Fibrosis	Chick, mouse	Pancreas		X	X		
Erythrocyte hemolysis	Rat, chick, man (premature infant)	Erythrocytes	X	X		X	
Plasma protein loss	Chick, turkey	Serum albumen		X	X		
Anemia	Monkey	Bone marrow		X			
Encephalomalacia	Chick	Cerebellum	X	X		X	
Exudative diathesis	Chick, turkey	Vascular system		X	X	X	
Kidney degeneration	Rat, mouse, monkey, mink	Kidney tubular epithelium	X	X	X		
Steatitis (ceroid)	Mink, pig, chick	Adipose tissue	X	X	X	X	
Depigmentation	Rat	Incisors	X	X		X	
Hair and feather loss	Chick, rat, horse, pig	Integument, hair, and feathers			X	?	
III. Nutritional myopathies							
Type A (nutritional muscular dystrophy)	Rabbit, guinea pig, monkey, duck, mouse, mink	Skeletal muscle		X		?	
Type B (white muscle disease)	Lamb, calf, kid	Skeletal and heart muscles		[a]	X[b]		
Type C	Turkey	Gizzard, heart		[a]			
Type D	Chicken	Skeletal muscle	[c]	X	X		X

[a] Not effective in diets severely deficient in selenium.
[b] When added to diets containing low levels of vitamin E.
[c] A low level (0.5%) of linoleic acid is necessary to produce dystrophy; higher levels did not increase vitamin E required for prevention.

research that has led to an understanding of the effects of selenium and its interactions with other nutrients in the prevention of these diseases, constitute the subject matter of this chapter.

2. Early Studies on Unknown Factors Interrelated with Vitamin E Nutrition

a. Studies on Dietary Necrotic Liver Degeneration in Rats. Dietary necrotic liver degeneration in rats was first described by Weichselbaum,[4] who attempted to maintain these animals on a cystine-free diet. This disease bears no relationship to fatty liver or cirrhosis. Daft *et al.*[5] showed that choline completely prevented cirrhosis of the liver and had no effect upon necrotic liver degeneration, whereas cystine completely prevented liver necrosis but did not prevent cirrhosis. These results demonstrated that cirrhosis and dietary liver necrosis are two separate and distinct entities.

Schwarz[6] demonstrated that dietary necrotic liver degeneration occurs only when the diet is deficient in both cystine and vitamin E. The significance of vitamin E as a protective agent against necrotic liver degeneration was first discovered by Schwarz in 1944.[7] His finding was confirmed by György and Goldblatt,[8] by Himsworth and Lindan,[9] and by Goettsch.[10]

Although György *et al.*[11] first showed that American yeasts prevented this disease, which occurred in rats fed a British brand of baker's yeast, these workers concluded that the "so-called dietary hepatic necrosis is due more to some unidentified 'toxic' factor than to pure deficiency" and believed that "tocopherol and cystine may act as 'detoxifying' agents." Schwarz[12] recognized that, in addition to either vitamin E or cystine, an unidentified nutritional factor was present in American brewer's yeasts which was capable of preventing dietary liver necrosis even in rats fed diets severely deficient in both vitamin E and cystine. Schwarz[12] termed this unidentified nutrient "factor 3."

b. Sources of Factor 3. Subsequent studies by Schwarz and his associates[13] showed that, in addition to American brewer's yeast, kidney, liver, and numerous other natural materials were good sources of factor 3. In contrast, torula yeast, produced on the sulfite waste liquor of the paper pulp industry, was a very poor source of factor 3 and thus was used as the protein source in semipurified diets to study factor 3 and to assay concentrates of it.

During the fractionation of factor 3 from acid hydrolysates of natural materials, Schwarz and Foltz[1] obtained two chemically similar substances with factor 3 activity. These were designated as α- and β-factors 3. α-Factor 3 was found to be water soluble, strongly anionic, and stable against oxidation, but sensitive to reducing agents. Dry ashing entirely eliminated factor

3 activity. Fractionation of α-factor 3 soon led to semicrystalline preparations, some of which developed a characteristic, "garlic-like" odor upon addition of alkali. This observation led to the discovery that selenium is an integral part of factor 3. Since it had been reported that the breath of cattle consuming high-selenium plants from seleniferous soils has a garlic-like odor, Schwarz and Foltz investigated the possibility that the factor 3 preparations might contain selenium. Analyses showed that the factor 3 activity of each concentrate was highly correlated with its selenium content. Inorganic selenium as sodium selenite was soon shown to be completely effective in preventing dietary liver necrosis. Schwarz and Foltz stated, therefore, that ". . . at least for the rat, factor 3 is not an essential organic dietary constituent, since it can be replaced by selenite." They went on to say:

The specific potencies of various organic and inorganic selenium compounds remain to be established. Most of the selenium in normal diets appears to be organically bound. The dose needed for prevention of dietary necrotic degeneration amounts to less than 1% of the chronic toxic dose, which has been placed at 300–400 μg/100 g of ration.[14]

In comparison with previously reported values for vitamin E and cystine,[13] selenium is exceedingly effective in the prevention of necrotic liver degeneration. When applied in the form of sodium selenite, it is approximately 500 times as active as vitamin E and 250,000 times as active as L-cystine. The element has been shown to be constantly present in tissues of higher animals.[15] It can be inferred from our results that selenium is an essential trace element.

*c. **Studies on Exudative Diathesis in Chicks.*** During the early 1950's, concurrently with the studies by Schwarz and his associates on dietary liver necrosis in rats, studies were being conducted by the author and others at Cornell University on an unidentified factor required for the prevention of exudative diathesis in chicks. An unknown factor in American brewer's yeast not present in torula yeast was found to spare the vitamin E requirement for the prevention of exudative diathesis in chicks.[16] The obvious probability that the unidentified factor required to prevent exudative diathesis in vitamin E-deficient chicks was identical with the unknown factor in yeast needed to prevent dietary liver necrosis in vitamin E-deficient rats was strengthened by Cornell studies showing that the potent factor 3 concentrates prepared by Schwarz were effective in preventing exudative diathesis in chicks, while all preparations found by Schwarz to be devoid of factor 3 activity also did not prevent exudative diathesis in vitamin E-deficient chicks.

Immediately after the discovery by Schwarz and Foltz[1] that selenium is the active principle in factor 3 concentrates, a collaborative study was undertaken at Cornell University and at the National Institutes of Health to determine the effect of selenium compounds upon exudative diathesis in chicks. The results showed that 0.1 ppm of selenium in sodium selenite completely prevented exudative diathesis in chicks, and also demonstrated that selenocystathionine can serve the same purpose. Subsequent studies have shown that selenocystathionine, selenomethionine, and selenocystine all are approximately as effective as selenium from selenite in preventing both dietary liver necrosis and exudative diathesis.

In completely independent studies, Patterson et al.,[17] using the torula yeast diet of Scott and his associates,[16] discovered that the unidentified factor required for prevention of exudative diathesis in chicks was present in the acid hydrolysate of hog kidney but that the ash of the acid-hydrolyzed kidney was inactive. Further work by Patterson et al.[17] showed that an ash of kidney made alkaline with calcium oxide before burning was active. This activity of an alkaline ash and inactivity of an acid ash suggested that the activity of the factor in kidney was due to an element which forms a volatile inorganic acid. Among the elements forming such volatile acids are arsenic, selenium, and tellurium. Subsequent studies by these workers showed that both arsenic and tellurium were inactive, but that selenium as sodium selenite was completely active in preventing exudative diathesis in chicks.

These studies, mutually confirming the effects of selenium in the prevention of exudative diathesis in chicks, were reported simultaneously at the 1957 meeting of the Poultry Science Association.[18,19]

3. Nutritional Importance of Selenium and Selenium-Deficiency Diseases in Laboratory and Farm Animals

a. Multiple Necrotic Degeneration. Dietary liver necrosis in the rat is described by Schwarz[20] as "a specialized case of the widespread systemic injury which involves other tissues as well." He goes on to say:

> In the mouse, for example, multiple necrotic degeneration ensues. Necrosis develops in the heart, the liver, the kidneys, and the peripheral muscles. In addition there is severe atrophy of the pancreas and degeneration of the testes. In multiple necrotic degeneration of the mouse the heart lesion predominates and develops several weeks before other gross changes can be seen. Its overall incidence is 91% as compared to 54% of the liver lesion.

> In the rat, the liver is apparently the site of predilection for massive necrosis. In this species the liver lesion develops very acutely, and kills the animal before many other changes appear. The acute necrosis of the liver is almost always the decisive pathological event which terminates the life of the animal.

The above observations prompted Schwarz and his associates to focus attention almost exclusively on the liver, describing the disease as dietary necrotic degeneration of this organ. They stated, however, that kidney lesions, in the form of degeneration of the intercorticomedullary zone, were detectable in about 87 % of all animals succumbing to liver necrosis; and occasionally muscle dystrophy was observed.

b. Necrotic Liver Degeneration. Schwarz[20] described dietary necrotic liver degeneration as consisting of two phases: (1) a latent phase, during which the liver appears essentially normal, and (2) a terminal phase. During the latent, prenecrotic phase, metabolic disturbances were detectable and degeneration of the endoplasmic reticulum and mitochondrial swelling were seen in electron micrographs. The metabolic abnormality was described as an inability of tissue slices and homogenates to maintain normal oxygen consumption *in vitro* for extended periods of time, and was attributed to the effects of a deficiency of factor 3 or selenium upon the enzyme lipoyldehydrogenase. The *in vitro* addition of vitamin E to homogenates protected the enzyme, but selenium was inactive *in vitro* and was not as effective as vitamin E even when included in the diet. It was suggested by Schwarz that selenium and vitamin E appear to be involved in different steps in intermediary metabolism.

In the absence of vitamin E or selenium the rats entered the "terminal phase" of necrotic liver degeneration at approximately 21–45 days on the deficient diets. At this time necrosis became grossly visible. It appeared primarily in the central portions of the lobes, but never as a very discrete central zone. The necrosis was not focal, but massive, invariably involving an area larger than an individual lobule. The usual necrosis was not related to areas of main blood supply. However, infarct-like changes often were seen. These had an irregular ramifying, serpentine pattern that corresponded to the branching of efferent veins. There appeared to be a narrow layer of preserved cells surrounding the portal veins. In seriously necrotic areas this zone was extremely narrow. Schwarz suggested that the better-nourished portal cells suffered less than the more peripheral cells of the liver.

In addition to frank necrosis a second outstanding feature of the gross picture of dietary liver necrosis was the finding of hemorrhagic areas which previous workers had grouped together with the necrosis as "hemorrhage and necrosis." Careful studies by Schwarz indicated that these changes did not constitute true hemorrhage but rather accumulation and congestion of blood within the sinusoids. The pattern suggested a block in the efferent part of the vascular system and possibly a back-tracking of blood from efferent veins into necrotic areas. In a report on the histological hepatic changes seen in dietary liver necrosis, Fite[21] described a diffuse change in the nucleus of the cells characterized by karyolysis and karyorrhexis. This change was consistently observed during the terminal phase of the disease. Fite also reported the existence of granules distributed evenly in the cytoplasm which had appeared to undergo some calcification during the latent phase of the disease. These "degenerative microbodies" were eosinophilic and were believed to possibly develop from degenerating mitochondria. The duration of the dramatic terminal phase of the disease varied from a few hours to several days. It always culminated in the sudden death of the animal. Schwarz and Mertz[22] reported that the pattern of symptoms found for the terminal phase of dietary liver necrosis closely resembled that seen in hepatectomized dogs. A severe hypoglycemia occurred, and lactic acid levels increased in the blood. After hepatectomy, as in the terminal phase of liver necrosis, intravenous glucose infusions prolonged life for a number of hours. The lack of hepatic function then caused death *via* some other mechanism not concerned with the maintenance of blood glucose levels.

The mechanism of action of inorganic or organic forms of selenium in the complete prevention of dietary liver necrosis remains to be determined.

c. **Studies with Chicks and Turkey Poults.**   Selenium metabolism in chicks and poults is intimately related to their nutritional status and to the metabolism of vitamin E, cystine, linoleic acid, and synthetic antioxidants.

Selenium spares the vitamin E requirement of the chick in two ways: (1) selenium is required to preserve the integrity of the pancreas, which in turn allows normal fat digestion, normal lipid-bile salt micelle formation, and thus normal vitamin E absorption;[23] and (2) selenium aids in some unknown way in the retention of vitamin E in blood plasma.[24] In turn, vitamin E reduces the selenium requirement of chicks, apparently by maintaining the body selenium in an active form or by preventing its loss from the body.

Vitamin E deficiency expresses itself in various ways in chicks and in turkey poults, depending on the levels of various other nutrients in the diet. Encephalomalacia in chicks, which responds to vitamin E and to synthetic antioxidants, is not prevented by selenium. Selenium spares the amount of vitamin E required to prevent nutritional muscular dystrophy in chicks but will not prevent dystrophy when added to diets severely deficient in both vitamin E and sulfur amino acids.[25] Thus, the selenium-responsive diseases of the chick are (1) atrophy and fibrosis of the pancreas, (2) exudative diathesis, and (3) poor growth and severe alopecia. Pancreatic degeneration and feather loss are prevented only by selenium, while exudative diathesis is prevented by either selenium or vitamin E.

Young poults have been reported to suffer from exudative diathesis,[26] but the chief symptoms of selenium deficiency in turkey poults are myopathies of the gizzard and the heart.[27,28]

d. **Exudative Diathesis.**   The term exudative diathesis was coined by Dam and Glavind[29] in 1940 to describe a severe edema apparently due to an increased permeability of the capillaries in vitamin E-deficient chicks. The disease occurs only in chicks receiving diets severely deficient in both selenium and vitamin E. Exudate fluids are prominently apparent under the skin covering the abdomen and also may appear as a "parrot-like" puffiness surrounding the mandibles; sometimes on the under side of the wings; and often draining down into the subcutaneous areas of the legs. The edematous fluid may be colorless, it may be straw colored, or it may have a purplish blue hue due to the presence of heme. Undoubtedly heme is present in the exudate fluids only when the vitamin E deficiency is sufficiently severe to result in a simultaneous occurrence of erythrocyte hemolysis. Goldstein and Scott[30] observed a markedly reduced level of blood proteins, particularly plasma albumin, in chicks suffering from exudative diathesis. They obtained some evidence of the presence of albumin in the tissue exudates. Whether the edema is due to a breakdown in capillary walls resulting in increased permeability of the plasma fluids, or to a decreased synthesis of blood proteins that results in a marked decrease in intravascular osmotic pressure, or both, remains to be determined. It has been postulated that exudative diathesis occurs because of peroxidation of the linoleic acid in the phospholipid, which, together with the mucopolysaccharides and proteins, makes up the cell structure of the capillary walls. However, this position proves vulnerable when it is known that synthetic antioxidants at levels which completely prevent encephalomalacia have no effect on exudative diathesis, and conversely that selenium at levels which completely prevent exudative diathesis has no effect on encephalomalacia.

The course of the disease, if it has not developed to its terminal stages, may be reversed within 48–72 hours by injection of either vitamin E or selenium. Many outbreaks of exudative diathesis which have occurred in commercial flocks of broilers being reared on feeds produced from low-selenium ingredients have been shown to respond almost immediately, with decreased mortality and disappearance of symptoms of exudative diathesis, to the administration of approximately 0.1 ppm of sodium selenite in the drinking water. Undoubtedly prevention and cure of this disease also could be achieved by supplementing the diet of broiler chicks with as little as 0.1 ppm of selenium as sodium selenite.

e. Selenium Deficiency and Pancreatic Fibrosis. A severe selenium deficiency in chicks produces atrophy and degeneration of the pancreas.[23] A description of this disease and details concerning its significance in the metabolism of chicks are given in Section XIII.A.4.

f. Selenium and Other Vitamin E-Deficiency Diseases in Chicks. Selenium has been reported to have some effect in preventing encephalomalacia, muscular dystrophy, erythrocyte hemolysis, and yellow fat disease in chicks receiving diets deficient but not completely devoid of vitamin E. In no instance has selenium been shown to prevent these diseases in the complete absence of vitamin E. Undoubtedly the effect of selenium upon these diseases stems from its action in improving the absorption of vitamin E, and helping in the retention of vitamin E in blood plasma.

g. Selenium and Turkey Nutrition.   Exudative diathesis has been described in poults by Creech and his associates.[26] The symptoms were similar to those described in chicks except that the anemia was found to be macrocytic rather than the microcytic type described by Scott and his associates[16] in chicks suffering from exudative diathesis. Furthermore, these workers did not show changes in the protein components of blood serum as determined electrophoretically in poults suffering from exudative diathesis. However, selenium-depleted poults had reduced A/G ratios, which could be significantly increased by supplementation of the basal diet with α-tocopheryl acetate or dried brewer's yeast (a source of selenium).

h. Myopathy of the Gizzard.   Gizzard myopathy in turkey poults was originally reported by Jungherr and Pappenheimer[31] to be the most characteristic, if not the only, vitamin E-deficiency symptom in turkeys. Since 1957, however, several groups of workers have shown that under appropriate conditions vitamin E-deficient poults also display nutritional muscular dystrophy (myopathy of the skeletal muscles), as well as severe myopathies of the smooth muscle (gizzard) and the cardiac muscle. Ferguson and his associates[32] reported nutritional myopathy of skeletal muscle in turkey poults receiving diets containing 10 % ethyl linoleate in which the methionine and arginine levels were varied. Walter and Jensen[28] reported nutritional myopathies of skeletal, gizzard, and heart muscles in 4-week-old poults that received a vitamin E-deficient, selenium-low diet containing torula yeast as the source of protein. These workers also noted the development of severe pericardial exudates and increased levels of glutamic oxaloacetic transaminase. They reported that the addition of sulfur-containing amino acids to the diet had no beneficial effect on these myopathies, whereas selenium as sodium selenite was completely effective in preventing all symptoms.

Interest in the selenium-responsive disease of poults was strengthened during the summer of 1964, when several commercial flocks of young turkeys in southwestern Ohio were found to grow poorly and suffer a high mortality rate, with no overt signs of disease or nutritional deficiency except for a severe hyaline degeneration of the gizzard musculature. Upon investigation it was found that both the corn and the soybean meal used in the turkey diets were produced locally in southern Ohio or Indiana, on soils found by Muth and Allaway[33] to be very low in selenium and to produce forages so deficient in this element that they failed to prevent white muscle disease in lambs and calves. Poults showing myopathy of the gizzard were found in four flocks totaling about 15,000 turkeys.

The case history of this outbreak was described by Bruins et al.[34] as follows:

All poults were of a late hatch from a breeding flock located in Northern Indiana. Histologic examination of the affected gizzards by Dr. Frank Mitchell of Pitman-Moore Laboratories, Indianapolis, Indiana, revealed "extensive acute coagulation necrosis and fibrous tissue replacement involving musculature of the gizzard." Mitchell observed that he had not previously encountered cases similar to this among turkeys; however, coagulation necrosis of muscle fibers is a principal pathologic feature of vitamin E deficiency as it occurs in sheep, cattle and other species. Similar histopathology also was observed in the myocardium of poults from one of the turkey flocks. Mortality at 6 weeks of age ranged from 14 to 30% in three of the four flocks, and 65% in the fourth flock which also was afflicted with chronic respiratory disease.

Scott and his associates[27] used both a selenium-low practical turkey starting diet and a semipurified basal diet to produce selenium deficiency in young turkey poults. They reported poor growth, mortality, and myopathies of the heart and gizzard in young poults receiving a practical diet containing all nutrients previously known to be required, except for supplemental vitamin E and methionine. The corn and soybean meal used in this diet were obtained from geographical areas where the soil is known to be low in selenium. Addition of vitamin E and methionine improved growth, but gizzard myopathy was not prevented and maximal growth was not achieved until the diet also was supplemented with at least 0.1 ppm of selenium as sodium selenite. Under the conditions of these experiments the selenium requirement in the practical-type diet depended to some extent on the amount of vitamin E or methionine supplementation; it ranged from approximately 0.18 ppm in the presence of supplemental vitamin E to approximately 0.28 ppm of selenium as sodium selenite in the absence of added vitamin E.

Although myopathy of the skeletal (pectoral) muscle was not observed grossly, necropsy studies revealed histological changes in skeletal muscle typical of Zenker's degeneration. The order of prominence of the "selenium-responsive" diseases of the young poult appears to be: first, myopathy of the smooth muscle (gizzard); second, myopathy of the myocardium; and third, myopathy of the skeletal muscle. Selenium appears to be the primary nutritional factor required. Vitamin E is of less importance, and sulfur amino acids are completely ineffective in the prevention of these myopathies in poults. Photomicrographs showing the myopathies of the gizzard and heart are presented in Figs. XIIIA-1 and XIIIA-2.

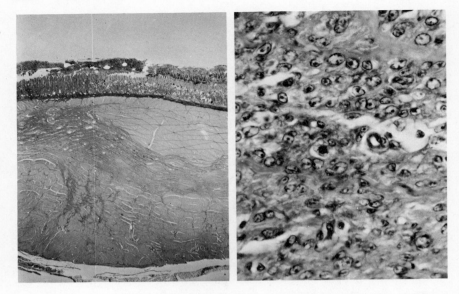

FIG. XIIIA-1   *Left:* Gizzard. Severe fibrosis, of haphazard distribution. Selenium-deficient poult, 4 weeks. Van Gieson, ×7.5.[27] *Right:* Pronounced proliferation of gizzard muscle nuclei. Nucleus just below center in mitosis. Selenium-deficient poult, 2 weeks. H & E, oil immersion, ×600.[27]

i.   Selenium Deficiency in Lambs and Calves.   A myopathy in young lambs generally referred to as "stiff lamb disease" was first recognized and described by Metzger and Hagen at Cornell University in 1927.[35] This disease was produced experimentally by feeding pregnant ewes a diet consisting of hay and cull beans obtained from certain areas of New York State. After the reports on the effects of inorganic selenium compounds on dietary liver necrosis in rats and on exudative diathesis in chicks, studies were initiated almost simultaneously by Proctor and his associates[36] of Cornell University and Muth *et al.*[37] of Oregon State University, to determine the effects of supplementation of the basal diets of pregnant ewes that previously had been shown to produce nutritional myopathy in lambs. These research groups reported almost complete prevention of nutritional myopathy in newborn lambs by addition of sodium selenite at a level of 0.1–1.0 part of selenium per million parts of total diet fed to the ewes.

This work led to a monumental study by Allaway and his colleagues of the U.S. Plant, Soil and Nutrition Laboratory at Ithaca, New York. These workers completely mapped the seleniferous and selenium-deficient areas of the United States, and showed that stiff lamb disease and white muscle

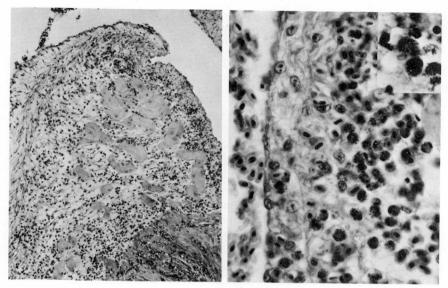

FIG. XIIIA-2 *Left:* Heart, left atrium. Acute Zenker's degeneration with muscle bundles separated by edema and hemorrhages. Selenium-deficient poult, 3 weeks. H & E, ×90.[27] *Right:* Heart, left atrium. Erythrocytes in left portion of field are outside epicardium. Extensive subepicardial hemorrhage and edema with admixture of heterophilic leukocytes. H & E, oil immersion, ×600.[27] Inserted in upper right corner: detail of heterophilic leukocytes. Selenium-deficient poult, 3 weeks. H & E, oil immersion, ×862.5.[27]

disease occur only in areas in which the soil and therefore the forage crops are severely deficient in selenium.[38,39]

The etiology and pathology of white muscle disease have been described in detail by Muth.[40,41] This disease commonly occurs in young calves or lambs born to dams fed a diet extremely low in selenium (approximately 0.02 mg Se/kg of dry matter, or less). The name "white muscle disease" has been applied to this selenium and vitamin E-deficiency disease because of the light color of the affected muscles, especially the myocardium.

j. Congenital White Muscle Disease. Affected lambs either are born dead or may die suddenly after exertion. Postmortem examination usually reveals clear fluid and fibrin strands in the body cavities and a congested liver. The heart shows a mild focal to an extensive diffuse grayish white discoloration of the subendocardial myocardium. This lesion extends for only approximately 1 mm into the myocardium and most commonly affects the right ventricle, although the other cavities may be involved. Skeletal musculature is rarely affected. Microscopically, the initial lesion consists of large areas of noninflammatory coagulative myonecrosis. This is superseded by lysis or calcification of the affected fibers and replacement by macrophages and fibroblasts.

k.   Delayed White Muscle Disease of Lambs and Calves.   A second form of white muscle disease occurs usually in lambs from 3 to 6 weeks of age but may manifest itself as late as 3 months of age. Only a few cases have been seen in calves between 1 and 4 months old. The affected animals walk with a stiff, stilted gait and an arched back. They move about with extreme difficulty, waste away, become prostrate, and die. In those with severe heart involvement exercise may bring about sudden death. Recently affected animals have been found to have a high serum glutamic oxaloacetic transaminase level and a high urinary creatine. Postmortem examination shows a bilateral symmetrical focal or diffuse yellowish gray discoloration of the skeletal muscles. Cardiac lesions, when present, are similar to those seen in the congenital form of white muscle disease. Histologically, the skeletal lesions consist of extensive areas of noninflammatory coagulative myonecrosis, followed by lysis or calcification of affected fibers and invasion by macrophages. Some muscle fiber regeneration is apparent.

l.   Selenium-Responsive Infertility in Ewes.   Andrews et al.[42] reported that the embryonic mortality which has occurred in ewes in the South Island of New Zealand for decades, and often runs as high as 75%, is prevented by administration of selenium to the ewe just before mating. In extreme cases lambing percentages were increased from 25% to 90%. Infertility due to selenium deficiency apparently does not affect cattle, since calving percentages for cattle grazing with sheep experiencing selenium-responsive infertility appear satisfactory.

m.   Selenium-Responsive Unthriftiness of Sheep and Cattle.   Andrews et al.[42] reported that selenium-responsive unthriftiness is probably the most widespread and economically important of all the selenium-responsive diseases of New Zealand livestock. Unthriftiness is characterized by subclinical inability to maintain optimum growth rate. Lambs may apparently thrive for several months and then show abruptly reduced weight gains; others stop eating, stop growing, lose weight, become dejected, and die. Unthriftiness in all ages of dairy and beef cattle occurs particularly in autumn and winter months and varies from a subclinical growth depression to a syndrome characterized by a sudden and rapidly progressive loss of condition, profuse diarrhea, and sometimes high mortality. Selenium has been shown by several workers to prevent this unthrifty condition in both sheep and cattle.

In the United States the usual observations in selenium-deficient ruminants have been confined to both congenital and delayed white muscle disease in lambs and calves. Few if any reports have been made concerning cases of reproductive failure or general unthriftiness which have been shown to respond to selenium supplementation. However, Wu *et al.*[43] reported impaired testicular function in selenium-deficient rats. Trinder *et al.*[44] in England investigated the effects of selenium therapy on the incidence of retained placenta in dairy cows. Parenteral administration of vitamin E and potassium selenate 1 month prepartum greatly reduced retention of the placenta, but potassium selenate alone was less effective than the combination of vitamin E and selenate.

n.   Liver Necrosis.   Todd and Krook[45] obtained evidence that fattening beef animals also develop liver necrosis (sawdust livers) when fed grain rations low in selenium.

o.   Selenium Deficiency in Swine.   When weanling pigs receive diets deficient in both selenium and vitamin E, necrotic liver degeneration or *hepatosis dietetica* occurs; this usually results in death unless the diet is supplemented with either vitamin E or selenium.[46] Some workers have reported both liver and cardiac myopathy (mulberry heart), as well as skeletal muscle degeneration in pigs fed a vitamin E-deficient diet also low in selenium.[47] These deficiency diseases are prevented by supplementing the diet with either vitamin E or selenium but not cystine. Orstadius *et al.*[48] found that pigs receiving a diet deficient in both vitamin E and selenium showed elevated glutamic oxalic transaminase (GOT) levels of the plasma, which were reduced by supplementing the diet with either vitamin E or selenium, but that maximum reduction of plasma GOT levels occurred when the diet was supplemented with both vitamin E and selenium.

Field cases of *hepatosis dietetica* have been observed in Michigan and Washington, areas known to be low in selenium, a well as in New Zealand. This disease has been reported by Hartley and Grant[49] to be due to a selenium deficiency. The pigs were found to show liver degeneration, generalized subcutaneous edema (similar to exudative diathesis in chicks), pale skeletal muscle (similar to white muscle disease in ruminants), accumulation of straw-colored fluid in the body cavity, and degeneration of the myocardium (mulberry heart). Losses from this disease have been controlled by the administration of selenium.

p.   Effects of Selenium on Hair and Feather Development.   Attempts have been made in the author's laboratory to determine the symptoms of a chronic submarginal selenium deficiency over a long period of time in chickens. The main result of such a deficiency appears to be an almost complete loss of feathers. The skin and feathers become rough and dry, and the birds gradually lose their feathers until they become almost completely nude.

This occurs with a level of selenium in the diet that is sufficiently high to prevent both pancreatic degeneration and exudative diathesis, and in chicks receiving sufficient antioxidant to prevent encephalomalacia and adequate methionine and cystine for normal growth and prevention of nutritional muscular dystrophy.

The specific effects of selenium on the growth of wool and hair have been investigated by Hartley and Grant[49] and by Slen et al.,[50] who noted that diet supplementation with subtoxic amounts of selenium resulted in higher total fleece weights and increased wool fiber thickness in sheep. In vitro treatment of wool fibers weakened by reduction of cystine cross linkages to cysteine residues with selenious acid caused them to be restrengthened, apparently because of formation of selenium-containing cross linkages.

Hair loss has been reported to be a sign of selenium deficiency in several mammals, including rats, pigs, and horses. The extent to which this represents a specific selenium deficiency, as opposed to an unknown interrelationship between selenium and some other nutrients, remains to be determined.

4. Proof That Selenium Is an Essential Nutrient

Selenium Deficiency and Pancreatic Fibrosis. Some recent experiments have been conducted with selenium-depleted chicks fed a crystalline amino acid diet adequate in all known nutrients, including high levels of vitamin E, but severely deficient in selenium.[23] The results showed that a specific selenium deficiency causes poor growth, poor feathering, and fibrotic degeneration of the pancreas. Death usually occurred after markedly decreased absorption of lipids, including vitamin E. The pancreatic degeneration results in a decrease in pancreatic and intestinal lipase, which causes a failure in fat digestion. Under these conditions bile flow is almost eliminated. With the absence of bile and of monoglycerides in the intestinal lumen, there is a failure of micelle formation, which in turn impairs the absorption of vitamin E. Addition of bile salts to the diet did not return fat digestion to normal levels and only temporarily enhanced vitamin E absorption. Addition to the basal diet of free fatty acids, monoglycerides, and bile salts improved the absorption of vitamin E and survival during the experimental period of 4 weeks, but did not prevent the degenerative changes in the pancreas. The selenium requirement for prevention of pancreatic degeneration was found to depend on the vitamin E level in the diet. With very high dietary vitamin E levels (100 IU/kg or more) as little as 0.01 mg Se as sodium selenite per kg of diet completely prevented pancreatic degeneration. However, when the vitamin E content of the diet was at more nearly normal levels (10–15 IU/kg), 0.02–0.04 mg Se/kg of diet was required.

It was observed in these experiments that exudative diathesis did not occur as long as some vitamin E was being absorbed. The disease occurred only after the absorption of vitamin E was decreased to very low levels by the pancreatic degeneration. With the administration of fatty acids, monoolein, and bile salts, which resulted in a marked improvement in vitamin E absorption, exudative diathesis did not occur, even though the pancreas under these dietary conditions continued to undergo severe atrophy and degeneration. As a result of these studies, the following scheme is proposed to depict the interrelationship between vitamin E and selenium in the prevention of both exudative diathesis and pancreatic degeneration:

1. When the diet contains adequate selenium, the integrity of the pancreas is preserved.

2. With the normal production of pancreatic lipase, fat is digested, and there is normal bile flow as well as normal formation of bile salt-lipid micelles.

3. Micelle formation then allows maximum absorption and good blood levels of vitamin E.

4. Either the vitamin E or the selenium in the diet will prevent exudative diathesis, but a good blood level of vitamin E also helps to preserve the selenium in the blood and tissues, thereby reducing the dietary level of selenium required to protect the pancreas.

5. Thus, vitamin E spares the selenium requirement, and, conversely, selenium enhances the absorption of α-tocopherol, thereby reducing the dietary requirement for vitamin E.

Recently, McCoy and Weswig[51] reported on the specific effects of selenium deficiency independent of vitamin E in young rats. Rats fed a low-selenium diet containing torula yeast supplemented with 60 mg d-α-tocopheryl acetate/kg grew and reproduced normally. However, their offspring were almost hairless, grew slowly, and failed to reproduce. Addition of selenium at a level of 0.1 ppm as sodium selenite restored the hair coat, growth, and reproductive capabilities. Selenium also appeared to be necessary for the prevention of eye discoloration.

5. Nutritional Interrelationships

a. Two Effects of Selenium on Vitamin E Metabolism.

The requirement for selenium for normal integrity of the pancreas is very low: approximately 0.01 ppm. Since most diets, both practical and purified, contain at least 0.02 ppm of this element, it appears highly unlikely that pancreatic fibrosis from selenium deficiency occurs under practical conditions.

The amount of selenium required for maximum retention of d-α-tocopherol, however, is greater than 0.1 ppm. Desai and Scott[24] showed, by chemical determination of plasma tocopherols and also by tracing the activities of tritiated tocopherols and ^{75}Se in serum and in various fractions of serum proteins in chicks receiving the nutrients alone and in combination, that the selenium and d-α-tocopherol activities followed each other very closely in the serum proteins. These studies indicated that vitamin E may be carried by a seleno-lipoprotein fraction associated with serum γ-globulin. Thus, one biological role of selenium appears to lie in a selenium-containing compound which acts as a carrier of vitamin E and which may function in the absorption, retention, and prevention of destruction of d-α-tocopherol, and perhaps in its transfer across cell membranes, thereby enhancing its biological activity in the blood and perhaps in the cells throughout the body.

Results of these studies indicated that the variations observed in the biological activities of the d- and l-epimers of α-tocopherol, originally thought to be due to differences in absorption, now appear to be brought about by differences in retention in blood plasma. Results showed that selenium improves the effectiveness of d-α-tocopherol much more than that of the l-epimer for the prevention of nutritional muscular dystrophy in chicks.[52]

b. Interrelationships with Zinc, Cadmium, Thallium, Mercury, and Arsenic. Parizek and Zahor[53] showed that subcutaneous injections of subtoxic doses of cadmium chloride will produce injury to the testis of the rat and other mammalian species. Evidence indicates that this injury is mediated through an effect of cadmium on the permeability of the capillary bed of the testis parenchyma, which leads to ischemic necrosis of the seminiferous tubules.[54] Although early work by Parizek[55] indicated that this cadmium injury is prevented by zinc, more recent experiments reported by Mason and Young[54] show that selenium dioxide is approximately 100 times as effective as zinc acetate, and that selenium is effective when injected simultaneously with the cadmium chloride, whereas zinc acetate prevents testis injury only when given at least 8 hours before the cadmium chloride injection.

Several interesting interrelationships have been discovered concerning the mutual effects of selenium, cadmium, mercury, thallium, and arsenic. These effects appear to be concerned with the influences of one element on the rate and mode of excretion of one or more of the others.

Arsenic counteracts some of the toxicity of high levels of selenium[56–58] and decreases the retention of the element in the livers of rats.[59] Levander and Baumann[60] showed that this effect was due to an influence of arsenic in increasing the removal of selenium from the liver into the bile. Ganther and Baumann[61] found that cadmium increases the retention of selenium in the liver by inhibiting the formation of volatile methylated selenium compounds. Levander and Argrett[62] showed that arsenic, mercury, and thallium also inhibit the pulmonary excretion of volatile selenium compounds. Parizek et al.[63] reported that injections of sodium selenite or the presence of selenite in the diet markedly increased the blood levels of cadmium and mercury in rats injected with these metals. Holmberg and Ferm[64] found that sodium selenite injections given simultaneously with cadmium sulfate or sodium arsenate provided significant protection against teratogenesis in embryos of pregnant hamsters.

The mode of action of these elements and the relationship of the various other elements to selenium metabolism are not known. However, Gunn and Gould[65] have shown that, like selenium, cysteine and BAL also protect against cadmium damage to the rat testis and have suggested that the effect of selenium may be due to its complexing with cadmium, thereby preventing the latter from reacting with sulfhydryl groups in the testicular blood vessels.

6. Quantitative Nutritional Requirements

Early studies on the prevention of dietary liver necrosis in rats showed that the quantitative requirement for selenium for this purpose was approximately 0.04 ppm when added to a torula yeast diet deficient in vitamin E. Studies with lambs and calves indicate that approximately 0.1 ppm of selenium is needed in the diet of the ewe in order to completely prevent stiff lambs. Although quantitative studies have not been reported for cattle, field results indicate that approximately 0.1 ppm of selenium in the diet of the pregnant cow will prevent white muscle disease in the calf. This level of selenium also appears to be adequate in the diet of the young lamb and the young calf. Chicks receiving a vitamin E-deficient diet require approximately 0.08 ppm of selenium, either as sodium selenite or as selenomethionine or selenocystine, for the prevention of exudative diathesis. The amount of selenium required to prevent pancreatic fibrosis is much lower, that is, approximately 0.02 ppm.

The studies with chicks have shown that the selenium requirement depends to a great extent on the vitamin E content of the diet. In diets containing the usual amounts of vitamin E present in normal practical feedstuffs (approximately 10–15 IU/kg of diet) the selenium requirement of chicks appears to be approximately 0.04 ppm.

The selenium requirement of turkeys is higher than that of chicks. In turkeys receiving a vitamin E-deficient diet, approximately 0.28 ppm Se is required to prevent gizzard and heart myopathy. A level of 0.18 ppm Se has been shown to be required even when the diet contains normal practical levels of vitamin E (15 IU/kg of diet).

Selenium Levels in Blood and Tissues. The selenium levels in the blood and tissues have been found to reflect the dietary intake up to levels of dietary selenium which are slightly above the nutritional requirement for this element. When graded levels of selenium as sodium selenite were fed to chicks and poults, the amounts of selenium found in the blood, muscle, skin, liver, and kidney increased until the diet contained about 0.4 ppm Se. As the dietary selenium increased above this level, the amount of selenium in the tissues tended to plateau. The results of these studies are presented in Tables XIIIA-2 and XIIIA-3.

An interesting finding in the course of these studies was that, whereas the selenium level of the tissues reached a plateau when sodium selenite was used as the source of dietary selenium, the tissue levels tended to increase above this plateau when the dietary selenium was present as natural selenium in wheat. It appears likely that, when selenium is ingested as inorganic selenite, the body can retain only that amount which is capable of reacting with certain selenium-binding sites in the animal body (possibly cystine and other sulfur compounds). Any excess, inorganic selenium over that which can be bound appears to be quickly excreted. However, when the dietary selenium is present as selenomethionine or selenocystine, these forms may be deposited directly into the tissue proteins of the animal and thus may accumulate considerably higher levels of selenium than can occur when inorganic forms of the element are present in the diet.

It is of interest that the selenium content of the blood and muscles of chicks is higher than that of turkeys at each level of dietary sodium selenite. The higher nutritional selenium requirement of turkeys as compared with chickens may be due to poorer retention of the element by turkeys. The fact that the kidney selenium levels in the turkeys were somewhat higher than those of chickens when sodium selenite was fed lends further support to this hypothesis.

The retention of dietary selenium has been studied using [75]Se.[66] These studies indicate that approximately 25–75 % of an oral dose is excreted within a few days. Selenium-deficient animals retain more selenium than do those which have been receiving a diet adequate in the element. Selenium is deposited at different rates in different tissues of the body; the pancreas, kidneys, and liver retain considerably more than the muscles, heart, or lungs. While simple-stomached animals tend to excrete excess selenium *via* the urine, ruminants pass more selenium into the feces than do nonruminants.

Moxon and Rhian[67] showed that rats changed from a high-selenium to a low-selenium diet excreted most of their body selenium within 2 weeks. Allaway *et al.*[68] found, however, that ewes previously receiving high-selenium diets (2.6 ppm) retained sufficient selenium so that, even when fed a low-selenium diet (0.03 ppm) for 10 months, they produced lambs that were free of white muscle disease and had more tissue selenium than was

Table XIIIA-2 Selenium content of blood and tissues of chicks

DIETARY TREATMENT	AVERAGE CHICK WEIGHT,[a] 4 weeks (g)	WHOLE BLOOD[b]		MUSCLE[c]		LIVER[c]		SKIN[c]		KIDNEY[c]	
		4 weeks (ppm)	8 weeks (ppm)	Dry Basis (ppm)	Wet Basis (ppm)	Dry Basis (ppm)	Wet Basis (ppm)	Dry Basis (ppm)	Wet Basis (ppm)	Dry Basis (ppm)	Wet Basis (ppm)
1. Basal (low-Se) diet (0.07 ppm Se)	616	0.082	0.147	0.23	0.061	0.74	0.25	0.17	0.09	1.31	0.39
2. Basal + 0.1 ppm Se (0.22 ppm Na$_2$SeO$_3$)	600	0.28	0.178	0.29	0.071	1.48	0.48	0.25	0.13	1.85	0.34
3. Basal + 0.2 ppm Se (0.44 ppm Na$_2$SeO$_3$)	617	0.188	0.188	0.40	0.103	1.58	0.53	0.35	0.16	2.50	0.80
4. Basal + 0.4 ppm Se (0.88 ppm Na$_2$SeO$_3$)	601	0.204	0.203	0.45	0.114	1.82	0.59	0.28	0.13	2.75	0.56
5. Basal + 0.6 ppm Se (1.32 ppm Na$_2$SeO$_3$)	628	0.230	0.227	0.50	0.126	2.28	0.80	0.43	0.18	2.81	0.62
6. Basal + 0.8 ppm Se (1.76 ppm Na$_2$SeO$_3$)	602	0.238	0.250	0.58	0.157	2.34	0.77	0.27	0.15	3.41	0.71
7. High-Se basal diet[d] (0.67 ppm Se)	619	0.327	0.338	1.20	0.293	2.37	0.80	0.51	0.25	3.83	1.08
8. High-Se basal diet + 0.2 ppm Se (0.44 ppm Na$_2$SeO$_3$)	610	0.390	0.368	1.63	0.423	2.88	1.05	0.62	0.29	4.27	0.98

[a] At end of experiment, 16–20 chicks per lot.

[b] Blood selenium values are averages of individual analyses on 10 chicks (5 male and 5 female) from each treatment at 4 weeks of age and 3 individual analyses (2 male and 1 female) at 8 weeks of age.

[c] Selenium values on muscle, liver, skin, and kidney are averages of individual analyses on 3 chicks (2 male and 1 female) per treatment, at 4 weeks of age.

[d] Selenium in this diet provided by high-selenium wheat.

Table XIIIA-3 Selenium content of blood and tissues of turkeys

DIETARY TREATMENT	AVERAGE POULT WEIGHT, 4 weeks[a] (g)	WHOLE BLOOD[b] (ppm)	MUSCLE[c] Dry Basis (ppm)	MUSCLE[c] Wet Basis (ppm)	LIVER[c] Dry Basis (ppm)	LIVER[c] Wet Basis (ppm)	SKIN[c] Dry Basis (ppm)	SKIN[c] Wet Basis (ppm)	KIDNEY[c] Dry Basis (ppm)	KIDNEY[c] Wet Basis (ppm)
1. Basal (low-Se) diet (0.07 ppm Se)	529	0.055	0.21	0.056	0.42	0.15	0.22	0.07	0.72	0.28
2. Basal + 0.1 ppm Se (0.22 ppm Na_2SeO_3)	550	0.082	0.27	0.07	0.96	0.33	0.36	0.14	2.00	0.63
3. Basal + 0.2 ppm Se (0.44 ppm Na_2SeO_3)	503	0.106	0.31	0.08	1.68	0.54	0.42	0.13	2.65	0.86
4. Basal + 0.4 ppm Se (0.88 ppm Na_2SeO_3)	514	0.116	0.37	0.11	1.84	0.56	0.41	0.12	2.90	0.80
5. Basal + 0.06 ppm Se (1.32 ppm Na_2SeO_3)	490	0.150	0.38	0.10	2.23	0.78	0.48	0.16	3.27	1.08
6. Basal + 0.8 ppm Se (1.76 ppm Na_2SeO_3)	540	0.150	0.43	0.10	2.75	0.94	0.48	0.17	3.42	1.23
7. High-Se diet (0.68 ppm Se)[d]	511	0.244	1.33	0.32	3.18	1.03	0.90	0.36	4.14	0.98
8. High-Se diet + 0.2 ppm Se (0.44 ppm Na_2SeO_3)	531	0.243	1.37	0.35	3.03	1.06	0.90	0.31	4.21	1.02

[a] At end of experiment, 14–20 turkeys per lot.

[b] Blood selenium values are averages of individual analyses on 10 poults from each treatment at 4 weeks of age.

[c] Selenium values on muscle, liver, skin, and kidney are averages of individual analyses on 3 poults (2 male and 1 female) at 4 weeks of age.

[d] Selenium in this diet provided by high-selenium wheat.

found in lambs of ewes that had been raised from birth on low-selenium diets.

7. Selenium Content of Foods and Feeds

a. Chemical Analyses. Upon discovery of the nutritional importance of selenium, it became desirable to determine the content of the element in foods and feeds. It was soon found that the macro method in use by the Association of Official Agricultural Chemists (AOAC) for detection of high levels of selenium in foods and feeds from seleniferous areas was much too cumbersome, and failed to detect quantitatively the extremely low levels present in many foods and feedstuffs. To determine the nutritional levels of selenium in foods and feeds, a highly satisfactory neutron activation method was developed by Leddicotte and Reynolds.[69] Although this method was very useful, it was expensive and could be conducted only by a handful of laboratories throughout the world. In 1964, Allaway and Cary[70] developed a combustion and fluorometric method that was an improvement upon the neutron activation method but still required much special equipment and allowed one to analyze only approximately 6–8 samples per day.

The wet-ashing method of Watkinson,[71] as modified by Olson,[72] has now become the tentatively accepted method of the AOAC. This method, which is very precise, can be used to assay 24–48 samples per operator in 48 hours. (See Chapter XVII for a review of methods for the determination of selenium.)

b. Selenium Content of Foods and Feeds. The selenium content of various foods and feeds is shown in Table XIIIA-4. Because the selenium content of a food or feedstuff depends largely on the selenium content of the soil in which the food was grown, it is necessary to show a wide range in selenium content for most foods and feeds. Those having very low selenium values usually were produced in the eastern or far western parts of the United States; those with high selenium values usually originated in the Midwest. One exception is the very high selenium values found in fish meals, particularly tuna meals. The source of this selenium is a mystery, since Oelschlaeger and Menke[74] have found sea water to contain only 0.0038 mg Se/l.

c. Biological Availability of Selenium from Various Feedstuffs. A number of studies have been conducted at Cornell University by the author and his co-workers to determine the biological availability of selenium in various feedstuffs. This was done by feeding graded levels of the feedstuff and comparing the amount of chemically determined selenium in the feed required to prevent exudative diathesis to the amount of pure sodium selenite required to prevent the disease. Having previously determined by chemical assay

Table XIIIA-4   Selenium content of foods and feedstuffs

FOODS

TYPE OF FOOD	SELENIUM CONTENT [μg Se/g (fresh basis[a])]	REF.
Animal Foods		
Meats: steak, ground beef, chicken, pork chops, lamb chops	0.22	74, 75
Livers: beef, chicken, pork, turkey	0.50	73, 75, 76
Milk products and eggs		
Cheese, American processed	0.09	75
", Camembert	0.06	74
", Swiss	0.10	75
Cream	0.005	75
Eggs, whole	0.10	73
", yolk	0.18	75
", white	0.05	75
", whole, dried	0.7	73, 74
Milk, whole, homogenized	0.012	75
", evaporated, canned	0.013	75
", whole, dried	0.14	74
", skimmed, dried		
Sample 1	0.096	75
Sample 2	0.24	75
Milk, human breast, dried	0.24	76
Seafoods		
Clams, crabs, lobster tails, oysters, scallops, shrimp, squid	0.60	75, 76
Fish (fresh water): carp, pike, trout, sunfish	0.50	73, 74, 75, 76
Fish fillets (salt water): cod, flounder, salmon, tuna	0.53	73, 74, 75
Fish flour	1.93	76
Cereal Foods		
Barley cereal	0.66	75
Bread, white	0.277	75
", whole wheat	0.665	75
Corn, sweet	0.004	75
Cornflakes	0.026	75
Flour, white	0.192	75
", whole wheat	0.634	75
Noodles, egg	0.625	75
Oat breakfast cereal (Cheerios)[b]	0.429	75
Oatmeal, quick cooking	0.110	75

Table XIIIA-4 (*continued*)

FOODS

TYPE OF FOOD	SELENIUM CONTENT [μg Se/g (fresh basis[a])]	REF.
Rice, polished	0.318	75
,, brown	0.388	75
Rice cereal, puffed	0.025	75, 76
Wheat cereal (Wheatena)[c]	0.241	75
Wheat cereal (Wheaties)[d]	0.100	75, 76
Fruits, Vegetables, and Prepared Foods		
Apples	0.005	75
Baby foods, strained	0.068	75
Bananas, peeled	0.010	75
Beans, red kidney (dried)	0.03	73
,, , green	0.006	75
Cabbage, white	0.023	75
Carrots	0.022	75
Cauliflower	0.007	75
Garlic	0.200	75, 76
Grapes (dry basis)	0.15	74
Lemons (dry basis)	0.12	74
Lentils (dry basis)	0.11	74
Lettuce	0.008	75
Mushrooms	0.135	75
Onions, white	0.015	75, 76
Oranges, peeled	0.013	75
Peaches, peeled	0.004	75
Pears, peeled	0.006	75
Peas (dry basis)	0.27	74
Peanuts, dried	0.38	74
Pineapple, canned	0.01	75
Potatoes, white	0.005	75
,, , sweet	0.007	75
Radishes	0.04	75
Sauerkraut (dry basis)	0.18	74
Sugar, white	0.003	75
,, , brown	0.011	75
Tomatoes, fresh	0.005	75
,, , canned	0.010	75
Turnips	0.007	75

Table XIIIA-4   (*continued*)

FEEDSTUFFS

FEED MATERIAL	SELENIUM CONTENT [μg Se/g (air-dried basis)]	REF.
Alfalfa meals and hays		
Eastern U.S.	0.10	73
Plains States, U.S.	0.38	73
Bakery by-product	0.40	73
Barley, eastern U.S. and Canada	0.10	73, 77
" , midwestern U.S. and Canada	0.30	73, 77
Blood meal	0.01	74
Brewer's dried grains	0.70	73
Cassava meal	0.10	74
Clover, red	0.11	74
Corn, yellow dent, eastern U.S.	0.025	73
" , Nebraska and South Dakota	0.38	73
" , Germany	0.11	74
" , Opaque-2	0.038	73
Corn distiller's dried solubles	0.50	73
Corn fermentation solubles	0.19	73
Corn gluten feed	0.20	73
Corn gluten meal, 60% protein	1.15	73
Cottonseed cake	0.06	74
Crab meal	1.3	73
Fish meals: anchovetta, herring,		
menhaden, sardine, whitefish	1.5	73
Fish meal: tuna	5.6	73
Fish solubles, dried	2.0	73
Hay, grass or trefoil	0.08	73, 74
Hominy feed	0.10	73
Limestone	0.17	74
Linseed meal	1.0	73
Meat and bone scraps, 50% protein	0.29	73
Oats	0.06	73, 77
Oyster shell	0.01	73
Palm kernel meal	0.12	74
Peanut meal	0.28	74
Phosphates, dicalcium	0.20	74
" , defluorinated rock	1.4	73
" , Curaphos	1.2	73
Potato meal	0.06	73
Poultry by-product meal	1.2	73
Rye	0.2	74

Table XIIIA-4 (*continued*)

FEEDSTUFFS

FEED MATERIAL	SELENIUM CONTENT [μg Se/g (air-dried basis)]	REF.
Shrimp meal	1.8	73
Soybean meal, midwestern U.S.	0.10	73
,, ,, , Nebraska	0.54	73
,, ,, , eastern U.S.	0.07	73
Water, tap, Ithaca, N.Y.	0.0	73
,, , ,, , Stuttgart, Germany	0.0016	74
,, , North Sea	0.0038	74
Wheat, Nebraska, South Dakota, western Canada	1.20	73, 77
Wheat, eastern U.S. and central Canada	0.06	73, 77
Wheat, durum, western Canada	2.22	77
Wheat bran, north-central U.S. and western Canada	0.85	73, 77
Wheat bran, durum, Canada	1.6	77
Wheat middling, north-central U.S.	0.6	73, 77
Wheat shorts, western Canada	0.6	73, 77
Whey, dried	0.08	73
Yeast, dried brewer's, U.S.	1.1	73
,, , ,, ,, , Germany	0.11	74
,, , torula	0.04	73

[a] Fresh basis except as indicated in table.

[b] Cheerios, General Mills, Minneapolis, Minn.

[c] Wheatena, Standard Milling Co., Kansas City, Mo.

[d] Wheaties, General Mills, Minneapolis, Minn.

the total selenium content of the feed, it was then possible to arrive at a figure for the biological availability of the selenium as a percentage of the total selenium in the sample.

It was found that the selenium in wheat, brewer's yeast, brewer's grains, and apparently most plant materials is highly available (85–100% availability). However, the selenium in tuna fish meal, poultry by-product meal, and other animal sources appears to be only 20–50% available.

Thus, the chemical assay provides a reasonably good assessment of the nutritional value of selenium in plant materials, but may fall far short of giving an accurate picture of the nutritional selenium value of animal products.

8.   Public Health Aspects of Selenium Fortification of Feeds and Foods

a.   The Need for Selenium Supplements.   White muscle disease in lambs and calves, myopathy of the gizzard and heart in turkeys, exudative diathesis in chicks, *hepatosis dietetica* and mulberry heart in swine, and sawdust livers in steers have caused widespread economic loss to livestock and poultry producers in the eastern and western parts of the United States and in many other areas of the world where the feeds being consumed by the animals have been produced on soils that are deficient in selenium. Economic losses have not been limited to those due to outright selenium-deficiency diseases, since it has been shown that borderline selenium deficiencies produce marked decreases in growth and greatly interfere with efficiency of feed utilization by animals.

The feeds for these animals therefore must be supplemented with enough selenium not only to prevent deficiency diseases but also to promote maximum growth and efficiency of feed utilization. The research described above shows that the maximum amount of selenium supplementation needed is 0.2 ppm, which is the level found to be required in some commerical turkey diets for prevention of myopathy of the heart and gizzard. Analyses conducted on a wide variety of feedstuffs and commerical feeds from various parts of the United States show that the selenium contents of these feeds range from a low of about 0.02 ppm in the selenium-deficient areas to a high of approximately 0.6 ppm in the selenium-rich areas of the Midwest. The universal use of a selenium concentrate to provide 0.2 ppm of supplemental selenium for all poultry and animal feeds would produce a range of selenium values in these feeds of approximately 0.25–0.85 ppm. The results of numerous experimental studies, of surveys, and of field experience show that levels of selenium below 1 ppm appear to be completely safe, and that selenium toxicity does not occur in any animal with selenium levels below 2 ppm. Furthermore, the results on tissue levels of chicks and turkeys presented earlier in this chapter show that when supplemental selenium is provided as sodium selenite very little of this selenium is retained if the selenium content of the animal body is above the minimum level required for optimum metabolism.

Although it is possible to provide selenium to large animals by periodic injection of selenium salts or by feeding a "heavy pellet" (usually compressed elemental selenium and iron filings), which dissolves slowly in the rumen and provides adequate selenium over a long period of time, these methods are not applicable for use with swine and poultry. Here the best method of administration of supplemental selenium is *via* the trace mineral supplement which is added to the complete feed.

b. Selenium Supplements. It is generally agreed by scientists working on selenium nutrition that a carefully prepared supplement is needed wherein the selenium content is adjusted so that approximately 1 pound of the supplement per ton of feed increases the selenium content of the feed by 0.1 ppm. Two pounds per ton of such a supplement could be used in diets for turkeys to provide the 0.2 ppm of selenium supplementation that is needed by turkeys in certain areas of the country. Such a selenium concentrate should contain 200 ppm Se as Na_2SeO_3 and should be prepared by a laboratory that has the facilities for conducting precise analyses on each batch of concentrate prepared for commercial use. The laboratory should be required to provide a certificate of analysis on each batch, showing that it contains 200 ppm Se as Na_2SeO_3. The preparation of these concentrates should adhere in every way to the Food and Drug Administration's "good manufacturing practices" regulations in regard to the master formula and the batch production records, the control procedures, packaging, labeling, quality control, and distribution records.

c. Possible Implications of Selenium Deficiency in Man. In 1968 Allaway and his associates[78] conducted an extensive study on the selenium content of whole blood from male donors to blood banks at different locations in the United States. The selenium concentrations ranged from 0.1 to 0.34 ppm with a mean value of 0.206 μg Se/ml of blood. The selenium content of these blood samples reflected to a remarkable degree the selenium content of the soil in the geographical areas from which the samples were taken. The lowest value found was 0.157 ppm Se as an average of 10 samples taken in Lima, Ohio, which is known to be a selenium-deficient area. The highest mean value obtained was 0.256 ppm Se on 20 samples of blood taken from Rapid City, South Dakota, a selenium-adequate area of the United States.

Comparison of the blood levels of chicks receiving the bare minimum amount of dietary selenium required for the prevention of exudative diathesis (0.1 ppm) shows that under these conditions the blood contains 0.128 ppm Se at 4 weeks of age and 0.178 ppm at 8 weeks. Indeed, in chicks receiving the basal low-selenium diet, those surviving to 8 weeks of age had an average blood selenium level of 0.147 ppm, which is approximately the same as the mean value for human male blood in Lima, Ohio.

This observation, coupled with the widespread incidence of white muscle disease in ruminants and cardiac myopathies in turkeys in Ohio and other areas of the northeastern United States, strongly hints at a possible selenium deficiency in the human population in this and perhaps other selenium-inadequate areas.

d. Selenium Supplementation of Human Diets. Schwarz[79] reported in 1965 that administration of selenium to two children with kwashiorkor promoted a growth response. Majaj and Hopkins[80] in 1966 found that administration of selenium as sodium selenite to three children with kwashiorkor appeared to produce a reticulocyte response. Burk and his associates[81] found that the blood of Guatemalan children suffering from kwashiorkor contained only about half as much selenium as did that of children who had recovered from the disorder. The untreated kwashiorkor patients showed an average blood selenium level of 0.08 ppm, while those who had received dietary treatment and had recovered had a value of 0.14 ppm, which ranks close to the blood selenium level found by Allaway and his associates in normal human males in Ohio.

9. Metabolic Role of Selenium

a. Nutritionally Active Forms of Selenium. Although the unknown "factor 3" of Schwarz was thought to be an organic compound, ample evidence now shows that selenium from sodium selenite and other inorganic compounds is at least as active as selenium from selenomethionine, selenocystine, and most other known organic forms of the element.

Schwarz and Fredga[82] assessed the biological potency of many organic selenium compounds in relation to sodium selenite. They found that in both mono- and diselenodicarboxylic acid series a minimum of potency was seen at chain length C_7 and a maximum at chain length C_{10} or C_{11}. Branching of the carbon chain reduced activity. Introduction of two methyl groups at the same carbon, forming a quaternary carbon atom, eliminated biological potency almost entirely. Tests of monoselenodicarboxylic acids of unsymmetrical structure showed that activity was not always predictable.

The most active substances appeared to be monoseleno-11,11'-di-*n*-undecanoic acid, diseleno-4,4'-di-*n*-valeric acid, and diseleno-11,11'-di-*n*-undecanamide. It was reported that these compounds had potencies which were 169, 138, and 180% that of selenite selenium, respectively. Whether these findings mean that these organic forms of selenium are more readily converted to a metabolic form than sodium selenite, or whether they are more stable *or better retained*[83] than selenite selenium, remains to be determined.

b. Possible Sites of Action and Nature of Metabolic Activity. Although many hypotheses have been proposed, there is no definitive evidence concerning the metabolic action of selenium. The possibility that some selenium-containing compound acts as a potent antioxidant[84] does not appear to explain its action, since few if any of the selenium-responsive diseases are prevented by synthetic antioxidants in the absolute absence of selenium. Also, the possibility that selenium is acting in a general way as a free-radical scavenger[85] does not seem to be compatible with its very specific effect in preventing degeneration of the pancreas in chicks and myopathies in turkeys, swine, and ruminants.

Its effect on such a wide variety of tissues, for example, in preventing liver necrosis, exudative diathesis, pancreatic degeneration, and various myopathies, indicates that selenium must be concerned with something very basic, such as protein synthesis or a similar metabolic mechanism. Discovery of the metabolic role of selenium may go far towards providing anwers to some of the most perplexing problems in animal and human metabolism, such as the causes of muscular dystrophy and perhaps other diseases.

REFERENCES

1. K. Schwarz and C. M. Foltz, *J. Amer. Chem. Soc.*, **79**, 3292 (1957).

2. J. N. Thompson and M. L. Scott, *J. Nutr.*, **97**, 335 (1969).

3. H. M. Evans and K. S. Bishop, *Amer. J. Physiol.*, **63**, 396 (1922).

4. T. E. Weichselbaum, *Quart. J. Exptl. Physiol.*, **25**, 363 (1935).

5. F. S. Daft, W. H. Sebrell, and R. D. Lillie, *Proc. Soc. Exptl. Biol. Med.*, **50**, 1 (1942).

6. K. Schwarz, *Z. Physiol. Chem.*, **283**, 187 (1948).

7. K. Schwarz, *Z. Physiol. Chem.*, **281**, 109 (1944).

8. P. György and H. Goldblatt, *J. Exptl. Med.*, **89**, 245 (1949).

9. H. P. Himsworth and O. Lindan, *Nature*, **163**, 30 (1949).

10. M. Goettsch, *Federation Proc.*, **9**, 177 (1950).

11. P. György, C. S. Rose, R. M. Tomarelli, and H. Goldblatt, *J. Nutr.*, **41**, 265 (1950).

12. K. Schwarz, *Proc. Soc. Exptl. Biol. Med.*, **78**, 852 (1951).

13. K. Schwarz, *Ann. N.Y. Acad. Sci.*, **57**, 878 (1954).

14. H. E. Munsell, G. M. Devaney, and M. H. Kennedy, *U.S. Dept. Agr. Tech. Bull.* **534** (1936).

15. E. J. Underwood, "Trace Elements," Academic Press, New York, N.Y., 1956, p. 2.

16. M. L. Scott, F. W. Hill, L. C. Norris, D. C. Dobson, and T. S. Nelson, *J. Nutr.*, **56**, 387 (1955).

17. E. L. Patterson, R. Milstrey, and E. L. R. Stokstad, *Proc. Soc. Exptl. Biol. Med.*, **95**, 617 (1957).

18. M. L. Scott, J. G. Bieri, G. M. Briggs, and K. Schwarz, *Poultry Sci.*, **36**, 1155 (1957) (abstract).

19. E. L. R. Stokstad, E. L. Patterson, and R. Milstrey, *Poultry Sci.*, **36**, 1160 (1957).

20. K. Schwarz, Symposium on Liver Function, *Amer. Inst. Biol. Sci. Publ.* **4**, Washington, D.C., 1958, p. 509.

21. G. L. Fite, *Ann. N.Y. Acad. Sci.*, **57**, 831 (1954).

22. K. Schwarz and W. Mertz, *Metabolism*, **8**, 79 (1959).

23. J. N. Thompson and M. L. Scott, *J. Nutr.*, **100**, 797 (1970).

24. I. D. Desai and M. L. Scott, *Arch. Biochem. Biophys.*, **110**, 309 (1965).

25. M. L. Scott, *Nutr. Abstr. Rev.*, **32**, 1 (1962).

26. B. C. Creech, G. L. Feldman, T. M. Ferguson, B. L. Reid, and J. R. Couch, *J. Nutr.*, **62,** 83 (1957)

27. M. L. Scott, G. Olson, L. Krook, and W. R. Brown, *J. Nutr.*, **91,** 573 (1967).

28. E. D. Walter and L. S. Jensen, *J. Nutr.*, **80,** 327 (1963).

29. H. Dam and J. Glavind, *Naturwissenschaften*, **28,** 207 (1940).

30. J. Goldstein and M. L. Scott, *J. Nutr.*, **60,** 349 (1956).

31. E. Jungherr and A. M. Pappenheimer, *Proc. Soc. Exptl. Biol. Med.*, **37,** 520 (1937).

32. T. M. Ferguson, E. M. Omar, and J. R. Couch, *Texas Rept. Biol. Med.*, **22,** 902 (1964).

33. O. H. Muth and W. H. Allaway, *J. Amer. Vet. Med. Assoc.*, **142,** 1379 (1963).

34. H. W. Bruins, L. E. Ousterhout, M. L. Scott, E. E. Cary, and W. H. Allaway, *Feedstuffs*, **38** (2), 66 (1966).

35. H. J. Metzger and W. A. Hagen, *Cornell Vet.*, **17,** 35 (1927).

36. J. F. Proctor, D. E. Hogue, and R. G. Warner, *J. Animal Sci.*, **17,** 1183 (1958).

37. O. H. Muth, J. E. Oldfield, L. F. Remmert, and J. R. Schubert, *Science*, **128,** 1090 (1958).

38. W. H. Allaway, E. E. Cary, and C. F. Ehlig, "Selenium in Biomedicine," O. H. Muth, Ed., AVI Publishing Co., Westport, Conn., 1967, p. 273.

39. W. H. Allaway, "Proceedings Semi-Annual Meetings, Nutrition Council, American Feed Manufacturers Association," Chicago, Ill., 1968, p. 27.

40. O. H. Muth, *J. Amer. Vet. Med. Assoc.*, **126,** 355 (1955).

41. O. H. Muth, *J. Amer. Vet. Med. Assoc.*, **142,** 272 (1963).

42. E. D. Andrews, W. J. Hartley, and A. B. Grant, *N. Zealand Vet. J.*, **16,** 3 (1968).

43. A. S. H. Wu, J. E. Oldfield, O. H. Muth, P. D. Whanger, and P. H. Weswig, *Proc. Western Sect., Amer. Soc. Animal Sci.*, **20,** 85 (1969).

44. N. Trinder, C. D. Woodhouse, and C. P. Renton, *Vet. Rec.*, **85,** 550 (1969).

45. G. C. Todd and L. Krook, "Proceedings Cornell Nutrition Conference for Feed Manufacturers," 1965, p. 33.

46. R. O. Eggert, E. Patterson, W. T. Akers, and E. L. R. Stokstad, *J. Animal Sci.*, **16,** 1037 (1957).

47. L. Pellegrini, Doctoral Thesis, University of Minnesota, St. Paul, Minn., 1958.

48. K. Orstadius, G. Nordstrom, and N. Lannek, *Cornell Vet.*, **53,** 60 (1963).

49. W. J. Hartley and A. B. Grant, *Federation Proc.*, **20,** 679 (1961).

50. S. B. Slen, A. S. Demiruren, and A. D. Smith, *Can. J. Animal Sci.*, **41,** 263 (1961).

51. K. E. M. McCoy and P. H. Weswig, *J. Nutr.*, **98,** 383 (1969).

52. M. L. Scott, *Federation Proc.*, **24**(4), 901 (1965).

53. J. Parizek and Z. Zahor, *Nature*, **177,** 1036 (1956).

54. K. E. Mason and J. O. Young, "Selenium in Biomedicine," O. H. Muth, Ed., AVI Publishing Co., Westport, Conn., 1967, p. 383.

55. J. Parizek, *J. Endocrinol.*, **15,** 56 (1957).

56. A. L. Moxon, *Science*, **88,** 81 (1938).

57. I. S. Palmer and C. W. Bonhorst, *J. Agr. Food Chem.*, **5,** 928 (1957).

58. N. T. Thapar, E. Guenther, C. W. Carlson, and O. E. Olson, *Poultry Sci.*, **48,** 1988 (1969).

59. A. L. Moxon and K. D. Du Bois, *J. Nutr.*, **18,** 447 (1939).

60. O. A. Levander and C. A. Baumann, *Toxicol. Appl. Pharmacol.*, **9**, 106 (1966).

61. H. E. Ganther and C. A. Baumann, *J. Nutr.*, **77**, 210 (1962).

62. O. A. Levander and L. C. Argrett, *Toxicol. Appl. Pharmacol.*, **14**, 308 (1969).

63. J. Pařízek, I. Beneš, A. Babický, J. Beneš, V. Procházková, and J. Lener, *Physiol. Bohemoslov.*, **18**, 105 (1969).

64. R. E. Holmberg and V. H. Ferm, *Arch. Environ. Health*, **18**, 873 (1969).

65. S. A. Gunn and J. C. Gould, "Selenium in Biomedicine," O. H. Muth, Ed., AVI Publishing Co., Westport, Conn., 1967, p. 395.

66. C. F. Ehlig, D. E. Hogue, W. H. Allaway, and D. J. Hamm, *J. Nutr.*, **92**, 121 (1967).

67. A. L. Moxon and M. Rhian, *Physiol. Rev.*, **23**, 305 (1943).

68. W. H. Allaway, D. P. Moore, J. E. Oldfield. and O. H. Muth, *J. Nutr.*, **88**, 411 (1966).

69. G. W. Leddicotte and S. A. Reynolds, *Nucleonics*, **8**, 62 (1951).

70. W. H. Allaway and E. E. Cary, *Anal. Chem.*, **36**, 1359 (1964).

71. J. H. Watkinson, "Selenium in Biomedicine," O. H. Muth, Ed., AVI Publishing Co., Westport, Conn., 1967, p. 97.

72. O. E. Olson, *J. Assoc. Offic. Agr. Chemists*, **52**, 627 (1969).

73. J. N. Thompson and M. L. Scott, "Proceedings Cornell Nutrition Conference for Feed Manufacturers," 1968, p. 121.

74. W. Oelschlaeger and K. H. Menke, *Z. Ernährungs.*, **9**, 208, 216 (1969).

75. V. C. Morris and O. A. Levander, *J. Nutr.*, **100**, 1383 (1970).

76. H. A. Schroeder, D. V. Frost, and J. J. Ballasa, *J. Chronic Dis.*, **23**, 227 (1970).

77. D. Arthur, University of Guelph, Ontario, Canada, personal communication, 1971.

78. W. H. Allaway, J. Kubota, F. Losee, and M. Roth, *Arch. Environ. Health*, **16**, 342 (1968).

79. K. Schwarz, *Lancet*, **1**, 1335 (1965).

80. A. S. Majaj and L. L. Hopkins, Jr., *Lancet*, **2**, 592 (1966).

81. R. F. J. Burk, Jr., W. N. Pearson, R. P. Wood, II, and F. Viteri, *Amer. J. Clin. Nutr.*, **20**, 723 (1967).

82. K. Schwarz and A. Fredga, *J. Biol. Chem.*, **244**, 2103 (1969).

83. M. K. Smith, B. B. Westfall, and E. F. Stohlman, *U.S. Public Health Rept.*, **53**, 119 (1938).

84. J. W. Hamilton and A. L. Tappel, *J. Nutr.*, **79**, 493 (1963).

85. A. L. Tappel, *Federation Proc.*, **24**, 73 (1965).

XIII Selenium compounds in nature and medicine

B. SELENIUM ASSIMILATION IN ANIMALS

JOHN L. MARTIN

Colorado State University
Department of Biochemistry
Fort Collins, Colorado
and
Metropolitan State College
Department of Chemistry
Denver, Colorado

1. Introduction and History

The ingestion of seleniferous plants by livestock, with the consequent development of well-defined disorders, has been a problem to stockmen and farmers for centuries. Perhaps the first report describing a disease syndrome resulting from the ingestion of seleniferous plants was that of Marco Polo[1] during his famous travels to the Orient. In 1295 Marco Polo recorded in his journal that beasts of burden were afflicted by a peculiar disorder when they fed on a particular plant which grew in western China near the border of Turkestan and Tibet. One of the symptoms he described was that the hoofs of the affected animals dropped off. Only recently has it become evident that this disorder is characteristic of a type of chronic selenosis.

In 1860 Dr. T. C. Madison,[2] an army surgeon stationed at Fort Randall, then located in the Nebraska Territory but now a part of South Dakota near the Nebraska border, published a sanitary report in which he described a fatal disease in horses grazing near the fort. This disease was of a chronic nature and began about 10 days after the horses were moved into the area. Losses continued for several months but ceased when other forage was provided. Again, one of the characteristic symptoms exhibited by the horses was extreme tenderness and inflammation of the feet. These symptoms were accompanied by loss of hair from the mane and tail. This account may not have represented the only instance in which cavalry horses suffered from selenium toxicosis. It has even been speculated that this ailment played an important role in the disaster suffered by General Custer and his troops at the Little Big Horn.[3] In this military engagement, a relief expedition failed to reach the beleaguered troops of General Custer in time to provide needed support. The officer in command of this expedition wrote in his official report that a peculiar sickness affected his horses and was responsible for the delay. This is a plausible explanation since the route taken by the relief expedition ran through seleniferous regions.

As the western part of this country became settled, farmers and stockmen experienced similar losses in all forms of livestock. The disease acquired the name "alkali disease" because of the belief that alkali (high salt) in the water and soil was responsible. In spite of the fact that the investigations of Larsen et al.[4,5] in 1912 and 1913 eliminated alkali water as the cause of the disease and that later selenium was definitely shown to be the causative agent, the ailment continued to be erroneously called "alkali disease."

Draize and Beath[6] reported a similar disease of a somewhat more acute nature occurring in range livestock in Wyoming. This disease was called "blind staggers" by the early settlers because some of the affected animals exhibited symptoms of central nervous system disturbances. Draize and Beath examined tissues from 100 affected cattle and sheep and determined that the gross and microscopic pathologies of blind staggers and alkali disease were very similar and probably had a common cause.

In 1929 Franke[7] began a series of investigations at the South Dakota Experiment Station, some of which were not published until 1934. This work established that grains and grasses grown in definite soil areas were toxic to animals, and that this toxicity was due to the presence of selenium in the grains and grasses.

An interbureau cooperative study of the problem was started by the U.S. government in 1931,[8] and at one of the interbureau committee meetings H. G. Knight, Chief of the Bureau of Chemistry and Soils, was reported to have suggested that selenium should be investigated as the possible ingredient in toxic grain. In 1933, W. O. Robinson[9] obtained samples of toxic wheat and after devising a method of analysis found 10–12 ppm Se in one sample and 5–6 ppm in another. Nontoxic wheat, on the other hand, was free of selenium.

2. Toxicity in Animals

The clinical signs and pathological changes produced by toxic quantities of selenium vary with the animal species, the age of the animal, the chemical form and quantity of selenium consumed or administered, and the nature of the diet exclusive of selenium.

In addition to acute selenosis, there are, according to Rosenfeld and Beath,[10] three types of chronic selenosis: (1) blind staggers, caused by organic selenium compounds, with or without small amounts of selenate, which are readily extractable with water from native selenium indicator plants; (2) alkali disease, produced in livestock having consumed plants or grain in which selenium is bound in the proteins, presumably as selenoamino acids, and is relatively insoluble in water; and (3) chronic selenosis, produced experimentally by the administration of selenate or selenite to livestock. The acute form is caused by the consumption of a sufficient quantity of selenium in a short period of time and produces severe signs of toxicosis. At the onset of acute poisoning, the movements and posture of the poisoned animal become abnormal. After walking unsteadily for a short distance, the animal is likely to stop and assume a characteristic stance with lowered head and drooped ears. Cyanosis of the mucous membranes, labored breathing, diuresis, and bloating are also usually seen. Complete prostration and apparent unconsciousness precede death. The duration of the illness varies from a few hours to several days.

The blind staggers type of chronic selenium poisoning occurs in cattle and sheep feeding on moderately toxic quantities of highly seleniferous weeds, such as certain species of the genera *Astragalus*, *Xylorhiza*, *Stanleya*, and *Oonopsis*. Affected cattle show signs of central nervous system disorder, such as aimless wandering and circling. They bump into, stumble over, or push against objects in their paths. Weakness and inappetence are also present. If recovery does not occur, affected cattle become paralyzed before death. The symptoms in sheep are less characteristic.

Alkali disease, the other type of chronic selenosis, occurs when livestock feed on seleniferous grains and grasses. The clinical signs in horses, cattle, and swine, reported by Franke et al.,[11] include retarded growth, emaciation, deformed hoofs, loss of hair, arthritis, and eventual death.

3. Deposition in Animal Tissues

Selenium is distributed by the circulatory system to the various organs of the body. The extent of its accumulation in the organs varies with the type of tissue, the level of selenium administered, and the individual susceptibility to selenium intoxication. Detoxifying organs tend to accumulate the highest quantities of selenium, and investigations on several animal species have resulted in general agreement that in both acute and chronic selenosis the highest concentration of selenium is found in the liver, followed in decreasing order by the kidneys, spleen, and lungs.[12–18] Myocardium, skeletal muscle, and brain contain only small amounts, and fat has practically no selenium. Signs of chronic selenosis begin to appear in animals when their blood selenium levels reach 2–4 ppm.[13,17,19] Much higher concentrations are observed in the blood of acutely poisoned animals and may reach 25 ppm.[12] Hair also accumulates selenium and is a good indicator for routine diagnosis of chronic selenium toxicity in cattle. Olson[20] pointed out that toxicity could be expected in animals when the hair contains over 10 ppm.

Values obtained by Glenn et al.[18] on eight sheep lethally poisoned by sodium selenate are presented as representative values. Mean values in parts per million on a wet weight basis are as follows: liver, 28.0; kidney, 6.9; lung, 3.0; spleen, 2.4; myocardium, 4.0; skeletal muscle, 0.6; and brain, 0.8. Considerable individual variation was observed in this study[21] since the values for the concentrations of selenium in the liver ranged between 15.2 and 49.8 ppm. Similar results were obtained by Morrow[22] in lambs acutely poisoned with selenite.

Other investigators have observed that in selenium toxicosis the kidney concentrates higher levels of selenium than does the liver.[23–26] However, in one of these investigations, Levander and Argrett[24] reported that there was a much higher percentage of radioselenium in the liver of rats 12 hours after injection of radioselenite than in the kidneys of the same animals.

The pattern of selenium accumulation is different in animals receiving very low levels of the element. For instance, it has been found that animals on very low levels of selenium concentrate more selenium in their kidneys than in any other organ.[27,28] In another study Allaway et al.[29] fed sheep alfalfa that had a selenium content of less than 0.01 ppm. The concentration of selenium in the kidneys of these sheep was 0.52 ppm, whereas that of the liver was only 0.4 ppm. However, in the same study other sheep were fed alfalfa containing 0.43–0.57 ppm Se. The level of selenium in the kidneys of these animals rose to 3.3 ppm, whereas that of the liver increased to over 4.0 ppm.

The quantity of selenium accumulated in tissues also depends on the chemical form in which the element is administered. For instance, Smith et al.[30] found that the concentrations of selenium in various tissues of rabbits fed seleniferous oats were many times those found in the tissues of rabbits fed comparable amounts of selenium as selenite. They found 7 times more selenium in the liver, 15 times more in the skin, and 29 times more in skeletal muscle, which comprises the bulk of the body protein. Similar findings were reported by Halverson et al.[31] in rats. Ehlig et al.[32] fed a sodium selenite-supplemented diet (1 ppm Se) to one group of lambs and a selenomethionine-supplemented diet (1 ppm Se) to another group. These investigators observed that all 13 tissues assayed from lambs on the selenomethionine-supplemented diet were significantly higher in selenium content than were the same tissues from lambs fed the selenite-supplemented diet. As would be expected, the greatest differences were found in highly proteinaceous tissues.

The mode of assimilation of selenium into testicular tissue has been shown to follow a very interesting pattern. In recently published data, Brown and Burk[32a] administered $^{75}SeO_3^{2-}$ (0.01 μg Se) to selenium-deficient rats fed a torula yeast diet for 5 weeks after weaning. The administered selenium was taken up rather slowly by the testes, but after 3 weeks 30% of the total body ^{75}Se was present in these organs. After 3 weeks there was a slow decline in this level of selenium. In rats fed a diet adequate in selenium (0.1 ppm) and administered $^{75}SeO_3^{2-}$, only about 7% of the total body ^{75}Se was found in the testes 3 weeks after $^{75}SeO_3^{2-}$ administration. This observation suggests that, once the requirement for selenium has been met by the testes, there is little need for concentrating more of it.

The deposition of selenium in tissues also depends on previous exposure of the animal to this element. Jaffé and Mondragon[33] recently observed that young rats born to mothers fed a moderately high selenium diet (seleniferous sesame press cake) and kept on this diet until after weaning showed a steady decrease in selenium content in their livers. On the other hand, selenium liver levels increased in stock rats fed this same diet. Hopkins et al.[34] have also demonstrated that previous selenium intake can influence the distribution of subsequent doses of the element.

4. Pathological Changes in Selenium Toxicosis

Extensive tissue damage results from both acute and chronic selenosis, and most investigators agree that the most severe and consistent degeneration occurs in the organs which concentrate selenium to the greatest extent, that is, liver and kidney. The characteristic gross pathological changes in animals suffering from alkali disease, blind staggers, and experimental selenosis are summarized by Rosenfeld and Beath.[35]

The observations by Glenn *et al.*[18] are somewhat contradictory to other reports. These investigators fed toxic quantities of sodium selenate to sheep and observed that most tissue damage occurred in the myocardium, with only occasional degenerative changes in the liver and kidneys. Because of the structural and functional similarities between cardiac and skeletal muscle, these researchers also compared the pathological changes in these tissues. They found that cardiac muscle, which accumulated more than twice as much selenium as did skeletal muscle, suffered much more severe degenerative changes.

5. Relative Toxicity of Selenium Compounds

Once it was established that the presence of selenium in certain plants was responsible for the devastating livestock losses suffered by ranchers in Wyoming and South Dakota, research was immediately undertaken to determine the toxicity of selenium in its various chemical forms. Early work by Franke and Potter[36] revealed that elemental selenium (gray allotrope) was virtually harmless, probably because of its low solubility and absorption. In this same study it was further found that inorganic selenium (*i.e.*, selenate and selenite) had about the same toxicity as an equivalent amount of selenium in natural grains.

Later work by Smith and Lillie[37] suggested that sodium selenite actually was slightly more toxic than an equivalent amount of selenium in wheat. In this same study these researchers warned that a daily intake of from 1.0 to 1.5 mg Se, as selenite, per kilogram of body weight was dangerously toxic.

Studies on the relative toxicities of selenite and selenate have revealed either that their toxicities are comparable[38] or that selenate is slightly less toxic than selenite.[39] In the latter study it was determined that the minimum lethal dose of selenite in rats was 3.25–3.5 mg Se/kg body weight, while that of selenate was 5.25–5.75 mg Se/kg body weight. However, Tinsley *et al.*[40] found that selenate produced a greater toxic response in rats than did an equivalent amount of selenite.

Se-Methylselenocysteine and selenocystathionine are the two most prominent selenium compounds found in selenium accumulator plants,[41–48] but unfortunately very limited research has been carried out on their toxicities. However, the toxicities of two common selenium accumulator plants have been investigated. Rosenfeld and Beath[49] fed steers *Astragalus pectinatus*, which supplied a daily dose of 2.2 mg Se/kg body weight. After receiving

6–16 doses, the steers developed blind staggers. The disease also developed in a steer that had been fed seven daily doses of *A. bisulcatus*, each dose supplying 3.4 mg Se/kg body weight. These data suggest that *Se*-methylselenocysteine and selenocystathionine possess toxicities comparable to that of selenite. However, additional controlled studies of the toxicities of these two selenoamino acids, separately and together, are needed.

Different species apparently have different susceptibilities to selenium intoxication. The minimum lethal doses in milligrams of selenium, as selenite, per kilogram of body weight, when administered orally, were determined on several species by Miller and Williams[13,50] and are as follows: 3.3 for horses, 11 for cattle, and 15 for pigs.

Dimethyl selenide is much less toxic than selenite. McConnell and Portman[51] determined that the LD_{50} of dimethyl selenide in rats was 1.6 g Se/kg body weight—approximately 1/500 the toxicity of an equivalent amount of selenium as selenite.

Hydrogen selenide, which has a very offensive odor, is probably the most toxic and irritating selenium compound known. When compared to SeO_2, which is toxic at 1.0 mg Se/m³ air,[52] hydrogen selenide is toxic at a level of 0.2 mg Se/m³ air, as established by the American Conference of Government Industrial Hygienists.

The toxicities of organoselenium compounds apparently depend on whether their sulfur analogs are intermediates in sulfur metabolism. For instance, the minimum lethal doses of selenocystine and selenomethionine in rats were determined by Klug *et al.*[53] to be 4.0 and 4.25 mg Se/kg body weight, respectively. Such levels of toxicity are very comparable to that of selenite. Moxon *et al.*[54] compared the toxic effects of D- and L-selenocystine in rats. Although the solubilities of these enantiomers are identical, the toxicity of the L-isomer was many times greater than that of its D-isomer. The toxicities of organoselenium compounds whose sulfur analogs are not intermediates in sulfur metabolism are less pronounced. For instance, the minimum lethal doses of *n*-propylseleninic acid, β-selenopropionic acid, and β,β'-diselenodipropionic acid in rats were shown by Moxon *et al.*[55] to be 20–25, 25–30, and 40 mg Se/kg body weight, respectively.

Another way of expressing the toxicity of selenium is to specify the quantity of the element in the diet that will produce the characteristic symptoms of selenium toxicity. Fitzhugh *et al.*[56] demonstrated that as little as 3 ppm Se in seleniferous grain produced toxic effects in rats, and 10 ppm killed most of the rats within 8 weeks. Similar studies carried out by Munsell and his colleagues[57] confirmed the results of Fitzhugh *et al.* More recently Witting and Horwitt[58] observed pronounced toxic effects in vitamin E-deficient rats fed a diet containing only 1 ppm Se as selenite.

6. Factors Affecting Toxicity

Knowing that selenium was the causative agent of alkali disease and blind staggers, researchers soon began to investigate ways in which selenium toxicosis could be prevented or at least minimized. Of course, the most obvious course of action would be to prevent livestock from grazing in seleniferous areas. Fortunately, the more highly seleniferous weeds are rather unpalatable to livestock, particularly to horses. Cattle show less ability to discriminate, and sheep the least. However, winter fat (*Eurotia lanata*) and salt-bushes (*Atriplex canescens* and *Atriplex nuttallii*) are quite palatable, and although normally free of selenium these plants can become sufficiently toxic in seleniferous soils to be hazardous to grazing livestock.

If it is known that selenium is present in a given ration, what can be done to minimize its toxicity? The quantity and the type of dietary protein present in the ration are important factors in reducing the toxicity of selenium. In 1939 Smith[59] observed that a high-casein diet significantly reduced selenium toxicity. Smith found that a diet having a selenium content of 10 ppm (from selenized wheat) and a protein level of 10 % was highly toxic to rats. However, when casein replaced an equivalent amount of starch, raising the percentage of protein to 30 %, scarcely any adverse effects were noted in rats fed such a ration.

In 1940 Gortner[60] investigated the protective effects of various proteins. He confirmed that casein provided good protection against selenosis; lactalbumin proved to be only fair in this respect, and edestin and gelatin exerted no protective action.

Just how casein exerts its protective effect against selenosis is still not clearly understood. According to Rosenfeld and Beath,[61] however, a true detoxification occurs, since all known manifestations—effusions, anemia, impairment of growth, and liver injury—are equally reduced.

Active methyl donors, such as methionine, betaine, and choline, provide considerable protection against chronic selenosis caused by selenite and selenate.[26,62-66] However, methionine was shown to be ineffective against chronic selenosis produced by seleniferous grain.[59,67,68]

The reason why methyl donors, particularly methionine, protect against selenosis produced by selenite and selenate but not against the same disease caused by seleniferous grains is better understood. McConnell and Portman[51] were able to identify dimethyl selenide in the breath of rats administered selenate. These researchers postulated that this conversion represented the method by which the selenate was detoxified. Earlier, Challenger *et al.*[69] pointed out that ionic selenium was essential for this methylation process. This apparent requirement of selenite for dimethyl selenide formation by reductive methylation may explain why methionine is effective against selenosis caused by inorganic selenium but not against selenosis caused by selenized grain, in which the selenium is probably present as selenoamino acids.

Halverson et al.[70] discovered that linseed oil meal provides even greater protection against chronic selenosis than does casein. The active principle, a nonprotein factor, reduces the incidence of liver injury caused by selenium but does not lower the concentration of selenium in that organ.[71] Furthermore, the protective action operates against selenosis produced by selenite and seleniferous corn.[70] Dialysis experiments by Levander et al.[25] revealed a difference in selenium binding between rat liver homogenates from animals fed linseed oil meal plus casein and homogenates from animals fed casein alone. On the basis of this observation, these investigators suggested that the selenium in the livers of rats fed linseed oil meal is bound in a less toxic form than that in rats fed casein alone.

Bromobenzene has been shown to increase the urinary excretion of selenium in dogs fed seleniferous corn and in steers grazing on a seleniferous range.[72] However, bromobenzene had no effect on the excretion of selenium when selenious acid was given to rabbits.[73] Presumably, bromobenzene conjugates with selenocysteine, forming the selenium analog of p-bromophenylmercapturic acid, which is then excreted in the urine. It is doubtful that monogastric animals, possibly excepting the cat family, can synthesize selenoamino acids from selenite.[74,75] This limitation would explain why bromobenzene had no effect on promoting the excretion of selenium in rabbits administered selenious acid.

Certain inorganic substances also can protect animals against selenosis. Early investigations by Moxon[76] demonstrated that chronic and acute selenoses produced by seleniferous grains were minimized by the administration of arsenite. Arsenite and arsenate were found to be equally effective in preventing the toxic action of selenium in seleniferous wheat, selenite, and selenocysteine.[15,77-82] Until recently, the exact mechanism by which arsenic counteracted selenium toxicity was somewhat obscure. Some researchers have been unable to detect any alteration in the distribution or excretion of selenium after arsenic treatment,[23,83] while others have reported that arsenic treatment decreases the amount of selenium deposited in the liver of experimental animals.[24,54,84,85] Ganther and Baumann[86] and, later, Levander and Baumann[87,88] reported that arsenic increased the gastrointestinal excretion of selenium in rats. They further found that this increased excretion was accompanied by a corresponding decrease in the retention of selenium in the livers of these rats. Moreover, Levander and Baumann[88] observed that gastrointestinal excretion of selenium was elevated because there was an increase in its clearance from the liver into the bile.

Although arsenite increases the biliary excretion of selenium, Kamstra and Bonhorst[89] found that it likewise prevents the formation of volatile selenium in animals injected with selenite. This observation was confirmed by Ganther and Baumann[86] and by Olson et al.[90] Later Ganther[91] was able to demonstrate that arsenite is a powerful inhibitor of dimethyl selenide synthesis in vitro, indicating that the action of arsenite on selenium volatilization in vivo is a direct inhibition of volatile selenium formation.

Sulfate has been shown to be an effective agent in protecting experimental animals against chronic selenosis brought on by selenate, but it is less effective or is ineffective against selenosis caused by selenite[92] or seleniferous wheat.[93] Previously, Bonhorst and Palmer[94] reported that the injection of magnesium sulfate to rats previously injected with a subtoxic dose of selenate decreased the concentration of selenium in the blood and liver of the animals. Ganther and Baumann[92] further observed that the urinary excretion of selenium in selenate-poisoned rats administered sulfate was three times greater than that in selenate-poisoned rats not administered sulfate. This increased urinary excretion of selenium was accompanied by consistent, although small, decreases in selenium retention in the blood, liver, kidney, and carcass. When selenate was substituted by selenite, the effect of sulfate on the urinary excretion of selenium was minimal.

Although vitamin E and selenium are related in the prevention of selenium-responsive diseases (cf. Chapter XIIIA), Morss and Olcott[95] reported that vitamin E alone apparently has no effect on reducing or increasing the toxicity of selenium in rats. However, Witting and Horwitt[58] observed that vitamin E-deficient rats were more susceptible to selenium intoxication than were control rats. Recently Levander and Morris[26] reported that a combination of methionine and vitamin E significantly reduced the severity of selenate-induced liver damage in rats. However, a deficiency in cobalt has been shown to increase the susceptibility of sheep to selenium toxicosis.[96]

7. Tissue Retention of Selenium

Even though the deposition of selenium in various animal tissues has been covered, nothing has been said in regard to how quickly selenium reaches the various tissues and how well, once in the tissues, it is retained. Selenium is absorbed rapidly and efficiently into the circulatory system regardless of its mode of administration. Lee et al.[97] have shown that, when selenium, as selenite, is added to human blood, it is rapidly taken up by the cells (50–70% within 1–2 minutes) and then released into the plasma so that most of the radioactivity is in the plasma within 15–20 minutes. From the blood the selenium is quickly distributed to the various tissues of the body. Heinrich and Kelsey[98] injected mice with radioselenite contained in a quantity of selenium that, if given in daily doses, could produce chronic selenosis. One hour after administration only 2.1 % of the original dose was found in the blood, 30 % in the liver, 7.3 % in the lungs, and 6.8 % in the kidneys.

The retention of selenium in the tissues depends on the quantity administered. Hopkins et al.[34] found that selenium-deficient rats administered microgram quantities of selenite (0.25 μg Se/kg body weight) retained 90 % of the administered dose after 24 hours. However, when the dose was increased to 10 μg/kg body weight, there was only a 65 % retention of the administered selenium during this same time period.

Blincoe[99] measured the rate of elimination of selenium in rats by measuring the decline in whole-body count in the animals administered a single dose of radioselenite. This investigator found that initially the rate of elimination was high (48 % in the first 24 hours). This rapid rate of elimination was followed by a much reduced rate. Blincoe further described the elimination of selenium as two first-order processes of widely differing rate constants ($t_{1/2} = 19$ hours, and $t_{1/2} = 318$ hours). Similar experiments carried out by Ewan et al.[100] confirmed the results of Blincoe; however, they found that the rate of selenium depletion from a fixed pool varied with selenium intake during the depletion period but was unaffected in rats fed diets containing linseed oil meal, high methionine, or high protein. The rate of selenium depletion was also unaffected in rats fed a selenium-deficient diet.

Later, Lopez et al.[101] administered $^{75}SeO_3^{2-}$ to lambs fed varying levels of dietary selenium. From 48 hours until termination of the experiment at 336 hours, the whole-body loss of selenium was described as a first-order rate constant which was inversely proportional to the level of dietary selenium.

Imbach and Sternberg[102] have found that the early phase of selenium elimination in rats injected with trace quantities of selenite could be further subdivided into a pulmonary phase with a $t_{1/2}$ of 3.5 hours and a urinary phase with a $t_{1/2}$ of 48 hours. After approximately 21 hours only 50 % of the original dose was retained.

As would be expected, there is a much longer retention time when organic selenium is administered. Lathrop et al.[103] found that, when human beings were orally administered microgram quantities of ^{75}Se-selenomethionine, a period of 80 days was required for the elimination of 50 % of the original dose. These investigators reported further that excretion of selenium by all excretory routes was characterized by a three-component curve. Eight per cent of the administered selenium was excreted during the first phase, which had a $t_{1/2}$ of 6.8 days; 64 % during the second phase, which had a $t_{1/2}$ of 73 days; and 28 % during the final phase, which had a $t_{1/2}$ of 325 days. However, it should be pointed out that selenomethionine is a protein amino acid. Greater retention would be expected with this amino acid than with the nonprotein selenoamino acids, such as those found in selenium accumulator plants.

8. Routes of Elimination

The major route of selenium excretion in most animals is *via* the kidney. In ruminant animals, particularly sheep, a considerable portion is also eliminated in the feces. In addition, depending on the quantity and chemical form of selenium administered, dietary factors, and the presence of certain additives, significant quantities of selenium can be eliminated through the lungs.

In the experiments of Heinrich and Kelsey[98] mice were given a single injection of radioselenite (0.1–0.7 mg Se per animal). After 24 hours 63% of the dose was excreted in the urine, 8% in the feces, and 6% in expired air. Similar studies were carried out by Ganther[104] on rats except that much smaller doses were administered (25 μg per animal). Again the major route of elimination was by way of the kidney, but after the first 24 hours only 29% of the administered selenium was excreted in the urine and 8% in the feces.

Urinary excretion also appears to be the preferred route of selenium elimination in chronic selenosis. Halverson *et al.*[93] fed rats a diet containing 5 ppm Se as selenite for a 2 week period. During this time 51% of the ingested selenium was excreted in the urine and 12% in the feces.

Gastrointestinal excretion apparently is an important route of selenium elimination in sheep. Cousins and Cairney[27] fed young sheep 5 mg Se as selenite, and after 72 hours 10% of the dose was found in the urine and 30% in the feces. Similar experiments by Butler and Peterson[105] confirmed these findings. However, Ewan *et al.*[106] found that the excretion pattern of selenium in young lambs administered selenate was similar to that in monogastric animals in that most of the administered selenium was excreted in the urine. This difference between the excretory pattern of sheep and that of young lambs was attributed to the fact that the latter probably did not have functional rumina.

Under certain conditions significant quantities of selenium are eliminated as volatile compounds. Long before selenium was recognized as the causative agent in blind staggers and alkali disease, Hofmeister[107] reported that the breath of dogs previously administered sodium selenite had a garlic-like odor. He further postulated that this volatile substance was dimethyl selenide. Final proof of the excretion of dimethyl selenide in animals came more than 50 years later when McConnell and Portman[51] isolated from the breath of selenate-treated rats a volatile compound which on analysis proved to be dimethyl selenide. Challenger[69] had previously shown that in fungi dimethyl selenide was formed by methylating sodium selenite. The chemical form of selenium administered is a determining factor in the quantity of volatile selenium formed. For instance, a larger percentage of pulmonary excretion of selenium follows the administration of selenite than follows the administration of selenate,[90,108] selenomethionine,[109] or seleniferous wheat.[90]

As the quantity of administered selenium increases, pulmonary excretion becomes an increasingly important excretory pathway. McConnell and Roth[109] found this to be true in rats administered either selenite or selenomethionine. On the other hand, Sternberg and Imbach[110] confirmed this only with selenite, as they were unable to detect any pulmonary excretion of selenium in rats administered selenomethionine. It is difficult to reconcile the differences between the observations of McConnell and Roth and those of Sternberg and Imbach. The former workers used male rats and administered the selenomethionine subcutaneously, whereas Sternberg and Imbach used female rats and administered the selenomethionine intravenously. Also, a question arises regarding the chemical purity of the preparations administered, particularly in the case of organic selenium compounds. Ganther et al.[111] found that volatilization was increased two- to threefold in rats by adding 5 ppm Se to the basal diet.

The inhibition of dimethyl selenide formation by arsenite has already been discussed.[86,89-91] Cadmium has likewise been shown by Ganther and Baumann[86] to be a powerful inhibitor of dimethyl selenide formation. These investigators also observed that the diet was an important factor in the elimination of volatile selenium. They noted that rats eliminated more than twice as much volatile selenium when fed a crude ration than when fed a nutritionally adequate purified diet. Mercury and thallium also inhibit the formation of volatile selenium.[24]

Phenobarbital, which has been shown to stimulate liver microsomal enzymes,[112] also increases the rate of pulmonary excretion of selenium in rats administered selenite.[113]

9. Metabolic Fate of Selenium Compounds

a. Inorganic Selenium. Most studies on the metabolic fate of selenium have involved the use of inorganic selenium (*i.e.*, selenite and selenate). It seems unlikely that animals, particularly livestock, would ever encounter selenate or selenite *per se* except experimentally. However, selenate and selenite salts are inexpensive, have a high degree of purity, and can be administered with precision in a variety of ways. Furthermore, as a chemical tool, selenite selenium has proved to be as toxic as any form of selenium except hydrogen selenide; selenate has generally been regarded as only slightly less toxic.[39] Organic selenium compounds, particularly those likely to be encountered by animals, are more expensive (especially if labeled), often less pure, sometimes have questionable authenticity, and are no more toxic than equivalent amounts of selenite. Understandably, therefore, researchers prefer using selenite and/or selenate to organic selenium compounds. This is not to say that metabolism studies using naturally occurring selenium compounds are unimportant. Quite the contrary, there is unquestioned need for such studies. Doubtless, when organic compounds become more readily available and possess unquestioned authenticity, significant investigations in this area of research will be performed.

Although living systems are incapable of oxidizing selenite to selenate, some animal tissues are quite capable of reducing selenate to selenite.[114] Now the question is: must selenate first be converted to selenite before becoming toxic? Potter and Elvehjem[115] suggested that this was the case after noting that succinic dehydrogenase was inhibited by selenite but not by selenate. Later Rosenfeld and Beath[114] observed that selenate at 75 ppm did not interfere with the normal motility of isolated guinea pig intestine but that selenite at only 8 ppm completely blocked such motility. From such data these workers concluded that (1) selenate was not converted to selenite in guinea pig intestine, although they observed that such a conversion, which was promoted enzymatically, occurred in beef liver, spleen, whole blood, and plasma; and (2) as far as intestinal motility of the guinea pig was concerned, selenate was nontoxic and had to be converted to selenite before becoming toxic. On the other hand, other research has indicated that selenate may be toxic *per se* without having to be first converted to selenite. Westfall and Smith[116] administered selenate intravenously to rabbits and reported that 85 % of the urinary selenium appeared as inorganic selenium, presumably selenate, indicating that the selenate had undergone little chemical change. Conversely, when the same quantity of selenite was similarly administered, considerable change took place since 80 % of the urinary selenium appeared in the ethereal and neutral sulfur fraction. If selenate can pass through the body without undergoing any appreciable chemical change, yet remain nearly as toxic as selenite, then it would appear that the toxicity of selenate is not entirely due to its conversion to selenite, and that the compound also probably follows a different metabolic route from that of selenite.

One area in which it was felt that selenate may follow a metabolic pathway not shared by selenite involved its possible incorporation into mucopolysaccharides in cartilage. It has been established that sulfate was readily incorporated into cartilage mucopolysaccharides,[117] and it had also been shown that selenium administered as selenate[118,119] or selenite[32,120] became incorporated in cartilagenous tissue. Cipera and Hidiroglou[121] have demonstrated that neither selenite nor selenate was incorporated into the mucopolysaccharides of rabbit cartilage. The selenium that was present in the cartilage was found in the protein and lipid fractions.

That both selenite and selenate are reduced to Se° and Se²⁻ is unquestioned. It is also believed that the limiting factor in the reduction of selenate to Se²⁻ is its reduction to selenite.[108] The question is: what are the chemical forms of Se° and Se²⁻? Although *in vitro* studies have indicated that mammalian tissues are capable of reducing selenite to the red allotrope of elemental selenium,[114,122] there is no evidence at present indicating that such a reduction occurs *in vivo*.[123] Ganther[124] has presented excellent evidence showing that selenite combines with biologically important sulfhydryl compounds such as cysteine, glutathione, and coenzyme A to form "selenotrisulfides" (RS–Se–SR),* in which the Se atom exists in the zero oxidation state. This reaction is depicted in equation 1.

$$(1) \qquad 4RSH + H_2SeO_3 \rightarrow RS\!-\!Se\!-\!SR + RSSR + 3H_2O$$

Such a reaction between selenite and sulfhydryl compounds was first suggested 30 years ago by Painter.[125] One way of balancing this oxidation-reduction reaction is to consider that the four sulfhydryl sulfurs, which are in an oxidation state of −2, are oxidized to disulfide sulfurs, which are in an oxidation state of −1. This change in oxidation state of the S atoms is balanced by the selenite selenium, which is at an oxidation state of +4, being reduced to selenium at an oxidation state of zero. However, according to Pauling's scale of electronegativities, sulfur is only slightly more electronegative than selenium (2.5 and 2.4, respectively). Therefore, it may be more appropriate to think of the −2 charge as being rather evenly distributed throughout the three atoms comprising the selenotrisulfide bridge, that is, each having an oxidation number of $-\frac{2}{3}$.†

The detoxification of selenite through the formation of dimethyl selenide, in which selenium is in an oxidation state of −2, has already been discussed. As early as 1942 McConnell[126] suggested that such formation occurred in minced rat liver. Later Ganther[91] presented more direct evidence showing that such synthesis takes place in mouse liver homogenates and requires GSH, NADPH + H⁺, and S-adenosylmethionine. It had previously been reported that fungi are capable of synthesizing dimethyl selenide from selenite.[69] Challenger[127] has proposed a reaction sequence for

---

* An alternative name for this class of compounds which has been suggested by Günther (Chapter I) is 1,3-dithia-2-selanes.

† Rules covering the oxidation states of sulfur in its various compounds are given in B. E. Douglas and D. H. McDaniel, "Concepts and Models of Inorganic Chemistry," Blaisdell Publishing Co., Waltham, Mass., 1968, p. 43.

dimethyl selenide formation in fungi; this is given in scheme 4 of Chapter XIIIE.

Sternberg *et al.*[113] have suggested that a similar sequence occurs in mammalian liver microsomes. It should be pointed out, however, that this mechanism was first proposed 20 years ago, and to date there has been no experimental verification of this mechanism either in microorganisms or in animals.

On the basis of the calculations of Schlenk[128] on the resting level of active methyl groups in liver, Sternberg *et al.*[113] calculated that the livers of rats administered toxic levels of selenium, as selenite, would have to renew this resting level every 7 minutes, that is, at a rate 18 times that observed in rats receiving low doses of selenite. These investigators suggested that such a rapid rate of renewal might well necessitate a diversion of methylation processes from other reactions to the reductive methylation of selenite in order to detoxify it when administered in large doses.

Researchers from two different laboratories have announced the isolation of a trimethylselenonium ion $[(CH_3)_3Se^+]$ in the urine of rats administered selenite.[129,130] In both cases the researchers determined that the trimethylselenonium ion was the major urinary detoxification metabolite of selenite. Palmer *et al.*[130] further suggested the possibility that trimethylselenonium ion might arise from the methylation of dimethyl selenide. To test this hypothesis these researchers administered arsenic on the assumption that this element, which lowers dimethyl selenide formation, should likewise inhibit the production of trimethylselenonium ion. Although arsenic did reduce the excretion of the latter, such reduction was not statistically significant. Palmer *et al.* concluded, therefore, that dimethyl selenide is not necessarily an intermediate in trimethylselenonium synthesis. Direct toxicity studies have been conducted only recently. Obermeyer *et al.*[130a] have shown that orally administered trimethylselenonium chloride was substantially less toxic to rats than selenite.

It has long been known that a considerable portion of tissue selenium is bound to proteins. In 1938, Smith *et al.*[30] administered seleniferous oats and inorganic selenium (selenite and selenate) to rabbits and found that in both cases most of the liver selenium was bound to proteins. Later, Westfall and Smith[131] reported that more than 90 % of the total selenium in the plasma of rabbits fed seleniferous grain was present in the protein portion. Subsequently, numerous investigators have shown that selenium administered as inorganic selenium becomes rather firmly attached to all types of proteins— from milk proteins[132,133] to keratins.[134,135] However, until recent times it was not known whether administered selenite or selenate became bound to proteins through their reacting with sulfhydryl groups or were reduced to selenoamino acids and subsequently incorporated into the proteins through peptide bonds. Although the biosynthesis of selenoamino acids from selenite in animal systems is postulated in the literature,[135–137] some investigators began to doubt that such synthesis could take place. Except perhaps in the cat family, there is no known pathway for the synthesis of sulfur amino acids from inorganic sulfur in monogastric animals. However, Ganther and Corcoran[138] have suggested that, even though most species lack the capability of reductive utilization of sulfur, the formation of reduced selenides (i.e., dimethyl selenide and trimethylselenonium ion) may lead to selenoamino acid biosynthesis. On the other hand, Schwarz[139,140] has maintained that selenoamino acids are not the naturally occurring forms of selenium in animal systems. In 1966, Hirooka and Galambos[141] failed to identify any selenoamino acids in serum protein hydrolysates of rats administered selenate.

In all studies that reported the biosynthesis of selenoamino acids from inorganic selenium, the sole criterion of identification of the selenoamino acids was based on the observation that ^{75}Se could be detected on chromatograms in the vicinity where their corresponding sulfur analogs were known to appear. However, Schwarz and Sweeney[142] observed that various sulfur compounds bind selenite ions, which in turn migrate with the sulfur compounds during chromatographic procedures. Therefore, the above criterion used in the identification of selenoamino acids became suspect. Later, Cummins and Martin[74] demonstrated that neither selenocystine nor selenomethionine was present in liver protein of rabbits fed a selenite-supplemented diet. The hydrolysates were enzymatically hydrolyzed to prevent the destruction of any selenoamino acids that may have been present in a sample. Urine samples from rabbits administered radioselenite were chromatographed by ion-exchange chromatography. The chromatograms revealed two radioactive peaks suggesting the presence of selenocystine and selenomethionine. However, on further analysis, it was determined that the selenium in both fractions complexed readily with diaminobenzidine, indicating that the element was present as selenite which was bound in some fashion to the sulfur amino acids in the urine sample. Imbach and Sternberg,[102] in their studies on rats administered selenite, have also presented evidence that such an association between selenite and sulfur amino acids exists.

The results of Cummins and Martin were confirmed by Jenkins.[75] In one phase of an investigation Jenkins crop-intubated one chick with Na$_2$-^{75}SeO$_3$ and another chick with a uniformly ^{14}C-labeled L-amino acid mixture. Serum proteins from both chicks were dialyzed for 17 hours at pH 11.5 at 25°. Such alkaline treatment brought about a release of one-third of the

[75]Se activity, but no such release was observed in the serum of the chick administered the ^{14}C-labeled amino acid mixture. This observation indicated that the release of selenium from the protein was not the result of protein hydrolysis, thus negating the possibility that peptide-bound seleno-amino acids had been released. Furthermore, Jenkins was unable to identify either selenocystine or selenomethionine in the alkaline dialysate.

In this investigation and a subsequent one[143] from Jenkins' laboratory it was postulated that the selenium was bound to the protein through a seleno-trisulfide type of bond similar to that first proposed by Painter.[125] Jenkins[75] then postulated a series of equations (scheme 2) illustrating how the protein-

$$
\begin{aligned}
&\text{(2)} \quad 2\text{PS—Se—SP} + 2\text{OH}^- \rightarrow 2\text{PS}^- + 2\text{PS—SeOH} \\
&\quad\quad 2\text{PS—SeOH} \rightarrow \text{PSH} + \text{PS—SeO}_2\text{H} + \text{Se}^\circ \\
&\quad\quad \underline{\text{PS—SeO}_2\text{H} + 2\text{OH}^- \rightarrow \text{PSH} + \text{SeO}_3^{2-} + \text{H}_2\text{O}} \\
&\quad\quad 2\text{PS—Se—SP} + 4\text{OH}^- \rightarrow 2\text{PS}^- + 2\text{PSH} + \text{Se}^\circ + \text{SeO}_3^{2-} + \text{H}_2\text{O} \\
&\quad\quad 2\text{PS—Se—SP} + 6\text{OH}^- \rightarrow 4\text{PS}^- + \text{Se}^\circ + \text{SeO}_3^{2-} + 3\text{H}_2\text{O}
\end{aligned}
$$

$$(\text{P} = \text{protein moiety})$$

selenotrisulfide complex could dissociate under alkaline conditions to re-form selenite, thus accounting for the release of selenium from proteins under alkaline conditions. The formation of Se° as shown here should not be overlooked. Although extracellular elemental selenium is biologically inert because of its limited cellular absorption,[36] intracellular elemental selenium may exhibit significant biological activity.

Ganther and Corcoran[138] have recently presented spectral evidence showing that reduced ribonuclease combines with selenite, forming a so-called selenotrisulfide type of product. Before proteins can react with selenite to form selenotrisulfides, the proteins would necessarily have to exist as free sulfhydryls (see equation 1). At the completion of protein synthesis, the sulfhydryls of most proteins are largely converted to disulfide groups. Indeed, Ganther and Corcoran found that reduced ribonuclease and not the native protein reacted with selenite to form a selenotrisulfide. Jenkins *et al.*[143] reported that selenite added to chick serum *in vitro* does not complex with the proteins. This observation led these investigators to postulate that selenium is incorporated into proteins during the final stages of protein synthesis.

Although protein selenotrisulfides dissociate readily under alkaline conditions to form selenite and colloidal selenium (see scheme 2), such dissociation would take place rather slowly at physiological pH's. Therefore, until the time when selenite becomes firmly bound to proteins as selenotrisulfides, a rather rapid rate of excretion could be expected, followed by a much reduced rate after the selenium becomes more firmly attached to proteins. This would explain the observations that animals administered selenite exhibit a rapid rate of selenium elimination, followed by a much reduced rate.[99,100]

McConnell et al.[144] were the first to demonstrate that selenium, soon after administration, is quite labile but rapidly becomes rather tightly bound to proteins. These investigators injected dogs with $^{75}SeCl_4$ and noted that initially the serum albumin contained most of the radioactivity, but after the first hour the selenium was transferred to the α_2- and β-globulin fractions. These workers suggested that the albumin-bound selenium was an anion-protein type of bond, but that by the time the selenium became bound to the α_2- and β-globulins, it was much more firmly bound, possibly as some type of covalent bond.

Similar results were obtained by Imbach and Sternberg,[102] who injected rats with trace amounts of selenite. However, these workers observed that, from 3 days until the termination of the experiment at 72 days, most of the selenium was bound to the γ-globulin fraction with somewhat less bound to the α-globulin fraction. They noted that only a small quantity of selenium was bound to the β-globulin fraction during this time period. Jenkins et al.[143] also noted that, from 24 to 173 hours after administration of $^{75}SeO_3{}^{2-}$ to chicks, 52–70% of the serum selenium was carried by the α_2- and γ-globulin fractions. Moreover, these investigators noted that during the first 12 hours after selenite administration the selenium remained in a somewhat labile phase.

Selenium also becomes bound to the serum lipoproteins in dogs and rats administered selenite[145] and in rats administered selenate.[141] In both investigations more of the selenium was bound to the α-lipoprotein fraction than to the β-fraction.

With the knowledge that selenite reacts with sulfhydryl groups, one would naturally investigate this type of reaction as a possible explanation for the toxicity of selenite. A likely starting point would be to examine the inhibitory effect of selenite on sulfhydryl-containing enzymes. Results from such investigations have not been as promising as originally hoped, although inhibition of these enzymes does occur. Klug et al.[146] demonstrated that the level of inhibition of several selenium-containing compounds on rat liver succinic dehydrogenase did not correlate with the relative toxicities of these compounds. Selenocystine was the most potent inhibitor, with no appreciable difference between the D- and L-isomers. Earlier it had been shown by Moxon et al.[54] that L-selenocystine was much more toxic than its D-isomer. Klug et al. found that sodium selenite was about 1/10 as potent an inhibitor as selenocystine, and selenomethionine only about 1/300 as potent. In rats the levels of toxicity of selenite, selenocystine, and selenomethionine are quite similar. These investigators also pointed out that, although rat liver succinic dehydrogenase was one of the most sensitive systems of selenite inhibitions *in vivo*, other as yet unidentified systems would have to be 25 times more sensitive to account for the low level at which selenium is toxic in animals. Tsen and Collier[147] also found that selenite was a relatively weak inhibitor of six sulfhydryl-containing enzymes. They found that selenite was a much poorer inhibitor of all enzymes than was either cupric chloride or *p*-chloromercuribenzoate. Tsen and Tappel[148] had previously demonstrated that selenite was an excellent catalyst for the oxidation of reduced glutathione—much better even than cupric chloride, the most active catalyst previously known for promoting this oxidation.[149] The role of selenite in the conversion of reduced glutathione to oxidized glutathione had been previously reported by other workers.[150-152] In each of these investigations selenite was shown as an oxidant, and it should be pointed out that selenious acid is an excellent oxidizing agent and is much more effective than its sulfur counterpart, sulfurous acid. In fact, the reaction between sulfurous acid and selenious acid was used in early quantitative studies to reduce selenious acid to elemental selenium (equation 3).

$$(3) \qquad H_2SeO_3 + 2H_2SO_3 \rightarrow Se + 2H_2SO_4 + H_2O$$

In the studies of Tsen and Tappel[148] selenite was presented as a true catalyst with molecular oxygen acting as the oxidant. These workers offered the equation shown in scheme 4 for the overall oxidation of reduced glutathione catalyzed by selenite. Later, Jenkins[75] demonstrated that oxygen

$$(4) \qquad \begin{array}{l} 4GSH + SeO_3{}^{2-} \rightarrow GSSG + GS\!-\!Se\!-\!SG + 2OH^- + H_2O \\ 2OH^- + GS\!-\!Se\!-\!SG + O_2 \rightarrow GSSG + SeO_3{}^{2-} + H_2O \\ \hline 4GSH + O_2 \rightarrow 2GSSG + 2H_2O \end{array}$$

was not required in the decomposition of selenotrisulfides by alkali.

Bearing these facts in mind and realizing that selenite is toxic in small quantities, Tsen and Tappel advanced the hypothesis that its toxicity might be due to its ability to catalyze the oxidation of such cofactors as glutathione, coenzyme A, and dihydrolipoic acid, thus resulting in disturbances in intermediary metabolism. In this connection, it has recently been shown that selenite and lipoic acid act synergistically in inhibiting cholesterol biosynthesis from acetate and mevalonic acid.[153] This effect was attributed in part to selenite-induced alterations of sulfhydryl enzymes and glutathione contained in the enzyme system.

Reports on the effects of selenium toxicity on tissue levels of glutathione have been conflicting. Klug *et al.*[154,155] stated that tissue levels of glutathione were reduced by selenium toxicity. Recently, however, Shtutman *et al.*[156] found that the injection of subtoxic quantities of selenite to rats considerably increased the conversion of methionine to glutathione in the livers of these animals. Martin and Spallholz[157] obtained similar results in chicks suffering from selenite intoxication. It would appear from these two studies that animals afflicted with selenium toxicity were endeavoring to synthesize glutathione at an increased rate in order to compensate for its inactivation brought about by its selenite-induced oxidation. Apparently, such inactivated glutathione has a slow turnover in the liver, for it tends to accumulate in this organ in chicks suffering from selenite intoxication.[157] It had previously been shown that increasing the level of reduced glutathione by dietary means will protect rats against selenite toxicity.[151]

From the foregoing comments it would appear that selenite exerts its toxic effects in several ways, all of which seem to involve sulfhydryl compounds, particularly its catalytic oxidation of such compounds. The toxicity of selenate depends, at least in part, on its reduction to selenite. Obviously, more research is required to answer all the nagging questions relating to the modes of toxicity of selenite and selenate.

The following statements seem appropriate in briefly summarizing the metabolic fates of selenite and selenate in animals, particularly as these compounds act as toxicants. Selenite rapidly finds its way into the general circulation, regardless of its mode of administration. In the general circulation it quickly, but loosely, combines with plasma albumin and, as such, is carried throughout the body. This phase has been described by Imbach and Sternberg[102] as the "carrier phase." Before selenite becomes more firmly attached to the various globular fractions, it can freely detach itself from the albumin and easily enter the various cells of the body. In the cells it exerts its toxic action by catalyzing the oxidation of important sulfhydryl metabolites, thus inactivating them. The body attempts to detoxify the selenite by reductive methylation, which occurs in the liver. Under most conditions the most important excretory pathway is *via* the kidneys, and the principal urinary excretory product is trimethylselenonium ion. When selenite is administered in doses approaching toxic levels, detoxification by the formation of volatile dimethyl selenide, which is eliminated by pulmonary excretion, assumes increasing importance. In animals having a functional rumen, fecal elimination is an important excretory pathway. After the selenium becomes more firmly attached to the plasma globular proteins, it is eliminated at a much slower rate by all excretory routes than that observed before such an attachment occurs.

Selenate, of course, would follow the same metabolic pathway after its reduction to selenite.

The mode of toxicity of hydrogen selenide, the most toxic selenium compound known, is probably very similar to that of hydrogen sulfide, which has been shown to be a powerful inhibitor of cytochrome oxidase.[158]

b. Organic Selenium. First, let us consider the fate of the selenoamino acids commonly found in selenium accumulator plants, namely *Se*-methylselenocysteine and selenocystathionine. The high purchase price of these selenoamino acids, particularly if tagged, makes their use in metabolic studies almost prohibitive. This limitation is indeed unfortunate, considering that these compounds are probably the ones most likely to be encountered by grazing livestock. Beath *et al.*[159] reported that approximately 340 mature sheep died within 24 hours after accidentally consuming *Astragalus bisulcatus*. This selenium accumulator plant contains large quantities of *Se*-methylselenocysteine and γ-glutamyl-*Se*-methylselenocysteine.[42–48] The toxicities of both the free amino acid and the dipeptide were shown to be quite extensive and about equal in magnitude.[48] Other *Astragalus* accumulators, such as *A. pectinatus* and *A. osterhoutii*, contain large quantities of both *Se*-methylselenocysteine and selenocystathionine.[46,47] *Stanleya pinnata*, another selenium accumulator, accumulates primarily selenocystathionine.[45,47] It has been suggested by Olson[160] that animals consuming selenium accumulator plants may exhale rather large quantities of volatile selenium compounds, in contrast to the very low amounts of volatile selenium eliminated by animals fed seleniferous grain.[90] If this is true, it would suggest that the nonprotein selenoamino acids, characteristic of the selenium accumulator plants, are oxidized to selenite or to some oxidation product biochemically similar to selenite, since either selenite or selenate has been proposed as the original precursor of volatile dimethyl selenide in molds[69] and in liver ribosomes.[113]

Much has been said about selenium compounds undergoing reduction reactions, but there has been little discussion of the possibility that selenoamino acids become oxidized in animal systems. Since it now appears that the conversion of selenite to selenoamino acids can at best occur to only a limited degree, to attribute the toxicity of selenite in such terms would seem rather futile. Moreover, it now appears that selenite is toxic *per se*, and most mechanisms proposed to explain the toxicity of selenium have as a starting point the involvement of selenite with sulfhydryl compounds. Also, any significant reduction of selenite which is likely to occur in animal systems serves primarily to produce reduction compounds far less toxic than selenite, and the oxidation of selenite in animal systems is out of the question.

Selenoamino acids probably undergo oxidation reactions similar to those undergone by their sulfur analogs to produce such oxidation products as selenocysteineseleninic acid, selenocysteic acid, and selenotaurine. Such reactions would not be surprising, for Shrift[161] has cataloged a number of enzyme systems that catalyze selenium metabolites quite as well as they do their sulfur analogs. The immediate environment of the Se atom in such compounds is similar to that of the Se atom in selenite in that O atoms are bound to it. Therefore, such compounds would be expected to have toxicities similar to that of selenite. Indeed, Mattii et al.[162] have recently found that the toxicological response of rats to selenotaurine was similar to their response to inorganic selenium compounds. However, at the present time there is little direct evidence to support the hypothesis that selenoamino acids are so oxidized. In 1968 Neethling et al.[163] reported the in vivo formation of selenotaurine in sheep administered selenomethionine. More recently, Martin and Gerlach[164] presented evidence that oxidation products, presumably ^{75}Se-selenocysteic acid and ^{75}Se-selenotaurine, are formed in the livers of chicks and mice after the administration of ^{75}Se-selenomethionine. It should be pointed out, however, that attempts to synthesize selenocysteic acid and selenotaurine, employing the conditions under which cysteic acid and taurine are synthesized, have, in general, proved quite unsucessful.[165–167] Apparently alkyl SeO_3H compounds either are difficult to synthesize or, once formed, are unstable. It may be, therefore, that such compounds cannot be normal metabolic end products of selenoamino acids. More research on the metabolic fates of the various selenoamino acids in animal systems is needed.

A quite different situation would be expected to exist in crop plants. In the past, plant researchers have referred to organic selenium compounds as the selenium compounds occurring in plants. Clearly, a distinction should be made. All existing evidence suggests that the selenoamino acids in crop plants are of the protein type, that is, selenocystine and selenomethionine,[168–172] whereas the selenoamino acids of accumulator plants are of the nonprotein type.

The most striking difference between the response of animals to the ingestion of accumulator plants and their response to the ingestion of selenized grain is that alkali disease can be produced only from selenized grain. This difference would suggest that perhaps it is the incorporation of the selenoamino acids from the selenized grain into the primary structure of the proteins of the affected animal that is responsible for the unique toxicity symptoms associated with alkali disease. Apparently, animals experience no difficulty in incorporating selenoamino acids into their proteins. Selenomethionine has been shown to be incorporated into all types of animal proteins, including pancreatic proteins,[173] egg albumin,[174] and even the protein of tarantula venom.[175] However, the incorporation of selenoamino acids into proteins would have to be rather extensive to account for their toxicity, since it has been shown that partial replacement of sulfur amino acids by their selenium analogs does not materially alter the biological activity of proteins. Also, Walter and du Vigneaud[176,177] have shown that the replacement by selenium of one or both of the S atoms of oxytocin does not seriously alter the biopotency of this hormone (cf. Chapter XIIB). For this reason one should not attribute the toxicity of the selenoamino acids primarily to their replacement of sulfur amino acids in proteins. Obviously, such replacement of sulfur by selenium would not occur with inorganic selenium or with the nonprotein selenoamino acids found in accumulator plants.

However, one would not expect all of the selenoamino acids from selenized grain to become part of the primary structure of body proteins. The selenoamino acids not incorporated might very well follow the oxidative processes experienced by their sulfur analogs, thereby forming toxic oxidation products similar or identical to those that may result from the oxidation of the nonprotein selenoamino acids.

Again, the final proof as to whether selenoamino acids are oxidized in a manner similar to that of their sulfur analogs has to await further investigation.

REFERENCES

1. Marco Polo, "The Travels of Marco Polo," revised from Marsden's translation and edited with introduction by Manual Komroff, Liveright, New York, N.Y., 1926, p. 81.
2. T. C. Madison, U.S. *Congr. 36th, 1st Session, Senate Exch. Doc.*, **52,** 37 (1860).
3. P. W. Selwood, "General Chemistry," Henry Holt, New York, N.Y., 1954, p. 236.
4. C. Larsen, W. White, and D. E. Bailey, *S. Dakota Agr. Exp. Sta. Bull.*, **132,** 220 (1912).
5. C. Larsen and D. E. Bailey, *S. Dakota Agr. Exp. Sta. Bull.*, **147,** 300 (1913).
6. J. H. Draize and O. A. Beath, *J. Amer. Vet. Med. Assoc.*, **86,** 753 (1935).
7. K. W. Franke, *J. Nutr.*, **8,** 597 (1934).
8. H. G. Byers, *U.S. Dept. Agr. Tech. Bull.*, **482,** 1 (1935).
9. W. O. Robinson, *J. Assoc. Offic. Agr. Chemists*, **16,** 423 (1933).
10. I. Rosenfeld and O. A. Beath, "Selenium," Academic Press, New York, N.Y., 1964, p. 145.
11. K. W. Franke, T. D. Rice, A. G. Johnson, and H. W. Schoening, *U.S. Dept. Agr. Circ.*, **320,** 1 (1934).
12. H. C. Dudley, *Amer. J. Hyg.*, 23, 169 (1936).
13. W. T. Miller and K. T. Williams, *J. Agr. Res.*, **61,** 353 (1940).
14. K. P. McConnell, *J. Biol. Chem.*, **141,** 427 (1941).
15. M. Rhian and A. L. Moxon, *J. Pharmacol. Exptl. Therap.*, **78,** 249 (1943).
16. I. Rosenfeld and O. A. Beath, *J. Nutr.*, **30,** 443 (1945).
17. D. D. Maag, J. S. Orsborn, and J. R. Clopton, *Amer. J. Vet. Res.*, **21,** 1049 (1960).
18. M. W. Glenn, J. L. Martin, and L. M. Cummins. *Amer. J. Vet. Res.*, **25,** 1495 (1964).
19. I. Rosenfeld and O. A. Beath, *Amer. J. Vet. Res.*, **7,** 52 (1946).
20. O. E. Olson, C. A. Dinkel, and L. D. Kamstra, *S. Dakota Farm Home Res.*, **6,** 12 (1954).
21. M. W. Glenn, "A Study of Sodium Selenate Toxicosis in Sheep," Dissertation, Colorado State University, Fort Collins, Colo., 1962.
22. D. A. Morrow, *J. Amer. Vet. Med. Assoc.*, **152,** 1625 (1968).

23. H. L. Klug, G. P. Lampson, and A. L. Moxon, *Proc. S. Dakota Acad. Sci.*, **29,** 57 (1950).

24. O. A. Levander and L. C. Argrett, *Toxicol. Appl. Pharmacol.*, **14,** 308 (1969).

25. O. A. Levander, M. L. Young, and S. A. Meeks, *Toxicol. Appl. Pharmacol.*, **16,** 79 (1970).

26. O. A. Levander and V. C. Morris, *J. Nutr.*, **100,** 1111 (1970).

27. F. B. Cousins and I. M. Cairney, *Australian J. Agr. Res.*, **12,** 927 (1961).

28. M. Hidiroglou, I. Hoffman, and K. J. Jenkins, *Can. J. Physiol. Pharmacol.*, **47,** 953 (1969).

29. W. H. Allaway, D. P. Moore, J. E. Oldfield, and O. H. Muth, *J. Nutr.*, **88,** 411 (1966).

30. M. I. Smith, B. B. Westfall, and E. F. Stohlman, *U.S. Public Health Rept.*, **53,** 1199 (1938).

31. A. W. Halverson, I. S. Palmer, and P. L. Guss, *Toxicol. Appl. Pharmacol.*, **9,** 477 (1966).

32. C. F. Ehlig, D. E. Hogue, W. H. Allaway, and D. J. Hamm, *J. Nutr.*, **92,** 121 (1967).

32a. D. G. Brown and R. F. Burk, *Federation Proc.*, **31,** 692 (1972).

33. W. G. Jaffé and M. C. Mondragon, *J. Nutr.*, **97,** 431 (1969).

34. L. L. Hopkins, Jr., A. L. Pope, and C. A. Baumann, *J. Nutr.*, **88,** 61 (1966).

35. I. Rosenfeld and O. A. Beath, "Selenium," Academic Press, New York, N.Y., 1964, p. 146.

36. K. W. Franke and V. R. Potter, *J. Nutr.*, **10,** 213 (1935).

37. M. I. Smith and R. D. Lillie, *U.S. Public Health Serv., Natl. Inst. Health Bull.*, **174,** 1 (1940).

38. M. I. Smith, E. F. Stohlman, and R. D. Lillie, *J. Pharmacol. Exptl. Therap.*, **60,** 449 (1937).

39. K. W. Franke and A. L. Moxon, *J. Pharmacol. Exptl. Therap.*, **58,** 454 (1936).

40. I. J. Tinsley, J. R. Harr, J. F. Bone, P. H. Heswig, and R. S. Yamamoto, "Selenium in Biomedicine," AVI Publishing Co., Westport, Conn., 1967, p. 150.

41. S. F. Trelease, A. A. DiSomma, and A. L. Jacobs, *Science*, **132,** 3427 (1960).

42. A. Shrift and T. K. Virupaksha, *Biochim. Biophys. Acta*, **100,** 65 (1965).

43. T. K. Virupaksha and A. Shrift, *Biochim. Biophys. Acta*, **107,** 69 (1965).

44. M. J. Horn and D. B. Jones, *J. Biol. Chem.*, **139,** 649 (1941).

45. T. K. Virupaksha and A. Shrift, *Biochim. Biophys. Acta*, **74,** 791 (1963).

46. J. L. Martin and M. L. Gerlach, *Anal. Biochem.*, **29,** 257 (1969).

47. J. L. Martin, A. Shrift, and M. L. Gerlach, *Phytochemistry*, **10,** 945 (1971).

48. S. N. Nigam and W. B. McConnell, *Biochim. Biophys. Acta*, **192,** 185 (1969).

49. I. Rosenfeld and O. A. Beath, "Selenium," Academic Press, New York, N.Y., 1964, p. 147.

50. W. T. Miller and K. T. Williams, *J. Agr. Res.*, **60,** 163 (1940).

51. K. P. McConnell and O. W. Portman, *Proc. Soc. Exptl. Biol. Med.*, **79,** 230 (1952).

52. F. A. Patty, "Industrial Hygiene and Toxicology," Vol. II, Interscience Publishers, New York, N.Y., 1949.

53. H. L. Klug, D. F. Petersen, and A. L. Moxon, *Proc. S. Dakota Acad. Sci.*, **28,** 117 (1949).

54. A. L. Moxon, K. P. DuBois, and R. L. Potter, *J. Pharmacol. Exptl. Therap.*, **72,** 184 (1941).

55. A. L. Moxon, H. D. Anderson, and E. P. Painter, *J. Pharmacol. Exptl. Therap.*, **63,** 357 (1938).

56. O. G. Fitzhugh, A. A. Nelson, and C. I. Bliss, *J. Pharmacol. Exptl. Therap.*, **80,** 289 (1944).

57. H. E. Munsell, G. M. DeVaney, and M. H. Kennedy, *U.S. Dept. Agr. Tech. Bull.*, **534,** 1 (1936).

58. L. A. Witting and M. K. Horwitt, *J. Nutr.*, **84,** 351 (1964).

59. M. I. Smith, *U.S. Public Health Rept.*, **54,** 1441 (1939).

60. R. A. Gortner, Jr., *J. Nutr.*, **19,** 105 (1940).

61. I. Rosenfeld and O. A. Beath, "Selenium," Academic Press, New York, N.Y., 1964, p. 185.

62. H. B. Lewis, J. Schultz, and R. A. Gortner, Jr., *J. Pharmacol. Exptl. Therap.*, **68,** 292 (1940).

63. E. A. Sellers, R. W. You, and C. C. Lucas, *Proc. Soc. Exptl. Biol. Med.*, **75,** 118 (1950).

64. K. P. McConnell, *Federation Proc.*, **11,** 255 (1952).

65. I. Rosenfeld and H. F. Eppson, *Amer. J. Vet. Res.*, **18,** 693 (1957).

66. O. E. Olson, C. W. Carlson, and E. Leitis, *S. Dakota Agr. Exp. Sta. Tech. Bull.*, **20,** 1 (1958).

67. H. L. Klug and R. H. Harshfield, *Proc. S. Dakota Acad. Sci.*, **28,** 99 (1949).

68. H. L. Klug, R. D. Harshfield, R. M. Pengra, and A. L. Moxon, *J. Nutr.*, **48,** 409 (1952).

69. F. Challenger, *Advan. Enzymol.*, **12,** 429 (1951).

70. A. W. Halverson, C. M. Hendrick, and O. E. Olson, *J. Nutr.*, **56,** 51 (1955).

71. O. E. Olson and A. W. Halverson, *Proc. S. Dakota Acad. Sci.*, **33,** 90 (1954).

72. A. L. Moxon, A. E. Schaefer, H. A. Lardy, K. P. DuBois, and O. E. Olson, *J. Biol. Chem.*, **132,** 785 (1940).

73. H. Sakurayama, *Shikoku Acta Med.*, **16,** 128 (1960).

74. L. M. Cummins and J. L. Martin, *Biochemistry*, **6,** 3162 (1967).

75. K. J. Jenkins, *Can. J. Biochem.*, **46,** 1417 (1968).

76. A. L. Moxon, *Science*, **88,** 81 (1938).

77. K. P. DuBois, A. L. Moxon, and O. E. Olson, *J. Nutr.*, **19,** 477 (1940).

78. A. L. Moxon, M. A. Rhian, H. D. Anderson, and O. E. Olson, *J. Animal Sci.*, **3,** 299 (1944).

79. A. L. Moxon, C. R. Paynter, and A. W. Halverson, *J. Pharmacol. Exptl. Therap.*, **84,** 115 (1945).

80. A. L. Moxon, C. W. Jensen, and C. R. Paynter, *Proc. S. Dakota Acad. Sci.*, **26,** 21 (1947).

81. C. W. Carlson, E. Guenthner, W. Kohlmeyer, and O. E. Olson, *Poultry Sci.*, **33,** 768 (1954).

82. N. T. Thapar, E. Guenthner, C. W. Carlson, and O. E. Olson, *Poultry Sci.*, **48,** 1988 (1969).

83. D. F. Petersen, H. L. Klug, R. D. Harshfield, and A. L. Moxon, *Proc. S. Dakota Acad. Sci.*, **29,** 123 (1950).

84. A. L. Moxon and K. P. DuBois, *J. Nutr.*, **18,** 447 (1939).

85. I. S. Palmer and C. W. Bonhorst, *J. Agr. Food Chem.*, **5,** 928 (1957).

86. H. E. Ganther and C. A. Baumann, *J. Nutr.*, **77,** 210 (1962).

87. O. A. Levander and C. A. Baumann, *Toxicol. Appl. Pharmacol.*, **9,** 98 (1966).

88. O. A. Levander and C. A. Baumann, *Toxicol. Appl. Pharmacol.*, **9,** 106 (1966).

89. L. D. Kamstra and C. W. Bonhorst, *Proc. S. Dakota Acad. Sci.*, **32,** 72 (1953).

90. O. E. Olson, B. M. Schulte, E. I. Whitehead and A. W. Halverson, *J. Agr. Food Chem.*, **11,** 531 (1963).

91. H. E. Ganther, *Biochemistry*, **5,** 1089 (1966).

92. H. E. Ganther and C. A. Baumann, *J. Nutr.*, **77,** 408 (1962).

93. A. W. Halverson, P. L. Guss and O. E. Olson, *J. Nutr.*, **77,** 459 (1962).

94. C. W. Bonhorst and I. S. Palmer, *J. Agr. Food Chem.*, **5,** 931 (1957).

95. S. G. Morss and H. S. Olcott, *Proc. Soc. Exptl. Biol. Med.*, **124,** 483 (1967).

96. M. R. Gardiner, *Australian Vet. J.*, **42,** 442 (1966).

97. M. Lee, A. Dong, and J. Yano, *Can. J. Biochem.*, **47,** 791 (1969).

98. M. A. Heinrich, Jr., and F. E. Kelsey, *J. Pharmacol. Exptl. Therap.*, **114,** 28 (1955).

99. C. Blincoe, *Nature*, **186,** 398 (1960).

100. R. C. Ewan, A. L. Pope, and C. A. Baumann, *J. Nutr.*, **97,** 547 (1967).

101. P. L. Lopez, R. L. Preston, and W. H. Pfander, *J. Nutr.*, **97,** 123 (1969).

102. A. Imbach and J. Sternberg, *Intern. J. Appl. Radiation Isotopes*, **18,** 545 (1967).

103. K. A. Lathrop, P. V. Harper and F. D. Malkinson, *Strahlentherapie, Sonderb.*, **67,** 436 (1968).

104. H. E. Ganther, *World Rev. Nutr. Dietetics*, **5,** 338 (1965).

105. G. W. Butler and P. J. Peterson, *New Zealand J. Agr. Res.*, **4,** 484 (1961).

106. R. C. Ewan, C. A. Baumann, and A. L. Pope, *J. Agr. Food Chem.*, **16,** 216 (1968).

107. F. Hofmeister, *Arch. Exptl. Pathol. Pharmakol.*, **33,** 198 (1894).

108. T. Hirooka and J. T. Galambos, *Biochim. Biophys. Acta*, **130,** 313 (1966).

109. K. P. McConnell and D. M. Roth, *Proc. Soc. Exptl. Biol. Med.*, **123,** 919 (1966).

110. J. Sternberg and A. Imbach, *Intern. J. Appl. Radiation Isotopes*, **18,** 557 (1967).

111. H. E. Ganther, O. A. Levander, and C. A. Baumann, *J. Nutr.*, **88,** 55 (1966).

112. J. C. Arcos, A. H. Conney, and N. P. Buu-Hoi, *J. Biol. Chem.*, **236,** 1291 (1961).

113. J. Sternberg, J. Brodeur, A. Imbach, and A. Mercier, *Intern. J. Appl. Radiation Isotopes*, **19,** 669 (1968).

114. I. Rosenfeld and O. A. Beath, *J. Biol. Chem.*, **172,** 333 (1948).

115. V. R. Potter and C. A. Elvehjem, *J. Biol. Chem.*, **117,** 341 (1937).

116. B. B. Westfall and M. I. Smith, *J. Pharmacol. Exptl. Therap.*, **72,** 245 (1941).

117. D. D. Dziewiatkowski, *J. Biol. Chem.*, **189,** 187 (1951).

118. R. D. Campo, P. A. Wengert, C. D. Tourtellotte, and M. A. Kirsch, *Biochim. Biophys. Acta*, **124,** 101 (1966).

119. R. D. Campo, C. D. Tourtellotte, and J. W. Ledrick, *Proc. Soc. Exptl. Biol. Med.*, **125,** 512 (1967).

120. I. Rosenfeld and H. F. Eppson, *Univ. Wyo. Agr. Exp. Sta. Bull.*, **414,** 53 (1964).

121. J. D. Cipera and M. Hidiroglou, *Can. J. Physiol. Pharmacol.*, **47,** 591 (1969).

122. I. Rosenfeld, *Proc. Soc. Exptl. Biol. Med.*, **109,** 624 (1962).

123. I. Rosenfeld and O. A. Beath, "Selenium," Academic Press, New York, N.Y., 1964, p. 347.

124. H. E. Ganther, *Biochemistry*, **7**, 2898 (1968).

125. E. P. Painter, *Chem. Rev.*, **28**, 179 (1941).

126. K. P. McConnell, *J. Biol. Chem.*, **145**, 55 (1942).

127. F. Challenger, "Aspects of the Organic Chemistry of Sulphur," Butterworths, London, 1959, p. 162.

128. F. Schlenk, "Transmethylation and Methionine Biosynthesis," S. K. Shapiro and F. Schlenk, Eds., University of Chicago Press, Chicago, 1965, p. 48.

129. J. L. Byard, *Arch. Biochem. Biophys.*, **130**, 556 (1969).

130. I. S. Palmer, D. D. Fischer, A. W. Halverson, and O. E. Olson, *Biochim. Biophys. Acta*, **177**, 336 (1969).

130a. B. D. Obermeyer, I. S. Palmer, O. E. Olson, and A. W. Halverson, *Toxicol. Appl. Pharmacol.*, **20**, 135 (1971).

131. B. B. Westfall and M. I. Smith, *U.S. Public Health Rept.*, **55**, 1575 (1940).

132. K. P. McConnell, *J. Biol. Chem.*, **173**, 653 (1948).

133. F. Kiermeier, and W. Wigand, *Z. Lebensm. Untersuch. Forsch.*, **139**, 205 (1969).

134. K. P. McConnell and A. E. Kreamer, *Proc. Soc. Exptl. Biol. Med.*, **105**, 170 (1960).

135. I. Rosenfeld, *Proc. Soc. Exptl. Biol. Med.*, **111**, 670 (1962).

136. K. P. McConnell and C. H. Wabnitz, *J. Biol. Chem.*, **226**, 765 (1957).

137. K. P. McConnell, A. E. Kreamer, and D. M. Roth, *J. Biol. Chem.*, **234**, 2932 (1959).

138. H. E. Ganther and C. Corcoran, *Biochemistry*, **8**, 2557 (1969).

139. K. Schwarz, *Federation Proc.*, **20**, 666 (1961).

140. K. Schwarz and C. M. Foltz, *J. Biol. Chem.*, **233**, 245 (1958).

141. T. Hirooka and J. T. Galambos, *Biochim. Biophys. Acta*, **130**, 321 (1966).

142. K. Schwarz and E. Sweeney, *Federation Proc.*, **23**, 421 (1964).

143. K. J. Jenkins, M. Hidiroglou, and J. F. Ryan, *Can. J. Physiol. Pharmacol.*, **47**, 459 (1969).

144. K. P. McConnell, C. H. Wabnitz, and D. M. Roth, *Texas Rept. Biol. Med.*, **18**, 438 (1960).

145. K. P. McConnell and R. S. Levy, *Nature*, **195**, 774 (1962).

146. H. L. Klug, A. L. Moxon, D. F. Petersen, and E. P. Painter, *J. Pharmacol. Exptl. Therap.*, **108**, 437 (1953).

147. C. C. Tsen and H. B. Collier, *Nature*, **183**, 1327 (1959).

148. C. C. Tsen and A. L. Tappel, *J. Biol. Chem.*, **233**, 1230 (1958).

149. C. Voegtlin, J. M. Johnson, and S. M. Rosenthal, *J. Biol. Chem.*, **93**, 435 (1931).

150. J. L. Svirbely, *Biochem. J.*, **32**, 467 (1938).

151. K. P. DuBois, M. Rhian, and A. L. Moxon, *Proc. S. Dakota Acad. Sci.*, **19**, 71 (1939).

152. H. L. Klug, D. F. Petersen, and A. L. Moxon, *Federation Proc.*, **10**, 209 (1951).

153. C. D. Eskelson and H. P. Jacobi, *Physiol. Chem. Phys.*, **1**, 487 (1969).

154. H. L. Klug, A. L. Moxon, and G. P. Lampson, *Proc. S. Dakota Acad. Sci.*, **29**, 16 (1950).

155. H. L. Klug, A. L. Moxon, and D. F. Petersen, *Proc. S. Dakota Acad. Sci.*, **29**, 38 (1950).

156. Ts. M. Shtutman, R. V. Chagovets, A. M. Sukharevskaya, and T. I. Kashleva, *Voprosy Pitan.*, **27**, 3 (1968).

157. J. L. Martin and J. E. Spallholz, *Federation Proc.*, **29,** 499 (1970).

158. E. C. Slater, *Biochem. J.*, **46,** 484 (1950).

159. O. A. Beath, J. H. Draize, and H. F. Eppson, *Wyo. Agr. Exp. Sta. Bull.*, **189,** 1 (1932).

160. O. E. Olson, "Selenium in Biomedicine," AVI Publishing Co., Westport, Conn., 1967, p. 305.

161. A. Shrift, *Federation Proc.*, **20,** 695 (1961).

162. M. Mattii, R. Badiello, and E. Cecchetti, *Radiobiol. Radioter. Fis. Med.*, **22,** 56 (1967).

163. L. P. Neethling, J. M. M. Brown, and O. J. de Wet, *J. S. African Vet. Med. Assoc.*, **39,** 25 (1968).

164. J. L. Martin and M. L. Gerlach, *Ann. N.Y. Acad. Sci.*, **192,** 193 (1972).

165. L. Pichat, M. Herbert, and M. Thiers, *Tetrahedron*, **12,** 1 (1960).

166. K. A. Caldwell and A. L. Tappel, *Biochemistry*, **3,** 1643 (1964).

167. D. L. Klayman, *J. Org. Chem.*, **30,** 2454 (1965).

168. K. W. Franke, *J. Nutr.*, **8,** 609 (1934).

169. E. P. Painter and K. W. Franke, *Cereal Chem.*, **13,** 172 (1936).

170. A. L. Smith, M.S. Thesis, South Dakota State College, Brookings, S. Dak., 1949.

171. P. J. Peterson and G. W. Butler, *Australian J. Biol. Sci.*, **15,** 126 (1962).

172. O. E. Olson, E. J. Novacek, E. I. Whitehead, and I. S. Palmer, *Phytochemistry*, **9,** 1181 (1970).

173. E. Hansson and M. Blau, *Biochem. Biophys. Res. Commun.*, **13,** 71 (1963).

174. A. Ochoa-Solano and C. Gitler, *J. Nutr.*, **94,** 243 (1968).

175. D. Lebez, Z. Maretić, F. Gubenšek, and J. Kristan, *Toxicon*, **5,** 261 (1968).

175a. D. B. Cowie and G. N. Cohen, *Biochim. Biophys. Acta*, **26,** 252 (1957).

176. R. Walter and V. du Vigneaud, *J. Amer. Chem. Soc.*, **87,** 4192 (1965).

177. R. Walter and V. du Vigneaud, *J. Amer. Chem. Soc.*, **88,** 1331 (1966).

XIII Selenium compounds in nature and medicine

C. SELENIUM AND HUMAN BIOLOGY

JAY R. SHAPIRO

Washington Hospital Center
Department of Endocrinology
Washington, D.C.

1. Introduction

It is difficult to assess whether selenium functions as a micronutrient for man: clearly, it is essential for health in many animal species.[1] We know that human tissues contain selenium, but if selenium has a specific metabolic role, it is not understood. This lack of data in regard to human beings contrasts with the abundant information about selenium metabolism and toxicity gleaned from numerous studies performed on a variety of animals. It has been noted that traces of selenite and molybdate are essential for formic dehydrogenase production in *Escherichia coli*.[2] Schwarz[2a] has proposed that selenium might influence mitochrondrial respiratory activity *via* the α-ketoglutarate oxidase system, while Bull and Oldfield[2b] found selenium to influence oxidation of pyruvate by rat liver preparations. However, selenium has not been demonstrated to act as an essential cofactor in mammalian enzyme systems. A comparison of the tissue distribution of selenium with that of other trace elements in organs from Glasgow residents has led Liebscher and Smith[3] to conclude that selenium, like copper and zinc but unlike arsenic, antimony, and mercury, must be an essential element for man.

The symptomatology of acute selenium toxicity is well described, but there are no data on the changes in blood chemistry that accompany the toxic state. Although chronic selenium toxicity was first described in man 40 years ago, the symptoms are, at best, poorly categorized. While the use of selenium in industry is increasing each decade, relatively few studies have correlated symptoms in exposed workers with changes in blood and urinary levels of selenium. Variations in selenium levels in disease states are hardly documented.

In view of the above uncertainties, a definitive description of selenium metabolism and toxicology as it pertains to man is not feasible. Instead, an attempt will be made to review the existing literature with the hope of providing a stimulus for future investigations.

2. Dietary Intake of Selenium in Man

It is not known what constitutes normal selenium balance in man. Only recently have reports appeared from Maryland,[4] Vermont,[5] the western United States,[6,7] and Germany[8,10] documenting the selenium content of both common foods and drinking water.[11] These data are summarized in Chapter XIIIA (Table XIIIA-4). It is estimated that in the average U.S. diet between 25 and 200 μg of selenium is consumed daily from various sources, such as mushrooms and garlic,[12] meat and fowl, fish and shellfish,[13-15] egg yolk,[16,17] milk, bread,[18] and cereals.[19-21] Table XIIIC-1 presents values for the selenium content of a typical diet over a 2-day period as measured in a Vermont hospital.[5] Because symptomatic selenium toxicity first appears in man when solid food contains greater than 5 mg Se/kg and water or milk contains greater than 0.05 mg Se/l, systemic toxicity due to dietary selenium must be extremely uncommon. On the other hand, beef and pork kidney, as well as many agricultural products in seleniferous areas

Table XIIIC-1 Selenium in 2-day hospital diet[a]

MEAL	WEIGHT (g)	DRY SUBSTANCE (%)	WET (µg/g)	DRY (µg/g)	TOTAL (µg)
Breakfast	518	18.6	0.059	0.318	30.6
Dinner	519	21–23	0.023	0.110	11.9
Supper	660	21.6	0.030	0.137	19.8
Totals and mean:	1697	20.5	0.037	0.178	62.3

[a] The selenium contents of the diets reported by Schroeder *et al.*[5] represent those of the northeastern United States, an area of the country known to be low in selenium as determined in agricultural products. Because of local variations in the selenium content of grain and dairy products, the average diet in parts of the midwestern and northwestern United States probably contains 2–3 times the amounts reported in the table.

such as Venezuela,[22,23] the northern United States, and Canada, contain significant amounts of selenium and if consumed to excess could cause toxicity. Little is known about the metabolism of different selenium compounds in the human diet, or the effect of modern techniques of food processing and cooking on selenium availability.[24] It is possible that inorganic selenium salts are, to a certain extent, volatilized during cooking. In frozen food dinners the mean concentration of selenium has been shown to decrease from 1.0 mg/g to 0.04 mg/g after cooking.[5]

Furthermore, as shown in a variety of animals, it is likely that other dietary constituents affect selenium metabolism in man. An increase in dietary protein[25] or sulfate[26] may decrease selenium retention. Elevated dietary selenium intake will saturate tissue stores and diminish the retention of any added selenium.[27,27a] Although crop levels of selenium vary widely, the rapid and extensive distribution of food supplies probably ensures a relatively constant selenium intake in the human diet. It is common practice for animals foraging in areas low in selenium to receive sodium selenite by injection so as to prevent deficiency. The addition of selenium to feed has been proposed.[28] Certainly, it seems necessary that we understand more about selenium metabolism before further manipulating the selenium content of meats or grains used for human consumption.

3. Tissue Content of Selenium

The quantity of selenium available from diets suggests that this element might be found widely distributed in human tissues. However, only fragmentary data exist upon which to draw any conclusions regarding the relationship between dietary intake, or degree of environmental exposure to selenium, and resulting tissue levels. Schroeder *et al.*[5] estimate that the adult body in the United States contains a total of approximately 14 mg of selenium.

Selenium is a constituent of normal teeth. Hadjimarkos and Bonhorst[29] have reported that the teeth from an ancient Greek contained 0.58 ppm Se, whereas modern teeth from the Athens area contained only 0.16 ppm. In Portland, Oregon, a "low-selenium area" in terms of the crop content of selenium, deciduous teeth were found to contain more selenium than permanent teeth.[30] In age groups ranging from under 20 to over 50 years, the selenium content of dentine ranged from 0.35 to 0.40 ppm, and enamel selenium varied from 0.43 to 1.60 ppm, but no age-related pattern was observed. Nixon and Myers[31] have found the enamel of English teeth to contain 0.21–2.08 ppm, while studies in Moscow have reported human jaw bone to contain 5.37 μg Se/g, teeth to contain 12.78 μg Se/g, and femur to contain 8.95 μg Se/g.[32] Hardwick and Martin[33] found 1–10 ppm in enamel, 10–100 ppm in dentine, and 1–10 ppm in dental plaque.

Japanese hair was reported to contain an average of 1.2 ppm selenium,[34] while mammalian hair in England contained 0.3–13 ppm.[35] In the northeastern United States, hair contained 0.57 μg Se/g,[5] but no relationship to age or hair color was evident. Bowen[35] did not find selenium present in skin, but Dickson and Tomlinson[36] reported that skin contained from 0.12 to 0.62 μg Se/g. Loss of selenium, along with other trace elements, in sweat has been documented.[37]

Ocular tissue selenium was measured because of the suggestion that it might play a role in photoreceptor reactions. Taussky and his co-workers[38] measured the selenium content of eyes obtained in New York City from individuals 60–80 years of age. A human eye weighing 6–8 g contained 5.98 μg Se, with the highest concentration in the iris (0.99 μg/g), while the sclera and cornea contained only trace amounts. In Maryland, a human retina was found to contain 0.04–0.41 ppm, sclera 0.15–0.26 ppm, and uvea 0.79–1.6 ppm; these low levels of selenium suggest that the element does not play a role in the visual process.[39] In fact, selenite and selenate will induce reversible damage when applied to frog retina in concentrations as low as 0.0001–0.1 %.[40]

Limited data on the large-organ content of selenium have been published in the Soviet Union,[41] Canada,[36] Scotland,[3] Vermont,[5] and California.[42] The marked variation from patient to patient found in these reports may be due to differences in analytical procedures or may be caused by differences in either selenium intake or excretion. Usually pancreas, kidney, and thyroid contain the highest concentrations of selenium. Liebscher and Smith[3] reported the liver selenium content of Glasgow residents to vary between 1.57 and 3.81 ppm, while in Canada[36] hepatic selenium content ranged from 0.18 μg/g to 0.66 μg/g of dry tissue.[36] In 10 individuals of varying ages, muscle selenium varied between 0.25 and 0.59 μg/g of whole tissue. Ermakov[41] reported considerable variation in tissue selenium in Russian children and adults dying of diverse causes. Table XIIIC-2 compares selenium levels in postmortem specimens from Vermont, Canada, and the Soviet Union

Analysis of blood cells and plasma from 254 Canadian residents revealed 22–24.7 μg Se/100 ml of washed cells, and plasma levels of 14.4 $\pm$ 2.86 μg Se/100 cc.[36] Plasma selenium content was observed to roughly parallel that of the formed elements in the same individual. The higher red cell concentration reflects the binding of selenium to hemoglobin. Bowen and Cawse[43] reported whole blood selenium in England to range from 26 to 37 μg/100 ml, with a mean value of 32 μg/100 ml.[43]

The blood selenium contents of inhabitants from a number of U.S. cities have been determined by Allaway and his co-workers[44] (cf. Table XIIIC-3). Mean whole blood selenium ranged from a low of 15.7 μg/100 ml in Lima, Ohio, to a high of 25.6 μg/100 ml in Rapid City, South Dakota. The overall mean value of 20.6 μg/100 ml was lower than the English values, but similar to those found in Canadian inhabitants. Brune and his co-workers[45] found the selenium contents of normal and uremic whole blood to be 0.12 μg/g and 0.11 μg/g, respectively.

Whole blood selenium in normal Thai children, age 12, was 20.2 μg/100 ml. Adult blood contained 14.4–14.7 μg Se/100 ml.[46]

Burk et al.[47] have reported a similarity of blood selenium levels in Guatemala and the southern United States; values ranged from 19 to 28 μg/100 ml, with a mean value of 22 μg/100 ml. Blood selenium in healthy children was similar to that in adults; and, as noted by Allaway et al.,[44] no marked geographical variation in blood selenium content was seen. This implies either that a homeostatic mechanism exists for maintaining uniform blood levels or that wide food distribution ensures a relatively constant selenium intake.

In Oregon, the placental selenium content was 0.18 ppm, while 0.12 ppm was measured in fetal cord blood, suggesting some storage capacity for selenium in the placenta.[48] Utilizing neutron activation analysis, Dawson et al.[49] found placental tissue from women in Texas to contain between 0.174 and 0.522 μg Se/g. Placental tissue from Guatemalan residents was found to contain less selenium than that from residents of Iowa.[49a]

The rate of selenium excretion in urine remains fairly constant from hour to hour, suggesting that selenium is excreted by the kidneys as a nonthreshold substance. In 1941, Sterner and Lidfeldt[50] reported urine selenium to range

Table XIIIC-2 Selenium content in human tissues (necropsy data) (μg Se/g whole tissue)

TYPE OF TISSUE	SEX: M AGE: 9 months LOCATION: Vermont[a]	"Infant" Canada[b]	F 15 years U.S.S.R.[c]	M 41 years Vermont[a]	M 68 years Vermont[a]	"Adult" Canada[b]	M 29 years U.S.S.R.[c]	M 64 years U.S.S.R.[c]
Liver	0.33	0.34	0.79	0.42	0.81	0.39	0.58	1.61
Lung	0.24	0.17	0.24	0.26	0.20	0.21	1.38	1.00
Heart	0.25	0.55	0.04	0.25	0.26	0.22	0.72	1.14
Kidney	0.70	0.92	0.80	0.75	1.84	0.63	2.71	1.61
Spleen	0.47		0.18	0.29	0.32		0.76	0.78
Bone (rib)	0.42		0.21				0.34	0.243
Muscle	0.18	0.37	0.16	0.17	0.36		0.28	2.26
Pancreas	0.34		0.32	0.29	0.27	0.27	0.26	1.00
Testes	0.15			0.36	0.38			
Brain	0.04	0.16		0.21		0.27		
Small intestine	0.18			0.32	0.12			
(Fat) intestine								
Fat				0.04				
Skin			0.12				0.28	1.24
Human milk								
Gonad		0.46				0.47		
Thyroid		0.64				1.24		

[a] Schroeder et al.[5]
[b] Dickson and Tomlinson.[36]
[c] Ermakov.[41]

Table XIIIC-3 Selenium concentrations in whole blood from 19 collection sites in the United States[a]

COLLECTION SITE	NO. OF SAMPLES	MEAN[b] (μg/100 ml)	STANDARD ERROR (μg/100 ml)	RANGE (μg/100 ml)
Rapid City, S.D.	20	25.6	3.6	20–34
Cheyenne, Wyo.	10	23.4	2.5	20–27
Spokane, Wash.	10	23.0	3.3	18–28
Fargo, N.D.	10	21.7	3.6	18–30
Little Rock, Ark.	10	20.1	2.8	15–24
Phoenix, Ariz.	10	19.7	3.4	15–25
Meridian, Miss.	10	19.5	4.4	14–27
Missoula, Mont.	20	19.4	3.1	12–25
El Paso, Tex.	10	19.2	3.4	15–26
Jacksonville, Fla.	10	18.8	3.0	15–26
Red Bluff, Calif.	10	18.2	4.5	12–27
Geneva, N.Y.	10	18.2	2.2	15–22
Billings, Mont.	10	18.0	4.9	10–27
Montpelier, Vt.	10	18.0	1.6	16–20
Lubbock, Tex.	10	17.8	2.9	12–24
Lafayette, La.	10	17.6	2.4	14–22
Canandaigua, N.Y.	10	17.6	2.0	14–20
Muncie, Ind.	10	15.8	2.4	11–20
Lima, Ohio	10	15.7	3.2	10–20
Summary	210	20.6		10–34

[a] The specimens used in this study were obtained from blood banks making collections in different cities and towns in the United States.[44]

[b] The mean for Rapid City, S.D., is significantly (0.05) greater than the means for all other cities, except Cheyenne, Spokane, and Fargo. The mean for Fargo, N.D., is significantly (0.05) greater than the means for Muncie, Ind., and Lima, Ohio, but not significantly different from the means for the other 16 location sites.

between 10 and 150 μg/l, with a mean of 42 μg/l in normal Rochester inhabitants. Glover[51] has found British adults to excrete 34 μg Se/l, a value that approaches the Rochester level when corrections are made for differences in specific gravity. In seleniferous regions of the United States urine values ranged from 100 to 2000 μg Se/l.[52] The urine of primary school children in a mountainous area of Japan contained 0.19–202 μg Se/l (mean, 0.86 μg/l), while in a seaside region urines contained 0.44–189 μg/l (mean, 115 μg/l).[52a] Although it was postulated that a seaside diet contained more selenium than the diet consumed by children living in the mountainous region, the wide range of values in both areas suggests the influence of factors other than diet alone.

4. Selenium Deficiency in Man

In animals, dietary selenium deficiency induces a number of pathologic changes such as diminished growth and increased mortality in quail,[53] pancreatic fibrosis[54,54a] and exudative diathesis in chicks,[55] hepatic necrosis in rats,[56,57] and white muscle disease in cattle and sheep.[58] The biochemical abnormality common to these various lesions remains unknown. Because of the general availability of selenium in human diets, it is only under the most drastic conditions of malnutrition, such as kwashiorkor, that selenium deficiency has been suspected. It must be stressed, however, that no survey studies of the selenium content of tissue have yet been done in man. The significance of blood levels is uncertain without data describing net metabolic balance.

In 1961, Schwarz[59] observed two infants with kwashiorkor who failed to gain weight after protein refeeding. Both infants had a prompt weight gain, however, when diets were supplemented with selenium in doses of 25–30 μg/day.

Majaj and Hopkins[60,61] reported similar results in three Jordanian children suffering from kwashiorkor and diarrhea who also initially failed to gain weight during 1 month of hospitalization. Initial selenium blood levels in these three children were 0.16, 0.13, and 0.06 μg/ml. The addition of 30–50 μg Se as selenite per day to the diets produced weight gains varying from 23 to 41 g/day. Prompt reticulocytosis also occurred, with the peak reticulocyte count observed in the first week of therapy. Slight elevation of erythrocyte count and hemoglobin followed therapy in two of the three children.

Burk and his co-workers[47] performed extensive investigations on children with kwashiorkor in Guatemala during 1965 and 1966. The mean selenium blood content was 0.08 μg/ml in untreated children, 0.14 μg/ml in children recovering from kwashiorkor, and 0.23 μg/ml in a group of nonmalnourished children (healthy normal level: 22 μg/ml). Although these children were not given supplementary selenium, gradual elevation of blood selenium content occurred during hospitalization over a 1 year period, suggesting that depleted tissue stores had gradually been replaced. In children with low selenium blood content, erythrocyte selenium uptake at 3 hours was 21%, compared to 12.7% observed in normal children. These results are qualitatively similar to those of Wright and Bell,[62] who reported elevated red cell ^{75}Se uptake in sheep fed selenium-deficient diets.

Levine and Olson[46] have found the mean whole blood selenium level to be

88 ± 38 ng/ml in a group of malnourished Thai children, while a mean value of 120 ± 36 ng/ml was observed in well-nourished children recovered from kwashiorkor. While plasma selenium was reduced in malnourished children (39 ± 10 ng/ml), red cell selenium levels (206 ± 138 ng/ml) did not differ from the control values, suggesting that in a state of selenium deficiency the red cell, and perhaps other tissue stores as well, compete with plasma for the limited supply available. It is not known whether selenium deficiency complicates the anemia or clinical symptoms seen in malabsorption diseases in man, and this is an area that should be studied further.

5. Selenium Toxicity in Man

During the 1960's, production of selenium from sulfide-rich ores was in excess of 1,300,000 lb, while consumption in the United States reached 900,000 lb. Selenium is used in the manufacture of colored glass, pigments, rectifiers, transistors, and lasers. The machinability of steel is improved by selenium, and the element is used as a catalyst and antioxidant in the production of rubber, fat products, and plastics. Selenium is essential to image reproduction on xerographic plates.

In addition to the risk of exposure to toxic selenium compounds in modern industry, many areas of the world are naturally seleniferous because of the presence of selenium-containing soils derived from certain geologic formations.[63] The addition of selenium to human food and water supplies could lead to chronic toxicity; but because the symptoms of this disorder are poorly defined, it is unlikely that even today the condition would be recognized.

a. Nonindustrial Selenium Toxicity. Chronic selenosis is said to have been responsible for loss of hair and nails and for birth defects in Colombia, South America, during the sixteenth century.[64] Even at that time it was appreciated that some relationship existed between the toxicity of plants and vegetables and the soil, an association which was proved in 1933, when Robinson[65] demonstrated the toxic factor in grain to be selenium. The toxicity of the seleniferous fruit of the monkey pod tree was described as early as 1763 by Jacquin.[66]

Water supplies, even in seleniferous areas of the western United States, have not generally been considered a potential source of selenium toxicity in man. However, a recent report by the Bureau of Water Hygiene suggests that excessive amounts of selenium may contaminate local water supplies in many areas of the United States.[67] One instance of toxicity due to selenium contained in an underground water source was described by Beath.[68] A

Colorado Ute Indian family was found to suffer from hair loss, weakened nails, and listlessness. Well water that they had used for 3 months was found to contain 9 ppm selenium.

Smith et al.[52] were the first to undertake a systematic epidemiologic study of chronic toxicity due to excessive dietary intake of selenium. These authors attempted to correlate various symptoms with diet selenium content and urinary selenium levels in 111 families living in South Dakota, Wyoming, and Nebraska, on farms where animals were known to be suffering from selenium toxicity ("alkali disease"). Several signs and symptoms were attributed to excessive selenium intake: bad teeth, jaundice, chloasma, vertigo, chronic gastrointestinal disease, dermatitis, nail changes, arthritis, edema, and fatigue. The food consumed by these families contained what was judged to be excessive amounts of selenium in meat, milk, and eggs, with some individuals found to ingest as much as 0.1–0.2 mg Se/kg body weight/day. Ninety-two per cent of the urines examined contained selenium, and 45% contained between 0.20 and 1.33 ppm. Although no precise correlation of symptoms to urinary excretion of selenium was apparent, individuals who excreted over 0.2 ppm had a greater number of complaints. In a subsequent field study Smith and Westfall[69] investigated selenium intake, urinary excretion, and symptomatology in 50 additional families living in highly seleniferous areas of South Dakota and Nebraska, but in general the findings were similar to those of the initial survey. Although there is reason to believe that the conclusions of the Smith and Westfall studies are valid, and that the cases described did represent chronic selenium toxicity, the totally non-specific nature of the illnesses does not allow the conclusion that a cause and effect (as regards selenium intake) relationship existed. These studies suggest, but do not firmly establish, that chronic selenium toxicity of dietary origin exists in man. Only when blood and tissue levels of selenium are measured in affected individuals, however, will such conclusions be valid. Hair and nails are tissues that could easily be sampled in future field studies.

Lemley[70] reported in detail the clinical history of a South Dakota rancher afflicted with acute and chronic dermatitis, rhinitis, and sinusitis, which were eventually attributed to excessive dietary selenium intake. In this patient and several others experiencing dizziness, fatigue, and dermatitis, elimination of dietary selenium was found to afford relief. Because of the report of Moxon et al.[71] showing bromobenzene to increase urinary excretion of selenium in cattle, treatment with bromobenzene was instituted, and this led to rapid diminution of systemic symptoms with clearing of the rash. In this case and 30 others reported by Lemley and Merryman[72] dermatitis variously described as follicular, postular, papular, or eczematoid, and initially diagnosed as a photosensitivity reaction, neurodermatitis, or contact dermatitis, was said to clear after therapy with bromobenzene.

Accidental poisoning due to the ingestion of selenium from various sources has been reported. In 1901, Tunnicliffe and Rosenheim[73] stated that selenium rather than arsenic was the cause of paresthesias, erythromelalgia, and melanosis which occurred during an epidemic of poisoning attributed to tainted beer in London. The fact that selenium was found to contaminate the sulfuric acid used in the manufacturing process led the authors to caution against future use of toxic materials in food manufacture, a warning that perhaps has not been entirely heeded during the past 70 years. For many years, hunters in South America have realized that ingestion of fruit of the monkey pod tree (*Lecythis ollaria*) could lead to nausea, vomiting, and generalized hair loss.[74] The toxic compound has recently been isolated and identified as selenocystathionine; it has been shown that the toxicity of selenocystathionine can be reversed by L-cystine.[75,76] While not reported to affect human beings, the flour of the Brazil nut has been shown to cause hepatic necrosis in rats.[77] In this instance the toxic compound has not yet been identified. The compounds γ-L-glutamyl-*Se*-methylseleno-L-cysteine and *Se*-methylselenocysteine, isolated from seeds and leaves, respectively, of *Astragalus bisulcatus*, are both toxic to mice.[78] Organic selenium compounds in species of *Lathyrus* grain used for feed have been thought to cause lathyrism in animals.[79]

The death of a child resulted from the ingestion of gun-bluing compound containing 1.8% selenious acid.[80] In this case, which is the only autopsy report describing the histopathology of acute selenium poisoning, there was fulminating peripheral vascular collapse, pulmonary edema, and coma. At autopsy the gastric mucosa was a brick-red color, and marked intestinal vascular congestion was observed. Garlic odor of the breath was present before death but was not detected in any organ postmortem. The lungs were diffusely hemorrhagic, congested, and edematous, but no specific renal or hepatic necrosis was described. Although selenium was identified in several tissues, the levels were not reported.

Buchan[81] cites a patient investigated by Cotter who was said to have developed nervousness, mental depression, metallic taste, vomiting, and pharyngitis secondary to the use of selenium red lipstick. Although selenium sulfide used in shampoo is relatively inert, a 46-year-old female using excessive amounts on abraded skin developed progressive generalized tremor, abdominal pain, metallic taste, and a garlic breath odor. These symptoms cleared when the use of the shampoo was discontinued.[82]

It is uncertain whether, as proposed by Hadjimarkos,[83,83a] an increase in the incidence of dental caries is one of the toxic effects of excessive dietary selenium intake. Studies of caries prevalence, urinary selenium, and diet selenium in children from three Oregon counties revealed urinary selenium to be twice as high in those children with high prevalence of caries.[84] Eggs and milk produced in the high-caries counties contained more selenium than those originating in counties where caries incidence was lower. An increased incidence of caries has been produced in rats fed selenium,[85] but the production of caries in animals by selenium feeding has not been consistently observed,[86] particularly after full tooth development has occurred.[87,88] Ludwig and Bibby[89] surveyed towns in the eastern United States for caries incidence and related this to selenium in local crops. Although caries prevalence was higher in high-selenium areas, the findings were of borderline significance. In certain instances, low caries incidence was recorded in regions considered to have high selenium contents in food and forage. However, enamel solubility in acid solutions has been shown to be increased by the addition of selenium.[83b] Eisenmann and Yeager[83c] reported the administration of selenium to inhibit mineralization of enamel and dentine in rats.

b. Industrial Selenium Toxicity. As previously noted, it is not entirely clear whether the nonspecific complaints described by Smith and Westfall[52,69] were due exclusively to excessive selenium intake. Reproducible and generally severe symptoms occur upon acute or subacute exposure to the various selenium compounds used industrially.[90] Selenium may be inhaled as fumes or dust, or absorbed through the skin or gastrointestinal tract. Marked irritation of the nasal, conjunctival, and tracheobronchial mucosa occurs rapidly, leading to cough, wheezing, dyspnea, chemical pneumonitis, and pulmonary edema. Low-grade fever may complicate the chemical pneumonitis. Abdominal pain, nausea, vomiting, and diarrhea ensue. Acute and chronic dermatitis of exposed or unexposed areas of the skin commonly occurs.[91]

Hepatic necrosis has not been observed following exposure of human beings to selenium, but detailed analyses of liver function have not been performed. Although myocarditis, known to occur in animals poisoned with inorganic selenates,[154] has not been reported in humans, myocardial failure could complicate severe pulmonary edema. Workers exposed to selenium have been noted to complain of nervousness, fatigue, depression, and pallor. A garlic odor of breath (and sweat) due to the pulmonary excretion of dimethyl selenide is one of the first signs of selenium absorption; a similar odor has been observed, however, after absorption of tellurium, because of dimethyl telluride excreted *via* the lungs. Metallic taste is commonly reported after selenium ingestion.

Nagai[92] observed hypochromic anemia and leukopenia in Japanese women and children exposed to selenium in a rectifier manufacturing plant. Depression of hemopoetic function caused by selenium has also been reported in fish,[93] chick embryo,[94] and ducks.[95]

Investigators have documented several specific acute and chronic lesions secondary to selenium poisoning in animals which have not thus far been described in humans.[96] In the rat, chronic therapy with selenium salts has produced electrocardiographic abnormalities of bradycardia and left axis shift.[97] Of particular interest because of the acute toxicity syndrome termed "blind staggers" is the neuronal degeneration of the cerebrum and cerebellum produced in cattle. Also, lytic and fibrotic myocardial lesions have been observed in sheep, and hepatic necrosis leading to cirrhosis in rats.[98] Erosion of the ends of weight-bearing bones in selenosed cattle and bone softening, with cartilaginous defects, have been noted in rats. Although selenium *per se* is probably responsible for these various lesions, Bel and his co-workers[99] claim that hepatic necrosis in selenium-fed rats is a consequence of gross dietary alterations rather than a specific effect of selenium.

No reports have yet described changes in various blood constituents in man as a consequence of acute or chronic selenium poisoning; however, selenate or selenite feeding has been found to increase serum cholesterol and aortic lipids in the rat,[100] while hypoglycemia has been produced in rabbits by the injection of selenite.[101] Tsuzuki *et al.*[101a] noted both decreased cholesterol levels and increased urinary protein in mice fed selenium.

c. **Toxicity of Commonly Used Compounds.** In 1929 Hamilton and Hardy[102] first reported selenium intoxication in an industrial setting. Sax[103] has listed over 20 industrial materials containing selenium which may be classified as potentially hazardous. Most commonly encountered among these are elemental selenium, selenic and selenious acid, selenium dioxide, selenium oxychloride and oxybromide. To date, "there are no clear indications of long-term systemic effects among industrial workers."[104] This statement has been confirmed by Holmquist,[105] who also found no serious industrial health problem to be caused by selenium when strict maintenance of hygienic standards was enforced.

i. *Elemental Selenium.* Elemental selenium in any of its allotropic forms is virtually insoluble in water and is apparently harmless when ingested. It has been found to pass unchanged through dog intestine and is probably not absorbed to any significant extent when applied to skin. Elemental selenium is capable of causing dermatitis in some, but not all exposed individuals. Molten elemental selenium has caused severe burns, and absorption of selenium through burned areas has been observed. When inhaled as dust or vaporized, elemental selenium is intensely irritating to nasal, tracheobronchial, and pulmonary tissues.[106,108]

Clinton[109] found that workers exposed to selenium fumes immediately developed conjunctivitis and rhinitis, followed by varying degrees of bronchitis or chemical bronchopneumonia. These symptoms were of varying severity but cleared within 3 days without residual effects. Similar findings were observed in a North Carolina plant where workers were exposed to selenium fumes.[110] Upper respiratory tract irritation was followed in 2–12 hours by bronchitis. Peribronchial infiltrates and atelectasis were seen on X-ray, but cleared within 1 week.

ii. Hydrogen Selenide. Hydrogen selenide, one of the most toxic substances known, is produced by the action of acids or water on inorganic selenides. The gas has an offensive odor but also has the property of inducing olfactory fatigue, so that at a concentration of 0.0001 mg/l the odor disappears, thereby increasing the danger of exposure. Dudley and Miller[111] found that 50% of guinea pigs were killed within 30 days of exposure to 0.001–0.004 mg hydrogen selenide per liter of air.[111] The threshold limit for exposure has been set by The American Conference of Governmental Industrial Hygienists as 0.05 ppm Se, or 0.2 mg Se/m³ air.[90]

Hydrogen selenide produces marked irritation of nasal mucous membranes with the production of large amounts of mucus. Pulmonary edema, bronchitis, and bronchopneumonia ensue.[111a] Lassitude, nausea, and vomiting, in addition to dizziness, follow exposure to the gas. Buchan[81] reported five cases of industrial selenosis resulting from exposure to hydrogen selenide in a concentration of less than 0.2 ppm, generated from selenious acid in etching ink. Exposed workers complained of nausea, vomiting, metallic taste, alliaceous breath, odor, dizziness, and extreme lassitude and fatigue. While five persons were symptomatic, fifteen others, also with garlic odor to their breath, had no subjective complaints.

iii. Dimethyl Selenide. Dimethyl selenide inhaled by persons working with animals injected with selenates has produced severe pharyngitis in some individuals and bronchitis in others.[112] The fact that pulmonary symptoms recurred after minimal re-exposure to selenate dust suggested that an allergic state had been produced.

iv. Selenium Oxychloride. This colorless to yellowish liquid may cause blistering upon contact. Dudley[113] found as little as 0.01 ml to result in death to rabbits and severe third-degree burns when applied to human skin. Examination of urine has demonstrated the absorption of selenium through the skin after the application of selenium oxychloride.

v. Inorganic Selenates and Selenites. In general, these compounds cause local irritation of skin upon contact. Duboir *et al.*[114] has reported positive skin reactions to potassium selenide and selenic anhydride. Porphryinuria is said to be associated with selenosis when it is the cause of skin lesions.[107] Halter[114a] reported porphyrinuria in a glass worker exposed to sodium selenite or selenium dust and assumed that a latent abnormality of porphyrin metabolism had been activated.

vi. Selenium Dioxide. Selenium dioxide is a white crystalline powder which forms selenious acid when in contact with water and is more irritating than sulfurous acid. This compound is not volatile under ambient conditions. Pringle[115] has described a fresh purpuric rash as a result of selenious acid burns.

Glover[116] has investigated cases of selenium dioxide intoxication in a British factory employing 5000 workers. The symptoms caused by selenium dioxide were in general similar to those due to other selenium compounds; a garlic odor of the breath and metallic taste were noted as two of the early signs of selenium absorption. Indigestion and mild epigastric pain were observed, and one patient experienced severe hematemesis. The sudden inhalation of selenium dioxide produced pulmonary edema. When the compound was in contact with the skin, an acute pruritic papular dermatitis or an urticarial rash occurred. Selenium dioxide penetrating under nail edges led to painful irritation, and conjunctivitis occurred if the substance entered the eyes.

Before the institution of environmental control, workers in this factory were found to excrete 30–35 mg Se/l of urine; after contamination was controlled, they excreted 0.08 mg Se/l. In this group the mortality rate from common diseases was not increased in comparison to that for a control population in spite of the chronic selenium exposure.

d. Therapy of Selenium Toxicity. No really effective nontoxic substance has been found that will protect human beings against acute selenium toxicity or mobilize tissue stores of selenium.

Studies in animals have shown that increased dietary protein will protect them against selenium toxicity.[117] This effect may be due to decreased absorption of amino acid-bound selenium, or to binding of inorganic selenium by products of protein hydrolysis. Casein added to the diet offsets selenium toxicity in rats.[118] Methionine has been found of value in preventing selenium-induced hepatic damage in rats as long as sufficient vitamin E or fat-soluble antioxidants are also present in the diet.[119] No beneficial effects have followed the feeding of cystine or cysteine to selenosed animals. Halverson and his co-workers[120] have reported that a nonprotein factor in linseed oil will increase the growth and survival of chronically selenosed rats. Levander *et al.*[121] confirmed the protective effect of linseed oil meal in rats, but found that this led to elevated selenium content in kidney and liver.

Reduced glutathione is reported to protect against selenium-induced embryonic malformation in chicks[122] and to decrease the toxicity of selenium in the rat.[123] Levander *et al.*[121,124] found glutathione to increase the dialyzability of selenium in the bile of rats. Dickson and Tappel[125] have proposed that both the toxic symptoms and the deficiency syndromes due to selenium are perhaps related to the glutathione/selenocystine ratios in tissues.

It has been recognized for many years that arsenic could protect animals fed otherwise toxic doses of selenium;[126] decreased hepatic selenium was associated with lessened toxicity in such animals.[127] Levander and Baumann[128] demonstrated that in animals treated with arsenic, decreased hepatic selenium was associated with increased gastrointestinal excretion of selenium. They subsequently demonstrated that arsenic lessened selenium toxicity by promoting the hepatic clearance of selenium into the bile. In arsenic-treated animals the selenium content in carcass and blood was decreased and volatile selenium excretion *via* the lungs was inhibited, while urinary excretion of selenium was unchanged.[129] Certain organic arsenicals have been utilized as growth-promoting agents in cattle. Hendrick *et al.*[130] studied the effects of arsanilic acid and 3-nitro-4-hydroxyarsonic acid in rats but found these compounds to exert only a mild protective effect in selenium toxicity, which varied with the dose and the method of administration.[130]

Bromobenzene was found by Moxon *et al.*[71] to mobilize tissue selenium in cattle and was used in cases of suspected human selenium toxicity by Lemley and Merryman.[70,72] Its hepatic toxicity, however, restricts consideration of its use in man. In addition, Westfall and Smith[131] did not find bromobenzene to increase selenium excretion in the rabbit, while McConnell *et al.*[132] reported that bromobenzene promoted the excretion of small amounts of selenium in the mercapturic acid fraction of dog urine.

British anti-Lewisite (BAL) increases rather than decreases selenium toxicity in animals.[133,134] Calcium or sodium ethylenediaminetetraacetic acid are of limited benefit in treating selenium toxicity in the rat if administered within 15 minutes of selenium ingestion, but their use in humans is probably limited both by this time factor and by their inherent renal toxicity.[135]

Penicillamine, fed to rats at a concentration of 0.1 %, has not proved effective in decreasing selenium toxicity in either acute or chronic cases,[136] but initial studies indicate that lower hepatic and renal tissue selenium is found after penicillamine therapy. To counteract the local irritative effects of selenium compounds on the skin, Glover[137] recommends the topical application of an aqueous solution or cream containing 10 % sodium thiosulfate.

*e. **Selenium as a Teratogen.*** Having observed the occurrence of embryonic chick monsters on farms where seleniferous grain was used as feed, Franke *et al.*[138] injected selenite and selenate into normal eggs and caused similarly deformed chicks. Wright and Mraz[139] observed 30–50 % of chick embryos injected with as much as 15 μg Se to be normal. No hatching occurred, however, if over 20 μg Se was injected, and at levels greater than 30 μg all embryos died within 24 hours. With lesser doses, dwarfness, shortened visceral cranium, shortened extremities, and digital fusion occurred, but the severity of deformities was not related to the amount of selenium injected.

Gruenwald[140] studied the embryos of eggs produced by selenium-fed hens and found focal tissue necrosis with subsequent retardation of development as an early manifestation of selenium toxicity. During segmentation in *Pleurodeles waltlii* eggs, the addition of 0.01 and 0.02 M selenium dioxide produced chromosomal alterations, leading to pseudospindle formation, and failure of chromosomal movement.[141] Telophasic transformation and cytodieresis were also inhibited in the presence of selenium.

On the other hand, a dose of 2 mg/kg sodium selenite injected into pregnant golden hamsters decreased the incidence of malformations and mortality in animals subsequently injected with cadmium sulfate.[142] Selenate also had a moderately protective effect on the malformed resorption rate induced by sodium arsenite in the hamster.

Sodium selenite is widely used in *Salmonella* culture media, so that long-term exposure of laboratory workers to selenium dust is possible. In one laboratory surveyed by Robertson,[143] ten females of child-bearing age were employed, eight of whom were exposed to selenium. Four certain pregnancies and one probable pregnancy occurred in these women, all of which aborted except one. A term infant was born with a club foot. Urinary selenium in these women was similar to that in a local control group. Inquiries at other laboratory centers handling selenite in this manner showed no definite relationship to fetal abnormalities. At another laboratory with three females handling selenite, one became pregnant and miscarried. Hadjimarkos[144] has pointed out that this situation demands further investigation, perhaps in females continually exposed in industry.

f. Is Selenium a Carcinogen? Considerable dispute exists as to whether or not selenium is carcinogenic.[144a] This matter is of more than theoretical interest because of the widespread use of selenium in industry, the administration of selenium to animals used for human consumption, and the recent suggestion that selenium be added to feed to prevent selenium deficiency in animals.[145] Miller[146] and Kraybill[147] both list selenium among tumorigenic and carcinogenic products occurring naturally in foods, but this has been disputed by Frost.[149]

Thyroid hyperplasia and loss of thyroid follicle colloid occurred in rats that received 0.05–0.1 % bis-(4-acetaminophenyl) selenium dihydroxide in their diets for 10 days.[150] Animals receiving this compound for 105 days developed multiple adenomas of the thyroid and adenomatous hyperplasia of the liver, but it is uncertain whether the selenium or a goitrogenic effect of compounds bearing acetaminophenyl groups was responsible for the adenomas.

In 1943, Nelson *et al.*[151,151a] reported the development of hepatic cell adenomas and "low-grade carcinoma" in female rats fed potassium ammonium sulfoselenide or seleniferous grain containing 5–10 ppm selenium for 18 months. Subacute hepatic necrosis leading to nodular cirrhosis was produced after only 3 months by this diet with the appearance of the presumed "hepatoma of low malignancy" in 11 of 53 rats; no metastases occurred. On the basis of these limited studies, selenium was listed as a carcinogen in The Food Additive Amendment of 1958 to the Food, Drug, and Cosmetic Act (U.S.A.).[151b]

Tscherkes and his co-workers[152] found that 10 of 23 rats fed 10 ppm of sodium selenate per kilogram of food in a 12% protein diet and surviving for 18–32 months developed neoplasms, of which three were carcinoma of the liver with metastatic lung lesions, and four were sarcomatous lesions of lymph nodes and mediastinum. In a subsequent study, 200 male rats were fed 0.43–0.86 mg Se/100 g diet, with differing amounts of casein, and with tocopherol, riboflavin, cystine, choline, and nicotinic acid added to the diets of some. Three animals receiving diets with 30% casein and 0.85 mg Se developed sarcomas, and one developed liver cirrhosis with nodules similar to the low-grade carcinoma described by Nelson. No spontaneous tumors were observed in control animals, whereas a total of four cancers and seven sarcomas were found in the experimental groups of 200 rats.[153] Tsuzuki *et al.*[101a] noted the appearance of cylindromatous dorsal neck tumors in a few selenium-fed mice, but the etiology of these is unclear.

Harr and his co-workers[154] did not find selenate to cause hepatic cancer in Wistar rats, although hyperplastic hepatic lesions occurred in rats on high-casein diets and in animals in which high-selenium intake was followed by a basal ration. Several authors have stated that there exists only incomplete evidence on which to base the statement that selenium is carcinogenic, and it is apparent that any change in hepatic histology may be dependent on other dietary factors besides the amounts of selenium added. Undoubtedly, species differences, sex, and age of animals used in selenium feeding studies all influence the appearance of histologic abnormalities. Detailed histologic investigation of animals dying of acute or chronic selenium intoxication has not, to date, revealed an increased incidence of malignancy, although hepatic necrosis and cirrhosis are common. This inconsistency raises the question of whether extraneous factors influenced the appearance of malignancy in the few studies cited above. Further investigation, perhaps utilizing animals other than rats or mice, is required before this question can be answered.

Selenium compounds as potential anticarcinogens are discussed in Chapter XIIID.

g. Selenium in the Atmosphere. In the vicinity of industrial plants that refine selenium-containing sulfide ores there may be heavy contamination by selenium compounds.[155] The National Air Pollution Control Administration estimated that 37 million tons of sulfur oxides, largely the result of the burning of fossil fuels, were emitted into the atmosphere over the United States in 1970. Scant evidence exists as to the extent of atmospheric contamination with selenium that may be expected to accompany such sulfur oxide pollution, or the health hazard that could result from this contamination. A recent report of air pollution surveillance systems lists selenium as one of the atmospheric pollutants being measured by the National Air Pollution Control Administration.[160]

Methods for the measurement of atmospheric selenium have been described by Tada[156] and Tabor and his co-workers.[157] Using neutron activation analysis, Hashimoto and Winchester[158] measured the selenium and sulfur contents of air, rain, and snow in Cambridge, Massachusetts, during 1964–1965. The average values of the Cambridge precipitation samples were 0.21 μg Se/l and 2.0 mg S/l, giving an average Se/S weight ratio of 1×10^{-4}. The selenium in Cambridge air samples ranged from 0.03 to 0.06 ng/100 m³, with a Se/S ratio similar to that in the precipitation. The Se/S ratio of 1×10^{-4} is also similar to that found in industrial fuels and sulfide ores, raising the possibility of pollution from these sources.

In 1970, Hashimoto and his co-workers[159] compared the Se/S ratio in Cambridge air to that in industrial fuel oils, rubber, and soil samples in Tokyo. Close agreement was found with the atmospheric Se/S ratio and that in various types of heavy petroleum and raw petroleum, suggesting that combustion of these materials was the source of polluting atmospheric selenium. Selenium is mobilized through the atmosphere to a lesser extent than sulfur by fuel combustion according to Weiss and co-workers[159a] who measured Se/S ratios in specimens of polar ice.

The toxicity of selenium dioxide to fish was reported by Weir and Hine.[161] The conditioned avoidance technique (light-shock conditioning system) was used to assess the effect of exposure for 48 hours to varying concentrations of the compound. When exposed to toxic concentrations of selenium dioxide (lethal concentration for 50% = 7.9–17 ppm) goldfish developed large hemorrhagic areas and other evidence of toxicity within 4 hours. The lowest concentration of selenium causing significant impairment of the conditioned response was 0.25 ppm, or 1/48 of the LD[50]. This concentration of selenium is similar to that found in river water draining heavily seleniferous soil. The lowest concentrations of other metals significantly impairing conditioned responses in 50 % of the fish were as follows: arsenic, 0.1 ppm; lead, 0.07 ppm; and mercury, 0.003 ppm. This study contradicts the notion that low concentrations of metals are harmless to aquatic life and also suggests that the potential neurotoxicity of selenium needs to be further investigated.

Allaway[28,162] has discussed in detail problems involved in the environmental control of selenium. He notes that, although adequate methods exist for the measurement of selenium in biological materials, there is a need for the expansion of analytical services and facilities for the measurement of selenium in industrial wastes and potable waters.

6. Clinical Use of Selenium-Containing Compounds

a. *Selenium Sulfide.* Selenium sulfide in the form of a 2.5 % suspension in detergents has been widely used as a shampoo for the control of seborrheic dermatitis. Ointments containing selenium sulfide combined with hydrocortisone acetate are also in use for the treatment of seborrheic rash on skin.

Henschler and Kirschner[163] found that water-insoluble selenium sulfide possesses very low systemic toxicity in comparison to soluble inorganic selenium compounds. Chronic oral administration of selenium sulfide in doses of 5, 25, or 125 mg/kg per day to mice for 7 weeks was tolerated without signs of toxicity, and only the highest dose resulted in significant accumulation of selenium in parenchymatous organs. The LD$_{50}$ for selenium sulfide in mice was 3.7 g/kg, compared with a LD$_{50}$ of 48 mg/kg for sodium selenite. Cummins and Kimura[164,164a] confirmed the low systemic toxicity of selenium sulfide.

One instance of systemic toxicity has been reported in a woman who developed progressive tremor, weakness, and anorexia after using selenium sulfide shampoo on open scalp lesions several times weekly for 8 months. The symptoms cleared rapidly after use of the shampoo was discontinued, but elevated urinary levels of selenium confirmed that absorption through the open skin had occurred.[82]

Slinger and Hubbard[165] found no skin sensitivity, as determined by patch testing, to occur after use of the shampoo and observed no increase in urinary selenium. Good therapeutic results were obtained in 81 % of treated cases of seborrheic dermatitis. Bereston[166] found 89 % of patients to have good results; however, excessive skin oiliness occurred in 31 % and orange tinting of gray hair was seen in 19 %, these side effects responding to washing with green soap.

Eisenberg[167] reported three cases of contact dermatitis following the use of selenium sulfide shampoo. Grover[168] observed diffuse hair loss in six patients, which ceased upon discontinuation of the shampoo. Alopecia due to local irritation from selenium sulfide shampoo has been noted by Sidi and Bourgeois-Spinasse.[169]

Archer and Luell[170] reported the occurrence of dysplastic and deformed hair roots after single applications of shampoo, but Maguire and Kligman,[171] in acute and chronic application studies in 20 patients, found neither root deformities nor changes in the rate of hair regrowth. Increased sebum secretion following selenium sulfide shampooing has been observed.[172]

Selenium sulfide and zinc pyrithione shampoos were found to be equally effective in treating seborrheic dermatitis. Oiliness of the scalp, however, remained after use of both preparations.[173]

Although 1.0% selenium sulfide ointment was found effective in treating seborrheic dermatitis of glabrous skin in 73% of 37 patients, a high incidence of irritation and positive patch-test reactions (21%) led to the substitution of a 0.5% ointment, which caused less irritation and was still effective.[174]

b. Selenomethionine.

i. Physical Characteristics and Metabolism.   In 1961, Blau and Bender[175] produced ^{75}Se-selenomethionine by a biosynthetic process and demonstrated its use for external pancreatic scanning.[175a] Within recent years selenium-labeled analogs of purines and pyrimidines, hormones, amino acids, and coenzymes have been synthesized and used in studying various biologic processes.[176,177] The fact that ^{75}Se, a γ-emitter, can be substituted for sulfur makes it suitable for conveniently studying the metabolism of sulfur-containing compounds in the body. The localization of ^{75}Se-substituted methionine in organs with a high capacity for synthesizing proteins underlies the usefulness of the compound as an agent for visualizing organs such as the pancreas or parathyroids by scintillation scanning. However, studies by Kovacs[178] and Holland *et al.*[179] have amply demonstrated that selenomethionine and methionine are not always metabolized in an identical manner. Awwad *et al.*[180] found the splenic uptake of methionine to be greater than that of selenomethionine, while the uptake of selenomethionine was more rapid into the pancreas and liver.

^{75}Se, which has a complex radiation spectrum consisting of 10 γ-photons, decays by electron capture, with a physical half-life of 120 days.[181] Lathrop *et al.*[182] have presented revised absorbed dose (rads/250 μC i.v.), estimates for ^{75}Se-selenomethionine as follows: blood, 2.2; kidneys, 5.0; liver, 6.3; ovaries, 1.2; pancreas, 3.0; spleen, 4.0; testes, 2.7; thyroid, 1.5; and total body, 2.0. The biological whole body disappearance half-times ($T_{1/2}$) of injected ^{75}Se-L-selenomethionine are the following: 13%, 0.55 day; 44%, 46 days; and 42%, 220 days. The publication by Lathrop *et al.*[182] presents extensive discussion of the dosage calculations, kinetics, and tissue distribution of ^{75}Se-L-selenomethionine in man.

Hine and Johnston[183] have recorded the following absorbed doses (millirads/microcurie absorbed) to various tissues after intravenous ^{75}Se-selenomethionine administration: total body, 7–9; blood, 8–10; gonad, 3–7; kidney, 10–12; lungs, 30; spleen, 15–20; thyroid, 5–7; and pancreas, 10–12.

The distribution of selenomethionine in the tissues of the rat has been studied by Hansson and Jacobsson[184] and by Awwad *et al.*[185] Diffuse tissue uptake of selenomethionine occurs with no single organ demonstrating preferential accumulation. The liver and pancreas generally have the highest initial concentration of isotope. Significant accumulation occurs also in gastrointestinal mucous membranes, kidney, salivary gland, and bone marrow, but lung and myocardial concentrations are low. In the dog pancreatic uptake is increased when the selenoamino acid is administered after a period of fasting.[186] Kupic *et al.*[187] observed that 30 minutes after intravenous administration of ^{75}Se-selenomethionine to a dog its pancreas accumulated 0.27%, its liver 0.088%, and its kidney 0.057% of the dose. Neither injection directly into aorta or celiac artery, nor the use of secretin or pancreozymin, significantly enhanced the pancreatic uptake of isotope. The organ distribution of ^{75}Se-selenomethionine has been studied at necropsy in three patients with cancer by Ben-Porath *et al.*[188] Three hours after injection the liver was found to contain the highest concentration of isotope, whereas at 3 and 52 days after injection the highest concentration was in the kidney. Relative uptake into tumor tissue varied with the type of tumor, suggesting that in some instances normal tissue may not be well differentiated from malignant tissue on scan (Table XIIIC-4). Values for the distribution of ^{75}Se-selenomethionine administered by various routes in the rat have been presented by Graham *et al.*[188a] In this investigation the ^{75}Se content of tissue was the same regardless of whether the selenomethionine was administered intravenously or orally.

Selenomethionine is rapidly absorbed after oral administration, peak blood levels in man being reached in 3 hours.[189] The transport system across the gut wall for selenomethionine is identical to that for methionine[190] and is an energy-requiring, aerobic pathway. Methionine, glucose, and galactose inhibit selenomethionine absorption, as does oubain, the latter effect indicating involvement of the sodium pump in transport of the amino acid.[191] Biliary excretion occurs promptly and at a constant rate (0.0054% of administered dose per hour), and fecal excretion increases gradually with 10% of the dose excreted in the dog at 24 hours, and 25% excreted by 7

Table XIIIC-4 Tissue concentration and distribution of ^{75}*Se-selenomethionine in man (3 cases)*[a]

	3 HOURS—COLON CANCER		3 DAYS—BRONCHIAL CANCER		53 DAYS—HEPATOMA	
	%Dose/ 100 g[b]	Total % dose	%Dose/ 100 g[b]	Total % dose	% Dose/ 100 g[b]	Total % dose
Brain			0.22	3.00		
Thyroid	0.34	0.04	0.50	0.10		
Myocardium	0.30	1.36	0.24	1.00		
Pericardium[c]	0.16	0.12	0.25	0.20		
Lung: normal tissue	0.50	2.30	0.34	11.20		
" : metastases	0.48					
" : primary tumor			0.38	2.10		
Aorta[c]			0.13	0.13		
Diaphragm: normal tissue[c]	0.36	0.82	0.15	0.30		
" : metastases	0.54					
Intestines and stomach[c]	0.40	7.00	0.36	6.40		
Liver: normal tissue	1.75	40.00	1.62	28.40	0.32	7.70
" : metastases	0.48					
" : primary tumor					0.29	
Spleen	0.70	1.00	0.90	0.90	0.22	1.30
Metastatic lymph node			0.33	0.10		
Pancreas: normal tissue	1.35	1.85	1.13	0.90	0.08	0.97
" : metastases	0.20					
Gall bladder[c]	0.45	0.10	0.70	0.14		
Bile			0.70			
Colon: normal tissue	0.22					
" : primary tumor	0.40					
Kidneys	1.10	3.20	1.80	6.30	0.40	1.0
Urinary bladder	0.12	0.18	0.20	0.30		
Testes	0.26	0.08	0.25	0.10		
Adrenal glands	0.80	0.15	0.67	0.10		
Prostate			0.40	0.10		
Skin[c]			0.09	5.50		
Skeletal muscle[c]			0.06	18.00		
Bone[c]			0.04	2.80		
Blood[c]			0.22	11.00		

[a] These data were obtained at necropsy by Ben-Porath *et al.*[188] Two hundred fifty microcuries of ^{75}Se-selenomethionine had been injected intravenously to each of three male patients with carcinoma, whose tissues were obtained 3 hours, 3 days, and 52 days after injection.

[b] Wet weight.

[c] Denotes that weight of the organ was estimated.

days. Urinary excretion of selenomethionine in the dog is less than fecal, 1 % of the dose being eliminated in the urine at day 1 with 15 % recovered in the urine after 12 weeks.[192] Total-body counting in human beings showed only 50 % of the dose eliminated by day 80, with 85 % excreted in the urine. Only 15 % of the administered ^{75}Se-selenomethionine leaves the body *via* the feces, and complete elimination of the dose in man can take as long as 3 years.[193] Unlike the situation with selenate and selenite, less than 1 % of administered ^{75}Se-L-selenomethionine is excreted *via* the lungs.[194]

When administered intravenously, selenomethionine is cleared from the vascular compartment into tissues over 30–45 minutes and reappears in plasma bound to serum proteins.[195,196] The activity of protein precursor pools in the liver after the injection of selenomethionine has been studied by Awwad *et al.*[197] However, it is doubtful that the liver can synthesize selenoproteins from precursor selenite.[189,199] In serum, selenomethionine is incorporated into albumin and several globulin fractions;[200] however, the amounts incorporated have varied widely from study to study. Penner[201] found in man that 7 hours after injection 10 % of the activity in blood was in the albumin fraction of plasma, 5 % in α_1-globulin, 20 % in α_2-globulin, 50 % in β-globulin, and 6 % in the γ-globulin fraction. A significant portion of the selenomethionine carried bound to plasma proteins can be released after sulfitolysis, indicating nonpeptide binding.[200] Diminished incorporation of selenomethionine into the plasma proteins of patients with hepatitis has been observed,[202,203] but no deviation from the normal patterns was evident in patients with cirrhosis or leukemia. There was increased selenomethionine incorporation into plasma globulin fractions in patients with multiple myeloma and rheumatoid arthritis.[201]

The patterns and the extent of incorporation of selenomethionine into erythrocytes, leukocytes, and blood platelets have been extensively studied in normal individuals and in patients with various hematologic diseases.[204,204a] Although it was initially thought that selenomethionine could function as a label in studying the life span of the formed elements in blood, studies by McIntyre *et al.*[205] cast doubt on this application. However, selenomethionine has been effectively utilized in the study of thrombopoesis.[206–208]

ii. Pancreatic Scanning with Selenomethionine. Selenomethionine has been widely used as an agent for external organ scanning. A detailed discussion of the reliability of scanning after the administration of ^{75}Se-selenomethionine is beyond the scope of this review, but the following comments will serve as an orientation to this topic, which is still in a developmental stage.

[75]Se-Selenomethionine is rapidly concentrated by tissues that actively synthesize or metabolize proteins. As such, it is concentrated by the

pancreas as it produces enzymes,[209,211] by the parathyroid as it synthesizes parathyroid hormone,[212] and by the liver. The uptake by liver frequently obscures pancreatic visualization. Therefore, a number of specialized techniques, such as simultaneous [198]Au hepatic scans,[212a] and isotope subtraction procedures have been developed to distinguish pancreatic from hepatic shadow[213,218] more effectively. Also, several workers have used either pancreatic secretogogs[219,220] or morphine[221] to increase the uptake of selenomethionine into the pancreas, or a prescan high-protein meal to increase the rate of digestive enzyme synthesis.[222] In most instances, however, these procedures have not been sufficiently effective to warrant their use. Direct instillation of isotope into the major vessels feeding the pancreas has also not proved feasible in increasing the visualization of the organ by scan. In the case of the parathyroid, direct instillation of ^{75}Se-selenomethionine into the thyrocervical vessels has increased visualization of these small glands, but the technique is technically difficult.[223]

Several clinical studies have been published which indicate that the selenomethionine pancreatic scan is clinically useful.[224-229] In general, with adequate technique, the pancreas can be visualized in 90% of the scans. If several variants of the normal in terms of location and configuration are taken into account, a pancreas that appears normal on scan is not likely to be the site of disease. With an abnormal scan (*i.e.*, one which does not fill, or contains a nonfunctioning area) it is difficult to differentiate between a benign cyst, pancreatitis, or carcinoma. The incidence of falsely negative and positive scans is sufficiently high so that the scan *per se* cannot be relied upon to establish a diagnosis of pancreatic disease. Studies in diabetics of various grades of severity have established that functional disorders of the pancreas may also produce abnormalities on scan.[230]

iii. Scanning of Parathyroid and Other Tissue with Selenomethionine. Selenomethionine scanning is of less value in the diagnosis of parathyroid than of pancreatic disease.[231] The thyroid gland also accumulates selenomethionine; and although this uptake can be blocked by the administration of thyroid hormone,[212] the ratio of isotope uptake by the parathyroid compared to that by surrounding tissues and blood is low, so that differentiation of the gland is poor.[232] Every published series of selenomethionine parathyroid scans contains an unacceptably large number that are falsely positive or falsely negative.[233,236] The use of sequential scans[237,238] or computer analysis of sequential scans has been reported to increase the accuracy of the procedure.[239]

Weinstein *et al.*[240,241] have observed that selenomethionine is concentrated by malignant lesions in the thyroid, whereas ^{131}I is not taken up into these tumors. Nonfunctioning benign thyroid nodules accumulate neither the selenomethionine nor the iodine, thus allowing differentiation of malignant from benign nodules. Preliminary studies appear to support the usefulness of this technique.

D'angio *et al.*[242] have found neuroblastoma tissue to concentrate ^{75}Se-selenomethionine and have successfully scanned brain lesions of this type in children. Masses of intra-abdominal lymphoma have been successfully localized with a high degree of accuracy by selenomethionine scans,[243,247] but only rarely have epidermoid or other malignant lesions been found to incorporate selenomethionine.[248]

[75]Se-Selenite has been utilized in the scanning of malignant and non-malignant lesions of bone and brain. Cavalieri and Scott[249] reported ^{75}Se-selenite to be more effective than [197]Hg-chloromerodrin or ^{99m}Tc-sodium pertechnetate in demonstrating vascular abnormalities of the brain. Ray *et al.*[250] found a high correlation between results of ^{75}Se-selenite bone scans and bone biopsy. The incorporation of ^{75}Se-selenite into tumor mucopolysaccharide has permitted external scanning of a malignant chondrosarcoma of the shoulder.[251]

In spite of early reports suggesting the interchangeability of ^{35}S-sulfate and ^{75}Se-selenate in the measurement of extracellular fluid space,[252] Nelp and Blumberg[253] found that in both man and dog the early plasma concentration of selenate is less than that of sulfate, leading to an erroneously enlarged volume of distribution. An additional artifact is introduced because of the return to the vascular space of protein-bound selenate 1 hour after injection. Thus it appears that selenate is of no value for the measurement of extracellular fluid space.

REFERENCES

1. H. E. Ganther, *World Rev. Nutr. Dietet.*, **5**, 338 (1965).

2. J. Pinsent, *Biochem. J.*, **57**, 10 (1954).

2a. K. Schwarz, *Federation Proc.*, **24**, 58 (1965).

2b. R. C. Bull and J. E. Oldfield, *J. Nutr.*, **91**, 237 (1967).

3. K. Liebscher and H. Smith, *Arch. Environ. Health*, **17**, 881 (1968).

4. V. C. Morris and O. A. Levander, *J. Nutr.*, **100**, 1383 (1970).

5. H. A. Schroeder, D. V. Frost, and J. J. Balassa, *J. Chronic Dis.*, **23**, 227 (1970).

6. W. B. Dye, E. Bretthauer, H. J. Seim, and C. Blincoe, *Anal. Chem.*, **35**, 1687 (1963).

7. G. Patrias and O. E. Olson, *Feedstuffs*, **41**, 32 (1967).

8. F. Kiermeier and W. Wigand, *Z. Lebensm. Untersuch. Forsch.*, **136**, 158 (1968).

9. W. Oelschlaeger and K. H. Menke, *Z. Ernährungswiss.*, **9**, 216 (1969).

10. W. Oelschlaeger, *Umschau*, **70**, 45 (1970).

11. H. Bostrom and P. O. Wester, *Acta Med. Scand.*, **181**, 465 (1967).

12. T. Goto and M. Fujino, *Eiyo To Shokuryo*, **20**, 311 (1967).

13. N. Ohta, N. Onuma, and K. Kiwasaki, *Nippon Kagaku Zasshi*, **81,** 920 (1960).

13a. R. R. Kifer, W. L. Payne and M. E. Ambrose, *Feedstuffs*, **41,** 24 (1969).

14. G. Lunde, *J. Sci. Food Agr.*, **21,** 242 (1970).

15. D. M. Hadjimarkos, *Lancet*, **1,** 605 (1965).

16. D. M. Hadjimarkos and C. W. Bornhorst, *Nature*, **202,** 296 (1961).

17. D. M. Hadjimarkos and C. W. Bonhorst, *J. Pediat.*, **59,** 256 (1961).

18. E. R. Morris and F. E. Greene, *Federation Proc.*, **29,** 500 (1970).

19. W. O. Robinson, *Ind. Eng. Chem.*, **28,** 736 (1936).

20. T. Thorvaldson and L. R. Johnson, *Can. J. Res.*, **18,** 138 (1940).

21. H. W. Lakin and H. G. Byers, *Cereal Chem.*, **18,** 73 (1941).

22. W. G. Jaffe, J. F. Chavez, and M. C. Mondragon, *Arch. Latinoamer. Nutr.*, **19,** 299 (1969).

23. W. G. Jaffe, J. F. Chavez, and M. C. Mondragon, *Arch. Latinoamer. Nutr.*, **17,** 59 (1967).

24. O. A. Levander, "Present Knowledge in Nutrition," The Nutrition Foundation, New York, N.Y., 1967, p. 138.

25. M. I. Smith and E. F. Stohlman, *J. Pharmacol. Exptl. Ther.*, **70,** 270 (1940).

26. H. E. Ganther and C. A. Baumann, *J. Nutr.*, **77,** 408 (1962).

27. M. L. Scott and A. H. Cantor, *Federation Proc.*, **30,** 237 (1971).

27a. W. G. Jaffe and M. C. Mondragon, *J. Nutr.*, **97,** 431 (1969).

28. W. H. Allaway, "Conference on Trace Substances in Environmental Health, II," D. D. Hemphill, Ed., University of Missouri, 1968, p. 181.

29. D. M. Hadjimarkos and C. W. Bonhorst, *Nature*, **193,** 177 (1962).

30. D. M. Hadjimarkos and C. W. Bonhorst, *Oral Surg.*, **12,** 113 (1959).

31. G. S. Nixon and V. B. Myers, *Caries Res.*, **4,** 179 (1970).

32. F. M. Mamedova, K. Nikolaeva, and E. A. Bozhevolinov, *Stomatologia*, **47,** 81 (1968).

33. J. L. Hardwick and C. V. Martin, *Helv. Odont. Acta*, **11,** 62 (1967).

34. T. Yugi, *Shikoku Acta Med.*, **15,** 1861 (1959).

35. H. J. M. Bowen, "Trace Elements in Biochemistry," Academic Press, London, 1966, p. 77.

36. R. C. Dickson and R. H. Tomlinson, *Clin. Chem. Acta*, **16,** 311 (1967).

37. C. F. Consolazio, R. A. Nelson, L. O. Matoush, R. C. Hughes, and P. Vrone, N.A.S.A. Accession No. N65-35035, Rept. No. AD 447382.

38. H. H. Taussky, A. Washington, and E. Zubillaga, *Nature*, **210,** 949 (1966).

39. G. D. Christian and M. Moritz, *Invest. Ophthalmol.*, **5,** 248 (1966).

40. H. Berger, *Acta Biol. Med. Ger.*, **19,** 405 (1967).

41. V. V. Ermakov, *Byull. Eksp. Biol. Med.*, **59,** 61 (1965).

42. J. M. Fuller, E. D. Beckman, M. Goldman, and L. K. Bustad, "Selenium in Biomedicine," O. H. Muth, Ed., AVI Publishing Co., Westport, Conn., 1967, p. 123.

43. H. J. M. Bowen and P. A. Cawse, *Analyst*, **88,** 721 (1963).

44. W. H. Allaway, J. Kubota, F. Losee, and M. Roth, *Arch. Environ. Health*, **16,** 342 (1968).

45. D. K. Brune, K. Samasahl, and P. O. Westev, *Clin. Chem. Acta*, **13,** 285 (1966).

46. R. J. Levine and R. E. Olson, *Proc. Soc. Exptl. Biol. Med.*, **134,** 1030 (1970).

47. R. F. J. Burk, W. N. Pearson, R. F. Wood, and F. Viter, *Amer. J. Clin. Nutr.*, **20,** 723 (1967).

48. D. M. Hadjimarkos, C. W. Bonhorst, and J. H. Mahice, *J. Pediat.*, **54,** 296 (1959).

49. E. B. Dawson, M. P. Menon, R. E. Wainerdi, and W. J. McGanity, *J. Nucl. Med.*, **9,** 160 (1968).

49a. D. H. Dayton, L. J. Filer, and C. Canosa, *Federation Proc.*, **28,** 488 (1969).

50. J. H. Sterner and V. Lidfeldt, *J. Pharmacol. Exptl. Ther.*, **73,** 205 (1941).

51. J. Glover, *Bull. Selenium-Tellurium Develop. Assoc.*, **8,** 5 (1969).

52. M. I. Smith, K. W. Franke, and B. B. Westfall, *U.S. Public Health Rept.*, **51,** 1496 (1936).

52a. H. Sakurayama, *Shikoku Igoku Zasshi*, **16,** 131 (1965).

53. L. S. Densen, *Proc. Soc. Exptl. Biol. Med.*, **128,** 970 (1968).

54. M. L. Scott and J. N. Thompson, *Federation Proc.*, **29,** 499 (1970).

54a. J. N. Thompson and M. L. Scott, *J. Nutr.*, **100,** 797 (1970).

55. K. Schwarz, J. G. Bieri, G. M. Briggs, and M. L. Scott, *Proc. Soc. Exptl. Biol. Med.*, **95,** 621 (1957).

56. K. Aterman, *Brit. J. Nutr.*, **13,** 38 (1959).

57. E. A. Porta, F. A. de la Iglesia, and W. S. Hartroft, *Lab. Invest.*, **18,** 283 (1968).

58. D. A. Egan, *Irish Vet. J.*, **20,** 182 (1966).

59. K. Schwarz, *Federation Proc.*, **20,** 666 (1961).

60. A. S. Majaj and L. L. Hopkins, Jr., *Lancet*, **2,** 592 (1966).

61. L. L. Hopkins, Jr. and A. S. Majaj, "Selenium in Biomedicine," O. H. Muth, Ed., AVI Publishing Co., Westport, Conn., 1967, p. 203.

62. P. L. Wright and M. C. Bell, *Federation Proc.*, **22,** 377 (1963).

63. I. Rosenfeld and O. A. Beath, "Selenium," Academic Press, New York, N.Y., 1964, p. 9.

64. S. T. Benavides and R. F. S. Mojica, *Publication 1T-3*, Instituto Geografico de Colombia, Bogota, Colombia, 1959.

65. W. O. Robinson, *J. Assoc. Offic. Anal. Chemists*, **16,** 423 (1933).

66. Jacquin, *Amer. Pict.*, 109 (1763); cited by F. Kerdel-Vegas, *J. Invest. Dermatol.* **42,** 91 (1964).

67. *Chem. Eng. News*, Aug. 24, 1970, p. 10.

68. O. A. Beath, *Sci. Newslett.*, **81,** 254 (1962).

69. M. I. Smith and B. B. Westfall, *U.S. Public Health Rept.*, **52,** 1375 (1937).

70. R. E. Lemley, *J. Lancet*, **60,** 528 (1940).

71. A. L. Moxon, A. E. Schaefer, H. A. Lardy, and K. P. Dubois, *J. Biol. Chem.*, **132,** 785 (1940).

72. R. E. Lemley and M. M. Merryman, *J. Lancet*, **61,** 435 (1941).

73. F. W. Tunnicliffe and O. Rosenheim, *Lancet*, **1,** 319 (1901).

74. F. Kerdel-Vegas, F. Wagner, P. B. Russell, N. H. Grant, H. E. Alburn, D. E. Clark, and J. A. Miller, *Nature*, **205,** 1186 (1965).

75. L. Aronow and F. Kerdel-Vegas, *Nature*, **205,** 1185 (1965).

76. G. J. Olivares, L. Aronow, and F. Kerdel-Vegas, *Acta Cient. Venez.*, **18,** 9 (1967).

77. J. F. Chavez, *Bol. Soc. Guim. Peru*, **32**, 195 (1966).

78. S. N. Nigam and W. B. McConnell, *Biochim. Biophys. Acta*, **192**, 185 (1969).

79. F. L. Zalkind and V. V. Ermakov, *Agrokhimiya*, 98 (1968).

80. R. F. Carter, *Med. J. Australia*, **1**, 525 (1966).

81. R. F. Buchan, *Occup. Med.*, **3**, 439 (1947).

82. J. W. Ranson, N. M. Scott, Jr., and E. C. Knoblock, *New Engl. J. Med.*, **264**, 384 (1961).

83. D. M. Hadjimarkos, *J. Pediat.*, **59**, 256 (1961).

83a. K. Schwarz, "Selenium in Biomedicine," O. H. Muth, Ed., AVI Publishing Co., Westport, Conn., 1967, p. 225.

83b. A. G. McLundie, J. B. Shepherd, and D. Mobbs, *Arch. Oral Biol.*, **13**, 1321 (1968).

83c. D. R. Eisenman and J. A. Yeager, *Arch. Oral Biol.*, **14**, 1045 (1969).

84. D. M. Hadjimarkos, *Caries Res.*, **3**, 14 (1969).

85. W. Buttner, *J. Dent. Res.* **42**, 453 (1963).

86. N. Muhleman and K. Konig, *Helv. Odont. Acta*, **8**, 79 (1964).

87. J. C. Muller and W. A. Shafer, *J. Dent. Res.*, **36**, 895 (1957).

88. W. Buttner, *Caries Res.*, **3**, 1 (1969).

89. T. G. Ludwig and B. Bibby, *Caries Res.*, **3**, 32 (1969).

90. E. A. Cerwenka, Jr., and W. C. Copper, *Arch. Environ. Health*, **3**, 189 (1961).

91. J. R. Glover, *Ind. Med. Surg.*, **39**, 50 (1970).

92. I. Nagai, *Igaku Kenkyu*, **29**, 1505 (1959).

93. M. Ellis, M. Motley, M. D. Ellis, and R. O. Jones, *Proc. Soc. Exptl. Biol. Med.*, **36**, 519 (1937).

94. G. Kury, L. Rev-Kury, and R. J. Crosby, *Toxicol. Appl. Pharmacol.*, **11**, 449 (1967).

95. R. H. Rigdon, R. H. Cross, and K. P. McConnell, *Arch. Pathol.*, **57**, 374 (1953).

96. A. L. Moxon and M. Rhian, *Physiol. Rev.*, **23**, 305 (1943).

97. S. Scarabicchi, M. Merlini, and D. Ribaldone, *Minerva Dietol.*, **8**, 222 (1968).

98. R. D. Lillie and M. I. Smith, *Amer. J. Pathol.*, **16**, 223 (1940).

98a. A. F. Stenn, *Arch. Pathol.*, **22**, 398 (1936).

99. A. Bel, J. Levrat, J. Nesmoz, and M. Girard, *J. Med. Lyon*, Suppl. 53–64 (1966).

100. H. A. Schroeder, *J. Nutr.*, **94**, 475 (1968).

101. V. E. Levine and R. A. Flaherty, *Proc. Soc. Exptl. Biol. Med.*, **24**, 251 (1926).

101a. H. T. Tsuzuki, K. Okawa, and T. Hosoya, *Yokohama Med. Bull.*, **11**, 368 (1960).

102. A. Hamilton and H. L. Hardy, "Industrial Toxicology," Harpers, New York, N.Y., 1949, p. 188.

103. N. I. Sax, "Dangerous Properties of Industrial Materials," Reinhold, New York, N.Y., 1968, p. 1085.

104. Editorial, *Bull. Selenium-Tellurium Devel. Assoc.*, 4 (1965).

105. I. Holmquist, International Symposium, European Selenium-Tellurium Comm., Stockholm, Apr. 16, 1969; cited in *Bull. Selenium-Tellurium Devel. Assoc.*, 2 (1969).

106. H. C. Dudley, *U.S. Public Health Rept.*, **53**, 281 (1938).

107. E. Browning, "Toxicity of Industrial Metals," Butterworths, London, 1961, p. 258.

108. H. C. Dudley, *Amer. J. Hyg.*, **23**, 181 (1936).

109. M. Clinton, Jr., *J. Indust. Hyg. Toxicol.*, **29,** 225 (1947).

110. H. M. Wilson, *N. Carolina Med. J.*, **23,** 74 (1962).

111. H. C. Dudley and J. W. Miller, *U.S. Public Health Rept.*, **52,** 1217 (1937).

111a. H. Synanski, *Deut. Med. Wochschr.*, **75,** 1730 (1950).

112. H. L. Motley, M. M. Ellis, and M. D. Ellis, *Amer. Med. Assoc.*, **109,** 1718 (1937).

113. H. C. Dudley, *U.S. Public Health Rept.*, **53,** 94 (1938).

114. M. Duboir, L. Pollet, and J. L. Herrenschmidt, *Bull. Soc. Fr. Dermat. Syph.*, **44,** 88 (1937).

114a. K. Halter, *Arch. Dermtol. Syphilol.*, **178,** 340 (1939).

115. P. Pringle, *Brit. J. Dermatol. Syphilis*, **54,** 54 (1942).

116. J. R. Glover, *Bull. Selenium-Tellurium Devel. Assoc.*, 8 (1969).

117. M. I. Smith, *U.S. Public Health Rept.*, **54,** 1441 (1939).

118. A. L. Moxon and M. Rhian, *Physiol. Rev.*, **23,** 305 (1943).

119. O. A. Levander and V. C. Morris, *J. Nutr.*, **100,** 1111 (1970).

120. A. W. Halverson, C. M. Hondrick, and O. E. Olson, *J. Nutr.*, **56,** 81 (1955).

121. O. A. Levander, M. L. Young, and S. A. Meeks, *Toxicol. Appl. Pharmacol.*, **16,** 79 (1970).

122. B. B. Westfall, Doctoral Dissertation, University of Missouri, Columbia, Mo., 1938.

123. K. P. DuBois, M. Rhian, and A. L. Moxon, *Proc. S. Dakota Acad. Sci.*, **19,** 71 (1939).

124. O. A. Levander and C. A. Baumann, *Toxical. Appl. Pharmacol.*, **9,** 106 (1966).

125. R. C. Dickson and A. L. Tappel, *Arch. Biochem. Biophys.*, **130,** 547 (1969).

126. A. L. Moxon and K. P. DuBois, *J. Nutr.*, **18,** 447 (1939).

127. I. S. Palmer and C. W. Bonhorst, *J. Agr. Food Chem.*, **5,** 928 (1957).

128. O. A. Levander and C. A. Baumann, *Toxicol. Appl. Pharmacol.*, **9,** 106 (1966).

129. O. A. Levander and C. A. Baumann, *Toxicol. Appl. Pharmacol.*, **9,** 98 (1966).

130. C. Hendrick, H. L. Klug, and O. E. Olson, *J. Nutr.*, **51,** 131 (1953).

131. B. B. Westfall and M. I. Smith, *J. Pharmacol. Exptl. Ther.*, **72,** 245 (1941).

132. K. P. McConnell, A. E. Kreamer, and D. M. Roth, *J. Biol. Chem.*, **234,** 2932 (1959)

133. H. A. Braun, L. M. Lusky, and H. O. Calvery, *J. Pharmacol. Exptl. Ther.*, **87,** 119 (1946).

134. J. B. Belogorsky and D. Slaughter, *Proc. Soc. Exptl. Biol. Med.*, **72,** 196 (1949).

135. K. I. Sivjakov and N. A. Braun, *Toxicol. Appl. Pharmacol.*, **1,** 602 (1959).

136. O. A. Levander, personal communication.

137. J. R. Glover, *Trans. Assoc. Ind. Med. Officers*, **4,** 94 (1959).

138. K. W. Franke, A. L. Moxon, W. E. Poley, and W. C. Tully, *Anat. Rec.*, **65,** 15 (1936).

139. P. L. Wright and F. R. Mraz, *Proc. Soc. Exptl. Biol. Med.*, **118,** 354 (1965).

140. P. Gruenwald, *Amer. J. Pathol.*, **34,** 77 (1958).

141. P. Sentein, *Chromosoma*, **17,** 336 (1965).

142. R. E. Holmberg, Jr., and V. H. Ferm, *Arch. Environ. Health*, **18,** 873 (1969).

143. D. S. F. Robertson, *Lancet*, **1,** 518 (1970).

144. D. M. Hadjimarkos, *Lancet*, **1,** 721 (1970).

144a. Editorial, *Nutr. Rev.*, **28,** 75 (1970).

145. H. C. Burns, *Vet. Med.*, **62,** 512 (1967).

146. J. A. Miller, N.A.S.–N.C.R. Publ. No. 1354, Washington, D.C., 1966, p. 24.

147. H. F. Kraybill, *Environ. Res.*, **2,** 231 (1969).

148. D. V. Frost, "Selenium in Biomedicine," O. H. Muth, Ed., AVI Publishing Co., Westport, Conn., 1967, p. 7.

149. D. V. Frost, *World Rev. Pest Control*, **9,** 6 (1970).

150. J. Seifter, W. E. Ehrlich, G. Hudyma, and G. Miller, *Science*, **103,** 762 (1946).

151. O. G. Fitzhugh, A. A. Nelson, and C. I. Bliss, *J. Pharmacol.*, **80,** 289 (1944).

151a. A. A. Nelson, O. G. Fitzhugh, and H. O. Calvery, *Cancer Res.*, **3,** 230 (1943).

151b. C. G. Durbin, "Selenium in Biomedicine," O. H. Muth, Ed., AVI Publishing Co., Westport, Conn., 1967, p. 419.

152. L. A. Tscherkes, M. N. Volgarev, and S. G. Aptekar, *Acta Univ. Intern. Contra Cancrum,* **19,** 632 (1963).

153. M. N. Volgarev and L. A. Tscherkes, "Selenium in Biomedicine," O. H. Muth, Ed., AVI Publishing Co., Westport, Conn., 1967, p. 179.

154. J. R. Harr, J. F. Bone, I. J. Tinsley, P. H. Weswig, and R. S. Yamamoto, "Selenium in Biomedicine," O. H. Muth, Ed., AVI Publishing Co., Westport, Conn., 1967, p. 153.

155. H. G. Byers, *Ind. Chem. Eng.*, **29,** 1200 (1937).

156. O. Tada, *Rept. Inst. Sci. Labour*, No. 69, 10 (1968).

157. E. C. Tabor, M. M. Braverman, H. E. Bumstead *et al.*, *Health Lab. Sci.*, **7,** Suppl. 96 (1970).

158. Y. Hashimoto and J. W. Winchester, *Environ. Sci. Tech.*, **1,** 338 (1967).

159. Y. Hashimoto, J. Y. Hwang, and S. Yanagisawa, *Environ. Sci. Tech.*, **4,** 157 (1970).

159a. H. V. Weiss, M. Koide, and E. D. Goldberg, *Science*, **172,** 261 (1971).

160. G. B. Morgan, G. Ozolins, and E. C. Tabor, *Science*, **170,** 289 (1970).

161. P. A. Weir and C. H. Hine, *Arch. Environ. Health*, **20,** 45 (1970).

162. W. H. Allaway, *Advan. Agron.*, **20,** 235 (1968).

163. D. Henschler and V. Kirschner, *Arch. Toxik.*, **24,** 341 (1969).

164. L. M. Cummins and E. T. Kimura, *Federation Proc.*, **29,** 743 (1970).

164a. L. M. Cummins and E. T. Kimura, *Toxicol. Appl. Pharmacol.*, **20,** 89 (1971).

165. W. N. Slinger and D. M. Hubbard, *Arch. Dermatol. Syphilol.*, **64,** 41 (1951).

166. E. S. Bereston, *J. Amer. Med. Assoc.*, **156,** 1246 (1954).

167. B. C. Eisenberg, *Arch. Dermatol.*, **72,** 71 (1955).

168. R. W. Grover, *J. Amer. Med. Assoc.*, **160,** 1397 (1956).

169. E. Sidi and M. Bourgeois-Spinasse, *Press Medicale*, **66,** 1767 (1958).

170. V. E. Archer and E. Luell, *J. Invest. Dermatol.*, **35,** 65 (1960).

171. H. C. Maguire and A. M. Kligman, *J. Invest. Dermatol.*, **39,** 469 (1962).

172. H. Goldschmidt, *Acta Dermatol.* (Stockholm), **48,** 489 (1968).

173. N. Orentreich, E. H. Taylor, R. A. Berger, and R. Auerbach, *J. Pharm. Sci.*, **58,** 1279 (1969).

174. S. Ayres III and S. Ayres, Jr., *Arch. Dermatol. Syphilol.*, **69,** 615 (1954).

175. M. Blau and M. A. Bender, *Biochim. Biophys. Acta*, **49,** 389 (1961).

175a. M. Blau, and M. A. Bender, *Prog. Med. Radioisotope Scanning*, USAEC, ORINS, October 1962, p. 263.

176. H. G. Mautner, *Radioactive Pharm., Proc. Symp., Oak Ridge, Tenn., 1965,* 409 (1966).

177. K. A. Scott, and H. G. Mautner, *Biochem. Pharmacol.,* **16,** 1903 (1967).

178. V. Kovacs, *Studii Cercetari Biochem.,* **12,** 367 (1969).

179. J. F. Holland, S. Peters, B. Bryant, and M. Blau. *J. Clin. Invest.,* **45,** 1024 (1966).

180. H. K. Awwad, E. J. Potchen, S. J. Adelstein, and J. P. Dealy, Jr., *Metabolism,* **15,** 370 (1966).

181. E. M. Smith, *Radioactive Pharm., Proc. Symp., Oak Ridge, Tenn., 1965,* 656 (1966).

182. K. A. Lathrop, R. E. Johnston, M. Blau, and E. O. Rothschild, "Radiation Dose to Humans from ^{75}Se-L-Selenomethionine," *J. Nucl. Med., Suppl. 6,* **13,** 10 (1972).

183. C. J. Hine and R. E. Johnston, *J. Nucl. Med.,* **11,** 468 (1970).

184. E. Hansson and S. O. Jacobsson, *Biochim. Biophys. Acta,* **115,** 285 (1966).

185. H. K. Awwad, E. J. Potchen, S. J. Adelstein, and J. B. Dealy, Jr., *Metabolism,* **15,** 370 (1966).

186. M. Blau and R. F. Manske, *J. Nucl. Med.,* **2,** 102 (1961).

187. E. A. Kupic, A. G. Kasenter, and C. D. Janney, *J. Nucl. Med.,* **12,** 35 (1971).

188. M. Ben-Porath, L. Case, and E. Kaplan, *J. Nucl. Med.,* **10,** 709 (1969).

188a. L. A. Graham, R. L. Veatch, and E. Kaplan, *J. Nucl. Med.,* **12,** 566 (1971).

189. R. P. Spencer and M. Blau, *Science,* **136,** 155 (1962).

190. A. Ochoa-Solano and C. Gitler, *J. Nutr.,* **94,** 249 (1968).

191. K. P. McConnell and G. J. Cho, *Amer. J. Physiol.,* **213,** 150 (1967).

192. G. E. van Goidsenhoven, A. F. Denk, B. A. Pfleger, and W. A. Knight, Jr., *Gastroenterology,* **53,** 403, (1967).

193. K. A. Lathrop, P. V. Harper, F. D. Malkinson, *Strahlentherapie,* **67,** 436 (1968).

194. J. Sternberg and A. Imbach, *Intern. J. Appl. Radiation Isotopes,* **18,** 217 (1967).

195. M. Blau and J. F. Holland, *Radioactive Pharm. Proc. Symp., Oak Ridge, Tenn., 1965,* 423 (1966).

196. W. H. Olendorf and M. J. Kitano, *J. Nucl. Med.,* **4,** 231 (1963).

197. H. K. Awwad, S. J. Adelstein, E. J. Potchen, and J. B. Dealy, Jr., *J. Biol Chem.,* **242,** 492 (1967).

198. L. M. Cummins and J. L. Martin, *Biochemistry,* **6,** 3162 (1967).

199. K. J. Jenkins, *Can. J. Biochem.,* **46,** 1417 (1968).

200. H. K. Awwad, E. J. Potchen, S. J. Adelstein, and J. B. Dealy, Jr., *Metab. Clin. Exptl.,* **15,** 626 (1966).

201. J. A. Penner, *Clin. Res.,* **12,** 277 (1964).

202. I. Szantay, S. Cotul, and M. Giduli, *Rev. Intern. Hepatol.,* **18,** 663 (1968).

203. O. Fodor, I. Szantay, and S. Cotul, *Nucl. Med.,* **7,** 130 (1968).

204. J. A. Penner, *J. Lab. Clin. Med.,* **67,** 427 (1966).

205. P. A. McIntyre, B. Evatt, B. A. Hodgkinson, and V. Schettel, *J. Lab. Clin. Med.,* **75,** 427 (1970).

206. B. L. Evatt and J. Levin, *J. Clin. Invest.,* **48,** 1615 (1969).

207. D. G. Pennington, *Brit. Med. J.,* **4,** 782 (1969).

208. P. Cohen, M. H. Cooley, and F. H. Gardner, *J. Clin. Invest.*, **44,** 1036 (1965).

209. J. E. Wheeler, *Proc. Soc. Exptl. Biol. Med.*, **70,** 187 (1949).

210. E. Hansson, *Acta Physiol. Scand., Suppl.*, **161,** 58 (1959)

211. E. Hansson, *Acta Physiol. Scand., Suppl.*, **213,** 59 (1963).

212. E. J. Potchen, *J. Nucl. Med.*, **4,** 480 (1963).

212a. S. Fink, M. Ben-Porath, B. Jacobson, G. D. Clayton, and E. Kaplan, *J. Nucl. Med.*, **10,** 78 (1969).

213. S. R. Reuter and J. J. Cohen, *Radiology*, **92,** 158 (1969).

214. M. Ben-Porath, G. Clayton, and E. Kaplan, *Trans. Amer. Nucl. Soc.*, **9,** 76 (1968).

215. E. Kaplan, S. Fink, M. Ben-Porath, and G. Clayton, *J. Nucl. Med.*, **9,** 330 (1968).

216. P. C. Blanquet, C. R. Beck, J. Fleury, and C. J. Palais, *J. Nucl. Med.*, **9,** 486 (1968).

217. S. Fink, M. Ben-Porath, B. Jacobson, G. Clayton, and E. Kaplan, *J. Nucl. Med.*, **10,** 78 (1969).

218. D. S. Fischer, E. Solario, W. Stewart, C. Cejas, and C. Kelly, *J. Nucl. Med.*, **10,** 459 (1969).

219. T. P. Haynie, K. C. Svoboda, and G. D. Zuidema, *J. Nucl. Med.*, **5,** 90 (1964).

220. D. L. Tabern, J. Kearney, and A. Dolbow, *J. Nucl. Med.*, **6,** 762 (1965).

221. A. Rodriquez-Antunez, *Cleveland Clinic Quart.*, **31,** 213 (1964).

222. D. B. Sodee, *Radiology*, **83,** 910 (1964).

223. A. Gottschalk, K. Ranninger, D. Paloyan, E. Paloyan, and P. V. Harper, *J. Nucl. Med.*, **7,** 374 (1966).

224. R. N. Melmed, J. E. Agnew, and I. A. D. Bouchier, *Quart J. Med.*, **37,** 607 (1968).

225. P. W. Brown, W. Sircus, A. A. Donaldson, C. W. A. Falconer, A. M. Smith, I. W. Dymock, and W. P. Small, *Lancet*, **1,** 160 (1968).

226. D. B. Sodee, Scientific Exhibit at the 12th Annual Meeting of the Society of Nuclear Medicine, June, 1965. Cited in H. N. Wagner, Jr., "Principles of Nuclear Medicine," W. B. Saunders, Philadelphia, Pa., 1968, p. 626.

227. A. Rodriguez-Antunez, E. J. Filson, B. H. Sullivan, and C. H. Brown, *Ann. Internal Med.*, **65,** 730 (1966).

228. J. A. Burdine and T. P. Haynie, *J. Amer. Med. Assoc.*, **194,** 979 (1965).

229. S. Landman, R. J. Polycyn, and A. Gottschalk, *J. Nucl. Med.*, **11,** 393 (1970).

230. D. H. Woodbury, and J. E. Yanez, *J. Nucl. Med.*, **10,** 453 (1969).

231. E. J. Potchen and D. B. Sodee, *J. Clin. Endocrinol. Metab.*, **24,** 1125 (1964).

232. J. R. Pickleman, E. Paloyan, K. A. Lathrop, and P. V. Harper, *J. Nucl. Med.*, **10,** 363 (1969).

233. M. G. McGoven, T. K. Bell, W. O. Soyann, S. S. A. Fenton, and S. D. Oreopoulos, *Brit. J. Radiol.*, **41,** 300 (1968).

234. N. C. Newton, M. G. Sumich, I. S. Jenkenson, *Med. J. Australia*, **2,** 205 (1968).

235. J. S. Garrow and R. Smith, *Brit. J. Radiol.*, **41,** 307 (1968).

236. W. D. Giulio and J. O. Morales, *J. Amer. Med. Assoc.*, **209,** 1873 (1969).

237. E. J. Potchen, "Principles of Nuclear Medicine," H. N. Wagner, Jr., Ed., W. B. Saunders, Philadelphia, Pa., 1968, p. 396

238. E. J. Potchen, H. G. Watts, and H. K. Awwad, *Radiol. Clin. N. A.*, **5,** 267 (1967).

239. E. O. Rothschild, A. Preciadosolis, D. Weber, and M. Powell, *J. Nucl. Med.*, **11,** 358 (1970).

240. M. B. Weinstein and W. M. Smoak, *J. Nucl. Med.*, **10,** 452 (1969).

241. M. B. Weinstein and F. Ashkar, *J. Nucl. Med.*, **11,** 376 (1970).

242. G. J. D'angio, M. Loken, and M. Nesbit, *Radiology*, **93,** 615 (1969).

243. R. P. Spencer, G. Montana, G. T. Scanlon, and O. R. Evans, *J. Nucl. Med.*, **8,** 197 (1967).

244. N. E. Herrera, R. Gonzalez, R. D. Schwartz, A. M. Diggs, and J. Belsky, *J. Nucl. Med.*, **6,** 792 (1965).

245. M. R. Halie, M. G. Woldring, M. A. Van Zanten, and H. O. Neiweg, *Ned. Tijdschr. Geneesk.*, **112,** 1687 (1968).

246. N. E. Herrera, R. D. Gonzalez, and R. N. Kranwinkel, *Lahey Clin. Found. Bull.*, **17,** 43 (1968).

247. W. L. Ashburn and D. G. McDonald, *J. Nucl. Med.*, **11,** 92 (1970).

248. R. R. Cavalieri, K. G. Scott, and E. Sairenji, *J. Nucl. Med.*, **7,** 197 (1966).

249. R. R. Cavalieri and K. G. Scott, *J. Amer. Med. Assoc.*, **206,** 591 (1968).

250. G. R. Ray, J. A. DeGrazia, and R. R. Cavalieri, *J. Nucl. Med.*, **11,** 354, (1970).

251. J. Esteban, D. Lasa, and S. Perez-Modrego, *Radiology*, **85,** 149 (1965).

252. S. N. Albert, C. A. Albert, E. F. Hirsch, and I. N. Brecher, *J. Nucl. Med.*, **7,** 290 (1966).

253. W. B. Nelp and F. A. Blumberg, *J. Nucl. Med.*, **6,** 822 (1965).

XIII Selenium compounds in nature and medicine

D. SELENIUM COMPOUNDS AS POTENTIAL CHEMOTHERAPEUTIC AGENTS

DANIEL L. KLAYMAN

Walter Reed Army Institute of Research
Division of Medicinal Chemistry
Washington, D.C.

1. Introduction

Anyone unfamiliar with the manner in which medicinal products are developed may very well wonder why effort has been expended to synthesize and test potential therapeutic agents containing a toxic element, selenium. Of course, the role of serendipity in drug design could be invoked as a justification for the study of these or any other compounds lacking an apparent basis for trial. A strong rationale does exist, however, for the incorporation of selenium into potential medicinal agents; this is predicated on the close resemblance of selenium to the biologically important element sulfur.

It is well established that some form of sulfur is needed by all living organisms. Among the vital biological roles in which sulfur compounds are involved are (a) the retention of configuration of the tertiary structure of certain proteins through intramolecular disulfide bonds; (b) the immune processes *via* thiol-disulfide interchange; (c) catalytic processes; (d) transport across cell membranes; (e) energy-transfer mechanisms; and (f) blood clotting.

Some of the organic sulfur compounds that have been isolated from natural sources include amino acids: cysteine, cystine, methionine, lanthionine, cystathionine, ergothioneine, taurine, djenkolic acid, and S-allylcysteine sulfoxide; a tripeptide: glutathione; vitamins: thiamine and biotin; hormones: insulin, vasopressin, and oxytoxin; and antibiotics: penicillin, cephalosporin, and lincomycin; as well as numerous enzymes. In the past few years, even compounds with trisulfide moieties have been isolated from plants.[1] Many of the natural products mentioned above enjoy extensive therapeutic usage. In addition, some of the most powerful *synthetic* drugs are also sulfur-containing. Among the best-known classes are sulfonamides (anti-infective agents); sulfonylureas (oral hypoglycemic agents); 2-thiouracils (antithyroid agents); phenothiazines (antihistamines, tranquilizers); thiobarbiturates (analgesics, anesthetics); dithiocarbamates (fungicides, treatment of chronic alcoholism); aromatic sulfones (antileprotic and antimalarial agents); BAL (antidote for heavy metal poisoning); 6-mercaptopurines (antileukemic agents); certain thiazoles (anticonvulsant agents); and saccharin (sweetener). Minor chemical variations in the structure of active drugs have been known to result in remarkable modifications in their biological properties; hence, the substitution of selenium for sulfur could, in theory, lead to a more effective agent.

The elements that constitute Group VIb of the periodic table (*i.e.*, oxygen, sulfur, selenium, and tellurium) possess the identical number of valence electrons. Of this group, only sulfur and selenium also approximate one another in size. Hence, this pair of elements may be considered to be isosteric, as originally defined by Langmuir[2] and as later redefined by Erlenmeyer.[3] The medicinal implications of the concept of isosterism have been reviewed by Schatz.[4] Many pairs of isosteric compounds form crystals having the same geometric structure, thus exhibiting isomorphism. The atomic weight of selenium itself was established in 1828, when it was observed that crystals of sodium sulfate and sodium selenate are isomorphous.[5] Although many isosteric pairs of sulfur and selenium compounds are isomorphous and form continuous mixed crystals (*cf.*, *e.g.*, Rheinboldt and Giesbrecht[6]), the two properties are not necessarily mutually dependent.

A comparison of the physical properties of the Group VIb elements (Table XIIID-1) indicates that sulfur and selenium resemble one another

Table XIIID-1   Comparison of the physical properties of oxygen, sulfur, selenium, and tellurium

ELEMENT	ELECTRONIC CONFIGURATION	PAULING ELECTRO-NEGATIVITY COEFFICIENTS	COVALENT RADII (Å)	IONIC RADII (Å) (COORDINATION NUMBER)	$\bar{D}_{X-X}$ (kcal/mole)
O	$1s^2 2s^2 2p^4$	3.5	0.73	1.40 (-2) 0.09 $(+6)$	33.2
S	$1s^2 2s^2 2p^6 3s^2 3p^4$	2.5	1.02	1.84 (-2) 0.29 $(+6)$	50.9
Se	$1s^2 2s^2 2p^6 3s^2 3p^6$ $3d^{10} 4s^2 4p^4$	2.4	1.16	1.98 (-2) 0.42 $(+6)$	44.0
Te	$1s^2 2s^2 2p^6 3s^2 3p^6$ $3d^{10} 4s^2 4p^6 4d^{10}$ $5s^2 5p^4$	2.1	1.35	2.21 (-2) 0.56 $(+6)$	33

more than do sulfur and oxygen. The Se atom is only slightly larger than the S atom and, being less electronegative, possesses somewhat more metallic character. The lower electronegativity and greater size of these two elements in relation to oxygen accounts, in part, for their lesser tendency to participate in hydrogen bonding than observed for oxygen. The Se–Se bond is weaker than the S–S bond but stronger than the O–O or Te–Te bond. The success that has been achieved in the replacement of oxygen by sulfur, notably in the thiobarbiturates, suggests that a sulfur-selenium exchange is not an illogical step to be taken in the search for improved medicinal agents. The pharmacological changes that occur upon any atomic substitution generally reflect the spatial and electronic differences of the two elements. Alterations in drug solubility, oil-water partition, polarity, therapeutic index, tolerance, basicity, stability, distribution, transport, and cross resistance are among the effects that may be seen in isosteric replacements.

Although a good case can be made for the substitution of selenium for oxygen or sulfur in chemotherapeutically active compounds, the toxic nature of most selenium compounds can be a serious obstacle. In general, compounds in which the Se atom is part of an actual or potential functional group capable of reacting with sulfhydryl groups (*e.g.*, –SeH, –SeCN, –SeSO$_3$H, –SeSe–) tend to be considerably more toxic than their sulfur isosteres (*cf.* Table XIIID-4, *e.g.*). High toxicity *per se* does not rule out a compound for drug use, but it frequently prevents the administration of an effective dose that is also safe.

On the other hand, when the Se atom is not readily accessible, as in a stable ring system such as selenophene, the toxicities of the sulfur and selenium analogs do not differ widely. In some instances, cyclic selenium compounds have been reported to be even less toxic than the comparable sulfur compounds (*vide infra*).

Although it is highly debatable whether selenium compounds are carcinogens[7] (*cf.* Chapter XIIIC), the element nevertheless suffers a bad reputation and has been listed among carcinogenic agents by the United States Food and Drug Administration in its Food Additive Amendment of 1958.[8] Thus, great suspicion is cast upon any application to use a selenium compound as a chemotherapeutic agent.

The only organic selenium compound in current medical usage is ^{75}Se-selenomethionine, which is of value in pancreatic scanning. The application of this compound is discussed in detail in Chapter XIIIC. The present chapter is concerned with other organic selenium compounds that have been considered as potential drugs.

2. Anti-infective Agents

a. Antibacterial Agents. Selenium isosteres of sulfur-containing antibacterial agents are, in some instances, as effective as their analogous sulfur compounds or even more effective. In contrast, inorganic selenium compounds, when tested at the same selenium concentration as the highly active organic selenium compounds, have been found to exhibit comparatively negligible antibacterial activity.[9]

Many of the selenium compounds studied as potential bactericides during World War II were modeled after the sulfonamide drugs[10–14] and mainly contain substituted 2-aminoselenazole or selenohydantoin[14] moieties linked to the sulfonyl group. Although no direct sulfur-selenium interchange to give the analogous selenonamides has been achieved, a first approximation may be found in bis(*p*-aminophenyl) selenide and diselenide.[9] The latter is a thousandfold more effective as an antibacterial agent than bis(*p*-aminophenyl) disulfide. Table XIIID-2 summarizes the antibacterial activity of various organoselenium compounds. Recently, methyl alkyl selenoxide hydronitrates [CH$_3$Se(=O)R·HNO$_3$] have been patented as germicides for inclusion in detergent formulations.[14a]

Table XIIID-2   *Antibacterial activity of organic selenium compounds*

TYPE OF COMPOUND	RESULTS REPORTED	REF.
$\left(H_2N-\!\!\!\bigcirc\!\!\!-Se-\right)_2$	Very effective against *Brucella paramelitensis* and *Staphylococcus aureus*.	9
$H_2N-\!\!\!\bigcirc\!\!\!-SeO_2H$	Active against *Br. paramelitensis* and *Staph. aureus*.	9
$H_2N-\!\!\!\bigcirc\!\!\!-SO_2\overset{H}{N}-\!\!\!\underset{N}{\overset{Se}{\diamond}}$	Showed same activity *in vitro* against *Pneumococcus* as sulfathiazole	11
$\left(\underset{R}{\bigcirc}-\right)_2 Se$ R = *o*- or *p*-NO$_2$	Inactive against *Streptococcus* infections.	15
$HO-\!\!\!\bigcirc\!\!\!-Se-\!\!\!\bigcirc$	Strong bactericidal activity.	16
$CH_3Se-\!\!\!\bigcirc\!\!\!-\overset{O}{\underset{}{C}}-\overset{\overset{O}{\overset{\Vert}{NHCCHCl_2}}}{\underset{}{CH}}-CH_2OH$	The selenium derivative of chloromycetin is a strong antibacterial agent, 10 times as active as the S analog.	17
(selenopurine structure)	Screened against 10 types of bacteria; exceeds the antibacterial activity of 6-mercaptopurine.	18
(1,2,5-selenadiazole diamide structure)	1,2,5-Selenadiazoles were found to exert marked *in vitro* inhibitory activity against more than 300 species of bacteria, yeasts, and fungi.	19–21

(Continued overleaf)

Table XIIID-2   (continued)

TYPE OF COMPOUND	RESULTS REPORTED	REF.

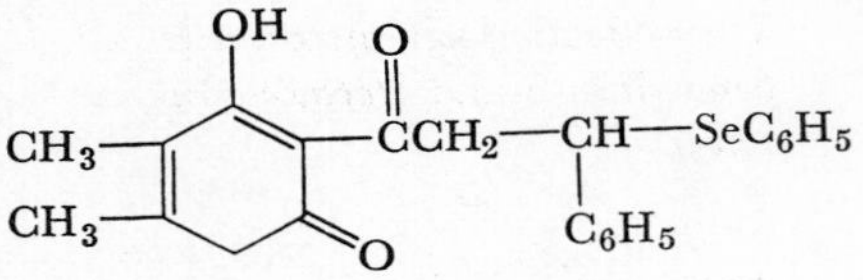

Bactericidal activity. — 22

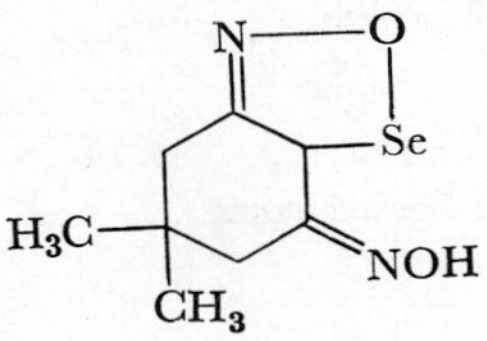

Inhibited the growth of *Myco-* — 23
bacterium tuberculosis in 1.95×10^{-5} *M* solution.

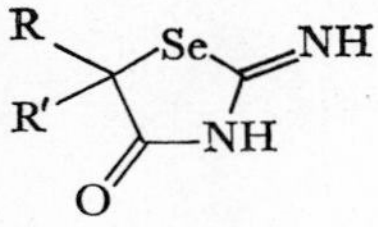

2-Iminoselenazolin-4-ones — 24
showed negligible activity
against *Staph. aureus, Strep.*
pyogenes, E. coli, and *Proteus*
vulgaris.

RSeO$_2$H

Bactericidal. — 25

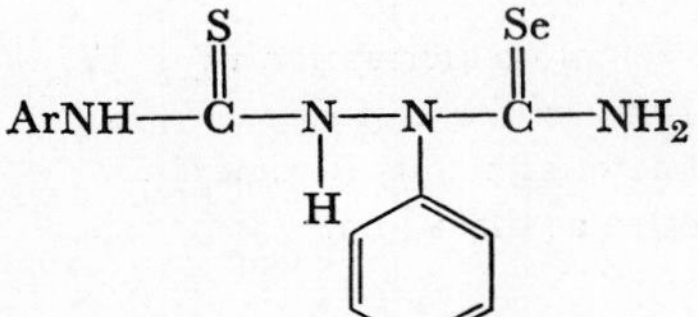

At a concentration of — 26a
100γ/ml the compounds of
this series have a weak effect on
Staphylococcus, Bacillus subtilis,
and *E. coli;* a moderate effect
on *Mycobacterium BCG, Myco-*
bacterium phlei, and *Mycobacterium*
smegmatis; and a strong effect
on *M. tuberculosis.*

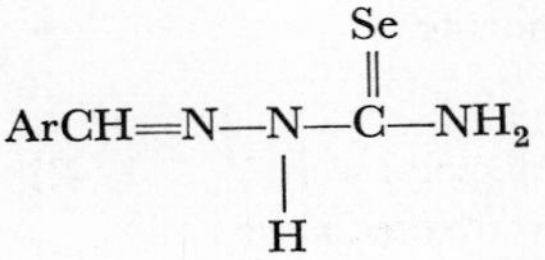

This series had a weak bacter- — 26b
iostatic effect at 100γ/ml
against *Staph. aureus, Bacillus*
subtilis, and *E. coli;* the compounds also inhibit growth in
Mycobacterium phlei and *Myco-*
bacterium smegmatis; all completely inhibit the growth of
M. tuberculosis R$_v$/47.

Selenium analogs of sulfur-containing antibiotics are unknown.* For example, the synthesis of "selenopenicillin," in which the S atom in the thiazolidine ring is replaced by a Se atom, has yet to be reported. Selenium and sulfur derivatives of chloromycetin have been prepared in which the p-nitro group of the antibiotic is replaced by Se-methyl and S-methyl groups (cf. Table XIIID-2).[17]

b. Antifungal Agents. Selenium-containing compounds that have been tested for their bactericidal properties have frequently been screened simultaneously as potential fungicides. Some aminoselenadiazoles have been found to possess both broad-spectrum antifungal and antibacterial activity (cf. Table XIIID-3).[20] Inorganic materials such as elemental selenium, hydrogen selenide, and selenite have been demonstrated to be poor fungicides.

n-Butyl-, n-hexyl-, and n-decyl selenocyanates were noted to be more fungicidal than the isosteric thiocyanates, but their "disgusting" odor was cited as a hindrance to their use.[27] A series of aralkyl and aryl selenocyanates, as well as alkyl esters of selenocyanatoacetic acid, was tested in conjunction with 450 organic sulfur compounds against ten organisms. Good fungicidal activity was reported for most of the selenium compounds.[28]

In comparing the antifungal activities of phenylselenourea and 2-phenyl-selenosemicarbazide with those of their sulfur and oxygen analogs, Mautner et al.[29] observed that the selenium compounds had 10–1000 times greater effectiveness on a molar basis than the sulfur compounds, while the oxygen analogs exhibited negligible activity. It was suggested that the superiority of the selenium compounds was derived from their ability to chelate with metal ions, thus forming complexes which, in themselves, may be active fungicides.

Thia- and selenacarbocyanine dyes, although moderately fungicidal, were less effective than the oxygen analogs in the test system used.[30]

The steroid 6β,19-seleno-Δ^{1,4}-cholestadiene-3-one (**1**) has been suggested for use as an agricultural fungicide.[32]

1

---

* It is not clear what relationship exists, if any, between selenium and the new antibiotic, selenomycin [C. Coronelli and J. Thiemann, Ger. Pat. 2,028,986; Chem. Abstr., **74,** 52132 (1971)].

Table XIIID-3   Antifungal activity of organic selenium compounds

TYPE OF COMPOUND	RESULTS REPORTED	REF.
	Neither the Se nor the S analog of the keto derivative of chloromycetin had activity against *Aspergillus niger;* however, the O analog was active.	17
	1,2,5-Selenadiazoles had marked broad-spectrum antifungal activity.	20,21
	These phenylselenosemicarbazide derivatives were tested against *Trichophyton metagrophytes, Botrytis cinerea, Monilia fructagena,* and *Penicillium notatum;* good antifungal activity was observed for most of the compounds.	29
	Selenocarbocyanines were fungitoxic against *Venturia inaequalis, Botrytis cinerea,* and *Fusarium bulbigenum.*	30,31

c. Antiparasitic Agents. Ehrlich and Bauer[33] in 1915 tested the dye 3,6-diaminoselenopyronine chloride (**2**) as a trypanocidal drug in mice. The

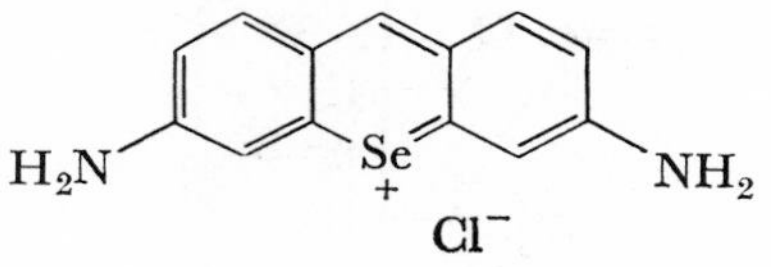

$$H_2N \quad \underset{\underset{Cl^-}{+}}{Se} \quad NH_2$$

2

compound produced only transitory healing effects, which were accompanied by marked edema.

Seven selenium compounds have been tested against strains of *Leptospira* and found to have activities in the range of 5 μg/ml *in vitro* and 5 mg/kg in hamsters.[34] Selenourea was found to exert greatest activity towards *L. australis*, *L. autumnalis*, *L. bataviae*, and *L. pomona*, with less towards *L. canicola*, *L. griptyphosa*, and *L. icterohaemorrhaiae*.

The compounds bis(*p*-hydroxyphenyl) selenide (**3**) and bis(*p*-hydroxyphenyl) diselenide (**4**) were found by Keimatsu and Yokota[35] to have

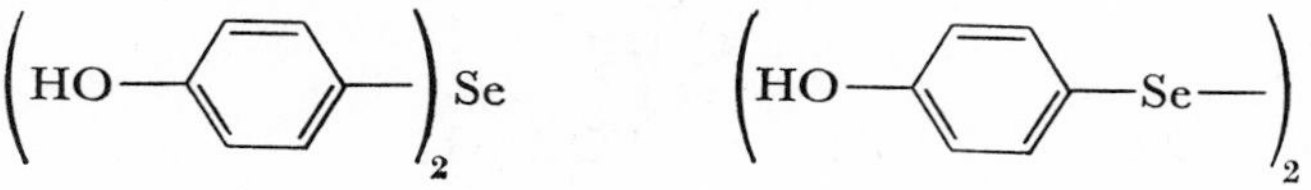

$$\left(HO - \bigodot - \right)_2 Se \qquad \left(HO - \bigodot - Se - \right)_2$$

3   **4**

150 times and 100 times, respectively, greater antiseptic power towards *Paramecium caudatum* than phenol. In a series of 4-substituted 2,6-diamino-pyridines, the 4-phenylseleno derivative (**5**) was among the most active

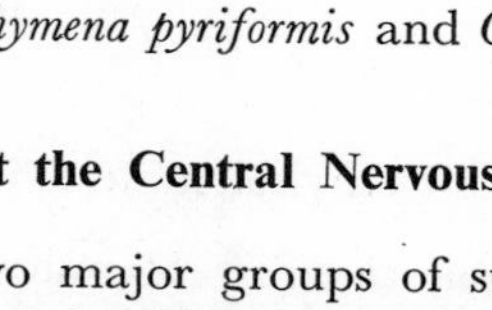

$$SeC_6H_5$$

$$H_2N \quad N \quad NH_2$$

5

against the protozoa *Tetrahymena pyriformis* and *Crithidia fasciculata*.[36]

3. Compounds That Affect the Central Nervous System

a. Hypnotics. Of the two major groups of sulfur-containing hypnotics (*i.e.*, the thiobarbiturates and the aliphatic disulfones) only the former are in current usage. The 2-thio-5,5-dialkyl barbiturates (*e.g.*, pentothal sodium), when administered intravenously in small doses, produce anesthesia of short duration (5–30 minutes). Mautner and Clayton[37] have reported the synthesis of 2-selenobarbituric acid (**6**, R = R′ = H), as well as of several

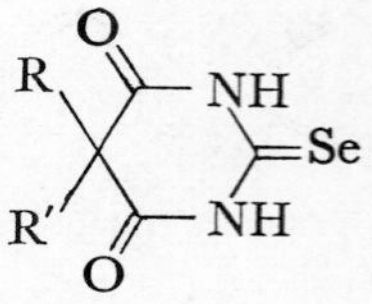

6

5-substituted and 5,5-disubstituted derivatives. On comparison with the sulfur and oxygen isosteres, their water-lipid partitions and found to more closely resemble those of the sulfur analogs. Apparently the 2-selenobarbiturates have not been tested physiologically.

The hypnotic activity of aliphatic disulfones has been known since the 1890's. The most active types are the ketone disulfones, exemplified by sulfonal (**7**); however, because of their toxicity they have been replaced by

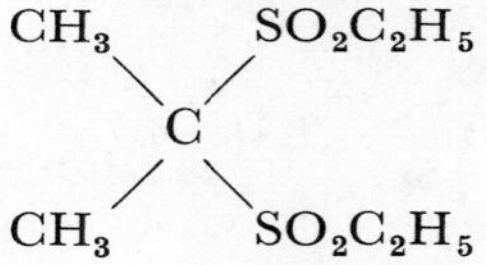

7

modern hypnotics. Shaw and Reid[38] attempted to prepare the selenium analog of sulfonal (**8**) by the nitric acid oxidation of the selenoketal shown,

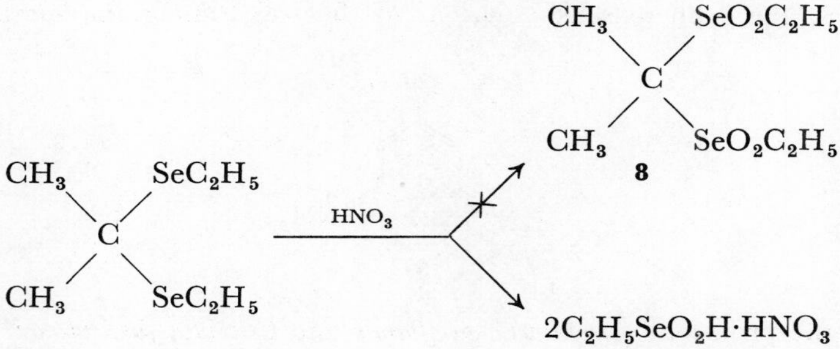

8

9

but the reaction gave instead ethylseleninic acid as its nitric acid salt (**9**).

b. Analgesics and Local Anesthetics. Hannig has been responsible in the past decade for the development of numerous selenium-containing analgesics. In 1963 he reported the synthesis of selenium derivatives of phenylbutazone (**10**), a potent antirheumatic, analgesic, and antipyretic, which

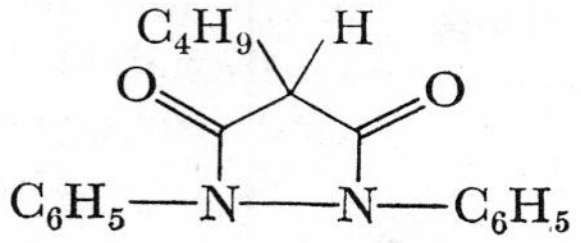

$$
\begin{array}{c}
\text{C}_4\text{H}_9 \quad \text{H} \\
\text{O} \qquad \text{O} \\
\text{C}_6\text{H}_5\text{—N} \quad \text{N—C}_6\text{H}_5
\end{array}
$$

10

were benzoylethylated at the 4-position (**11**).[39] Selenium analogs of the local

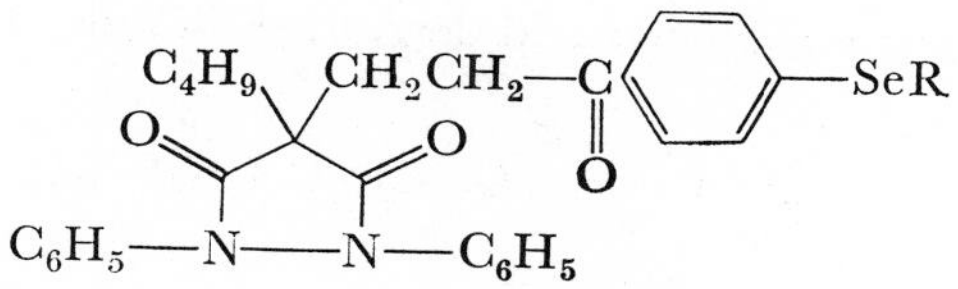

$$
\begin{array}{c}
\text{C}_4\text{H}_9 \quad \text{CH}_2\text{CH}_2\text{—C} \text{—SeR} \\
\text{O} \qquad \text{O} \qquad \text{O} \\
\text{C}_6\text{H}_5\text{—N} \quad \text{N—C}_6\text{H}_5
\end{array}
$$

11

anesthetic Falicaine (**12**, β-piperidino-*p*-propoxypropiophenone) were also

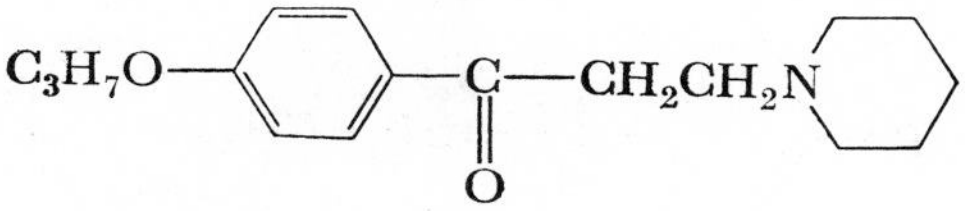

$$
\text{C}_3\text{H}_7\text{O—} \bigcirc \text{—C—CH}_2\text{CH}_2\text{N} \bigcirc
$$
$$
\text{O}
$$

12

synthesized; in these, the propoxy group was replaced by alkylseleno (RSe—)
groups in which R ranged from methyl through *n*-hexyl (**13**).[40] The selenium

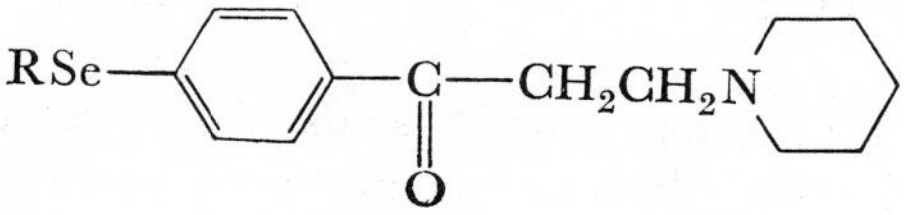

$$
\text{RSe—} \bigcirc \text{—C—CH}_2\text{CH}_2\text{N} \bigcirc
$$
$$
\text{O}
$$

13

analogs were less toxic than Falicaine. The results of topical anesthesia
experiments on the rabbit cornea showed that activity increased as R was
extended. The butyl, pentyl, and hexyl homologs were more active than
Falicaine and less toxic. Additional selenium-containing analogs of **12** were
also made in which the piperidine ring was expanded to a seven-membered
hexamethylenimino heterocycle (**14**).[41] Several compounds of the latter

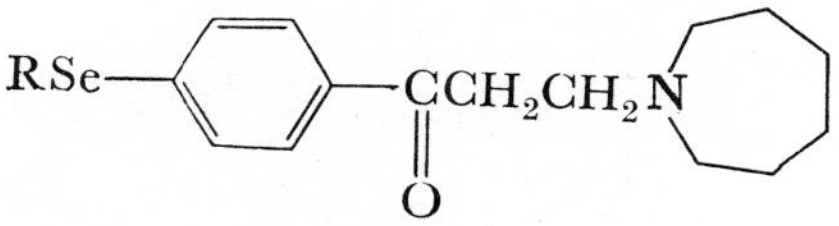

$$
\text{RSe—} \bigcirc \text{—CCH}_2\text{CH}_2\text{N} \bigcirc
$$
$$
\text{O}
$$

14

type have been found to be capable of inhibiting the incorporation of ^{14}C-leucine into lymphocytes from peritoneal exudate of the rat.[41a] Acetophenone-like variants of **14**, having the general structure **15**,[42] were also prepared,

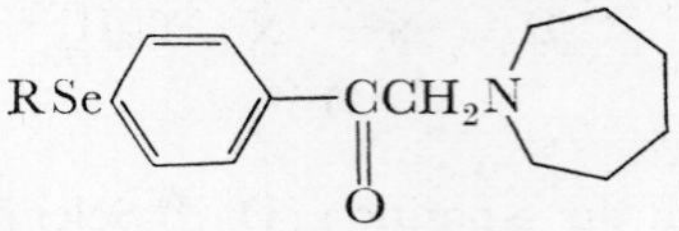

$$RSe-\underset{\underset{\displaystyle O}{\|}}{C}CH_2N\bigcirc$$

15

as well as 4-alkylselenomandelic acid derivatives of type **16**.[42]

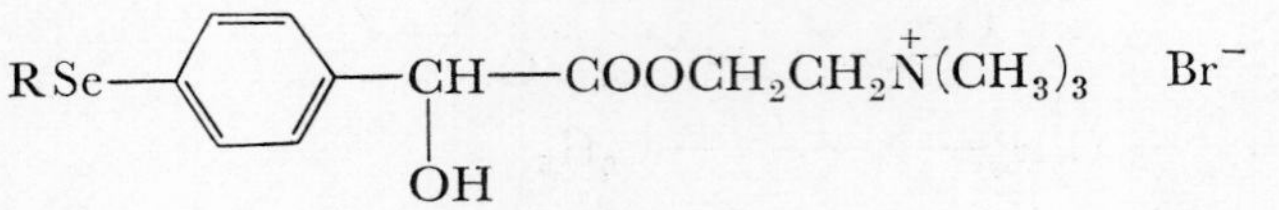

$$RSe-\underset{\underset{\displaystyle OH}{|}}{C}H-COOCH_2CH_2\overset{+}{N}(CH_3)_3 \quad Br^-$$

16

Bekemeier *et al.*[43] studied some *p*-alkylseleno-α,β-hexamethyleniminopro-piophenones (**17**) and noted slightly greater toxicity with these compounds

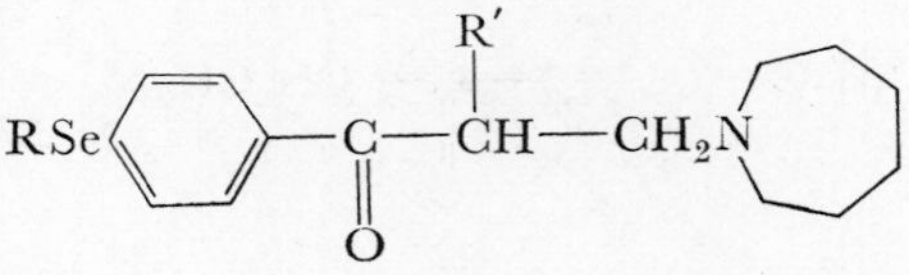

$$RSe-\underset{\underset{\displaystyle O}{\|}}{C}-\underset{\underset{\displaystyle R'}{|}}{C}H-CH_2N\bigcirc$$

17

than with the oxygen analog but less toxicity than with Falicaine (**12**). In general, the derivatives have a lesser local anesthetic effect than Falicaine but were reported to be effective against electroshock and/or pentylenetetrazole convulsions.

Many esters of substituted benzoic acids (*e.g.*, procaine) and esters of 4-alkylthiobenzoic acids show excellent local anesthetic properties.[44] Hannig and Zeibandt[45] prepared basic esters of 4-(alkylseleno)benzoic acids (**18**) and 3-[4-(alkylseleno)benzoyl]propionic acids (**19**) but have not reported test results.

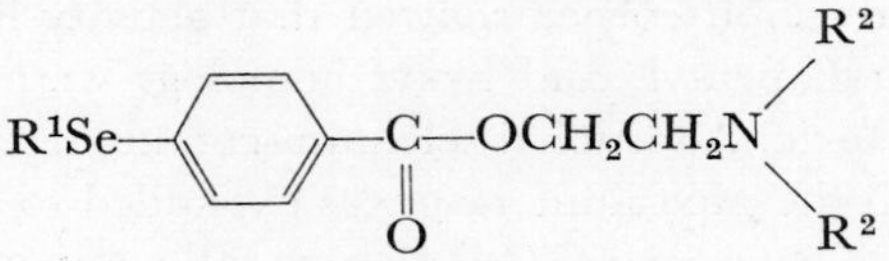

$$R^1Se-\underset{\underset{\displaystyle O}{\|}}{C}-OCH_2CH_2N\underset{\displaystyle R^2}{\overset{\displaystyle R^2}{<}}$$

18

$$R^1Se-\underset{\underset{\displaystyle O}{\|}}{C}-CH_2CH_2-\underset{\underset{\displaystyle O}{\|}}{C}-OCH_2CH_2N\underset{\displaystyle R^2}{\overset{\displaystyle R^2}{<}}$$

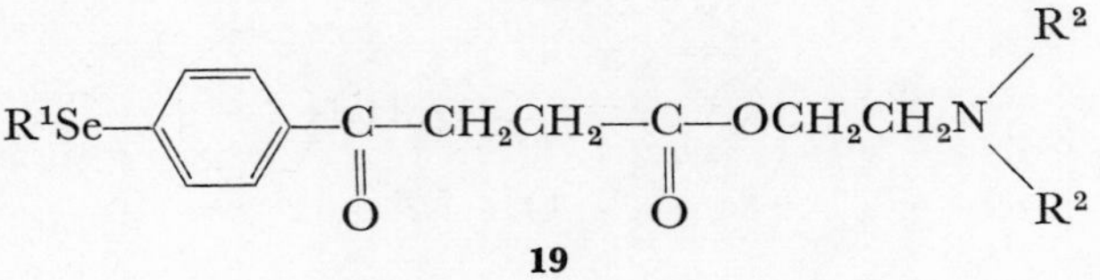

19

c. Tranquilizing Drugs. During the past 20 years a large number of phenothiazine derivatives, such as promazine (**20**) and chlorpromazine (**21**), have found use as psychopharmacologic agents as well as antiemetics, antihistamines, anthelmintics, and antimalarials.

Phenoselenazines, first prepared by Müller *et al.*,[46] have been reported by Glassman and his co-workers[47] to have lower psychopharmacologic potency, but otherwise to retain the characteristics of promazine and chlorpromazine in regard to their cardiovascular, adenolytic, antihistaminic, and hypoglycemic actions. Craig *et al.*[48-50] found that 10-substituted phenoselenazines are effective as central nervous system depressants. Cordella and Sparatore[51,52] reported the synthesis of 2,10-disubstituted phenoselenazines but gave no pharmacologic data.

Protiva and his group[52] have developed tricyclic psychotropic drugs that are derivatives of 6,11-dihydrodibenzo[b,e]thiepin. These sulfur-containing compounds are reported to be effective antagonists of reserpine and to have good antidepressant activities. The 6,11-dihydrodibenzo[b,e]selenepins (**22**, R = H, C_2H_5; and **23**), prepared by methods used to make the sulfur

analogs, were found to resemble the above-mentioned thiepins pharmacologically.[54] Compound **23** had particularly high antireserpine activity. Šindelar *et al.*[55] have also prepared selenium analogs of substituted thioxanthenes, the latter being a class of compounds which had been reported earlier to posssess central nervous system depressant and antiemetic properties.[56,57] The compound *cis*-2-chloro-9-(3-dimethylaminopropylidene)-selenoxanthene (**24**) showed the greatest central nervous system depressant activity but was less effective than its sulfur analog.[55] Related compounds exhibited antihistamine, antireserpine, and low central nervous system depressant activity.

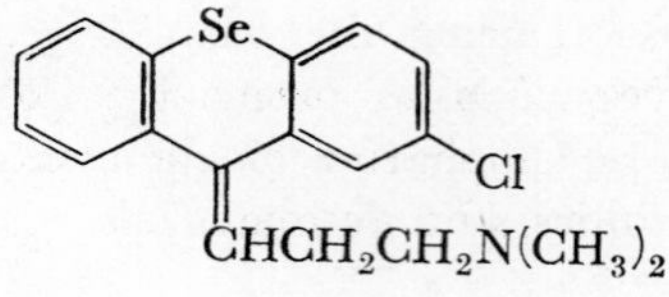

24

The 8-unsubstituted or 8-chloro-10-(4-methylpiperazino)-10,11-dihydro-dibenzo[b,f]selenepins (**25**) showed less central nervous system depressant

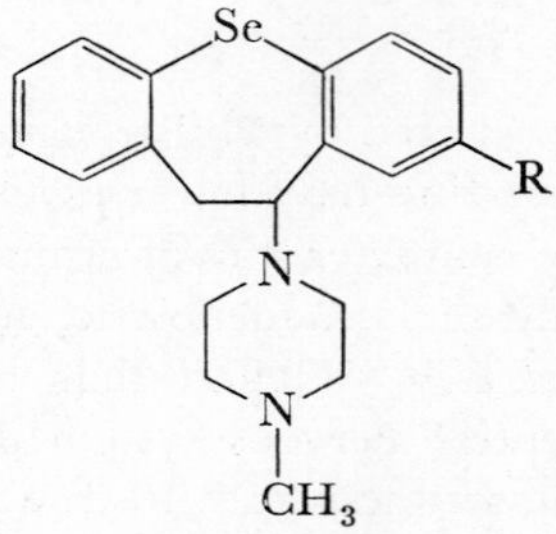

(R = H, Cl)

25

activity than their corresponding sulfur analogs,[58] which are reported to be active neuroleptics.[59] Interestingly, the oral toxicity in mice of the selenium analog of the 8-chloro derivative was lower than that of the sulfur analog. Enamine analogs of **25** in which the selenium-containing ring is unsaturated have been found to be highly active central nervous system depressants.[59a]

In the Soviet Union there has been considerable interest in studying the pharmacological properties of selenophene derivatives of the type shown in structure **26**. These compounds, for the most part, were developed by the

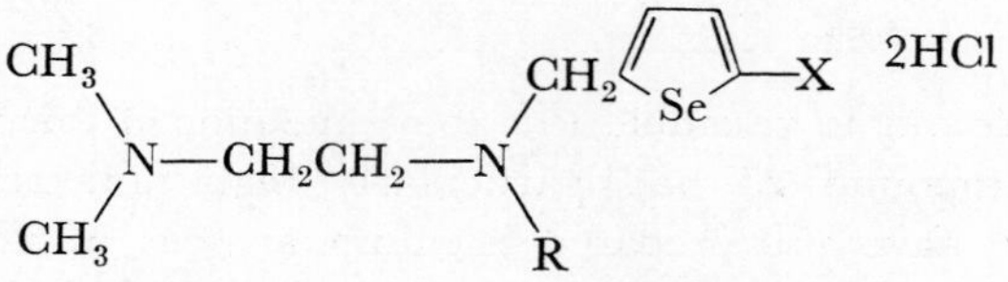

26

late Yu. K. Yur'ev and by N. N. Magdesieva of Moscow State University, who have specialized in the investigation of the chemistry of selenophene. The compounds Se-5 (**26**, R = α-pyridyl, X = Cl) and Se-6 (**26**, R = α-pyridyl, X = Br) have been examined for their neurotropic and psychotropic activity.[60] The toxicity of this class is relatively low; for example,

LD_{50} in mice, for Se-1 (**26**, R $=$ C_6H_5, X $=$ H) $=$ 160 mg/kg; for Se-5 $=$ 340 mg/kg; and for Se-6 $=$ 215 mg/kg.[61] The effect of Se-6 on the pregnancy of rats[62] and on their estrous cycle[63] and the influence of Se-5 on frog heart[64] have also been studied. In addition, selenophene derivatives have been investigated for their antihistamine activity (*vide infra*).

4. Compounds That Affect the Autonomic Nervous System

The compound *o*-carboxybenzeneseleninic acid (**27**) was reported by Limongi

$$\text{—SeO}_2\text{H}$$
$$\text{—COOH}$$

27

and his co-workers to exhibit adrenergic[65] and cholinergic[66] blocking activity. It can reverse the pressor effect of epinephrine in the dog, can inhibit the action of ephedrine on the cat's nictitating membrane, and can protect mice from lethal doses of epinephrine up to 48 hours after administration.

Chemical mediation in the transmission of nerve impulses is believed to involve (1) the liberation by autonomic nerve endings of acetylcholine (**28**)

28

and its reversible combination with receptor sites; (2) the breakdown by acetylcholinesterase of some of the acetylcholine into choline and acetic acid; and (3) the resynthesis of the acetylcholine. Selenium and sulfur analogs of acetylcholine, in which the chalcogen replaces either or both of the choline oxygen atoms,* have been extensively studied by Mautner and his associates. Günther and Mautner[67a,b] have prepared selenocholine iodide (**29**, R $=$ H)[67b] and derivatives (**29**, R $=$ Ac, Bz, PrCo), as well as seleno-choline diselenide and some *N*-methyl and *N,N*-dimethyl analogs. These compounds have been investigated in various test systems, such as the guinea pig ileum,[68] frog rectus abdominis,[68] phrenic nerve-stimulated rat diaphragm,[69] and isolated single-cell *Electrophorus* electroplax preparation.[71,72a,b]

---

* The closest analog of acetylcholine in which the N atom is replaced by a chalcogen is trimethylsulfonium iodide. The latter was found to have lower muscarinic activity than tetramethylammonium iodide [R. Hunt and R. R. Renshaw, *J. Pharmacol. Exptl. Therap.*, **25**, 315 (1925)]. The related selenium isolog does not appear to have been investigated for this property.

$$CH_3 \overset{\underset{\displaystyle CH_3}{|}}{\overset{+}{\underset{|}{N}}} -CH_2CH_2SeR$$
$$\underset{CH_3}{}$$

29

When 2-dimethylaminoethyl benzoate (**30**, A = O) was compared with

$$(CH_3)_2NCH_2CH_2-A-\overset{\displaystyle O}{\overset{\|}{C}}-C_6H_5$$
$$(A = O, S, Se)$$

30

its sulfur and selenium isologs for the ability to block the electrical activity of squid axons,[72a,b] the sulfur isolog was found to have about 10 times and the selenium isolog about 30 times the potency of the oxygen isolog. Replacement of the carbonyl group with a thiocarbonyl gave agents with enhanced activity. All compounds of the series were also tested at the synaptic junctions of the electroplax. Their relative activities were again in the order Se > S > O, suggesting a similarity in the receptor sites.[70,71]

The hydrolysis and aminolysis of the benzoyl derivatives of choline, thiocholine, and selenocholine and of their 2-dimethylaminoethyl analogs as they are influenced by neighboring groups have also been investigated over a wide pH range.[73] It was observed that hydrolysis of the esters occurred at similar rates, but that aminolysis at pH 8 took place at a considerably higher rate for the selenol ester than for the thiol ester while benzoylcholine was unreactive.

The carbonyl oxygen atom of benzoylcholine and its sulfur and selenium derivatives has been replaced by a thiocarbonyl function (*cf.* **30a**) and studied

$$CH_3 \overset{\underset{\displaystyle CH_3}{|}}{\overset{+}{\underset{|}{N}}} -CH_2CH_2-A-\overset{\displaystyle S}{\overset{\|}{C}}-C_6H_5$$

$$(A = O, S, Se; R = H, CH_3)$$

30a

in the single-cell electroplax preparation,[74] as well as in squid giant axons.[72a] Again, the selenium isologs showed the greatest electrical inhibitory activity in both the synaptic and the axonal preparations.

Numerous sulfur-containing derivatives of phosphoric, phosphonic, and pyrophosphoric acids have been found to be active inhibitors of cholinesterase.[74a,74b] Åkerfeldt and Fagerlind[74c] have synthesized four related selenium organophosphorus compounds that exhibit potent inhibition of cholinesterase when tested in human erythrocyte enzyme. The most toxic was O-ethyl Se-(2-diethylaminoethyl)ethylphosphonoselenoate (Selenophos, **30b**) which has an extremely low LD_{50} of 0.021 mg/kg when administered

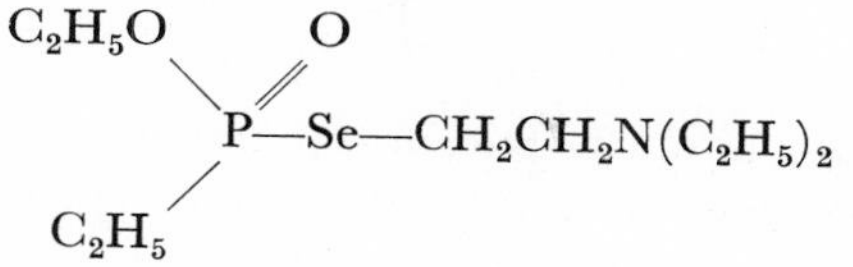

$$C_2H_5O \quad O$$
$$\diagdown \quad \diagup\!\!\!\diagup$$
$$P\!\!-\!\!Se\!-\!CH_2CH_2N(C_2H_5)_2$$
$$\diagup$$
$$C_2H_5$$

30b

to mice subcutaneously. In a subsequent study, Hoskin and co-workers[74d] found that Selenophos, when applied for 30 minutes in concentrations as high as 5×10^{-3} M, did not block conduction in the squid giant axon unless the latter was pretreated with cotton mouth moccasin venom. Then, **30b** caused a marked and irreversible reduction in the action potential.

5. Compounds That Affect the Circulatory System

The compound known as Se-5 (**26**, R = α-pyridyl, X = Cl) was found to be the most potent of four selenophene derivatives studied in a test for anti-arrhythmic activity on isolated rat atria.[75] Not only was Se-5 effective at a lower dose than procaine amide but it also had a longer duration of action.

1-Phenyl-2-aminoethanols of the type shown in structure **31**, where

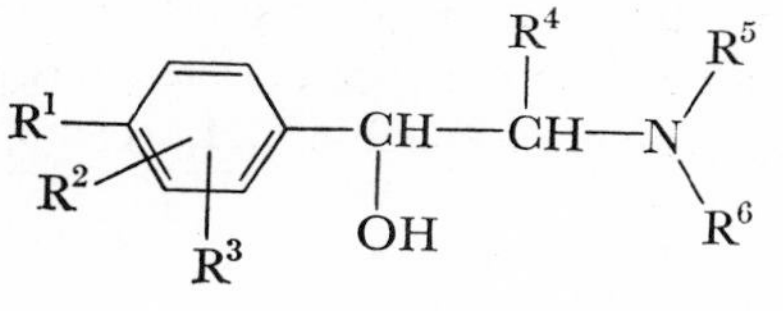

$$R^4 \qquad R^5$$
$$R^1 \qquad | \qquad |$$
$$R^2 \qquad CH\!-\!CH\!-\!N$$
$$R^3 \qquad | \qquad |$$
$$OH \qquad R^6$$

31

R^1 may be RSe or $RSeO_2$, have peripheral vasodilating activity and have been suggested for use in the treatment of cardiac arrhythmias.[76]

6. Anti-inflammatory Compounds

The anti-inflammatory properties of low concentrations of γ,γ'-diseleno-divaleric acid, benzylselenovaleric acid, and selenocystine were observed *in vitro* by Roberts.[77] The compounds β,β'-diselenodipropionic acid and diphenyl selenide, however, were ineffective.

Phillips *et al.*,[78] in assessing compounds for anti-inflammatory activity, found that 42 of them were more active than acetylsalicyclic acid. Among the active compounds was 5,5'-selenobissalicyclic acid (**32**), which was observed

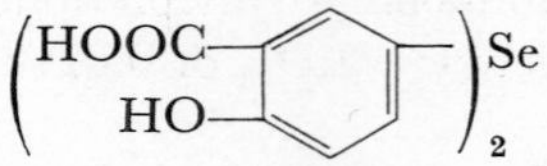

$$\left(\begin{array}{c} \text{HOOC}\!-\!\!\!\!\!\\ \text{HO}\!-\!\!\!\!\! \end{array} \right)_2 \text{Se}$$

32

to be active in *in vitro* but not *in vivo* anti-inflammatory assays.

Penicillamine (**33**) has been studied for the treatment of rheumatoid

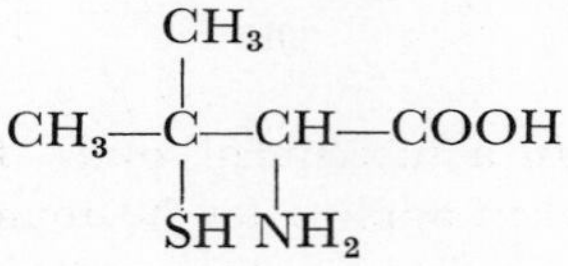

$$\begin{array}{c} \text{CH}_3 \\ | \\ \text{CH}_3\!-\!\text{C}\!-\!\text{CH}\!-\!\text{COOH} \\ | \quad | \\ \text{SH} \;\; \text{NH}_2 \end{array}$$

33

arthritis.[79] Its action may stem from its ability to chelate copper, the mechanism held responsible for its great efficacy in the treatment of chronic copper toxicity found in Wilson's disease. It would be of interest to compare its effectiveness with that of the recently synthesized selenopenicillamine,[79a] which would also be expected to be a strong chelating agent.

7. Antihistamines

The two important classes of sulfur-containing antihistamines are the 10-substituted phenothiazines (exemplified by the powerful drug Promethazine) and the ethylenediamine derivatives substituted by a 2-thenyl moiety. Only one selenium isostere of the former class has been examined: 10-(3-dimethylaminopropyl)phenoselenazine (**34**, selenopromazine).[80] Rosell

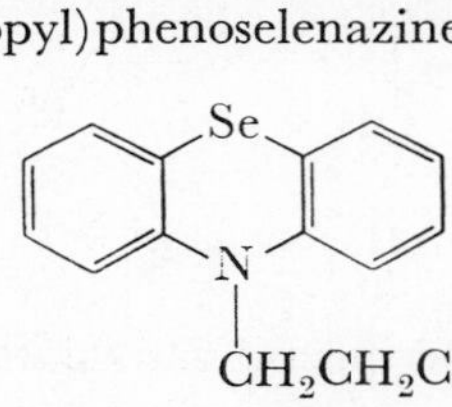

$$\text{CH}_2\text{CH}_2\text{CH}_2\text{N(CH}_3)_2$$

34

and Axelrod,[80] in studying the uptake of noradrenalin-^{3}H by rat heart, noted that **34** has greater antihistaminic activity than promazine and anti-adrenalin action of about the same order of magnitude..

The second group of sulfur-containing antihistamines may be exemplified by methapyrilene (**35**, Thenylene), whose activity is increased and toxicity lowered by chlorination of the thiophene nucleus.[81] As has been mentioned earlier, selenium isosteres of methapyrilene-type compounds (**26**) have

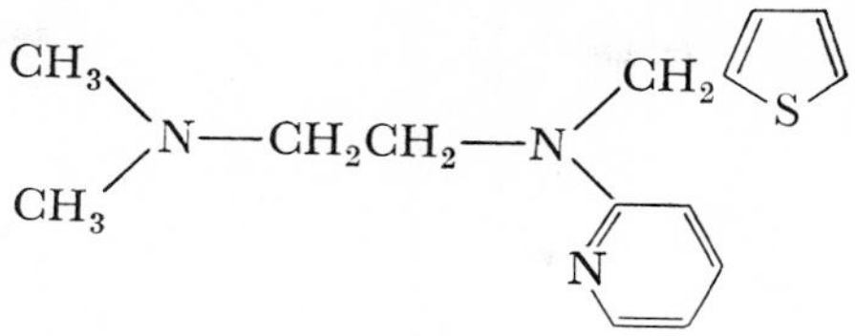

35

occupied the attention of numerous investigators in the Soviet Union because of their diverse properties. The compounds Se-4 (**26**, R = α-pyridyl, X = H),[82-85] Se-5 (**26**, R = α-pyridyl, X = Cl), Se-6 (**26**, R = α-pyridyl, X = Br)[82,83,86] and Se-1 (**26**, R = phenyl, X = H) were reported to be 2–5 times as active antihistamines as their sulfur analogs, and the duration of their action was 12 times longer.[87] All were found to inhibit deamination of tyramine and dopamine.[88] In addition, Se-4 showed preventive action against traumatic shock at 1–5 mg/kg in dogs.[89]

The compound Se-5 showed the highest activity against histamine-induced cardiac impairment in the guinea pig[82] and has been recommended for use in heart surgery;[90] Se-6 was found to be useful in the treatment of toxemia induced in pregnant rabbits by the administration of placental extracts from women with late toxicosis.[91] Such rabbits bore fetuses with fewer abnormalities if given Se-6.[92]

8. Anticancer Agents

Some types of cancer, a class of diseases marked by abnormal growth of certain tissues, have shown good response to treatment by numerous purine and pyrimidine antagonists. The most useful, thus far, in the chemotherapy of certain leukemias has been the powerful drug 6-mercaptopurine (**36**,

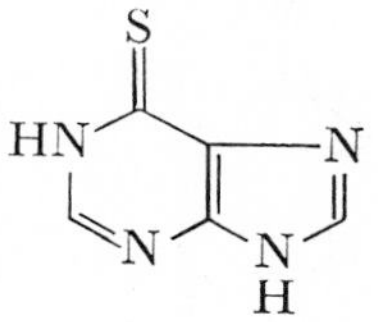

36

6-MP), discovered by Elion *et al.*[93] Mautner[94] first prepared its selenium isostere, 6-selenopurine (**37**, 6-SP), which was found[95] to inhibit mouse leukemia L1210 as effectively as 6-MP. When administered to mice having leukemia L5178Y or sarcoma 180, however, it was not as effective as 6-MP and was more toxic. The incorporation of formate and, to a lesser degree, adenine in the Ehrlich ascites tumor cell system is inhibited by

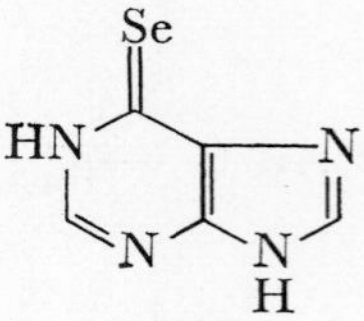

37

6-SP and 6-MP.[18] Later studies[96] showed that 6-SP, unlike 6-MP, is unstable at body temperature and has a half-life of only 6 hours, making it a less satisfactory compound from the stability standpoint. The S- or Se-methylation of 6-MP or 6-SP diminished the antitumor activity.[95] Mautner and his co-workers[97] have also synthesized selenoguanine (**38**), selenocytosine (**39**), its 5-methyl derivative, and diselenothymine (**40**). Thioguanine and its

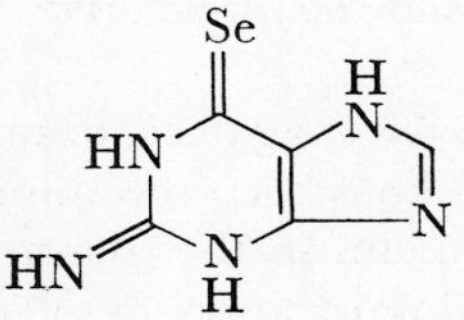

38 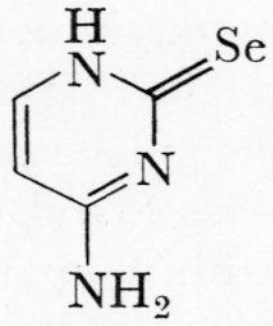   **39** 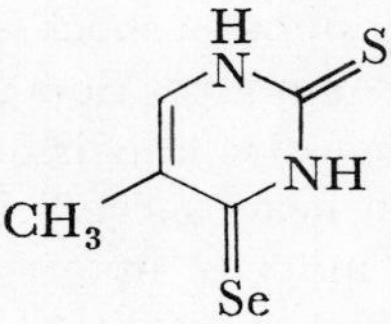   **40**

selenium analog, **38**, showed comparable antitumor activities against mouse neoplasms.

Chu[98] has reported that 6-selenoguanosine (**41**), independently synthesized by Townsend and Milne,[99] had activity approximately equal to

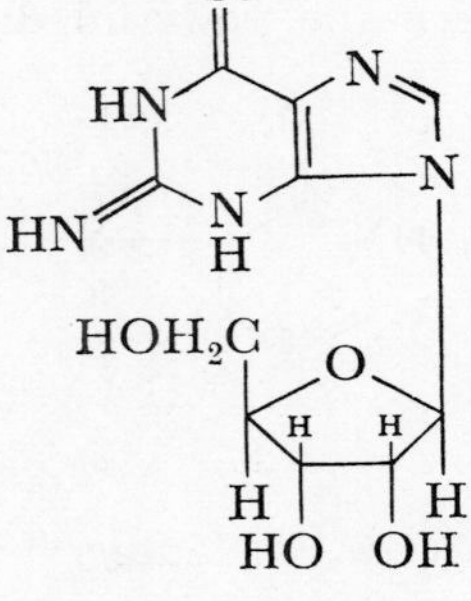

41

thioguanine in mice bearing L5178Y lymphomas. The Se-methyl derivatives of selenoguanosine and selenoguanine were less active than thioguanine. The toxicities of several other Se-methyl derivatives of 6-selenopurines were

observed to be 4–6 times greater than those of the selenopurines themselves.[100] It was suggested that the *in vivo* formation of methaneselenol might
be responsible for this manifestation. Carr *et al.*[101] have synthesized 8-
selenapurines (**42**) as potential purine antimetabolites, but the results have
been disappointing in trials against sarcoma 180.

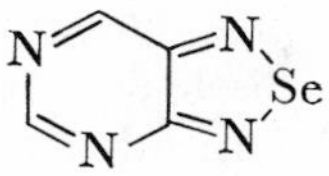

42

Some 9-ethoxycarbonyl derivatives of 2-amino-6-selenopurine and their
sulfur isosteres were found to be virtually inactive as antitumor agents.[102]

Shealy and his co-workers[20,21] have reported that some 4-amino-1,2,5-
selenadiazoles (**43**) are highly cytotoxic to KB cells in culture. The related

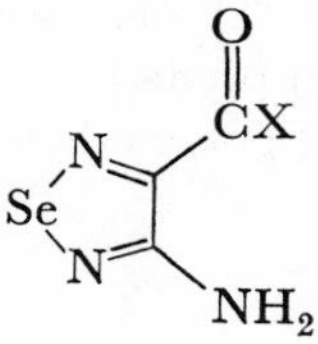

43

pyrimido[4,5-c][2,1,3]selenadiazoles were found to be ineffective against
mouse leukemia L1210.[103] Other selenium-containing heterocycles have also
been tested. Takeda *et al.*[104] reported that, of ten selenium compounds
screened, compound **44** was the most active against Ehrlich ascites

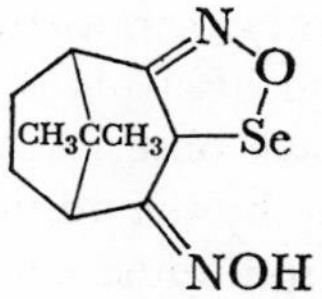

44

tumor cells, to the extent of being more effective than nitrogen mustard
N-oxide. Lack of activity was reported by other workers, however, for
2-phenylbenzoselenazole-4'-arsonic acid[105] and 3-methyl-2-benzoselenazo-
lium iodide.[106]

Inasmuch as leukemic leukocytes have been shown to require L-cysteine
and L-cystine for continued cell growth, Weisberger and Suhrland[107]

investigated selenocystine as an agent that might arrest their multiplication. The compound was administered orally to two patients with acute leukemia and to two patients with chronic myeloid leukemia in a daily average dose of 100 mg. The result was a rapid decrease in the total leukocyte count and in spleen size. One patient, resistant to 6-MP, appeared to reacquire sensitivity. Side effects, notably nausea and severe vomiting, made it impossible to administer the selenocystine long enough to ascertain whether a remission could be obtained. Diphenyl diselenide, given to one patient, failed to decrease the leukocyte count.[107] In other studies it was found that selenocystine inhibits the incorporation of ^{35}S-L-cystine by rat Murphy lymphosarcoma tumor cells *in vitro* and *in vivo*[108] and is capable, as is selenocystine, of decreasing by *ca.* one-half the incorporation of the ^{35}S-L-cystine into leukemic leukocytes.[109] Diphenyl diselenide and inorganic selenium compounds were found to be ineffective in the latter system.

Because of the antileukemic activity obtained with N-hydroxyurea, Adamson[110] tested a series of urea-related compounds, including selenourea, against mouse leukemia L1210. The latter compound did not increase the survival time of the animals. In preliminary tests in mice and rats, Senning and Sørensen[111] found that α,ω-polymethylenebisselenouronium dibromides (**45**) were carcinostatic.

$$
\begin{array}{cc}
\overset{\displaystyle +NH_2}{\underset{\displaystyle \parallel}{}} & \overset{\displaystyle +NH_2}{\underset{\displaystyle \parallel}{}} \\
H_2N-C-Se-(CH_2)_x-Se-C-NH_2 \cdot 2Br^- \\
(x = 1-6)
\end{array}
$$

45

Steroids bearing various selenium moieties in the 3-position were tested against Dunning leukemia at 10 mg/kg;[112] however, the survival time of the test animals was not significantly increased. Matti[15] tested some *o*- and *p*-substituted diphenyl mono- and diselenides but found no activity.

Clayton and Baumann[113] have observed a reduction in the incidence of induced hepatic tumors in mice by the addition of 5 ppm of sodium selenite to the animals' diets, whereas sodium selenide, when fed to mice, was noted by Shamberger[114–116] to decrease the number of chemically induced skin tumors. This work has been confirmed by Riley.[117]

Shamberger and Frost[118] have compared sex- and age-adjusted cancer death rates for certain states in the United States and provinces in Canada with the selenium contents of their forage crops and the selenium blood levels. Where the average forage crop selenium concentration was $\geqslant 0.06$ ppm, the cancer death rates for 1965 were found to be lower than in localities where the level was $\leqslant 0.05$ ppm. The inverse relationship between selenium blood levels and cancer death rates suggested that an adequate selenium intake resulted in a diminished cancer death rate. Frost[119] has proposed that selenium may function in an anticarcinogenic manner, through either its protective association with SH groups or its possible role in the regulation of protein synthesis.

9. Antiradiation Agents

The development of a drug that, when given before exposure to lethal ionizing radiation, prolongs the life of the exposed organisms has been the goal of numerous groups throughout the world almost since the beginning of the atomic age. Virtually all effective antiradiation agents contain sulfur and are variants of 2-aminoethanethiol, $H_2NCH_2CH_2SH$.

The first selenium-containing compound tested for radioprotective activity was selenophenol. When administered to mice i.p. (0.2 or 0.4 mg/g) in olive oil 10 minutes before X-irradiation, this compound was found by Bacq and his co-workers[120] to be ineffective.

Shimazu and Tappel[121,122] compared the *in vitro* radioprotective abilities of the selenium-containing amino acids DL-selenomethionine and DL-selenocystine with those of the known radioprotectants L-cystine, 2-amino-ethylthiopseudourea dihydrochloride, and 2-aminoethanethiol hydrochloride. The selenoamino acids were found to protect various amino acids and the enzymes yeast alcoholic dehydrogenase and ribonuclease against γ-rays more effectively than methionine and the other sulfur radioprotectants. The high resistance of selenomethionine to radiation was attributed to its ability to form stable radical intermediates which can combine with free H atoms or electrons to return the compound to its original state. It was proposed that the behavior of selenoamino acids as antioxidants may be a clue to the biological role of selenium. Schwarz[123] challenged this suggestion by noting that the selenium levels tested by Shimazu and Tappel were in considerable excess over those found in normal tissue.

In related experiments, Dickson and Tappel[124] noted that selenocystine and selenomethionine bind reversibly to the sulfhydryl enzymes papain and glyceraldehyde-3-phosphate dehydrogenase to protect them from oxidative inactivation. The compounds 2-aminoethaneselenol (selenocysteamine) hydrochloride,[125] bis(2-aminoethyl) diselenide (selenocystamine) dihydrochloride,[125–127] 2-aminoethaneselenosulfuric acid,[67a,125] and 2-aminoethyl-selenopseudourea dihydrobromide[128] are analogs of well-known mammalian radioprotective agents which have been synthesized and screened for their antiradiation activity. As can be seen from Table XIIID-4, the selenium compounds are considerably more toxic than their sulfur counterparts and are devoid of activity.

Table XIIID-4   A comparison of the ability of selenium and sulfur compounds to protect mice from lethal ionizing radiation

COMPOUND (X = Se, S)	LD$_{50}$ (mg/kg)		DRUG DOSE (mg/kg)[a]		% SURVIVAL (30 day)	
	Se	S	Se	S	Se	S
H$_2$NCH$_2$CH$_2$—X—H[b]	10	250	5	150	0	80
(H$_2$NCH$_2$CH$_2$—X—)$_2$	11[c]	~375[d]	7.5[c]	~355[d]	0[c]	100[d]
H$_2$NCH$_2$CH$_2$—X—SO$_3$H[b]	17.5	450	10	350	0	73
H$_2$NCH$_2$CH$_2$—X—$\overset{\overset{\displaystyle NH}{\|}}{C}$—NH$_2$	38[c]	690[e]	25[c]	250[e]	0[c]	88[e]

[a] Intraperitoneal administration 15–30 minutes before exposure.
[b] Klayman et al.[129]
[c] Grenan.[130]
[d] Recalculated values: Doherty et al.[131]
[e] Doherty and Burnett.[132]

The metal-complexing ability of selenocysteamine has been investigated because of its relationship to cysteamine.[133,134]

Selenourea was reported to be superior to cysteine as a radioprotectant for rats irradiated at sublethal (600 R), lethal (750 R) and supralethal (950 R) doses.[135] However, in another species (*i.e.*, in mice), Grenan[130] found that selenourea lacked any radioprotective activity at the lethal level (1000 R) when administered at a dose of 200 mg/kg. The distribution of ^{75}Se after 6 hours to 14 days from labeled selenourea in rat organs has been determined after i.p. administration.[136] Selenourea has been reported to protect amino acids from radiation damage in solution[137] by serving as an efficient free-radical scavenger.[138] In pulse radiolysis studies, selenourea was found to be superior to thiourea and urea in competing for free-radicals formed by the radiolysis of water.[139]

Breccia *et al.*[140] evaluated selenomethionine, colloidal selenium, selenoxanthene (**46**), selenoxanthone (**47**), and selenochromone (**48**) in rats exposed to 600, 750, or 900 R. These substances were reported to have activity similar to that of cysteine and, in some instances, activity superior to it.

46   47   48

The ability of various selenium heterocycles to protect ATP from losing orthophosphate on irradiation was studied by Brucker and Bulka.[141] Only 2-amino-4,5-dimethylselenazole hydrochloride (**49**) exhibited any radio-

49

protective activity. Other 2-aminoselenazoles, as well as selenosemicarbazide and acetoneselenosemicarbazone, not only failed to protect the spores of the mushroom *Phycomyces blakesleeanus* from the effects of radiation but also exhibited a strong radiation-sensitizing effect.[141a] A series of 2-substituted selenazolidines has been synthesized as potential antiradiation agents; however, test results have not been reported as yet.[142]

One theory of radioprotection holds that sulfur-containing antiradiation agents form unsymmetrical disulfides (RSSR') with protein sulfhydryl groups. Thus, oxidation, resulting from the radiolysis of water in the tissues, is prevented, and the protein sulfhydryl groups are restored to their normal state, probably by some enzymatic process. Roy *et al.*[143] have remarked that, if thioselenates (RSSeR') occur in living organisms, their great reactivity must severely limit the duration of their existence.

Sodium selenate, when administered to rats 55–60 minutes postirradiation (800 R) at 4.2–4.6 mg/kg intraperitoneally, followed by an additional 80% subcutaneously, resulted in the survival of all the animals.[144]

10. Steroids

The first selenium-containing steroids, made inadvertently by what was supposed to have been a selenium dioxide oxidation of the nucleus,[139] have not been tested for their medicinal properties.[32]

Segaloff and Gabbard[112,146] synthesized a series of steroids related to dehydroepiandrosterone and pregnenolone in which the 3β-hydroxy group is replaced by a selenium- or sulfur-containing moiety. The selenols prepared were converted to the stable benzoate esters, since the former compounds are prone to air oxidation. The diselenide forms were also compared with their analogous disulfides and tested for their androgenicity, antiandrogenicity, progestogenicity, antiprogestogenicity, and inhibiting effect on ovulation and implantation. The selenium and sulfur isosters of dehydroepiandrosterone and pregnenolone resembled one another and were moderately androgenic but less active than the parent steroids. The diselenide,

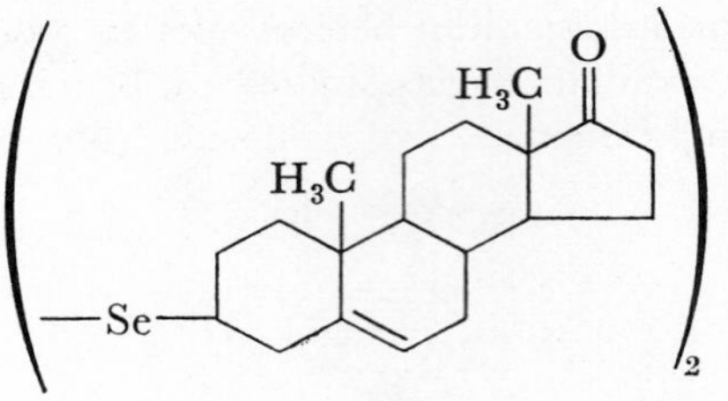

50

$3\beta,3\beta'$-diselenobisandrost-5-en-17-one (**50**), effectively inhibited the endo-metrial proliferation caused by progesterone in the rabbit but was ineffective in inhibiting ovulation and implantation in the rat.

The synthesis of 3α- and 3β-selenyl benzoate derivatives of cholestane and androstanone has recently been reported.[147] The androgenic activity of steroids of type **51**, where $Z = $ O, S, Se, or Te, was found by Wolff and

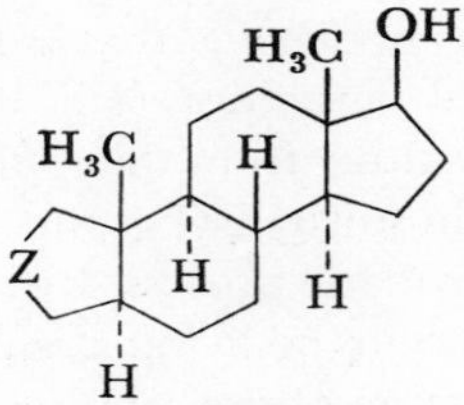

51

Zanati[148] to be good when the heteroatom was sulfur, selenium, or tellurium but nonexistent when oxygen. The authors concluded from this observation that the *steric* rather than *electronic* requirements at C(2) and/or C(3) of androgens govern their activity.

11. Selenocoenzyme A

Coenzyme A (CoA), a metabolic intermediate, was discovered by Lipmann[149] in 1945 in the course of investigating the ATP-dependent acetylation of sulfanilamide. From *S*-acetyl-CoA (so-called "active acetate") the acetyl group is readily transferred to a suitable acceptor such as an aromatic amine or an amino acid. The cofactor, which has been isolated from pigeon liver extracts and many microorganisms, is composed of adenosine, phosphoric acid, and pantetheine [*N*-(pantothenyl-β-aminoethanethiol] moieties. The acyl portion of the latter is related to pantothenic acid, a vitamin whose isolation was reported by Williams *et al.*[150] in 1938 and which is a biogenetic precursor of CoA. The disulfide form of pantetheine, pantethine, and its

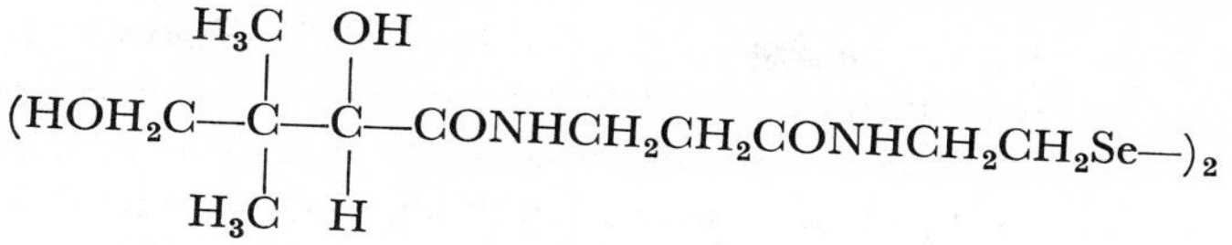
H3C OH
(HOH2C—C—C—CONHCH2CH2CONHCH2CH2Se—)2
H3C H

52

selenium isostere, selenopantethine (**52**), were demonstrated to be biologically equivalent on a molar basis in *Lactobacillus helveticus*,[151] an organism that has a requirement for pantethine and is believed to be capable of converting it to CoA.

The complex synthesis of selenocoenzyme A (Se-CoA, **53**) was achieved by

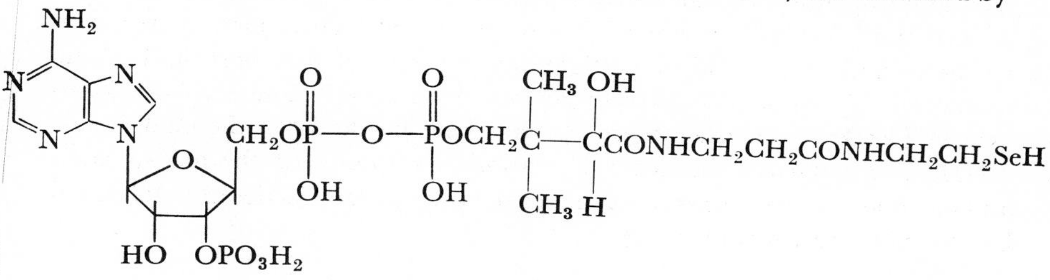
NH2
N
N
N
N
O
O
O
CH3 OH
CH2OP—O—POCH2C—CCONHCH2CH2CONHCH2CH2SeH
OH
OH
CH3 H
HO OPO3H2

53

Günther and Mautner[152] in 1965. The compound was obtained initially as the diselenide in admixture with isoselenocoenzyme A, the ribose-2-phosphate. Reductive acylation with 2-dimethylaminoethyl selenobenzoate[152] resulted in selenobenzoate esters of the CoA analogs, which could then be separated by ion-exchange chromatography on ECTEOLA-cellulose. Aminolysis of the separated products finally yielded Se-CoA in the selenol form and isoselenocoenzyme A, which are subject to rapid air oxidation to the symmetrical diselenides.

As the diselenide, Se-CoA had no CoA activity in several enzyme systems. However, in the presence of dithiothreitol, a reagent that reduces diselenides to selenols under physiological conditions,[153] Se-CoA was fully functional[154] in the acetylcoenzyme A synthetase (E.C.6.2.1.1) system. Although enzyme binding constants for the sulfur and selenium analogs are identical, Se-CoA has about one-third the rate of acyl acceptance under identical kinetic conditions. *In vitro* Se-CoA functions as a partial competitive agonist of coenzyme A.

12. Selenium-Containing Carbohydrates

The history of selenium-containing sugars goes back to 1917, when Schneider and Wrede[155] reported the preparation of selenoisotrehalose, a disaccharide containing two glucose moieties linked by a Se atom. Although the taste of the selenium analog was similar to that of the oxygen sugar,[155] the new species was not subject to hydrolysis by enzymes or microorganisms.[156] Against all expectations it also appeared to be nontoxic to mammals.[156] After subcutaneous injection of selenoisotrehalose into rabbits (600 mg/kg), urinary excretion in 24 hours accounted for 85 % of the unchanged material.

Syntheses were rapidly extended to a diglucosyl diselenide, to seleno-digalactose,[158] and to several other species,[159–161] but then interest evidently flagged and no additional selenium-containing carbohydrates were reported until 1950. At that time Bonner and Robinson[162,163] synthesized phenyl-seleno-β-D-glucopyranoside, the first example of this class.

Selenium-containing carbohydrates were neglected thereafter for another decade until Wagner et al.[164–177] started an extensive, ongoing series of synthetic and mechanistic investigations. Reports of this work and of the syntheses of 1-seleno-1-deoxy-D-glucose[178] and of 5-seleno-5-deoxy-D-xylose[179] appear to constitute the entire recent literature in the field. Table XIIID-5 summarizes the general relationships found for the acid-, base-, and enzyme-catalyzed scission of arylchalcogen-β-D-glucosides as determined by Wagner and his co-workers.

Table XIIID-5   Relative reactivities for scission of arylchalcogen-β-D-glucosides

Y = O, S, Se

R	REAGENT	REACTION ORDER	REF.
(phenyl)	OH⁻	Se ≫ S > O	173
	H⁺	O ≫ S > Se	174
	Enzymatic[a]	O S[b] Se[b]	171
(p-NO₂-phenyl)	OH⁻	Se > O > S	173
	H⁺	O ≫ Se > S	174
	Enzymatic[a]	O > Se > S	172
(o-NO₂-phenyl)	OH⁻	Se > O > S	172
	H⁺	O ≫ Se > S	172
	Enzymatic[a]	O ≫ S Se[b]	172

[a] Almond emulsin, that is, a preparation of β-glucosidase, pH 5 buffer.
[b] Sulfur or selenium isologs did not react under these conditions.

In the single case in which the enzyme was found to be active for the complete series, the rates closely followed the order observed for acid hydrolysis. Hence, a protonation step does appear to be implicated for the enzyme-substrate interaction. The possible role of the nonreacting selenoglycoside species as competitive enzyme inhibitors has, apparently, not been investigated.

13. Miscellaneous

Selenosaccharin (**54**), synthesized in 1912 by Lesser and Weiss,[180] was reported to have a weakly astringent, rather than a sweet, taste. This may

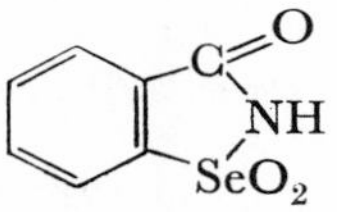

54

be due to the greater ionizability of its imide proton in comparison to that of saccharin.

The potentially useful selenium-containing amino acids and peptides are reviewed in Chapter XII. A drug formulation that includes selenium- and sulfur-containing amino acids and antioxidants has been proposed by Passwater[181] as a means to "significantly retard aging and improve the quality of life."

14. Conclusion

As can be seen from the foregoing, organic selenium compounds have not enjoyed much practical usage in chemotherapy, for only a few of them offer such advantages over existing agents as to warrant more than the briefest of clinical trials. Selenium compounds have served, however, as important tools in the elucidation of the mechanism of action of sulfur-containing chemotherapeutic agents and have been used to good advantage when radiolabeled. Thus ^{75}Se-selenomethionine has proved to be valuable for human pancreatic and parathyroid scanning for the detection of lymphomas (*cf*. Chapter XIIIC). In all probability, further use will be made of radiolabeled selenium compounds for the study of binding sites, biological pathways, and reaction mechanisms.

REFERENCES

1. K. Morita and S. Kobayashi, *Tetrahedron Lett.*, 573 (1966); D. Brewer, R. Rahman, S. Safe, and A. Taylor, *Chem. Commun.*, 1571 (1968).

2. I. Langmuir, *J. Amer. Chem. Soc.*, **41**, 1543 (1919).

3. H. Erlenmeyer, *Bull. Soc. Chim. Biol.*, **30**, 792 (1948).

4. V. B. Schatz, "Medicinal Chemistry," 2nd ed., A. Burger, Ed., Interscience Publishers, New York, N.Y., 1960, p. 72.

5. A. F. Wells, "Structural Inorganic Chemistry," Oxford University Press, London, 1950, p. 146.

6. H. Rheinboldt and E. Giesbrecht, *Justus Liebigs Ann. Chem.*, **568**, 197 (1950).

7. J. R. Harr, J. F. Bone, I. J. Tinsley, P. H. Weswig, and R. S. Yamamoto, "Selenium in Biomedicine," O. H. Muth, Ed., AVI Publishing Co., Westport, Conn., 1967, p. 153.

8. C. G. Durbin, "Selenium in Biomedicine," O. H. Muth, Ed., AVI Publishing Co., Westport, Conn., 1967, p. 417.

9. H. N. Green and F. Bielschowsky, *Brit. J. Exptl. Pathol.*, **23**, 13 (1942).

10. H. J. Backer and J. De Jonge, *Rec. Trav. Chim. Pays-Bas*, **60**, 495 (1941).

11. K. A. Jensen and K. Schmith, *Dansk Tidskr. Farm.*, **15**, 197 (1941); *Chem. Abstr.*, **38**, 1483[1] (1944).

12. A. R. Frisk, *Acta Med. Scand., Suppl.*, **142**, 1 (1943); *Chem. Abstr.*, **38**, 4692[6] (1944).

13. P. C. Guha and A. N. Roy, *Current Sci.*, **12**, 150 (1943).

14. A. N. Roy and P. C. Guha, *J. Ind. Chem. Soc.*, **22**, 82 (1945).

14a. H. M. Priestly, U.S. Pat. 3,642,909 (1972); *Chem. Abstr.*, **76**, 112693 (1972).

15. J. Matti, *Bull. Soc. Chim. Fr.*, [5], **7**, 617 (1940).

16. S. Keimatsu, K. Yokota, and I. Satoda, *J. Pharm. Soc. Japan*, **53**, 994 (1933).

17. J. Supniewski, S. Misztal, and J. Krupinska, *Bull. Acad. Polon. Sci., Classe II*, **2**, 153 (1954).

18. H. G. Mautner, *Biochem. Pharmcol.*, **1**, 169 (1959).

19. D. E. Hunt and R. F. Pittillo, *Antimicrobial Agents Chemother.*, 551 (1966).

20. Y. F. Shealy, J. D. Clayton, G. J. Dixon, E. A. Dulmadge, R. F. Pittillo, and D. E. Hunt, *Biochem. Pharmacol.*, **15**, 1610 (1966).

21. Y. F. Shealy and J. D. Clayton, *J. Heterocycl. Chem.*, **4**, 96 (1967).

22. S. J. Kuhn and J. S. McIntyre, U.S. Pat. 3,493,586; *Chem. Abstr.*, **72**, 90289 (1970); S. J. Kuhn and J. S. McIntyre, U.S. Pat. 3,489,774; *Chem. Abstr.*, **72**, 100517 (1970).

23. K. Takeda, R. Kitagawa, T. Konaka, R. Noguchi, H. Nishimura, N. Shimoaka, N. Kimura, K. Kaida, K. Sugiyama, H. Ilada, and T. Maki, *Ann. Rept. Shionogi Res. Lab.*, **2**, 12 (1952).

24. A. M. Comrie, D. Dingwall, and J. B. Stenlake, *J. Pharm. Pharmacol.*, **16**, 268 (1964).

25. Aktiebolag Astra, Brit. Pat. 1,174,753; *Chem. Abstr.*, **72**, 59060 (1970).

26a. K. Bednarz, *Dissertationes Pharm.*, **9**, 249 (1957); *Chem. Abstr.*, **52**, 8083f (1958).

26b. K. Bednarz, *Dissertationes Pharm.*, **10**, 93 (1958).

27. W. E. Weaver and W. M. Whaley, *J. Amer. Chem. Soc.*, **68**, 2115 (1946).

28. T. Zsolnai, *Biochem. Pharmacol.*, **11**, 271 (1962).

29. H. G. Mautner, W. D. Kumler, Y. Okano, and R. Pratt, *Antibiot. Chemother.*, **6**, 51 (1956).

30. M. Pianka and J. C. Hall, *J. Sci. Food Agr.*, **8**, 432 (1957).

31. M. Pianka and J. C. Hall, *J. Sci. Food Agr.*, **10**, 385 (1959).

32. K. G. Florey, U.S. Pat. 2,917,507.

33. P. Ehrlich and H. Bauer, *Ber.*, **48**, 502 (1915).

34. F. C. Goble, E. A. Konopka, and H. C. Zoganas, *Antimicrobial Agents Chemother.*, **7**, 531 (1967).

35. S. Keimatsu and K. Yokota, *J. Pharm. Soc. Japan*, **51**, 605 (1931).

36. D. G. Markees, V. C. Dewey, and G. W. Kidder, *J. Med. Chem.*, **11**, 126 (1968).

37. H. G. Mautner and E. M. Clayton, *J. Amer. Chem. Soc.*, **81**, 6270 (1959).

38. E. H. Shaw, Jr. and E. E. Reid, *J. Amer. Chem. Soc.*, **48**, 520 (1926).

39. E. Hannig, *Arch. Pharm.*, **296**, 441 (1963).

40. E. Hannig, *Pharmazie*, **19**, 201 (1964).

41. E. Hannig and H. Ziebandt, *Pharm. Zentralhalle*, **104**, 301 (1965).

41a. R. Hirschelmann and H. Bekemeier, *Acta Biol. Med. Germ.*, **25**, 41 (1970).

42. E. Hannig and H. Ziebandt, *Pharmazie*, **23**, 552 (1968).

43. H. Bekemeier, W. Schmollack, H.-J. Feicht, and K.-H. Lemnitzer, *Pharmazie*, in press.

44. J. J. Donleavy and J. English, Jr., *J. Amer. Chem. Soc.*, **62**, 220 (1940); C. Rohmann and T. Wischniewski, *Arch. Pharm.* (Weinheim), **292**, 787 (1959).

45. E. Hannig and H. Ziebandt, *Pharmazie*, **22**, 626 (1967).

46. P. Müller, N. P. Buu-Hoi, and R. Rips, *J. Org. Chem.*, **24**, 37 (1959).

47. J. M. Glassman, A. J. Begany, H. H. Pless, G. M. Hudyma, and J. Seifter, *Federation Proc.*, **19**, 280 (1960).

48. P. N. Craig, U.S. Pat. 3,043,839; *Chem. Abstr.*, **57**, 16636 (1962).

49. P. N. Craig, Brit. Pat. 814,065; *Chem. Abstr.*, **54**, 1566 (1960).

50. M. P. Olmsted, P. N. Craig, J. J. Lafferty, A. M. Pavloff, and C. L. Zirkle, *J. Org. Chem.*, **26**, 1901 (1961).

51. G. Cordella and F. Sparatore, *Farmaco, Ed. Sci.*, **20**, 446 (1965).

52. G. Cordella and F. Sparatore, *Ann. Chim.* (Rome), **56**, 182 (1966).

53. M. Protiva, M. Rajsner, E. Adlerova, V. Seidlova, and Z. J. Vejdelek, *Collection Czech. Chem. Commun.*, **29**, 2161 (1964).

54. K. Šindelar, J. Metyšová, E. Svatek, and M. Protiva, *Collection Czech. Chem. Commun.*, **34**, 2122 (1969).

55. K. Šindelar, E. Svatek, J. Metysová, J. Metyš, and M. Protiva, *Collection Czech. Chem. Commun.*, **34**, 3792 (1969).

56. P. V. Petersen and I. M. Nielsen, "Medicinal Chemistry," Vol. 4, Academic Press, New York, N. Y., 1964, p. 301.

57. J. O. Jílek, M. Rajšner, J. Pomykácek, and M. Protiva, *Cesk. Farm.*, **14**, 294 (1965); *Chem. Abstr.*, **65**, 2211f (1966).

58. K. Šindelar, J. Metyšová, and M. Protiva, *Collection Czech. Chem. Commun.*, **34**, 3801 (1969).

59. K. Šindelar, J. Metyšová, J. Metyš, and M. Protiva, *Naturwissenschaften*, **56**, 374 (1969).

59a. J. Jilek, K. Šindelar, J. Metyšová, J. Metyš, J. Pomykacek, and M. Protiva, *Collection Czech. Chem. Commun.*, **35**, 3721 (1970).

60. N. I. Kapitonov, *Farmakol. Toksikol. Prep. Selena, Mater. Simp.*, 90 (1967); *Chem. Abstr.*, **70**, 86057 (1969).

61. L. F. Chernysheva and A. N. Kudrin, *Farmakol. Toksikol. Prep. Selena, Mater. Simp.*, 50 (1967); *Chem. Abstr.*, **70**, 66599 (1969).

62. N. E. Kuznetsova, *Farmakol. Toksikol. Prep. Selena, Mater. Simp.*, 75 (1967); *Chem. Abstr.*, **70**, 66602 (1969).

63. V. M. Lotis and N. E. Kuznetsova, *Farmakol. Toksikol. Prep. Selena, Mater. Simp.*, 68 (1967); *Chem. Abstr.*, **70**, 66601 (1969).

64. N. F. Zhdanova, *Farmakol. Toksikol. Prep. Selena, Mater. Simp.*, 56 (1967); *Chem. Abstr.*, **70**, 66600 (1969).

65. J. P. Limongi, *Arch. Intern. Pharmacodyn.* **103**, 160 (1955).

66. C. E. Corbett, J. P. Limongi, and A. O. Ramos, *Arch. Intern. Pharmacodyn.*, **111**, 245 (1957).

67a. W. H. H. Günther and H. G. Mautner, *J. Med. Chem.*, **7**, 229 (1964).

67b. W. H. H. Günther and H. G. Mautner, *J. Med. Chem.*, **8**, 845 (1965).

68. K. A. Scott and H. G. Mautner, *Biochem. Pharmacol.*, **13**, 907 (1964).

69. K. A. Scott and H. G. Mautner, *Biochem. Pharmacol.*, **16**, 1903 (1967).

70. H. G. Mautner, E. Bartels, and G. D. Webb, *Biochem. Pharmacol.*, **15**, 187 (1966).

71. G. D. Webb and H. G. Mautner, *Biochem. Pharmacol.*, **15**, 2105 (1966).

72a. P. Rosenberg and H. G. Mautner, *Science*, **155**, 1569 (1967).

72b. P. Rosenberg, H. G. Mautner, and D. Nachmansohn, *Proc. Natl. Acad. Sci.*, **55**, 835 (1966).

73. S.-H. Chu and H. G. Mautner, *J. Org. Chem.*, **31**, 308 (1966).

74. S.-H. Chu and H. G. Mautner, *J. Med. Chem.*, **11**, 446 (1968).

74a. E. Heilbronn-Wilkström, *Svensk Kem. Tidskr.*, **77**, 598 (1965).

74b. J. E. Gearien, "Medicinal Chemistry," 3rd ed., A. Burger, Ed., John Wiley, New York, N.Y., 1970, p. 1309.

74c. S. Åkerfeldt and L. Fagerlind, *J. Med. Chem.*, **10**, 115 (1967).

74d. F. C. G. Hoskin, L. T. Kremzner, and P. Rosenberg, *Biochem. Pharmacol.*, **18**, 1727 (1969).

75. A. N. Kudrin and Ya. I. Zaidler, *Farmakol. Toksikol.*, **31**, 41 (1968).

76. N. P. Buu-Hoi, G. Lambelin, J. Roba, G. Jacques, and C. Gillet, Belg. Pat. 739,678; *Chem. Abstr.*, **73**, 109490 (1970).

77. M. E. Roberts, *Toxicol. Appl. Pharmacol.*, **5**, 500 (1963).

78. B. M. Phillips, L. F. Sancilio, and E. Kurchacova, *J. Pharm. Pharmacol.*, **19**, 696 (1967).

79. I. A. Jaffe, *Arthritis Rheumat.*, **8**, 1064 (1965); I. A. Jaffe and R. W. Smith, *Arthritis Rheumat.*, **11**, 585 (1968).

79a. C. Draguet and M. Renson, *Bull. Soc. Chim. Belges*, **81**, 303 (1972).

80. S. Rosell and J. Axelrod, *Experientia*, **19**, 318 (1963).

81. D. T. Witiak, "Medicinal Chemistry," 3rd ed., A. Burger, Ed., John Wiley, New York, N. Y., 1970, p. 1647.

82. L. F. Chernysheva and S. V. Eliseeva, *Farmakol. Toksikol.*, **29**, 679 (1966); *Chem. Abstr.*, **66**, 45,284 (1967).

83. A. I. Radostina, *Vliyanie Nekototykh Fiz. i Biol. Faktorov na Organizm, Materialy Nauchn. Konf., Univ. Druzhby Narodov*, 106 (1965); *Chem. Abstr.*, **65**, 11208 (1966).

84. A. N. Kudrin, L. F. Chernysheva, M. A. Gal'bershtam and Yu. K. Yur'ev, *Farmakol. Toksikol.*, **27**, 692 (1964).

85. A. N. Kudrin, L. F. Chernysheva, O. D. Kolyutskaya, V. S. Gigauri, and G. S. Ovchinnikov, *Vopr. Anestziol. i Reanimatsii, Stavropol.*, *Sb.*, 163 (1964); *Chem. Abstr.*, **63**, 18906 (1965).

86. A. I. Radostina, *Tr. Univ. Druzby Nar.*, **38**, 83 (1968); *Chem. Abstr.*, **71**, 121,999 (1969).

87. L. F. Chernysheva and M. A. Gal'bershtam, *Farmakol. Toksikol. Prep. Selena, Mater. Simp.*, 44 (1967); *Chem. Abstr.*, **70**, 55925 (1969).

88. L. A. Romanova, V. Z. Gorkin, A. N. Kudrin, and L. F. Chernysheva, *Byul. Eksp. Biol. Med.*, **66**, 39 (1968); *Chem. Abstr.*, **70**, 45968 (1969).

89. L. F. Chernysheva, A. N. Kudrin, and V. Gigauri, *Vop. Farmakol. Regul. Deyatel. Serdtsa, Mater. Simp.*, 110 (1969); *Chem. Abstr.*, **73**, 129437 (1970).

90. I. G. Bodkov, *Vop. Farmakol. Regul. Deyatel. Serdtsa, Mater. Simp.*, 106 (1969); *Chem. Abstr.*, **73**, 129436 (1970).

91. V. M. Lotis, A. N. Kudrin, V. V. Korzhova, and L. F. Chernysheva, *Farmakol. Toksikol. Prep. Selena, Mater. Simp.*, 64 (1967); *Chem. Abstr.*, **71**, 79510 (1969).

92. V. V. Korzhova and E. I. Smirnova, *Vop. Farmakol. Regul. Deyatel. Serdtsa, Mater. Simp.*, 119 (1969); *Chem. Abstr.*, **73**, 129440 (1970).

93. G. B. Elion, E. Burgi, and G. H. Hitchings, *J. Amer. Chem. Soc.*, 74, 411 (1952); *cf.* also "Symposium on 6-Mercaptopurine," *Ann. N. Y. Acad. Sci.*, **60**, 183 (1954).

94. H. G. Mautner, *J. Amer. Chem. Soc.*, **78**, 5292 (1956).

95. H. G. Mautner and J. J. Jaffe, *Cancer Res.*, **18**, 294 (1958).

96. J. J. Jaffe and H. G. Mautner, *Cancer Res.*, **20**, 381 (1960).

97. H. G. Mautner, S.-H. Chu, J. J. Jaffe, and A. C. Sartorelli, *J. Med. Chem.*, **6**, 36 (1963).

98. S.-H. Chu, *J. Med. Chem.*, **14**, 254 (1971); **15**, 1088 (1972).

99. L. B. Townsend and G. H. Milne, *J. Heterocycl. Chem.*, **7**, 753 (1970).

100. F. Bergmann and M. Rashi, *Israel J. Chem.*, **7**, 63 (1969).

101. A. Carr, E. Sawicki, and F. E. Ray, *J. Org. Chem.*, **23**, 1940 (1958); A. A. Carr, Jr., *Dissertation Abstr.*, **18**, 1613 (1958).

102. E. Dyer and C. E. Minnier, *J. Med. Chem.*, **11**, 1232 (1968).

103. H. Endo, K. Sato, and T. Kawasaki, *Sci. Rept. Res. Inst., Tohoku Univ., Ser. C*, **11**, 201 (1963); *Chem. Abstr.*, **60**, 3390c (1964).

104. K. Takeda, H. Nishimura, N. Simaoka, R. Noguchi, and K. Nakajima, *Ann. Rept. Shionogi Res. Lab.*, **5**, 1 (1955).

105. M. T. Bogert and A. Stull, *J. Amer. Chem. Soc.*, **49**, 2011 (1927).

106. A. G. Mueller and J. P. Phillips, *J. Med. Chem.*, **10**, 110 (1967).

107. A. S. Weisberger and L. G. Suhrland, *Blood*, **11**, 19 (1956).

108. A. S. Weisberger and L. G. Suhrland, *Blood*, **11**, 11 (1956).

109. A. S. Weisberger, L. G. Suhrland, and J. Seifter, *Blood*, **11**, 1 (1956).

110. R. H. Adamson, *Proc. Soc. Exptl. Biol. Med.*, **119**, 456 (1965).

111. A. Senning and O. N. Sørensen, *Acta Chem. Scand.*, **20**, 1445 (1966).

112. A. Segaloff and R. B. Gabbard, *Steroids*, **5**, 219 (1965).

113. C. C. Clayton and C. A. Baumann, *Cancer Res.*, **9**, 595 (1949).

114. R. J. Shamberger, *J. Natl. Cancer Inst.*, **44,** 931 (1970).

115. R. J. Shamberger, *Proc. Amer. Assoc. Cancer Res.*, **10,** 79 (1969).

116. R. J. Shamberger and G. Rudolph, *Experientia*, **22,** 116 (1966).

117. J. F. Riley, *Experientia*, **24,** 1237 (1968).

118. R. J. Shamberger and D. V. Frost, *Can. Med. Assoc. J.*, **100,** 682 (1969).

119. D. V. Frost, *World Rev. Pest Control*, **9,** 6 (1970).

120. Z. M. Bacq, C. Onkelinx, and G. Barac, *C. R. Soc. Biol.*, **157,** 899 (1963).

121. F. Shimazu and A. L. Tappel, *Radiation Res.*, **23,** 210 (1964).

122. F. Shimazu and A. L. Tappel, *Science*, **143,** 369 (1964).

123. K. Schwarz, *Federation Proc.*, **24,** 58 (1965).

124. R. C. Dickson and A. L. Tappel, *Arch. Biochem. Biophys.*, **131,** 100 (1969).

125. D. L. Klayman, *J. Org. Chem.*, **30,** 2454 (1965).

126. V. Coblentz, *Ber.*, **24,** 2131 (1891).

127. W. H. H. Günther and H. G. Mautner, *J. Amer. Chem. Soc.*, **82,** 2762 (1960).

128. S.-H. Chu and H. G. Mautner, *J. Org. Chem.*, **27,** 2899 (1962).

129. D. L. Klayman, M. M. Grenan, and D. P. Jacobus, *J. Med. Chem.*, **12,** 510 (1969).

130. M. M. Grenan, Walter Reed Army Institute of Research, personal communication.

131. D. G. Doherty, W. T. Burnett, Jr., and R. Shapira, *Radiation Res.*, **7,** 13 (1957).

132. D. G. Doherty and W. T. Burnett, Jr., *Proc. Soc. Exptl. Biol. Med.*, **89,** 312 (1955).

133. H. Tanaka, H. Sakurai, and A. Yokoyama, *Chem. Pharm. Bull.* (Tokyo), **18,** 1015 (1970).

134. A. Yokoyama, H. Sakurai, and H. Tanaka, *Chem. Pharm. Bull.* (Tokyo), **18,** 1021 (1970).

135. R. Badiello, A. Trenta, M. Mattii, and S. Moretti, *Med. Nucl. Radiobiol. Lat.*, **10,** 1 (1967).

136. A. Breccia, A. Trenta, R. Badiello, S. Moretti, and M. Mattii, *Experientia*, **22,** 475 (1966).

137. R. Badiello and A. Breccia, "Radioprotection and Sensitization," H. L. Moroson, and M. Quintiliani, Eds. Taylor and Francis, London, 1970, p. 103.

138. R. Badiello and E. M. Fielden, "Radiation and Sensitization," H. L. Moroson, and M. Quintiliani, Eds., Taylor and Francis, London, 1970, p. 109.

139. R. Badiello and E. M. Fielden, *Intern. J. Radiation Biol.*, **17,** 1 (1970).

140. A. Breccia, R. Badiello, A. Trenta, and M. Mattii, *Radiation Res.*, **38,** 483 (1969).

141. W. Brucker and E. Bulka, *Stud. Biophys.*, **1,** 253 (1966).

141a. W. Brucker and H.-G. Rohde, *Pharmazie*, **23,** 310 (1968).

142. T. Sadeh and G. Shtacher, *Abstr. IV⁰ Congrès Intern. de Radiobiolgie et de Physico-Chemie des Rayonnements, Evian, 29 June–4 July, 1970*, Abstr. 733.

143. J. Roy, I. L. Schwartz, and R. Walter, *J. Org. Chem.*, **35,** 2840 (1970).

144. Z. M. Hollo and S. Zlatarov, *Naturwissenschaften*, **47,** 328 (1960).

145. K. Florey and A. R. Restivo, *J. Org. Chem.*, **22,** 406 (1957).

146. A. Segaloff and R. B. Gabbard, U.S. Pat. 3,372,173.

147. S. M. Hiscock, D. A. Swann, and J. H. Turnbull, *J. Chem. Soc.*, D, 1310 (1970).

148. M. E. Wolff and G. Zanati, *Experientia*, **26,** 1115 (1970).

149. F. Lipmann, *J. Biol. Chem.*, **160**, 173 (1945).

150. R. J. Williams, J. H. Truesdail, H. H. Weinstock, Jr., E. Rohrmann, C. M. Lyman, and C. H. McBurney, *J. Amer. Chem. Soc.*, **60**, 2719 (1938).

151. W. H. H. Günther and H. G. Mautner, *J. Amer. Chem. Soc.*, **82**, 2762 (1960).

152. W. H. H. Günther and H. G. Mautner, *J. Amer. Chem. Soc.*, **87**, 2708 (1965).

153. W. H. Günther, *J. Org. Chem.*, **32**, 3931 (1967).

154. W. H. H. Günther and H. G. Mautner, personal communication.

155. W. Schneider and F. Wrede, *Ber.*, **50**, 793 (1917).

156. F. Wrede, *Biochem. Z.*, **83**, 96 (1917).

157. F. Wrede, *Ber.*, **52**, 1756 (1919).

158. W. Schneider and A. Beuther, *Ber.*, **52**, 2135 (1919).

159. F. Wrede, *Hoppe-Seylers Z. Physiol. Chem.*, **112**, 1 (1920).

160. F. Wrede, *Hoppe-Seylers Z. Physiol. Chem.*, **115**, 284 (1921).

161. F. Wrede and W. Zimmermann, *Hoppe-Seylers Z. Physiol. Chem.*, **148**, 65 (1925).

162. W. A. Bonner and A. Robinson, *J. Amer. Chem. Soc.*, **72**, 354 (1950).

163. W. A. Bonner and A. Robinson, *J. Amer. Chem. Soc.*, **72**, 356 (1950).

164. G. Wagner and G. Lehmann, *Pharm. Zentralhalle*, **100**, 160 (1961).

165. G. Wagner and P. Nuhn. *Arch. Pharm.*, **296**, 374 (1963).

166. G. Wagner and P. Nuhn, *Z. Chem.*, **3**, 64 (1963).

167. G. Wagner, E. Fickweiler, P. Nuhn, and H. Pischel, *Z. Chem.*, **3**, 62 (1963).

168. G. Wagner and P. Nuhn, *Arch. Pharm.*, **297**, 81 (1964).

169. G. Wagner and P. Nuhn, *Arch. Pharm.*, **297**, 461 (1964).

170. G. Wagner and P. Nuhn, *Pharm. Zentralhalle*, **104**, 328 (1964).

171. G. Wagner and R. Metzner, *Naturwissenschaften*, **52**, 61 (1965).

172. G. Wagner and R. Metzner, *Pharmazie*, **20**, 752 (1965).

173. G. Wagner and P. Nuhn, *Arch. Pharm.*, **298**, 686 (1965).

174. G. Wagner and P. Nuhn, *Arch. Pharm.*, **298**, 692 (1965).

175. G. Wagner and G. Valz, *Pharmazie*, **22**, 548 (1967).

176. P. Nuhn and G. Wagner, *Arch. Pharm.*, **301**, 186 (1968).

177. H. Frenzel, P. Nuhn and G. Wagner, *Arch. Pharm.*, **302**, 62 (1969).

178. J. Kocourek, J. Kleňha and V. Jiráček, *Chem. Ind.* (London), 1397 (1963).

179. T. van Es and R. L. Whistler, *Tetrahedron*, **23**, 2849 (1967).

180. R. Lesser and R. Weiss, *Ber.*, **45**, 1835 (1912).

181. R. J. Passwater, quoted in *Chem. Eng. News*, Oct. 26, 1970, p. 17.

GENERAL REFERENCES

D. Dingwall, *J. Pharm. Pharmacol.*, **14**, 765 (1962).

H. G. Mautner, *Radioactive Pharm.*, *Proc. Symp.*, Oak Ridge, Tenn., *1965*, 409 (1966).

A. Burger, Ed., "Medicinal Chemistry," 3rd ed., John Wiley, New York, N. Y., 1970.

XIII Selenium compounds in nature and medicine

E. METABOLISM OF SELENIUM BY PLANTS AND MICROORGANISMS

ALEX SHRIFT

State University of New York at Binghamton
Department of Biological Sciences
Binghamton, New York

1. Higher Plants

a. Historical Background. Many years elapsed between the discovery by Berzelius and Gahn in 1817 of the element selenium and the first reports of its biological effects on plants. It was not until the late 1880's, a period of intensive research into the nutritional requirements of plants, that a few of the many investigators, such as Cameron[1] in 1880 and Knop[2] in 1884–85, were curious enough to test the effect of this element on plant growth. Cameron had the insight to realize that "the analogy between sulfur and selenium suggests that selenium may wholly or partly replace sulfur as a constituent of vegetables." This concept pervaded much of the subsequent research with plants, as well as with other organisms, and still influences the thinking in the field.

Whereas Cameron failed to detect any injury to his leguminous plants by potassium selenate, Knop reported that 0.01 % selenic and selenious acids were toxic to corn plants. Cameron's experiment was rather crude; as described in his own words:

A sod was taken from a field in which a crop of the so-called artificial grasses (which are chiefly leguminous plants, and not grasses at all) was just peeping over the ground. The sod was two feet in depth, three feet in length, and one foot wide. . . . During four weeks the total quantity of potassium selenate applied to the plants amounted to twenty grammes . . . .

In Knop's experiments, better conceived than those of Cameron, defined nutrient solutions were used, and his results could be corroborated and extended to other species by later investigators[3–8] in the first quarter of this century.

It was not until the early 1930's, however, that a hitherto unsuspected biological role for selenium was recognized. At that time range plants, long thought to be the cause of certain livestock disorders in many western regions of the United States, were discovered to contain selenium.[9–11] Though these disorders had been known for many decades,* their cause had remained elusive. Attention eventually had narrowed to certain native plants and to grains. In 1933 selenium was found in toxic grains by Robinson;[12] the following year Beath and his co-workers[13] detected it in toxic native plants. These investigators were actually not the first to report that plants could absorb selenium under natural conditions; but the earlier findings had remained obscure,[14] and some were questionable.[15–18] Intensive research by several western laboratories and by the U.S. Department of Agriculture soon established that selenium was a constituent, not of just a few plants, but of many range plants and grains. When these seleniferous plant materials were fed to experimental animals, the syndromes seen on the range were duplicated. The occurrence of selenium at high concentrations

---

* The symptoms in horses described in 1860 by Madison, a U.S. Army surgeon, and even by Marco Polo during his thirteenth century travels, are thought to have been the same as the disorders caused by selenium.[10,11]

(hundreds or thousands of parts per million) was found to be restricted to a few groups of plants, yet other species sometimes contained levels of selenium considerably higher than the usual few parts per million. Continued research has established that seleniferous vegetation occurs in many other parts of the world as well.

At the other extreme, regions exist in which the selenium level of vegetation is so low that a selenium-deficiency disease of sheep known as white muscle disease is prevalent. Analyses of plants from these nonseleniferous regions throughout the world show selenium contents less than 0.1 ppm and as low as 0.02 ppm, a level close to the sensitivity limits of current analytical techniques. Determination of whether lower concentrations of selenium exist in plants awaits further refinement of assay methods.

b. Selenium Indicator Plants.

i. Primary and Secondary Indicators.

When selenium was implicated as a toxic agent, a survey of western range plants was undertaken to determine their geographical distribution, their geological associations, and the extent to which they absorbed selenium.[9-11,13,20-58] Although the studies lasted from the 1930's into the 1940's, it became apparent quite early that certain species were more reliable than others as indicators of toxic regions. These species are called indicator species; their occurrence always shows the presence of soils from which selenium can be accumulated. Such soils have become known as seleniferous soils even though they may contain only as much selenium as other soils that do not support seleniferous vegetation, or even less. The map shown in Fig. XIIIE-1 illustrates how widespread are these seleniferous regions; they extend from Mexico north through the United States into Canada.

Species from four genera of plants, representing three different families, are notoriously toxic in North America and have been classified as primary indicators (Table XIIIE-1).[10,11,59,60] Other plants also absorb toxic levels of selenium from seleniferous soils; however, they are often found growing in nonseleniferous soils (Table XIIIE-2), whereas the primary indicators are always associated with seleniferous soils.[10,11,59,60]

In other parts of the world, seleniferous vegetation has been found in Australia,[61] Ireland,[78,90] Israel,[62,63] South Africa,[64] and Venezuela.[79,96] A great variety of species from many genera and families are seleniferous in these regions, but, surprisingly, *Astragalus* species, which are widespread throughout the world, have not been implicated. Although the seleniferous species accumulate sufficient selenium to be toxic, their levels of this element, with two exceptions, are usually lower than the levels found in primary accumulators. One exception, *Neptunia amplexicaulis*, a legume, accumulates several thousand parts per million of selenium and is responsible for selenium

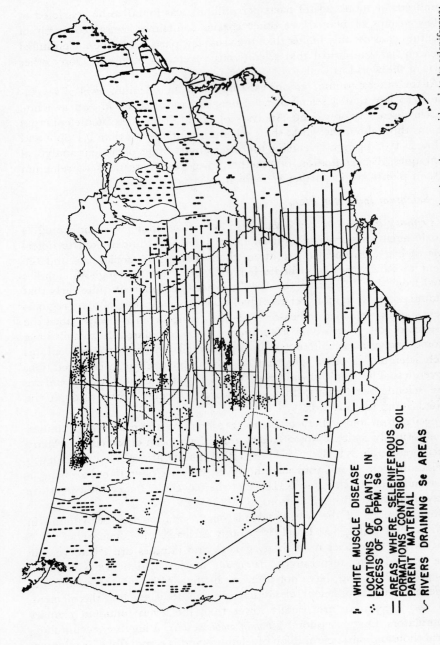

FIG. XIIIE-1   Map of the United States, showing (1) distribution of seleniferous vegetation in regions considered to be seleniferous; and (2) distribution of white muscle disease of sheep in regions considered to be nonseleniferous. From Muth and Allaway.[19]

· WHITE MUSCLE DISEASE
· LOCATIONS OF PLANTS IN
** EXCESS OF 50 PPM. Se**

☰ AREAS WHERE SELENIFEROUS
** FORMATIONS CONTRIBUTE TO SOIL**
** PARENT MATERIAL**

∿ RIVERS DRAINING Se AREAS

Table XIIIE-1 Primary selenium indicators[a]

GENUS	FAMILY	COMMON NAME
Astragalus	Leguminosae	Milk vetch
Xylorhiza (Machaeranthera)	Compositae	Woody aster
Oonopsis (Haplopappus)	Compositae	Goldenweed
Stanleya	Cruciferae	Prince's plume

[a] From Trelease and Beath,[10] Rosenfeld and Beath,[11] and Munz and Keck.[59]

Table XIIIE-2 Secondary selenium indicators[a]

GENUS	FAMILY	COMMON NAME
Aster	Compositae	Wild aster
Atriplex	Chenopodiaceae	Saltbush
Castilleja	Scrophulariaceae	Paint-brush
Comandra	Santalaceae	Bastard toad-flax
Grayia	Chenopodiaceae	
Grindelia	Compositae	Gum plant
Gutierrezia	Compositae	Snakeweed, matchweed
Machaeranthera	Compositae	
Sideranthus	Compositae	

[a] From Trelease and Beath,[10] Rosenfeld and Beath,[11] and Munz and Keck.[59]

toxicity to animals in Australia.[61] The other exception is *Lecythis ollaria*, a South American tree that bears a nut reported to contain as much as 18,000 ppm selenium; in Venezuela, ingestion of these nuts has caused severe toxicity symptoms and death in human beings.[79,207]

ii. Astragalus Species Known as Indicators. Of the primary indicators, the large and extensive genus *Astragalus* has received the most attention; the latest taxonomic revision of the genus by Barneby[60] establishes over 500 species and varieties in North America alone. The striking feature about this large genus of plants is that so few of the species, only about 25, have evolved a mechanism for extracting and accumulating extremely high levels of selenium from seleniferous soils. These species, taken from the revision of the genus by Barneby, are listed in Table XIIIE-3.

Barneby recognized that the ability to accumulate selenium is a useful taxonomic criterion and therefore grouped together all selenium-accumulating species into two taxonomic units. Table XIIIE-3 also gives as a comparison the species names used by Trelease and Beath,[10] whose classification

Table XIIIE-3   Astragalus species known to accumulate selenium

BARNEBY[60]	TRELEASE AND BEATH,[10] ROSENFELD AND BEATH[11]
A. flavus var. *flavus*	*A. confertiflorus* var. *flaviflorus* (Kuntze) Jones
A. flavus var. *candicans* Gray	*A. confertiflorus* Gray
A. flavus var. *argillosus* (Jones)	*A. argillosus* Jones
A. sophoroides	*A. confertiflorus* Gray
A. moencoppensis Jones	*A. moencoppensis* Jones
A. albulus Woot. & St.	*A. albulus* Woot. & Stand.
A. bisulcatus var. *bisulcatus*	*A. diholcos* (Rydb.) Tidestrom
	A. bisulcatus (Hook.) Gray
A. bisulcatus var. *haydenianus* (Gray)	*A. haydenianus* Gray
A. bisulcatus var. *nevadensis* (Jones) Barneby	*A. haydenianus* Gray
A. racemosus var. *racemosus*	*A. racemosus* Pursh
A. racemosus var. *longisetus*	*A. racemosus* Pursh
A. racemosus var. *treleasei*	*A. racemosus* Pursh
A. oocalycis Jones	*A. oocalycis* Jones
A. pectinatus Dougl.	*A. pectinatus* Doubl.
A. nelsonianus	*A. pectinatus* var. *platyphyllus* Jones
A. grayi Parry	*A. grayi* Parry
A. toanus Jones	*A. toanus* Jones
A. linifolius Osterh.	*A. toanus* Jones
A. rafaelensis Jones	*A. toanus* Jones
A. saurinus Barneby	*A. toanus* Jones
A. osterhouti Jones	*A. osterhouti* Jones
A. woodruffi Jones[a]	
A. preussii var. *Preussii*	*A. preussii* Gray
A. preussii var. *laxiflorus*	*A. preussii* Gray
A. eastwoodae	*A. eastwoodae* Jones
A. crotalariae (Bth) Gray	*A. crotalariae* (Benth.) Gray
A. mokiacensis Gray[b]	
A. beathii	*A. beathii* Porter
A. praelongus var. *praelongus*	*A. pattersoni* var. *praelongus* (Sheld.) Jones; *A. recedens* (Jones)
A. praelongus var. *ellisiae* (Rydb.) Barneby	*A. ellisiae* Porter
A. praelongus var. *lonchopus* Barneby	
A. pattersoni Gray	*A. pattersoni* Gray
A. sabulosus Jones	*A. sabulosus* Jones

[a] Has not been tested for ability to accumulate selenium; has odor typical of seleniferous specimens of other species.

[b] Grouped by Barneby with other seleniferous species because of morphological affinities; nothing known of its ability to accumulate selenium.

of the species into six sections was based primarily on that of Jones.[65] Barneby's revision has the perspective of modern botanical taxonomy, as well as the advantage of extensive collections from a number of herbaria. The reader is referred to his treatise for a thorough analysis of North American species. A number of these have eight chromosomes or multiples of eight; this relationship to Old World species, as well as the geographical distribution of *Astragalus*, is used as evidence to suggest that they originated in Asia and migrated into North America across the Bering land-bridge as long ago as the Pleistocene period.[60,66,67]

The identification of species in so large a genus requires great expertise; the beginner is referred to the books on selenium by Trelease and Beath[10] and by Rosenfeld and Beath[11] for useful diagrams and taxonomic keys.

iii. Selenium Content of Indicator Species under Natural Conditions. Countless analyses made in the 1930's clearly established differences in the abilities of plants to absorb selenium from soils. Accumulators often were found with selenium contents in the thousands of parts per million, although occasional specimens contained only a few parts per million.[10,11] A wide sampling becomes necessary, therefore, if a species is to be properly labeled as an accumulator. Secondary selenium indicators accumulate only moderate amounts of selenium, in the hundreds of parts per million; these plants are never found with a selenium content as high as the highest content in *Astragalus* accumulators.

Most of the other plants growing in seleniferous regions accumulate only a few parts per million of selenium. Included in this group are grasses and grains such as corn and wheat and a number of *Astragalus* species that are commonly found in nonseleniferous regions. The dichotomy within the genus is well illustrated by these *Astragalus* species, which contain only a few parts per million of selenium even though they grow adjacent to accumulator species with selenium contents of 400–5560 ppm.[9,34,42,114]

iv. Factors That Influence Selenium Content of Indicators under Natural Conditions. The reasons for the low selenium values in some specimens of *Astragalus* accumulators have not been fully determined. Suggested possibilities include (1) the selenium level in the soil was extremely low; (2) the specimens were collected at different stages of growth; (3) the selenium volatilized between the time of collection and the time of analysis; (4) genetic variability exists within individuals of the particular species. The first three factors have been shown to influence selenium content, but in no case has the absence of selenium from some specimens been proved to be due to any of these factors. Genetic variability within a species has not been studied. All of the factors bear on the question of selenium as a micronutrient for accumulator species, a role not universally accepted (see reviews by Shrift[68] and Beeson[69]). If selenium is a micronutrient, it becomes important to know why some specimens apparently have little or no selenium. If selenium is not a micronutrient, the low values could reflect nothing more than one of the four conditions mentioned above.

Soil type. Soils in regions of the United States known to be seleniferous are not uniform in their selenium content;[11] geological changes apparently left a spotty distribution.[10,11,58,70,71] Accordingly, several investigators believe that the best way to determine whether a soil should be considered seleniferous is to analyze the vegetation growing in it.[48] Even so, the amount of soil selenium does not necessarily correlate with the amount of selenium found in accumulators.[24] Some soils contain much more selenium than others, yet accumulators found on these soils may contain less of the element than plants of the same species found growing on soils with less selenium.[29,72]

One correlation that has emerged exists between the amount of soluble selenium present in the soil and the amount of selenium accumulated.[48,49,72,73] In western seleniferous soils there is appreciably more soluble selenium than in nonseleniferous soils, in which the selenium apparently is tightly bound to soil colloids. In seleniferous soils, on the other hand, the soil environment is alkaline, so that selenium tends to be oxidized to selenate, a reaction not likely to occur in the acid environment of nonseleniferous soils.[53,72] Organic selenium compounds, derived from the decay of seleniferous vegetation, also contribute to selenium absorption of accumulators, a factor considered to be highly important by Beath and his co-workers.[10,11]

Climate. Although the nature of the growing season can be expected to influence the selenium content of accumulators, few studies have been made with this factor in mind. Moreover, those that have been made are contradictory. There is a report of increased selenium content in vegetation with high rainfall,[74] but another study found the opposite correlation.[9]

Stage of growth. To determine accurately the variation in selenium content as a species develops from seedling stage to seed stage, it would be best to take periodic samplings from a uniform population of plants. But since accumulators in the field are not uniform, one can only rely on specimens that appear to be at the different stages of growth as the seasons progress. With this type of sampling, changes in selenium content have been noted, the variations being dependent on the species. *Astragalus bisulcatus*, for example, shows a continuous drop in selenium concentration from the pre-bloom stage in June to the past-seeding stage in winter, whereas *Oonopsis condensata*, another accumulator, does not reach a maximum selenium concentration until its full bloom stage.[25]

Part of the plant. The selenium contents in different parts of a plant vary with the species, the stage of development, and the regrowth in certain species. In general, the roots of *Astragalus* species, but not of other genera such as *Oonopsis* and *Stanleya*, contain less selenium than do aerial portions. At fruiting and seeding, selenium appears to be redistributed. In some species a regrowth occurs during the fall after the plants have bloomed, and the new leaves contain considerably more selenium than do the old ones.[11]

Volatilization. Loss of selenium by volatilization received early recognition when plant tissues decreased in selenium during the ashing of samples in preparation for analysis.[21] Later it was established that appreciable losses occurred even when green tissues were air dried. As much as 66% of the selenium is recorded to have been lost in this way.[25]

c. Selenium Levels in Forage, Crop, and Other Plants.

When selenium was first recognized as a toxic constituent in certain range plants, there was concern that other plants used by animals and man might absorb toxic levels of the element. Extensive surveys established, however, that forage and crop plants in seleniferous regions, with the exceptions discussed below, usually absorbed too little selenium to be considered dangerous. Generally, the levels of selenium in grasses and vegetables were only a few parts per million; in nonseleniferous regions throughout the world the selenium contents of plants can be as low as 0.01–0.02 ppm.[75–77]

Occasionally, however, specimens of forage and crop plants collected in seleniferous regions are found to assay high in selenium and therefore are potentially dangerous to animals (toxic levels are approximately 4–5 ppm or more). Among the grasses of South Dakota, for example, *Agropyron smithii* (western wheat grass), though not an accumulator, consistently absorbs more selenium than other species; hence, it has been suggested that this plant be used as an indicator of seleniferous regions.[46,47]

Among crop plants, crucifers such as cabbage and mustard accumulate much more selenium from seleniferous soils than do other vegetables. These plants are also high in sulfur, and it has been suggested that absorption of the two elements may be correlated.[11,78,81–83] Grains and forage crops, though usually low in selenium, sometimes assay high; samples of wheat, corn, and even alfalfa (which is rarely toxic) are known with extraordinarily high values, though they never reach the thousands of parts per million found in *Astragalus* and other indicator species.

The causes of these high values in wheat and other grains are incompletely understood. There is apparently no relation to the total selenium content of the soil, for some fields in which the selenium level is higher than in others support crops with lower selenium contents.[43,57] But the concentration of water-soluble selenium, a factor that appears to influence the amount of absorption by indicator species, seems also to affect the level in native grasses and crops. Such a correlation has been established for *Agropyron smithii*.[49,72,73] There also seems to exist a correlation between the chemical form of the selenium in the soil and the amount of the element found in vegetation. Beath observed early in his studies that proximity to seleniferous *Astragalus* increased the selenium content of grasses. The high levels were attributed to a leaching from the seleniferous vetches of organic selenium compounds which were readily available to plants ordinarily low in selenium.[24]

Other factors known to affect the selenium levels of indicator species can also be expected to influence the selenium contents of nonaccumulator plants. Hence, type of soil,[44,81,84,97,102,298] stage of development,[84-86] part of the plant,[84,112,113] and volatilization[87-89] must all be considered when studying the selenium levels in nonaccumulators. Volatile selenium, often lost during drying and storage, is also released by living plants; it is apparently taken up by plants in small amounts,[181] but the extent to which it contributes to the total selenium in a plant has not been ascertained.

Grasses in certain regions of Ireland sometimes contain as much as 5 ppm selenium, a value much higher than that in most grasses from seleniferous soils in the United States. The Irish soils responsible for these toxic grasses differ in a number of respects from the seleniferous soils in the United States. It is not uncommon to find between 5 and 100 ppm selenium, and in some soils in Ireland the element has accumulated to concentrations as high as 1200 ppm. This accumulation is believed to have occurred when glacial beds gradually filled with seleniferous materials eroding from surrounding shales and limestones.[78,90] One characteristic of these soils is their very high content of organic material, unlike the situation in the seleniferous soils of the United States, which are thought to have originated in the weathering of seleniferous rocks under arid and semiarid conditions.[10,11,58]

Occasionally, an unsuspected plant concentrates selenium. For instance, the mushroom *Amanita muscaria* accumulates 100–600 times more of the element than do the leaves of surrounding herbs and trees.[80] Another unusual example, mentioned previously, is the level of 18,200 ppm selenium in seeds of *Lecythis ollaria*, a nut-bearing tree that grows in Venezuela and other regions of Central and South America.[79] In Venezuela the toxic trees grow on alluvial plains where the seleniferous soils probably arose by deposition of seleniferous materials carried there by the many rivers emptying into this coastline. Other plants from this region also are known to contain toxic levels of selenium.[96]

d. Association of Selenium with Other Minerals.

High levels of several minerals have been reported in seleniferous soils and in seleniferous vegetation. Such an association has been found between molybdenum and selenium in soils from Ireland,[78,90] Wales and England,[91,92] and the Colorado Plateau in the United States.[93] High molybdenum is also found in seleniferous vegetation,[78,90,93] especially in members of the Leguminosae and Cruciferae.[78] Uranium and vanadium often accompany high concentrations of selenium, and the occurrence of several indicator species of *Astragalus* has in fact been used successfully by geobotanists as a means of botanical prospecting for uranium.[93–95]

e. Experimental Uptake of Selenium Compounds. Potential economic losses from seleniferous crops, forages, and indicator species, and in more recent years from low levels of selenium and the accompanying selenium-responsive diseases,[77] aroused considerable scientific interest in the absorption of selenium compounds by plants. Field experiments established that a great many factors, including type of soil, form of selenium in the soil, and species of plant, had to be considered when uptake was evaluated. By manipulation of experimental growing conditions the selenium content of "nonseleniferous" plants could be varied, and the extraordinarily high levels of selenium found in indicators could sometimes be reached. In nonseleniferous regions, where the selenium content of plants often is as low as 0.01–0.02 ppm, it has been found possible to raise the selenium levels sufficiently to prevent white muscle disease in sheep and other selenium-responsive diseases without giving rise to toxic vegetation.[77]

i. Soil Type. The colloid and organic content of soils plays an important role in the availability of selenium, especially selenite, which is readily bound.[53,97–103] Selenate, less readily bound, can be made increasingly toxic by varying the soil composition.[81,84] A decrease in sand content or in percentage of soil particles, for example, can increase the selenium concentration of such crop plants as clover, barley, and mustard.[84]

ii. Form of Selenium. Some forms of selenium are more readily available to plants than others. Crops sown on soils known to be seleniferous usually absorb no more than a few parts per million of selenium; but if mulch made from highly seleniferous indicator plants is incorporated into these soils before sowing, the crop plants then absorb enough selenium to render them toxic to animals.[13,21,24]

The soil plot experiments have been corroborated by comparable experiments under more clearly defined conditions. Experiments with corn and *Astragalus racemosus* grown in a defined culture solution under greenhouse conditions clearly showed that the selenium in extracts prepared from *A. bisulcatus* was superior to selenite as a source of selenium.[104,105] Comparable results were obtained with a variety of garden vegetables grown in soil under greenhouse conditions;[106] however, with other crops no general difference in the uptake of selenate or organic selenium was found,[107] and with a variety of range plants selenate proved a better source of selenium than organic selenium.[108] The selenoamino acid now known to predominate in these extracts is *Se*-methylselenocysteine;[109,110] it is uncertain, however, whether this was the form that entered the plants, because there was no attempt to eliminate microbial contamination in these relatively long-term experiments. These experiments reinforce the conclusion drawn from field experiments: that the return of highly seleniferous plants to soil, either naturally or by plowing under, serves to make the soil more seleniferous and thereby provides selenium to crops that would otherwise remain nontoxic.[10,11]

Selenate, is usually absorbed more readily than is selenite;[81,84,111,112] this difference can perhaps be attributed in part to the relative ease with which selenite is adsorbed to colloids or reduced to elemental selenium. There is also likely to be a difference in permeability at the membranes of the root cells.[114] Sulfate, which exerts little effect on selenite uptake, will, however, influence the absorption of selenate.

The elemental form of selenium has been considered unavailable to plants.[81] But when *A. bisulcatus* and *A. pectinatus* were grown for several months with elemental selenium, they contained 1150 ppm Se, and wheat contained 6.3 ppm.[25] Other plants, as well, are reported to contain selenium, though usually in small amounts, after having been grown with the elemental form.[112,115–118] In all these studies the experimental design has not ruled out the possibility of a microbial or nonbiological transformation of selenium into available forms.

iii. Enhancement of Selenium Uptake.   A variety of substances enhance selenite uptake and toxicity. In studies with several crop plants, proteins, peptones, and several amino acids increased the selenite uptake appreciably;[104,105] the results are difficult to evaluate, however, since bacteria might have contributed to the effect. But even when enhancement is observed with axenic microbial cultures, it has no ready explanation.[119–122]

iv. Antagonism.   Because of the resemblance in chemical and physical properties between the S and Se atoms, a relationship clearly recognized by Cameron as early as 1880, most studies that involve the effects of substances on selenium uptake and toxicity have been concerned with antagonism by sulfur compounds, particularly sulfate. Table XIIIE-4 summarizes a number of such investigations with plants and microorganisms. The ability of other sulfur metabolites to counteract selenate toxicity in a noncompetitive manner is usually interpreted to mean that the selenate ion is assimilated into a series of analogs, each of which contributes to the overall toxic effect, and that sulfur compounds provided externally counteract the effects of the selenium analogs produced internally.

Unlike the competition between selenate and sulfate, no competitive antagonism between selenite and any sulfur compound has ever been

Table XIIIE-4   *Sulfur-selenium antagonism*

SELENIUM COMPOUND	SULFUR ANTAGONISTS	ORGANISM	REF.
Selenate	Sulfate, elemental sulfur	Wheat	81, 82, 111, 123-127
	Sulfate	Barley roots	129, 130
	Sulfate	*Astragalus* roots	114
	Sulfate	*Astragalus* callus cultures	131
	Sulfate	*Crassula argentea, Thuidium tamariscinum, Vallisnaria gigantea*, barley	126, 127
	Sulfate, L-methionine, D-methionine	*Chlorella vulgaris*	132
	Sulfate	*Scenedesmus* sp.	128
	Sulfate, cysteine, L-methionine, thiamine	*Saccharomyces cerevisiae*	133–138
	Cysteine, glutathione	*Escherichia coli*	139
	Sulfate, cysteine, cystine, homocystine, methionine, glutathione	*Aspergillus niger*	140
	Sulfate	*Penicillium* spp., *Aspergillus nidulans*	141, 142
	Sulfate, methionine, homocysteine	*Neurospora crassa*	143
	Sulfate, sulfite	*Desulfovibrio desulfuricans*	144–146
Selenite	Sulfate	Wheat	111, 147, 148
	Sulfate	Buckwheat	147
	Sulfate	Tobacco, soybean	149
	Sulfate	*Astragalus racemosus*	150
	Sulfate, cysteine, methionine	*Escherichia coli*	151, 152
	Cystine	*Salmonella typhimurium, Escherichia coli, Citrobacter freundii*	230
	Pentathionate ($K_2S_5O_6$)	*Salmonella choleraesuis*	230

(Continued overleaf)

Table XIIIE-4   (*continued*)

SELENIUM COMPOUND	SULFUR ANTAGONISTS	ORGANISM	REF.
Seleno-methionine	L-Methionine, D-methionine	*Chlorella vulgaris*	153
	DL-Methionine	*Escherichia coli*	152
	L-Methionine	*Pityrosporum ovale*	154
	L-Methionine	Hamster intestine	156
	L-Methionine	*Penicillium chrysogenum*	155
Selenocystine	L-Cystine	Leukemic leukocytes, lymphosarcoma tumor cells	157, 158

demonstrated. Some sulfur compounds, especially sulfate, counteract the effects of low concentrations of selenite, but they are ineffective at higher concentrations.[111,147–150] Unfortunately, selenite reacts nonmetabolically with sulfhydryl compounds[159–163] so that any effect must be carefully evaluated to distinguish between a metabolic antagonism and a nonenzymatic reaction with selenite. Such a reaction conceivably might decrease concentrations of the ion to levels relatively less effective.

From observations of sulfur-selenium antagonism, it would be logical to predict that selenium uptake by crops and forage plants should be depressed by the application of sulfur compounds to seleniferous soils. Such an antagonism has been shown with soils under greenhouse conditions;[81,123–125,164] but when elemental sulfur or gypsum ($CaSO_4$) was added to field plots, the effects were negligible.[25,44,74] The problem may have been that the selenium in the soil was structurally and physiologically unrelated to the form of sulfur added.

Nonsulfur compounds such as phosphate,[133,165–167] arsenate,[133,165,166] arsenite,[165,166] and histidine[151] counteract selenite and selenate toxicity. How these substances act is unknown. Phosphate has even been reported to enhance selenite toxicity.[120]

v.   Active Transport of Selenium Compounds.   A great many organisms concentrate selenium at levels above the one initially present in the external medium, though the degree of concentration is less than that typical of accumulators. Of the various mechanisms generally recognized for the movement of substances across cell membranes, an active transport for selenium compounds has been inferred from accumulation data and from data with respiratory inhibitors.[132,153] These types of experiments alone are inconclusive, however, because they overlook the possibility that selenium compounds may enter by diffusion, followed by a continuous assimilation into other compounds *via* energy-dependent reactions. Such a process could simulate active transport.

An active transport mechanism is favored by the many studies that show a competitive antagonism between sulfur and selenium analogs. The existence of a transport site for which two molecules compete is usually deduced from experiments in which the ratio between the concentrations of the two molecules determines the degree to which each is absorbed. Such ratio dependence was seen first by Hurd-Karrer[82,111,124,125] in her extensive studies with wheat, and then by many others who studied selenium uptake and toxic effects on growth. If the ratio of the competitors is kept the same despite changes in the absolute levels, the growth response or the amount of selenium (or sulfur) absorbed will remain constant. Changing the ratio will vary the growth response or absorption.

A more precise analysis, the Michaelis-Menten analysis, usually applied to enzyme action, has also been used successfully to assess ion transport by plant roots.[129,130] Interaction between the two ions was found to be competitive; the data indicated that the two ions were bound to the cell membrane at a common binding site (variously termed carrier or permease) with the same affinity. A common binding site for sulfate and selenate has also been found in several fungi.[141,142]

If binding by the carrier, however, were the only factor for transport into the cell, the internal ratio of sulfur to selenium would approximate the external ratio. Earlier plant studies by Hurd-Karrer[125] had demonstrated, however, that the internal S/Se ratio often was much greater than the external ratio. Additional factors, therefore, must be instrumental in the uptake of these two ions by the intact plant.

One of the criteria for active transport requires that a substance accumulate unchanged. Excised roots of *Astragalus* species were found to accumulate selenate in this way.[114] By the end of 1 hour a gram of root tissue from one species had accumulated 0.4 μmole of selenium from a solution that contained 0.01 μmole/ml, about a fortyfold concentration. Uptake was inhibited by sulfate, by the respiratory inhibitors azide and dinitrophenol, and by a zero degree temperature. Most of the selenium was extractable in the form of selenate.[168] Comparable studies with selenomethionine also show an active transport in the alga *Chlorella vulgaris*[169] and in hamster intestine cells.[156]

The transport of selenite, however, is more complicated. When this ion is supplied to various organisms, selenium accumulates and uptake is prevented by respiratory inhibitors.[114,169] But a high proportion of the extractable selenium, as studied in excised *Astragalus* roots, was no longer selenite but had been converted to other forms.[168,170] At present is is unknown whether selenite enters excised *Astragalus* roots by a carrier mechanism comparable to that for the active transport of many other substances, or whether it diffuses in continuously and is then rapidly converted to different forms, either metabolically, by energy-dependent reactions, or in part nonmetabolically.

f. Influence of Selenium on Growth and Development.

i. Toxic Effects.   In nature, many nonaccumulators absorb relatively high concentrations of selenium under certain conditions, with no apparent effect on growth. In the laboratory, however, low concentrations of selenium compounds inhibit the growth of a great many plants. The degree of inhibition by selenate will depend, as was discussed earlier, in large measure on the presence of sulfur compounds—in particular, sulfate, which acts as a competitive antagonist.

ii. Selenium as a Micronutrient for Indicator Species.   Many facts indicate a micronutrient role for selenium in accumulator species. The evidence in favor of such a micronutrient role, though not conclusive, is summarized in the following paragraphs.

The 25 species of *Astragalus* and the species from other genera classified as primary accumulators are always found growing where soils contain available forms of selenium. Nonaccumulator species from the same genus, when found adjacent to an accumulator, contain only a few parts per million of selenium. Other plants that accumulate moderate amounts of selenium from these soils, the secondary accumulators, often can be located on non-seleniferous soils and presumably, therefore, have no need for the element. The extremely high concentrations of selenium accumulated and assimilated by primary indicators seem to signify a special need for the element, although why individual specimens sometimes assay low for selenium is still to be explained.

In greenhouse experiments, Trelease and Trelease[150,171] demonstrated that extremely small amounts of selenium stimulated the growth of accumulators, whereas nonaccumulators were severely inhibited by the same concentrations. Figure XIIIE-2 illustrates the type of response obtained when the plants were grown in solution culture. A somewhat greater response occurred in sand culture, where the amount of stimulation reached 2.8 times that of the minus selenium control.[171] Figure XIIIE-3 shows the stimulation by selenium of another species, *A. beathii*.[10] Nine accumulator species (*A. racemosus, pattersoni, beathii, bisulcatus, confertiflorus, grayi, osterhouti, pectinatus,* and *preussii*) are reported to have been so stimulated, and 6 nonaccumulator species (*A. crassicarpus,** asclepiadoides, canadensis, drummondii, lonchocarpus,*

---

* In the original publication by Trelease and Trelease[171] this nonaccumulator was referred to as *A. crassicarpus;* in two subsequent reviews the species was called *A. succulentus.*[10,11] Barneby lists *A. succulentus* under *A. crassicarpus* var. *crassicarpus.*[60]

FIG. XIIIE-2   Opposite effects of selenium (as selenite) on two species of *Astragalus* grown in solution cultures. *Above: A. racemosus; below: A. crassicarpus.* Selenium concentrations in solutions (left to right): 0, 0.33, 1, 3, 9 ppm. From Trelease and Trelease.[171]

and *lentiginosus*) to have been severely poisoned.[10,11] The divergence shown by these experiments parallels the divergence seen in the field. A germination test devised by Trelease[172] also differentiated between the two types of *Astragalus* species. In this test, 20 ppm selenium, as selenite, allowed 15 species of accumulators to germinate readily, whereas this level of selenite severely inhibited 11 nonaccumulator species.

Assimilation of selenium by accumulators differs in a number of respects from assimilation by nonaccumulators.[68] A distinctive type of metabolism in accumulator plants could be associated with a micronutrient role. Furthermore, differences in the way in which sulfur and selenium are metabolized (see Section E.1f) also may indicate a special role for selenium.

Frequently, biochemical and physiological needs found in one group of organisms are also present in others. On a comparative basis, therefore, a

FIG. XIIIE-3   Influence of selenium on the growth of *Astragalus beathii* in sand culture. Plants on the left received selenium in the culture solution; those on the right did not. From Trelease and Beath.[10]

micronutrient role for selenium in plants could be expected since such a role has been established for several animals.[173] Although a requirement for selenium has not yet been demonstrated in microorganisms, selenite apparently is needed for the induction of formic dehydrogenase, if not for growth, by several enteric microorganisms,[174] and it improves the growth of several yellow-green algae.[252]

The evidence as a whole seems to favor selenium as a micronutrient for the selenium accumulator species, but the growth experiments of Trelease and Trelease fall somewhat short of the criteria enumerated by Arnon and Stout[175] for the establishment of an element as a micronutrient. According to these criteria an element is not essential unless (*a*) a deficiency makes it impossible for the plant to complete its life cycle; (*b*) the deficiency is specific to that element; and (*c*) the element participates directly in the nutrition of the plant.

Although the experiments of Trelease and Trelease[171] show a striking growth stimulation, a criterion frequently used in micronutrient studies, the plants were not carried through the reproductive stages of the life cycle.

Nevertheless, the growth stimulation seems to have been specific for selenium, since the medium used in these studies contained all trace elements currently recognized as micronutrients in higher plants. The criterion of metabolic participation seems to have been met by the observation that selenium and sulfur sometimes are not metabolized in the same way.

A micronutrient interpretation has been questioned, however, by Broyer and Johnson,[175a] who grew *A. crotalariae*, an accumulator, in a medium carefully freed of traces of selenium (see the review by Shrift[68]). They believe that the medium used by Trelease and Trelease contained levels of phosphate which were toxic to *Astragalus* and that the improved growth seen with the accumulators was a manifestation of an antagonism between phosphate and selenate or selenite. The complete inhibition of nonaccumulators by added selenite, observed by the Treleases, is attributed to the absence in these species of a detoxification mechanism, which is suggested as responsible for the tolerance of accumulators to high levels of selenium.[176]

iii. Selenium as a Micronutrient for Other Plants. The evidence for selenium as a micronutrient in other plants is mainly negative. Although there are several references to small amounts of growth stimulation in plants after the addition of selenium compounds to the culture medium, the amount of stimulation was slight, and, moreover, no attempts were made to meet criteria for essentiality.[4,7,25,111,178–180]

The major evidence against a general micronutrient role was provided by Broyer *et al.*[181] in an investigation specifically designed to test this hypothesis. Alfalfa and subterranean clover were grown successfully without deficiency symptoms in a medium composed of specially purified salts which contained less than 1×10^{-8} g-atom Se/mole (0.79 μg Se/mole). No growth improvement was seen after the addition of selenium. The authors estimate that, if these plants require selenium, the level is probably less than 0.001 μg-atom/g of dry weight (0.079 ppm), a level close to that found in many crops that appear normal (about 0.02 ppm).

g. Metabolism of Selenium Compounds.

i. Properties of Selenium Compounds. The resemblance between sulfur and selenium has been mentioned many times. But it is also important, in order to understand the assimilation of selenium and the varied effects caused by selenium compounds in many organisms, to be aware of the dissimilarities. The four selenium compounds most widely used in biological and biochemical studies are selenate, selenite, selenomethionine, and selenocyst(e)ine.

Selenate. Selenate is the most highly oxidized form of selenium. Chemically and biologically this compound is an analog of sulfate. The structural similarity between the selenate and sulfate ions seems to be the basis both for the competitive antagonism so often been observed between the two ions in growth studies with plants and microorganisms, and for the ability of organisms to assimilate selenate into analogs of sulfur metabolites. The two ions differ, however, in their oxidation potentials;[182] the free energy of formation is much less for selenate than for sulfate. Selenate, therefore, is the stronger oxidizing agent and consequently is more readily reduced than is sulfate. It will react with sulfur dioxide, though not as readily as will selenite, to give elemental selenium.

Selenite. Selenite has a lower free energy of formation than does selenate or sulfate[182] and consequently is the stronger oxidizing agent. It is easily reduced by sulfur dioxide but requires strong oxidizing agents such as hydrogen peroxide or potassium permanganate to be oxidized to selenate. It is apparent, therefore, that the sulfite and selenite ions, even though their empirical formulas suggest some degree of identity, are not equivalent chemically; they should not be expected to act biologically as a pair of competitive antagonists.

The instability of the selenite ion and the ease with which it oxidizes sulfhydryl compounds are factors to remember when the effect of this compound on growth is under study, when transport into cells is being considered, or when cells are fractionated and selenium compounds isolated. Selenite reacts with sulfhydryl metabolites such as cysteine and glutathione to give rise to compounds with the general formula R–S–Se–S–R.[159–163] Because such nonbiological transformations of selenite can accompany biological transformations, it may sometimes be difficult to assess the degree to which selenite is enzymatically converted.

Selenocyst(e)ine. This analog of cysteine predictably shares some of the properties of the sulfur amino acid. However, despite limited research, enough important differences between the two compounds are known to warrant their discussion.[183] The pK of the selenol group is much lower than that of the sulfhydryl group; at physiological pH's, therefore, the selenoamino acid would be expected to exist mainly as the anionic selenide, whereas the cysteine molecule would still be mainly in the protonated sulfhydryl form. Higher half-wave potentials obtained by polarography show that selenol groups are more readily oxidized to the corresponding diselenide than are sulfhydryl groups to disulfide. Proteins into which selenocysteine had been incorporated would, therefore, be chemically and biologically different from their sulfhydryl counterparts.

Because selenocysteine itself is labile under the concentrated acid conditions often used for the hydrolysis of proteins,[183] reports of selenocysteine in proteins analyzed by this technique must be studied with care. The incorporation of selenocysteine into a polypeptide chain, however, seems to confer some stability on this compound during acid hydrolysis. Analysis of an acid hydrolysate of simple selenopolypeptides, such as 1-selenooxytocin or 6-selenooxytocin, showed destruction of only 60% of the selenocystine residues; the remaining selenoamino acid could be identified when the hydrolysate was analyzed by ion-exchange column chromatography.[184]

Another hindrance to the identification of selenocyst(e)ine is the possible formation during the isolation procedure of mixed selenium-selenium or selenium-sulfur artifacts through the reduction of diselenides by thiols or selenols. For example, selenocystine reacts easily with cysteine by a diselenide-sulfhydryl interchange to give 2,7-diamino-4-thia-5-selenaoctanedioic acid (equation 1).[184]

(1) $\text{HOOC—CH—CH}_2\text{—SH}$ +
$$\quad\quad\quad\; |$$
$$\quad\quad\quad NH_2$$

$\text{HOOC—CH—CH}_2\text{—Se—Se—CH}_2\text{—CH—COOH} \longrightarrow$
$$\quad\quad\quad\; | \quad\quad\quad\quad\quad\quad\quad\quad\quad\; |$$
$$\quad\quad\quad NH_2 \quad\quad\quad\quad\quad\quad\quad\quad NH_2$$

$\text{HOOC—CH—CH}_2\text{—S—Se—CH}_2\text{—CH—COOH}$
$$\quad\quad\quad\quad\quad\; | \quad\quad\quad\quad\quad\quad\quad\quad\quad\; |$$
$$\quad\quad\quad\quad\quad NH_2 \quad\quad\quad\quad\quad\quad\quad\quad NH_2$$

Selenomethionine. At the biochemical level, as discussed in Section E.2.e, several enzymes handle selenomethionine as if it were the normal substrate; it also competes with methionine in several absorption systems in a typical way. The two amino acids differ, however, in several properties. Selenomethionine proved highly effective in protecting amino acids from radiation damage; in itself, it was more stable to the γ-radiation than was methionine. Selenomethionine also protected several enzymes from radiation damage more completely than did methionine.[185] Selenomethionine has been reported as stable to acid hydrolysis;[186] other investigators, however, found that it decomposed completely under these conditions.[187] The stability of selenomethionine during release from a polypeptide chain has not been investigated. It would be helpful to know whether selenomethionine is more stable under these circumstances, as has been determined for selenocysteine.[184]

One physical property of selenomethionine that may bear on its biological properties is its solubility in demineralized water; it is about one-third as soluble as methionine. This difference in hydrophobic character between the two amino acids has been suggested as affecting the hydrophobic interactions of proteins in which methionine has been replaced by selenomethionine.[187]

Identification of selenium compounds of biological origin. Two major problems arise during the isolation of seleno compounds from biological materials: (*a*) artifacts can develop from the interaction of selenium compounds with other substances and can be mistaken for naturally occurring substances;[188] (*b*) the lability of selenium compounds can cause them to decompose partially or completely. Gentle methods such as enzymatic digestion are preferred, therefore, for protein hydrolysis.[188] Criteria for the identification of selenium compounds have been outlined elsewhere.[169] Ion-exchange column chromatography is the most desirable method and should eventually supersede paper chromatography. Several satisfactory column methods have now been developed to separate selenoamino acids from their sulfur analogs.[184,189-191]

ii. Organic Selenium Compounds in Plants.

Classical biochemical methods of analysis, such as differential solubility, steam distillation, dialysis, and precipitation with selective agents, served to separate gross fractions and thereby to show that selenium can exist in organic forms.[192] Despite the considerable work done since then with grains, forages, and grasses, selenium compounds in these plants are still inadequately identified. We do know that much naturally occurring selenium is associated with protein fractions,[10,11] and there is increasing evidence of selenomethionine and selenocysteine in the proteins of these plants.[193,203,204] Although some organic selenium compounds in *Astragalus* and *Stanleya* have been identified, the nature of the compounds in other primary indicators and in secondary indicators (see Tables XIIIE-1 and 2) is virtually unknown.

Table XIIIE-5 lists a number of selenium compounds reported in higher plants. Most of these compounds have been identified by modern chromatographic methods. No attempt is made here to evaluate the data; in order to ascertain whether sufficient criteria were used in the identification of the compounds each paper should be consulted individually.

Murti[212a] has reviewed the area of naturally occurring sulfur- and selenium-containing amino acids.

iii. Biochemical Distinctions between Indicator and Nonindicator Astragalus Species.

Se-Methylselenocysteine, the first compound listed in Table XIIIE-5, is a nonprotein amino acid consistently found in indicator species of *Astragalus*.[109,110,194,198,199] Although it occurs also in some nonindicator species, the amounts are always lower.[110,194] Nonindicator species of *Astragalus* apparently synthesize relatively large amounts of *Se*-methylselenomethionine; this compound, however, has not always been found in these species,[194] perhaps because it is labile.

The distinction between the two groups of *Astragalus* holds also for sulfur compounds in seeds of a large number of *Astragalus* species collected from throughout the world.[213] Among approximately 120 species analyzed, only the 10 North American accumulators contained S-methylcysteine and glutamyl-S-methylcysteine; the remaining species, all nonaccumulators, lacked these compounds. Another distinction is revealed by the presence of large amounts of selenocystathionine in several accumulator species of *Astragalus*,[190,194,194a] in *Stanleya pinnata*,[194,206] and in *Neptunia amplexicaulis* and *Morinde reticulata*, two Australian accumulators.[176,176a,177] Several nonaccumulator species of *Astragalus* also synthesize this compound, but only in trace quantities.[194]

One other difference emerges from a study of the metabolism of seleno-methionine by a species from each of the two groups. Both species methylated the amino acid to give Se-methylselenomethionine, but only the accumulator converted it to selenohomocystine and Se-methylselenocysteine.[200]

It has been suggested that in accumulator species many of the selenium compounds function in a detoxification mechanism.[110,176,196] Although the function of these compounds remains speculative, their identification and the study of their metabolism in *Astragalus* have provided new taxonomic information and have illustrated that the physiological divergence that evolved in this genus of higher plants is associated with a biochemical divergence as well.

iv. Selenium and Sulfur Metabolism—Similarities. The selenium compounds listed in Table XIIIE-5 are analogs of sulfur compounds, a number of which are metabolites in plants and microorganisms.[214] Many of these compounds have been identified in plants supplied with radioactive sulfate, selenite, or selenate. When *Astragalus* species, for example, were supplied with these inorganic ions, radioactive Se-methylselenocysteine or S-methyl-cysteine was identified among the radioactive components.[109,110] This kind of evidence could mean that the Se atom is metabolized along the same route as sulfur, though separate routes leading to the same end product are also quite possible.

Many gaps exist in our knowledge of the intermediate steps by which selenate and even sulfate are assimilated. Often, great reliance is placed on comparative biochemistry; it is assumed that metabolic steps in one group of organisms occur also in other groups. Quite often such assumptions prove correct, but at times other pathways are discovered and the assumption of biochemical unity is not borne out.

The first step in sulfate assimilation in both plants and microorganisms is the reaction of the ion with ATP to yield adenosine-5'-phosphosulfate (equation 2). ATP sulfurylase has been characterized in both plants and

$$(2) \qquad ATP + SO_4{}^{2-} \xrightleftharpoons{\text{ATP-sulfurylase}} APS + PP$$

microorganisms;[214–216] the enzyme, prepared from yeast and several bacteria,

Table XIIIE-5 Selenium compounds reported in higher plants

SELENIUM COMPOUND	FORMULA	PLANT	REF
Se-Methylselenocysteine	$CH_3-Se-CH_2-CH-COOH$, with NH_2	*Astragalus* spp.	109, 110, 190, 194–196
		Oonopsis condensata, Stanleya pinnata	197–200
Selenocysteineseleninic acid	$HO_2Se-CH_2-CH-COOH$, with NH_2	*Trifolium pratense, T. repens, Lolium perenne*	193
Se-Propenylselenocysteine selenoxide	$CH_3-CH=CH-Se-CH_2-CH-COOH$, with O (double bond to Se) and NH_2	*Allium cepa*	201
Selenohomocystine	$HOOC-CH-CH_2-CH_2-Se-Se-CH_2-CH_2-CH-COOH$, with NH_2 and NH_2	*Astragalus crotalariae*	200
Selenocystine	$HOOC-CH-CH_2-Se-Se-CH_2-CH-COOH$, with NH_2 and NH_2	*Zea mays*
Trifolium pratense, T. repens, Lolium perenne
Allium cepa
Spirodela oligorrhiza
Brome grass	202
193
201
115
203
γ-L-Glutamyl-*Se*-methyl-seleno-L-cysteine	$HOOC-CH-CH_2-CH_2-CONH-CH-CH_2-Se-CH_3$, with NH_2 and $COOH$	*Astragalus bisulcatus*	195, 196

Compound	Structure	Species	References
Selenocystathionine	$HOOC-CH-CH_2-CH_2-Se-CH_2-CH-COOH$ (with NH_2 on each α-carbon)	*Astragalus* spp. *Stanleya pinnata* *Lecythis ollaria* *Neptunia amplexicaulis* *Morinda reticulata* *Oxytropis* spp.	190, 194, 194a, 205 194, 206 79, 207–210 176, 176a 177 194
Selenomethionine	$CH_3-Se-CH_2-CH_2-CH-COOH$ (with NH_2)	*Allium cepa* *Trifolium pratense*, *T. repens*, *Lolium perenne* *Spirodela oligorrhiza* Brome grass Wheat	201 193 115 203 204
Selenomethionine selenoxide	$CH_3-\overset{O}{\underset{}{Se}}-CH_2-CH_2-CH-COOH$ (with NH_2)	*Trifolium pratense*, *T. repens*, *Lolium perenne*	193
Se-Methylselenomethionine	$CH_3-\overset{+}{Se}(CH_3)-CH_2-CH_2-CH-COOH$ (with NH_2)	*Astragalus* spp. *Trifolium pratense*, *T. repens*, *Lolium perenne*	110 193
Dimethyl diselenide	$CH_3-Se-Se-CH_3$	*Astragalus racemosus*	211
Selenopeptides		*Astragalus* spp.	109, 110, 200
Seleno wax		*Stanleya bipinnata*	212

also converts selenate to APSe.[217,219] Although selenate prevents the formation of APS by the enzyme prepared from several higher plants,[215] APSe has not been described from these sources. Its existence in plants, however, is very likely, because the activation of sulfate by ATP-sulfurylase is so well established; otherwise another mechanism for the assimilation of selenate to amino acids would have to be assumed.

The second step in sulfate assimilation is generally recognized to be a reaction between APS and another molecule of ATP catalyzed by the enzyme APS-kinase to give 3′-phosphoadenosine-5′-phosphosulfate (equation 3). The occurrence of this enzyme in microbial cells is well estab-

$$(3) \qquad APS + ATP \underset{}{\overset{APS\text{-}kinase}{\rightleftharpoons}} PAPS + ADP$$

lished,[214] but its existence in higher plants is unconfirmed[214,215] and has even been questioned.[215] Efforts to demonstrate PAPSe with a yeast APS-kinase have been unsuccessful;[218] either PAPSe was too unstable under the experimental conditions used or the enzyme was unable to convert APSe to

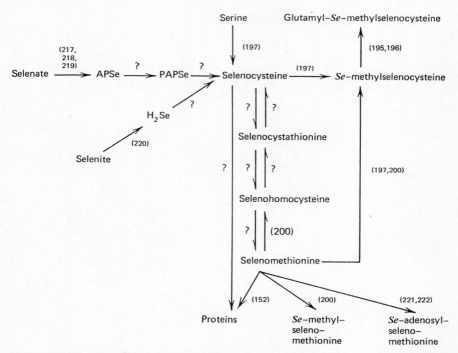

FIG. XIIIE-4   Assimilation of selenate and selenite into selenoamino acids. The question marks indicate steps which have not been proved.

PAPSe. Knowledge of the remaining steps for the assimilation of selenate or selenite is equally uncertain. There is evidence,[197] based on the use of radioactive precursors, that serine serves as intermediate in the synthesis of both cysteine and selenocysteine in *Astragalus bisulcatus*, and that methylation to *S*-methylcysteine and *Se*-methylselenocysteine occurs from the methyl group of methionine, but the mechanism for insertion of the Se atom is virtually unknown. Figure XIIIE-4 gives a scheme for the biosynthesis of selenoamino acids, some of the steps being based on those known from sulfur metabolism in plants and microorganisms.

v. Selenium and Sulfur Metabolism—Differences. Early indications of the metabolic nonequivalence of sulfur and selenium analogs were noted by Hurd-Karrer[125] in experiments with wheat; the internal ratio of sulfur to selenium was not the same as the external ratio that had been supplied to the plants. Sulfate, apparently, was favored in the absorption process. The work of Painter and Franke[223] showed that the ratio of the two elements also changed in different parts of cereal plants, as well as in different protein fractions from the same part of the plant.

Nonequivalence is sometimes found when structural analogs are compared biochemically. The reasons are still conjectural. The reactions listed in Table XIIIE-6 are ones for which selenium compounds were sought but not demonstrated. The failure to detect compounds such as choline selenate, flavonoid selenates, a plant selenolipid, and selenoglutathione can be interpreted, pending further research, to mean that certain branches of sulfur metabolism are closed to selenium.

Reactions in which seleno compounds are made, but not, apparently, their sulfur analogs, are also known. *Astragalus vasei* is a nonaccumulator that synthesized *Se*-methylselenocysteine from selenite;[110] when the leaves were supplied with ^{35}S-sulfate, however, *S*-methylcysteine was not detected. Nor was the sulfur amino acid found in seeds of this species, though it occurred in those of other nonaccumulators.[213] Another *Astragalus* species that metabolized sulfur and selenium differently is *A. crotalariae*, an accumulator. ^{35}S-Methionine administered to excised leaves remained largely unchanged, though small amounts were converted to *S*-methylmethionine. ^{75}Se-Selenomethionine, however, was converted to *Se*-methylselenomethionine, selenohomocystine, and *Se*-methylselenocysteine.[200]

An unusual reaction with selenium for which a comparable reaction with sulfur is unknown has been reported for elemental selenium. The element in this form is considered to be inert. It has been incorporated into soils and has found its way into plants, but the interpretation of these results is complicated by the possible action of soil microorganisms. In experiments with *Spirodela oligorrhiza*, a species of duckweed, however, ^{75}Se-labeled elemental

Table XIIIE-6   Metabolic reactions involving sulfur for which comparable reactions with selenium have been sought but not demonstrated

REACTION	ORGANISM	REF.
$SO_4^{2-} \rightarrow$ choline sulfate	*Aspergillus niger*	224
	Zea mays	
	Hordeum vulgare	
	Triticum aestivum	
	Helianthus annuus	
$SO_4^{2-} \rightarrow$ flavonoid sulfates	*Zostra marina*	224
$SO_4^{2-} \rightarrow$ plant sulfolipid	*Euglena gracilis*	224
	Chlorella sp.	
	Brassica oleracea var. *acephala*	
	Phaseolus vulgaris	
	Triticum aestivum	
$SO_4^{2-} \rightarrow$ chondroitin sulfate	*Bos tauros* (calf)	225
	Oryctolagus cuniculus (rabbit)	226
$SO_4^{2-} \rightarrow$ PAPS	*Saccharomyces cerevisiae*	218
$SO_4^{2-} \rightarrow$ glutathione	*Escherichia coli*	152
	Astragalus spp.	109, 110
$SO_4^{2-} \rightarrow$ sulfur amino acids in proteins	*Neptunia amplexicaulis*	176, 193
$SO_4^{2-} \rightarrow H_2S$	*Desulfovibrio desulfuricans*	144, 145
$SO_4^{2-} \rightarrow$ cysteine in β-galactosidase	*Escherichia coli*	287

selenium, incorporated into an axenic culture, was converted in an appreciable amount to selenomethionine, and in a much smaller amount to selenocystine; in addition there was some oxidation to selenate and selenite.[115] Such a transformation of elemental sulfur by plants is unknown, though it can be attacked by a variety of microorganisms. In the work with *Spirodela* there is a possibility, as pointed out by the authors, that the elemental selenium might have been atmospherically oxidized to an assimilable form.

2. Microorganisms

a. Historical Background. Little metabolic research with selenium in microorganisms took place during the nineteenth century. The few papers that appeared all described the ability of microorganisms to reduce selenite to elemental red selenium. In the first three decades of the twentieth century, however, at a time when many bacteriologists were occupied with the reducing properties of microorganisms, as well as with a search for chemicals that provided specific antipathogen action, selenium compounds began to receive increasing attention. Reduction of selenite to elemental selenium proved to be widespread among microorganisms, and some bacteria and fungi were found that reduced selenite or selenate to volatile, organic selenium compounds. Generally selenite was found to be more inhibitory, than selenate, but bacteria were discovered that tolerated toxic levels of these and other selenium compounds. A thorough review of the subject has been given by Levine.[227]

Research in the next few decades uncovered a sulfur-selenium antagonism, first observed with higher plants. The effects of selenite, it was learned, could be enhanced or modified by changes in the components (other than sulfur) of the nutrient growth medium. The reduction of selenite, originally thought to be nonmetabolic, was discovered to be enzymatically mediated. More startling, it proved possible, with little or no effect on growth, to replace a sulfur nutrient with its selenium analog. It is now known that many isolated enzyme systems can utilize sulfur and selenium analogs interchangeably. What was traditionally thought to be the mechanism of selenium toxicity, namely, a general interference with the enzymes involved in sulfur assimilation has proved to be more complex.

b.　Influence of Selenium Compounds on Growth

i. Species Differences. In their response to selenium compounds,[119,132,227-232] as to many other inhibitory substances, microorganisms show great species and strain variation. *Chlorella vulgaris*, for example, was inhibited by low concentrations of selenomethionine,[153] whereas another species, *Chlorella pyrenoidosa*, tolerated concentrations of this selenoamino acid up to 1×10^{-3} *M*.[233] Even strains derived from the same parent may show very different growth responses. For example, selenite is far more toxic to a wild type of *Escherichia coli*, inducible for β-galactosidase, than it is to a constitutive strain of this bacterium.[234]

ii. Antagonisms. Although toxic effects of selenium compounds were known for many years, it was not until after the pioneering work with higher plants[82,111,124] that concepts of antagonism by sulfur compounds were applied to studies with microorganisms (*cf.* Table XIIIE-4). Competitive antagonism between selenate and sulfate, first described by Fels and Cheldelin for yeast,[135] is known in a variety of other microorganisms.[128,132,140,142,144,145] Sulfur compounds such as methionine, cysteine, and thiamine also antagonized selenate toxicity but not in a competitive way.[132,134,136,139,140,143] As new selenium compounds such as selenomethionine became available and were found to be toxic, their sulfur analogs also were shown to exert a competitive effect.[153]

This type of antagonism undoubtedly results from resemblances between sulfur and selenium compounds. But compounds not containing sulfur, such as arsenic compounds,[165] phosphate,[133,165–167] and histidine[151] also antagonize selenium to varying degrees. These types of antagonism are more difficult to explain.

iii. Enhancement of Toxicity. Selenite toxicity can be increased by several substances. One of these substances, phosphate, has recently been described as antagonistic to selenium toxicity.[169] Likewise, sulfur compounds, almost always antagonistic to selenium toxicity, enhanced selenite toxicity to several strains of *E. coli*.[121,122] The choice of organism, as well as the experimental conditions, undoubtedly had much to do with these conflicting results.

iv. Nutritional Effects. It is possible to modify selenium toxicity by changing the carbon source.[234,235] When grown with lactose as such a source, a situation under which the cells are induced to synthesize enzymes for the breakdown of this carbohydrate, *E. coli* was highly sensitive to selenite; with succinate, however, a compound for which enzymes are already present, a concentration of selenite that was highly toxic to the lactose-grown cells had little effect. These results do not reflect antagonisms in the familiar competitive sense, but they illustrate the fact that organic components of a medium, other than sulfur compounds, can modify selenium toxicity.

v. Tolerance and Adaptation. Sometimes an organism is found that grows in a medium high in selenium. Currently, however, *Salmonella* is the only genus of bacteria the majority of whose species have this capacity. The initial observation of this tolerance was made by Handel and Theororascu (cited by Guth[236]) as early as 1916. Guth,[236] who extended and confirmed their findings, suggested that selenite be used in selective media for the isolation of typhoid bacilli. It was not until 1936, however, that a medium of this type was more fully developed by Leifson[119] and later recommended by others.[231] The medium consists of a basal nutrient broth to which is added 0.4% $NaHSeO_3$ (0.026 M). Many *Salmonella* species grow well in this medium,[237] whereas the growth of other enteric bacteria, such as *E. coli*, is strongly inhibited.

Occasionally other organisms are found that tolerate high selenium levels. In 1916 Brenner[238] isolated two types of bacteria from the mud of Kiel harbor in Germany. One, which he named *Micrococcus selenicus*, developed deep red colonies but failed to grow unless selenium was added. The best growth was obtained with a mixture of Na_2Se (impure) and Na_2SeO_3. Selenium, however, was not required for growth; Brenner found that he could replace the selenium mixtures with dyes such as indigo carmine, litmus, or methylene blue, or even with sodium thiosulfate. It appears that the selenium mixture functioned to maintain the oxidation-reduction conditions needed by this organism for growth.

Another organism isolated from a medium high in selenium was described by Lipman and Waksman[239] in 1923. At that time interest centered on the ability of certain bacteria to grow autotrophically through oxidation of inorganic substances and coupling of the energy, released during the oxidation, to synthetic reactions. From soil to which elemental selenium had been added Lipman and Waksman reported the isolation of a rod-shaped bacterium able to oxidize selenium to selenic acid, much as *Thiobacillus thiooxidans* derives energy from the oxidation of elemental sulfur to sulfuric acid.

It is sometimes possible to adapt microorganisms that are not naturally tolerant of selenium compounds to concentrations that were originally toxic. Table XIIIE-7 summarizes a number of growth studies that illustrate adaptive responses by a variety of bacteria, fungi, and algae; most of these experiments do not distinguish between adaptive responses that involve the selection of mutants and adaptive responses that involve most or all of the inoculum cells.

vi. Replacement of Sulfur-Containing Nutrients by Selenium Analogs. No organism can exist without a supply of sulfur in some form. Partial replacement of sulfur nutrients by selenium compounds, however, is known. A methionine-requiring mutant of *E. coli*, for example, was subcultured for more than 100 generations with selenomethionine in place of methionine and with sulfate to supply the other sulfur needs of these cells. Growth, though somewhat slower than with methionine, was exponential; the enzymes necessary for growth and division were apparently functioning despite their altered state.[152,246] Strain differences, however, appear to influence the extent to which methionine can be replaced. In one strain of *E. coli* a similar replacement was possible,[247] but in another strain selenomethionine supported growth for only about five generations.[248]

In *Chlorella vulgaris* selenomethionine inhibited division, but growth of the cell continued up to a point where division resumed. During this period of cell enlargement, methionine, derived from sulfate, no longer was incorporated into cell proteins. Presumably, selenomethionine replaced the normal sulfur metabolite and gave rise to altered proteins, some of which were unable to function in division.[153,249]

Replacement without loss of function is again seen with the precursor of coenzyme A, pantethine.[250] This sulfur-containing cofactor is required for the growth of *Lactobacillus helveticus*, but the selenium analog replaces it, mole for mole.

Table XIIIE-7   Adaptations to selenium compounds

ORGANISM	CHARACTERISTICS OF ADAPTATION	SELENIUM COMPOUND	REF.
Bacillus coli-communis, *Streptococcus pyogenes-aureus*	No growth in high concentrations at 24 hours; growth at 72 hours.	Selenite, Selenate, Selenocyanate	227
Escherichia coli	Lag followed by renewed growth (0.027 M, 0.4%).	Selenite	119
Escherichia coli, Proteus vulgaris, Salmonella thompson	Viable counts drop and then increase.	Selenite	232
Escherichia coli	Trained to grow with $4 \times 10^{-2}\,M\,K_2SeO_4$ and $2 \times 10^{-4}\,M$ K_2SO_4; adaptation permanent.	Selenate	240
Candida albicans	Colonies from selenite plates subcultured with $10^{-2}\,M$ selenite; adaptation permanent.	Selenite	167, 241
Anacystis nidulans	Trained to grow in 20 mg/100 ml; resistance lost after subculture without selenate.	Selenate	242
Chlorella vulgaris	Resistance correlated with decreased permeability to methionine and to Se analog; adaptation permanent.	Seleno-methionine	153, 243 244, 245
Chlamydomonas reinhardi, *Chlorella pyrenoidosa*	Growth curves indicate adaptive response.	Seleno-methionine	233

An attempt to replace sulfate with selenate in a strain of *E. coli* gave a highly reduced yield of cells per gram of glucose consumed during growth;[251] the growth that occurred seems to have been at the expense of sulfur, which was found to be a contaminant in the selenate. Most of the selenate used during growth was incorporated into protein; but when the protein was examined by enzyme hydrolysis, only selenomethionine, but no selenocysteine, could be detected.

vii.   Selenium as a Micronutrient.   Sodium selenate, in work cited by Levine,[227] increased the development of molds. No further details were provided. In addition to this mention, there is another report of growth stimulation caused by selenium. Three species of *Chrysochromulina*, members of a group of yellow-green algae known as chrysomonads, responded to the addition of 10 ng % to 1 μg % selenium as H_2SeO_3.[252] The dose response curve was uneven, but the values were significantly greater than those for non-selenium controls. There has been no further work with microorganisms, comparable in scope to the experiments with *Astragalus*,[150,171] to see whether selenium stimulates growth or is required for it. However, the experiments of Pinsent,[174] which showed that cells of *E. coli* needed selenite to produce the enzyme formic dehydrogenase, can perhaps serve as a guide to an approximate functional level of selenium. She found that enzyme activity was induced during growth of this organism in a highly purified medium only if ferrous, molybdate, and selenite ions were present. Growth in the absence of molybdate and selenite, however, was normal. Since the salts in the growth medium had been purified to the extent that at least 10^{-8} M selenite had to be added for enzyme induction, it may be presumed that a selenium requirement for the growth of *E. coli*, if one exists, will probably be less than 10^{-8} M.

The isolation of a *Thiobacillus*-like bacterium that oxidized elemental selenium to selenate[239] may have represented a case of selenium requirement. But beyond this brief published report, no additional details were forthcoming.

c. **Uptake of Selenium Compounds.**   Absorption studies show that the total amount of selenium within the cells of microorganisms can be higher than the initial external concentration, that respiratory inhibitors reduce uptake, and that structural analogs exhibit a competitive antagonism.[132,153,167,241] Results of this kind suggest an active transport but require, in addition, evidence that the accumulated selenium is the same as the selenium originally provided to the cells. Such an active transport for selenomethionine has been demonstrated in the alga *Chlorella vulgaris*, in which the selenoamino acid predominated in cell extracts.[169] There is also evidence of an active transport mechanism in *Penicillium* and *Aspergillus* species.[141,142] The criteria of inhibition by respiratory inhibitors and structural analogs were met; but in addition it was shown that temperature and pH, as well as a number of nutritional changes which caused repression or derepression of the sulfate permease, all had the same effect on the selenate, thiosulfate, and molybdate ions. In these fungi, therefore, a common permease for the four ions exists.

d. **Influence of Selenium on Development and Form.**   Selenium compounds are known to selectively influence some aspect of cell metabolism without being entirely toxic. As a result, the cell continues to function but in an altered way; either the form of the cell changes or the cell fails to divide but continues to grow. Selenate, for example, inhibited development of the cap-like, apical structure of *Acetabularia*, a large, single-celled alga that attains a height of 4–6 cm.[254] Through its interference with sulfur metabolism, selenate is thought to have hindered the production and function of morphogenetic substances whose synthesis is nuclear controlled.

Another example of selective metabolic inhibition is the effect of selenomethionine on the division of *Chlorella vulgaris*.[153] Cell growth was not inhibited and the growing cells remained metabolically active despite the synthesis of altered proteins.[249] The giant cells eventually adapted, resumed division, and gave rise to progeny normal in size and resistant to the growth-uncoupling effect of the selenomethionine.[243,244] During this adaptation period, permeability changed so that methionine (and presumably selenomethionine) was excluded. The profound changes within the cell, induced by the selenium analog in some way involved sulfur metabolism, since sulfur starvation restored both sensitivity to the selenium analog and active transport of methionine.[245] The sulfur pathway seemed to control the functioning or the synthesis of the transport system for the methionine molecule and, by inference, the transport of the selenomethionine molecule as well.

The growth uncoupling effect is apparently restricted to *Chlorella vulgaris;* other microorganisms, including *Chlorella pyrenoidosa*, failed to respond to selenomethionine in the same way. The effect seems specific for selenomethionine as well; a variety of methionine structural analogs are known, but none of them elicited the same response as selenomethionine.[255] (An isolated study reports that another selenium compound, selenite, arrested division of *Salmonella heidelberg* but allowed the cells to elongate.[256]) The implication of these results is that the S atom in methionine is critical for cell division of this one strain of *Chlorella*.

Selenium compounds can exert other surprising effects. One unusual response was shown by a filamentous strain of the fungus *Candida albicans*.[241] The filamentous habit of growth is caused by a genetic block to some step or steps of the division mechanism. In some way selenite counteracted the effect of this block and allowed cell division to resume so that the filaments were transformed into the more typical yeast-like single cells. The response was not restricted to this mutant but was found also in other fungal species. Cysteine also changed filaments to single cells, but the relation between the action of the sulfur amino acid and that of selenite has not been worked out.

e. Metabolism of Selenium Compounds

i. Organic Selenium Compounds in Microorganisms.   A number of organic selenium compounds reported in microorganisms is listed in Table XIIIE-8. The first one, APSe, was discussed in Section E.1.f in connection with similarities between selenium and sulfur metabolism in higher plants.

Table XIIIE-8   Selenium compounds reported in microorganisms or from enzyme reactions

COMPOUND	ORGANISM	REF.
Adenosine phospho-selenate (APSe)	*Saccharomyces cerevisiae*	217, 218
	Desulfovibrio desulfuricans	219
	Clostridium nigrificans	
Selenomethionine	*Saccharomyces cerevisiae*	257
	Candida albicans	258, 261
	Escherichia coli	188, 232, 287
	Ewe rumen bacteria[a]	262
Se-Adenosylseleno-methionine	*Saccharomyces cerevisiae*	259, 260
Selenocystine	*Saccharomyces cerevisiae*	257
	Candida albicans	258, 261
	Escherichia coli, Proteus vulgaris, Salmonella thompson	232
Dimethyl selenide	*Scopulariopsis brevicaulis*	264, 265
	Aspergillus niger	
	Penicillium notatum	

[a] But see Paulson *et al.*[263] for a comparable experiment in which selenomethionine could not be detected.

The existence of selenomethionine in microorganisms is no longer in question. It is a fairly stable compound, and its chromatographic properties are sufficiently different from those of other selenium compounds to allow its ready identification. *Se*-Adenosylselenomethionine, the analog of *S*-adenosylmethionine, which is an important intermediate in the metabolism of methionine, has also been authenticated. It has been isolated from a reaction mixture in which selenomethionine was incubated with a methionine-activating enzyme derived from yeast.

The occurrence of selenocysteine, one of the most troublesome seleno-metabolites to identify, is cause for disagreement because it is labile, and also because artifacts are known to arise between sulfhydryl and selenium compounds. Reports of this compound should be read with these possibilities kept in mind.

Certain fungi convert selenite and selenate to dimethyl selenide, a volatile compound. This compound has also been detected in exhalations from rats injected with selenate[266,267] and in mouse liver extracts incubated with selenite.[268] In fungi, methionine serves as the best methyl donor. The theoretical mechanism shown in scheme 4 for the reduction and methylation has been suggested.[264]

$$H_2SeO_3 \longrightarrow H^+ + \; :\!Se\!\!\begin{smallmatrix} O^- \\ \diagup \\ \diagdown \\ O \end{smallmatrix}\!\!OH \xrightarrow{\; CH_3^+ \;}$$

(4)

Ion

$$CH_3Se\!\!\begin{smallmatrix} O \\ \diagup \\ \diagdown \\ O \end{smallmatrix}\!\!OH \xrightarrow[\text{and reduction}]{\text{ionization}} CH_3Se\!\!\begin{smallmatrix} O^- \\ \diagup \\ \diagdown \\ O \end{smallmatrix}\!\!: \xrightarrow{\; CH_3^+ \;}$$

Methaneselenonic   Ion of Methaneseleninic
Acid   Acid

$$(CH_3)_2Se\!\!\begin{smallmatrix} O \\ \diagup \\ \diagdown \\ O \end{smallmatrix} \xrightarrow{\text{reduction}} (CH_3)_2\ddot{S}e:$$

Dimethyl   Dimethyl
Selenone   Selenide

ii. Selenium Compounds as Substrates for Isolated Enzymes. Isolated enzyme systems that normally convert sulfur metabolites convert their selenium analogs equally well. The enzymes listed in Table XIIIE-9 represent various stages of sulfur metabolism, from the first step in sulfate assimilation to the terminal sequences in which sulfur amino acids are readied for incorporation into polypeptides. Also included is an esterase capable of cleaving thiol- and selenolesters, both of which are artificial substrates.

The interchangeability of methionine and selenomethionine in protein synthesis is illustrated by an experiment in which the two amino acids were used separately as substrates for an *E. coli* methionyl-tRNA synthetase that attaches methionine to a transfer RNA molecule before polypeptide synthesis.[276] The two "substrate" molecules were competitors with almost the same affinity for the enzyme ($K_m = 7.0 \times 10^{-6}$ for methionine; $K_m = 1.1 \times 10^{-5}$ for selenomethionine); utilization of both molecules was almost equal.

Such lines of evidence raise serious doubt about what has generally been held to be the mechanism of selenium toxicity, namely, that selenium analogs interfere with sulfur metabolism by forming complexes with the active sites of enzymes and thereby inactivating them. Rather, the work with isolated enzymes shows that selenium analogs are readily metabolized and enter terminal products such as proteins. It is possible, of course, that the toxicity of selenite may result in part from its strong oxidizing properties

Table XIIIF-9   Utilization of selenium analogs by enzyme systems

ENZYME	SOURCE	REACTION	REF.
ATP: sulfate adenylyltransferase	*Saccharomyces cerevisiae*	$ATP + SeO_4^{2-} \rightarrow$ adenosine phosphoselenate $+ PP$	217, 218, 219
NADPH sulfite and nitrite reductase	*Escherichia coli*	$SeO_3^{2-} \xrightarrow{NADPH} H_2Se$	220
ATP: L-methionine S-adenosyltransferase	*Saccharomyces cerevisiae* and rabbit liver	$ATP +$ selenomethionine $\rightarrow$ Se-adenosylseleno-methionine $+ PP + P$	221, 222, 269
Methyltransferase	Rat liver		270
	Pig liver	Se-Adenosylselenomethionine $+$ guanidinoacetic acid $\rightarrow$ creatine	221
	Rat liver microsomes	Se-Adenosylselenomethionine $+$ microsomes $\rightarrow$ choline	271
	Rat liver microsomes	Se-Adenosylselenomethionine $+$ 2-methoxy-ethanthiol $\rightarrow$ O,S-dimethylmercaptoethanol	271
	Rat ventral prostate	Se-Adenosylselenomethionine $+$ putrescine $\rightarrow$ spermidine	272
L-Methionine: tRNA ligase (AMP)	*Escherichia coli*	$ATP +$ selenomethionine $\rightarrow$ AMP-seleno-methionine adenylate $+ PP$	273
	Sarcina lutea		274
	Rat lens	$ATP +$ selenomethionine $+$ tRNA $\rightarrow$	275
	Escherichia coli	L-selenomethionyl-tRNA $+ AMP + PP$	276
Acetylcholinesterase	Rat spinal cord	Acetylselenocholine $\rightarrow$ cholineselenol	277

and that enzymes may some day be found which are inactivated by a selenium antimetabolite. But for analogs such as selenate, the mechanism of toxicity requires reinterpretation. One explanation is that Se atoms replace essential sulfur groupings during protein synthesis so that altered, malfunctioning enzymes arise.[110,176]

iii.   Selenopolypeptides.   The effect on metabolic activity of substituting selenium for sulfur has been investigated with several polypeptides. Putidaredoxin,[278,279] an iron-sulfur enzyme derived from *Pseudomonas putida*, hydroxylates methylene C(5) of (+)-camphor; it is distinguished by the presence of two acid-labile atoms of sulfur. Replacement of these two S atoms led to several changes in physical properties which indicated that the Se atoms did not occupy precisely the same position in the protein as the two acid-labile S atoms. Of significance was the absence of any pronounced effect on enzyme activity. Comparable results were obtained with adrenodoxin, an iron-sulfur protein from pig and beef adrenals.[279] The adrenal selenoproteins were as active as the iron-sulfur protein in a deoxycorticosterone-11-β-hydroxylase assay.

Parsley ferredoxin, a similar protein isolated from chloroplasts, also contains two Fe and two S atoms. Replacement of the sulfur by selenium caused some changes in spectral properties and oxidation-reduction properties and a 20% reduction in its ability to reduce cytochrome-c. The selenium homolog proved to be less stable than the natural substance.[279a,279b]

Oxytocin, an animal hormone, contains two cysteine residues joined by a disulfide bridge. Selenium analogs of this polypeptide have been synthesized with either the S(1) or the S(6) atom, or both, replaced by a Se atom.[280–286] Physiological potencies varied with the test and with the analog, but activity was never completely lost (*cf.* Chapter XIIB).

Sometimes it is possible to substitute an amino acid analog for a natural amino acid in the proteins of an organism and to isolate functional, though somewhat altered, enzymes. An attempt to replace sulfate with selenate in a strain of *E. coli* although it led to poor growth, yielded enough of a purified selenoenzyme, β-galactosidase, to allow comparison with the normal enzyme obtained from sulfate-grown cells.[287] Approximately 80 of the 150 methionine residues in this large protein (mol. wt. about 540,000) had been replaced by selenomethionine, but none of the 80 cystine residues had been replaced by selenocystine. Despite the high degree of replacement of methionine, the final purified material was as active as the normal enzyme.

In another approach to this problem, a different strain of *E. coli* was grown with selenomethionine; 70–75% of the methionine in the purified β-galactosidase was replaced under these experimental conditions. Again the sulfur and selenium enzymes were equally active.[247]

From these few experiments with selenoproteins it appears that a major replacement of methionine by selenomethionine does not change the catalytic activity, but that replacement of cysteine by selenocysteine seems to have some effect. The stability of the polypeptides also appears to decrease. The hemiseleno isologs of oxytocin, for example, tended to form aggregates of higher molecular weight to a greater extent than did the disulfide molecules.[282] The seleno-β-galactosidase was less stable to several agents such as heat and urea, though it renatured more readily than did the sulfur protein.[287] In addition, significantly less of the selenoenzyme was produced in selenate-grown cells than of the normal enzyme in sulfate-grown cells. A relatively large proportion of small, inactive proteins, higher in selenomethionine than the larger β-galactosidase and having the characteristics of the subunits that constitute the active enzyme, was detected.

The findings discussed above, admittedly limited, indicate that a substitution of selenium for sulfur, particularly selenocystine for cystine, can give rise to malfunctioning selenoenzymes. The instability of selenoenzymes when present within the milieu of the cell, effectively lowering enzyme levels and metabolic rates, also seems to be factor in selenium toxicity.

iv. Reduction of Selenite to Elemental Selenium. The ability to convert selenite to elemental selenium, known almost from the first biological work with selenium (see the review by Levine[227]) is characteristic of many organisms. The distinguishing red color was often attributed to the action of such strong reducing agents as ascorbic acid or glutathione, known to carry out the reduction *in vitro*, although a metabolic reduction sometimes was indicated.[8] In recent years, a number of investigators have described a metabolic reduction of selenite by microorganisms. Table XIIIE-10 summarizes several characteristics of selenite reduction both by intact cells and by cell-free extracts.

Although strain differences exist, reduction is generally favored by anaerobic conditions. Flavin compounds are sometimes needed, an indication that the reduction of selenite is coupled to the electron flow mechanism of the cell. Ties to other phases of cell metabolism may also exist, as seen from reduction by certain methionine-requiring mutants of the fungus *Neurospora*. Enzyme activity has been found to exist mainly in $18,000-25,000 \times g$ supernatants, an indication that the enzyme is not particulate. There is evidence for a divalent Se^{2+} compound as an intermediate, as well as for a further reduction of the elemental selenium to a Se^{2-} compound in some organisms. The intermediate steps in the reduction of selenite remain to be elucidated.

v. Microbial Oxidation of Selenium Compounds. Direct microbial oxidation of reduced forms of selenium has received limited investigation. *Aspergillus niger*, a fungus, is reported to have oxidized selenite to selenate,[295]

Table XIIIE-10 Reduction of selenite to elemental selenium

ORGANISM	CHARACTERISTICS OF REDUCTION			
	INTACT CELLS	REF.	CELL-FREE EXTRACTS	REF.
Streptococcus faecalis N83	Reduces high concentrations under aerobic and anaerobic conditions; flavin compounds needed; ascorbate enhances reduction	288, 289	Oxygen sensitive, heat stable; FAD or riboflavin for maximum reduction; inhibited by $HgCl_2$, iodoacetamide, and other inhibitors; $K_m = 0.00135$.	288, 289
Streptococcus faecium K6A	Reduces low concentrations under anaerobic but not aerobic conditions; no effect of flavin compounds; ascorbate enhances reduction	288, 289	Oxygen sensitive, heat stable; FAD or riboflavin for maximum reduction; inhibited by $HgCl_2$, iodoacetamide, and other inhibitors; $K_m = 0.0020$; sensitive to cell rupture; activity restored by cysteine.	288, 289
Candida albicans RM 806	Reduction inhibited by dinitrophenol, reversed by FMN; stimulated by anaerobiosis.	167, 290, 253, 291	Enzyme activity in 18,000 $\times$ g supernatant; loss of activity by dialysis restored by boiled yeast extract or by combination of glucose-6-phosphate, TPN, GSSG, and menadione.	253, 291

Salmonella heidelberg	Red amorphous granules in cells identified as Se by X-ray analysis; intermediate in reduction is a divalent form of Se^{2+}	256
Neurospora crassa	Methionine-requiring mutants blocked after homocysteine unable to reduce selenite in absence of methionine; methionine-requiring mutants blocked before homocysteine reduced selenite with or without necessary growth factors.	292
Escherichia coli	Reduction depends on growth; reducing system not induced but is present in normal cells	293
Micrococcus lactilyticus	Enzyme activity in $25{,}000 \times g$ supernatant; H_2 used in reduction; two steps: (a) selenite to elemental Se, (b) elemental Se to H_2Se; other ions such as tellurite and uranyl acetate also reduced.	294

and two bacterial species are said to have converted elemental selenium to selenate.[239,296] Unfortunately, these interesting but brief accounts were not followed by more extensive study. There are a number of indirect bits of evidence for microbial oxidations;[25,87,106-108,115,297,298] these experiments are ambiguous, however, because the possibility of oxidation by multicellular organisms or even nonbiological oxidation was not excluded.

3. Biological Cycling of Selenium

Estimates of the amount of selenium in the earth's crust range from 0.03 to 0.8 ppm;[72] in the oceans, the average selenium content is approximately 0.09 ppm.[58] These concentrations, lower than those of many other elements, nevertheless reveal that, in total quantity, appreciable selenium exists in the biosphere. Few attempts have been made to quantitate the amount of selenium removed by plants from soils under natural conditions; however, according to one rough calculation,[29] *Xylorhiza*, a primary accumulator, removed a ton of selenium from 2000 acres of seleniferous soil in a single season's growth.* Decay of this and similar seleniferous plant material returns considerable selenium, often changed in form, to the soil. Volatilization of soil selenium can also be expected to release some selenium to the atmosphere.[300,301]

The role of plants in cycling of this type has been recognized by several investigators.[46,302] It has been suggested that plants cycled selenium as far back as the early Cretaceous period, and perhaps even as early as the Paleozoic era.[23,24,33,46] Evidence for this hypothesis derives from the high selenium content in carboniferous shales, as much as 150 ppm in some specimens.[23] It has even been suggested that selenium accumulator plants may have existed so abundantly in the Cretaceous that they were responsible for the extinction of dinosaurs, massive consumers of herbage.[303]

The cycling of selenium can also be discussed in terms of biochemical valence changes, brought about by plants, microorganisms, and animals. The relative contribution made by each group of organisms has not yet been evaluated, but plant accumulator species probably provide a key share. As shown in the cycle depicted by Fig. XIIIE-5, plants, bacteria, and fungi are able to assimilate selenate and selenite. Animals injected with selenate are known to exhale dimethyl selenide,[266,267] but there is the recurrent question of whether the reduction was carried out by the animal cell or through the action of organisms in the gastrointestinal tract. Selenite, however, is enzymatically reduced to dimethyl selenide by mouse liver extracts,[268]

* By comparison, 2000 acres of land plants fix about 1×10^3 tons of carbon each year.[299] See ref. 75 for a hypothetical estimate of the daily flux of selenium in a low-selenium soil-plant-animal system on 1 acre of land.

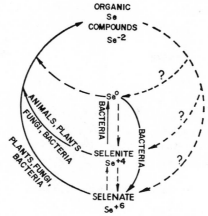

FIG. XIIIE-5   Biological selenium cycles.

and the trimethylselenonium ion has been identified as an excretory product in the urine of rats.[304–306] Bacteria also reduce selenite, but the end product is elemental selenium.

Microorganisms can logically be expected to play a leading part in a selenium cycle, much as they do in the carbon, nitrogen, and sulfur cycles. Their involvement in the reductive part of the cycle is established, but their full participation in the remainder of the cycle either requires corroboration (indicated by a broken line in Fig. XIIIE-5) or has yet to be established (indicated by dotted lines and question marks). The contribution of non-biological oxidations to the cycle also remains to be evaluated. Although we know how selenium enters the cycle, considerable research is needed to determine how the element is returned to its initial oxidized state.

REFERENCES

1. C. A. Cameron, *Sci. Proc. Roy. Dublin Soc.*, **2**, 231 (1880).

2. W. Knop, *Ber. Verhandl. K. Sachs. Ges. Wiss. Math.-phys. Classe*, 39 (1884–1885).

3. M. Awschalom, *Rev. Fac. Agron. La Plata, Ser. 2*, **14**, 122 (1921).

4. V. E. Levine, *Amer. J. Bot.*, **12**, 82 (1925).

5. M. J. Stoklasa, *C. R. Acad. Sci., Paris*, **174**, 1075 (1922).

6. M. J. Stoklasa, *C. R. Acad. Sci., Paris*, **174**, 1256 (1922).

7. M. J. Stoklasa, *Biochem. Z.*, **130**, 604 (1922).

8. B. Turina, *Biochem. Z.*, **129**, 507 (1922).

9. H. G. Byers, *U.S. Dept. Agr. Tech. Bull.*, **482**, 1 (1935).

10. S. F. Trelease and O. A. Beath, "Selenium," published by the authors, New York, N.Y., 1949.

11. I. Rosenfeld and O. A. Beath, "Selenium," Academic Press, New York, N.Y., 1964.

12. W. O. Robinson, *J. Assoc. Offic. Agr. Chemists*, **16,** 423 (1933).

13. O. A. Beath, J. H. Draize, H. F. Eppson, C. S. Gilbert, and O. C. McCreary, *J. Amer. Pharm. Assoc.*, **23,** 94 (1934).

14. F. Taboury, *C. R. Acad. Sci., Paris*, **195,** 171 (1932).

15. T. Gassmann, *Hoppe-Seylers Z. Physiol. Chem.*, **98,** 182 (1916–17).

16. T. Gassmann, *Hoppe-Seylers Z. Physiol. Chem.*, **108,** 38 (1919–20).

17. R. Fritsch, *Hoppe-Seylers Z. Physiol. Chem.*, **104,** 59 (1919).

18. R. Fritsch, *Hoppe-Seylers Z. Physiol. Chem.*, **109,** 186 (1920).

19. O. H. Muth and W. H. Allaway, *J. Amer. Vet. Med. Assoc.*, **142,** 1379 (1963).

20. O. A. Beath, J. H. Draize, and C. S. Gilbert, *Wyoming Agr. Exp. Sta. Bull.*, **200,** 1 (1934).

21. O. A. Beath, H. F. Eppson, and C. S. Gilbert, *Wyoming Agr. Exp. Sta. Bull.*, **206,** 1 (1935).

22. O. A. Beath, *Science*, **83,** 104 (1936).

23. O. A. Beath and C. S. Gilbert, *Science*, **84,** 484 (1936).

24. O. A. Beath, *Wyoming Agr. Exp. Sta. Bull.*, **221,** 29 (1937).

25. O. A. Beath, H. F. Eppson, and C. S. Gilbert, *J. Amer. Pharm. Assoc.*, **26,** 394 (1937).

26. O. A. Beath, C. S. Gilbert, and H. F. Eppson, *Amer. J. Bot.*, **24,** 96 (1937).

27. O. A. Beath, *Amer. J. Bot.*, **26,** 729 (1939).

28. O. A. Beath, H. F. Eppson, C. S. Gilbert, and W. B. Bradley, *Wyoming Agr. Exp. Sta. Bull.*, **231,** 1 (1939).

29. O. A. Beath, C. S. Gilbert, and H. F. Eppson, *Amer. J. Bot.*, **26,** 257 (1939).

30. O. A. Beath, C. S. Gilbert, and H. F. Eppson, *Amer. J. Bot.*, **26,** 296 (1939).

31. O. A. Beath, C. S. Gilbert, and H. F. Eppson, *Amer. J. Bot.*, **27,** 564 (1940).

32. O. A. Beath, C. S. Gilbert, and H. F. Eppson, *Amer. J. Bot.*, **28,** 887 (1941).

33. O. A. Beath, *Amer. J. Bot.*, **30,** 698 (1943).

34. H. G. Byers, *U.S. Dept. Agr. Tech. Bull.*, **530,** 1 (1936).

35. H. G. Byers, K. T. Williams, and H. W. Lakin, *Ind. Eng. Chem.*, **28,** 821 (1936).

36. H. G. Byers, *Ind. Eng. Chem.*, **29,** 1200 (1937).

37. H. G. Byers, J. T. Miller, K. T. Williams, and H. W. Lakin, *U.S. Dept. Agr. Tech. Bull.*, **601,** 1 (1938).

38. H. G. Byers and H. W. Lakin, *Can. J. Res.*, **17,** 364 (1939).

39. S. H. Knight, *Wyoming Agr. Exp. Sta. Bull.*, **221,** 3 (1937).

40. H. W. Lakin and H. G. Byers, *U.S. Dept. Agr. Tech. Bull.*, **783,** 1 (1941).

41. H. W. Lakin and F. J. Hermann, *Amer. J. Bot.*, **27,** 245 (1940).

42. H. W. Lakin and H. G. Byers, *U.S. Dept. Agr. Tech. Bull.*, **950,** 1 (1948).

43. J. T. Miller and H. G. Byers, *J. Agr. Res.*, **55,** 59 (1937).

44. A. L. Moxon, *S. Dakota Agr. Exp. Sta. Bull.*, **311,** 1 (1937).

45. A. L. Moxon, O. E. Olson, W. V. Searight, and K. M. Sandals, *Amer. J. Bot*, **25,** 794 (1938).

46. A. L. Moxon, O. E. Olson, and W. V. Searight, *S. Dakota Agr. Exp. Sta. Tech. Bull.*, **2,** 1 (1939).

47. O. E. Olson, D. F. Jornlin, and A. L. Moxon, *J. Amer. Soc. Agron.*, **34**, 607 (1942).

48. O. E. Olson, D. F. Jornlin, and A. L. Moxon, *Soil Sci.*, **53**, 365 (1942).

49. O. E. Olson, E. I. Whitehead, and A. L. Moxon, *Soil Sci.*, **54**, 47 (1942).

50. W. V. Searight and A. L. Moxon, *S. Dakota Agr. Exp. Sta. Tech. Bull*, **5**, 1 (1945).

51. W. V. Searight, A. L. Moxon, E. I. Whitehead, and F. G. Viets, Jr., *Proc. S. Dakota Acad. Sci.*, **26**, 87 (1946–1947).

52. K. T. Williams and H. G. Byers, *Ind. Eng. Chem., Anal. Ed.*, **6**, 296 (1934).

53. K. T. Williams and H. G. Byers, *Ind. Eng. Chem.*, **28**, 912 (1936).

54. K. T. Williams, *J. Assoc. Offic. Agr. Chemists*, **20**, 225 (1937).

55. K. T. Williams, *U.S. Dept. Agr. Yearbook*, 830 (1938).

56. K. T. Williams, H. W. Lakin, and H. G. Byers, *U.S. Dept. Agr. Tech. Bull.*, **702**, 1 (1940).

57. K. T. Williams, H. W. Lakin, and H. G. Byers, *U.S. Dept. Agr. Tech. Bull.*, **758**, 1 (1941).

58. H. W. Lakin and D. F. Davidson, "Selenium in Biomedicine," O. H. Muth, Ed., AVI Publishing Co., Westport, Conn., 1967, p. 27.

59. P. A. Munz and D. D. Keck, "A California Flora," University of California Press, Berkeley, Calif., 1959.

60. R. C. Barneby, "Atlas of North American Astragalus," Memoirs of the New York Botanical Garden, Vol. 13, Parts I and II, 1964.

61. C. W. R. McRay and I. S. H. Hurwood, *Queensland J. Agr. Res.*, **20**, 475 (1963).

62. S. Ravikovitch and M. Margolin, *Israel Agr. Res. Sta., Rehovot 1957 Series 145-E*, **7**, 41 (1957).

63. S. Ravikovitch and M. Margolin, *Israel Sci. Soc. Conv. Proc. 2, Biol. Geol. Bull. 6B*, 265 (1957).

64. J. M. M. Brown and P. J. DeWet, *Onderstepoort J. Vet. Res.*, **29**, 111 (1962).

65. M. E. Jones, "Revision of North American Species of *Astragalus*," published by the author, Salt Lake City, Utah, 1923.

66. G. F. Ledingham, *Can. J. Genet. Cytol.*, **2**, 119 (1960).

67. G. F. Ledingham and B. M. Rever, *Can. J. Genet. Cytol.*, **5**, 18 (1963).

68. A. Shrift, *Ann. Rev. Plant Physiol.*, **20**, 475 (1969).

69. K. C. Beeson, "Selenium in Agriculture," M. S. Anderson, H. W. Lakin, K. C. Beeson, F. F. Smith, and E. Thacker, Eds., *U.S. Dept. Agr. Handbook*, **200**, 34 (1961).

70. W. H. Allaway and J. F. Hodgson, *J. Animal Sci.*, **23**, 271 (1964).

71. W. H. Allaway and E. E. Cary, *Feedstuffs*, **38**, 62 (1966).

72. H. W. Lakin, "Selenium in Agriculture," M. S. Anderson, H. W. Lakin, K. C. Beeson, F. F. Smith, and E. Thacker, Eds., *U.S. Dept. Agr. Handbook*, **200**, 27 (1961).

73. O. E. Olson and A. L. Moxon, *Soil Sci.*, **47**, 305 (1939).

74. K. W. Franke and E. P. Painter, *Ind. Eng. Chem.*, **29**, 591 (1937).

75. W. H. Allaway, E. E. Cary, and C. F. Ehlig, "Selenium in Biomedicine," O. H. Muth, Ed., AVI. Publishing Co., Westport, Conn., 1967, p. 270.

76. J. Kubota, W. H. Allaway, D. L. Carter, E. E. Cary, and V. A. Lazar, *J. Agr. Food Chem.*, **15**, 448 (1967).

77. E. D. Andrews, W. J. Hartley and A. B. Grant. *N. Zealand Vet. J.*, **16**, 3 (1968).

78. G. A. Fleming, *Soil Sci.*, **94**, 28 (1962).

79. F. Kerdel-Vegas, *Separata Revista Dermatol. Venezolana*, **4**, 109 (1964).

80. J. H. Watkinson, *Nature*, **202,** 1239 (1964).

81. A. M. Hurd-Karrer, *J. Agr. Res.*, **50,** 413 (1935).

82. A. M. Hurd-Karrer, "Annual Report of the Board of Regents of the Smithsonian Institution," U.S. Government Printing Office, Washington, D.C., Publ. No. 3348, 289 (1936).

83. A. M. Hurd-Karrer, *J. Agr. Res.*, **54,** 601 (1937).

84. B. Bisbjerg and G. Gissel-Nielsen, *Plant and Soil*, **31,** 287 (1969).

85. J. C. Lane and G. A. Fleming, *Trans. Comm. II and IV, Intern. Soc. Soil Sci., Aberdeen*, 289 (1966).

86. J. R. Lessard, M. Hidiroglou, R. B. Carson, and P. Dermine, *Can. J. Plant Sci.*, **48,** 581 (1968).

87. C. J. Asher, C. S. Evans, and C. M. Johnson, *Australian J. Biol. Sci..* **20,** 737 (1967).

88. B. G. Lewis, C. M. Johnson, and C. C. Delwiche, *Agr. Food Chem.*, **14,** 638 (1966).

89. G. Gissel-Nielsen, *Plant and Soil*, **32,** 242 (1970).

90. G. A. Fleming and T. Walsh, *Proc. Roy. Irish Acad.*, **58,** 151 (1957).

91. J. S. Webb and W. J. Atkinson, *Nature*, **208,** 1056 (1965).

92. J. S. Webb, I. Thornton, and K. Fletcher, *Nature*, **211,** 327 (1966).

93. H. L. Cannon, *U.S. Geol. Survey Bull. No. 1085-A*, 1 (1960).

94. H. L. Cannon, *U.S. Geol. Survey Bull. No. 1030-M*, 399 (1957).

95. H. L. Cannon, *Science*, **132,** 591 (1960).

96. D. S. Ortiz and A. Carrasquero, *Agron. Trop. (Marcay, Venezuela)*, **18,** 369 (1968).

97. E. E. Cary, G. A. Wieczorek, and W. H. Allaway, *Proc. Soil Sci. Soc. Amer.*, **31,** 21 (1967).

98. P. L. Gile, H. W. Lakin, and H. G. Byers, *J. Agr. Res.*, **57,** 1 (1938).

99. P. L. Gile and H. W. Lakin, *J. Agr. Res.*, **63,** 559 (1941).

100. O. E. Olson, *Proc. S. Dakota Acad. Sci.*, **19,** 22 (1939).

101. L. W. Strock, *Amer. J. Pharm.*, **107,** 144 (1935).

102. E. B. Davies and J. H. Watkinson, *N. Zealand J. Agr. Res.*, **9,** 641 (1966).

103. J. H. Watkinson and E. B. Davies, *N. Zealand J. Agr. Res.*, **10,** 116 (1967).

104. S. F. Trelease and A. A. DiSomma, *Amer. J. Bot.*, **31,** 544 (1944).

105. S. F. Trelease and S. S. Greenfield, *Amer. J. Bot.*, **31,** 630 (1944).

106. J. W. Hamilton and O. A. Beath, *Agr. Food Chem.*, **12,** 371 (1964).

107. J. W. Hamilton and O. A. Beath, *Agron. J.*, **55,** 528 (1963).

108. J. W. Hamilton and O. A. Beath, *Range Management*, **16,** 261 (1963).

109. A. Shrift and T. K. Virupaksha, *Biochim. Biophys. Acta.* **100,** 65 (1965).

110. T. K. Virupaksha and A. Shrift, *Biochim. Biophys. Acta.*, **107,** 69 (1965).

111. A. M. Hurd-Karrer, *Amer. J. Bot.*, **24,** 720 (1937).

112. G. Gissel-Nielsen and B. Bisbjerg, *Plant and Soil*, **32,** 382 (1970).

113. J. H. Watkinson and E. B. Davies, *N. Zealand J. Agr. Res.*, **10,** 122 (1967).

114. J. M. Ulrich and A. Shrift, *Plant Physiol.*, **43,** 14 (1968).

115. G. W. Butler and P. J. Peterson, *Australian J. Biol. Sci.*, **20,** 77 (1967).

116. G. W. Butler and P. J. Peterson, *N. Zealand J. Agr. Res.*, **4,** 484 (1961).

117. P. J. Peterson and D. J. Spedding, *N. Zealand J. Agr. Res.*, **6,** 13 (1963).

118. P. J. Peterson and G. W. Butler, *Nature*, **212,** 961 (1966).

119. E. Leifson, *Amer. J. Hyg.*, **24,** 423 (1936).

120. W. R. North and M. T. Bartram, *Appl. Microbiol.*, **1,** 130 (1953).

121. J. Scala and H. H. Williams, *Arch. Biochem. Biophys.*, **99,** 363 (1962).

122. J. Scala and H. H. Williams, *Arch. Biochem. Biophys.*, **101,** 319 (1963).

123. A. M. Hurd-Karrer, *Science*, **78,** 560 (1933).

124. A. M. Hurd-Karrer, *J. Agr. Res.*, **49,** 343 (1934).

125. A. M. Hurd-Karrer, *Amer. J. Bot.*, **25,** 666 (1938).

126. A. Kylin, *Physiol. Plantar.*, **13,** 366 (1960).

127. A. Kylin, *Botan. Notiser*, **113,** 49 (1960).

128. A. Kylin, *Physiol. Plantar.*, **20,** 139 (1967).

129. E. Epstein, *Plant Physiol.*, **30,** 529 (1955).

130. J. E. Leggett and E. Epstein, *Plant Physiol.*, **31,** 222 (1956).

131. N. K. Ziebur and A. Shrift, *Plant Physiol.*, **47,** 545 (1971).

132. A. Shrift, *Amer. J. Bot.*, **41,** 223 (1954).

133. C. W. Bonhorst and I. S. Palmer, *Agr. Food Chem.*, **5,** 931 (1957).

134. I. G. Fels and V. H. Cheldelin, *J. Biol. Chem.*, **176,** 819 (1948).

135. I. G. Fels and V. H. Cheldelin, *Arch. Biochem.*, **22,** 402 (1949).

136. I. G. Fels and V. H. Cheldelin, *J. Biol. Chem.*, **185,** 803 (1950).

137. G. A. Maw, *Folia Microbiol.*, **8,** 325 (1963).

138. G. A. Maw, *Pure Appl. Chem.*, **7,** 655 (1963).

139. I. G. Fels and V. H. Cheldelin, *Arch. Biochem.*, **22,** 323 (1949).

140. G. S. Weissman and S. F. Trelease, *Amer. J. Bot.*, **42,** 489 (1955).

141. J. W. Tweedie and I. H. Segal, *Biochim. Biophys. Acta*, **196,** 95 (1970).

142. L. A. Yamamoto and I. H. Segal, *Arch. Biochem. Biophys.*, **114,** 523 (1966).

143. V. R. Widstrom, *Proc. S. Dakota Acad. Sci.*, **40,** 208 (1961).

144. J. Postgate, *Nature*, **164,** 670 (1949).

145. J. Postgate, *J. Gen. Microbiol.*, **6,** 128 (1952).

146. C. Furusaka, *Nature*, **192,** 427 (1961).

147. A. L. Martin, *Amer. J. Bot.*, **23,** 471 (1936).

148. A. L. Martin, *Amer. J. Bot.*, **24,** 198 (1937).

149. A. L. Martin and S. F. Trelease, *Amer. J. Bot.*, **25,** 380 (1938).

150. S. F. Trelease and H. M. Trelease, *Amer. J. Bot.*, **25,** 372 (1938).

151. J. Opienska-Blauth and H. Iwanowski, *Acta Microbiol. Polon.*, **1,** 273 (1952).

152. D. B. Cowie and G. N. Cohen, *Biochim. Biophys. Acta*, **26,** 252 (1957).

153. A. Shrift, *Amer. J. Bot.*, **41,** 345 (1954).

154. J. Brotherton, *J. Gen. Microbiol.*, **49,** 393 (1967).

155. P. V. Benko, T. C. Wood, and I. H. Segel, *Arch. Biochem. Biophys.*, **122,** 783 (1967).

156. K. P. McConnell and G. J. Cho, *Amer. J. Physiol.*, **208,** 1191 (1965).

157. A. S. Weisberger, L. G. Suhrland, and J. Seifter, *J. Hematol.*, **11,** 1 (1956).

158. A. S. Weisberger and L. G. Suhrland, *J. Hematol.*, **11,** 11 (1956).

159. H. L. Klug and D. F. Petersen, *Proc. S. Dakota Acad. Sci.*, **28,** 87 (1949).

160. D. F. Petersen, *Proc. S. Dakota Acad. Sci.*, **30,** 53 (1951).

161. C. C. Tsen and A. L. Tappel, *J. Biol. Chem.*, **233,** 1230 (1958).

162. H. E. Ganther, *Biochemistry*, **7,** 2898 (1968).

163. H. E. Ganther and C. Corcoran, *Biochemistry*, **8,** 2557 (1969).

164. A. M. Hurd-Karrer, *J. Agr. Res.*, **52,** 933 (1936).

165. C. W. Bonhorst, *Agr. Food Chem.*, **3,** 700 (1955).

166. M. C. Mahl and E. I. Whitehead, *Proc. S. Dakota Acad. Sci.*, **40,** 93 (1961).

167. G. Falcone and W. J. Nickerson, *Giorn. Microbiol.*, **8,** 129 (1960).

168. A. Shrift and J. M. Ulrich, *Plant Physiol.*, **44,** 893 (1969).

169. A. Shrift, "Selenium in Biomedicine," O. H. Muth, Ed., AVI Publishing Co., West-port, Conn., 1967, p. 241.

170. I. Rosenfeld, *Univ. Wyoming Agr. Exp. Sta. Bull.*, **385,** 31 (1962).

171. S. F. Trelease and H. M. Trelease, *Amer. J. Bot.*, **26,** 530 (1939).

172. S. F. Trelease, *Science*, **95,** 656 (1942).

173. J. N. Thompson and M. L. Scott, *J. Nutr.*, **97,** 335 (1969).

174. J. Pinsent, *Biochem. J.*, **57,** 10 (1954).

175. D. I. Arnon and P. R. Stout, *Plant Physiol.*, **14,** 371 (1939).

175a. T. C. Broyer and C. S. Johnston, personal communication.

176. P. J. Peterson and G. W. Butler, *Nature*, **213,** 599 (1967).

176a. P. J. Peterson and P. J. Robinson, *Phytochem.*, **11,** 1837 (1972).

177. P. J. Peterson and G. W. Butler, *Australian J. Biol. Sci.*, **24,** 175 (1971).

178. A. T. Perkins and H. H. King, *J. Amer. Soc. Agron.*, **30,** 664 (1938).

179. G. W. Stanford and O. E. Olson, *Proc. S. Dakota Acad. Sci.*, **19,** 25 (1939).

180. K. von Scharrer and W. Schropp, *Z. Pflanzenernaehr. Dueng. Bodenk.*, **50,** 187 (1950).

181. T. C. Broyer, D. C. Lee, and C. J. Asher, *Plant Physiol.*, **41,** 1425 (1966).

182. W. M. Latimer, "The Oxidation States of the Elements and Their Potentials in Aqueous Solutions," Prentice-Hall, Englewood Cliffs, N.J., 1938.

183. R. E. Huber and R. S. Criddle, *Arch. Biochem. Biophys.*, **122,** 164 (1967).

184. R. Walter, D. H. Schlesinger, and I. L. Schwartz, *Anal. Biochem.*, **27,** 231 (1969).

185. A. L. Tappel and K. A. Caldwell, "Selenium in Biomedicine," O. H. Muth, Ed., AVI Publishing Co., Westport, Conn., 1967, p. 345.

186. E. I. Whitehead, C. M. Hendrick, and F. M. Moyer, *Proc. S. Dakota Acad. Sci.*, **34,** 52 (1955).

187. L. Shepherd and R. E. Huber, *Can. J. Biochem.*, **47,** 877 (1969).

188. T. Tuve and H. H. Williams, *J. Biol. Chem.*, **236,** 597 (1961).

189. J. V. Benson and J. A. Patterson, *Anal. Biochem.*, **29,** 130 (1969).

190. J. L. Martin and M. L. Gerlach, *Anal. Biochem.*, **29,** 257 (1969).

191. K. P. McConnell and C. H. Wabnitz, *Biochim. Biophys. Acta*, **86,** 182 (1964).

192. O. A. Beath and H. F. Eppson, *Univ. Wyo. Agr. Exp. Sta. Bull.*, **278,** 1 (1947).

193. P. J. Peterson and G. W. Butler, *Australian J. Biol. Sci.*, **15,** 126 (1962).

194. J. L. Martin, A. Shrift, and M. L. Gerlach, *Phytochem.*, **10,** 945 (1971).

194a. S. N. Nigam and W. B. McConnell, *Phytochem.*, **11,** 377 (1972).

195. S. N. Nigam, Jan-I Tu, and W. B. McConnell, *Phytochem.*, **8,** 1161 (1969).

196. S. N. Nigam and W. B. McConnell, *Biochim. Biophys. Acta*, **192**, 185 (1969).

197. D. M. Chen, S. N. Nigam, and W. B. McConnell, *Can. J. Biochem.*, **48**, 1278 (1970).

198. A. Shrift and T. K. Virupaksha, *Biochim. Biophys. Acta*, **71**, 483 (1963).

199. S. F. Trelease, A. A. DiSomma, and A. L. Jacobs, *Science*, **132**, 618 (1960).

200. T. K. Virupaksha, A. Shrift, and H. Tarver, *Biochim. Biophys. Acta*, **130**, 45 (1966).

201. C. G. Spare and A. I. Virtanen, *Acta Chem. Scand.*, **18**, 280 (1964).

202. A. L. Jacobs, "The Isolation and Identification of a Selenoamino Acid from Corn," Doctoral Thesis, Columbia University, 1963.

203. K. J. Jenkins and M. Hidiroglou, *Can. J. Biochem.*, **45**, 1027 (1967).

204. O. E. Olson, E. J. Novacek, E. I. Whitehead, and I. S. Palmer, *Phytochemistry*, **9**, 1181 (1970).

205. M. J. Horn and D. B. Jones, *J. Biol. Chem.*, **139**, 645 (1941).

206. T. K. Virupaksha and A. Shrift, *Biochim. Biophys. Acta*, **74**, 791 (1963).

207. L. Aronow and F. Kerdel-Vegas, *Separata Revista Dermatol. Venezolana*, **4**, 109 (1964).

208. L. Aronow and F. Kerdel-Vegas, *Nature*, **205**, 1185 (1965).

209. F. Kerdel-Vegas, F. Wagner, P. B. Russell, N. H. Grant, H. E. Alburn, D. E. Clark, and J. A. Miller, *Nature*, **205**, 1186 (1965).

210. G. J. Olivares, L. Aronow, and F. Kerdel-Vegas, *Acta Cient. Venezolana*, **18**, 9 (1967).

211. C. S. Evans, C. J. Asher, and C. M. Johnson, *Australian J. Biol. Sci.*, **21**, 13 (1968).

212. R. J. McColloch, J. W. Hamilton, and S. K. Brown, *Biochem. Biophys. Res. Commun.*, **11**, 7 (1963).

212a. V. V. S. Murti, "Advancing Frontiers in the Chemistry of Natural Products," Hindustan Publishing Corp., Delhi, India, 1965, p. 157.

213. P. M. Dunnill and L. Fowden, *Phytochemistry*, **6**, 1659 (1967).

214. J. F. Thompson, *Ann. Rev. Plant Physiol.*, **18**, 59 (1967).

215. R. J. Ellis, *Planta*, **88**, 34 (1969).

216. C. A. Adams and R. W. Rinne, *Plant Physiol.*, **44**, 1241 (1969).

217. L. G. Wilson and R. S. Bandurski, *Arch. Biochem. Biophys.*, **62**, 503 (1956).

218. L. G. Wilson and R. S. Bandurski, *J. Biol. Chem.*, **233**, 975 (1958).

219. J. M. Akagi and L. L. Campbell, *J. Bacteriol.*, **84**, 1194 (1962).

220. J. D. Kemp, D. E. Atkinson, A. Ehret, and R. A. Lazzarini, *J. Biol. Chem.*, **238**, 3466 (1963).

221. S. H. Mudd and G. L. Cantoni, *Nature*, **180**, 1052 (1957).

222. J. Skupin, *Acta Biochim. Polon.*, **9**, 253 (1962).

223. E. P. Painter and K. W. Franke, *Amer. J. Bot.*, **27**, 336 (1940).

224. P. Nissen and A. A. Benson, *Biochim. Biophys. Acta*, **82**, 400 (1964).

225. R. D. Campo, P. A. Wengert, Jr., C. D. Tourtellotte, and M. A. Kirsch, *Biochim. Biophys. Acta*, **124**, 101 (1966).

226. J. D. Cipera and M. Hidiroglou, *Can. J. Physiol. Pharmacol.*, **47**, 591 (1969).

227. V. E. Levine, *J. Bacteriol.*, **10**, 217 (1925).

228. G. Joachimoglu, *Biochem. Z.*, **107**, 300 (1926).

229. J. Sternberg and A. Mercier, *Intern. J. Appl. Radiation Isotopes*, **15**, 587 (1964).

230. H. G. Smith, *J. Gen. Microbiol.*, **21**, 61 (1959).

231. M. A. Gohar, *J. Trop. Med. Hyg.*, **46,** 29 (1943).

232. K. F. Weiss, J. C. Ayres, and A. A. Kraft, *J. Bacteriol.*, **90,** 857 (1965).

233. A. Shrift, *Plant Physiol.*, **35,** 510 (1960).

234. J. Scala, P. Ulbrich, and H. H. Williams, *Arch. Biochem. Biophys.*, **107,** 132 (1964).

235. T. Abe and R. Nakaya, *Japan. J. Bacteriol.*, **6,** 463 (1951); *Chem. Abstr.*, **48,** 9457 (1954).

236. F. Guth, *Centralbl. Bakteriol.*, *Abt. 1*, **77,** 487 (1916).

237. G. J. Banwart and J. C. Ayres, *Appl. Microbiol.*, **1–2,** 296 (1953–54).

238. W. Brenner, *Jahrb. Wiss. Bot.*, **57,** 95 (1916).

239. J. G. Lipman and S. A. Waksman, *Science*, **57,** 60 (1923).

240. A. Shrift and E. Kelly, *Nature*, **195,** 732 (1962).

241. W. J. Nickerson, W. A. Taber, and G. Falcone, *Can. J. Microbiol.*, **2,** 575 (1956).

242. H. D. Kumar, *Plant Cell Physiol.*, **5,** 465 (1964).

243. A. Shrift, J. Nevyas, and S. Turndorf, *Plant Physiol.*, **36,** 502 (1961).

244. A. Shrift, J. Nevyas, and S. Turndorf, *Plant Physiol.*, **36,** 509 (1961).

245. A. Shrift and M. Sproul, *Biochim. Biophys. Acta*, **71,** 332 (1963).

246. G. N. Cohen and D. B. Cowie, *C. R. Acad. Sci., Paris*, **244,** 680 (1957).

247. E. H. Coch and R. C. Greene, *Biochim. Biophys. Acta*, **230,** 223 (1971).

248. M. Wu and J. T. Wachsman, *J. Bacteriol.*, **104,** 1393 (1970).

249. A. Shrift, *Plant Physiol.*, **34,** 505 (1959).

250. H. G. Mautner and W. H. H. Günther, *Biochim. Biophys. Acta*, **36,** 561 (1959).

251. R. E. Huber, I. H. Segal, and R. S. Criddle, *Biochim. Biophys. Acta*, **141,** 573 (1967).

252. I. J. Pintner and L. Provasoli, *Bull. Misaki Marine Biol. Inst., Kyoto Univ.*, **12,** 25 (1968).

253. G. Falcone and W. J. Nickerson, *J. Bacteriol.*, **85,** 754 (1963).

254. G. Werz, *Planta*, **57,** 250 (1961).

255. A. Shrift, *Plant Physiol.*, **35,** 510 (1960).

256. R. G. L. McCready, J. N. Campbell, and J. I. Payne, *Can. J. Microbiol.*, **12,** 703 (1966).

257. M. Blau, *Biochim. Biophys. Acta*, **49,** 389 (1961).

258. J. Hedegaard, G. Falcone, and S. Calabro, *C. R. Séances Soc. Biol.*, **157,** 280 (1963).

259. S. H. Mudd and G. L. Cantoni, *Nature*, **180,** 1052 (1957).

260. J. Skupin, *Acta Biochim. Polon.*, **9,** 253 (1962).

261. G. Falcone and V. Giambanco, *Nature*, **213,** 396 (1967).

262. M. Hidiroglou, D. P. Heaney, and K. J. Jenkins, *Can. J. Physiol. Pharmacol.*, **46,** 229 (1968).

263. G. D. Paulson, C. A. Baumann, and A. L. Pope, *J. Animal. Sci.*, **27,** 497 (1968).

264. F. Challenger, *Advan. Enzymol.*, **12,** 429 (1951).

265. F. Challenger, *Endeavour*, **12,** 173 (1953).

266. K. P. McConnell, *J. Biol. Chem.*, **145,** 55 (1942).

267. K. P. McConnell and O. W. Portman, *J. Biol. Chem.*, **195,** 277 (1952).

268. H. E. Ganther, *Biochemistry*, **5,** 1089 (1966).

269. R. C. Greene, *Biochemistry*, **8,** 2255 (1969).

270. F. Pan and H. Tarver, *Arch. Biochem. Biophys.*, **119,** 429 (1967).

271. J. Bremer and Y. Natori, *Biochim. Biophys. Acta,* **44,** 367 (1960).

272. A. E. Pegg, *Biochim. Biophys. Acta,* **177,** 361 (1969).

273. B. Nisman and M. L. Hirsch, *Ann. Inst. Pasteur,* **95,** 615 (1958).

274. G. A. Hahn and J. W. Brown, *Biochim. Biophys. Acta,* **146,** 264 (1967).

275. C. A. Weller and M. Green, *Exptl. Eye Res.,* **8,** 84 (1969).

276. J. L. Hoffman, K. P. McConnell, and D. R. Carpenter, *Biochim. Biophys. Acta,* **199,** 531 (1970).

277. A. Kokko, H. G. Mautner, and R. J. Barrnett, *J. Histochem. Cytochem.,* **17,** 625 (1969).

278. J. C. M. Tsibris, M. J. Namtyedt, and I. C. Gunsalus, *Biochem. Biophys. Res. Commun.,* **30,** 323 (1968).

279. W. H. Orme-Johnson, R. E. Hansen, H. Beinert, J. C. M. Tsibris, R. C. Bartholomaus, and I. C. Gunsalus, *Proc. Natl. Acad. Sci.,* **60,** 368 (1968).

279a. J. A. Fee and G. Palmer, *Biochim. Biophys. Acta,* **245,** 175 (1971).

279b. J. A. Fee, S. G. Mayhew, and G. Palmer, *Biochim. Biophys. Acta,* **245,** 196 (1971).

280. R. Walter and V. du Vigneaud, *J. Amer. Chem. Soc.,* **87,** 4192 (1965).

281. R. Walter and V. du Vigneaud, *J. Amer. Chem. Soc.,* **88,** 1331 (1966).

282. R. Walter and W. Y. Chan, *J. Amer. Chem. Soc.,* **89,** 3892 (1967).

283. R. Walter, W. Gordon, I. L. Schwartz, F. Quadrifoglio, and D. W. Urry, *Proceedings of the 9th European Peptide Symposium,* E. Ericas, Ed., North-Holland Publishing Co., Amsterdam, 1968, p. 50.

284. D. W. Urry, F. Quadrifoglio, R. Walter, and I. L. Schwartz, *Proc. Natl. Acad. Sci.,* **60,** 967 (1968).

285. W. Gordon, R. T. Havran, I. L. Schwartz, and R. Walter, *Proc. Natl. Acad. Sci.,* **60,** 1353 (1968).

286. C. C. Chiu, I. L. Schwartz, and R. Walter, *Science,* **163,** 925 (1969).

287. R. E. Huber and R. S. Criddle, *Biochim. Biophys. Acta,* **141,** 587 (1967).

288. R. C. Tilton, H. B. Gunnar, and W. Litsky, *Can. J. Microbiol.,* **13,** 1175 (1967).

289. R. C. Tilton, H. B. Gunnar, and W. Litsky, *Can. J. Microbiol.,* **13,** 1183 (1967).

290. V. Giambanco and G. Falcone, *Rivista Inst. Sieroterapico Ital.,* **41,** 219 (1966).

291. W. J. Nickerson and G. Falcone, *J. Bacteriol.,* **85,** 763 (1963).

292. M. Zalokar, *Arch. Biochem. Biophys.,* **44,** 330 (1953).

293. G. S. Ahluwalia, Y. R. Saxena, and H. H. Williams, *Arch. Biochem. Biophys.,* **124,** 79 (1968).

294. C. A. Woolfolk and H. R. Whiteley, *J. Bacteriol.,* **84,** 647 (1962).

295. M. L. Bird, F. Challenger, P. T. Charlton, and J. O. Smith, *Biochem. J.,* **43,** 78 (1948).

296. D. I. Sapozhnikov, *Mikrobiologia (USSR),* **6,** 643 (1937).

297. H. R. Geering, E. E. Cary, L. H. P. Jones, and W. H. Allaway, *Proc. Soil Sci. Soc. Amer.,* **32,** 35 (1968).

298. E. E. Cary and W. H. Allaway, *Proc. Soil Sci. Soc. Amer.,* **33,** 571 (1969).

299. E. I. Rabinowitch, "Photosynthesis," Vol. 1, Interscience Publishers, New York, N.Y., 1945, p. 7.

300. T. J. Ganje and E. I. Whitehead, *Proc. S. Dakota Acad. Sci.,* **37,** 81 (1958).

301. G. M. Abu-Erreish, E. I. Whitehead, and O. E. Olson, *Soil Sci.,* **106,** 415 (1968).

302. A. Shrift, *Nature*, **201,** 1304 (1964).

303. N. C. Koch, *J. Paleontol.*, **41,** 970 (1967).

304. J. L. Byard, *Arch. Biochem. Biophys.*, **130,** 556 (1969).

305. I. S. Palmer, D. D. Fischer, A. W. Halverson, and O. E. Olson, *Biochim. Biophys. Acta,* **177,** 336 (1969).

306. I. S. Palmer, R. P. Gunsalus, A. W. Halverson, and O. E. Olson, *Biochim. Biophys. Acta,* **208,** 260 (1970).

XIV Selenium-containing polymers

LUIGI MORTILLARO and **MARIO RUSSO**

Montecatini Edison Research Center
Milan, Italy

Some data are reported in the literature dealing with selenium-containing polymers, but the only polymer that has been extensively studied and characterized is poly(selenomethylene). These polymers should be attractive as semiconducting materials because of their selenium content and the ease with which the plastic materials can be molded.

The general routes followed for synthesizing linear organic selenium-containing polymers involve:

1. Polycondensation between alkali selenides and dihaloalkanes.
2. Hydrolysis of alkanediselenocyanates in alkaline medium.
3. Ionic polymerization in bulk or in solution of cyclic oligomers.
4. Solid-state radiation polymerization of cyclic oligomers.

Method 1 is the most general route. By this method cyclization and polymerization reactions may occur at the same time, but it is possible to drive the reaction by selecting the appropriate experimental conditions. As expected, high dilution of the reactants favors cyclization. Method 2 was followed originally to obtain cyclic 1,2-diselenides, but the low stability of these compounds leads to polymerization with the formation of very low-molecular-weight diselenide polymers. These depolymerize on heating to cyclic monomers, which are stable only in solution. Method 3 leads to polymerization of cyclic oligomers for which the polyaddition may take place. High-molecular-weight polymers are obtained in good yield. The main application of method 4 is to study the kinetics and mechanism of solid-state topotactic reactions, particularly when they lead to oriented polymer crystals.

A. POLY(SELENOMETHYLENE)

The preparation of "selenoformaldehyde" or of its polymeric compounds from aqueous formaldehyde and hydrogen selenide has been investigated by several authors. In 1915 Vanino and Schinner[1] found that on passing hydrogen selenide into aqueous solutions of formaldehyde in dilute hydrochloric acid a pasty solid, together with a small quantity of an oil, is obtained. The nature of these compounds has not been investigated. When the reaction was carried out in strong hydrochloric acid, a solid product was recovered. By crystallization of this product a "selenoformaldehyde" was obtained. Substantially the same results as those of Vanino and Schinner[1] were reported by Brandt and Van Valkenburgh[2] in 1929. In 1950 Bridger and Pittman[3] found that this product is a cyclic trimer, the structure of which was confirmed in 1965 by Mortillaro et al.[4] from X-ray analysis.

In 1965 Mortillaro et al.[5] ascertained that the reaction actually yields two products: (a) the previously described 1,3,5-triselenane,[4] and (b) the linear polymer of selenoformaldehyde, namely poly(selenomethylene) (1). The

$$(-CH_2-Se-)_n$$
1

mechanism of formation of these compounds has been reported by Russo and Credali.[6] In the first step, a reaction between hydrogen selenide and hydrated formaldehyde (methylene glycol) takes place with formation of methylene selenoglycol in equilibrium with methylene glycol through an interchange of hydrogen selenide and water (equation 1). In the second step, which is

$$(1) \qquad H_2C\diagup\substack{OH \\ \diagdown OH} + H_2Se \rightleftharpoons H_2C\diagup\substack{SeH \\ \diagdown OH} + H_2O$$

catalyzed by hydrochloric acid, there is elimination of water from methylene selenoglycol, giving the trimer and the polymer (equation 2).

(2) 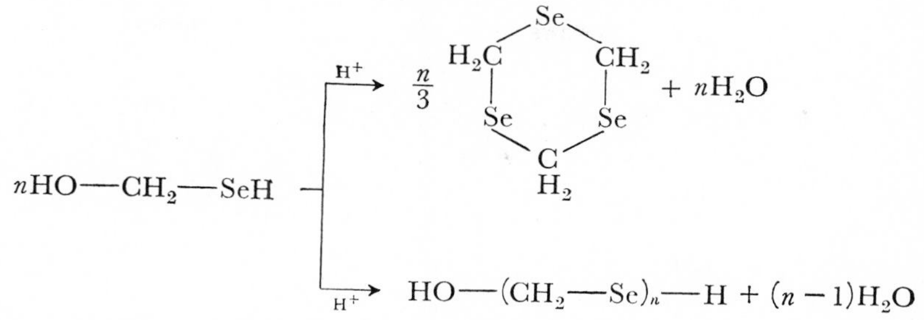 nHO—CH$_2$—SeH

$$\xrightarrow{\text{H}^+} \frac{n}{3} \quad \begin{array}{c} \text{Se} \\ \text{H}_2\text{C} \quad\quad \text{CH}_2 \\ | \quad\quad\quad | \\ \text{Se} \quad\quad \text{Se} \\ \text{C} \\ \text{H}_2 \end{array} \quad + n\text{H}_2\text{O}$$

$$\xrightarrow{\text{H}^+} \text{HO—(CH}_2\text{—Se)}_n\text{—H} + (n-1)\text{H}_2\text{O}$$

The poly(selenomethylene) obtained from the CH$_2$O–H$_2$O–H$_2$Se system appears as a white powder with a selenium content lower than that calculated for a fully –CH$_2$–Se– structure. It is crystalline, is highly insoluble in common solvents, and begins to melt at 133–139° with partial decomposition. By annealing the polymer at 125°, Mortillaro et al.[5] found that it undergoes at first a limited weight loss, after which it remains quite stable; the melting range of the annealed product rises to 174–178° and the selenium content reaches the required value for the –CH$_2$–Se– structure. According to Russo and Credali,[6] the annealing causes a loss of oxymethylenic units at the extremity of the chains and/or a condensation between the selenomethylenic residues. The latter hypothesis is justified by the increase of the melting range.

High-molecular-weight poly(selenomethylene) (yield 75 %, m.p. 185–190°) was obtained by Mortillaro et al.[5] by bulk polymerization with cationic catalyst (235°, BF$_3$-etherate) of 1,3,5-triselenane. According to these authors, the polyaddition probably proceeds via a mechanism similar to that reported for the polymerization of 1,3,5-trioxane[7] and of 1,3,5-trithiane[8] (scheme 3).

(3) 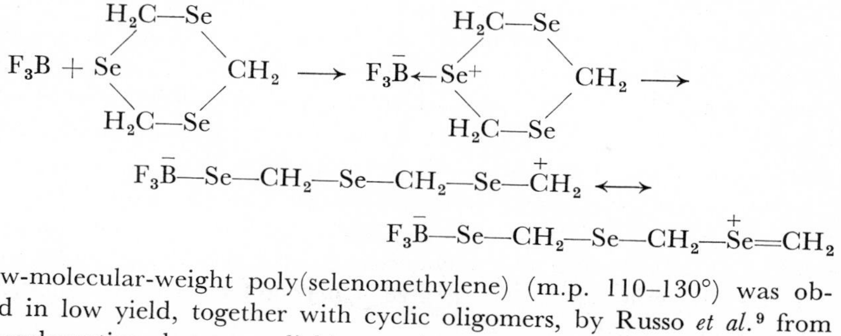 $\text{F}_3\text{B} + \begin{array}{c} \text{H}_2\text{C—Se} \\ \diagup \quad\quad \diagdown \\ \text{Se} \quad\quad\quad \text{CH}_2 \\ \diagdown \quad\quad \diagup \\ \text{H}_2\text{C—Se} \end{array} \longrightarrow \begin{array}{c} \text{H}_2\text{C—Se} \\ \diagup \quad\quad \diagdown \\ \text{F}_3\bar{\text{B}}{\leftarrow}\text{Se}^+ \quad\quad \text{CH}_2 \\ \diagdown \quad\quad \diagup \\ \text{H}_2\text{C—Se} \end{array} \longrightarrow$

$\text{F}_3\bar{\text{B}}\text{—Se—CH}_2\text{—Se—CH}_2\text{—Se—}\overset{+}{\text{C}}\text{H}_2 \longleftrightarrow$

$\text{F}_3\bar{\text{B}}\text{—Se—CH}_2\text{—Se—CH}_2\text{—}\overset{+}{\text{Se}}\text{=CH}_2$

Low-molecular-weight poly(selenomethylene) (m.p. 110–130°) was obtained in low yield, together with cyclic oligomers, by Russo et al.[9] from polycondensation between dichloromethane and sodium selenide in a water-methanol medium (scheme 4). This poly(selenomethylene) shows an

$$(4) \quad n\mathrm{CH_2Cl_2} + n\mathrm{Na_2Se} \xrightarrow[\substack{N_2 \\ \text{refl., 5 hr}}]{\text{aq. } CH_3OH}$$

$$\frac{n}{4} \quad
\begin{array}{c}
\mathrm{Se\!-\!CH_2} \\
\mathrm{H_2C} \qquad \mathrm{Se} \\
\mathrm{Se} \qquad \mathrm{CH_2} \\
\mathrm{H_2C\!-\!Se}
\end{array} \qquad 4\%$$

$$(\!-\!\mathrm{CH_2\!-\!Se}\!-\!)_n \qquad 7\%$$

$$\frac{n}{3} \quad
\begin{array}{c}
\mathrm{Se} \\
\mathrm{H_2C} \qquad \mathrm{CH_2} \\
\mathrm{Se} \qquad \mathrm{Se} \\
\mathrm{C} \\
\mathrm{H_2}
\end{array} \qquad 30\%$$

increase in its melting range (to 150–165°) when melted under vacuum, thus confirming its low molecular weight and the ease of further polycondensation of selenomethylenic residues.

Schmidt et al.[10] in 1966, by reaction of dichloromethane with hydrogen selenide in a sealed tube at room temperature and in the presence of triethylamine, obtained a yellow powder polymer (yield 32%, m.p. ~160° dec.), very probably with a selenomethylenic structure.

Russo et al.[9] obtained high-molecular-weight poly(selenomethylene) (yield 53%, m.p. 187–195°) by bulk polymerization of 1,3,5,7-tetraselenocane with BF_3-etherate as catalyst at 100°. In 1967 Prince and Bremer[11] prepared high-molecular-weight poly(selenomethylene) (m.p. 225°) in 76% yield (without cyclic oligomers) from reaction 4 (24 hours at 77°), using dibromomethane as the dihalo compound and ethyl acetate as the reaction medium. In their paper, which has been criticized by Mortillaro et al.,[12] Prince and Bremer reported that the poly(selenomethylene) they prepared had an electrical conductivity of 6.72×10^{-4} ohm⁻¹-cm⁻¹ at 25°. Since a polymeric material exhibiting such a conductivity would be of great interest for possible applications, Sandrolini et al.[13] investigated the d.c. conductivity of poly(selenomethylene) prepared by different methods. They found that, in contrast with the statements of Prince and Bremer, none of the poly(selenomethylenes) examined showed conductivity values exceeding 7.5×10^{-13} ohm⁻¹-cm⁻¹. During this work, Sandrolini et al.[13] found that 1,3,5-triselenane polymerizes in the solid state with cationic catalysts (BF_3-etherate at 180°) to give orthorhombic poly(selenomethylene) in quantitative yield.

The reaction of methylene radicals with selenium was also studied by several authors. The formation of "selenoformaldehyde" by the reaction of methylene radicals with selenium mirror was investigated by Rice and Glasebrook[14] in 1934 and by Pearson et al.[15] in 1938. Pearson stated that selenoformaldehyde exists first as a gas and polymerizes to two solid forms, melting at 62–68° and 205–215°. However, in 1968 Williams and Dunbar[16] proved by mass spectrometry that the product of the reaction of methylene radicals with a selenium mirror is not a polymeric substance, but 1,3,5-triselenane.

Carazzolo and Mammi[17] obtained poly(selenomethylene) by exposure of crystals of 1,3,5-triselenane to a γ-radiation dose of 64 Mrad at room temperature. Irradiated crystals were subsequently annealed at various temperatures in the range 60–120°. The sample annealed at 100° and below was shown to be unchanged triselenane, but the sample annealed at 120° proved to have been partially polymerized (about 20%) to poly(selenomethylene).

Poly(selenomethylene) Structure

Structural investigations of poly(selenomethylene) by X-ray analysis were carried out by Carazzolo et al.,[17–19] who found that, depending on the polymerization conditions, two crystalline modifications were obtained. In particular, by polycondensation methods[5,9] and by bulk polymerization of triselenane,[5] hexagonal poly(selenomethylene)[18,19] was formed. In contrast, bulk polymerization of 1,3,5,7-tetraselenocane[9] and solid-state radiation[17] or cationic[13] polymerization of 1,3,5-triselenane yielded orthorhombic poly(selenomethylene).[17,20] The two crystalline modifications of poly(selenomethylene) give rise to a monotropic dimorphic system with orthorhombic → hexagonal transition temperature near the melting range of hexagonal poly(selenomethylene).

It is very interesting to note that the same monotropic dimorphism observed in poly(selenomethylene) appears also in the isologous poly(oxymethylene),[21,22] whereas only one crystalline modification (hexagonal) of poly(thiomethylene) has been described.[23,24] The structures of poly(oxy-), poly(thio-), and poly(selenomethylenes) have been compared by Carazzolo and Mammi.[25] Table XIV-1 reports some data relative to the structures of these polymers, and Fig. XIV-1 shows the chain top views.

Carazzolo and Mammi[17] have proved that the solid-state radiation polymerization of 1,3,5-triselenane to orthorhombic poly(selenomethylene) takes place in two directions, parallel to the two [hh0] and [hh̄0] diagonal axes in the a-b plane of the trimer orthorhombic unit cell (Fig. XIV-2).

Hexagonal poly(selenomethylene) i.r. absorption data were reported by Prince and Bremer,[11] but were quite different from those given by Mortillaro et al.[12]

Table XIV-1   Some data relative to the structures of hexagonal poly(oxymethylene) (HPOM), orthorhombic poly(oxymethylene) (OPOM), hexagonal poly(selenomethylene) (HPSeM), orthorhombic poly(selenomethylene) (OPSeM), and hexagonal poly(thiomethylene) (HPTM)

PARAMETER	HPOM	OPOM	HPSeM	OPSeM	HPTM
Unit cell parameters					
$\quad a$ (Å)	4.47	4.47	5.22	5.37	5.07
$\quad b$ (Å)	4.47	7.66	5.22	9.03	5.07
$\quad c$ (Å)	56.00	3.56	46.25	4.27	36.52
Melting temperature (°C)	176–193	...	185–200	...	240–260
Transition temperature					
$\quad$ (°C) to hexagonal	...	60	...	>170	...
Density (g/cm³)					
$\quad$ Calculated	1.492	1.533	2.968	2.985	1.60
$\quad$ Observed	a	a	2.82	...	1.52
Chains crossing the unit					
$\quad$ cell	1	2	1	2	1
Monomeric units					
$\quad$ (—CH₂O— or					
$\quad$ —CH₂Se— or					
$\quad$ —CH₂S—) contained					
$\quad$ in the c identity period	29	2	21	2	17
Turns of the helix contained					
$\quad$ in the c identity					
$\quad$ period	16	1	11	1	9
Monomeric units per					
$\quad$ turn of the helix	1.813	2	1.909	2	1.889
Monomeric unit					
$\quad$ periodicity (Å)	1.931	1.780	2.203	2.135	2.148
Pitch of the helix (Å)	3.500	3.56	4.205	4.27	4.058
Distance of a C atom					
$\quad$ from the helix axis (Å)	0.68	0.79		1.08	0.94
Distance of an O or Se or					
S atom			1.08ᵇ		
$\quad$ from the helix axis (Å)	0.68	0.79		1.19	1.03
Bond length C—O					
$\quad$ or C—Se or C—S (Å)	1.430	1.430	1.930	1.930	1.815
Bond angle					
$\quad$ C—O—C or C—Se—C					
$\quad\quad$ or C—S—C	110°45′	112°50′		104° 0′	104°
$\quad$ O—C—O or			105°59′ᵇ		
$\quad$ Se—C—Se or S—C—S	110°45′	112°50′		111°50′	110°
Internal rotation angle	76°14′	63°50′	63°11′	57°59′	65°59′

ᵃ Observed densities of HPOM and OPOM may vary with the type of samples; at the limit, values have been measured that are practically equal to the calculated ones.[26]

ᵇ Mean value.

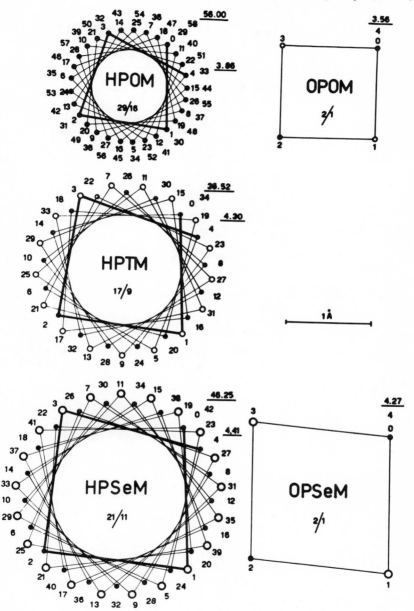

FIG. XIV-1 Top view of hexagonal poly(oxymethylene) (HPOM), orthorhombic poly(oxymethylene) (OPOM), hexagonal poly(selenomethylene) (HPSeM), orthorhombic poly(selenomethylene) (OPSeM), and hexagonal poly(thiomethylene) (HPTM) chains.[25]

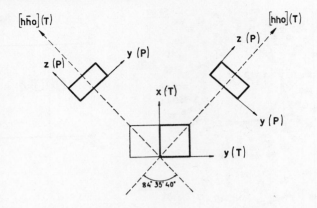

(T)   Triselenane
(P)   Orthorhombic Polyselenomethylene

FIG. XIV-2   Relative orientation of the two twinned orthorhombic poly(selenomethylene) unit cells with respect to the triselenane orthorhombic unit cell.[17]

B. COPOLYMERS WITH SELENOFORMALDEHYDE UNITS

Russo *et al.*,[27] by reaction of α,α'-dichlorodimethyl ether with sodium selenide in anhydrous methanol or in a water-methanol medium, obtained 1,3,5-oxadiselenane (**2**), together with a copolymer (yield 26%, m.p. 80–95°) with formaldehyde and selenoformaldehyde units (equation 5).

(5)

$$ClCH_2{-}O{-}CH_2Cl + Na_2Se \xrightarrow[\substack{N_2 \\ \text{refl., 2 hr}}]{\text{aq. } CH_3OH}$$

$$(-CH_2{-}O{-}CH_2{-}Se{-})_n$$
3

The ratio CH_2Se/CH_2O in the copolymer (**3**) was found to equal 1.

The same authors, by bulk polymerization of **2** with BF_3-etherate as catalyst, obtained the corresponding copolymer (**4**) with CH_2Se/CH_2O ratio = 2 in 70% yield (m.p. 193–198°) (equation 6). The reported X-ray

(6) $$2 \xrightarrow[\substack{N_2,100°,1\,\text{hr}}]{BF_3\text{-etherate}} [-CH_2{-}O{-}(-CH_2{-}Se{-})_2{-}]_n$$

4

powder spectra of both copolymers show that the more intense reflection corresponds to an interplanar spacing intermediate between that of hexagonal poly(oxymethylene)[21] and that of hexagonal poly(selenomethylene).[5]

C. POLY(SELENOTRIMETHYLENE)

Morgan and Burstall,[28] during the preparation of selenetane by reaction of 1,3-dibromopropane with sodium selenide, obtained a pale yellow, soap-like substance (m.p. 38–40°). The elemental analysis data are in agreement with a selenotrimethylenic composition, whereas molecular weight determinations suggest that the average degree of polymerization is about 6. A poly(selenotrimethylenic) structure (5) appears reasonable, with $n \geqslant 6$.

$$(-CH_2-CH_2-CH_2-Se-)_n$$
$$5$$

On heating 5, propylene gas was evolved, and the residue was probably low-molecular-weight poly[(diseleno)trimethylene]. According to Morgan and Burstall,[28] 1,5-diselenocane tetranitrate and tetrachloride may be obtained from 5 (scheme 7).

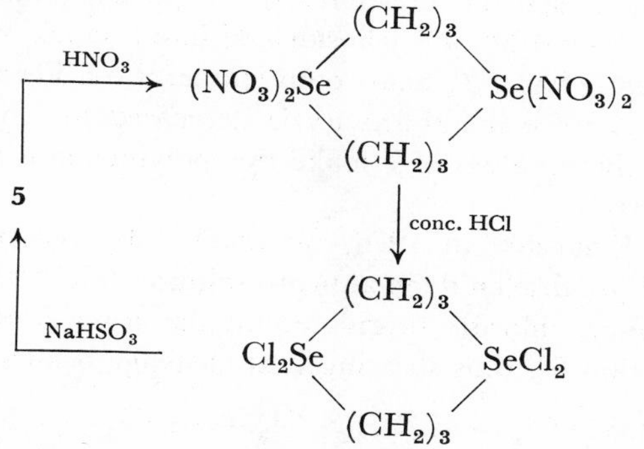

Cozzens and Harvey[29] found that the photochemical and thermal polymerization of selenetane occurs through ring cleavage at the C–Se bond. By Raman and e.p.r. studies a regular head-to-tail poly(selenotrimethylenic) structure (5) was suggested. Molecular oxygen plays an important role in the polymerization kinetics, since degassed selenetane, in contrast to that which is not degassed, polymerizes rapidly even under room illumination.

D. POLY(SELENOHEXAMETHYLENE)

Morgan and Burstall[28] obtained, by the reaction of 1,6-dibromohexane with sodium selenide, a white, waxy, microcrystalline powder (m.p. 36–37°)

for which analytical data agree with a poly(selenohexamethylenic) structure (6) with an n value of approximately 12. With 6 there were obtained small fractions of selenepane and 1,8-diselenacyclotetradecane. On heating, 6

$$(-CH_2-CH_2-CH_2-CH_2-CH_2-CH_2-Se-)_n$$
6

melted at 36–37°, showed no further change until at about 200°, and decomposed at about 220° into 2-methylselenane.

E. POLY[(DISELENO)METHYLENE]

In 1967, Prince and Bremer[31] found that under anhydrous conditions formaldehyde and sodium selenide react to give poly[(diseleno)methylene] (7) in high yield (87%). Two forms of 7 have been shown to exist. The first (a)

$$CH_2O + Na_2Se \xrightarrow[N_2,\,24\,hr]{CH_3COOC_2H_5} (-CH_2-Se_2-)_n$$
7

is a crystalline, red-brown solid easily molded at 70° and 5000 psi. Its density is 3.4 g/cm³, and its specific resistivity was reported as 3×10^5 ohm-cm. When heated to above 120°, a is transformed into a black, rubbery polymer (b) which apparently has a T_g near room temperature. Form b has the same density as a, but its specific resistivity is decreased to 1.6×10^4 ohm-cm. The specific resistivity values of 7 make this polymer interesting as a semiconducting material.

Paetzold and Knaust,[32] in 1970, obtained 7 by reaction of dibromomethane and sodium diselenide in aqueous solution (yield 98%). Compound 7 reacts with sulfuryl chloride to give equimolar quantities of $ClCH_2SeCl_3$ and $SeCl_4$ (equation 8), thus showing that cleavage occurs at Se–Se bonds

(8) $7 + 4nSO_2Cl_2 \rightarrow nClCH_2SeCl_3 + nSeCl_4 + 4nSO_2$

and at one Se–C bond every two monomeric units.

F. POLY[(DISELENO)TRIMETHYLENE]

The hydrolysis of 1,3-propanediselenocyanate with alcoholic potassium hydroxide yields a yellow powder (m.p. 54–59°) that was regarded as 1,2-diselenolane by Morgan and Burstall[28] and as its cyclic dimer by Hagelberg.[33] The same product was obtained by Morgan and Burstall[28] by heating poly(selenotrimethylene). Bergson and Claeson[34] found in 1957 that this product was insoluble at room temperature in the common spectroscopic solvents, but when heated to about 60° went into solution. The solution showed an absorption maximum at about 440 nm, characteristic of 1,2-diselenolane ring systems.[35] After standing several days at room temperature the yellow powder slowly precipitated from solution. Bergson and Claeson[34] suggested that the solution contains 1,2-diselenolane monomer, whereas the yellow powder is a dimer or a low-molecular-weight polymer (**8**), which depolymerizes when heated. This idea is confirmed by the

$$
\begin{array}{c}
CH_2\!-\!SeCN \\
\diagup \\
H_2C \\
\diagdown \\
CH_2\!-\!SeCN
\end{array}
\xrightarrow[\text{air}]{C_2H_5OH-KOH}
(\!-\!CH_2\!-\!CH_2\!-\!CH_2\!-\!Se\!-\!Se\!-\!)_n
$$

8

$$
\Delta \Big\uparrow -nCH_2\!=\!CH\!-\!CH_3
$$

$$
(\!-\!CH_2\!-\!CH_2\!-\!CH_2\!-\!Se\!-\!)_{2n}
$$

molecular weight value found, which is higher than that of diselenolane monomer, and is in agreement with a shoulder found in its u.v. spectrum.[34]

G. POLY[(DISELENO)TETRAMETHYLENE]

By air oxidation of 1,4-butanediselenocyanate in an alkaline medium, Morgan and Burstall[36] obtained a yellow powder (m.p. 41–42°) regarded as 1,2-diselenane. According to Bergson,[37] in analogy to the findings for 1,2-diselenolane (see Section F), this product is not 1,2-diselenane, but its low-molecular-weight polymer (**9**).

$$
\begin{array}{l}
CH_2\!-\!CH_2\!-\!SeCN \\
| \\
CH_2\!-\!CH_2\!-\!SeCN
\end{array}
\xrightarrow[\text{air}]{C_2H_5OH-KOH}
$$

$$
(\!-\!CH_2\!-\!CH_2\!-\!CH_2\!-\!CH_2\!-\!Se\!-\!Se\!-\!)_n
$$

9

Compound **9** in suitable solvents depolymerized on heating and gave solutions in which 1,2-diselenane monomer was identified by u.v. absorption

$$
\xrightarrow[\text{warm}]{HNO_3-H_2O} \qquad\qquad \mathbf{9} \qquad\qquad \xrightarrow{\Delta}
$$

$$
\Big\downarrow Br_2
$$

$$
\begin{array}{l}
CH_2\!-\!CH_2\!-\!SeO_2H\!\cdot\!HNO_3 \\
| \\
CH_2\!-\!CH_2\!-\!SeO_2H\!\cdot\!HNO_3
\end{array}
\qquad
\begin{array}{c}
H_2C\!-\!\!-\!\!-\!CH_2 \\
|\qquad\quad| \\
H_2C\qquad CH_2 \\
\diagdown\quad\diagup \\
Se \\
Br_2
\end{array}
\qquad
\begin{array}{c}
H_2C\!-\!\!-\!\!-\!CH_2 \\
|\qquad\quad| \\
H_2C\qquad CH_2 \\
\diagdown\quad\diagup \\
Se
\end{array}
$$

measurements[37,38] (maximum at 365 nm). Morgan and Burstall[36] described the reactions of **9**. When strongly heated it decomposed to give selenolane. It reacted with bromine in chloroform or with excess bromine to give 1,1-dibromoselenolane. Treatment of **9** or of 1,4-butanediselenocyanate with warm nitric acid yielded colorless crystals of 1,4-tetramethylene-diseleninic acid dinitrate.

Brown et al.[39] found that freshly prepared solutions of **9** in chloroform gave an u.v. spectrum with one absorption maximum at 308 nm, characteristic of Se–Se noncyclic bond. On allowing these solutions to stand at 60°, the peak at 308 nm disappeared and the characteristic maximum of Se–Se bond in the 1,2-diselenane (364 nm) appeared. Thermal depolymerization in solution in the absence of light was increased by raising the temperature. Although in diffuse sunlight the peak at 308 nm was replaced by one at 364 nm, in very strong sunlight the latter peak was not observed at all. These facts suggest that thermal depolymerization gives 1,2-diselenane, whereas two photochemical processes may occur, yielding, respectively, 1,2-diselenane and products containing no Se–Se bonds.

Depolymerization proceeds by a free-radical mechanism, both photochemically and thermally. The rate of the purely thermal depolymerization of **9** in chloroform at 60° in the dark and in the presence of air follows first-order kinetics in polymeric diselenide concentration, with a rate constant of 2.75×10^{-5} sec^{-1}. The photochemical depolymerization at 25° in chloroform solution, with light predominantly of 253.7 and 366 nm, was also first order in polymeric diselenide concentration with a rate constant of 8.96×10^{-5} sec^{-1}.

Brown et al.[39] studied also the effects of **9** and its depolymerization product, 1,2-diselenane, on the thermal polymerization of styrene, methyl methacrylate, vinyl acetate, and acrylonitrile at 60° in the presence and the absence of azobisisobutyronitrile and on the direct photopolymerization of these vinyl monomers at 25°.

H. POLY[(DISELENO)PENTAMETHYLENE]

Morgan and Burstall,[40] by treating 1,5-pentanediselenocyanate with alcoholic potassium hydroxide, obtained a yellow uncrystallizable gum that, from elemental analysis data, was regarded as 1,2-diselenepane monomer. It is reasonable to state, in analogy with what has been found for 1,2-diselenocycloalkane five- and six-membered rings, that this compound is not 1,2-diselenepane but its low-molecular-weight polymer (**10**) (equation 9). This product was quite stable; but, when heated strongly, it decomposed, giving selenane. On reaction of **10** with bromine, 1,1-dibromoselenane was obtained.

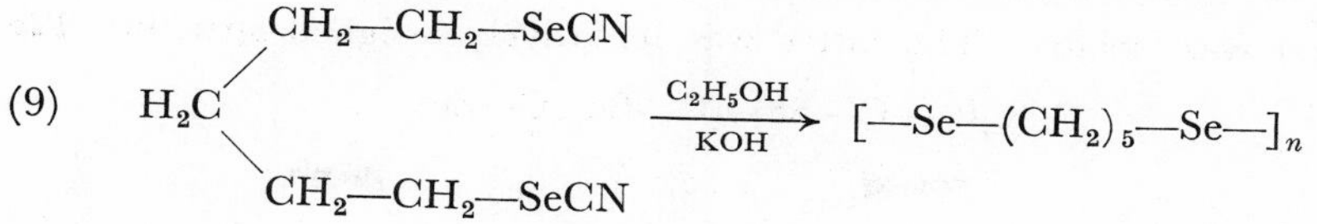

(9) H_2C
$\Big\langle$
$\begin{array}{c} CH_2-CH_2-SeCN \\ \\ CH_2-CH_2-SeCN \end{array}$
$\xrightarrow[\text{KOH}]{\text{C}_2\text{H}_5\text{OH}}$ $[-Se-(CH_2)_5-Se-]_n$

10

I. POLY[(DISELENO)HEXAMETHYLENE]

Morgan and Burstall,[30] by treating 1,6-hexanediselenocyanate with a solution of sodium in alcohol, obtained a yellow solid, regarded as polymer (**11**).

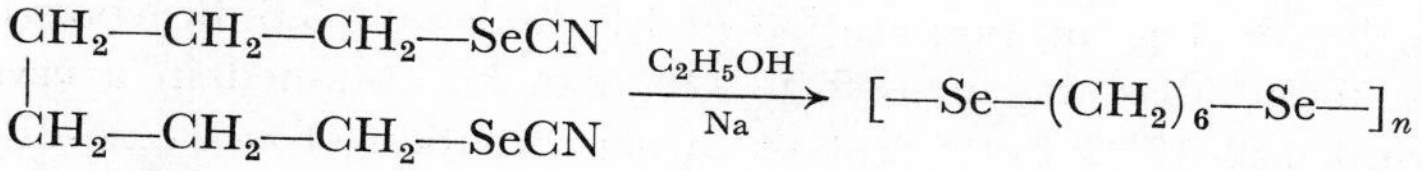

$\begin{array}{c} CH_2-CH_2-CH_2-SeCN \\ | \\ CH_2-CH_2-CH_2-SeCN \end{array}$ $\xrightarrow[\text{Na}]{\text{C}_2\text{H}_5\text{OH}}$ $[-Se-(CH_2)_6-Se-]_n$

11

When this polymer was heated, it melted rather indefinitely at 40° and showed no further change until about 250°, when it decomposed with loss of selenium and formation of 2-methylselenane.

J. POLY(CARBON DISELENIDE)

Grimm and Metzer[41] and Ives et al.[42] observed that carbon diselenide spontaneously precipitates a black polymeric solid at the rate of about 1 % per month at room temperature. In 1968 Brown and Whalley,[43] investigating the structure of this product by i.r. spectroscopy over the range of 4000–50 cm^{-1}, found three bands which they attributed to a polymeric structure. According to Brown and Whalley, a strong band at 885 cm^{-1} was associated with the C=Se stretching vibration, and the weaker band at 775 cm^{-1} can perhaps be associated with the C–Se stretching vibration. It seems very likely, therefore, that the product has essentially a poly(carbon diselenide) structure (**12**). Like poly(carbon disulfide), **12** is amorphous to X-ray diffraction.[43]

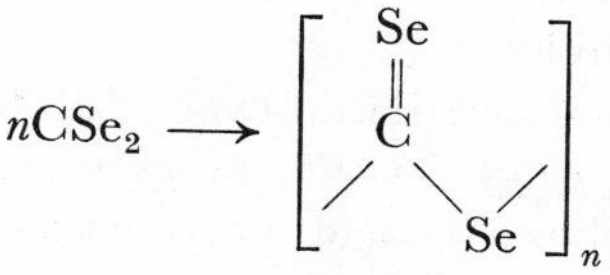

$$nCSe_2 \longrightarrow \left[\begin{array}{c} Se \\ \| \\ C \\ \diagup \ \diagdown \\ \qquad Se \end{array} \right]_n$$

12

Jensen and Nielsen[43a] have formed adducts of CSe_2 and trialkylphosphines of the type $R_3P^+CSe_2^-$. The tripropylphosphine compound was transformed on standing into a brown-red semisolid, from which a dark red crystalline material was isolated. The latter was tentatively assigned structure **12a**

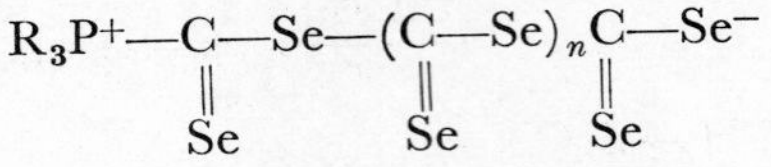

$$R_3P^+\!\!-\!\!C\!\!-\!\!Se\!\!-\!\!(C\!\!-\!\!Se)_n C\!\!-\!\!Se^-$$
$$\quad\; \underset{Se}{\|} \qquad \underset{Se}{\|} \;\; \underset{Se}{\|}$$

12a

on the basis of its elemental analysis and the ease with which CSe_2 polymerizes.

K. MISCELLANEOUS SELENIUM-CONTAINING POLYMERS

Fredga,[44,45] during the preparation of 1,2-diselenane-3,6-dicarboxylic acid isomers, found that the meso form (**13**) can be obtained in a crystalline state (m.p. 172–176°) by hydrolysis in acid solution of *meso*-α,α'-di(selenocyanato)adipic acid salt. However, **13** is readily converted to a polymer (scheme 10).

(10)

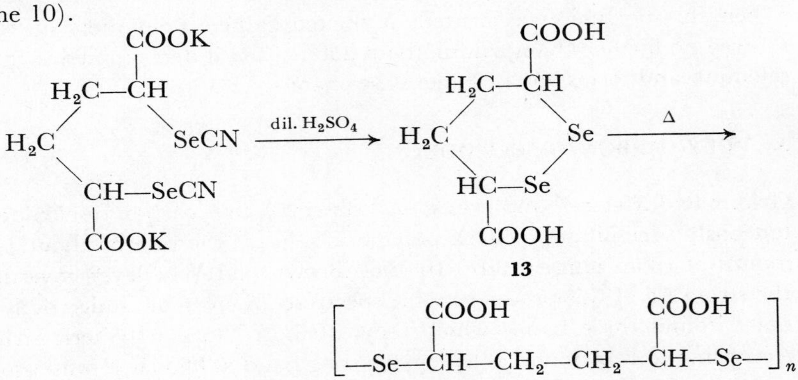

13

Fredga and Styrman[46] found that the reaction product obtained from *meso*-dibromopimelic acid and alkali diselenide was a mixture of selenane-2,6-dicarboxylic acid (*meso* form) and an amorphous diselenide, in all probability a polymer chain. The acid hydrolysis of di(selenocyanato) derivatives of both the *rac*- and *meso*-dibromopimelic acids gives rise in each case to polymeric diselenides.

The same ease of polymerization of cyclic 1,2-diselenides was also found in Se-S bond-containing rings. In fact, Bergson and Biezais[47] reported that 1-thia-2-selenolane-4-carboxylic acid polymerizes just above its melting point (105.5–107.5°) to polymeric materials for which alternative structures **14** and **15** have been proposed (equation 11).

Bergson,[48] during the synthesis of 1-thia-2-selenolane-4-aminomethyl hydrochloride (**16**), found some insoluble polymeric material whose chloroform solution was found to be rather unstable and which after only 1 day

$$\left[\begin{array}{c} COOH \\ | \\ -CH_2-CH-CH_2-Se-S- \end{array} \right]_n$$

14

(11)
$$\begin{array}{c} Se-CH_2 \\ | \qquad\quad \diagdown \\ \qquad\qquad CH-COOH \\ | \qquad\quad \diagup \\ S-CH_2 \end{array} \xrightarrow{\text{heat} > \text{m.p.}}$$

or

$$\left[\begin{array}{cc} COOH & COOH \\ | & | \\ -CH_2-CH-CH_2-Se-Se-CH_2-CH-CH_2-S-S- \end{array} \right]_n$$

15

deposited a polymeric precipitate. The u.v. spectrum of an aqueous solution of **16** showed a peak at 380 nm which was not as pronounced as that characteristic of 1-thia-2-selenolane-4-carboxylic acid[47] (386 nm). According to Bergson,[48] this might be due to a partial polymerization of **16**. When heated, **16** polymerized at about 120° and then melted again at about 180°.[48]

$$\begin{array}{c} Se-CH_2 \\ | \qquad\quad \diagdown \\ \qquad\qquad CH-CH_2-NH_3{}^+Cl^- \\ | \qquad\quad \diagup \\ S-CH_2 \end{array}$$

16

Simon and Heintz[49] reported that, during the preparation of 1,3-dioxa-2-selenepane-2-oxide, heating caused some polymerization of this selenacycloalkane. No further data were given.

A patent[50] described the preparation of a selenium-containing polymer with formula **17**, obtained by reaction of 16 parts of sodium sulfotriselenide with 6 parts of 1,2-dichloroethane and 4.5 parts of water at 40–50° for 20 minutes (equation 12). According to the patent, the resulting resin is tough

(12)

$$nClCH_2-CH_2Cl + nNa_2SSe_3 \xrightarrow{H_2O, 40-50°}$$
$$(-S-Se-Se-Se-CH_2-CH_2-)_n + 2nNaCl$$

17

but sufficiently plastic to work on a rubber mill. It is reddish in color, soluble in CS_2, insoluble in benzene and it softens when heated to 70°.

In 1965 a patent[51] claimed the preparation of linear copolymers formed by copolymerization of 1,5-diolefins with SeO_2. For example, a linear copolymer (**18**) was obtained by reaction of isopropenyl allyl ether with selenium dioxide in water, in the presence of azobisisobutyronitrile as initiator, at 50–60° for 10 hours. Compound **18** had the appearance of a rubber gum stock.

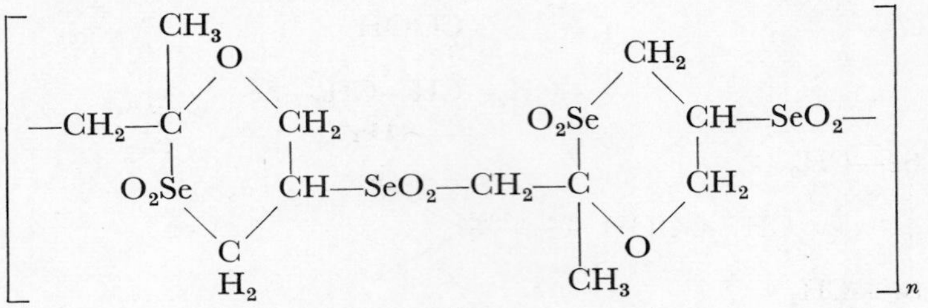

18

Kroll and Bolton[52] investigated the condensation (at temperatures below 150°) of bis(2-hydroxyethyl) selenide and bis(2-aminoethyl) selenide with a variety of polyfunctional isocyanates, epoxides, carboxylic acids, acid chlorides, esters, and anhydrides. These reactions yielded polymeric materials that varied from oils to brittle solids.

Zundel and Metzger[53-56] studied the hydration and the intermolecular interaction of selenium-containing polyelectrolytes. For this purpose, polystyreneselenonic acids (**19**) were prepared by reaction of H_2SeO_4 on polystyrene membranes in the presence of Ag_2SeO_4 (equation 13). The i.r.

(13)

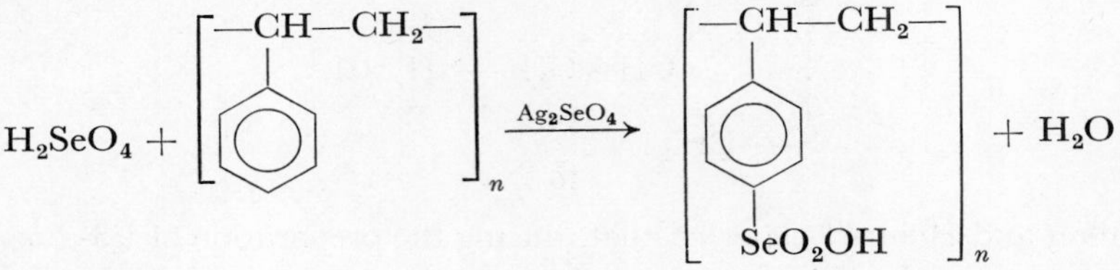

spectra of the compounds obtained in this way indicate the presence of *para* $-SeO_2OH$ groups, characterized by bands at 965, 911, and 725 cm^{-1}. Compounds of type **19** are quantitatively reduced to polystyreneseleninic acids (**20**) by treatment with concentrated HCl (equation 14).[53] The i.r. spectrum

(14)

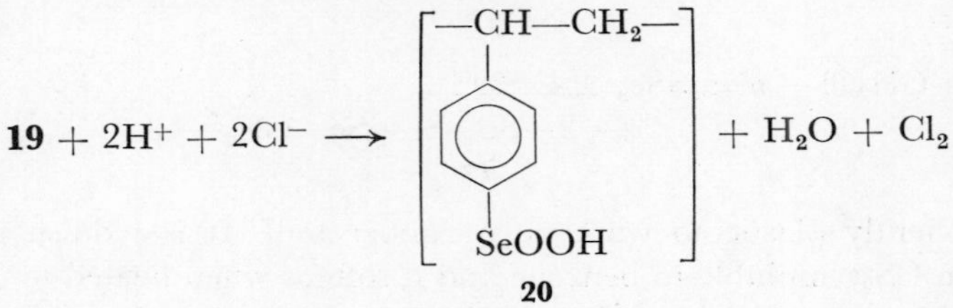

of **20** shows characteristic bands at 839 and 658 cm^{-1}. The i.r. spectrum has also revealed the existence in **19** and **20** of strong intermolecular interactions originating from hydrogen bonds, and has indicated how these polyelectrolytes tend to associate with water molecules to give a hydrated lattice with a definite structure.[54,56] Analogous studies were carried out on different salts of **19** and **20**.[53,56]

Günther and Salzman[57] have recently reported that benzylic and aromatic polydiselenides can be produced fairly readily *via* cyclic oligomers by a thermal polymerization (equation 15). Benzylic compounds in this series

(15) 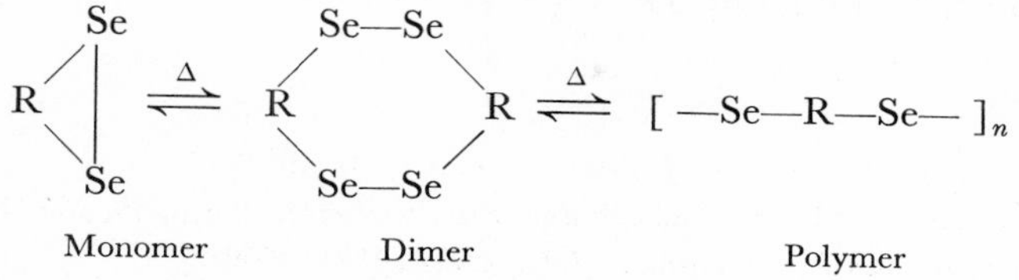   Monomer ⇌ Dimer ⇌ Polymer $[—Se—R—Se—]_n$

were generated by nucleophilic displacement reactions on the corresponding halides with aqueous potassium selenosulfate, followed by alkaline hydrolysis of the corresponding seleno-Bunte salts (*cf.* Chapter IVB.1.h). Poly(α,α-diseleno-*m*-xylene) (**21**) was then obtained as a yellow amorphous resin with $T_g \sim 22°$ and $T_m \sim 112–115°$ by heating the corresponding crystalline dimer (**22**) above its m.p. of 136–138°.[57]

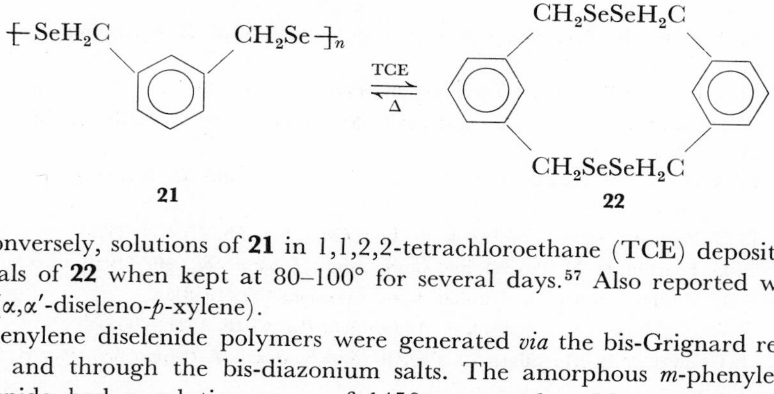

21   **22**

Conversely, solutions of **21** in 1,1,2,2-tetrachloroethane (TCE) deposited crystals of **22** when kept at 80–100° for several days.[57] Also reported was poly(α,α'-diseleno-*p*-xylene).

Phenylene diselenide polymers were generated *via* the bis-Grignard reagents and through the bis-diazonium salts. The amorphous *m*-phenylene diselenide had a solution m.w. of 1450; repeated melting and cooling established T_g 63° for the solid.[57] *p*-Phenylene diselenide (**23**) exists as a crystalline tetramer, solution m.w. 930, T_m 252–256°. Rapidly cooled melts had T_g 77–78° and no crystal m.p.[57]

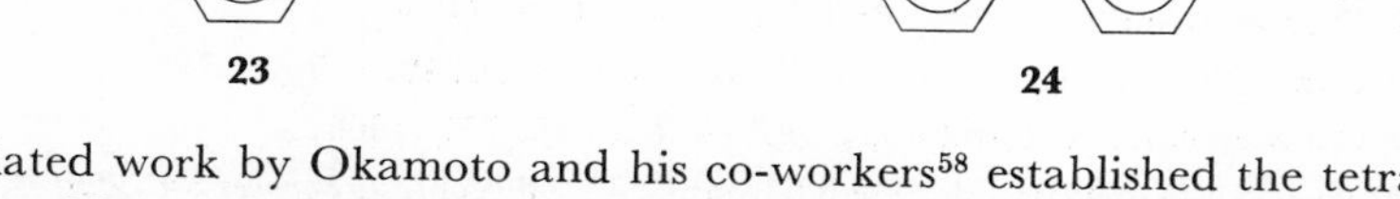

23   **24**

Related work by Okamoto and his co-workers[58] established the tetramer nature of **23** (m.p. 248°) by mass spectrometry. These workers also prepared poly(4,4'-biphenylene diselenide) (**24**) in two forms, m.p. 145° and m.p. 190°. The mass spectrogram indicated **24** (m.p. 145°) to be a cyclic dimer.

It appears remarkable that **21** and **23** and related species are able to incorporate additional Se atoms into the chain through a simple melt "alloying" process, thus giving products whose structures are intermediate between organic and inorganic polymers.[57,59]

REFERENCES

1. L. Vanino and A. Schinner, *J. Prakt. Chem.*, **91**, 116 (1915).

2. W. E. Bradt and M. Van Valkenburgh, *Proc. Indiana Acad. Sci.*, **39**, 165 (1929).

3. H. J. Bridger and R. W. Pittman, *J. Chem. Soc.*, 1371 (1950).

4. L. Mortillaro, L. Credali, M. Mammi, and G. Valle, *J. Chem. Soc.*, 807 (1965).

5. L. Mortillaro, L. Credali, M. Russo, and C. De Checchi, *J. Polymer Sci.*, *Part B*, **3**, 581 (1965).

6. M. Russo and L. Credali, *J. Macromol. Sci.* (*Chem.*), **1A**, 387 (1967).

7. V. Kern and V. Jaacks, *J. Polymer Sci.*, **48**, 399 (1960).

8. E. Gipstein, E. Wellisch, and O. J. Sweeting, *J. Polymer Sci.*, *Part B*, **1**, 239 (1963).

9. M. Russo, L. Mortillaro, L. Credali, and C. De Checchi, *J. Polymer Sci.*, *Part A1*, **4**, 248 (1966).

10. M. Schmidt, K. Blaettner, P. Kochendörfer, and H. Ruf, *Z. Naturforsch.*, *Part B*, **21**, 622 (1966).

11. M. Prince and B. Bremer, *J. Polymer Sci.*, *Part B*, **5**, 843 (1967).

12. L. Mortillaro, M. Russo, G. Carazzolo, M. Mammi, and G. Valle, *J. Polymer Sci.*, *Part B*, **6**, 431 (1968).

13. F. Sandrolini, P. Manaresi, S. Pietra, L. Mortillaro, and M. Russo, *J. Polymer Sci.*, *Part B*, **8**, 749 (1970).

14. F. O. Rice and A. L. Glasebrook, *J. Amer. Chem. Soc.*, **56**, 2381 (1934).

15. T. G. Pearson, R. H. Purcell, and G. S. Saigh, *J. Chem. Soc.*, 409 (1938).

16. F. D. Williams and F. X. Dunbar, *Chem. Commun.*, 459 (1968).

17. G. Carazzolo and M. Mammi, *J. Polymer Sci.*, *Part C*, **16**, 1521 (1967).

18. G. Carazzolo, L. Mortillaro, L. Credali, and S. Bezzi, *J. Polymer Sci.*, *Part B*, **3**, 997 (1965).

19. G. Carazzolo and G. Valle, *J. Polymer Sci.*, *Part A-3*, 4013 (1965).

20. G. Carazzolo and M. Mammi, *Makromol. Chem.*, **100**, 28 (1967).

21. G. Carazzolo and M. Mammi, *J. Polymer Sci.*, *Part A-1*, 965 (1963).

22. L. Mortillaro, G. Galiazzo, and S. Bezzi, *Chim. Ind.* (Milan), **46**, 139 (1964).

23. G. Carazzolo, L. Mortillaro, L. Credali, and S. Bezzi, *Chim. Ind.* (Milan), **46**, 1484 (1964).

24. G. Carazzolo and G. Valle, *Makromol. Chem.*, **90**, 66 (1966).

25. G. Carazzolo and M. Mammi, *Plast. Mod. Elastomeres*, **19**, 165 (1967).

26. G. Carazzolo and G. Putti, *Chim. Ind.* (Milan), **45**, 771 (1963).

27. M. Russo, L. Mortillaro, L. Credali, and C. De Checchi, *J. Polymer Sci.*, *Part B*, **4**, 167 (1966).

28. G. T. Morgan and F. H. Burstall, *J. Chem. Soc.*, 1497 (1930).

29. R. F. Cozzens and A. B. Harvey, *J. Polymer Sci., Part A-2*, **8,** 1279 (1970).

30. G. T. Morgan and F. H. Burstall, *J. Chem. Soc.*, 173 (1931).

31. M. Prince and B. Bremer, *J. Polymer Sci., Part B*, **5,** 847 (1967).

32. R. Paetzold and D. Knaust, *Z. Chem.*, **10,** 269 (1970).

33. L. Hagelberg, *Ber.*, **23,** 1083 (1890).

34. G. Bergson and G. Claeson, *Acta Chem. Scand.*, **11,** 911 (1957).

35. G. Bergson, *Ark. Kemi*, **19,** 195 (1962).

36. G. T. Morgan and F. H. Burstall, *J. Chem. Soc.*, 1096 (1929).

37. G. Bergson, *Ark. Kemi*, **13,** 11 (1958).

38. G. Bergson, G. Claeson, and L. Schotte, *Acta Chem. Scand.*, **16,** 1159 (1962).

39. J. R. Brown, G. P. Gillman, and M. H. George, *J. Polymer Sci., Part A-1*, **5,** 903 (1967).

40. G. T. Morgan and F. H. Burstall, *J. Chem. Soc.*, 2197 (1929).

41. H. G. Grimm and H. Metzger, *Ber.*, **69,** 1356 (1936).

42. D. J. G. Ives, R. W. Pittman, and W. Wardlaw, *J. Chem. Soc.*, 1080 (1947).

43. A. J. Brown and E. Whalley, *Inorg. Chem.*, **7,** 1254 (1968).

43a. K. A. Jensen and P. H. Nielsen, *Acta Chem. Scand.*, **17,** 549 (1963).

44. A. Fredga, *Ark. Kemi, Mineral. Geol.*, **11B, No. 15,** 1 (1933).

45. A. Fredga, *Uppsala Univ. Arsskrift*, **No. 5,** 125 (1935).

46. A. Fredga and K. Styrman, *Ark. Kemi*, **14,** 461 (1959).

47. G. Bergson and A. Biezais, *Ark. Kemi*, **18,** 143 (1961).

48. G. Bergson, *Ark. Kemi*, **19,** 75 (1962).

49. A. Simon and G. Heintz, *Naturwissenschaften*, **47,** 468 (1960).

50. U.S. Pat. 2,587,805 (1952).

51. Brit. Pat. 1,013,230 (1965).

52. H. Kroll and E. F. Bolton, *J. Appl. Polymer Sci.*, **14,** 2319 (1970).

53. G. Zundel, *Z. Naturforsch., Part B*, **23,** 119 (1968).

54. G. Zundel and H. Metzger, *Spectrochim. Acta*, **23A,** 759 (1967).

55. G. Zundel and H. Metzger, *Z. Phys. Chem.* (Leipzig), **240,** 50 (1969).

56. G. Zundel, *Angew. Chem.* (*Intern. Ed.*), **8,** 499 (1969).

57. W. H. H. Günther and M. N. Salzman, *Ann. N.Y. Acad. Sci.*, **192,** 25 (1972).

58. Y. Okamoto, T. Yano, and R. Homsany, *Ann. N.Y. Acad. Sci.*, **192,** 60 (1972).

59. W. H. H. Günther, U.S. Pat. 3,671,467 (1972).

XV Physicochemical investigations of selenium compounds

A. INFRARED SPECTRA OF ORGANIC SELENIUM COMPOUNDS

K. A. JENSEN, L. HENRIKSEN, and P. H. NIELSEN

University of Copenhagen
Chemical Laboratory II
The H. C. Ørsted Institute
Copenhagen, Denmark

1. Introduction

Infrared spectroscopy is a comparatively recent tool in the investigation of organic selenium compounds, few reports being found before 1965. Today, however, complete vibrational assignments have been published for many of the simpler compounds, and these have proved very useful for the interpretation of the infrared spectra of more complicated structures.

Unlike bonds to oxygen, bonds to selenium seldom give rise to infrared absorptions of high intensity. Moreover, most of these absorptions are found below 1100 cm^{-1}. The region from 1100 cm^{-1} down to around 600 cm^{-1} is usually rather crowded because of the presence of absorptions from common structural elements. This makes an assignment of the observed absorptions difficult and, unless selection rules are imposed by molecular symmetry, allows for the occurrence of highly mixed vibrations. On the other hand, the low frequencies encountered for vibrations of bonds to selenium should place the deformation modes of these bonds in the less crowded region below 400 cm^{-1}. Absorptions due to such modes, therefore, might prove to be of analytical value.

When a series of simple, analogous organic chalcogen compounds is compared, the following gradual changes influencing the vibrational spectra occur in the sequence oxygen, sulfur, selenium, tellurium: (1) the mass of the heteroatom increases; (2) the bonds to this atom usually become weaker; and (3) the molecular geometry is changed so that the bond angle at the heteroatom decreases and also the angles at a neighboring C atom are influenced.

The first type of change produces the same effects as an isotopic substitution. The effects of the second type of change can, as an approximation, be considered to be restricted to the absorption, to which vibrations of the bonds to the heteroatom contribute significantly. Since the two effects usually work in the same direction, a lowering of such frequencies can be expected in the sequence oxygen, sulfur, selenium, tellurium. However, the change in force constants varies from case to case. The third factor, the change in molecular geometry, together with the change in force constants, in some cases accounts for the dependence of vibrations of alkyl groups on the chalcogen to which they are bound.

In addition to the usual aids to the interpretation of vibrational spectra, such as Raman spectroscopy, evaluation of line contours of the gas-phase infrared spectra, and isotopic substitution, particularly deuteriation, a further tool is available for the evaluation of the spectra of organic selenium compounds, namely, comparison with the spectra of corresponding sulfur compounds. Among the compounds which contain a chalcogen incorporated in a π-electron system there is a radical difference between the oxygen and sulfur compounds, whereas the sulfur and selenium compounds behave similarly. When sulfur is substituted by selenium, often only a few absorptions, which can be attributed to vibrations of the sulfur- or selenium-containing groups, are shifted significantly, leaving the rest of the spectrum virtually unchanged. Accordingly the exchange of sulfur by selenium, or *vice versa*, may be as useful as an isotopic substitution and can serve analogous purposes. Because of this analogy the method may conveniently be termed elemental substitution.

The result of an elemental substitution of selenium with sulfur is always a shift towards higher wave numbers of the absorptions due to bonds to the chalcogen atoms. The magnitude of this shift is usually 50–150 cm^{-1} (*cf.* Tables XVA-2–4). Larger shifts are found for selenium-oxygen compounds (Table XVA-5).

It should be emphasized that elemental substitution may significantly influence several vibrations of the group attached directly to the chalcogen atom. When, for example, methylseleno and methylthio compounds are compared, there are only small differences between the C–H stretching frequencies and the asymmetric CH$_3$ deformation frequencies, but substantial differences exist between the symmetric deformation modes and between the rocking modes. The direction of the displacements is towards lower wavenumbers in going from sulfur to selenium, and the magnitude commonly exceeds 50 cm^{-1}. This introduces a hazard to the use of elemental substitution for the interpretation of the spectra of compounds containing alkylseleno groups, since the absorptions due to these groups may be confused with those due to C–Se or C–S bonds. As a typical example, the absorptions assigned to CH$_3$ deformation and rocking for dimethyl sulfide and dimethyl selenide are shown in Table XVA-1.

The following discussion of the infrared spectra of organic selenium compounds has been restricted mainly to vibrations of bonds to selenium. Tellurium compounds have been included so far as they are known, and data for corresponding sulfur compounds have been cited for comparison.

Table XVA-1   Wave numbers for the observed CH_3 deformation and rocking modes for dimethyl chalcogenides (C_{2v} model)[a]

					$\rho(CH_3)$	
	$\delta_{as}(CH_3)$	$\delta_s(CH_3)$		In-Plane	Out-of-Plane	In-Plane
Compound	$A_1 + B_1 + B_2$	A_1[b]	B_1	A_1	B_2	B_1
Me_2S	1434	1331	1308	1027	972	906
Me_2Se	1423	1284	1263	954	908	857
Me_2Te	1415	1227	1208	875	837	790

[a] From Allkins and Hendra.[7]
[b] From Raman spectrum.

Abbreviations.   To simplify formulas the following abbreviations have been used: Me = methyl, Et = ethyl, Pr = propyl, Pr^i = isopropyl, Ph = phenyl. Chalcogen atoms are denoted by Y, and halogen atoms by X.

Vibrational Modes

ν = stretching (also used for quoting an absorption frequency)
δ = angle deformation $\Big\}$ in-plane bend
ρ = rocking
ω = wagging $\Big\}$ out-of-plane bend
t = twisting
τ = torsion around a single bond
Δ = deformation, out-of-plane
s = symmetric; as = asymmetric (or antisymmetric)

To avoid ambiguities, in most cases vibrational modes have been designated in a slightly more elaborate way than usual; for example, the symmetric and asymmetric C–Se stretching modes of a dialkyl selenide are designated by ν(C–Se–C), not simply by νCSe.

2.   Selenols and Tellurols

The infrared spectra of selenols were first investigated by Sharghi and Lalezari.[1] Seven aliphatic and aromatic selenols were examined and were found to exhibit the Se–H band at 2280–2300 cm^{-1}.

A more complete investigation was performed by Harvey and Wilson,[2,3] who examined the infrared and Raman spectra of methaneselenol and its mono-, tri-, and tetradeuterio analogs in gaseous as well as in liquid and solid states. These authors reported complete vibrational assignments supported by a normal coordinate analysis.

The Se–H stretching vibration was found at 2330 cm^{-1} for CH$_3$SeH, at 2323 cm^{-1} for CD$_3$SeH, and at 1680 cm^{-1} for CH$_3$SeD and CD$_3$SeD (gas phase). Solid-phase infrared absorptions and liquid-phase Raman lines were found at lower wave numbers. The liquid-phase data for ν(Se–H) agree closely with data reported for higher selenols[1,4] (Mangini et al.,[5] however, quote the value 2364 cm^{-1} for benzeneselenol in C$_2$Cl$_4$).

The C–Se–H deformation band of MeSeH was found at 712 cm^{-1}. Finally, the C–Se stretching band was found at 590 cm^{-1}.

The ν(Se–H) band is very slightly influenced by the state of aggregation, by the concentration of solutions of the selenols, or by the nature of the solvent. Hence the tendency of selenols to form intermolecular hydrogen bonds seems to be very small.[1]

The Raman spectrum of methanetellurol[6] shows, as expected, a near correspondence to that of methaneselenol. The Raman lines of 2016, 516, and 608 cm^{-1} were assigned to Te–H stretch, Te–C stretch, and C–Te–H bend, respectively.

See Table XVA-2 for principal absorption bands of thiols, selenols, and tellurols.

Table XVA-2   Wave numbers (cm^{-1}) of infrared absorption bands assigned to Y—H and C—Y absorptions (Y = S, Se, Te)[a]

ABSORPTION BAND	S	Se	Te	COMPOUNDS[b]
ν(Y—H)	2550–2605	2280–2330	2016	R—Y—H
δ(C—Y—H)	802	712	608	R—Y—H
ν(C—Y)[c]	710	590	516	R—Y—H
		⎧ 550–610	528	R—Y—R
	600–750	⎨ 530–625		R—YO—R
		⎪		R—YO$_2$—R
		⎩ 507–584	507	R—Y—R
ν(Y—Y)	505–540	286–293	188	R—Y—Y—R
δ(C—Y—C)	284	233	198	R—Y—R
δ(Y—Y—C)	240 (A), 272 (B)	177	170	R—Y—Y—R

[a] Sharghi and Lalezari,[1] Harvey and Wilson,[2–3] Merijanian et al.,[4] Greenwood and Hunter,[15] Bahr and Fowles,[16] Hendra and Sadasivan,[17] Anysley,[18] Green and Harvey,[37] Bergson,[38] Allum et al.,[39] and Paetzold and Bochmann[186] (Se compounds); Sink and Harvey[6,43] and Allkins and Hendra[7] (Te compounds).

[b] If only one value is given, it refers to the methyl compound (R = CH$_3$).

[c] The symmetric and asymmetric stretching bands often coincide.

3. Selenides and Tellurides

The infrared and Raman spectra of dimethyl selenide and dimethyl telluride have been thoroughly investigated by Allkins and Hendra,[7] who reported complete vibrational assignments. Allkins and Hendra[8] have also investigated the infrared and Raman spectra of hexadeuteriodimethyl selenide.

The structures of Me_2O,[9] Me_2S,[10] and Me_2Se,[11,12] have been proved to possess C_{2v} symmetry by microwave absorption spectra and electron diffraction results. The spectra of Me_2Te could be interpreted also on this assumption.

Recently, the results of Allkins and Hendra have been used for vibrational analyses of Me_2Se and Me_2Te[13,14] The Raman spectrum[7] in the $\nu(C-H)$ region exhibited only three bands in the case of Me_2Se and Me_2Te, but five for Me_2S and six for Me_2O. This is taken as an indication that the bonding about the C atoms approaches a true tetrahedral disposition as one descends Group VI.

In the region of skeletal vibrations, liquid Me_2Se exhibited two bands at 604 and 589 cm^{-1}, the corresponding bands of Me_2S being found at 742 and 692 cm^{-1}. These two bands were assigned to $\nu_{as}(C-Se-C)$ and $\nu_s(C-Se-C)$, respectively.[7] The polarization of the Raman lines and the vapor-phase absorption envelopes confirmed that the lower frequency vibration is the symmetric one. In the case of dimethyl telluride, only one band, at 528 cm^{-1}, was observed in this region, presumably because the symmetric and asymmetric C–Te–C stretching vibrations are accidentally coincident.

A band near 580 cm^{-1}, which may be assigned as C–Se–C stretching, has been observed also in several other selenides.[15-18]

Methyltellurium compounds of various types (Me_2Te, Me_2TeI_2, Me_3TeI, MePhTe, etc.) all exhibited the C–Te stretching band of the CH_3–Te groups within the narrow range 521–525 cm^{-1}. For the corresponding ethyl derivatives the corresponding band was found near 500 cm^{-1}.[19]

The deformation vibration of the C–Y–C grouping gives rise to Raman lines at 284, 237, and 198 cm^{-1} for Me_2S, Me_2Se, and Me_2Te, respectively.[7]

Allkins and Hendra[20] have further examined platinum and palladium complexes of the type trans-$[MX_2(Me_2Y)_2]$, where M = Pt or Pd, X = halogen, and Y = S, Se, or Te. The bands due to vibrations in which the metal atom participates were identified in the region below 300 cm^{-1}.

Whereas the asymmetric deformation vibration of the methyl group is only slightly influenced by the attached chalcogen, there is a conspicuous lowering of the frequencies of the CH_3 symmetric deformation vibration and the CH_3 rocking vibrations with diminishing electronegativity of the chalcogen.[7] This effect has also been found by comparing the infrared spectra of more complicated sulfur and selenium compounds.[21]

Paetzold et al.[22] have pointed out that the intensity of a ν(C–Se) band is remarkably dependent on the formal charge of the Se atom. This band is very strong in selenoxides and compounds like $(CH_3)_2Se(OH)Cl$ and $(CH_3)_2SeCl_2$, but much weaker in selenious acid and the ions $[(CH_3)_2-Se(OH)]^+$ and $[CH_3Se(OH)_2]^+$.

Harvey et al.[23,24] have performed a vibrational analysis of the vibrational spectrum of selenetane (trimethylene selenide) and have assigned the 24 normal vibrations according to C_s symmetry. The normal coordinate analysis supports the assignment of a band at 410 cm^{-1} to the symmetrical C–Se stretching vibration; the frequency of this mode is more than 100 wave numbers lower than in analogous selenides such as dimethyl selenide. The band is no longer found at such low frequencies in polymerized selenetane. Therefore, the frequency shift is largely the result of the incorporation of C–Se–C into the strained ring system. A ν_{as}(C–Se–C) band was found at 563 cm^{-1} in the Raman spectrum of selenetane.

The infrared spectrum of poly(methylene selenide) has been reported by Prince and Bremer[25] and by Mortillaro et al.,[26] although not in the region below 600 cm^{-1}, where the C–Se–C vibrations are to be expected. The two reports are quite different.

Petrov et al.[27] have investigated a series of compounds of the type $R–Y–C{\equiv}C–CH{=}CH_2$ (Y = S, Se, Te) and have found that the chalcogen induces a shift of ν(C$\equiv$C) towards lower frequencies, the effect being most pronounced where Y = Te. A similar effect has been found by Radchenko et al.[28] in compounds of the type $R–Y–C{\equiv}C–R'$, where R' = 2,3-diphenyl-isoxazolidin-5-yl or 3-phenylisoxazolin-5-yl.

The Hg(II)-catalyzed addition of H_2O to the triple bond of compounds $Me–C{\equiv}C–SeR$ proceeds with the formation of esters, $MeCH_2COSeR$, rather than of the isomeric ketones, $MeCOCH_2SeR$, as shown by the infrared spectra of the reaction products.[29]

See Table XVA-2 for group frequencies of sulfides, selenides, and tellurides.

Coordination Compounds of Selenides. The work by Allkins and Hendra[20] has already been mentioned. Greenwood and Hunter[30,31] investigated coordination compounds of Ni(II), Pd(II), and Pt(II) with the ligand $Pr^i–Se–CH_2–CH_2–Se–Pr^i$ and showed that the ligand does not form a bridge between two metal atoms but rather functions as a bidentate chelating ligand.

Palladium(II) and platinum(II) form complexes with unsaturated selenides[32] such as $CH_2{=}CHCH_2CH_2{-}Se{-}Bu$. The stretching frequency of the M–Se bond was found near 220 cm^{-1}. The infrared spectra of Pd(II) and Pt(II) complexes of unsaturated sulfides[33] show similar behavior.

Baker and Fowles[34] have investigated the infrared spectra of metal complexes of thioxane and selenoxane and found that the metal ion is bonded to sulfur or selenium.

Infrared and far infrared spectra of 1,4-dithiane and 1,4-diselenane and their iodine complexes have been studied by Hendra and Sadasivan.[35] Hayward and Hendra[36] investigated the iodoform complexes of 1,4-dithiane and 1,4-diselenane.

4. Diselenides and Ditellurides

In an investigation of the infrared and Raman spectra of $CH_3SeSeCH_3$ and $CD_3SeSeCD_3$, Green and Harvey[37] have observed all 24 fundamental vibrations. On the basis of these observations and the fact that both polarized and nonpolarized Raman lines were observed it was concluded that dimethyl diselenide had C_2 symmetry with an optimal dihedral angle of 82°. The fundamental vibrations were assigned and reproduced reasonably well by a normal coordinate analysis.

The C–Se stretching band was observed at a similar wave number (577 cm^{-1}) as in selenides. Because of the small degree of coupling between the two C–Se vibrations usually only one C–Se stretching band is observed [designated in the following as ν(C–Se)]. The vibrations ν(Se–Se) and δ(Se–Se–C) were observed at 286 and 177 cm^{-1}, respectively.

Bergson[38] has investigated a series of dialkyl diselenides and 1,2-diselenolanes and assigned ν(C–Se) to an absorption in the 507–570 cm^{-1} range.

Allum et al.[39] also investigated dibutyl diselenide. A Raman line of dibutyl diselenide at 293 cm^{-1} was assigned to Se–Se stretching. This is not far from the ν(Br–Br) band at 317 cm^{-1}; in comparison, ν(S–S) is 505–540 cm^{-1} and ν(Cl–Cl) is 537 cm^{-1}. The absorptions in the spectrum of dibutyl diselenide at 631 and 542 cm^{-1}, respectively, were assigned to ν(C–Se) of trans and gauche conformations.

Infrared spectra of compounds containing both selenium and sulfur or tellurium and sulfur in a chain have been investigated in only a few cases. The Se–S stretching band of 1,2-thiaselenolane-4-carboxylic acid has tentatively been assigned to an absorption at 345 cm^{-1}.[40] For tellurium compounds of the type R–S–Te–S–R, where R = $CHMeCO_2H$, CMe_2CO_2H, or $CH_2CH_2CO_2H$, the ν(S–Te–S) absorptions are found in the 197–244 cm^{-1} range.[41]

The symmetric Te–Se stretching mode for Te(II) complexes of thioureas occurs at ca. 230 cm^{-1}.[42]

Sink and Harvey[43] have investigated the infrared and Raman spectra of dimethyl ditelluride and its hexadeuterio analog and have assigned ν(C–Te) at 507 c m^{-1}, ν(Te–Te) at 188 cm^{-1}, and δ(Te–Te–C) at 170 cm^{-1}.

See Table XVA-2 for the principal absorption bands of disulfides, diselenides, and ditellurides.

5. Derivatives of Selenocarbonic and Selenocarboxylic Acids

a. Introduction. Measurements of the electric dipole moments of compounds containing the C=S group[44] show that the polarity of this group varies considerably, and accordingly the ν(C=S) absorption band would be expected to be found within a wider range. The "purest" C-S double bond is found in thioacetone, for which ν(C=S) is 1269 cm^{-1}.[45] This is expected to represent approximately the upper limit for the ν(C=S) vibration, except for compounds with the grouping =C=S.

The often-repeated assertion that ν(C=S) for thioamides and thioureas gives rise to an absorption as high as *ca.* 1500 cm^{-1} [*cf.* also the assignment of ν(C—Se) as 1552 cm^{-1} for 2-selenouracil][46] is obviously quite unwarranted. The infrared bands of thioamides with the highest contribution from ν(C=S) are found below 800 cm^{-1}.[47]

Since sulfur and selenium do not differ much in electronegativity, the exchange of sulfur by selenium has in many cases only a minor influence on the force constants of the bonds of a larger organic molecule, so that to a first approximation only the vibrations of groups containing selenium or directly attached to selenium will be affected. In fact, "selenation" can be used to identify vibrations which have a more pronounced ν(C=S) or ν(C=Se) nature.[47]

Coordination compounds of the monoanion of diselenoacetylacetone can be prepared indirectly (*cf.* Chapter XVI). However, as in acetylacetonates, delocalization of electrons causes the C-Se double bonds to be equivalent, and, in fact, the observed C-Se stretching frequencies of these compounds (~ 600 cm^{-1}) are indicative of a C-Se single bond.[48]

The nearest approach to a C-Se double bond is realized in the few known and very unstable esters of selonecarboxylic acids, RC(=Se)YR' (Y = O, S, or Se). Even in these cases, because of electron delocalization the C-Se π-bond order is less than 1.

The CS$_2$ and CSe$_2$ groups, when occurring in compounds of proper symmetry, give rise to an antisymmetric ν_{as}(CYY) and a symmetric ν_s(CYY) stretching frequency. However, in compounds containing these groups in surroundings without the proper symmetry, as, for example, –CSSMe and CSeSeMe, vibrations occur of a similar type, properly described as out-of-phase and in-phase combinations, but often termed, in short, $\nu_{as}(CYY)$ and $\nu_{s}(CYY)$. If the coupling is small, one may prefer to speak of $\nu(C{=}X)$ and $\nu(C{-}X)$ $(Y = S, Se)$. In the literature the latter terms are often used even in the case of metal dithiocarboxylates and diselenocarboxylates, which are then formulated with a localized negative charge. In practice this distinction is not important, because empirically $\nu_{as}(YY)$ of salts and out-of-phase combinations of $\nu(C{=}Y)$ and $\nu(C{-}Y)$ of thione and selone esters are found in the same range.

b.   Carbon Diselenide and Selenocarbonates.

i.   Carbon Diselenide.   The fundamental frequencies of carbon diselenide and of the mixed compounds SCSe and SCTe were measured by Wentink[49,50] and compared with the values calculated by a normal coordinate analysis. Slightly modified values have been reported by other workers.[51,52] In the infrared spectrum of CSe_2 only two strong infrared bands are observed. The asymmetric stretching mode gives rise to an absorption at 1303 cm^{-1} in gaseous CSe_2 and at 1267 cm^{-1} in liquid CSe_2, whereas the bending absorption in both cases is near 310 cm^{-1}. These values may be compared with the values 1532 and 397 cm^{-1} for gaseous CS_2 and 1510 cm^{-1} for liquid CS_2. In the Raman spectrum of CSe_2 an intense line at 364 cm^{-1} is assigned to the symmetric stretching mode.

In a coordination compound of carbon diselenide,[53] $Pt(Ph_3P)_2(CSe_2)$, the asymmetric stretching vibration has been shifted from 1270 (CSe_2 in CCl_4) to 995 cm^{-1}. Poly(carbon diselenide)[54] has a strong band at 885 cm^{-1}.

ii.   Selenocarbonates.   The triselenocarbonate ion has D_{3h} symmetry, resulting in four fundamental vibrations:

$\nu_{s}(CSe_3)$	total symmetric C–Se stretching
$\Delta(CSe)$	out-of-plane deformation
$\nu_{as}(CSe_3)$	asymmetric C–Se stretching
$\delta(CSe_3)$	asymmetric CSe_3 deformation

The observed wave numbers[55] are listed below and compared with the corresponding values for the trithiocarbonate ion:

	$\nu_{s}(CSe_3)$	$\Delta(CSe)$	$\nu_{as}(CSe_3)$	$\delta(CSe_3)$
$CS_3{}^{2-}$	520	505	905	320 cm^{-1}
$CSe_3{}^{2-}$	316	420	802	185 cm^{-1}

The infrared spectrum of the selenodithiocarbonate ion has been studied by Seidel.[56]

Analyses of the spectra of dialkyl trithiocarbonates have indicated that two strong bands, near 1050 and 850 cm^{-1}, are characteristic of these compounds.[55,57,58] These bands were tentatively assigned to $\nu(C{=}S)$ and

ν_{as}(S–C–S), respectively. In addition, two weaker bands near 700 cm^{-1} in the spectra of dialkyl trithiocarbonates with lower alkyl groups have been assigned to ν(S–C$_{sp^3}$) and two bands near 500 cm^{-1} to ν_s(S–C–S) and Δ(C–S). This pattern is reproduced in the spectra of the corresponding triselenocarbonates,[59] which exhibit strong absorptions at *ca.* 880 and 750 cm^{-1}, and one or two weaker absorptions at *ca.* 600 cm^{-1}. For Me$_2$CS$_3$ a solvent-dependent doubling of the ν(C=S) and ν_{as}(S–C–S) absorptions has been observed and ascribed to the presence of conformers.[58] A similar doubling has also been observed for Me$_2$CSe$_3$ (856 + 883 cm^{-1} and 764 + 720 cm^{-1}) and for the mixed selenothiocarbonates, (MeSe)$_2$CS and (MeSe)(MeS)CS.[59]

The infrared spectra of a series of thiocarbonates and selenocarbonates of the following types:

RO—CS—SCH$_2$COOH   RS—CS—SCH$_2$COOH
RO—CSe—SeCH$_2$COOH   RS—CSe—SeCH$_2$COOH

have been studied by Jensen and Anthoni.[60] The compounds with a C=S group all have a strong band near 1050 cm^{-1}. The corresponding selenium compounds have instead a strong band near 940 cm^{-1}. Therefore, these bands were assigned essentially to ν(C=S) and ν(C=Se), respectively.

The infrared spectrum of potassium O-ethyl diselenocarbonate has been reported[60] to have a band originating mainly from C-Se stretching at 942 cm^{-1}, corresponding to a band at 1055 cm^{-1} of potassium O-ethyl dithiocarbonate.

c. ***Diselenocarbamates.*** The CSe$_2$ group attached to nitrogen is expected to give rise to six bands, approximately described as:

ν_{as}(CSeSe)	asymmetric CSe$_2$ stretch
ν_s(CSeSe)	symmetric CSe$_2$ stretch
δ(CSeSe)	CSe$_2$ deformation
ρ(CSeSe)	CSe$_2$ rock
ω(CSeSe)	CSe$_2$ wag
τ(CSeSe)	CSe$_2$ twist

A recent vibrational analysis of the dimethyl diselenocarbamate ion[61] has confirmed the identification of four of these modes in the spectrum of potassium dimethyl diselenocarbamate; ρ(CSeSe) and τ(CSeSe) have not been observed. The ν_{as}(CSeSe) vibration occurs [coupled with ν_{as}(CNC)] as a strong band at 872 cm^{-1}, and [coupled with ρ(CNC)] as another strong band at 391 cm^{-1}. The position of the first of these bands corresponds closely to the position of ν_{as}(CSeSe) in the above-mentioned compounds. The ν_s(CSeSe) vibration couples with δ(CNC) to give a rather weak band at 351 cm^{-1}. Finally, ω(CSeSe) and δ(CSeSe) occur at 536 and 224 cm^{-1}, respectively.

The infrared spectra of diselenocarbamato complexes show related features[62-65] and have recently been assigned on the basis of a normal coordinate treatment.[66]

Phosphoniodiselenoformates, R$_3\overset{+}{P}$CSeSe$^-$, have a strong band near 900 cm^{-1}, assigned to ν_{as}(CSe$_2$).[67] This corresponds to a strong band near 1050 cm^{-1} in the spectra of analogous sulfur compounds.[68]

d. Diselenocarbazates.

On the basis of infrared spectroscopic studies of 3,3-dimethyldithiocarbazic acid and 3,3-dimethyldiselenocarbazic acid Anthoni[69] concluded that these acids exist principally in the dipolar form, Me$_2\overset{+}{H}$N–NH–CYY$^-$. This was also shown to be the case with the unsubstituted acids[70] and with most other N-alkyl derivatives.[71,72] However, trimethyldiselenocarbazic acid assumed the nonpolar structure [ν(SeH) 2286 cm^{-1}].[72]

Investigations of the infrared spectra of unsubstituted and methyl-substituted dithio- and diselenocarbazic acids,[72] their salts, and their coordination compounds[73] [with Ni(II) and Cr(III)] have shown that intense bands at 960–1040 cm^{-1} and 590–690 cm^{-1} in the spectra of the sulfur compounds have their counterparts at 860–930 cm^{-1} and 490–615 cm^{-1} in the spectra of the selenium compounds. It is concluded, therefore, that these bands should be due mainly to asymmetric and symmetric skeletal stretching of the CS$_2$ and the CSe$_2$ group, respectively.

Diselenides obtained by oxidation of diselenocarbazic acids are either nonpolar (**1**) or dipolar (**2**). Infrared and ^{1}H n.m.r. spectroscopic investigations of these compounds[72] have shown the existence of both nonpolar and dipolar forms.

$$
\begin{array}{ccc}
\text{R}^1\text{R}^2\text{N—NR}^3\text{—C}{=}\text{Se} & & \text{R}^1\text{R}^2\overset{+}{\text{H}}\text{N—N}{=}\text{C—Se}^- \\
\mid & & \mid \\
\text{Se} & & \text{Se} \\
\mid & \longrightarrow & \mid \\
\text{Se} & & \text{Se} \\
\mid & & \mid \\
\text{R}^1\text{R}^2\text{N—NR}^3\text{—C}{=}\text{Se} & & \text{R}^1\text{R}^2\overset{+}{\text{H}}\text{N—N}{=}\text{C—Se}^- \\
\mathbf{1} \quad \text{R}^3 = \text{H} & & \mathbf{2}
\end{array}
$$

Like the dithio- and diselenocarbamates, the ionic dithio- and diseleno-carbazates have an N–C(Y$_2$) stretching band near 1480 cm^{-1}, which is shifted somewhat towards higher wave numbers in the coordination compounds.[73]

e. Cyanimidodiselenocarbonates and 1,1-Ethylenediselenolates. The infrared spectra of derivatives of cyanimidodiselenocarbonic acid (**3**) and the related 2,2-dicyanoethylene-1,1-diselenol (**4**), together with derivatives of several other 1,1-ethylenediselenols, have been investigated by Jensen and Henriksen.[74] The alkali-metal salts of **3** and **4** exhibit an intense infrared band at 870 and at 845 cm^{-1}, respectively, assigned to ν_{as}(CSeSe). A similar

$$\text{NC—N}{=}\text{C(SeH)}_2 \qquad \text{(NC)}_2\text{C}{=}\text{C(SeH)}_2$$

3   **4**

band can be recognized in a great variety of derivatives of these diselenols containing the following groups:

$$=\text{C}\big\langle \begin{smallmatrix} \text{SeR} \\ \text{SeR} \end{smallmatrix} \qquad =\text{C}\big\langle \begin{smallmatrix} \text{Se—} \\ \text{Se—} \end{smallmatrix} \qquad =\text{C}\big\langle \begin{smallmatrix} \text{Se—} \\ \text{Se—} \end{smallmatrix}$$

The analogous sulfur compounds have the corresponding band at wave numbers *ca.* 100 cm^{-1} higher.[75]

The spectra of coordination compounds of **4** have been reported.[76]

f. Esters of Monoseleno- and Diselenocarboxylic Acids. Collard-Charon and Renson[77] have investigated the infrared spectra of *O*-methyl thiobenzoate, C_6H_5–CS–OCH_3, and *O*-methyl selenobenzoate, C_6H_5–CSe–OCH_3, and have assigned the stretching vibrations of the C–S and C–Se double bonds to the same frequency (1227 cm^{-1}). In view of the difference in mass, this result is very improbable. A comparison of the infrared spectra of types R–CS–SMe, R–CS–SeMe, R–CSe–SMe, and R–CSe–SeMe[78] has shown that, apart from changes in intensity, the main difference between the spectra of Ph–CS–SMe and Ph–CSe–SMe is that an intense band at 1046 cm^{-1} in the spectrum of the sulfur compound is replaced by an intense band at 982 cm^{-1} in the spectrum of the selenium compound. These results have been corroborated by a study of carboxymethyl esters of various aromatic dithio- and selenothiocarboxylic acids.

Comparison of the spectra of aliphatic dithio-, selenothio-, and diselenoesters showed that a more extensive coupling of ν(C=S) or ν(C=Se) with other vibrations occurs. A normal coordinate analysis of the dithioacetate ion[79] has shown that two infrared bands, at 1141 and 865 cm^{-1}, of this ion result from a coupling of ν_{as}(CSS) with ρ(CH$_3$). In accordance herewith, the spectra of the thioneacetates, Me–CS–SMe and Me–CS–SeMe, differ from those of the seloneacetates, Me–CSe–SMe and Me–CSe–SeMe, both in the 1100–1200 cm^{-1} region and in the 800–900 cm^{-1} region. A band near 840 cm^{-1} was assigned to ν_{as}(CSSe), a band at 870 cm^{-1} to ν_{as}(CSS), and a band at 780 cm^{-1} to ν_{as}(CSeSe).

The spectra of selenothio analogs of propionic, pivalic, and phenylacetic acids show greater differences when $C{=}S$ is exchanged by $C{=}Se$, but here again the differences are located in the 800–1200 cm^{-1} region.

The carbonyl group frequency of thiocarboxylic S-esters is lower by *ca.* 40 cm^{-1} than that of carboxylic esters. It might be expected, therefore, that the carbonyl group stretching frequency would be still lower for seleno-carboxylic Se-esters, but actually Renson and Draguet[80] found it to be higher than for the corresponding sulfur compounds. It has been suggested that this effect is due to the greater π-inductive effect of sulfur.[81]

See Table XVA-3 for the principal stretching bands of CS_2 and CSe_2 compounds.

Table XVA-3 Infrared absorption bands (cm^{-1}) assigned principally to stretching vibrations of CY_2 groups (Y = S or Se)

APPROXIMATE DESCRIPTION	S	Se	COMPOUNDS	REF.
$\nu(C{=}Y)$	1532 (g)	1303 (g)	Carbon disulfide and	49, 51
	1510 (l)	1267 (l)	diselenide	
	1220–1260		Thioketones	
	1000–1200	800–1000	Dithioic and diselenoic esters; dialkyl trithio- and triselenocarbonates	58–60, 78
	700–800	600–700	Thio- and selenoamides, -hydrazides, and -ureas[a]	47
$\nu_{as}\left(=C{<}^{YR}_{YR}\right)$	800–900	700–800	Ketene thioacetals and selenoacetals; dialkyl trithiocarbonates and triselenocarbonates	58, 59, 74, 75
$\nu_{as}\left(-C{<}^{Y}_{Y}\ominus\right)$	900–1050	800–950	Ionic dithioates, dithio-carbamates, and diselenocarbamates	61, 63, 64, 79
$\nu_{as}\left(=C{<}^{Y\ominus}_{Y\ominus}\right)$	850–980	750–870	Ionic ethylene-1,1-di-thiolates and -1,1-di-selenolates, cyanimido-dithiocarbonates, and -diselenocarbonates	74, 75

[a] Absorptions at higher wave numbers may also contain a contribution from $\nu(C{=}Y)$.

g. Selenoamides and Selenohydrazides. In a study of the infrared spectra of thioamides and selenoamides Jensen and Nielsen[47] found that it was possible to locate various characteristic bands, termed the A, B, . . . , G bands, each of which behaved in a characteristic way when the compounds were deuteriated, S- or Se-alkylated, or transformed into metal complexes. Two of the bands, the B band at 1400–1600 cm^{-1} and the G band at 600–800 cm^{-1}, were thought to originate mainly in the stretching vibrations of the grouping $Y{=}C{-}N< \leftrightarrow \overset{-}{Y}{-}C{=}\overset{+}{N}<$; the B band being due principally to the $\nu(CN)$ vibration and the G band to the $\nu(YC)$ vibration. In accord with this, the spectra of thio- and selenoamides, on S- or Se-alkylation or transformation into metal complexes with the metal ion bound to sulfur or selenium, exhibit a shift of the B band towards higher wave numbers and a shift of the G band towards lower wave numbers. The influence of deuteriation is, in most cases, negligible. It was concluded, therefore, that the bands having most pronounced C–S or C–Se character should be sought for in the region of C-S and C-Se single-bond vibrations, that is, at 700–800 cm^{-1} for thioamides and thioureas and at wave numbers 30–100 cm^{-1} lower for selenoamides and selenoureas.

However, coupling of C–S or C–Se vibrations with other vibrations may be expected. The D band, which is medium strong in the 1000–1200 cm^{-1} region, sometimes reveals that C–S or C–Se vibrations may contribute to it. For some compounds (*e.g.*, thioacetamide and thiourea and their selenium analogs) the D band is shifted towards lower frequencies on S- or Se-alkylation. Collard-Charon and Renson[82] have assigned a band near 1200 cm^{-1} in the spectra of selenoamides, selenoureas, and selenosemicarbazides to $\nu(C{=}Se)$; but the changes induced by selenation are much too small to justify this assignment, and the band in question can probably in all cases be classified as the D band.

Comparison of the spectra of a large number of thiosemicarbazides and selenosemicarbazides,[83] thiocarboxylic hydrazides and selenocarboxylic hydrazides,[84] and thiourethanes and selenourethanes[85] has shown that they can be interpreted in a manner similar to that used for the spectra of thioamides and selenoamides.

Wiles and Suprunchuk[87] have examined the infrared spectra of a series of aldehyde and ketone thiosemicarbazones and selenosemicarbazones.

Hallam and Jones[88] have compared the infrared spectra of a series of thio- and selenolactams, $(CH_2)_n \overset{\displaystyle \frown CY}{\underset{\displaystyle \smile NH}{\big|}}$, where Y = S or Se, and n = 3, 4, 5, 7, and 11. In general, the results support the rules established by Jensen and Nielsen. However, in the $n = 3$ and 4 compounds the $\nu(C{=}S)$ and $\nu(C{=}Se)$ frequencies seem to be reasonably well localized, at about 1115 (S) and 1085 (Se) cm^{-1}.

The Raman and infrared spectra of thiourea and selenourea complexes of Pd(II) and Pt(II) have been reported,[89] and the sulfur-metal and selenium-metal stretching modes have been assigned. Infrared spectra of thiourea and selenourea complexes of Co(II), Ni(II), and Cd(II) thiocyanates have been reported by Domiano et al.[90]

The diselenobisformamidinium cation formed by the oxidation of selenourea has been subjected to X-ray crystallographic and infrared spectroscopic studies by Chiesi et al.[91] The infrared spectra of this ion and of the corresponding sulfur compound[92] show greater differences than are usual for corresponding sulfur and selenium compounds. Among the more prominent bands, however, there is fair agreement.

h. Selenocyanates and Isoselenocyanates. From a number of publications[93-95] on the infrared spectra of selenocyanates it is evident that the C–N stretching vibration should give rise to an absorption near 2150 cm⁻¹. Aynsley et al.[96] have investigated both the infrared and the Raman spectra of methyl and phenyl selenocyanates and have given complete assignments. The following refer to vibrations involving the Se atom:

		CH_3SeCN	C_6H_5SeCN
Se—C(N)	stretch	519 cm⁻¹	523 cm⁻¹
Se—CH₃	stretch	576 cm⁻¹	
Se—C≡N	in-plane bend	393 cm⁻¹	390 cm⁻¹
Se—C≡N	out-of-plane bend	358 cm⁻¹	349 cm⁻¹
C—Se—C	bend	168 cm⁻¹	

The two stretching frequencies of the NCSe group, whether it is nitrogen-bonded or selenium-bonded, are the out-of-phase and the in-phase combinations of $\nu(CN)$ and $\nu(CSe)$, which will be designated as $\nu_{as}(NCSe)$ and $\nu_s(NCSe)$, respectively. In cases where the coupling is expected to be small, the terms $\nu(CN)$ and $\nu(CSe)$ are used.

By analogy with the infrared spectra of thiocyanates and isothiocyanates, isoselenocyanates should differ from the selenocyanates in the contour and intensity of the $\nu_{as}(NCSe)$ or $\nu(CN)$ band. This has been confirmed by Tarantelli and Pecile[97] and by Bulka et al.[98] The latter found that a series of aromatic selenocyanates exhibited a medium strong and sharp band near 2160 cm⁻¹, whereas the isomeric isoselenocyanates showed a very strong and broad band with two submaxima in the range 2000–2200 cm⁻¹. This difference is similar to that found between the spectra of thiocyanates and isothiocyanates.[99,100]

A great number of coordination compounds with either nitrogen-bonded (Se=C=N–) or selenium-bonded (N≡C–Se–) selenocyanate groups have been prepared (cf. Chapter XVI). They may be distinguished either by the much higher intensity of the absorption in the region around 2000 cm^{-1} in the case of *N*-bonded SeCN groups or by the occurrence of the ν_s(NCSe) absorption at a higher wave number (605–674 cm^{-1}) for the *N*-bonded ligand than that of the ν(CSe) absorption for the *Se*-bonded ligand (520–543 cm^{-1}).

Isoselenocyanates with the N atom bound to elements other than carbon apparently do not exhibit a satellite to the ν_{as}(NCSe) band.[101–103] For the compounds R$_3$Si–NCSe, R$_3$Ge–NCSe, and R$_3$Sn–NCSe, ν_{as}(NCSe) is found near 2050 cm^{-1}. An infrared band of these compounds near 800 cm^{-1} has been assigned to ν_s(NCSe).[102] However, this assignment seems doubtful because a similar band of isoselenocyanato complexes is considered to be the first overtone of the NCSe bending frequency at 370–430 cm^{-1}.[105]

Selenocyanates of types R$_2$Tl–SeCN, R$_3$Pb–SeCN, and R$_2$Pb(SeCN)$_2$ have been studied by Aynsley *et al.*[104]

Gabrio and Barnikow[106] have investigated the infrared spectra of a series of isoselenocyanatophosphates, (RO)$_2$P(O)NCSe, and isoselenocyanatothiophosphates, (RO)$_2$P(S)NCSe. They found the asymmetric NCSe stretching band in the region 1960–1975 cm^{-1}, that is, considerably lower than for the above-mentioned isoselenocyanates.

6. Heterocyclic Selenium Compounds

a. Selenophene Derivatives. Early work on the Raman spectrum of selenophene[107] seemed to indicate that the selenophene molecule was not planar. The planarity of selenophene has been confirmed, however, by its microwave spectrum.[108] The infrared and Raman spectroscopic investigation of selenophene by Trombetti and Zauli,[109] in which 21 fundamentals were assigned, could also be interpreted on the basis of planar selenophene. Strong bands (liq.) at 1424, 1349, 824, and 760 cm^{-1} are assigned to ring vibrations. Nevertheless, there still seems to be disagreement concerning the absorption bands of selenophene,[109–111] possibly because of difficulties in obtaining a rigorously pure sample.

Tsukerman *et al.*[112] have investigated the infrared spectra of a series of chalcones containing the selenophene ring and ascribed three bands near 1430, 1330, and 1237 cm^{-1} to the selenophene group.

b. Selenazole Derivatives. Bassignana *et al.*[113] have investigated the infrared spectra of several selenazoles, benzoselenazoles, and selenazolines. A band in the range 1537–1570 cm^{-1} was assigned to the –N=C–Se– grouping. The benzoselenazoles have an additional band at 1590–1610 cm^{-1}. The selenazolines have an absorption band at 1650–1680 cm^{-1} assigned to ν(C=N).

c. 1,2,4-Selenadiazole Derivatives. Goerdeler *et al.*[114] have studied the infrared spectra of 5-amino-1,2,4-selenadiazoles and assigned a band in the 1525–1540 cm^{-1} range to a ring vibration.

d. 1,2,5-Selenadiazole Derivatives. A vibrational analysis of 1,2,5-selenadiazole by Benedetti and Bertini[115] has resulted in the identification of all 15 fundamental frequencies, 9 of which are due to ring vibrations.

Infrared absorptions have been reported for several derivatives of 1,2,5-selenadiazole[116–118] and 2,1,3-benzoselenadiazole.[119,120]

e. 1,3,4-Selenadiazole. A detailed vibrational analysis of this recently discovered compound has not yet been published. Infrared absorptions have been reported,[121] however, and its microwave spectrum has been investigated.[122]

f. 1,2-Diselenolylium Ions. Assignments for the in-plane vibrations of the 3,5-dimethyl-1,2-dithiolylium ion have been obtained by a normal coordinate analysis.[123] By adoption of a cyclic and planar structure for the corresponding diselenolylium ion vibrational assignments were obtained by the same normal-coordinate treatment. A weak band near 600 cm^{-1} was assigned to the C–Se stretching vibration, and an absorption band at 320 cm^{-1} to a coupled mode containing the Se–Se stretching vibration.

Infrared absorption of the 3,5-diamino-1,2-diselenolylium ion has been reported.[124] The infrared spectrum is very similar to that of the corresponding sulfur compound.[125]

7. Selenium Compounds of Group IV Elements

The vibrational spectra of disilyl selenide,[126] digermyl selenide,[127,128] disilyl telluride,[129] and digermyl telluride,[128] as well as of the corresponding sulfur compounds,[126,130] have been reported.

For bis(trimethylstannyl) selenide the values 224 and 238 cm^{-1} have been reported for the symmetric and the asymmetric Sn–Se stretch.[137] Coordination of $(Me_3Sn)_2Se$ to the metal carbonyl group $M(CO)_5$, where M = Cr, Mo, or W, induces a shift of the stretching vibrations to lower wave numbers and a greater separation of the symmetric and the asymmetric stretch.[132] Schumann and Schmidt[133] have given a short review of the infrared spectra of compounds of the type $Ph_3M–Y–MPh_3$, where M = Ge, Sn, or Pb, and Y = S, Se, or Te.

Also, Si–C, Ge–C, and Sn–C vibrations of selenides and tellurides containing triethylsilyl, triethylgermyl, or triethylstannyl groups have been reported.[134,135]

8. Selenium Compounds of Group V Elements

a. Nitrogen Compounds. As mentioned in Section A.9.d on selenoxides, selenium compounds with the grouping $R_2N–Se$ exhibit an infrared band near 540 cm^{-1} which has been assigned to Se–N stretching. For dimeric seleninic methylimide, $(OSeNMe)_2$, three infrared absorptions (at 625, 735, and 757 cm^{-1}) have been assigned to $\nu(N–Se–N)$.[136]

b. Phosphorus Compounds. Phosphine selenides exhibit the P=Se stretching frequency within a very wide range. Triphenylphosphine selenide has a strong band at 560–562 cm^{-1} which is absent from the spectra of triphenylphosphine sulfide and, therefore, was assigned[137,138] to $\nu(P=Se)$. This conclusion is supported by a shift of this absorption to 530–540 cm^{-1} in the spectra of metal complexes of triphenylphosphine selenide.[139–141] For tris(cyclopropyl)phosphine selenide $\nu(P=Se)$ seems to be considerably higher (685 cm^{-1}).[141a]

However, aliphatic selenides exhibit the $\nu(P=Se)$ absorption at considerably lower wave numbers,[137,142] (*e.g.*, at 422 cm^{-1} for Et$_3$PSe), this assignment being supported by Raman evidence[142] and by a shift of this absorption to lower frequencies in metal complexes of trimethylphosphine selenide.[143] A similar difference is found between the wave numbers of $\nu(P=S)$ of triphenylphosphine sulfide (637 cm^{-1}) and triethylphosphine sulfide (539 cm^{-1}). Durig *et al.*[142] explained the low value in the latter case by the coupling of $\nu(P=S)$ with $\nu(PC_3)$.

X-ray structural evidence suggests intermolecular association through double-bonded selenium in the compounds $[Et_2P(=Se)Se]_2Se$ and $[Et_2P(=Se)Se]_2Te$.[144,145] This may in part be responsible for the low value ($\sim$450 cm^{-1}) of $\nu(P=Se)$ found for these compounds.

Chittenden and Thomas[146] and Schmidtpeter and Brecht[146a] suggested that both $\nu(P=S)$ and $\nu(P=Se)$ should give rise to two bands, separated by approximately 150 cm^{-1}. It seems more probable that the band at higher wave number arises from a skeletal stretching vibration.[147]

A strong absorption in the 500–600 cm^{-1} range, which is probably caused by P=Se stretching, has been observed in the spectra of derivatives (esters, salts, amides, acid chlorides) of selenophosphoric,[148–153] selenophosphonic,[155–156a] and selenophosphinic acids.[157] Selenothiophosphates of the type $(EtO)_2P(=Se)SR$ exhibit a strong band at 590 cm^{-1}, whereas the isomers, $(EtO)_2P(=S)SeR$, have a strong band at 650–670 cm^{-1}, attributed to $\nu(P=S)$.[158] Compounds of the latter type should also give rise to $\nu(P–SeR)$ absorption, but this has not been identified with certainty. By analogy with the corresponding sulfur compounds this absorption is expected at somewhat lower wave numbers than $\nu(P=Se)$. Thus, it seems improbable that the assignment of ν(P–SePh) to an absorption in the 1000–1075 cm^{-1} range[159] can be correct.

Ionic *O,O'*-dialkyl diselenophosphates[148,154] and dialkyldiselenophosphinates[157] exhibit two separate bands which may be assigned to the symmetric and asymmetric stretching vibrations of the $\begin{array}{c} \diagup \diagup^{Se} \\ P \\ \diagup \diagdown_{Se} \end{array} \Big\} \ominus$ group.

Present evidence suggests an overall characteristic absorption range of 420–600 cm^{-1} for ν(P=Se).

Infrared spectra of phosphorus-sulfur compounds have been studied very extensively (*cf.* a review by Corbridge[160]). Many of the correlations for the sulfur compounds will undoubtedly prove to be valid also for the corresponding selenium compounds.

Trialkylphosphine tellurides have been studied by Zingaro *et al.*[161,162]

c. Arsenic Compounds. A series of trialkylarsine selenides has been prepared by Zingaro and Merijanian.[163] The ν(As=Se) stretching frequency is assigned to an absorption in the 330–360 cm^{-1} range, in most cases observed as a doublet.

Some attempts to prepare triphenylarsine selenide[163,164] have been unsuccessful. However, it was synthesized by Jensen and Nielsen,[138] who suggested that ν(As=Se) for this compound might be coincident with a phenyl group absorption near 460 cm^{-1}.

d. Antimony Compounds. Trialkylstibine selenides[165] exhibit ν(Sb=Se) as an intense singlet band around 300 cm^{-1}.[162]

For comparison the M=Y stretching frequencies of various chalcogenides of type R$_3$MY are given in Table XVA-4. Chremos and Zingaro[166] have shown that ν(M=Y) of these compounds depends on the total mass of the molecule and the masses and electronegativities of the respective atoms, M and Y.

Table XVA-4   M=Y *stretching frequencies* (cm^{-1}) *of chalcogenides,* R$_3$MY[a]
R = alkyl;[b] M = P, As, or Sb; Y = O, S, Se, or Te

M	O	S	Se	Te
P	1143–76	552–599	421–511	400–467
As	885–903	473–487	331–358	
Sb	650–78	438–40	272–300	

[a] Data from Chremos and Zingaro.[162]

[b] For R = phenyl or cyclohexyl the upper limits may be 15–100 cm^{-1} higher.

9. Selenium-Oxygen Compounds*

The very extensive investigations of Paetzold and his co-workers during recent years have furnished much valuable information on Se–O vibrations in compounds such as selenates, selenites, selenones, and selenoxides (*cf.* Table XVA-5).

Table XVA-5   Infrared absorption bands of Se—O *and* Se—N *groups*[a,b]

VIBRATIONAL MODE	WAVE NUMBER (cm^{-1})	COMPOUNDS
ν_{as}(O—Se—O)	1010–1040	Selenates
	910–920	Selenones
ν_s(O—Se—O)	930–960	Selenates
	880–890	Selenones
ν(Se=O)	930–1005	Seleninyl halides
	930	Selenites
	880–890	Selenious amides
	850–900	Seleninic acids, their esters and anhydrides
	800–840	Selenoxides
ν(Se—O—C)	600–700	Selenates, seleninates
ν(Se—OH)	680–700	Seleninic acids
ν_{as}(Se—O—Se)	600–700	Seleninic anhydrides, diselenates
ν_s(Se—O—Se)	500–560	Seleninic anhydrides, diselenates
δ(O—Se—O)	390–420	Selenones
δ(Se—O—Se)	170–230	Seleninic anhydrides, diselenates
ν(N—Se—N)[c] or ν(Se—N)	540–590	Dialkylamino compounds: $(R_2N)_2SO$, $(R_2N)_2Se$, $(R_2N)SeOCl$, and $(R_2N)_2SeR_2'$

[a] Data from Paetzold *et al.*,[167–170,172,173,176–182,185–195] Krejči *et al.*,[171] Astin *et al.*,[174] Prideaux and Green,[175] Bredereck *et al.*,[183] Mock and McCausland,[184] Jensen and Krishnan,[196] and Tanaka and Kamitani.[197]

[b] The counterparts of corresponding sulfur compounds are at wave numbers 200–300 cm^{-1} higher.

[c] The symmetric and asymmetric stretching bands usually coincide.

a.   Selenates and Selenonates. Paetzold and Amoulong[167] examined the infrared and Raman spectra of dimethyl selenate, $(MeO)_2SeO_2$, and dimethyl diselenate, $(MeO)_2Se_2O_5$. In both compounds two very strong bands are observed in the infrared spectra near 1010 and 936 cm^{-1} and are assigned to the asymmetric and symmetric stretching vibrations of the SeO_2 groups.

* The vibrations of bonds between selenium and other Group VI elements (sulfur and tellurium) were treated in Section A.4.

The Se–OC stretching band is located near 650 cm^{-1}. The diselenate exhibits two additional bands at 687 cm^{-1} and 555 cm^{-1}, which are assigned to the asymmetric and symmetric stretching vibrations of the Se–O–Se group. On the basis of the Raman spectrum, the Se–O–Se deformation occurs at 229 cm^{-1}.

Similar results have been found with methyl and ethyl fluoroselenate,[168] bis(trimethylsilyl) selenate,[169] and methaneselenonic and benzeneselenonic acids.[170]

The infrared spectra of dialkyl selenates have also been investigated by Krejči et al.[171]

b. Selenones. Paetzold and Bochmann[172] have prepared selenones, R_2SeO_2, with R = methyl, ethyl, propyl, and butyl. The very strong stretching bands of the SeO_2 group are found at wave numbers 50–100 cm^{-1} lower than those for the corresponding bands in the selenates.[173]

c. Selenites and Seleninic Acids. The nature of the reaction product of SeO_2 with methanol has been the subject of some controversy.[174,175] Simon and Paetzold[176] have shown that methyl hydrogen selenite is formed and have given a complete vibrational analysis of the product on the basis of infrared and Raman spectra. The spectra of alkyl selenite ions have also been investigated.[177] In addition, Simon and Paetzold[178] have studied the infrared spectrum of dimethyl selenite and have obtained assignments of the absorptions by comparison with the assignments for dialkyl sulfites. Furthermore, Paetzold and Knaust[169] have carried out a vibrational analysis of bis(trimethylsilyl) selenite.

Other papers of Paetzold and his co-workers deal with the infrared spectra of methane- and ethaneseleninic acids;[179] their acidium ions, that is, the ions $RSe(OH)_2^+$; and esters of fluoro- and chloroseleninic acids.[180]

The anhydrides of seleninic acids were shown by Paetzold and his associates[181,182] to be true anhydrides, R–Se(O)–O–Se(O)–R, in contrast to the "sulfinic anhydrides," which, according to Bredereck et al.,[183] are sulfinylsulfones with a S-S bond, R–S(O)–SO$_2$–R.

The adduct of butadiene and selenium dioxide is rapidly transformed into a product that has been shown, by infrared spectroscopy, to be a cyclic seleninic ester.[184]

d. Selenoxides. Dimethyl selenoxide and diethyl selenoxide have been subjected to detailed infrared and Raman spectroscopic studies.[185,186] The Se=O stretching band is a strong band near 800 cm^{-1}. The symmetric and asymmetric C–Se–C stretching vibrations coincide in the case of the methyl compound.

The frequency ascribed to the ν(Se=O) vibration covers a wide range, from *ca.* 800 to 1000 cm^{-1}. Paetzold[187] has pointed out that the frequency decreases with the electronegativity of X in compounds of type $X_2Se=O$, as shown in the following table:

X:	F	Cl	Br	OMe	NMe$_2$	Ph	Me	Et
$\nu(Se=O)$	1005	948	930	929	877	838	820	818

It is assumed that there is no appreciable coupling between the stretching vibrations of $Se=O$ and $Se-X$, so that $\nu(Se=O)$ can be used as a measure of the S-O double-bond strength in various X_2SeO molecules. A linear relationship was established between valence force constants, calculated from $S=O$ and $Se=O$ stretching frequencies, for compounds of types X_2SO and X_2SeO.[188] Accordingly, the $Se=O$ force constant of a compound X_2SeO can be calculated from that of the corresponding sulfur compound and *vice versa*. A linear relationship has further been established between valence force constants and lengths of Se-O double bonds.[189]

Infrared studies of the halides Ph_2SeCl_2 and Ph_2SeBr_2, and of the alkoxides $Ph_2Se(OR)_2$ (R = Me or Et) and $Me_2Se(OMe)_2$, indicate that they have a trigonal-bipyramidal structure.[190] This is probably true also for the hydrochloride of dimethyl selenoxide in the crystalline state.[185] The corresponding perchlorate, in the crystalline state, is a true hydroxy-selenonium salt, containing the cation $Me_2Se(OH)^+$. The crystalline chloride also exhibits a broad band near 275 cm^{-1}, assigned to $\nu(Se-Cl)$.

The infrared and Raman spectra of aqueous solutions of selenoxides indicate that they do not form dihydroxy compounds of type $R_2Se(OH)_2$.[185]

Paetzold and Rönsch[191,192] have also studied the infrared and Raman spectra of selenious amides $(R_2N)_2SeO$, selenium(II) bis(dimethylamide) $(Me_2N)_2Se$, and $Me_2N-SeO-Cl$. All three types exhibit one SeN$_2$ or SeN stretching band near 540 cm^{-1}. In this region $(Me_2N)_2SeO$ exhibits two bands which are so widely separated (539 and 587 cm^{-1}) that it is concluded that they originate from rotational isomers rather than from $\nu_s(N-Se-N)$ and $\nu_{as}(N-Se-N)$, respectively.

Several metal complexes have been prepared from diphenyl selenoxide[193] and dimethyl selenoxide.[194-197] For all these complexes, a lowering of the $Se=O$ stretching frequency was observed, indicating that coordination occurs, in all cases, to the O atom.

10. Selenium-Halogen Compounds

Dialkyl sulfides, selenides, and tellurides form well-defined 1:1 adducts with halogens of the type R_2YX_2 (Y = S, Se, Te; X = Cl, Br, Te). Three types of structure need to be considered: the ionic type, $R_2YX^+X^-$; the covalent type, R_2YX_2, with two equivalent Y-X bonds; and the charge-transfer type, with one Y-X and one X-X bond. Although some cations of the first type are known, their formation appears to require the presence of a complex anion.[198,199]

Wynne[200] interpreted his infrared results on dialkylselenium difluorides in terms of a covalent structure. Hayward and Hendra[201] have similarly interpreted the infrared spectra of several selenium and tellurium compounds to be indicative of type Me_2YX_2 and have assigned X–Y stretching frequencies. These fall in the same region as the vibrations of selenium and tellurium tetrahalides.[202–205] The infrared spectrum of Me_2SeI_2 could be interpreted[201] in terms of the charge-transfer type of spectrum.

REFERENCES

1. N. Sharghi and I. Lalezari, *Spectrochim. Acta*, **20**, 237 (1964).

2. A. B. Harvey and M. K. Wilson, *Inorg. Nucl. Chem. Lett.*, **1**, 101 (1965).

3. A. B. Harvey and M. K. Wilson, *J. Chem. Phys.*, **45**, 678 (1966).

4. A. Merijanian, R. A. Zingaro, L. S. Sagan, and K. J. Irgolic, *Spectrochim. Acta*, **25A**, 1160 (1966).

5. A. Mangini, A. Trombetti, and C. Zauli, *J. Chem. Soc.*, A, 153 (1967).

6. C. W. Sink and A. B. Harvey, *Chem. Commun.*, 1023 (1969).

7. J. R. Allkins and P. J. Hendra, *Spectrochim. Acta*, **22**, 2075 (1966).

8. J. R. Allkins and P. J. Hendra, *Spectrochim. Acta*, **23A**, 1671 (1967).

9. U. Blukis, P. H. Kasai, and R. J. Myers, *J. Chem. Phys.*, **38**, 2753 (1963).

10. L. Pierce and M. Hayashi, *J. Chem. Phys.*, **35**, 479 (1961).

11. E. Goldish, K. Hedberg, R. E. Marsh, and V. Schomaker, *J. Amer. Chem. Soc.*, **77**, 2948 (1955).

12. J. F. Beecher, *J. Mol. Spectry.*, **21**, 414 (1966).

13. J. Shiro, M. Okaku, M. Hayashi, and H. Murata, *Bull. Chem. Soc. Japan*, **43**, 612 (1970).

14. J. M. Freeman and T. Henshall, *J. Mol. Struct.*, **1**, 31 (1967).

15. N. N. Greenwood and G. Hunter, *J. Chem. Soc.*, A, 929 (1969).

16. K. W. Bahr and G. W. A. Fowles, *J. Chem. Soc.*, A, 801 (1968).

17. P. J. Hendra and N. Sadasivan, *Spectrochim. Acta*, **21**, 1127 (1965).

18. E. E. Aynsley, *Chem. Ind.* (London), 379 (1966).

19. H. P. Fritz and H. Keller, *Chem. Ber.*, **94**, 1524 (1961).

20. J. R. Allkins and P. J. Hendra, *J. Chem. Soc.*, A, 1325 (1967).

21. K. A. Jensen and L. Henriksen, *Acta Chem. Scand.*, **24**, 3213 (1970).

22. R. Paetzold, U. Lindner, G. Bochmann, and P. Reich, *Z. Anorg. Allg. Chem.*, **352**, 295, 305 (1967).

23. A. B. Harvey, J. R. Durig, and A. C. Morrissey, *J. Chem. Phys.*, **47**, 4864 (1967).

24. A. B. Harvey, J. R. Durig, and A. C. Morrissey, *J. Chem. Phys.*, **50**, 4949 (1969).

25. M. Prince and B. Bremer, *J. Polymer Sci.*, Part B, **5**, 843 (1967).

26. L. Mortillaro, M. Russo, G. Carazzolo, M. Mammi, and G. Valle, *J. Polymer Sci.*, Part B, **6**, 431 (1968).

27. A. A. Petrov, S. I. Radchenko, K. S. Mingaleva, I. G. Savich, and V. B. Lebedev, *Zh. Obshch. Khim.*, **34**, 1899 (1964).

28. S. I. Radchenko, V. N. Chistokletov, and A. A. Petrov, *Zh. Obshch. Khim.*, **35**, 1735 (1965).

29. Yu. A. Boiko, B. S. Kupin, and A. A. Petrov, *Zh. Org. Khim.*, **5**, 1553 (1969).

30. N. N. Greenwood and G. Hunter, *J. Chem. Soc.*, *A*, 1520 (1967).

31. N. N. Greenwood and G. Hunter, *J. Chem. Soc.*, *A*, 929 (1969).

32. D. C. Goodall, *J. Chem. Soc.*, *A*, 890 (1969).

33. D. C. Goodall, *J. Chem. Soc.*, *A*, 887 (1968).

34. K. L. Baker and G. W. A. Fowles, *J. Chem. Soc.*, *A*, 801 (1968).

35. P. J. Hendra and N. Sadasivan, *Spectrochim. Acta*, **21**, 1127 (1965).

36. G. C. Hayward and P. J. Hendra, *Spectrochim. Acta*, **23A**, 1937 (1967).

37. W. H. Green and A. B. Harvey, *J. Chem. Phys.*, **49**, 3586 (1968).

38. G. Bergson, *Ark. Kemi*, **13**, 11 (1959).

39. K. G. Allum, J. A. Creighton, J. H. S. Green, G. J. Meinkoff, and L. J. S. Prince, *Spectrochim. Acta*, **24A**, 927 (1968).

40. G. Bergson and A. Biezais, *Ark. Kemi*, **18**, 143 (1962).

41. E. R. Clark and A. J. Collett, *J. Chem. Soc.*, *A*, 2129 (1969).

42. P. J. Hendra and Z. Jović, *J. Chem. Soc.*, *A*, 911 (1968).

43. C. W. Sink and A. B. Harvey, *J. Mol. Struct.*, **4**, 203 (1969).

44. A. Lüttringhaus and J. Grohmann, *Z. Naturforsch.*, *Part B*, **106**, 365 (1955).

45. C. Andrieu, M. Demuynck, and Y. Mollier, *C.R. Acad. Sci.*, *Paris, Ser. B.*, **265**, 127 (1967).

46. K. Sy and J. Oiry, *Bull. Soc. Chim. Fr.*, 331 (1969).

47. K. A. Jensen and P. H. Nielsen, *Acta Chem. Scand.*, **20**, 597 (1966).

48. C. G. Barraclough, R. L. Martin, and I. M. Stewart, *Australian J. Chem.*, **22**, 891 (1969).

49. T. Wentink, Jr., *J. Chem. Phys.*, **29**, 188 (1958).

50. T. Wentink, Jr., *J. Chem. Phys.*, **30**, 105 (1959).

51. G. W. King and K. Srikasmewaran, *J. Mol. Spectry.*, **29**, 491 (1969).

52. G. Gattow and M. Dräger, *Z. Anorg. Allg. Chem.*, **343**, 11 (1966).

53. K. A. Jensen and E. Huge-Jensen, *Acta Chem. Scand.*, to be published.

54. A. J. Brown and E. Whalley, *Inorg. Chem.*, **7**, 1254 (1968).

55. A. Müller, G. Gattow, and H. Seidel, *Z. Anorg. Allg. Chem.*, **347**, 24 (1966).

56. H. Seidel, *Naturwissenschaften*, **52**, 539 (1965).

57. B. Krebs and A. Müller, *Z. Anorg. Allg. Chem.*, **348**, 107 (1966).

58. K. Herzog, E. Steger, P. Rosmus, S. Scheithauer, and R. Mayer, *J. Mol. Struct.*, **3**, 339 (1969).

59. M. Dräger and G. Gattow, *Chem. Ber.*, **104**, 1429 (1971).

60. K. A. Jensen and U. Anthoni, *Acta Chem. Scand.*, **24**, 2055 (1970).

61. K. A. Jensen, B. M. Dahl, P. H. Nielsen, and G. Borch, *Acta Chem. Scand.*, **25**, 2039 (1971).

62. K. A. Jensen, J. B. Carlsen, A. Holm, and P. H. Nielsen, *Acta Chem. Scand.*, **17**, 550 (1963).

63. K. A. Jensen and V. Krishnan, *Acta Chem. Scand.*, **24,** 1088 (1970).

64. A. T. Pilipenko and N. V. Mel'nikova, *Zh. Neorg. Khim.*, **14,** 1843 (1969).

65. G. Durgaprasad, D. N. Sathyanarayana, and C. C. Patel, *Can. J. Chem.*, **47,** 631 (1969).

66. K. A. Jensen, B. M. Dahl, P. H. Nielsen, and G. Borch, *Acta Chem. Scand.*, **26,** in press (1972).

67. K. A. Jensen and P. H. Nielsen, *Acta Chem. Scand.*, **17,** 549 (1963).

68. K. A. Jensen and P. H. Nielsen, *Acta Chem. Scand.*, **17,** 547 (1963); K. A. Jensen, O. Dahl, and L. Engels-Henriksen, *Acta Chem. Scand.*, **24,** 1179 (1970).

69. U. Anthoni, *Acta Chem. Scand.*, **20,** 2742 (1966).

70. U. Anthoni, B. M. Dahl, C. Larsen, and P. H. Nielsen, *Acta Chem. Scand.*, **23,** 1061 (1969).

71. U. Anthoni, C. Larsen, and P. H. Nielsen, *Acta Chem. Scand.*, **21,** 2571 (1967).

72. U. Anthoni, B. M. Dahl, C. Larsen, and P. H. Nielsen, *Acta Chem. Scand.*, **24,** 959 (1970).

73. K. A. Jensen and B. M. Dahl, *Acta Chem. Scand.*, to be published.

74. K. A. Jensen and L. Henriksen, *Acta Chem. Scand.*, **24,** 3213 (1970).

75. K. A. Jensen and L. Henriksen, *Acta Chem. Scand.*, **22,** 1107 (1968).

76. K. A. Jensen and V. Krishnan, *Acta Chem. Scand.*, **24,** 1092 (1970).

77. C. Collard-Charon and M. Renson, *Bull. Soc. Chim. Belges*, **71,** 563 (1962).

78. K. A. Jensen, H. Mygind, and P. H. Nielsen, *Acta Chem. Scand.*, to be published.

79. K. A. Jensen, H. Mygind, P. H. Nielsen, and G. Borch, *Acta Chem. Scand.*, **24,** 1492 (1970).

80. M. Renson and C. Draguet, *Bull. Soc. Chim. Belges*, **71,** 260 (1962).

81. A. J. Collings, P. F. Jackson, and K. J. Morgan, *J. Chem. Soc.*, B, 581 (1970)

82. C. Collard-Charon and M. Renson, *Bull. Soc. Chim. Belges*, **72,** 149, 291, 304 (1969).

83. K. A. Jensen, E. Binderup, and C. Larsen, *Acta Chem. Scand.*, to be published.

84. K. A. Jensen and P. H. Nielsen, *Acta Chem. Scand.*, to be published.

85. K. A. Jensen and V. Krishnan, *Acta Chem. Scand.*, to be published.

86. D. M. Wiles, B. A. Gingras, and T. Suprunchuk, *Can. J. Chem.*, **45,** 469 (1967).

87. D. M. Wiles and T. Suprunchuk, *Can. J. Chem.*, **45,** 2258 (1967).

88. H. E. Hallam and C. M. Jones, *Spectrochim. Acta*, **25A,** 1791 (1969).

89. P. J. Hendra and Z. Jović, *Spectrochim. Acta*, **24A,** 1713 (1969).

90. P. Domiano, A. G. Manfredotti, G. Grossoni, M. Nardelli, and M. E. V. Tani, *Acta Cryst.*, **25B,** 591 (1969).

91. A. Chiesi, G. Grossoni, M. Nardelli, and M. E. Vidoni, *Chem. Commun.*, 404 (1969).

92. K. A. Jensen and A. Holm, *Acta Chem. Scand.*, **18,** 570 (1964).

93. L.-B. Agenäs, *Ark. Kemi*, **23,** 155 (1965); **28,** 145 (1968); **30,** 417 (1969).

94. M. A. A. Bey and S. K. Hasan, *Pakistan J. Sci. Ind. Res.*, **7,** 220 (1964).

95. N. Welcman and M. Wulf, *Israel J. Chem.*, **6,** 37 (1968).

96. E. E. Aynsley, N. N. Greenwood, and M. J. Sprague, *J. Chem. Soc.*, 2395 (1965).

97. T. Tarantelli and C. Pecile, *Ann. Chim.* (Rome), **52,** 75 (1962).

98. E. Bulka, K.-D. Ahlers, and E. Tucek, *Chem. Ber.*, **100,** 1367 (1967).

99. U. Mazzuccato, A. Foffani, A. Iliceto, and G. Svegliado, "Advances in Molecular Spectroscopy," Pergamon Press, London, 1962, p. 861.

100. N. S. Ham and J. A. Willis, *Spectrochim. Acta*, **16**, 279 (1960).

101. J. S. Thayer, *Inorg. Chem.*, **7**, 2599 (1968); *J. Organometal. Chem.*, **9**, P30 (1967).

102. H. Bürger and U. Goetze, *J. Organometal. Chem.*, **10**, 380 (1967).

103. H. Böhland and E. Niemann, *Z. Chem.*, **8**, 191 (1968).

104. E. E. Aynsley, N. N. Greenwood, G. Hunter, and M. J. Sprague, *J. Chem. Soc., A*, 1344 (1966).

105. A. Turco, C. Pecile, and M. Nicolini, *J. Chem. Soc.*, 3008 (1962).

106. T. Gabrio and G. Barnikow, *Z. Chem.*, **9**, 183 (1969).

107. H. Gerding, G. Milazzo, and H. H. K. Rossmark, *Rec. Trav. Chim. Pays-Bas*, **72**, 957 (1953).

108. R. D. Brown, F. R. Burden, and P. D. Godfrey, *J. Mol. Spectry.*, **25**, 415 (1968).

109. A. Trombetti and C. Zauli, *J. Chem. Soc., A*, 1106 (1967).

110. Ye. G. Treshchova, D. Ekkhardt, and Yu. K. Yur'ev, *Zh. Fiz. Khim.*, **38**, 159 (1964).

111. G. Milazzo, *Gazz. Chim. Ital.*, **98**, 1511 (1968).

112. S. V. Tsukerman, V. D. Orlov, Yu. S. Rozum, and V. F. Lavrushin, *Khim. Geterotsikl. Soedin.*, **4**, 623 (1969).

113. P. Bassignana, C. Cogrossi, M. Gandino, and P. Merli, *Spectrochim. Acta*, **21**, 605 (1965).

114. J. Goerdeler, D. Gross, and H. Klinke, *Chem. Ber.*, **96**, 1289 (1963).

115. E. Benedetti and V. Bertini, *Spectrochim. Acta*, **24A**, 57 (1968).

116. V. Bertini, *Gazz. Chim. Ital.*, **97**, 1870 (1967).

117. Y. F. Shealy and J. D. Clayton, *J. Heterocycl. Chem.*, **4**, 96 (1967).

118. V. A. Pozdyshev, Z. V. Todres-Selektor and L. S. Efros, *Zh. Obshch. Khim*, **30**. 2551 (1960).

119. C. W. Bird and G. W. H. Cheeseman, *Tetrahedron*, **20**, 1701 (1964).

120. P. G Pesin, *Khim. Geterotsikl. Soedin.*, **4**, 235 (1969).

121. R. V. Kendall and R. A. Olofson, *J. Org. Chem.*, **35**, 806 (1970).

122. D. M. Levine, W. D. Krugh, and L. P. Gold, *J. Mol. Spectry.*, **30**, 459 (1969).

123. O. Siimann and J. Fresco, *Inorg. Chem.*, **9**, 294 (1970).

124. K. A. Jensen and U. Henriksen (née Svanholm), *Acta Chem. Scand.*, **21**, 1991 (1967).

125. K. A. Jensen, H. R. Baccaro, and O. Buchardt, *Acta Chem. Scand.*, **17**, 163 (1963).

126. E. A. V. Ebsworth, R. Taylor, and L. A. Woodward, *Trans. Faraday Soc.*, **55**, 211 (1959).

127. J. E. Drake and C. Riddle, *J. Chem. Soc., A*, 1573 (1969).

128. S. Cradock, E. A. V. Ebsworth, and D. W. H. Rankin, *J. Chem. Soc., A*, 1628 (1969).

129. H. Bürger and U. Götze, *Inorg. Nucl. Chem. Lett.*, **3**, 549 (1967).

130. T. D. Goldfarb and S. Sujushi, *J. Amer. Chem. Soc.*, **86**, 1679 (1964).

131. H. Kriegsmann, H. Hoffmann, and H. Geissler, *Z. Anorg. Allg. Chem.*, **341**, 24 (1965).

132. H. Schumann and R. Weiss, *Angew. Chem.*, **82**, 256 (1970).

133. H. Schumann and M. Schmidt, *J. Organometal. Chem.*, **3**, 485 (1965).

134. A. N. Egorochkin, N. S. Vyazankin, M. N. Bochkarev, and S. Ya. Khorshev, *Zh. Obshch. Khim.*, **37**, 1165 (1967).

135. A. N. Egorochkin, S. Ya. Khorshev, N. S. Vyazankin, M. N. Bochkarev, O. A. Kruglaya, and G. S. Semchikova, *Zh. Obshch. Khim.*, **37**, 2308 (1967).

136. J. Goubeau and U. Weser, *Z. Anorg. Allg. Chem.*, **319**, 276 (1963).

137. R. A. Zingaro, *Inorg. Chem.*, **2**, 192 (1963).

138. K. A. Jensen and P. H. Nielsen, *Acta Chem. Scand.*, **17**, 1875 (1963).

139. M. G. King and G. P. McQuillan, *J. Chem. Soc.*, 898 (1967).

140. P. Nicpon and D. W. Meek, *Chem. Commun.*, 398 (1966).

141. D. Brown, J. Hill, and C. E. F. Rickard, *J. Less-Common Met.*, **20**, 57 (1970).

141a. A. H. Cowley and J. L. Mills, *J. Amer. Chem. Soc.*, **91**, 2915 (1969).

142. J. R. Durig, J. S. DiYorio, and D. W. Wertz, *J. Mol. Spectry.*, **28**, 444 (1968).

143. A. M. Brodie, G. A. Rodley, and C. J. Wilkins, *J. Chem. Soc.*, A, 2927 (1969).

144. S. Husebye, *Acta Chem. Scand.*, **20**, 2007 (1966).

145. S. Husebye and G. Helland-Madsen, *Acta Chem. Scand.*, **23**, 1398 (1969).

146. R. A. Chittenden and L. C. Thomas, *Spectrochim. Acta*, **20**, 1679 (1964).

146a. A. Schmidtpeter and H. Brecht, *Z. Naturforsch.*, *Part B*, **24**, 179 (1969).

147. J. R. Durig, D. W. Wertz, B. R. Mitchell, F. Bloch, and J. M. Greene, *J. Phys. Chem.*, **71**, 3815 (1967).

148. M. V. Kudchaker, R. A. Zingaro, and K. J. Irgolic, *Can. J. Chem.*, **46**, 1415 (1968).

149. R. G. Melton and R. A. Zingaro, *Can. J. Chem.*, **46**, 1425 (1968).

150. V. V. Turkevich and I. F. Viblyi, *Zh. Prikl. Spectrosk.*, **4**, 77 (1966).

151. I. A. Nuretdinov and N. P. Grechkin, *Izv. Akad. Nauk SSSR, Ser. Khim.*, 2831 (1968).

152. F. Herail and V. Viossat, *C.R. Acad. Sci., Paris*, **259**, 4629 (1964).

152a. I. A. Nuretdinov, N. P. Grechkin, N. A. Buina, and L. K. Nikonorova, *Izv. Akad. Nauk SSSR, Ser. Khim.*, 1535 (1969).

152b. I. A. Nuretdinov, L. K. Nikonorova, and N. P. Grechkin, *Zh. Obshch. Khim.*, **39**, 2265 (1969).

152c. L. K. Nikonorova, N. P. Grechkin, and I. A. Nuretdinov, *Izv. Akad. Nauk SSSR, Ser. Khim.*, 464 (1969).

153. L. C. Thomas and R. A. Chittenden, *Chem. Ind.* (London), 1913 (1961).

154. V. Krishnan and R. A. Zingaro, *Inorg. Chem.*, **8**, 2337 (1969).

155. J. Quinchon, M. Le Sech and E. Gryszkiewicz-Trochimowski, *Bull. Soc. Chim. Fr.*, 735 (1961).

156. E. Gryszkiewicz-Trochimowski, J. Quinchon, O. Gryszkiewicz-Trochimowski and M. Le Sech, *Bull. Soc. Chim. Fr.*, 739 (1961).

156a. C. Krawiecki, J. Michalski, R. A. Y. Jones, and A. R. Katritzky, *Rocz. Chem.*, **43**, 869 (1969).

157. W. Kuchen and H. Hertel, *Angew. Chem.*, **81**, 127 (1969).

158. N. I. Zemlyanski, N. M. Chernaya, V. V. Turkevich, and V. I. Krasnoshchek, *Zh. Obshch. Khim.*, **36**, 1240 (1966).

159. N. Petragnani, V. G. Toscano, and M. de Moura Campos, *Chem. Ber.*, **101**, 3070 (1968).

160. D. E. C. Corbridge, "Topics in Phosphorus Chemistry," Vol. 6, Interscience Publishers, New York, N.Y. 1969, p. 235.

161. R. A. Zingaro, B. H. Steeves, and K. J. Irgolic, *J. Organometal. Chem.*, **4**, 320 (1965).

162. G. N. Chremos and R. A. Zingaro, *J. Organometal. Chem.*, **22**, 637 (1970).

163. R. A. Zingaro and A. Merijanian, *J. Organometal. Chem.*, **3**, 580 (1964).

164. P. Nicpon and D. W. Meek, *Inorg. Chem.*, **5**, 1297 (1966).

165. R.A. Zingaro and A. Marijanian, *J. Organometal. Chem.* **1**, 369 (1964).

166. G. N. Chremos and R. A. Zingaro, *J. Organometal. Chem.*, **22**, 647 (1970).

167. R. Paetzold and H. Amoulong, *Z. Chem.*, **6**, 29 (1966).

168. R. Paetzold, R. Kurze, and G. Engelhardt, *Z. Anorg. Allg. Chem.*, **353**, 62 (1967).

169. R. Paetzold and D. Knaust, *Z. Anorg. Allg. Chem.*, **368**, 196 (1969).

170. R. Paetzold and D. Lienig, cited in refs. 181 and 187.

171. J. Krejči, L. Zborilová, and I. Horsák, *Collection Czech. Chem. Commun.*, **32**, 3468 (1967).

172. R. Paetzold and G. Bochmann, *Z. Anorg. Allg. Chem.*, **360**, 293 (1968).

173. R. Paetzold and G. Bochmann, *Spectrochim. Acta*, **26A**, 391 (1970).

174. S. Astin, L. de V. Moulds, and H. L. Riley, *J. Chem. Soc.*, 901 (1935).

175. E. B. R. Prideaux and G. Green, *J. Phys. Chem.*, **28**, 1273 (1924).

176. A. Simon and R. Paetzold, *Z. Anorg. Allg. Chem.*, **303**, 72 (1960).

177. R. Paetzold and E. Rönsch, *Z. Anorg. Allg. Chem.*, **315**, 64 (1962).

178. A. Simon and R. Paetzold, *Z. Anorg. Allg. Chem.*, **303**, 79 (1960).

179. R. Paetzold, H.-D. Schumann, and A. Simon, *Z. Anorg. Allg. Chem.*, **305**, 78, 88, 98 (1960).

180. R. Paetzold and K. Aurich, *Z. Anorg. Allg. Chem.*, **317**, 156 (1962).

181. R. Paetzold, S. Borek, and E. Wolfram, *Z. Anorg. Allg. Chem.*, **353**, 53 (1967).

182. R. Paetzold and D. Lienig, *Z. Anorg. Allg. Chem.*, **335**, 289 (1965).

183. H. Bredereck, A. Wagner, H. Beck, and R. J. Klein, *Chem. Ber.*, **93**, 2736 (1960).

184. W. L. Mock, and J. H. McCausland, *Tetrahedron Lett.*, 391 (1968).

185. R. Paetzold, U. Linder, G. Bochmann, and P. Reich, *Z. Anorg. Allg. Chem.*, **352**, 295 (1967).

186. R. Paetzold and G. Bochmann, *Spectrochim. Acta*, **26A**, 391 (1970).

187. R. Paetzold, *Z. Chem.*, **4**, 321 (1964).

188. R. Paetzold, *Spectrochim. Acta*, **24A**, 717 (1968).

189. R. Paetzold, *Spectrochim. Acta*, **26A**, 577 (1970).

190. R. Paetzold and U. Lindner, *Z. Anorg. Allg. Chem.*, **350**, 295 (1967).

191. R. Paetzold and E. Rönsch, *Z. Anorg. Allg. Chem.*, **338**, 22 (1965).

192. R. Paetzold and E. Rönsch, *Spectrochim. Acta*, **26A**, 569 (1970).

193. R. Paetzold and P. Vordank, *Z. Anorg. Allg. Chem.*, **347**, 294 (1966).

194. R. Paetzold and G. Bochmann, *Z. Chem.*, **8**, 308 (1968).

195. R. Paetzold and G. Bochmann, *Z. Anorg. Allg. Chem.*, **368**, 202 (1969).

196. K. A. Jensen and V. Krishnan, *Acta Chem. Scand.*, **21**, 1988 (1967).

197. T. Tanaka and T. Kamitani, *Inorg. Chim. Acta*, **2**, 175 (1968).

198. K. J. Wynne and J. W. George, *J. Amer. Chem. Soc.*, **87**, 4750 (1965).

199. R. J. Gillespie and A. Whitla, *Can. J. Chem.*, **48**, 657 (1970).

200. K. J. Wynne, *Inorg. Chem.*, **9**, 299 (1970).

201. G. C. Hayward and P. J. Hendra, *J. Chem. Soc.*, *A*, 1710 (1969).

202. D. M. Adams and P. J. Lock, *J. Chem. Soc.*, *A*, 145 (1967).

203. G. C. Hayward and P. J. Hendra, *J. Chem. Soc.*, *A*, 643 (1967).

204. N. N. Greenwood, A. C. Sarma, and B. P. Straughan, *J. Chem. Soc.*, *A*, 1561 (1968).

205. N. N. Greenwood, B. P. Straughan, and A. E. Wilson, *J. Chem. Soc.*, *A*, 2209 (1968).

206. K. J. Wynne and J. W. George, *J. Amer. Chem. Soc.*, **91**, 1649 (1969).

XV Physicochemical investigations of selenium compounds

B. ULTRAVIOLET AND VISIBLE LIGHT SPECTROSCOPY

JAMES E. KUDER

Xerox Corporation
Chemistry Research Laboratory
Webster, New York

1. Introduction

Literature reports of electronic absorption spectra generally belong to one of three categories: theoretical studies relating structure to electronic properties, papers dealing with the determination of structure by analogy with the spectra of related compounds, and articles on the characterization of compounds along with their melting points and elemental analyses. Of these, the third type is perhaps the least ambiguous, while the first is the most challenging to the chemist. This chapter is a survey of the spectra of organoselenium compounds and includes examples of each of the three types. Emphasis has been given to papers that show structure-property relationships and to those that illustrate the trends seen in the spectra of compounds containing the chalcogen elements. Qualitatively, it may be said that these trends are dependent both on the decreasing ionization potential of the heteroatom† and on the decreasing overlap between the orbitals of the chalcogen and adjacent carbon atoms. In many cases these effects have been discussed in terms of molecular orbital (MO) theory, which in its simplest form successfully accounts for the observed trends and in its more advanced forms is capable for reproducing the position and intensity of absorption bands with remarkable accuracy.

2. Selenols

Although aliphatic selenols appear to have no distinct ultraviolet absorption maxima, a relatively intense (log $\varepsilon \sim 3.7$) absorption peak in the range 243–253 nm is seen for the aliphatic selenol anion.[2] This absorption peak appears to correspond in position and intensity to the absorption of aliphatic mercaptide ions in the 230–240 nm region.[3] No explanation of the nature of this electronic transition has been advanced, and, in general, the absorption spectra of both aliphatic and aromatic selenols might be expected to be difficult to obtain because of the facile oxidation to diselenides which occurs on exposure to air.

Table XVB-1   Absorption spectra of phenol, thiophenol, and selenophenol[a]

COMPOUND	$\lambda_{\max}$ (nm) (log ε)		SOLVENT
Phenol	272.5 (3.24)	219 (3.74)	Ethanol
Thiophenol	279 (2.75)	238 (3.93)	Ethanol
Selenophenol	262 (3.22)[b]	241 (3.82)	Light petroleum

[a] From Chierici and Passerini.[4]

[b] Inflection.

The spectrum of selenophenol in ethanol reported by Chierici and Passerini[4] was shown by Kiss and Muth[5] to represent, in fact, a mixture of selenophenol and diphenyl diselenide. The latter authors discuss the spectra of phenol, thiophenol, and selenophenol (*cf.* Table XVB-1) in terms of inductive effects, which operate in the order OH > SH > SeH, and mesomeric effects, decreasing in the reverse order.[5] The interpretation of the

† The first ionization potentials[1] (in electron volts) of the Group VIB elements are as follows: O, 13.6; S, 10.4; Se, 9.8; Te, 9.0.

spectra of these compounds is essentially the same as that given for phenyl alkyl selenides in the next section.

3.   Selenides

Few reports of the absorption spectra of alkyl selenides appear in the literature. Diethyl selenide is reported[6] to have a maximum of 250 nm (log ε 1.70), while the corresponding telluride exhibits maxima at 235 (3.84), 290 (1.38), and 360 (1.26) nm. It has been proposed[7] that the Se atom in β-ketoselenides increases the intensity of the carbonyl n-π^* absorption band by an order of magnitude, so that this band, which is normally buried under the tail of the more intense π-π^* band of the carbonyl group, becomes visible.

Considerably more work has been done with the phenyl alkyl selenides, whose absorption spectra are more readily accessible. According to Bowden and Braude,[8] the short-wavelength band is due to a transition to an excited state involving increased contributions from *para*-dipolar structures such as **1**, in which a pair of electrons migrates from the heteroatom into the benzene ring, while the longer-wavelength band is thought to result from transitions involving homopolar excited structures such as **2**. The same assignments were made by Kiss and Muth.[5]

In contrast, Chierici and Passerini[4] considered the band at 250 nm to be a perturbed benzenoid transition involving only the six π-electrons of the benzene ring, and the long-wavelength band to be due to the eight-electron chromophore consisting of the selenium 4-p-electrons in conjugation with the benzene ring. More recently, Mangini *et al.*[12] concluded that the long-wavelength band is related to the $A_{1g} \rightarrow B_{2u}$ absorption seen in benzene at 255 nm, and found that the shift in frequency in such monosubstituted benzenes is linearly related to the function S/I, where S is the overlap integral between the heteroatom and the ring carbon atom and I is the ionization potential of the substituent. The short-wavelength band involves much larger perturbations by the substituent, so that the analogy with transitions in benzene itself breaks down.

The absorption spectra of 1- and 2-methyl naphthyl ethers, sulfides, and selenides have been reported by Gasco *et al.*[13] The use of semiempirical MO

Table XVB-2   Absorption spectra in ethanol of compounds PhXEt[a]

COMPOUND	λ_{max} (nm) (log ε)	
PhOEt	219.5 (3.98)	272 (3.28)
PhSEt	205 (4.31), 255 (3.93)	278 (3.04)
PhSeEt	250 (3.69)	269 (3.50)
PhTeEt	225 (4.17)	270 (3.60)

[a] Data of Bowden and Braude.[8]

theory by these authors to explain the effects of substituent and position of substitution on absorption wavelength and dipole moments appears to be successful.

The absorption spectra of diphenyl ether, sulfide, and selenide are given in Table XVB-3. As for the phenyl alkyl chalcogenides, a variety of spectral interpretations have been advanced.

Table XVB-3   Absorption spectra of diphenyl ether, sulfide, and selenide in ethanol[a]

COMPOUND	λ_{max} (nm) (log ε)		
PhOPh	271 (3.31)	224 (4.05)	
PhSPh	275 (3.75)	250 (4.05)	206 (4.39)
PhSePh	280 (3.63)[b]	255 (3.99)	

[a] Given by Kiss and Muth.[14]
[b] Inflection.

Koch[15] interpreted the long-wavelength band of diphenyl sulfide as involving conjugation between the S atom and one benzene ring, analogous to the low-energy transition seen in phenyl alkyl sulfides. The band at 250 nm was ascribed to a transition involving both phenyl rings conjugated *via* the pair of p-electrons on the connecting heteroatom.

Mangini and Passerini[16] concluded, by analogy with the spectra of phenyl alkyl sulfides, that it is not necessary to consider conjugation extending over both phenyl rings in diphenyl sulfide and selenide.[4] Instead, the long-wavelength band is ascribed to a transition involving the S (or Se) atom in conjugation with one phenyl ring at a time, while the short-wavelength band is attributed to a perturbed benzenoid transition involving the six electrons of the benzene chromophore. The same arguments hold for a series of phenyl pyridyl selenides prepared by Chierici and Passerini.[17]

Kiss and Muth[14] are in agreement that conjugation in diphenyl selenide (or sulfide) does not extend over the entire molecule, but they reverse the assignment given by Mangini and Passerini to the two absorption bands. In fact, it is not possible to decide on the extent of conjugation from a comparison of the spectra of diphenyl selenide and phenyl alkyl selenides. Molecular orbital calculations[18,19] have shown that the lowest empty orbital for PhXR compounds lies at the same level as for PhXPh and that the highest occupied MO is only slightly lower in PhXR than in PhXPh. (This slight difference would be canceled if twisting of the phenyl rings in PhXPh is considered.) Thus, the presence of a single π-system in diphenyl selenide is unrelated to any shift of the long-wavelength bands relative to those observed in the phenyl alkyl selenides.

4.   Diselenides

a.   Aliphatic.   The observation by Fredga[20] in 1936 that among acids with the general formula $HO_2CCH_2Se–(CH_2)_n–SeCH_2CO_2H$ those with $n \neq 0$ were colorless, whereas those with $n = 0$, as well as other diselenides, were yellow, appears to be the first recognition of the diselenide group as a distinct chromophore. The nature of this long-wavelength absorption band and of the analogous band in disulfides has been discussed by Bergson in terms of a simple LCAO-MO treatment.[21,22] According to this model, the bonds formed by divalent Se atoms involve mainly p-orbitals, leaving each Se atom with two lone pairs of electrons occupying an s- and a p-orbital. Linear combination of the two lone-pair p-orbitals gives rise to doubly occupied π- and π^*-orbitals whose energies are dependent on the dihedral angle (and hence on the overlap between the constituent atomic orbitals) of the diselenide group. The lone-pair s-electrons are more tightly bound and thus presumably of little importance in bonding. The longest-wavelength absorption band, then, involves an electronic transition from the π^*-orbital to a σ^*-orbital whose energy is independent of the diselenide dihedral angle. It should be noted that Bergson's model accounts both for the preferred ground-state dihedral angle of nearly 90° for open-chain diselenides (and disulfides) and for the bathochromic shift observed in cyclic diselenides, which have their dihedral angle reduced to less than 90°.† Thus, the open-chain diselenides with a dihedral angle of 90° have absorption maxima near 312 nm,[24] the six-membered ring diselenides with a dihedral angle of 56° absorb at 365 nm,[25] and the five-membered cyclic diselenides with a dihedral angle of 0–20° have absorption maxima near 440 nm.[21,26]

The effect of electron-withdrawing substituents is to increase the effective electronegativity of atoms of the diselenide chromophore and thus increase

---

† It may be noted that the same general features of Bergson's model are reproduced in an advanced CNDO-MO treatment.[23]

the energy of the lowest electronic transition.[22,26] Thus, in the series $HO_2C(CH_2)_n$–SeSe–$(CH_2)_nCO_2H$, as n goes from 4 to 1, the absorption maximum shifts from 311 to 300 nm. Similarly, the replacement of an H atom in dimethyl diselenide (λ max 316) by Cl to afford bis(chloromethyl) diselenide results in a hypsochromic shift of 11 nm.[21]

The inductive effect of the *n*-butyl group should be about the same as that of the *tert*-butyl group, and yet di-*n*-butyl diselenide (λ max 312 nm) absorbs 26 nm to the red of the di-*tert*-butyl compound. This red shift, explained by Bergson[21,27] as being due to hyperconjugation, is readily seen in the series RSeSeR, where R = Me, Et, *i*-Pr, *tert*-Bu and also in the series R = CH_2CO_2H, $CH(CH_3)CO_2H$, $C(CH_3)_2CO_2H$. The transition energy has been found[10] to vary linearly with $\sqrt{k}$, where k = the number of H

Table XVB-4   Typical absorption maxima in the series RXXR

COMPOUND	λ_{max} (nm)	ELECTRONEGATIVITY OF ATOM X (32)	COVALENT DIAMETER OF ATOM X (Å) (29)
CH_3SSCH_3	255 (6, 29)[a]	2.5	3.93
$C_2H_5SSC_2H_5$	252 (29)		
$CH_3SeSeCH_3$	316 (29)	2.4	4.42
$C_2H_5SeSeC_2H_5$	312 (6, 29)		
$C_2H_5TeTeC_2H_5$	395 (6)	2.1	4.99
n-$C_4H_9TeTeC_4H_9$-n	399 (33)		

[a] References are given in parentheses.

atoms attached to the α-C atom. The hyperconjugative effect is also seen in the case of aliphatic disulfides,[29] although the disulfide maximum is located at shorter wavelengths than is the corresponding diselenide band.

In the LCAO-MO model for electronic absorption in the chalcogen series, the change in wavelengths is assumed to vary with the Coulomb integral, which in turn is often taken[30] as being proportional to the electronegativity of the atom in question. The same series has also been treated by the free-electron model,[6] in which the absorption maximum is found to vary with the square of the covalent diameter of the heteroatom. As seen in Table XVB-4, both approaches lead to the same result.

Finally, the saturated aliphatic thioselenenates, R–S–Se–R, must be mentioned.[31] Compounds **3** (λ max 285 nm) and **4** (λ max 386 nm) each

$$C_2H_5—SeS—C_2H_5$$

3

```
      Se—CH₂
       |       \
       |        CH—CO₂H
       |       /
      S—CH₂
```

4

show only one absorption peak in the ultraviolet. The S–Se group therefore acts as a single chromophore, and thioselenenates exhibit the same dependence on dihedral angle as the corresponding diselenides and disulfides.

b. Aromatic. The absorption spectrum of diphenyl diselenide[14,34,35] consists of a broad band at 332 nm (log ε 3.01), a slight inflection at 276 nm (log ε 3.43), and an intense band at 241 nm (log ε 4.31). Diphenyl ditelluride shows peaks at 407 nm (log ε 3.97) and 250 nm (log ε 4.40).[36,37] In contrast to the situation for the aliphatic compounds, no attempt has been made to interpret the spectrum and the favored conformation of these relatively large, nonplanar[38] molecules on the basis of MO theory. Assignments of the various transitions have been made on the basis of analogy, however.[14,35] The 332 nm band is ascribed to the excitation of the –SeSe– chromophore, since a similar band appears in the spectra of the aliphatic diselenides, and since an analogous band due to the –SS– chromophore appears as an inflection at 310 nm (log ε 3.11) in the spectrum of diphenyl disulfide. It might be pointed out, however, that the diselenide band is somewhat more intense in diphenyl diselenide than in the aliphatic diselenides. In addition, the aliphatic diselenides absorb some 60 nm to the red of the corresponding disulfides, whereas the difference is only 20 nm in the case of the two diphenyl compounds. The inflection at 276 nm is presumably due to a perturbed benzenoid transition, since similar absorption occurs with selenophenol (278 nm) and diphenyl selenide (inflection at 280 nm). The peak at 241 nm is at the same position as the K band of selenophenol (241 nm) and is believed to have the same origin. In other words, the excited state associated with this transition may be represented by the following resonance structures:

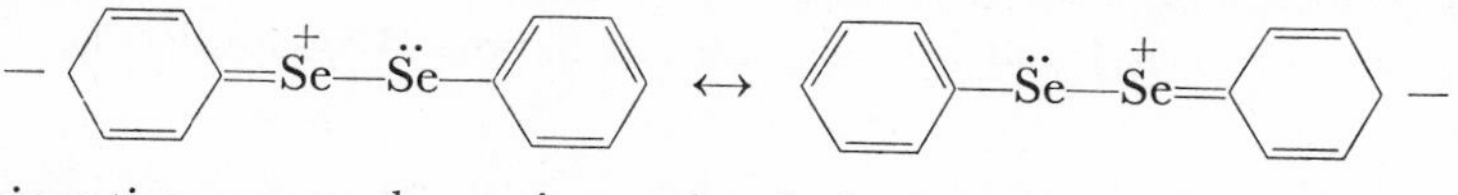

Conjugation across the entire molecule has not been invoked for any of the three transitions involved in the spectrum of diphenyl diselenide. Campaigne has shown[39] that the absorption spectra of unsymmetrical diaryl disulfides may be represented by one-half the sum of the corresponding two symmetrical disulfides and that no transmission of electronic effects across the –SS– bond can be detected. Similarly, Bergson and Wold[31] found that the absorption curve of the thioselenenate **5** closely fits the arithmetic mean of the absorption curves of diselenide **6** and disulfide **7**, and

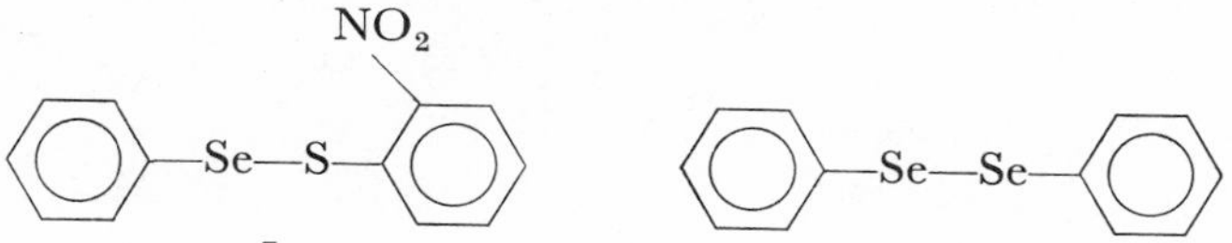

7

that therefore the Se-S bond (as well as the S-S and Se-Se bonds) acts as a π-electron insulator. Furthermore, the structure of diphenyl diselenide, as determined by X-ray analysis,[38] indicates two mutually orthogonal π-systems.

In addition to the parent compound, a number of substituted diphenyl diselenides[11,14,40] have had their electron spectra reported, along with those of the diindolyl diselenide series.[41]

Finally, it has been known for some time that diselenides exhibit thermochromic behavior,[42] and the suggestion has been made that this is due to dissociation into free radicals. Recent work has shown, however, that diselenide thermochromism is caused by broadening of an absorption band which lies on the edge of the visible region.[43]

5. Compounds Containing the Selenoamide and Selenourea Groups

The absorption spectra of a number of compounds containing a selenourea or selenium amide moiety have been investigated by Mautner and his co-workers because of their biological as well as their theoretical interest. Among the compounds examined are the isologous series related to phenylurea (**8**),[44] uracil (**9**),[45] thymine (**10**),[45] barbituric acid (**11**),[46] hypoxanthine (**12**),[45] (X = O, S, Se) and the benzimidazolinones (**13**), studied by El'tsov *et al.*[47,48] (X = O, S, Se, Te). The bathochromic shift observed in this series of compounds was discussed by Mautner and his colleagues[44-46,49] in terms of resonance structures, the ground state presumably being represented as a resonance hybrid of **14** and **15**, while the excited state resembles **15**. The batho-

$$
\begin{array}{cc}
\overset{\displaystyle X}{\underset{\displaystyle }{\|}} & \overset{\displaystyle X^-}{\underset{\displaystyle }{|}} \\
-\text{NH}-\text{C}- & -\overset{+}{\text{N}}\text{H}=\text{C}- \\
\mathbf{14} & \mathbf{15}
\end{array}
$$

chromic shift in the series oxygen, sulfur, selenium is then explained either by destabilization of the ground state due to increased difficulty of π-bond formation with increasing interatomic distance or by stabilization of the excited state due to increased ability of the chalcogen atom to accept a negative charge (possibly by octet expansion). In either case, the result is to decrease the energy difference between the two states and to shift the absorption band to a longer wavelength. The increasing importance of **15** to the ground state is supported by the observed trends in dipole moment[44] and acid dissociation constants.[45]

Mautner and Bergson[50] considered the effect of the replacement of oxygen by sulfur and sulfur by selenium in an LCAO-MO treatment of several pyrimidines and purines. The position of the two long-wavelength bands was successfully interpreted by this method, as was also the magnitude of change in absorption maximum, depending on which O atom was being replaced. A consideration of total π-electron energies of the ground and excited states shows, however, that it is impossible to interpret the observed bathochromic shift in terms of either a destabilization of the ground state or a stabilization of the excited state.

In a similar study, Azman and his co-workers[51] demonstrated that in the series acetamide, thioacetamide, and selenoacetamide there is simultaneous lowering of the lowest empty MO and raising of the highest filled MO in going from oxygen to sulfur to selenium, resulting in a bathochromic shift. In contrast to Mautner's initial suggestion, these workers showed that there is a decrease in the dipole moment upon electronic excitation. Similar results were found in the series urea (λ_{max} 186 nm), thiourea (λ_{max} 236 nm), and selenourea (λ_{max} 260 nm).[51]

The semicarbazides (**16**)[44,52] and semicarbazones (**17**)[52] are related series

$$
\begin{array}{cc}
\overset{\displaystyle X}{\underset{\displaystyle }{\|}} & \overset{\displaystyle X}{\underset{\displaystyle }{\|}} \\
\text{NH}_2\text{CNNH}_2 & \text{NH}_2\text{CNHN}=\text{C} \overset{\displaystyle \diagup R}{\underset{\displaystyle \diagdown R}{}} \\
\underset{\displaystyle \text{Ph}}{\overset{\displaystyle |}{}} & \\
\mathbf{16} & \mathbf{17}
\end{array}
$$

Table XVB-5   Absorption maxima of some semicarbazides and semicarbazones[a,b]

COMPOUND	λ_{max} (nm) (log ε)
2-Phenylsemicarbazide	242 (4.00)
2-Phenylthiosemicarbazide	255 (3.91)
2-Phenylselenosemicarbazide	271 (3.86)
Selenosemicarbazide	259 (3.94)
Butyraldehyde selenosemicarbazone	287–288 (4.23)
Crotonaldehyde selenosemicarbazone	313 (4.40)
Benzaldehyde selenosemicarbazone	327–328 (4.40), 276 (3.96)
Cinnamaldehyde selenosemicarbazone	346 (4.61), 294 (4.24)

[a] From Mautner and Kumler[44] and Huls and Renson.[52]

[b] Ethanol solution.

of compounds which show the trends expected in the Group VI elements. The effect of increasing conjugation on the long-wavelength band is clearly seen in Table XVB-5, which lists the absorption maxima of a number of selenosemicarbazones.[52] In this case, the effect of introducing a benzene ring into the conjugated system is the same as the introduction of 1.5 ethylene units. Huls and Renson[52] have also related the positions of the absorption maxima in a series of substituted benzaldehyde selenosemicarbazones to the electron-donating ability of the substituent.

6. Selenochromones and Related Compounds

The chromanones (**18**, X = O, S, Se) constitute another series in which is seen the shift to longer wavelength with increasing atomic weight.[53] Thus the longest-wavelength band of chromanone (**18**, X = O) is at 321 nm (log ε 3.62), that of thiachromanone (**18**, X = S) at 346 nm (3.59), and that of selenochromanone at 353 nm (3.54). The shifts in the absorption bands are considerably greater between the oxygen and sulfur compounds than between the sulfur and selenium analogs, and are accompanied by only small variations in intensity.

The same trend is seen with the chromones (**19**) and the xanthones (**20**).[54] Selenoflavone (2-phenylselenochromone) resembles isoselenoflavone

18   **19**   **20**

(3-phenylselenochromone) in the positions of its longest-wavelength band (354 and 352 nm, respectively), but is distinguished by the positions of the shorter-wavelength maxima (274 and 258 nm, respectively).[55]

A comparison of the spectra of **19** and **20** in concentrated H_2SO_4 with those of the corresponding chromylium (**21**) and xanthylium (**22**) cations

reveals a close resemblance in position and structure of the absorption bands.[54] This is taken as evidence that the protonation of chromone and xanthone occurs at the carbonyl oxygen to afford the hydroxyl chromylium and xanthylium cations, respectively. The spectra of a number of seleno-chromylium perchlorates have been reported by Renson *et al.*;[56,57] these closely resemble the spectra of the corresponding sulfur analogs. That the selenocarbonyl group may also function as a proton acceptor has been shown by studies of the ultraviolet and visible spectra of 2,6-dimethyl-4-seleno-γ-pyrone (**23**) in various solvents.[58,59]

7. Selenophene

The vapor-phase spectrum of selenophene, as determined by Milazzo and his co-workers,[60-62] consists of band systems at 266–268, 242–252, and 197–209 nm, and, overlaid on a continuous absorption, three distinct bands at 186.5, 182.8, and 172.1 nm. The existence of three separate band systems in selenophene is presumed to correspond to the bands at 204–220, 221–245, and 221–260 nm (the last two overlapping) seen in the vapor spectrum of thiophene (*cf.* Table XVB-6).[63-64] Two bands have been reported for the vapor spectrum of furan, with most intense absorption at 191 and 205 nm.[64] Milazzo and Paolini[60] concluded, from a comparison of spectra, that selenophene has less aromatic character than thiophene. This decrease in aromatic character is seen also in the position of the short-wavelength band, which lies intermediate between the corresponding bands in furan (0–0 band at 211 nm)[65] and cyclopentadiene (0–0 band at 198 nm).[65] Trombetti and Zauli[66] were unable to see the first two absorption bands reported by Milazzo, and instead report a continuous absorption in the region 271–210 nm, with a maximum at 245 nm (log ε 3.74) and a broad shoulder at 232 nm (log ε 3.62). In cyclohexane the maximum and the shoulder are at 250 (log ε 3.84) and 238 nm (log ε 3.72), respectively, in agreement with the solution spectra reported by Chierici and Pappalardo.[67] In addition, some low-intensity peaks not seen by Milazzo were reported in the high-energy portion of the spectrum. The peak at 245 nm is presumably due to a $\pi \rightarrow \pi^*$ transition, which, when compared to the spectra of thiophene and furan, seems related to the ionization potential of the heteroatom.[69] The shoulder at 232 nm may be related to the absorption that aliphatic selenium compounds display in this region.[21]

Table XVB-6   Solution spectra of five-membered heterocycles

COMPOUND	λ_{max} (nm)	LOG ε	SOLVENT	REF.
Furan	208[a]	3.9	Ethanol	69
Thiophene	235	3.65	Hexane	70
Selenophene	250	3.84	Cyclohexane	66
Tellurophene	280	3.62	Cyclohexane	71

[a] Shoulder.

A number of attempts have been made to provide a theoretical description of the spectra of five-membered heterocycles, but the agreement in this regard appears to be no better than for the spectra themselves. The MO treatments of thiophene (and thus, by implication, selenophene) by Mulliken[72] and by Longuet-Higgins[73] differ chiefly in the manner in which the vacant d-orbital on the heteroatom is allowed to interact with the π-system, but neither is completely satisfactory from a theoretical standpoint.[74]

Fabian *et al.*[75] have treated the isologous series furan, thiophene, seleno-phene, and tellurophene by a variation of the Pariser-Parr-Pople MO method, which neglects d-orbital participation but includes configuration interaction. This treatment successfully accounts for the position of the long-wavelength band in these compounds without the introduction of undue assumptions. The authors point out that the bathochromic shift seen in the series is a result both of the decreasing overlap between the p-orbitals of the heteroatom and the adjacent C atom and of the increasingly lower ionization potential of the heteroatom.

The spectra of numerous derivatives of selenophene have been reported, particularly by Yur'ev and his co-workers[76-80] and by Chierici *et al.*[81-83] The effect of substituents on the spectra of various selenophene, thiophene, and furan derivatives led Chierici to conclude that the aromaticity of the ring systems decreases in the order selenophene $\geq$ thiophene $>$ furan. However, Braye *et al.*[84] claim that the spectra and chemical properties of the tetraphenyl derivatives of furan, thiophene, selenophene, and tellurophene show that all of these ring systems have similar aromaticities. The syntheses and spectra of tetrachloroselenophene and tetrachlorotellurophene have also been reported.[85] Hammett σ-constants have been derived for a number of heterocyclic ring systems, including selenophene, by the study of wavelength dependence in analogs of chalcone.[86]

The spectrum of selenanaphthene[87] (benzo[b]selenophene, **24**) is quite

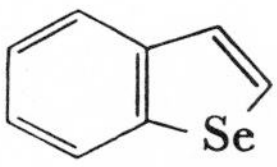

24 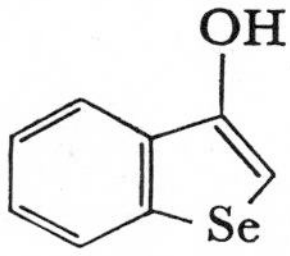   25 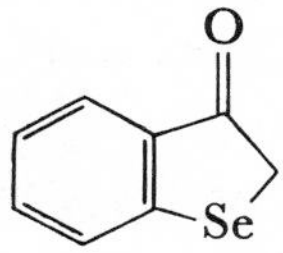   26

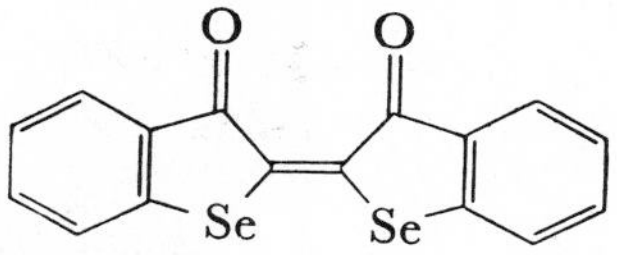

27

similar to that of the corresponding sulfur compound, the greatest difference being seen in the position of the longest-wavelength band, which lies at 304 nm as compared to 297.5 nm for thianaphthene. Substitution of a methyl group in the 2-position results in bathochromic shifts of 1–2 nm. In contrast to this, the long-wavelength band of 3-hydroxyselenanaphthene (**25**) is red-shifted in either neutral or acid solution by over 60 nm relative to the parent compound.[88] It is suggested that the compound may exist in the keto form (**26**), but the possibility that this dramatic shift may be the result of having an electron-releasing substituent at the 3-position of this fused 5,6-ring system cannot be overlooked.

A simple MO treatment of the spectrum of the parent compound has been reported.[89] Selenoindigo (**27**), the oxidation product of **25**, shows two intense (log ε 4.1) absorption bands, typical of the indigo structure, one in the visible and one in the ultraviolet region.[88, 90] Both cis- (λ max 485 nm) and trans-selenoindigo (λ max 562) are known and are photochemically interconvertible.[91] The fusion of a second benzene ring to form dibenzoselenophene (**28**) results in a further bathochromic shift of the spectrum. The spectra of the dibenzo five-ring heterocycles of Group VI come increasingly to resemble one another[92–94] and also to resemble the isoelectronic compound phenanthrene.[95] Thus, the band at 316–326 nm in dibenzoselenophene is

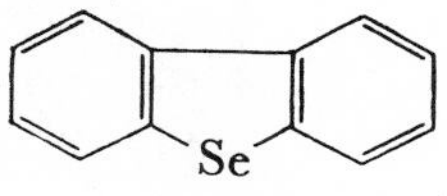

28

identified as the ^1L_b band, while the bands at 260–286 and 238 nm are identified as the ^1L_a and ^1B_b bands, respectively.[93,96-98] It is of interest to note in this context that, when LCAO-MO calculations were done on dibenzothiophene, the correlation of spectra and reduction potentials was successful when the $3d$ orbitals on sulfur were neglected, but was unsuccessful when the model of Longuet-Higgins was used.[99] The spectra of a number of substituted dibenzoselenophenes have also been reported.[96,100,101]

8. Other Heterocyclics

The synthesis and absorption spectra of 1,2,5-selenadiazole (**29**) (λ max 285 nm)[102,103] and 1,3,4-selenadiazole (**30**) (λ max 232 nm)[104] have been reported. The sulfur analogs of **29** and **30** exhibit maxima at 253 and 211 nm, respectively, while the oxygen analogs show end absorption only.

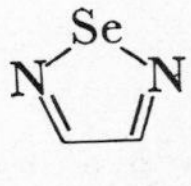

29    30 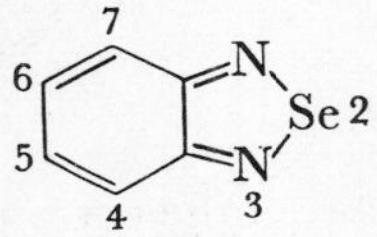   31

2,1,3-Benzoselenadiazole (**31**), also called piazselenole,[105] has been known for some time, and its properties have been well characterized by numerous workers. Its ground state seems to be adequately represented by the o-quinoid structure shown, as indicated by both chemical[106,107] and X-ray[108] data. Its absorption spectrum[106,107] in ethanol consists of bands at 230–233 nm (log ε 3.66) and 331 nm (log ε 4.22), with an inflection at 360 nm (log ε 3.32). In benzene solution the band at 331 nm is red-shifted to 334 nm,[109] suggesting a decrease in molecular polarity upon excitation analogous to that observed in azulene.[110]

The effect of substituents on the long-wavelength absorption band[106,109,111,112] is proportional to the electron-donating ability of the substituent and has been related to the Hammett $\sigma_m + \sigma_p$ constants.[109] Vibrational analyses of the near-ultraviolet spectra of 2,1,3-benzoselena-diazole and the related oxygen and sulfur compounds[113] have shown that the long-wavelength band consists of two π-π^* absorptions, and involves a decrease in moment of inertia about the long axis of the molecule.

The spectra of 2,1,3-benzoselenadiazole and a number of its derivatives have been determined in acid solution and show increasing red shifts in going from the base to the monoprotonated to the diprotonated salts, because of a successive lowering of the excited state by protonation.[114] Comparison of spectra reveals that, with the 5-amino compound, protonation occurs first on the ring N atom, while the second proton adds to the amino nitrogen. The 4-amino derivative is protonated first at the amino group and second at the ring nitrogen. The use of 2,1,3-benzoselenadiazole and its analogs in the detection of selenium is described in Chapter XVII and therefore need not be discussed here.

Other selenadiazoles whose spectra have been reported include the two isomeric naphthoselenadiazoles and their respective radical anions.[115] The spectra of 1,2,5-selenadiazolo(3,4-b)pyridine (**32**) and 1,2,5-selena-diazolo(3,4,-c)pyridine (**33**) (λ_{max}^{EtOH} 332 and 320 nm, respectively) closely

| 32 | 33 | 34 |

resemble the spectrum of 2,1,3-benzoselenadiazole in position and intensity of the long-wavelength π-π* band.[116] The selenadiazolopyrimidines (**34, 35**, and **36**)[117,118] have been synthesized because of their resemblance to

| 35 | 36 |

purine. Compounds **35** and **36** are cleaved by various reagents to afford 4-amino-3-carboxy-1,2,5-selenadiazole derivatives whose long-wavelength absorption is located 40–60 nm to the red of the unsubstituted compound.

The absorption maximum of isoselenazole (**37**) is reported to lie at 268 nm.[119] The spectrum of the isomeric compound, selenazole (**38**), does not appear to have been reported, but the spectrum of benzoselenazole[120] was recorded some 30 years ago.

| 37 | 38 |

39

Tautomerism in 2-aminoselenazolin-4-ones (**39**) has been investigated by the spectroscopic comparison of model compounds.[121,122] It was found that the amino form predominates when R = H, but that the imino tautomer is favored when R = phenyl.

Other selenium heterocyclic compounds whose spectra have been reported include various azomethine dyes[123–125] with a selenazole ring incorporated in the molecule; cyanine dyes;[126,127] 1-thiol-3-selenol-2-thione,[128] a novel compound for which a mesoionic structure may be drawn; and compound **40,** which is described as possessing a "no-bond resonance" system.[129]

40

9. Miscellaneous Compounds

Limitations of space do not permit detailed discussions of the spectra of all the various organic compounds containing the element selenium. Details may be obtained by consulting the original literature pertaining to the spectra of aromatic selenocyanates,[41,130] diarylselenium dihalides,[135–137] selenoxides,[138–141] selenenyl halides,[142] and selenium-containing esters.[143–145]

REFERENCES

1. L. Pauling, "The Nature of the Chemical Bond," 3rd ed., Cornell University Press, Ithaca, N.Y., 1960, p. 57.

2. W. H. H. Günther, *J. Org. Chem.*, **32,** 3931 (1967).

3. L. H. Noda, S. A. Kudy, and H. A. Lardy, *J. Amer. Chem. Soc.*, **75,** 913 (1953).

4. L. Chierici and R. Passerini, *Atti Accad. Naz. Lincei, Rend., Classe Sci. Fis. Mat. Nat.*, **14,** 99 (1953).

5. A. I. Kiss and B. R. Muth, *Acta Chim. Acad. Sci. Hung.*, **22,** 396 (1960).

6. G. M. Bogolyubov and Yu. N. Shlyk, *Zh. Obshch. Khim.*, **39,** 1759 (1969).

7. G. Bergson and A.-L. Delin, *Ark. Kemi*, **18,** 441 (1961).

8. K. Bowden and E. A. Braude, *J. Chem. Soc.*, 1068 (1952).

9. L. Chierici and R. Passerini, *Atti. Accad. Naz. Lincei, Rend., Classe Sci. Fiz. Mat. Nat.* **15,** 69 (1953).

10. J. W. Baker, G. F. C. Barrett, and W. T. Tweed, *J. Chem. Soc.*, 2831 (1952).

11. A. I. Kiss and B. R. Muth, *J. Chem. Phys.*, **23,** 2187 (1955).

12. A. Mangini, A. Trombetti, and C. Zauli, *J. Chem. Soc, B,* 153 (1967).

13. A. Gasco, G. DiModica, and E. Barni, *Ann. Chim.* (Rome), **58**, 385 (1968).

14. A. I. Kiss and B. R. Muth, *Acta Chim. Acad. Sci. Hung.*, **24**, 231 (1960).

15. H. P. Koch, *J. Chem. Soc.*, 387 (1949).

16. A. Mangini and R. Passerini, *J. Chem. Soc.*, 1168 (1952).

17. L. Chierici and R. Passerini, *Ric. Sci.*, **25**, 2316 (1955).

18. W. W. Robertson and F. A. Matsen, *J. Amer. Chem. Soc.*, **72**, 5250 (1950).

19. L. I. Lagutskaya and V. I. Danilova, *Zh. Strukt. Khim.*, **6**, 591 (1965).

20. A. Fredga, *Svensk Kem. Tidskr.*, **48**, 91 (1936).

21. G. Bergson, *Ark. Kemi*, **13**, 11 (1958).

22. G. Bergson, G. Claeson, and L. Schotte, *Acta Chem. Scand.*, **16**, 1159 (1962).

23. J. Linderberg and J. Michl, *J. Amer. Chem. Soc.*, **92**, 2619 (1970).

24. L. Pauling, *Proc. Nat. Acad. Sci. U.S.*, **35**, 495 (1949).

25. O. Foss, private communication quoted by G. Bergson, *Ark. Kemi*, **13**, 11 (1948).

26. G. Bergson, *Ark. Kemi*, **19**, 195 (1962).

27. G. Bergson, *Ark. Kemi*, **9**, 121 (1955).

28. D. S. Margolis and R. W. Pittman, *J. Chem. Soc.*, 799 (1947).

29. G. Bergson, *Nord. Kemikermode Aarhus*, **9**, 47 (1946).

30. A. Streitwieser, Jr., "Molecular Orbital Theory for Organic Chemists," John Wiley, New York, N.Y., 1961, p. 117.

31. G. Bergson and S. Wold, *Ark. Kemi*, **19**, 215 (1962).

32. L. Pauling, "The Nature of the Chemical Bond," 3rd ed., Cornell University Press, Ithaca, N.Y., 1960, p. 93.

33. G. Bergson, *Acta Chem. Scand.*, **11**, 571 (1957).

34. M. Chaix, *Bull. Soc. Chim. Fr.*, **53**, 700 (1933).

35. L. Chierici and R. Passerini, *Boll. Sci. Fac. Chim. Ind. Bologna*, **12**, 131 (1954).

36. W. V. Farrar, *Research* (London), **4**, 177 (1951).

37. A. J. Parker, *Acta Chem. Scand.*, **16**, 855 (1962).

38. R. E. Marsh, *Acta Cryst.*, **5**, 458 (1952).

39. E. Campaigne, J. Tsurugi, and W. W. Meyer, *J. Org. Chem.*, **26**, 2486 (1961).

40. L. Chierici and R. Passerini, *Boll. Sci. Fac. Chim. Ind. Bologna*, **11**, 102 (1953); **12**, 131 (1954).

41. L.-B. Agenäs, *Ark. Kemi*, **23**, 155 (1964).

42. T. W. Campbell, H. G. Walker, and G. M. Coppinger, *Chem. Rev.*, **50**, 279 (1952).

43. J. E. Kuder and M. A. Lardon, *Ann. N.Y. Acad. Sci.*, **192**, 147 (1972).

44. H. G. Mautner and W. D. Kumler, *J. Amer. Chem. Soc.*, **78**, 97 (1956).

45. H. G. Mautner, *J. Amer. Chem. Soc.*, **78**, 5292 (1956).

46. H. G. Mautner and E. M. Clayton, *J. Amer. Chem. Soc.*, **81**, 6270 (1959).

47. M. Z. Girshovich and A. V. El'tsov, *Zh. Obshch. Khim.*, **39**, 941 (1969).

48. A. V. El'tsov, Kh. L. Muravich-Aleksandr, and I. Yu. Tsereteli, *Zh. Org. Khim.*, **4**, 110 (1968).

49. H. G. Mautner, S.-H. Chu, and C. M. Lee, *J. Org. Chem.*, **27**, 3671 (1962).

50. H. G. Mautner and G. Bergson, *Acta Chem. Scand.*, **17**, 1694 (1963).

51. A. Azman, M. Drofenik, D. Hadzi, and B. Lukman, *J. Mol. Struct.*, **1**, 181 (1968).

52. R. Huls and M. Renson, *Bull. Soc. Chim. Belges*, **66**, 55 (1957).

53. J. Gosselck, *Angew. Chem. (Intern. Ed.)*, **2**, 660 (1963).

54. I. Degani, R. Fochi, and G. Spunta, *Boll. Sci. Fac. Chim. Ind. Bologna*, **26**, 3 (1968).

55. A. Ruwet and M. Renson, *Bull. Soc. Chim. Belges*, **75**, 260 (1966).

56. J. M. Danze and M. Renson, *Bull. Soc. Chim. Belges*, **75**, 169 (1966).

57. M. Renson and P. Pirson, *Bull. Soc. Chem. Belges*, **75**, 456 (1966).

58. M. Rolla and P. Franzosini, *Gazz. Chim. Ital.*, **88**, 837 (1958).

59. P. Franzosini, *Gazz. Chim. Ital.*, **88**, 1109 (1958).

60. G. Milazzo and L. Paolini, *Gazz. Chim. Ital.*, **82**, 576 (1952).

61. G. Milazzo and E. Miescher, *Gazz. Chim. Ital.*, **83**, 782 (1953).

62. G. Milazzo, *Gazz. Chim. Ital.*, **98**, 1511 (1968).

63. G. Milazzo, *Gazz. Chim. Ital.*, **78**, 835 (1948).

64. G. Milazzo, *Gazz. Chim. Ital.*, **83**, 392 (1953).

65. W. C. Price and A. D. Walsh, *Proc. Roy. Soc., Ser. A*, **179**, 201 (1941).

66. A. Trombetti and C. Zauli, *J. Chem. Soc., A*, 1106 (1967).

67. L. Chierici and G. Pappalardo, *Gazz. Chim. Ital.*, **88**, 453 (1958).

68. A. Mangini and C. Zauli, *J. Chem. Soc.*, 2210 (1960).

69. D. G. Manly and E. D. Amstutz, *J. Org. Chem.*, **22**, 323 (1957).

70. G. Leandri, A. Mangini, F. Montanari, and R. Passerini, *Gazz. Chim. Ital.*, **85**, 769 (1955).

71. W. Mack, *Angew. Chem. (Intern. Ed.)*, **5**, 896 (1966).

72. R. S. Mulliken, *J. Chem. Phys.*, **7**, 339 (1939).

73. H. C. Longuet-Higgins, *Trans. Faraday Soc.*, **45**, 173 (1949).

74. H. H. Jaffe and M. Orchin, "Theory and Application of Ultraviolet Spectroscopy," John Wiley, New York, N.Y., 1962, p. 349.

75. J. Fabian, A. Mehlhorn, and R. Zahradnik, *Theoret. Chim. Acta*, **12**, 247 (1968).

76. Yu. K. Yur'ev and N. N. Mezentsova, *Zh. Obshch. Khim.*, **31**, 1449 (1961).

77. Yu. K. Yur'ev, N. N. Magdesieva, and V. V. Titov, *Zh. Obshch. Khim.*, **33**, 1156 (1963).

78. Yu. K. Yur'ev, N. K. Sadovaya, and Ye. A. Grekova, *Zh. Obshch. Khim.*, **34**, 847 (1964).

79. Ye. G. Treshchova, D. Ekkhardt, and Yu. K. Yur'ev, *Zh. Fiz. Khim.*, **38**, 295 (1964).

80. Yu. K. Yur'ev, N. N. Magdesieva, and A. T. Monakhova, *Zh. Org. Khim.*, **1**, 1094 (1965).

81. L. Chierici and G. Pappalardo, *Gazz. Chim. Ital.*, **89**, 1900 (1959).

82. L. Chierici and G. Pappalardo, *Gazz. Chim. Ital.*, **90**, 69 (1960).

83. A. Bellotti and L. Chierici, *Gazz. Chim. Ital.*, **90**, 1125 (1960).

84. E. H. Braye, W. Hübel, and I. Caplier, *J. Amer. Chem. Soc.*, **83**, 4406 (1961).

85. W. Mack, *Angew. Chem. (Intern. Ed.)*, **4**, 245 (1965).

86. S. V. Tsukerman, V. D. Orlov, and V. F. Lavrushin, *Khim. Geterotsikl. Soedin.*, 67 (1969).

87. B. R. Muth and A. I. Kiss, *J. Org. Chem.*, **21**, 276 (1956).

88. A. I. Kiss and B. R. Muth, *Acta Chim. Acad. Sci. Hung.*, **11**, 57 (1957).

89. M. Baiwir, *Bull. Soc. Roy. Sci. Liege*, **38**, 708 (1969).

90. G. Marondel, *Chem. Ber.*, **93**, 2834 (1960).

91. D. L. Ross, J. Blanc, and F. J. Matticoli, *J. Amer. Chem. Soc.*, **92**, 5751 (1970).

92. E. Sawicki and F. E. Ray, *J. Amer. Chem. Soc.*, **75,** 2519 (1953).

93. A. Mangini and R. Passerini, *Gazz. Chim. Ital.*, **84,** 606 (1954).

94. R. Passerini and G. Purrello, *Ann. Chim.* (Rome), **48,** 738 (1958).

95. H. Christol, M. Mousseron, and R. Salle, *Bull. Soc. Chim. Fr.*, 556 (1958).

96. E. Sawicki, *J. Org. Chem.*, **19,** 1163 (1954).

97. A. Cerniani, R. Passerini, and G. Righi, *Boll. Sci. Fac. Chim. Ind. Bologna*, **12,** 75 (1954).

98. A. Cerniani and R. Passerini, *Boll. Sci. Fac. Chim. Ind. Bologna*, **12,** 65 (1954).

99. R. Gerdil and E. A. C. Lucken, *J. Amer. Chem. Soc.*, **88,** 733 (1966).

100. E. Sawicki, *J. Amer. Chem. Soc.*, **77,** 957 (1955).

101. N. Marciano, *Ric. Sci.*, **30,** 743 (1960).

102. L. M. Weinstock, P. Davis, D. M. Mulvey, and J. C. Schaeffer, *Angew. Chem.* (*Intern. Ed.*), **6,** 364 (1967).

103. V. Bertini, *Angew. Chem.* (*Intern. Ed.*), **6,** 563 (1967).

104. R. V. Kendall and R. A. Olofson, *J. Org. Chem.*, **35,** 806 (1970).

105. O. Hinsberg, *Ber.*, **22,** 862 (1889).

106. E. Sawicki and A. Carr, *J. Org. Chem.*, **22,** 503 (1957).

107. L. S. Efros and Z. V. Todres-Selektor, *Zh. Obshch. Khim.*, **27,** 983 (1957).

108. V. Lazzati, *Acta Cryst.*, **4,** 193 (1951).

109. M. Goto and K. Toei, *Talanta*, **12,** 124 (1965).

110. E. Heilbronner, "Non-benzenoid Aromatic Compounds," D. Ginsberg, Ed., Interscience Publishers, New York, N.Y. 1959, p. 244.

111. L. S. Efros and Z. V. Todres-Selektor, *Zh. Obshch. Khim.*, **27,** 3127 (1957).

112. E. Sawicki and A. Carr, *J. Org. Chem.*, **23,** 610 (1958).

113. J. M. Hollas and R. A. Wright, *Spectrochim. Acta*, **25A,** 1211 (1969).

114. E. Sawicki and A. Carr, *J. Org. Chem.*, **22,** 507 (1957).

115. J. Fajer, *J. Phys. Chem.*, **69,** 1773 (1965).

116. N. M. D. Brown and P. Bladon, *Tetrahedron*, **24,** 6577 (1968).

117. A. Carr, E. Sawicki, and F. E. Ray, *J. Org. Chem.*, **23,** 1940 (1958).

118. Y. F. Shealy and J. D. Clayton, *J. Heterocycl. Chem.*, **4,** 96 (1967).

119. F. Wille, A. Ascherl, G. Kaupp, and L. Capeller, *Angew. Chem.* (*Intern. Ed.*), **1,** 553 (1962).

120. O. Behaghel and E. Schneider, *Ber.*, **69,** 88 (1936).

121. A. M. Comrie, D. Dingwall, and J. B. Stenlake, *J. Chem. Soc.*, 5713 (1963).

122. J. F. Giudicelli, J. Menin, and H. Najer, *Bull. Soc. Chim. Fr.*, 1099 (1968).

123. E. Bulka, H. G. Patzwaldt, F. K. Peper, and H. Beyer, *Chem. Ber.*, **94,** 2759 (1961).

124. E. Bulka, M. Morner, and H. Beyer, *Chem. Ber.*, **94,** 2763 (1961).

125. E. Bulka, W. Dietz, H. G. Patzwaldt, and H. Beyer, *Chem. Ber.*, **96,** 1996 (1963).

126. T. Tani and S. Kikuchi, *Photogr. Sci. Eng.*, **11,** 129 (1967).

127. A. Leifer, D. Bonis, M. Boedner, P. Dougherty, A. J. Fusco, M. Koral, and J. E. LuValle, *Appl. Spectroscopy*, **21,** 71 (1967).

128. R. Mayer and A. K. Müller, *Z. Chem.*, **4,** 384 (1964).

129. J. H. van den Hende and E. Klingsberg, *J. Amer. Chem. Soc.*, **88,** 5045 (1966).

130. G. Cordella and R. Passerini, *Boll. Sci. Fac. Chim. Ind. Bologna*, **13,** 38 (1955).

131. L. Chierici and R. Passerini, *Boll. Sci. Fac. Chim. Ind. Bologna*, **12,** 138 (1954).

132. T. M. Lowry, R. R. Goldstein, and F. L. Gilbert, *J. Chem. Soc.*, 307 (1928).

133. F. L. Gilbert and T. M. Lowry, *J. Chem. Soc.*, 2658, 3179 (1928).

134. T. M. Lowry and F. L. Gilbert, *J. Chem. Soc.*, 2076 (1929).

135. M. A. A. Beg and A. R. Shaikh, *Tetrahedron*, **22,** 653 (1966).

136. J. D. McCullough and D. Mulvey, *J. Amer. Chem. Soc.*, **81,** 1291 (1959).

137. J. D. McCullough and D. Mulvey, *J. Phys. Chem.*, **64,** 264 (1960).

138. H. P. Koch, *J. Chem. Soc.*, 2892 (1950).

139. A. Cerniani, *Ric. Sci.*, **26,** 2100 (1956).

140. J. Grundnes and P. Klaeboe, *Acta Chem. Scand.*, **18,** 2022 (1964).

141. D. N. Jones, D. Mundy, and R. D. Whitehouse, *Chem. Commun.*, 86 (1970).

142. G. Bergson and G. Nordström, *Ark. Kemi*, **17,** 569 (1961).

143. H. G. Mautner, S.-H. Chu, and W. H. H. Günther, *J. Amer. Chem. Soc.*, **85,** 3458 (1963).

144. W. H. H. Günther and H. G. Mautner, *J. Med. Chem.*, **8,** 845 (1965).

145. M. Renson and R. Collienne, *Bull. Soc. Chem. Belges*, **73,** 419 (1964).

XV Physicochemical investigations of selenium compounds

C. CHIROPTICAL PROPERTIES OF SELENIUM COMPOUNDS

GÜNTHER SNATZKE

Institute of Organic Chemistry
University of Bonn
Bonn, Germany

MÁRTON KAJTÁR

Institute of Organic Chemistry
Eötvös University
Budapest, Hungary

1. Introduction

The terms "chiroptical"[1] and "chiral-optical"[2] refer to spectroscopic properties that depend on the chirality of the structure being investigated. Literature data range from simple measurements of the optical rotatory power at a single wavelength to fully documented optical rotatory dispersion (o.r.d.) and circular dichroism (c.d.) spectra. An attempt has been made to cite all relevant references. The discussion has been restricted, however, to species for which meaningful o.r.d. or c.d. data are available. Measurement techniques, sources of instrumentation, theoretical interpretations, and the terminology pertaining to chiroptical properties are given in several recent books.[3-7]

2. Compounds with Chiral Selenium Atoms

*a. **Selenonium Salts.***   Optical activity requires the presence of an element of chirality (center, axis, or plane[8]) in a molecule. A free-electron pair in a compound with three ligands attached to a Se atom can stabilize the configuration sufficiently so that resolution into enantiomers is possible, as is well known for the corresponding sulfur analogs. Thus, both optically active diastereomers of carboxymethyl methyl phenyl selenonium (+)-bromocamphor sulfonate [**1**, X = (+)-bromocamphor sulfonate (*cf.* Table XVC-1)] have been prepared[9] (the absolute configurations are unknown). The chloroplatinates of the two enantiomers could be obtained in the usual manner; reaction with HgI_2-KI leads, however, to complete racemization.[9]

On the other hand, no such difficulty arose when the triiodomercurates from the two enantiomeric 2-p-chlorophenacyl-2-selenaisochromanium bromocamphor sulfonates (**2**, *cf.* Table XVC-1) were prepared.[10] The molecular rotations $[M]_D$ for the picrates increase here in the following series: sulfonium < selenonium < telluronium compound[10] (−242, −533, more negative than −753, respectively), which is in agreement with the increasing polarizability in going from sulfur to tellurium (*cf.* the values of the distortion polarization for SF_6, SeF_6, and TeF_6: 15.7, 18.5, and 22.7, respectively[11]).

*b. **Selenoxides.***   In analogy to sulfoxides, selenoxides should also be resolvable into enantiomers. Although earlier efforts aimed at resolving some acyclic[12] and cyclic[10] selenoxides were not successful, n.m.r. studies done on racemic selenoxides[13,14] proved the Se atom to be a center of chirality in such compounds. The n.m.r. spectrum of benzyl phenyl selenoxide shows an AB-quartet for the two benzylic protons in achiral solvents ($CDCl_3$, water, *etc.*). Such an anisochrony is possible only for a chiral molecule, that is, the Se atom must be a center of chirality. By this means the two benzylic protons become diastereotopic to each other, as found.[8S]

Recently, however, two diastereomeric steroidal selenoxides have been prepared by Jones and his colleagues.[15] By these studies the earlier hypothesis[13,16] that traces of water may be the reason for the ready racemization of selenoxides, by forming symmetrical hydrates, was proven to be incorrect. The absolute configurations of the two 6β-phenylseleninyl-5α-cholestanes (**3**, *cf.* Table XVC-2) were determined from their different rates of decomposition by heat. The (S)-compound (**4**) gives a negative, the (R)-isomer a positive, Cotton effect around 270 nm, that is, within the benzene ^1L_b band. The signs of the Cotton effects are the same as for the corresponding sulfoxides of identical configurations at the chiral heteroatom (determined by the same method of rate analysis during thermal decomposition[15]); the

Table XVC-1   *Chiroptical data of chiral selenonium salts*

COMPOUND	$[M]_D{}^a$	REF.

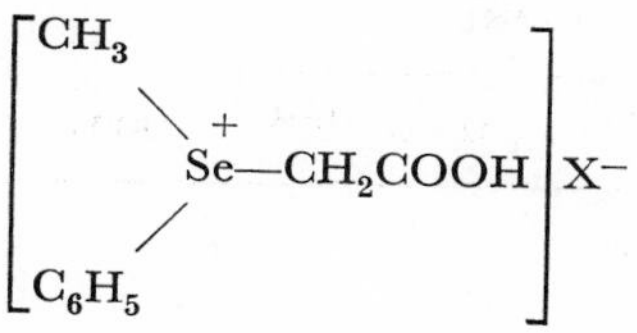

1

$X = (+)$-bromocamphor sulfonate

$(+)$-selenonium $(+)$-sulfonate	$+330.8$ (W)[b]	9
$(-)$-selenonium $(+)$-sulfonate	$+209.6$ (W)[c]	9
Calculated for the enantiomeric cations[d]	$+60.8$ (W)	9
	-60.4 (W)	
$X = (PtCl_6)/2$ (two enantiomers)	$+27.5$ (A)[e]	9
	-27.2 (A)[e]	

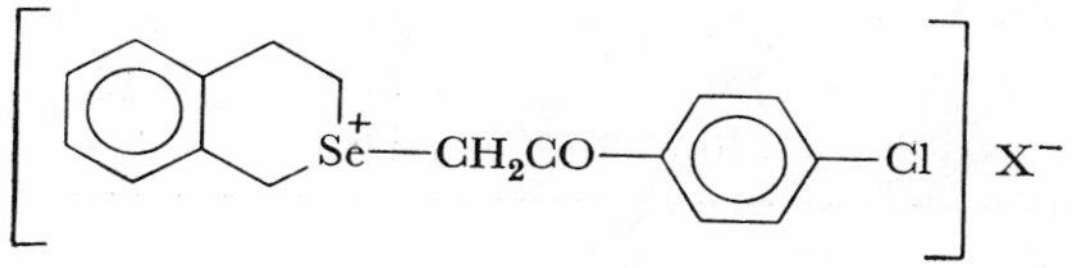

2

$X =$ bromocamphor sulfonate

$(+)$-selenonium $(-)$-sulfonate	$+291$ (M)	10
$(-)$-selenonium $(+)$-sulfonate	-293 (M)	10
	-271 (E)	
$X =$ picrate (two enantiomers)	$+504$ (A)	10
	-533 (A)	10
$X = HgI_3{}^-$	-535 (A)	10

[a] Letters in parentheses refer to solvents: A = acetone, E = ethanol, M = methanol, W = water.

[b] In the original paper[9] $[\alpha]_D$ $+61.26$ (W) is given.

[c] In the original paper[9] $[\alpha]_D$ $+38.81$ (W) is given.

[d] The molecular rotation of $(+)$-bromocamphor sulfonate ion was taken as $[M]_D$ $+270.0$ (W).[9]

[e] In the original paper[9] $[\alpha]_D$ $+6.34$ (W) and $[\alpha]_D$ -6.25 (W), respectively, are given, and the molecular rotations are calculated for the chloroplatinates containing 2 moles of the chiral cation (mol. wt. 868.4): $[M]_D$ $+55.0$ (A) and -54.3 (A), respectively.

Table XVC-2 Chiroptical data of chiral selenoxides

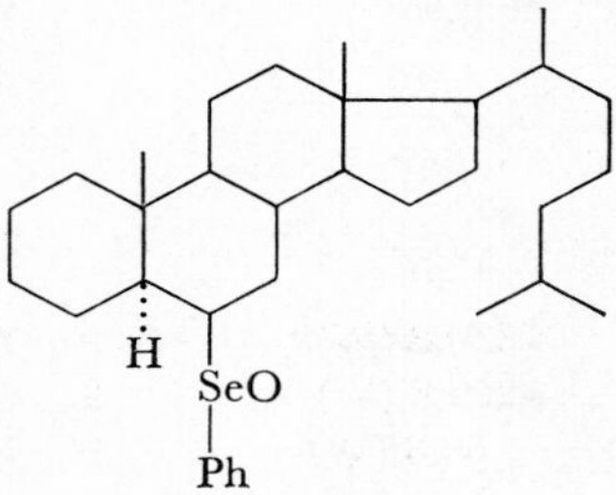

COMPOUND	OPTICAL ROTATORY DISPERSION EXTREMA (IN HEXANE)		REF.
	λ (nm)	$[M] \times 10^{-4}$	

3

(*R*)-selenoxide	290	$+1.2$	15
	253	-6.0	
(*S*)-selenoxide	281	-1.6	15
	246	$+3.8$	

amplitudes are, however, somewhat smaller (Fig. XVC-1). In both series the extrema appear at longer wavelengths for the (*R*)-isomers. Analogous chiroptical properties for steroidal 6β-benzyl selenoxides and 4β-phenyl selenoxides and their sulfur analogs are cited in the same paper.[15]

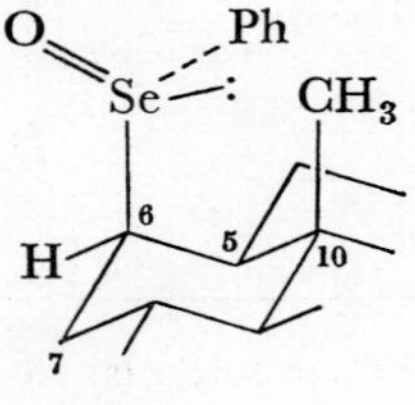

4*

c. Diselenides. Like peroxides and disulfides, diselenides are chiral compounds. However, their configurational stability is low,[17] so that at room temperature both possible helical conformers will be in rapid equilibrium. Only by incorporation of the diselenide moiety into a ring system is rotation around the Se-Se bond prevented, and a strong intrinsic Cotton effect to be

* In the (R)-isomer, the Ph group and the lone-electron pair are interchanged.

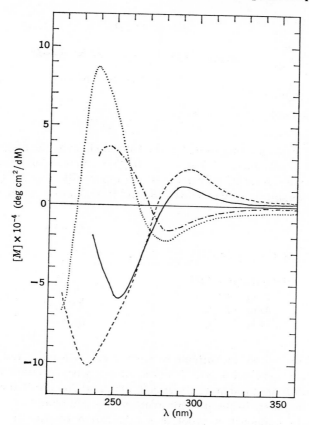

FIG. XVC-1 Optical rotatory dispersion of 6β-(R)-phenylseleninyl- (————), 6β-(S)-phenylseleninyl- (—·—·—·), 6β-(R)-phenylsulfinyl- (- - - - -), and 6β-(S)-phenylsulfinyl- (· · · · · ·) 5α-cholestanes in hexane.

expected. If centers of chirality are close to the Se_2 grouping, one of the two possible helices may be prevalent, even in acyclic compounds, because diastereomerism and stronger Cotton effects within the Se_2 absorption bands may again be found. If the M- and the P-helix[8] have equal populations, different achiral perturbations for the two diastereomers will still lead to optical activity for the Se_2 chromophore, but much smaller rotational strengths are expected in these cases. This is seen from a comparison of the diselenides (+)-1,2-diselenane-3,6-*trans*-dicarboxylic acid (5) and L-(−)-selenocystine (6). In the cyclic compound the conformation around the Se-Se bond is fixed because of the equatorial arrangement of the two carboxylic

COOH
|
[ring structure with Se—Se]
|
COOH

5 (*trans*)

COOH   COOH
| |
H_2N—C—H H_2N—C—H
| |
CH_2—Se—Se—CH_2

6

groups.[18] The apparent amplitude* of the Cotton effect at the longest wavelengths in the o.r.d. curve of **5** is $a = +87$,[19,20] whereas for selenocystine (though in another solvent) only $a = -0.1$ is reported[19] (Fig. XVC-2). For the three diselenobiscarboxylic acids **7**, **8**, and **9**, no Cotton effect was seen

SeCH(Q)COOH
|
SeCH(Q)COOH
    7: Q = CH_3
    8: Q = C_2H_5
    9: Q = C_6H_5

COOH
|
[ring structure with S—S]
|
COOH

10 (*trans*)

down to 270 nm (these measurements were done with older instrumentation).[19] The c.d. curve of **5** (Fig. XVC-3) reveals with certainty that three absorption bands are present at about 350, 280, and 250 nm;[20] these can be seen only as shoulders in the u.v. spectrum of this compound.

The hitherto undescribed absolute configuration of the (+)-1,2-diselenane-3,6-*trans*-dicarboxylic acid (**5**) can be deduced from a comparison with the corresponding dithiane-3,6-dicarboxylic acid (**10**). That both compounds have the same absolute configuration was determined earlier.[21,22] Chiral disulfides also give rise to several Cotton effects which appear, however, at shorter wavelengths than the corresponding bands of the diselenides (*e.g.*, the c.d. band at longest wavelength is found at 280 nm for **10**; *cf.* Fig. XVC-3), and the signs of the corresponding Cotton effects are equal in both series.[19,20] The correlation between the chirality of the S–S chromophore and the signs of its Cotton effects has been determined by Carmack and Neubert:[23] a negative torsion angle of the disulfide grouping corresponds to a negative Cotton effect at about 290 nm and a positive one at about 240 nm.

---

* The distance between the first two extrema in the o.r.d. curve is not a real amplitude, since two Cotton effects merge together, as can clearly be seen from the c.d. spectrum (*cf.* Fig. XVC-3).[20]

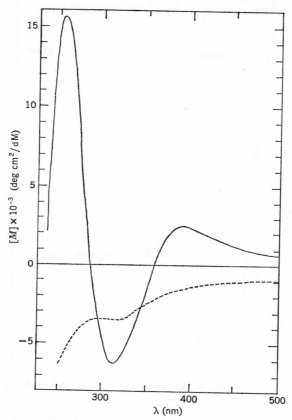

FIG. XVC-2   Optical rotatory dispersion of (+)-(*R, R*)-1,2-diselenane-3,6-*trans*-dicarbox-ylic acid (**5**) in dioxane (————) and of L-(−)-selenocystine (**6**) in 0.1 *N* hydrochloric acid (- - - - -).

From this follows the (*R,R*)-configuration for the (+)-1,2-dithiane-3,6-*trans*-dicarboxylic acid (**10**) and, therefore, also for the configurationally identical (+)-1,2-diselenane-3,6-*trans*-dicarboxylic acid (**5**).

The disulfide bridge of diaminooxytocin has been displaced by a diselenide

$$(+)-(R, R)\text{--}\mathbf{5}\colon \ X = Se$$
$$(+)-(R, R)\text{--}\mathbf{10}\colon X = S$$

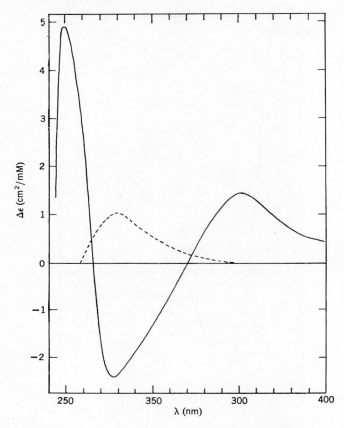

FIG. XVC-3   Circular dichroism of (+)-(R, R)-1,2-diselenane-3,6-*trans*-dicarboxylic acid
(**5**) (————) and of (+)-(R, R)-1,2-dithiane-3,6-*trans*-dicarboxylic acid (**10**) (- - - - -) in
dioxane.

bridge and by the two possible –S–Se– groupings.[24–26] The correct assignment of the c.d. bands of these compounds is difficult, however, because of
the presence of the tyrosine Cotton effects.[27] The c.d. bands assigned to the
–Se–Se– chromophore become stronger on temperature enhancement, and
this unusual behavior has been ascribed to a change of the torsion angle of
this moiety.[27]

The chiroptical properties of these diselenides are collected in Table
XVC-3. Additional diselenides have been prepared in different classes of
compounds; for these only the specific rotations at the Na_D line have,
however, been reported.[24,26,28,32–37,50]

Table XVC-3 Chiroptical data of chiral diselenides and selenosulfides

COMPOUND	$[\alpha]_D^a$	OPTICAL ROTATORY DISPERSION EXTREMA[a]				CIRCULAR DICHROISM		
		REF.	λ (nm)	$[M] \times 10^{-2}$	REF.	λ (nm)	$\Delta\varepsilon$	REF.
SeCH(Q)COOH | SeCH(Q)COOH								
Q = CH₃; (R, R)	+266.4 (HCl) +254.0 (W) −33.5 (NW) +105.8 (E) +129.5 (EA) +201.8 (C)	29	270!	+97.8[b] (M)	19			
(S, S)	−253.6 (W)	29						
Q = C₂H₅; (R, R)	+292.5 (W)	29	290!	+129[b] (M)	19			
(S, S)	−311 (HCl) −293 (W) +57 (NW) −149.4 (E) −177.9 (EA) −209.1 (C)	29						
Q = C₆H₅[c]	+446.1 (W)	29	290!	+238[b] (M)	19			

Table XVC-3 (Continued)

COMPOUND	$[\alpha]_D{}^a$	REF.	OPTICAL ROTATORY DISPERSION EXTREMA^a			CIRCULAR DICHROISM^a		
			λ (nm)	$[M] \times 10^{-2}$	REF.	λ (nm)	$\Delta\varepsilon$	REF.
trans; (R, R)	+351 (NW)	30	378.5	+26.2	19	351	+1.43	20
	−244 (HCl)		315	−60.1		277	−2.39	
	+226.5 (DE)			(M)		249	+4.90	
	+178.5 (E)		390	+25.0	20		(D)	
	+170.5 (A)		313	−62.0				
			255	+156.0				
				(D)				
(S, S)	−351 (NW)	30						
$\left[-SeCH_2-\overset{\displaystyle NH_2}{\underset{\displaystyle \mid}{CH}}-COOH \right]_2$ L	−183 (0.1 N HCl)	31	320	−34.6	19			
			290	−34.5				
			(0.1 N HCl)					

X—CH₂CH₂CO → Tyr → Ile^d
                                    ↓
Y—CH₂CHNH ← Asn ← Gln

CO → Pro → Leu → GlyNH₂

X	Y							
Se	Se	−51.0 (Ac)	25				290	+0.36	27

Se	S	−46.4 (Ac)	26	270i	+0.27
				241	−0.84
				225	+2.70
				197	−18.78
			27	(W)	
				300	−0.12
				280	−0.09
				257	+0.33
				235	−0.72
				198	−18.78
				(W)	
S	Se	−54.0 (Ac)	24	290	+0.14
			26	280	+0.09
				260	+0.14
				237	−0.24
				227	+0.60
			27	197	−18.78
				(W)	

a Letters in parentheses refer to solvents: A = acetone, Ac = acetic acid, C = chloroform, D = dioxane, DE = diethyl ether, E = ethanol, EA = ethyl acetate, HCl = dilute hydrochloric acid, M = methanol, NW = neutralized water solution, W = water.

b Positive plain curve; "i" indicates that first extremum not reached.

c Configuration unknown.

d All amino acids are of L-configuration. For amino acid symbols, see IUPAC-IUB Combined Commission on Biochemical Nomenclature.[38] The arrows show the usual direction of peptide bonds.

3. Compounds without Chiral Selenium Atoms

Many other optically active selenium-containing compounds have been prepared in which the selenium replaces either oxygen (in the sugar[34,36,37,65—77] and amino acid series[39,63,64]) or sulfur (amino acids[26,49—62]). In these the selenium may be bound to an asymmetric C atom, but in no case is it a center of chirality. For most of them only the rotation at the

Table XVC-4   Chiroptical data of amino acid selenoaryl esters

COMPOUND[a]	OPTICAL ROTATORY DISPERSION EXTREMA (IN t-BUTYL ALCOHOL)			CIRCULAR DICHROISM (IN ETHANOL)		
	λ (nm)	$[M] \times 10^{-3}$	REF.	λ (nm)	$\Delta\varepsilon$	REF.
Cys(Bzl)—SePh·HBr	308	+11.30	39			
	265	−14.86				
	239	−20.42				
	226	−0.33				
Glu(OMe)—SePh·HBr	304	+6.17	39			
	262	−4.10				
Leu—SePh·HBr	303.5	+7.46	39			
	262.5	−8.50				
Phe—SePh·HBr	304	+11.87	39	284	+4.43	40
	263	−2.89		248	+1.37	
	249	+0.40		221	+9.1	
	225	+34.95		205	+7.0	
Leu—SeNap·HBr	311	+9.74	39			
	270	−3.13				
Phe—SeNap·HBr	312	+10.45	39			
	269.5	−7.56				
	236	−15.80				
Ac—Leu—SePh[b]	310.5	−5.96	39			
	281	−0.04				
Ac—Phe—SePh[b]	312	−4.91	39			
	276	+3.94				
Ac—Leu—SeNap[b]	307	−5.73	39			
	269	+0.25				
Phe—Gly—SePh·HBr	308	−0.31	39			
	253	+7.78				

[a] All amino acids are of L-configuration. For amino acid symbols see IUPAC-IUB Combined Commission on Biochemical Nomenclature.[38] Ac = acetyl, Bzl = benzyl, Me = methyl, Ph = phenyl, Nap = α-naphthyl.

[b] The $[\alpha]_D$ value is described in Table XVC-5.

Na$_D$ line has been determined. The Cotton effects (by o.r.d.) have been measured[39] (Table XVC-4) only for some selenophenyl and selenonaphthyl esters of L-amino acid derivatives. A positive Cotton effect appears at about 284 nm for the phenyl compounds and at about 291 nm for the naphthyl compounds. This band shows a bathochromic shift and sign inversion by acetylation at the amino group. It is characteristic for the C(=O)Se-aryl moiety and is assumed[39] to be mainly of $n_{C=O} \rightarrow \pi^*$ origin. A second (negative) Cotton effect, which is not always separated distinctly from the

Table XVC-5   *Chiroptical data of miscellaneous chiral selenium compounds*

COMPOUND		$[\alpha]_D$[a]	REF.	OPTICAL ROTATORY DISPERSION EXTREMA[a]		
				λ(nm)	$[M]$	REF.
11						
Q = H;	(R)	+66 (NaOH)	41	316	+39,000	41
		+37 (M)		281	−31,000	
		+44 (E)			(E)	
		+46 (A)				
	(S)	−44 (E)	41			
Q = CH$_3$;	(R)	+63.5 (A)	41	327	+35,000	41
				281	−31,000	
					(E)	
12						
(L, L)				606	+1,200	43
				533	−11,000	
				461	+10,000	
				413	+11,000	
				351	+4,600	
					(W)	

[a] Letters in parentheses refer to solvents: A = acetone, E = ethanol, M = methanol, NaOH = dilute sodium hydroxide solution, W = water; en = ethylenediamine.

first one, was described as appearing at about 225–245 nm for some of these compounds and was believed to be due to a transition involving the unshared electrons of the Se atom.[39] A reinvestigation of the c.d. of L-phenylalanine selenophenyl ester hydrobromide revealed at least four Cotton effects, all of positive signs, that is, at 284, 248, 221, and 205 nm (Fig. XVC-4).[40] Thus, the "second" Cotton effect seen in some of the o.r.d. spectra is only apparently negative because of overlapping of several different o.r.d. extrema.

A number of other simple, optically active compounds containing selenium have been prepared and their specific rotations at the Na_D line reported.[28,29,78–84] Cotton effects were measured in only two cases. 2,2'-Dinitrobisselenophen-3,3'-yl-4,4'-dicarboxylic acid (**11**, Q = H, *cf.* Table XVC-5), whose chirality is due to a restricted rotation around the pivot bond,

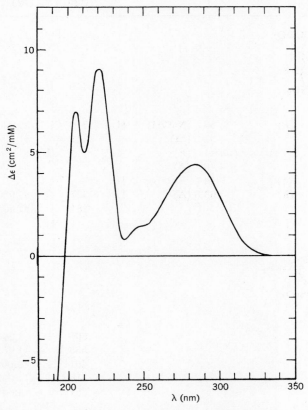

FIG. XVC-4   Circular dichroism of L-phenylalanine selenophenyl ester hydrobromide in ethanol.

has been resolved, and the chirality of the $(+)$-compound has been determined by the quasi-racemate method to be (R).[41] Its o.r.d. curve resembles that of the sulfur analog of known absolute configuration,[42] thus independently proving the assigned (R)-chirality (cf. Table XVC-5).

The o.r.d. of a selenato-bridged cobalt complex (**12**, cf. Table XVC-5) has been described[43]; it is very similar to the o.r.d.'s of the corresponding sulfato- and phosphato-bridged compounds.

Several optically active complexes show strong changes in their c.d. curves when selenite ions are added to their aqueous solutions;[44-47] e.g., the $^1A_2$ component of the $^1T_{1g}$ c.d. band is enhanced in the case of trisdiamine complexes, and the ^1E_a component is diminished. This method, which gives good results in the case of bidentate ligands, is unreliable, however, with polydentate ligands.[48]

REFERENCES

1. V. Prelog, *Proc. Koninkl. Ned. Acad. Wetenschap.*, *Ser. B*, **71**, 108 (1968).

2. U. Weiss, *Experientia*, **24**, 1088 (1968).

3. T. M. Lowry, "Optical Rotatory Power" (republication), Dover, New York, N.Y., 1964.

4. C. Djerassi, "Optical Rotatory Dispersion," McGraw-Hill, New York, N.Y., 1960.

5. P. Crabbé, "Optical Rotatory Dispersion and Circular Dichroism in Organic Chemistry," Holden-Day, San Francisco, Calif., 1965. [French edition: Gauthier-Villars, Paris, 1968.]

6. L. Velluz, M. Legrand, and M. Grosjean, "Optical Circular Dichroism," Verlag Chemie, Weinheim, 1965.

7. G. Snatzke, Ed., "Optical Rotatory Dispersion and Circular Dichroism in Organic Chemistry," Heyden, London, 1967.

8. R. S. Cahn, C. Ingold, and V. Prelog, *Angew. Chem.*, **78**, 413 (1966); *Angew. Chem. (Intern. Ed.)*, **5**, 385 (1966).

9. W. J. Pope and A. Neville, *J. Chem. Soc.*, **81**, 1552 (1902).

10. F. G. Holliman and F. G. Mann, *J. Chem. Soc.*, 37 (1945).

11. R. Linke, *Z. Phys. Chem.*, **48B**, 193 (1941).

12. W. R. Gaythwaite, J. Kenyon, and H. Phillips, *J. Chem. Soc.*, 2280 (1928).

13. M. Oki and H. Iwamura, *Tetrahedron Lett.*, 2917 (1966).

14. M. Cinquini, S. Colonna, and D. Landini, *Boll. Sci. Fac. Chim. Ind. Bologna*, **27**, 207 (1969).

15. D. N. Jones, D. Mundy, and R. D. Whitehouse, *J. Chem. Soc.*, *D*, 86 (1970).

16. T. W. Campbell, H. G. Walker, and G. M. Coppinger, *Chem. Rev.*, **50**, 279 (1952).

17. H. Kessler and W. Rundel, *Chem. Ber.*, **101**, 3350 (1968).

18. O. Foss, K. Johnsen, and T. Reistad, *Acta Chem. Scand.*, **18**, 2345 (1964).

19. C. Djerassi, A. Fredga, and B. Sjöberg, *Acta Chem. Scand.*, **15**, 417 (1961).

20. C. Djerassi, H. Wolf, and E. Bunnenberg, *J. Amer. Chem. Soc.*, **84**, 4552 (1962).

21. A. Fredga, *Ark. Kemi, Mineral. Geol.*, **12A**, No. 27 (1938).

22. A. Rosenberg and L. Schotte, *Ark. Kemi*, **8**, 143 (1955).

23. M. Carmack and L. A. Neubert, *J. Amer. Chem. Soc.*, **89**, 7134 (1967).

24. R. Walter and V. du Vigneaud, *J. Amer. Chem. Soc.*, **87**, 4192 (1965).

25. R. Walter and V. du Vigneaud, *J. Amer. Chem. Soc.*, **88**, 1331 (1966).

26. R. Walter and W. Y. Chan, *J. Amer. Chem. Soc.*, **89**, 3892 (1967).

27. D. W. Urry, F. Quadrifoglio, R. Walter, and I. L. Schwartz, *Proc. Natl. Acad. Sci. U.S.*, **60**, 967 (1968).

28. J. Kenyon, H. Phillips, and V. P. Pittman, *J. Chem. Soc.*, 1072 (1935).

29. A. Fredga, *Uppsala Univ. Årsskr.*, No. 5 (1935); *Chem. Abstr.*, **29**, 7281 (1935).

30. A. Fredga, *Ark. Kemi, Mineral. Geol.*, **11B**, No. 15 (1933).

31. A. Fredga, *Svensk Kem. Tidskr.*, **49**, 124 (1937).

32. J. Roy, I. L. Schwartz, and R. Walter, *J. Org. Chem.*, **35**, 2840 (1970).

33. W. H. H. Günther and H. G. Mautner, *J. Amer. Chem. Soc.*, **87**, 2708 (1965).

34. C. Wagner and P. Nuhn, *Arch. Pharm.* (Weinheim), **297**, 461 (1964).

35. F. Wrede, *Ber.*, **52**, 1756 (1919).

36. F. Wrede, *Hoppe-Seylers Z. Physiol. Chem.*, **115**, 284 (1921).

37. T. van Es and R. L. Whistler, *Tetrahedron*, **23**, 2849 (1967).

38. IUPAC-IUB Combined Commission on Biochemical Nomenclature, *J. Biol. Chem.*, **241**, 527 (1966).

39. K. Bláha, I. Frič, and H.-D. Jakubke, *Collection Czech. Chem. Commun.*, **32**, 558 (1967).

40. K. Bláha, I. Frič, G. Snatzke, and M. Kajtár, unpublished results.

41. C. Dell'Erba, D. Spinelli, G. Garbarino, and G. Leandri, *J. Heterocycl. Chem.*, **5,** 45 (1968).

42. S. Gronowitz and P. Gustavson, *Ark. Kemi*, **20**, 289 (1963).

43. K. Garbett and R. D. Gillard, *J. Chem. Soc.*, A, 1725 (1968).

44. S. F. Mason and B. J. Norman, *Proc. Chem. Soc.*, 339 (1964).

45. S. F. Mason and B. J. Norman, *Chem. Commun.*, 73 (1965).

46. R. Larsson, S. F. Mason, and B. J. Norman, *J. Chem. Soc.*, A, 301 (1966).

47. S. F. Mason and B. J. Norman, *J. Chem. Soc.*, A, 307 (1966).

48. R. D. Gillard and P. R. Mitchell, *Structure and Bonding*, **7**, 45 (1970).

49. S. N. Nigam and W. B. McConnell, *Biochim. Biophys. Acta*, **192**, 185 (1969).

50. J. Roy, W. Gordon, I. L. Schwartz, and R. Walter, *J. Org. Chem.*, **35**, 510 (1970).

51. W. Frank, *Hoppe-Seylers Z. Physiol. Chem.*, **339**, 202 (1964).

52. D. Theodoropoulos, I. L. Schwartz, and R. Walter, *Biochemistry*, **6**, 3927 (1967).

53. G. Zdansky, *Ark. Kemi*, **29**, 443 (1968).

54. D. Theodoropoulos, I. L. Schwartz, and R. Walter, *Tetrahedron Lett.*, 2411 (1967).

55. W. Frank, *Hoppe-Seylers Z. Physiol. Chem.*, **339**, 214 (1964).

56. W. Frank, *Hoppe-Seylers Z. Physiol. Chem.*, **339**, 222 (1964).

57. H.-D. Jakubke, J. Fischer, K. Jošt, and J. Rudinger, *Collection Czech. Chem. Commun.*, **33,** 3910 (1968).

58. F. Pan and H. Tarver, *Arch. Biochem. Biophys.*, **119**, 429 (1967).

59. G. Zdansky, *Ark. Kemi*, **29**, 437 (1968).

60. C. S. Pande, J. Rudick, and R. Walter, *J. Org. Chem.*, **35**, 1440 (1970).

61. G. Zdansky, *Ark. Kemi*, **29**, 449 (1968).

62. F. Kerdel-Vegas, F. Wagner, P. B. Russell, N. H. Grant, H. E. Alburn, D. E. Clark, and J. A. Miller, *Nature*, **205**, 1186 (1965).

63. H.-D. Jakubke, *Chem. Ber.*, **97**, 2861 (1964).

64. H.-D. Jakubke, *Justus Liebigs Ann. Chem.*, **682**, 244 (1965).

65. J. Kocourek, J. Kleňha, and V. Jiráček, *Chem. Ind.* (London), 1397 (1963).

66. G. Wagner and P. Nuhn, *Z. Chem.*, **3**, 64 (1963).

67. H. Frenzel, P. Nuhn, and G. Wagner, *Arch. Pharm.* (Weinheim), **302**, 62 (1969).

68. W. A. Bonner and A. Robinson, *J. Amer. Chem. Soc.*, **72**, 354 (1950).

69. G. Wagner and G. Lehmann, *Pharm. Zentralhalle*, **100**, 160 (1961).

70. G. Wagner and P. Nuhn, *Pharm. Zentralhalle*, **104**, 328 (1965).

71. G. Wagner and P. Nuhn, *Arch. Pharm.* (Weinheim), **269**, 374 (1963).

72. G. Wagner and P. Nuhn, *Arch. Pharm.* (Weinheim), **297**, 81 (1964).

73. G. Wagner and G. Valz, *Pharmazie*, **22**, 548 (1967).

74. W. Schneider and F. Wrede, *Ber.*, **50**, 793 (1917).

75. F. Wrede, *Hoppe-Seylers Z. Physiol. Chem.*, **112**, 1 (1920).

76. W. Schneider and A. Beuther, *Ber.*, **52**, 2135 (1919).

77. F. Wrede and W. Zimmermann, *Hoppe-Seylers Z. Physiol. Chem.*, **148**, 65 (1925).

78. A. Fredga, *J. Prakt. Chem.*, **123**, 129 (1929).

79. A. Fredga, *J. Prakt. Chem.*, **121**, 56 (1929).

80. W. Gerrard, J. Kenyon, and H. Phillips, *J. Chem. Soc.*, 153 (1937).

81. A. Fredga, *Ark. Kemi, Mineral. Geol.*, **17A**, No. 17 (1944).

82. A. Fredga, *Ark. Kemi, Mineral. Geol.*, **11B**, No. 43 (1934).

83. A. Fredga, *J. Prakt. Chem.*, **130**, 180 (1931).

84. H. J. Backer and W. van Dam, *Rec. Trav. Chim. Pays-Bas*, **49**, 482 (1930).

85. K. Mislow and M. Raban, "Topics in Stereochemistry," N. L. Allinger and E. L. Eliel, Eds., Vol. 1, Interscience, New York, N.Y., 1967, p. 1.

XV Physicochemical investigations of selenium compounds

D. PROTON MAGNETIC RESONANCE SPECTROSCOPY

ULLA SVANHOLM

University of Copenhagen
Chemical Laboratory II
The H. C. Ørsted Institute
Copenhagen, Denmark

1. Introduction

Nuclear magnetic resonance is one of the most useful techniques in the study of structural problems. The proton has previously been by far the most frequently used nucleus, and only a few n.m.r. investigations of other nuclei have been reported in the field of organoselenium compounds. Both chemical shifts and spin-spin coupling constants are of analytical use in the study of the structure as well as the conformation of selenium compounds, and the information has originated mainly in the following areas.

Because of the magnetic ^{77}Se isotope the ^{1}H n.m.r. spectra of selenium compounds often offer additional information to that provided by spectra of, for example, corresponding sulfur compounds. The natural abundance (7.5%) of ^{77}Se $(I = \frac{1}{2})$ is usually sufficient to allow observation of satellite spectra which readily identify protons one, two, or three bonds apart from Se atoms.

^{1}H Chemical shifts are dependent on both magnetic and electric fields and are very sensitive to the nature and spatial orientation of substituents. Chemical shifts of related compounds may be compared in terms of contributions from magnetically anisotropic atoms or bonds and inductive and mesometric effects.

Coupling constants are dependent on electric fields, but independent of magnetic fields, and are sensitive to conformation, bond angles, and bond lengths, and to the electronegativity of substituents. The sign and the magnitude of spin-spin coupling constants are dependent on the position in the periodic table of the nuclei involved. It should be noted, however, that coupling constants and, in particular, chemical shifts also depend on external factors such as solvent and temperature.

Nuclear magnetic resonance spectroscopy is also useful in the determination of energy barriers for intra- or intermolecular exchange processes. The rate constants may be found from the line shape of the signals from the exchanging protons if the interconversion is slow on the n.m.r. time scale.

In view of the vast number of ^{1}H n.m.r. spectra of organoselenium compounds reported in the literature, it has not been possible to present all published information explicitly. The aim of this chapter is to present the ^{1}H n.m.r. parameters of a great variety of selenium compounds; no detailed interpretation of chemical shifts or coupling constants will be attempted. The spectral parameters are tabulated and discussed according to type of compound. Included are comparisons with related compounds (especially oxygen and sulfur compounds) and brief discussions on correlations between chemical shifts and electronegativity and diamagnetic anisotropy, the origin and sign of spin-spin coupling constants, and correlations between coupling constants and electronegativity, geometry, and bond order.

2. Compounds Containing Selenium(II)-Carbon Single Bonds

Some characteristic proton chemical shifts and coupling constants of selenols, selenides, diselenides (and compounds with Se-Si, Ge, or Sn bonds) are given in Fig. XVD-1 and Table XVD-2, respectively.

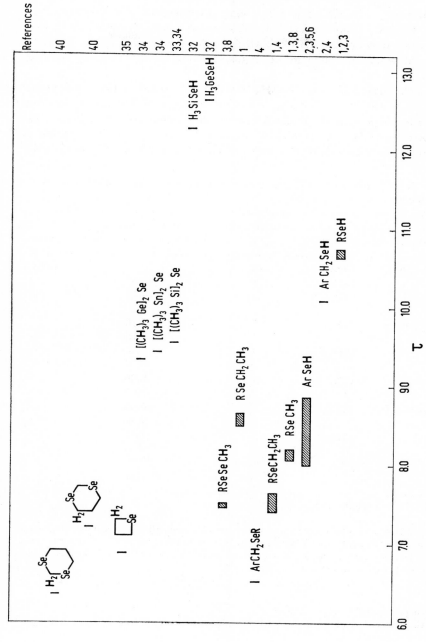

FIG. XVD-1 Proton chemical shifts (τ-values) of compounds with Se(II)-C single bonds, and compounds with Se-Si, Ge, or Sn bonds. R = H or alkyl, Ar = aryl.

a. Selenols, Acyclic Selenides, and Diselenides. The signals from
R–Se–**H*** (R = alkyl or aralkyl) are always† found at higher field than
that from tetramethyl silane (TMS). A downfield shift of *ca*. 2 ppm is
observed when the alkyl group is displaced by an aryl moiety (*cf*. Fig.
XVD-1).

The chemical shift of R–Se–**CH₃** (Fig. XVD-1) is approximately the same
for R = H and R = alkyl and shows no pronounced dependence on concentration or solvent (*cf*. Table XVD-1). Many attempts[7b] have been made
to correlate $(CH_3)_nX$ chemical shifts with the electronegativity of the X
group. It has been found that a plot of chemical shifts *vs*. electronegativity

*Table XVD-1   The dependence of the chemical shifts of selenols and selenides on solvent and
concentration*

COMPOUND	SOLVENT	CONCENTRATION (%)	CHEMICAL SHIFTS (τ) OF INDICATED PROTONS	REF.
CH_3CH_2SeH	None	100	10.67	2
	C_6H_6	20	10.63	1
	CCl_4	12	10.73	1
$(CH_3)_2Se$	C_6H_6	10	8.18	
		5	8.18	
	C_6H_{12}[a]	10	8.10	1
		5	8.10	
	CCl_4	10	8.08	
		5	8.07	
$(CH_3CH_2)_2Se$	C_6H_6	11	7.46	
		5	7.46	
	C_6H_{12}[a]	10	7.41	1
		5	7.43	
	CCl_4	11	7.48	
		7	7.47	

[a] C_6H_{12} = cyclohexane.

* Bold-face type indicates the nuclei under consideration. All chemical shifts are given in
τ-values, that is, relative to TMS (used mainly as internal reference). Chemical shifts,
originally given relative to benzene or water (used mainly as external references), are
converted to somewhat uncertain τ-values by[7a]

$$\delta_{ext}^{C_6H_6} = \delta_{ext}^{TMS} + 7.27 \quad \text{and} \quad \delta_{ext}^{H_2O} = \delta_{ext}^{TMS} + 4.78$$

Coupling constants are given in hertz.
† All evidence is against the chemical shifts given at lower field by Lalezari and Sharghi.[5]

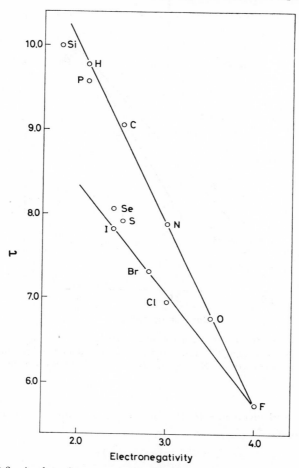

FIG. XVD-2 A plot of proton chemical shifts (τ-values) of compounds (dilute CCl_4 solutions) with the general formula $(CH_3)_nX$ *vs.* Pauling electronegativity of X. The chemical shifts, except for X = Se and P, are from Appendix B of ref. 7. The data for $(CH_3)_2Se$ are from the work of Mila and Laurent,[1] and for $(CH_3)_3P$ from the work of McCoy and Allred.[9]

yields a fairly linear curve for X = Group IV and V elements and one with a different slope for X = sulfur and Group VII elements. The chemical shift of $(CH_3)_2Se$ follows the relationship for X = S and Group VII elements (*cf.* Fig. XVD-2). A plot of τ_{CH_3}-τ_{CH_2} *vs.* X-electronegativity for ethyl derivatives, $(CH_3CH_2)_nX$, shows the same tendency (*cf.* Fig. XVD-3), and Dailey and Shoolery[12] have derived the relationship between chemical shift

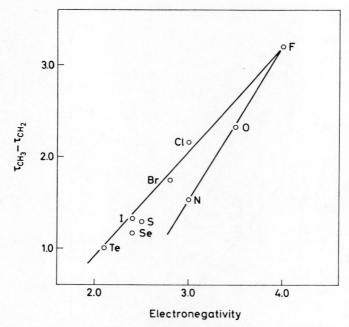

FIG. XVD-3   A plot of the internal chemical shift difference $(\tau_{CH_3} - \tau_{CH_2})$ in ethyl deriv-atives, $(CH_3CH_2)_nX$, *vs.* Pauling electronegativity of X. The data are from the work of Narasimhan and Rogers,[10] except for X = Se and Te, which are from the work of Breuninger *et al.*[11]

differences (Δ) and electronegativity (x) shown in equation 1 from a plot where X = halogen.*

(1) $x = 0.695 \ (ppm) + 1.71$

(2) $x = 2 + 0.564Δ \ (ppm)$

Mila and Laurent[1] have used equation 1 and a similar equation (2), derived by Müller,[13] to calculate the effective electronegativity of selenium in CH_3CH_2SeH and $(CH_3CH_2)_2Se$. Equation 2 is derived by treating ethyl derivatives of elements of Groups IV, V, VI, and VII as one series and yields, as expected on the basis of Fig. XVD-3, a higher value than equation 1; equation 1 gave 2.50 and equation 2 gave 2.64 for selenium in diethyl selenide.

* The corresponding equation[10] for elements of Groups IV and V is $x = 0.64Δ + 2.07$. Oxygen follows this relationship.

The deviation of Groups VI and VII elements (except oxygen and fluorine) from the relationship for Groups IV and V elements has been interpreted[7b,c] as an effect of the magnetic susceptibility of the C-X bonds in the former series. The net effect[7c] of the diamagnetic anisotropy of the C-X bonds (X = S, Se, Te, Cl, Br, I) on the α-protons is deshielding (*i.e.*, the shielding constants are reduced more than might be expected from the electronegativity of X). For an axially symmetrical group the change in σ from the magnetic susceptibility (χ) may be found from McConnell's equation (equation 3):[7c]

$$(3) \qquad \Delta\sigma = \frac{\chi_{\parallel} - \chi_{\perp}}{3R^3} (1 - 3\cos^2\theta)$$

where R is the distance of the proton from the anisotropic center, and θ is the angle between the direction of χ and the line connecting the center with the proton.

Gabdrakhmanov *et al.*[14] have used equation 3 to make a rough estimate of the diamagnetic anisotropy of the Se–C bonds ($\Delta\chi_{C-Se}$, where $\Delta\chi = \Delta_{\parallel} - \chi_{\perp}$) in $(CH_3)_2C{=}CCH_3CH_2SePh$. Calculations using the chemical shift differences between the signals from the methyl groups, assuming the Se atom to be isotropic, yielded an average value of $\Delta\chi_{C-Se} = 12.5 \times 10^{-6}$ cm³/ mole. As a comparison, calculations of $\Delta\chi_{C-C}$ from chemical shift differences in alkanes yield values of the magnitude 4.5×10^{-6} cm³/mole.[7c]

It may be inferred that the dominant effects of selenium on the shielding constants of neighbor protons in alkyl derivatives are the inductive effect and the effect of the diamagnetic anisotropy of C-Se bonds. However, other effects, which have received less attention, should not be neglected, for example, paramagnetic contributions from the heavy Se atom and effects of selenium on electric fields.

McFarlane and Nash[15] have found that the *Se*-CH₃ groups of $[(CH_3)_2SeCH(CH_3)Ph]^+Br^-$ are magnetically nonequivalent (the chemical shift difference is 0.19 ppm in $CDCl_3$).

The introduction of another Se atom (diselenides) causes a downfield shift of the α-proton of *ca.* 0.6 ppm (Fig. XVD-1) and a small increase in $^1J(Se^{13}C–H)$ (*cf.* Table XVD-2). The chemical shift dependence on the number of Se atoms has been used by Grant and van Wazer[16] to find the product distribution in a mixture of polyselenides.

Kessler and Rundel[17] have found that the *m*-protons and the *o-tert*-butyl protons of bis(2,4,6-tri-*tert*-butylphenyl) diselenide are magnetically non-equivalent and have calculated $\Delta G_c^{\ddagger} = 12.6$ kcal/mole (at coalescence) for the intramolecular exchange process from the temperature dependence of the ¹H n.m.r. spectrum.

Table XVD-2 Proton coupling constants (Hz) in compounds with Se(II)-C single bonds

COMPOUND TYPE[a]	^2J(H—C—H)	^3J(H—C—C—H)	^3J(H—C—Se—H)	^1J(^{77}Se—H)	^2J(^{77}Se$\cdots$H)	^3J(^{77}Se$\cdots$H)	^1J(^{13}C—H)[b]	REF.
RSeH		7.2	6.4–7.0	41.2–44.7	9.4–12.8		142.7	1–4
ArSeH			57.5					3
RSeR		6.9–7.5			10.2–11.2	10.8	139.0–140.3[c]	1, 11, 20–22
(CH₃)₃Se⁺I⁻					9.3		145.8	19
RSeSeR					11.9	2.3[d]	141.7	3, 21
Selenetane	−7.91	6.47 (trans) 8.70 (cis)						35
Selenane-4,4-d_2		8.46 (trans) 3.09 (cis)						38

a R = alkyl, Ar = aryl.
b The values are for ^1J(Se—^{13}**C**—**H**).
c The values for corresponding O and Te compounds are 140 and 139.7–140.7 Hz, respectively.
d The value is for ^3J(^{77}**Se**—Se—**C**—**H**).

A survey of other compounds with Se(II)-C single bonds is given in Table XVD-3. The chemical shifts of the compounds with selenium as part of a conjugated system do not follow the electronegativity of the heteroatom (*i.e.*, the electronegativity O > S > Se > Te would give the τ-values O < S < Se < Te). In this type of compound the reverse (τ: O > S > Se > Te) is observed, indicating that mesomeric effects may be of some importance.

A relationship between $^3J(CH_3CH_2X)$ and the electronegativity of X has also been deduced;[7d] this is given approximately by equation 4.

$$^3J = 7.9 - 0.7 \, \Delta x \quad (\Delta x = x_X - x_H) \tag{4}$$

Pople and Santry[18a] have shown, on the basis of a LCAO-MO treatment, that the Fermi contact term usually is the dominant one in nuclear spin-spin coupling constants. For comparison of heteronuclear coupling constants, it is convenient to use the reduced coupling constants, K, defined as in equation 5.[18]

$$K_{AB} = \frac{2\pi}{\hbar \gamma_A \gamma_B} J_{AB} \tag{5}$$

A simplified expression for the Fermi contact term for coupling between two directly bonded nuclei, A and B, is given in equation 6, where $s_A{}^2(0)$

$$K_{AB} \propto (\Delta E)^{-1} s_A{}^2(0) s_B{}^2(0) P^2_{s_A s_B} \tag{6}$$

is the density of the valence s orbital at the A nucleus, $P_{s_A s_B}$ is the $s_A - s_B$ bond order, and ΔE is the average excitation energy from the ground state to the triplet states.[18a]

For comparison of $^{13}C-H$ coupling constants of related molecules, it is usually assumed that $^1J(^{13}C-H)$ is proportional to the s-character of the carbon bonding orbital [*i.e.*, that $s_C{}^2(0)$, $s_H{}^2(0)$, and ΔE are constant, and that the s character of the bonding hydrogen orbital is unity]. Values of $J(^{13}C-H)$ in selenols, selenides, and diselenides are given in Table XVD-2. However, it has been shown[19] that ΔE is not constant in a series of substituted methanes, and care must be taken in deducing the amount of s-character of the carbon orbital in the C–H bonds from these data.

Equation 6 predicts that the coupling constants for directly bonded nuclei are always positive, but this is not true when one of the nuclei is a Group V, VI, or VII element.* The natural abundance (7.5%) of ^{77}Se

---

* Pople and Santry[18a] have shown (by MO treatment) that the use of a mean excitation energy in equation 6 is not valid for elements when the valence s atomic orbitals have appreciably lower energy than the p-orbitals. To explain the variation in sign and magnitude of the coupling constants when one of the nuclei is ^{31}P or ^{77}Se, Jameson[23] has used a model which includes, besides the direct Fermi contact term (≥ 0), a term (positive when s-orbitals, and negative when p-orbitals, are used as bonding orbitals) from the polarization of the s-electrons in the core.

Table XVD-3 Proton chemical shifts of miscellaneous compounds with Se(II)-C *single bonds*[a]

| COMPOUND | CHEMICAL SHIFTS (τ) OF INDICATED PROTONS | | | |
	(A)	(B)	(C)	REF.
$(PhSe)_2CH_2$	5.98			24
$(PhSe)_3CH$	4.58			24
	7.80	9.05	9.48	25
$PhCH_2CSeCH_3$ (B) (A)	6.27	7.96		26
$PhC\!\!=\!\!CSeCH_3$	8.07			26
$PhSeCH_2C\!\!\equiv\!\!CH$ (B) (A)	6.58	7.91		27, 28
$C_3H_7SeCH\!\!=\!\!CHC\!\!\equiv\!\!CH$	6.71[b]			29
$CH_2\!\!=\!\!CHC\!\!\equiv\!\!CSeCH_3$	7.74			30
$PhSeCH\!\!=\!\!C\!\!=\!\!CH_2$	4.04			31
	5.72			43
	7.30	8.10		44
	7.34	5.95		45
	6.68	7.35		43
	8.58			46

[a] Most of the chemical shifts are from recorded spectra of dilute CCl_4 solutions. The table is not exhaustive; only one example of each type of compound is included.

[b] Solvent dependent.

$(I = \frac{1}{2})$ is usually sufficient to detect the satellites in a normal proton spectrum. Selenols and selenides, especially methyl and ethyl derivatives, have, therefore, been of interest in investigations[3,4,11,20-23] of the dependence of the sign of spin-spin coupling constants on the position of the nuclei in the periodic table. The sign of $^1J(^{77}Se-Y)$ depends on the nature of Y; that is, $^1J(^{77}Se-H)$ is positive but small, $^1J(^{77}Se-^{13}C)$ is negative, and $^1J(^{77}Se-^{77}Se)$ is reported[3] to have a small, positive value.*

In substituted ethanes, $(CH_3CH_2)_nX$, all $^3K(X-C-C-H)$ values are positive, independently of the position of X in the periodic table, whereas the sign[11,18] of $^2K(X-C-H)$ is dependent on the nature of X. In other words, $^2K(X-C-H)$ is negative for X = Group IV elements (opposite sign of ^1K) and small, negative for Group V elements (P), but positive for X = Groups VI and VII elements (Se, Te, and F).

b. Selenols and Selenides with Selenium-Silicon, Germanium, or Tin Bonds.

In the selenols[32] (H_3XSeH, X = Group IV elements), the τ-values of the SeH proton increase in descending Group IV (τ: C < Si < Ge; cf. Fig. XVD-1), as expected from the electronegativity of X. In the selenides[33,34] $[(CH_3)_3XSeX(CH_3)_3$, X = Group IV elements], the chemical shift sequence for the methyl protons is τ: Ge < Sn < Si (cf. Fig. XVD-1), indicating the following trend in electronegativity: Ge > Sn > Si. The apparent low electronegativity of silicon in comparison to the other Group IV elements is in agreement with the results obtained for $(CH_3)_4X$ (X = C, Si, Ge, Sn). The $^1J(^{13}C-H)$ values increase with increasing atomic number of X. The exchange of selenium with oxygen or sulfur for X = Si, Ge, or Sn yields, in all cases, τ: Se < S < O.

c. Cyclic Selenides

i. Selenetane.[35]

The chemical shift of the α-methylene protons and the coupling constants are given in Fig. XVD-1 and Table XVD-2, respectively. The methylene protons are magnetically nonequivalent; they appear at room temperature with the same chemical shift but have different coupling constants to the other protons in the molecule. The assignment of J_{cis} and J_{trans} is based on the assumption that $J_{cis} > J_{trans}$ in cyclobutane derivatives.

ii. Selenane and Diselenane (cf. Figure XVD-1 and Table XVD-2).

Valence bond calculations[36] have predicted that $^3J(H \cdots H)$, across formally

---

* High-resolution n.m.r. spectroscopy does not give absolute signs; in most cases,[3,20,21] the relative signs have been determined from double resonance experiments, and the absolute signs are deduced under the assumption that $^1J(^{13}C-H) > 0$. The values of $^1J(^{125}Te-H)$ and $^1J(^{125}Te-^{13}C)$ are opposite in sign to the corresponding ^{77}Se-coupling constant, but the reduced coupling constants have the same sign since γ_H, γ_{13C}, $\gamma_{77Se} > 0$ and $\gamma_{125Te} < 0$.

single bonds, is dependent on the H–C–C–H dihedral angle, ϕ (for normal angles and bond distances approximately given by equation 7), dependent

(7) $$^3J = A + B \cos\phi + C \cos 2\phi$$

on the electronegativity of the substituent (given approximately by equation 4), and dependent on the C–C bond length and C-C-H bond angles. where $A \sim 4$ Hz, $B \sim -0.5$ Hz, and $C \sim 4.5$ Hz.

In the chair forms of rapidly inverting cyclohexane derivatives, two different ^3J-values (equations 8 and 9) may be observed.

(8) $$^3J_{trans} = \tfrac{1}{2}(J_{aa} + J_{ee})$$

(9) $$^3J_{cis} = \tfrac{1}{2}(J_{ae} + J_{ea})$$

Lambert[37] assumes that the C–C bond distances and the C-C-H bond angles are invariant in a series of cyclohexane derivatives and that the electronegativity of the heteroatoms has nearly the same effect on J_{cis} and J_{trans} in rapidly inverting systems; he suggests that the ratio R (defined in equation 10) be used as a measure of the deviation from a pure chair

(10) $$R = \frac{J_{trans}}{J_{cis}}$$

conformation. For perfect chair conformations, it is inferred, from experimental data, that $R = 2$. Lambert et al.[38] have found that $R = 2.74$ for selenane-4,4-d_2 (protonation of this ring system yielded a compound with $^+$Se-**H** in an axial position). 1,4-Diselenane gave[37] $R = 3.49$. The values $R > 2$ for these compounds have been explained[37] by reduced C–Se–C bond angles, which push the Se atom further up and out of the plane of C(2), C(3), C(5), and C(6), resulting in a decrease of ϕ_{aa} and ϕ_{ee} and an increase in ϕ_{ae}. Comparison with oxygen, sulfur, and tellurium analogs[37,38] reveals that R increases in the following order: O (close to 2) $<$ S $<$ Se $\leq$ Te, in accordance with C–X–C bond angles [e.g., X = O: 110°, S: 105°, and Se: 98° in $(CH_3)_2X$].[39]

Greens et al.[40] have calculated $\Delta G_c^{\ddagger} = 8.2$ kcal/mole for inversion in 1,3-diselenane from the temperature dependence of the ^1H n.m.r. spectrum (coalescence temperature, T_c, is 178°K).

Lehn and Riddell[41] have found $\Delta G_c^{\ddagger} = 14.4$ kcal/mole ($T_c = 288°K$) for the intramolecular exchange process of 3,7-dimethyl-1,5-diselena-3,7-diazacyclooctane from the temperature dependence of the ^1H n.m.r. spectrum.

Olsson and Almqvist[42] have investigated the ^1H n.m.r. spectra of polyselenaadamantanes. For example, $\tau = 7.36$ and $|^3J(^{77}Se \cdots H)| = 19.1$ Hz have been found for tetramethylhexaselenaadamantane.

Chemical shifts of miscellaneous cyclic selenides are given in Table XVD-3.

3. Compounds Containing Carbon-Selenium Double Bonds

The energy barriers for internal rotation about C–N amide, thioamide, or selenoamide bonds (*e.g.*, **1**) are sufficiently high to allow observation of two different sets of signals for protons in positions *A* and *B* (*i.e.*, *s-cis* and *s-trans* to X, respectively).

$$
\begin{array}{ccc}
\underset{CH_3}{\overset{X}{\diagdown}}C-N\overset{CH_3\ (A)}{\underset{CH_3\ (B)}{\diagup}} & \leftrightarrow & \underset{CH_3}{\overset{X^-}{\diagdown}}C=\overset{+}{N}\overset{CH_3\ (A)}{\underset{CH_3\ (B)}{\diagup}}
\end{array}
$$

1

It has been shown[47] that the rotational barriers are higher in thioamides than in the corresponding amides and, more recently,[48,49] that the corresponding selenoamides have still higher barriers. This is in agreement with the decreasing tendency, $O > S > Se$, to participate in π_p double bonds (*cf.* **1**). It has been inferred[50a–d] that the diamagnetic anisotropy of the thioamide moiety is greater than that of the amide moiety, that the effect on proton shielding constants is greatest for protons *s-cis* to X, and that the net effect on *s-cis-N*-methyl protons is shielding for *N*-methyl amides, but deshielding for *N*-methylthioacetamide and higher thioamides. The data on selenoamides are limited to two investigations,[48,49] but apparently the anisotropy of the magnetic susceptibility of the selenoamide moiety exceeds that of the thioamide moiety by a small amount, as revealed by a greater chemical shift difference between *s-cis* and *s-trans* protons in the selenoamides. (However, great care must be taken in comparisons of this type because of contributions from, *e.g.*, electric fields.) Chemical shifts and thermodynamic parameters for selenoamides, together with chemical shifts for a few other compounds having C–Se double bonds, are listed in Table XVD-4.

4. Miscellaneous Selenium Compounds

Proton chemical shifts of a variety of selenium compounds are given in Fig. XVD-4 and Tables XVD-5 and XVD-6.

a. Selenium(IV) Halides and Their Boron Trihalide Adducts[8,56]

Proton chemical shifts of $(CH_3)_2Se$ and $(CH_3)_{4-n}SeX_n$ (X = F, Cl, Br) and of their BCl_3 adducts are shown, in graphical representation, in Fig. XVD-4. The shifts to lower field with increasing *Cl*-substitution suggest that the dominant effect in this series is inductive (however, the linear relationship is probably fortuitous[8] because of differences in solvation and state of aggregation). Comparison of chemical shifts of $(CH_3)_2SeX_2$, τ: F > Br ~ Cl, indicates

Table XVD-4   Proton chemical shifts of compounds with C–Se *double bonds, and Arrhenius energies of activation for selenoamides*[a]

COMPOUND	CHEMICAL SHIFTS (τ) OF INDICATED PROTONS		SOLVENT[b]	Ea (kcal/mole)	REF.
	(A)	(B)			
Se⎓C—N, CH₃ (A), CH₃, CH₃ (B)	6.67 (6.72)	7.11 (6.97)	$o\text{-}C_6H_4Cl_2$ $(o\text{-}C_6H_4Cl_2)$	26.0 (21.0)	49 (53)
Se⎓C—N, CH₃ (A), Ph, CH₃ (B)	6.57 6.71 (6.70)	7.27 7.36 (7.27)	$o\text{-}C_6H_4Cl_2$ C_6H_5Cl (C_6H_5Cl)	19.5 21.1 (15.4)	49 48 (48)
H₂NCH₃N—C(Se)(SeCH₃)	*s-cis* 6.12 (6.13)	*s-trans* 6.45 (6.37)	CDCl₃ (CDCl₃)		51 (51)
CH₃HNCNCH₃NH₂ (Se) (B)(A)	6.22 (6.35)	6.82 (6.85)	CDCl₃ (CDCl₃)		54 (54)
H(—N), CH₃—N, N—CH₃ (A), (B), Se	6.10 (6.17)	6.30 (6.38)	CDCl₃ (CDCl₃)		52 (52)
[NH₂NCH₃C(Se)Se]₂	6.48 (6.40)		DMSO-d_6 (DMSO-d_6)		55 (55)
Se=C(SeCH₃)(SeCH₃)	7.27		CCl₄		43

[a] The table is not exhaustive. Except for the selenoamides, the data are limited to one example of each compound type. Values in parentheses are for the corresponding sulfur compounds.

[b] The chemical shifts and energy barriers are strongly solvent dependent for these types of compound.

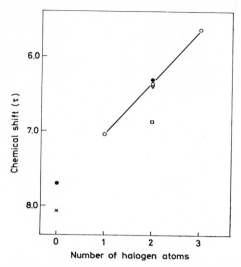

FIG. XVD-4 A plot of chemical shifts (τ-values) of $(CH_3)_2Se$, $(CH_3)_{4-n}SeX_n$, and their BCl_3 adducts *vs.* the number of halogen atoms. x: $(CH_3)_2Se$; o: $X = Cl$; $\triangledown$: $X = Br$; $\square$ $X = F$; $\bullet$: $(CH_3)_2Se\cdot BCl_3$ and $(CH_3)_2SeCl_2\cdot BCl_3$. The data, except for $X = F$, are from the work of Wynne and George.[8] The data for $(CH_3)_2SeF_2$ are from the work of Wynne and Puckett.[56]

Table XVD-5 Proton chemical shifts and coupling constants of miscellaneous selenium compounds[a]

COMPOUND	CHEMICAL SHIFTS (τ) OF INDICATED PROTONS	COUPLING CONSTANTS[b] (Hz) FOR INDICATED NUCLEI	REF.
$[(CH_3)_3Si]_2N-Se-CH_3$	7.72		57
$D_3\overset{+}{N}CH_2CH_2SeO^-$	6.99		58
$FSeO_2OCH_3$	5.60	$^3J(^{77}SeOCH) = 29.2$	59
$PhSeC(\mathbf{H_A})\mathbf{H_B}Ph$ (with O double bond)	(A) 5.80; (B) 6.02	$^2J_{AB} = 11.3$	60
(structure with O=Se, H_X, H_A, H_B, H_Y)	(A) 5.16; (B) 6.24; (X) 8.59; (Y) 7.80;	$^1J_{AB} = 12; ^3J_{AY} = 12;$ $^3J_{AX} = 3.0; ^1J_{XY} = 13;$ $^3J_{BX} = 3.0; ^3J_{BY} = 4.6$	61
(B) $(CH_3)_2PSeCH_3$ (A)	(A) 8.13 (B) 8.66	$^1J(Se^{13}C-H) = 140.0$ $^2J(^{77}Se\cdots H_A) = 10.7$ $^3J(^{77}Se\cdots H_B) = 8.1$	62

Table XVD-5   (*continued*)

COMPOUND	CHEMICAL SHIFTS (τ) OF INDICATED PROTONS	COUPLING CONSTANTS[b] (Hz) FOR INDICATED NUCLEI	REF.
(B) $(CH_3)_2P(S)SeCH_3$ (A)	(A) 7.85 (B) 7.98	$^1J(Se^{13}C—H) = 145.0$ $^2J(^{77}Se \cdots H_A) = 10.8$ $^3J(^{77}Se \cdots H_B) = 3.9$	62
$\overset{\text{Se}}{\overset{\|}{PhP}}(OCH_3)_2$	6.40	$^3J(POCH) = 14.0$	63
$\overset{\text{O}}{\overset{\|}{PhP}}(OCH_3)(SeCH_3)$ (A) (B)	(A) 6.29; (B) 8.07	$^3J(POCH) = 12.1$ $^3J(PSeCH) = 13.1$	63
$(CH_3O)_3P{=}Se$	6.50	$^3J(HCOP) = 14$	64
$[(CH_3CH_2O)_2\overset{\text{Se}}{\overset{\|}{P}}]_2Se_3$	5.88	$^3J(POCH) = 10.5$	65
$(A)H_2$ (S, Se, P, S, CH_3(B))	(A) 7.62; (B) 8.13	$^2J(PCH) = 12.5$	66
$[(CH_3)_2N]_3P(Se)CH_3I$ (A) (B)	(A) 7.08; (B) 7.71	$^3J(PNCH) = 11.4$ $J(P \cdots H_B) = 12.3$	67
NCSe, H, SeCN	2.33		68
SeCN, H, N, H (indole)	2.31		69
H, NCSe, N, H (indoline)	3.47		70

[a] The table is not exhaustive; it contains only examples.
[b] Numerical values.

Table XVD-6 Proton chemical shifts, coupling constants, and thermodynamic parameters for metal complexes with selenium-containing ligands (Values of corresponding sulfur compounds are given in parentheses.)

COMPOUND	CHEMICAL SHIFTS (τ)		COUPLING CONSTANTS (Hz)	SOLVENT	T_c (°K)	$\Delta G_c^{\neq}$ (kcal/mole)	REF.
	(A)	(B)					
$(CH_3Se)_2PtCl_2$	cis: 5.51 trans: 5.31	5.94 5.92	cis: $^2J(^{77}Se\cdots H) = 10.0$; $^3J(^{195}Pt\cdots H) = 41.8(49.5)$; trans: $^2J(^{77}Se\cdots H) = 9.5$; $^3J(^{195}Pt\cdots H) = 36.6(41.0)$	CH_2Cl_2 CH_2Cl_2			73 (71) 73 (71)
$(PhCSe)_2PtCl_2$ H (A) H (B)	5.35ª	5.92	$^2J_{AB} = -11.3(-13.0)$; $^3J(^{195}Pt\cdots H) \sim 38(43.0)$ $^2J_{AB} = -11.7$; $^3J(^{195}Pt\cdots H) = 32.2(37.0)$	Nitrobenzene-d_5 (CDCl₃) Nitrobenzene-d_5 (CDCl₃)	>373 (307.5) 364 (<300)	>19 (16.6) 18.0 (<15)	72 (71) 72 (71)
$(PhCSe)_2PdCl_2$ H (A) H (B)	5.42 (5.69)	6.01 (6.08)	$^2J_{AB} = -12.0$ $^2J_{AB} = -12.0(-12.9)$	Nitrobenzene-d_5 CDCl₃ (CDCl₃)	328.5 321.9 (233)	16.0 15.9 (11.4)	72 72 (72)
$C_2H_4(SeCH_3)_2Re(CO)_3Cl$ (A) (B)	7.45	7.65		Benzene			74

ª Only one isomer.

that mesomeric effects are of some importance, especially in the fluorine compound. The larger shift towards lower field for $(CH_3)_2Se$ than for $(CH_3)_2SeCl_2$ by BCl_3 adduct formation has been explained[8] as due to the fact that $(CH_3)_2Se \cdot BCl_3$ is essentially undissociated in solution, and $(CH_3)_2SeCl_2 \cdot BCl_3$ essentially dissociated.

b. Selenoxides. Oki and Iwamura[60] have found that the methylene protons of benzyl phenyl selenoxide are magnetically nonequivalent even at 100° when water is excluded, indicating that the Se atom is an asymmetric center and that water facilitates the interconversion. Chemical shifts and coupling constants for the magnetically nonequivalent methylene protons are given in Table XVD-5.

c. Cyclic Selenites. Samitov[61] has analyzed the ${}^1H$ n.m.r. spectra of propane-1,3-diol selenite and 2,2-dimethylpropane-1,3-diol selenite and has found that the axial proton, in a methylene group, is less shielded than the equatorial proton (contrary to the case in many other cyclohexane derivatives where the equatorial protons are less shielded). It has been suggested[61] that the axial protons on $C(1)$ and $C(3)$ are less shielded because the anisotropic $Se{=}O$ group is in an axial position. Chemical shifts and coupling constants for propane-1,3-diol selenite are given in Table XVD-5.

d. Metal Complexes with Selenium-Containing Ligands. Nuclear magnetic resonance studies[71] indicate that the S atom in sulfide-platinum and sulfide-palladium complexes $[(RS)_2MX_2$: R = alkyl; M = Pt, Pd; X = halogen] is pyramidal. A similar temperature dependence of the ${}^1H$ n.m.r. spectra of selenide-platinum and selenide-palladium complexes (and the tellurium analogs) has been observed.[72]

The signals from the methylene protons in bis(dibenzylselenide)dichloroplatinum(II)[72] form an AB system (66%, corresponding to the abundance of platinum isotopes with $I = 0$) and the AB part of an ABX system (34%, corresponding to the abundance of ${}^{195}Pt$, $I = \frac{1}{2}$; only a few lines of the ABX system corresponding to the 7.5% ${}^{77}Se$ could be detected). It has been found[71-73] that $3J({}^{195}\mathbf{PtSeCH})$ is positive, greater for *cis* than for *trans* complexes, and greater for sulfur than for selenium compounds. Furthermore, *cis* complexes have higher energy barriers[71,72] to inversion than *trans* complexes, and the barrier is higher at selenium than at sulfur. Proton chemical shifts, coupling constants, and $\Delta G_c^{\neq}$ values for inversion of selenide-platinum and selenide-palladium complexes are given in Table XVD-6.

5. Heteroaromatic Selenium Compounds

In most Group VI aromatic heterocycles, the chemical shifts of the ring protons are found in the order τ: Se < S < O, opposite to that expected from the electronegativity of these elements. It may be inferred that the dominant effects in heteroaromatic compounds apparently are mesomeric effects and diamagnetic anisotropy. A comparison of the chemical shifts of Group VI heteroaromatic compounds is shown in Fig. XVD-5. The chemical shifts and coupling constants given in Tables XVD-7 to XVD-9 are, in most cases, refined parameters obtained by comparison of experimental spectra with computer-simulated ones. All of the H $\cdots$ H coupling constants listed in these tables are assumed to have the same sign (positive) on the basis of comparison with calculated spectra.

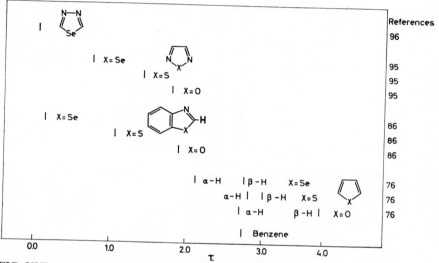

FIG. XVD-5 Proton chemical shifts (τ-values) of hetero-aromatic compounds containing Group VI elements. The chemical shift of benzene has been included for comparison.

a. Selenophenes. Proton chemical shifts and H $\cdots$ H coupling constants are given in Table XVD-7 [$^2J(^{77}Se \cdots H_2) = 48$ Hz and $^3J(^{77}Se \cdots H_3) = 9.5$ Hz].

In the series furan, thiophene, and selenophene (**2**), the chemical shifts of the β-protons are found in the order τ: Se < S < O (*cf.* Fig. XVD-5). This may be rationalized by a decrease, O > S > Se, in the contribution of structure **3** in connection with an increase in the contribution of **5** and an increase, O < S < Se, in the deshielding effect of the diamagnetic anisotropy of the ring systems.

(X = O, S, Se)

2   3   4   5

Table XVD-7 Proton chemical shifts and coupling constants of selenophenes

SUBSTITUENTS	SOLVENT	CHEMICAL SHIFTS[a] (τ)				PROTON COUPLING CONSTANTS[b] (Hz)				REF.
		H_2	H_3	H_4	H_5	$J(H_2 \cdots H_5)$	$J(H_3 \cdots H_4)$	$J(H_3 \cdots H_5)$	$J(H_4 \cdots H_5)$	
—H	None	2.30[c]	2.88	2.88	2.30	2.47	3.56	1.05	5.35	75
	TMS[d]	2.12	2.77	2.77	2.12	2.34	3.74	1.46	5.40	76
2-Cl	TMS[d]		3.06	3.07	2.40		3.87	1.42	6.04	76
2-Br	TMS[d]		3.04	3.10	2.46		3.88	1.34	6.05	76
2,5-diCl	TMS[d]		3.30	3.30			4.26			76
2,5-diBr	TMS[d]		3.13	3.13			4.17			76
			3.05	3.05						78[e]
2,5-diI	Acetone		2.73	2.73						75
2,5-diCHO			1.9	1.9						79
2,5-diCN			2.15	2.15						78[e]
2-CH$_3$,5-Ph	CCl$_4$		3.15	2.84						80

Compound					Ref.
2,3-diBr	2.9	2.12		6	77, 78
2,3-diCHO	2.14	1.60		5	78
2,3-diCN	2.42	1.54		5	77, 78
2,4-diBr	2.80	2.24	2		77, 78
2,4-diCHO	1.54	0.69	1		77, 78
2,4-diCN	1.95	1.04	1		77, 78
3,4-diBr	2.32	2.32			78[e]
3,4-diCHO	0.99	0.99			77, 78
3,4-diCN	1.27	1.27			78[e]
2,3,4-triCHO		0.6			78, 81

[a] The table is not exhaustive; it lists only the chemical shifts of the ring protons. The signals from the formyl protons are found at $\tau \sim 0$.

[b] The table lists only the coupling constants for the ring protons. The value of $J[\mathrm{H_5 \cdots C(2)CHO}]$ is 1 Hz. For comparison, the values of $J(\mathrm{H \cdots H})$ for the thiophene are as follows: $J_{25} = 2.84$, $J_{34} = 3.50$, $J_{35} = 1.04$, and $J_{45} = 4.90$.[76]

[c] The chemical shift difference between α- and β-protons is solvent dependent: 47.7, 39.8, 44.1, and 52.9 Hz (at 60 MHz) in $(CH_3)_2CO$, CCl_4, CH_3CN, and DMSO, respectively.[75]

[d] Extrapolated to infinite dilution.

[e] In Morel et al.[77] the values for 2,5-substituted compounds are interchanged with those of 3,4-substituted compounds.

Table XVD-8 Proton chemical shifts and coupling constants for the ring protons in benzoselenazoles

SUBSTITUENT	SOLVENT	CHEMICAL SHIFT (τ) OF PROTONS ON					$J(\text{H} \cdots \text{H})$ COUPLING CONSTANTS (Hz)						$^1J(^{13}\text{C—H})$ FOR		REF.
		C(2)[a]	C(4)	C(5)	C(6)	C(7)	J_{45}	J_{46}	J_{47}	J_{56}	J_{57}	J_{67}	2-H	2-CH$_3$	
-H	CCl$_4$	0.14											214.0[b]		86
	CF$_3$COOH	-1.11											216.0		86
2-CH$_3$	Acetone[c]	7.18	2.08	2.55	2.71	1.98	8.12	1.32	0.53	7.28	1.26	7.98			87
	Cyclo-hexane[c]	7.27	2.13	2.72	2.85	2.29									87
	CCl$_4$	7.18												130.5	86, 88
	CF$_3$COOH	6.73												134.8	86
2-CH$_3$-6-NH$_2$	CCl$_4$	7.28														88[d]
2-CH$_3$-6-NO$_2$	CCl$_4$	7.12														88
2-CH$_3$-5-Cl	CCl$_4$	7.20														88

[a] The values are for the methyl protons when the compound contains a methyl group in the 2-position.
[b] The value is for the neat substance. The values for the corresponding O and S compounds are 213.5 Hz and 211.0 Hz, respectively.[86] The value of $^3J(^{77}\text{SeCCH}_3)$ is 7.7 Hz.[87]
[c] Extrapolated to infinite dilution.
[d] The table includes only selected examples from this work.

The chemical shifts of the α-protons of **2** are found in the order τ: Se < O < S. The inductive effect may be of some importance for the α-protons, that is, responsible for τ: O < S. The greater chemical shift difference between α- and β-protons for selenophene than for thiophene (cf. Fig. XVD-5) may be caused by an increase, S < Se, in the deshielding of the α-protons from paramagnetic contributions of the heteroatom. The chemical shifts of the ring protons in selenophenes are sensitive to the electronic properties of the substituents (cf. Table XVD-7), but, as in thiophene, no linear correlation with Hammett's σ-values exists (a correlation between chemical shifts and σ-values is more fruitful for benzenes).

The coupling constants are of similar magnitude in selenophene and thiophene. Heffernan and Humffray[75] have concluded, on the basis of a comparison of chemical shifts and H $\cdots$ H coupling constants in thiophene and selenophene, that selenophene, like thiophene, is probably a planar molecule (this opinion has since been supported by a microwave investigation[82] of selenophene).

The ^{1}H n.m.r. spectra[83,84] of β-diketones containing the selenophene ring reveal, for α-unsubstituted compounds, that the enolic forms are favored in the presence of the selenophene moiety. ^{1}H n.m.r. spectroscopy has, furthermore, revealed that 3-hydroxybenzo[b]selenophene[85] exists exclusively in the ketonic form (contrary to the corresponding sulfur compound).

b. Benzoselenazoles. Chemical shifts and coupling constants for benzoselenazoles are listed in Table XVD-8.

In the benzazole series, the chemical shift of the proton on C(2) is found in the order τ: Se < S < O (cf. Fig. XVD-5). The same order has been found[87] for the chemical shifts of the benzoid ring protons, with the effect of the heteroatoms decreasing in the order $H_7 > H_4 > H_6 > H_5$, which may indicate transmission of electronic effects through the N atom. Di Modica et al.[88] have correlated the chemical shifts of the 2-methyl protons of substituted 2-methyl benzoselenazoles with Hammett's substituent constant, σ. The best linear correlation was obtained when σ_m was used for substituents in the 5-position and σ_p for substituents in the 6-position; this offers further indication of the transmission of electronic effects through the N atom.

c. Benzo-2,1,3-selenadiazoles. Extensive investigations of the chemical reactivity of benzo-2,1,3-selenadiazoles, supported by an X-ray investigation, indicate that these compounds have basically an ortho-quinoid structure (cf. Brown and Bladon[89] and references cited therein). This theory is in agreement with the magnitude of the coupling constants (cf. Table XVD-9) observed for benzo-2,1,3-selenadiazoles. The values of $J(H_4 \cdots H_5)$ and $J(H_6 \cdots H_7)$ are greater than the value of $J(H_5 \cdots H_6)$, indicating that the C_4-C_5 and C_6-C_7 bonds are shorter than the C_5-C_6 bond. Brown and

Table XVD-9 *Proton chemical shifts and coupling constants of benzo-2,1,3-selenadiazoles*[a]

SUBSTITUENT	SOLVENT	CHEMICAL SHIFT (τ) OF PROTONS ON				COUPLING CONSTANTS (Hz)						REF.
		C(4)	C(5)	C(6)	C(7)	$J(H_4 \cdots H_5)$	$J(H_4 \cdots H_6)$	$J(H_4 \cdots H_7)$	$J(H_5 \cdots H_6)$	$J(H_5 \cdots H_7)$	$J(H_6 \cdots H_7)$	
—H	CDCl₃	2.20[b]	2.64	2.64	2.20	9.09[c]	1.20	0.94	6.51	1.20	9.09	89
	Acetone	2.16	2.50	2.50	2.16	9.10	1.18	0.92	6.40	1.18	9.10	91
	Cyclohexane	2.30	2.76	2.76	2.30							91
5-CH₃[d]	CDCl₃	2.42		2.80	2.36		1.54	1.02			9.40	89
5-CH₃O	CDCl₃	3.05		2.88	2.40		2.20	0.04			10.47	89
5-(CH₃)₂N	CDCl₃	3.31		2.67	2.40		2.65	0.33			9.82	89
5-Cl	CDCl₃	2.20		2.64	2.29		2.05	0.84			9.36	89
5,6-diCH₃	CDCl₃	2.55	2.55									89
5,6-diCl	CDCl₃	2.02	2.02									89

[a] Chemical shifts of benzo-2,1,3-selenadiazoles have also been discussed by Fedin and Todres.[92]

[b] The values for the corresponding O and S compounds[89] are as follows: O: $\tau_4 = \tau_7 = 2.18$, $\tau_5 = \tau_6 = 2.56$; S: $\tau_4 = \tau_7 = 2.11$, $\tau_5 = \tau_6 = 2.57$.

[c] The values (Hz) for the corresponding O and S compounds[89] are as follows: O: $J_{45} = J_{67} = 9.21$, $J_{46} = J_{57} = 0.99$, $J_{47} = 1.23$, $J_{56} = 6.61$; S: $J_{45} = J_{67} = 8.95$, $J_{46} = J_{57} = 1.35$, $J_{47} = 0.76$, $J_{56} = 6.63$.

[d] The numerical value of $J(H_4 \cdots CH_3)$ is 1.1 Hz.

Table XVD-10 Proton chemical shifts and coupling constants for miscellaneous selena-heteroaromatic compounds

COMPOUND	SOLVENT	CHEMICAL SHIFTS (τ) OF INDICATED PROTONS			COUPLING CONSTANTS (Hz)	REF.
		(A)	(B)	(C)		
	$CH_3CN/$ $HClO_4$	-1.08	1.17	1.09	$J_{23} = 11.3; J_{24} = 2.1$ $J_{26} = 2.5; J_{34} = 9.1$	93
	None CCl$_4$	0.6 0.76	7.47 7.40		$^1J(^{13}C-H) = 187.3$ $^3J(^{77}Se\cdots H) = 27.9$	94 95
	None CCl$_4$	0.96 1.08	7.47 7.40		$^1J(^{13}C-H_A) = 184.3; {}^1J(^{13}C-H_B) = 129.0$ $^3J(^{77}Se\cdots H_A) = 28.8$	94 95
	CCl$_4$	7.50				95
	None CCl$_4$	-0.89 0.04			$^1J(^{13}C-H) = 214.3; {}^2J(^{77}Se\cdots H) = 55.3$	96

Table XVD-10 (*Continued*)

COMPOUNDS	SOLVENT	CHEMICAL SHIFTS (τ) OF INDICATED PROTONS			COUPLING CONSTANTS (Hz)	REF.
		(A)	(B)	(C)		
CH$_3$NH—structure **(B)**	DMSO-d_6	0.73	7.10[a]		$^2J(^{77}\mathrm{Se}\cdots\mathrm{H_A}) = 57.0$	97, 98[b]
	CDCl$_3$	0.83	6.87		$^2J(^{77}\mathrm{Se}\cdots\mathrm{H_A}) = 58.5$	
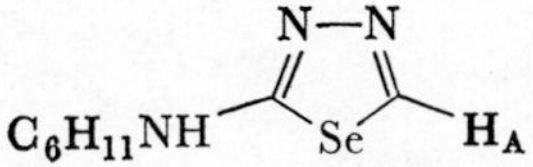 **C$_6$H$_{11}$NH**—structure	DMSO-d_6	0.8			$^2J(^{77}\mathrm{Se}\cdots\mathrm{H_A}) = 56.0$ $^1J(^{13}\mathrm{C}{-}\mathrm{H_A}) = 213.11$	98
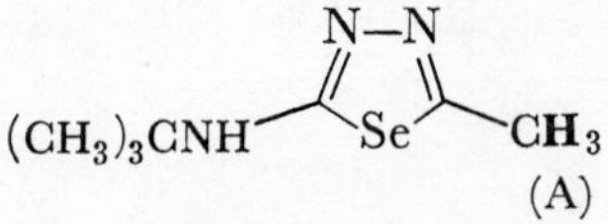 (CH$_3$)$_3$CNH—structure (A)	DMSO-d_6	7.52			$^3J(^{77}\mathrm{Se}\cdots\mathrm{H_A}) = 7.0$ $^1J(^{13}\mathrm{C}{-}\mathrm{H_A}) = 129.56$	98
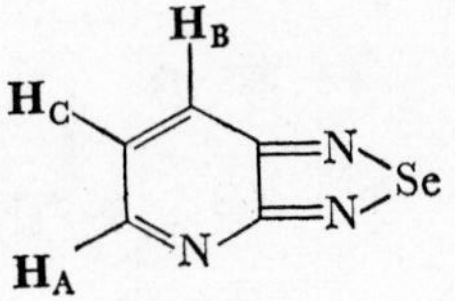 structure (H$_B$, H$_C$, H$_A$)	CDCl$_3$	0.87	1.77	2.55	$J_{BC} = 9.11;\ J_{AB} = 1.82$ $J_{AC} = 3.35$	99
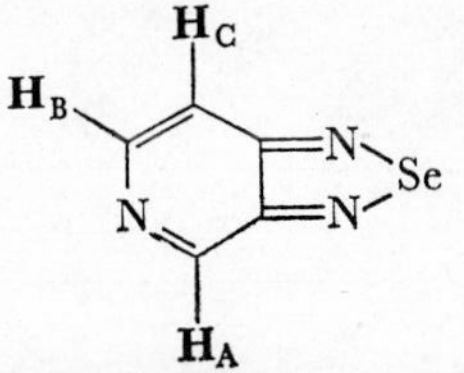 structure (H$_C$, H$_B$, H$_A$)	CDCl$_3$	0.53	1.52	2.23	$J_{AC} = 1.19;\ J_{AB} = 0.02$ $J_{BC} = 6.66$	99

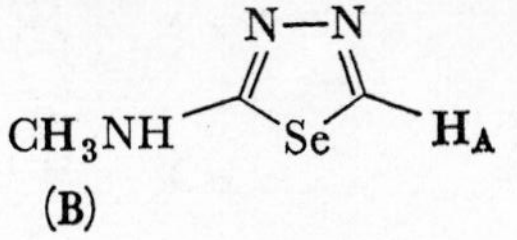

[a] In DMSO-d_6, the signal is observed as a doublet with spacing 4.6 Hz, because of coupling with the N—H proton (temperature *ca.* 40°). In CDCl$_3$, at the same temperature, the signal is observed as a singlet (too fast exchange of the N—H proton).

[b] Only examples of the investigated 2-amino-1,3,4-selenadiazoles are given in the table.

Bladon[89] have, on the basis of the $H \cdots H$ coupling constants, estimated the π-bond order P, of the C-C bonds from equation 11[90] and have found

$$(11) \qquad J(H_n \cdots H_m) = 12.7P(C_n - C_m) - 1.1$$

that $P_{45} = P_{67}$ is greater than P_{56}.

Proton chemical shifts of benzo-2,1,3-selenadiazoles are given in Table XVD-9. In the series benzo-2,1,3-oxa-, benzo-2,1,3-thia-, and benzo-2,1,3-selenadiazole, the chemical shifts of the ring protons are found in the order τ: S < O < Se; that is, it appears that neither mesomeric nor inductive effects are dominant (although the influence of the heteroatoms is small in this series). Structural changes, as well as changes in magnetic anisotropy, may have some responsibility for the trends observed in this series.[91,92]

Proton chemical shifts and coupling constants for miscellaneous hetero-aromatic selenium compounds are given in Table XVD-10. The chemical shifts of the selenapyrylium salt,[93] the 1,2,5-selenadiazoles,[94,95] the 2-amino-1,3,4-selenadiazoles,[98] and their corresponding oxygen and sulfur analogs reveal, in all cases, the order τ: Se < S < O, and it may be concluded that mesomeric effects are more important than inductive ones in hetero-aromatic selenium compounds.

REFERENCES

1. J.-P. Mila and J.-P. Laurent, *Bull. Soc. Chim. Fr.*, 2735 (1967).

2. A. Merijanian, R. A. Zingaro, L. S. Sagan, and K. J. Irgolic, *Spectrochim. Acta*, **25A,** 1160 (1969).

3. W. McFarlane, *J. Chem. Soc., A*, 670 (1969).

4. W. McFarlane, *Chem. Commun.*, 963 (1967).

5. I. Lalezari and N. Sharghi, *Spectrochim. Acta*, **23A,** 1948 (1967).

6. C. G. Pierpont, B. J. Corden, and R. Eisenberg, *Chem. Commun.*, 401 (1969).

7. J. W. Emsley, J. Feeney, and L. H. Sutcliffe, "High Resolution Nuclear Magnetic Resonance Spectroscopy, "Vols. 1 and 2, Pergamon Press, New York, N.Y., 1965 and 1966: (a) p. 263; (b) p. 666; (c) p. 130; (d) p. 678; (e) p. 1011; (f) p. 685.

8. K. J. Wynne and J. W. George, *J. Amer. Chem. Soc.*, **91,** 1649 (1969).

9. C. R. McCoy and A. L. Allred, *J. Inorg. Nucl. Chem.*, **25,** 1219 (1963).

10. P. T. Narasimhan and M. T. Rogers, *J. Amer. Chem. Soc.*, **82,** 5983 (1960).

11. V. Breuninger, H. Dreeskamp, and G. Pfisterer, *Ber. Bunsenges. Phys. Chem.*, **70,** 613 (1966).

12. B. P. Dailey and J. N. Shoolery, *J. Amer. Chem. Soc.*, **77,** 3977 (1965).

13. J. C. Müller, *Bull. Soc. Chim. Fr.*, 1815 (1964).

14. F. G. Gabdrakhmanov, Yu. Yu. Samitov, and E. G. Kataev, *Zh. Obshch. Khim.*, **37,** 761 (1967).

15. W. McFarlane and J. A. Nash, *Chem. Commun.*, 524 (1969).

16. D. Grant and J. R. van Wazer, *J. Amer. Chem. Soc.*, **86,** 3012 (1964).

17. H. Kessler and W. Rundel, *Chem. Ber.*, **101,** 3350 (1968).

18a. J. A. Pople and D. P. Santry, *Mol. Phys.*, **8**, 1 (1963).

18b. G. E. Maciel, J. W. McIver, Jr., N. S. Ostlund, and J. A. Pople, *J. Amer. Chem. Soc.*, **92**, 1 (1970).

19. N. Cyr and T. J. R. Cyr, *J. Chem. Phys.*, **47**, 3082 (1967).

20. H. Dreeskamp and G. Pfisterer, *Mol. Phys.*, **14**, 295 (1968).

21. W. McFarlane, *Mol. Phys.*, **12**, 243 (1967).

22. G. Klose, *Ann. Phys.*, **8**, 220 (1961).

23. C. J. Jameson, *J. Amer. Chem. Soc.*, **91**, 6232 (1969).

24. D. Seebach and N. Peleties, *Angew. Chem.*, **81**, 465 (1969).

25. U. Schöllkopf and H. Küppers, *Tetrahedron Lett.*, 105 (1963).

26. Yu. A. Boiko, B. S. Kupin, and A. A. Petrov, *Zh. Org. Khim.*, **4**, 1355 (1968).

27. M.-P. Simonnin, *Bull. Soc. Chim. Fr.*, 1774 (1966).

28. M.-P. Simonnin, *C.R. Acad. Sci.*, *Paris*, **257**, 1075 (1963).

29. N. V. Elsakov and A. A. Petrov, *Opt. Spectry.* (USSR), **16**, 434 (1964).

30. A. A. Petrov, S. I. Radchenko, K. S. Mingaleva, I. G. Savich, and V. B. Lebedev, *Zh. Obshch. Khim.*, **34**, 1899 (1964).

31. M.-P. Simonnin and G. Pourcelot, *C.R. Acad. Sci.*, *Paris*, **262**, 1279 (1966).

32. C. Glidewell, D. W. H. Rankin, and G. M. Sheldrick, *Trans. Faraday Soc.*, **65**, 1409 (1969).

33. H. Schmidbaur, *J. Amer. Chem. Soc.*, **85**, 2336 (1963).

34. H. Schmidbaur and I. Ruidisch, *Inorg. Chem.*, **3**, 599 (1964).

35. W. B. Moniz, *J. Phys. Chem.*, **73**, 1124 (1969).

36. M. Karplus, *J. Amer. Chem. Soc.*, **85**, 2870 (1963), and references cited therein.

37. J. B. Lambert, *J. Amer. Chem. Soc.*, **89**, 1836 (1967).

38. J. B. Lambert, R. G. Keske, and D. K. Weary, *J. Amer. Chem. Soc.*, **89**, 5921 (1967).

39. R. J. Gillespie, *J. Amer. Chem. Soc.*, **82**, 5978 (1960).

40. A. Geens, G. Swallens, and M. Anteunis, *Chem. Commun.*, 439 (1969).

41. J. M. Lehn and F. G. Riddell, *Chem. Commun.*, 439 (1969).

42. K. Olsson and S.-O. Almqvist, *Acta Chem. Scand.*, **23**, 3271 (1969).

43. L. Henriksen, *Acta Chem. Scand.*, **21**, 1981 (1967).

44. P. Mazerolles, J. Dubac, and M. Lebre, *J. Organometal. Chem.*, **12**, 143 (1968).

45. K. L. Baker and G. W. A. Fowles, *J. Chem. Soc.*, *A*, 801 (1968).

46. M. Wieber and G. Schwarzmann, *Monatsh. Chem.*, **100**, 74 (1969).

47. R. C. Neuman, Jr., D. N. Roak, and V. Jonas, *J. Amer. Chem. Soc.*, **89**, 3412 (1967), and references cited therein.

48. G. Schwenker and H. Rosswag, *Tetrahedron Lett.*, 4237 (1967).

49. K. A. Jensen and J. Sandström, *Acta Chem. Scand.*, **23**, 1911 (1969).

50a. P. L. Southwick, J. A. Fitzgerald, and G. E. Milliman, *Tetrahedron Lett.*, 1247 (1965).

50b. H. Paulsen and K. Todt, *Chem. Ber.*, **100**, 3385 (1967).

50c. W. Walter and G. Maerten, *Justus Liebigs Ann. Chem.*, **712**, 58 (1968).

50d. W. Walter, E. Schaumann, and H. Paulsen, *Justus Liebigs Ann. Chem.*, **727**, 61 (1969).

51. B. M. Dahl, *Acta Chem. Scand.*, to be published.

52. U. Svanholm, *Acta Chem. Scand.*, in press.

53. J. Sandström, *J. Phys. Chem.*, **71**, 2318 (1967).

54. K. A. Jensen, O. Buchardt, N. G. Carlsen, M. G. Ettlinger, and U. Svanholm, *Acta Chem. Scand.*, to be published.

55. U. Anthoni, B. M. Dahl, C. Larsen, and P. H. Nielsen, *Acta Chem. Scand.*, **24**, 959 (1970).

56. K. J. Wynne and J. Puckett, *Chem. Commun.*, 1532 (1968).

57. O. J. Scherer and J. Wokulat, *Z. Anorg. Allg. Chem.*, **357**, 92 (1968).

58. D. L. Klayman and J. W. Lown, *J. Org. Chem.*, **31**, 3396 (1966).

59. R. Paetzold, R. Kurze, and G. Engelhardt, *Z. Anorg. Allg. Chem.*, **353**, 62 (1967).

60. M. Oki and H. Iwamura, *Tetrahedron Lett.*, 2917 (1966).

61. Yu. Yu. Samitov, *Dokl. Akad. Nauk SSSR*, **164**, 347 (1965).

62. W. McFarlane and J. A. Nash, *Chem. Commun.*, 913 (1969).

63. G. Mavel, R. Mankowski-Favelier, and N. T. Thuong, *J. Chim. Phys.*, **64**, 1692 (1967).

64. L. Élégant, J.-F. Gal, and M. Azzaro, *Bull. Soc. Chim. Fr.*, 4273 (1969).

65. M. V. Kudchadker, R. A. Zingaro, and K. J. Irgolic, *Can. J. Chem.*, **46**, 1415 (1968).

66. M. Wieber and H. U. Werther, *Monatsh. Chem.*, **99**, 1153 (1968).

67. I. A. Nuretdinov, N. A. Buina, and N. P. Grechkin, *Izv. Akad. Nauk SSSR, Ser. Khim.* 169 (1969).

68. L.-B. Agenäs and S.-O. Almqvist, *Ark. Kemi*, **30**, 433 (1969).

69. L.-B. Agenäs, *Ark. Kemi*, **23**, 154 (1964).

70. L.-B. Agenäs, *Ark. Kemi*, **31**, 159 (1969).

71. P. Haake and P. C. Turley, *J. Amer. Chem. Soc.*, **89**, 4611, 4617 (1967), and references cited therein.

72. K. A. Jensen, P. Lindgreen, and U. Svanholm, *Acta Chem. Scand.*, to be published.

73. W. McFarlane, *Chem. Commun.*, 755 (1968).

74. E. W. Abel and G. V. Hutson, *J. Inorg. Nucl. Chem.*, **31**, 3333 (1969).

75. M. L. Heffernan and A. A. Humffray, *Mol. Phys.*, **7**, 527 (1964).

76. J. M. Read, Jr., C. T. Mathis, and J. H. Goldstein, *Spectrochim. Acta*, **21**, 85 (1965).

77. J. Morel, C. Paulmier, and P. Pastour, *C.R. Acad. Sci., Paris*, **266**, 1300 (1968).

78. C. Paulmier, J. Morel, P. Pastour, and D. Semard, *Bull. Soc. Chim. Fr.*, 2511 (1969).

79. C. Paulmier and P. Pastour, *Bull. Soc. Chim. Fr.*, 4021 (1966).

80. K. E. Schulte, J. U. Reisch, and D. Bergenthal, *Chem. Ber.*, **101**, 1540 (1968).

81. J. Morel, C. Paulmier, and P. Pastour, *C.R. Acad. Sci., Paris*, **267**, 1842 (1968).

82. N. M. Pozdeev, O. B. Akulinin, A. A. Shapkin, and N. N. Magdesieva, *Dokl. Akad. Nauk SSSR*, **185**, 384 (1969).

83. N. N. Magdesieva, V. V. Titov, V. F. Bystrov, V. P. Lezina, and Yu. K. Yur'ev, *Zh. Strukt. Khim.*, **6**, 402 (1965).

84. Yu. K. Yur'ev, N. N. Magdesieva, and A. T. Monakhova, *Khim. Geterotsikl. Soedin.*, 650 (1968).

85. G. Grandolini, A. Ricci, N. P. Buu-Hoi, and F. Périn, *J. Heterocycl. Chem.*, **5**, 133 (1968)

86. E. Barni, G. Di Modica, and A. Gasco, *J. Heterocycl. Chem.*, **4**, 139 (1967).

87. F. L. Tobiason and J. H. Goldstein, *Spectrochim. Acta*, **23A**, 1385 (1967).

88. G. Di Modica, E. Barni, and A. Gasco, *J. Heterocycl. Chem.*, **2**, 457 (1965).

89. N. M. D. Brown and P. Bladon, *Spectrochim. Acta*, **24A**, 1869 (1968).

90. N. Jonathan, S. Gordon, and B. P. Dailey, *J. Chem. Phys.*, **36,** 2443 (1962).

91. F. L. Tobiason and J. H. Goldstein, *Spectrochim. Acta*, **25A,** 1027 (1969).

92. E. I. Fedin and Z. V. Todres, *Khim. Geterotsikl. Soedin.*, 416 (1968).

93. I. Degani, F. Taddei, and C. Vincenzi, *Boll. Sci. Fac. Chim. Ind. Bologna*, **25,** 61 (1967).

94. P. Bucci, V. Bertini, G. Ceccarelli, and A. de Munno, *Chem. Phys. Lett.*, **1,** 473 (1967).

95. V. Bertini, *Gazz. Chim. Ital.*, **97,** 1870 (1967).

96. R. V. Kendall and R. A. Olofson, *J. Org. Chem.*, **35,** 806 (1970).

97. K. A. Jensen, *Z. Chem.*, **9,** 121 (1969).

98. U. Svanholm, *Acta Chem. Scand.*, **26,** 459 (1972).

99. N. M. D. Brown and P. Bladon, *Tetrahedron*, **24,** 6577 (1968).

XV Physicochemical investigations of selenium compounds

E. SELENIUM-77 NUCLEAR MAGNETIC RESONANCE

MARCEL A. LARDON

Xerox Corporation
Chemistry Research Laboratory
Webster, New York

1. Introduction

Selenium contains several naturally occurring isotopes, one of which is suitable for n.m.r. experiments. ^{77}Se possesses a nuclear spin of $\frac{1}{2}$, and the magnetic moment associated with this spin has been determined by Dharmatti and Weaver[1] to be $\mu = +0.53326 \pm 0.00005$. This isotope occurs in low natural abundance (7.5 %) and has a low sensitivity relative to that of protons (0.693 %). The combination of these two effects causes the detectability of this nucleus to be 5.2×10^{-4} relative to that of the proton and accounts for the difficulties in studying this nucleus by magnetic resonance techniques. The literature on n.m.r. studies of ^{77}Se is sparse.

Two different approaches to the determination of ^{77}Se n.m.r. spectra are reported. Birchall and his co-workers[2] and Lardon[3] used the single-sideband technique, whereas McFarlane[4] took advantage of the high sensitivity of the proton and performed a heteronuclear double resonance experiment to measure ^{77}Se chemical shifts and ^{77}Se-^{77}Se coupling constants.

2. ^{77}Se Chemical Shift

a.   General Considerations.   The chemical shift (δ) is generally defined as the difference between the shielding coefficients of the sample (σ) and of reference (σ_R) compounds (equation 1). Quantum-mechanical expressions

$$(1) \qquad\qquad \delta = \sigma - \sigma_R$$

for the shielding coefficient for simple molecules were derived by Lamb[5] and by Ramsey,[6] who treated the molecular system as a whole. This type of approach is applicable for small molecules only. In more complex systems, the shielding coefficient for atom A within the molecule is usually written as the sum of a number of contributions:[7]

$$(2) \qquad\qquad \sigma_A = \sigma_A^{dia} + \sigma_A^{para} + \sum_{A \neq B} \sigma_{AB} + \sigma_A^{deloc}$$

The first term of equation 2, σ_A^{dia}, depends on the electron density around the nucleus of A and therefore on the electronegativity of any substituent on A. It is normally the dominant term for the chemical shift of protons.

The second term, σ_A^{para}, arises from paramagnetic currents induced on atom A. It is zero only for nuclei with a spherical distribution of the electrons and consequently is quite influential in nuclei with valence p-electrons (*e.g.*, ^{19}F, ^{31}P).

The third term, $\sum_{A \neq B} \sigma_{AB}$, is referred to as the neighbor anisotropy effect and represents the influence of induced local currents on nuclei other than A. Only nuclei having an anisotropic susceptibility tensor contribute to this term.

The last term, σ_A^{deloc}, arises from the contribution of electrons delocalized over several bonds and is particularly important in aromatic hydrocarbons.

The Se atom contains two $4s$ and four $4p$ electrons, and chemical bonds involving these valence electrons are likely to result in nonspherical electron distributions. It can be expected, therefore, that the paramagnetic contribution to the shielding coefficient will be the dominant factor. Selenium nuclei with an electron distribution of predominantly spherical symmetry will have their resonance at high magnetic fields compared to nuclei surrounded by electrons arranged in a different symmetry.

Ramsey[6] gives the following expression for the paramagnetic term of equation 2:

$$(3) \qquad \sigma_A^{para} \propto \sum_n{}' \frac{1}{E_n - E_0} \left[\left(0 \left| \sum_k m_{zk} \right| n \right) \right]$$

Deviations from spherical symmetry of the electron cloud—and, therefore, large paramagnetic shifts—occur in molecules having low-lying electronic excited states (E_n) which mix with the ground state (E_0) to yield nonzero matrix elements of the magnetic moment operator (m_z). The paramagnetic term may be of particular importance in organic diselenides, which exhibit a long-wavelength electronic absorption band associated with the –Se–Se– part of the molecule, and the predominant contribution of this term to the shielding coefficient is borne out by the large spread in chemical shift of a number of selenium-containing compounds described in the following section.

b. Experimental Data. The chemical shift of ^{77}Se has been determined and published for a number of compounds. Walchli[8] reported the chemical shifts of Se(IV) and Se(VI) in H_2SeO_3 and H_2SeO_4, respectively, and his value of 1504 ± 40 ppm downfield from external liquid H_2Se for the chemical shift of $H_2SeO_4(aq.)$ was confirmed by Birchall et al.[2] However, the value of 1560 ± 80 ppm for the hexavalent selenium in $H_2SeO_4(aq.)$ was not reproduced by other authors. Birchall, in particular, pointed out that hexavalent selenium is generally more shielded than tetravalent selenium. Lardon[9] confirmed Birchall's statement and found that the Se(VI) resonance in $H_2SeO_4(aq.)$ occurred 273 ± 10 ppm upfield from that of Se(IV) in $H_2SeO_3(aq.)$. McFarlane has reported data on ter- and quinquevalent phosphorus compounds containing selenium[10] and on organic selenium compounds[11] and concludes that the ^{77}Se chemical shifts qualitatively follow a pattern similar to ^{31}P chemical shifts. Thus, replacement of a methyl substitutent on selenium by a phenyl group causes a shift to a lower field, whereas replacement by a directly bonded proton causes a shift to higher field. The completely hydrogenated selenium (H_2Se) experiences the largest shielding.

Lardon[3] has investigated a number of organic selenides and diselenides. Whenever a diselenide was compared with its corresponding monoselenide, a downfield shift of the selenium resonance was encountered. This shift was attributed to the influence of the low-energy electronic diselenide band on the paramagnetic contribution to the shielding coefficient (equation 3).

All the consistent results on the chemical shifts reported to date are compiled in Table XVE-1. In comparing Birchall's and Lardon's data, it was assumed that the reference used by Lardon (H_2SeO_3, aq.) had a chemical shift of 197 ppm from external selenium oxychloride ($SeOCl_2$). McFarlane's

Table XVE-1   [77]Se *chemical shifts*

COMPOUND	PHYSICAL STATE	δ (ppm from $SeOCl_2$)[a]	REF.
$SeOBr_2$	Liquid	−80.3	2
$SeOCl_2$	Liquid	0	2
SeOFCl	$SeOF_2/SeOCl_2$ mixture	0.4	2
H_2SeCl_6	Aqueous solution	28.0	2
$SeOF_2$	Liquid	100.8	2
H_2SeO_3	Saturated solution in water	196.6	2
Se_2Cl_2	Liquid	204.5	2
Na_2SeO_3	Saturated solution in water	226.1	2
Se_2Br_2	Liquid	304.9	2
$SeCl_4$[b]	Saturated solution in DMF	325.2	2
$SeF_4 \cdot BF_3$	Liquid (m.p. 46°)	356.8	2
SeF_4	Liquid	386.8	2
$SeF_4 \cdot SO_3$	Liquid (m.p. 70°)	421.8	2
K_2SeO_4	Saturated solution in water	454.8	2
$HSeO_3Cl$	Solution in SO_2	475.8	2
H_2SeO_4	Slightly aqueous	478.1	2
$HSeO_3F$	Impure liquid containing H_2SeO_4	478.0	2
SeO_3	Saturated solution in phosphorus oxychloride	521.8	2
SeO_2F_2	Liquid	531.2	2
SeO_3	Liquid (m.p. 120°)	535.3	2
$(CF_3)_2Se$	Liquid	784.6	2
SeF_6	Liquid	868.7	2
$(CF_3Se)_2$	Liquid	950.9	2
$(CH_3)_2P^{III}SeCH_3$	Liquid	999	10
PhSeSePh	Liquid, 60°	1019	3
Ph$\underline{Se}$SeCH$_3$	Liquid	1040	11
PhSePh	Liquid	1077	3
$(CH_3)_2P^{V}(S)SeCH_3$	Liquid	1137	10
$(CF_3Se)_2Hg$	Solution in methanol	1145	2
$C_2H_5SeSeC_2H_5$	Liquid	1164	3
PhSe$\underline{Se}$CH$_3$	Liquid	1191	11
$CH_3SeSeCH_3$	Liquid	1213	3
$CF_3SeHgCl$[c]	Solution in methanol	1215	2
$C_2H_5SeC_2H_5$	Liquid	1262	3
PhSeH	Liquid	1333	11
		1343	3
CH_3SeCH_3	Liquid	1479	3
CH_3SeH	Liquid	1600	11
H_2Se	Liquid	1705	2

[a] Positive entry designates a shift to high field.

[b] The signal was very weak because of low solubility and could only be seen by using the broad-line technique.

[c] The signal was weak and was observed by the rapid-passage technique.

data were converted to Birchall's scale by assuming a value of 1213 ppm for dimethyl diselenide.

The ^{77}Se chemical shift values available so far indicate that the paramagnetic contribution to the shielding coefficient is the dominant factor for this nucleus. The electron cloud surrounding the selenium nucleus is about three times as large as that of the proton.[12] This makes the selenium nucleus sensitive to small changes in its chemical environment. The wide range of values found for the chemical shift of ^{77}Se suggests that this parameter in itself is of considerable analytical significance.

3. Spin-Spin Coupling with ^{77}Se

Coupling constants for the spin-spin interaction of ^{77}Se with other nuclei are theoretically important because they allow the testing of quantitative theories of chemical bonding.[13] Coupling with protons was reviewed in Chapter XVD. We will now consider coupling between ^{77}Se and nuclei other than protons.

Any coupling between ^{77}Se and another nucleus leads to low-intensity sidebands in the resonance signal of the other nucleus; these are often difficult to see by conventional recording techniques[2] but can be observed in enhanced spectra.[14] Heteronuclear double resonance and tickling techniques[15] make it possible to obtain both the magnitude and the sign of the coupling constant, with the conventional assumption that the coupling constant for $^{13}C-^{1}H$ has to be taken as positive. McFarlane[4,16,17] and Dreeskamp and Pfisterer[18] used these techniques successfully to determine $^{1}H-^{77}Se$, $^{13}C-^{77}Se$, $^{31}P-^{77}Se$, and $^{77}Se-^{77}Se$ coupling constants. The results are compiled in Table XVD-2. More data on coupling between ^{77}Se and protons can be found in Chapter XVD on proton n.m.r.

4. Conclusions

The chemical shift of ^{77}Se appears to cover a range of the order of 1800 ppm, according to the experimental data available so far. The main contribution to the shielding coefficient arises from Ramsey's paramagnetic term and is small for molecules in which the electron distribution around the selenium nucleus has nearly spherical symmetry (e.g., H_2Se). In most cases, however, deviations from this symmetry occur and result in large negative (paramagnetic) shifts of the ^{77}Se resonance. Chemical shift differences of the order of several hundred parts per million are observed even within the range of organic selenium compounds, thus making the ^{77}Se chemical shift a very valuable analytical tool.

Table XVE-2   Coupling constants (Hz) for spin-spin interaction involving ^{77}Se

COMPOUND	COUPLING	J	REF.
H_2Se	$^{1}H—^{77}Se$	63.4 ± 0.5	2
$PhCH_2SeH$	$^{1}H—^{77}Se$	$+42.2$	17
	$^{1}H—C—^{77}Se$	$+12.8$	17
$(CH_3)_2Se$	$^{13}C—^{77}Se$	-62 ± 1	16
		-62.0 ± 0.5	18
$(CH_3Se)_2$	$^{13}C—^{77}Se$	-75 ± 1	16
CH_3SeH	$^{13}C—^{77}Se$	-48 ± 10	18
$SeOF_2$	$^{19}F—^{77}Se$	837.1 ± 0.6	2
$SeOFCl$	$^{19}F—^{77}Se$	647.5 ± 2.0	2
SeO_2F_2	$^{19}F—^{77}Se$	1583.9 ± 0.6	2
SeF_6	$^{19}F—^{77}Se$	1420.9 ± 0.5	2
$HSeO_3F$	$^{19}F—^{77}Se$	1453.6 ± 0.7	2
$(CF_3)_2Se$	$^{19}F—C—^{77}Se$	11.1 ± 0.2	2
$(CF_3Se)_2$	$^{19}F—C—^{77}Se$	6.4 ± 0.2	2
$(CF_3Se)_2Hg$	$^{19}F—C—^{77}Se$	39.5 ± 0.6	2
$CF_3SeHgCl$	$^{19}F—C—^{77}Se$	37.4 ± 0.5	2
$(CH_3)_2PSeCH_3$	$^{31}P—^{77}Se$	-205 ± 6	10
$(CH_3)_2P(S)SeCH_3$	$^{31}P—^{77}Se$	-341 ± 4	10
$(CH_3Se)_2$	$^{77}Se—^{77}Se$	*ca.* $+20$	4
$PhSeSeCH_3$	$^{77}Se—^{77}Se$	$+22 \pm 4$	4

To date only simple liquids have been studied, mainly because of sensitivity problems. A drastic improvement in instrument sensitivity is needed before the technique can be applied to investigate compounds that do not melt within the temperature range accessible to the n.m.r. probe (up to $+200°$) but can be dissolved in suitable solvents. Such gains in sensitivity would also render solvent-solute interactions with organic selenium compounds available for investigation. This would be of particular interest, for instance, with selenols, where hydrogen bonding effects are likely to be reflected in the selenium n.m.r. parameters.

Double resonance techniques such as the INDOR method[4] and proton noise decoupling experiments have been shown to lead to significant improvements in the signal/noise ratio and have recently been applied to ^{77}Se n.m.r. studies.[19]

Fourier transform spectroscopy combined with spectrum accumulation increases the signal/noise ratio by about 2 orders of magnitude for a given recording time and has been successfully applied to n.m.r. studies on isotopes of low natural abundance, such as ^{13}C. Advantage of this technique will certainly be taken in future selenium n.m.r. investigations.

Pulsed n.m.r. experiments with ^{77}Se would provide another means to study this nucleus and would yield either wide-line or even high-resolution data on solid selenium compounds.

REFERENCES

1. S. S. Dharmatti and H. E. Weaver, Jr., *Phys. Rev.*; **86,** 259 (1952).

2. T. Birchall, R. J. Gillespie, and S. L. Vekris, *Can. J. Chem.,* **43,** 1672 (1965).

3. M. Lardon, *J. Amer. Chem. Soc.,* **92,** 5063 (1970).

4. W. McFarlane, *J. Chem. Soc., A,* 670 (1969).

5. W. E. Lamb, *Phys. Rev.,* **60,** 817 (1941).

6. N. F. Ramsey, *Phys. Rev.,* **78,** 699 (1950); **86,** 243 (1952).

7. J. W. Emsley, J. Feeney, and L. H. Sutcliffe, "High Resolution Nuclear Magnetic Resonance Spectroscopy," Vol. 1, Pergamon Press, New York, N.Y., 1965, p. 130.

8. H. E. Walchli, *Phys. Rev.,* **90,** 331 (1953).

9. M. Lardon, unpublished results.

10. W. McFarlane and J. A. Nash, *Chem. Commun.,* 913 (1969).

11. W. McFarlane, *J. Chem. Soc., A,* 670 (1969).

12. M. L. Huggins, *J. Amer. Chem. Soc.,* **75,** 4126 (1953). Atomic radii: Se, 1.17 Å; H, 0.38 Å.

13. J. A. Pople and D. P. Santry, *Mol. Phys.,* **8,** 1 (1964).

14. R. A. Dwek, R. E. Richards, D. Taylor, G. J. Penney, and G. M. Sheldrick, *J. Chem. Soc., A,* 935 (1969).

15a. W. McFarlane, *Chem. Brit.,* **5,** 142 (1969).

15b. W. McFarlane, "Annual Review of NMR Spectroscopy," E. F. Mooney, Ed., Vol. 1, Academic Press, New York, N.Y., 1968, p. 135.

16. W. McFarlane, *Mol. Phys.,* **12,** 243 (1967).

17. W. McFarlane, *Chem. Commun.,* 963 (1967).

18. H. Dreeskamp and G. Pfisterer, *Mol. Phys.,* **14,** 295 (1968).

19. W. McFarlane and R. J. Wood, *J. Chem. Soc., Dalton Trans.,* 1397 (1972).

XV Physicochemical investigations of selenium compounds

F. ORGANOSELENIUM RADICALS

HENRY J. SHINE

Texas Tech University
Department of Chemistry
Lubbock, Texas

1. Introduction

The reactions of organosulfur radicals are well authenticated and documented.[1] Authentic reactions of organoselenium radicals, on the contrary, are hard to find in the chemical literature; and, in comparison with the situation for sulfur analogs, data on the chemistry of organoselenium radicals may be described as almost nonexistent. Certainly, radicals containing selenium are known. The SeH and SeO radicals[2] as well as the SeF radicals[3] have been made in the gas phase, and their e.s.r. spectra have been recorded. Also, as will be presented in more detail later, the e.s.r. spectra of a small number of radicals and anion radicals containing selenium have been published. But, on the whole, very little is known about the preparation and chemistry of organoselenium radicals.

Conventional methods of making organosulfur radicals include the abstraction of a H atom from thiols and the thermal and photolytic scissions of disulfides. One would have thought that the literature would contain analogous reactions of selenium compounds. For the most part, however, this is not so. The literature offers examples of reactions of selenols and diselenides, but in no case has it been clearly shown that organoselenium radicals participate. The uncertainty about some of these reactions will be brought out in Sections F.2 and F.3, that is, the sections that cover particular additions to unsaturated systems and particular reactions of diselenides.

2. Additions to Unsaturated Systems

As far as I am aware, no unquestionable example is to be found in the literature of the free-radical addition of a selenol to an unsaturated molecule. Selenophenols have been made to undergo both base-catalyzed and so-called uncatalyzed addition to unsaturated acids and to alkynes. For example, in the absence of a base, phenylselenol adds to maleic acid to give phenylselenosuccinic acid (97% yield), and to maleic anhydride in 56% yield. It adds to fumaric acid to give phenylselenosuccinic acid in 68% yield.[4] Additions to alkynes take place in absolute ether at room temperature in yields[5] of 60–80% (equation 1). In none of these cases of so-called

$$(1) \qquad ArC{\equiv}CH + Ar'SeH \rightarrow ArCH{=}CHSeAr'$$

uncatalyzed reactions is there any real evidence of free-radical participation. Additions to 1,3-dienes in the absence of solvent (equation 2)[6] and to chalcones in ethanol (equation 3)[7] have been reported.

$$(2) \quad ArSeH + CH_2{=}CH{-}CH{=}CH_2 \rightarrow ArSeCH_2{-}CH{=}CH{-}CH_3$$

$$(3) \qquad ArSeH + Ar'CH{=}CH{-}\overset{\overset{\displaystyle O}{\|}}{C}Ar'' \longrightarrow Ar'{-}\underset{\underset{\displaystyle SeAr}{|}}{C}H{-}CH_2\overset{\overset{\displaystyle O}{\|}}{C}Ar''$$

One may wonder whether these reactions also occur under genuine free-radical conditions, and one may even be tempted to think that some of these reported additions are free-radical reactions. But there is no evidence to support such a possibility, and mechanistically the reactions are probably either simple nucleophilic additions or autocatalytically acidic.

Uncatalyzed addition of phenylselenol to epoxyolefins has, indeed, been described as being probably free radical.[8] For example, in the presence of its sodium salt, phenylselenol added to the epoxide ring of 3,4-epoxy-1-butene (equation 4). In the absence of the catalyst, however, a second prod-

$$(4) \quad C_6H_5SeH + CH_2{=}CH{-}CH\underset{\displaystyle O}{\diagdown\!\!\diagup}CH_2 \longrightarrow$$

$$CH_2{=}CH{-}\underset{\displaystyle \underset{OH}{|}}{CH}{-}CH_2SeC_6H_5$$

uct was also formed by 1,4-addition (equation 5) and is thought to involve

$$(5) \quad C_6H_5SeH + CH_2{=}CH{-}CH\underset{\displaystyle O}{\diagdown\!\!\diagup}CH \longrightarrow$$

$$C_6H_5SeCH_2{-}CH{=}CH{-}CH_2OH$$

free radicals.[8]

Further work in the Soviet Union indicates a free-radical addition, initiated by u.v. light, of some organometalloid selenols.[9] These are shown in equation 6. Irradiation of sealed, evacuated ampoules was carried out for

$$(6) \qquad Et_3MSeH + CH_2{=}CHR \rightarrow Et_3MSeCH_2CH_2R$$

$$[M = Si, R = C_6H_5 \ (90\%); \ C_4H_9 \ (72\%)]$$

$$[M = Ge, R = C_6H_5 \ (77\%); \ CO_2Et \ (52\%)]$$

lengthy periods (15–72 hours), and high yields of adduct were obtained.

There *is* an addition to olefins that involves the *formation* of an organoselenium radical, and this occurs in the addition of a Se atom. Flash photolysis of carbon diselenide in the u.v. region leads to decomposition and the formation of Se atoms in the triplet state: $Se(4^3P)$.[10–12] In the absence of an olefin the Se atoms dimerize. In the presence of an olefin an episelenide is formed (equation 7–9). Addition occurs in the triplet state and is deduced,

$$(7) \qquad\qquad CSe_2 + h\nu \rightarrow CSe + Se$$

$$(8) \qquad\qquad 2Se \rightarrow Se_2$$

$$(9) \qquad Se + RCH{=}CHR \longrightarrow RCH\underset{\displaystyle Se}{\diagdown\!\!\diagup}CHR$$

from most of the data in Table XVF-1, assembled by Callear and Tyerman[12] from several sources, to be electrophilic in nature. Table XVF-1 shows that the rate of addition increases when electron-donating methyl groups are attached to the olefinic center, and brings out the analogies among the several atoms listed. The reason for the discrepancy regarding electrophilic character in the addition to the last three compounds is not known.

Table XVF-1   Relative rates of atom additions to unsaturated compounds[a]

UNSATURATED COMPOUND	$O(2^3P)$	$S(3^3P)$	$Se(4^3P)$	Br
Ethene	1.0	1.0	1.0	1.0
Propene	5.8	7.8	2.6	17.7
Butene-1	5.8	11	7.1	22.6
cis-Butene-2	24	18	23.9	95.0
trans-Butene-2	28	23	56	98.5
2-Methylpropene	25	56	44.7	384
1,3-Butadiene	24	100	114	...
Pentene-1	...	11	5.0	...
Vinyl chloride	...	1.3	1.3	...
Acrylonitrile	...	...	3.8	...
Allyl chloride	...	...	2.3	...

[a] Data from Callear and Tyerman.[12]

Addition of $Se(4^3P)$ to olefins is stereospecific.[11,12] The same type of addition is observed with triplet sulfur, $S(3^3P)$, and contrasts with the non-stereospecific addition of triplet oxygen, $O(2^3P)$, and electronically analogous triplet carbenes and nitrenes. This situation has been discussed by Callear and Tyerman[11] and by Leppin and Gollnick,[13] who list two phenomena as the possible causes of stereospecificity of the $S(3^3P)$ and $Se(4^3P)$ additions. These heavy atoms are thought to slow down rotation around the central bond of the triplet diradical formed in the first step of the addition, allowing time for the triplet-singlet conversion necessary for ring closure. At the same time spin-orbit coupling with the heavy atomic nucleus also enhances spin inversion, which must precede ring closure. Leppin and Gollnick also consider, however, that participation of Walsh orbitals in the addition steps may be the controlling factor in the stereospecificity.

Of interest to the scope of this type of reaction is the relatively recent, first report of the addition of triplet tellurium to olefins.[14] The atom $Te(5^3P)$ is formed by the flash photolysis of dimethyl telluride in the gas phase. Addition of $Te(5^3P)$ appears to be slower and more selective than addition of triplet sulfur and selenium (Table XVF-2).

Whereas flash photolysis of carbon diselenide leads to $Se(4^3P)$, photolysis of carbonyl selenide gives singlet selenium,[15] $Se(4^1D_2)$. Photolysis in the presence of saturated hydrocarbons is accompanied by the insertion of selenium into C-H bonds, as well as by some H-atom abstraction, leading to the HSe radical. Ethene was converted into the episelenide; in contrast, $S(3^1D_2)$ gave both thiacyclopropane and vinyl thiol in equal amounts.[16]

Finally, in connection with stereospecific additions in selenium chemistry,

Table XVF-2 Rates of addition of $S(^3P)$, $Se(^3P)$, *and* $Te(^3P)$ *to ethene and propene*[a]

COMPOUND	S	Se	Te
C_2H_4[b]	7×10^8	1×10^8	2×10^7
C_3H_6/C_2H_4	6.8[c]	3.5[c]	10[c]

[a] Data from Connor *et al.* [14]
[b] In liters per mole per second.
[c] Ratio of rates for the two alkenes.

mention can be made of the addition of phenylselenylcarbene to *cis-* and *trans*-butene-2. The addition is stereospecific (equations 10 and 11), control

(10)

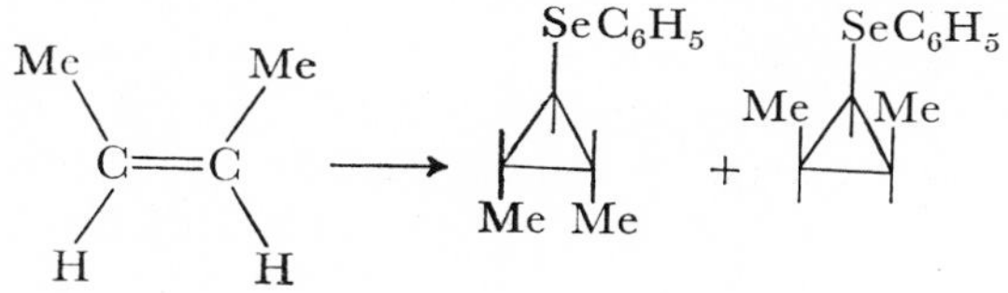

(11)

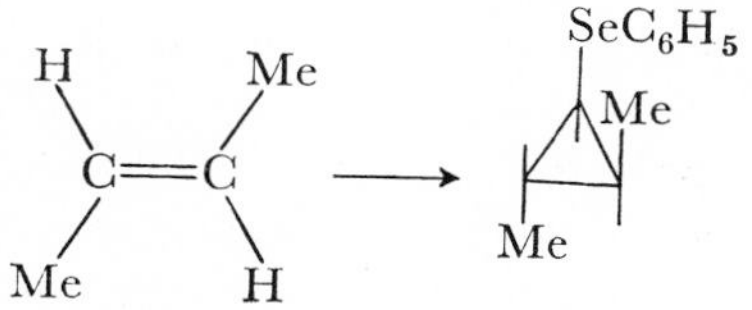

being attributed to stabilization of the sp^2 configuration of carbene carbon by *p*-orbital overlap with selenium.[17]

3. Reactions of Diselenides

Diaryl diselenides exhibit thermochromism in solution, and this phenomenon has been attributed to dissociation. For example, the color of a solution of diphenyl diselenide in xylene increases when the solution is heated and decreases when it is cooled again.[18] A similar, reversible color change from yellow to orange occurs in toluene solution.[19] Unsymmetrical diaryl diselenides, ArSeSeAr', are also thermochromic.[20] Kuder and Lardon[20a] have shown recently, however, that the thermochromism of diphenyl diselenide in the temperature range 77–388°K does not involve dissociation but is due solely to broadening of the long wavelength diselenide absorption band (λ_{max} 332 nm) as the temperature is raised. The oscillator strength of this transition was constant over the indicated temperature range. [1]H and [77]Se Nuclear magnetic resonance spectra showed no evidence of new species, nor did e.s.r. studies suggest the presence of free radicals in equilibrium with diselenides.

When a diaryl diselenide is strongly heated, selenium is ejected and a diaryl selenide is formed. It is thought[18] that this reaction involves arylseleno radicals (equation 12–14). For the case of dibenzyl diselenide,

(12) $ArSeSeAr \rightarrow 2ArSe\cdot$

(13) $ArSe\cdot \rightarrow Ar\cdot + Se$

(14) $Ar\cdot + ArSe\cdot \rightarrow Ar_2Se$

Lardon[20b] has shown by [1]H n.m.r. that thermal decomposition at 150–170° leads to a complex mixture of products, including selenium, dibenzyl selenide, and various dibenzyl polyselenides. Longer exposure to 225° also yielded substantial quantities of toluene and of 1,2-diphenylethane. Günther and Salzman[20c] prepared a series of benzylic and aromatic diselenide polymers by thermal conversion of the corresponding cyclic dimers (*cf.* Chapter XIV). It was found[20d] that these species also readily incorporated elemental selenium into the polymer chain upon rapid melting of the components at 225–240°. Again, participation of selenium radicals is strongly indicated for these thermal reactions, but no rigorous proof exists at this time. Equations 12–14 represent a simplified picture of a complex reaction scheme.

Along this line, it should also be noted that diaryl ditellurides undergo the same types of reactions when heated.[21] Diphenyl ditelluride itself appears to be 30% dissociated, according to molecular weight measurements in camphor at 160°.[21] In contrast, molecular weight measurements with other diaryl ditellurides in benzene indicated no dissociation at all, although the solutions were red.[22,23] Of added interest is the fact that a solution of diphenyl ditelluride in ethanol became red on warming at 80°, and after 24 hours at room temperature had absorbed oxygen from the air to deposit a colorless, gelatinous, peroxide-like substance.[21] All of these results suggest that the ditellurides and, by analogy, the diselenides dissociate into radicals when heated.

5,6-Dimethylbenzo-1,2,3-selenadiazole forms 2,3,7,8-tetramethylselenanthrene in 60% yield when boiled gently in amyl alcohol (equation 15).[24]

This reaction may well result from the dimerization of a selenoaryl diradical formed by the loss of nitrogen from the diazole.

Work with dialkyl diselenides has been confined mostly to cyclic compounds: 1,2-diselenane (**1**), 1,2-diselenolane (**2**), and derivatives of **2**.

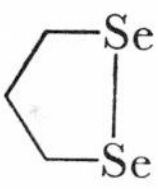

Cyclic diselenides are readily made from the corresponding open-chain diselenocyanates. The reaction is probably free radical, as indicated by the preparation of **1** and **2** themselves. These compounds are obtained initially as polymers, not as the monomers **1** and **2**. For example, the alkaline hydrolysis of 1,4-tetramethylene diselenocyanate in the presence of air leads to poly-1,6-diselenahexamethylene (equation 16, $n = 4$).[25] Hydrolysis of 1,3-trimethylene diselenocyanate leads to poly-1,5-diselenapentamethylene (equation 16, $n = 3$).[26] These polymers are obtained as yellow powders,

$$(16) \qquad xNCSe(CH_2)_nSeCN \rightarrow [Se(CH_2)_nSe]_x$$

insoluble in most cold organic solvents. They dissolve on warming and reappear slowly from the cooled solution. The warm solutions are thought to contain the monomer. The molecular weight of **1** in bromoform was initially 516. After the solution was kept at 60° and 24 hours in the dark, the molecular weight was 210, as compared with 214 for the monomer.

The change from polymer to monomer has also been followed spectroscopically.[25] Polymeric **1** in chloroform has an absorption maximum at 308 nm, characteristic of linear diselenides. When the solution was kept at 60° the 308 nm band was replaced by one at 364 nm, attributed to cyclic diselenides. The molecular weight of **2** in camphor was 240 (monomer 200). Thus, although **1** and **2** exist in solution, it is unlikely that either of them has ever been isolated as the pure monomer.

Some derivatives of **2** have been isolated as monomers by Bergson.[27,28] A particularly interesting example is 2,3-diselenaspiro[4,5]decane.[28] This compound is prepared by heating a solution of cyclohexane-1,1-dimethylbenzene sulfonate and potassium selenocyanate. If we accept the formation of the diselenocyanate as occurring first, the subsequent reactions may involve free radicals by either path (*a*) or path (*b*) (scheme 17).

The overall situation with the cyclic diselenides seems to be that the monomers and polymers are interconvertible in solution and that the interconversions may go through organoselenium radicals. Radicals have been detected by e.s.r. in experiments with poly-**1** in chloroform solution.[25] The experiments were not reproducible, however, and the *g*-values (2.000)

(17)

were far too low for RSe· radicals, which should have g-values near 2.02. Therefore, although free-radical participation is the most reasonable explanation of the reactions leading to the formation and depolymerization of poly-**1** and poly-**2**, no proof has been provided as yet.

The effects of poly-**1** and **1** on vinyl polymerizations are also not clearly interpretable in terms of the participation of organoselenium radicals.[25] Poly-**1** dissolved in styrene, methyl methacrylate, vinyl acetate, and acrylonitrile depolymerizes into **1** at 60°. Yet, if radicals are formed, they do not initiate vinyl polymerization. In fact, the rate of thermal polymerization of styrene is decreased, and that of methyl methacrylate only slightly increased. At the same time the molecular weight of the polymer obtained from these two monomers is lowered by the presence of the diselenide. Furthermore, although vinyl acetate does not polymerize at 60° in either the presence or the absence of diselenide, its polymerization at 60°, initiated by bisazoisobutyronitrile, is severely inhibited.

The several results obtained suggest that organoselenium radicals, if present at 60°, do not add to the vinyl monomers. Indeed, the results may well indicate that organoselenium radicals are *not* initially present. The retardation effects may result from termination reactions of vinyl-compound radicals with the diselenide (either mono or poly) (equation 18). In that

$$(18) \qquad R'· + RSeSeR \rightarrow RSeR' + RSe·$$

case, of course, the newly formed RSe· radical would not feature in chain initiation, because rate retardations are observed.

In line with the possibility shown in equation 18 is the finding that polystyrene and poly(methyl methacrylate) prepared at 60° in the presence of bisazoisobutyronitrile and poly-**1** contained some selenium.[25]

Polarographic reduction of diphenyl diselenide[29] and a number of dialkyl diselenides (both linear and cyclic)[30,31] has been carried out by Nygård *et al.* Although these reductions must involve one-electron transfers at some stage, the initial reaction in each case is cleavage of the diselenide bond by mercury (equations 19, 20).[29]

$$(19) \qquad RSeSeR + 2Hg \rightarrow 2RSeHg$$

$$(20) \qquad RSeHg + H^+ + e^- \rightarrow RSeH + Hg$$

4. Organoselenium Radicals in Oxidation Reactions

Organoselenium compounds have been used as antioxidants in biological reactions. Antioxidant action is believed to involve organoselenium radicals, but again direct evidence for these has never been obtained. For example, small amounts of selenoamino acids in the diet of chicks were found to inhibit lipid peroxidation in the chick liver. Inhibition was also found in *in vitro* studies. The thought is that stable selenyl-type radicals are formed which divert free-radical peroxidation reactions from their liver-damaging pathways.[32] In a related study, rats, chicks, and ewes were fed a diet containing sodium selenite and selenate. Lipid-free, selenium-containing tissue was obtained from the kidneys and livers of the animals, and was found to be 50–500 times as effective as α-tocopherol in inhibiting the uptake of oxygen by linoleic acid. It is believed that organoselenium compounds are formed from the diet of inorganic selenium, and that these compounds interrupt peroxide reactions, partly by free-radical pathways, as in, for example, equation 21.[33]

$$(21) \qquad ROO\cdot + R'SeH \rightarrow ROOH + R'Se\cdot$$

These interpretations are entirely speculative. Denison and Condit[34] have found that dialkyl selenides are 10–50 times more effective than dialkyl sulfides in inhibiting the oxidation of lubricating oils. In this case the interpretation is that oxygen transfer occurs, interrupting the peroxide chain (equation 22).

$$(22) \qquad R_2Se + R'O_2\cdot \rightarrow R_2SeO + R'O\cdot$$

Very little is known about the reactions of organoselenium compounds with peroxides. It has been found that a dialkyl diselenide reacts with *tert*-butyl hydroperoxide in benzene *at room temperature* to give a seleninic anhydride (equation 23).[35] This reaction is nonradical in nature. When the

$$(23) \qquad RSeSeR + 3t\text{-}BuOOH \longrightarrow \underset{\overset{\|}{O}}{RSe}\text{—}O\text{—}\underset{\overset{\|}{O}}{SeR} + 3t\text{-}BuOH$$

reaction is carried out at 75°, the products are ultimately SeO_2, O_2 and *tert*-butyl alcohol. The proposal[36] is that at 75° the first-formed seleninic anhydride reacts with *tert*-butyl hydroperoxide to form a perester, which then undergoes homolytic decomposition (equations 24, 25).

$$(24) \qquad \underset{\underset{O}{\|}}{RSe}—O—\underset{\underset{O}{\|}}{SeR} + t\text{-BuOOH} \longrightarrow \underset{\underset{O}{\|}}{RSeOO}t\text{-Bu}$$

$$\overset{O}{\overset{\|}{RSeO\cdot}} + t\text{-BuO}\cdot$$

$$(25) \qquad \underset{\underset{O}{\|}}{RSeOO}t\text{-Bu} \Big\langle$$

$$RSe\cdot \ + t\text{-BuOO}\cdot$$

The fate of the organoselenium radicals is not known. Since SeO_2 is formed, it is possible that the selenoxyl-type radicals decompose (equation 26), while the oxyselenyl-type radicals may be reconverted into the seleninic

$$(26) \qquad \overset{O}{\overset{\|}{RSeO\cdot}} \longrightarrow R\cdot + SeO_2$$

anhydride (equations 27, 28).[36]

$$(27) \qquad 2RSe\cdot \longrightarrow \underset{\underset{O}{\|}}{RSe}—O—\underset{\underset{O}{\|}}{SeR}$$

$$(28) \qquad \underset{\underset{O}{\|}}{RSeOSeR} \overset{oxn}{\longrightarrow} \underset{\underset{O}{\|}}{RSe}—O—\underset{\underset{O}{\|}}{SeR}$$

An understanding of the role of selenium in biology has also been sought in radiation-protection studies. It has been found, for example, that seleno-methionine and selenocystine are more effective than methionine and cystine in preventing X-radiation damage to amino acids.[37,38] Nothing definite is known, however, about *how* protection is afforded. Among the possibilities considered[38] are the ease of ionization of the selenide (equation 29) and abstraction of hydrogen from an adjacent C–H group (equation 30).

$$(29) \qquad RSeR \rightarrow R\overset{\cdot +}{Se}R$$

$$(30) \qquad RCH_2SeR \rightarrow R\overset{\cdot}{C}HSeR \leftrightarrow RCH\!\!=\!\!\overset{\cdot}{Se}R$$

Stabilization of the selena-alkyl radical by selenium is proposed as a factor in favor of α-hydrogen abstraction.[38] The last reaction is looked upon as a way of repairing damage sites in the host, that is, by transferring a H atom to a site at which a radical had been formed. These ideas are attractive, but, as seems to be the case in most aspects of organoselenium radical chemistry, there is no direct evidence supporting them.

Irradiation of selenides and diselenides with 2537 Å light at 77°K gave alkyl and organoselenyl radicals, characterized by e.s.r. spectroscopy.[39] For example, dibenzyl diselenide and dibenzyl selenide each lost an α-H atom and gave radicals with a g-value of 2.003, whereas benzylselenol, octadecyl selenide, and dodecyl selenide each gave, in part, organoselenyl radicals with a g-value near 2.09 (equations 31, 32).

(31) $C_6H_5CH_2SeSeCH_2C_6H_5 \rightarrow C_6H_5CHSeSeCH_2C_6H_5$

(32) $C_{12}H_{26}SeC_{12}H_{26} \rightarrow C_{12}H_{26}Se\cdot + R\cdot$

Results of this kind are helpful to some extent in trying to understand the part played by selenium compounds in biological reactions and as antioxidants. At the same time, though, there is a danger that the interpretations of results like these, or their extension into supposedly related systems, may be misleading. *Definite* knowledge of the roles played by organoselenium radicals is still extremely meager.

5. Organoselenium Radicals in Hydrogen Transfer Reactions

House and Orchin[40] have noted that, in certain aromatizations of hydro-aromatics with selenium, less hydrogen selenide is evolved than corresponds to the amount of aromatic obtained. The selenium was shown to act as a hydrogen transfer agent. For example, the conversion of guaiene (**3**) into guaiazulene (**4**) (equation 33) was enhanced by the presence of oleic acid,

(33)

3   **4**

and stearic acid was formed. Aromatizations like this are thought to involve organoselenium radicals.[41] In this connection, 2,2′-dinaphthyl diselenide was used to aromatize guaiene at 200°. A possible reaction sequence was formulated as in equations 34–37.[41] In these, equation 36 represents the hydro-

(34) $ArSeSeAr \rightarrow 2ArSe\cdot$

(35) $ArSe\cdot + 3 \rightarrow ArSeH + 3\cdot$

(36) $2ArSeH + 3 \rightarrow \rightarrow ArSeSeAr + HH\text{–}3$

(37) $3\cdot + 3 \rightarrow \rightarrow 4 + \text{saturate}$

genation of the isopropenyl group of **3**, while equation 37 represents several steps ending in **4**.

In a somewhat similar way, diphenyl diselenide participates in the photocyclization of stilbenes.[42] The irradiation of stilbene and diphenyl diselenide with a 200 watt Hanovia mercury lamp gave phenanthrene in 21 % yield. Picene was obtained from di-1-naphthylethylene in 100 % yield (equation 38). These dehydrogenatation cyclizations are thought to be initiated by the

(38)

ArSe· radicals formed photolytically.[42] Schmidt et al.[43] have shown that diphenyl diselenide undergoes photolytic scission. The solid, green $C_6H_5Se·$, formed at 77°K, was found to be somewhat more stable toward dimerization than the $C_6H_5S·$ radical.

6. Anion and Cation Radicals Containing Selenium

A small number of heterocyclic ion radicals containing selenium have been made. In these cases there is no doubt that a selenium-containing radical was at hand. The objectives of research in this field are mostly either measuring e.s.r. splittings by naturally abundant ^{77}Se or establishing the extent of p- or d-orbital involvement in electron delocalization.

Questions of p- and d-orbital involvement have been raised chiefly with selenadiazoles. A common technique[44,45] has been to make the anion radicals of 2,1,3-benzoxa-, 2,1,3-benzothia-, and 2,1,3-benzoselenadiazole (compounds 5–7), and to compare experimental coupling constants with those

5, X = O;   **6**, X = S;   **7**, X = Se

calculated for models, using p- and d-orbitals. Behind this technique is the reasoning that d-orbital participation is more likely to be experienced in **7** than in **6**, and is not available in **5**. Therefore, if a p-orbital model accounted for the experimental results with **6** and **7** as suitably as with **5**, there would be no particular advantage in invoking d-orbital participation. The question of

Table XVF-3 Experimental and calculated coupling constants for the anion radicals of compounds **5**, **6**, *and* **7**

COMPOUND	a_N	$a_{4,7}$ Exptl.	Calcd.[a]	$a_{5,6}$ Exptl.	Calcd.[a]	REF.
5	5.24	3.33	3.41	2.02	2.01	44
	5.61	3.51		2.19		45
6	5.26	2.63	2.64	1.53	1.52	44
	5.18	2.59		1.59		45
7	5.97	1.99		1.99		44
	5.79	2.48		1.65		45

[a] As presented by Urberg and Kaiser.[46]

d-orbital participation is a persistent one in Group VI chemistry, and the availability of these anion radicals provides a convenient test system.

Two sets of results with **5–7** show that there is no particular advantage in using a *d*-orbital model. The data are assembled in Table XVF-3.

Fajer and his co-workers[47] have also found that a *p*-orbital model is satisfactory for the anion radical of perfluoro-2,1,3-benzoselenadiazole ($a_F = 3.56$ and 4.34 G; $a_N = 6.08$ G).

Analogous calculations for the anion radicals of dibenzofuran (**8**), dibenzothiophene (**9**), and dibenzoselenophene (**10**) led to the conclusion that

8, X = O;   **9**, X = S;   **10**, X = Se

a *p*-orbital model for **9** (and presumably, by analogy, for **10**) was better than a *d*-orbital one.[48]

It is noticeable in these examples that there are no calculations for the selenium-containing anion radicals. The reason is that necessary suitable parameters (*e.g.*, overlap integrals) are not known. This fact brings up another point about the calculations for sulfur compounds: there is uncertainty about how sophisticated these calculations should be. Some workers have used the Hückel-MO theory, and others the SCF-MO theory. Also, the appropriate model for *d*-orbital participation is not known; some workers[48] have used the Longuet-Higgins model, and others[45] a modified Longuet-Higgins. One gets the impression that the *d*-orbital-model calculations cannot have much significance as long as the uncertainty about types of *d*-orbital model persists.

Strom and Norton[49] have made a detailed analysis of the e.s.r. spectra of some semidione radicals containing C_6H_5S and C_6H_5Se substituents, and have concluded that $d\pi-p\pi$ bonding within the substituents best explains the results. Certainly these radicals are not thiyl and selenyl radicals, but the claim for explaining the effect of the C_6H_5Se substituent makes one wonder what the situation really is in selenyl and more closely related selenium-containing radicals.

The methods of reduction of these heterocycles are varied. Potassium metal and solvent dimethoxyethane,[45,48] sodium and tetrahydrofuran,[47] electron donation from another anion, such as that of dihydroanthracene,[44] and polarography[45] have all been used. Recently, anion radicals of 4- and 5-substituted 2,1,3-benzoselenadiazoles were made polarographically in DMF by Todres et al.[50a] The half-wave potentials for reversible one-electron reduction were shown to be capable of correlation with Hammett σ constants.

The Soviet investigators have also found that, among the diazole anion radicals, the ease of removing the extra electron with oxygen goes in the order $6 > 5 > 7$. In contrast, the ease of protonating the anion radicals follows the order $7 > 5 > 6$. Protonation occurs during cathodic reduction in aqueous solvents and is followed by further reduction. The final result of reduction in aqueous solvents is formation of o-phenylenediamine and H_2X.[50b] The Soviet group[50c] has shown also, in studies of redox equilibria between the cyclooctatetraene dianion and the heterocycles **6** and **7**, that the anion radical of **7** is more stable than that of **6**. Anion radicals of 5,6-dimethyl-2,1,3-selenadiazole and of the two naphtho-2,1,3-selenadiazoles have also been made, using sodium in THF.[51]

Work with organoselenium cation radicals is fragmentary. Anodic oxidation in acetonitrile has established the oxidation potentials for one-electron and some two-electron removals in heterocycles related to dibenzo-dioxine (Table XVF-4). Oxidation is easier when selenium replaces sulfur and oxygen.[52]

Table XVF-4   Oxidation potentials for group VI heterocycles in acetonitrile[a]

COMPOUND	E_1	E_2
Dibenzodioxine	0.991	...
Phenoxathiine	0.825	1.32
Phenoxaselenine	0.751	1.31
Phenoxatellurine	0.366	...
Thianthrene	0.865	1.19
Selenanthrene	0.832	...

[a] Data from Barry et al.[52]

Lamotte and Berthier[53] prepared the cation radicals of phenoxaselenine and selenanthrene in concentrated sulfuric acid and obtained a broad, single-line e.s.r. spectrum in each case. The respective g-values and line widths were 2.0225 and 5.8 G, 2.0315 and 9.3 G.

Somewhat more detailed spectroscopic data for the cation radicals of selenanthrene (11) and phenoselenazine (12) have been obtained by Shine and Abler.[54] The absorption and e.s.r. spectra ($g = 2.0314$) of the selenanthrene cation radical in concentrated sulfuric acid are given in Figs. XVF-1 and XVF-2. The two small wing peaks in the e.s.r. spectrum are believed to be caused by naturally abundant ^{77}Se ($m = \frac{1}{2}$), with splitting 32.2 G. The relative intensities of the minor and major peaks (7%) are in accord with the natural ^{77}Se abundance of 7.6%. Selenanthrene is like its sulfur analog in forming a stable cation radical in sulfuric acid. Phenoselenazine is also like its sulfur analog.[55] The cation radical is stable in 60% sulfuric acid, but has only a broad, four-line e.s.r. spectrum ($g = 2.0206$) (Figs. XVF-3 and XVF-4). The cation radical is formed rapidly in concentrated sulfuric acid but is converted, in that solution, into the dication, and the change can be followed nicely in the ultraviolet and the visible. Clean isosbestic points are to be seen (Fig. XVF-5).

Absorption maxima for 11 and 12 and their ions are given in Table XVF-5. The maxima are similar to those of the corresponding sulfur compounds.[55]

Tsujino[56a] has prepared the cation radical of phenoselenazine (12) by oxidation with iodine in DMSO. One of the products isolated was the 3,10'-dimer (13) in 65% yield, which is represented as arising from the dimerization of the cation radical (equation 39).

In a second publication, however, Tsujino[56b] shows the dimer as coming from the neutral radical (i.e., deprotonated cation radical). The puzzling thing about this is that the dimer is formed more readily in concentrated than in dilute sulfuric acid. The dimerization reaction is also observed with phenothiazine. My own intuition about these reactions says that the monomer dication may be involved, rather than the cation radical.

An e.s.r. splitting from naturally occurring ^{77}Se was mentioned above.

(39)

13

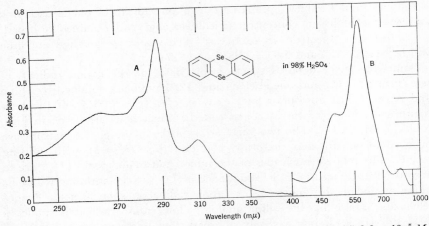

FIG. XVF-1 Absorption spectra of selenanthrene in 96% sulfuric acid. (*A*) $3.6 \times 10^{-5} M$; (*B*) $1 \times 10^{-4} M$.

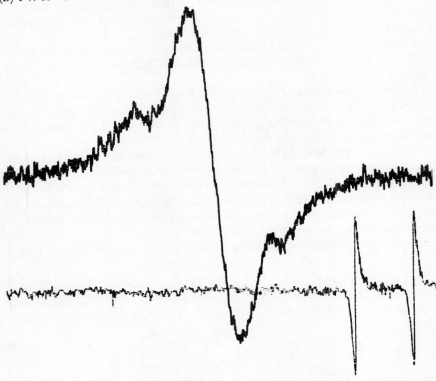

FIG. XVF-2 The e.s.r. spectrum of a solution of selenanthrene in 96% sulfuric acid. The field increases from left to right. The superimposed lines are of a potassium nitrosodisulfonate spectrum, splitting 13 G, recorded simultaneously with the cation radical spectrum but displaced slightly upfield by a pen differential.

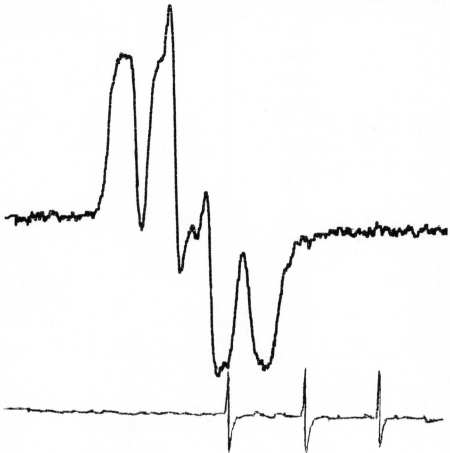

FIG. XVF-3   The e.s.r. spectrum of a $1 \times 10^{-3}\,M$ solution of phenoselenazine in 60% sulfuric acid after 1 hr. The field increases from the left of the spectrum to the right. The superimposed lines are of a potassium nitrosodisulfonate spectrum, of splitting 13 G, recorded simultaneously with the cation radical spectrum but displaced slightly upfield by a pen differential.

Müller and his co-workers[57] have obtained ^{77}Se splittings in some selenium-containing 2,6-di-*tert*-butylphenoxy radicals. Splittings of 10.5–11.1 G were seen in the e.s.r. spectra of radicals **14**, R = *i*-Pr and *t*-Bu. Splitting was not seen when R was Me or phenyl.

·O—⬡—SeR

14

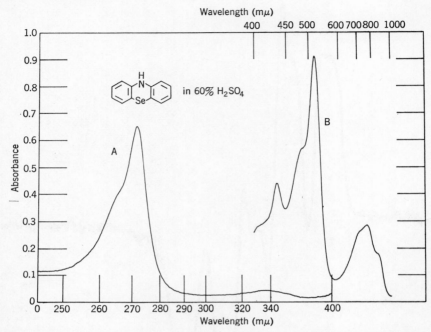

FIG. XVF-4   The ultraviolet (lower scale) and visible (upper scale) spectra of solutions of phenoselenazine in 60% sulfuric acid. (A) 1×10^{-5} M; (B) 1×10^{-4} M. Each solution was 1 day old.

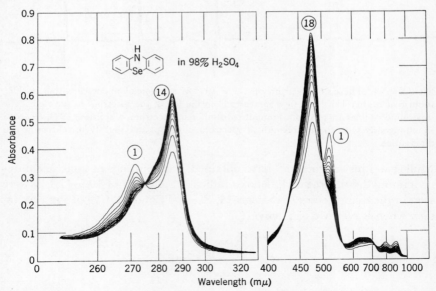

FIG. XVF-5   Changes with time in the spectra of solutions of phenoselenazine in 96% sulfuric acid. Ultraviolet: 1×10^{-5} M, between 2 and 35 min; visible: 6×10^{-5} M between 2 and 30 min.

Table XVF-5 Absorption maxima for selenanthrene (11), phenoselenazine (12), and their ions[a]

COMPOUND	SOLVENT	MAXIMA (nm)	$(\varepsilon \times 10^{-4})$	
11	Cyclohexane	254		
11·+	96% H_2SO_4	281 (0.436)	287 (1.88)	308 (0.69)
12	95% EtOH	257 (3.74)	316 (0.55)	
12·+	60% H_2SO_4	272 (6.54)	522 (0.91)	
12^{2+}	96% H_2SO_4	286 (6.08)	476 (1.37)	

[a] Data from Shine and Abler.[54]

7. Cation Radicals in Mass Spectrometry

Under conventional usage a mass spectrometer is the richest source known of information regarding gas-phase, cation-radical chemistry. The first electron-impact ionization produces the cation radical of the parent molecule. Thereafter fragmentations and rearrangements follow, and these are reactions of the parent and other cations. Several investigations of organoselenium compounds have been carried out by mass spectrometry. Some of these are particularly interesting because they allow comparison with the corresponding oxygen and sulfur compounds. Some of the differences in behavior that have been observed are remarkable.

Table XVF-6 Relative abundancies of ions obtained from C_6H_5—X—C_6H_5 (X = O, S, Se) *in the mass spectrometer*

ION	RELATIVE ABUNDANCE		
	X = O[a]	X = S[b]	X = Se[c]
$C_{12}H_{10}X$	100	100	44.0
$C_{12}H_9X$	17.9	49.0	
$C_{12}H_8X$		26.0	
$C_{12}H_{10}$		1.1	100
$C_{12}H_9$		2.6	
$C_{12}H_8$		5.6	
$C_{11}H_{10}$	15.4	1.1	
$C_{11}H_9$	28.4	2.3	
C_6H_5	60.0	11.0	98.0

[a] Natalis and Franklin.[58]

[b] Wszolek et al.[59]

[c] Bergman.[60]

Diaryl selenides split out selenium easily in the mass spectrometer, forming the diaryl cation radical. This is in contrast with diaryl ethers. Data for diphenyl ether, diphenyl sulfide, and diphenyl selenide in Table XVF-6 illustrate this point. The selenide forms the biphenyl ion easily, whereas the ether forms it not at all. The sulfide loses atomic sulfur to a small extent. The behavior of the selenide can be attributed to the weak C–Se bond, but the $C_{12}H_{10}$ ion must be formed by a rearrangement accompanying C–Se bond scission; otherwise the parent ion would only fragment.

Selenium is also ejected in stages from diaryl diselenides (equation 40).[61]

$$(40) \qquad C_6H_5SeSeC_6H_5 \cdot^+ \rightarrow C_6H_5SeC_6H_5 \cdot^+ \rightarrow C_{12}H_{10} \cdot^+$$

Similar reactions occur with diphenyl disulfide.[62]

Since organoselenocyanates are easily made, some of them have also been examined in the mass spectrometer. The parent cation radicals lose selenium and nitrile radical in stages, and thereafter the hydrocarbon cation undergoes its customary fragmentations.[63] (Also see Chapter XVG.)

REFERENCES

1a. W. A. Pryor, "Free Radicals," McGraw-Hill, New York, N.Y., 1966.

1b. W. A. Pryor, "Mechanisms of Sulfur Reactions," McGraw-Hill, New York, N.Y., 1962.

1c. M. J. Janssen, Ed., "Organosulfur Chemistry," Interscience Publishers, New York, N.Y., 1967.

2. A. Carrington, G. N. Currie, P. N. Dyer, D. H. Levy, and T. A. Miller, *Chem. Commun.*, 641 (1967); A. Carrington, *Proc. Roy. Soc., A*, **302,** 291 (1967).

3. A. Carrington, G. N. Currie, T. A. Miller, and D. H. Levy, *J. Chem. Phys.*, **50,** 2726 (1969).

4. E. G. Kataev and F. G. Gabdrakhmanov, *J. Gen. Chem. USSR*, **37,** 725 (1967).

5. J. Gosselck and E. Wolters, *Z. Naturforsch., Part B*, **17,** 131 (1962).

6. F. G. Gabdrakhmanov, Yu. Yu. Samitov, and E. G. Kataev, *J. Gen. Chem. USSR*, **37,** 715 (1967).

7. H. Gilman and L. F. Cason, *J. Amer. Chem. Soc.*, **73,** 1074 (1951).

8. G. I. Zaitseva and V. M. Al'bitskaya, *J. Org. Chem. USSR*, **4,** 1987 (1968).

9. N. S. Vyazankin, M. N. Bochkarev, and L. P. Maiorova, *J. Gen. Chem. USSR*, **39,** 444 (1969).

10. A. B. Callear and W. J. R. Tyerman, *Nature*, **202,** 1326 (1964); A. B. Callear and W. J. R. Tyerman, *Proc. Chem. Soc.*, 296 (1964); A. B. Callear and W. J. R. Tyerman, *Trans. Faraday Soc.*, **61,** 2395 (1965).

11. A. B. Callear and W. J. R. Tyerman, *Trans. Faraday Soc.*, **62,** 371 (1966).

12. A. B. Callear and W. J. R. Tyerman, *Trans. Faraday Soc.*, **62,** 2760 (1966).

13. E. Leppin and K. Gollnick, *Tetrahedron Lett.*, 3819 (1969).

14. J. Connor, G. Greig, and O. P. Strausz, *J. Amer. Chem. Soc.*, **91,** 5695 (1969).

15. W. J. R. Tyerman, W. B. O'Callaghan, P. Kebarle, O. P. Strausz, and H. E. Gunning, *J. Amer. Chem. Soc.*, **88,** 4277 (1966).

16. H. E. Gunning and O. P. Strausz, *Advan. Photochem.*, **4,** 143 (1966).

17. U. Schöllkopf and H. Küppers, *Tetrahedron Lett.*, 105 (1963).

18. T. W. Campbell, H. G. Walker, and G. M. Coppinger, *Chem. Rev.*, **50,** 279 (1952).

19. A. Schönberg, *Ber.*, **65,** 289 (1932).

20. H. Rheinboldt and E. Giesbrecht, *Chem. Ber.*, **85,** 357 (1952).

20a. J. E. Kuder and M. A. Lardon, *Ann. N.Y. Acad. Sci.*, **192,** 147 (1972).

20b. M. A. Lardon, *Ann. N.Y. Acad. Sci.*, **192,** 132 (1972).

20c. W. H. H. Günther and M. N. Salzman, *Ann. N.Y. Acad. Sci.*, **192,** 25 (1972).

20d. W. H. H. Günther, U.S. Pat. 3,671,467 (1972).

21. W. V. Farrar, *Research* (London), **4,** 177 (1951).

22. G. T. Morgan and H. D. K. Drew, *J. Chem. Soc.*, 2307 (1925).

23. G. T. Morgan and R. E. Kellet, *J. Chem. Soc.*, 1080 (1926).

24. S. Keimatsu, I. Satoda, and T. Kobayashi, *J. Pharm. Soc. Japan*, **56,** 869 (1936); *Chem. Abstr.*, **33,** 155 (1939).

25. J. R. Brown, G. P. Gillman, and M. H. George, *J. Polymer Sci.*, Part A, **5,** 903 (1967).

26. G. Bergson and G. Claeson, *Acta Chem. Scand.*, **11,** 911 (1957).

27. G. Bergson, *Ark. Kemi*, **19,** 195 (1962).

28. G. Bergson and A. Biezais, *Ark. Kemi*, **22,** 475 (1964).

29. B. Nygård, *Acta Chem. Scand.*, **20,** 1710 (1966).

30. B. Nygård, J. Olofsson, and G. Bergson, *Ark. Kemi*, **28,** 41 (1967).

31. B. Nygård, *Ark. Kemi*, **27,** 425 (1967) and earlier papers. A full discussion is given by Nygård in *Acta Univ. Upsaliensis, Abstr. Uppsala Dissertations Sci.*, 1967.

32. H. Zalkin, A. L. Tappel, and J. P. Jordan, *Arch. Biochem. Biophys.*, **91,** 117 (1960).

33. J. W. Hamilton and A. L. Tappel, *J. Nutr.*, **79,** 493 (1963).

34. G. H. Denison and P. C. Condit, *Ind. Eng. Chem.*, **41,** 944 (1949).

35. G. Ayrey, D. Barnard, and D. T. Woodbridge, *J. Chem. Soc.*, 2089 (1962).

36. D. T. Woodbridge, *J. Chem. Soc.*, B, 50 (1966).

37. F. Shimazu and A. L. Tappel, *Science*, **143,** 369 (1964).

38. F. Shimazu and A. L. Tappel, *Radiation Res.*, **23,** 210 (1964).

39. J. J. Windle, A. K. Wiersma, and A. L. Tappel, *J. Chem. Phys.*, **41,** 1996 (1964).

40. W. T. House and M. Orchin, *J. Amer. Chem. Soc.*, **82,** 639 (1960).

41. H. A. Silverwood and M. Orchin, *J. Org. Chem.*, **27,** 3401 (1962).

42. E. J. Levi and M. Orchin, *J. Org. Chem.*, **31,** 4302 (1966).

43. U. Schmidt, A. Müller, and K. Markau, *Chem. Ber.*, **97,** 405 (1964).

44. E. T. Strom and G. A. Russell, *J. Amer. Chem. Soc.*, **87,** 3326 (1965).

45. N. M. Atherton, J. N. Ockwell, and R. Dietz, *J. Chem. Soc.*, A, 771 (1967).

46. M. M. Urberg and E. T. Kaiser, "Radical Ions," E. T. Kaiser and L. Kevan, Eds., Interscience Publishers, New York, N.Y., 1968, p. 305.

47. J. Fajer, B. H. J. Bielski, and R. H. Felton, *J. Phys. Chem.*, **72,** 1281 (1968).

48. R. Gerdil and E. A. C. Lucken, *J. Amer. Chem. Soc.*, **87,** 213 (1965).

49. E. T. Strom and J. R. Norton, *J. Amer. Chem. Soc.*, **92,** 2327 (1970).

50a. Z. V. Todres, V. Sh. Tsveniashvili, S. I. Zhdanov, and D. N. Kursanov, *Proc. Acad. Sci. USSR*, **181,** 715 (1968).

50b. V. Sh. Tsveniashvili, Z. V. Todres, and S. I. Zhdanov, *J. Gen. Chem. USSR*, **38,** 1835, 1840 (1968).

50c. Z. V. Todres, Yu. I. Lyakhovetskii, and D. N. Kursanov, *Bull. Acad. Sci. USSR, Div. Chem. Sci.*, 1349 (1969).

51. J. Fajer, *J. Phys. Chem.*, **69,** 1773 (1965).

52. C. Barry, G. Cauquis, and M. Maurey, *Bull. Soc. Chim. Fr.*, 2510 (1968).

53. B. Lamotte and G. Berthier, *J. Chim. Phys.*, **63,** 369 (1966).

54. H. J. Shine and R. J. Abler, Jr., unpublished work.

55. H. J. Shine and E. E. Mach, *J. Org. Chem.*, **30,** 2130 (1965).

56a. Y. Tsujino, *Tetrahedron Lett.*, 763 (1969).

56b. Y. Tsujino, *Nippon Kagaku Zasshi*, **90,** 809 (1969). I am indebted to Prof. Kimio Ohno for the translation during his visit to Texas Tech University.

57. E. Müller, H. B. Stegmann, and K. Scheffler, *Justus Liebigs Ann. Chem.*, **657,** 5 (1962).

58. P. Natalis and J. L. Franklin, *J. Phys. Chem.*, **69,** 2943 (1965).

59. P. C. Wszolek, F. W. McLafferty, and J. H. Brewster, *Org. Mass Spectrom.*, **1,** 127 (1968).

60. J. Bergman, *Acta Chem. Scand.*, **22,** 1883 (1968).

61. L.-B. Agenäs, *Acta Chem. Scand.*, **22,** 1763 (1968).

62. J. Ø. Madsen, C. Nolde, S. O. Lawesson, G. Schroll, J. H. Bowie, and D. H. Williams, *Tetrahedron Lett.*, 4377 (1965).

63. L.-B. Agenäs, *Ark. Kemi*, **31,** 45 (1969), and earlier publications.

XV Physicochemical investigations of selenium compounds

G. MASS SPECTROMETRY

LARS-BÖRGE AGENÄS

Uppsala University
Institute of Chemistry
Uppsala, Sweden

1. Introduction

Although a great variety of organosulfur compounds have been studied by mass spectrometry and the results have been reviewed,[1,2] comparatively few mass spectra of the analogous organoselenium compounds have so far been reported. The occurrence of three stable natural isotopes of sulfur does not give rise to any significant complexity of the mass spectra of organosulfur compounds. Six stable natural isotopes of selenium (Fig. XVG-1), however, make the mass spectra of organoselenium compounds very rich in peaks,

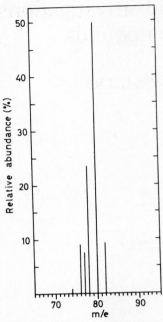

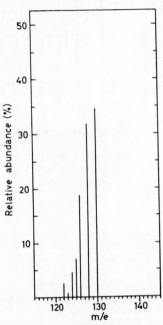

FIG. XVG-1   Graph representing the natural abundances of stable selenium isotopes (^{74}Se 0.87; ^{76}Se 9.02; ^{77}Se 7.58; ^{78}Se 23.52; ^{80}Se 49.82; ^{82}Se 9.19).

FIG. XVG-2   Graph representing the natural abundances of stable tellurium isotopes (^{120}Te 0.09; ^{122}Te 2.46; ^{123}Te 0.87; ^{124}Te 4.61; ^{125}Te 6.99; ^{126}Te 18.71; ^{128}Te 31.79; ^{130}Te 34.48).

often rendering the complete analysis of the spectra very difficult. This problem is still more pronounced for the few mass spectra recorded for organotellurium compounds, because there are eight stable natural isotopes of tellurium (Fig. XVG-2). Thus, organoselenium and organotellurium compounds are, in general, less suitable as model substances for mass spectrometric studies. However, selenium- and tellurium-containing fragments are often easily recognized in mass spectra from the very characteristic groups of peaks resulting from the typical distribution of natural isotopes (Fig. XVG-3), thus making it possible to study the fragmentation patterns involving the Se and Te atoms present in the molecule. For mass spectra of

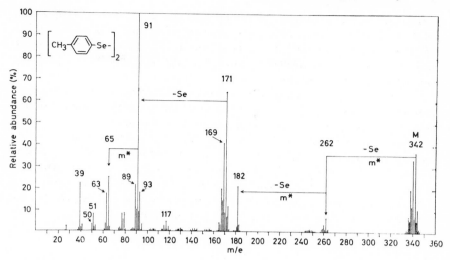

FIG. XVG-3 Mass spectrum of 4,4′-dimethyldiphenyl diselenide, recorded with the ion source at 20°. The ionization energy was 70 eV.[9]

selenium compounds, the peak arising from the most abundant ^{80}Se isotope is in general chosen to represent a selenium-containing fragment. In this respect, extensive application of mass spectrometry as an analytical tool for selenium and tellurium chemists is to be expected.

Mass spectra of organoselenium compounds are most conveniently obtained by use of the direct inlet technique. Many times an uncontrolled change of structure occurs when organoselenium compounds are subjected to gas chromatographic treatment. This fact makes it necessary, as a rule, to purify the samples very carefully before mass spectrometric investigations, even if fractionation to some degree is to be expected because of differences in the rate of volatilization between the components in a mixture of substances.

From the mass spectrometric investigations of organoselenium compounds so far published, two main stability categories can be observed.

1. *The organoselenium compound is stable towards heating.* The selenium-containing groups of the molecule usually split off in a very characteristic way from the molecular ion upon electron impact. Sometimes other groups, less stable towards electron impact than the selenium-containing ones, will be lost simultaneously, giving rise to a more complex spectrum, because of alternative fragmentation paths. Frequently, low-molecular-weight selenium-containing fragments are observed, providing evidence for the high stability of certain C–Se bonds. Peaks arising from such fragments usually are of a

comparatively low abundance, making them less useful from an analytical point of view. Selenium-containing impurities, often not observable by other methods, are easily seen in mass spectra. For instance, diselenides give rise to easily observed groups of peaks centered at m/e $M + 80$, where M is the molecular weight of the monoselenide.

2. *The organoselenium compound is labile towards heating.* Some organoselenium compounds are unstable even towards very moderate heating, and the result is a separation of elemental selenium from the sample. In such cases, the initial vaporization of the sample gives rise to decomposition before ionization. The mass spectrum in such a case shows no molecular ion or one of very low abundance. The fragmentation observed is consequently in accord with the resulting selenium-free molecule. Selenium-containing fragments of an organic nature are of very low abundance. The peaks in the spectrum are always overlapped by a characteristic series of peaks from Se_n $(n = 1–8)$ fragments.[3–5]

2. Mass Spectra of Aliphatic Selenium Compounds

a. *Dialkyl Selenides.*   The mass spectrum for diethyl selenide has been reported by Bogolyubov *et al.*[6] Evidence was found in the spectrum for a main fragmentation pattern (scheme 1) in accord with what has proved to be

$$[C_2H_5Se]^+ \xleftarrow{-C_2H_5} [C_2H_5SeC_2H_5]^{+\cdot} \xrightarrow{-C_2H_4} [C_2H_5SeH]^+$$

$$m/e\ 109 \qquad\qquad m/e\ 138 \qquad\qquad m/e\ 110$$

(1)

$$\downarrow{-C_2H_5} \quad \searrow{-C_2H_4} \qquad\qquad\qquad\qquad\qquad \downarrow{-C_2H_4}$$

$$[Se] \qquad [HSe] \qquad\qquad\qquad\qquad\qquad [H_2Se]$$

$$m/e\ 80 \qquad m/e\ 81 \qquad\qquad\qquad\qquad\qquad m/e\ 82$$

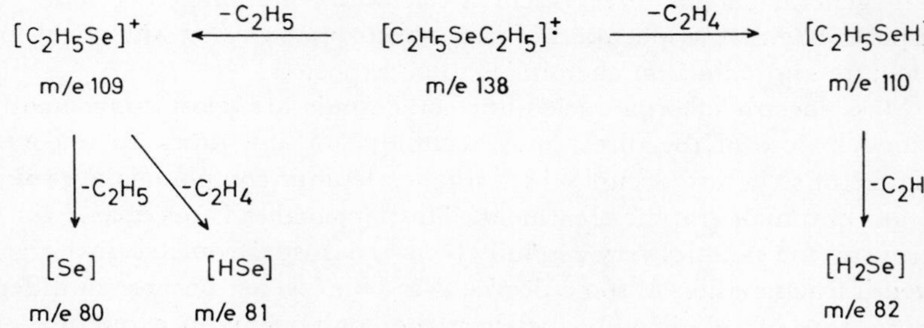

characteristic for dialkyl sulfides.[7] In addition, prominent fragments at m/e 29, 28, and 27 were observed to occur from $[C_2H_5]^+$ $[C_4H_4]^{+\cdot}$, and $[C_2H_3]^+$ ions.

A similar fragmentation pattern, deduced from the spectrum for 2-ethylseleno-2-hexene-4-yne, was reported by Gronowitz and Frejd.[8]

The mass spectra of dibenzyl selenide and 2-carboxydibenzyl selenide have been reported by Agenäs.[9] In the case of dibenzyl selenide, prominent peaks were observed, suggesting the fragmentation pattern in scheme 2. The mass spectrum recorded for the unsymmetrical 2-carboxydibenzyl selenide displayed a fragmentation pattern similar to that in scheme 2. The more complex spectrum observed is a consequence of the alternative losses of the two different groups attached to the Se atom. The presence of the carboxylic

(2) $\left[\text{(C}_6\text{H}_5)-CH_2-Se-CH_2-(C}_6\text{H}_5) \right]^{+\cdot}$ $\xrightarrow{-HSe}$ $(C_6H_5)-CH_2-$ (+)

m/e 262   m/e 181

$\downarrow -C_6H_5CH_2\cdot$ $\downarrow -C_7H_6$

$C_7H_7Se^{\oplus}$ $\xrightarrow{-Se}$ (+)

m/e 171   m/e 91

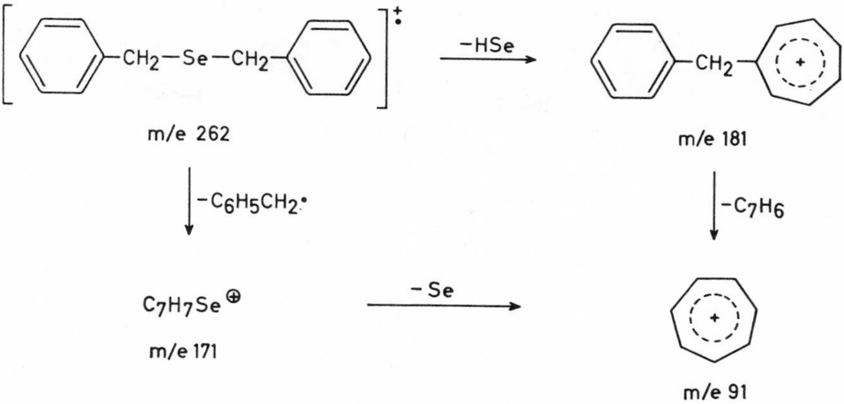

group gives further fragmentation steps in accord with what has been found for carboxylic acids.[10] The latter fragmentation steps do not, however, influence the pattern as regards the selenide group.

The mass spectrum recorded for 3-indolylmethyl benzyl selenide by Agenäs[11] was similarly complex in character as a consequence of the unsymmetrical molecule. The fragmentation pattern was observed also for the sulfur analog.[11] One interesting difference between the mass spectra of the two analogous compounds is that, whereas the molecular ion of 3-indolylmethyl benzyl selenide readily splits off a benzylseleno group as well as benzylselenol, in the sulfur analog the loss of a benzylthio group predominates.

The analogous fragmentation pattern was observed also in the mass spectrum of the symmetrical analog di(3-indolyl)methyl selenide,[11] which was found to fragment similarly in many respects to dibenzyl selenide upon electron impact. The indolylic groups of di(3-indolyl)methyl selenide, however, give rise to other fragmentation paths, in addition to those described in scheme 2, by which the Se atom can be split off.

The mass spectra recorded for nine homologous selenodialkanoic acids, reported by Agenäs and Lindgren,[12] give evidence of fragmentation patterns similar to the one for diethyl selenide which is characterized by successive cleavages of the two C–Se bonds. Fragmentation steps for common carboxylic acids, such as the expulsion of water molecules,[13] the carboxyl group, or a formic acid molecule[14–16] from the molecular ion, were also observed, giving rise to more complex spectra. Thus, in the case of selenodiacetic acid, the fragmentation pattern in scheme 3 was found to occur. Metastable peaks† were observed to support some fragmentation steps. Further selenium-containing fragments were observed at m/e 109, 107, 95, 94, 93, and 80,

† An asterisk is used in scheme 3 and elsewhere in this chapter to designate the observation of metastable ions.

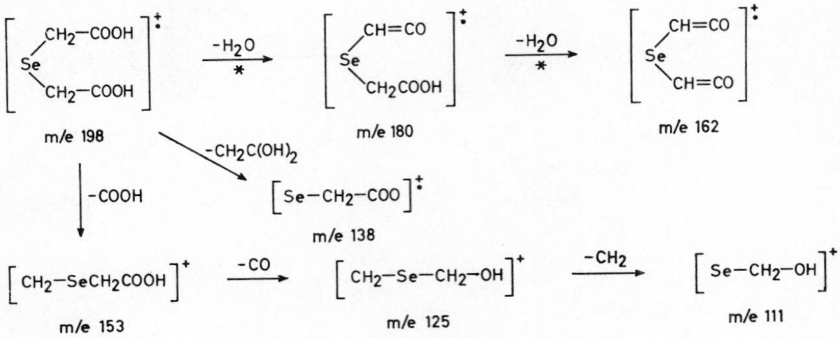

(3)

$$\left[\begin{array}{c} CH_2-COOH \\ Se \\ CH_2-COOH \end{array} \right]^{\ddagger} \xrightarrow[*]{-H_2O} \left[\begin{array}{c} CH=CO \\ Se \\ CH_2COOH \end{array} \right]^{\ddagger} \xrightarrow[*]{-H_2O} \left[\begin{array}{c} CH=CO \\ Se \\ CH=CO \end{array} \right]^{\ddagger}$$

m/e 198 m/e 180 m/e 162

$\xrightarrow{-CH_2C(OH)_2}$ $\left[Se-CH_2-COO \right]^{\ddagger}$ m/e 138

$\downarrow -COOH$

$\left[CH_2-SeCH_2COOH \right]^+ \xrightarrow{-CO} \left[CH_2-Se-CH_2-OH \right]^+ \xrightarrow{-CH_2} \left[Se-CH_2-OH \right]^+$

m/e 153 m/e 125 m/e 111

corresponding to $[C_2H_5Se]^+$, $[C_2H_3Se]^+$, $[CH_3Se]^+$, $[CH_2Se]^{\ddagger}$, $[CHSe]^+$, and $[Se]^{\ddagger}$ fragment ions.

In the mass spectra recorded for the higher homologs, there was found a somewhat different fragmentation pattern. Scheme 4, deduced for 3,3′-selenodipropionic acid, is representative. Other selenium-containing

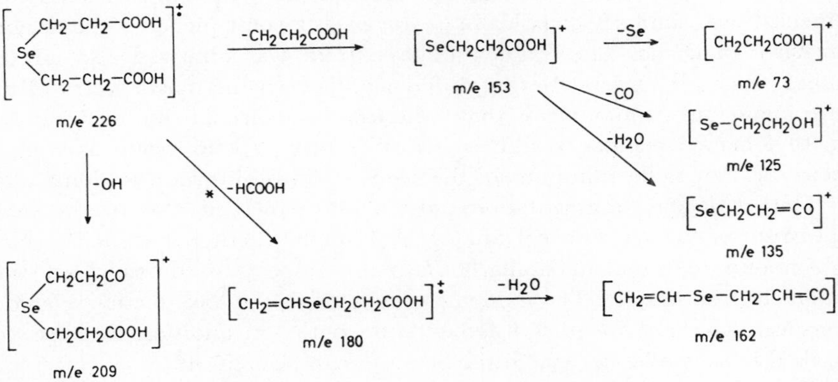

(4)

$$\left[\begin{array}{c} CH_2-CH_2-COOH \\ Se \\ CH_2-CH_2-COOH \end{array} \right]^{\ddagger} \xrightarrow{-CH_2CH_2COOH} \left[SeCH_2CH_2COOH \right]^+ \xrightarrow{-Se} \left[CH_2CH_2COOH \right]^+$$

m/e 226 m/e 153 m/e 73

$\xrightarrow{-CO}$ $\left[Se-CH_2CH_2OH \right]^+$ m/e 125

$\xrightarrow{-H_2O}$ $\left[SeCH_2CH_2=CO \right]^+$ m/e 135

$\downarrow -OH$ $\xrightarrow[*]{-HCOOH}$

$\left[\begin{array}{c} CH_2CH_2CO \\ Se \\ CH_2CH_2COOH \end{array} \right]^+$ $\left[CH_2=CHSeCH_2CH_2COOH \right]^{\ddagger} \xrightarrow{-H_2O} \left[CH_2=CH-Se-CH_2-CH=CO \right]^{\ddagger}$

m/e 209 m/e 180 m/e 162

fragments were observed at m/e 108, 107, 95, 93, 81, and 80, corresponding to $[C_2H_4Se]^{\ddagger}$, $[C_2H_3Se]^+$, $[CH_3Se]^+$, $[CHSe]^+$, $[HSe]^+$, and $[Se]^{\ddagger}$ fragments.[12]

In the mass spectra of the higher homologs, 4,4′-selenodibutyric acid, 5,5′-selenodivaleric acid, 6,6′-selenodihexanoic acid, 7,7′-selenodiheptanoic acid, 8,8′-selenodioctanoic acid, 9,9′-selenodinonanoic acid, and 10,10′-selenodidecanoic acid, the molecular ions were found to simultaneously split off an unsaturated carboxylic acid molecule and a water molecule. This may be explained by McLafferty-type rearrangements, as illustrated for 4,4′-selenodibutyric acid in equation 5 and for 7,7′-selenodiheptanoic acid in scheme 6.[12]

(5)

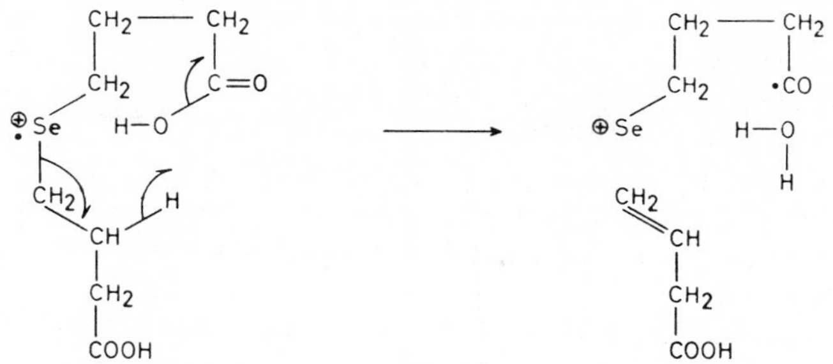

(6) 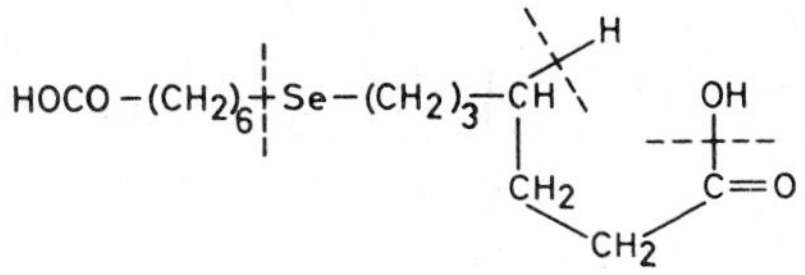 HOCO−(CH₂)₆−Se−(CH₂)₃−CH ...

The mass spectrum of selenomethionine has been reported by Svec and Junk,[17] and the fragmentation observed was compared with that found for methionine. Good accord between the two spectra was evident, the fragmentation patterns being similar to those for selenodialkanoic acids.[12]

Some benzylseleno-substituted, branched carboxylic acids have been investigated by mass spectrometry by Agenäs.[18] The fragmentation patterns were found to be alike in some respects to those reported for straight-chain selenodialkanoic acids.[12] Thus, the spectrum recorded for 3-benzylseleno-isovaleric acid gave evidence of the fragmentation pattern described in scheme 7.

The mass spectra of 4-benzylseleno-2,2-dimethylbutyric acid, 4-benzyl-seleno-3,3-dimethylbutyric acid, and 5-benzylseleno-3,3-dimethylvaleric

(7)

acid suggested fragmentation patterns similar to the one for 3-benzylseleno-isovaleric acid, but in these cases the two C-Se bonds are alternatively broken to form fragment ions rather than a benzylselenol fragment, as described for 3-benzylselenoisovaleric acid in scheme 7. This type of cleavage is expected in these cases, along with β-splittings of the carbon chains. Similar fragmentations were observed in the mass spectrum of 4-benzylseleno-2,3-dimethyl-2-butenyl acetate, which was reported by Mock and McCausland.[19]

In the mass spectra of β-(3-indolyl)ethylselenoacetic acid, β-(3-indolyl)-ethylseleno-β-propionic acid, β-(3-indolyl)ethylseleno-γ-butyric acid, and β-(3-indolyl)ethylseleno-δ-valeric acid, reported by Agenäs,[20] fragmentation behavior comparable to that found for 3-benzylselenoisovaleric acid was observed. The C-Se bonds of the molecular ion in these compounds do not have equal stability, and indolylethylselenium ions or indolylethaneselenol fragments were exclusively formed; in a succeeding step these split off the Se atom to give very stable indolylic ions.

b. Dialkyl Diselenides. The mass spectrum of diethyl diselenide has been reported by Bogolyubov et al.[6] The fragmentation pattern in scheme 8

$$[SeSeC_2H_5]^+ \xleftarrow{-C_2H_5} [-SeC_2H_5]_2^+ \xrightarrow{-C_2H_4} [HSeSeC_2H_5]^{+}$$

m/e 189   m/e 218   m/e 190

(8) $\downarrow_{-C_2H_5}$ $\downarrow_{-SeC_2H_5}$ $\downarrow_{-C_2H_4}$

$$[SeSe]^{+} \quad\quad [SeC_2H_5]^+ \quad\quad [HSeSeH]^{+}$$

m/e 160   m/e 109   m/e 162

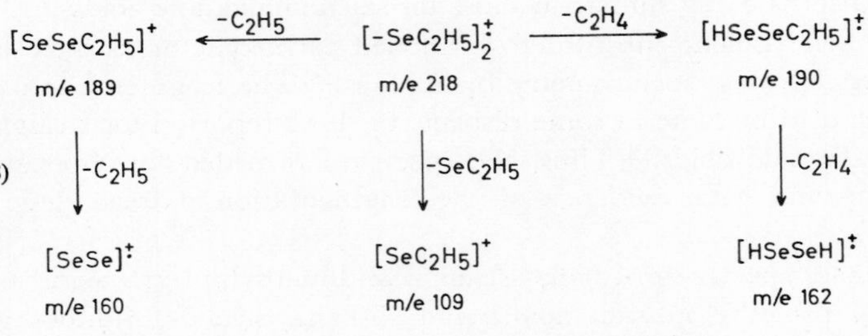

follows from this spectrum and is similar to what has been seen for aliphatic disulfides.[21] In addition, there were other selenium-containing fragments at m/e 161, 110, 108, 82, 81, and 80, corresponding, respectively, to $[HSe_2]^+$, $[C_2H_5SeH]^+$, $[C_2H_4Se]^+$, $[H_2Se]^+$, $[HSe]^+$, and $[Se]^{+}$ ions. This type of fragmentation was also found for dibutyl diselenide.[22]

In the mass spectrum recorded for dibenzyl diselenide by Agenäs,[9] peaks were observed in accord with a fragmentation pattern similar to that for the monoselenide analog. Successive losses of the two Se atoms are favored by the formation of very stable fragments with the tropylium ion structure (scheme 9). An analogous fragmentation occurs with dibenzyl disulfide.[21]

In the mass spectrum of di(3-indolyl)methyl diselenide determined by Agenäs,[11] the absence of peaks corresponding to the molecular ion indicated a low thermal stability of the sample. This observation is in accord with results from the synthesis of di(3-indolyl)methyl diselenide. The occurrence

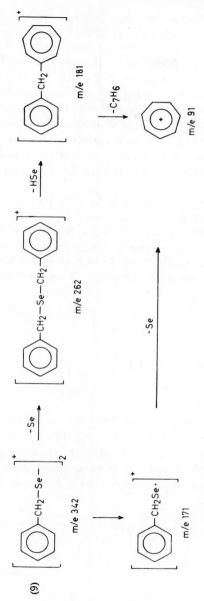

of peaks corresponding to Se_n fragments is further evidence of thermal decomposition of the sample before ionization. The corresponding disulfide[11] was found to be more stable towards heat and gave a spectrum similar to that of dibenzyl disulfide, while the spectrum for di(3-indolyl)methyl diselenide was almost identical with that of the analogous monoselenide. In the mass spectrum recorded for β,β'-di(3-indolyl)ethyl diselenide,[20] evidence was found for its higher thermal stability. Low-intensity peaks corresponding to the molecular ion and other selenium-containing fragments are indicative of the fragmentation pattern in scheme 10. A selenium-containing fragment of low abundance at m/e 368 corresponds to the loss of one Se atom from the molecular ion to form the analogous monoselenide fragment.

The mass spectra of a series of dihydroxydialkyl diselenides have been reported by Agenäs.[22] In the spectra for 3,3'-dihydroxydipropyl diselenide, 4,4'-dihydroxydibutyl diselenide, 4,4'-dihydroxy-2,2'-dibutyl diselenide, and 3,3'-dihydroxydiisobutyl diselenide prominent peaks disclosed similar fragmentation patterns, as illustrated for 3,3'-dihydroxydipropyl diselenide in scheme 11. Other selenium-containing fragments were observed at m/e

(11)

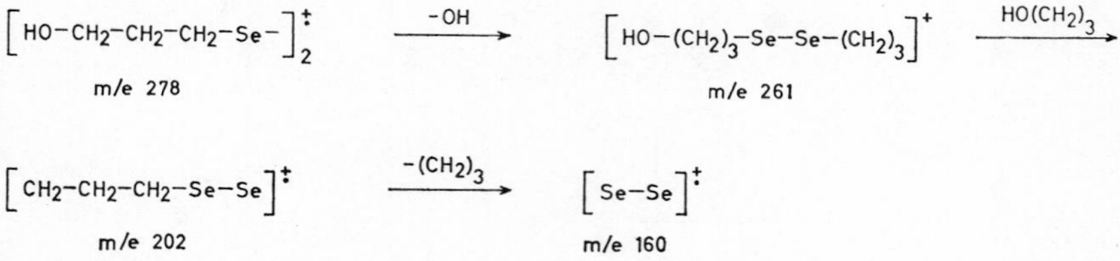

137, 122, 121, 109, 107, 95, and 93 giving evidence for the formation of $[\frac{1}{2}M-H_2]^+$, $[C_3H_6Se]^{+\cdot}$, $[C_3H_5Se]^+$, $[C_2H_5Se]^+$, $[C_2H_3Se]^+$, $[CH_3Se]^+$, and $[CHSe]^+$ fragment ions. The spectrum recorded for 3,3'-dihydroxydi-(2,2-dimethyl)propyl diselenide[22] showed peaks from selenium-containing fragments corresponding to another type of fragmentation pattern (scheme 12).

(12)

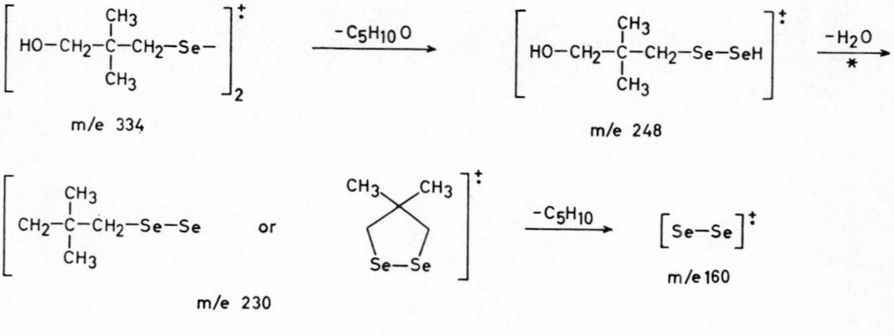

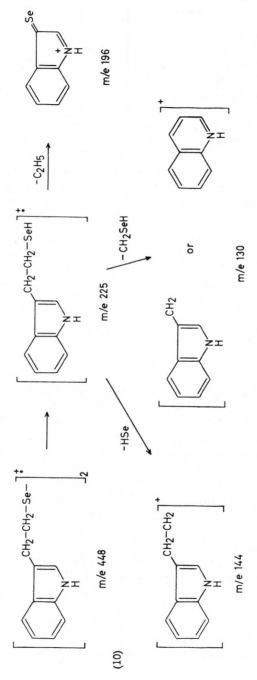

m/e 196

m/e 225

m/e 130

m/e 448

m/e 144

−C₂H₅

−CH₂SeH

−HSe

or

(10)

The mass spectra of 5,5'-diacetoxydipentyl diselenide and 5,5'-diacetoxy-2,2'-dipentyl diselenide have been compared.[22] These compounds were found to fragment similarly to 3,3'-dihydroxydipropyl selenide. Initially, one acetoxy group is split off from the molecular ion, and in a succeeding step the one C-Se bond is broken. The resulting selenium-containing fragment at m/e 230 will split off either the remaining alkyl group or an HSe moiety.

The mass spectra of nine homologous, symmetrical diselenodialkanoic acids have been reported by Agenäs and Lindgren.[12] In many respects the spectra indicated fragmentation related to what was described for diethyl diselenide in scheme 8; in addition, fragmentation steps in accord with those characteristic for carboxylic acids[10] were found. Thus, in the mass spectrum of diselenodiacetic acid, prominent peaks from selenium-containing fragments were observed to support the pattern given in scheme 13. Addi-

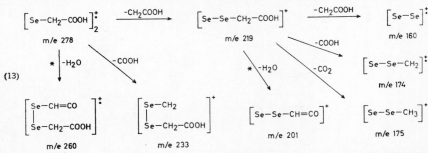

tional selenium-containing fragments were observed at m/e 140, 122, 111, 95, 94, 93, and 80, indicative of the formation of $[SeCH_2C(OH)_2]^+$, $[SeCH_2CO]^+$, $[SeCH_2OH]^+$, $[CH_3Se]^+$, $[CH_2Se]^+$, $[CHSe]^+$, and $[Se]^+$ fragment ions.

In the spectra recorded for the higher homologs, evidence was found for other types of fragmentation patterns. For instance, in the spectrum of 3,3'-diselenodipropionic acid the fragmentation depicted in scheme 14 was observed. Further selenium-containing fragments were noted at m/e 135, 125, 108, 107, 93, 81, and 80, giving support for the formation of $[SeCH_2CH=C=O]^+$, $[SeCH_2CH_2OH]^+$, $[C_2H_4Se]^+$, $[C_2H_3Se]^+$, $[CHSe]^+$, $[HSe]^+$, and $[Se]^+$ ions.

The diselenides 4,4'-diselenodibutyric acid, 5,5'-diselenodivaleric acid, 6,6'-diselenodihexanoic acid, 7,7'-diselenodiheptanoic acid, 8,8'-diseleno-dioctanoic acid, 9,9'-diselenodinonanoic acid, and 10,10'-diselenodidecanoic acid were found to cleave similarly to the straight-chain selenodialkanoic acids to which they correspond. The molecular ions of these samples simultaneously split off an unsaturated carboxylic acid molecule and a water molecule; this doubtlessly occurs by a McLafferty-type rearrangement, as illustrated for 4,4'-diselenodibutyric acid in equation 15.

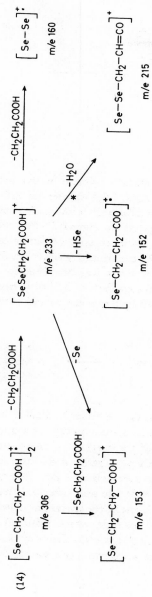

$$(15)$$

The analogous fragmentation pattern of the branched carboxylic acids 3,3′-diselenodiisovaleric acid, 4,4′-diselenodi(2,2-dimethyl)butyric acid, 4,4′-diselenodi(3,3-dimethyl)butyric acid, and 5,5′-diselenodi-(3,3-dimethyl)-valeric acid was investigated by Agenäs.[18] For the last three compounds the molecular ions were found to split off an unsaturated carboxylic acid molecule and a water molecule simultaneously. In the case of 3,3′-diselenodi-isovaleric acid, contrary to what was described for 3,3′-diselenodipropionic acid, the molecular ion expelled a 3,3′-dimethylacrylic acid molecule; this is reasonably explained by a rearrangement of the type given in equation 16.

$$(16)$$

$$m/e\ 362 \longrightarrow m/e\ 262$$

c. **Selenonium Compounds.** Although dialkyl selenides and diselenides have been studied by mass spectrometry by various investigators and the fragmentation patterns of these types of compounds are known to follow certain general paths, only one compound of the selenonium type has so far been investigated by mass spectrometry. Byard[23] reported the mass spectrum for trimethylselenonium chloride and observed that peaks corresponding to the molecular ion were absent in the spectrum. The sample decomposes upon heating to yield dimethyl selenide, which splits off one methyl group to give a methylselenium fragment at m/e 95. This explanation is further supported by the base peak of the spectrum, originating at m/e 50 from a methyl chloride fragment.

3. Mass Spectra of Aromatic Selenium Compounds

a. **Aryl Selenocyanates.** A number of selenocyanato-substituted aromatic compounds have been investigated by mass spectrometry, and the results obtained have been summarized by Agenäs.[24] The investigation was started by studying phenyl selenocyanate and 4-methylphenyl selenocyanate.[25] The spectra of these compounds were found to contain easily identified groups of peaks corresponding to selenium-containing fragments, suggesting a two-step expulsion of the selenocyanato group along two alternative paths, as illustrated for phenyl selenocyanate in scheme 17. Additional selenium-containing fragments were observed at m/e 156, 117, 93, and 80 corresponding

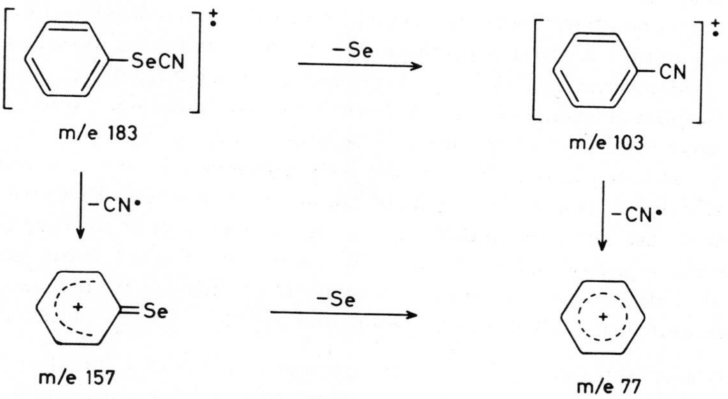

to $[C_6H_4Se]^{+\cdot}$, $[C_3HSe]^+$, $[CHSe]^+$, and $[Se]^{+\cdot}$ fragment ions. These fragments indicate a possible rupture of the aromatic ring before the Se atom had split off.

The spectrum recorded for 4-methylphenyl selenocyanate gave evidence for the same fragmentation pattern. Similarly, the spectra for 3-seleno-cyanatoindole, 1-methyl-3-selenocyanatoindole, and 2-methyl-3-seleno-cyanatoindole, reported by Agenäs,[22] showed peaks in accord with the pattern described in scheme 17. This result was compared with what was found for the sulfur analogs phenyl thiocyanate and 3-thiocyanatoindole, which upon electron impact fragmented along similar paths.[26] The mass spectra for 2-methyl-3-selenocyanatopyrrole and 2,5-dimethyl-3-seleno-cyanatopyrrole reported by Agenäs and Lindgren,[27] as well as the spectrum recorded for 5-selenocyanatoindole by Agenäs,[28] also indicated the same type of fragmentation.

In the spectrum of 5-selenocyanatoindoline,[28] however, peaks were found in accord with an alternative fragmentation path, by which a HSe group is split off from the molecular ion. The H atom, no doubt, splits off from the saturated five-membered ring. It was also observed that, after the loss of a Se atom or a cyanide group from the molecular ion of 5-seleno-cyanatoindoline, the succeeding fragmentation step was a rearrangement whereby a hydrogen cyanide molecule or a HSe fragment was expelled and a very stable indole fragment ion was formed.

In another investigation, Agenäs recorded[29] the mass spectra for 4-aminophenyl selenocyanate, 2-methyl-4-aminophenyl selenocyanate, and 3-methyl-4-aminophenyl selenocyanate; these showed fragmentation patterns in marked agreement with the one observed for 5-selenocyanatoindoline. In the spectrum for 4-aminophenyl selenocyanate, however, peaks for aniline[30,31] were seen, giving a complex spectrum very difficult to analyze. This spectrum was further compared with the spectra recorded for N-methyl-4-aminophenyl selenocyanate, N-ethyl-4-aminophenyl selenocyanate, N-propyl-4-aminophenyl selenocyanate, N-butyl-4-aminophenyl selenocyanate, N-sec-butyl-4-aminophenyl selenocyanate, N,N-dimethyl-4-aminophenyl selenocyanate, N-methyl-N-ethyl-4-aminophenyl selenocyanate, N,N-diethyl-4-aminophenyl selenocyanate, N,N-dipropyl-4-aminophenyl selenocyanate, and N,N-dibutyl-4-aminophenyl selenocyanate. In this series of spectra, groups of peaks were observed, indicating fragmentations similar to those proposed for 4-aminophenyl selenocyanate. By extension of alkyl substitution, the spectra were found to be more markedly characterized by prominent fragmentations of the alkyl groups before the expulsion of the Se atom.

b. Diaryl Selenides. Only a few aromatic selenides have thus far been investigated by mass spectrometry. Bergman[32] has determined the spectra for 3,3'-diindolyl selenide and 3,3'-di(1-methylindolyl) selenide. In these spectra, prominent peaks were displayed which apparently resulted from the loss of the Se atom from the molecular ion and were due to a biindolyl fragment (scheme 18).

(18)

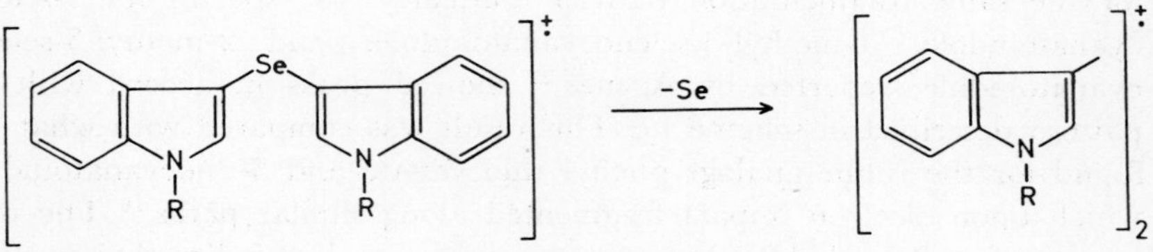

An alternative fragmentation path also occurs, by which one indolyl group is split off from the molecular ion. The spectra for 3,3'-diindolyl selenide and 3,3'-di(1 methylindolyl) selenide were compared with the spectrum recorded for diphenyl selenide, in which peaks were found in agreement with an analogous fragmentation pattern.

Korček et al.[33] reported the mass spectrum for 2,2'-dihydroxy-3,3'-di-tert-butyl-5,5'-dimethyldiphenyl selenide. The spectrum obtained was not analyzed, however, in regard to the fission of the selenide group. One C-Se bond of this compound is broken initially upon electron impact, simultaneously with a hydrogen transfer from the one hydroxyl group to the selenium-free fragment formed.

c. Diaryl Diselenides. In the mass spectra recorded for diphenyl diselenide and 4,4'-dimethyldiphenyl diselenide by Agenäs,[9] characteristic series of peaks were observed, indicative of the successive loss of the two Se atoms from the molecular ion or, alternatively, the splitting of the molecular ion into halves. These observations lead to the fragmentation pattern shown in scheme 19 for diphenyl diselenide. Further selenium-

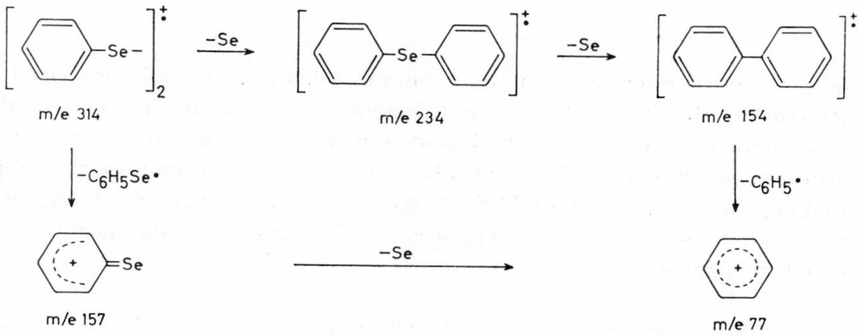

containing fragments were observed at m/e 117 and 93, arising from $[C_3HSe]^+$ and $[CHSe]^+$ ions formed by splitting of the aromatic ring.

Other aromatic diselenides, 3,3'-diindolyl diselenide, 3,3'-di(1-methyl-indolyl) diselenide, and 3,3'-di(2-methylindolyl) diselenide, investigated by Agenäs,[25] gave mass spectra with peaks in accord with a fragmentation pattern analogous to that described in scheme 19. Moreover, the same pattern was found for 5,5'-diindolinyl diselenide and 5,5'-diindolyl diselenide.[34] A similar fragmentation pattern has also been found for aromatic disulfides.[21]

d. Diaryl Selenoxides. Aromatic as well as aliphatic sulfoxides have been studied by mass spectrometry, and characteristic fragmentation patterns were observed.[35] Thus, rearrangements were found to occur within the sulfoxide group, by which alkyloxy as well as alkylthio fragments could be formed from the molecular ion. The aromatic analogs fragmented similarly upon electron impact. So far, only two aromatic selenoxides have been investigated by mass spectrometry. Rebane[36] found that diphenyl selenoxide and 4,4'-dimethyldiphenyl selenoxide are fragmented similarly to the analogous sulfur compounds, as shown for diphenyl selenoxide in scheme 20.

e. Areneselenonic Acids and Derivatives. Few compounds containing an –SeO$_2$– group have so far been subjected to electron impact. The mass spectra

$$[(C_6H_5)_2 Se]^{\ddagger} \xleftarrow{-O} [(C_6H_5)_2 SeO]^{\ddagger} \longrightarrow [C_6H_5O-SeC_6H_5]^{\ddagger} \xrightarrow{*} [C_6H_5O]^{+}$$

m/e 234   m/e 93

(20) $* \Big|$ -Se $* \Big|$ -Se $[C_6H_5Se]^{+}$ $* \Big|$ -CO

m/e 157

$$[C_6H_5-C_6H_5]^{\ddagger} \qquad [(C_6H_5)_2O]^{\ddagger} \qquad \Big|$$ -Se $[C_5H_5]^{+}$

m/e 154   m/e 170 $[C_6H_5]^{+}$ m/e 65

m/e 77

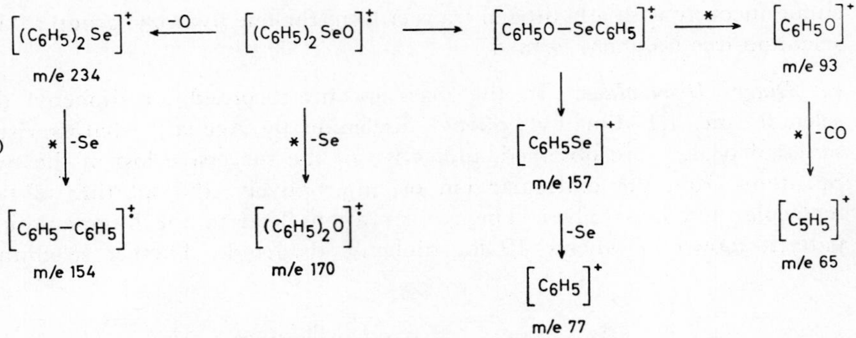

of two methyl esters of aromatic selenonic acids have been investigated by Rebane,[37] who found that methyl 4-chlorobenzeneselenonate and methyl 4-bromobenzeneselenonate are fragmented similarly, in some respects, to aromatic selenoxides. The molecular ions were thus found to rearrange, making the expulsion of a CH_3SeO_2 group possible. Other prominent peaks were also observed, giving evidence for the fragmentation pattern indicated in scheme 21.

$$[4-Cl-C_6H_4-SeO_3CH_3]^{\ddagger} \longrightarrow [4-Cl-C_6H_4-O-SeO_2CH_3]^{\ddagger} \xrightarrow{-CH_3SeO_2} [4-Cl-C_6H_4O]^{+}$$

m/e 254   m/e 254   m/e 127

$-CH_3SeO_3$

$[ClC_6H_4]^{+}$

m/e 111

(21) $\Big|$ -CH_3O $\Big\uparrow$ -Se $* \Big|$ -CO

$[4-Cl-C_6H_4Se]^{+}$

m/e 191

$$[4-Cl-C_6H_4SeO_2]^{+} \xrightarrow{-O} [4-Cl-C_6H_4SeO]^{+} \xrightarrow[*]{-CO} [ClC_5H_4Se]^{+} \xrightarrow{-Se} [ClC_5H_4]^{+}$$

m/e 223   m/e 207   m/e 179   m/e 99

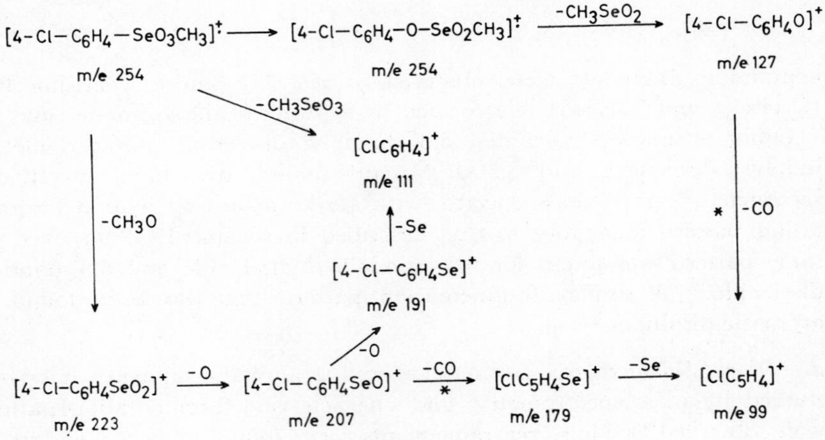

4. Mass Spectra of Heterocyclic Selenium Compounds

a. Cycloaliphatic Selenium Compounds. Cyclic sulfides of different ring sizes have been investigated by mass spectrometry, and their characteristic fragmentation patterns are well known.[38] Among their selenium analogs, thus far only five-membered rings have been subjected to electron impact. For instance, the parent compound selenolane has been investigated by mass spectrometry by Duffield *et al.*,[39] and its spectrum compared with the spectra for thiolane and tellurolane. In order to make a proper assignment to the peaks observed in the spectra, 2,2,5,5-tetradeuteriothiolane, -selenolane, and -tellurolane were also investigated by mass spectrometry. It was found that selenolane is fragmented similarly to the sulfur analog; however, interesting differences in the fragmentation patterns were observed. Whereas thiolane will readily split off an ethylene molecule from the molecular ion to form the most abundant fragment of the spectrum at m/e 60, the base peak in the spectrum of selenolane was found to occur from the expulsion of a SeH fragment from the molecular ion. Other prominent peaks in the spectrum of selenolane provided evidence for the fragmentation pattern in scheme 22. The expulsion of an ethylene molecule from the molec-

$$m/e\ 108 \qquad m/e\ 136 \qquad m/e\ 55$$

(22)

$-C_2H_5 \qquad -C_3H_6$

$$m/e\ 107 \qquad [Se{=}CH_2]^{\ddagger} \quad m/e\ 94$$

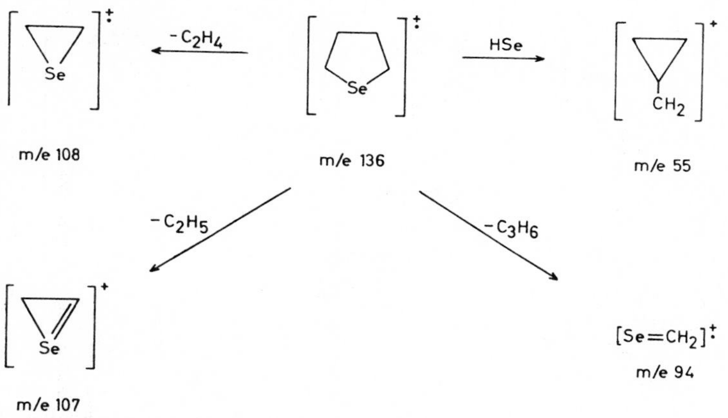

ular ion occurs also in this case, but gives an ion of considerably less abundance at m/e 108. The resulting fragment is followed by another at m/e 107 of related structure. Low-molecular-weight selenium-containing fragments were observed at m/e 94, 81, and 80, corresponding to $[CH_2Se]^{\ddagger}$, $[HSe]^+$, and $[Se]^{\ddagger}$ ions.

In the mass spectrum recorded for tellurolane the most abundant fragment was similarly at m/e 55, apparently formed by the expulsion of a TeH fragment from the molecular ion. The observation of other tellurium-containing fragments, of very low abundance, indicated a fragmentation pattern resembling that of selenolane.

In an investigation of selenium-containing monosaccharides, van Es and Whistler[40] used mass spectrometry in an attempt to determine the structures of two isomeric D-threo-3,4-dihydroxyselenolane-2-dimethyl acetals, supposedly the C_2-anomers. The recorded spectra were not analyzed completely. From the reported peaks, the observation of a selenophene ion occurring at m/e 132 is of importance.

b. Selenophene and Related Aromatic Compounds. Selenophene and alkyl-substituted selenophenes have not as yet been investigated by mass spectrometry. However, Chizhov *et al.*[41] have recorded the mass spectra for a number of alkylseleno-substituted selenophenes and related compounds. In this investigation, 2-methylselenoselenophene was studied in comparison with 2-methylselenofuran and 2-methylselenothiophene, as well as 2-methylthioselenophene and 3-methylthioselenophene. The mass spectrum recorded for 2-methylselenoselenophene was found to contain prominent peaks, giving evidence for a fragmentation pattern very similar to the patterns of 2-methylthiothiophene, 3-methylselenothiophene, and 2-methylthioseleno-phene, as described in scheme 23. In the spectra for these three compounds,

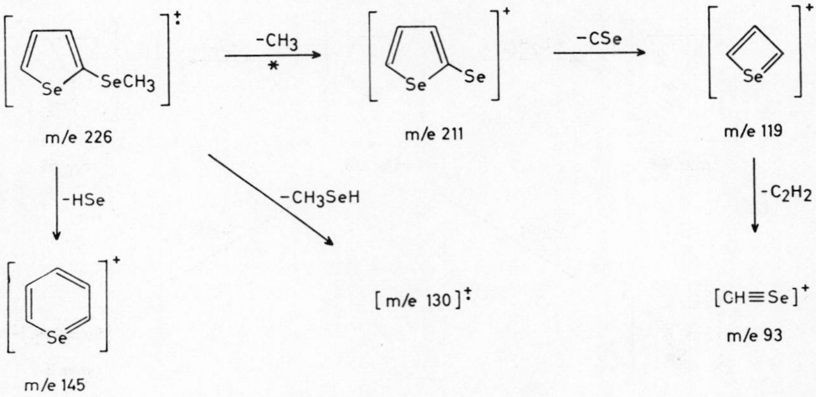

peaks were observed at *m/e* 71 and 45, corresponding to sulfur-containing fragments and analogous to those found at *m/e* 119 and 93 in the spectrum of 2-methylselenoselenophene.

The mass spectrum of 2-methylselenofuran indicated a fragmentation pattern different in some respects from that of 2-methylselenoselenophene, but analogous to what was observed for 2-methylthiofuran. The fragmentations related to the spectrum of 2-methylselenofuran are described in scheme 24.

When the mass spectrum recorded for 2-methyl-5-methylselenothiophene was compared with that for 2-methyl-5-methoxythiophene, analogous fragmentation patterns were noted. Peaks in accord with fragmentations analogous to those in scheme 24 were observed in these spectra. Similarly, the spectra of the two isomers, 2-butylselenothiophene and 3-butylseleno-thiophene, were indicative of fragmentations similar to those in scheme 24.

Furthermore, the mass spectrum for 2-methylthio-5-methylselenothio-phene was compared with that for its 5-*d₃*-methylseleno analog. It was found that the H₃C-Se bond is broken more readily than the H₃C-S bond. Similar

(24)

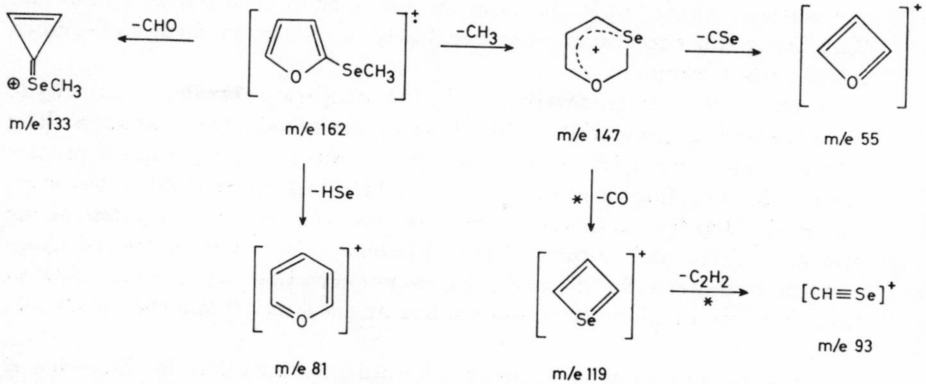

fragmentation patterns were also observed for 2-propylthio-5-methylseleno-thiophene, 2-methylthio-5-propylselenothiophene, and 2-methylthio-5-propylselenofuran.

In an investigation of benzo[b]selenophenes, Buu-Hoi et al.[42] found that these compounds are much less stable to electron impact than the analogous benzo[b]thiophenes. In the mass spectrum of the parent compound benzo[b]selenophene, evidence was found that selenium is expelled from the molecular ion much more readily than sulfur was observed to be split off from the analogous benzo[b]thiophene. The characteristic fragmentations of benzo[b]selenophene are given in scheme 25. The loss of selenium from the

(25)

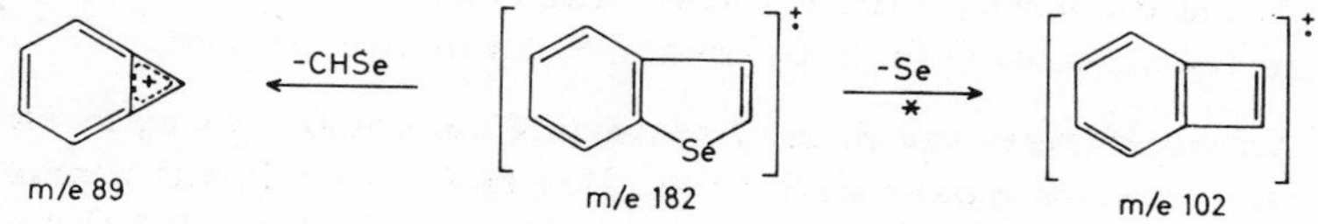

molecular ion will reasonably give a benzocyclobutadiene fragment. The alternative fragmentation step is certainly the expulsion of a CHSe group to form a $[C_7H_5]^+$ ion, as shown by a peak at m/e 89 in the spectrum,[42] rather than a $[C_7H_6]^+$ fragment at m/e 90. The fragment at m/e 89 may be either a benzyne analog to a tropylium ion or, as in scheme 25, a benzocyclopropanium ion. From a comparison between the fragmentations for benzo[b]-selenophene and the sulfur analog, it was of interest to observe that the molecular ion of benzo[b]selenophene does not expel a C_2H_2 fragment, whereas in the sulfur case this step is of considerable importance.

In the mass spectra of the homologous compounds 2-methylbenzo[b]-selenophene and 3-methylbenzo[b]selenophene, evidence was found for fragmentation steps similar to those described for benzo[b]selenophene. The molecular ion was consequently found to split off one H atom, and through a simultaneous rearrangement a stable selenochromenylium ion was formed, which lost its Se atom to form a benzocyclopentadienium ion. This latter ion successively lost two C_2H_2 molecules to form well-known selenium-free fragments.

Further, the fragmentations of 2-benzo[b]selenophenecarbaldehyde were studied in comparison with those of the nonaldehydic analogs. The introduction of the carbaldehyde group into the molecule causes a greater variety in the fragmentation pattern. The loss of selenium, however, occurred along fragmentation paths observed also for benzo[b]selenophene and 2-methyl- and 3-methylbenzo[b]selenophene. The proposed fragmentation pattern for 2-benzo[b]selenophenecarbaldehyde seems not to agree in every detail with the mass spectrum published[42] and the metastable peaks given.

The mass spectrum of benzo[b]tellurophene, recorded by Buu-Hoi *et al.*,[43] indicated a loss of tellurium from the molecular ion, whereby a benzo-cyclobutadiene fragment was formed, analogous to what was found for benzo[b]selenophene. However, there was no evidence for the loss of tellurium along alternative paths—an important difference from the selenium analog.

In an investigation of some benzoselenopheno[2,3-b]benzoselenophenes, Elmaleh *et al.*[44] determined the mass spectra of the parent compound, as well as its 3,8-dimethyl, 3,8-di-*tert*-butyl, and 3,8-dichloro derivatives. The mass spectrum for benzoselenopheno[2,3-b]benzoselenophene showed successive expulsions of the two Se atoms from the molecular ion by paths similar to the one found for benzo[b]selenophene. When substituents are present in the molecules, these are split off before the Se atoms are expelled.

c. ***Selenochromane and Related Aromatic Compounds.***   In an investigation of some selenochromane derivatives, Buu-Hoi *et al.*[45] found characteristic fragmentation patterns similar to those described for benzo[b]selenophenes. Thus, 4-selenochromanone was found to fragment as shown in scheme 26.

(26)

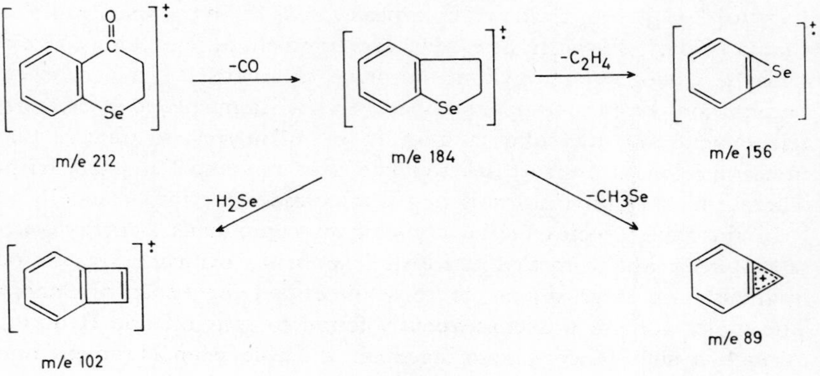

The molecular ion of 4-selenochromanone loses a carbon monoxide molecule, and the resulting ion alternatively splits off a hydrogen selenide and an ethylene molecule or a methylseleno fragment. In the mass spectrum of 4-hydroxyselenocoumarin, peaks were found in accord with a like fragmentation pattern. Similarly, 3,3'-methylenebis(4-hydroxyselenocoumarin) was found to be labile towards electron impact, and no peaks corresponding to its molecular ion were observed. In that case the molecule is split into parts to form either a 4-hydroxyselenocoumarin fragment or a 3-methylene-4-selenochromanone fragment, which is further fragmented in a manner akin to that shown in scheme 26.

The mass spectrum of selenoxanthone has been reported by Buu-Hoi et al.;[42] peaks were observed in accord with the alternative expulsions of a Se atom or a carbon monoxide molecule from the molecular ion. In this way, a biphenylene fragment is formed along two alternative paths, as expected from the xanthone structure.

d. Selenazoles and Other Aromatic Compounds. In a short communication, Buu-Hoi et al.[46] reported the mass spectra of benzo-2,1,3-selenadiazole, 5-methylbenzo-2,1,3-selenadiazole, 4,5,6-trimethylbenzo-2,1,3-selenadiazole, and phenanthro[1',2']-2,1,3-selenadiazole. Although the spectra were not analyzed with respect to the fragmentation pattern, it was observed that the heteroatoms were split off from the molecular ion, thus forming ions of the benzyne type. Later, Buu-Hoi et al.[47] published the mass spectra of benzo-2,1,3-selenadiazole and phenanthro[1',2']-2,1,3-selenadiazole, giving evidence for the expulsion of HCN and selenium from the molecular ions.

In 1966, Klayman and Milne[48] reported the mass spectrum of 2-amino-2-selenazoline and compared it with the spectrum of the sulfur analog. The investigated compounds were observed to fragment similarly, and the alternative expulsions of ethylene or cyanamide from the molecular ions were found to be characteristic. The Se atom may be lost as a CH_2Se moiety. Important peaks were reported in correspondence with hydrogen losses from the molecular ion, in addition to those occurring from the principal fragmentation steps.

Kendall and Olofson[49] determined the mass spectra of 1,3,4-selenadiazole and 2,5-dideuterio-1,3,4-selenadiazole. The spectra were not analyzed but provided full evidence for the successive losses of two hydrogen cyanide molecules from the molecular ion, by which a Se atom was formed.

The mass spectra of benzoselenino[4,3-b]indole, benzo[g][1]benzoselenino[4,3-b]indole, and benzo[e][1]benzoselenino[4,3-b]indole were reported by Buu-Hoi et al.[50] The complex spectra obtained were, however, not analyzed. Deutsch et al.[51] have reported the mass spectrum of 3,7-dimethyl-6-selenopurine without analyzing it with respect to the fragmentation pattern.

In an investigation of various related oxygen-containing heterocyclic systems, Calder *et al.*[52] recorded the mass spectra of phenoxaselenine and phenoxatellurine and compared the fragmentations observed with those for dibenzo-1,4-dioxin and phenoxathiine. These four analogs were found to behave similarly upon electron impact, and the molecular ions alternatively expel a S, Se, or Te atom or a CHO group, as described for phenoxaselenine in scheme 27.

(27)

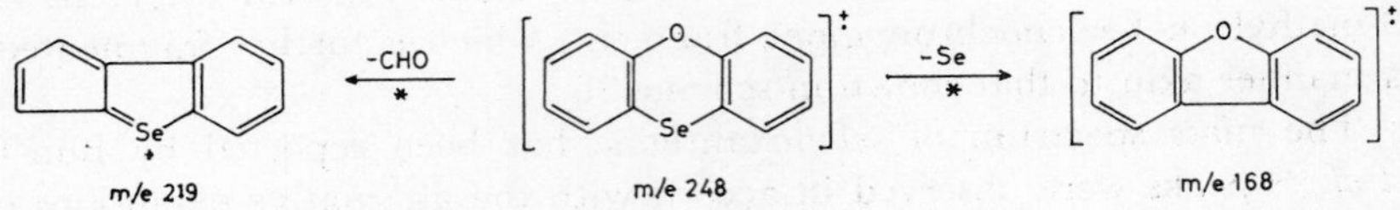

m/e 219   m/e 248   m/e 168

The mass spectra of the analogous dibromides 10,10-dibromophenoxaselenine and 10,10-dibromophenoxatellurine were also recorded. In these spectra, prominent peaks were observed in accord with the successive losses of the two Br atoms from the molecular ions, whereby phenoxaselenine and phenoxatellurine fragments, respectively, were formed. The loss of selenium or tellurium was similarly observed from the latter fragments, while no peak

(28)

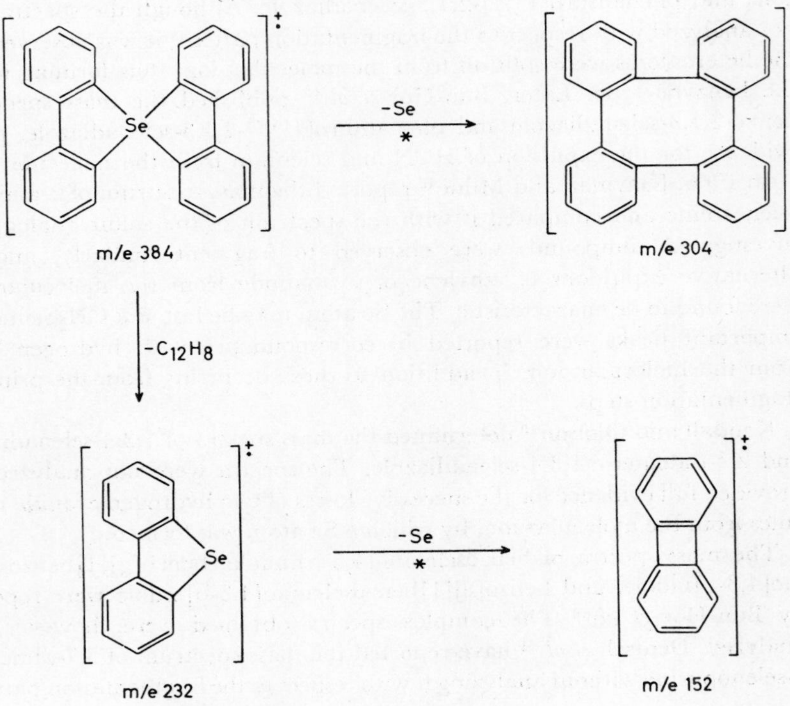

m/e 384   m/e 304

$-C_{12}H_8$

m/e 232   m/e 152

in correspondence with the alternative loss of a CHO group was found in the spectra.

e. Spiro Compounds Containing Selenium.

Few spiro compounds containing a central selenium or tellurium atom have been described. Hellwinkel and Fahrbach[53,54] prepared three analogous compounds of this type and reported their mass spectra. In the spectra for 9,9'-spirobi(9-selenafluorene),[53] 9,9'-spirobi(3,7-dimethyl-9-selenafluorene),[53] and 9,9'-spirobi(9-tellurafluorene),[54] prominent peaks were observed suggesting comparable fragmentation patterns. Varying the temperature of the ion source changed the spectra considerably, giving reason to assume that splitting of the molecules occurs partly from thermal decomposition. The observed fragmentations are given in scheme 28 for 9,9'-spirobi(9-selenafluorene).

REFERENCES

1. H. Budzikiewicz, C. Djerassi, and D. H. Williams, "Mass Spectrometry of Organic Compounds," Holden-Day, San Francisco, Calif., 1967, pp. 276, 552, 625, 634.

2. J. H. Beynon, R. A. Saunders, and A. E. Williams, "The Mass Spectra of Organic Molecules," Elsevier Publishing Co., Amsterdam, 1968, p. 352.

3. H. Fujisaki, J. B. Westmore, and A. W. Tichner, *Can. J. Chem.*, **44**, 3063 (1966).

4. J. Berkowitz and W. A. Chupka, *J. Chem. Phys.*, **45**, 4289 (1966).

5. R. Yamdagni and R. F. Porter, *J. Electrochem. Soc. Japan*, **115**, 601 (1968).

6. G. M. Bogolyubov, N. N. Grishin, and A. A. Petrov, *Zh. Obshch. Khim.*, **39**, 2244 (1969).

7. E. J. Levy and W. A. Stahl, *Anal. Chem.*, **33**, 707 (1961).

8. S. Gronowitz and T. Frejd, *Acta Chem. Scand.*, **23**, 2540 (1969).

9. L.-B. Agenäs, *Acta Chem. Scand.*, **22**, 1763 (1968).

10. H. Budzikiewicz, C. Djerassi, and D. H. Williams, "Mass Spectrometry of Organic Compounds," Holden-Day, San Francisco, Calif., 1967, p. 214.

11. L.-B. Agenäs, *Ark. Kemi*, **31**, 31 (1969).

12. L.-B. Agenäs and B. Lindgren, *Ark. Kemi*, **30**, 529 (1969).

13. S. Meyerson and L. C. Leitch, *J. Amer. Chem. Soc.*, **88**, 56 (1966).

14. G. P. Happ and D. W. Stewart, *J. Amer. Chem. Soc.*, **74**, 4404 (1952).

15. R. I. Reed and W. K. Reid, *J. Chem. Soc.*, 5933 (1963).

16. R. G. Cooks and D. H. Williams, *Chem. Commun.*, 51 (1967).

17. H. J. Svec and G. A. Junk, *J. Amer. Chem. Soc.*, **89**, 790 (1967).

18. L.-B. Agenäs, *Ark. Kemi*, **30**, 549 (1969).

19. W. L. Mock and J. H. McCausland, *Tetrahedron Lett.*, 391 (1968).

20. L.-B. Agenäs, *Ark. Kemi*, **30**, 471 (1969).

21. J. H. Bowie, S.-O. Lawesson, J. Ø. Madsen, C. Nolde, G. Schroll, and D. H. Williams, *J. Chem. Soc.*, B, 946 (1966).

22. L.-B. Agenäs, *Ark. Kemi*, **30**, 497 (1969).

23. J. L. Byard, *Arch. Biochem. Biophys.*, **130**, 556 (1969).

24. L.-B. Agenäs, *Acta Univ. Upsaliensis, Abstr. Uppsala Dissertations Sci.*, **132,** 18 (1969).

25. L.-B. Agenäs, *Acta Chem. Scand.*, **22,** 1773 (1968).

26. L.-B. Agenäs, *Ark. Kemi*, **30,** 465 (1969).

27. L.-B. Agenäs and B. Lindgren, *Ark. Kemi*, **29,** 479 (1969).

28. L.-B. Agenäs, *Ark. Kemi*, **31,** 159 (1969).

29. L.-B. Agenäs, *Ark. Kemi*, **31,** 45 (1969).

30. J. Momigny, *Bull. Soc. Roy. Sci. Liège*, **22,** 541 (1953).

31. P. N. Rylander, S. Meyerson, E. L. Eliel, and J. D. McCollum, *J. Amer. Chem. Soc.*, **85,** 2723 (1963).

32. J. Bergman, *Acta Chem. Scand.*, **22,** 1883 (1968).

33. Š. Korček, Š. Holotík, J. Leško, and V. Veselý, *Chem. Zwesti*, **23,** 281 (1969).

34. L.-B. Agenäs, *Ark. Kemi*, **31,** 159 (1969).

35. H. Budzikiewicz, C. Djerassi, and D. H. Williams, "Mass Spectrometry of Organic Compounds," Holden-Day, San Francisco, Calif., 1967, p. 552.

36. E. Rebane, *Acta Chem. Scand.*, **24,** 717 (1970).

37. E. Rebane, *Acta Chem. Scand.*, **23,** 1817 (1969).

38. H. Budzikiewicz, C. Djerassi, and D. H. Williams, "Mass Spetrometry of Organic Compounds," Holden-Day, San Francisco, Calif., 1967, p. 284.

39. A. M. Duffield, H. Budzikiewicz, and C. Djerassi, *J. Amer. Chem. Soc.*, **87,** 2920 (1965).

40. T. van Es and R. L. Whistler, *Tetrahedron*, **23,** 2849 (1967).

41. O. S. Chizhov, B. M. Zolotarev, A. N. Sukiasian, V. P. Litvinov, and Ya. L. Goldfarb, *Org. Mass Spectrom.*, **3,** 1379 (1970).

42. N. P. Buu-Hoi, M. Mangane, M. Renson, and L. Christiaens, *J. Chem. Soc., B*, 971 (1969).

43. N. P. Buu-Hoi, M. Mangane, M. Renson, and J. L. Piette, *J. Heterocycl. Chem.*, **7,** 219 (1970).

44. D. Elmaleh, S. Patai, and Z. Rappoport, *J. Chem. Soc., C*, 939 (1970).

45. N. P. Buu-Hoi, M. Mangane, O. Périn-Roussel, M. Renson, A. Ruwet, and M. Maréchal, *J. Heterocycl. Chem.*, **6,** 825 (1969).

46. N. P. Buu-Hoi, P. Jacquignon, and M. Mangane, *Chem. Commun.*, 624 (1965).

47. N. P. Buu-Hoi, P. Jacquignon, O. Roussel-Périn, F. Périn, and M. Mangane, *J. Heterocycl. Chem.*, **4,** 415 (1967).

48. D. L. Klayman and G. W. A. Milne, *J. Org. Chem.*, **31,** 2349 (1966).

49. R. V. Kendall and R. A. Olofson, *J. Org. Chem.*, **35,** 806 (1970).

50. N. P. Buu-Hoi, A. Croisy, P. Jacquignon, M. Renson, and A. Ruwet, *J. Chem. Soc., C*, 1058 (1970).

51. J. Deutsch, Z. Neiman, and F. Bergmann, *Org. Mass Spectrom.*, **3,** 1219 (1970).

52. I. C. Calder, R. B. Johns, and J. M. Desmarchelier, *Org. Mass Spectrom.*, **4,** 121 (1970).

53. D. Hellwinkel and G. Fahrbach, *Justus Liebigs Ann. Chem.*, **715,** 68 (1968).

54. D. Hellwinkel and G. Fahrbach, *Justus Liebigs Ann. Chem.*, **712,** 1 (1968).

XV Physicochemical investigations of selenium compounds

H. X-RAY DIFFRACTION

ISABELLA L. KARLE and **JEROME KARLE**

Laboratory for the Structure of Matter
U.S. Naval Research Laboratory
Washington, D.C.

1. Introduction

An X-ray diffraction analysis of a single crystal or an electron diffraction analysis of a vapor can yield various types of structural information about the material being studied. For example, such investigations can establish the structural formula. As an interesting illustration, it has been shown by X-ray analysis that β-dimethyltellurium diiodide[1] exists as the two ions, $[(CH_3)_3Te]^+$ and $[CH_3TeI_4]^-$. Also the geometry of the molecule can be determined, answering such questions as whether the configuration of the four bonds to the Se atom in diphenylselenium dibromide[2] is a square plane, square pyramid, or trigonal bipyramid or whether 1,4-diselenane·$2I_2$ may or may not be isostructural with 1,4-dithiane·$2I_2$.[3,4] Precise bond lengths and angles can be determined, as well as the range of values that certain types of linkages, such as the Se-Cl bond, assume. Finally, the packing of molecules or ions in a crystal affords detailed information concerning the coordination around Se and Te atoms, the nearest intermolecular approaches, and the possible existence of hydrogen bonds involving Se atoms, as in 2,4-diselenouracil.[5]

In diffraction experiments, the presence of a heavy atom such as Se or Te in the midst of much lighter atoms such as C and N will dominate the scattering, and, in general, the positions of the light atoms will not be found with great precision. Consequently, the accuracy of Se–C bond lengths and the values of C–Se–C angles will suffer from a lack of knowledge of the precise coordinates of the C atoms. In more recent investigations, the intensity data from single-crystal diffraction experiments have been collected by automatic diffractometers rather than by visual estimation, the intensities of all possible diffraction spots rather than just the zonal reflections have been recorded, and the accessibility of high-speed computers has enabled the investigator to make extensive least-squares refinements to obtain the best possible atomic parameters from the available data. Consequently, the bond lengths and angles reported in more recent investigations have a higher accuracy than those published earlier. Nevertheless, the overall geometry of the molecules and the molecular packing appear to be correct in even the very early papers.

Elemental selenium and tellurium are semiconductors, intermediate between nonmetals and metals. As part of organic molecules many of their properties are similar to those of the analogous sulfur compounds; in fact, it has been found that many sulfur, selenium, and tellurium analogs are isomorphous and, if not isomorphous, at least isostructural. In 1956 a review by Abrahams[6] on "The Stereochemistry of Sub-Group VIB of the Periodic Table" indicated some of the similarities and differences between sulfur, selenium, and tellurium compounds. In the intervening years to the present, the crystal structures of a considerable number of organoselenium and organotellurium compounds have been determined and the results will be summarized here. Stereodiagrams have been prepared to illustrate the different configurations assumed by various selenium and tellurium compounds.

2. Selenium-Carbon and Tellurium-Carbon Bonds: Lengths and Angles

The crystal structures of several dozen organoselenium compounds have been determined, thus making it possible to draw some conclusions about the Se–C bonds and C–Se–C angles. Even though it is difficult to obtain high precision in regard to the Se–C bond lengths, as mentioned previously, the values for different types of Se–C bonds cluster within fairly limited bounds. Table XVH-1 summarizes the values obtained in the various investigations.

Table XVH-1 Se—C *bond lengths and* C—Se—C *bond angles*

COMPOUND	Se—C (Å)	C—Se—C	REF.
(a) Se—C (aliphatic)			
CH_3SeCH_3	1.98 ± 0.01	$98° \pm 10°$	7
$CH_3SeSeCH_3$	1.953 ± 0.005		40
Se⌒Se (ring)	$\begin{cases} 1.99 \pm 0.04 \\ 2.04 \pm 0.06 \end{cases}$	$98° \pm 3.5°$	8
Se—Se—Se (ring)	$\begin{cases} 1.945 \pm 0.015 \\ 1.947 \pm 0.015 \\ 1.967 \pm 0.015 \end{cases}$	$\begin{cases} 96.8° \pm 1° \\ 100.3° \pm 1° \end{cases}$	8a
I—I—Se ⌬ Se—I—I	$\begin{cases} 1.95 \pm 0.024 \\ 1.98 \pm 0.024 \end{cases}$	$101° \pm 2°$	3
S ⌬ Se(Br)(Br)	$\begin{cases} 2.01 \pm 0.03 \\ 2.02 \pm 0.03 \end{cases}$	$105° \pm 2°$	9
O ⌬ Se—I—Cl	1.98 ± 0.03	$96° \pm 2°$	10
O ⌬ Se—I—I	$\begin{cases} 1.93 \pm 0.03 \\ 1.98 \pm 0.03 \end{cases}$	$94° \pm 2°$	11
⌬ Se—I—I	1.96 ± 0.025	$93° \pm 2°$	12
$=NH^+$ / Se =NH_2	$\begin{cases} (1.88 \pm 0.03)^a \\ 1.99 \pm 0.03 \end{cases}$	$86° \pm 2°$	13
$(CH_3)_3Se^+$	$\begin{cases} 1.96 \pm 0.02 \\ 1.95 \pm 0.02 \end{cases}$	$\begin{cases} 98° \pm 1° \\ 99° \pm 1° \end{cases}$	14
$(CH_3)_2\overset{+}{N}HCH_2CH_2Se\overset{O}{C}C_6H_5$	$\begin{cases} 1.945 \pm 0.005 \\ 1.945 \pm 0.005 \end{cases}$	$96.4° \pm 0.3°$	15
$(CH_3)_2CH$—Se—Pd(Cl)(Cl)—Se—$CH(CH_3)_2$	$\begin{cases} 1.98 \\ 2.03 \pm 0.05 \\ 2.04 \\ 2.10 \end{cases}$	$\begin{cases} 98.7° \pm 2.2° \\ 100.4° \pm 2.0° \end{cases}$	15a
$(C_6H_5)_2CHSeSeCH(C_6H_5)_2$	1.97 ± 0.01		16
HOOC ⌬ COOH (Se—Se)	1.97 ± 0.02		17

Table XVH-1   (*Continued*)

COMPOUND	Se—C (Å)	C—Se—C	REF.
(b) Se—C (aromatic)			
	$\begin{cases}1.92 \pm 0.045 \\ 1.93 \pm 0.045\end{cases}$	$106° \pm 2°$	18
	1.93 ± 0.03	$106.5° \pm 1°$	19
	1.95 ± 0.03	$108° \pm 1°$	19
	1.93 ± 0.05		20
	$\begin{cases}1.92 \pm 0.10 \\ 1.94 \pm 0.10\end{cases}$		21
	$\begin{cases}1.92 \pm 0.02 \\ (1.84 \pm 0.02)^a\end{cases}$	$94.4° \pm 0.9°$	21a
(c) Se—C (unsaturated heterocycles)			
	$\begin{cases}1.75 \\ 1.97\end{cases}$	$80°$	22
	$\begin{cases}1.87 \pm 0.01 \\ 1.93 \pm 0.02\end{cases}$	$86° \pm 1°$	23
—COOH	$\begin{cases}1.85 \pm 0.01 \\ 1.87 \pm 0.01\end{cases}$	$87° \pm 0.5°$	24
	1.899 ± 0.005	$86.6° \pm 0.2°$	25
(d) Se=C	1.86 ± 0.03		26

Table XVH-1   (*Continued*)

COMPOUND	Se—C (Å)	C—Se—C	REF.
	1.87 ± 0.03		27
	1.82 ± 0.01		28
	a 1.89 ± 0.05 b 1.99 ± 0.05		5
(*e*) Se—C (seleninic acids)			
$NH_2CH_2CH_2SeOOH$	1.87 ± 0.02		29
	1.90 ± 0.02		30
	1.85 ± 0.07		31
	1.99 ± 0.04		31a

[a] Se-C distance for bond adjacent to C-N

The standard deviations listed are those set by the individual investigators. Most of the bond lengths between selenium and an aliphatic C atom fall between 1.95 and 1.99 Å, while those between selenium and a C atom in an aromatic ring are near 1.93 Å. When selenium is part of an unsaturated five-membered ring, the Se-C bond length becomes even shorter and averages near 1.88 Å. In selenourea and two derivatives of this compound, the C–Se double bond ranges from 1.82 to 1.87 Å. An anomaly arises in 2,4-diselenouracil, where the C–Se double bonds are 1.89 and 1.99 (±0.05) Å. Shefter *et al.*[5] suggest that the high values may be due to the existence of a polar form (**1**) which is consistent with dipole moment, u.v., and ionization measurements.

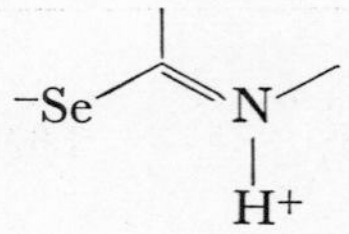

1

A short length, 1.88 Å, is reported for the 2-amino-1,3-selenazol-2-inium ion (**2**) for the Se–C bond adjacent to the NH_2 group.[13] Some unsaturation

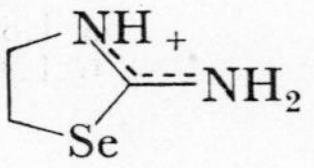

2

appears to exist in this bond as well as in the adjacent C–N bonds. In 1,4-diselenocyanatobenzene,[21a] the Se–C_{ar} bond length of 1.92 Å is consistent with all the values observed for a selenium atom attached to an aromatic ring; however, the Se–CN bond is only 1.84 Å which again suggests that the unsaturation in the CN bond affects the length of the adjacent Se–C bond.

The values for the C–Se–C angle also show some regularity, depending on the type of molecular linkage. When the Se atom is between two aliphatic C atoms and when it is not constrained by a five-membered ring, the C–Se–C angle is generally somewhat less than 100°. On the other hand, when the Se atom is between two aromatic rings, the C–Se–C angle is near 107°. In planar, five-membered rings, the value for the angle is near 87°. For comparison, the HSeH bond angle is 91°.[67]

Tellurium-carbon bond lengths are summarized in Table XVH-2. There are not as many examples for tellurium compounds as there are for selenium compounds, and the accuracy with which Te-C bond lengths can be determined is less than that for the Se-C bond lengths. The reported bond lengths range from 2.01 to 2.18 Å, and there seems to be no correlation between the bond lengths and the types of molecules. The C–Te–C angles are near 100° or a little less.

3. Conformations

Organoselenium and tellurium compounds, like their sulfur analogs, assume various configurations and coordinations. In many instances they are isomorphous with the corresponding sulfur analogs; in others they are strikingly dissimilar.

Table XVH-2 Te—C *bond lengths and* C—Te—C *angles*

COMPOUND	Te—C (Å)	C—Te—C	REF.
(a) Te—C (aliphatic)			
$CH_3-\underset{Cl}{\overset{Cl}{Te}}-CH_3$	$\begin{cases}2.08 \pm 0.03 \\ 2.10 \pm 0.03\end{cases}$	$98° \pm 1°$	32
[cyclic S—Te(I)(I) ring]	$\begin{cases}2.13 \\ 2.18\end{cases}$	$100°$	33
$(CH_3)_3Te^+$	$\left.\begin{matrix}2.01 \\ 2.08 \\ 2.13\end{matrix}\right\} \pm 0.06$	$\left.\begin{matrix}91° \\ 95° \\ 97°\end{matrix}\right\} \pm 2°$	1
$CH_3TeI_4^-$	2.15 ± 0.06		1
(b) Te—C (aromatic)			
$CH_3-\!\!\!\bigcirc\!\!\!-Te-\!\!\!\bigcirc\!\!\!-CH_3$	2.05 ± 0.05	$101° \pm 3°$	34
$Cl-\!\!\!\bigcirc\!\!\!-\underset{I}{\overset{I}{Te}}-\!\!\!\bigcirc\!\!\!-Cl$	$\begin{cases}2.13 \pm 0.02 \\ 2.10 \pm 0.02\end{cases}$	$101° \pm 1°$	35
$\bigcirc\!\!-\underset{Br}{\overset{Br}{Te}}-\!\!\bigcirc$	2.18 ± 0.03	$94° \pm 1.5°$	36
$\bigcirc\!\!-\underset{S=C(NH_2)_2}{\overset{Cl}{Te}}$	2.12 ± 0.06		37
$\bigcirc\!\!-\underset{S\cdots C(NH_2)_2}{\overset{S\cdots C(NH_2)_2}{Te^+}}$	2.11 ± 0.04		38
$Cl-\!\!\!\bigcirc\!\!\!-Te-Te-\!\!\!\bigcirc\!\!\!-Cl$	$\begin{cases}2.10 \pm 0.20 \\ 2.16 \pm 0.20\end{cases}$		21

The structures of several diselenides and of one ditelluride that is iso-morphous with the corresponding diselenide analog have been studied, and the results are summarized in Table XVH-3. A stereodiagram of bis(di-phenylmethyl) diselenide[16] is shown in Fig. XVH-1. The particular feature of interest is the peroxide[41] conformation (3) assumed by the CSeSeC moiety.

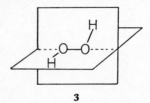

FIG. XVH-1   A stereodiagram of bis(diphenylmethyl) diselenide[16] illustrating the peroxide conformation of the CSeSeC moiety. The diagram should be seen with a three-dimensional viewer for printed stereophotographs. The stereodiagrams in this chapter were made by means of a computer program prepared by C. K. Johnson of Oak Ridge National Laboratory.

3

The dihedral angle CSeSe/SeSeC ranges from 74° to 87° in the compounds listed in Table XVH-3 except for 1,2-diselenane-3,6-dicarboxylic acid,[17] in which the conformation of the CSeSeC grouping is constrained by ring closure. (The ring has the chair conformation with the carboxyl groups attached in equatorial positions.) The only ditelluride compound studied has a CTeTe/TeTeC dihedral angle of 72°.[21] In disulfides such as tetraethylthiuram disulfide,[42] N,N'-diglycyl-L-cystine·2H$_2$O,[43] and L-cystine·HCl,[44] the dihedral angles for CSS/SSC have been found to be 96–101°. A conclusion based on the results from the few compounds studied thus far with the –SS–, –SeSe–, and –TeTe– linkages is that the dihedral angle decreases from the sulfur to the tellurium compounds.

The peroxide conformation is also evident in a molecule that has a single Se atom. In Fig. XVH-2, which shows the configuration of 2-dimethyl-aminoethyl selenolbenzoate,[15] one can see that the CCSe/CSeC angle is

Table XVH-3 Diselenides and ditellurides

COMPOUND	Se—Se (Å) (Te—Te)	Se—Se—C (Te—Te—C)	DIHEDRAL ANGLE	REF.
NCSeSeSeCN	2.33 ± 0.03	95°	94°	39
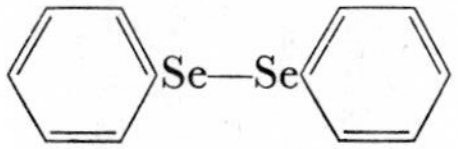 (Se—Se diphenyl)	2.29 ± 0.01	{105° ± 2° / 107° ± 2°}	82° ± 3°	20
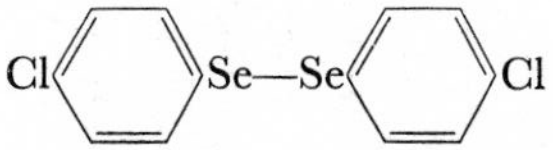 (Cl—phenyl Se—Se phenyl—Cl)	2.33 ± 0.015	{100° / 102°}	74.5°	21
$(C_6H_5)_2CH$—Se—Se—$CH(C_6H_5)_2$	2.285 ± 0.005	100° ± 0.5°	82°	16
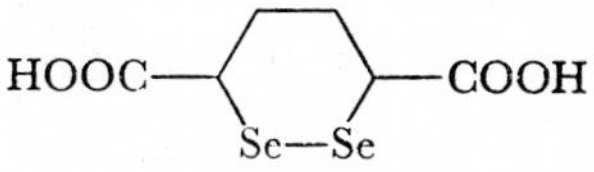 HOOC—…—COOH (Se—Se ring)	2.32 ± 0.015	96° ± 0.8°	56° ± 1.5°[a]	17
CH_3Se—$SeCH_3$	2.325 ± 0.003	98.9°	87.5°	40
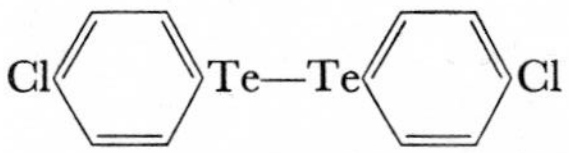 Cl—phenyl Te—Te phenyl—Cl	2.70 ± 0.01	{95° / 93°}	72°	21

[a] Dihedral angle constrained by ring closure.

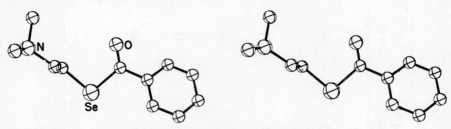

FIG. XVH-2   Conformation of 2-dimethylaminoethyl selenolbenzoate.[15]

near 90°. Thus far no structure studies have been made on linear aliphatic molecules containing selenium in the chain except for CH_3SeCH_3. Such an investigation would show whether the peroxide conformation is forced by the presence of one Se atom when not constrained by ring closure or other structural restraints.

A complex formed with $PdCl_2$ and 1,2-bis(isopropylseleno)ethane[15a] is shown in Fig. XVH-3. The $PdCl_2Se_2$ moiety is square planar with Pd-Cl

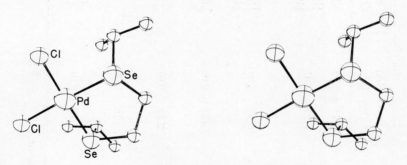

FIG. XVH-3   The complex of $PdCl_2$ with 1,2-bis(isopropylseleno)ethane.[15a]

bonds of 2.32 Å and Pd-Se bonds near 2.38 Å. Both of the isopropyl groups lie on the same side of the CSePdSeC chelate ring.

Five-membered rings which are saturated or partially unsaturated,[12,13] containing selenium, have so far been found to be nonplanar. Figure XVH-4 shows the 2-amino-1,3-selenazol-2-inium ion (**2**), where the ring has the envelope conformation with four atoms coplanar and the C atom not adjacent to either of the N atoms 0.43 Å below the plane,[13] similar to the conformation of cyclopentane and five-membered aliphatic heterocyclic rings. Six-membered rings containing Se atoms and aliphatic C atoms have the chair conformation,[8,8a] as illustrated in several halogen derivatives in Figs. XVH-5a and XVH-7.

FIG. XVH-4   Envelope conformation of five-membered ring containing selenium.[13]

When 1,4-diselenane, 1-thia-4-selenane, or 1-oxa-4-selenane is combined with Cl_2, Br_2, I_2, or ICl, crystalline adducts are formed. Figure XVH-5a illustrates the configuration of a typical adduct[3] between selenium in a saturated ring and I_2, while the adduct[4] between 1,4-dithiane and I_2 is shown in Fig. XVH-5b. This pair of molecules illustrates a difference between the Se and the S atoms, in that the halogen atoms are attached to Se atoms in *axial* positions whereas they are attached to S atoms in *equatorial*

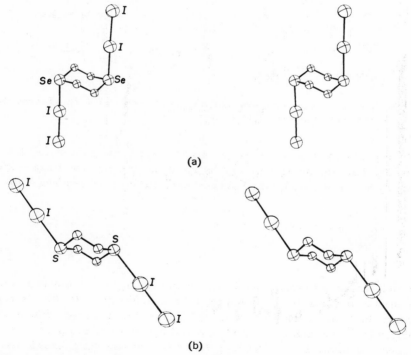

(a)

(b)

FIG. XVH-5   (a) 1,4-Diselenane·$2I_2$ with the Se–I bonds axial to the ring.[3] (b) 1,4-Dithiane·$2I_2$ with the S–I bonds equatorial to the ring.[4]

positions. The molecular complexes 1,4-diselenane·diiodoacetylene (**4**) and 1,4-dithiane·diiodoacetylene (**5**), while not adducts but infinite chains, show the same distinction in the coordination to the Se and S atoms; that is, I atoms are in *axial* positions with respect to the Se atoms and in *equatorial* positions with respect to the S atoms.[45] The molecular complex 1,4-diselenane·2CHI$_3$ forms a two-dimensional lattice,[45a] as illustrated in Fig. XVH-6. In this instance, each Se atom is coordinated to two I atoms, each in the *axial* position. The Se · · · I separations are 3.47 and 3.51 Å.

FIG. XVH-6   A two-dimensional network formed by the coordination of two I atoms from CHI$_3$ molecules to each Se atom of diselenane.[45a]

Other substances with configurations very similar to that of 1,4-diselenane·2I$_2$, but containing only one Se atom in the ring, are 1-oxa-4-selenane·I$_2$ (**6**),[11] 1-oxa-4-selenane·ICl (**7**),[10] and tetrahydroselenophene·I$_2$(**8**).[12]

In each case the halogen atoms form a chain with an I atom attached to the Se atom in the axial position, and the I–Se–C angles are 100–107°. In contrast to the I$_2$ compounds, when Br$_2$ or Cl$_2$ is used to make the adducts, the halogen molecules dissociate into atoms and two Se-halogen axial bonds are formed,[9] as shown in Fig. XVH-7. The 1,4-diselenane tetrachloride has a similar configuration with two axial Se-Cl bonds on each Se atom.[46] It is interesting to note that the only tellurium compound of this type that has been

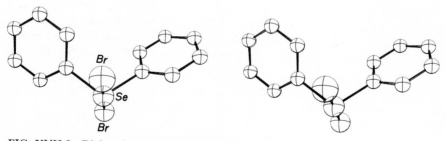

FIG. XVH-7   1-Thia-4-selenane-4,4-dibromide.[9]

studied, 1-thia-4-tellurane-4,4-diiodide (**9**),[33] is isostructural with 1-thia-4-selenane-4,4-dibromide (Fig. XVH-7), rather than with the selenane-iodine type adduct (*e.g.*, **6**).

$$S \diagup\diagdown Te \diagup^{I}_{\diagdown I}$$

9

The configurations about the Se and Te atoms, as shown in Fig. XVH-7, persist even when the selenium or tellurium is not part of a ring that constrains the C–Se–C or C–Te–C angle. For example, diphenylselenium dibromide,[2] shown in Fig. XVH-8, has the four bonds to the Se atom in an

FIG. XVH-8   Diphenylselenium dibromide,[2] showing the four bonds to the Se atom in a trigonal-bipyramid arrangement.

approximate trigonal-bipyramid arrangement, with the two Se-Br bonds directed up and down and the two Se-C bonds in equatorial positions. Similar configurations are assumed by diphenylselenium dichloride,[47] di-*p*-tolylselenium dichloride and dibromide,[19] diphenyltellurium dibromide,[36] and di-*p*-chlorodiphenyltellurium diiodide.[35]

The tellurium compound α-$(CH_3)_2TeCl_2$ exists in a configuration similar to that of the diphenyl compounds mentioned above (Fig. XVH-8). The compound called β-$(CH_3)_2TeI_2$ exists as the two ions $[(CH_3)_3Te]^+$ and $[CH_3TeI_4]^-$,[1] which are illustrated in Fig. XVH-9. The $(CH_3)_3Te^+$ cation has a pyramidal shape with a mean C–Te–C angle of 95° and is very similar to the $(CH_3)_3Se^+$ cation in trimethylselenonium iodide,[14] where the C–Se–C

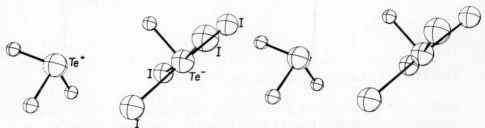

FIG. XVH-9   Configurations of the two ions in β-$(CH_3)_2TeI_2$,[1] the pyramidal, $[(CH_3)_3Te]^+$, and the octahedral (with one site vacant), $[CH_3TeI_4]^-$.

angles are near 98°. In the $CH_3TeI_4^-$ anion, the four I atoms are in a square-planar configuration with the methyl group occupying one of the octahedral sites and the other octahedral site vacant. All angles are near 90°. The closest approaches of atoms from neighboring ions in the crystal are such as to complete an approximate octahedral coordination about the Te atoms in both $(CH_3)_3Te^+$ and $CH_3TeI_4^-$.

The only example of a structure of an organoselenium compound with five bonds attached to the Se atom is the adduct formed from $SeOCl_2$ and two pyridine molecules[48] (Fig. XVH-10). The two Se–Cl bonds and the two Se–N bonds are in a *trans* square-planar configuration. All the bond angles about the Se atom are close to 90° except for one O–Se–Cl angle, which is

FIG. XVH-10   The incomplete octahedral configuration about a Se atom having five bonds in the $SeOCl_2 \cdot 2C_5H_5N$ adduct.[48]

100°. In the direction of the vacant site of the incomplete octahedral configuration about the Se atom, there is a Cl atom from a neighboring molecule at a distance of 3.65 Å. The double-bond length for Se-O is 1.59 Å, whereas the Se-N bond lengths are 2.20 Å, a rather high value compared to Se-C bond lengths.

Bond lengths between selenium and tellurium and halogen atoms are listed in Table XVH-4. It is immediately apparent that the lengths for each type of bond vary over a wide range of values. In addition, the sums of commonly accepted bond radii are much smaller than the observed bond lengths in these halogen compounds.

The structures of five compounds in which selenium is part of a five-membered aromatic ring have been studied: the four compounds listed under part (c) of Table XVH-1 and 3,4-benzo-1,2,5-selenadiazole (10).[53]

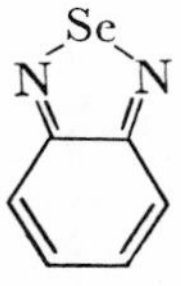

10

In each of these crystals, the molecules are essentially planar. However, when selenium replaces a middle ring atom in anthracene, as in selenanthrene (11)[54] or dibenzo[c,d]phenoselenazine (12),[55] the molecule is

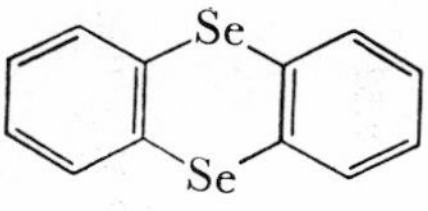

11

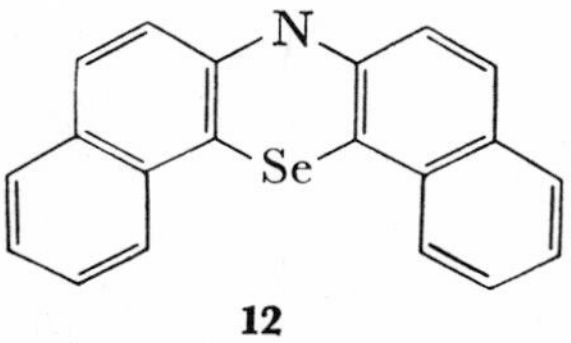

12

folded with dihedral angles reported at 127° and 150°, respectively. Hosoya[56] has suggested that in anthracene, if the middle ring atoms are replaced by C, N, or O, the molecule remains planar, but becomes folded if any one of the middle atoms is S, Se, or Te.

The stereochemistry of selenium in seleninic acids has been established by structure investigations of benzeneseleninic acid (13),[30] p-chlorobenzeneseleninic acid,[31] 2-aminoethylseleninic acid,[29] and trans-ethanediseleninic anhydride (14).[31a] The stereoconfigurations of the first two compounds are almost identical.[30] The three bonds to the Se atom form a pyramid with C-Se-O angles near 99°, and the O-Se-O angle at 103.5°. The two Se-O bond lengths are 1.71 and 1.76 Å. In the 2-aminoethylseleninic acid,[29]

Table XVH-4 Selenium and tellurium bonds to halogen atoms[a,b]

COMPOUND	Se—Cl	Se—Br	Se—I	Te—Cl	Te—Br	Te—I	REF.
(Se ring, Cl)	2.24 ± 0.02						46
C_6H_5—Se—C_6H_5 (Cl)	2.30						47
$CH_3C_6H_4$—Se—$C_6H_4CH_3$ (Cl)	2.38 ± 0.02						19
$SeOCl_2 \cdot 2C_5H_5N$	$\begin{cases} 2.39 \pm 0.15 \\ 2.57 \pm 0.15 \end{cases}$						48
C_6H_5—Se—C_6H_5 (Br)		2.52					2
$CH_3C_6H_4$—Se—$C_6H_4CH_3$ (Br)		2.55 ± 0.02					19
(Se ring, Br)		2.547 ± 0.005					9

Compound				Ref.
![Se—ICl dioxane] Se—ICl	2.630 ± 0.005			10
Se—I—I (dioxane)	2.755 ± 0.004			11
Se—I—I (tetrahydro)	2.762 ± 0.005			12
I—I⋯Se⋯Se⋯I—I	2.829 ± 0.004			3
$(CH_3)_2TeCl_2$		$\begin{cases}2.48 \pm 0.01\\2.54 \pm 0.01\end{cases}$		32
$trans\text{-}TeCl_4 \cdot 2(CH_3)_4(tu)$		$\begin{cases}2.520 \pm 0.008\\2.536 \pm 0.008\end{cases}$		50
$C_6H_5\overset{Cl}{-}Te-(tu)$		3.00 ± 0.015		37
$cis\text{-}TeCl_2 \cdot 2(tu)$		2.92 ± 0.015		49
$C_6H_5\overset{Br}{\underset{Br}{-}}TeC_6H_5$			$2.682 \pm 0 \cdot 003$	36
$trans\text{-}TeBr_4 \cdot 2(CH_3)_4(tu)$			$\begin{cases}2.686 \pm 0.007\\2.707 \pm 0.007\end{cases}$	50

Table XVH-4 (*Continued*)

COMPOUND	Se—Cl	Se—Br	Se—I	Te—Cl	Te—Br	Te—I	REF.
trans-TeBr$_2$·2 ethylene(tu)					2.78 ± 0.01		51
cis-TeBr$_2$·2(tu)					3.05 ± 0.01		49
C$_6$H$_5$—Te—(tu) (with Br, I)					3.11 ± 0.01		37
ClC$_6$H$_4$TeC$_6$H$_4$Cl (with I)						{2.922 ± 0.002, 2.947 ± 0.002}	35
CH$_3$TeI$_4^-$						{2.84, 2.89 ± 0.007, 2.95, 2.98}	1
trans-TeI$_2$·2 ethylene(tu)						2.97 ± 0.01	51
(Te ring with 2 I, S)						{2.85, 2.98}	33
Sum of bond radii	2.16	2.31	2.50	2.36	2.51	2.70	52

a Thiourea ≡ (tu).
b Bond lengths given in Å.

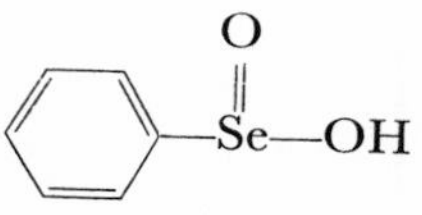

13

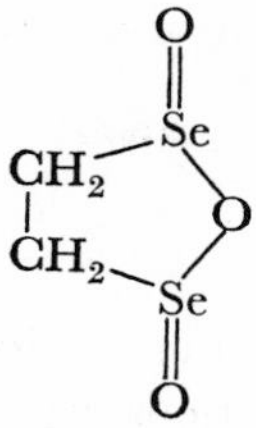

14

illustrated in Fig. XVH-11, the configuration about the Se atom is very similar to that in benzeneseleninic acid, the main difference being in the Se-O bond lengths, which are 1.65 and 1.78 Å. It is interesting to note the orientation of the aminoethyl chain, which assumes a *gauche* conformation with respect to both O atoms. 2-Aminoethylseleninic acid crystallizes as the hydrochloride. In the crystal, octahedral coordination about the Se atom is

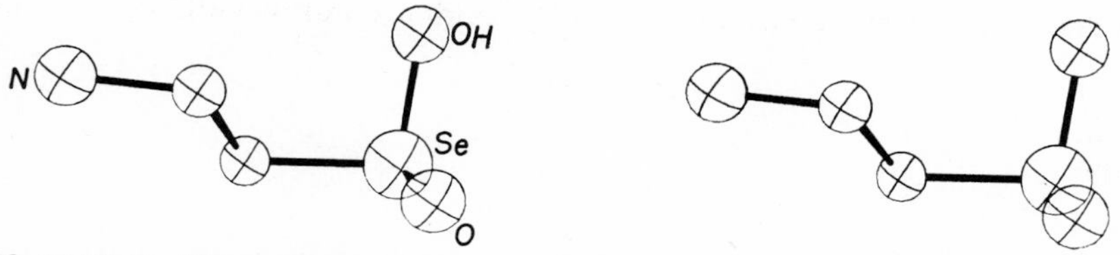

FIG. XVH-11   The configuration of 2-aminoethylseleninic acid,[29] illustrating the pyramidal bond arrangement to the Se atom.

completed by two neighboring Cl⁻ ions and one Se atom at distances of 3.26–3.48 Å. *Trans*-Ethanediseleninic anhydride,[31a] illustrated in Fig. XVH-12, also has pyramidal Se atoms with C–Se–O angles of 91° and 107° and an O–Se–O angle of 106°. The two Se-O bond lengths are 1.61 ± 0.03 Å and 1.82 ± 0.02 Å. Oxygen atoms from two neighboring molecules complete

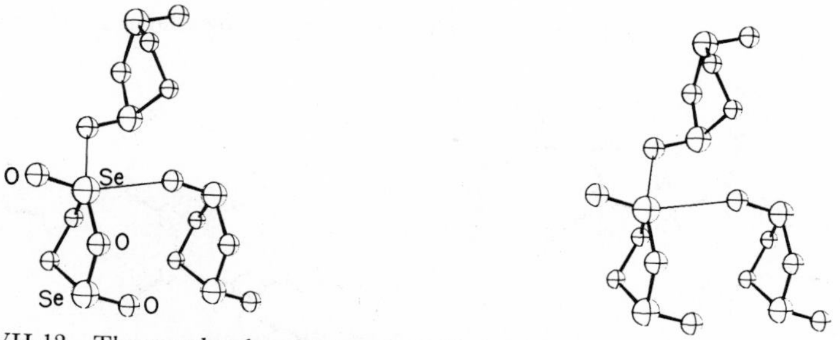

FIG. XVH-12   Three molecules of *trans*-ethanediseleninic acid anhydride, showing the five-fold coordination about the Se atom.[31a]

a fivefold coordination about the Se atom with intermolecular Se · · · O
distances of 2.70 and 3.11 Å.

4. Hydrogen Bonds

Selenoureas are analogs of the thioureas and ureas. Hydrogen bonds between
S and HN exist in various crystalline derivatives of thiourea, for example,
trimethylenethiourea,[57] allylthiourea,[58] and 1-thiocarbamoylimidazolidine-
2-thione.[58a] A similar hydrogen bonding scheme has been demonstrated for
selenourea,[26] where infinite chains of molecules are formed, as shown in Fig.
XVH-13. The Se · · · HN distance is 3.51 Å. 1-Benzoyl-3-phenyl-2-seleno-
urea,[28] whose stereoconfiguration is shown in Fig. XVH-14, exists as a
dimer formed by two Se · · · HN bonds of 3.83 Å across a center of symmetry
in the crystal. On the other hand, 1-acetyl-3-phenyl-2-selenourea[27] does not
form any hydrogen bonds involving the Se atom. Each molecule of 2,4-
diselenouracil[5] forms a pair of Se · · · HN bonds around two different centers

FIG. XVH-13   The hydrogen bonding scheme in selenourea.[26]

FIG. XVH-14   1-Benzoyl-3-phenyl-2-selenourea.[28]

of symmetry with lengths of 3.47 and 3.75 Å. Although the locations of the H atoms were not established in any of these determinations, the assumption that hydrogen bonds are formed appears to be very plausible. For example, the 2,4-diselenouracil is isomorphous with its sulfur analog, 2,4-dithiouracil, in which the H atoms were located in proper positions for hydrogen bond formation.

5. Selenium–Sulfur and Tellurium–Sulfur Linkages

Divalent tellurium is bonded to sulfur in the organothiosulfonates and thiophosphates. In these compounds the SSTeSS (or PSTeSP) chain is unbranched and nonplanar; see Fig. XVH-15. It assumes the same form as the

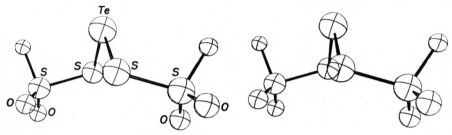

FIG. XVH-15   The configuration of Te[SS(CH$_3$)O$_2$]$_2$.

trans-pentathionic chain.[59] The bond lengths and angles for TeS and STeS are listed in Table XVH-5.

Another example of a chain resembling the *trans*-pentathionic chain is afforded by the group of atoms C–S–Se–S–C in selenium dibenzenesulfinate,[59a] illustrated in Fig. XVH-16, even though, in this case, the S atoms are hexavalent rather than divalent. The Se–S distance is 2.20 ± 0.03 Å, and the S–Se–S angle is 105 ± 2°. The S–Se–S angle is smaller than the S–S–S angle in the isomorphous sulfur compound[59b] and larger than the S–Te–S angles listed in Table XVH-5.

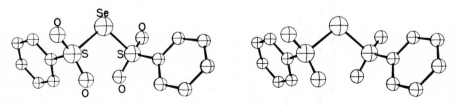

FIG. XVH-16   The configuration of selenium dibenzenesulfinate.[59a]

Table XVH-5 Tellurium-sulfur bond lengths, angles, and coordinations

COMPOUND	Te—S (Å)	S—Te—S	COORDINATION ABOUT Te	REF.
(a) Thiosulfonates				
$CH_3SSTeSSCH_3$	2.35 ± 0.03 2.36 ± 0.03	100° ± 3°	} Unbranched, nonplanar SSTeSS chain; *trans* form of pentathionic chain.	60
(tolyl) SSTeSS (tolyl)	2.41 ± 0.04	96° ± 3°		61
(phenyl) SSTeSS (phenyl)	2.41 ± 0.03	97° ± 3°		62
(b) Dithiophosphates				
$(CH_3O)_2PSTeSP(OCH_3)_2$	2.44 ± 0.01	98.3° ± 0.3°	*Trans* form of pentathionic chain for PSTeSP; square-planar coordination with two neighboring =S.	63

(c) Thiourea[a] adducts

Compound	Distance (Å)	Angle	Geometry	Ref.
TeCl·(tu)	2.50 ± 0.015		Square-planar; one position vacant.	37
Te⁺·2(tu)	2.61 ± 0.015 2.74 ± 0.015	188° ± 1°	Trans square-planar; one position vacant.	38
TeCl₂·2(tu) TeBr₂·2(tu)	2.48 ± 0.02	93° ± 1°	Cis square-planar	49
TeBr₂·2 ethylene(tu) TeI₂·2 ethylene(tu)	2.69 ± 0.015	180°	Trans square-planar	51
CH₃SSTeSSCH₃·2(tu)	2.67 ± 0.015 2.68 ± 0.015	90.6° 180°	Trans square-planar	64
Te²⁺·4(tu)	2.69 ± 0.01	90°	Square-planar	65
TeBr₄·2(CH₃)₄(tu) TeCl₄·2(CH₃)₄(tu)	2.71 ± 0.01 2.70 ± 0.01	180°	Octahedral; halogens square planar; S trans.	50
Te₂⁴⁺·6(tu)	2.47 ± 0.015 cis 2.53 ± 0.015 2.86 ± 0.015 3.02 ± 0.015 bridge	83–95°	Square planar.	66

[a] Thiourea ≡ (tu).

Many different thiourea adducts are formed with tellurium compounds or ions. Table XVH-5 lists the adducts whose structures have been determined. In all these adducts the square-planar coordination about the Te atom occurs. Figure XVH-17 illustrates the tribonded Te atom in the phenyl bisthiourea tellurium(II) ion,[38] in which the bonds to Te assume the square-planar configuration with one position vacant. A Cl^- ion is directed towards the vacant site in the square-planar configuration at a distance of

FIG. XVH-17   The tribonded ion $C_6H_5Te^+·2$ thiourea and a neighboring Cl^-, exhibiting a square-planar configuration about the Te atom.

3.61 Å from the tellurium. A similar configuration is assumed by the tribonded Te atom in benzenetellurenyl chloride·thiourea (**15**),[37] where the

15

vacant site in the square-planar configuration is filled by a Cl atom from a neighboring molecule at a Te $\cdots$ Cl distance of 3.71 Å. The six-bonded $TeBr_4·2$ thiourea[50] and the isomorphous chlorine compound have an octahedral configuration with the four halogen atoms in the square-planar configuration, and the two S atoms *trans* with respect to each other. The square-planar configuration persists in the more complex ion, $Te_2^{4+}·$ 6 thiourea,[66] illustrated in Fig. XVH-18, where the two Te atoms share the S atoms from two of the thiourea moieties. In this ion, the two Te atoms and the six S atoms are coplanar.

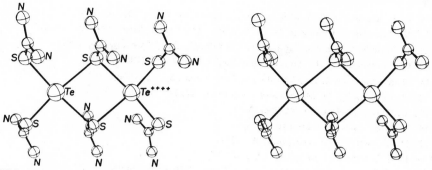

FIG. XVH-18 The configuration of the $Te_2^{4+} \cdot 6$ thiourea ion, illustrating the square-planar configuration about each Te atom.

An inspection of the Te-S bond lengths in Table XVH-5 shows that the values range from 2.35 to 3.02 Å. The Te-S bond length appears to be dependent on the number of bonds to tellurium. A comparison of

$$CH_3S(O_2)STeSS(O_2)CH_3$$

(Fig. XVH-15)[60] and *trans*-$CH_3S(O_2)STeSS(O_2)CH_3 \cdot 2$ thiourea,[64] where the four Te-S bonds are in a square-planar configuration, shows that in the latter compound, which has four Te-S bonds instead of the two in the former compound, the Te-S bond length is longer by 0.32 Å. The Te-S bond length also seems to be a function of whether the thiourea groups are *cis* or *trans* to each other. In *cis*-$TeBr_2 \cdot 2$ thiourea[49] the Te–S distance is 2.48 Å, whereas in *trans*-$TeBr_2 \cdot 2$ ethylenethiourea[51] the Te–S distance is 2.69 Å. In the compounds listed in Table XVH-5, all those with only two bonds to tellurium and those with thiourea groups *cis* to each other, that is, having S–Te–S angles of 90–100°, have Te-S bond lengths less than 2.50 Å, whereas those with thiourea groups *trans* to each other have Te-S bond lengths greater than 2.60 Å.

REFERENCES

1. F. Einstein, J. Trotter, and C. Williston, *J. Chem. Soc.*, A, 2018 (1967).
2. J. D. McCullough and G. Hamburger, *J. Amer. Chem. Soc.*, **63,** 803 (1941).
3. G. Y. Chao and J. D. McCullough, *Acta Cryst.*, **14,** 940 (1961).
4. G. Y. Chao and J. D. McCullough, *Acta Cryst.* **13,** 727 (1960).
5. E. Shefter, M. N. G. James, and H. G. Mautner, *J. Pharm. Sci.*, **55,** 643 (1966).
6. S. C. Abrahams, *Quart. Rev.* (London), **10,** 407 (1956).
7. E. Goldish, K. Hedberg, R. E. Marsh, and V. Schomaker, *J. Amer. Chem. Soc.*, **77,** 2948 (1955).

8. R. E. Marsh and J. D. McCullough, *J. Amer. Chem. Soc.*, **73,** 1106 (1951).

8a. M. Mammi, G. Carazzolo, G. Valle, and A. Del Pra, *Z. Krist.*, **127,** 401 (1968).

9. L. Battelle, C. Knobler, and J. D. McCullough, *Inorg. Chem.*, **6,** 958 (1967).

10. C. Knobler and J. D. McCullough, *Inorg. Chem.*, **7,** 365 (1968).

11. H. Maddox and J. D. McCullough, *Inorg. Chem.*, **5,** 522 (1966).

12. H. Hope and J. D. McCullough, *Acta Cryst.*, **17,** 712 (1964).

13. L. Karle, *Angew. Chem.*, **80,** 793 (1968).

14. H. Hope, *Acta Cryst.* **20,** 610 (1966).

15. D. D. Dexter, *Acta Cryst.*, **B28,** 49 (1972).

15a. H. J. Whitfield, *J. Chem. Soc.*, *A*, 113 (1970).

16. H. T. Palmer and R. A. Palmer, *Acta Cryst.*, **25B,** 1090 (1969).

17. O. Foss, K. Johnsen and T. Reistad, *Acta Chem. Scand.*, **18,** 2345 (1964).

18. W. R. Blackmore and S. C. Abrahams, *Acta Cryst.*, **8,** 323 (1955).

19. J. D. McCullough and R. E. Marsh, *Acta Cryst.*, **3,** 41 (1950).

20. R. E. Marsh, *Acta Cryst.*, **5,** 458 (1952).

21. F. H. Kruse, R. E. Marsh, and J. D. McCullough, *Acta Cryst.*, **10,** 201 (1957).

21a. W. S. McDonald and L. D. Pettit, *J. Chem. Soc.*, *A*, 2044 (1970).

22. H. von Eller, *C. R. Acad. Sci., Paris*, **239,** 1043 (1954).

23. A. Chiesi Villa, M. Nardelli, and C. Palmieri, *Acta Cryst.*, **25B,** 1374 (1969).

24. M. Nardelli, G. Fava, and G. Giraldi, *Acta Cryst.*, **15,** 737 (1962).

25. H. Hope, C. Knobler, and J. D. McCullough, *Acta Cryst.*, **26B,** 628 (1970).

26. J. S. Rutherford and C. Calvo, *Z. Krist.*, **128,** 229 (1969).

27. M. Perez-Rodriguez and A. Lopez-Castro, *Acta Cryst.*, **25B,** 532 (1969).

28. H. Hope, *Acta Cryst.*, **18,** 259 (1965).

29. I. L. Karle and J. A. Estlin, *Z. Krist.*, **129,** 147 (1969).

30. J. H. Bryden and J. D. McCullough, *Acta Cryst.*, **7,** 833 (1954).

31. J. H. Bryden and J. D. McCullough, *Acta Cryst.*, **9,** 528 (1956).

31a. E. S. Gould and B. Post, *J. Amer. Chem. Soc.*, **78,** 5161 (1956).

32. G. D. Christofferson, R. A. Sparks, and J. D. McCullough, *Acta Cryst.*, **11,** 782 (1958).

33. C. Knobler, J. D. McCullough, and H. Hope, *Abstr.* H3, Amer. Cryst. Assoc. Meeting, Seattle, Wash., March 1969.

34. W. R. Blackmore and S. C. Abrahams, *Acta Cryst.*, **8,** 317 (1955).

35. G. Y. Chao and J. D. McCullough, *Acta Cryst.*, **15,** 887 (1962).

36. G. D. Christofferson and J. D. McCullough, *Acta Cryst.*, **11,** 239 (1958).

37. O. Foss and S. Husebye, *Acta Chem. Scand.*, **20,** 132 (1966).

38. O. Foss and K. Marøy, *Acta Chem. Scand.*, **20,** 123 (1966).

39. O. Aksnes and O. Foss, *Acta Chem. Scand.*, **8,** 1787 (1954).

40. C. George, P. D'Antonio, A. Lowrey, and J. Karle, *J. Chem. Phys.*, **55,** 1071 (1971).

41. S. C. Abrahams, R. L. Collin, and W. N. Lipscomb, *Acta Cryst.*, **4,** 15 (1951).

42. I. L. Karle, J. A. Estlin and K. Britts, *Acta Cryst.*, **22,** 273 (1967).

43. H. L. Yakel and E. W. Hughes, *Acta Cryst.*, **7,** 291 (1954).

44. L. K. Steinrauf, J. Peterson, and L. H. Jensen, *J. Amer. Chem. Soc.*, **80,** 3835 (1958).

45. O. Holmesland and C. Rømming, *Acta Chem. Scand.*, **20,** 2601 (1966).

45a. T. Bjorvatten, *Acta Chem. Scand.* **17,** 2292 (1963).

46. A. Amendola, E. S. Gould, and B. Post, *Inorg. Chem.*, **3,** 1199 (1964).

47. J. D. McCullough and G. Hamburger, *J. Amer. Chem. Soc.*, **64,** 508 (1942).

48. I. Lindquist and G. Nahringbauer, *Acta Cryst.*, **12,** 638 (1959).

49. O. Foss, K. Johnsen, K. Maartmann-Moe, and K. Marøy, *Acta Chem. Scand.*, **20,** 113 (1966).

50. S. Husebye and J. W. George, *Inorg. Chem.*, **8,** 313 (1969).

51. O. Foss, H. Kjøge, and K. Marøy, *Acta Chem. Scand.*, **19,** 2349 (1964).

52. L. Pauling, "The Nature of the Chemical Bond," 3rd ed., Cornell University Press, Ithaca, N.Y., 1960, p. 224.

53. V. Luzzati, *Acta Cryst.*, **4,** 193 (1951).

54. R. G. Wood and G. Williams, *Nature*, **150,** 321 (1942).

55. S. Hosoya and S. Satake, *J. Phys. Soc. Japan*, **21,** 1611 (1966).

56. S. Hosoya, *Acta Cryst.*, **16,** 310 (1963).

57. H. W. Dias and M. R. Truter, *Acta Cryst.*, **17,** 937 (1964).

58. K. S. Dragonette and I. L. Karle, *Acta Cryst.*, **19,** 978 (1965).

58a. G. Valle, G. Cojazzi, V. Busetti, and M. Mammi, *Acta Cryst.*, **B26,** 468 (1970).

59. O. Foss, *Acta Chem. Scand.*, **7,** 1221 (1953).

59a. S. Furberg and P. Öyum, *Acta Chem. Scand.*, **8,** 42 (1954).

59b. A. McL. Mathieson and J. M. Robertson, *J. Chem. Soc.*, 724 (1949).

60. O. Foss and E. H. Vihovde, *Acta Chem. Scand.*, **8,** 1032 (1954).

61. O. Foss and P. Öyum, *Acta Chem. Scand.*, **9,** 1014 (1955).

62. P. Öyum and O. Foss, *Acta Chem. Scand.*, **9,** 1012 (1955).

63. S. Husebye, *Acta Chem. Scand.*, **20,** 24 (1966).

64. O. Foss, K. Marøy, and S. Husebye, *Acta Chem. Scand.*, **19,** 2361 (1965).

65. K. Fossheim, O. Foss, A. Scheie, and S. Solheimsnes, *Acta Chem. Scand.*, **19,** 2336 (1965).

66. O. Foss and S. Hauge, *Acta Chem. Scand.*, **19,** 2395 (1965).

67. E. D. Palik, *J. Mol. Spectrosc.* **3,** 259 (1959).

XVI Coordination compounds with organic selenium- and tellurium-containing ligands

K. A. JENSEN

University of Copenhagen
Chemical Laboratory II
The H. C. Ørsted Institute
Copenhagen, Denmark

C. K. JØRGENSEN

University of Geneva
Department of Chemistry
Geneva, Switzerland

A. INTRODUCTION

Coordination compounds with sulfur-containing ligands have been intensively studied during recent years (see the reviews by Livingstone[1] and

Jørgensen[2]), especially because of the theoretical significance connected with concepts such as noninnocent (or suspect) ligands, collectively oxidized ligands, and delocalized ground states.[2-6] As a consequence of the great interest in sulfur ligands an increasing number of publications have also appeared on coordination compounds with selenium- and tellurium-containing ligands. One of the purposes of extending the study of sulfur ligands to selenium and tellurium ligands is to increase knowledge of the chemical behavior of Group VI elements (oxygen, sulfur, selenium, tellurium) in their complexes, which have been somewhat neglected compared to those of Group V (nitrogen, phosphorus, arsenic, antimony) and Group VII (fluorine, chlorine, bromine, iodine).

For a comparison of the chemical behavior of ligands with these various elements, knowledge of spectrochemical parameters such as optical electronegativity and the nephelauxetic effect is essential. A recent review of the spectra and electronic structures of complexes with sulfur-containing ligands contains a comparison with selenium-containing ligands.[2] Of great importance are also the bond distances and i.r. bands of selenium- and tellurium-containing complexes.

The chemical properties of selenium ligands are very similar to those of the corresponding sulfur compounds. For tellurium the similarity is restricted mainly to the tellurols and organic tellurides. Tellurium analogs of carboxylic acids, amides, *etc.*, are unknown and possibly unstable.

B. SELENIUM AND TELLURIUM AS CENTRAL ATOMS

This chapter centers around complexes with selenium-containing ligands. It should be pointed out, however, that selenium and especially tellurium may also function as central atoms. A review of such compounds, mainly with inorganic ligands, has been published by Ryabchikov and Nazarenko.[7] Very few compounds have been reported with organic ligands bound to a Se central atom. Selenium(II) dithiocarbamates,[8] *O*-alkyl dithiocarbonates (xanthates),[9] dithiophosphinates,[10] and dithiophosphates[10] have been prepared, but these are probably comparable to the polythionates rather than to dithioate complexes, that is, only one S atom of the dithioate group is coordinated to selenium. The reported existence of a selenium complex of dithizone[11] has been alleged to be erroneous,[12] but this question is still under discussion.[13,14,14a]

Several workers have reported that $SeCl_4$ forms 1,2-addition compounds with pyridine or thiourea.[15-19] A recent X-ray investigation of $SeCl_4 \cdot 2py$ has shown that it is isomorphous with $SnCl_4 \cdot 2py$ and accordingly is an octahedrally coordinated *trans* compound[19] and not, as was earlier concluded from conductivity measurements,[16] an ionic compound, $[SeCl_3py_2]Cl$, with pentacoordinate selenium (*cf.* Katsaros and George[17]). With bidentate ligands like bipyridine, $SeCl_4$ forms complexes of the type $[SeCl_2L_2]Cl_2$, which without doubt also contain hexacoordinate selenium.[18] However, oxodichlorobis(pyridine)selenium(IV), $SeOCl_2 \cdot 2py$, has been shown by X-ray analysis[20] to contain pentacoordinate selenium with the group *trans*-$[SeCl_2py_2]$ in the base of a square pyramid and the O atom in the apical position.

Besides these compounds and the selenonium salts the only authentic examples of selenium complexes with organic ligands coordinated to a central Se atom seem to be the alkoxide complexes of the type $[SeO(OR)_3]^-$, formed by addition of NaOR to solutions of selenious esters.[21]

Because of its more metallic nature, tellurium would be expected to be more likely to function as a central atom. Typical coordination compounds of the types $[TeX_2tu_2]$ and $[Tetu_4]X_2$ are formed from Te(II) halides and thiourea * or substituted thioureas (tu). These complexes have been studied very extensively by Foss and his co-workers (one paper[24] contains references to earlier publications), who have shown by X-ray structure determinations that they have a square-planar or distorted square-planar structure. It is very interesting that this structure also applies to tellurium(II) xanthate,[25] in which the $(EtO)CS_2^-$ ion functions as a bidentate ligand so that the Te atom becomes surrounded by four S atoms in a nearly planar arrangement. A similar structure has recently been found for tellurium (II) *N,N*-diethyl-dithiocarbamate.[25a] Similar chelates are not formed by dithiophosphates or dithiophosphinates. However, the molecules are joined together by weak intermolecular bonds from a Te atom to S atoms of two neighboring molecules, so that again there is a tendency to square-planar coordination around tellurium.[26]

The i.r., Raman,[27] and Mössbauer spectra[27a] of Te(II)-thiourea complexes have also been studied.

Tellurium tetrahalides also form complexes with thioureas.[28,29] An X-ray crystallographic analysis of tetrachlorobis(tetramethylthiourea)tellurium(IV) has shown it to have a *trans*-octahedral structure.[29] With tetramethylthiourea, trimethyltellurium halides form 1:1 adducts for which a structure with pentacoordinate tellurium has been proposed.[30]

The numerous organometallic derivatives of tellurium, as well as complexes or Lewis adducts of tellurium tetrahalides or telluric acid with amines,[31-36] polyols,[37-40] and other organic compounds, will not be treated in this chapter.

---

* The reaction of thiourea with tellurium compounds was discovered by Falciola[22] and gained some analytical application (for references see a paper by Vřeštál[23]).

C. TYPES OF SELENIUM- AND TELLURIUM-CONTAINING LIGANDS

1. Selenolate and Tellurolate Ions

Several examples of organometallic selenolates with an SeR group bound to germanium, tin, *etc.*, are known, but these cannot strictly be considered coordination compounds but are rather to be compared to dialkyl selenides. However, some organoselenium derivatives with alkylseleno bridges are known. Abel *et al.*[41] prepared $[Mn(CO)_4SeR]_2$ and $[Re(CO)_4SeR]_2$ by reaction of phenylselenotrimethyltin with bromopentacarbonylmanganese and chloropentacarbonylrhenium, respectively. Since the analogous reactions with phenylthiotrimethyltin produce trimers under all experimental circumstances, the tin-selenium reactions seemingly involve a different mechanism from that involved in the tin-sulfur reaction. However, the trimer $[Re(CO)_3SePh]_3$ was formed by thermolysis of the dimer. Manganese compounds of the type $[Mn(CO)_4SeR]_2$ are also formed from hydridopentacarbonylmanganese and diselenides.[42] Analogous mixed carbonyl-phosphine complexes and the C–O stretching frequencies of binuclear complexes of the form $[LM(CO)_4]_2$ (M = Mn, Tc, Re; L = RS, RSe, RTe, R_2P, R_2As) have also been studied by Abel *et al.*[43,44] The reaction of $Mn_2(CO)_{10}$ with $F_3Se-P(CF_3)_2$ yields $Mn_2(CO)_8(F_3Se)(F_3C)_2P$, probably with $SeCF_3$ and $P(CF_3)_2$ bridging groups.[45] The corresponding arsenic compound has also been prepared.[46]

Tetracarbonyliron and dicarbonyldinitrosyliron react with diethyl diselenide to form the bridged compounds $(CO)_3Fe(SeEt)_2Fe(CO)_3$ and $(NO)_2Fe(SeEt)_2Fe(NO)_2$, respectively.[47,48] It is remarkable that the latter differs from the corresponding sulfur compound by having a dipole moment of zero. The analogous tellurium compounds are also known.[49]

Selenium- and tellurium-bridged iron carbonyls of the type

$$(CO)_3Fe(YR)_2Fe(CO)_3$$

(Y = Se or Te; R = C_6H_5 or C_6F_5) have been studied by Kostiner *et al.*[50,51] Mössbauer, i.r., n.m.r., and mass spectra are given.

Cyclopentadienyltitanium compounds with selenium ligands have been investigated by Koepf *et al.*[52] Cyclopentadienyltitanium and cyclopentadienylzirconium dihalides react with selenols to form RSe-bridged complexes. The dimer $[(C_5H_5)_2Ti]_2$ reacts with diphenyl diselenide to give bis(cyclopentadienyl)bis(benzeneselenolato)titanium, $(C_5H_5)_2Ti(SePh)_2$.[53] By irradiation of a solution of this compound and molybdenum hexacarbonyl in THF a phenylseleno-bridged Ti–Mo complex is formed.[54] An unusual compound with the formula $(C_5H_5)_2TiSe_5$ was obtained by Koepf *et al.*[55] and, on the basis of n.m.r. spectral evidence, formulated with a $TiSe_5$ six-membered ring.

Diselenides and ditellurides react with tricarbonyl(cyclopentadienyl)-molybdenum dimer to give $(C_5H_5)Mo(CO)_3Se(or\ Te)R$.[56]

Recently Andrä[57] has described the preparation of several metal benzene-selenolates from $Al(SePh)_3$ and metal halides. The aluminum compound was prepared from tris(isobutyl)aluminum and benzeneselenol. With $[NiX_2(Bu_3P)_2]$, $Al(SePh)_3$ reacted to give bis(tributylphosphine)bis(benzene-selenolato)nickel(II).

Only a few other nonbridged selenolato complexes are known. Bis-(trifluoromethylseleno)mercury(II) coordinates to anions to form

$$[(F_3CSe)_2HgX_2]^{2-}$$

and to 1 or 2 moles of triphenylphospine.[58] Ethyl(ethylseleno)beryllium adds 2 moles of pyridine.[59]

Benzeneselenol adds to carbene pentacarbonyl complexes of chromium, molybdenum, and tungsten to form complexes with coordinated selenide.[60]

Apart from the diselenolates, which represent a special type to be discussed in Section XVI.C.8, only a few chelating selenol ligands have been investigated. Quinoline-8-selenol[61] forms strongly colored and slightly soluble complexes similar to those of quinoline-8-thiol,[62,63] and chelates derived from (2-hydroselenophenyl)diphenylphosphine can be prepared by demethylation of the corresponding methylseleno compounds.[106] A recent addition to these examples are the anils of o-aminoselenophenol.[63a] From qualitative observations it is evident that hydroselenoacetic acid, 2-aminoethane-selenol, selenocysteine, and similar compounds resemble the corresponding sulfur compounds in their behavior towards metal ions. A recent quantitative investigation of the complex formation of selenocysteamine (2-aminoethane-selenol)[64,64a] has shown, however, that the nickel complexes differ in a remarkable way from those of 2-aminoethanethiol.[62,65] The red cation $[(\overset{+}{H_3}NCH_2CH_2Se-)_2Ni]$ is stable within wider pH ranges, and a chelate compound could not be isolated.

2.   Selenides and Tellurides

Unidentate dialkyl selenides resemble dialkyl sulfides in that they do not coordinate strongly to metals, apart from Pd(II), Pt(II), and a few others. Complexes analogous to the platinum compounds $[PtX_2(R_2S)_2]$ have been prepared from a few selenides and also tellurides and with palladium instead of platinum. Their electric dipole moments were measured in 1935 and found to be 2.2–2.4 D for the *trans* compounds.[66] This result was rather surprising at that time, because one would have expected the centrosymmetric conformation to be most stable. It is interesting that a recent X-ray structural analysis of *trans*-$[PdCl_2(Et_2Se)_2]$ has shown that this conformation is in fact stabilized in the solid state.[67] In solution, however, there is a rapid inversion, producing a certain population of *trans* conformers with dipole moments different from zero. It has been found that the *cis* complexes have higher barriers to inversion than the *trans* complexes and that the barrier is higher on going up in the series S < Se < Te.[68,69]

The stability of chloro-bridged Pt(II) compounds of the type $Pt_2L_2Cl_4$ (L = R_2S, R_2Se, R_2Te) follows the order $R_2S > R_2Te > R_2Se$, that is, the selenide complexes are the least stable.[70] This sequence seems to be specific, however, for platinum complexes since the corresponding Pd(II) complexes exhibit the reverse order for R_2Se and R_2Te. This behavior has been explained as being due to the relative sizes of the orbitals used for σ-bonds,[71] or to an abnormally weak π-bond from platinum to selenium.

An abnormal character of the Pt-Te bond is further indicated by a study of the vibrational spectra of complexes of the type $[PtX_2(Me_2Y)_2]$ (Y = S, Se, Te).[72] The frequencies of the metal chalcogen vibrations strongly suggest the following relation of bond orders: M–S > M–Se < M–Te.

Other investigations that are relevant to this question are those of Chatt[73-75] and Aires[72a] and their co-workers on the transmission of inductive or mesomeric effects across a Pt or Pd atom. These effects were studied by measuring the N–H stretching frequencies of complexes of the type *trans*-[Pt(or Pd)Cl_2-(amine)L][73] the ligand field splitting of complexes of the type *trans*-$[PtCl^3(piperidine)L]$,[74] and the pyridine proton chemical shifts in complexes of the type *trans*-$[PtCl_2pyL]$.[75] These results suggest that considerable π-bonding is involved in the bonds between certain ligands, especially phosphines, tellurides or ethylene, and platinum. In the Pd(II) complexes the effects transmitted across the Pd atom are mainly electrostatic. The degree of donation of electrons to platinum from different donor atoms does not determine the qualitatively observed general stabilities of their complexes. Thus, the complexes of dialkyl selenides and tellurides are less stable than corresponding dialkyl sulfide complexes.

An organometallic platinum compound, *trans*-$[Pt(SeEt_2)_2ClPh]$, has been prepared by treatment of the corresponding dichloro compound with phenyllithium.[109] A series of corresponding organopalladium complexes has recently been prepared.[109a] The rate of replacement of chloride by bromide, iodide, NO_2^-, N_3^-, SCN^-, or $SC(NH_2)_2$ has been determined spectrophotometrically.[109]

The difference between compounds of type a and type b metals (*cf.* Ahrland *et al.*[110]) is illustrated by the following examples. Dimethyl selenide and dimethyl telluride form unstable complexes with AgI and with Hg(II) halides. In both cases the stability order is Te > Se > S.[72] This is the reverse order of the stabilities of addition compounds with Me_3Al, whereas Me_3Ga assumes an intermediate position with the order Se > S = Te.[76]

Adducts of dialkyl selenides with $TiCl_4$,[77] WF_6,[78] and boron halides[79–81] have also been reported.

Complexes of 1,4-selenoxane and 1,4-diselenane have been studied by i.r. spectroscopy.[82,83] The selenoxane complexes exhibit an unchanged C–O–C stretching frequency, indicating that coordination does not occur to oxygen. In the complexes of 1,4-diselenane the ligand probably is bidentate, functioning as a bridge between different metal atoms with the formation of polymeric structures.

Tetrahydroselenophene forms a complex with $HgCl_2$.[84]

The addition compounds of selenides with iodine are of the charge-transfer type and will not be discussed here. They have been studied especially by McCullough and his co-workers (see refs. 85 and 86 and references given therein).

Some examples of metal carbonyls containing selenides or tellurides are also known.[87–95]

Compounds with two selenide groups an appropriate distance apart may function as chelate ligands. The formation of complexes between Pt(II) or Pd(II) halides and 1,3-bis(ethylseleno)propane was described by Fritzmann in 1912.[96] Recently Aynsley et al.[97] reported the preparation of other bidentate selenide ligands of the type $RSe(CH_2)_nSeR$ and their complex formation with a variety of transition halides. An attempt was made to increase the solubility of the complexes by varying the terminal alkyl group; as a result the ligand 1,2-bis(isopropylseleno)ethane, Pr^i–Se–CH_2CH_2–Se–Pr^i, was prepared, and its complexes with Pd(II) and Pt(II) halides were investigated.[98] The ligand forms complexes of the type MX_2L, which crystallize as monomers from acetone and as dimers from chloroform. In the monomers the ligand functions as a chelating group, but in the dimers two of the ligands form bridges between two metal atoms, as indicated by the fact that the ^1H n.m.r. spectra of the dimers are completely different from those of the monomers. A halogen-bridged dimeric system should give a ^1H n.m.r. spectrum very similar to that of the monomer. This conclusion has been supported by an X-ray structure analysis.[99] In the monomer the two isopropyl groups both lie on the same side of the $PdCl_2S_2$ plane.

Pluščec and Westland[100] have investigated Pd(II) and Pt(II) complexes of the ligand 1,3-bis(phenylseleno)propane. The halides are of the type MX_2L, but, on treating them with silver nitrate and adding additional selenide, nitrates of the type $[ML_2](NO_3)_2$ are formed. According to conductivity measurements in nonaqueous solutions, these complexes exhibit ion association, which was interpreted as implying the formation of five-coordinated complexes. The molar conductivities indicated an increase in ion association when sulfur was replaced by selenium in the palladium complexes, but the platinum complexes showed the reverse effect. The authors explain this difference by supposing an abnormally weak π-bond from platinum to selenium. This conclusion is supported by the i.r. spectra, the ligand-metal frequencies being decreased to a greater extent on replacement of sulfur by selenium in the platinum complexes than in the corresponding palladium compounds. As mentioned earlier, Chatt and Venanzi[70,71] found the order of stability to be $R_2S > R_2Te > R_2Se$ for platinum complexes of a similar type.

Dyer and Meek[101] have synthesized the quadridentate ligand tris(2-methylselenophenyl)phosphine and investigated its nickel complexes. This ligand forms pentacoordinate complexes of the types $[NiLX]ClO_4$, in which the fifth ligand is an anion, and $[NiLL'](ClO_4)_2$, in which L' is a neutral molecule.

These complexes are intensely colored and exhibit two ligand field bands characteristic of the trigonal-bipyramidal structure. The intensity of the first band is very high, and comparison with the corresponding sulfur ligand[102,103] shows that the order of increasing intensities is S (~ 1100) $<$ Se (~ 1700). This sequence indicates that the Ni-Se bond in these complexes involves a greater metal electron delocalization than the analogous Ni-S bond.

The related ligands (2-methylselenophenyl)diphenylphosphine and bis-(2-methylselenophenyl)phenylphosphine similarly form pentacoordinate complexes with Co(II) and Ni(II).[104,105] The first also forms four- and six-coordinated Ni(II) complexes. Meek has shown[106] that the complexes of this ligand easily undergo demethylation on heating in solution, forming complexes of the corresponding selenolate.

Uhlig et al.[107] have prepared complexes of the terdentate ligand bis-[2-(2-pyridyl)ethyl] selenide. It forms Cu(II) compounds of the type $CuLX_2$, possibly with a trigonal bipyramidal structure.

As mentioned in connection with the discussion of the i.r. spectra of selenides (Chapter XVA, Section 3), Goodall[108] has investigated complexes of Pd(II) and Pt(II) with unsaturated selenides, which function as chelating ligands, the metal being bound also to the double bond.

Quantitative comparisons of the stability constants of silver complexes in aqueous solutions with the unidentate ligands p-R–Se–$C_6H_4SO_3^-$ (R = Et or Ph)[111] and with the bidentate (phenylseleno)acetic acids (RC_6H_4–Se–CH_2COOH, with various substitutents in the o-, m-, or p-position)[112] have shown that the selenide complexes are more stable than the corresponding sulfide complexes. Selenium may therefore be classed as definitely "softer" or more "class b" (*cf.* ref. 110) than sulfur and much more so than oxygen (the silver complex of phenoxyacetic acid has a very low stability). Pettit *et al.*[112] considered the trends in the stability constants of the silver complexes of twenty substituted (phenylseleno)acetic acids in terms of the Hammett σ-functions for the substituents to evaluate the extent of d_π-bonding in the complexes. They concluded that d_π-bonding, if present, can play only a minor role in the stability trends found. The formation of silver complexes from saturated and unsaturated alkylselenoacetic acids has been recently discussed.[112a]

The stabilities of complexes of various divalent metal ions with the terdentate anions of selenodi(acetic acid), $Se(CH_2COOH)_2$, and of *cis*-tetrahydroselenophene-2,5-dicarboxylic acid have been determined by Suzuki *et al.*[113,114] According to their results, the stability constants of the complexes of these metal ions are always smaller for the selenium ligands investigated than for the corresponding sulfur ligands, even for the borderline "class b" ions (Cu^{2+}, Ni^{2+}, Co^{2+}), which show greater affinity to sulfur than to oxygen.

3. Heteroaromatic Selenium Compounds

A palladium complex of 2-(2-pyridyl)benzo[b]selenophene has been investigated with a view to its possible analytical application.[115] The ligand functions as a chelate ligand, forming a complex with the formula $[PdLCl_2]$, whereas the isomer 2-(4-pyridyl)benzo[b]selenophene functions as an unidentate ligand (presumably nitrogen-bonded), forming the complex $[PdL_2Cl_2]$. The u.v.-visible spectrum of the first complex differs from that of the ligand by the presence of an absorption band at 400 nm. The establishment of a metal bond to a Se atom that is part of an aromatic system is somewhat surprising and deserves to be more closely examined.*

It might be mentioned that benzo-2,1,3-selenadiazole forms a very slightly soluble palladium complex.[119,120] In this case, however, the metal atom is probably bound to one of the N atoms.

4. Selenoxides

Complexes of diphenyl selenoxide with various metal chlorides have been described by Paetzold and Vordank.[121]

A Rh(III) complex of diphenyl selenoxide has been proposed for the photometric determination of Rh(III).[122] Jensen and Krishnan[122] prepared

---

* The complexes of 1-(selenophene-2-yl)-1,3-butanedione with rare earth metals[116,117] and the complexes of 2,5-dibenzoyl-3,4-dihydroxythiophene[118] are undoubtedly normal β-diketone complexes.

several complexes of dimethyl selenoxide (DMSeO) from metal chlorides and DMSeO in ethanolic solution. The compounds were obtained as precipitates, which in most cases contained 2 moles of DMSeO per mole of metal chloride. According to this composition the compounds must contain coordinated halogen ions, and this inference was also indicated by their far-infrared spectra.

For all the DMSeO complexes a lowering of the Se–O stretching frequency from that of free DMSeO indicates that bonding of the metal ions occurs *via* oxygen. This inference applies also to the Pd(II) and Pt(II) complexes,[122] in contrast to the corresponding DMSO complexes, for which sulfur bonding has been inferred.[123]

Using metal perchlorates and an excess of DMSeO, Paetzold and Bochmann[124] succeeded in preparing complex ions with only DMSeO (4–6 moles) coordinated to the metal ion; no coordination of the perchlorate ion occurred, as shown by the presence of the unmodified absorption bands of the perchlorate ion in the i.r. spectra. These complexes are of the high-spin type, and from both their electronic and their i.r. spectra it may be concluded that the complexes of Cr(III), Mn(II), Fe(III), Ni(II), Zn(II), Cd(II), and Mg(II) are strictly octahedral, whereas $[Cu(DMSeO)_4]^{2+}$ is planar and $[Co(DMSeO)_6]^{3+}$ has a distorted octahedral structure.

Values of the Racah parameter B and the nephelauxetic effect slightly higher than those for DMSO complexes were calculated from the electronic spectra of hexa(dimethyl selenoxide) complexes of Mn(II), Fe(III), and Ni(II). Accordingly, DMSeO has a slightly lower d-electron delocalizing effect than DMSO.

The Dq values (calculated according to Jørgensen[126]) are in agreement with the sequence $H_2O > DMSeO > DMSO$ in the spectrochemical series, indicating that DMSeO has a higher ligand field strength than DMSO because of the higher polarity of the SeO bond.[127]

Tanaka and Kamitani[128] have studied tin complexes of DMSeO of the types $[SnCl_4(Me_2SeO)_2]$ and $[R_2SnCl_2(Me_2SeO)_2]$ (R = Me, Et, Ph). Infrared data suggest octahedral configuration with *trans*-alkyl, *cis*-halogen, and *cis*-Me$_2$SeO configurations for the dialkyldichlorotin complexes.

5. Diselenoketoens

Although monomeric selenoketones are unstable, a nickel complex of diselenoacetylacetone (4-hydroseleno-3-pentene-2-selone),

$$CH_3—CSeH=CH—C(=Se)—CH_3,$$

can be prepared directly by bubbling hydrogen selenide through an ethanolic solution of nickel chloride and acetylacetone.[129] The complex separates as red-brown crystals, which are soluble in chloroform, and its composition and physical properties—diamagnetism and i.r. spectrum—suggest that it has a square-planar configuration with equivalent Se atoms:

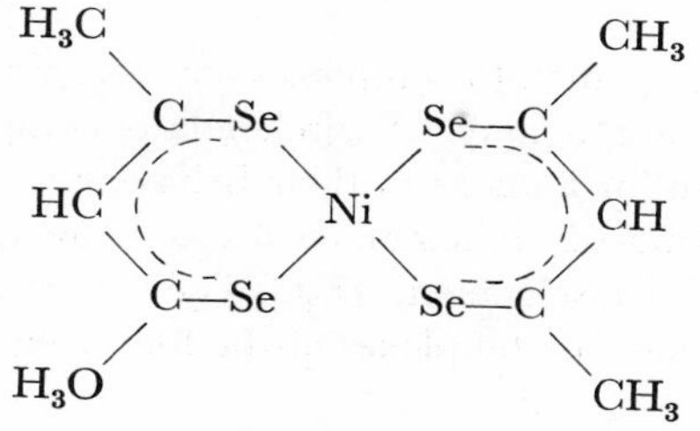

A compound obtained in a similar manner from cobalt(II) chloride has the formula $(C_5H_7Se_2)_2CoCl_4$.[130,131] The identity of its i.r. spectrum with that of 3,5-dimethyl-1,2-diselenolylium iodide showed, however, that it was not a diselenoacetylacetonate but instead was 1,3-dimethyl-1,2-diseleno-lylium tetrachlorocobaltate(II).

6. Carbon Diselenide

Tris(triphenylphosphine)platinum(0) forms a well-defined, green complex with the composition $[Pt(Ph_3P)_3CSe_2]$. A rhodium complex with the composition $[Rh(Ph_3P)_2Cl(CSe_2)_2]$ and a nickel complex with the composition $[Ni(Ph_3P)_2(CSe_2)_2]$ were also isolated.[132] The latter complexes formally contain two molecules of carbon diselenide, but these are probably combined into a tetraselenooxalate ion. This question is at present under investigation.

7. *O*-Alkyl Diselenocarbonates ("Selenoxanthates")

O-Alkyl diselenocarbonates form colored precipitates with several transition-metal ions.[133] Bis(*O*-ethyl diselenocarbonato)nickel(II) has been prepared by Jensen and Krishnan[134] and by Sgamelotti *et al.*[135] The solid compound is almost black but dissolves with a dark red color in chloroform. On addition of pyridine the solution instantly turns yellow. Although the adduct could not be isolated, the solution spectrum[135] has the typical pattern of the pseudooctahedral Ni(II) complexes, indicating the formation of the compound $[Ni(EtOCSe_2)_2py_2]$.

The main difference between the i.r. spectra of $[Ni(EtOCSe_2)_2]$ and $[Ni(EtOCS_2)_2]$ is that the sulfur compound has a strong band at 1025 cm^{-1}, whereas the selenium compound has a similar band at 935 cm^{-1}. The latter band is therefore assigned to the $\nu_{as}(C-Se_2)$ stretching vibration.[136]

8.   Diselenolates

1,1-Dithiolato and 1,2-dithiolato complexes have been intensively studied.[2–6] Analogous 1,1-diselenolato and 1,2-diselenolato complexes do not differ essentially from the sulfur analogs in their behavior.

Simple ionic ethylene-1,2-diselenolates are unknown, but the heterocycle bis(trifluoromethyl)-1,2-diseleneten, $(F_3C)_2C_2Se_2$, reacts with some metal carbonyls[137] to give neutral complexes of the following type:

(M = Ni, Mo, W; n = 2 or 3)

These compounds may be considered formally to be 1,2-diselenolato complexes of M^{4+} or M^{6+} ions. Like the analogous sulfur compounds, they undergo redox reactions to give a series of complexes related by one-electron transfer reactions. They are easily reduced, for example, by hydrazine, to uni- or bivalent anions that can be isolated as tetraphenylarsonium or tetrabutyl-ammonium salts. Only anionic complexes could be isolated with M = Cu, Fe, V, or Co.

Complexes of the 2,2-dicyanoethylene-1,1-diselenolate and cyanimido-diselenocarbonate ions can be prepared directly from the corresponding alkali-metal salts.[138] The bis- and tris-1,1-diselenolatometallates of several transition-metal ions were isolated as tetraphenylphosphonium salts. The double-bond frequency of the 1,1-diselenolate ion is increased 40–100 cm^{-1} on coordination;[139] this shift is explained like that of the diselenocarbamato complexes. The electronic spectra of the 2,2-dicyanoethylene-1,1-diseleno-lato complexes are very similar to those of the corresponding dithiolato complexes.[139]

Nickel, cobalt, and copper complexes of the mixed sulfur-selenium ligand 2-selenolatobenzenethiolate have been described by Pierpont et al.[140] The half-wave potential for reduction of these complexes, and their electronic and e.s.r. spectra, show great similarities to those of the analogous 1,2-benzene-dithiolato complexes.

9. Diselenocarbamates

N,N-Diethyl diselenocarbamates of Cu(II) and Zn(II) were first prepared by Barnard and Woodbridge.[141] Copper(II) complexes of other dialkyl-diselenocarbamic acids have also been reported.[142,143] Sodium N,N-dimethyl diselenocarbamate has been studied as an analytical reagent for a great number of metal ions.[144]

Tanaka et al.[145–146a] have studied selenothiocarbamato complexes of the type [Me$_2$Sn(SeSCNMe$_2$)$_2$], as well as similar diselenocarbamato complexes. Their i.r. and n.m.r. spectra suggest chelation of selenothiocarbamate and diselenocarbamate groups and a distorted octahedral configuration with *trans*-alkyl groups. The n.m.r. spectra indicate that the difference in the coordination power towards the Sn atom is hardly discernible between the diselenocarbamate and the dithiocarbamate ion in

$$[Me_2Sn(Se_2CNMe_2)(S_2CNMe_2)].\text{[146]}$$

Several transition-metal complexes were prepared by Jensen and Krishnan[134] and Furlani et al.[147] The crystal structure of bis(diethyldiselenocarbamato)nickel(II) has been determined.[148] Contrary to the statement in this paper, both Ni-Se bond lengths in each diselenocarbamato group are 2.32 Å and are identical within the experimental error.[148a,148b]

According to their electronic spectra,[147,149] the spectrochemical difference between dithiocarbamates and diselenocarbamates is very small. The diselenocarbamates have only slightly lower ligand field strengths than the corresponding dithiocarbamates, and the optical electronegativity, $\chi_{opt} = 2.65$, is almost insignificantly lower than the value for the dithiocarbamates ($\chi_{opt} = 2.7$).

In contrast to the anomalous (threshold) magnetic behavior of Fe(III) dithiocarbamates,[150,151] [Fe(Et$_2$NCSe$_2$)$_3$] is definitely a low-spin complex.[152] This is due to the pronounced nephelauxetic effect of the selenium ligand, which favors magnetic cross-over to the low-spin d^5 ground state. Dithiocarbamates have a marked tendency to form Fe(III) complexes of the type [FeL$_2$Cl] with approximately square-pyramidal structure.[153,154] Diselenocarbamates show a similar behavior, as demonstrated by the preparation of the complex [Fe(Et$_2$NCSe$_2$)$_2$Cl].[152]

The i.r. spectra of diselenocarbamato complexes exhibit the ν(C–N) absorption band at higher frequencies than the spectra for the alkali-metal salts. The shift is dependent on the coordination number and is similar to the shift observed for dithiocarbamates. The nitrosyl complexes

$$[Fe(Et_2NCSe_2)_2NO] \quad \text{and} \quad [Co(Et_2NCSe_2)_2NO]$$

exhibit an N–O stretching band at almost the same wave numbers as the corresponding dithiocarbamates.[134]

The i.r. spectra of dimethyl diselenocarbamate complexes have been investigated by Jensen and Krishnan[155] and by Pilipenko and Mel'nikova.[156] The far-infrared spectra of several dimethyl diselenocarbamate complexes have been studied by Jensen and Andersen.[157] Comparison with the spectra of the corresponding dithiocarbamates indicates that an absorption band at 870 cm^{-1} is due mainly to asymmetric C–Se$_2$ stretching. This inference has been confirmed by a normal coordinate analysis of [Ni{(CH$_3$)$_2$NCSe$_2$}$_2$] and [Ni{(CD$_3$)$_2$NCSe$_2$}$_2$].[158] Durgaprasad et al.[159] concluded from a normal coordinate analysis that the ν(C–Se$_2$) of dimethyl diselenocarbamates should be at 940 cm^{-1}, but actually they exhibit no absorption in this region.

The natural abundance of ^{77}Se (7.5%) with spin $= \frac{1}{2}$ makes it possible to determine the delocalization of unpaired electrons in selenium ligands by e.s.r. spectroscopy, and e.s.r. investigations of diselenocarbamates of Cu(II),[160–162c] as well as of VO^{2+},[163] have been reported. The e.s.r. spectra of bis(dibutyldiselenocarbamato)copper(II) agree with what is expected for a planar complex of approximately D_{2h} symmetry with one unpaired electron in an antibonding orbital.[162] A ^{59}Co n.m.r. study of chelates of Co(III) includes the dimethyl diselenocarbamate.[164]

10.   Diselenocarbazates

Anthoni[165] described bis(3,3-dimethyldiselenocarbazato)nickel(II) as a yellow-brown microcrystalline product. Undoubtedly it is similar in structure to the dithiocarbamates, that is, CSe$_2$ functions as a chelate group. It is soluble in strong bases, forming the anion [(Me$_2$N–N=CSe$_2$)$_2$Ni]$^{2-}$.

Various transition-metal complexes of unsubstituted and methyl-substituted diselenocarbazate ions have been prepared.[166] They are also probably analogous to diselenocarbamates. However, when the first N atom carries no substituent or only a methyl group, there is also the possibility of the formation of complexes with coordination to nitrogen, as in the case of selenosemicarbazides. The esters of mono- and diselenocarbazic acids form complexes of the latter type. These are mentioned in Section XVI.C.12 on selenohydrazides.

11.   Selenoureas

N,N'-Disubstituted selenoureas have been shown to function as unidentate, Se-bonded ligands in Pd(II) and Pt(II) complexes, which are similar to the corresponding thiourea complexes but show an enhanced tendency to 5-coordination.[167,168]

Complexes of unsubstituted selenourea with Cu(I), Ag(I), and Au(I),[169,169a] Co(II), and Ni(II),[169b] Tl(I),[169c] and with Pt(II),[170] have been described. Like thiourea, selenourea (su) reacts with a cobalt salt and dimethylglyoxime (H₂dmg) to form the complex ion *trans*-[Co(Hdmg)₂su₂]⁺, which in acid solution is easily transformed into *trans*-[CoX(Hdmg)₂su].[171] Selenourea complexes of Mo(V) have been studied by e.s.r. spectra.[171a]

Selenourea and thiourea form strongly colored osmium complexes.[172] Titrations have been interpreted to indicate that Os(VIII) is reduced to Os(III), which binds 8 moles of selenourea. A coordination number of 8 for Os(III) is, however, very unlikely.

Domiano *et al.*[173] have concluded, from X-ray powder data and crystal structures, that bis(selenourea)metal(II) thiocyanates, [M(su)₂(NCS)₂] (M = Co, Ni, Cd), are isostructural with the corresponding thiourea complexes. They have a polymeric structure consisting of chains of octahedra joined by two selenourea molecules, which form bridges between two metal atoms. The structure of [CdCl₂su₂] is also analogous to that of the corresponding thiourea complex.[174]

The Raman and i.r. spectra of selenourea complexes of Pd(II) and Pt(II) have been reported, and the selenium-metal stretching modes have been assigned.[175]

Goddard *et al.*[176,176a] have determined the stability constants of metal complexes of selenourea and selenosemicarbazides, as well as of thiourea and thiosemicarbazides. As would be expected, the order of affinity towards a "class b" metal like Hg(II) is Se > S ≫ O.

In this connection, it may be mentioned that selenoantipyrine forms complexes with SnCl₄ or TiCl₄.[177] On the basis of unconvincing evidence (shift of an i.r. band near 1300 cm⁻¹) it was concluded that coordination occurs through selenium.

12. Selenohydrazides

Thiosemicarbazides, mono- and dithiocarbazates, and thiocarboxylic hydrazides form complexes in which the metal atom is coordinated to the S atom and one of the N atoms (of the neutral or ionized ligand), forming a five-membered chelate ring. It was shown in 1936 that 4-phenylselenosemicarbazide forms similar complexes,[178] and an investigation of several substituted selenosemicarbazides, *O*-alkyl selenocarbazates, and selenocarboxylic hydrazides has shown that these are very similar to the corresponding sulfur compounds.[179] The inner-complex Ni(II) compounds of thiobenzhydrazide and 1-substituted thiosemicarbazides are oxidized by atmospheric oxygen to form intensely colored (dark blue) compounds containing collectively oxidized ligands.[180,181] They may be derived formally from (and prepared from) the corresponding azo and Ni(0) compounds. The corresponding selenium compounds, such as selenopivaloylhydrazine, show similar behavior.[181]

Complexes of unsubstituted selenosemicarbazide have been studied by Ablov et al.,[182–183a] who have also investigated complexes of salicylaldehyde selenosemicarbazone,[184–184d] selenosemicarbazones + dimethylglyoxime,[185] selenosemicarbazide + dimethylglyoxime,[188] and 2,3-butanedione oxime selenosemicarbazone.[186,187] Metal complexes of selenocarbonohydrazide have been studied by Linke and Turley.[189]

As mentioned previously, Goddard et al.[176] have determined the stability constants of complexes of selenosemicarbazide.

The selenium analog of dithizone has been prepared by Ramakrishna and Irving[190] and shown to form strongly colored, extractable complexes with a great variety of metal ions and thus to behave like dithizone. The diselenide also forms extractable colored complexes but is more selective.

13. Selenocyanate Ion

Ambidentate ligands have attracted the interest of many coordination chemists during recent years (see the reviews by Burmeister,[191] Jørgensen,[2] and Norbury and Sinha[192]). In the case of a coordinated thiocyanate ion the question arises whether it is attached to the metal atom *via* nitrogen or *via* sulfur, or whether it functions as a bridging group. All three modes can be realized, and it has even been possible to prepare complexes containing an *S*-bonded and an *N*-bonded thiocyanate group in the same molecule.[193,194] The same problems arise with a coordinated selenocyanate ion. Selenocyanate complexes have been quite extensively investigated (for the literature before 1964 see the review by Golub and Skopenko[195]). Structural assignments have been based mainly on i.r. spectroscopy.[196–204] The fundamentals of the selenocyanate ion[205] appear at 2070 cm^{-1} (mainly C–N stretching), 558 (mainly C–Se stretching) and 416 and 424 cm^{-1} (components of the bending mode). The ν(C–N) band is shifted to higher frequencies on *Se*-coordination (selenocyanate complexes), and the ν(C–Se) band is shifted to higher frequencies on *N*-coordination (isoselenocyanate complexes). The following ranges (cm^{-1}) have been suggested:[204]

	ν(C—N)	ν(C—Se)
M—NCSe	2025–2105	605–614
M—SeCN	2080–2124	520–543
M—SeCN—M	2135–2154	528–529

Accordingly, there is some overlap for ν(C–N); however, there is a pronounced difference in the integrated intensities of this band, which is up to an order of magnitude higher for the isoselenocyanates.[201,206] The position of the ν(C–Se) band seems to be quite conclusive, though this band may be covered if the complex contains other ligands or organic cations.

At a first approximation "a-type" metals[110] form isoselenocyanato complexes, whereas "b-type" metals form selenocyanato complexes. Thus, in complexes of Zn(II),[199–201] Cd(II),[201,201a] In(III),[207] Y(III),[201] lanthanides (La, Ce, Pr, Nd),[209–212b] Mo(III),[213,214] UO₂,[215] Ti(III),[216] V(III) and V(IV),[214,217,218] Cr(III),[219–222,227] Mn(II),[199,201,213,223,224] Fe(II) and Fe(III),[199,201,225,226] Co(II),[196,197,199,201,202,220,224,227,244] and Ni(II),[198,199,201,224,228–230,244] Zr(IV) and Hf(IV),[230a] Nb(V) and Ta(V),[230b] the selenocyanate ion is N-bonded, and in complexes of Rh(III),[201,213] Pd(II),[201,208,213,227,231–232] Pt(II),[196,201,202,213,227,233] Ag(I),[196] Au(III),[213] Hg(II),[196,197,202,213] Pb(II),[202] diarylthallium(III),[233a] and Bi(III)[234–236] it is Se-bonded or appears as a bridging group. Bridging occurs also in a Cd complex, [Cd₂(SeCN)₆]²⁻, studied by Burmeister and Williams,[201] whereas [Cd(NCSe)₆]⁴⁻ is wholly an isoselenocyanate complex.

The complexes MHg(SeCN)₄ (M = Co, Zn, Cd) and M[Hg(SeCN)₃]₂ (M = Cu, Pb) contain SeCN bridging groups with mercury bound to selenium.[196,237–242] The nickel complex [Nipy₄(NCSe)₂] is an isoselenocyanate, whereas the complex [Nipy₄(NCSe)₂] is polymeric with NCSe bridging groups.[228,229] A complex of nickel(II) selenocyanate with 6 moles of pyridine has been found to consist of [Nipy₄(NCSe)₂] with clathrate pyridine molecules.[230]

A bridged structure is also suggested for dioxane-containing magnesium selenocyanate,[243] whereas the corresponding calcium, strontium, and barium selenocyanates are wholly ionic.

The foregoing conclusions have been supported by X-ray analyses, although complete structure determinations have been performed for only a few compounds. One example is [Ni(dmf)₄(NCSe)₂] (dmf = dimethylformamide), which has been shown to have an octahedral *trans* configuration with the Ni–N distances = 2.05 Å.[244]

The complex [Co(CN)₅(NCSe)]³⁻ is an isoselenocyanate,[227] whereas the corresponding thiocyanate has been found to exhibit linkage isomerism and can be isolated as salts with the SCN group either N- or S-bonded. The ammines [Co(am)₅(NCSe)]²⁺ and [Cr(am)₅(NCSe)]²⁺ are analogous to the corresponding thiocyanates.[220] Vanadium forms a complex ion,

$$[V(NCS)_3(NCSe)_3]^{3-},$$

containing both thiocyanate and selenocyanate groups, all of which are N-bonded.[245]

The i.r. spectra of the Cu(II) selenocyanates [Cu(en)₂](SeCN)₂ and [Cu(am)₄](SeCN)₂ suggest Se-bonded or, more probably, ionic SeCN groups.[198,202] The corresponding nickel complexes are hexacoordinated with NCSe coordinated through nitrogen.

Cyclopentadienyl and carbonyl complexes apparently follow the same trend. The titanium cyclopentadienyl complex, $(C_5H_5)_2Ti(NCSe)_2$, is an isoselenocyanate,[246,247] and the same probably applies also to the corresponding vanadium complex.[248] In complexes of the type

$$(C_5H_5)M(CO)_2(SeCN)$$

the SeCN group is *Se*-bonded when M = Fe, Mo, or W, but *N*-bonded when M = Cr.[249] The binuclear manganese complex

$$[Mn_2(CO)_6Cl_2(NCSe)_2]^{2-}$$

(isolated as the tetraphenylarsonium salt) contains *N*-bonded terminal selenocyanate groups and chloro bridges.[250] In the corresponding tetra-selenocyanato complex two of the selenocyanate groups function as bridges.

However, the bonding mode of a coordinated thiocyanate ion is solvent dependent,[251] and Burmeister and Gysling[252] have succeeded in rearranging a *Se*-bonded palladium selenocyanate to the *N*-bonded isomer simply by dissolving it in dimethylformamide at room temperature. Rate studies of the isomerization in various solvents supported the assumption of a dissociative process, the first-order rate of isomerization decreasing as the dielectric constant of the solvent was decreased.[253]

Jennings and Wojcicki[254] also succeeded in obtaining two isomeric cyclopentadienyl complexes with the compositions

$$[(C_5H_5)Fe(CO)(Ph_3P)(NCSe)] \quad \text{and} \quad [(C_5H_5)Fe(CO)(Ph_3P)(SeCN)].$$

From the thiocyanates it is also known that the coordination of the thio-cyanate group to certain metals may be modified by the presence of other ligands or by the nature of the counter-ion used. These problems have not as yet been extensively investigated for selenocyanates, but at present it seems to be more difficult to change the bonding mode of a selenocyanate ion than of a thiocyanate ion.[208]

However, although Co(II), Fe(II), and Zn(II) normally form *N*-bonded selenocyanates, coordination through selenium has been suggested on the basis of i.r. spectroscopic evidence for the dimethylglyoximato complexes $[Fe(Hdmg)_2(SeCN)_2]^-$ and $[Co(Hdmg)_2(SeCN)_2]^-$,[255] as well as for the zinc complex $[Zn(bipy)_2(SeCN)_2]$.[256] An X-ray study of the Co(III) complex $NH_4[Co(SeCN)_2(Hdmg)_2] \cdot 3H_2O$ indicates *Se*-bonded seleno-cyanate ions,[257] and it is concluded that this applies also to the complex $[Co(SeCN)(dmg)_2(CH_3CSNH_2)]$.[258]

König *et al.*[259] have presented some evidence for the presence of both *Se*- and *N*-bonded selenocyanate in an iron complex with the empirical formula

$Fe_3(bipy)_7(NCSe)_6$. For the 1,10-phenanthroline complex $[Fe(phen)_2-(SeCN)_2]$, the variation of its magnetic moment with temperature, as well as i.r. data, indicates a change from a bridged Se-bonded structure at low temperature to a N-bonded structure at room temperature.[260]

According to a vibrational analysis, the complex zinc ions $[Zn(NCY)_4]^{2-}$ ($Y = O$, S, Se) have a tetrahedral structure.[200] An octahedral structure has been proved by X-ray analysis for the ion $[Pt(SeCN)_6]^{2-}$.[261] For some compounds of the types $[ML_4(NCSe)_2]$ and $[M(L-L)_2(NCSe)_2]$ ($M = $ Mn, Fe, Co, Ni, Zn; L = DMF,[202,223,244] N-methylformamide,[261a] dioxane,[223,243] pyridine,[226] water;[262] L–L = bipyridyl,[223,256] 1,10-phenanthroline,[215,225,256] 4-aminopyridine[261b]), octahedral $trans$ configuration is suggested by X-ray powder data. The complex

$$[Cr(py)_3(NCSe)_3]$$

has been isolated in two forms, red and yellow, which may be the anticipated fac and mer isomers.[221] The bidentate ligand diphenyl(2-diphenylarsinophenyl)phosphine forms a 5-coordinate Ni(II) complex,

$$[NiL_2(NCSe)]ClO_4,$$

in which the isoselenocyanate group is at the apex of a square pyramid.[263] As indicated by rate studies a 5-coordinate selenocyanato complex is formed by the addition of $SeCN^-$ to the cation $[PdL_2]^{2+}$, where L = o-phenylenebis(dimethylarsine).[264]

The electronic spectra of selenocyanates have been studied extensively;[2,196–207,214,231,265–267b] –NCSe is slightly higher than –NCS in the spectrochemical series, whereas –SeCN is lower than –SCN. According to Schmidtke,[265] the optical electronegativity is 2.8 for –SeCN and 2.6 for –NCSe (–SCN 2.9 and –NCS 2.65). The magnetic properties of isoselenocyanato complexes have been discussed by Salzmann and Schmidtke.[268]

The stabilities of selenocyanate complexes of Ag(I), Cu(I), Pb(II), Cd(II), Hg(II), Ni(II), and Co(II) have been studied in aqueous solutions, and of Th(IV), Ni(II), and Co(II) in nonaqueous solutions (acetonitrile, DMF) by conductometric, spectrophotometric, and polarographic methods.[269–80] The stabilities of Se-bonded selenocyanates are greater than those of N-bonded selenocyanates. Łodzińska[281] has compared the stability constants of selenocyanate complexes with those of halide complexes, and Czakis-Sulikowska[282] has investigated mixed halide-selenocyanate complexes.

The mechanism of substitution of the $SeCN^-$ of selenocyanate complexes with other anions or water has been investigated for a few special compounds.[283–286b]

Finally, it should be mentioned that an organic selenocyanate may function as a ligand. Goodall[287] investigated a number of metal complexes of bis-(selenocyanato)ethane, $NCSe-CH_2CH_2-SeCN$, and studied their i.r. spectra. It was concluded that the binding of the metal atoms [Pd(II), Pt(II), Rh(III), Ir(III)] occurred by means of the Se atoms, and that the ligand has to be in the *gauche* conformation to function as a chelating ligand.

14.   Phosphine Selenides

Several triphenylphosphine selenide complexes have been reported.[288–290,292] A shift of the P–Se stretching vibration from 560 cm^{-1} to 530–540 cm^{-1} indicates that coordination occurs to the Se atom.

Complexes of the types $[M(Me_3PSe)_4](ClO_4)_2$ and $[M(Me_3PSe)_2X_2]$ (M = Zn, Ni, Co; X = Cl, Br, I) have been prepared from trimethylphosphine selenide.[291] Their structure was discussed on the basis of electronic and i.r. spectra, magnetism, and X-ray powder diagrams.

15.   Diselenophosphate and Diselenophosphinate Ions

Complexes of O,O'-diethyl dithiophosphate have been investigated extensively.[293–296] Some recent studies of the corresponding O,O'-diethyl diselenophosphate complexes make a comparison possible.[297–299a]

The electronic spectra of the inner complexes of Cr(III), Rh(III), and Ir(III) with the O,O-diethyl diselenophosphate ion (dsep$^-$) were studied by Jørgensen,[296] who discussed the particularly strong nephelauxetic effect of this ion, β_{35} being as low as 0.37. The nephelauxetic effect in Cr(dsep)$_3$ is twice as large on (γ_5, γ_3) interactions as on (γ_5, γ_5) interactions, suggesting a very extensive delocalization of the σ-antibonding γ_3-subshell on the neighboring Se atoms. The strong absorption bands in the near-u.v. occur at a wave number *ca.* 3000 cm^{-1} lower in M(dsep)$_3$ than in the analogous dithiophosphates. This difference indicates that the optical electronegativity, χ_{opt}, of dsep$^-$ is 2.6, that is, 0.1 lower than for the dithiophosphates.

In a study of the diselenophosphate complexes of Tl(I), Pb(II), Sn(IV), As(III), Sb(III), Bi(III), and In(III) Krishnan and Zingaro[298] found that the electronic absorptions fall very close to those observed for the bromide complexes. The i.r. spectrum of potassium O,O'-diethyl diselenophosphate exhibits two bands at 584 and 620 cm^{-1}, ascribed respectively to the asymmetric and the symmetric stretching vibration of the $>PSe_2^-$ group. In the complexes listed (except Tl) these two bands are found at lower wave numbers, indicating coordination of the metal to both Se atoms.

Complexes of the O,O'-diphenyl selenothiophosphate ion have also been prepared.[299,300]

Complexes of dialkyl and diaryl selenothiophosphinate and diselenophosphinate ions have been prepared by Kuchen and Knop.[301,302] The selenothiophosphinate complexes of $Zn(II)$ and $Co(II)$ exhibit association in benzene or chloroform. This association has been attributed to the formation of ligand bridges.[303] The electronic spectra of various diselenophosphinato complexes have been discussed.[304–304b]

D. COVALENT BONDING IN TRANSITION-METAL COMPLEXES

Many crystalline selenides are either metallic (even when stoichiometric) or semiconductors having a low energy gap.[305] Such materials are black, and hence it was of great theoretical interest when the absorption spectra of inorganic chromophores[306] such as $Cr(III)Se_6$ could be studied in monomeric complexes like the diselenophosphates[296] and diselenocarbamates.[147,149] In more recent treatments[126,307] of the ligand field theory emphasis is given to the effects of interelectronic repulsion in the partly filled shell. It is worthwhile noting that the nephelauxetic ratio β_{35} still is as high as 0.4 in the $Cr(III)$ complexes because full covalency with half of the density of the d-like electrons localized on the central atom would decrease β_{35} to 0.25. It is possible to evaluate fractional charges of most central atoms,[3,308] in the interval from $+1$ to $+2$, though they seem to be $+0.6$ in $Cr(III)Se_6$. The nephelauxetic effect is more pronounced in selenium-containing complexes than in most others. However, indirect evidence is available[3] that phosphines and arsines produce an even stronger delocalization of the partly filled shell.

The absorption spectra of selenium-containing complexes show that there is much less difference between S and Se as ligating atoms than between O and S. This is particularly true for the electron transfer spectra,[309,310] where the difference between χ_{opt} for Cl^- (3.0) and Br^- (2.8) is larger than for S and Se. The stability constants of complexes of "class b" metals with selenium donors also manifest the slight difference between sulfur and selenium ligands. The stability constants show that selenium is "softer" than sulfur, in accordance with the Se atom being larger and more polarizable than a S atom. However, this chemical difference, as well as the slightly different position in the spectrochemical series, does not require the assumption of more extensive π back-donation in the selenium-containing ligands. On the whole, the d-group complexes of selenium-containing ligands are not as covalent as previously assumed, though they are less electrovalent than most other complexes.

Compounds with a Se atom forming a bridge between two atoms of the same or two different metal ions have been less extensively investigated than corresponding sulfur compounds.[2] Quite interesting magnetochemical effects have been reported by Zobrist[311] in the case of $[Cr(SCH_2CH_2NH_2)_3]$ and its adducts with divalent metal ions in which M–S–Cr bridges are present. There is little doubt that similar behavior could be detected for comparable selenium-containing ligands.

REFERENCES

1. S. E. Livingstone, *Quart. Rev.* (London), **19,** 386 (1965).

2. C. K. Jørgensen, *Inorg. Chim. Acta, Rev.,* **2,** 65 (1968).

3. C. K. Jørgensen, "Oxidation Numbers and Oxidation States," Springer Verlag, Berlin, 1969.

4. J. A. McCleverty, *Progr. Inorg. Chem.,* **10,** 49 (1968).

5. D. Coucouvanis, *Progr. Inorg. Chem.,* **11,** 233 (1970).

6. G. N. Schrauzer, *Accounts Chem. Res.,* **2,** 72 (1969).

7. D. I. Ryabchikov and I. I. Nazarenko, *Usp. Khim.,* **33,** 108 (1964).

8. O. Foss, *Acta Chem. Scand.,* **3,** 435 (1949).

9. O. Foss, *Acta Chem. Scand.,* **3,** 1385 (1949).

10. S. Husebye, *Acta Chem. Scand.,* **19,** 1045 (1965).

11. J. Starý and J. Růžička, *Talanta,* **15,** 505 (1968).

12. R. S. Ramakrishna and H. M. N. H. Irving, *Chem. Commun.,* 1356 (1969), *Anal. Chim. Acta,* **49,** 9 (1970).

13. J. Starý and J. Marek, *Chem. Commun.,* 519 (1970).

14. H. M. N. H. Irving, *Chem. Commun.,* 519 (1970).

14a. J. Starý, J. Marek, K. Kratzer, and F. Sebesta, *Anal. Chim. Acta,* **57,** 393 (1971).

15. V. G. Tronev and A. N. Grigorovich, *Zh. Neorg. Khim.,* **2,** 2400 (1957).

16. A. W. Cordes and T. V. Hughes, *Inorg. Chem.,* **3,** 1640 (1964).

17. N. Katsaros and J. W. George, *J. Inorg. Nucl. Chem.,* **31,** 3503 (1969).

18. D. A. Couch, P. S. Elmes, J. E. Ferguson, M. L. Greenfield, and C. J. Wilkins, *J. Chem. Soc., A,* 1813 (1967).

19. I. R. Beattie, M. Milne, M. Webster, H. E. Blayden, P. J. Jones, R. C. G. Killean, and J. L. Lawrence, *J. Chem. Soc., A,* 482 (1969).

20. I. Lindquist and G. Nahringbauer, *Acta Cryst.,* **12,** 638 (1959).

21. R. Paetzold and K. Aurich, *Z. Chem.,* **6,** 152 (1966).

22. P. Falciola, *Ann. Chim. Appl.,* **17,** 359 (1927).

23. J. Vřeštál, *Collection Czech. Chem. Commun.,* **25,** 443 (1960).

24. O. Foss, K. Johnson, K. Maartmann-Moe, and K. Marøy, *Acta Chem. Scand.,* **20,** 113 (1966).

25. S. Husebye, *Acta Chem. Scand.,* **21,** 42 (1967).

25a. C. Fabiani, R. Spagna, A. Vaciago, and L. Zambonelli, *Acta Crystallogr., Sect. B,* **27,** 1499 (1971).

26. S. Husebye, *Acta Chem. Scand.,* **19,** 1045 (1965); **20,** 24, 2007 (1966); **23,** 1389 (1969).

27. J. P. Hendra and Z. Jović, *J. Chem. Soc., A,* 735 (1967); 911 (1968).

27a. A. Yu. Aleksandrov, A. M. Babeshkin, V. I. Gol'danskii, S. P. Ionov, V. A. Lebedev, and R. A. Lebedev, *Zh. Strukt. Khim.*, **12**, 328 (1971).

28. O. Foss and W. Johannesen, *Acta Chem. Scand.*, **15**, 1939 (1961).

29. S. Husebye and J. W. George, *Inorg. Chem.*, **8**, 313 (1969).

30. K. J. Wynne and P. S. Pearson, *Chem. Commun.*, 556 (1970); *Inorg. Chem.*, **10**, 2735 (1971).

31. S. Prasad and B. L. Khandelwal, *J. Indian Chem. Soc.*, **38**, 837 (1961); **39**, 67, 112 (1962).

32. E. Montignie, *Z. Anorg. Allg. Chem.*, **307**, 109 (1961).

33. E. E. Aynsley and A. C. Hazell, *Chem. Ind.* (London), 611 (1963).

34. N. N. Greenwood, B. P. Straughan, and A. E. Wilson, *J. Chem. Soc.*, A, 2209 (1968).

35. N. N. Greenwood, A. C. Sarma, and B. P. Straughan, *J. Chem. Soc.*, A, 1561 (1968).

36. R. C. Paul, K. K. Paul, and K. C. Malhotra, *Australian J. Chem.*, **22**, 847 (1969).

37. P. J. Antikainen and P. J. Mälkönen, *Suomen Kemistilehti, B*, **32**, 23 (1959); P. J. Antikainen and G. Lundgren, *Suomen Kemistilehti, B*, **32**, 175 (1959).

38. R. G. Mehrotra and S. N. Mathur, *J. Indian Chem. Soc.*, **42**, 749 (1965).

39. H. R. Ellison, J. O. Edwards, and E. A. Healy, *J. Amer. Chem. Soc.*, **84**, 1820 (1962).

40. H. R. Ellison, J. O. Edwards, and L. Nyberg, *J. Amer. Chem. Soc.*, **84**, 1824 (1962).

41. E. W. Abel, B. C. Crosse, and G. V. Hutson, *J. Chem. Soc.*, A, 2014 (1967); *Chem. Ind.* (London), 238 (1966).

42. N. Welcman and I. Rot, *J. Chem. Soc.*, 7515 (1965).

43. E. W. Abel, A. M. Atkins, B. C. Crosse, and G. V. Hutson, *J. Chem. Soc.*, A, 687 (1968).

44. E. W. Abel, J. Dalton, I. Paul, J. G. Smith, and F. G. A. Stone, *J. Chem. Soc.*, A, 1203 (1968).

45. J. Grobe, *Z. Anorg. Allg. Chem.*, **331**, 63 (1964).

46. J. Grobe, J. E. Helgerud, and H. Stierand, *Z. Anorg. Allg. Chem.*, **371**, 123 (1969).

47. W. Hieber and W. Beck, *Z. Anorg. Allg. Chem.*, **305**, 265 (1960).

48. W. Hieber and W. Beck, *Z. Anorg. Allg. Chem.*, **305**, 274 (1960).

49. W. Hieber and R. Kramolowsky, *Z. Anorg. Allg. Chem.*, **321**, 94 (1963).

50. E. Kostiner, M. L. N. Reddy, D. S. Urch, and A. G. Massey, *J. Organometal. Chem.*, **15**, 383 (1968).

51. E. S. Kostiner and A. G. Massey, *J. Organometal. Chem.*, **19**, 233 (1969).

52. H. Koepf, B. Block, and M. Schmidt, *Z. Naturforsch., Part B*, **22**, 1077 (1967).

53. H. Koepf and B. Block, *Z. Naturforsch., Part B*, **23**, 1536 (1968).

54. H. Koepf and K. H. Raethlein, *Angew. Chem.*, **81**, 1001 (1969).

55. H. Koepf, B. Block and M. Schmidt, *Chem. Ber.*, **101**, 272 (1968).

56. E. W. Tillay, E. D. Schermer, and W. H. Baddley, *Inorg. Chem.*, **7**, 1925 (1968).

57. K. Andrä, *Z. Anorg. Allg. Chem.*, **373**, 209 (1970).

58. H. J. Clase and E. A. V. Ebsworth, *J. Chem. Soc.*, 2002 (1951).

59. G. E. Coates and A. H. Fischwich, *J. Chem. Soc.*, A, 635 (1968).

60. E. O. Fischer and V. Kiener, *Angew. Chem.*, **79**, 982 (1967).

61. E. Sekido, Q. Fernando, and H. Freiser, *Anal. Chem.*, **35**, 1550 (1963); **37**, 1556 (1965).

62. K. A. Jensen, *Z. Anorg. Allg. Chem.*, **229**, 265 (1936).

63. A. Corsini, Q. Fernando, and H. Freiser, *Anal. Chem.*, **35,** 1424 (1963).

63a. A. D. Garnovskii, L. S. Minkina, L. V. Sakhashchik, V. P. Kurbatov, and O. A. Osipov, *Zh. Obshch. Khim.*, **41,** 1884 (1971).

64. A. Yokoyama, H. Sakurai, and H. Tanaka, *Chem. Pharm. Bull.*, **18,** 1021 (1970).

64a. H. Sakurai, A. Yokoyama, and H. Hisashi, *Chem. Pharm. Bull.*, **19,** 1270 (1971).

65. D. C. Jicka and D. H. Busch, *Inorg. Chem.*, **1,** 872 (1962).

66. K. A. Jensen, *Z. Anorg. Allg. Chem.*, **225,** 97, 115 (1935).

67. P. E. Skakke and S. E. Rasmussen, *Acta Chem. Scand.*, **24,** 2634 (1970).

68. K. A. Jensen, P. Lindgreen, and U. Svanholm, article to be published.

69. W. McFarlane, *Chem. Commun.*, 755 (1968).

70. J. Chatt and L. M. Venanzi, *J. Chem. Soc.*, 2787 (1955).

71. J. Chatt and L. M. Venanzi, *J. Chem. Soc.*, 2351 (1957).

72. J. R. Allkins and P. J. Hendra, *J. Chem. Soc.*, A, 1325 (1967).

72a. B. E. Aires, J. E. Fergusson, D. T. Howarth, and J. M. Miller, *J. Chem. Soc.*, A, 1144 (1971).

73. J. Chatt, L. A. Duncanson, and L. M. Venanzi, *J. Chem. Soc.*, 4461 (1955); 3203 (1958).

74. J. Chatt, G. A. Gamlen and L. E. Orgel, *J. Chem. Soc.*, 1047 (1959).

75. J. Chatt and A. D. Westland, *J. Chem. Soc.*, A, 88 (1968).

76. G. E. Coates, *J. Chem. Soc.*, 2003 (1951).

77. A. D. Westland and L. Westland, *Can. J. Chem.*, **43,** 426 (1965).

78. A. M. Noble and J. M. Winfield, *Inorg. Nucl. Chem. Lett.*, **4,** 339 (1968).

79. K. J. Wynne and J. W. George, *J. Amer. Chem. Soc.*, **87,** 4750 (1965).

80. J. Le Calve and J. Lascombe, *Spectrochim. Acta*, **24A,** 736 (1968).

81. A. A. Palko and J. R. Durig, *J. Chem. Phys.*, **46,** 2297 (1967).

82. K. L. Baker and G. W. A. Fowles, *J. Chem. Soc.*, A, 801 (1968).

83. P. J. Hendra and N. Sadasivan, *J. Chem. Soc.*, 2063 (1965).

84. G. T. Morgan and F. H. Burstall, *J. Chem. Soc.*, 1096 (1929).

85. J. D. McCullough and A. Brunner, *Inorg. Chem.*, **6,** 1251 (1967).

86. H. Maddox and J. D. McCullough, *Inorg. Chem.*, **5,** 522 (1966).

87. W. Hieber and K. Wollmann, *Chem. Ber.*, **95,** 1552 (1962).

88. W. Hieber and T. Kruck, *Chem. Ber.*, **95,** 2027 (1962).

89. T. Kruck and M. Höfler, *Chem. Ber.*, **96,** 3035 (1963).

90. W. Hieber and A. Zeidler, *Z. Anorg. Allg. Chem.*, **329,** 92 (1964).

91. W. Hieber and F. Stanner, *Chem. Ber.*, **102,** 2930 (1969).

92. W. Hieber, W. Opavsky, and W. Rohm, *Chem. Ber.*, **101,** 2244 (1968).

93. H. Schumann and R. Weiss, *Angew. Chem.*, **82,** 256 (1970).

94. L. W. Houk and G. R. Dobson, *Inorg. Chem.*, **5,** 2119 (1966).

95. E. W. Abel and G. V. Hutson, *J. Inorg. Nucl. Chem.*, **31,** 3333 (1969).

96. E. Fritzmann, *Z. Anorg. Allg. Chem.*, **73,** 239 (1912); **133,** 119 (1924).

97. E. E. Aynsley, N. N. Greenwood, and J. B. Leach, *Chem. Ind.* (London), 379 (1966).

98. N. N. Greenwood and G. Hunter, *J. Chem. Soc.*, A, 1520 (1967); 929 (1969).

99. H. J. Whitfield, *J. Chem. Soc., A*, 113 (1970).

100. J. Pluščec and A. D. Westland, *J. Chem. Soc.*, 5371 (1965).

101. G. Dyer and D. W. Meek, *Inorg. Chem.*, **6**, 149 (1967).

102. G. Dyer and D. W. Meek, *Inorg. Chem.*, **4**, 1398 (1965).

103. G. S. Benner, W. E. Hatfield, and D. W. Meek, *Inorg. Chem.*, **3**, 1544 (1964).

104. G. Dyer and D. W. Meek, *J. Amer. Chem. Soc.*, **89**, 3983 (1967).

105. M. O. Workman, G. Dyer, and D. W. Meek, *Inorg. Chem.*, **6**, 1543 (1967).

106. D. W. Meek, *Inorg. Nucl. Chem. Lett.*, **5**, 235 (1969).

107. E. Uhlig, B. Borek, and H. Glänzer, *Z. Anorg. Allg. Chem.*, **348**, 189 (1966).

108. D. C. Goodall, *J. Chem. Soc., A*, 890 (1969).

109. R. Pietropaolo, S. Sergi, and G. Gaetano, *Ric. Sci.*, **38**, 195 (1968).

109a. S. Sergi, F. Faraone, L. Silvestro, and R. Pietropaolo, *J. Organometal. Chem.*, **33**, 403 (1971).

110. S. Ahrland, J. Chatt, and N. R. Davies, *Quart. Rev.* (London), **12**, 265 (1958).

111. S. Ahrland, J. Chatt, N. R. Davies, and A. Williams, *J. Chem. Soc.*, 264 (1958).

112. L. D. Pettit, C. Sherrington, and R. J. Whewell, *J. Chem. Soc., A*, 2204 (1968).

112a. D. S. Barnes, G. J. Ford, L. D. Pettit, and C. Sherrington, *J. Chem. Soc., A*, 2883 (1971).

113. K. Suzuki and K. Yamasaki, *J. Inorg. Nucl. Chem.*, **28**, 473 (1966).

114. K. Suzuki, I. Nakano, and K. Yamasaki, *J. Inorg. Nucl. Chem.*, **30**, 545 (1968).

115. L. S. Bark and D. Brandon, *Talanta*, **14**, 759 (1967).

116. A. I. Byrke, N. N. Magdesieva, L. I. Martynenko, and V. I. Spitsyn, *Dokl. Akad. Nauk SSSR*, **170**, 593 (1966).

117. V. M. Peshkova, I. P. Efimov, and N. N. Magdesieva, *Zh. Anal. Khim.*, **21**, 499 (1966).

118. M. Tomašković, A. Balenović-Solter, and Z. Štefanac, *Mikrochim. Acta*, 1156 (1969).

119. M. Ziegler and O. Glemser, *Chem. Ber.*, **91**, 2889 (1958); *Z. Anal. Chem.*, **146**, 29 (1955).

120. T. G. Bunting and C. E. Meloan, *Anal. Chem.*, **40**, 435 (1968).

121. R. Paetzold and P. Vordank, *Z. Anorg. Allg. Chem.*, **347**, 294 (1966).

122. K. A. Jensen and V. Krishnan, *Acta Chem. Scand.*, **21**, 1988 (1967).

123. F. A. Cotton and R. Francis, *J. Amer. Chem. Soc.*, **82**, 2986 (1960).

124. R. Paetzold and G. Bochmann, *Z. Chem.*, **8**, 308 (1968); *Z. Anorg. Allg. Chem.*, **368**, 202 (1969); **385**, 256 (1971).

125. M. Ziegler and H. Schröder, *Mikrochim. Acta*, 782 (1967).

126. C. K. Jørgensen, "Absorption Spectra and Chemical Bonding in Complexes," Pergamon Press, Oxford, 1962.

127. R. Paetzold, *Z. Chem.*, **4**, 321 (1964).

128. T. Tanaka and T. Kamitani, *Inorg. Chim. Acta*, **2**, 175 (1968).

129. C. G. Barraclough, R. L. Martin, and I. M. Stewart, *Australian J. Chem.*, **22**, 891 (1969).

130. G. A. Heath, I. M. Stewart, and R. L. Martin, *Inorg. Nucl. Chem. Lett.*, **5**, 169 (1969).

131. G. A. Heath, I. M. Stewart, and R. L. Martin, *Australian J. Chem.*, **22**, 83 (1969).

132. K. A. Jensen and E. Huge-Jensen, *Acta Chem. Scand.*, to be published.

133. A. Rosenbaum, H. Kirchberg, and E. Leibnitz, *J. Prakt. Chem.*, **19**, 1 (1963).

134. K. A. Jensen and V. Krishnan, *Acta Chem. Scand.*, **21**, 2904 (1967).

135. A. Sgamelotti, P. Porta, and E. Cervone, *Ric. Sci.*, **38**, 1223 (1968).

136. K. A. Jensen and U. Anthoni, *Acta Chem. Scand.*, **24**, 2055 (1970).

137. A. Davison snd E. T. Shawl, *Chem. Commun.*, 670 (1967); *Inorg. Chem.*, **9**, 1820 (1970)

138. K. A. Jensen and V. Krishnan, *Acta Chem. Scand.*, **24**, 1090 (1970).

139. K. A. Jensen and V. Krishnan, *Acta Chem. Scand.*, **24**, 1092 (1970).

140. C. G. Pierpont, B. J. Corden, and R. Eisenberg, *Chem. Commun.*, 401 (1969).

141. D. Barnard and D. T. Woodbridge, *J. Chem. Soc.*, 2922 (1961).

142. B. Lorenz and E. Hoyer, *Z. Chem.*, **8**, 230 (1968).

143. H. C. Brinkhoff, J. A. Cras, J. J. Steggerda, and J. Willemse, *Rec. Trav. Chim. Pays-Bas*, **88**, 633 (1969).

144. A. I. Busev and Kh. K. Kirspuu, *Tartu Riikliku Ülikooli Toimetised*, No. 219, 215 (1969).

145. T. Kamitani and T. Tanaka, *Inorg. Nucl. Chem. Lett.*, **6**, 91 (1970).

146. T. Kamitani, H. Yamamoto, and T. Tanaka, *J. Inorg. Nucl. Chem.*, **32**, 2621 (1970).

146a. T. Tanaka and N. Sonoda, *Inorg. Chem.*, **10**, 2337 (1971).

147. C. Furlani, E. Cervone, and F. D. Camassei, *Inorg. Chem.*, **7**, 265 (1968).

148. M. Bonamico and G. Dessy, *Chem. Commun.*, 1114 (1967).

148a. M. Bonamico, private communication.

148b. M. Bonamico and G. Dessy, *J. Chem. Soc.*, A, 264 (1971); P. T. Beurskens and J. A. Cras, *J. Cryst. Mol. Struct.*, **1**, 63 (1971).

149. K. A. Jensen, V. Krishnan, and C. K. Jørgensen, *Acta Chem. Scand.*, **24**, 743 (1970).

150. A. H. White, E. Kokot, R. Roper, H. Waterman, and R. L. Martin, *Australian J. Chem.*, **17**, 294 (1964).

151. A. H. Ewald, R. L. Martin, I. G. Ross, and A. H. White, *Proc. Roy. Soc.*, A, **280**, 235 (1964).

152. E. Cervone, F. D. Camassei, M. L. Luciani, and C. Furlani, *J. Inorg. Nucl. Chem.*, **31**, 1101 (1969).

153. B. F. Hoskins, R. L. Martin, and A. H. White, *Nature*, **211**, 627 (1966).

154. R. L. Martin and A. H. White, *Inorg. Chem.*, **6**, 713 (1967).

155. K. A. Jensen and V. Krishnan, *Acta Chem. Scand.*, **24**, 1088 (1970).

156. A. T. Pilipenko and N. V. Mel'nikova, *Zh. Neorg. Khim.*, **14**, 462 (1969).

157. K. A. Jensen and F. A. Andersen, *Acta Chem. Scand.*, to be published.

158. K. A. Jensen, B. M. Dahl, P. H. Nielsen, and G. Borch, *Acta Chem. Scand.*, in press.

159. G. Durgaprasad, D. N. Sathyanarayana, and C. C. Patel, *Can. J. Chem.*, **47**, 631 (1969).

160. G. M. Larin, P. M. Solozhenkin, N. I. Kopitsya, and Kh. K. Kirspuu, *Izv. Akad. Nauk SSSR, Ser. Khim.*, 968 (1969).

161. T. Ramasubba Reddy and R. Srinivasan, *Proc. Intern. Conf. Spectry.*, Bombay, 412 (1967).

162. J. G. M. van Rens, C. P. Keijzers, and H. van Willigen, *J. Chem. Phys.*, **52**, 2858 (1970).

162a. R. Kirmse, B. Lorenz, E. Hoyer, and W. Windsch, *Z. Chem.*, **10**, 305 (1970).

162b. R. Kirmse, B. Lorenz, E. Hoyer, and W. Windsch, *Z. Chem.*, **10**, 305 (1970).

162c. R. Kirmse, B. Lorenz, W. Windsch, and E. Hoyer, *Z. Anorg. Allg. Chem.*, **384,** 160 (1971).

163. G. M. Larin, P. M. Solozhenkin, N. I. Kopitsya, and Kh. K. Kirspuu, *Dokl. Akad. Nauk Tadzh. SSR*, **12,** 32 (1969).

164. R. L. Martin and A. H. White, *Nature*, **223,** 394 (1969).

165. U. Anthoni, *Acta Chem. Scand.*, **20,** 2742 (1966).

166. K. A. Jensen and B. M. Dahl, *Acta Chem. Scand.*, to be published.

167. C. Furlani and T. Tarantelli, *Inorg. Nucl. Chem. Lett.*, **2,** 391 (1966).

168. T. Tarantelli and C. Furlani, *J. Chem. Soc., A*, 1717 (1968).

169. V. L. Varand, V. M. Shul'man, and E. V. Khlystunova, *Izv. Akad. Nauk SSSR, Ser. Khim.*, 450 (1970).

169a. V. M. Shul'man, V. L. Varand, E. V. Khlystunova, and V. E. Fedorov, *Izv. Akad. Nauk SSSR, Ser. Khim.*, 915 (1971).

169b. E. V. Khlystunova, I. M. Kheremisina, V. L. Varand, and V. M. Shul'man, *Izv. Akad. Nauk SSSR, Ser. Khim.*, 1551 (1971).

169c. V. M. Shul'man, L. N. Makeeva, E. V. Khlystunova, V. L. Varand, and V. E. Fedorov, *Izv. Akad. Nauk SSSR, Ser. Khim.*, 1552 (1971).

170. V. M. Shul'man and L. I. Tyuleneva, *Izv. Akad. Nauk SSSR, Ser. Khim.*, 1189 (1970).

171. N. N. Proskina, A. V. Ablov, and V. N. Shafranskii, *Zh. Neorg. Khim.*, **14,** 3034 (1969).

171a. I. N. Marov, V. K. Belyaeva, Yu. N. Dubrov, and A. N. Ermakov, *Zh. Neorg. Khim.*, **16,** 154 (1971).

172. A. T. Pilipenko and I. P. Sereda, *Zh. Anal. Khim.*, **13,** 3 (1958); *Zh. Neorg. Khim.*, **6,** 413 (1961).

173. P. Domiano, A. G. Manfredotti, G. Grossoni, M. Nardelli, and M. E. V. Tani, *Acta Cryst.*, **25B,** 591 (1969).

174. N. N. Proskina, S. M. Chulskaya, G. F. Volodina, and A. V. Ablov, *Zh. Strukt. Khim.*, **9,** 1095 (1968).

175. P. J. Hendra and Z. Jović, *Spectrochim. Acta*, **24A,** 1713 (1968).

176. D. R. Goddard, B. D. Lodam, S. O. Ajayi, and M. J. Campbell, *J. Chem. Soc., A,* 506 (1969).

176a. D. R. Goddard and S. O. Ajayi, *J. Chem. Soc., A,* 2673 (1971).

177. V. T. Panyushkin, A. D. Garnovskii, I. I. Grandberg, O. A. Osipov, V. I. Minkin, V. S. Troitskaya, and V. G. Vinokurov, *Zh. Obshch. Khim.*, **38,** 1154 (1968).

178. K. A. Jensen and E. Frederiksen, *Z. Anorg. Allg. Chem.*, **230,** 31 (1936).

179. K. A. Jensen *et al.*, article to be published.

180. K. A. Jensen and J. F. Miquel, *Acta Chem. Scand.*, **6,** 189 (1952).

181. K. A. Jensen, K. Bechgaard, and C. Th. Pedersen, *Acta Chem. Scand.*, **22,** 3341 (1968).

182. A. V. Ablov, N. V. Gerbeleu, and A. M. Romanov, *Zh. Neorg. Khim.*, **14,** 981 (1969).

183. A. M. Romanov, A. V. Ablov, and N. V. Gerbeleu, *Zh. Neorg. Khim.*, **14,** 381 (1969).

183a. E. V. Suntsov, A. M. Romanov, A. V. Ablov, and N. V. Gerbeleu, *Zh. Strukt. Khim.*, **11,** 437 (1971).

184. A. V. Ablov, N. V. Gerbeleu, and A. M. Romanov, *Zh. Neorg. Khim.*, **13,** 3034 (1968).

184a. A. V. Ablov, N. V. Gerbeleu, and N. Y. Negryatse, *Zh. Neorg. Khim.*, **16,** 1069 (1971).

1044   **K. A. Jensen and C. K. Jørgensen**

184b. A. V. Ablov, N. V. Gerbeleu, A. M. Romanov, and V. M. Vlad, *Zh. Neorg. Khim.*, **16**, 1357 (1971).

184c. B. T. Oloi, N. V. Gerbeleu, and A. V. Ablov, *Zh. Neorg. Khim.*, **16**, 2888 (1971).

184d. K. I. Turta, A. V. Ablov, V. I. Fol'danskii, N. V. Gerbeleu, and R. A. Stukan, *Dokl. Akad. Nauk SSSR*, **196**, 1383 (1971).

185. A. V. Ablov, N. V. Gerbeleu, and A. M. Romanov, *Zh. Neorg. Khim.*, **13**, 787 (1968).

186. A. V. Ablov, N. V. Gerbeleu, and N. Y. Negryatse, *Zh. Neorg. Khim.*, **14**, 986 (1969).

187. A. V. Ablov, N. V. Gerbeleu, and N. Y. Negryatse, *Zh. Neorg. Khim.*, **15**, 119 (1970).

188. A. V. Ablov, N. V. Gerebeleu, and A. M. Romanov, *Zh. Neorg. Khim.*, **14**, 3317 (1969).

189. K. H. Linke and R. Turley, *Z. Naturforsch., Part B*, **24**, 821 (1969).

190. R. S. Ramakrishna and H. M. N. H. Irving, *Anal. Chim. Acta*, **48**, 251 (1969); *Chem. Ind.* (London), 325 (1969).

191. J. L. Burmeister, *Coord. Chem. Rev.*, **1**, 205 (1966); **3**, 225 (1968).

192. A. H. Norbury and A. I. P. Sinha, *Quart. Rev.* (London), **24**, 69 (1970).

193. G. R. Clark, G. J. Palenik, and D. W. Meek, *J. Amer. Chem. Soc.*, **92**, 1077 (1970).

194. D. W. Meek, P. E. Nicpon, and V. I. Meek, *J. Amer. Chem. Soc.*, **92**, 5351 (1970).

195. A. M. Golub and V. V. Skopenko, *Usp. Khim.*, **34**, 2098 (1965).

196. A. Turco, C. Pecile, and M. Nicolini, *J. Chem. Soc.*, 3008 (1962).

197. F. A. Cotton, D. M. L. Goodgame, M. Goodgame, and T. E. Haas, *Inorg. Chem.*, **1**, 565 (1962).

198. M. E. Farago and J. M. James. *Inorg. Chem.*, **4**, 1706 (1965).

199. D. Forster and D. M. L. Goodgame, *Inorg. Chem.*, **4**, 1712 (1965).

200. D. Forster and W. de W. Horrocks, Jr., *Inorg. Chem.*, **6**, 339 (1967).

201. J. L. Burmeister and L. E. Williams, *Inorg. Chem.*, **5**, 1113 (1966).

201a. G. V. Tsintsadze, Yu. Ya. Kharitonov, A. Yu. Tsivadze, A. M. Golub, and A. S. Managadze, *Zh. Neorg. Khim.*, **15**, 2336 (1970).

202. Yu. Ya. Kharitonov and G. V. Tsintsadze, *Zh. Neorg. Khim.*, **10**, 35, 1191 (1965).

203. Yu. Ya. Kharitonov, G. V. Tsintsadze, and M. A. Porai-Koshits, *Dokl. Akad. Nauk SSSR*, **160**, 1351 (1965).

204. Yu. Ya. Kharitonov and V. V. Skopenko, *Zh. Neorg. Khim.*, **10**, 1803 (1965).

205. H. W. Morgan, *J. Inorg. Nucl. Chem.*, **16**, 367 (1961).

206. C. Pecile, *Inorg. Chem.*, **5**, 210 (1966).

207. V. V. Skopenko, V. F. Mikitchenko, and G. V. Tsintsadze, *Zh. Neorg. Khim.*, **14**, 1790 (1969).

208. J. L. Burmeister and H. J. Gysling, *Inorg. Chim. Acta*, **1**, 100 (1967).

209. J. L. Burmeister, S. D. Patterson, and E. A. Deardorff, *Inorg. Chim. Acta*, **3**, 105 (1969)

210. J. L. Burmeister and E. A. Deardorff, *Inorg. Chim. Acta*, **4**, 97 (1970).

211. G. V. Tsintsadze, A. M. Golub, and M. V. Kopa, *Zh. Neorg. Khim.*, **14**, 2743 (1969).

212. A. M. Golub, M. V. Kopa, and O. E. Andreichenko, *Zh. Obshch. Khim.*, **39**, 1382 (1969).

212a. A. M. Golub, M. V. Kopa, V. V. Skopenko, and G. V. Tsintsadze, *Z. Anorg. Allg. Chem.*, **375**, 302 (1970).

212b. A. M. Golub, M. V. Kopa, and G. V. Tsintsadze, *Zh. Obshch. Khim.*, **41**, 15 (1971).

213. H. H. Schmidtke and D. Garthoff, *Helv. Chim. Acta*, **50,** 1631 (1967).

214. H. H. Schmidtke and D. Garthoff, *Z. Naturforsch., Part A*, **24,** 126 (1969).

215. V. V. Skopenko and E. I. Ivanova, *Ukr. Khim. Zh.*, **36,** 16 (1970).

216. J. L. Burmeister and L. E. Williams, *J. Inorg. Nucl. Chem.*, **29,** 839 (1969).

217. V. V. Skopenko and E. I. Ivanova, *Zh. Neorg. Khim.*, **14,** 742 (1969).

218. V. V. Skopenko, E. I. Ivanova, and G. V. Tsintsadze, *Ukr. Khim. Zh.*, **34,** 1000 (1968).

219. K. Michelsen, *Acta Chem. Scand.*, **17,** 1811 (1963).

220. N. V. Duffy and F. G. Kossel, *Inorg. Nucl. Chem. Lett.*, **5,** 519 (1969).

221. G. V. Tsintsadze, V. V. Skopenko, and A. I. Brusilovets, *Soobshch. Akad. Nauk Gruz.*, **50,** 109 (1968).

222. A. I. Brusilovets, V. V. Skopenko, and G. V. Tsintsadze, *Zh. Neorg. Khim.*, **14,** 467 (1969).

223. P. Stancheva, V. V. Skopenko, and G. V. Tsintsadze, *Ukr. Khim. Zh.*, **35,** 166 (1969).

224. I. G. Shevchenko and G. V. Tsintsadze, *Zh. Neorg. Khim.*, **9,** 2675 (1964).

225. V. V. Skopenko, A. I. Brusilovets, and G. V. Tsintsadze, *Ukr. Khim. Zh.*, **35,** 489 (1969).

226. G. V. Tsintsadze, *Soobshch. Akad. Nauk Gruz.*, **57,** 57 (1970).

227. J. L. Burmeister and M. Y. Al-Janabi, *Inorg. Chem.*, **4,** 962 (1965).

228. S. M. Nelson, *Proc. Chem. Soc.*, 372 (1961).

229. S. M. Nelson and T. M. Shepherd, *Inorg. Chem.*, **4,** 813 (1965).

230. A. H. Norbury, E. A. Ryder, and R. F. Williams, *J. Chem. Soc., A*, 1439 (1967).

230a. A. M. Golub, T. P. Lishko, and V. V. Skopenko, *Ukr. Khim. Zh.*, **37,** 835 (1971).

230b. T. M. Brown and B. L. Bush, *J. Less-Common Metals*, **25,** 397 (1971).

231. H. H. Schmidtke, *J. Inorg. Nucl. Chem.*, **28,** 1735 (1966).

232. A. M. Golub, G. B. Pomerants, and S. A. Ivanova, *Zh. Neorg. Khim.*, **14,** 2826 (1969).

233. A. Sabatini and I. Bertini, *Inorg. Chem.*, **4,** 959 (1965).

233a. T. N. Srivastava and K. K. Bajpai, *J. Organometal. Chem.*, **31,** 1 (1971).

234. A. M. Golub, V. V. Skopenko, and A. Z. Zhumabaev, *Zh. Neorg. Khim.*, **14,** 2995 (1969).

235. V. V. Skopenko and A. Z. Zhumabaev, *Ukr. Khim. Zh.*, **35,** 428 (1969); **37,** 528 (1971).

236. V. V. Skopenko, A. Z. Zhumabaev, and G. V. Tsintsadze, *Ukr. Khim. Zh.*, **36,** 329 (1970).

237. E. Frasson, A. Turco, and C. Panattoni, *Gazz. Chim. Ital.*, **91,** 750 (1961).

238. R. M. Alasaniya, V. V. Skopenko, and G. V. Tsintsadze, *Tr. Gruz. Politekh. Inst.*, No. 7, 21 (1967).

239. R. M. Alasaniya, G. V. Tsintsadze, and V. V. Skopenko, *Tr. Gruz. Politekh. Inst.*, No. 3, 28 (1968).

240. G. V. Tsintsadze, A. E. Shvelashvili, and V. V. Skopenko, *Tr. Gruz. Politekh. Inst.*, No. 1, 29 (1967).

241. G. V. Tsintsadze, A. E. Shvelashvili and V. V. Skopenko, *Tr. Gruz. Politekh. Inst.*, No. 4, 53 (1967).

242. G. V. Tsintsadze, V. V. Skopenko, and A. E. Shvelashvili, *Soobshch. Akad. Nauk Gruz.*, **41,** 337 (1966).

243. V. V. Skopenko, G. V. Tsintsadze, R. M. Alasaniya, and L. V. Glushchenko, *Ukr. Khim. Zh.*, **35**, 1317 (1969).

244. G. V. Tsintsadze, M. A. Porai-Koshits, and A. S. Antsyskina, *Zh. Strukt. Khim.*, **8**, 296 (1967).

245. H. Böhland and P. Malitzke, *Z. Anorg. Allg. Chem.*, **350**, 70 (1967).

246. J. L. Burmeister, E. A. Deardorff, and C. E. Van Dyke, *Inorg. Chem.*, **8**, 170 (1969).

247. J. L. Burmeister, E. A. Deardorff, A. Jensen, and V. H. Christensen, *Inorg. Chem.*, **9**, 58 (1970).

248. G. Doyle and R. S. Tobias, *Inorg. Chem.*, **7**, 2479 (1968).

249. M. A. Jennings and A. Wojcicki, *J. Organometal. Chem.*, **14**, 231 (1968).

250. M. F. Farona, L. M. Frazee, and N. J. Bremer, *J. Organometal. Chem.*, **19**, 225 (1969).

251. J. L. Burmeister, R. L. Hassel, and R. J. Phelan, *Chem. Commun.*, 679 (1970).

252. J. L. Burmeister and H. J. Gysling, *Chem. Commun.*, 543 (1967).

253. J. L. Burmeister, H. J. Gysling, and J. C. Lim, *J. Amer. Chem. Soc.*, **91**, 44 (1969).

254. M. A. Jennings and A. Wojcicki, *Inorg. Chim. Acta*, **3**, 335 (1969).

255. K. Burger and B. Pinter, *Magy. Kem. Folyoirat*, **73**, 209 (1967).

256. R. M. Alasaniya, V. V. Skopenko, and G. V. Tsintsadze, *Ukr. Khim. Zh.*, **35**, 568 (1969).

257. A. V. Ablov and I. D. Samus', *Dokl. Akad. Nauk SSSR*, **146**, 1071 (1962).

258. A. V. Ablov and V. N. Shafranskii, *Zh. Neorg. Khim.*, **9**, 585 (1964).

259. E. König, K. Madeja, and W. H. Böhmer, *J. Amer. Chem. Soc.*, **91**, 4582 (1969).

260. W. A. Baker, Jr., and H. M. Bobonich, *Inorg. Chem.*, **3**, 1184 (1964).

261. G. V. Tsintsadze, V. V. Skopenko, and A. E. Shvelashvili, *Tr. Gruz. Politekh. Inst.*, No. 1, 19 (1968).

261a. G. V. Tsintsadze, V. V. Skopenko, and L. B. Kereselidze, *Soobshch. Akad. Nauk Gruz.*, **61**, 53 (1971).

261b. J. de O. Cabral and M. F. Cabral, *Rev. Port. Quim.*, **12**, 85 (1970).

262. G. V. Tsintsadze, A. M. Mumulashvili, and L. P. Demchenko, *Zh. Neorg. Khim.*, **15**, 276 (1970).

263. T. D. Dubois and D. W. Meek, *Inorg. Chem.*, **6**, 1395 (1967).

264. R. Ettore, A. Peloso, and G. Dolcetti, *Gazz. Chim. Ital.*, **97**, 968 (1967).

265. H. H. Schmidtke, *Ber. Bunsenges. Phys. Chem.*, **71**, 1138 (1967).

266. P. Day, *Inorg. Chem.*, **5**, 1619 (1966).

267. F. Pruchnik and S. Wajda, *Rocz. Chem.*, **44**, 933 (1970).

267a. F. Pruchnik, S. Wajda, and E. Kwaskowska-Cheć, *Rocz. Chem.*, **45**, 537 (1971).

267b. S. Wajda, F. Pruchnik, and E. Kwaskowska-Cheć, *Bull. Acad. Pol. Sci.*, *Ser. Sci. Chim.*, **19**, 141 (1971).

268. J. J. Salzmann and H. H. Schmidtke, *Inorg. Chim. Acta*, **3**, 207 (1969).

269. V. F. Toropova, *Zh. Neorg. Khim.*, **1**, 243 (1956).

270. A. M. Golub and G. B. Pomerants, *Zh. Neorg. Khim.*, **4**, 769 (1959).

271. A. M. Golub, *Zh. Neorg. Khim.*, **4**, 1577 (1959).

272. A. M. Golub and V. V. Skopenko, *Zh. Neorg. Khim.*, **6**, 140 (1961).

273. A. M. Golub and V. V. Skopenko, *Zh. Neorg. Khim.*, **7**, 1012 (1962).

274. A. M. Golub and V. V. Skopenko, *Zh. Neorg. Khim.*, **7**, 1265 (1962).

275. A. M. Golub, V. V. Skopenko, and G. B. Pomerants, *Zh. Neorg. Khim.*, **10**, 344 (1965)

276. A. M. Golub and V. A. Kalibabchuk, *Zh. Neorg. Khim.*, **12**, 2370 (1967).

277. V. V. Skopenko and A. I. Brusilovets, *Ukr. Khim. Zh.*, **30**, 24 (1964).

278. V. V. Skopenko and A. I. Brusilovets, *Ukr. Khim. Zh.*, **34**, 1210 (1968).

279. V. V. Skopenko, R. M. Alasaniya, and L. V. Glushchenko, *Ukr. Khim. Zh.*, **36**, 16 (1970).

280. A. A. Humffray, A. M. Bond, and J. S. Forrest, *J. Electroanal. Chem.*, **15**, 67 (1967).

281. A. Łodzińska, *Rocz. Chem.*, **41**, 1007 (1967).

282. D. M. Czakis-Sulikowska, *Rocz. Chem.*, **39**, 1161 (1965); **40**, 1821 (1966).

283. J. L. Burmeister and J. C. Lim, *Chem. Commun.*, 1154 (1969).

284. A. Peloso, R. Ettore, and G. Dolcetti, *Inorg. Chim. Acta*, **1**, 307 (1967).

285. A. Peloso, G. Dolcetti, and R. Ettore, *Gazz. Chim. Ital.*, **97**, 1507 (1967). *Inorg. Chim. Acta*, **1**, 403 (1967).

286. G. Schiavon and C. Paradisi, *Ric. Sci.*, **37**, 952 (1967).

286a. C. Varhelyi, I. Zsako, and Z. Finta, *Stus. Univ. Babes-Bolyai, Ser. Chem.*, **15**, 81 (1970).

286b. C. Varhelyi, I. Zsako, and Z. Finta, *Monatsh. Chem.*, **101**, 1013 (1970).

287. D. C. Goodall, *J. Inorg. Nucl. Chem.*, **30**, 1269, 2483 (1968).

288. E. Bannister and F. A. Cotton, *J. Chem. Soc.*, 1959 (1960).

289. M. G. King and G. P. McQuillan, *J. Chem. Soc.*, 898 (1967).

290. P. Nicpon and D. W. Meek, *Chem. Commun.*, 398 (1966).

291. A. M. Brodie, G. A. Rodley, and C. J. Wilkins, *J. Chem. Soc., A*, 2927 (1969).

292. D. Brown, J. Hill, and C. E. F. Rickard, *J. Less-Common Met.*, **20**, 57 (1970).

293. C. K. Jørgensen, *J. Inorg. Nucl. Chem.*, **24**, 1571 (1962).

294. S. Chaston, S. E. Livingstone, T. N. Lockyer, V. A. Pickles, and J. M. Shanon, *Australian J. Chem.*, **18**, 673 (1965).

295. D. M. Adams and J. B. Cornell, *J. Chem. Soc. A.*, 1299 (1968).

296. C. K. Jørgensen, *Mol. Phys.*, **5**, 485 (1962).

297. M. V. Kudchaker, R. A. Zingaro, and K. J. Irgolic, *Can. J. Chem.*, **46**, 1415 (1968).

298. V. Krishnan and R. A. Zingaro, *Inorg. Chem.*, **8**, 2337 (1969).

299. S. V. Larionov and L. A. Il'ina, *Zh. Obshch. Khim.*, **39**, 1587 (1969).

299a. S. V. Larionov and L. A. Il'ina, *Zh. Obshch. Khim.*, **41**, 762 (1971).

300. L. A. Il'ina, N. I. Zemlyanski, S. V. Larionov, and N. M. Chernaya, *Izv. Akad. Nauk SSSR, Ser. Khim.*, 198 (1969).

301. W. Kuchen and B. Knop, *Angew. Chem.*, **76**, 496 (1964).

302. W. Kuchen and B. Knop, *Angew. Chem.*, **77**, 259 (1965).

303. W. Kuchen and H. Hertel, *Angew. Chem.*, **81**, 127 (1969).

304. F. Galsbøl, *Proc. 13. Nord. Kemikermøde, Copenhagen*, 54 (1968).

304a. H. Hertel and W. Kuchen, *Chem. Ber.*, **104**, 1735, 1740 (1971).

304b. A. Müller, P. Christophliemk, and V. V. Krishna Rao, *Chem. Ber.*, **104**, 1905 (1971).

305. F. Hulliger, *Structure and Bonding*, **4**, 83 (1968).

306. C. K. Jørgensen, "Inorganic Complexes," Academic Press, London, 1963.

307. C. K. Jørgensen, "Modern Aspects of Ligand Field Theory," North Holland Publishing Co., Amsterdam, 1971.

308. C. K. Jørgensen, *Helv. Chim. Acta, Fasc. Extraord. Alfred Werner*, 131 (1967).

309. C. K. Jørgensen, "Halogen Chemistry," Vol. 1, V. Gutmann, Ed., Academic Press, London, 1967, p. 265.

310. C. K. Jørgensen, *Progr. Inorg. Chem.*, **12,** 101 (1970).

311. J. Zobrist, Thesis, ETH, Zürich, 1970.

XVII Analytical methods

JOSEPH F. ALICINO and JOHN A. KOWALD

Squibb Institute for Medical Research
New Brunswick, New Jersey

A. INTRODUCTION

With the growing importance of selenium chemistry comes the need for precise methods of analysis for this element. This need is reflected by the astounding increase in the number of publications dealing with organic selenium chemistry. More than one hundred papers on selenium analysis were published in the past decade, whereas only seven references were cited for the microanalysis of selenium in "Quantitative Organic Microanalysis" by Steyermark, published in 1961.[150] In addition, analytical methods for the determination of selenium have been reviewed by Bagnall[4a] and Rosenfeld and Beath.[136a] The current interest in selenium and its compounds is especially prevalent in the medical, biological, pharmacological, and toxicological disciplines now concerned with air and water pollution.

The purpose of this chapter is to help the chemist select the method best suited to solve a particular analytical problem arising during an investigation in selenium chemistry.

For any method of analysis, the sample must be prepared in such a way that the desired element or compound can be easily and precisely measured. This preparation generally takes the form of two separate and distinct steps: (1) decomposition of the material and (2) measurement of the decomposition product, with or without a separation step.

B. SEPARATION AND ISOLATION

1. Decomposition, Dissolution, and Preliminary Treatment of Sample

a. Fusion. One of the earliest decomposition methods, and one which is still in use today, is the fusion method, in which the sample is fused with an inorganic material to destroy the organic matter. The three most common approaches are *carbonate* fusion, utilizing the mixture $Na_2CO_3 + KNO_3$ or $Na_2CO_3 + NaNO_3$; *alkali* fusion with NaOH; and *peroxide* fusion with Na_2O_2. The fusion mixtures will generally produce the hexavalent form of selenium if a sufficient amount of oxidizing agent is present. When other elements, such as boron, silicon, aluminum, tungsten, and molybdenum, are present, these will solubilize along with the selenate salt after extraction with water. Metals such as iron, copper, and nickel will be present as insoluble oxides or carbonates and are removed by filtration or centrifugation.

The carbonate fusions are carried out in platinum crucibles, whereas the alkali and peroxide fusions require nickel or iron crucibles. The ratio of fusion mixture to sample is generally about 10 to 1, and in the carbonate fusion the ratio of carbonate to oxidizing agent must be about 5 to 1.

b. Combustion. Combustion of the sample may also be accomplished by heating at high temperature (so-called tube combustion) in a quartz or Vycor combustion tube. The sample is best weighed in a porcelain boat, because of the fouling effect of selenium in one made of platinum. A sample may be combusted by burning in a McCullough[110] oxyhydrogen flame. Selenium dioxide is formed and determined iodometrically.

A very simple type of combustion, and one that is ideally suited for the destruction of most organic compounds, is accomplished by the method of Schöniger,[141] generally referred to as "closed oxygen flask" combustion. The application of this method to the analysis of selenium compounds is discussed in Section XVIID.2.a.

c. Wet Digestion. This type of decomposition is variously referred to as wet digestion, wet ashing, or mineralization. McNulty[113] gave a detailed procedure for the wet ashing of material such as soil, nutrient media, or tissues. He used ammonium metavanadate in nitric acid for digestion of the sample, followed by the application of nitric, sulfuric, and perchloric acids, in that order. Finally, a temperature of 210° was utilized for complete digestion. These are rather drastic conditions, but were found necessary for the large sample (10–50 g) used.

On a smaller scale, Kahane and Korach[88] digested 5-50 mg of selenium-containing compounds in 1.5 ml nitric and 2 ml perchloric acid. A water trap was used to absorb small amounts of selenium that might be vaporized during heating. The water in the trap was combined with the digest for subsequent treatment.

Gould[70] decomposed organoselenium compounds by digestion with 3 ml of concentrated sulfuric acid, followed by smaller portions of fuming nitric acid until digestion was complete. He lists 22 organoselenium compounds, the selenium content of which ranges from 10 to 30%.

2. Selective Separation

a. Interfering Elements. A number of elements, metallic and otherwise, that may interfere in various methods must be removed, depending on the decomposition technique or subsequent treatment. Elements mentioned by Gould[70] (bromine and mercury), as well as arsenic, antimony, tin, and tellurium, can interfere. In certain iodometric methods and gravimetric finishes, sulfur can also interfere; this element will be considered later in Section XVIIC.5.

b. Volatilization and Distillation Methods. In 1896, Gooch and Pierce[63] used the reaction $H_2SeO_3 + 4HBr \rightarrow SeBr_4 + 3H_2O$ to separate selenium from other materials. The interfering elements are left behind on distillation with hydrobromic acid, and the distillate contains selenium dibromide and bromine. On the addition of water, H_2SeO_3 and HBr are formed and selenium may be determined in a variety of ways. McNulty et al.[114] used the separation technique to remove selenium from copper and organic material and then determined selenium iodometrically after destroying the bromine. The same type of separation, on a larger scale, was described by Trelease and Beath.[159]

c. Precipitation and Coprecipitation Procedures. Ferric hydroxide has been used to coprecipitate H_2SeO_3 at pH 8.0. This procedure is best used to concentrate a small amount of selenium from large amounts of material like copper alloys.[83]

d. Ion-Exchange Procedures. A cation-exchange procedure was used by Yoshino[174] in the accumulation of selenite/selenate from sea water.

The separation of methionine and selenomethionine was accomplished by Martin and Cummins[108] by an ion-exchange technique. Martin and Gerlach[109] separated the important selenoamino acids selenocystine, Se-methylselenocysteine, selenomethionine, and selenocystathionine, using a Bio-Rad resin of the Aminex A-4 type and employing modifications of the procedures described by Spackman et al.[148]

Benson and Patterson[14] described an accelerated ion-exchange chromatographic procedure for the quantitative analysis of selenocystine and selenomethionine, using a spherical highly cross-linked cation-exchange resin, type PA-35, which is widely employed for the chromatography of amino acids and peptides. By this method, the appearance of selenocystine required 45 minutes, and of selenomethionine 90 minutes, using identical buffer, temperature, and buffer flow rate.

e. Solvent Extraction Methods. Iron, copper, and several other heavy metals can be separated from selenium by extracting with chilled chloroform and 6% aqueous cupferron solution.

Jordanov and Futekov[87] recommend acetophenone and 1–2 M H_2SO_4 for the extraction of selenium into chloroform. These authors have studied various other ketones as well.

f. Paper, Thin-Layer, and Gas Chromatography.

Combining paper chromatography with an electrophoretic separation, Peterson and Butler[134] showed that sulfur and selenium amino acids can be identified.

By coupling the techniques of paper chromatography and neutron activation analysis, Barak and Swanberg[11] perfected a method for quantitating selenomethionine in the presence of methionine in sources such as serum, liver, or muscle. The analogs were isolated from the tissues and chromatographed on paper; the methionine area, which included the selenomethionine, was located with ninhydrin. This area was cut from the paper, placed in a Tygon tube, and irradiated for 20 seconds in a Triga Mark I atomic reactor, using a rapid-transit pneumatic rabbit system and a neutron flux of 1.1×10^{11} n/cm²/sec. After a delay of 20 seconds, the sample was analyzed in an RIDL 400 channel γ-ray spectrometer, using a 1 minute live time count. This provided a spectrum from 0 to 1 MeV, which includes the peak due to ⁷⁷ᵐSe, a 17.5 second isotope of selenium with peak emission at 0.160 MeV. By a direct comparison using selenium standards, the selenium content and the micromoles of selenomethionine in the sample were calculated. The lower limit of detectability of this method was reported to be 0.1 μM of selenomethionine per gram of tissue or per milliliter of serum. The two isotopes most likely to interfere with the neutron activation analysis of ⁷⁷ᵐSe are ⁴⁶ᵐSc, derived from scandium in natural sources, and ¹⁹O, which arises from irradiated samples containing water. When dried ninhydrin-sprayed spots were used in the neutron activation analysis for selenium, the ¹⁹O interference was eliminated. Scandium was found to remain at the origin of the chromatogram and created no problem in this technique.

Millar[118] developed a solvent mixture for the separation of methionine from selenomethionine, using thin-layer chromatography on silica gel G plates. Spencer et al.[149] reported on 5 solvent systems for characterizing radioactive ⁷⁵Se-selenocystine on cellulose thin-layer chromatographic plates. These authors used isobutanol-HCl-H_2O, H_2O-butanol-acetic acid, H_2O saturated with phenol, butanol-acetic acid-H_2O, and 2-propanol-ammonia-H_2O.

Evans and Johnson[51] used gas chromatography for the separation and identification of the following alkylselenium compounds: dimethyl, diethyl, and di-n-propyl selenides; diethyl and di-n-propyl diselenides; and ethyl selenocyanate. All compounds were fully characterized on three columns. The best separation was achieved with a polymetaphenyl ether packing.

The g.l.c. separation of organic selenides, sulfides, and ethers was reported by Benes and Prochazkova.[13] The selenides investigated included: ^{14}C-dimethyl selenide, dimethyl selenide, diethyl selenide, dipropyl selenide, di-n-butyl selenide, phenylselenol, phenyl methyl selenide, and phenyl ethyl selenide. The chromatographic data obtained for the selenides were compared with those for the corresponding sulfides and ethers under identical conditions, using 20% squalene as well as 20% dinonylphthalate as stationary phases at temperatures of 90 and 110°. The relative elution volumes for each group are given, using di-n-propyl selenide and the corresponding sulfide and ether as the reference elution volumes.

Caldwell and Tappel[23] separated the silylated derivatives of the selenium analogs of cysteine and methionine on a 2% SE-30 column on Anakron SD.

g.   Ring Oven Techniques.   Selenium, in the order of parts per billion, is determined semiquantitatively by the ring oven technique, which uses 2,3-diaminonaphthalene[48] under u.v. light. Traces of selenium in water can be also determined by this technique. After separation by means of thiourea, the selenium is compared with standards. West and Cimerman[167] used 3,3'-diaminobenzidine for determining selenium in air pollution studies.

h.   Reduction Methods.

i.   Hydrogen Sulfide and Ammonium Sulfide.   Taimni *et al.*[152-154] proposed a separation technique employing precipitation with H_2S, which removes the copper and arsenic group metals. The conditions employed are critical, however, and are difficult to achieve.

ii.   Sulfur Dioxide.   Sulfur dioxide is a widely used reductant. However, the procedure[66,67] that reduces selenium(IV) and selenium(VI) to the elemental state is not recommended for microgram quantities of selenium. After the sample has been treated with a mixture of perchloric and nitric acids, the nitric acid is removed by boiling. The acid digest is cooled and diluted with 9 N hydrochloric acid, and the precipitation of selenium is completed by the addition of sulfur dioxide. Selenium is filtered, washed, and weighed on a fritted glass crucible. De Salas[139] uses hydroxylamine hydrochloride in combination with sulfur dioxide to ensure the complete reduction of any selenious or selenic acid that may be present.

iii.   Stannous Chloride.   Stannous chloride is a more efficient reducing agent than sulfur dioxide and is recommended when the selenium concentration is less than 1 mg/100 ml of solution. This procedure employs a method of digestion similar to that used for sulfur dioxide reduction. To the digest containing concentrated hydrochloric acid, 5 g of solid $SnCl_2$ is added and the solution is heated to just below the boiling point for 15 minutes. After slow cooling of the solution, the precipitate is filtered through a Silas crucible, washed, and weighed. Wiberly *et al.*[169] used only 5 ml of 0.6 M $SnCl_2$ in 25 ml of 3 N hydrochloric acid for samples containing 2–10 mg of selenium. Noakes[121] used hydrazine hydrochloride, in addition to $SnCl_2$, to precipitate selenium.

iv. Thiourea. A qualitative test for as little as 0.1 μg of selenium is claimed by Treadwell and Hall,[158] who placed some dry thiourea on a filter paper and moistened it with a drop of the solution to be tested. An orange to red coloration indicates the presence of selenium if large amounts of copper and nitrates are absent; a yellow coloration signifies the presence of tellurium. Tellurium forms a soluble complex with thiourea.[86] Thiourea has been used for a quantitative determination of selenium by Deshmukh and Sankaranarayanan.[45]

v. Other Reducing Agents. The use of various other reducing agents has been suggested as a means of separating selenium from tellurium. Simon and Grim[147] employed ferrous sulfate in the presence of Complexon III at pH 10 to reduce the tellurite ion. Sodium hypophosphite was used by McKenna and Templeton[112] and Noakes[121] to precipitate tellurium. Reduction of tellurium by means of chromous ion was accomplished by Goto *et al.*[68] Hydrazine sulfate was used by Mellor[116] to precipitate tellurium in a hydrochloric acid solution. Hydroxylamine, hydrazine hydrate, phenyl-hydrazine, and hypophosphorous acid have also served to reduce telluric acid quantitatively.[96] Simon and Grim,[146] as well as Rudra and Rudra,[138] used ascorbic acid to reduce selenite and selenate to selenium. Fidler[57] employed glucose in slightly alkaline solution for this purpose.

Anger and Fisher[3a] selectively reduced aliphatic and aromatic disulfides and diselenides with Raney nickel, using nickel-aluminum alloys (50/50) in hydrochloric acid-alcohol to yield thiols or selenols. The resulting thiols or selenols are then detected through the iodine-azide reaction. This permits a selective detection of disulfides in the absence of other compounds that react positively with the iodine-azide solution (thiols, thiones, mustard oils, thio acids, and derivatives). Some of the examples and the detectable quantities reported are as follows:

COMPOUND	QUANTITY (μg)
Cystine	0.005
p,p'-(Dimethylamino)diphenyl disulfide	0.1
p,p'-Dinitrodiphenyl disulfide	0.2
o,o'-Dinitrodiphenyl disulfide	0.01
Dibenzyl diselenide	1

C. DETECTION AND IDENTIFICATION OF ELEMENTAL SELENIUM AND SELENIUM OXIDES

1. Chromatography

Burstall *et al.*[21] distinguished selenium from tellurium by the orange to red color of the former and the black color of the latter after reduction with stannous chloride. Weatherley[165] detected and was able to estimate from 5 to 45% of selenium in silicates by a chromatographic system using 40% hydrofluoric acid for development. Thiourea solution served for detection in this case.

A spot detection system using $SnCl_2$ was applied by Bighi and Mantovani.[17] The solvents employed were pyridine, *n*-propanol, and water which enabled the authors to distinguish H_2SeO_3, H_2SeO_4, $H_2SeS_2O_6$, and $Na_2SeS_4O_6$ by their R_f-values.

2. Spectrography

Selenium does not show very sensitive lines in the range of ordinary spectrographs, and its detection is impeded by its high volatility. Gatterer[61] utilized high-vacuum discharge tubes that were under the influence of a high-frequency electromagnetic field. It is claimed that as little as 0.001% selenium can be detected by this means. Another investigator, Feldman,[56] tried to detect selenium at 2413.5 Å and found the matrix to have a profound effect on the sensitivity, especially in compounds and alloys that tended to repress the volatility of selenium. Mitchell[119] and Milbourn[117] also investigated spectrographic methods.

3. Methylene Blue Procedures

Feigl and West[55] used the unique ability of elemental selenium to catalyze the reducing power of Na_2S solution on methylene blue. Elemental selenium instantly decolorizes the dye, but other elements take 15–20 minutes. Goto *et al.*[65] also studied this reaction and showed that concentration and temperature determined the limits of interference of other elements.

4. Thiourea Procedures

These techniques are given on page 1055.

5. Precipitation as Barium Selenate

The barium ion can be used in a detection method inasmuch as barium selenite is soluble in dilute acid whereas barium selenate is not. A strong oxidizing agent, such as $KMnO_4$, will convert selenium from the tetravalent to the hexavalent form, which may be precipitated with barium chloride. However, the presence of sulfite, thiosulfate, or sulfate invalidates this test and must be taken into consideration.

6.   Complex in Concentrated Sulfuric Acid

An intensely green solution is produced when elemental selenium is heated strongly in concentrated sulfuric acid, according to equation 1.

$$(1) \qquad Se + H_2SO_4 \rightarrow SeSO_3 + H_2O$$

Red selenium is precipitated upon the addition of water (equation 2).

$$(2) \qquad SeSO_3 + H_2O \rightarrow Se + H_2SO_4$$

The qualitative detection is accomplished by heating the selenium compound with concentrated H_2SO_4 in a dry test tube that had previously been heated for 15 minutes in an oil bath at 170–180°. A green color indicates the presence of selenium if tellurium is absent. (See the recent paper by Barr et al.[12a])

7.   Diaminobenzidine Test

Gillis and Hoste[62] used this reagent as a very sensitive test for selenium. For qualitative purposes 1 ppm of selenium, as the oxide, when treated with 3 drops of a 2.5% aqueous solution of diaminobenzidine, will produce a yellow precipitate. The test is valid in the presence of about 40 common cations. Cheng[27] claims interference from some oxidizing agents and some colored salts. Additional details concerning the quantitative use of this important reagent for selenium will be found on pages 1066 and 1070.

D.   QUANTITATIVE METHODS

Many of the methods discussed in Sections XVIIB and XVIIC may be regarded as quantitative, in whole or in part. Some of these methods will now be discussed in greater detail as they apply to the assay of selenium and its compounds.

1.   Gravimetric Methods

a.   Reduction Procedures

i.   Thiourea.   The method of Deshmukh and Sankaranarayanan[45] converts most forms of selenium except HSe^- to the elemental state.

ii. Sulfur Dioxide. Following a peroxide fusion, Shaw and Reid[145] precipitated selenium with sulfur dioxide for the assay of ethylseleninic acid hydronitrate, *p*-nitrophenyl selenocyanate, and similar compounds. Satisfactory results were also obtained by these investigators using potassium iodide as reducing agent.

The procedure of Goto and Kakita[66] involves treatment of the sample with perchloric acid and nitric acid, if required. The nitric acid is fumed off, the solution is diluted with 9 *N* HCl, and sulfur dioxide is added to precipitate elemental selenium.

De Salas[139] recommends that, if the selenium is in the form of selenious or selenic acid, hydroxylamine hydrochloride solution be added to complete the precipitation. Goto and Ogawa[67] claim that selenium may be quantitatively precipitated from a mixture of sulfuric and hydrochloric acids.

iii. Glucose. Selenate and selenite can be reduced to Se° with glucose from a weakly alkaline solution.[57]

b. Precipitation of Selenious Acid with Mercuric Nitrate. An excess of $Hg(NO_3)_2$ is used to precipitate selenate from solution in the pH range 4–10 as an amorphous, white solid having the composition $HgSeO_3$.[46] Avoidance of strong acids and ammonium hydroxide is necessary, because the precipitate is soluble in these media. Silver interferes by forming Ag_2SeO_3.

c. Precipitation as Selenium Sulfide. The separation of selenium from other elements is accomplished by precipitating with H_2S from a cool acidic solution. Yellow ammonium sulfide dissolves selenium and partially separates it from other elements. The filtrate, when strongly acidified and reprecipitated, forms Se + 2S, which is collected on a filter and calculated as SeS_2. For solutions containing only selenic acid, the selenium can be precipitated by H_2S from 12 *N* hydrochloric acid at room temperature.[123] Arsenic and tellurium interfere, as do large amounts of tin and antimony. Hillebrand *et al.*[81] claim interference from copper also.

d. Miscellaneous Methods. Very strong reducing agents are not required when the selenate ion is absent, and their use should be avoided as they increase the chances that coprecipitation and occlusion will occur. Ideally, gravimetric reagents that are completely destroyed or are very soluble should be employed. Hydroxylamine and sulfur dioxide are, therefore, to be recommended, but not the use of $SnCl_2$ and of phosphorous or hypophosphorous acids, in gravimetric procedures.

2. Titrimetric Methods

a. Iodometry. Iodometric methods can be recommended as sensitive and accurate when dealing with organic compounds. Inorganic ions (Cu, Fe) interfere by oxidizing the iodide ion, but in materials lacking these metals iodometric methods are preferred. Strong oxidizing agents, if present, also interfere, but most of these can be removed. In general, there are three approaches to the iodometric estimation of selenium, two of which deal with the determination of selenites or selenious acid and the third with selenate ions.

i. By Hydrochloric Acid Reduction—Method I. With a moderate excess of KI solution, iodine is liberated according to reaction 3.

$$(3) \qquad H_2SeO_3 + 4HCl + 4KI \rightarrow Se + 2I_2 + 4KCl + 3H_2O$$

The liberated iodine is titrated with standard thiosulfate solution, as above, and the equivalence is Se $\rightleftharpoons$ 4I. This method has the distinct advantage of requiring only one standard solution; however, precipitated red selenium formed in this reaction may present difficulties in determining the starch-iodide end point. The precipitated selenium tends to occlude some iodine, but with low concentrations of selenium this difficulty is minimal. Table XVII-1 shows the results obtained by Alicino[1] on two typical selenium-containing organic compounds and attests to the precision and accuracy of the method.

Table XVII-1   Analysis of selenium compounds by Method I

Selenourea

CALCULATED	FOUND
64.24	64.39, 64.02, 64.12, 64.33

Selenomethionine

CALCULATED	FOUND
40.37	40.18, 40.53, 40.29, 40.32

The method used is essentially the modified method of Meier and Shaltiel.[115] The procedure is as follows.

Three to five milligrams of sample is weighed and combusted in a 500 ml oxygen flask[141] containing 10–20 ml of water plus a few drops of 30% hydrogen peroxide. After 15–20 minutes (to allow the products of combustion to be absorbed), the stopper and the platinum basket are rinsed with water. Five milliliters of concentrated HCl is added to the flask and the solution is boiled on the hot plate for 30 minutes, with water added, if necessary. After cooling, the volume is brought up to about 50 ml. Approximately 100 mg of solid KI is added to the flask, thus liberating the iodine, and after 5 minutes the iodine is titrated with 0.01 N standard thiosulfate solution. Starch is added near the end point, and the deep purple color changes to red at the end point.

With a little practice, the end point is easily discernible. If difficulty is experienced, Green and Turley[73] suggest the addition of small increments of 1% KI solution to the solution containing the starch indicator. As the starch iodide color is dispelled, more KI is added until, finally, the blue color persists at the end point. In this manner, little or no red coloration due to the presence of selenium is encountered.[97]

Modifications of this method by McNulty et al.[114] and Norris and Fay[122] have been published, but these procedures are more time consuming.

ii. *By Liberation of Iodine from Potassium Iodide—Method II.*   Another iodometric method for selenium depends on reaction 4a. The chlorine formed then reacts with iodide to liberate iodine, according to equation 4b. The iodine is titrated in the usual way with thiosulfate.

(4a) $$H_2SeO_4 + 2HCl \rightarrow H_2SeO_3 + Cl_2 + H_2O$$

(4b) $$2KI + Cl_2 \rightarrow I_2 + 2KCl$$

The chlorine obtained according to equation 4a is distilled into a dilute solution of KI maintained at room temperature. This procedure is also slow, is subject to air oxidation, and the completeness of the reaction is uncertain. Moreover, the chemical factor is half the equivalence obtained for Method I.

Rowley and Swift[137] claim that an improvement can be made by excluding oxygen from the system. Other iodometric methods devised by Deshmukh and Sant[47] and by Schulek and Koros[142] are more tedious and offer no advantage over the simpler methods mentioned above.

Fredga[59] has summarized the iodometric properties of organoselenium acids containing carboxyl groups as follows.

(*a*) Mercury complexes in mineral acid solution consume one atom of iodine per selenium equivalent with formation of the diselenides (equation 5).

Selenols behave in a similar fashion. In neutral or alkaline solutions more iodine is consumed [cf. (c)].

$$(5) \qquad \begin{array}{c} R\text{—}Se \\ \diagdown \\ \phantom{R\text{—}Se}Hg + I_2 + 2KI \\ \diagup \\ R\text{—}Se \end{array} \longrightarrow \begin{array}{c} R\text{—}Se \\ | \\ R\text{—}Se \end{array} + K_2HgI_4$$

(b) Hydrogen selenide or alkali selenides are oxidized by iodine with the liberation of red selenium.

(c) Diselenide dicarboxylic acids do not react in acid media, but in alkaline bicarbonate solutions 6 equivalents of iodine are consumed per mole with the formation of seleninic acids (equation 6). Generally this reaction takes

$$(6) \qquad \begin{array}{c} R\text{—}Se \\ | \\ R\text{—}Se \end{array} + 3I_2 + 4H_2O \longrightarrow 2RSeO_2H + 6HI$$

place rapidly, but with cyclic diselenodicarboxylic acids it slows down towards the end. After a waiting period of 5 minutes, the correct analytical value is obtained.

(d) Selenodicarboxylic acids, which do not react with iodine in acid media, consume 2 equivalents of iodine per mole of acid in bicarbonate solutions with the formation of hydrated selenoxides (equation 7). This

$$(7) \qquad \begin{array}{c} R \\ \diagdown \\ Se \\ \diagup \\ R \end{array} + I_2 + H_2O \longrightarrow \begin{array}{c} R \\ \diagdown \\ SeO \\ \diagup \\ R \end{array} + 2HI$$

reaction proceeds less smoothly than the preceding one, and only certain acids give good data for equivalent weights.

(e) Seleninic acids readily oxidize hydriodic acid in acid media with the liberation of 3 equivalents of iodine, which is titratable with thiosulfate.

(f) Selenoxide-dicarboxylic acids oxidize hydriodic acid in acid media with the liberation of 2 equivalents of iodine per mole of acid. This reaction also occurs rapidly and gives satisfactory analytical values.

(g) Dibromides of selenodicarboxylic acids are hydrolyzed to selenoxides by water and behave like the compounds in (f).

(h) Diiodides of selenodicarboxylic acids are solubilized with great difficulty in water. In potassium iodide solutions, they behave as if they were in a mixture of iodine and selenodicarboxylic acid.

(i) Selenotin compounds do not seem to react quantitatively by means of iodometry.

(*j*) Selenious acid oxidizes HI with the formation of red selenium and iodine. The red selenium is best observed after decolorization of the iodine with thiosulfate.

For the above titrations the iodine solutions were standardized against thiosulfate, and the thiosulfate was standardized against bromate. It was found, however, that the iodine solution could be standardized satisfactorily against recrystallized and well-dried diselenodiglycolic acid according to equation 6.

Fredga[58] has also studied the behavior of diselenodicarboxylic acids towards mercury salts. He has shown that selenols can be oxidized with iodine in acid media to diselenides and may be titrated as such. The method is even useful for the mercury salts, where iodide ion is present in excess and complexes with mercury, as shown in equation 8. In a method developed

$$(8) \qquad 2R—Se—HgCl + I_2 + 6I^- \rightarrow RSe—SeR + 2HgI_4^{2-} + 2Cl^-$$

by Bengtsson[13a] the selenium in organoselenium compounds is quantitatively converted to selenious acid by a wet digestion method described by Fredga.[58] An amperometric end-point detection with polarized electrodes under completely oxygen-free conditions is used. The amperometric circuit is described by Wernimont and Hopkinson.[166]

A rapid, simple, and accurate method for the determination of selenium derived from organic selenium compounds was evolved by Barcza[12] in which the entire process from combustion to the final iodometric titration may be carried out in the same flask. The organic selenium compound, containing 0.4–4.0 mg of selenium, is burned in a Schöniger flask. In this procedure the selenium dioxide formed by combustion is dissolved in water to form selenious acid. This is reduced to elemental selenium with ascorbic acid, and the selenium is then converted with cyanide into selenocyanate. Excess cyanide is removed, and the selenocyanate is measured by iodometry as cyanogen bromide. Organic selenium compounds assayed and tabulated by this method included anilinium hydrogen selenite, piazselenole, dipiazselenole, and 5-(3,4-diaminophenyl)-2,1,3-benzoselenadiazole.

iii. By Excess Thiosulfate—Method III.   Selenious acid is reduced by an excess of sodium thiosulfate solution (equation 9).[122] The excess $Na_2S_2O_3$

$$(9) \quad H_2SeO_3 + 4Na_2S_2O_3 + 4HCl \rightarrow$$

$$Na_2SeS_4O_6 + Na_2S_4O_6 + 4NaCl + 3H_2O$$

is then back-titrated in the usual way with standard iodine solution (equation 10). Therefore, 1 ml 0.1 N $Na_2S_2O_3$ ≃ 1 ml 0.1 N iodine ≃ 0.001974 g

$$(10) \qquad 2Na_2S_2O_3 + I_2 \rightarrow Na_2S_4O_6 + 2NaI$$

selenium. This determination is done at room temperature in a solution containing 10–20% hydrochloric acid. The conventional starch-iodide end point is sharp.

b. Determination by Thiourea. A method, already described as a gravimetric procedure, is suggested by Deshmukh and Sankaranarayanan.[45] Excess standard thiourea solution is back-titrated with a standard solution of ceric sulfate. Here, again, this procedure lacks the directness of the iodometric methods.

c. Argentometric Titration. The determination of selenium as selenite, by a method similar to the Volhard method for chloride, is described by Hahn and Viohl.[76] This procedure lacks the ease and sensitivity of the other volumetric methods described here.

d. Potassium Permanganate Oxidation. Although not as sensitive as the iodometric methods, permanganate titration of selenium (equation 11)

$$(11) \quad 2KMnO_4 + 5H_2SeO_3 + 3H_2SO_4 \rightarrow$$

$$5H_2SeO_4 + 2MnSO_4 + 3H_2O + K_2SO_4$$

can be used if only moderate amounts of iron and copper are present. Nitric and halogen acids interfere and must be removed. The excess standard $KMnO_4$ is back-titrated with ferrous ammonium sulfate to the disappearance of the purple permanganate color. One milliliter of 0.1 N $KMnO_4$ $\approx$ 1 ml 0.1 N FAS $\approx$ 0.003948 g Se. A large excess of phosphate ion is required to complex the Mn^{2+} and thus prevent its reaction with excess MnO_4^-.

A detailed procedure is described by Barabas and Cooper.[10] An alkaline permanganate oxidation is reported by Issa *et al.*[85]

3. Electrochemical Methods

Three electroanalytical methods for the determination of selenium are described below: (*a*) potentiometric, (*b*) polarographic, and (*c*) coulometric. These techniques are used to determine inorganic forms of selenium. Organic selenium compounds must, therefore, be appropriately decomposed and dissolved (as described in Section XVIIB.1) in order to convert the selenium to an inorganic form.

a. Potentiometry. Kolthoff *et al.*[96] have summarized the potentiometric methods for the determination of selenium previously reported by Kolthoff and Furman.[97]

More recently, a cyanometric titration for selenium in nonaqueous solutions was reported by Erdey *et al.*[50] Elemental selenium is determined by dissolving it in an excess of a standard solution of potassium cyanide in isopropyl alcohol and back-titrating the excess potentiometrically with a standard solution of sulfur in benzene-acetone. A visual end point is possible with bromothymol blue as indicator because the inflection point of the potentiometric curve coincides with the color transition of bromothymol blue.

Potentiometric methods for the determination of selenium(IV) were also reported by Barabas and Bennett,[9] who developed a one-vessel technique, and by Issa et al.[85]

Hahn and Bartels[77] developed a method for the potentiometric determination of micro quantities of selenium in which the selenium present as selenite is reduced with hydrazine sulfate, in the presence of standardized silver nitrate solution, and precipitated as Ag_2Se. The unused quantity of silver nitrate is measured by potentiometric titration with potassium iodide, and the selenium content is calculated from the amount of silver consumed.

b. Polarography.   The polarographic behavior of selenite was originally investigated by Schwaer and Suchy.[143] Lingane and Niedrach[105,106] subsequently studied the basic polarographic chemistry of the various oxidation states of selenium. Polarography has since become a useful tool for the assay of selenium, with a sensitivity and a rapidity comparable to those of spectrophotometric methods of analysis. Watkinson[162] has reviewed the polarographic contributions of Faulkner et al.,[53] Bock et al.,[18] Betteridge,[16] and Christian et al.[28-30] It was shown that, after the elimination of interferences with dithizone and isolation of the element by precipitation or by extraction with 2,3-diaminobenzidine, quantities of selenium as small as 0.2 μg can be assayed polarographically. Williams and Haskett[171] examined the conditions under which selenium(IV) and selenium(VI) are reduced by sulfur dioxide and were able to make quantitative polarographic determinations of selenium(IV) in the presence of selenium(VI). Atomic absorption determinations confirmed these results.

Among those who have contributed to the polarography of organic selenium compounds are Stricks and Mueller,[151] who studied the polarography of two potential antiradiation drugs, 2-aminoethanethiosulfuric acid and 2-aminoethaneselenosulfuric acid, and Griffin,[74] who looked into the selenium-diaminobenzidine complex. The latter investigator studied the parameters for the single-sweep polarographic analysis of the selenium-diaminobenzidine complex and developed a technique for the determination of selenium with a sensitivity of 1 ppb of the element. This polarographic technique offers sensitivity equal to that of existing spectrofluorimetric methods or even greater, and should be at least as fast. Application of this method to the analysis of selenium in biological samples is being developed.[74]

The polarographic investigation of organic compounds has been extensively reported by Nygård.[124–129] In one of these investigations, Nygård[129] comprehensively studied the polarographic behavior of diselenodiacetic acid. He showed that $(HOOCCH_2Se)_2$ and $HSeCH_2COOH$ form a reversible system at the dropping mercury electrode. The reversibility of the following electrode reaction (equation 12) was shown by a number of methods. The dissociation

$$(12) \qquad HgSeCH_2COOH + H^+ + e \rightleftharpoons HSeCH_2COOH + Hg$$

constant of the SeH group in $HSeCH_2COOH$ is 1.0×10^{-7}.

These experiments were extended by Nygård[128] to ω,ω'-diseleno-dialkanoic acids such as $HO_2C(CH_2)_nSeSe(CH_2)_nCOOH$, where $n = 2$ to 4, and to their hydroseleno compounds, which were prepared electrolytically. These compounds also behave as reversible redox systems at the dropping mercury electrode, giving continuous anodic and cathodic waves with magnitudes depending on the relative concentrations of reductant and oxidant. Diffusion coefficients and dissociation constants were calculated for these diselenodialkanoic acids. A study of the instantaneous current of single drops showed an initial delay in current, which was attributed to a currentless formation of a compact surface layer of molecules resulting from a fission of the Se-Se double bond and to the formation of organic mercury compounds in the surface layer.

The polarography of selenocystine and selenocysteine was investigated by oscillopolarography and a.c. polarography.[127] The polarographic properties of the two compounds were compared with those of their sulfur analogs, cystine and cysteine.

Nygård[126] studied the polarographic reduction mechanisms for the five-membered cyclic diselenides, 4,4-bis(hydroxymethyl)-1,2-diselenolane and 1,2-diselenolane-3-valeric acid, and the six-membered cyclic diselenides, 1,2-diselenane-3,6-dicarboxylic acid and 1,2-diselenane-4,5-dicarboxylic acid. Information about the electrode processes for these compounds was gained from: (1) d.c. polarography curves, (2) electrocapillary curves, (3) oscillopolarography dE/dt $vs.$ E curves, (4) instantaneous current-time functions. The polarographic differences between these compounds and the analogous five- and six-membered cyclic disulfides were pointed out.

A polarographic study of some inhibiting effects of five-membered cyclic diselenides and disulfides was made by Nygård $et\ al.$[124] Three five-membered cyclic selenium and sulfur compounds containing a Se-Se, S-S, and Se-S bond, respectively, were investigated. The relative inhibiting action of these groups upon the reduction of the peroxydisulfate ion $(S_2O_8)^{2-}$ was demonstrated in the following compounds: 1,2-diselenolane-4-carboxylic acid (**1**), 1,2-dithiolane-4-carboxylic acid (**2**), and 1-thia-2-selenacyclopentane-4-carboxylic acid (**3**).

$$\text{1}\qquad\text{2}\qquad\text{3}$$

The polarographic reduction of diphenyl diselenide was studied by Nygård[125] and compared with that of the corresponding disulfide.

c. Coulometry. Although Taylor and Smith[157] state that coulometric titrations are the most accurate and precise of all chemical analytical methods, very few applications of coulometry have been reported in selenium analysis. One example of the coulometric determination of selenium, illustrated by Lingane,[104] is the method developed by Rowley and Swift.[137]

4. Spectroscopy

a. Colorimetric and Spectrophotometric Methods.

i. 3,3′-Diaminobenzidine (DAB). One of the aromatic *o*-diamine reagents for the photometric determination of selenium is 3,3′-diaminobenzidine (DAB). Selenium(IV) forms a yellow and intensely colored piazselenole with DAB, as shown in equation 13.

(13)

Cheng[27] showed that the piazselenole was developed in acid medium between pH values of 2 and 3. Optimum results were obtained by extracting the piazselenole into toluene at pH 6. Under these conditions the color had an absorption maximum at 420 nm and a molar extinction coefficient of 19,900. Ethylenediaminetetraacetic acid was used to prevent interferences from iron, copper, and other metals. With excess DAB, Parker and Harvey[133] showed that the monopiazselenole is formed.

Handley and Johnson[80] applied Cheng's method to the analysis of samples of plant material containing added selenium. Amounts as low as 0.25 μg Se in 1.0 g samples of plant material were determined by this procedure.

Gutenmann and Lisk[75] determined the selenium content in 1 g samples of oats by oxygen flask combustion, followed by the DAB spectrophotometric method at 420 nm. The results compared favorably with determinations made by the iodometric method of Klein.[94] As little as 0.25 ppm Se in 2 g samples is detectable by this method.

Kelleher and Johnson[90] modified Cheng's procedure and extended the sensitivity of the method to a few hundredths of a part per million of selenium in organic samples. This is in the range previously achieved only by neutron activation analysis and by fluorimetric analysis. An isotope-dilution technique was incorporated into this micro method to check the selenium losses during the various stages of the analysis. Gorsuch[64] also carried out extensive radiochemical investigations on the recovery of trace amounts of selenium and other elements in organic and biological materials.

A technique for the detection and quantitative estimation of air-borne selenium particulates in air pollution studies was developed by West and Cimerman.[167] The method is based on ring oven techniques, coupled with the use of 3,3'-diaminobenzidine.

Cummins et al.[40,41] simplified their earlier DAB procedure for the determination of selenium in animal tissues. They found that maximum color development of the Se-DAB complex can occur in less than 20 minutes (plateau 10–20 minutes) if the reaction mixture is heated to 60° before toluene extraction. A single sample can be assayed in considerably less than an hour, at a rate of 15 samples in about 2 hours. Assay results are comparable to those obtained by neutron activation analysis.

ii. 2,3-Diaminonaphthalene. Although 2,3-diaminonaphthalene (DAN), which forms a bright red piazselenole, is most widely used as a fluorimetric reagent for selenium, it has also served for the colorimetric determination of selenium(IV) by Lott et al.[107]

For the determination of traces of selenium in pharmaceuticals, Klein et al.[95] used the colorimetric DAN procedure after combustion of the sample in an oxygen flask. The red piazselenole has an absorption maximum at 380 nm and is linear with concentration in the range of 1–12 μg Se, as the piazselenole, in 5 ml of cyclohexane. The procedure was applied to 69 compounds that will be official in U.S.P. XVIII and N.F. XIII. All these compounds either contain sulfur or require the use of sulfur or selenium in their manufacture.

iii. o-Phenylenediamine and Derivatives. Hinsberg[82] depicted the reaction of o-phenylenediamine with selenium(IV) as shown in equation 14. Within

(14)

$$+ H_2SeO_3 \longrightarrow \text{(Se)} + 3H_2O$$

Piazselenole

the last 10 years, a number of analytical methods for the determination of selenium have been reported that use o-phenylenediamine and its derivatives.

Ariyoshi *et al.*[4] developed a u.v. spectrophotometric method for the determination of trace amounts of selenium with o-phenylenediamine. Tanaka and Kawashima[155] found that, in addition to o-phenylenediamine, 4-substituted o-phenylenediamines react with selenious acid in acid solution to form benzoselenadiazoles which can be extracted into toluene. Goto and Tōei[69] studied other 4-substituted 1,2-phenylenediamine reagents and found that 3,4-dichloro-1,2-phenylenediamine appears to be the most sensitive for the spectrophotometric determination of selenium.

iv.   Phenylhydrazines. Feigl and Demant[54] first described the use of asymmetric diphenylhydrazine for detecting selenium(IV). Kirkbright and Yoe[92] developed a spectrophotometric method for the determination of microgram quantities of selenium(IV), based on the oxidation of phenylhydrazine p-sulfonic acid by selenious acid. The diazonium oxidation product is coupled with 1-naphthylamine, forming an azo dyestuff whose color is measured at 520 nm. Equations 15 and 16 depict the reactions involved in the determination of selenium in acidic solution.

The colorimetric procedure described by Kirkbright and Yoe[92] was applied with a few modifications by Kowald[98] for the determination of selenium in selenomethionine. Selenomethionine containing 5 μg and 10 μg Se was assayed by the colorimetric procedure after combustion in a Schöniger oxygen flask. The products of combustion were absorbed in 0.15% sodium oxalate containing 0.3 M ammonium hydroxide, as described by Taussky.[156] The colorimetric reactions were then carried out in the 500 ml

Schöniger flasks used for combustion. The absorbancies at 520 nm for the 5 μg and 10 μg combustion Se samples, after correction for the paper blank, were in excellent agreement with the values obtained by reacting selenious acid standards containing 5 μg and 10 μg Se(IV).

Another example of the use of the diazotization-coupling reaction for the determination of selenium has been reported by Osburn et al.[131a] This new procedure, which is sensitive in the submicrogram range of selenium(IV), is based upon the oxidation of hydroxylamine hydrochloride to nitrous acid by selenious acid, followed by the diazotization of sulfanilamide by the nitrite produced and subsequent coupling of the diazonium salt with N-(1-naphthyl)-ethylenediamine dihydrochloride. Absorbancy measurements are made at 544 nm. The system obeys Beer's law in the range of 0.01–0.20 mg Se(IV)/l. The method is simple and sensitive ($\varepsilon = 193,000$) and has been applied for the determination of selenium in air pollution studies of smoke samples collected with Telematic 150A air samplers (Unico, Fall River, Mass.).

v. 4,5-Diamino-6-thiopyrimidine. Chan[25] described the use of 4,5-diamino-6-thiopyrimidine (4,5-D6TP) for the spectrophotometric determination of microgram quantities of selenium.

vi. Dianthrimides. 2,2'-Dianthrimide has been investigated by Langmyhr and Dahl[100] as an analytical reagent for selenium(IV) and compared with 1,1'-dianthrimidine.[101]

vii. 1,4,6,11-Tetraazanaphthacene (TAN). The catalytic effect of elemental selenium on the reduction of methylene blue has been studied in qualitative and quantitative analyses for this element by Feigl and West,[55] Goto et al.,[65] and, more recently, by West and Ramakrishna.[168] Kawashima and Tanaka[89] described details for the determination of submicrogram amounts of selenium(IV) by means of the catalytic reduction of TAN, which is prepared from 2,3-diaminophenazine (DAP) by reaction with glyoxal (equation 17).

(17)

| 2,3-Diaminophenazine (DAP) | Glyoxal | 1,4,6,11-Tetraazanaphthacene (TAN) |

In the presence of hypophosphorous acid, TAN is reduced to a blue compound, 1,6-dihydro-1,4,6,11-tetraazanaphthacene (equation 18). This reduction is catalyzed by submicrogram amounts (*i.e.*, less than 0.15 μg)

(18) 1,4,6,11-Tetraazanaphthacene (TAN)
($\lambda_{max} = 478$ nm)

1,6-Dihydro-1,4,6,11-tetraazanaphthacene
(DHTAN)
($\lambda_{max} = 570, 600$ nm)

of selenium(IV). The method is remarkably simple and, in the absence of interfering ions, is carried out as a one-tube reaction.

viii. Sulfur-Containing Reagents. Busev[22] studied the reactions of selenious acid with diethylphosphorodithioate, 2-mercaptobenzimidazole, N-mercaptoacetyl-p-anisidine, and N-mercaptoacetyl-p-toluidine. Diethylphosphorodithioate and 2-mercaptobenzimidazole react with selenious acid in a 4:1 molar ratio, and N-mercaptoacetyl-p-toluidine reacts with it in a 2:1 molar ratio.

Kirkbright and Ng[91] reported the use of thioglycolic acid for the spectrophotometric determination of selenium in the presence of tellurium.

Bera and Chakrabartty[15] further developed the use of 2-mercaptobenzimidazole for the spectrophotometric determination of 1–7 μg/ml Se(IV). The yellow color is stable in strong acid solution.

Cresser and West[38] have developed a rapid and sensitive method for the determination of selenium(IV) with a 2-mercaptobenzoic acid.

xi. Cyclohexanone. The use of cyclohexanone as a reagent for the spectrophotometric determination of selenium(IV) was described by Cresser and West.[37] By this method, 2–100 μg Se(IV) can be determined directly in the aqueous phase by formation of the ketone-chloro complex with cyclohexanone in about 7 M hydrochloric acid heated for 5 minutes at 100°. Under the conditions employed, selenium(VI) is rapidly reduced to selenium(IV). Hence, the initial oxidation state of selenium is not important, an unusual advantage for a spectrophotometric procedure.

x. ESCA. A series of selenium compounds has been studied through the use of the recently developed technique, ESCA (electron spectroscopy for chemical analysis). The electron binding energies of the $3d$ and $3p_{3/2}$ levels in Se were ascertained; also, the binding energy and the charge on the Se atom were correlated.[107a]

b. Spectrofluorimetric Methods.

i. 3,3'-Diaminobenzidine. Cousins[36] has been credited with discovering that the yellow piazselenole formed by selenium(IV) with 3,3'-diaminobenzidine (DAB) is strongly fluorescent. The excitation and emission wavelengths are 425 and 565 nm, respectively, as reported by Dye *et al.*[49]

Grant[71] developed a spectrofluorimetric assay permitting the determination of 0.01 μg Se, with a standard deviation of 0.001 μg. Watkinson[164] and Taussky et al.[156] have used DAB for the fluorimetric determination of traces of selenium, particularly in plant and animal materials. Watkinson[164] removed interfering fluorescent substances, which are not quenched by EDTA, by completely extracting the digested selenium from strong hydrochloric acid into a mixture of ethylene chloride and carbon tetrachloride, using the zinc complex of toluene-3,4-dithiol described by Clark.[31,32] According to Serlin,[144] DAB is suspect as a possible carcinogen and should be handled with extreme caution.

ii.   2,3-Diaminonaphthalene (DAN).   The most widely used fluorimetric reagent for selenium is 2,3-diaminonaphthalene (DAN), which was introduced by Parker and Harvey.[132] The reagent forms the bright red, extremely fluorescent 4,5-benzopiazselenole (**4**). The fluorescence emission spectrum

4

of 4,5-benzopiazselenole was shown by Parker and Harvey[132] to have a maximum at 521 nm with an excitation wavelength at 366 nm. The sensitivity of DAN is in the nanogram range of 0.002 μg Se.

Traces of fluorescent compounds from the oxygen flask combustion were eliminated by Watkinson[163] before measurement by oxidation with nitric and perchloric acids, and by Cukor et al.[39] by use of persulfate.

The versatility of DAN as a selenium reagent was described by Lott et al.,[107] who used it for macro, micro, and submicro determinations of selenium. Milligram quantities of selenium were determined gravimetrically after precipitation by means of DAN. The use of an organic precipitant for the gravimetric determination of selenium has not been previously reported. For microgram quantities of selenium, DAN was employed spectrophotometrically, whereas submicrogram amounts were determined fluorimetrically. For the fluorimetric determinations Lott et al.[107] used an Aminco-Bowman Spectrofluorimeter with an excitation wavelength of 390 nm and a fluorescent wavelength of 540 nm.

Cukor et al.[39] applied isotopic dilution analysis to the fluorimetric determination of selenium in plant materials. The samples were prepared by oxygen flask combustion, and their selenium content was determined fluorimetrically with DAN. The loss of selenium throughout the analytical procedure was compensated for by the incorporation of radioactive selenium into the analytical scheme. In a study of the effect of foreign ions in the procedure, radioactive selenium was used to correct for the interferences of the 20 metallic ions tested. Ion exchange cannot be used in fluorimetry because traces of the resin monomer are carried along in the procedure and the fluorescence of the monomer cannot be filtered out. In this procedure, EDTA was employed as a masking agent.

Recent applications of DAN have been made almost entirely in the fluorimetric determination of selenium in biological materials, both plant and animals. Ewan et al.[52] used DAN fluorimetry to analyze animal tissues containing organically bound [75]Se. The recovery of organically bound ^{75}Se was 95 % when carried through their entire procedure.

Submicrogram quantities of selenium were determined, using DAN fluorimetry, by Hoffman et al.,[84] Hall and Gupta,[78] Watkinson,[163] and Olson.[131] The method of Watkinson[163] was comprehensively studied by Olson.[131] The modified Watkinson assay procedure for selenium was applied to various plant materials and compared with assays according to Klein,[93] Cheng,[27] Allaway and Cary,[3] and Cummins et al.[40] and with neutron activation analysis. The results of these collaborative studies and of two statistical tests for ruggedness, conducted as described by Youden,[175] have been reported. The method described by Olson[131] is recommended for adoption by the Committee on Recommendations of Referees[34] of the Association of Official Analytical Chemists, as an official method for the determination of selenium in plants.

c.   X-ray Fluorescence. One advantage of X-ray fluorescence, as an analytical technique, is that the sample is not destroyed. Another advantage is that, unlike the situation in emission spectroscopy, very few lines are obtained for each element, thereby decreasing interferences from other elements. Any element with an atomic number higher than that of magnesium can be readily assayed.

Unfortunately, the intensity of the few X-ray emission lines of selenium is very low, thus limiting their analytical value. Nevertheless, Olson and Shell[130] developed a procedure for the simultaneous determination of selenium and mercury in organic compounds. This X-ray fluorescence method gave results that were accurate to ± 1 ppm in the range of 2–40 ppm for both selenium and mercury. A 600 mg sample is required, but the method is rapid and nondestructive. The average time for each analysis, including sample preparation and X-ray fluorescence analysis, was 45 minutes.

Handley[79] used X-ray fluorescence to determine selenium in plant materials containing 7.73–946.0 ppm.

5.   Atomic Absorption and Fluorescence Spectroscopy

The use of atomic absorption spectroscopy was first suggested by Walsh,[161] who introduced hollow-cathode lamps to analysis. The measurement of selenium by atomic absorption was introduced by Allan,[2] who reported on the detection limits of 28 elements in the air-acetylene flame. The rapid progress in selenium analysis by atomic absorption spectroscopy during the short span between 1962 and 1966 has been reviewed by Watkinson.[162] Watkinson reported on Allan's development of the tin selenide cathode and the flame-in-tube method that gave a sensitivity of 0.015 ppm Se. The work of Rann and Hambly[135] led to electrodeless discharge in vaporized selenium. In 1966 Walsh[160] improved the hollow-cathode lamp by developing a high-intensity version for selenium analysis. In this lamp, one electrical discharge is used to produce atomic vapor and a second functions independently to excite the vapor. Excitation is accomplished at low energies, without line broadening, thus bringing about a marked increase in the intensity of the resonance lines. In 1968 Chakrabarti[24] reported on these findings regarding cationic and anionic interferences, the selenium sensitivities possible at different wavelengths in air-acetylene and air-hydrogen flames, and the effect of organic solvents on the atomic absorption of selenium. From his findings, a sensitive extraction procedure was developed.

Atomic fluorescence spectroscopy is another form of flame spectroscopy, which has been applied to selenium analysis by Winefordner and Vickers[173] and by Winefordner and Staab.[173] The radiation from a light source is used to excite the atoms to fluorescence, and the intensity of the fluorescence serves as a measure of the concentration of the particular element. Dagnall et al.[42,43] described in detail the production and operation of microwave-excited electrodeless discharge tubes for use as sources in atomic fluorescence, as well as in atomic absorption spectroscopy. Selenium can be determined down to 0.25 ppm by measuring the atomic fluorescence signals at 2040 Å. The oxidation state of the selenium is unimportant, and preliminary extraction into organic solvents is unnecessary. The authors report that the selenium sensitivities of these atomic fluorescence methods are about ten times greater than those of atomic absorption methods.

A new atomic absorption accessory recently developed by the Perkin–Elmer Corporation[84a] is a highly sensitive selenium sampling system which works with their atomic absorption spectrophotometer. It is equipped with a suitable hollow cathode lamp and a three-slot burner head. The detection limit for selenium with this system is reported to be 0.0015 μg/ml. In this system, the sample of selenium is treated chemically to convert the selenium present to the gaseous hydride. The collected gas is then passed to an argon-hydrogen flame where the atomic absorption is measured and logged on a suitable recorder.

6. Neutron Activation

Much of the literature of neutron activation analysis in the last 10 years has been concerned with either the determination of trace impurities in high-purity selenium[5-8,170] or the determination of trace selenium in various inorganic materials.[33,35,44,60,120,136]

Surprisingly few papers describe in detail the estimation of selenium in biological materials by thermal neutron activation analysis. Bowen and Cawse[20] pointed out that only three radionuclides of selenium produced by irradiation with slow neutrons have sufficiently high specific activity to be useful for analytical purposes. Data on these three radionuclides, taken from the paper by Bowen and Cawse,[20] are shown in Table XVII-2.

Table XVII-2   Radionuclides produced by a thermal neutron flux of 10^{12} neutrons/cm²/sec on selenium

RADIONUCLIDE	SPECIFIC ACTIVITY OF SELENIUM AFTER ACTIVA- TION FOR ONE HALF- LIFE (mCi/g)	HALF-LIFE	MAXIMUM BETA ENERGY (MeV)	GAMMA ENERGIES (MeV)
^{75}Se	25	120 days	...	0.27, 0.14, others
^{77m}Se	97	17.5 sec	...	0.16
^{81}Se	25	18.6 min	1.60	...
All others	3	...	...	...

Some analysts make use of the longer-lived ^{75}Se, but its radioactivation requires a long neutron irradiation (half-life 120 days). In practice, the irradiation periods are much shorter, resulting in 5–10% of the specific activity shown in the table, which permits complete or partial radiochemical separation of ^{75}Se from other sources of radioactivity.

Grant *et al.*[72] made the separations by distillation of the selenium with HBr, followed by precipitation of the selenium. The radiochemical purity was established by spectrometry. The smallest amount of selenium measurable by this method was about 0.01 μg.

Samsahl[140] described a radiochemical method for the determination of selenium in the presence of arsenic, bromine, mercury, and antimony in neutron-irradiated biological material. The radionuclides ^{75}Se, ^{76}As, ^{82}Br, ^{197}Hg, ^{203}Hg, ^{122}Sb, and ^{124}Sb are separated by chemical means and counted by γ-ray spectrometry. Quantitative results are obtained by comparison with standards similarly treated. The irradiation is carried out in 1–2 days with a thermal neutron flux of 2×10^{13} n/cm²/sec. The high radioactivity of the samples is allowed to decay for 2–3 days before starting the chemical separation.

Bowen and Cawse[20] described a radiochemical separation based on the

β-emitter ^{81}Se. The 18.6 minute half-life was sufficiently long to permit an efficient radiochemical separation and to start counting within two half-lives (36 minutes) after removal of six samples from the reactor, with two analysts performing the chemical manipulations. The sensitivity of this method was 0.005 μg, and selenium was satisfactorily separated from arsenic, bromine, manganese, sodium, and zinc. The selenium contents of fertilizers, tomato tissues, and human blood were measured by the process.

In 1969 Lampert et al.[99] used ^{77m}Se in a rapid, nondestructive method for the determination of selenium in biological samples. The method is free of any chemical procedures. The isotope ^{77m}Se has a half-life of approximately 17.5 seconds and a γ-ray energy of 160 keV. The results were highly dependent on the value of the ^{77m}Se half-life used in the computation. The best value was selected by comparing results for different values of half-life with those from a colorimetric determination of identical samples reported by Cummins et al.[41] A value of 17.5 seconds was finally selected on the basis of these results. This value agreed with three of the four values listed by Lederer et al.[102] The selenium was determined in kidney and liver samples from rats that had been chronically fed toxic levels of selenium to test the effects of heavy metals on selenium metabolism in animals with chronic selenosis.

In an assessment of the accuracy and precision of neutron activation analysis, Bowen[19] reported the results obtained from 29 laboratories on the analytical determination of 40 elements, including selenium, in standard kale powder. Neutron activation and fluorescence analysis techniques were used for selenium. The results showed that fluorimetry gave slightly lower results than did neutron activation analysis.

7. Radiochemical Methods and Isotope Dilution

Reference has been made to the method of Kelleher and Johnson,[90] which uses combined spectrophotometric and isotope dilution techniques for the determination of selenium in organic matter. The method is a micro modification of a spectrophotometric method that employs DAB as the selenium(IV) reagent and incorporates an isotope dilution technique utilizing radioactive selenious acid solutions containing ^{75}Se to compensate for losses that occur during the analysis.

Cukor et al.[39] also incorporated radioactive selenious acid solutions containing ^{75}Se for isotope dilution studies in a fluorimetric method that employs DAN as the selenium(IV) reagent. Their findings are described in Section XVIID.4.b on spectrofluorimetric methods. A description of the determination of organically bound ^{75}Se by Ewan et al.[52] is included in the same section.

A rapid, selective separation of the nuclides ^{75}Se, [74]As, and [71]Ge, in appropriate carrier solutions, has been reported by McGee *et al.*[111]

Lewis *et al.*[103] studied the release of volatile selenium compounds in two plant species, one selenium-accumulating and the other nonaccumulating. Trace amounts of ^{75}Se were incorporated with the nutrient solutions used to culture these plants. Charcoal proved most effective in trapping the ^{75}Se activity released by the plants.

The determination of selenium in sea water, silicates, and marine organisms was the subject of a thorough study by Chau and Riley.[26] Isotope dilution methods, using [75]Se (as selenite), were used in developing the multistage analytical procedures.

REFERENCES

1. J. F. Alicino, unpublished data.

2. J. E. Allan, *Spectrochim. Acta*, **18**, 259 (1962).

3. W. H. Allaway and E. E. Cary, *Anal. Chem.*, **36**, 1359 (1964).

3a. V. Anger and G. Fisher, *Microchim. Acta*, 501 (1962).

4. H. Aritoshi, M. Kiniwa, and K. Tōei, *Talanta*, **5**, 112 (1960).

4a. K. W. Bagnall, "The Chemistry of Selenium, Tellurium, and Polonium," Elsevier Publishing Co., New York, N.Y., 1966, Chap. 2.

5. C. Ballaux, R. Dams, and J. Hoste, *Anal. Chim. Acta*, **45**, 337 (1969).

6. C. Ballaux, R. Dams, and J. Hoste, *Anal. Chim. Acta*, **43**, 1 (1968).

7. C. Ballaux, R. Dams, and J. Hoste, *Anal. Chim. Acta*, **41**, 147 (1968).

8. C. Ballaux, R. Dams, and J. Hoste, *Anal. Chim. Acta*, **37**, 164 (1967).

9. S. Barabas and P. W. Bennett, *Anal. Chem.*, **35**, 135 (1963).

10. S. Barabas and W. C. Cooper, *Anal. Chem.*, **28**, 129 (1956).

11. A. J. Barak and S. C. Swanberg, *J. Chromatog.*, **31**, 282 (1967).

12. L. Barcza, *Acta Chim. Acad. Sci. Hung.*, **47**, 137 (1966); *Chem. Abstr.*, **64**, 18401b (1966).

12a. J. Barr, R. J. Gillespie, R. Kapoor, and K. C. Malhotra, *Can. J. Chem.*, **46**, 149 (1968).

13. J. Benes and V. Prochazkova, *J. Chromatog.*, **29**, 239 (1967).

13a. T. A. Bengtsson, Uppsala University, Uppsala, Sweden, private communication.

14. J. V. Benson, Jr. and J. A. Patterson, *Anal. Biochem.*, **29**, 130 (1969).

15. B. C. Bera and M. M. Chakrabartty, *Analyst* (London), **93**, 50 (1968).

16. D. Betteridge, *At. Energy Res. Estab. (Gt. Brit.)*, Rept. R4881; *Anal. Abstr.*, **12**, 6592 (1969).

17. C. Bighi and I. Montovani, *Bull. Sci. Fac. Chim. Ind. Univ. Bologna*, **13**, 102 (1955); *Chem. Abstr.*, **50**, 8388b (1956).

18. R. Bock, D. Jacob, M. Fariwar, and K. Rankenfeld, *Z. Anal. Chem.*, **200**, 81 (1964).

19. H. J. M. Bowen, *Analyst* (London), **92**, 124 (1967).

20. H. J. M. Bowen and P. A. Cawse, *Analyst* (London), **88**, 721 (1963).

21. F. H. Burstall, G. R. Davies, R. P. Linstead, and R. A. Wells, *J. Chem. Soc.*, 516 (1950).

22. A. I. Busev, *Talanta*, **11**, 485 (1964).

23. K. A. Caldwell and A. L. Tappel, *J. Chromatog.*, **32**, 635 (1968).

24. C. L. Chakrabarti, *Anal. Chim. Acta*, **42**, 379 (1968).

25. F. L. Chan, *Talanta*, **11**, 1019 (1964).

26. Y. K. Chau and J. P. Riley, *Anal. Chim. Acta*, **33**, 36 (1965).

27. K. L. Cheng, *Anal. Chem.*, **28**, 1738 (1956).

28. G. D. Christian, E. C. Knoblock, and W. C. Purdy, *Anal. Chem.*, **37**, 425 (1965).

29. G. D. Christian, E. C. Knoblock, and W. C. Purdy, *J. Assoc. Offic. Agr. Chemists*, **48**, 877 (1965).

30. G. D. Christian, E. C. Knoblock, and W. C. Purdy, *Anal. Chem.*, **35**, 1128 (1963).

31. R. E. D. Clark, *Analyst* (London), **82**, 182 (1957).

32. R. E. D. Clark, *Analyst* (London), **83**, 396 (1958).

33. R. F. Coleman, *Analyst* (London), **86**, 39 (1961).

34. Committee on Recommendations of Referees, *J. Assoc. Offic. Anal. Chemists*, **52**, 321 (1969).

35. F. J. Conrad and B. T. Kenna, *Anal. Chem.*, **39**, 1001 (1967).

36. F. B. Cousins, *Australian J. Exptl. Biol. Med. Sci.*, **38**, 11 (1960).

37. M. S. Cresser and T. S. West, *Talanta*, **16**, 416 (1969).

38. M. S. Cresser and T. S. West, *Analyst* (London), **93**, 595 (1968).

39. P. Cukor, J. Walzcyk, and P. F. Lott, *Anal. Chim. Acta*, **30**, 473 (1964).

40. L. M. Cummins, J. L. Martin, and D. D. Maag, *Anal. Chem.*, **37**, 430 (1965).

41. L. M. Cummins, J. L. Martin, G. W. Maag, and D. D. Maag, *Anal. Chem.*, **36**, 382 (1964).

42. R. M. Dagnall, K. C. Thompson, and T. S. West, *Talanta*, **14**, 551 (1967).

43. R. M. Dagnall, K. C. Thompson, and T. S. West, *Talanta*, **14**, 557 (1967).

44. R. Dams and J. Hoste, *Anal. Chim. Acta*, **41**, 205 (1968).

45. G. S. Deshmukh and K. M. Sankaranarayanan, *J. Sci. Res. Banaras Hindu Univ.*, **3**, 5 (1952–3); *Chem. Abstr.*, **48**, 6320c (1954).

46. G. S. Deshmukh and K. M. Sankaranarayanan *J. Indian Chem. Soc.*, **29**, 527 (1952); *Chem. Abstr.* **47**, 2634f (1953).

47. G. S. Deshmukh and B. R. Sant, *Analyst*, **77**, 272 (1952).

48. D. W. Dickey, J. H. Wiersma, R. G. Barnekow, Jr., and P. F. Lott, *Mikrochim. Acta*, 605 (1969).

49. W. B. Dye, E. Bretthauer, H. J. Seim, and C. Blincoe, *Anal. Chem.*, **35**, 1687 (1963).

50. L. Erdey, O. Gimesi, and G. Rady, *Talanta*, **11**, 461 (1964).

51. C. S. Evans and C. M. Johnson, *J. Chromatog.*, **21**, 202 (1966).

52. R. C. Ewan, C. A. Baumann, and A. L. Pope, *J. Agr. Food Chem.*, **16**, 212 (1968).

53. A. G. Faulkner, E. C. Knoblock, and W. C. Purdy, *Clin. Chem.*, **7**, 22 (1961).

54. F. Feigl and V. Demant, *Mikrochim. Acta*, 322 (1937).

55. F. Feigl and P. W. West, *Anal. Chem.*, **19**, 351 (1947).

56. C. Feldman, *J. Opt. Soc. Amer.*, **35**, 180 (1945).

57. J. Fidler, *Chem. Listy*, **46**, 221 (1952); *Chem. Abstr.*, **46**, 11029c (1952).

58. A. Fredga, *Ber.*, **71**, 286 (1938).

59. A. Fredga, *Uppsala Univ. Arsskr.*, **5**, 16 (1935).

60. I. Fujii, T. Inouye, H. Muto, and K. Onodera, *Analyst* (London), **94**, 189 (1969).

61. A. Gatterer, *Mikrochem. ver. Mikrochim. Acta*, **36/37,** 476 (1951); *Chem. Abstr.* **45,** 5065h (1951).

62. J. Gillis and J. Hoste, *Compt. Rend. 27e Congr. Intern. Chim. Ind.* (Brussels) (1954); *Chem. Abstr.*, **50,** 8374f (1956).

63. F. A. Gooch and A. W. Pierce, *Amer. J. Sci.*, **1,** 181 (1896).

64. T. T. Gorsuch, *Analyst* (London), **84,** 135 (1959).

65. H. Goto, T. Hirayama, and S. Ikeda, *J. Chem. Soc. Japan, Pure Chem. Sect.*, **73,** 652 (1952).

66. H. Goto and Y. Kakita, *Sci. Rep. Res. Inst. Tohuku Univ.*, Ser. A, **4,** 121 (1952); *Chem. Abstr.*, **47,** 2635c (1953).

67. H. Goto and T. Ogawa, *Sci. Rept. Res. Inst. Tohuku Univ.*, Ser. A, **4,** 121 (1952); *Chem. Abstr.*, **47,** 3745b (1953).

68. H. Goto, Y. Kakita, and S. Suzuki, *Nippon Kinzoku Gakkai-Shi*, **B15,** 617 (1951); *Chem. Abstr.*, **47,** 12115b (1953).

69. M. Goto and K. Tōei, *Talanta*, **12,** 124 (1965).

70. E. S. Gould, *Anal. Chem.*, **23,** 1502 (1951).

71. A. B. Grant, *New Zealand J. Sci.*, **6,** 577 (1963).

72. C. A. Grant, B. Thafrielin, and R. Christell, *Acta Pharmacol. Toxicol.*, **18,** 285 (1961).

73. T. E. Green and M. Turley, in "Treatise on Analytical Chemistry," I. M. Kolthoff, P. J. Elving, and E. B. Sandell, Part II, Vol. 7, Sect. A, Interscience Publishers, New York, N.Y. 1961, p. 176.

74. D. A. Griffin, *Anal. Chem.*, **41,** 462 (1969).

75. W. H. Gutenmann and D. J. Lisk, *J. Agr. Food Chem.*, **9,** 488 (1961).

76. H. Hahn and U. Viohl, *Z. Anal. Chem.* **149,** 40 (1955); *Chem. Abstr.*, **50,** 5453g (1956).

77. H. Hahn and H. Bartels, *Mikrochim. Acta*, 250 (1961).

78. R. J. Hall and P. L. Gupta, *Analyst* (London), **94,** 292 (1969).

79. R. Handley, *Anal. Chem.*, **32,** 1719 (1960).

80. R. Handley and L. M. Johnson, *Anal. Chem.*, **31,** 2105 (1959).

81. W. F. Hillebrand, G. E. F. Lundell, H. A. Bright, and H. I. Hoffman, "Applied Inorganic Analysis," 2nd ed., John Wiley, New York, N.Y., 1953, p. 327.

82. O. Hinsberg, *Ber.*, **22,** 2895 (1889).

83. S. Hirano, M. Suzuki and T. Noguchi, *J. Chem. Soc. Japan, Ind. Chem. Sect.*, **55,** 514 (1952); *Chem. Abstr.*, **48,** 3845a (1954).

84. I. Hoffman, R. J. Westerby, and M. Hidiroglou, *J. Assoc. Offic. Anal. Chemists*, **51,** 1039 (1968).

84a. *Instrument News* (Perkin–Elmer Corp.), **22,** 3 (1972).

85. I. M. Issa, S. A. Eid, and R. M. Issa, *Anal. Chim. Acta*, **11,** 275 (1954).

86. A. Jílek, J. Vřešťál, and J. Havíř, *Chem. Zvesti*, **10,** 110 (1956); *Chem. Abstr.*, **50,** 8388d (1956).

87. N. Jordanov and L. Futekov, *Talanta*, **15,** 850 (1968).

88. E. Kahane and S. Korach, *Microchim. Acta*, **36/37,** 781 (1951).

89. T. Kawashima and M. Tanaka, *Anal. Chim. Acta*, **40,** 137 (1968).

90. W. J. Kelleher and M. J. Johnson, *Anal. Chem.*, **33,** 1429 (1961).

91. G. F. Kirkbright and W. K. Ng, *Anal. Chim. Acta*, **35,** 116 (1966).

92. G. F. Kirkbright and J. H. Yoe, *Anal. Chem.*, **35**, 809 (1963).

93. A. K. Klein, *J. Assoc. Offic. Agr. Chemists*, **26**, 346 (1943).

94. A. K. Klein, *J. Assoc. Offic. Agr. Chemists*, **24**, 363 (1941).

95. H. R. Klein, R. H. King, and W. J. Mader, *J. Pharm. Sci.*, **58**, 1524 (1969).

96. I. M. Kolthoff, P. J. Elving, and E. B. Sandell, "Treatise on Analytical Chemistry," Part II, Vol. 7, Sect. A, Interscience Publishers, New York, N.Y. 1961.

97. I. M. Kolthoff, and N. J. Furman, "Potentiometric Titrations," 2nd ed., John Wiley, New York, N.Y., 1931.

98. J. A. Kowald, unpublished data.

99. J. L. Lampert, P. Arthur, and T. E. Moore, *Anal. Chem.*, **23**, 1101 (1951).

100. F. J. Langmyhr and I. Dahl, *Anal. Chim. Acta*, **29**, 377 (1963).

101. F. J. Langmyhr and S. H. Omang, *Anal. Chim. Acta*, **23**, 565 (1960).

102. C. M. Lederer, J. M. Hollander, and I. Perlman, "Table of Isotopes," John Wiley, New York, N.Y., 1968.

103. B. G. Lewis, C. M. Johnson, and C. C. Delwiche, *J. Agr. Food Chem.*, **14**, 638 (1966).

104. J. J. Lingane, "Electroanalytical Chemistry," 2nd ed., Interscience Publishers, New York, N.Y., 1958, p. 553.

105. J. J. Lingane and L. W. Niedrach, *J. Amer. Chem. Soc.*, **71**, 196 (1949).

106. J. J. Lingane and L. W. Niedrach, *J. Amer. Chem. Soc.*, **70**, 4115 (1948).

107. P. F. Lott, P. Cukor, G. Moriber, and J. Solga, *Anal. Chem.*, **35**, 1159 (1963).

107a. G. Malmsten, I. Thoren, S. Hogberg, J. E. Bergmark, S. E. Karlsson, and E. Rebane, *Phys. Scr.*, **3**, 96 (1971); *Chem. Abstr.*, **75**, 92788 (1971).

108. J. L. Martin and L. M. Cummins, *Anal. Biochem.*, **15**, 530 (1966).

109. J. L. Martin and M. L. Gerlach, *Federation Proc.*, **27**, 417 (1968).

110. J. D. McCullough, T. W. Campbell, and N. J. Krilanovich, *Ind. Eng. Chem., Anal. Ed.*, **18**, 638 (1964).

111. T. McGee, J. Lynch, and G. G. J. Boswell, *Talanta*, **15**, 1437 (1968).

112. F. E. McKenna and D. H. Templeton, *Natl. Nucl. Energy Ser.*, Div. VIII, **1**, 303 (1950); *Chem. Abstr.*, **45**, 1909a (1951).

113. J. S. McNulty, *Anal. Chem.*, **19**, 809 (1947).

114. J. S. McNulty, B. J. Center, and R. M. MacIntosh, *Anal. Chem.*, **23**, 123 (1951).

115. E. Meier and N. Shaltiel, *Mikrochim. Acta*, 580 (1960).

116. J. W. Mellor, "A Comprehensive Treatise on Inorganic and Theoretical Chemistry," Vol. II, Longmans Green, New York, N.Y., 1931, p. 1.

117. M. Milbourn, *J. Soc. Chem. Ind.*, **56**, 205 (1937).

118. K. R. Millar, *J. Chromatog.*, **21**, 344 (1966).

119. R. W. Mitchell, *Commonwealth Bur. Soil. Sci. (Gt. Brit.), Tech. Commun.*, **44**, 115 (1948).

120. D. F. C. Morris and R. A. Killick, *Talanta*, **10**, 279 (1963).

121. F. D. L. Noakes, *Analyst*, **76**, 542 (1951).

122. J. F. Norris and H. Fay, *Amer. Chem. J.*, **20**, 278 (1898).

123. A. A. Noyes, and W. C. Bray, "A System of Qualitative Analysis for the Rare Elements," Macmillan, New York, N.Y., 1927, p. 330.

124. B. Nygård, J. Olofsson, and G. Bergson, *Ark. Kemi*, **28**, 41 (1967).

125. B. Nygård, *Acta Chem. Scand.*, **20**, 1710 (1966).

126. B. Nygård, *Ark. Kemi*, **27**, 405 (1967).

127. B. Nygård, *Ark. Kemi*, **27**, 341 (1967); *Chem. Abstr.*, **67**, 121806j (1967).

128. B. Nygård, *Ark. Kemi*, **27**, 325 (1967); *Chem. Abstr.*, **68**, 8761y (1968).

129. N. Nygård, *Acta Chem. Scand.*, **15**, 1039 (1961).

130. E. C. Olson and J. W. Shell, *Anal. Chim. Acta*, **23**, 219 (1960).

131. O. E. Olson, *J. Assoc. Offic. Anal. Chemists*, **52**, 627 (1969).

131a. R. L. Osburn, A. D. Shendrikar, and P. W. West, *Anal. Chem.*, **43**, 594 (1971).

132. C. A. Parker and L. G. Harvey, *Analyst* (London), **87**, 588 (1962).

133. C. A. Parker and L. G. Harvey, *Analyst* (London), **86**, 54 (1961).

134. P. J. Peterson and G. W. Butler, *J. Chromatog.*, **8**, 70 (1962).

135. C. S. Rann and A. N. Hambly, *Anal. Chim. Acta*, **32**, 346 (1965).

136. H. Rausch and A. Salamon, *Talanta*, **15**, 975 (1968).

136a. I. Rosenfeld and O. A. Beath, "Selenium" Academic Press, New York, N.Y. 1964, Chap. IX.

137. K. Rowley and E. H. Swift, *Anal. Chem.*, **27**, 818 (1955).

138. M. N. Rudra and S. Rudra, *Current Sci.*, **21**, 229 (1952); *Chem. Abstr.*, **47**, 3735g (1953).

139. S. M. de Salas; *Rev. Obras. Sauit. Naciou* (Buenos Aires), **20**, 264 (1947); *Chem. Abstr.* **42**, 2039h (1948).

140. K. Samsahl, *Anal. Chem.*, **39**, 1480 (1967).

141. W. Schöniger, *Mikrochim. Acta*, 123 (1955).

142. E. Schulek and E. Koros, *Z. Anal. Chem.*, **139**, 20 (1953); *Chem. Abstr.*, **47**, 9853h (1953).

143. L. Schwaer and K. Suchy, *Collection Czech. Chem. Commun.*, **7**, 25 (1935).

144. I. Serlin, *Anal. Chem.*, **35**, 2221 (1963).

145. E. H. Shaw, Jr. and E. E. Reid, *J. Amer. Chem. Soc.*, **49**, 2330 (1927).

146. V. Simon and V. Grim, *Chem. Listy*, **48**, 1774 (1954); *Chem. Abstr.*, **49**, 4453b (1955).

147. V. Simon and V. Grim, *Chem. Listy*, **48**, 1415 (1954); *Chem. Abstr.*, **49**, 775i (1955).

148. D. H. Spackman, W. H. Stein, and S. Moore, *Anal. Chem.*, **30**, 1191 (1958).

149. R. P. Spencer, K. R. Brody, W. H. H. Günther, and H. G. Mautner, *J. Chromatog.*, **21**, 342 (1966).

150. A. Steyermark, "Quantitative Organic Microanalysis," 2nd ed., Academic Press, New York, N.Y., 1961, p. 143.

151. W. Stricks and R. G. Mueller, *Anal. Chem.*, **36**, 40 (1964).

152. I. K. Taimni and R. P. Agarwal, *Anal. Chim. Acta*, **9**, 116 (1953); *Chem. Abstr.*, **48**, 79a (1954).

153. I. K. Taimni, and R. P. Agarwal, *Proc. Natl. Acad. Sci. India*, **20A**, 48 (1951); *Chem. Abstr.*, **48**, 11239i (1954).

154. I. K. Taimni and M. N. Srivastava, *Anal. Chim. Acta*, **15**, 517 (1956).

155. M. Tanaka and T. Kawashima, *Talanta*, **12**, 211 (1965).

156. H. H. Taussky, A. Washington, E. Zubillaga, and A. T. Milhorat, *Microchem. J.*, **10**, 470 (1966).

157. J. K. Taylor and S. W. Smith, *J. Res. Natl. Bur. Stand.*, **63A**, 153 (1959).

158. F. P. Treadwell and W. T. Hall, "Analytical Chemistry," Vol. I, 9th ed., John Wiley, New York, 1948, p. 110.

159. S. F. Trelease and O. A. Beath, "Selenium," 1st ed., published by the authors at the Champlain Printers, Burlington, Vt., 1949.

160. A. Walsh, *J. New Zealand Inst. Chem.*, **30**, 7 (1966).

161. A. Walsh, *Spectrochim. Acta*, **7**, 108 (1955).

162. J. H. Watkinson "Selenium in Biomedicine," O. H. Muth, Ed., AVI Publishing Co., Westport, Conn., 1967, p. 107.

163. J. H. Watkinson, *Anal. Chem.*, **38**, 92 (1966).

164. J. H. Watkinson, *Anal. Chem.*, **32**, 981 (1960).

165. E. G. Weatherley, *Analyst* (London), **81**, 404 (1956).

166. G. Wernimont and F. J. Hopkinson, *Anal. Chem.*, **12**, 308 (1940).

167. P. W. West and C. Cimerman, *Anal. Chem.*, **36**, 2013 (1964).

168. P. W. West and T. V. Ramakrishna, *Anal. Chem.*, **40**, 966 (1968).

169. S. E. Wiberley, L. Bassett, A. M. Burrill, and H. Lyng, *Anal. Chem.*, **25**, 1586 (1953).

170. A. I. Williams, *Analyst* (London), **86**, 172, (1961).

171. L. R. Williams and P. R. Haskett, *Anal. Chem.*, **41**, 1138 (1969).

172. J. D. Winefordner and R. A. Staab, *Anal. Chem.*, **36**, 165 (1964).

173. J. D. Winefordner and T. J. Vickers, *Anal. Chem.*, **36**, 161 (1964).

174. Y. Yoshino, *J. Chem. Soc. Japan, Pure Chem. Sect.*, **71**, 577 (1950); *Chem. Abstr.*, **45**, 6537i (1951).

175. W. J. Youden, "Statistical Techniques for Collaborative Tests," Association of Official Analytical Chemists, Washington, D.C., 1967.

Author Index

Roman numerals and letters in parens identify chapters and Arabic numerals that follow give reference numbers.

Subject Index

from hydrazine selenocyanate, 291
I.R., 849
reaction with silver ion, 294
U.V., 873
Selenosis, acute, 665
chronic, 665
Selenosteroids, *see* Seleno steroids
Selenosugars, *see* Seleno sugars
Selenosulfates, 94, 151
from ethylenimine, 151
from halides, 151, 419
from lactones, 152
hydrolysis, 50, 94, 152
oxidation, 94
reactions, 152
see also Seleno Bunte salts
Selenosulfides, chiroptical data (table), 893;
    see also Selenenyl sulfides
Selenosulfuric acid, 50
Selenotaurine, 143, 684
toxicity, 685
4-Seleno-1-thia-γ-pyrones, 259
P-Selenothioacids, 346ff
Selenothiocarbamates, 295
Selenothiocarbamic esters, 295
Selenothiocarbonates, I.R., 845
Selenothiocarboxylic acids, I.R., 847
Selenothiocarboxylic esters, 266
I.R., 847
Selenothiophosphates, 346ff
I.R., 853
Selenothiophosphinate ions, as complexing
    ligands, 1037
Selenothiophthenes, 534
Selenothiopyrophosphates, 351
Selenothymine, 500
U.V., 872
Selenotrisulfides, 677, 680, 682
Selenouracil, 500
U.V., 872
2-Selenouracil, 500
I.R., 843
U.V., 872
X-ray, 503
4-Selenouracil, U.V., 504
Selenourea, 38, 38, 42, 56, 175, 478, 499
antiparasitic properties, 735
complexes, I.R., 850
hydrogen bonds, 1008
in leukemia, 748
iodometry, 1059
pulse radiolysis, 750
radiation protection by, 750
^{75}Se, 281
U.V., 872, 873
X-ray, 992, 993

Selenoureas, 270, 274, 281
acylated, 282
alkylation, 284, 285
analytical applications, 288, 289
as complexing ligands, 1030
as vulcanization accelerators, 289
clathrates with hydrocarbons, 288
cyclization reactions, 286
decomposition, 288
from alkylthiopseudoureas, 283
from carbodiimides, 281
from carbon diselenide, 59, 283
from cyanamides, 281
from hydrogen selenide, 281
from isocyanides, 282
from isoselenocyanates, 282
from thioureas, 283
from ureas, 284
hydrolysis, 288
I.R., 848, 849
in photography, 289
metal complexes, 287, 288
nomenclature, 4
oxidation, 287
phosphorus substituted, 282
radioactive, 281
selenazoles from, 286, 471, 488
selenium removal from, 288
selenium-75, 281
tetrasubstituted, 284
X-ray, 1008
Selenourethanes, I. R., 849
Selenovaleraldehyde, 247
1,4-Selenoxane, metal complexes, 1023; *see*
    also 1-Oxa-4-selane
1,4-Selenoxane-4,4-diiodide, 392
X-ray, 991, 1000, 1005
Selenoxanthates, as complexing ligands,
    1027
Selenoxanthen-9-ol, 558
Selenoxanthenes, 523, 554, 558
central nervous system depressants, 739
radiation protection by, 750
Selenoxanthenols, 558
Selenoxanthogenacetic acid, 296
Selenoxanthone, 235, 554, 556, 558
E.S.R., 557
mass spectrum, 557, 985
p*K*, 550, 557
radiation protection by, 750
Selenoxanthone-10-oxide, 557
Selenoxanthonecarboxylic acid, 555
Selenoxanthones, 554ff
Selenoxanthydrol, 230
Selenoxanthylium perchlorate, 230, 555
Selenoxanthylium salts, 559